AF696515

HUMAN LANGUAGE

HUMAN LANGUAGE

FROM GENES AND BRAINS TO BEHAVIOR

Peter Hagoort
Editor-in-chief

Section Editors: Christian F. Beckmann
Simon E. Fisher
Peter Hagoort
Evan Kidd
Stephen C. Levinson
James M. McQueen
Antje S. Meyer
David Poeppel
Caroline F. Rowland
Constance Scharff
Carel ten Cate
Ivan Toni
Willem Zuidema

THE MIT PRESS
CAMBRIDGE, MASSACHUSETTS
LONDON, ENGLAND

© 2019 The Massachusetts Institute of Technology

All rights reserved. No part of this book may be reproduced in any form by any electronic or mechanical means (including photocopying, recording, or information storage and retrieval) without permission in writing from the publisher.

This book was set in ITC New Baskerville Std by Westchester Publishing Services. Printed and bound in the United States of America.

Library of Congress Cataloging-in-Publication Data

Names: Hagoort, Peter, editor.
Title: Human language : from genes and brains to behavior / edited by Peter Hagoort.
Description: Cambridge, MA : The MIT Press, [2019] | Includes bibliographical references and index.
Identifiers: LCCN 2018050123 | ISBN 9780262042635 (hardcover : alk. paper)
Subjects: LCSH: Language and culture. | Anthropological linguistics. | Language acquisition. | Neurolinguistics.
Classification: LCC P35 .H76 2019 | DDC 401/.9--dc23
LC record available at https://lccn.loc.gov/2018050123

10 9 8 7 6 5 4 3 2 1

CONTENTS

VII THE GENETICS OF LANGUAGE
SIMON E. FISHER

VIII ANIMAL MODELS OF LANGUAGE
CAREL TEN CATE AND CONSTANCE SCHARFF

1 Introduction

PETER HAGOORT

Language is not only one of the most complex cognitive functions that we command, it is also the aspect of the mind that makes us uniquely human. Most of our culture clearly depends on the human ability to communicate ideas and construct artifacts with the help of natural language. To a large extent, language is constitutive for human interaction and cooperation. At the same time, the diversity in form and content of existing languages is shaped by the interactional settings in which they are grounded. One of the remarkable features of linguistic phenotypes is that they come in very different forms, at all levels of organization. The sound repertoires of the more than 6,000 languages that are still around today vary widely, as do their grammatical structures, and the meanings that their lexical items code for. For instance, some languages have a sound repertoire of only a dozen phonemes, whereas others have more than 100; some languages have a very elaborate system of morphological markers, whereas others are very limited in their morphological inventory; some languages make semantic distinctions in one domain, others in another domain. Furthermore, sign languages are expressed by movements of hands and face, whereas spoken languages are expressed by movements of the vocal tract. In addition to the variability in the world's languages, there is individual variation in language skills within the population of any given language community. Some people command only a limited vocabulary and simple sentence structures, whereas others are polyglots who can speak multiple languages fluently or do simultaneous translation between languages.

Despite these differences between languages and individual language skills, most children master their mother tongue (spoken or signed) within about the first four years of life and with little or no explicit instruction. They command their native language at a remarkable level of complexity and computational sophistication well before they are capable of lacing their shoes or performing even simple mathematical operations such as addition and subtraction. Moreover, they are internally driven to communicate, exhibiting "proto-conversation" in gesture and vocalization, and passive comprehension well before they can produce any words. This implies that the human brain exhibits a language readiness not found in the brains of other species. What makes for the language readiness of human brains, which genetic instructions contribute to building such a brain, and how does that language capacity build on the other systems of perception, action, memory, and cognitive control?

Insights into the organization of the uniquely human system for language (including the *universality* and *variability* of the human language faculty) will not be gained if we study language at just one level of description and explanation. We will only make progress if we build on the insights from multiple levels of organization. For instance, there are both genetic and cultural contributions to language variability. In addition, we need to combine insights about different cognitive systems that are in continuous interaction with language and that codetermine the structure and content of the utterances that speakers produce and listeners understand. In recent decades, language has been increasingly studied at multiple levels, including neurobiological, psychological, and linguistic levels of description and explanation. The aim of this book is to bring together the knowledge acquired in many different fields (genetic/genomic, neurobiological, psychological, linguistic, computational, and animal research) about the organization of the uniquely human faculty for language.

Ways to investigate this complex cognitive capacity were traditionally restricted to observational and behavioral methods in healthy people and neuropsychological patients with a language disorder. In recent decades, this picture has changed dramatically. Partly due to technological developments and partly as a result of developments in other fields of research, methods to study language and communication have seen a vast increase in number and level of sophistication. Due to the technological progress in computing power, we are now able to build computational models of language processing that are much more advanced than ever before. Thanks to developments in neuroimaging and genetic sequencing, we are able to study the neural basis and the genetic underpinnings of the language-ready brain in an unprecedented manner. These developments, however, come at a price. To be able to appreciate research findings or actively and fruitfully participate in research on human language, one has to be acutely aware of the developments in the multiple specialized fields that are currently involved in understanding the foundations of human language skills. This is why the need arose for a volume that brings together and summarizes the insights from different angles. The most logical way to do this was to divide up this massive volume into eight parts, each with its own focus and with specialists as section editors. All parts contain state-of-the-art review chapters, and end with a chapter in which the section editors summarize the key issues and point at fruitful future directions for their branch of language research.

Part I on cognitive architectures specifies blueprints for the functional organization of the major language skills, including language comprehension, speaking, and reading. Moreover, it discusses the multimodal nature of human language use and the role of the visual modality in the use of co-speech gestures and sign language.

Part II focuses on learning and developmental trajectories. Children acquire the multiple aspects of language within the first few years of life and beyond, mostly without formal instruction and driven by a need for communication. This part summarizes our current understanding of the many steps and neurocognitive mechanism required to process, encode, and store incoming speech in monolingual and multilingual environments.

Part III discusses the preconditions, evolutionary and otherwise, for advanced communication by means of natural language. Interactional needs and drives might have shaped the coding machinery of a full-fledged language, but they are not dependent on language, and have their own organizational principles, from cognitive as well as neuronal perspectives. The interactional engine is both a precursor of and precondition for a human language system.

Part IV centers on recent developments in modeling language. Enhanced computational power and the availability of large language corpora, as well as the increasing inspiration from knowledge of the brain when modeling data, have together resulted in an enormous boost to attempts at modeling human language processing. We are now at a stage when even the modeling of neurophysiological mechanisms implementing language skills is on the horizon.

Part V discusses the insights from studies that exploit relatively recent ways of measuring brain activity triggered by different forms of cognitive processing. Today a toolkit of cognitive neuroscience methods is available that allows investigations of language in action in the living human brain at an unprecedented scale. Studies using techniques such as functional MRI, electroencephalography/magnetoencephalography, transcranial magnetic stimulation, electrocorticography, and others, enable a linkage between brain and language that goes well beyond the possibilities provided by neuropsychological patients with a lesion in language-relevant brain areas.

Part VI reviews the structural components of the neural infrastructure that makes our brain language-ready. Insights from cyto- and receptor architectonics and from the connectomics of the human brain are discussed. Evidence is provided in favor of important adaptations to the requirements of analyzing and producing speech in areas of auditory and motor cortex, respectively. In addition, neural plasticity in congenitally blind and deaf people offers an inroad into the issue of neuronal degrees of freedom for cognitive computations, in particular for language.

Part VII focuses on the role of genetic factors. The rapid and important developments in molecular methods of the last few decades have made it possible to start deciphering how the genome contributes to building and maintaining a language-ready brain. Genes that have been implicated in developmental forms of language disorders, reading disabilities, and speech motor problems offer novel windows into the biology of language. Investigations of these genes in a range of model systems are deepening our understanding of how different levels of organization (genetic, neurobiological, cognitive) are connected.

Finally, part VIII reviews insights from studying relevant behaviors in nonhuman animals, such as song in birds and vocalizations in primates, to help illuminate possible evolutionary building blocks for speech and language. Although language in the forms that we know seems uniquely human, it is perhaps best conceptualized as a multicomponent system. At least some of the components have analogies and (deep) homologies with behavioral and cognitive features observed in other species. Auditory-guided vocal learning, sequencing abilities, and patterning of sounds are key examples of language-relevant abilities that can be found elsewhere in the animal kingdom. Studies of such traits in animal models, including songbirds and nonhuman primates, might be instructive for the question of how language came into being in our evolutionary history.

Collectively this volume gives a state-of-the-art overview of our current understanding of human language in all its facets. It shows that developing such an understanding depends on interactions between different fields of research and expertise. As for communication between speakers of different languages, cross talk between experts of different fields is not without the risk of misunderstanding and miscommunication. In these cases, real and substantive interaction happens only with the motivation to listen to each other and to develop a common creole. The purpose of this book is to contribute to the establishment of a common language in interaction between the different fields that are relevant for the scientific endeavor of the language sciences.

This book would not have been possible without a substantial grant from the Dutch government to a consortium of researchers under the title "Language in Interaction." This grant enabled us to organize a two-week summer school for the PhD candidates and postdoctoral researchers of the consortium in the summer of 2016. The contributors to this volume came to an inspiring location near Nijmegen, the oldest city in the Netherlands, to present their ideas as a preparation for the chapters of the book. They profited from the lively and insightful comments of a group of very talented students and postdoctoral researchers interested in language research. I also thank the MIT Press for its enthusiasm and support during the transformation of chapters into print. Sander Berends, the program officer of the Language in Interaction grant was instrumental in the organization of the summer school. My biggest thanks I owe to Carolin Lorenz whose unlimited energy and efficiency guided this project in all its phases toward the outcome that you have in front of you.

I COGNITIVE ARCHITECTURES

SECTION EDITORS: JAMES M. MCQUEEN AND ANTJE S. MEYER

2 Mental Representations for Language

RAY JACKENDOFF

1. What Is a Mental Representation?

Before discussing specifically linguistic representations—phonological, syntactic, and semantic/conceptual structures—it is worth briefly considering mental representations in general.[1] The most productive way to think about mental representations is as data structures instantiated in the brain, which collectively create our sense of "the real world"—as well as that of imagined and possible worlds. The brain has no direct access to what the world is like: as stressed already by the early 20th-century gestalt psychologists (Koffka, 1935; Wertheimer, 1923/1999), not to mention (on some readings) Kant, the mind/brain has to *construct* the world of experience. Different species, with different brains, construct different worlds of experience (Dennett, 1996). The focus of such a *constructivist* approach is discovering the structure and organization of mental representations themselves, in terms of which the world of experience is constructed. In particular, words and sentences are part of reality for humans—but not for other species, except possibly in the most limited sense—because only humans have the capacity for the appropriate sorts of mental representations built into their brain.

The mental representations for different domains of cognition are not one-size-fits-all. Representations suitable for encoding verb agreement are not going to be suitable for encoding visual experience, music, or morality. Hence a fundamental question for a theory of mental representation is what kinds (or *levels*) of representation there are. For each posited level, the theory must characterize the space of possible representations—its repertoire of structures. In particular, phonemes and syntactic structures are basically digital and algebraic in character, while mental representations for higher level vision are most likely geometrical or topological (Kosslyn, 1980; Kosslyn & Koenig, 1995; Marr, 1982).

Despite such differences, many levels of representation share *hierarchical structure* as an organizing principle. For instance, individual speech sounds are grouped into syllables, which are grouped into words, which are grouped into phrases, which are grouped into sentences. In vision there is also hierarchical grouping, but over two- and three-dimensional spaces. For instance, objects can be parsed into parts, which themselves have parts, and so on (Biederman, 1987; Marr, 1982). Hierarchical grouping is also an indispensable aspect of music cognition (Jackendoff & Lerdahl, 2006; Lerdahl & Jackendoff, 1983), and it arguably plays a role in the structure of complex action (Jackendoff, 2007a) and in the comprehension of visually presented narrative (Cohn, 2013; Cohn, Jackendoff, Holcomb, & Kuperberg, 2014).

If there are multiple levels of representation, there must be a way for them to interact computationally. For instance, there has to be an informational pathway between visual and linguistic representations. If there were not, one could not talk about what one sees. And reading requires a different and rather eccentric informational conduit between the visual system and linguistic representations (Dehaene, 2009; Wolf, 2007; Andrews & Reichle, chapter 5 of this volume).

Similarly, hand-eye coordination and imitation require connections between visual and motor representations. And singing or playing music in an ensemble requires connections among auditory representations (the music one is hearing, both self-produced and from the other members of the group), visual representations (reading the sheet music and following the conductor), and motor control (singing or playing one's instrument). Finally, for a more unusual case, synesthetes evidently have connections between various disparate representations (e.g., numbers and colors) that most of us lack.

Fodor (1983) proposed that low-level perceptual processes (including syntactic parsing) constitute "mental modules" and are "informationally encapsulated," that is, they are impervious to input from other modules. For instance, perceiving a speaker's spatial location or hair color has no effect on the perceived phonology of this individual's utterance. However, the cross-modal examples just enumerated show that Fodor's formulation is too rigid. A more forgiving assessment is that levels of mental representation cannot interact *indiscriminately*: there must be principles that specify, for any level of representation, which other levels it connects with, and how richly. For instance, although language and vision interact, not every visual feature can actually be transmitted to the language faculty, for instance the

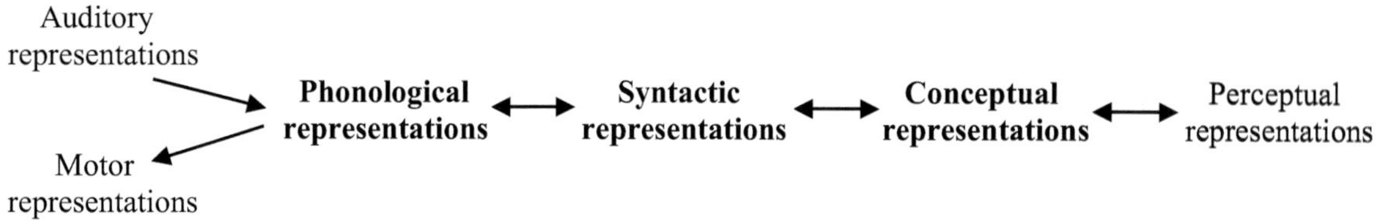

FIGURE 2.1 Levels of mental representation involved in language, with their interfaces.

exact shape or precise color of a particular leaf (Landau & Jackendoff, 1993). And conversely, nonphysical concepts expressed in language, such as kinship and obligations, have no visual counterpart.

Similarly, one can look at one object and handle another and judge whether they have the same shape. This suggests a relatively close connection between visual and haptic representations of shape. However, only visual representations encode color, and only haptic representations (directly) encode temperature. Moreover, hapsis is confined to objects within reach, while this is obviously not the case for vision. In short, these two levels, like language and vision, do not communicate everything to each other.

It follows that a theory of mental representations must delineate not only the individual levels of representation, but also the connections (or *interfaces*) between them. An interface is characterized in terms of a correspondence between aspects of structure in two or more levels of representation. To the degree that two levels of representation do *not* have established correspondences, they are informationally encapsulated from each other in Fodor's sense.

One more general point: Following in the spirit of Marr (1982), a theory of linguistic representations ideally ought to lend itself to postulated processing mechanisms and neural implementation. At the same time, theories of processing and neural computation must be rich enough to support one's postulated representations. Should the theories come into conflict, the theory of representations should not be immune to evidence from processing. But by the same token, neither should a theory of processing or neural computation be immune to evidence from what an adequate theory of linguistic representation requires.[2]

2. Architecture

In these terms, language is to be conceived of fundamentally as a mapping between two sorts of mental representation: sound structure (phonology) and meaning (semantics or *conceptual structure*), with syntactic structure to help mediate the connection. Phonology in addition interfaces with the auditory system, so that one can perceive speech, and with the motor system, so that one can produce speech. Conceptual structure is the mental representation in terms of which one makes inferences and in terms of which language ultimately interfaces with the perceptual systems, so that one can for instance talk about what one sees. The overall architecture is sketched in figure 2.1, following the Parallel Architecture proposed in Jackendoff (2002). The parts in bold and their interfaces are those directly pertinent to the language faculty.[3]

Figure 2.1 depicts the three major components of language as equals linked by interfaces. All the interfaces are bidirectional, with the possible exception of those between phonology on one hand and auditory and motor representations on the other. As mentioned, literacy adds a special interface from visual representations to phonology; sign languages (Özyürek & Woll, chapter 6 of this volume) substitute visually and gesturally based phonology for the auditory/vocal phonology in figure 2.1.[4]

Where do words fit into this picture? Conventionally, words are thought of as listed in a *mental lexicon*, and they are combined into sentences by a component of *grammatical rules*. Theories differ in the character of their grammatical rules. In particular, some approaches keep grammar and lexicon separate, while others assimilate rules of grammar into the lexicon, an issue to which we return in section 6. First, however, we ask what the mental representation of a word is and what the mental representation of a sentence is.

3. The Mental Representation of Words

According to just about every theory, a word is a linkage of components from all three structures: phonology, syntax, and semantics. A customary notation for the word *sheep*, for example, is shown in (1). (There are many variants on this notation, but they all say essentially the same thing.)

(1) Phonology: /ʃip/$_1$
Syntax: Noun$_1$
Semantics: SHEEP$_1$

Conventionally, the three structures are enclosed in large square brackets. For reasons to be clear in the next section, I notate their connection by cosubscripting them instead. The cosubscripts can be thought of as marking the ends of association lines, and the association between these three structures can be thought of as a small component of the interface among the three levels of linguistic representation.[5] We now look more closely at the three components in turn.

3.1. Phonological Structure in Words Phonological structure is notated in (1) in terms of the phonetic alphabet; each symbol represents an individual speech sound. However, speech sounds are actually composite, made up of a matrix of *distinctive features.* Figure 2.2 zooms in on the structure of /ʃip/, now broken into distinctive features. These features define a similarity space among speech sounds: speech sounds with similar features tend to pattern together within languages as well as crosslinguistically.[6] There is considerable discussion in the literature on the repertoire of features, in an effort to make them crosslinguistically valid (Chomsky & Halle, 1968; Clements, 1985; Clements & Ridouane, 2011; Jakobson, Fant, & Halle, 1952; Ladefoged, 2007); figure 2.2 is a more or less generic version.

Figure 2.2 also notates the syllable structure of *sheep* as a hierarchical tree structure. The vowel constitutes the *syllabic nucleus.* It groups with the final segment, the *coda,* to form the *rhyme.* This groups with the initial segment, the *onset,* to form the entire syllable.

In addition to the features that identify speech segments and syllabic structure, phonological structure represents stress patterns. For instance, the verb *permít* is stressed on the second syllable, while the noun *pérmit,* with the same sequence of segments, is stressed on the first syllable; the word *òrganizátion* has secondary stress on the first syllable and primary stress on the fourth. In tone languages such as Mandarin and many West African languages, the pitch contour is also part of the phonological representation of words (Yip, 2002).

In accordance with figure 2.1, the phonological representation of a word maps to some sort of *gestural score* (Browman & Goldstein, 1986) that directs the timing of movements executed by articulators (lips, tongue tip, tongue blade, velum, larynx) to produce speech. It also maps to some sort of acoustic phonetic representation that enables segments to be identified in speech recognition.

While gestural and acoustic representations are continuous and gradient, phonological representations are discrete; they might be thought of as a sort of digitization of sound. This digitization is crucial to the architecture of language. It is what makes possible the large vocabulary of natural languages, through combination and recombination of a finite repertoire of phonological segments.

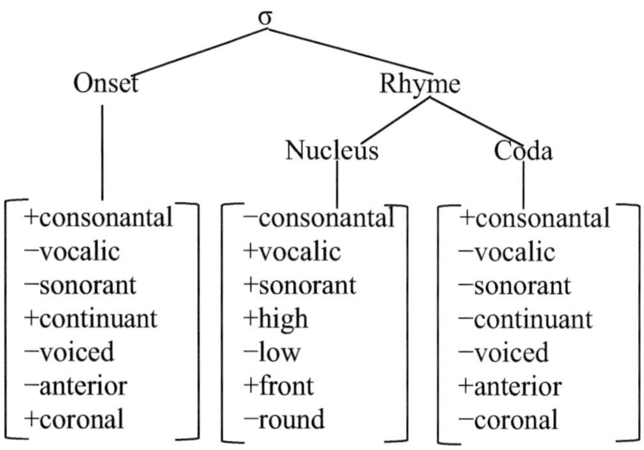

Figure 2.2 Syllabic structure and feature decomposition of the word *sheep.*

3.2. Syntactic Structure in Words With a few exceptions such as *hello* and *ouch,* all words belong to a syntactic category such as noun, verb, adjective, preposition, determiner, and auxiliary. These categories have no significance for words in isolation; rather, they represent affordances for combinations with other words. For instance, in English, a noun may be preceded by a determiner and an adjective and followed by a relative clause, the whole forming a noun phrase (NP). Depending on the language, the syntactic structure of a noun may also include *inflectional features* such as number, grammatical gender, and case; a verb may include features such as tense, aspect, and person, as well as number and gender agreement with the subject and/or object.

A word can also have internal *morphosyntactic* structure. For example, the compound noun *football* is built out of two nouns, as in (2a), and the noun *baker* is built out of a verb plus a suffix (2b). Morphosyntactic structure can also be more complex, as in (2c).

(2) a. [$_N$ [$_N$ foot] [$_N$ ball]]
b. [$_N$ [$_V$ bake] [$_{suffix}$ er]]
c. [$_N$ [$_A$ [$_{prefix}$ in] [$_A$ [$_V$ comprehens] [$_{suffix}$ ibil]]] [$_{suffix}$ ity]]

We return to syntactic structure in the next section.

3.3. Conceptual Structure in Words The meaning of the word *sheep* is notated in (1) simply as *SHEEP,* following a commonplace practice. This is to be taken as a convenient abbreviation for a formidable amount of mental structure involved in a concept, much of which is not well understood. Ultimately, in order for language to be used to express thought and in order to make inferences based on linguistic input, the full richness of meaning must be drawn upon, hence the term *conceptual structure.*[7]

To convey some of this richness: The conceptual structure of the word *sheep* has to at least classify sheep as animate physical objects, as animals, as mammals, and as domestic animals. Already in this classification we can see the limitations of any sort of feature or vector representation—regardless of whether the features in question are binary or gradient and whether they are stipulated or derived from statistical analysis of a corpus.[8] For instance, a feature such as [+domestic animal] tells us exactly nothing about what we know: that domestication is a human institution; that sheep are typically kept in herds; what a herd is and how it is maintained; that the reason for keeping sheep is to get meat, wool, and perhaps cheese; and so on. Going further, we have to know about wool, how it is used to make clothing, and so on. The connections go on indefinitely.

In addition, the concept expressed by *sheep* has to make possible the identification of sheep in the world—the *referential* function of concepts. Again, this cannot be done in terms of a system of features or the statistics of a corpus. Rather, the word must be associated with a mental representation of what a sheep looks like, derived from the visual system, perhaps along the lines of Marr's (1982) 3D model, but likely much richer. It is only by means of such a representation that one can accomplish standard experimental tasks such as picture naming, let alone use language to speak about the perceived world.

This latter sort of representation has sometimes been referred to as something like "an image of a stereotypical instance" (e.g., Putnam, 1976), and it is typically thought of as a sort of "picture in the head." But a picture is too precise: it has to represent an image of a sheep from a particular angle and in a particular posture. Moreover, not all visually identifiable concepts lend themselves to imagery. Consider the word *end.* Objects such as ropes, sticks, beds, queues, rivers, and oval tables all have ends, as do events such as speeches, songs, and rainstorms. There is no conceivable image of a stereotypical instance for such disparate cases. What makes all these entities susceptible to having ends is that they can be idealized as a bounded linear structure in space or time, of which the end constitutes the boundary. In other words, the mental representation of the appearance of objects must allow a considerable degree of abstraction and structuring, well beyond a bitmap-like image (Jackendoff, 1991, 2012a; Marr, 1982).

In addition, many concepts, unlike *SHEEP* and *END,* are not perceptually grounded. Yet we can refer to such entities, for instance the *similarity* of two items (in any modality or even across modalities), the *value* of a house, the *points* in a game, the *purpose* of an action, and the *right* to free speech (Jackendoff, 2007a). These abstract concepts require an ontology of mental representations in which similarity, values, points, purposes, and rights are part of the world of our experience. Some of these concepts lend themselves to treatment in terms of metaphor (Lakoff & Johnson, 1980) or embodiment (Varela, Thompson, & Rosch, 1991); for others it is less clear (Jackendoff, 2007a).

A further difficulty in pinning down the conceptual structures of words is that they typically lack sharp edges. Classic examples include the distinctions between color words (at what point does it stop being red and become orange?), the so-called *sorites paradox* (how little hair does one need in order to be considered bald?), and "family resemblance" concepts (Wittgenstein, 1953). A good example of family resemblance comes from Lakoff (1987): the term *X's mother* can be used to refer to the woman whose genes X has inherited, the woman who gave birth to X, or the woman who has raised X. A stereotypical mother satisfies all these criteria, but modern technology allows these functions to be performed by three different people. In such a case, which one is X's "real mother"? On the face of it, there is no fact of the matter. A conceptual representation for the word has to be able to encode such collections of conditions, which, once one begins to look for them, turn out to be ubiquitous in the vocabulary (Jackendoff, 2012b). Still more complexity will be illustrated in section 4.

Little of this detail matters to syntax. For instance, all that syntax cares about *SHEEP* is that it denotes a singular animate physical object. In general, it turns out to be difficult if not impossible to draw a line between the parts of meaning that pertain directly to language (semantics) from the full complexity of information the word conveys (pragmatics or world knowledge) (Bolinger, 1965; Jackendoff, 1983, 2002; also psycholinguistically: Hagoort, Hald, Bastiaansen, & Petersson, 2004). Thus an important issue is what aspects of conceptual structure do make a difference to syntax—that is, how rich the interface between them is.

Language does not consist simply of isolated words. Language is what it is because words can be combined into phrases and sentences. So next we look at issues in the representation of combinatoriality.

4. The Mental Representation of Sentences

Figure 2.3 sketches the mental representation of an utterly simple sentence of English, *Mike bought a bright blue bike.* All three component structures display combinatorial structure. The notation here is that of the Parallel Architecture (Jackendoff, 2002), but the same basic ideas appear in just about all treatments.

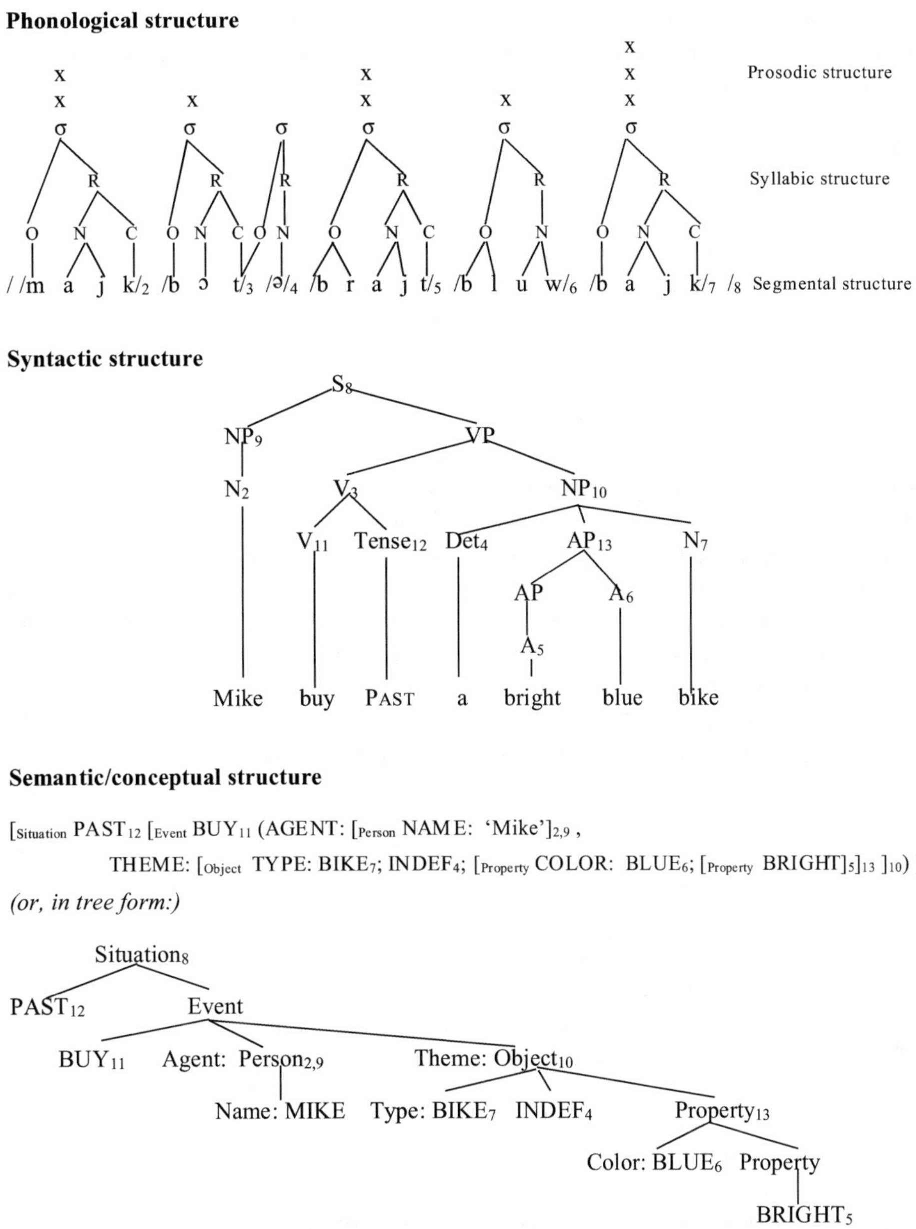

FIGURE 2.3 The structure of *Mike bought a bright blue bike.*

A first thing to notice is that each word is distributed across the three structures. *Mike,* for instance, has a representation in phonology (*/majk/*), in syntax (N), and in semantics ($[_{\text{Person}}$ NAME: "Mike"]). The three pieces are bound together by the co-index 2, just as in *sheep.* Hence, again, the word serves as part of the interface among the three structures.[9] The same is true of all the other parts of the sentence—though as we will see in a moment, the co-indexation is not entirely uniform.

Next we look at the individual structures in turn: The phonological structure shows the segmental and syllabic structure. The only thing we need to add here is that segments in certain positions may be *ambisyllabic,* that is, they belong to two syllables at once. A possible example is the */t/* in *bought a.*

Above the syllabic structure is the *metrical grid,* which notates degrees of relative stress in terms of relative number of *X*s. Each of the words in figure 2.3 receives a stress, with the exception of the unstressed indefinite article *a.* The last word of the sentence receives the greatest stress. On the pronunciation represented in figure 2.3, the remaining words fall into an alternating stress pattern. This configuration can be altered by emphasis or contrast, as in *Mike bought a bright blue bike, not a red one.*

The syntactic structure in figure 2.3 is a fairly conventional tree structure. Three things of note: First, with one exception, the words of the phonological structure, set off by slashes, are co-indexed with corresponding parts of speech in the syntactic tree (co-indices 2–7). In addition, the entire sentence is co-indexed with the entire phonological string (co-index 8).

Second, the exception in co-indexing is the phonological string */bɔt/* (*bought*), which corresponds to two constituents of the syntax: the verb *buy* and the past tense. With regular verbs, the past tense is expressed by a separate piece of phonology, such as the /t/ in *walk<u>ed</u>*. In contrast, with the irregular verb *buy*, the past tense and the verb itself are folded into a single idiosyncratic piece of phonology. Hence the phonological string is co-indexed with the combination of *buy* and PAST, each of which has its own co-indexation to the semantics.

Third, the second NP contains the adjective phrase (AP) *bright blue*. The modifier in this AP, *bright*, has the affordance of itself being modified, as in *a stunningly bright blue bike*. For this reason it is notated in figure 2.3 as being in its own AP. Hence the larger AP contains a smaller one. This is a hint of the depths of hierarchical embedding (or *recursion*) that syntactic structure can achieve effortlessly. For example, in sentence (3) (from an actual interview with an innkeeper), several NPs contain smaller NPs, and several clauses contain smaller clauses.

(3) $[_S$ $[_S$ $[_S$We have a young child] and $[_S$ our chef has twins]] so $[_S$ we know $[_S$ how difficult it is $[_S$ to find $[_{NP}$ a first rate restaurant $[_S$ that doesn't shudder $[_S$ when you show up at the door with kids in tow]]]]]]]

Turning to the conceptual structure: Figure 2.3 notates it in a fairly conventional labeled bracketing à la predicate logic (though hardly as complex as the structures posited in formal semantics), and also in a tree structure that is somewhat easier to decode. Informally, it says that there is a situation in the past that consists of an event of buying. The buying involves two characters, an Agent and a Theme. The Agent is a person named Mike. The Theme (the thing being bought) is a physical object belonging to the category of bikes, and this one happens to be blue, and not only that, but a shade of blue that has the property of being bright.

All the words of the sentence are co-indexed with the appropriate pieces of this expression. Note however that the verb *buy* and the past tense are linked to separate constituents of the conceptual structure (co-indices 11 and 12), whereas their combination is linked to the single phonological word *bought* (co-index 3). Also, since the word *Mike* constitutes an NP all by itself, both it and the entire NP are linked to the same piece of conceptual structure (co-indices 2 and 9).

Every part of this conceptual structure raises the same sorts of difficulties we encountered with *sheep* in section 3. Here I want to dwell on the internal structure of *BUY*, which is especially complex. What constitutes an act of buying? Figure 2.3 shows two of the characters in such an action—the *arguments* of *BUY*. These arguments are labeled in figure 2.3 by their roles in the action—so-called *thematic roles*. The Agent is the individual carrying out the action; the Theme is the individual undergoing change. Other thematic roles, with other verbs, include Patient (the individual affected by the action), Goal (the final point of motion or change), Source (the initial point of motion or change), Experiencer (the individual having an experience such as a perception, thought, or emotion), Stimulus (the entity causing an experience), and so on.

It is not clear that there is a discrete list of thematic roles, or whether they are merely convenient descriptive labels (Dowty, 1991; Jackendoff 1990). But despite their somewhat uncertain theoretical status, they are widely used to categorize the arguments of verbs. In particular, one important generalization over thematic roles is that crosslinguistically, Agents are invariably linked to the subject position of sentences (except in passives).

However, *BUY* tells us more about this event than simply that it has two characters, an Agent and a Theme. We know that at the beginning of the event, the Agent was not in possession of the Theme, and at the end of the event, he was. So far, this is the same as an event of getting or obtaining. Part of what makes the event *buying* is that there is another individual (the Source) who is in possession of the Theme at the beginning of the event and is not in possession of it at the end—that is, buying involves a change of possession. However, this is still not enough to distinguish buying from giving (or taking). What makes this event go beyond giving is that the transfer of possession is accompanied by a transfer of something else (we might call it the *Secondary Theme*) from the Source. If the Secondary Theme is another object, the event is *trading*; if it is money, finally the event is *buying*. These other characters—the Source and the Secondary Theme—can be expressed optionally in the syntax and phonology:

(4) Mike bought a bright blue bike <u>from Ike</u> (Source) <u>for 50 euros</u> (Secondary Theme).

But even if they are not expressed, the verb *buy* tells us that these *implicit arguments* are part of the event being described, and a more accurate version of the conceptual structure in figure 2.3 would reflect this.

There is yet another level of complexity. The two transfers are interdependent; together they constitute a jointly agreed-upon action by Mike and Ike, in which each of them judges that the transaction is to his own benefit. So we are very quickly off into questions of what benefits are and what constitutes joint agreement (or joint intention: Clark, 1996; Bratman, 1987; Gilbert, 1989). Moreover, in order to understand transfer of possession, we have to understand what possession is (Jackendoff, 2007a, chapters 8–10).

This hugely complex notion of a transaction underlies a collection of verbs in English. The sentences in (5) describe essentially the same transaction as (4), but with different characters highlighted. And the sentences in (6) describe different sorts of transactions.

(5) a. Ike sold a bright blue bike to Mike.
   b. Mike paid Ike 50 euros for a bright blue bike.
   c. That bright blue bike cost Mike 50 euros.

(6) a. Jill traded her bike to Bill (for a toaster).
   b. Jill rented a bike (from Bill) (for 3 euros).
   c. Jill hired Bill to fix her bike (for 10 euros).

This suggests that human conceptualization encompasses complex scenarios that can be expressed from multiple perspectives. Another simple example is the contrast between *chasing* and *fleeing*—the same event from the point of view of different characters. Even a single word such as *bride* evokes a whole social scenario of matrimonial ritual, mutual obligation, and presumed sexual fidelity. This insight is at the basis of Frame Semantics (Fillmore, Lee-Goldman, & Rhodes, 2012), which seeks to identify a wide range of such scenarios. It also comes into play in Minsky's (1975) notion of frames (such as birthday parties), Schank and Abelson's (1975) scripts (such as what a customer does in a restaurant), and Goffman's (1974) wide-ranging Frame Analysis. How such frames are to be formalized is a wide-open question at present.

Going beyond figure 2.3, a further aspect of conceptual structure is *information structure* (Lambrecht, 1994). The sentences in (7) all depict the same event as (4), using the same words, but with different emphasis. The differences can be conveyed by shifting stress (7a) or by special focusing constructions such as the cleft in (7b) and the pseudocleft in (7c).

(7) a. Mike bought a bright *blue* bike. (not some other color)
   b. It was *Mike* who bought a bright blue bike. (not someone else)
   c. What Mike did was *buy a bright blue bike.* (as opposed to doing something else)

Information structure phenomena include the marking of topic, focus, and common ground and the division of an utterance into discourse-old versus discourse-new information. Notice that this organization cuts across the marking of thematic roles and modifiers illustrated in figure 2.3: any of the entities or properties can be focused, as seen in (7). For information structure to be expressed, there has to be a mapping that relates it to stress and intonation and to syntactic structure such as the special constructions in (7b, c) and/or special morphological markers (e.g., *wa* in Japanese).

Another important element of linguistic structure is *anaphora*: the use of pronouns, reflexives, and other expressions to refer to someone or something mentioned previously. Some varieties of anaphoric elements appear in (8).

(8) a. Joe adores <u>himself</u>. [reflexive: himself = Joe]
   b. Joe thinks that Fred adores <u>him</u>. [pronoun: him = Joe or someone else]
   c. Mike bought a bright blue bike, and Sally bought a red <u>one</u>. [*one*-anaphora: one = bike]
   d. Mike bought a blue bike, and Sally did too. [VP-ellipsis: did = bought a blue bike]
   e. Sally likes soup, and Fred fish. [Gapping: second clause = Fred likes fish]
   f. *Speaker A:* Ozzie drinks scotch. *Speaker B:* No, vodka. [Bare Argument Ellipsis: vodka = Ozzie drinks vodka]

A major issue in the field is how to delimit the syntactic configuration in which anaphoric elements can find their antecedents (so-called *binding theory*, not to be confused with the neuroscientific notion of binding). In brief, with a few exceptions, the antecedent of a reflexive is always in the same clause (so-called *Principle A*; Chomsky, 1981), and the antecedent of a pronoun is always in a different clause (*Principle B*).

(9) a. Sally thinks that Joe adores himself/*herself. [himself = Joe; herself ≠ Sally]
   b. Sally thinks that Joe adores him/her. [him ≠ Joe; her = Sally or someone else]

The problems are all in the exceptions and in the behavior of such elements crosslinguistically (Culicover & Jackendoff, 2005; Davis, 2009; Hale, 1992).

Another important syntactic phenomenon, which involves elements of both information structure and anaphora, is *unbounded dependency*, in which an element at the beginning of a clause is understood as if it occupied a "gap" position somewhere interior to the clause. Example (10) illustrates a number of such constructions in English. The underlined phrases are "understood" in the position marked with ^. For instance, in (10a), Sarah is understood to have seen some movie, that is, the semantic role of *movie* is the same as in *Sarah saw that movie.*

(10) a. Which movie does Susan imagine that Sarah saw ^ last night? [*wh*-direct question]
b. John was wondering who Sarah decided she would go to the movies with ^ on Sunday. [Indirect question]
c. I didn't like the movie which you said that everyone was talking about ^ the other day. [Relative clause]
d. You may take whichever sandwich you find ^ on the table over there. [Free relative]
e. That movie, I wouldn't recommend that anyone ought to consider taking their kids to ^. [Topicalization]

The gap ^ can be inside a subordinate clause, or even deeply embedded within multiple subordinate clauses, as in (10e); this is why these configurations are called unbounded dependencies.

These constructions place strong constraints on the structural positions that a gap can occupy with respect to its antecedent, namely the constituent at the front of the clause. For example, the antecedent cannot be outside of a conjoined construction (11a), a relative clause (11b), an indirect question (11c), or a subject (11d) that the gap is inside of. The examples in (11) all involve direct wh-questions, but the same thing happens with all the constructions in (10).

(11) a. **What* did Beth eat peanut butter and ^ for dinner?
(cf. Beth ate peanut butter and lobster for dinner.)
b. **Who* does Sam know a girl who is in love with ^?
(cf. Sam knows a girl who is in love with Max.)
c. **Who* does Betty know which professor flunked ^?
(cf. Betty knows which professor flunked Harriet.)
d. **Who* does a good friend of ^ go to MIT?
(cf. A good friend of Suzanne goes to MIT.)

Accounting for these constraints, different versions of which are widely attested crosslinguistically (e.g., Phillips, 2013), has been a preoccupation of linguistic research since Ross (1967/1986). The main effort has been to treat them as properties of syntactic structure (Chomsky, 1977), but other approaches have chipped away at the phenomena, ascribing some cases to semantics (Erteschik-Shir, 1997; Kuno & Takami, 1993; Van Valin 1998) and some to difficulties in processing (Culicover, 2013; Hofmeister & Sag, 2010; Kluender, 1998; Phillips, 2013).

5. *What Are the Principles behind Each Level?*

The central question of linguistic theory is where all these structures come from. The answer must account for the hallmark of human language: the ability to combine and recombine items from the lexicon into an unending flood of novel utterances—as Humboldt (1836/1999, p. 91) famously put it, "the infinite use of finite means." In order to combine lexical items nonrandomly, it is necessary that there be a set of principles—a *mental grammar*—that determines how lexical items can be assembled into larger structures such as figure 2.3.

5.1. Derivational Approaches There have been two basic approaches to the nature of the grammar. The mainstream approach (see Chomsky, 1957, 1965, 1981, 1995; Chomsky & Halle, 1968 for different versions) can be thought of as derivational, algorithmic, or procedural. For instance, in pre-1995 syntactic theory in this vein, the derivation of figure 2.3 begins with the initial symbol S (sentence), and expands it into NP plus VP (verb phrase), then the VP is expanded into V plus NP, and so on. These expansions are accomplished by *phrase structure rules*, applying successively to create the syntactic structure in figure 2.3. Words are then selected from the lexicon to insert into the tree at terminal nodes (N [noun], V [verb], A [adjective], Det [determiner], and so on).

In the more recent *Minimalist Program* (Chomsky 1995, 2000), syntactic structure is generated bottom-up instead of top-down. Items are taken out of the lexicon and combined by means of a "Merge" operation, which joins two constituents into a single constituent. For instance, in figure 2.3, *bright* and *blue* are merged into the constituent *bright blue*, which is merged with *bike* to form *bright blue bike*, and so on.

The derivation typically does not stop there. In a sentence containing an unbounded dependency, such as *What did Mike buy?*, *movement rules* move *what* from the gap position in "underlying" structure to the front of the clause, leaving behind a "trace"—an unpronounced pronoun (Chomsky, 1977).[10] Following the movement of *what*, other adjustments are made to the form of the sentence to derive the surface form. At certain points in the derivation, the syntactic structure is "sent" to phonology and semantics.

In this approach, ellipsis also involves an underlying structure. For instance, in VP-ellipsis (8d), the underlying form of the second clause, *Sally did too*, has the full syntactic structure of *bought a blue bike*, but this structure is deleted (or fails to map into phonology) in the course of the derivation (Merchant, 2001). Similar derivational approaches are found in phonology, stemming from Chomsky and Halle (1968), and in morphology (Hale & Keyser, 2002; Distributed Morphology: Halle & Marantz, 1993; Siddiqi, 2019).

This approach is difficult to apply to language processing. One surely does not produce a sentence by building a syntactic tree (either top-down or bottom-up), moving its constituents around, then sending the result off to semantics and phonology—thus deciding at the end of this process what one wants to say and how to pronounce it. For this reason, a central tenet of the Chomskyan approach has been that the rules of linguistic theory describe "competence," and that it is somewhat obscure how the rules per se play a role in "performance" (Ferreira, 2005; Phillips & Lau, 2004; Poeppel & Embick, 2005).

5.2. CONSTRAINT-BASED AND CONSTRUCTIONAL APPROACHES A second major approach to principles of grammar replaces procedural, step-by-step rules with declarative templates, variously called *constraints* (Head-driven Phrase Structure Grammar: Pollard & Sag, 1994; Lexical-Functional grammar: Bresnan, 2001), *constructions* (Construction Grammar: Croft, 2001; Goldberg, 1995; Hoffmann & Trousdale, 2013), or *schemas* (Construction Morphology: Booij, 2010; Parallel Architecture: Jackendoff & Audring, 2019). These templates can be applied simultaneously to a structure to license its well-formedness, or they can be used to build structure by "clipping" pieces together in any order that is convenient. For example, these approaches have templates or "treelets" along the lines of (12a, b, c) (abstracting away from notational differences among the theories).

(12) a. b. c.

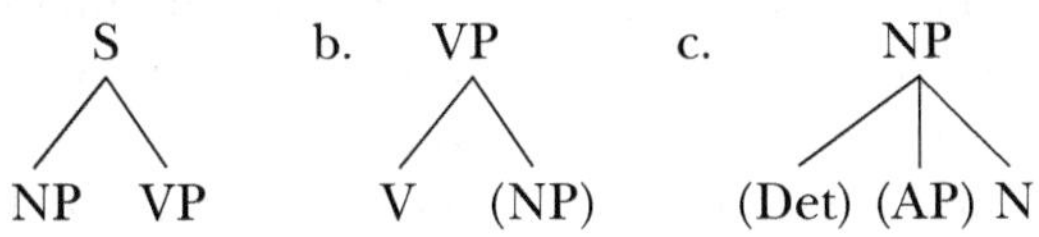

It can be readily seen how these pieces can be fit together into the tree structure in figure 2.3. The order in which they are combined is irrelevant.[11] For instance, a determiner *the* at the beginning of an utterance can be clipped onto treelet (12c) to form the beginning of an NP:

(13)

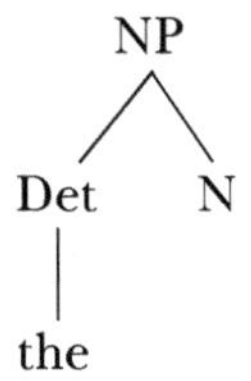

This can be clipped into (12a) to form the beginning of an S:

(14)

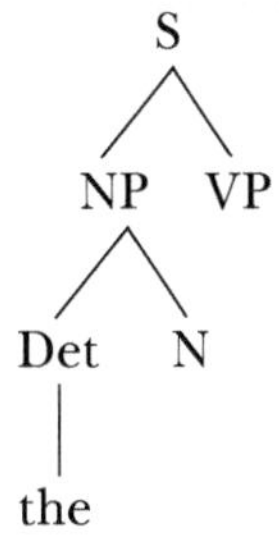

The S in turn has a VP onto which can be clipped treelet (12b), and words can be clipped into these trees in any order. Moreover, there is no need to *derive* phonology and semantics from syntax; they can be treated as quasi-independent but *correlated* with syntax (Jackendoff, 2007b).

Thus these approaches lend themselves better to theories of processing, in that the construction of sentence structure in comprehension can proceed from phonology to semantics, and in production the other way around. Syntactic structure in particular can be built from the top down, the bottom up, and/or left to right, all of which are illustrated in this example.

These theories still must account for those phenomena that the procedural theory treats as movement. The techniques are somewhat different from one approach to the next, but the general idea is that among the templates for the language is a characterization of the surface form of such constructions, along the lines of (15). No movement is involved. Rather, the gap is simply occupied by an unpronounced pronoun (a "trace" or another formal device), labeled *t*, that links this syntactic position with the constituent at the front (Culicover & Jackendoff, 2005; Pollard & Sag, 1994; Sag, 1997). The dashed line in (15) indicates that the trace is *somewhere* beneath the VP, but not necessarily directly beneath it. The constraints illustrated in (11), to the extent that they are truly syntactic, are implemented as conditions on exactly where in the tree the trace is allowed to be in relation to its antecedent.

(15)

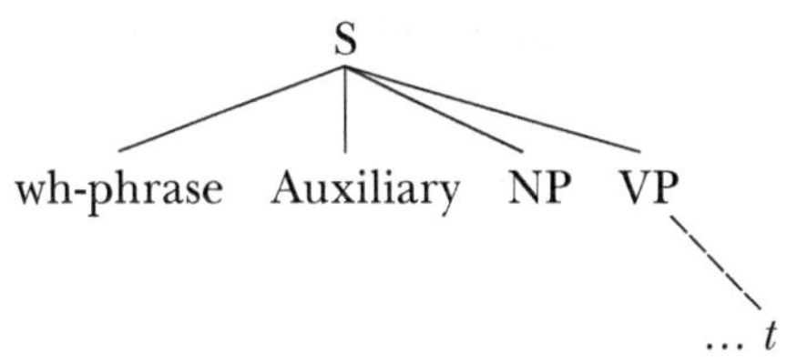

As for ellipsis, these theories avoid "underlying forms" in syntax that spell out the understood semantics. For instance in (8d) (*Mike bought a blue bike, and Sally did too*), the semantics of *bought a blue bike* is present in the conceptual structure of the second clause, but it has no syntactic counterpart. Rather, the elliptical construction is treated like a pronoun—a minimal piece of syntax that receives its interpretation by virtue of standing in some relation to an antecedent, in this case the first clause. For details see Culicover & Jackendoff (2005), Jackendoff & Culicover (2012).

6. Mismatches between Levels and a View of the Lexicon

An important advantage of constructional approaches is that they are tolerant of mismatches between levels. We saw in section 4 a mismatch between phonology and syntax, where the single word *bought* expresses both the verb *buy* and the past tense. Such mismatches are rampant in inflectional morphology. For instance, the German word *deinem* 'your', expresses at once possession of something that is singular, in dative case, and either masculine or neuter gender (the contribution of *-em*); moreover, the possessor is second person singular (the contribution of *dein-*)—a considerable amount of morphosyntactic information packed into a single phonological word.

Similarly, language is full of mismatches between syntax and semantics. The simplest such phenomena are idioms such as *spill the beans* and *a breath of fresh air.* These have canonical syntax but a meaning that is at best metaphorical. Hence idioms have to be stored as a whole, as lexical items with phrasal syntactic structure. A derivational theory, where only simple words are inserted at the bottom of syntactic trees, is difficult to reconcile with such stored phrasal expressions. In contrast, a constructional theory can store phrasal items in the lexicon and clip them into composed structure in exactly the same way as words.

Another class of syntax-semantics mismatches are phenomena in which the meaning of an expression is more than the meanings of its parts—so-called *enriched composition.* A classic case (Talmy, [1978] 2000) is *aspectual coercion,* in which there is a sense of repetition without any word that expresses it (16a); another is *reference transfer* (Nunberg, 1979), in which an NP refers not to its normal referent but to an individual associated with that referent (16b).

(16) a. The light flashed until dawn. (multiple flashes)
b. [One waitress to another:] The ham sandwich wants some more coffee. (*ham sandwich*= "person with the ham sandwich")

These options are easy to incorporate into constructional theories as pieces of semantics that need not have a syntactic counterpart; they are not so readily formulated in terms of syntactic deletions (Culicover & Jackendoff, 2005; Jackendoff, 1997).

Another important feature, shared by Construction Grammar and Parallel Architecture, is that they recognize schemas (or *constructions*) that specify an idiosyncratic interpretation for a stipulated syntactic configuration. Thus there are pieces of meaning that come not from particular words but rather from the configuration as a whole. Three examples appear in (17).

(17) a. Bill won't eat clams, let alone lobster. (Fillmore, Kay, & O'Connor, 1988)
b. The more you eat, the fatter you get. (Culicover & Jackendoff, 2005)
c. Sam sneezed the handkerchief off the table. (Goldberg, 1995)

The cumulative effect of all these mismatches is to blur the traditional distinction between words and rules of grammar. On one hand, as suggested in section 3, words can be regarded as interface rules that connect the three levels of representation. On the other hand, rules of grammar, stated as templates/constructions/schemas, can also be regarded as lexical items—pieces of stored linguistic structure that can be clipped into larger structures. The lexicon is therefore heterogeneous. First, treelets like (12) and (15) specify syntactic structure without any accompanying semantics. Second, idioms are listed items with internal syntactic structure, as though generated by rule, but their interpretation is not (entirely) derived from their constituent words. Third, constructions like those illustrated in (17) specify a template for composite syntactic structure and for a special interpretation, but either this interpretation is signaled by some minor element (*let alone* in (17a) and *the...-er, the...-er* in (17b)) or it is entirely implicit in the structure, as in (17c).

From the perspective of constructionist theories, then, a radical conclusion emerges from these observations: there are not really separate components of words and rules. A lexical item is more "word-like" to the extent that it specifies its full content, as in *sheep.* It is more "rule-like" to the extent that it has open slots, as in the treelets in (12) and (15). Between the two extremes are all sorts of conventionalized structures such as idioms, enriched compositions, and the constructions illustrated in (17), which specify pieces of phonology, syntactic structure, and meaning but also have slots (or variables) to be filled by other material. Thus in effect the grammar is *within* the lexicon.[12]

7. The Challenge for Psycholinguistics and Neuroscience

I want to close by outlining the challenges that this very superficial survey of linguistic representations poses for psycholinguistics and the neuroscience of language.

All of the many linguistic theories in the literature take it as a given that the human language capacity involves the employment of structures along the lines discussed in sections 3 and 4—and that are often far more complex. These structures are justified by a wide range of facts and generalizations about the languages of the world. The notations may differ from theory to theory, but they are all motivated by the same insights about language.

Constructionist theories, unlike derivational theories, lend themselves to implementation in theories of processing, thanks to (a) the free order of assembly, (b) the acknowledgment of mismatches among levels, and (c) the treatment of rules of grammar as lexical items in their own right. The challenge to theories of processing is to try to meet linguistic theory halfway, for instance, striving to be totally explicit about what levels of representation are being invoked at any particular moment and how these levels interact. One might hope, for instance, that if the linguistic theory is correct, it might predict that more complex syntax-semantics correspondences would result in more intense processing. And there is promising evidence that this prediction is correct for the Parallel Architecture (e.g., Paczynski, Jackendoff, & Kuperberg, 2014; Wittenberg et al., 2014; Wittenberg & Piñango, 2011).

In addition, the structures posited by linguistic theory must somehow be instantiated in the brain. Jackendoff (2002, section 3.5) lays out some of the challenges on this front. In particular, linguistic structures cannot be reduced to vectors or probability distributions. Gallistel and King (2009) make a similar argument for mental capacities far removed from language, e.g., dead reckoning.) At the time of this writing, virtually nothing is known about how neural architecture can support mental representations of this sort, especially hierarchical structures of the sort in figure 2.3, let alone the detailed connections between the disparate representations. There are at least three separate issues:

1. How are these structures instantiated in memory?
2. How are novel structures built in real time from smaller structures stored in memory?
3. How do the structures in memory get there; in other words, how are new structures added to memory?

These are basic questions of linguistic theory, translated into neural terms.

Note that discovering the spatial and temporal localization of linguistic activity in the brain, while valuable, does not completely answer these questions. It is indeed useful to find out *where* or *when* the structures are stored, built, or learned, as revealed by brain imaging. The far deeper challenge is to discover how the neural circuitry works in detail, such that it instantiates these structures and the operations on them. Even if this challenge is nowhere near being met at present, it is important to keep it in mind. Sufficiently sophisticated computational modeling may be the source of insight. In any event, I cannot stress enough that the complexity of linguistic structure must not be sacrificed.

NOTES

1. This is necessarily a personal view of a large and complex field; my work (Jackendoff 2002, 2011, 2012b) deals with these issues in much greater detail. I am grateful to Jenny Audring, Peter Culicover, Antje Meyer, James McQueen, and an anonymous reviewer for useful comments on an earlier version of this chapter.
2. Unfortunately, such interplay is sometimes discouraged in linguistics. Chomsky's well-known distinction between "competence" and "performance" (Chomsky, 1965; Chomsky & Lasnik, 1993) has the effect of insulating linguistic theory from experimental evidence; and many other linguistic theories are implicitly or explicitly indifferent about the relation of linguistic representations to the mind/brain. Linguistic theory has indeed uncovered many complex phenomena that must be accounted for in theories of processing and acquisition, and much research in psycholinguistics addresses such phenomena. However, the influence tends to be only one-way.
3. Mental functions are often characterized in terms of flow diagrams in which one box—or brain area—has an arrow leading to another box or brain area. The intended interpretation is that information is transmitted from one functional unit to another. (A typical example appears in Roelofs and Ferreira [chapter 4 of this volume].) However, this conception is problematic: the information processed by one component, such as semantics, is useless to another component, such as syntax, which traffics only in its own types of representations. A more accurate way to understand the arrows in figure 2.1 is that they stand for interfaces that establish a potentially complex mapping between one representational space and another.
4. This is by no means a consensual view in the literature. Mainstream generative grammar (Chomsky 1965, 1981, 1995) assumes the primacy of syntactic structure, while the other two structures are derived from syntax, in effect treating the arrows emanating from syntax as unidirectional. For a different view, Cognitive Grammar and its variants (Langacker, 1987, Lakoff, 1987) claim to derive all allegedly syntactic phenomena from semantics, in effect eliminating an independent syntax.
5. Notice that, strictly speaking, there is no independent "lexical level" here, as often invoked in psycholinguistics; a canonical "lexical item" such as a word is an association of structures in the three domains.
6. Sign languages use a set of distinctive features having to do with gesture: handshape, location, and motion of the

signing hand, use of the nondominant hand, and facial and postural cues (Özyürek & Woll, chapter 6 of this volume).

7. This characterization deviates from perhaps the predominant school of thought in semantics, growing out of the philosophical tradition, namely *formal* or *truth-conditional* or *model-theoretic* semantics (e.g., Heim & Kratzer, 1998; Portner & Partee, 2002). This approach seeks to explicate linguistic meaning directly in terms of reference to the world or to possible worlds, or to set-theoretic models of these, without the mediation of the minds of language users. It therefore makes no contact with psychological or cognitive concerns; it is not a theory of mental representations. To be sure, phenomena explored in this tradition, such as quantification and modality, have not been without interest to psycholinguists. Again, however, the influence is largely only one-way. (See Jackendoff [1983, 1990, 2002] for critiques of truth-conditional semantics.) Formal semantics amplifies the notation *SHEEP* to λx.sheep(x), 'is a sheep', and the factors mentioned in the next paragraph might be treated as truth-conditions or meaning postulates applying to this predicate.
8. In particular, it should be clear that meaning cannot be mentally encoded in terms of statistics of co-occurrence in a corpus, as in Latent Semantic Analysis (Landauer & Dumais, 1997), for instance. The co-occurrence of words in a corpus is a *symptom* of their meaning, and for certain purposes it may function as a proxy for the meaning, but it is not the meaning itself.
9. In the discussion of *sheep* in section 3, I noted that traditional notation encloses the three components together in large square brackets. Figure 2.3 shows why I abandon it here in favor of co-indexation: once the word takes part in larger structures, it becomes quite inconvenient to keep the components together in brackets.
10. In the Minimalist Program, movement is treated as a special case of Merge.
11. An in-between approach is Tree-Adjoining Grammar (Joshi & Schabes, 1997), which builds structure from "treelets" but maintains an ordering of rule application. Another in-between approach, primarily in phonology, is Optimality Theory (Prince & Smolensky, 2004), which applies constraints simultaneously to determine an "output" from an "input." The operation of clipping templates together is more precisely termed *unification,* a sort of Boolean union defined over structures rather than sets (Shieber, 1986). For instance, the unification of the string ABCD with the string CDEF is ABCDEF, with one instance of the parts they have in common and including all the parts that each has uniquely.
12. Most variants of Construction Grammar go further and claim that all lexical items are complexes of form (syntactic and phonological structure) and function (in present terms, conceptual structure). This requires that even basic syntactic templates like (12) are associated with an interpretation. Construction Morphology and Parallel Architecture are somewhat more flexible, in that they allow for lexical items that have no semantics—not only syntactic templates such as (12), but also, for instance, meaningless case-marking (nominative and accusative) and words such as the *do* of *do*-support (as in *What* did *Joe buy?*), which is present only to give the tense something to attach to.

REFERENCES

Biederman, I. (1987). Recognition-by-components: A theory of human image understanding. *Psychological Review, 94,* 115–147.

Bolinger, D. (1965). The atomization of meaning. *Language, 41,* 555–573.

Booij, G. (2010). *Construction morphology.* Oxford: Oxford University Press.

Bratman, M. (1987). *Faces of intention.* Cambridge: Cambridge University Press.

Bresnan, J. (2001). *Lexical functional syntax.* Oxford: Blackwell.

Browman, C., & Goldstein, L. (1986). Towards an articulatory phonology. *Phonology Yearbook, 3,* 219–252.

Chomsky, N. (1957). *Syntactic structures.* The Hague: Mouton.

Chomsky, N. (1965). *Aspects of the theory of syntax.* Cambridge, MA: MIT Press.

Chomsky, N. (1977). On wh-movement. In P. W. Culicover, T. Wasow, & A. Akmajian (Eds.), *Formal syntax* (pp. 71–132). New York: Academic Press.

Chomsky, N. (1981). *Lectures on government and binding.* Dordrecht: Foris.

Chomsky, N. (1995). *The minimalist program.* Cambridge, MA: MIT Press.

Chomsky, N. (2000). *New horizons in the study of language and mind.* Cambridge: Cambridge University Press.

Chomsky, N., & Halle, M. (1968). *The sound pattern of English.* New York: Harper & Row.

Chomsky, N., & Lasnik, H. (1993). The theory of principles and parameters. In J. Jacobs, A. von Stechow, W. Sternefeld, & T. Venneman (Eds.), *Syntax: An international handbook of contemporary research* (pp. 506–569). Berlin: Walter de Gruyter.

Clark, H. H. (1996). *Using Language.* Cambridge: Cambridge University Press.

Clements, G. (1985). The geometry of phonological features. *Phonology Yearbook, 2,* 225–252.

Clements, G., & Ridouane, R. (Eds.). (2011). *Where do phonological features come from?* Amsterdam: Benjamins.

Cohn, N. (2013). *The visual language of comics.* London: Bloomsbury.

Cohn, N., Jackendoff, R., Holcomb, P., & Kuperberg, G. (2014). The grammar of visual narrative: Neural evidence for constituent structure in sequential image comprehension. *Neuropsychologia, 64,* 63–70.

Croft, W. (2001). *Radical construction grammar.* Oxford: Oxford University Press.

Culicover, P. W. (2013). *Grammar and complexity.* Oxford: Oxford University Press.

Culicover, P. W., & Jackendoff, R. (2005). *Simpler syntax.* Oxford: Oxford University Press.

Davis, H. (2009). Cross-linguistic variation in anaphoric dependencies: Evidence from the Pacific Northwest. *Natural Language and Linguistic Theory, 27,* 1–43.

Dehaene, S. (2009). *Reading in the brain.* New York: Penguin Viking.

Dennett, D. C. (1996). *Kinds of minds.* New York: Basic Books.

Dowty, D. (1991). Thematic proto-roles and argument selection. *Language, 67,* 547–619.

Erteschik-Shir, N. (1997). *The dynamics of focus structure.* Cambridge: Cambridge University Press.

Ferreira, F. (2005). Psycholinguistics, formal grammars, and cognitive science. *Linguistic Review, 22,* 365–380.

Fillmore, C., Kay, P., & O'Connor, C. (1988). Regularity and idiomaticity in grammatical constructions: The case of *let alone*. *Language, 64*, 501–539.

Fillmore, C., Lee-Goldman, R. & Rhodes, R. (2012). The FrameNet constructicon. In H. Boas, & I. Sag (Eds.), *Sign-based construction grammar* (pp. 309–372). Stanford: CSLI.

Fodor, J. A. (1983). *Modularity of mind*. Cambridge, MA: MIT Press.

Gallistel, C. R., & King, A. (2009). *Memory and the computational brain: Why cognitive science will transform neuroscience*. Malden, MA: Wiley-Blackwell.

Gilbert, M. (1989). *On social facts*. Princeton, NJ: Princeton University Press.

Goffman, E. (1974). *Frame analysis*. Cambridge, MA: Harvard University Press.

Goldberg, A. (1995). *Constructions: A construction grammar approach to argument structure*. Chicago: University of Chicago Press.

Hagoort, P., Hald, L., Bastiaansen, M., & Petersson, K. M. (2004). Integration of word meaning and world knowledge in language comprehension. *Science, 304*, 438–441.

Hale, K. (1992). Subject obviation, switch reference, and control. In R. K. Larson, S. Iatridou, U. Lahiri, & J. Higginbotham (Eds.), *Control and grammar* (pp. 51–78). Dordrecht: Kluwer Academic.

Hale, K., & Keyser, S. J. (2002). *Prolegomena to a theory of argument structure*. Cambridge, MA: MIT Press.

Halle, M., & Marantz, A. (1993). Distributed morphology. In K. Hale & S. J. Keyser (Eds.), *The view from building 20* (pp. 111–176). Cambridge, MA: MIT Press.

Heim, I., & Kratzer, A. (1998). *Semantics in generative grammar*. Oxford: Blackwell.

Hoffmann, T., & Trousdale, G. (2013). *The Oxford handbook of construction grammar*. Oxford: Oxford University Press.

Hofmeister, P., & Sag, I. (2010). Cognitive constraints and island effects. *Language, 86*, 366–415.

Humboldt, Wilhelm von. (1999 [1836]). *On the diversity of human language construction and its influence on the mental development of the human species* (Originally published as *Über die Verschiedenheit des menschlichen Sprachbaus und seinen Einfluss auf die geistige Entwicklung des Menschengeschlechts*). Michael Losonsky (Ed.). Cambridge: Cambridge University Press.

Jackendoff, R. (1983). *Semantics and cognition*. Cambridge, MA: MIT Press.

Jackendoff, R. (1990). *Semantic structures*. Cambridge, MA: MIT Press.

Jackendoff, R. (1991). Parts and boundaries. *Cognition, 41*, 9–45.

Jackendoff, R. (1997). *The architecture of the language faculty*. Cambridge, MA: MIT Press.

Jackendoff, R. (2002). *Foundations of language*. Oxford: Oxford University Press.

Jackendoff, R. (2007a). *Language, consciousness, culture*. Cambridge, MA: MIT Press.

Jackendoff, R. (2007b). A parallel architecture perspective on language processing. *Brain Research, 1146*, 2–22.

Jackendoff, R. (2011). What is the human language faculty? Two views. *Language, 87*, 586–624.

Jackendoff, R. (2012a). Language as a source of evidence for theories of spatial representation. *Perception* [Special issue], *41*(9), 1128–1152 (commemorating 30th anniversary of David Marr's *Vision*).

Jackendoff, R. (2012b). *A user's guide to thought and meaning*. Oxford: Oxford University Press.

Jackendoff, R., & Audring, J. (2019). *The texture of the lexicon*. Oxford: Oxford University Press.

Jackendoff, R., & Culicover, P. W. (2012) Same-except: A domain-general cognitive relation and how language expresses it. *Language, 88*, 305–340.

Jackendoff, R., & Lerdahl, F. (2006). The capacity for music: What is it, and what's special about it? *Cognition, 100*, 33–72.

Jakobson, R., Fant, G., & Halle, M. (1952). *Preliminaries to speech analysis: The distinctive features and their correlates*. Cambridge, MA: MIT Press.

Joshi, A., & Schabes, Y. (1997). Tree-adjoining grammars. In G. Rosenberg & A. Salomaa (Eds.), *Handbook of formal languages* (pp. 69–124). Berlin: Springer-Verlag.

Kluender, R. (1998). On the distinction between strong and weak islands: A processing perspective. In P. Culicover & L. McNally (Eds.), *Syntax and semantics Series: Vol. 29. The limits of syntax* (pp. 241–279). San Diego: Academic Press.

Koffka, K. (1935). *Principles of gestalt psychology*. New York: Harcourt Brace & World.

Kosslyn, S. (1980). *Image and mind*. Cambridge, MA: Harvard University Press.

Kosslyn, S., & Koenig, O. (1995). *Wet mind: The new cognitive neuroscience*. New York: Free Press.

Kuno, S., & Takami, K.-I. (1993). *Grammar and discourse principles: Functional syntax and GB theory*. Chicago: University of Chicago Press.

Ladefoged, P. (2007). Articulatory features for describing lexical distinctions. *Language, 83*, 161–180.

Lakoff, G. (1987). *Women, fire, and dangerous things*. Chicago: University of Chicago Press.

Lakoff, G., & Johnson, M. (1980). *Metaphors we live by*. Chicago: University of Chicago Press.

Lambrecht, K. (1994). *Information structure and sentence form*. Cambridge: Cambridge University Press.

Landau, B., & Jackendoff, R. (1993). "What" and "where" in spatial language and spatial cognition. *Behavioral and Brain Sciences, 16*, 217–238.

Landauer, T. K., & Dumais, S. T. (1997). A solution to Plato's problem: The latent semantic analysis theory of the acquisition, induction, and representation of knowledge. *Psychological Review, 104*, 211–240.

Langacker, R. (1987). *Foundations of cognitive grammar* (Vol. 1). Stanford, CA: Stanford University Press.

Lerdahl, F., & Jackendoff, R. (1983). *A generative theory of tonal music*. Cambridge, MA: MIT Press.

Marr, D. (1982). *Vision*. San Francisco: Freeman.

Merchant, J. (2001). *The syntax of silence*. Oxford: Oxford University Press.

Minsky, M. (1975). A framework for representing knowledge. In P. H. Winston (Ed.), *The psychology of computer vision* (pp. 211–277). New York: McGraw-Hill.

Nunberg, G. (1979). The non-uniqueness of semantic solutions: Polysemy. *Linguistics and Philosophy, 3*, 143–184.

Paczynski, M., Jackendoff, R., & Kuperberg, G. (2014). When events change their nature: The neurocognitive mechanisms underlying aspectual coercion. *Journal of Cognitive Neuroscience, 26*, 1905–1917.

Phillips, C. (2013). On the nature of island constraints I: Processing and reductionist accounts. In J. Sprouse & N. Hornstein (Eds.), *Experimental syntax and island effects* (pp. 64–108). Cambridge: Cambridge University Press.

Phillips, C., & Lau, E. (2004). Foundational issues [Review of the book *Foundations of language*, by R. Jackendoff]. *Journal of Linguistics, 40,* 1–21.

Poeppel, D., & Embick, D. (2005). Defining the relation between linguistics and neuroscience. In A. Cutler (Ed.), *Twenty-first century psycholinguistics: Four cornerstones* (pp. 103–118). Hillsdale, NJ: Lawrence Erlbaum.

Pollard, C., & Sag, I. (1994). *Head-driven phrase structure grammar.* Stanford: CSLI.

Portner, P., & Partee, B. (Eds.). (2002). *Formal semantics: The essential readings.* Oxford: Blackwell.

Prince, A., & Smolensky, P. (2004). *Optimality theory: Constraint interaction in generative grammar.* Malden, MA: Blackwell.

Putnam, H. (1976). The meaning of "meaning." In K. Gunderson (Ed.), *Language, mind, and knowledge* (pp. 131–193). Minneapolis: University of Minnesota Press.

Ross, J. R. (1967). *Constraints on variables in syntax* (Doctoral dissertation, Massachusetts Institute of Technology). Published as *Infinite syntax.* Norwood, NJ: Ablex Publishing Corporation, 1986.

Sag, I. (1997). English relative clause constructions. *Journal of Linguistics, 33,* 431–484.

Schank, R., & Abelson, R. (1975). *Scripts, plans, goals, and knowledge.* Hillsdale, NJ: Erlbaum.

Shieber, S. (1986). *An introduction to unification-based approaches to grammar.* Stanford: CSLI.

Siddiqi, D. (2019). Distributed morphology. In J. Audring & F. Masini (Eds.), *The Oxford handbook of morphological theory.* Oxford: Oxford University Press.

Talmy, L. (2000). The relation of grammar to cognition: A synopsis (Rev. version). In L. Talmy (Ed.), *Toward a cognitive semantics* (pp. 21–96). Cambridge, MA: MIT Press. (Originally published in *Theoretical issues in natural language processing 2* (pp.14–24), by D. Waltz (Ed.), 1978.) New York: Association for Computing Machinery.)

Van Valin, R. (1998). The acquisition of wh-questions and the mechanisms of language acquisition. In M. Tomasello (Ed.), *The new psychology of language: Cognitive and functional approaches to language structure* (pp. 221–249). Hillsdale, NJ: Erlbaum.

Varela, F., Thompson, E., & Rosch, E. (1991). *The embodied mind.* Cambridge, MA: MIT Press.

Wertheimer, M. (1999). Laws of organization in perceptual forms. In W. Ellis, (Ed.), *A source book of gestalt psychology* (pp. 71–88). London: Routledge & Kegan Paul. (Originally published as Untersuchungen zur Lehre von der Gestalt II. *Psychologische Forschung, 4,* 301–350.)

Wittenberg, E., & Piñango, M. (2011). Processing light verb constructions. *Mental Lexicon, 6,* 393–413.

Wittenberg, E., Jackendoff, R., Kuperberg, G., Paczynski, M., Snedeker, J., & Wiese, H. (2014). The processing and representation of light verb constructions. In A. Bachrach, I. Roy, & L. Stockall (Eds.), *Structuring the argument: Multidisciplinary research on verb argument structure* (pp. 61–80). *Language Faculty and Beyond Series.* Amsterdam: John Benjamins.

Wittgenstein, L. (1953). *Philosophical investigations.* Oxford: Blackwell.

Wolf, M. (2007). *Proust and the squid: The story and science of the reading brain.* New York: HarperCollins.

Yip, M. (2002). *Tone.* Cambridge: Cambridge University Press.

3 Language Comprehension: Insights from Research on Spoken Language

DELPHINE DAHAN AND FERNANDA FERREIRA

Language is a system that gives physical form to meaning by combining discrete elements. In order to use language, people must learn the specific ways in which their language expresses meaning and recruit this knowledge when expressing the meanings they wish to convey or when retrieving those of others. This chapter discusses the architecture and the working of the machinery that enables listeners to derive the meaning of the utterances they hear.

Mainly for practical reasons, research in psycholinguistics has tended to focus on written language. But with the advent of speech technology and the increased availability of tools to edit and analyze speech, as well as to control the presentation of spoken stimuli, work in the area of spoken language comprehension has taken off. The goal of this chapter is to review how this research has advanced our knowledge of how people comprehend the utterances they are exposed to. Indeed, speech differs from print in a few but significant ways. Some of these differences have brought unique insight to existing debates, while others have led to the discovery of phenomena that pertain to speech specifically.

1. Prosody and the Architecture of the Language Processor

The comprehension of a sentence requires the listener to derive several linguistic structures. Each one operates over specific categories, is guided by distinct rules or principles constraining how categories combine, and makes a unique contribution to the meaning of the sentence. In the context of speech comprehension, the process has been described as a "conversion of acoustic input into meaning" (Cutler & Clifton, 1999, p. 123), and is traditionally described as involving separate (i.e., modular) and hierarchically organized stages (see Cutler & Clifton, 1999; Hagoort & Poeppel, 2013, for reviews largely rooted in the traditional approach). According to such a view, each stage takes as input the outcome of the one that precedes it in order to generate the relevant structure. The data reduction that this architecture entails optimizes the efficiency of the computations taking place at each stage. Before discussing this approach and its alternatives, we briefly describe the nature of the structures generated at each of the stages.

The putative first step takes the acoustic signal as input and infers the phonological structure of the utterance. One aspect of the structure consists of a sequence of sound categories or phonemes. These categories capture the functionally relevant contrasts in the sound system of the language and generalize across contexts, speakers, or speaking genres. In addition to this linearly organized phonemic string, the phonological structure represents the hierarchically organized phrasing and prominence structure of these phonemic elements organized into a prosodic structure.

The sequence of phonemes, in conjunction with its prosodic structure, may then be segmented out to retrieve the morphosyntactic elements and their relationships. Retrieval of the syntactic properties associated with each of these elements, along with the order in which they appear, gives rise to a structure that captures the groupings of the elements into larger constituents and the relationships that these constituents entertain with one another in a hierarchically organized fashion, in other words, in the syntactic structure of the sentence. Closely related but not identical to the syntactic structure, the semantic structure marks the meaning-based relationships between the morphosyntactic elements, based on both their syntactic relationships and the semantic properties associated with each element. Thus, the semantic structure relies on the constituency captured in the syntactic structure.

In this traditional model, the meaning of the sentence, represented in its semantic structure, is then integrated into the discourse and its broader context to yield its information structure. The information structure marks which parts of the sentence meaning are highlighted and brought to the foreground within the broader discourse and which ones are known or presupposed and left to the background. Information structure also captures some attitudes of the speaker attributed to himself or herself or to the hearer regarding parts of the sentence's meaning. For instance, Steedman (2014) pointed out that the sentence *You put my trousers in the microwave* may assert the proposition, in effect marking it as part of the knowledge common to both interlocutors, but it may also be used to convey disbelief by marking the proposition as not becoming part of that common knowledge (see also Kurumada, Brown, Bibyk, Pontillo, & Tanenhaus, 2014).

The traditional, modular and serial, architecture of the language comprehension system just described has been under attack for several decades (e.g., Rumelhart, McClelland, & PDP Research Group, 1986) and we refer the reader to other sources for a report on the debate (e.g., Joanisse & McClelland, 2015). Our goal here is to describe the theoretical contributions that research on spoken language comprehension, and prosody in particular, has made to the debate.

Prosodic representations capture the utterance's phrasing and prominence. For the most part, research on written language has not been concerned with prosody under the assumption that it plays little or no role during reading. Although there is now evidence that people spontaneously generate a prosodic structure to sentences that they read (Breen, 2014; Jun, 2010; Stolterfoht, Friederici, Alter, & Steube, 2007; Swets, Desmet, Hambrick, & Ferreira, 2007), it remains the case that an utterance makes its prosody immediately available to listeners, whereas a written sentence does not. Access to prosodic information is significant because the prosodic structure of a sentence is intimately related to, and constrained by, its phonological, syntactic, and information structures (e.g., Shattuck-Hufnagel & Turk, 1996; see Dahan, 2015, for a recent review). This relationship is best illustrated by the fact that prosody can reveal aspects of these structures. For example, the phonemically ambiguous phrases *crisis turnip* and *cry sister nip* can be distinguished from each other based on the prosodic rendering that the speaker adopts because a different prosodic phrasing is associated with each of the two sequences of morphosyntactic elements (e.g., Dilley, Mattys, & Vinke, 2010). Likewise, the sentences *Raoul murdered the man with a gun* and *I asked the teacher who left* have more than one literal meaning and speakers may be able to reveal which one they intend through the sentence's prosody (Schafer, Carlson, Clifton, & Frazier, 2000; Schafer, Speer, & Warren, 2005; Snedeker, & Trueswell, 2003). For instance, the presence of a prosodic boundary before *with a gun* increases the likelihood that listeners interpret *with a gun* as referring to the instrument of the murder, while the absence of such boundary biases them to interpret *with a gun* as a modifier of the noun phrase *the man*. These two interpretations correspond to syntactic structures that differ, among other things, in the way *the man* and *with a gun* are syntactically related. Numerous studies have shown that listeners are exquisitely sensitive to the prosodic cues that differentiate interpretations across the many contrasts that have been examined (e.g., Wagner & Watson, 2010).

The influence of prosody on multiple linguistic structures is a challenge for the traditional approach because the strictly sequential ordering of the stages that it assumes limits the dependencies (or interfaces) between structures it can account for. For instance, the traditional approach captures the fact that the syntactic structure depends on identifying the morphosyntactic elements of the sentence, and their identity is largely constrained on the retrieval of phonemic categories; likewise, the information structure is often viewed as assigning information status to the syntactic constituents that have been established at an earlier stage. However, research on prosody has revealed that interfaces between structures cannot be fully captured by the linear ordering that the traditional approach embodies. We describe three kinds of challenges that this research has raised.

First, there is ample evidence that elements of the phonological structure affect the establishment of both the syntactic structure and that of the informational structure. For instance, the informational structure of a sentence is constrained by the presence or absence of prominence on some constituents, elements that are encoded in the phonological structure (e.g., Steedman, 2014). For instance, the utterance *I would love some COFFEE* (with an accent on *coffee*) is ill-fitted to the preceding question *Would you like some coffee?* but very appropriate following the question *What would you like to drink?*. The two contexts differ in the information status of *coffee*: The first context mentions it explicitly; by contrast, *coffee* is what is being inquired about in the second context. It is unclear how that distinction is maintained if the mapping of the phonological structure onto the information structure is mediated solely through the syntactic structure because the syntactic structure, as traditionally formalized, lacks the relevant categories.

Another challenge to sequential ordering arises from cases that suggest the influence of a late-built structure on an earlier one. Consider, for example, the ambiguous sentence *I asked the pretty little girl who is cold*. In one interpretation of the sentence, *who* is a relative pronoun and *who is cold* a relative clause that specifies which girl is under discussion. In another interpretation, *who* is a wh-word and *who is cold* an embedded question and the object of the asking. As shown by Schafer et al. (2000), the presence of an accent on *who* biases listeners toward an embedded question interpretation whereas its absence biases them toward a relative clause one. This bias, the authors persuasively argued, reflects the fact that in English, focused elements often receive a pitch accent. Because a questioned constituent is by definition in focus, listeners are more likely to interpret *who* as a wh-word standing for a questioned constituent if it carries a pitch accent than if it does not. Importantly here, each interpretation is associated with a different syntactic structure. The example illustrates the fact that the syntactic structure can depend on the information structure (itself informed by the sentence's prominence pattern and encoded in its phonological structure). This dependency cannot be accounted for if the establishment of syntactic structure strictly precedes that of the information structure. The dependencies between these two structures appear to go both ways. Further evidence of bidirectionality between structures comes from demonstrations that listeners are more likely to hear the presence of a major prosodic boundary in locations where such a boundary is licensed by the syntax, even when the acoustic correlates for the prosodic boundary are weak or absent (Buxó-Lugo & Watson, 2016; Cole, Mo, & Baek, 2010). Thus, the prosodic structure influences and is influenced by the syntactic structure.

Finally, there is mounting evidence for interdependencies between structures such that the way a given portion of the sentence is accounted for in one structure affects the way it is accounted for in another structure whose establishment allegedly precedes the former. Examples of such interdependencies come from Clifton, Carlson, and Frazier (2006). Their study examined the role that the presence or absence of a prosodic phrase boundary, encoded in the phonological structure of the sentence, has on its syntactic structure. As predicted by the sequential and unidirectional approach to sentence interpretation, Clifton et al. showed that in sentences such as *Pat or Jake and Lee convinced the bank president to extend the mortgage*, the presence of a prosodic boundary after *Pat* leads people to assume that either one person (i.e., Pat) or two people (i.e., Jake and Lee) talked to the banker; by contrast, a prosodic boundary after *Jake* leads people to assume that two people talked to the bank, Lee and someone who was either Pat or Jake. An account of this effect within the traditional approach has been offered by Watson and Gibson's (2005) *antiattachment* hypothesis, according to which the presence of the prosodic boundary is interpreted as "a strong cue not to attach the upcoming word to the last potential attachment site before the boundary" (p. 285).

However, Clifton et al. (2006) also showed that when the sentence was modified to include longer noun phrases to refer to the individuals, as in *Patricia Jones or Jacqueline Frazier and Letitia Connolly convinced the bank president to extend the mortgage*, the effect of the pause location on people's interpretation was significantly reduced. As proposed by the authors, listeners appear to consider multiple influences that operate on the production of prosodic phrase boundaries, one originating from the syntactic structure and one associated with the phonological structure and the need to produce phrases of a manageable length. The more the presence of a pause is attributed to the phonological factor, that is, the length of the phrases, the less it is attributed to the syntactic factor. Such a trade-off relationship is difficult to account for in the traditional architecture because the stages are linearly organized and operate on different categories, with syntactic constituency assumed to be impervious to the constituents' phonological contents.

Another illustration of interdependency across structures is offered in Brown, Salverda, Gunlogson, and Tanenhaus (2015). Their investigation built on Salverda, Dahan, and McQueen (2003), who showed that a syllable is more likely to be interpreted as a monosyllabic word (e.g., pump) rather than the initial syllable of a polysyllabic word (e.g., pumpkin) when that syllable is long than when it is short. This effect is attributed to listeners' knowledge that syllable lengthening is an acoustic correlate of a prosodic phrase boundary and such boundaries are necessarily aligned with word boundaries. But syllable lengthening is an acoustic correlate of other prosodic phenomena, including the presence of a pitch accent. Brown et al. showed that the tendency to interpret syllable lengthening as marking prosodic phrasing diminished when the context was conducive to interpreting it instead as marking a (contrastive) pitch accent. Because the domain of pitch accents is the stressed syllable, not the prosodic phrase, an accent can fall on a monosyllabic word as well as on the first syllable of a polysyllabic word (provided that the syllable is stressed). In their study, listeners' attribution of the lengthening of a stressed syllable to a pitch accent rather than to a prosodic phrase boundary was revealed by a diminished tendency to favor the monosyllabic word when the syllable was lengthened than when it was shortened.

How might a processing model of language comprehension capture interdependencies between linguistic levels? The notion of a unidirectional flow of information between hierarchically organized levels of analysis is difficult to maintain. One possibility is to view the phonological structure, which encompasses both the segmental and prosodic structures, as a representation of the sentence that listeners try to explain by deciding what the other structures need to be in order to motivate this phonological structure. Thus, comprehending a sentence is viewed as a data-explanation process, where listeners aim to select the meaning, represented by all the linguistic structures simultaneously, that best explains the data (i.e., the utterance they hear). There is no data reduction or seriality in the process. It is generative because it aims to discover or infer the underlying structures that generated the data. Such a model provides a natural way to capture the interdependencies between the various linguistic structures that listeners aim to establish for a given utterance. Indeed, an explanation favored on one such structure in order to "explain away" the realization of a sentence may compete with a hypothesis at another level if both claim to account for the same portion or characteristic of the phonetic signal.

This approach is not new: Stevens and Halle's (1967) *analysis-by-synthesis* belonged to this class of models. However, it has recently gained popularity in cognitive science. As discussed by Clark (2013) within the realm of perception and action, generative models explain how people build an interpretation of the world from input to their senses by selecting a set of hypotheses that best explain the sensory signal. These hypotheses are based on the observer's knowledge of what the signal would be like if it were generated by each hypothesis. Hypotheses represent causal structures that may have generated the signal. This view follows Helmholtz's (1860/1962) view of perception as a process of knowledge-driven and probabilistic inference. Bayesian models of perception and cognition embody this approach (e.g., Chater & Manning, 2006; Norris & McQueen, 2008). The interactive activation model, originally proposed by Rumelhart and McClelland (1982) and revised over the years, offers one implementation of these ideas (see McClelland, Mirman, Bolger, & Khaitan, 2014, for a recent review, and Griffiths, Chater, Kemp, Perfors, & Tenenbaum, 2010, for a discussion on the similarities and differences between Bayesian and connectionist models).

Generative models are being increasingly recognized in the psycholinguistic community. An example pertaining to the perception of prosody was offered by Frazier, Carlson, and Clifton (2006) and their *rational speaker* hypothesis: listeners use what they know to be the constraints on a speaker's production of prosodic structures to infer the underlying causes of the prosodic rendering of a sentence. This approach helps explain the fact that a prosodic phrase boundary is interpreted as the marking of the closure of a syntactic constituent more if the phonological phrase before the boundary is short (i.e., containing few words or syllables) than if the phonological phrase is long. According to the rational speaker hypothesis, listeners attribute the prosodic phrase boundary to the need to break a long syntactic phrase into more manageable constituents when uttering the sentence, rather than to the marking of syntactic structure. While the scope of the rational speaker hypothesis has remained limited, the underlying principles are very general and in line with the generative model approach.

In summary, the traditional approach to spoken language comprehension that set the stage for current models assumes a sequential, stage-by-stage processing architecture. As just reviewed, the unidirectionality with which information flows in such an architecture is at odds with numerous empirical findings. Generative models offer a viable alternative: in such models, listeners simultaneously identify the linguistic structures that best account for the phonetic and phonological characteristics of an utterance, with trading relationships between structures playing a role in explaining the sensory data.

2. *Speaker- and Context-Specific Adaptation and Plasticity in Speech Perception*

In the first section, we discussed prosody, an aspect that research on written language has largely ignored. In this section, we discuss another aspect that speech, but not print, conveys: the voice of the talker and the numerous aspects pertaining to his or her current emotions and attitudes, all encoded in much the same acoustic dimensions as those used to convey phonetic, and thus linguistic, information. This research, as we will describe, has radically changed our understanding of how listeners establish the phonological elements of an utterance. Interestingly, the type of modeling that scholars have gravitated toward in order to capture the new findings takes a data-explanation perspective, the same framework as the one we discussed in section 1.

As was pointed out, speech conveys much more than the words that a talker chooses to utter. However, comprehension relies in large part on retrieving the linguistic elements that compose an utterance. Because these elements were assumed to be abstract, context-independent categories, speaker normalization was long assumed to be the process by which these stable phonetic units could be retrieved. The goal was to remove the impact of speaker specificity from the acoustic signal (see Johnson, 2005, for a review). However, seminal work by Pisoni and colleagues (e.g., Goldinger, Pisoni, & Logan, 1991; Nygaard & Pisoni, 1998; Nygaard, Sommers, & Pisoni, 1994, 1995) showed that experience with a given talker facilitates the processing of subsequent speech from the same talker. This finding suggests two things. First, the retrieval of the phonological units of a sentence is dependent on the acoustic details with which these units are produced. Second, because the retrieval of these units becomes more effective with experience with a specific speaker, some form of learning takes place during speech perception. A large body of research has since documented listeners' remarkable plasticity in accommodating speaker-specific pronunciations (e.g., Clarke & Garrett, 2004; Eisner & McQueen, 2005; Kraljic & Samuel, 2006; Magnuson & Nusbaum, 2007; Maye, Aslin, & Tanenhaus, 2008; Norris, McQueen, & Cutler, 2003).

What changes, exactly, as a result of exposure to the utterances of a given speaker? In a feed-forward model that transforms the acoustic signal by reducing it to a sequence of context-independent phonological categories, learning could only affect the normalization process, in other words, the computations that carry out the data reduction. However, Dahan, Drucker, and Scarborough (2008) provided compelling evidence that the categories themselves are changed: Listeners first heard a talker produce words such as *bag* with a vowel that has moved toward and approaches that of *beg*, a feature found in some American English dialects and limited to *g* contexts. Another group of listeners heard standard pronunciations of *bag*. When listeners later heard words such as *back* with standard vowels produced by the same talker, which version of *bag* they had heard before affected their recognition of *back*: Listeners who had heard the nonstandard *bag* words were faster at identifying *back* as *back* than listeners who had heard standard *bag* words. If exposure to *bag* with a nonstandard vowel altered the normalization process, one would not expect to see an impact of this learning on the interpretation of words with the standard vowel such as *back*. If, on the other hand, people modified how they expect words such as *bag* to sound based on their experience with their nonstandard vowels, *bag* is no longer a plausible interpretation to consider based on hearing the beginning of the word *back* with a different vowel. Thus, it appears that people can modify their expectations of how words (or perhaps smaller linguistic categories such as vowels) are expected to sound when uttered by that talker.

What kinds of architecture may capture such learning? Here Bayesian modeling is a framework that lends itself to this kind of approach, and Kleinschmidt and Jaeger (2015) recently offered an elegant and computationally explicit model following this formalism. Bayesian models of speech perception describe the process as one in which people infer the phonetic (and, at a higher level, lexical) units of an utterance by selecting the hypotheses that explain the data best. Hypotheses can be conceptualized in listeners' minds as probability distributions over acoustic dimensions. Distributions are mathematical tools to capture the variability with which a given phonetic unit is acoustically realized within, as well as across, speakers and contexts: Values that this unit can take along these dimensions correspond to probability distributions. In this formalism, listeners adapt to a given speaker or dialect by establishing speaker- or dialect-specific distributions and relying on them when interpreting subsequent speech from the same source.

While listeners benefit from prior familiarity with talkers, we know that they are also perfectly able to understand speech from a speaker of their language they have no prior experience hearing, provided that the talker's idiolect is similar enough to that of known talkers. Thus, as pointed out by Kleinschmidt and Jaeger, speech perception is a process requiring listeners to recognize speech from a familiar talker, to generalize this acquired knowledge to similar talkers, but also not to generalize, and instead to adapt to novel speakers who differ significantly from familiar ones. What prompts listeners to establish and apply speaker-specific distributions, as opposed to resorting to "default" distributions? Kleinschmidt and Jaeger proposed that adaptation is triggered when the default distribution associated with a hypothesis favored by additional contextual cues predicts the observed data with low probability. This so-called *error signal* is what allows such models to *update their beliefs*, that is, to learn and adapt (see Norris, McQueen, & Cutler, 2016, for a similar view).

The data explanation approach naturally explains a host of phenomena that point to the flexibility with which listeners adjust to the context when inferring the phonological units present in a given utterance. For instance, Dilley and Pitt (2010; see also Baese-Berk et al., 2014) showed that whether people perceived the presence of a short function word in a phrase (e.g., *leisure and time* vs. *leisure time*) depends on the *perceived* duration of the fragment of speech comprising the syllable that precedes the potential function word site and the function word itself, if produced. If the fragment is perceived as too long to be a single syllable, given the speech rate, a function word is posited. Conversely, if the fragment is perceived as too short to include a function word, its presence goes largely undetected. In the light of this work, one must conclude that the segmental representation of an utterance is only partially determined by the spectral cues present in the signal; equally important are the temporal characteristics of the utterance to which listeners try to attribute lexical content. Furthermore, listeners do not hypothesize specific segments or phonemes solely based on the support for them in the signal per se. Instead, listeners infer the presence of a specific element based on their beliefs that the element *in its context* would have resulted in the signal. This finding is reminiscent of the evidence that people adjust their expectations of what a word should sound like given what they know about who said it.

Is there any evidence for plasticity and context-specific adaptation at higher linguistic levels? These ideas are receiving increased attention because they can be viewed as natural extensions of the now widely accepted view that much of the syntactic knowledge that people recruit when establishing the structure of sentences reflects the statistics of the language they have been exposed to (e.g., MacDonald, Pearlmutter, & Seidenberg, 1994). These statistics can be viewed as probability distributions associated with structures. It is thus natural to ask how contextually dependent these statistics may be.

There is some evidence that listeners can learn to comprehend and produce structures that are initially judged to be ungrammatical (Kaschak & Glenberg, 2000; Luka & Barsalou, 2005). Furthermore, evidence for such learning is not limited to unusual structures. Syntactic priming, in other words, the tendency for people to use a structure that they have recently used or heard used, is a phenomenon that illustrates such learning (e.g., Bock, 1986; see Pickering & Ferreira, 2008, for a review). Studies that capitalize on syntactic ambiguity (i.e., where more than one structure for the sentence is licensed, given the incomplete fragment of the sentence available to the listener or the reader) showed that repeated exposure to a given syntactic structure can also tip people to anticipate this structure more than they would otherwise (Fine & Jaeger, 2013; Thothathiri & Snedeker, 2008; Wells, Christiansen, Race, Acheson, & MacDonald, 2009), or to process the repeated structure more rapidly (e.g., Myslín & Levy, 2016). These findings suggest that people adjust the probability distributions associated with structures to reflect their experience most closely when this experience is contextually relevant.

Overall, there is a wide consensus that the representations that listeners bring to bear to their interpretation of speech are flexible and can be temporarily adjusted to contexts. Next, we discuss how these representations are dynamically recruited in order to comprehend speech incrementally.

3. Incremental Processing of Spoken Language

Contrary to written language, spoken language is fleeting and the sensory information does not remain available to the hearer for later inspection. Furthermore, listeners do not have control over the rate at which that information becomes available. As discussed by Christiansen and Chater (2016), these constraints are some of the reasons why listeners do not wait to have heard the complete sentence to build its structures and meaning; rather, they utilize partial, incomplete information to project these structures. For instance, the first sounds of an utterance may result in positing the presence of the fricative */th/* followed by the reduced vowel sometimes called *schwa*, and this unstressed syllable may be immediately interpreted as indicating the presence of the definite determiner *the*, which itself leads to the projection (or anticipation) of a noun phrase. Incremental processing of speech was perhaps best demonstrated by the work of Marslen-Wilson and colleagues (Marslen-Wilson, 1975; Marslen-Wilson & Tyler, 1980), and later confirmed through different paradigms (e.g., Allopenna, Magnuson, & Tanenhaus, 1998).

Furthermore, because elements of a sentence can entertain long-distance relationships with one another, the spoken signal must be encoded in a way that enables later inspection and manipulation in some type of short-term memory buffer. As argued by Christiansen and Chater (2016), the immediate interpretation and organization of the sentence in terms of constituents at multiple levels of linguistic analysis (i.e., the process they call *chunking*) enables this encoding.

Just as it has been proposed in other areas of cognition (Clark, 2013), human listeners are viewed as prediction engines who learn the detailed properties of their environments and use this knowledge to anticipate and predict what can follow. While this "educated guessing" appears naturally helpful before any information has been gained, it is also critical in order to interpret what is being heard because sensory cues and their neural encoding are inherently noisy and can therefore provide only probabilistic information regarding the objects of perception people try to retrieve. Integrating multiple sources of information together (i.e., each of the constraints that the context provides together with the spoken signal) helps compensate for the uncertainty associated with each individual source.

Three questions have been addressed with regard to incremental processing: First, what knowledge do listeners base their predictions on? We know that people interpret partial information within the context in which it is encountered, and this rich set of knowledge constrains people's predictions. This was illustrated by

Altmann and Kamide's (1999) study showing that people can anticipate which type of objects (or their properties) will be mentioned as the upcoming theme of a verb upon hearing a fragment of a sentence that includes the verb such as *The boy will eat the....* The prediction goes beyond the verb and incorporates semantic properties of the subject and the agent of the verb, as subsequently shown by Kamide, Altmann, and Haywood (2003). Thus, upon hearing *the girl will ride,* people are more likely to anticipate a carousel as the upcoming theme of *ride* than a motorcycle, while the opposite pattern is observed following *the man will ride.* The specificity of the anticipations suggests that people anticipate what may be coming at multiple levels of representation.

The second question concerns what happens when partial information is compatible with multiple continuations. An incomplete sequence is compatible with and predicts many possible continuations. Are all continuations considered and entertained? This question has been the subject of much debate. In the field of spoken-word recognition, a consensus prevails: Multiple candidates are considered simultaneously, proportionally to their goodness of fit with the sensory information available (Luce & Pisoni, 1998; Marslen-Wilson, 1990; Marslen-Wilson, Moss, & van Halen, 1996). As more information is accumulated, the fit changes and so does the support for each candidate. There has been little concern for some form of limitation in the resources associated with multiple-candidate consideration, perhaps because the process is assumed to involve passive retrieval of content-addressable information from long-term memory.

In the field of syntactic parsing, most of the work on this question has been done on written language. We nonetheless discuss it here because the theoretical positions that have been proposed may apply to the spoken modality as well. As described above, if people interpret a sentence in an incremental fashion, this means that they project a structure based on partial information. This structure is often viewed as being built and actively maintained in working memory. Because the capacity of working memory is limited, resource limitation has played a major role in constraining theoretical approaches (see Clifton & Staub, 2008, for a review). According to serial approaches, only one parse can be pursued; ambiguity is resolved by choosing which structure to maintain, although the criteria for choosing have varied (e.g., Frazier, 1979; Gibson, 1998). Parallel approaches, on the other hand, do not subscribe to the notion of resource limitation; instead, in a way that is reminiscent to assumptions pertaining to word recognition, these approaches posit that multiple parses are held in memory, each one associated with some likelihood that it is the parse intended by the speaker (e.g., MacDonald et al., 1994; McRae, Spivey-Knowlton, & Tanenhaus, 1998). Note that serial accounts do assume that all (or most) parses are established in order to be evaluated relative to one another. Thus, the divergence between the serial and parallel views rests on how long multiple parses can be held in working memory.

Distinguishing between these two approaches empirically has proven challenging (e.g., Green & Mitchell, 2006; Sturt, 2007; Van Gompel, Pickering, Pearson, & Liversedge, 2005). Levy (2008) proposed a model of syntactic-processing difficulty that blurs the distinction even further: In this Bayesian model, grounded in parallel probabilistic ambiguity resolution, the difficulty of processing a new word is equivalent to the degree to which the set of parses (which, in the model's formalism, is the probability distribution of structures) that are compatible with the sentence before the new word is known differs from the probability distribution of the sentence after the word is known (i.e., a measure of "surprisal"). Importantly, Levy showed that this quantity is formally equivalent to the negative probability of the word given the sequence of words that preceded it, a quantity that may be difficult to empirically distinguish from what serial models predict if only one analysis were constructed and entertained at a time.

The third question that incremental processing raises concerns what happens when the most plausible structure (whether the only one pursued or among alternatives) is disconfirmed by later-arriving information, a phenomenon that has been extensively studied in the processing of *garden path* sentences. Although this research has largely focused on the conditions under which garden path effects emerge, important insight has come from studies examining the consequences of these effects on sentence comprehension (Christianson, Hollingworth, Halliwell, & Ferreira, 2001; Ferreira, 2003; Slattery, Sturt, Christianson, Yoshida, & Ferreira, 2013): When asked about the events described in the sentences they read, people seem to hold true propositions derived from both the original and revised structures. This finding suggests that people do not always revise the linguistic representation of the sentence based on subsequent information. Ferreira proposed that in many cases, people represent the meaning of sentences in terms of "good enough" representations, in other words, representations that may not be the most veridical rendition of the sentence but that a constellation of factors has rendered most plausible. A study by van Alphen and van Berkum (2010) reached a similar conclusion: people retrieved the meaning of a monosyllabic word (e.g., *pain*) upon hearing a longer word for which it forms the last syllable (e.g., *champagne*) if the semantic context preceding the longer word was more congruent with the short than the long word (e.g., *The patient asked the nurse when the champagne [would be cold enough to be served]*), even though that parsing would leave some portion of the utterance (i.e., the first syllable of *champagne*) unaccounted for.

Garden path effects have also been examined in the realm of speech perception and word recognition. The influence of subsequent information on the analysis of earlier portions of speech, a so-called *right-context* effect, has been known since Warren (1970): For instance, when the first consonant of a syllable ending in ... *ite* was replaced by a cough, the sound that people reported hearing (e.g., *b* or *k*) depended on which word the following context made more plausible. The perceptual basis of the effect, as well as its generality, was later confirmed (see Dahan, 2010, and references therein). Influence of the subsequent context can also be observed when the portion that affects the interpretation of the ambiguous sound comes a few syllables after it (McMurray, Tanenhaus, & Aslin, 2009) or even after a few words (Connine, Blasko, & Hall, 1991; Szostak & Pitt, 2013), as in phrases like *the tent/dent in the forest/fender.*

Retroactive effects of this sort can be found even with clearly articulated, undistorted speech. Hearing the pronunciation of a word (*carrot*) can elicit the consideration of a rhyming alternative (*parrot*), as reflected in listeners' transient fixations to the picture of the rhyming word (Allopenna et al., 1998). This result cannot be attributed to the occasional misperception of the initial sound because in all cases, people eventually identified the referent correctly. Instead, the rhyming portion of the spoken word *carrot* affects the interpretation of its initial sound. Everything else being equal, the initial sound of *carrot* is more likely to be /*p*/ (due to the influence of *parrot*) than the initial of sound of *candle* is (because there is no such word as *pandle*). This is, in effect, evidence that the later portion of a spoken stimulus can affect the interpretation of an earlier portion.

The existence of right-context effects raises the question of what kind of representation people hold in memory that allows them to revise their interpretation given later information. One possibility is that listeners make a series of discrete decisions regarding the presence of phonological (and, more generally, linguistic) elements in the speech signal (Christiansen & Chater, 2016). Such decisions may be subject to revisions if subsequent information renders prior decisions less plausible. This approach assumes that people work toward retrieving the combination of discrete phonological elements that the speaker produced. Because of the arbitrary mapping of sound onto meaning, the goodness of match between a sound and its best-fitting category or categories may interfere with extracting a semantic interpretation of the utterance. This approach is akin to the serial parser discussed earlier. Alternatively, people maintain uncertainty over time, and contextual information acts as an additional source of constraints that, together, modulate probabilistic evidence for lexical candidates (see Bicknell, Jaeger, & Tanenhaus, 2016, for helpful discussion). As was pointed out, the integration of multiple sources of information may reduce uncertainty and thereby render speech perception more robust to noise and variability. This approach is similar to many of the parallel-parser models.

Although the jury is still out, current evidence favors the latter, probabilistic approach: The impact of subsequent context on the identification of an earlier ambiguous sound varies as a function of the phonetic details of the ambiguous sound: It is greatest when the sound is most ambiguous (Connine et al., 1991; McMurray et al., 2009; Szostak & Pitt, 2013). For the integration to show sensitivity to the phonetic details of the sound requires that people have maintained more than the discrete category it best predicts; instead, people appear to maintain the probabilistic information (the uncertainty) associated with past sensory input for several words (e.g., Szostak & Pitt, 2013; Zellou & Dahan, 2019). For how long and why this representation is maintained are questions that future work must elucidate.

4. *Processing of Disfluent Speech*

In this section, we discuss a feature unique to spoken language, the presence of disfluencies, and their consequences on comprehension. Spoken and written languages often differ in their register. Indeed, written language has usually benefitted from some editing and the author's formulation is meant to reveal the intended meaning as effectively as possible. Spoken language, on the other hand, is often generated on the fly: speakers frequently interrupt, backtrack, and correct their utterances, and listeners must interpret these repairs as best they can. Research is illuminating how listeners cope with, as well as benefit from, disfluencies.

Disfluencies are important not just because they exist and we evidently can deal with them, but also because they reside at the intersection of a number of issues researchers have investigated in psycholinguistics over the last 50 years: incremental processing, revision of erroneous interpretations, integration of different sources of information, and prediction. If language is processed incrementally, then it is to be expected that a disfluency in the input is a piece of language that the listener will process and typically integrate into the ongoing representation of the utterance. Identifying a word or stretch of words as an error on the part of the speaker requires that information at the lexical, syntactic, prosodic, and semantic levels be integrated to support the inference that part of the input does not belong in the ultimate representation for the sentence.

Studies of the comprehension of disfluent utterances have mainly investigated fillers such as *uh*, *um*, and *er*, and these investigations demonstrate that listeners tend to expect speakers to refer to something novel (Arnold, Tanenhaus, Altmann, & Fagnano, 2004; Barr & Seyfeddinipur, 2010) or unfamiliar (Arnold, Kam, & Tanenhaus, 2007) when *uh* precedes a word, compared with a fluent production. This effect is seen even in children as young as two years of age (Kidd, White, & Aslin, 2011). Listeners show a reduced N400 to a less predictable word when that word is preceded by *er* (Corley, MacGregor, & Donaldson, 2007) but not when preceded by a repetition such as *burnt my my tongue* (MacGregor, Corley, & Donaldson, 2010). These findings indicate that fillers lead listeners to infer that the speaker is having trouble formulating her utterance and therefore lead to the expectation of something new, unusual, or unpredictable. In addition to the significance they can have on what listeners expect to hear next, our research has demonstrated that *uh* and *um* utterances can influence syntactic parsing (Bailey & Ferreira, 2003; Ferreira & Bailey, 2004). Previous work had shown that it is more difficult for comprehenders to recover from a garden path the longer they have been committed to the incorrect analysis, as measured in number of words (Ferreira & Henderson, 1991). Our disfluency experiments demonstrated that it did not matter whether words or disfluencies were used to lengthen the garden path. They also showed that listeners used the presence of an *uh* or *um* to help resolve a garden path, because listeners tend to assume that fillers preceded a more syntactically difficult constituent. Fillers also seem to direct attention to subsequent material, facilitating its recall (Fraundorf & Watson, 2011).

A more complex challenge arises from disfluencies including self-repairs (*turn left I mean turn right at the statue*). One model of how self-repairs are interpreted (Ferreira, Lau, & Bailey, 2004; Lau & Ferreira, 2005) assumes that the process is backward-looking: the mechanism begins from the point of breakdown and then moves backward through the input to arrive at the speaker's intended meaning. Many models designed to automatically recognize self-repairs in speech are also essentially backward-looking: the properties of the repair and the interruption point trigger nonmonotonic operations designed to locate the misspoken word (henceforth, the reparandum) and then delete the content. For example, Zwarts, Johnson, and Dale (2010) developed an automatic disfluency recognition system that is nonmonotonic and succeeds in recognizing a self-repair disfluency about five words after the repair has occurred. This level of performance is superior to many other automatic systems, but Zwarts et al. acknowledged that a more robust computational model would identify self-repairs much earlier.

An alternative approach takes advantage of the evidence for prediction during language comprehension and assumes that listeners can anticipate that a repair will occur and even what its content will be. This approach to the processing of spoken language containing disfluencies brings together two models that have been independently proposed and empirically tested: (i) the noisy channel model of language processing (Gibson, Bergen, & Piantadosi, 2013) and (ii) models of focus that assume that contrastive prosody signals a listener to generate a set of candidates that stand in semantic contrast to some critical item. Putting these together, the idea is that listeners treat each word in the input probabilistically and therefore as something that potentially will be repaired. Listeners predict the repair by creating a contrast set similar to the type of contrast set generated during the processing of contrastive focus, in which it has been shown that reference to, for example, a "large pitcher" causes the comprehender to infer the existence of a set of pitchers, one of which is large (Sedivy, 2002). Recent evidence supports these assumptions (Lowder & Ferreira, 2016): listeners displayed a greater tendency to predict the upcoming repair when the speaker signaled that they had made an error (e.g., … *some salt uh I mean* …), compared to their tendency to predict the second conjunct when the speaker signaled the presence of a coordination structure (e.g., … *some salt and also* …). A follow-up experiment replicated this finding and also examined prediction during the processing of contrastive focus (e.g., … *not some salt but rather* …). Results showed that listeners use information about the negated noun phrase to begin generating predictions about the alternate noun phrase—a pattern that resembles listeners' tendency to generate predictions during the processing of repair disfluencies. Furthermore, as soon as listeners have evidence that a portion of the input was spoken in error, they begin to generate hypotheses for likely repairs. If listeners receive a sequence that is implausible (*put the milk in the stove* …), they begin the process of repairing the input even before there is any explicit, bottom-up evidence of any error. These findings confirm listeners' ability to process linguistic input rapidly and integrate linguistic representations formed online with knowledge and expectations generated on the basis of linguistic experience.

To summarize, research on the comprehension of spoken language containing disfluencies reinforces the conclusions that have emerged from more traditional studies of fluent speech. Listeners incrementally process the input, leading them to integrate content that may later need to be removed from the sentence representation due to speaker error. Listeners use both backward-looking (revision) and forward-looking (anticipation) mechanisms to establish an interpretation that corresponds to the speaker model they have built and that captures the speaker's communicative intentions. Listeners also capitalize on the information value associated with filled pauses, editing terms, and other disfluencies to draw inferences about the speaker's goals, thus enabling them to build an interpretation that captures the speaker's intended meaning.

5. *Relationship between Comprehension and Production*

There is a growing interest in addressing the nature of the relationship between speech production and speech comprehension (e.g., Dell & Chang, 2014; MacDonald, 2013) and some of the recent developments in research on language comprehension are relevant to this endeavor. Here, we discuss what we take as the two theoretical perspectives put forward on the topic.

At various points in this review, we have described research that adopts a view according to which listeners do not decode spoken language, but instead make inferences about what the speaker intended to say. These inferences apply to both the spoken signal that listeners have heard and what may come next. In this section, we discuss two theoretical approaches that capture this general idea. One approach, which this review has discussed on multiple occasions, can be referred to as the noisy channel model, originally proposed by Shannon (1948) and embodied in many of the Bayesian models of language processing (see Gibson et al., 2013). Here, listeners infer what the speaker is likely to have said by combining two sources of information: prior beliefs about what may and may not be said in the present context, and the likelihood that each possible hypothesis would have given rise to the acoustic signal. The other approach, which we call the interactive model of dialogue, assumes a direct connection between production and perception.

Noisy channel models refer to a family of models that assume that processing takes place over a noisy, and therefore variable, communication channel. As alluded to earlier, speech always contains some degree of ambiguity; furthermore, the neural encoding of this signal is noisy. Thus, a rational comprehender whose goal is to recover the intention behind an utterance can make the best guess by combining the sensory data with any prior knowledge of what is likely or unlikely to be said. This idea is the foundation for many models of language processing (Levy, 2008; MacDonald, 2013). The information that is accessed to establish the priors can range from biases related to structural forms all the way to beliefs concerning speaker characteristics (e.g., van Berkum, van den Brink, Tesink, Kos, & Hagoort, 2008).

Noisy channel models imply a tight, albeit indirect, connection between listeners and speakers because the listener's goal is to infer the speaker's communicative intention. Models of interactive dialogue assume an even closer connection, one in which interlocutors closely collaborate during dialogue, allowing them to efficiently communicate (e.g., Clark, 1996). As anyone who has attempted to transcribe a real-world conversation can attest, the language that people produce in conversations is quite different from the way people express their thoughts in writing: Conversational turns overlap, interruptions occur, many utterances are mere fragments, and errors are frequent. But although transcription might be difficult, conversation is typically quite easy. One reason that conversation is easy despite the apparent challenges, Pickering and Garrod (2004) argued, is that interlocutors tend to align at every level of linguistic representation. This insight is captured in the interactive alignment model (Pickering & Garrod, 2004). Interlocutors converge on the same referring expressions; they even come to imitate each other's accents and speech styles (Giles, Coupland, & Coupland, 1991). With respect to syntax, the interactive alignment model views the phenomenon of syntactic priming as a reflection of this alignment tendency: speaker and listener come to use the same grammatical forms, even when they communicate across different languages (Hartsuiker & Pickering, 2008). This alignment skill is supposed to arise from people's tendency to model each other's minds (see Pickering & Clark, 2014). For the interlocutors' representations to be compatible, they must be aligned—the representations must be stated over similar vocabularies. This means, in effect, that the machinery involved in the real-time comprehension of an utterance and its working are those involved in the production of the same utterance. This is a stronger claim than that made by Bayesian approaches such as the noisy channel model and one that future research will actively examine.

REFERENCES

Allopenna, P. D., Magnuson, J. S., & Tanenhaus, M. K. (1998). Tracking the time course of spoken word recognition using eye movements: Evidence for continuous mapping models. *Journal of Memory and Language, 38*(38), 419–439.

Altmann, G. T. M., & Kamide, Y. (1999). Incremental interpretation at verbs: Restricting the domain of subsequent reference. *Cognition, 73*(3), 247–264.

Arnold, J. E., Kam, C. L. H., & Tanenhaus, M. K. (2007). If you say thee uh you are describing something hard: The on-line attribution of disfluency during reference comprehension. *Journal of Experimental Psychology: Learning, Memory, and Cognition, 33*(5), 914–930.

Arnold, J. E., Tanenhaus, M. K., Altmann, R. J., & Fagnano, M. (2004). The old and thee, uh, new: Disfluency and reference resolution. *Psychological Science, 15*(9), 578–582.

Baese-Berk, M. M., Heffner, C. C., Dilley, L. C., Pitt, M. A., Morrill, T. H., & McAuley, J. D. (2014). Long-term temporal tracking of speech rate affects spoken-word recognition. *Psychological Science, 25*, 1546–1553.

Bailey, K. G. D., & Ferreira, F. (2003). Disfluencies affect the parsing of garden-path sentences. *Journal of Memory and Language, 49*(2), 183–200.

Barr, D. J., & Seyfeddinipur, M. (2010). The role of fillers in listener attributions for speaker disfluency. *Language and Cognitive Processes, 25*(4), 441–455.

Bicknell, K., Jaeger, T. F., & Tanenhaus, M. K. (2016). Now or ... later: Perceptual data are not immediately forgotten during language processing. *Brain and Behavioral Sciences, 39*, 23–24.

Bock, J. K. (1986). Meaning, sound, and syntax: Lexical priming in sentence production. *Journal of Experimental Psychology: Learning, Memory, and Cognition, 12*(4), 575–586.

Breen, M. (2014). Empirical investigations of the role of implicit prosody in sentence processing. *Linguistics and Language Compass, 8*(2), 37–50.

Brown, M., Salverda, A. P., Gunlogson, C., & Tanenhaus, M. K. (2015). Interpreting prosodic cues in discourse context. *Language, Cognition and Neuroscience, 30*(1–2), 149–166.

Buxó-Lugo, A., & Watson, D. G. (2016). Evidence for the influence of syntax on prosodic parsing. *Journal of Memory and Language, 90*, 1–13.

Chater, N., & Manning, C. D. (2006). Probabilistic models of language processing and acquisition. *Trends in Cognitive Sciences, 10*(7), 335–344.

Christiansen, M. H., & Chater, N. (2016). The now-or-never bottleneck: A fundamental constraint on language. *Behavioral and Brain Sciences, 39*, 1–72.

Christianson, K., Hollingworth, A., Halliwell, J. F., & Ferreira, F. (2001). Thematic-roles assigned along the garden path linger. *Cognitive Psychology, 42*, 368–407.

Clark, A. (2013). Are we predictive engines? Perils, prospects, and the puzzle of the porous perceiver. *Behavioral and Brain Sciences, 36*(3), 233–253.

Clark, H. H. (1996). *Using language.* Cambridge: Cambridge University Press.

Clarke, C. M., & Garrett, M. F. (2004). Rapid adaptation to foreign-accented English. *Journal of the Acoustical Society of America, 116*(6), 3647–3658.

Clifton, C., Carlson, K., & Frazier, L. (2006). Tracking the what and why of speakers' choices: Prosodic boundaries and the length of constituents. *Psychonomic Bulletin and Review, 13*, 854–861.

Clifton, C., & Staub, A. (2008). Parallelism and competition in syntactic ambiguity resolution. *Language and Linguistics Compass, 2*(2), 234–250.

Cole, J., Mo, Y., & Baek, S. (2010). The role of syntactic structure in guiding prosody perception with ordinary listeners and everyday speech. *Language and Cognitive Processes, 25*(7–9), 1141–1177.

Connine, C. M., Blasko, D. G., & Hall, M. (1991). Effects of subsequent sentence context in auditory word recognition: Temporal and linguistic constraints. *Journal of Memory and Language, 30*(2), 234–250.

Corley, M., MacGregor, L. J., & Donaldson, D. I. (2007). It's the way that you, er, say it: Hesitations in speech affect language comprehension. *Cognition, 105*(3), 658–668.

Cutler, A., & Clifton, C. Jr. (1999). Comprehending spoken language: A blueprint of the listener. In C. M. Brown & P. Hagoort (Eds.), *The neurocognition of language* (pp. 123–166). Oxford: Oxford University Press.

Dahan, D. (2010). The time course of interpretation in speech comprehension. *Current Directions in Psychological Science, 19*, 121–126.

Dahan, D. (2015). Prosody and language processing. *Wiley Interdisciplinary Reviews: Cognitive Science, 6*, 441–452.

Dahan, D., Drucker, S. J., & Scarborough, R. A. (2008). Talker adaptation in speech perception: Adjusting the signal or the representations? *Cognition, 108*, 710–718.

Dell, G. S., & Chang, F. (2014). The P-chain: Relating sentence production and its disorders to comprehension and acquisition. *Philosophical Transactions of the Royal Society B, 369*, 20120394.

Dilley, L. C., Mattys, S. L., & Vinke, L. (2010). Potent prosody: Comparing the effects of distal prosody, proximal prosody, and semantic context on word segmentation. *Journal of Memory and Language, 63*, 274–294.

Dilley, L. C., & Pitt, M. (2010). Altering context speech rate can cause words to appear or disappear. *Psychological Science, 21*(11), 1664–1670.

Eisner, F., & McQueen, J. M. (2005). The specificity of perceptual learning in speech processing. *Attention, Perception, and Psychophysics, 67*(2), 224–238.

Ferreira, F. (2003). The misinterpretation of noncanonical sentences. *Cognitive Psychology, 47*(2), 164–203.

Ferreira, F., & Bailey, K. G. D. (2004). Disfluencies and human language comprehension. *Trends in Cognitive Sciences, 8*(5), 231–237.

Ferreira, F., & Henderson, J. M. (1991). Recovery from misanalyses of garden path sentences. *Journal of Memory and Language, 30*(6), 725–745.

Ferreira, F., Lau, E. F., & Bailey, K. G. D. (2004). Disfluencies, language comprehension, and tree adjoining grammars. *Cognitive Science, 28*(5), 721–749.

Fine, A. B., & Jaeger, T. F. (2013). Evidence for implicit learning in syntactic comprehension. *Cognitive Science, 37*(3), 578–591.

Fraundorf, S. H., & Watson, D. G. (2011). The disfluent discourse: Effects of filled pauses on recall. *Journal of Memory and Language, 65*(2), 161–175.

Frazier, L. (1979). *On comprehending sentences: Syntactic parsing strategies* (Doctoral dissertation). University of Connecticut. Available from ProQuest Dissertations & Theses Global. (302925499). Retrieved from https://search.proquest.com/docview/302925499?accountid=10552. (Order No. 7914150).

Frazier, L., Carlson, K., & Clifton, C. (2006). Prosodic phrasing is central to language comprehension. *Trends in Cognitive Sciences, 10*(6), 244–249.

Gibson, E. (1998). Linguistic complexity: Locality of syntactic dependencies. *Cognition, 68*(1), 1–76.

Gibson, E., Bergen, L., & Piantadosi, S. T. (2013). Rational integration of noisy evidence and prior semantic expectations in sentence interpretation. *Proceedings of the National Academy of Sciences of the United States of America, 110*(20), 8051–8056.

Giles, H., Coupland, N., & Coupland, J. (1991). Accommodation theory: Communication, context, and consequence. In H. Giles, J. Coupland, & N. Coupland (Eds.), *Contexts of accommodation* (pp. 1–68). Cambridge: Cambridge University Press.

Goldinger, S. D., Pisoni, D. B., & Logan, J. S. (1991). On the nature of talker variability effects on recall of spoken word lists. *Journal of Experimental Psychology: Learning, Memory, and Cognition, 17*(1), 152–162.

Green, M. J., & Mitchell, D. C. (2006). Absence of real evidence against competition during syntactic ambiguity resolution. *Journal of Memory and Language, 55*(1), 1–17.

Griffiths, T. L., Chater, N., Kemp, C., Perfors, A., & Tenenbaum, J. B. (2010). Probabilistic models of cognition: Exploring representations and inductive biases. *Trends in Cognitive Sciences, 14*, 357–364.

Hagoort, P., & Poeppel, D. (2013). The infrastructure of the language-ready brain. In M. A. Arbib (Ed.), *Language, music, and the brain: A mysterious relationship* (pp. 233–255). Cambridge, MA: MIT Press

Hartsuiker, R. J., & Pickering, M. J. (2008). Language integration in bilingual sentence production. *Acta Psychologica, 128*(3), 479–489.

Helmholtz, H. von. (1962). *Introduction to Physiological Optics*, by J. P. C. Southall, Ed., New York: Dover. (Translation and reprint of *Handbuch der physiologischen optik*. 1860, Leipzig: Leopold Vos.)

Joanisse, M. F., & McClelland, J. L. (2015). Connectionist perspectives on language learning, representation and processing. *Wiley Interdisciplinary Reviews: Cognitive Science, 6*, 235–247.

Johnson, K. (2005). Speaker normalization in speech perception. In D. B. Pisoni & R. E. Remez (Eds.), *The handbook of speech perception* (pp. 363–389). Oxford: Blackwell.

Jun, S.-A. (2010). The implicit prosody hypothesis and overt prosody in English. *Language and Cognitive Processes, 25*(7–9), 1201–1233.

Kamide, Y., Altmann, G. T., & Haywood, S. L. (2003). The time-course of prediction in incremental sentence processing: Evidence from anticipatory eye movements. *Journal of Memory and Language, 49*, 133–156.

Kaschak, M. P., & Glenberg, A. M. (2000). Constructing meaning: The role of affordances and grammatical constructions in sentence comprehension. *Journal of Memory and Language, 43*(3), 508–529.

Kidd, C., White, K. S., & Aslin, R. N. (2011). Toddlers use speech disfluencies to predict speakers' referential intentions. *Developmental Science, 14*(4), 925–934.

Kleinschmidt, D. F., & Jaeger, T. F. (2015). Robust speech perception: recognize the familiar, generalize to the similar, and adapt to the novel. *Psychological Review, 122*(2), 148–203.

Kraljic, T., & Samuel, A. G. (2006). Generalization in perceptual learning for speech. *Psychonomic Bulletin and Review, 13*(2), 262–268.

Kurumada, C., Brown, M., Bibyk, S., Pontillo, D., & Tanenhaus, M. K. (2014). Is it or isn't it: Listeners make rapid use of prosody to infer speaker meanings. *Cognition, 133*, 335–342.

Lau, E. F., & Ferreira, F. (2005). Lingering effects of disfluent material on comprehension of garden path sentences. *Language and Cognitive Processes, 20*(5), 633–666.

Levy, R. (2008). Expectation-based syntactic comprehension. *Cognition, 106*(3), 1126–1177.

Lowder, M. W., & Ferreira, F. (2016). Prediction in the processing of repair disfluencies. *Language, Cognition, and Neuroscience, 31*(1), 73–79.

Luce, P. A., & Pisoni, D. B. (1998). Recognizing spoken words: The neighborhood activation model. *Ear and Hearing, 19*, 1–36.

Luka, B. J., & Barsalou, L. W. (2005). Structural facilitation: Mere exposure effects for grammatical acceptability as evidence for syntactic priming in comprehension. *Journal of Memory and Language, 52*(3), 436–459.

MacDonald, M. C. (2013). How language production shapes language form and comprehension. *Frontiers in Psychology, 4*(226). doi: 10.3389/fpsyg.2013.00226.

MacDonald, M. C., Pearlmutter, N. J., & Seidenberg, M. S. (1994). The lexical nature of syntactic ambiguity resolution. *Psychological Review, 101*(4), 676–703.

MacGregor, L. J., Corley, M., & Donaldson, D. I. (2010). Listening to the sound of silence: Disfluent silent pauses in speech have consequences for listeners. *Neuropsychologia, 48*(14), 3982–3992.

Magnuson, J. S., & Nusbaum, H. C. (2007). Acoustic differences, listener expectations, and the perceptual accommodation of talker variability. *Journal of Experimental Psychology: Human Perception and Performance, 33*(2), 391–409.

Marslen-Wilson, W. D. (1975). Sentence perception as an interactive parallel process. *Science, 189*(4198), 226–228.

Marslen-Wilson, W. D. (1990). Activation, competition, and frequency in lexical access. In G. T. M. Altmann (Ed.), *Cognitive models of speech processing: Psycholinguistic and computational perspectives* (pp. 148–172). Cambridge, MA: MIT Press.

Marslen-Wilson, W. D., Moss, H. E., & van Halen, S. (1996). Perceptual distance and competition in lexical access. *Journal of Experimental Psychology: Human Perception and Performance, 22*, 1376–1392.

Marslen-Wilson, W. D., & Tyler, L. K. (1980). The temporal structure of spoken language understanding. *Cognition, 8*(1), 1–71.

Maye, J., Aslin, R. N., & Tanenhaus, M. K. (2008). The weckud wetch of the wast: Lexical adaptation to a novel accent. *Cognitive Science, 32*(3), 543–562.

McClelland, J. L., Mirman, D., Bolger, D. J., & Khaitan, P. (2014). Interactive activation and mutual constraint satisfaction in perception and cognition. *Cognitive Science, 38*, 1139–1189.

McMurray, B., Tanenhaus, M. K., & Aslin, R. N. (2009). Within-category VOT affects recovery from "lexical" garden-paths: Evidence against phoneme-level inhibition. *Journal of Memory and Language, 60*(1), 65–91.

McRae, K., Spivey-Knowlton, M. J., & Tanenhaus, M. K. (1998). Modeling the influence of thematic fit (and other

constraints) in on-line sentence comprehension. *Journal of Memory and Language, 38*(3), 283–312.
Myslín, M., & Levy, R. (2016). Comprehension priming as rational expectation for repetition: Evidence from syntactic processing. *Cognition, 147,* 29–56.
Norris, D., & McQueen, J. M. (2008) Shortlist B: A Bayesian model of continuous speech recognition. *Psychological Review, 115*(2), 357–395.
Norris, D., McQueen, J. M., & Cutler, A. (2003). Perceptual learning in speech. *Cognitive Psychology, 47*(2), 204–238.
Norris, D., McQueen, J. M., & Cutler, A. (2016). Prediction, Bayesian inference and feedback in speech recognition. *Language, Cognition, and Neuroscience, 31*(1), 4–18.
Nygaard, L. C., & Pisoni, D. B. (1998). Talker-specific learning in speech perception. *Perception and Psychophysics, 60*(3), 355–376.
Nygaard, L. C., Sommers, M. S., & Pisoni, D. B. (1994). Speech perception as a talker-contingent process. *Psychological Science, 5*(1), 42–46.
Nygaard, L. C., Sommers, M. S., & Pisoni, D. B. (1995). Effects of stimulus variability on perception and representation of spoken words in memory. *Perception and Psychophysics, 57*(7), 989–1001.
Pickering, M. J., & Clark, A. (2014). Getting ahead: Forward models and their place in cognitive architecture. *Trends in Cognitive Sciences, 18*(9), 451–456.
Pickering, M. J., & Ferreira, V. S. (2008). Structural priming: A critical review. *Psychological Bulletin, 134*(3), 427.
Pickering, M. J., & Garrod, S. (2004). Toward a mechanistic psychology of dialogue. *Behavioral and Brain Sciences, 27*(02), 169–190.
Rumelhart, D. E., & McClelland, J. L. (1982). An interactive activation model of context effects in letter perception: II. The contextual enhancement effect and some tests and extensions of the model. *Psychological Review, 89*(1), 60–94.
Rumelhart, D. E., McClelland, J. L., & PDP Research Group. (1986). *Parallel distributed processing: Explorations in the microstructure of cognition* (Vols. 1–2). Cambridge, MA: MIT Press.
Salverda, A. P., Dahan, D., & McQueen, J. M. (2003). The role of prosodic boundaries in the resolution of lexical embedding in speech comprehension. *Cognition, 90,* 51–89.
Schafer, A. J., Carlson, K., Clifton, H., & Frazier, L. (2000). Focus and the interpretation of pitch accent: Disambiguating embedded questions. *Language and Speech, 43*(1), 75–105.
Schafer, A. J., Speer. S. R., & Warren, P. (2005). Prosodic influences on the production and comprehension of syntactic ambiguity in a game-based conversation task. In J. C. Trueswell & M. K. Tanenhaus (Eds.), *Approaches to studying world-situated language use* (pp. 209–225). Cambridge, MA: MIT Press.
Sedivy, J. C. (2002). Invoking discourse-based contrast sets and resolving syntactic ambiguities. *Journal of Memory and Language, 46*(2), 341–370.
Shannon, C. (1948). A mathematical theory of communication. *Bell Systems Technical Journal, 27,* 623–656.
Shattuck-Hufnagel, S., & Turk, A. E (1996). A prosody tutorial for investigators of auditory sentence processing. *Journal of Psycholinguistic Research, 25,* 193–247.
Slattery, T. J., Sturt, P., Christianson, K., Yoshida, M., & Ferreira, F. (2013). Lingering misinterpretations of garden path sentences arise from competing syntactic representations. *Journal of Memory and Language, 69*(2), 104–120.
Snedeker, J., & Trueswell J. (2003). Using prosody to avoid ambiguity: Effects of speaker awareness and referential context. *Journal of Memory and Language, 48,* 103–130.
Steedman M. (2014). The surface-compositional semantics of English intonation. *Language, 90,* 2–57.
Stevens, K. N., & Halle, M. (1967). Remarks on analysis by synthesis and distinctive features. In W. Wathen-Dunn (Ed.), *Models for the perception of speech and visual form* (pp. 88–102). Cambridge, MA: MIT Press.
Stolterfoht, B., Friederici, A. D., Alter, K., & Steube, A. (2007). Processing focus structure and implicit prosody during reading: Differential ERP effects. *Cognition, 104*(3), 565–590.
Sturt, P. (2007). Semantic re-interpretation and garden path recovery. *Cognition, 105*(2), 477–488.
Swets, B., Desmet, T., Hambrick, D. Z., & Ferreira, F. (2007). The role of working memory in syntactic ambiguity resolution: a psychometric approach. *Journal of Experimental Psychology: General, 136*(1), 64–81.
Szostak, C. M., & Pitt, M. A. (2013). The prolonged influence of subsequent context on spoken word recognition. *Attention, Perception, and Psychophysics, 75*(7), 1533–1546.
Thothathiri, M., & Snedeker, J. (2008). Syntactic priming during language comprehension in three-and four--year-old children. *Journal of Memory and Language, 58*(2), 188–213.
van Alphen, P. M., & van Berkum, J. J. (2010). Is there pain in champagne? Semantic involvement of words within words during sense-making. *Journal of Cognitive Neuroscience, 22*(11), 2618–2626.
van Berkum, J. J., van den Brink, D., Tesink, C. M., Kos, M., & Hagoort, P. (2008). The neural integration of speaker and message. *Journal of Cognitive Neuroscience, 20*(4), 580–591.
Van Gompel, R. P., Pickering, M. J., Pearson, J., & Liversedge, S. P. (2005). Evidence against competition during syntactic ambiguity resolution. *Journal of Memory and Language, 52*(2), 284–307.
Wagner, M., & Watson, D. G. (2010). Experimental and theoretical advances in prosody: A review. *Language and Cognitive Processes, 25,* 905–945.
Warren, R. M. (1970). Perceptual restoration of missing speech sounds. *Science, 167*(3917), 392–393.
Watson, D., & Gibson, E. (2005). Intonational phrasing and constituency in language production and comprehension. *Studia Linguistica, 59,* 279–300.
Wells, J. B., Christiansen, M. H., Race, D. S., Acheson, D. J., & MacDonald, M. C. (2009). Experience and sentence processing: Statistical learning and relative clause comprehension. *Cognitive Psychology, 58*(2), 250–271.
Zellou, G., & Dahan, D. (2019). *Listeners maintain phonological uncertainty over time: the case of vowel nasality in English.* Unpublished manuscript.
Zwarts, S., Johnson, M., & Dale, R. (2010, August). Detecting speech repairs incrementally using a noisy channel approach. *Proceedings of the 23rd International Conference on Computational Linguistics (Coding 2010),* Beijing, pp. 1371–1378. Retrieved from http://www.aclweb.org/anthology/C10-1154.

4 The Architecture of Speaking

ARDI ROELOFS AND VICTOR S. FERREIRA

Theorizing about the architecture of speaking has a long history (see Levelt, 2013), ranging from the classic work of Wernicke (1874), Lichtheim (1885), and Wundt (1900) in the 19th century to that of Garrett (1975) and Levelt (1989) in the modern era. Although Wernicke gave his cognitive account a neuroanatomical basis using lesion-deficit analyses, later architectures were often mainly based on behavioral evidence (but see Geschwind, 1970). Since the 1990s, structural and functional neuroimaging has increasingly informed theoretical views on the cognitive architecture of speaking (see Shallice & Cooper, 2011, for a defense of this approach; see Kemmerer, 2015, for a literature review). In this chapter, we outline a neuroanatomically grounded cognitive architecture for speaking that takes account of modern neuroimaging insights.

Speaking involves multiple component processes and representations. We describe current views on these processes and representations and discuss debates along the way. It is generally assumed that three major types of processes underlie speaking: conceptualization, formulation, and articulation (Garrett, 1975; Levelt, 1989). Conceptualization processes generate messages, which are conceptual representations to be verbally expressed. Formulation processes use a message to retrieve appropriate words and to build a syntactic structure, which is used to generate a phonetic plan for the utterance. Articulation processes execute the phonetic plan, which yields overt speech. During conceptualization, formulation, and articulation, a speaker monitors progress on these processes. Figure 4.1 outlines the major component processes and representations, largely following Levelt (1989). We first briefly describe these processes and representations and discuss them in greater detail in the remainder of this chapter.

Conceptualization processes generate a message, for example, based on visually perceived scenes (see Konopka & Brown-Schmidt, 2014, for a review). Assume that your friend's cat is lost and you discover that the animal is hidden under the kitchen table. To say this to your friend requires the construction of a message making this information explicit, for example, BE(CAT, UNDER(TABLE)). The intended utterance *the cat is under the table* is useful if your friend doesn't know this fact yet (which is in your model of the addressee) and you are with her in the kitchen (as specified by your situation model; otherwise, you should make the exact table explicit by saying *kitchen table*). A controversial and unresolved issue is whether a message is specified in terms of nondecomposed lexical concepts (e.g., Roelofs, 1992), such as CAT and TABLE, or in terms of the conceptual features of these concepts (e.g., Bierwisch & Schreuder, 1992; Dell, Schwartz, Martin, Saffran, & Gagnon, 1997), such as FURRY, MEOWING, PURRING, FELINE, ANIMAL. Either way, messages are input to formulation processes.

According to a prominent view, the formulation of sentences is (for the most part) lexically driven (for reviews, see Ferreira, 2010; Bock & Ferreira, 2014). That is, concepts in the message activate corresponding words in memory, which make available the syntactic and morphophonological information that is needed for constructing a syntactic structure and phonetic plan. During syntactic encoding, syntactic constituents are assembled (e.g., the noun phrase *the cat*) and grammatical functions are assigned (e.g., *the cat* is made the subject of the sentence). During morphophonological encoding, the sound structure of the utterance is constructed, including its prosody. Some theorists assume that formulation processes draw on distinct declarative and procedural types of knowledge (e.g., Levelt, 1989), whereas other theorists do not make this distinction (e.g., Chang & Fitz, 2014). Declarative knowledge includes knowing that the word *cat* is a noun, and procedural knowledge includes knowing how to construct a word form and a noun phrase. According to a prominent theory (Levelt, Roelofs, & Meyer, 1999), a particular type of representation in declarative memory, called a lemma, links the conceptual, syntactic, and phonological representations of a word. For example, the lemma of the word *cat* links conceptual information (e.g., the lexical concept CAT, linked to other concepts such as FURRY, MEOWING, ANIMAL), syntactic properties such as word class and syntactic frame (e.g., specifying that a noun is the head of a noun phrase), morphological and phonological information (e.g., the morpheme <cat> and the phonemes /k/, /æ/, and /t/), and phonetic information (e.g., a motor program for [kæt]). In this theory, the phonetic plan makes explicit

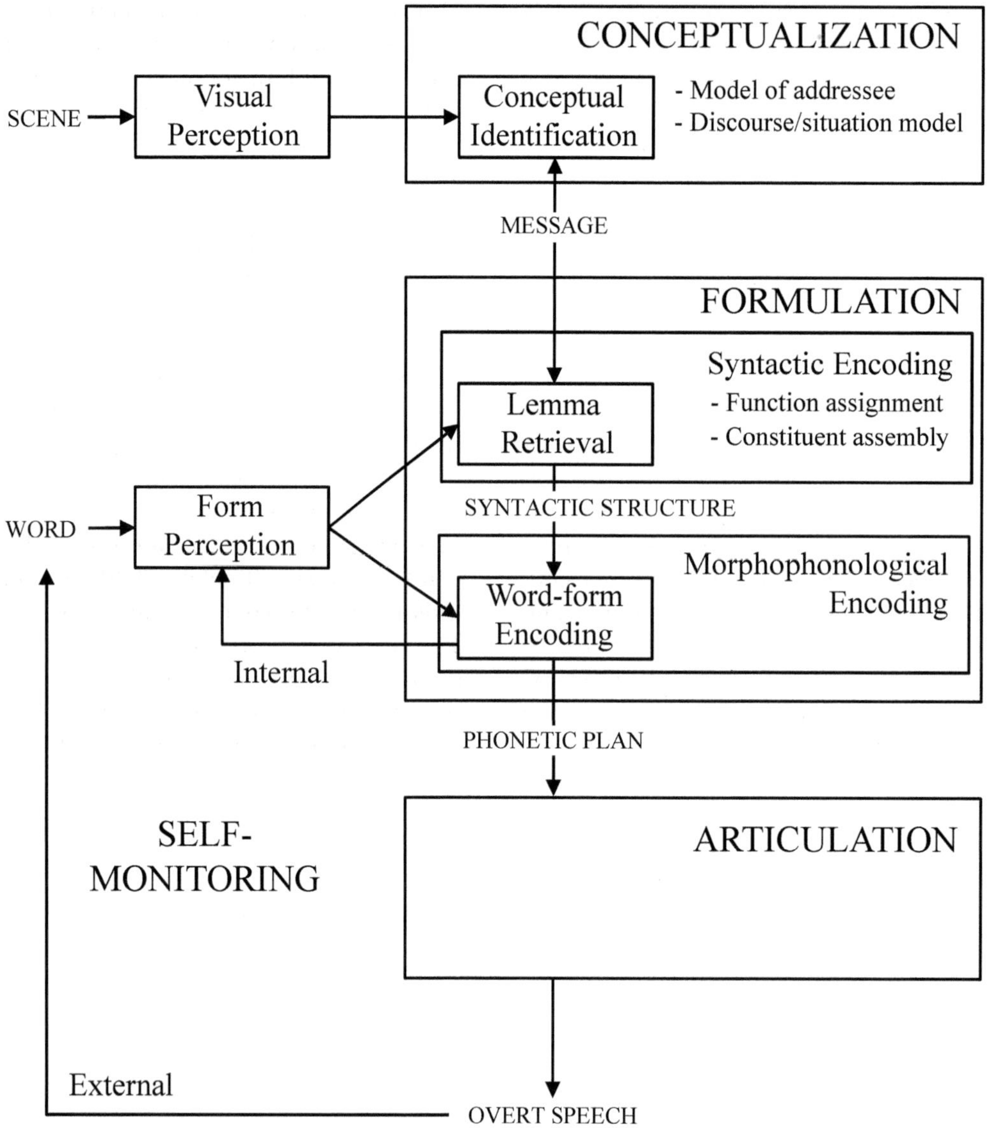

FIGURE 4.1 Cognitive architecture for speaking.

the motor programs for syllables, which are executed by the articulation process.

Through self-monitoring, a speaker assesses whether formulation and articulation are consistent with intent, and if not, this may trigger a correction (e.g., *the dog... uh... cat is under the table*). Errors may not only be detected when speakers hear their own overt speech, but also already before articulation onset. According to a comprehension-based account, the latter is accomplished by internal feedback through the speech perception system (Levelt, 1989; Levelt et al., 1999), whereas according to a production-based account, errors are detected inside the production system, for example, by monitoring for conflict (Nozari, Dell, & Schwartz, 2011).

In the remainder of this chapter, we provide further details on this cognitive architecture for speaking. We discuss key evidence from behavioral studies measuring response time and accuracy, eye tracking, electrophysiological recordings, hemodynamic neuroimaging, tractography, disorders, and computational modeling. Converging evidence from multiple methods is discussed with the idea that if evidence from multiple sources is in agreement, the conclusion can be strong even if each method has its weaknesses (cf. Shallice & Cooper, 2011). Moreover, some major past and current controversies are evaluated.

1. Word Production

For word-level processes in speaking, a number of computational models, which agree but also differ in several respects, have been developed. Differences concern the number and nature of the representation levels and the information flow between the levels. For example, the WEAVER++ model of Levelt et al. (1999) assumes concept, lemma, morpheme, phoneme, and motor program levels, whereas the model of Dell et al. (1997) assumes concept, lexical, and phoneme levels only (but see Dell, 1986). We use the more extensive model of Levelt et al. as a framework and discuss empirical evidence for and against the levels assumed by this model. The main processing stages in the model are lemma retrieval and word-form encoding, which is further subdivided into morphological, phonological, and phonetic encoding. Next, we discuss evidence on the nature of processing within each stage (i.e., whether lexical selection is a competitive process) and the information flow between levels (i.e., whether this is discrete, cascading, or interactive). Moreover, we discuss the relation between word planning and word comprehension, which supports self-monitoring in the model. The WEAVER++ model does not say much about phonetic encoding and articulation, which is addressed by several other models in the literature (e.g., Guenther & Vladusich, 2012; Hickok, 2012). We briefly discuss the most comprehensive of these models, namely the DIVA model of Guenther and colleagues, which fits well with WEAVER++.

1.1. Lemma Retrieval and Word-form Encoding Figure 4.2 outlines a recent extension of WEAVER++, called WEAVER++/ARC, which synthesizes behavioral psycholinguistic, functional neuroimaging, tractography, and aphasiological evidence (Roelofs, 2014). The model makes a distinction between a declarative lexical network and a procedural rule system (cf. Levelt et al., 1999), following Levelt (1989), Ullman (2004), and others. There is considerable cognitive neuroscience evidence for distinct declarative and procedural memory systems in the brain (see, e.g., Eichenbaum, 2012, for review). The declarative lexical network is thought to be underpinned by temporal and inferior frontal areas of the human brain (including Wernicke's and Broca's areas), whereas the procedural system is assumed to be underpinned by, among others, the basal ganglia, thalamus, frontal cortex (including Broca's area), and cerebellum. The lexical network is accessed by spreading activation while procedural if-then rules select nodes among the activated lexical information depending on the task demands specified in working memory. Activation spreads continuously from level to level (Roelofs, 2008), whereby each node sends a proportion of its activation to connected nodes. After a message concept is selected, an if-then rule enhances its activation (thought to be achieved through a ventral white matter pathway running from frontal cortex, including Broca's area, to anterior-ventral temporal cortex, possibly including the inferior frontal-occipital fasciculus) to achieve speeded and accurate lemma retrieval and word-form encoding operations.

The lexical network consists of lexical concepts (e.g., CAT) thought to be represented in anterior-ventral temporal cortex, lemmas (e.g., *cat*) in the middle section of the left middle temporal gyrus (MTG), input phonemes (e.g., /k/, /æ/, and /t/) and input and output lexical forms or morphemes (e.g., <cat>) in left

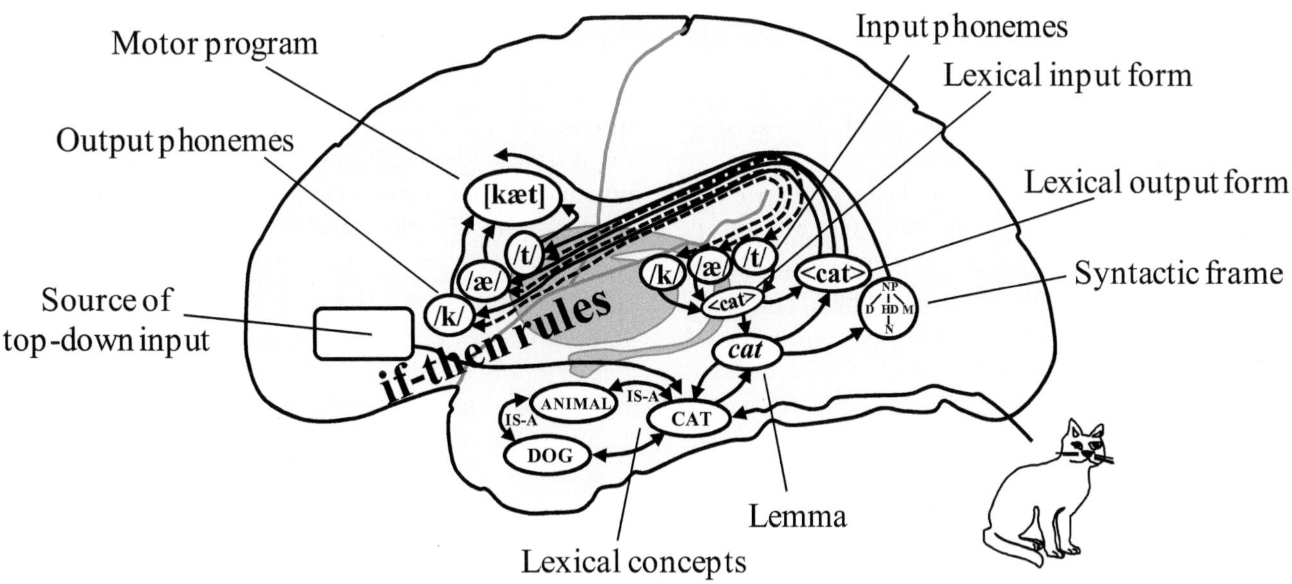

Figure 4.2 Schematic of the WEAVER++/ARC model of spoken word production and its relation with comprehension and repetition. D = determiner; HD = head; M = modifier; N = noun; NP = noun phrase.

posterior superior temporal gyrus (STG) and posterior MTG (Wernicke's area), output phonemes (e.g., /k/, /æ/, and /t/) in left posterior inferior frontal gyrus (Broca's area), and syllable motor programs (e.g., [kæt]) in ventral precentral gyrus. The functional localizations are based on an extensive meta-analysis of neuroimaging studies performed by Indefrey and Levelt (Indefrey & Levelt, 2004; Indefrey, 2011), with converging evidence from lesion-deficit analyses in aphasia (e.g., Schwartz, Faseyitan, Kim, & Coslett, 2012; Schwartz et al., 2009). Lexical concepts are assumed to be part of a hub of supramodal conceptual representations (in the anterior temporal lobes), which integrate modality-specific features that are represented in widespread brain areas for perception and action (Lambon Ralph, 2014; Patterson, Nestor, & Rogers, 2007; see Kemmerer, 2015, for a review). Lemmas specify the grammatical properties of words, which are thought to be represented in left posterior STG and MTG (Hagoort & Indefrey, 2014). These properties are propagated to Broca's area for the encoding of phrases and sentences (see section 2.1 for further discussion), as suggested by evidence from neuropsychology (Sapolsky et al., 2010; Wilson et al., 2010) and neuroimaging (Haller, Radue, Erb, Grodd, & Kircher, 2005; Indefrey et al., 2001). For example, the lemma of *cat* specifies that the word is a noun (for languages such as Dutch, lemmas also specify grammatical gender) and also that is the head of a noun phrase that may include a determiner and modifiers. Lemmas also allow for the specification of abstract morphosyntactic parameters, such as number (singular, plural) for nouns and number, person (first, second, third), and tense (past, present) for verbs, so that the appropriate lexical output forms may be retrieved (i.e., one or more morphemes, such as singular <cat>).

The connections from lexical output forms to output phonemes are thought to be underpinned by a dorsal white matter pathway in the brain made up by the left arcuate fasciculus. Also, connections from input to output phonemes, serving repetition (i.e., saying heard words or pseudowords), are assumed to be underpinned by this pathway. In vivo tractography and fiber microdissection of postmortem human brains suggest that there are two distinct subtracts within the left arcuate fasciculus (Brauer, Anwander, Perani, & Friederici, 2013; Fernandez-Miranda et al., 2015; Glasser & Rilling, 2008). One subtract interconnects left posterior MTG and Broca's area, and the other subtract interconnects left posterior STG and Broca's area. The MTG pathway is assumed to map lexical output forms for lexical-semantic representations onto output phonemes, whereas the STG pathway maps input phonemes onto output phonemes. Output phonemes also activate input phonemes, followed by activation of lexical input forms, lemmas, and lexical concepts, which supports internal self-monitoring in the WEAVER++/ARC model (Levelt et al., 1999; Roelofs, 2004b).

In word production, activation traverses from lexical concepts to syllable motor programs. If-then rules select lexical concepts, lemmas, output lexical forms (i.e., morphemes), and output phonemes; syllabify the phonemes; and assign a stress pattern across syllables in polysyllabic words. The resulting phonological word representation is used to select the corresponding syllable motor programs, which are stored in a syllabary (Levelt et al., 1999). Suppose, for example, that the plural form of *cat* needs to be produced. The lemma of the word *cat* will be retrieved for the lexical concept CAT plus a plural specification. Next, during morphological encoding, the lemma with the number parameter set to *plural* will be mapped onto the stem <cat> and the plural suffix <-s>. During phonological encoding, phonemes are selected for these morphemes, namely /k/, /æ/, /t/ for the stem and /s/ (rather than /z/) for the suffix. The phonemes are then syllabified (and for polysyllabic words, a stress pattern is assigned), yielding a phonological word representation. Sahin, Pinker, Cash, Schomer, and Halgren (2009) obtained evidence from intracranial electrophysiology that morphological and phonological encoding are spatially and temporally distinct processes within Broca's area. During phonetic encoding, the constructed phonological representation is used to select corresponding syllable motor programs (i.e., [kæts]). Evidence that these programs are retrieved from a syllabary comes from syllable frequency effects, which have been obtained for several languages (e.g., Cholin, Dell, & Levelt, 2011, for a review). Electrophysiological evidence suggests that the syllable frequency effect arises about 100–200 ms before articulation onset (Bürki, Pellet, & Laganaro, 2015), supporting the syllabary hypothesis.

Several of the processing and representational assumptions made by the WEAVER++ model are debated in the literature. Major controversies concern the distinction between lemmas and lexical forms (e.g., Caramazza & Miozzo, 1997), the activation flow between levels (e.g., Rapp & Goldrick, 2000), the nature of lexical selection (e.g., Dhooge & Hartsuiker, 2010), and the functional role of the arcuate fasciculus (e.g., Ueno, Saito, Rogers, & Lambon Ralph, 2011). We address these controversies in turn.

1.2. Levels of Representation Classic evidence for the distinction between lemmas and lexical forms comes from speech errors, in particular, the existence of two types of morpheme exchanges (Dell, 1986;

Garrett, 1980, 1988). The properties of these exchanges provide support for a lemma level with abstract morphosyntactic parameters that is distinct from a morphological level with concrete lexical form representations. For example, in "how many *pie*s does it take to make an *apple*?" (from Garrett, 1988), the interacting stems belong to the same syntactic category (i.e., noun) and come from distinct phrases. Note that this concerns a morpheme error, since the plurality of *apple* is stranded, that is, it is realized on *pie*. The distributional properties of these morpheme exchanges are similar to those of whole-word exchanges, such as the exchange of *roof* and *list* in "we completely forgot to add the *list* to the *roof*" (from Garrett, 1980). This suggests that these morpheme errors and whole-word errors occur at the same level of processing, namely when lemmas in a developing syntactic structure trade places. By contrast, the exchanging morphemes in an error such as "*slice*ly *thinn*ed" (from Stemberger, 1985) belong to different syntactic categories (adjective and verb) and come from the same phrase, which is also characteristic of phoneme exchanges, such as "she is a real *r*ack *p*at" for "pack rat" (from Garrett, 1988). This suggests that this second type of morpheme error and phoneme errors occur at the same level of processing, namely the level at which morphemes and phonemes are retrieved and the morphophonological form of the utterance is constructed. The errors occur when morphemes in a developing morphophonological structure trade places.

However, Caramazza and Miozzo (1997) argued against lemmas based on evidence from aphasic patients indicating that semantic errors and syntactic class deficits may be restricted to speaking or writing. For example, some patients make semantic substitution errors (e.g., *dog* for *cat*) in spoken but not in written naming, while the reverse holds true for other patients. Also, patients may have difficulty saying but not writing verbs, or vice versa. According to Caramazza and Miozzo, this suggests that conceptually driven and syntactically constrained lexical selection concerns modality-specific lexical forms rather than modality-neutral lemmas. On this account, brain damage in patients may differentially affect these conceptually and syntactically specified written and spoken forms. However, on an alternative account (Roelofs, Meyer, & Levelt, 1998), the brain damage may differentially affect the connections between lemmas and spoken or written forms (but see Caramazza & Miozzo, 1998, for a rejoinder).

1.3. Activation Flow Another issue concerns the activation flow between the network levels. According to one view (Levelt et al., 1999), activation spreads from a lemma to its associated word form only upon selection of the lemma. Moreover, within the form network, activation spreads forward from morphemes to phonemes and syllable programs, but not backward. Others (Dell, 1986; Dell et al., 1997; Rapp & Goldrick, 2000) hold that activation spreads forward from lemmas to forms and also backward from forms to lemmas.

Accumulating evidence has suggested that activation of word forms does not depend on lemma selection, contrary to what Levelt et al. (1999) maintained. For example, Morsella and Miozzo (2002) observed that in naming a target picture in green superimposed on a distractor picture in red, the naming time is shorter when the picture names share part of their form (e.g., *cat* and *cap*) than when they do not (e.g., *cat* and *pen*). This suggests that activation cascades from concepts (CAT, CAP, PEN) to lemmas (*cat*, *cap*, *pen*) and corresponding forms (/k/, /æ/, /t/, /p/, /e/, /n/), which speeds up the encoding of the target form (*cat*) when the distractor picture activates overlapping phonemes (i.e., /k/ and /æ/). Based on these and other findings, the discreteness assumption that was originally part of WEAVER++ (i.e., Levelt et al., 1999) has been dropped by Roelofs (2008).

Classic evidence that activation spreads backward comes from speech error biases (Dell et al., 1997), such as the statistical overrepresentation of mixed semantic-phonological errors (e.g., *rat* for *cat* vs. *dog* for *cat*) and the statistical overrepresentation of word errors among phonological errors (i.e., lexical bias). These error biases suggest that activation spreads back from forms to lemmas, yielding phonological influences on lemma retrieval and lexical influences on phoneme selection. However, it has remained unclear whether activation spreads back directly from output phonemes to output lexical forms and lemmas (Dell et al., 1997; Rapp & Goldrick, 2000) or via the speech perception network (i.e., via the connections between output and input phonemes). Moreover, it has been argued that the error biases are, at least partly, due to self-monitoring (e.g., Levelt et al., 1999). We refer to Rapp and Goldrick (2000, 2004) and Roelofs (2004a, 2004b) for an extensive discussion.

In meta-analyses of magnetoencephalography (MEG) studies on word production, Indefrey and Levelt (2004) and Munding, Dubarry, and Alario (2016) observed that it takes several hundred milliseconds for activity to progress from occipital-temporal via middle temporal to inferior frontal and precentral areas in picture naming. This evidence supports WEAVER++/ARC and the models of Dell et al. (1997) and Rapp and Goldrick (2000). We refer to Strijkers and Costa (2016) for a different view, which holds that lexical-semantic and phonological representations are activated in parallel in temporal and frontal cortex (but see Indefrey, 2016, for a critical discussion).

1.4. Selection by Competition Researchers also have found no agreement on how lemma selection occurs, see Abdel Rahman and Melinger (2009) and Navarrete, Del Prato, Peressotti, and Mahon (2014) for reviews of work in this area and multiple perspectives about the selection issue. WEAVER++ implements a competition account (e.g., Roelofs, 1992), which holds that lemmas compete for selection. Evidence for competition comes from semantic interference in picture naming. In the picture-word interference paradigm, speakers name pictures while trying to ignore spoken or written distractor words. For example, they say "cat" to a pictured cat combined with the semantically related word *dog* or the unrelated word *pen*. The picture naming time is typically longer on semantically related than on unrelated trials. According to the competition account, the lemma of a semantically related distractor word (*dog*) receives activation from the target picture (cat) and is therefore a stronger competitor to the picture name than the lemma of an unrelated distractor word (*pen*), which is not activated by the picture. However, other researchers have argued that lemma selection does not happen by competition (e.g., Finkbeiner & Caramazza, 2006). According to this noncompetition account, a word is selected if its activation exceeds some threshold, but selection is assumed to be independent of the activation state of other words. Semantic interference is assumed to arise after word planning, reflecting the exclusion of an articulatory response to the distractor word from an output buffer. However, using MEG, Piai, Roelofs, Jensen, Schoffelen, and Bonnefond (2014) observed that the semantic interference effect in picture naming occurs in the time window of word planning, well before articulation onset. Moreover, the semantic interference effect arises in the left MTG, where lemmas are thought to be stored.

1.5. Dorsal versus Ventral Pathway Controversy also exists about the white matter pathway that mediates the mapping of concepts onto motor-related form representations. According to the classic view (e.g., Geschwind, 1970), the arcuate fasciculus is the critical white matter fiber tract. This view has been adopted by WEAVER++/ARC (Roelofs, 2014). In the model, lemmas are linked to lexical output forms, which are thought to be connected by the arcuate fasciculus to output phonemes that are linked to motor programs (see figure 4.2). Dell, Schwartz, Nozari, Faseyitan, and Coslett (2013) also adhere to a dorsal pathway view. However, according to an alternative account, computationally implemented in the Lichtheim 2 model (Ueno et al., 2011), concepts are mapped onto motor-related form representations primarily by a ventral pathway underpinned by the left uncinate fasciculus and fiber tracts passing through the left extreme capsule, which include the inferior fronto-occipital fasciculus and a branch of the middle longitudinal fasciculus. Computer simulations revealed that both Lichtheim 2 and WEAVER++/ARC can account for the typical patterns of impaired and spared language performance associated with classic abrupt-onset and progressive aphasias (Roelofs, 2014; Ueno et al., 2011). However, WEAVER++/ARC but not Lichtheim 2 accounts for the evidence that damage to the dorsal but not the ventral pathway yields impaired production performance (Marchina et al., 2011; Wang, Marchina, Norton, Wan, & Schlaug, 2013).

1.6. Articulation In line with WEAVER++/ARC, the DIVA model (e.g., Guenther & Vladusich, 2012) assumes that motor programs for syllables play an important role in speech motor control. DIVA assumes that access codes for these programs (cf. the syllable program nodes in WEAVER++/ARC) are stored in left posterior inferior frontal gyrus and premotor cortex. These access codes activate representations that specify the articulatory tasks to be performed in pronouncing the syllables (such as closing the lips), thought to be represented in bilateral motor cortex. Initiation of the execution of the motor programs is controlled by the basal ganglia, whereas the temporal coordination of the motor execution is controlled by the cerebellum. Important for learning the motor programs or executing them under special circumstances (e.g., speaking while eating), the access codes for syllables also generate predictions about the auditory and somatosensory effects of producing them. These predictions may be compared with the actually perceived effects, and discrepancies may be used to adjust the motor programs.

2. Sentence Production

Sentences are collections of words, but producing a sentence involves much more than retrieving words (as described by the models) and stringing them one after another. This is because whereas individual words have the function of conveying specific entities (e.g., *cat*), states (*mangy*), and actions (*eat*), sentences have the function of conveying entire events (e.g., *the mangy cat is eating*). This greater expressive function brings with it the need for a more complex set of production mechanisms.

2.1. Stages of Sentence Production Like word production, sentence production is often viewed as proceeding through a sequence of processing stages, and ones that correspond to those involved in word production (see figure 4.1). A number of accounts have been put forth of this type (e.g., Bock & Levelt, 1994; Ferreira & Slevc, 2007). Here, we describe an account based largely on an influential proposal by Garrett (1975; see also Garrett, 1982, 1988), adapted to some extent by Ferreira and Slevc.

Any act of production begins with the speaker determining his or her intent for the utterance, what is often termed a *speech act.* Then, a process that is sometimes termed *message encoding* has the function of determining the features of meaning the speaker must express to be able to fulfill his or her intent. The message that serves as the input to sentence production includes as one of its components the same conceptual information that is used for word production. This component of a message can be termed *semantic meaning*—the who, what, and whom participating in an event. But additionally, the message for sentence production must represent *relational meaning*—who did what to whom in the event. Finally, the message must also include *perspective meaning*—what part of the event is more important and should be foregrounded, and what part is less important and should be backgrounded.

From this multifaceted message (consisting of semantic, relational, and perspective meaning), sentence production proceeds through two separate though interdependent processing streams. One stream, corresponding to the mechanisms of word production, has the task of retrieving the content words that are to be produced in the sentence. This *content stream* can be seen as taking the semantic meaning that is represented in the multifaceted message and engaging the process of lemma retrieval, that is, the retrieval of representations of the individual words to be spoken in the sentence, including those words' grammatical properties (e.g., form class, and if needed, grammatical gender), as described in section 1.1.

Meanwhile, the other stream has the task of determining the structural characteristics of the sentence to be produced. The first step within this *structure stream,* which operates in tandem and interdependently with lemma retrieval within the content stream, is often termed *function assignment* (Bock & Levelt, 1994; Bock, 1995; Ferreira & Slevc, 2007). Function assignment uses the relational and perspective components of the multifaceted message to determine the grammatical functions that will be needed in a sentence. If the relational meaning represented in the discourse corresponds to an intransitive sentence such as *the cat sleeps,* function assignment must formulate grammatical functions for the subject and verb; for a transitive sentence like *the cat eats the food,* function assignment must additionally formulate a direct object; for a ditransitive sentence like *the cat bats the toy to its owner,* function assignment must additionally formulate an indirect object. Meanwhile, as lemma retrieval mechanisms retrieve lemmas for the appropriate content words, those lemmas must become *bound* to the corresponding grammatical functions, so that the subject noun is spoken as the grammatical subject, the object noun as the direct object, and so forth (and if these bindings become crossed up, a speech error of the type reviewed in section 1.2 results).

As lemmas are retrieved in the content stream and functions are assigned in the structure stream, the next stages of sentence production can unfold. On the content side, the above-described process proceeds: lemma retrieval feeds to word-form encoding, wherein the morphophonological specification of the words to be produced in the sentence is determined. On the structure side, function assignment proceeds to a process often termed *constituent assembly* (Ferreira & Slevc, 2007) or *positional processing* (Bock, 1995; Bock & Levelt, 1994; Garrett, 1975). Constituent assembly processes take the grammatical functions that were determined at function assignment and specify a hierarchical structure that represents the relations among the grammatical functions in a manner that allows that structure to be linearized into a sequence of words. To do so, constituent assembly processes must consult the language-internal grammatical rules or conventions that dictate how grammatical functions are to be linguistically encoded. As the parallel word-form encoding process (within the content stream) specifies the morphophonological features of words to be produced in the sentence, those features must be coordinated with (or bound to) the hierarchical structures that are built by constituent assembly processes, so that the syllables that compose the words of the sentence can be arranged into the required linear sequence. As the syllabic sequence is determined, the *prosody* of the sentence is set by prosodic mechanisms (see Ferreira, 1993; Levelt, 1989), which use the syntactic, morphological, and phonemic content of the to-be-produced sequence, along with independent prosodic principles, to determine the "rhythm and melody" of the to-be-produced sentence (for a review, see Shattuck-Hufnagel, 2014).

Although not commonly considered in traditional theories of sentence production, function assignment and constituent assembly must also take into account the perspective meaning that is represented as part of the multifaceted message. Perspective meaning determines what is often termed the *information structure* of sentences (Lambrecht, 1994). Perspective meaning indicates that some aspects of the event represented in the message form the focus or topic to be expressed, whereas other aspects correspond to added information or a comment on the topic. Perspective meaning also represents the *discourse* properties of an event—that the participants in a conversation already know some aspects of the event to be expressed—*given* information—whereas other information is *new* information (as relevant to the model of the listener). The topic-comment and given-new aspects represented within the perspective meaning can then determine both which grammatical functions are used to express a sentence and the order that a particular set of grammatical functions is expressed with.

There is as of yet limited evidence regarding the neural basis of syntactic encoding. Evidence suggests that a particular form of aphasia due to neurodegenerative disease, progressive nonfluent/agrammatic aphasia, is associated with syntactic deficits, and often involves thinning of ventrolateral prefrontal regions, especially the pars opercularis (Sapolsky et al., 2010; Wilson et al., 2010; for review, see Kemmerer, 2015). These observations converge with brain-imaging evidence, suggesting that sentence production (as opposed to, for example, producing word lists) shows activity in and around Broca's area (e.g., Indefrey et al., 2001; Indefrey, Hellwig, Herzog, Seitz, & Hagoort, 2004). In a functional MRI study on healthy participants producing and comprehending sentences, Segaert, Menenti, Weber, Petersson, and Hagoort (2012) observed that repetition of syntactic structure (e.g., actives, passives), both within and between production and comprehension, reduced activity in Broca's area as well as in left posterior MTG. According to Segaert et al., syntactic frames are retrieved from posterior MTG and used to construct sentences in Broca's area in production and comprehension. In section 1.1 (see also figure 4.2), we indicated that Broca's area is involved in the top-down control of word retrieval and in word-form encoding, including morphological and phonological encoding. Here, we see that Broca's area is involved in syntactic encoding as well. Tractography evidence suggests that the arcuate fasciculus mediates the mapping of syntactic information between posterior MTG and Broca's area (e.g., Wilson et al., 2011), see also figure 4.2.

Evidence from stroke patients supports the distinction between function assignment and constituent assembly (see Kemmerer, 2015, for a review). Neuropsychological evidence points to the possibility that function assignment may specifically be impaired, as at least some patients have shown language deficits where they tend to bind the wrong referents to grammatical functions (e.g., saying *the boy hits the ball* to describe a situation where a boy is hit by a ball; e.g., Schwartz, Linebarger, & Saffran, 1985). Evidence concerning the neural locus of constituent assembly comes from Thothathiri, Schwartz, and Thompson-Schill (2010). They tested stroke patients in a task that induced interference that is highly likely localized at a positional level of processing (i.e., constituent assembly). Patients saw a pair of pictures that they named in a conjoined phrase (*The eye and the pencil*). In critical trials, one of the pictures appeared in the ordinal position opposite of where it had appeared on the preceding two trials, inducing interference in ordering—positional interference. Two patients had unusual difficulty on such interference trials, and both had damage to a small amount of brain tissue at the junction of Brodmann's areas 44 and 6. This points to the possibility that this brain region plays a critical role in the ordering process that operates at the level of positional processing.

2.2. An Example of the Sentence Production Stages in Action Imagine that a speaker wished to inform his or her listener of a particular fact about the best-known comet in our solar system, leading ultimately to the utterance, *Halley's comet orbits the sun every 75 years.* This act forms the beginning of the sentence-production process, namely, forming the speaker's intent for the utterance—here, to inform the listener of a particular astronomical fact. This intent then leads sentence-production mechanisms to specify a multifaceted message consisting of semantic, relational, and perspective meaning. The semantic meaning represents that the speaker needs to describe objects and actions, including particular celestial bodies (those we refer to as *Halley's comet* and *sun*), a particular duration (75 years), and a particular action (*orbiting*). The relational meaning represents that the comet orbits the sun (and not the other way around), and that the duration refers to a property of the orbiting process. Perspective meaning represents that, for example, the identity of the celestial bodies is foregrounded and the duration is added information, which will lead to *Halley's comet orbits the sun every 75 years* rather than *Every 75 years, Halley's comet orbits the sun.*

This multifaceted message serves as input to the processes within the content and structure streams. Lemma retrieval retrieves the content words in the sentence, including the lemmas corresponding to *comet* and *sun*, representing them as singular nouns, and *orbit,* representing it as a verb. Function assignment retrieves a grammatical subject function for *Halley's comet* and a direct object function for *the sun*, as well as an adjunct function to express the information about the duration of the orbiting action. Lemma retrieval and function assignment must coordinate to bind lemmas to their corresponding functions, so that the lemmas of *Halley's comet* are assigned to the subject function and not the direct object function. As lemmas are retrieved, word-form encoding processes can determine the morphological structure of a complex form such as *orbits*, representing that it consists of a base verb form (*orbit*) plus a singular affix (*-s*). In tandem, constituent assembly processes determine hierarchical relations that allow *the sun* to be the direct object of *orbit* and *Halley's comet* as the subject of the entire verb phrase *orbits the sun*, as well as (based on the represented perspective meaning) the arrangement of the adjunct phrase *every 75 years* after the main clause rather than before. As the hierarchical structure is specified, constituent assembly determines the linear sequence of constituents, which must be specified before the parallel word-form encoding processes can construct syllables (because, for example, the third syllable of *Halley's comet* may be encoded as /skɑ/, with the /s/ from *Halley's* grouped into the same syllable as the /kɑ/ from *comet*). With the constituent structure and syllabic (and phonemic) content of the sentence determined, prosodic mechanisms can then compute pitch modulations, as well as the durations of individual phonemes and pauses between phonemes to determine the ultimate form the utterance will take before it is sent to articulatory processes.

2.3. Beyond Staged Models of Sentence Production The classic staged model of sentence production has provided a valuable service to the field since the modern era of sentence-production research began (in the 1970s, mainly from work published in Fromkin, 1971; Garrett, 1975; Shattuck-Hufnagel, 1979). It has organized much of the empirical work that has been undertaken in the decades since, and aspects of the theoretical framework remain viable today. However, some recent research efforts have moved the field to questions that are to some extent orthogonal to the claims of the classic framework. Here, we describe some of these efforts.

First, one of the most prominent phenomena in the field of sentence production, often termed *syntactic priming*, is not particularly usefully explained within the classic model. Syntactic priming, first reported by Levelt and Kelter (1982) and Bock (1986b), refers to the tendency whereby speakers will produce current sentences (often called *target* sentences) with the same syntactic structure as they encountered in previous sentences (often called *prime* sentences). For example, if a speaker hears or says a prepositional dative sentence like *the governess made [a pot of tea] [for the princess]*, she or he is likely to describe a subsequent picture with another prepositional dative like *the girl is handing [the paintbrush] [to the boy]*, but instead, if the speaker hears or says a double-object dative like *the governess made [the princess] [a pot of tea]*, she or he is likely to describe a picture also with a double object, like *the girl is handing [the boy] [a paintbrush]* (Bock, 1986b).

Syntactic priming has proven extremely useful for exploring the basis of the production of sentence structure (for reviews, see Mahowald, James, Futrell, & Gibson, 2016; Pickering & Ferreira, 2008), which is squarely within the processing domain of the structure stream we described. The difficulty is that syntactic priming effects crosscut the distinctions posited within the structural stream by the classic model. For example, syntactic alternations that unambiguously involve differences in function assignment, such as the transitive alternation (wherein the active, the agent argument is assigned the grammatical subject and the patient the direct object, but in the passive, the reverse is true), cause structural priming effects (e.g., Bock, 1986b), suggesting that priming may operate at the level of function assignment. But syntactic alternations that do not involve differences in function assignment, such as the production of the optional complementizer *that* (as in "The coach knew [that] his starting quarterback missed practice"; Ferreira, 2003), also cause structural priming, suggesting that priming also operates at the level of constituent assembly. It is possible that priming operates at both levels, but to the extent that it does, the distinction posited by the classic model is not useful. Even more seriously, some evidence fails to support a prediction of the classic model: Syntactic alternations that have the same set of grammatical functions do not prime each other relative to other structures that do have such distinctions. For example, a *shifted* prepositional dative like *The driver showed to the mechanic the torn overalls* does not prime a regular prepositional dative like *The patient showed his injury to the doctor* any more than a neutral prime sentence like *The racing driver fainted*, even though the shifted preposition has the same grammatical function assignments as the (straight) prepositional dative does (Pickering, Branigan, & McLean, 2002). This raises the question of why distinctions in grammatical function assignment sometimes cause syntactic priming (as in the transitive alternation) and sometimes do not.

In fact, syntactic priming has played a part in promoting a different emphasis within sentence production, one that has influenced language production (and the cognitive sciences outside of language research) more broadly: A focus on how linguistic knowledge is subject to learning and tuning as a function of ongoing experience, even in adulthood (in the sentence-production domain, see Chang, Dell, & Bock, 2006; in the lexical domain, see Oppenheim, Dell, & Schwartz, 2010; in the syllabic domain, see Dell, Reed, Adams, & Meyer, 2000; for comprehension effects, see Gaskell & Dumay, 2003, Leach & Samuel, 2007, and Norris, McQueen, & Cutler, 2003). This is a second research effort that is less usefully explored within the classic sentence-production approach, as the approach does not make claims about learning and tuning effects. Instead, syntactic priming (as well as many of the other noted learning and tuning effects) has been used to develop rich computational frameworks for understanding how we acquire and then adapt the mechanisms that are used to convert multifaceted prelinguistic messages into grammatical sequences of words—frameworks that do not appeal to the distinctions made within the structural stream of the classic model (e.g., Chang et al., 2006).

The exploration of learning and tuning effects in adult language production (and comprehension) has led to a third research effort that is less usefully explored within the classic framework: An emphasis on the rational or optimal nature of linguistic behavior (see Kleinschmidt & Jaeger, 2015; Jaeger, 2010; for a broad framework for such analyses, see Clark, 2013). In an important way, this approach represents a different emphasis for developing our understanding of sentence-production phenomena (as well as cognitive phenomena more generally). In particular, the emphasis in most approaches to sentence production (including the classic approach described) has been on the representations and processes that give rise to sentence-production behavior (what is sometimes called the *algorithmic* or *representational* level of analysis; Marr, 1982). Instead, recent research efforts that focus on the rational or optimal nature of linguistic behavior more strongly emphasize the *design* properties of the system (what is sometimes called the *computational* level of analysis; Marr).

The most prominent of these rational approaches to sentence-production behavior is *uniform-information density* (Jaeger, 2010). The idea underlying uniform-information density is that each overt feature of a sentence carries more or less information for an addressee, both in terms of meaning and in terms of grammatical and relational structure. Information-theoretically, it is optimal if information is transmitted through a channel (i.e., a sentence that is spoken by a speaker for an addressee) as uniformly as possible. "Peaks" of information—highly informative short stretches of language—can overwhelm processing systems (both production and comprehension), whereas information troughs—weakly informative longer stretches of language—underutilize the channel's bandwidth. According to uniform-information density, the communicative process operates optimally if such peaks and troughs can be smoothed out—and so that is what the sentence-formulation process does, either by adding content to stretch an information peak, or omitting content to condense an information trough.

For example, an optional *that* in a sentence like *The coach knew (that) his starting quarterback missed practice* carries useful information in that it signals that the material following the *that* is likely to be a complete clause (*his starting quarterback missed practice*) rather than a simple noun phrase (as in *The coach knew his starting quarterback well*). This raises a puzzle: If the optional *that* carries useful information, why is it optional? According to the uniform-information density framework, more information is not always better. In particular, if properties of a sentence other than the optional *that* already successfully indicate that a complete clause is upcoming, then the *that* can be unnecessary (from an information-theoretical perspective), and can even "overload" the information-processing channel. For example, if the main verb of a sentence is often followed by a complete clause, then the *that* is not as useful, and speakers should be more likely to omit it (for evidence of just such an effect, see Ferreira & Schotter, 2013; Jaeger, 2010).

Critically, the uniform-information density analysis is (strictly speaking) agnostic with respect to the specific representations or processes that give rise to behaviors that fit with the framework (though it is more compatible with some hypothetical mechanisms than others). One possibility is that the production system explicitly represents the information value of the different features of a sentence as it is produced, and decision-making processes within sentence production consult these information values to maintain uniform-information density. Or, uniform-information behaviors may emerge more implicitly, based on correlated cognitive factors that are otherwise independently needed for production. For example, a well-documented effect in production is the *accessibility effect* (see Bock, 1982, 1986a). Words that are easier to retrieve from memory (more accessible words) tend to be positioned in sentences sooner, whereas words that are harder to retrieve from memory (less accessible words) tend to be positioned later. If higher-information words tend to be harder to retrieve from memory (a reasonable conjecture) and so will tend to be mentioned later in a sentence, and production mechanisms include other material (such as the optional *that*) to allow such later mention (Ferreira & Dell, 2000), then the information-density effect emerges. But important for present purposes, the precise nature of the relevant cognitive mechanisms is not as critical as the design principle that is explicated by the framework in the first place. This is the sense in which rational- or optimality-motivated approaches to sentence production represent a shift in the field—but one that can complement the traditional approach rather than replace it.

In the remainder, we return to a description of the cognitive mechanisms underlying speaking and their neural basis. We discuss the relation between speech production and comprehension, self-monitoring, and the attention demands of speaking.

3. Interfacing Systems

3.1. RELATION TO SPEECH COMPREHENSION According to Levelt et al. (1999), lemmas are shared between production and comprehension (see figure 4.2). Lemmas were localized to the middle section of the left MTG in a meta-analysis of neuroimaging studies on spoken word production by Indefrey and Levelt (2004). This evidence from normal healthy speakers has been corroborated by evidence from voxel-based lesion-deficit analyses concerning production and comprehension deficits in individuals with poststroke aphasia. The Montreal Neurological Institute (MNI) coordinates (a common brain atlas), of the middle MTG area identified by Indefrey and Levelt, are $y = -6$ to -39 mm. We provide MNI coordinates to make clear that the "anterior" temporal area for lemmas identified by lesion-deficit analyses on production (Schwartz et al., 2009) actually overlaps with the "posterior" temporal area identified for comprehension (Dronkers, Wilkins, Van Valin, Redfern, & Jaeger, 2004). Schwartz et al. (2009) found that damage to the left middle MTG (centered at MNI $y = -18$ mm), taken to be part of the "left anterior temporal lobe," was most highly associated with semantic error rate in picture naming (see also Walker et al., 2011). Moreover, Baldo, Arévalo, Patterson, and Dronkers (2013) observed that performance on the Boston Naming Test was also most critically dependent on damage to the left mid-posterior MTG (centered at MNI $y = -24$ mm). A voxel-based lesion-deficit analysis of spoken sentence comprehension by Dronkers et al. (2004) indicated that performance was most affected by damage to the left mid-posterior MTG (approximately MNI $y = -23$ to -63 mm), which they referred to as the "posterior middle temporal gyrus." Patients with damage to this area also performed worse on tests of single-word comprehension and picture naming than patients whose lesions spared the area. In investigating resting functional and structural connectivity of the MTG region identified by Dronkers et al. (2004), Turken and Dronkers (2011) observed that its anterior part (MNI $y = -18$ to -44 mm) showed the strongest connectivity with other parts of the sentence comprehension network. To conclude, neuroimaging evidence from healthy adult individuals as well as voxel-based lesion-deficit analyses suggests that, during comprehension and production, lemmas located in the left middle MTG mediate between conceptual, syntactic, and phonological structures in the neighboring cortex.

However, the evidence on lemmas is not universally accepted. On some accounts, there are no lemmas, and the left middle MTG is assumed to play no role in language comprehension and production (Ueno et al., 2011). According to others (Hickock & Poeppel, 2007), a lexical interface exists in the brain but it resides bilaterally in the posterior MTG and posterior inferior temporal sulcus or gyrus.

3.2. SELF-MONITORING During planning and articulation, a speaker monitors progress and initiates a correction if needed. Errors may not only be detected in overt speech, but also already before articulation onset, which suggests monitoring of inner speech. Internal monitoring explains why error-to-interruption times may be as short as 100–150 ms (Blackmer & Mitton, 1991), which seems too short for error detection based on auditory perception. Lackner and Tuller (1979) elicited speech errors with or without masking noise and observed that errors were detected faster in the condition with noise. This suggests that internal monitoring takes place, which leads to faster error detection than external monitoring. Using an error elicitation procedure, Motley, Camden, and Baars (1982) observed that phoneme exchanges that led to taboo words (e.g., *shit head* for *hit shed*) were made less often than normal exchanges (e.g., *bad mack* for *mad back*), but they led to an elevated galvanic skin response, suggesting that the taboo words were internally detected and suppressed.

According to a comprehension-based account, internal monitoring is accomplished by internal feedback through the speech perception system and a comparison of production and comprehension representations (Levelt, 1989; Levelt et al., 1999). In contrast, according to a production-based account, internal monitoring is based on information inside the production system, such as conflict (Nozari et al., 2011). A comprehension-based account predicts perception-specific effects, like the perceptual uniqueness-point effect, in the monitoring of internal speech. Özdemir, Roelofs, and Levelt (2007) tested this prediction using internal phoneme monitoring and picture naming tasks. They observed an effect of the perceptual uniqueness point of a word in internal phoneme monitoring but not in picture naming, which supports comprehension-based internal monitoring. However, self-monitoring and comprehension abilities may be uncorrelated in aphasia (Nozari et al., 2011). Moreover, the abilities may double dissociate. Some aphasic patients show poor error detection in spite of intact comprehension (e.g., Marshall, Robson, Pring, & Chiat, 1998), while other patients show the reverse (Marshall, Rappaport, & Garcia-Bunuel, 1985). However, under the comprehension-based account, good comprehension is neither necessary nor sufficient for error detection. Patients may have poor comprehension due to deficient processing of external speech, which may leave internal monitoring intact (perhaps as in Marshall et al., 1985). Also, patients may have good comprehension ability, but problems with error detection because of the comparison process required (perhaps as in Marshall et al., 1998).

According to the production-based account of self-monitoring proposed by Nozari et al. (2011), errors are detected by monitoring for conflict at lemma and phoneme levels within the production system. The conflict monitor is thought to be located in the anterior cingulate cortex, following a prominent theory about human performance monitoring of Botvinick, Braver, Barch, Carter, and Cohen (2001). Yeung, Botvinick, and Cohen (2004) provided evidence from a flanker task with manual responding that the N2 and error-related negativity (ERN) components of the event-related brain potential reflect response conflict and arise in the anterior cingulate cortex. Nozari et al. showed that their proposal explains the double dissociation between comprehension and self-monitoring in aphasia, and in particular, the observation that production but not comprehension parameters predict monitoring performance in patients. However, the account is challenged by neuroimaging evidence that anterior cingulate cortex activity reflects regulation rather than conflict monitoring (e.g., Aarts, Roelofs, & Van Turennout, 2008; Roelofs, Van Turennout, & Coles, 2006), and evidence that the magnitudes of the ERN and N2 are inversely related to the amount of conflict in manual responding (Burle, Roger, Allain, Vidal & Hasbroucq, 2008) and picture naming (Shao, Roelofs, Acheson, & Meyer, 2014).

3.3. Attention Demands Although speaking is one of our most highly exercised psychomotor skills, evidence suggests that it can have detrimental effects on unrelated tasks such as driving (Kubose et al., 2006) and vice versa (Strayer & Drews, 2007). In particular, the selection of lemmas and lexical forms precludes simultaneous selection of responses in an unrelated task (Ferreira & Pashler, 2002). These findings suggest that speaking does not happen completely automatically, but requires some form of attention.

According to an influential theory proposed by Posner and colleagues (Posner, 2012, for a recent review), attention consists of alerting, orienting, and executive control. Alerting is the ability to achieve and maintain alertness, either briefly or sustained over an extended period of time. Orienting is the ability to shift the locus of processing toward a particular spatial position, either with or without corresponding eye movements. Executive control is the ability to regulate other processes to remain goal directed in the face of distraction. According to Miyake and colleagues (Miyake et al., 2000), executive control can be decomposed into updating (the ability to actively manipulate the contents of working memory), inhibiting (the ability to lower the activation of unwanted responses), and shifting (the ability to rapidly switch back and forth between tasks or mental sets). Evidence suggests that all these attentional abilities contribute to speaking. Speakers have to maintain alertness in producing utterances, and as a consequence, their sustained attention ability influences the speed of naming and describing pictures (Jongman, Roelofs, & Meyer, 2015). Also, speakers need to orient to perceptual information until the phonological form of the corresponding utterance has been encoded (e.g., Meyer, Sleiderink, & Levelt, 1998). For example, using eye tracking, it has been observed that speakers gaze longer at pictures with disyllabic than monosyllabic names (Meyer, Roelofs, & Levelt, 2003). Speakers must update the contents of working memory while engaging in conceptual and linguistic processes, and as a consequence, updating ability determines the speed of picture naming (Piai & Roelofs, 2013; Shao, Roelofs, & Meyer, 2012) and picture description (Sikora, Roelofs, Hermans, & Knoors, 2016). Moreover, the inhibiting ability is required to suppress incorrect names that are coactivated during lexical selection (Shao et al., 2014). Furthermore, the shifting ability is needed to switch between planning one type of phrase to another (Sikora, Roelofs, Hermans, & Knoors, 2016). Evidence from event-related brain potentials suggests that the shifting and inhibiting abilities are engaged before the updating ability in speaking (Sikora, Roelofs, & Hermans, 2016).

4. Conclusion

In the 1970s and 1980s, accounts of the cognitive architecture of speaking (i.e., Garrett, 1975; Levelt, 1989) were mainly based on behavioral evidence, including analyses of speech errors and reaction times. During the past three decades, new accounts have been advanced building on this classic work. These accounts synthesize evidence from reaction times and errors as well as electrophysiological and hemodynamic neuroimaging, tractography, and lesion-deficit analyses. However, many controversies endure and there remains a lot to be learned. Whereas much is known about the neural basis of word production, comparable evidence on sentence production is lacking. Correspondingly, whereas comprehensive computationally implemented accounts of word production have been developed, comparable accounts of sentence production are lacking. We expect future cognitive neuroscience research and modeling to make progress in these areas.

REFERENCES

Aarts, E., Roelofs, A., & Van Turennout, M. (2008). Anticipatory activity in anterior cingulate cortex can be independent of conflict and error likelihood. *Journal of Neuroscience, 28,* 4671–4678.

Abdel Rahman, R., & Melinger, A. (2009). Semantic context effects in language production: A swinging lexical network proposal and a review. *Language and Cognitive Processes, 24,* 713–734.

Baldo, J. V., Arévalo, A., Patterson, J. P., & Dronkers, N. F. (2013). Grey and white matter correlates of picture naming: Evidence from a voxel-based lesion analysis of the Boston Naming Test. *Cortex, 49,* 658–667.

Bierwisch, M., & Schreuder, R. (1992). From concepts to lexical items. *Cognition, 42,* 23–60.

Blackmer, E. R., & Mitton, J. L. (1991). Theories of monitoring and the timing of repairs in spontaneous speech. *Cognition, 39,* 173–194.

Bock, J. K. (1982). Toward a cognitive psychology of syntax: Information processing contributions to sentence formulation. *Psychological Review, 89,* 1–47.

Bock, J. K. (1986a). Meaning, sound, and syntax: Lexical priming in sentence production. *Journal of Experimental Psychology: Learning, Memory, and Cognition, 12,* 575–586.

Bock, J. K. (1986b). Syntactic persistence in language production. *Cognitive Psychology, 18,* 355–387.

Bock, J. K. (1995). Sentence production: From mind to mouth. In J. L. Miller & P. D. Eimas (Eds.), *Handbook of perception and cognition. Vol 11. Speech, language, and communication* (pp. 181–216). Orlando, FL: Academic Press.

Bock, J. K., & Ferreira, V. S. (2014). Syntactically speaking. In M. Goldrick, V. Ferreira, & M. Miozzo (Eds.), *The Oxford handbook of language production* (pp. 21–46). Oxford: Oxford University Press.

Bock, J. K., & Levelt, W. J. M. (1994). Language production: Grammatical encoding. In M. A. Gernsbacher (Ed.), *Handbook of psycholinguistics* (pp. 945–984). San Diego: Academic Press.

Botvinick, M. M., Braver, T. S., Barch, D. M., Carter, C. S., & Cohen, J. D. (2001). Conflict monitoring and cognitive control. *Psychological Review, 108,* 624–652.

Brauer, J., Anwander, A., Perani, D., & Friederici, A. D. (2013). Dorsal and ventral pathways in language development. *Brain and Language, 127,* 289–295.

Bürki, A., Pellet, P., & Laganaro, M. (2015). Do speakers have access to a mental syllabary? ERP comparison of high frequency and novel syllable production. *Brain and Language, 150,* 90–102.

Burle, B., Roger, C., Allain, S., Vidal, F., & Hasbroucq, T. (2008). Error negativity does not reflect conflict: A reappraisal of conflict monitoring and anterior cingulate activity. *Journal of Cognitive Neuroscience, 20,* 1637–1655.

Caramazza, A., & Miozzo, M. (1997). The relation between syntactic and phonological knowledge in lexical access: Evidence from the "tip-of-the-tongue" phenomenon. *Cognition, 64,* 309–343.

Caramazza, A., & Miozzo, M. (1998). More is not always better: A response to Roelofs, Meyer, and Levelt. *Cognition, 69,* 231–241.

Chang, F., Dell, G. S., & Bock, K. (2006). Becoming syntactic. *Psychological Review, 113,* 234–272.

Chang, F., & Fitz, H. (2014). Computational models of sentence production: A dual-path approach. In M. Goldrick, V. Ferreira, & M. Miozzo (Eds.), *The Oxford handbook of language production* (pp. 70–87). Oxford: Oxford University Press.

Cholin, J., Dell, G. S., & Levelt, W. J. M. (2011). Planning and articulation in incremental word production: Syllable-frequency effects in English. *Journal of Experimental Psychology: Learning, Memory, and Cognition, 37,* 109–122.

Clark, A. (2013). Whatever next? Predictive brains, situated agents, and the future of cognitive science. *Behavioral and Brain Sciences, 36,* 181–204.

Dell, G. S. (1986). A spreading-activation theory of retrieval in sentence production. *Psychological Review, 93,* 283–321.

Dell, G. S., Reed, K. D., Adams, D. R., & Meyer, A. S. (2000). Speech errors, phonotactic constraints, and implicit learning: A study of the role of experience in language production. *Journal of Experimental Psychology: Learning, Memory, and Cognition, 26,* 1355–1367.

Dell, G. S., Schwartz, M. F., Martin, N., Saffran, E. M., & Gagnon, D. A. (1997). Lexical access in aphasic and nonaphasic speakers. *Psychological Review, 104,* 801–838.

Dell, G. S., Schwartz, M. F., Nozari, N., Faseyitan, O., & Coslett, H. B. (2013). Voxel-based lesion-parameter mapping: Identifying the neural correlates of a computational model of word production. *Cognition, 128,* 380–396.

Dhooge, E., & Hartsuiker, R. J. (2010). The distractor frequency effect in picture-word interference: Evidence for response exclusion. *Journal of Experimental Psychology: Learning, Memory, and Cognition, 36,* 878–891.

Dronkers, N. F., Wilkins, D. P., Van Valin, R. D., Redfern, B. B., & Jaeger, J. J. (2004). Lesion analysis of the brain areas involved in language comprehension. *Cognition, 92,* 145–177.

Eichenbaum, H. (2012). *The cognitive neuroscience of memory: An introduction* (2nd ed.). Oxford: Oxford University Press.

Fernandez-Miranda, J. C., Wang, Y., Pathak, S., Stefaneau, L., Verstynen, T., & Yeh, F.-C. (2015). Asymmetry, connectivity, and segmentation of the arcuate fascicle in the human brain. *Brain Structure and Function, 220,* 1665–1680.

Ferreira, F. (1993). Creation of prosody during sentence production. *Psychological Review, 100,* 233–253.

Ferreira, V. S. (2003). The persistence of optional complementizer mention: Why saying a "that" is not saying "that" at all. *Journal of Memory and Language, 48*, 379–398.

Ferreira, V. S. (2010). Language production. *Wiley Interdisciplinary Reviews: Cognitive Science, 1*, 834–844.

Ferreira, V. S., & Dell, G. S. (2000). Effect of ambiguity and lexical availability on syntactic and lexical production. *Cognitive Psychology, 40*, 296–340.

Ferreira, V. S., & Pashler, H. (2002). Central bottleneck influences on the processing stages of word production. *Journal of Experimental Psychology: Learning, Memory, and Cognition, 28*, 1187–1199.

Ferreira, V. S., & Schotter, E. R. (2013). Do verb bias effects on sentence production reflect sensitivity to comprehension or production factors? *Quarterly Journal of Experimental Psychology, 66*, 1548–1571.

Ferreira, V. S., & Slevc, L. R. (2007). Grammatical encoding. In G. Gaskell (Ed.), *Oxford handbook of psycholinguistics* (pp. 453–470). Oxford: Oxford University Press.

Finkbeiner, M., & Caramazza, A. (2006). Now you see it, now you don't: On turning semantic interference into facilitation in a Stroop-like task. *Cortex, 42*, 790–796.

Fromkin, V. A. (1971). The non-anomalous nature of anomalous utterances. *Language, 47*, 27–52.

Garrett, M. F. (1975). The analysis of sentence production. In G. H. Bower (Ed.), *The psychology of learning and motivation* (Vol. 9, pp. 133–177). New York: Academic Press.

Garrett, M. F. (1980). Levels of processing in sentence production. In B. Butterworth (Ed.), *Language production* (Vol. 1, pp. 177–220). London: Academic Press.

Garrett, M. F. (1982). Production of speech: Observations from normal and pathological language use. In A. Ellis (Ed.), *Normality and pathology in cognitive functions* (pp. 19–76). London: Academic Press.

Garrett, M. F. (1988). Processes in language production. In F. J. Newmeyer (Ed.), *Linguistics: The Cambridge survey* (Vol. 3, pp. 69–96). Cambridge, MA: Harvard University Press.

Gaskell, M. G., & Dumay, N. (2003). Lexical competition and the acquisition of novel words. *Cognition, 89*, 105–132.

Geschwind, N. (1970). The organization of language and the brain. *Science, 170*, 940–944.

Glasser, M. F., & Rilling, J. K. (2008). DTI tractography of the human brain's language pathways. *Cerebral Cortex, 18*, 2471–2482.

Guenther, F. H., & Vladusich, T. (2012). A neural theory of speech acquisition and production. *Journal of Neurolinguistics, 25*, 408–422.

Hagoort, P., & Indefrey, P. (2014). The neurobiology of language beyond single words. *Annual Review of Neuroscience, 37*, 347–362.

Haller, S., Radue, E. W., Erb, M., Grodd, W., & Kircher, T. (2005). Overt sentence production in event-related fMRI. *Neuropsychologia, 43*, 807–814.

Hickok, G. (2012). Computational neuroanatomy of speech production. *Nature Reviews Neuroscience, 13*, 135–145.

Hickok, G., & Poeppel, D. (2007). The cortical organization of speech processing. *Nature Reviews Neuroscience, 8*, 393–402.

Indefrey, P. (2011). The spatial and temporal signatures of word production components: A critical update. *Frontiers in Psychology, 2*, article 255.

Indefrey, P. (2016). On putative shortcomings and dangerous avenues: Response to Strijkers & Costa. *Language, Cognition and Neuroscience, 31*, 517–520.

Indefrey, P., Brown, C. M., Hellwig, F., Amunts, K., Herzog, H., Seitz, R. J., & Hagoort, P. (2001). A neural correlate of syntactic encoding during speech production. *Proceedings of the National Academy of Sciences, 98*, 5933–5936.

Indefrey, P., Hellwig, F., Herzog, H., Seitz, R. J., & Hagoort, P. (2004). Neural responses to the production and comprehension of syntax in identical utterances. *Brain and Language, 89*, 312–319.

Indefrey, P., & Levelt, W. J. M. (2004). The spatial and temporal signatures of word production components. *Cognition, 92*, 101–144.

Jaeger, T. F. (2010). Redundancy and reduction: Speakers manage syntactic information density. *Cognitive Psychology, 61*, 23–62.

Jongman, S. R., Roelofs, A., & Meyer, A. S. (2015). Sustained attention in language production: An individual differences investigation. *Quarterly Journal of Experimental Psychology, 68*, 710–730.

Kemmerer, D. (2015). *Cognitive neuroscience of language.* New York: Psychology Press.

Kleinschmidt, D. F., & Jaeger, T. F. (2015). Robust speech perception: Recognize the familiar, generalize to the similar, and adapt to the novel. *Psychological Review, 122*, 148–203.

Konopka, A. E., & Brown-Schmidt, S. (2014). Message encoding. In M. Goldrick, V. Ferreira, & M. Miozzo (Eds.), *The Oxford handbook of language production* (pp. 3–20). Oxford: Oxford University Press.

Kubose, T. T., Bock, K., Dell, G. S., Garnsey, S. M., Kramer, A. F., & Mayhugh, J. (2006). The effects of speech production and speech comprehension on simulated driving performance. *Applied Cognitive Psychology, 20*, 43–63.

Lackner, J. R., & Tuller, B. H. (1979). Role of efference monitoring in the detection of self-produced speech errors. In W. E. Cooper & E. C. T. Walker (Eds.), *Sentence processing* (pp. 281–294). Hillsdale, NJ: Erlbaum.

Lambon Ralph, M. A. (2014). Neurocognitive insights on conceptual knowledge and its breakdown. *Philosophical Transactions of the Royal Society B, 369*, 20120392.

Lambrecht, K. (1994). *Information structure and sentence form.* Cambridge: Cambridge University Press.

Leach, L., & Samuel, A. G. (2007). Lexical configuration and lexical engagement: When adults learn new words. *Cognitive Psychology, 55*, 306–353.

Levelt, W. J. M. (1989). *Speaking: From intention to articulation.* Cambridge, MA: MIT Press.

Levelt, W. J. M. (2013). *A history of psycholinguistics: The pre-Chomskyan era.* Oxford: Oxford University Press.

Levelt, W. J. M., & Kelter, S. (1982). Surface form and memory in question answering. *Cognitive Psychology, 14*, 78–106.

Levelt, W. J. M., Roelofs, A., & Meyer, A. S. (1999). A theory of lexical access in speech production. *Behavioral and Brain Sciences, 22*, 1–38.

Lichtheim, L. (1885). On aphasia. *Brain, 7*, 433–484.

Mahowald, K., James, A., Futrell, R., & Gibson, E. (2016). A meta-analysis of syntactic priming in language production. *Journal of Memory and Language, 91*, 5–27.

Marchina, S., Zhu, L. L., Norton, A., Zipse, L., Wan, C. Y., & Schlaug, G. (2011). Impairment of speech production predicted by lesion load of the left arcuate fasciculus. *Stroke, 42*, 2251–2256.

Marr, D. (1982). *Vision: A computational investigation into the human representation and processing of visual information.* San Francisco: Freeman & Co.

Marshall, J., Robson, J., Pring, T., & Chiat, S. (1998). Why does monitoring fail in jargon aphasia? Comprehension, judgment, and therapy evidence. *Brain and Language, 63*, 79–107.

Marshall, R. C., Rappaport, B. Z., & Garcia-Bunuel, L. (1985). Self-monitoring behavior in a case of severe auditory agnosia with aphasia. *Brain and Language, 24*, 297–313.

Meyer, A. S., Roelofs, A., & Levelt, W. J. M. (2003). Word length effects in object naming: The role of a response criterion. *Journal of Memory and Language, 48*, 131–147.

Meyer, A. S., Sleiderink, A. M., & Levelt, W. J. M. (1998). Viewing and naming objects: Eye movements during noun phrase production. *Cognition, 66*, B25–B33.

Miyake, A., Friedman, N. P., Emerson, M. J., Witzki, A. H., Howerter, A., & Wager, T. (2000). The unity and diversity of executive functions and their contributions to complex "frontal lobe" tasks: A latent variable analysis. *Cognitive Psychology, 41*, 49–100.

Morsella, E., & Miozzo, M. (2002). Evidence for a cascade model of lexical access in speech production. *Journal of Experimental Psychology: Learning, Memory, and Cognition, 28*, 555–563.

Motley, M. T., Camden, C. T., & Baars, B. J. (1982). Covert formulation and editing of anomalies in speech production: Evidence from experimentally elicited slips of the tongue. *Journal of Verbal Learning and Verbal Behavior, 21*, 578–594.

Munding, D., Dubarry, A.-S., & Alario, F.-X. (2016). On the cortical dynamics of word production: A review of the MEG evidence. *Language, Cognition and Neuroscience, 31*, 441–462.

Navarrete, E., Del Prato, P., Peressotti, F., & Mahon, B. Z. (2014). Lexical selection is not by competition: Evidence from the blocked naming paradigm. *Journal of Memory and Language, 76*, 253–272.

Norris, D., McQueen, J. M., & Cutler, A. (2003). Perceptual learning in speech. *Cognitive Psychology, 47*, 204–238.

Nozari, N., Dell, G. S., & Schwartz, M. F. (2011). Is comprehension necessary for error detection? A conflict-based account of monitoring in speech production. *Cognitive Psychology, 63*, 1–33.

Oppenheim, G. M., Dell, G. S., & Schwartz, M. F. (2010). The dark side of incremental learning: A model of cumulative semantic interference during lexical access in speech production. *Cognition, 114*, 227–252.

Özdemir, R., Roelofs, A., & Levelt, W. J. M. (2007). Perceptual uniqueness point effects in monitoring internal speech. *Cognition, 105*, 457–465.

Patterson, K., Nestor, P. J., & Rogers, T. T. (2007). Where do you know what you know? The representation of semantic knowledge in the human brain. *Nature Reviews Neuroscience, 8*, 976–987.

Piai, V., & Roelofs, A. (2013). Working memory capacity and dual-task interference in picture naming. *Acta Psychologica, 142*, 332–342.

Piai, V., Roelofs, A., Jensen, O., Schoffelen, J.-M., & Bonnefond, M. (2014). Distinct patterns of brain activity characterise lexical activation and competition in spoken word production. *PLOS ONE, 9*, e88674.

Pickering, M. J., Branigan, H. P., & McLean, J. F. (2002). Constituent structure is formulated in one stage. *Journal of Memory and Language, 46*, 586–605.

Pickering, M. J., & Ferreira, V. S. (2008). Structural priming: A critical review. *Psychological Bulletin, 134*, 427–459.

Posner, M. I. (2012). *Attention in a social world.* Oxford: Oxford University Press.

Rapp, B., & Goldrick, M. (2000). Discreteness and interactivity in spoken word production. *Psychological Review, 107*, 460–499.

Rapp, B., & Goldrick, M. (2004). Feedback by any other name is still interactivity: A reply to Roelofs' comment on Rapp & Goldrick (2000). *Psychological Review, 111*, 573–578.

Roelofs, A. (1992). A spreading-activation theory of lemma retrieval in speaking. *Cognition, 42*, 107–142.

Roelofs, A. (2004a). Comprehension-based versus production-internal feedback in planning spoken words: A rejoinder to Rapp and Goldrick (2004). *Psychological Review, 111*, 579–580.

Roelofs, A. (2004b). Error biases in spoken word planning and monitoring by aphasic and nonaphasic speakers: Comment on Rapp and Goldrick (2000). *Psychological Review, 111*, 561–572.

Roelofs, A. (2008). Tracing attention and the activation flow in spoken word planning using eye movements. *Journal of Experimental Psychology: Learning, Memory, and Cognition, 34*, 353–368.

Roelofs, A. (2014). A dorsal-pathway account of aphasic language production: The WEAVER++/ARC model. *Cortex, 59*, 33–48.

Roelofs, A., Meyer, A. S., & Levelt, W. J. M. (1998). A case for the lemma-lexeme distinction in models of speaking: Comment on Caramazza and Miozzo (1997). *Cognition, 69*, 219–230.

Roelofs, A., Van Turennout, M., & Coles, M. G. H. (2006). Anterior cingulate cortex activity can be independent of response conflict in Stroop-like tasks. *Proceedings of the National Academy of Sciences USA, 103*, 13884–13889.

Sahin, N. T., Pinker, S., Cash, S. S., Schomer, D., & Halgren, E. (2009). Sequential processing of lexical, grammatical, and phonological information within Broca's area. *Science, 326*, 445–449.

Sapolsky, D., Bakkour, A., Negreira, A., Nalipinski, P., Weintraub, S., Mesulam, M. M., ... Dickerson, B. C. (2010). Cortical neuroanatomic correlates of symptom severity in primary progressive aphasia. *Neurology, 75*, 358–366.

Schwartz, M. F., Faseyitan, O., Kim, J., & Coslett, H. B. (2012). The dorsal stream contribution to phonological retrieval in object naming. *Brain, 135*, 3799–3814.

Schwartz, M. F., Kimberg, D. Y., Walker, G. M., Faseyitan, O., Brecher, A., Dell, G. S., & Coslett, H. B. (2009). Anterior temporal involvement in semantic word retrieval: Voxel-based lesion-symptom mapping evidence from aphasia. *Brain, 132*, 3411–3427.

Schwartz, M. F., Linebarger, M. C., & Saffran, E. M. (1985). The status of the syntactic deficit theory of agrammatism. In M. L. Kean (Ed.) *Agrammatism* (pp. 83–124). New York: Academic Press.

Segaert, K., Menenti, L., Weber, K., Petersson, K. M., & Hagoort, P. (2012). Shared syntax in language production and language comprehension—an fMRI study. *Cerebral Cortex, 22*, 1662–1670.

Shallice, T., & Cooper, R. (2011). *The organisation of mind.* Oxford: Oxford University Press.

Shao, Z., Roelofs, A., Acheson, D. J., & Meyer, A. S. (2014). Electrophysiological evidence that inhibition supports lexical selection in picture naming. *Brain Research, 1586*, 130–142.

Shao, Z., Roelofs, A., & Meyer, A. S. (2012). Sources of individual differences in the speed of naming objects and

actions: The contribution of executive control. *Quarterly Journal of Experimental Psychology, 65*, 1927–1944.
Shattuck-Hufnagel, S. (1979). Speech errors as evidence for a serial-ordering mechanism in sentence production. In W. E. Cooper & E. C. T. Walker (Eds.), *Sentence processing: Psycholinguistic studies presented to Merrill Garrett* (pp. 295–342). Hillsdale, NJ: Erlbaum.
Shattuck-Hufnagel, S. (2014). Phrase-level phonological and phonetic phenomena. In M. Goldrick, V. Ferreira, & M. Miozzo (Eds.), *The Oxford handbook of language production* (pp. 259–274). Oxford: Oxford University Press.
Sikora, K., Roelofs, A., & Hermans, D. (2016). Electrophysiology of executive control in spoken noun-phrase production: Dynamics of updating, inhibiting, and shifting. *Neuropsychologia, 84*, 44–53.
Sikora, K., Roelofs, A., Hermans, D., & Knoors, H. (2016). Executive control in spoken noun-phrase production: Contributions of updating, inhibiting, and shifting. *Quarterly Journal of Experimental Psychology, 69*, 1719–1740.
Stemberger, J. P. (1985). An interactive activation model of language production. In A. Ellis (Ed.), *Progress in the psychology of language* (Vol. 1, pp. 143–186). London: Erlbaum.
Strayer, D. L., & Drews, F. A. (2007). Cell-phone-induced driver distraction. *Current Directions in Psychological Science, 16*, 128–131.
Strijkers, K., & Costa, A. (2016). The cortical dynamics of speaking: Present shortcomings and future avenues. *Language, Cognition and Neuroscience, 31*, 484–503.
Thothathiri, M., Schwartz, M. F., & Thompson-Schill, S. L. (2010). Selection for position: The role of left ventrolateral prefrontal cortex in sequencing language. *Brain and Language, 113*, 28–38.
Turken, A. U., & Dronkers, N. F. (2011). The neural architecture of the language comprehension network: Converging evidence from lesion and connectivity analyses. *Frontiers in Systems Neuroscience, 5*, article 1.
Ueno, T., Saito, S., Rogers, T. T., & Lambon Ralph, M. A. (2011). Lichtheim 2: Synthesizing aphasia and the neural basis of language in a neurocomputational model of the dual dorsal-ventral language pathways. *Neuron, 72*, 385–396.
Ullman, M. T. (2004). Contributions of memory circuits to language: The declarative/procedural model. *Cognition, 92*, 231–270.
Walker, G. M., Schwartz, M. F., Kimberg, D. Y., Faseyitan, O., Brecher, A., Dell, G. S., & Coslett, H. B. (2011). Support for anterior temporal involvement in semantic error production in aphasia: New evidence from VLSM. *Brain and Language, 117*, 110–122.
Wang, J., Marchina, S., Norton, A. C., Wan, C. Y., & Schlaug, G. (2013). Predicting speech fluency and naming abilities in aphasic patients. *Frontiers in Human Neuroscience, 7*, article 831.
Wernicke, C. (1874). *Der aphasische Symptomencomplex. Eine psychologische Studie auf anatomischer Basis.* Breslau, Germany: Cohn und Weigert.
Wilson, S. M., Galantucci, S., Tartaglia, M. C., Rising, K., Patterson, D. K., Henry, M. L., ... Gorno-Tempini, M. L. (2011). Syntactic processing depends on dorsal language tracts. *Neuron, 72*, 397–403.
Wilson, S. M., Henry, M. L., Besbris, M., Ogar, J. M., Dronkers, N. F., Jarrold, W., ... Gorno-Tempini, M. L. (2010). Connected speech production in three variants of primary progressive aphasia. *Brain, 133*, 2069–2088.
Wundt, W. (1900). *Die Sprache.* Leipzig: Verlag von Wilhelm Engelmann.
Yeung, N., Botvinick, M. M., & Cohen, J. D. (2004). The neural basis of error detection: Conflict monitoring and the error-related negativity. *Psychological Review, 111*, 931–959.

5 The Cognitive Architecture of Reading: The Organization of an Acquired Skill

SALLY ANDREWS AND ERIK D. REICHLE

Reading is a highly complex skill that is unique to humans. Written language allows us to convey, transmit, and store the contents of our collective knowledge. This cultural tool has transformed and extended human's cognitive abilities (Menary & Kirschhoff, 2014). At the level of the individual, reading is an essential educational and vocational skill in modern societies.

Critically, reading is a culturally engineered skill. Written scripts were only invented around 6,000 years ago and literacy has been a widespread human capacity for less than 200 years (Robinson, 2001). Reading is therefore too recent a cultural invention to be the product of evolution. Paralleling its recent development by our species, learning to read is a late, gradual, and laborious process for individual children. This contrast with the natural, implicit, biologically determined mechanisms underlying acquisition of spoken language led Gough and Hillenger (1980) to describe learning to read as "an unnatural act" that requires explicit instruction.

These differences in the difficulty and timing of spoken and written language acquisition influence the cognitive architecture of the reading system. By the time children learn to read, they are already proficient at producing and comprehending speech. Written language therefore needs to be integrated into the system established for spoken language. By learning to map written words to their spoken forms, children can exploit the knowledge and processes developed to produce and understand speech. Reading builds on biologically determined capacities for spoken language, but requires new metacognitive insights into the symbolic structure of language that are specific to reading: "children are wired for sound but orthography must be painstakingly bolted on" (Pinker, 1997, p. ix). Learning to read, even in adulthood, leads to changes in the organization of visual cortex (Dehaene et al., 2010). Reading therefore provides unique insights into how the cognitive architecture of the mind is shaped by experience.

1. The Functional Architecture of Skilled Reading: Modular or Interactive?

The basic structural components of the cognitive architecture of reading are relatively well-accepted. The reading system requires a store of written word forms (an *orthographic lexicon*), mechanisms for mapping from print to units in this lexicon, and procedures for integrating information extracted from print with existing lexical and semantic knowledge in *long-term memory* (LTM). The central theoretical questions about the architecture of the language system concern how these various components are organized and the time course of their operation during online reading.

The *simple model of reading* (Hoover & Gough, 1990) that has played an influential role in theories of both reading development (e.g., Oakhill & Cain, 2012) and skilled reading (e.g., Perfetti, 2007) assumes that word recognition and comprehension processes rely on independent, but potentially interacting, cognitive structures. The strongest version of this approach implicates a *staged processing architecture* (Staub, 2011) in which word identification precedes comprehension: The words in a text are identified one-by-one, based on bottom-up perceptual features, with the outputs of this lexical system then providing the input to comprehension processes, which integrate this information with existing knowledge to construct the meaning of the text. A precise mathematical specification of such a word-by-word theory of the reading process is instantiated in the *E-Z Reader* model of eye movements in reading (Reichle, Pollatsek, Fisher, & Rayner, 1998) which accounts for a large body of very detailed evidence about readers' eye movements during reading (see Reichle, 2011).

Such staged accounts of the reading process are consistent with *modular* views of the architecture of the skilled reading system that assume that word-recognition processes operate *automatically* and *autonomously*, encapsulated from effects of semantic or syntactic context

(Fodor, 1983; Forster, 1979). The modular framework is supported by recent neuroimaging evidence that the same highly localized, left posterior region of visual cortex responds selectively to written text in skilled readers across languages as diverse as English and Chinese. The invariance of this *visual word-form area* across both individuals and scripts implicates a specialized cortical system that is tuned by experience with written words (Dehaene, 2009). Learning to read is associated with increasingly selective activation of the word-form area by words (compared to faces and objects), even when literacy was acquired in adulthood, indicating that experience with written language can shape cortical organization across the life span (Dehaene et al., 2010). This specialized perceptual system, with its tight connections to temporal and frontal regions implicated in phonological and semantic processing, aligns well with the staged functional architecture assumed by modular theories.

According to this modular view, word identification is central to skilled reading. But reading obviously requires much more than the identification of individual words. The meanings associated with words and phrases must be retrieved from LTM and assembled into sentence and discourse representations while comprehension is monitored to ensure sufficient understanding of the material to meet the reader's goals (e.g., van den Broek, Lorch, Linderholm, & Gustafson, 2001). The simple view of reading assumes that a modular architecture provides an effective cognitive solution to achieving these goals because, if readers can automatically retrieve the phonological and semantic codes for written words, then comprehension can exploit the mechanisms used to understand speech. As we will elaborate, this solution is consistent with many general cognitive constraints on learning and performance.

However, the modular framework is not universally accepted. Indeed, many cognitive and educational theorists advocate the completely opposite view of skilled reading. The seminal educational exemplar is Goodman's (1967) "psycholinguistic guessing game" view of reading, which argues that skilled readers use contextually driven processes to reduce the need for detailed perceptual analysis of words. The early influence of contextual knowledge in skilled reading is also emphasized by probabilistic models from cognitive psychology that challenge the distinction between lexical and comprehension processes (e.g., Jurafsky, 1996) or assume that lexical and contextual factors jointly influence processing (e.g., Demberg & Keller, 2008). Context also plays a critical role in recent accounts highlighting the role of expectancies (Federmeier, 2007) and predictive coding mechanisms (Clark, 2013), which equate expertise with generative, hypothesis-driven processes rather than efficient perceptual analysis.

The aim of this chapter is to consider *why* the reading system might develop a functionally modular architecture by focusing on the fundamental constraints on its organization and operation: the critical bottlenecks and limitations that have shaped the organization of the reading system. This approach is grounded in the "emergentist" view (McClelland et al., 2010) that the cognitive architecture of the reading system is dynamically shaped by the intersecting constraints of the mind, the brain, the language, and the writing system. We will analyze why a modular structural organization—leading to a staged processing architecture—may emerge as a relatively universal solution to the constraints imposed by the interaction between human cognitive capacities and the demands of translating written text into meaning. Figure 5.1 provides a sketch of the broad architectural and processing assumptions underlying our arguments and some of the structures and processes that we discuss. Because of the brevity of this chapter, we present our arguments using a few key examples and theories of reading that have been selected to illustrate why a modular organization emerges as an efficient, effective solution to addressing physiological and computational bottlenecks and constraints on the cognitive systems that support reading.

2. *Constraint 1: Mapping from Text to Lexical Knowledge*

Just as the development of written representations of speech required new cultural inventions by our ancestors, the successful acquisition of reading by an individual requires novel meta-cognitive insights that were not necessary to acquire spoken language. A critical challenge faced by the developing reader is to learn to map from written text to existing knowledge about spoken language. This requires them to "crack the code" that links written words to phonology and meaning and develop a body of knowledge and a set of procedures for retrieving that knowledge, which operate effectively under the perceptual and cognitive constraints associated with comprehending connected text.

Although pictographs were common in early scripts (e.g., Egyptian hieroglyphs; Robinson, 2001), all modern writing systems are *glottographic* (Pulgram, 1976): They represent the *words* of the language through a sequence of symbols representing the spoken units of the language (Kessler & Treiman, 2015). Through this ingenious, general-purpose solution, writing systems inherited the *productivity* that is a universal feature of human languages to support "creation of an infinity of

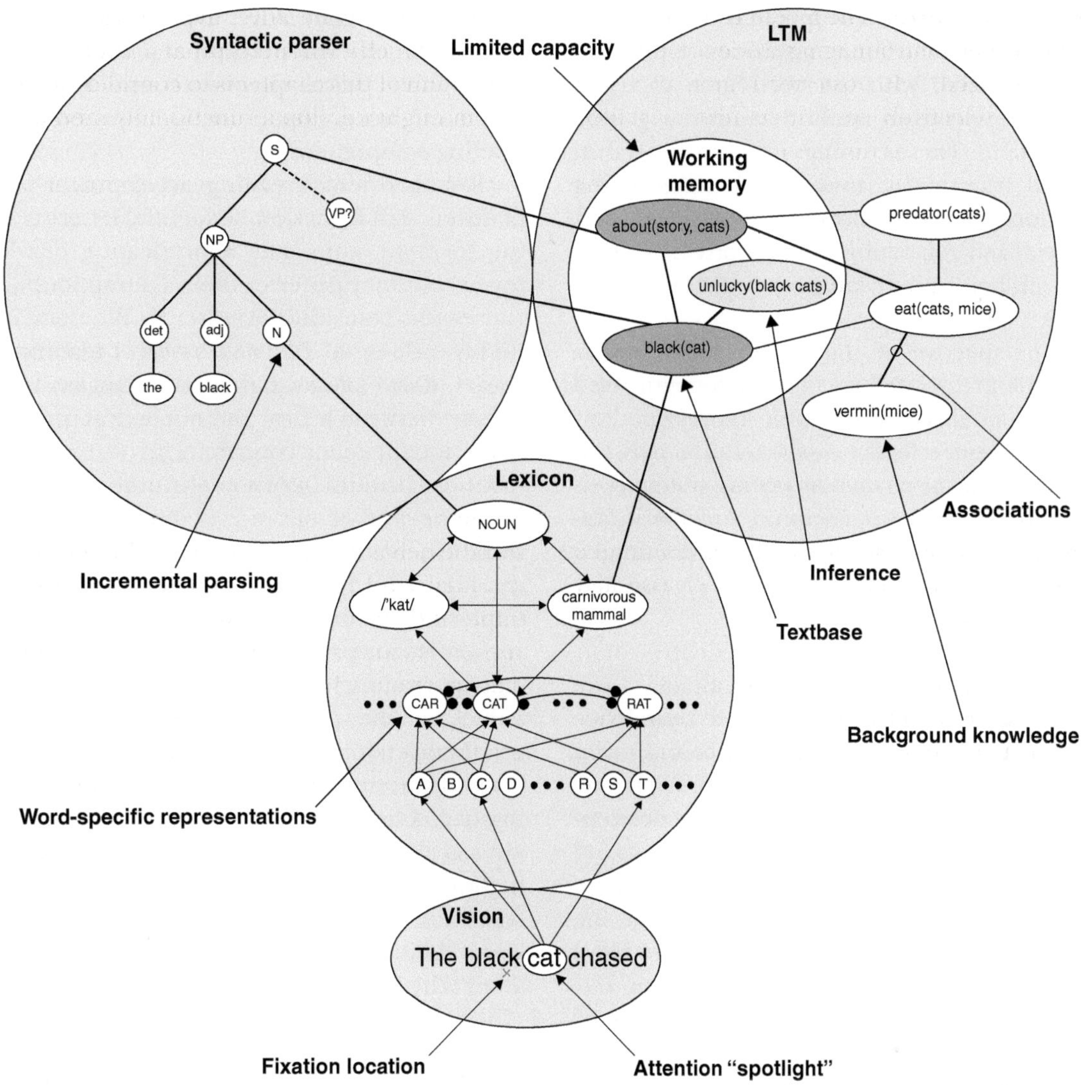

FIGURE 5.1 A tentative sketch of the cognitive architecture of reading, with key components labeled in bold. Many details of the architecture have not been shown for the purpose of clarity (e.g., sublexical units corresponding to intermediate sublexical units such graphemes and morphemes) and some of the implied theoretical assumptions (e.g., the role of lexical semantics in the textbase) are admittedly speculative. The shading of the propositions in working memory indicates their relative activation levels (e.g., the inferred proposition about black cats being unlucky is less strongly activated than the two propositions of the textbase). adj = adjective; det = determiner; N = noun; NP = noun phrase; S = subject; VP = verb phrase.

messages (sentences) from finite means (syllables, phonemes)" (Perfetti & Harris, 2013, p. 297). The cognitive advantage of coding language *form* rather than meaning is that it eliminates the need to store an infinitely large number of symbol-meaning associations. However, it carries the cost that discovering the principles linking these symbol systems to language imposes new, difficult cognitive demands.

Some of these demands vary across writing systems as a function of the "division of labor" among phonology, morphology/semantics (Harm & Seidenberg, 2004), the "grain size" of the units that link orthography to phonology (Ziegler & Goswami, 2005), and the transparency and consistency of the mapping between them (Katz & Frost, 1992). Alphabetic scripts vary from "shallow" orthographies, such as Finnish and Welsh, where letters directly correspond to sounds, to "deep" orthographies such as English, where letter-sound mapping is inconsistent and context-dependent. The gulf between alphabetic and nonalphabetic scripts is even greater. Brain imaging studies of word recognition across different writing systems (e.g., Bolger, Perfetti, & Schneider, 2005; Tan, Laird, Li, & Fox, 2005) provide evidence of script-specific variation in brain activity suggesting that "reading processes and the neural structures that support them accommodate to specific visual and structural features of writing systems" (Perfetti & Harris, 2013, p. 302). However, the data also reveal substantial similarities across scripts in the patterns of activation within both the left hemisphere occipitotemporal areas associated with the word-form area and premotor areas involved in reading handwriting (e.g., Rueckl et al., 2015). These findings implicate universal developmental trajectories involved in strengthening the connections between the posterior areas that become specialized for orthographic word identification and the left hemisphere language areas (Perfetti & Harris, 2013).

From the perspective of the modular framework, these universals are related to the development of a lexicon of high-quality orthographic representations that support efficient, effective discrimination between similar words within the particular writing system (Perfetti, 2007). This chapter will focus on alphabetic languages, and particularly English, which has dominated research and theory (Share, 2008), but briefly consider implications for other writing systems.

2.1. Establishing an Orthographic Lexicon Core phonological skills, such as letter-sound knowledge and phonemic awareness, are a critical constraint on early reading development in English (Frith & Happe, 1998). These skills are required to support the development of *alphabetic insight* into abstract spelling-sound relationships (Byrne et al., 2008). But to discriminate between the 10,000 or more words that compose the lexicon of the typical adolescent, readers must move beyond simple letter-sound associations to develop *consolidated alphabetic representations* of words defined by stable, multiletter, context-dependent graphophonic connections that amalgamate the spelling of a word with its pronunciation (Ehri, 2015). Extracting spelling-sound regularities at multiple grain sizes is particularly important for English because it is only quasi-regular at the grapheme-phoneme level (Ziegler & Goswami, 2005). Learning letter-sound associations takes less time in more transparent alphabetic orthographies, such as Spanish and German (Share, 2008), and is less sensitive to phonemic awareness (Ziegler et al., 2010). However, reading practice still plays an important role in developing the consolidated representations required for fluent word identification (Ehri, 2015).

The modular framework assumes that achieving such fluency depends on establishing a hierarchically structured orthographic lexicon of precise, well-specified word representations that can be rapidly and automatically retrieved by printed word forms to yield access to associated phonological, semantic, and syntactic knowledge (e.g., Compton, Miller, Elleman & Steacy, 2014). Perfetti and Hart (2001) coined the term *lexical quality* to capture the combination of precise, word-specific orthographic knowledge and tight interconnections with other lexical attributes, and they elaborated how high-quality lexical representations contribute to the automatic word identification required for effective reading comprehension.

Most theories of reading development predict that children shift from slow, sequential letter-sound decoding to rapid, automatic identification of whole word forms, but they differ in their assumptions about what causes this transition (Vaessen & Blomert, 2010). The highly influential *dual route model* of reading (e.g., Coltheart, Rastle, Perry, Langdon & Ziegler, 2001) distinguishes between lexical and nonlexical procedures that make independent contributions to reading development. Difficulties in one or the other route can lead to different profiles of reading impairment that parallel those observed in brain-damaged patients with acquired dyslexia (e.g., Marshall & Newcombe, 1973): selective impairment in development of the lexical route leads to *surface dyslexic* pattern, characterized by specific difficulty in reading irregular words (e.g., *yacht*), while difficulties in the nonlexical procedure for translating graphemes to phonemes is associated with a *phonological dyslexic* pattern, defined by specific difficulties in decoding (e.g., Castles & Coltheart, 1993). In contrast to this view, attributing fluent reading to the development of high-quality lexical representations implies that the transition from phonological decoding to automatic word identification is a continuous process that operates at the item level. This view is consistent with Ehri's (2015) consolidation account and Share's (1995) self-teaching hypothesis which both assume that, through practice at decoding words, children develop increasingly efficient recoding strategies that are applied more effectively the more frequently a word is encountered.

Support for such item-based accounts is provided by Vaessen and Blomert's (2010) evidence that the predictive power of measures of decoding decreased across first to sixth grades, while the contribution of *Rapid Automatized Naming (RAN)*, which measures automatic recognition of word-specific orthographic patterns (e.g., Clarke, Hulme, & Snowling, 2005; Georgiou, Parrila, Kirby, & Stephenson, 2008), gradually increased, particularly for high-frequency words. This item-specific shift in the relative contribution of decoding and RAN was argued to contradict models assuming independent processing routes and to "point to the development of one (and only one) reading network for all types of words in which processing load or type of processing may differ depending on the stimulus properties and amount of reading experience" (Vaessen & Blomert, 2010, p. 226). This evidence that word frequency modulates the developmental relationship between reading fluency and the automatic word identification indexed by RAN supports the claim by Perfetti (2007) that the quality of lexical representations varies both within and between individuals due to frequency of reading experience. The critical source of variability at both levels is "the knowledge a reader has about words, specific lexical representations" (p. 359).

Precise word-specific knowledge is also central to the numerous computational models of skilled visual word identification. In the seminal *Interactive Activation (IA)* (McClelland & Rumelhart, 1981) and *Dual-Route Cascaded (DRC)*(Coltheart et al., 2001) models, these word-specific representations and their full set of interconnections to letter- and feature-level representations are prespecified by the models, so the question of how readers acquire such a rich body of knowledge is ignored.

The issue of learning is directly addressed in the connectionist *Parallel Distributed Processing (PDP)*, or *Triangle*, model (Seidenberg & McClelland, 1989) and its progeny (e.g., Plaut, McClelland, Seidenberg, & Paterson, 1996; Harm & Seidenberg, 2004), which learn associations between orthographic, phonological, and semantic knowledge through training, although the rates of learning in these models do not scale well with that of children (e.g., Nation, Angell, & Castles, 2007). The distributed representational assumptions of these models are incompatible with assumptions of fixed, localist orthographic codes (e.g., Seidenberg & Plaut, 2006), but word-specific knowledge can instead be identified with network "attractors" that capture the covariation between the patterns of orthographic and semantic activity that defines lexical identity (Woollams, 2015).

Although these models provide very different accounts of word identification, they all posit (a) orthographic codes, representing individual letters/graphemes, that have been abstracted from visual features for the purpose of accessing word-specific knowledge and (b) a parallel activation process in which visual input is filtered through a network of orthographic and phonological associations to select the best matching word. The models differ in the precise assumptions about the levels of orthographic and phonological knowledge that are represented and how they are organized. However, in many contexts, they make identical predictions, and the models all simulate major empirical phenomena (e.g., letter-sound typicality), although sometimes for different reasons (e.g., whether effects of spelling-sound typicality is due to regularity or consistency; Andrews, 1982). The models account for similar, relatively modest proportions of variance in empirical data (Perry, Ziegler, & Zorzi, 2007), suggesting that they may fail to capture critical aspects of human performance—such as the decision processes that have been added to some more recent models (e.g., Davis, 2010; Dufau, Grainger, & Ziegler, 2012). The difficulty in empirically discriminating among the different models suggests that the detailed internal wiring of the lexical module may be of little relevance to understanding task performance in many contexts. What is critical is their shared assumption that skilled readers have established a well-structured body of lexical knowledge that allows direct, automatic retrieval of the semantic and phonological features of written words.

2.2. Selecting the Right Word From the perspective of the general architectural issues that are the focus of this chapter, the models also share a common constraint that influences the online mapping from print to lexical knowledge once an orthographic lexicon has been established. The dynamics of parallel activation models mean that, even in models with fully specified word representations, a written word will potentially activate multiple lexical candidates. The mechanisms for selecting the best-matching candidate are therefore critical to accurate lexical retrieval. Within the IA architecture that forms the core of many models (e.g., Coltheart et al., 2001; Davis, 2010; Grainger & Jacobs, 1996), this depends on competition between word-level representations, while in the PDP models it emerges from distributed subsymbolic orthographic-phonological connections established during training. Functionally, both frameworks assume that selecting the correct word depends on a competition between words that partially match the visual input that is biased toward higher frequency words. Selecting the correct lexical representation depends on balancing the degree of match with the stimulus input against "top-down" probabilistic knowledge and precisely mapping the perceptual input onto the orthographic lexicon.

The *masked-priming paradigm* (Forster & Davis, 1984) has been a major source of insight into this early interaction between bottom-up and top-down constraints. It has provided clear evidence that a briefly presented, orthographically related stimulus, of which the reader is unaware, can influence responses to a subsequent target word (see Kinoshita & Lupker, 2003, for reviews). When the masked prime contains all the letters of the target in a transposed order (e.g., *jugde, JUDGE*) priming can be as strong as for an identical prime (e.g., *judge, JUDGE*; Perea & Lupker, 2003), suggesting that readers have difficulty resolving the position or order of letters (Gomez, Ratcliff, & Perea, 2008; Norris & Kinoshita, 2012). Nonword neighbors also facilitate target processing (e.g., *bontrast, CONTRAST*), but word neighbors (e.g., *able*, *AXLE*) can yield interference (Davis & Lupker, 2006; Segui & Grainger, 1990), particularly in more proficient spellers (Andrews & Lo, 2012).

A number of well-specified models of early *orthographic coding* have been developed to explain the complex pattern of priming effects obtained using variants of the masked-priming paradigm (see Grainger, 2008). These models offer different solutions to the difficulty in distinguishing between similar words. Some propose additional levels of representations coding multiletter units (e.g., *open bigrams*; Grainger & van Heuven, 2003; subsyllabic segments; Taft & Krebs-Lazendic, 2013), while others assume that individual letters in a word are processed sequentially so that the relative timing of their processing provides evidence about their order (Davis, 2010; Whitney, 2001). As we will elaborate in section 3.1, this latter assumption aligns with the more general computational benefits of serial processing in addressing cognitive constraints on reading.

The broader implications of these findings in the present context is that they demonstrate that successful lexical retrieval is constrained by both perceptual factors related to resolving the position or order of letters within words (e.g., Gomez et al., 2008) and competition at the lexical level between words that partially match the stimulus (e.g., Davis, 2010). These interacting constraints converge with the representational constraints discussed earlier to underscore the central role of a lexicon of precise word-specific representations in rapidly selecting the best-matching lexical candidate. As we will elaborate in section 4, this automatic word-identification process supports effective deployment of readers' limited attentional resources to integrative comprehension processes and aligns with the comprehension processes developed to understand speech.

2.3. Maintaining Multiple Words Another basic constraint on the early processes involved in mapping from written text to meaning that emerges from consideration of the models of isolated word identification relates to how many lexical representations can be simultaneously active in the orthographic lexicon. As noted, lexical competition is central to lexical selection in many models (e.g., Davis, 2010). Presenting another word changes the inputs to the system and, therefore, the pattern of lexical activation, so most IA implementations assume that its processing state is reset for each new stimulus. Thus, as highlighted by Forster (2013), parallel activation models have no mechanism for maintaining activation of more than one word at a time.

This reflects a general problem with "display board models of the mind" (Kahneman & Treisman, 1984, p. 41) that equate perceptual experience with a succession of states of activation of units in conceptual LTM. Models that rely entirely on selective activation of semipermanent structures in LTM are incomplete because they provide no mechanism for representing the perceptual unity and continuity of objects and events—in the context of reading, there is no way of maintaining activation of previous words when processing of the next word begins, let alone coding information about the order of words. Kahneman and Treisman argued that there must be another form of mental representation—a "perceptual analogue of episodic memory" (p. 54)—that specifies the sensory properties of the eliciting stimulus and binds them to information about conceptual identity retrieved from LTM. In the context of sentence processing, this function may be served by the *textbase* or propositional representation of the literal sentence structure assumed in models of text and discourse comprehension. We will return to these issues in sections 3 and 4, where we discuss how serial lexical processing aligns with working memory and comprehension processes.

3. Constraint 2: Extracting and Using Information from Text

The second broad set of constraints that fundamentally affect reading are those that delimit the "bandwidth" of cognitive processing—the capacity limits imposed by attention and perceptuomotor processes that constrain the speed and efficiency of online reading. This section examines how these factors influence the other cognitive systems that are involved in reading.

3.1. The Spotlight of Visual Attention *Visual attention* has been likened to an internal "spotlight" that enhances the processing of whatever is in its focus (Johnson & Dark, 1986). In the context of reading, attention is required for the identification of words on the printed page (e.g., Kahneman & Treisman, 1984; Lachter, Forster, & Ruthruff, 2004), although both precisely *why* this is necessary and *how* it actually supports reading remain open questions (cf., Radach & Kennedy, 2013; Reichle, Liversedge, Pollatsek, & Rayner, 2009).

Computational models of reading make very different theoretical assumptions about the role of attention in reading. To illustrate this point, consider models that explain how the perceptual, cognitive, and motor processes involved in reading influence readers' eye movements. The two most successful of these models, *E-Z Reader* (Reichle et al., 1998; Reichle, Pollatsek, & Rayner, 2012) and *SWIFT* (Engbert, Nuthmann, Richter,

& Kliegl, 2005), make very different assumptions about how attention is allocated during reading and thus its role. According to the E-Z Reader model, attention is allocated in a strictly serial manner, to one word at any given time (Reichle, 2011). In contrast, the SWIFT model posits that attention is allocated to support the concurrent lexical processing of multiple words (Engbert & Kliegl, 2011). This latter model equates visual attention with a flexible "zoom-lens," rather than a fixed spotlight, that can be varied in accordance with task demands and reader goals (Schad & Engbert, 2012). This difference has ramifications for the functional role ascribed to attention: By the serial account, this function is restricted to one word at a time, whereas the parallel account assumes it can be simultaneously applied to three or four words. What, then, might this function be?

Attention appears to be necessary to convert the visual features of letters (or characters in nonalphabetic languages such as Chinese) into abstract orthographic codes that can be used to access associated phonological and semantic information. One function of attention may therefore be to *bind* the visual features of a word together to form a cue to retrieve lexical information from memory. Computational models of isolated word identification such as the IA, DRC, and PDP models begin with precise letter codes tied to specific within-word positions, finessing questions about how these features are extracted. This limitation has been addressed in recent *orthographic coding* models that provide detailed accounts of how perceptual constraints influence the extraction of features from written words (Davis, 2010; Gomez et al., 2008; Whitney, 2001). As noted, the solutions offered by some of these models hinge on an assumption of serial letter processing; for example, in the *SOLAR* model by Davis (2010), visual attention is applied sequentially to individual letters to convert the temporal sequence into a *spatial code* that preserves the order of letters in the word. This implies that attention may be focused more narrowly than individual words, perhaps even letter-by-letter. More generally, it suggests that, in addition to there being some inherent computational advantage to delimiting what is processed at any given time (e.g., representing the relative positions of letters within words), visual attention may also serve a more general functional role in segregating input to extract the information required for subsequent processing.

3.2. Coordinating the Mind with the Eyes An additional source of constraint on early visual processing derives from perceptuomotor demands of reading. Readers' eyes move across the lines of a text in a series of *fixations* that average 200–250 ms duration, interspersed by *saccades*, averaging about 8 characters in length, during which no useful information is extracted (Matin, 1974). Thus, at some level, reading requires integrating the information obtained from these temporally and spatially distributed "snapshots." Critically, only 7–8 characters within foveal vision can be resolved with sufficient acuity to extract the high-spatial-frequency information required to identify letters and words (McConkie & Rayner, 1975).

However, there is also clear evidence that information extracted from outside the fovea (i.e., from the *parafovea*) plays a critical role in reading. For example, the speed and accuracy of comprehension is dramatically reduced if the information available to the reader is artificially limited to foveal vision (Rayner, 1975, 1998, 2009). This is consistent with a distinction between "what" and "where" processing in online reading: foveal processing is critical to extracting the fine-grained information necessary to access the orthographic lexicon, whereas the coarse information extracted from parafoveal and peripheral vision is used to control where to move the eyes (Rayner & Pollatsek, 1981).

There is also evidence that preview of parafoveal words directly facilitates lexical retrieval when the word is later fixated (e.g., Reingold, Reichle, Glaholt, & Sheridan, 2012). In addition to the *preview benefit* provided by an identical parafoveal preview, benefits have been demonstrated for orthographically and/or phonologically related previews (see Schotter, Angele, & Rayner, 2012, for review). Evidence is also accumulating for semantic preview benefits (Schotter, 2013), although the precise source of these benefits remains a focus of research (e.g., Veldre & Andrews, 2016). These preview benefits indicate that readers extract at least partial information from words in the parafovea before they have directly fixated them.

Thus, information extracted from both foveal and parafoveal words can contribute to lexical retrieval. The interaction between these sources of information is constrained by the timing parameters of the oculomotor system. There is a delay of 150–175 ms between the programming of a saccade and its execution (Becker & Jürgens, 1979; Reichle & Reingold, 2013). To optimize reading speed, readers therefore need to initiate the programming of a saccade before completing retrieval of the currently fixated word.

To address these constraints, the E-Z Reader model distinguishes between two stages of lexical retrieval: the first "familiarity" stage provides preliminary evidence that lexical retrieval is imminent, which triggers planning and programming of a saccade to the next word, and is followed immediately by the second stage of "full lexical access." When lexical access completes before the planned saccade has been executed, the model assumes that covert attention shifts to the next word, allowing its processing to begin before it is directly fixated. This accounts for preview benefits and the occasional skipping of words that occurs when the parafoveal word is identified quickly enough to allow reprogramming of the planned saccade (Reichle & Drieghe, 2013). This *decoupling* of attention from the fixation location allows E-Z Reader to account for the parafoveal influences on lexical processing while maintaining the assumption that word identification is a serial process: Lexical retrieval of word *n+1* can only begin after word *n* has been identified, but attention can shift to word *n+1* as soon as processing of word *n* completes, even if the eyes have not yet moved to that location. Consistent with this assumption, lexical attributes of the fixated word are the best predictors of fixation time (Rayner, 1998), suggesting that an early index of the progress of lexical processing for a fixated word triggers a saccade to a new location (Morrison, 1984; Reichle & Laurent, 2006).

Thus, in E-Z Reader visual attention is a central constraint that is required for lexical retrieval and causes serial word-by-word identification. Rather than limiting comprehension, the serial processing imposed by these attentional constraints benefits comprehension by maintaining information about the order of words that is critical to assigning meaning, particularly in word-order–dependent languages such as English (Pollatsek & Rayner, 1999). Thus, a more general function of visual attention may be to "package" information in a manner that aligns with the requirements of higher-level comprehension processes.

Evidence consistent with this possibility is provided by a recent comparison of eye movements during reading of very carefully matched texts in Finnish, English, and Chinese—writing systems they vary in terms of their visual density, orthographic depth, and other characteristics (Liversedge et al., 2016). The data showed robust cross-language differences in the distribution and timing of initial fixations on the text that were predicted by measures related to the visual density of the different scripts. However, the informational density (indexed by measures of word frequency, word length, and the number of words in the sentences) accounted for similar proportions of common variance providing evidence for the "universality of representation and process in reading" (Liversedge et al., 2016, p. 1). Similar conclusions derive from a comparison of cross-language neuroimaging data by Rueckl et al. (2015) that revealed very similar reading circuits for English, Spanish, Hebrew, and Chinese. Thus, script-specific properties determine how attention is initially allocated to extract relevant lexical information, but the behavioral and neural processes that operate on that information appear to be general across writing systems.

4. Constraint 3: Mapping from Lexical Representations to Comprehension

Comprehension of sentences and text requires readers to integrate the meanings of the individual words that compose phrases and/or sentences and to then combine the meaning of these successive phrases and sentences with the global context of the passage, or *discourse*, and the readers' world knowledge and reading goals.

As noted, serial word-by-word processing reduces the computational demands of initial postlexical integration by maintaining word order, which is critical for syntactic processing, without adding assumptions about memory buffers or reconstruction processes (e.g., Pollatsek & Rayner, 1999). The language systems responsible for generating linguistic representations can then integrate words incrementally, in the same way they are processed in speech, by inserting the higher-level information derived from each word into a syntactic "scaffolding" from which the meaning of a sentence is built. This also solves the problem of how previously identified words are maintained when processing of the next word is initiated: the lexical output for the previous word is immediately integrated into the developing representation of the sentence, or textbase (as it is referred to in sentence- and discourse-processing models), which can then be used to construct the higher-order meaning of the text, known as the *discourse model* or *situation model* (Gernsbacher & Kaschak, 2013).

Consistent with this intuitive conceptualization of text processing, many sentence-processing models assume that lexical processing, and the attentional processes that support it, operate serially (e.g., Elman, 1991; Gibson, 1998; Lewis & Vasishth, 2005; Van Dyke & Lewis, 2003). This auxiliary assumption aligns with the earlier discussion of how domain-general constraints on perceptual processing contribute to the emergence of a functionally modular reading architecture reflected in a serial, staged processing strategy.

4.1. Mechanisms of Text and Discourse Processing These assumptions are exemplified in the *Construction-Integration* (CI) model of discourse processing (Kintsch, 1988, 1998; Kintsch & van Dijk, 1978), which is arguably the most successful and influential discourse model as demonstrated by the fact that many competing accounts have borrowed one or more of its theoretical assumptions (e.g., Goldman & Varma, 1995;

Myers & O'Brien, 1998; van den Broek, Risden, Fletcher, & Thurlow, 1996).

The defining feature of the CI model is its assumption that text comprehension involves two sequential stages that operate iteratively on phrase or clause chunks. In the first, the *construction* stage, the textbase encoded in working memory activates semantically related information from LTM to form an associative network of propositions. Information from LTM is retrieved quickly and automatically, but the number of propositions that can be simultaneously processed is constrained by the capacity limits of working memory. The second, the *integration* stage, then incorporates the new information into the developing discourse model. This process is slow and resource-demanding because it requires filtering out irrelevant information activated during the construction process to create a coherent representation of the discourse that integrates relevant information from the textbase with associations and inferences drawn from semantic memory.

As we will briefly elaborate, alternative "one-stage" models of discourse processing have been proposed that rely only on the bottom-up memory processes of the construction stage (e.g., Myers & O'Brien, 1998; Sanford & Garrod, 2005), or only on the active, strategic processes of the integration stage (e.g., Graesser, Singer, & Trabasso, 1994). Before evaluating the relative merits of the two-stage framework and its alignment with the modular reading architecture, we will briefly consider the constraints on comprehension processes imposed by working memory.

4.2. Working Memory and Comprehension *Working memory* has been conceptualized in many ways, ranging from Baddeley and Hitch's (1974) original conception of interrelated, specialized memory buffers for holding and manipulating information, and its successors (e.g., Baddeley, 2012), to views that propose a domain-general "global workspace" (Baars, 1997), or that identify working memory with the active contents of LTM (Cowan, 2005). It is beyond the scope of this chapter to consider the implications of these differences for understanding the architecture of the reading system (Miyake & Shah, 1999), but working memory is clearly a critical constraint on comprehension because it limits the number of mental representations that can be actively maintained and manipulated at any time. This has a number of consequences for the maintenance, construction, and integration of the mental representations required to understand a text.

One of the most widely cited illustrations of how working memory constraints influence the construction of representations of the individual phrases and sentences that comprise a text is provided by sentences containing embedded clauses, such as:

(1) The mouse the cat chased escaped.
(2) The mouse the cat the man owned chased escaped.

Although (1) is constructed so that a subordinate clause (i.e., *the cat chased*) is embedded within a main clause (i.e., *The mouse escaped*), the sentence itself is not difficult to understand. This is less true of (2), which contains two embedded clauses. Here, the linguistic operations required to link each of the entities to their respective roles (e.g., to determine who escaped) tax the limits of working memory, making the sentence difficult to understand.

Reducing working memory load is a central operating principle in many models of both spoken and written discourse processing. For example, the parsing principles of the seminal *garden-path model* of sentence processing (e.g., Frazier, 1987; Frazier & Clifton, 1996) generate phrase structures that minimize working memory demands on both processing and integration. Working memory capacity is captured more formally within various computational models of sentence processing as an important parameter that delimits the complexity of the phrase structures that can be generated and the likelihood of making comprehension errors (e.g., Gibson, 1998; Jurafsky, 1996; Lewis & Vasishth, 2005).

The central function of working memory in the CI model of discourse processing (Kintsch & van Dijk, 1978) is to serve as a *buffer* in which the propositional representations of individual phrases and/or sentences in the textbase are maintained and integrated through the formation and strengthening of associative links both within the textbase and with propositions retrieved from semantic LTM. The capacity constraints on working memory delimit the time that propositions can remain concurrently active in working memory, because each newly encoded proposition displaces one that is currently active. These activation dynamics determine the strengths of the associative links between the propositions and thus the propensity to both remember new propositions and relate them to preexisting knowledge.

To illustrate this point, consider the following short narrative:

(3) The black cat chased the mouse into the corner. After finishing its meal, the cat slept contentedly.

Understanding the meaning of even this very simple text requires a number of *inferences* about relationships that are unstated in the actual sentences; for example, that the agent denoted by the possessive pronoun *its* in the second sentence refers to the cat and that the meal consisted of the mouse. These inferences also require the use of many kinds of knowledge, including knowledge of language and writing conventions and *world knowledge* of the dietary habits of cats. Importantly, both the ability to link the representations of the two sentences and to generate inferences is a direct function of working memory capacity. A larger working memory capacity allows the propositions of individual phrases and/or sentences to remain coactive in working memory longer because they will be less rapidly displaced by other, more recently encoded propositions. This, in turn, strengthens the associations between coactive propositions (e.g., from the pronoun *its* back to *black cat*) and any relevant information retrieved from LTM (e.g., that cats eat mice), thereby facilitating both comprehension and memory of the text.

The CI model assumes that, although working memory *capacity* imposes limits on the initial construction stage, the memory retrieval processes that construct the initial propositional network operate automatically. However, working memory *resources* are required for the second integration stage of discourse processing in which newly encoded propositions are incorporated into the discourse model. This stage requires *elaborative inferences* that add detail or embellish the literal text (e.g., the cat ate the mouse). Readers must also connect new propositions to their discourse representation and update or modify links to maintain *local coherence* (McKoon & Ratcliff, 1992). Some models assume that readers also retrieve semantic and world knowledge to construct a *globally coherent* discourse model (Graesser et al., 1994). Demonstrating the contribution of working memory to these integrative processes, larger working memory capacity is associated with more elaborative inferences (Barreyro, Cevasco, Burin, & Marotto, 2012), more complete discourse models of text (Dutke & von Hecker, 2011), and better comprehension, even when intelligence is controlled (Christopher et al., 2012).

Thus, within the CI framework, the distinction between the initial, bottom-up construction stage that automatically and incrementally retrieves both relevant and irrelevant information from LTM, followed by a resource-demanding integration process that selectively filters the products of the initial stage, is driven by the constraints on working memory. The model thus effectively meshes the perceptual and linguistic constraints on extracting information from text and mapping it to lexical knowledge, on one hand, with the constraints imposed by domain-general limits on working memory capacity and resources, on the other.

5. Modularity and Discourse Processing: Why a Modular Architecture Works for Reading

The CI model provides a framework for conceptualizing the relationship between lexical retrieval and comprehension processes that is highly compatible with the modular reading architecture underpinning this chapter. The model's assumption that the initial construction stage is bottom-up and autonomous is compatible with the modular conception of an encapsulated lexical retrieval system that is insensitive to context, as described in section 1. It also aligns with the serial, incremental processing dictated by the constraints on visual attention and perceptuomotor coordination discussed in section 2. Importantly, the model assumes that semantic and world knowledge have an early influence on the formation of the propositional network constructed from the textbase because they have shaped the structure of the lexical and semantic knowledge that is retrieved by the words of the text. Indeed, it assumes that the effects of such knowledge occur before any influence of the current sentence or discourse context, which only affect the later integration stage (Cook & Myers, 2004). The effects of active strategic processes are also confined to integration.

The validity of these claims is the focus of current empirical research and theoretical debate (see Gernsbacher & Kaschak, 2013; O'Brien & Cook, 2015, for reviews). Much of the explanatory power of the CI framework lies in its inclusion of two stages with different properties. Other models of discourse processing have proposed potentially more parsimonious one-stage alternatives. For example, one class of *constructionist* models attributes all comprehension to active, capacity-limited, structure-building processes (Gernsbacher, 1990; Graesser et al., 1994) that are strategically guided by the reader's goals to achieve local and global coherence. Another class of models proposes the opposite view—that all discourse processing can be explained by automatic, bottom-up memory processes (e.g., Sanford & Garrod, 2005). For example, *resonance* models (e.g., Myers & O'Brien, 1998) assume that the propositions currently active in working memory function as retrieval cues that resonate with all relevant information in LTM (e.g., Hintzman, 1988) and select the most highly activated information to enter working memory. A critical feature of this resonance retrieval process is that, like the construction stage of the CI model, it is "dumb"—it has the characteristics of automatic, encapsulated, context-independent processing that define modularity (Fodor, 1983). These models provide a clearly specified account of how a simple, passive activation mechanism (i.e., resonance) can yield early effects of semantic and world knowledge without any contribution from strategic processes (O'Brien & Cook, 2015).

Distinguishing between strategy-based and memory-based models of discourse processing is difficult. Both processes are almost certainly required for a complete theory, which is consistent with the hybrid account proposed by the CI model. However, questions remain about how the two forms of processing interact during online reading (O'Brien & Cook, 2015). The passive, bottom-up, memory-based construction process assumed by the resonance model appears inefficient because it activates irrelevant information that must be subsequently filtered out and inhibited. However, assuming that comprehension relies entirely on active, resource-demanding processes appears incompatible with the constraints on working memory. Achieving the optimal balance between these two modes of processing depends on the relative resource demands imposed by the need to filter the output of bottom-up memory processes versus the demands of implementing active, strategic comprehension processes. Determining the factors influencing this balance is a major focus of current research on text processing (Gernsbacher & Kaschak, 2013; Graesser & Forsyth, 2013; Van Dyke, Johns, & Kukona, 2014).

The general architectural principles focused on in this chapter provide an important perspective on this debate. The patterns of performance demonstrated in studies of online reading arise from interactions between the structural architecture of the reading system, which is shaped by experience, and the temporal and cognitive constraints imposed by domain-general limitations (e.g., visual and attentional processes). Consideration of these intersecting constraints provides clues about the factors that influence the emergence of a functionally modular architecture. For example, the information-processing bottleneck is much more severe for the lower-level, "front end" systems involved in "gathering the raw materials" for comprehension (Gernsbacher & Kaschak, 2013). A general solution that seems to be adopted to address the perceptual, motor, and cognitive constraints on these processes is to limit the focus of attention to a single chunk of information at any point in time (i.e., serial processing). In addition to complying with limits on visual attention, sequential processing contributes to preserving the order of information (e.g., letters within words, words within sentences) that aligns with both the structure of the linguistic codes stored in memory and the linguistic processes required for comprehension. The success of these processes depends on the development of a well-structured orthographic lexicon containing precisely specified word forms that are tightly bound to their associated semantic and phonological codes to support the rapid integration of a word's lexical attributes into the textbase of the sentence after it has been identified.

However, in contrast to the structured, sequential processing that defines the lower-level stages of reading, the higher-level systems that are responsible for constructing more elaborate representations of a text, using both the meanings of sentences and prior knowledge, seem to depend on the capacity for highly parallel processing, like that captured by the resonance mechanism underpinning memory models of discourse processing (Myers & O'Brien, 1998). Thus, a modular architecture in which working memory serves as a "gateway" that constrains the early information extraction processes to generate a structured, sequential output (e.g., preserving the literal order of the words in the text) both limits the amount of information that needs to be encoded and operated on by working memory and represents it in a form that mirrors spoken language. This therefore allows reading and speech comprehension to rely on a common set of discourse processes.

5.1. Reading Is All About Words A central assumption of the modular framework is that lexical knowledge plays a critical role in reading. According to the staged processing architecture that we have described, words are the units that are sequentially extracted from printed text, mapped to structured lexical representations, and then incrementally incorporated into the textbase that provides the input to comprehension processes.

This view is challenged by recent neurally inspired probabilistic theories, both connectionist models (e.g., McClelland et al., 2010) and Bayesian frameworks (e.g., Norris, 2006), that reject the notion of a "fixed vocabulary of representational units" (McClelland et al., p. 349) in favor of graded, probabilistic patterns of activation that are dynamically modulated by task demands and context, potentially including proactive, anticipatory processes (e.g., Bar, 2007; Schacter, Addis, & Buckner, 2007). Such probabilistic accounts emphasize the benefits of plasticity and flexibility of mental representation to account for our capacity to rapidly construct new knowledge. However, McClelland et al. acknowledged that there may be a point at which the advantages of "imposing a specific structural form on knowledge outweigh the disadvantages" (p. 354), suggesting that expertise in some domains may be defined by the development of structured representations.

According to the modular framework outlined in this chapter, reading exemplifies a skill in which expertise depends on the development of structured, word-specific lexical knowledge. This view predicts that, in paradigms that capture the initial stages of lexical processing and that are uncontaminated by later construction and integration processes, higher-proficiency readers will be *less* sensitive to context than will lower-proficiency readers (Andrews, 2015). Direct evidence for this view is provided by recent investigations of individual differences in preview effects on eye movements during sentence reading, showing that higher-proficiency readers are faster at extracting lexical information from parafoveal words (Veldre & Andrews, 2015), but *less* sensitive to the contextual plausibility of parafoveal previews than average readers (Veldre & Andrews, 2016). These data are consistent with the modular framework's prediction that optimally efficient reading depends on high-quality, word-specific lexical representations (Andrews, 2015; Perfetti, 2007).

Individual differences will be an increasingly important focus of future reading research. Both studies of the trajectory of reading development across the life span (e.g., Blythe, 2014; Rayner, Yang, Schuett, & Slattery, 2014) and investigations of individual differences among skilled readers (e.g., Andrews, 2015; Kuperman & Van Dyke, 2011; Yap, Tse, & Balota, 2009) and readers of different scripts (e.g., Frost, 2012; Rueckl et al., 2015) provide a vehicle for tracking how universal and script-specific factors shape the organization of the reading system.

The central role of word-level analysis in reading is also driven by visual, attentional, and perceptuomotor constraints on reading that limit the amount of information that can be processed at any point in time and encourage serial processing strategies. The E-Z Reader model's success in simulating a broad range of data from both skilled and developing readers (e.g., Reichle, 2011; Reichle et al., 1998; Reichle et al., 2013) demonstrates that combining a modular lexical system that processes words serially with relatively simple, principled assumptions about the constraints imposed by attention, saccadic planning, and postlexical processing provides a thorough account of complex patterns of eye-movement data, including the influence of higher-level language processing on sentence reading (Reichle et al., 2009). The model provides a valuable heuristic for further evaluating the implications of a modular, staged processing architecture for understanding the cognitive processes in reading.

A central challenge for future computational models of reading is to describe how the various components of reading that are well described by existing models (e.g., word identification) "fit together" to explain the overall architecture of reading. Computational modeling has substantially advanced our understanding of many components of reading. But there have been very few attempts to combine two or more of these models to specify how the components interact (e.g., how lexical processing influences sentence processing). An important goal for future researchers is to develop more comprehensive models of the reading architecture that integrate models of lexical and LTM knowledge and retrieval with models of the online reading process to provide direct insight into how the basic components of reading interact and thus enhance our understanding of the reading architecture (see Reichle, 2019).

6. Conclusion

We have overviewed the perceptual, cognitive, and motor processes that are involved in reading and discussed how the constraints imposed by these processes delimit the way in which the cognitive architecture of reading emerges through learning and experience to allow skilled readers to rapidly convert ink marks on printed pages into lexical representations that can be used by the comprehension systems that evolved for spoken language. Although the brevity of the chapter prevented a detailed discussion of these constraints, we hope that the broad sketch that we have provided demonstrates how a modular reading architecture—characterized by discrete stages of processing—might facilitate the acquisition of "the most remarkable specific performance that civilization has learned in all its history" (Huey, 1908, p. 6)—that of reading.

REFERENCES

Andrews, S. (1982). Phonological recoding: Is the regularity effect consistent? *Memory and Cognition, 10,* 565–575.

Andrews, S. (2015). Individual differences among skilled readers: The role of lexical quality. In A. Pollatsek & R. Treiman (Eds.), *The Oxford handbook of reading* (pp. 129–148), Oxford: Oxford University Press.

Andrews, S., & Lo, S. (2012). Not all skilled readers have cracked the code: Individual differences in masked form priming. *Journal of Experimental Psychology: Learning, Memory, and Cognition, 38,* 152–163.

Baars, B. J. (1997). Some essential differences between consciousness and attention, perception, and working memory. *Consciousness and Cognition, 6,* 363–371.

Baddeley, A. D. (2012). Working memory: Theories, methods and controversies. *Annual Review of Psychology, 63,* 1–29.

Baddeley, A. D., & Hitch, G. J. (1974). Working memory. In G. H. Bower (Ed.), *Recent advances in learning and motivation* (Vol. 8, pp. 47–89). New York: Academic Press.

Bar, M. (2007). The proactive brain: Using analogies and associations to generate predictions. *Trends in Cognitive Sciences, 11,* 280–289.

Barreyro, J., Cevasco, J., Burin, D., & Marotto, C. (2012). Working memory capacity and individual differences in the making of reinstatement and elaborative inferences. *Journal of Spanish Psychology, 15,* 471–479.

Becker, W., & Jürgens, R. (1979). An analysis of the saccadic system by means of double step stimuli. *Vision Research, 19,* 967–983.

Blythe, H. I. (2014). Developmental change in eye movements and visual information encoding associated with learning to read. *Current Directions in Psychological Science, 23,* 201–207.

Bolger, D. J., Perfetti, C. A., & Schneider, W. (2005). A cross-cultural effect on the brain revisited: Universal structures plus writing system variation. *Journal of Human Brain Mapping, 25,* 92–104.

Byrne, B., Olson, R. K., Hulslander, J., Wadsworth, S., Defries, J. C., & Samuelsson, S. (2008). A behavior-genetic analysis of orthographic learning, spelling and decoding. *Journal of Research in Reading, 31,* 8–21.

Castles, A., & Coltheart, M. (1993). Varieties of developmental dyslexia. *Cognition, 47,* 149–180.

Christopher, M. E., Miyake, A., Keenan, J., Pennington, B., De Fries, J., Wadsworh, S., . . . Olson, R. K. (2012). Predicting word reading and comprehension with executive function and speed measures across development: A latent variable analysis. *Journal of Experimental Psychology: General, 141,* 470–488.

Clark, A. (2013). Whatever next? Predictive brains, situated agents and the future of cognitive science. *Behavioral and Brain Sciences, 36,* 181–204.

Clarke, P., Hulme, C., & Snowling, M. (2005). Individual differences in RAN and reading: A response time analysis. *Journal of Research in Reading, 28,* 73–86.

Coltheart, M., Rastle, K., Perry, C., Langdon, R., & Ziegler, J. (2001). DRC: A dual route cascaded model of visual word recognition and reading aloud. *Psychological Review, 108,* 204–256.

Compton, D. L., Miller, A. C., Elleman, A. M., & Steacy, L. M. (2014). Have we forsaken reading theory in the name of "quick fix" interventions for children with reading disability? *Scientific Studies of Reading, 18,* 55–73.

Cook, A. E., & Myers, J. L. (2004). Processing discourse rules in scripted narratives: The influence of context and world knowledge. *Journal of Memory and Language, 50,* 268–288.

Cowan, N. (2005). *Working memory capacity.* Hove, UK: Psychology Press.

Davis, C. J. (2010). The spatial coding model of visual word identification. *Psychological Review, 117,* 713–758.

Davis, C. J., & Lupker, S. J. (2006). Masked inhibitory priming in English: Evidence for lexical inhibition. *Journal of Experimental Psychology: Human Perception and Performance, 32,* 668–687.

Dehaene, S. (2009). *Reading in the brain.* New York: Penguin Viking.

Dehaene, S., Pegado, F., Braga, L., Ventura, P., Filho, G., Jobert, A., . . . Cohen, L. (2010). How learning to read changes the cortical networks for vision and language. *Science, 3,* 1359–1364.

Demberg, V., & Keller, F. (2008). Data from eye tracking corpora and theories of syntactic processing. *Cognition, 109,* 193–210.

Dufau, S., Grainger, J., & Ziegler, J. C. (2012). How to say "no" to a nonword: A leaky accumulator model of lexical decision. *Journal of Experimental Psychology: Learning, Memory and Cognition, 38,* 1117–1128.

Dutke, S., & von Hecker, U. (2011). Comprehending ambiguous texts: A high reading span helps to constrain the situation model. *Journal of Cognitive Psychology, 23,* 227–242.

Ehri, L. (2015). How children learn to read words. In A. Pollatsek & R. Treiman (Eds.), *The Oxford handbook of reading* (pp. 293–310), Oxford: Oxford University Press.

Elman, J. L. (1991). Distributed representations, simple recurrent networks, and grammatical structure. *Machine Learning, 7,* 195–225.

Engbert, R., & Kliegl, R. (2011). Parallel graded attention models of reading. In S. P. Liversedge, I. D. Gilchrist, & S. Everling (Eds.), *Oxford handbook of reading* (pp. 787–800). Oxford: Oxford University Press.

Engbert, R., Nuthmann, A., Richter, E., & Kliegl, R. (2005). SWIFT: A dynamical model of saccade generation during reading. *Psychological Review, 112,* 777–813.

Federmeier, K. D. (2007). Thinking ahead: The role and roots of prediction in language comprehension. *Psychophysiology, 44,* 491–505.

Fodor, J. A. (1983). *Modularity of mind.* Cambridge, MA: MIT Press.

Forster, K. I. (1979). Levels of processing and the structure of the language processor. In W. Cooper & E. Walker (Eds.), *Sentence processing.* Hillsdale, NJ: Erlbaum.

Forster, K. I. (2013). How many words can we read at once? More intervenor effects in masked priming. *Journal of Memory and Language, 69,* 563–573.

Forster, K. I., & Davis, C. (1984). Repetition priming and frequency attenuation in lexical access. *Journal of Experimental Psychology: Learning, Memory and Cognition, 10,* 680–698.

Frazier, L. (1987). Sentence processing: A tutorial review. In M. Coltheart (Ed.), *Attention and performance XII: The psychology of reading* (pp. 559–586). London: Erlbaum.

Frazier, L., & Clifton, C., Jr. (1996). *Construal.* Cambridge, MA: MIT Press.

Frith, U., & Happe, F. (1998). Why specific developmental disorders are not specific: On-line and developmental effects in autism and dyslexia. *Developmental Science,* 1, 267–272.

Frost, R. (2012). Towards a universal model of reading. *Behavioral and Brain Sciences, 35,* 263–279.

Georgiou, G. K., Parrila, R., Kirby, J. R., & Stephenson, K. (2008). Rapid naming components and their relationship with phonological awareness, orthographic knowledge, speed of processing and different reading outcomes. *Scientific Studies of Reading, 12,* 325–350.

Gernsbacher, M. A. (1990). *Language comprehension as structure building.* Hillsdale, NJ: Erlbaum.

Gernsbacher, M. A., & Kaschak, M. P. (2013). Text comprehension. In D. Reisberg (Ed.), *The Oxford handbook of cognitive psychology* (pp. 462–474). Oxford: Oxford University Press.

Gibson, E. (1998). Linguistic complexity: Locality of syntactic dependencies. *Cognition, 68,* 1–76.

Goldman, S. R., & Varma, S. (1995). CAPping the construction-integration model of discourse representation. In C. Weaver,

S. Mannes, & C. Fletcher (Eds.), *Discourse comprehension: Essays in honor of Walter Kintsch* (pp. 337–358). Hillsdale, NJ: Lawrence Erlbaum.

Gomez, P., Ratcliff, R., & Perea, M. (2008). The overlap model: A model of letter position coding. *Psychological Review, 115*, 577–600.

Goodman, K. (1967). Reading: A psycholinguistic guess game. *Journal of the Reading Specialist, 6*, 126–135.

Gough, P. B., & Hillenger, M. L. (1980). Learning to read: An unnatural act. *Bulletin of the Orton Society, 30*, 179–196.

Graesser, A. C., & Forsyth, C. M. (2013). Discourse comprehension. In D. Reisberg (Ed.), *The Oxford handbook of cognitive psychology* (pp. 475–491). Oxford: Oxford University Press.

Graesser, A. C., Singer, M., & Trabasso, T. (1994). Constructing inferences during narrative text comprehension. *Psychological Review, 101*, 371–395.

Grainger, J. (2008). Cracking the orthographic code: An introduction. *Language and Cognitive Processes, 23*, 1–35.

Grainger, J., & Jacobs, A. M. (1996). Orthographic processing in visual word recognition: A multiple read-out model. *Psychological Review, 103*, 518–565.

Grainger, J., & van Heuven, W. (2003). Modeling letter position coding in printed word perception. In P. Bonin (Ed.), *The mental lexicon* (pp. 1–23). New York: Nova Science.

Harm, M. W., & Seidenberg, M. S. (2004). Computing the meanings of words in reading: Cooperative division of labor between visual and phonological processes. *Psychological Review, 111*, 662–720.

Hintzman, D. L. (1988). Judgements of frequency and recognition memory in a multiple-trace memory model. *Psychological Review, 95*, 528–551.

Hoover, W. A., & Gough, P.B. (1990). The simple view of reading. *Reading and writing, 2*, 127–160.

Huey, E. B. (1908). *The psychology and pedagogy of reading.* Cambridge, MA: MIT Press.

Johnson, W. A., & Dark, V. J. (1986). Selective attention. *Annual Review of Psychology, 37*, 43–75.

Jurafsky, D. (1996). A probabilistic model of lexical and syntactic access and disambiguation. *Cognitive Science, 20*, 137–194.

Kahneman, D., & Treisman, A. (1984). Changing views of attention and automaticity. In R. Parasuraman & D. R. Davies (Eds.), *Varieties of attention* (pp. 29–61). New York: Academic Press.

Katz, L., & Frost, R. (1992). The reading process is different for different orthographies: The orthographic depth hypothesis. *Advances in Psychology, 94*, 67–84.

Kessler, B., & Treiman, R. (2015). Writing systems: Their properties and implications for reading. In A. Pollatsek & R. Treiman (Eds.), *The Oxford handbook of reading* (pp. 10–25), Oxford: Oxford University Press.

Kinoshita, S., & Lupker, S. (Eds.) (2003), *Masked priming: The state of the art.* New York: Psychology Press.

Kintsch, W. (1988). The role of knowledge in discourse comprehension: A construction-integration model. *Psychological Review, 95*, 163–182.

Kintsch, W. (1998). *Comprehension: A paradigm for cognition.* Cambridge: Cambridge University Press.

Kintsch, W., & van Dijk, T. A. (1978). Toward a model of text comprehension and production. *Psychological Review, 85*, 363–394.

Kuperman, V., & Van Dyke, J.A. (2011). Effects of individual differences in verbal skills on eye-movement patterns during sentence reading. *Journal of Memory and Language, 65*, 42–73.

Lachter, J., Forster, K. I., & Rothruff, E. (2004). Forty-five years after Broadbent: Still no identification without awareness. *Psychological Review, 111*, 888–913.

Lewis, R. L., & Vasishth, S. (2005). An activation-based model of sentence processing as skilled memory retrieval. *Cognitive Science, 29*, 375–419.

Liversedge, S. P., Drieghe, D., Li, X., Yan, G., Bai, X., & Hyona, J. (2016). Universality in eye movements and reading: A trilingual investigation. *Cognition, 147*, 1–20.

Marshall, J. C., & Newcombe, F. (1973). Patterns of paralexia: A psycholinguistic approach. *Journal of Psycholinguistic Research, 2*, 175–199.

Matin, E. (1974). Saccadic suppression: A review. *Psychological Bulletin, 81*, 899–917.

McClelland, J., Botvinick, M., Noelle, D., Plaut, D. C., Rogers, T., Seidenberg, M., & Smith, L. (2010). Letting structure emerge: Connectionist and dynamical systems approaches to cognition. *Trends in Cognitive Sciences, 14*, 348–356.

McClelland, J. L., & Rumelhart, D. E. (1981). An interactive activation model of context effects in letter perception: Part 1. An account of basic findings. *Psychological Review, 88*, 375–407.

McConkie, G. W., & Rayner, K. (1975). The span of the effective stimulus during a fixation in reading. *Perception and Psychophysics, 17*, 578–586.

McKoon, G., & Ratcliff, R. (1992). Inference during reading. *Psychological Review, 99*, 440–466.

Menary, R., & Kirschoff, M. (2014). Cognitive transformations and extended expertise. *Educational Philosophy and Theory, 46*, 610–623.

Miyake, A., & Shah, P. (1999). *Models of working memory: Mechanisms of active maintenance and executive control.* Cambridge: Cambridge University Press.

Morrison, R. E. (1984). Manipulation of stimulus onset delay in reading: Evidence for parallel programming of saccades. *Journal of Experimental Psychology: Human Perception and Performance, 10*, 667–682.

Myers, J. L., & O'Brien, E. O. (1998). Accessing the discourse representation during reading. *Discourse Processes, 26*, 131–157.

Nation, K., Angell, P., & Castles, A. (2007). Orthographic learning via self-teaching in children learning to read English: Effects of exposure, durability, and context. *Journal of Experimental Child Psychology, 96*, 71–84.

Norris, D. (2006). The Bayesian reader: Explaining word recognition as an optimal Bayesian decision process. *Psychological Review, 113*, 327–357.

Norris, D., & Kinoshita, S. (2012). Reading through a noisy channel: Why there's nothing special about the perception of orthography. *Psychological Review*, 119, 517–545.

Oakhill, J., & Cain, K. (2012). The precursors of reading ability in young readers: Evidence from a four-year longitudinal study. *Scientific Studies of Reading, 16*, 91–121.

O'Brien, E. J., & Cook, A. E. (2015). Models of discourse comprehension. In A. Pollatsek & R. Treiman (Eds.), *The Oxford handbook of reading* (pp. 217–231). Oxford: Oxford University Press.

Perea, M., & Lupker, S. (2003). Transposed-letter confusability effects in masked form priming. In S. Kinoshita & S. Lupker (Eds.), *Masked priming: The state of the art* (pp. 97–120). Hove, UK: Psychology Press.

Perfetti, C. A. (2007). Reading ability: Lexical quality to comprehension. *Scientific Studies of Reading, 11,* 357–383.

Perfetti, C. A., & Harris, L. (2013). Universal processes are modulated by language and writing system. *Language Learning and Development, 9,* 296–316.

Perfetti, C., & Hart, L. (2001). The lexical basis of comprehension skill. In D. Gorfein (Ed.), *On the consequences of meaning selection: Perspectives on resolving lexical ambiguity* (pp. 67–86). Washington, DC: American Psychological Association.

Perry, C., Ziegler, J. C., & Zorzi, M. (2007). Nested incremental modeling in the development of computational theories: The CDP+ model of reading aloud. *Psychological Review, 114,* 273–315.

Pinker, S. (1997). Foreword. In D. McGuinness, *Why our children can't read and what we can do about it* (pp. ix–x). New York: Free Press.

Plaut, D., McClelland, J. L., Seidenberg, M. S., & Paterson, K. (1996). Understanding normal and impaired word reading: Computational principles in quasi-regular domains. *Psychological Review, 103,* 56–115.

Pollatsek, A., & Rayner, K. (1999). Is covert attention really unnecessary? *Behavioral and Brain Sciences, 22,* 695–696.

Pulgram, E. (1976). Typologies of writing systems. In E. Pulgram & W. Haas (Eds.), *Writing without letters* (pp. 1–28). Manchester: Manchester University Press.

Radach, R., & Kennedy, A. (2013). Eye movements in reading: Some theoretical context. *Quarterly Journal of Experimental Psychology, 66,* 429–452.

Rayner, K. (1975). The perceptual span and peripheral cues in reading. *Cognitive Psychology, 7,* 65–81.

Rayner, K. (1998). Eye movements in reading: 20 years of research. *Psychological Bulletin, 124,* 372–422.

Rayner, K. (2009). Eye movements and attention in reading, scene perception, and visual search. *Quarterly Journal of Experimental Psychology, 62,* 1457–1506.

Rayner, K., & Pollatsek, A. (1981). Eye movement control during reading: Evidence for direct control. *Quarterly Journal of Experimental Psychology, 33A,* 351–373.

Rayner, K., Yang, J., Schuett, S., & Slattery, T. J. (2014). The effect of foveal and parafoveal masks on the eye movements of older and younger readers. *Psychology and Aging, 29,* 205–212.

Reichle, E. D. (2011). Serial attention models of reading. In S. P. Liversedge, I. D. Gilchrist, & S. Everling (Eds.), *Oxford handbook of reading* (pp. 767–786). Oxford: Oxford University Press.

Reichle, E. D. (2019). *Computational models of reading: A handbook.* Oxford: Oxford University Press.

Reichle, E. D., & Drieghe, D. (2013). Using E-Z Reader to examine word skipping during reading. *Journal of Experimental Psychology: Learning, Memory, and Cognition, 39,* 1311–1320.

Reichle, E. D., & Laurent, P. A. (2006). Using reinforcement learning to understand the emergence of "intelligent" eye-movement behavior during reading. *Psychological Review, 113,* 390–408.

Reichle, E. D., Liversedge, S. P., Drieghe, D., Blythe, H. I., Joseph, H. S. S. L., White, S. J., & Rayner, K. (2013). Using E-Z Reader to examine the concurrent development of eye-movement control and reading skill. *Developmental Review, 33,* 110–149.

Reichle, E. D., Liversedge, S. P., Pollatsek, A., & Rayner, K. (2009). Encoding multiple words simultaneously in reading is implausible. *Trends in Cognitive Sciences, 13,* 115–119.

Reichle, E. D., Pollatsek, A., Fisher, D. L., & Rayner, K. (1998). Toward a model of eye movement control in reading. *Psychological Review, 105,* 125–157.

Reichle, E. D., Pollatsek, A., & Rayner, K. (2012). Using E-Z Reader to simulate eye movements in non-reading tasks: A unified framework for understanding the eye-mind link. *Psychological Review, 119,* 155–185.

Reichle, E. D., & Reingold, E. M. (2013). Neurophysiological constraints on the eye-mind link. *Frontiers in Human Neuroscience, 7,* article 361.

Reingold, E. M., Reichle, E. D., Glaholt, M. G., & Sheridan, H. (2012). Direct lexical control of eye movements in reading: Evidence from a survival analysis of fixation durations. *Cognitive Psychology, 65,* 177–206.

Robinson, A. (2001). *The story of writing.* London: Thames & Hudson.

Rueckl, J. G., Paz-Alonso, P. M., Molfese, P. J., Kuo, W. J., Bick, A., Frost, S. J., ... Lee, J. R. (2015). Universal brain signature of proficient reading: Evidence from four contrasting language, *Proceedings of the National Academy of Science, 112,* 15510–15515.

Sanford, A. J., & Garrod, S. C. (2005). Memory-based approaches and beyond. *Discourse Processes, 39,* 205–224.

Schacter, D. L., Addis, D. R., & Buckner, R. L. (2007). Remembering the past to imagine the future: The prospective brain. *Nature Reviews Neuroscience, 8,* 657–661.

Schad, D. J., & Engbert, R. (2012). The zoom lens of attention: Simulating shuffled versus normal text reading using the SWIFT model. *Visual Cognition, 20,* 391–421.

Schotter, E. R. (2013). Synonyms provide semantic preview benefit in English. *Journal of Memory and Language, 69,* 619–633.

Schotter, E. R., Angele, B., & Rayner, K. (2012). Parafoveal processing in reading. *Attention, Perception and Psychophysics, 74,* 5–35.

Segui, J., & Grainger, J. (1990). Priming word recognition with orthographic neighbors: Effects of relative prime-target frequency. *Journal of Experimental Psychology: Human Perception and Performance, 16,* 65–76.

Seidenberg, M. S., & McClelland, J. L. (1989). A distributed developmental model of word recognition and naming. *Psychological Review, 96,* 523–568.

Seidenberg, M. S., & Plaut, D. C. (2006). Progress in understanding word reading: Data fitting versus theory building. In S. Andrews, (Ed.), *From inkmarks to ideas: Current issues in lexical processing* (pp. 25–49). Hove, UK: Psychology Press.

Share, D. L. (1995). Phonological recoding and self-teaching: Sine qua non of reading acquisition. *Cognition, 55,* 151–218.

Share, D. L. (2008). On the Anglocentricities of current reading research and practice: The perils of overreliance on an "outlier" orthography. *Psychological Bulletin, 134,* 584–615.

Staub, A. (2011). Word recognition and syntactic attachment in reading: Evidence for a staged architecture. *Journal of Experimental Psychology: General, 140,* 407–433.

Taft, M., & Krebs-Lazendic, L. (2013). The role of orthographic syllable structure in assigning letters to their position in visual word recognition. *Journal of Memory and Language, 68*, 85–97.

Tan, L. H., Laird, A., Li, K., & Fox, P. T. (2005). Neuroanatomical correlates of phonological processing of Chinese characters and alphabetic words. *Human Brain Mapping, 25*, 83–91.

Vaessen, A., & Blomert, L. (2010). Long-term cognitive dynamics of fluent reading development. *Journal of Experimental Child Psychology, 105*, 213–231.

van den Broek, P., Lorch, R. E. J., Linderholm, T., & Gustafson, M. (2001). The effects of readers' goals on inference generation and memory for texts. *Memory and Cognition, 29*, 1081–1087.

van den Broek, P., Risden, K., Fletcher, C. R., & Thurlow, R. (1996). A "landscape" view of reading: Fluctuating patterns of activation and the construction of a memory representation. In B. K. Britton & A. C. Graesser (Eds.), *Models of understanding text* (pp. 165–187). Mahwah, NJ: Lawrence Erlbaum.

Van Dyke, J. A., Johns, C. L., & Kukona, A. (2014). Low working memory is only spuriously related to poor reading comprehension. *Cognition, 131*, 373–403.

Van Dyke, J. A., & Lewis, R. L. (2003). Distinguishing effects of structure and decay on attachment and repair: A cue-based parsing account of recovery from misanalyzed ambiguities. *Journal of Memory and Language, 49*, 285–316.

Veldre, A., & Andrews, S. (2015). Parafoveal lexical activation is modulated by the precision of skilled readers' lexical representations. *Journal of Experimental Psychology: Human Perception and Performance, 41*, 219–232.

Veldre, A., & Andrews, S. (2016). Is semantic preview benefit due to relatedness or plausibility? *Journal of Experimental Psychology: Human Perception and Performance, 42*, 939–952.

Whitney, C. (2001). How the brain encodes the order of letters in a printed word: The SERIOL model and selective literature review. *Psychonomic Bulletin and Review, 8*, 221–243.

Woollams, A. M. (2015). Lexical is as lexical does: Computational approaches to lexical representation. *Language, Cognition and Neuroscience, 30*, 395–408.

Yap, M. J., Tse, C. S., & Balota, D. A. (2009). Individual differences in the joint effects of semantic priming and word frequency revealed by RT distributional analyses: The role of lexical integrity. *Journal of Memory and Language, 61*, 303–325.

Ziegler, J., Bertrand, D., Toth, D., Csepe, V., Reis, A., Faisca, L., ... Blomert, L. (2010). Orthographic depth and its impact on universal predictors of reading: A cross-language investigation. *Psychological Science, 21*, 551–559.

Ziegler, J., & Goswami, U. (2005). Reading acquisition, developmental dyslexia and skilled reading across language: A psycholinguistic grain size theory. *Psychological Review, 131*, 3–29.

6 Language in the Visual Modality: Co-speech Gesture and Sign Language

ASLI ÖZYÜREK AND BENCIE WOLL

As humans, our ability to communicate and use language is instantiated not only in the vocal modality but also in the visual modality. The main examples of this are sign languages and (co-speech) gestures used in spoken languages. Sign languages, the natural languages of Deaf[1] communities, use systematic and conventionalized movements of the hands, face, and body for linguistic expression (Brentari, 2010; Emmorey, 2002; Klima & Bellugi, 1979; Stokoe, 1960). Co-speech gestures, though nonlinguistic, are produced and perceived in tight semantic and temporal integration with speech (Kendon, 2004; McNeill, 1992, 2005). Thus, language—in its primary face-to-face context (as is the case both phylogenetically and ontogenetically)—is a multimodal phenomenon (Kendon, 2014; Vigliocco, Perniss, & Vinson, 2014). Expression in the visual modality appears to be an intrinsic feature of human communication. As such, our models of language need to take these visual modes of communication into account and provide a unified framework for how the semiotic and expressive resources of the visual modality are recruited in both spoken and sign languages and what the consequences of this recruitment are for cognitive architecture and processing of language. Most research on language, however, has focused on spoken or written language and has rarely considered the visual context in which it is embedded as a means of understanding our linguistic capacity.

The aim of the current chapter is to outline what the expressive resources of language look like both in spoken and sign languages, and what their roles are in communication, in cognition, and specifically in language processing. This will be set in the context of an exploration of the cognitive and neural architecture of language, addressing such issues as commonalities and contrasts in the brain's network for auditory and visual components of human communication, taking gestures in spoken languages as well as sign languages into account.

Even though historically, co-speech gestures and signs have been studied separately and belong to different traditions of literature, recently there have been attempts to achieve a joint perspective that seeks to understand the role of the visual modality in language in general by studying both gesture and sign (e.g., Goldin-Meadow & Brentari, 2015; Perniss, Özyürek, & Morgan, 2015). Studies comparing hearing speakers' gestures with systems used in emerging sign languages and *homesign* systems (e.g., Goldin-Meadow, 2003; Senghas, Kita, & Özyürek, 2004) have shown that as gestures move toward sign language, idiosyncratic gestures used with speech are replaced by conventionalized expressions, and linguistic properties increase (McNeill, 1992). Unlike co-speech gestures, sign languages are complete linguistic systems exhibiting linguistic structure and language-specific constraints at the phonological, morphological, lexical, grammatical, and discourse levels.

Even though sign languages differ from gestures in significant ways, there is now a clearer understanding that the pragmatic, semantic, and cognitive functions employed by co-speech gestures during the use of spoken languages are also evident in the use of sign languages. Furthermore there are similarities arising from the iconic and indexical properties afforded by the visual modality and these properties may be difficult to express within the auditory affordances of the speech channel. The joint perspective that we will adopt here may shed new light on how communication in the visual modality reflects modality-specific as well as modality-independent aspects of our language capacity and on the extent to which a common cognitive and neural architecture underpins linguistic and nonlinguistic communication across modalities.

1. Visual Modality in Spoken Language: Co-speech Gestures

Kendon (2004) defined *gestures* as visible actions of the hand, body, and face that are intentionally used to communicate and are expressed together with the verbal utterance.

Gestures are universal in the sense that all speaking communities around the world are known to produce gestures, even though the communicative and social value of gesturing and thus the frequency of its use may differ among cultures or show variation across individuals (Chu, Meyer, Foulkes, & Kita, 2014; Kita, 2009). Whenever individuals speak, gesture—a unique human ability starting around the first year of life—is involved. The link between gesturing and speaking also appears innate in the sense that it does not require people to see other people gesturing—congenitally blind people also gesture while speaking (Iverson & Goldin-Meadow, 1998; Özçalışkan, Lucero, & Goldin-Meadow, 2016).

In spite of the close links of gesture to language use, most grammatical theories and linguistic descriptions omit gestural specifications. Recognition of the relation of gesture to language has been even more recent (Kendon, 1986; McNeill, 1992) than the recognition that sign languages are on a par with spoken languages (e.g., Klima & Bellugi, 1979; Stokoe, 1960). One of the reasons for this is that linguistic theories have generally taken what can be spoken or written as their main domain of investigation and have been mostly occupied with aspects of language that denote things arbitrarily and categorically (e.g., words, phrases, sentences).

Less attention has been allocated to expressions of other aspects of communication such as *indicating* and *depicting*. Gestures are frequently found to serve the latter functions. For *indicating*, speakers use pointing gestures to place and locate referents in the shared communicative context and for *depicting*, they use so-called representational or iconic gestures to represent virtual objects and events in the gesture space around them (Clark, 2016). In many face-to-face contexts, without *indicating* and *depicting*, successful communication may be hard to achieve.

1.1. Forms and Functions of Gestures in Language Co-speech gestures manifest themselves in different form and meaning pairings as well as in different semiotic types and functions during communication (Clark, 1996; Kendon, 2004; McNeill, 1992, 2005). While some gestures, such as representational gestures, abstract points, and beats, occur as accompaniments to speech, other gestures, such as emblems or interactional gestures, can replace or complement speech in an utterance or can be used without speech, as will be discussed further.[2] Gestures allow speakers to be co-expressive and create *composite signals* (Clark, 1996) by making use of the specific representational capacities of each modality (visual and auditory).

Different forms of gestures fulfill different semantic and communicative functions when used with speech. For example, in so-called *emblems* there is an arbitrary relationship between their form and the meaning they convey, and they serve very similar functions to lexicalized words. On the other hand, *representational* gestures (also referred to as *iconic* gestures) bear a more visually motivated relation between their form and the referent, action, or event they represent. For example, a stirring hand movement accompanying a spoken utterance about cooking bears a resemblance in form to the actual act of stirring. Even though such gestures are visually motivated, the meaning they convey relies heavily on the speech they accompany. Experimental studies have shown that in the absence of speech, the meaning of these gestures is highly ambiguous and not at all transparent from their form (Krauss, Dushay, Chen, & Rauscher, 1995). When these gestures occur, they almost always overlap with semantically relevant speech—which supports the disambiguation of their meaning: speech and such gestures form a co-expressive ensemble.

Representational gestures vary in terms of their semiotic characteristics—that is, in the way they can represent objects, actions, or events—revealing modality-specific ways to convey or depict information such as the different visual perspectives of speakers to events, size, three-dimensional characteristics, shapes, relative spatial relations among objects (Debreslioska, Özyürek, Gullberg, & Perniss, 2013; McNeill, 1992; Müller, 2009; Tversky, 2011). Their meaning comes from the holistic representation of the image they represent rather than from a combinatorial representation of individual meaning units such as those we see in spoken languages. As such they mostly fulfill the *depictive* aspects of communication—reenacting objects and events talked about in a "visible" way in the shared space between the speaker and the interlocutor.

Points are co-expressive accompaniments of demonstrative forms and pronouns in discourse, specifying referents, places, and locations (for a review, see Peeters & Özyürek, 2016). Their form is not informative, but the direction of the point links the referent to the object/space, fulfilling indexical aspects of reference. Such pointing gestures can either be oriented toward concrete objects (when targeting objects or places in the here-and-now of the participants' discourse) or point to meaningful abstract locations in the gesture space in front of the speaker. Pointing to abstract gesture space allows speakers to express coherent relationships among the referents that figure in their discourse by locating them in the gesture space (e.g., McNeill, Cassell, & Levy, 1993; Perniss & Özyürek, 2015).

Finally, *beats* (meaningless repetitive hand movements) can be used to emphasize parts of speech at the information structure level to express focus on certain parts of speech. The so-called *interactional* gestures (e.g., a gesture for *I don't know*) are used to regulate different aspects of dialogic interaction (e.g., expressing stance, turn taking) between the interlocutors (Bavelas, Chovil, Lawrie, & Wade, 1992).

Thus when all types of gestures are considered, we see that they have very similar functions to lexical, semantic, pragmatic, discourse, and interactional features of spoken languages—albeit conveyed in a different format and thus allowing speakers to also convey aspects of messages (e.g., iconic, indexical) that cannot be conveyed through the affordances and the linguistic structures of spoken language (arbitrary, categorical). In this chapter we will mostly focus on representational and pointing gestures, discuss whether and how they are integrated with the spoken language in their use, and consider their roles in language processing and their neural correlates.

1.2. Role of Gesture in Language Processing The production of co-speech gestures is closely linked to the production of the linguistic message conveyed in speech. This is evident especially when we consider the close timing between gestures and speech during production and the co-expressive meaning they convey. With respect to timing, a co-speech gesture is produced along with the relevant part of speech and together they express a communicative act. People do not gesture the entire time they are speaking. Nor is it the case that each and every gesture is accompanied by speech. The important point, rather, is that when people produce co-speech gestures, those gestures are almost always temporally aligned in some meaningful way with a spoken utterance. With respect to meaning, gesture and speech have been argued to share an underlying conceptual message (Bernardis & Gentilucci, 2006; McNeill, 1992). In this sense, gesture and speech are considered to be co-expressive, although the contributions of these communicative channels may be supplementary to, or redundant with, one another, and the representational formats of speech and gesture differ (de Ruiter, Bangerter, & Dings, 2012; McNeill, 1992), in other words, speech is categorical and discrete, whereas gesture is continuous and analog.[3]

More evidence that speech and gesture are tightly linked comes from studies demonstrating that the timing and co-expressive meaning alignment between speech and gesture vary systematically between typologically different languages (Defina, 2016; Floyd, 2016; Gu, Mol, Hoetjes, & Swerts, 2017; Kita & Özyürek, 2003). One demonstration of this phenomenon is how people speak and gesture about motion events across languages. Languages vary in how they linguistically encode the path and manner of motion events (Talmy, 1985), and the gestures that speakers of a given language use have been shown to reflect this variation. Speakers of Japanese and Turkish (*contra* English) are unlikely to tightly package information about both path and manner within a linguistic clause (e.g., rather than *she runs down the stairs* they will say *she runs and goes down the stairs*). Likewise, when gesturing about motion events, speakers of Japanese and Turkish (*contra* English) are unlikely to encode both path and manner within a single gesture and prefer either to represent only one element or split the two elements into separate expressive gestures (Kita & Özyürek, 2003). In English, however, since it is grammatically possible to express manner and path within one linguistic clause (*she ran down the stairs*), speakers are more likely to express both components in a single gesture. That is, gestures in each language seem to be shaped by the syntactic and semantic packaging of information at the clause level. This is confirmed by recent evidence from blind speakers of Turkish and English, who show similar patterns of speech and gesture to those of sighted speakers (Özçalışkan et al., 2016). Recently, Defina (2016) showed that speakers of Avatime (an indigenous spoken language of Australia), which has serial verb constructions, also package two or more semantic elements into a single gesture.

A second area where gesture may vary in relation to linguistic structure is in the expression of spatial frames of reference. Speakers of languages (e.g., Guugu Yimidhirr) that preferentially express spatial relationships using cardinal directions (e.g., east, west, north, and south) rather than egocentric ones (e.g., left, right), also tend to express cardinal relationships in their gestures (Haviland, 1993). Gu et al. (2017) also showed that Chinese speakers' time metaphors based along the vertical dimension are also reflected in their gestures, unlike English speakers' speech and gestures that reflect time along a horizontal axis. Finally, in some languages, gestures may consistently express semantic information not expressed in speech. In a recent study, Floyd (2016) showed that speakers of the Brazilian indigenous language Nheengatú use pointing gestures "adverbially" to indicate time. While speech in Nheengatú gives information about time in general terms (e.g., morning,

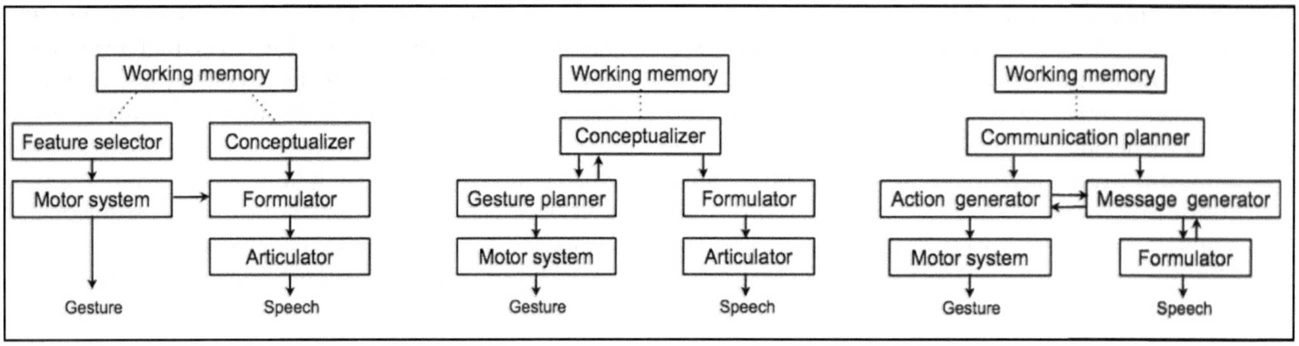

FIGURE 6.1 Adapted schematic overview of different models in relation to speech and gesture production (from left to right: Morrel-Samuels & Krauss, 1992; de Ruiter, 2000; Kita & Özyürek, 2003) (taken from Wagner, Malisz, & Kopp [2014], reprinted with permission).

noon, evening), celestial pointing gestures indicate more specific times of the day (e.g., 10 in the morning). Speakers' judgments about these composite utterances have also shown that they rely on gestures for further specification of time. Floyd argued that there is no *a priori* reason for linguistic properties not to develop in the visual practices (i.e., gestures) that accompany spoken language.

The studies presented show that speech and gesture (at least for representational and pointing gestures) are orchestrated in different ways in different languages, and that gestures are informed by the lexical, syntactic, and pragmatic possibilities of different languages. This poses interesting questions about how speakers coordinate speech and gesture during production to achieve the semantic and temporal alignment they co-express—also considering crosslinguistic variation in this respect. This brings us to language production models proposed so far that take into account the close links between speech and gesture.

The speech and gesture production models suggested to date have either viewed gestures as representing an independent but parallel expressive channel (de Ruiter, 2000; Krauss, Chen, & Gottesman, 2000) or have assumed that speech and gesture interact during the formulation of the linguistic message at different levels (Kita & Özyürek, 2003; McNeill, 1992). These models also differ in terms of whether they consider gesture to be part of the communicatively intended message (de Ruiter, 2000; Kita & Özyürek, 2003) or to be produced independently (Krauss et al., 2000).[4]

According to Krauss and colleagues, gestures are generated from images in working memory, which might help to prime lexical items cross-modally. They are not necessarily assumed to be communicative. In de Ruiter's (2000) model, gestures are generated from conceptualizations that are intended to be part of the communicative message, but during production the information to be conveyed in both modalities is split and expressed through independent channels. In contrast, according to more interactionist models, for example that of McNeill (1992, 2005), gesture and speech are derived from an initial single unit, which McNeill refers to as a *growth point*, composed of both types of representations, imagistic and linguistic. Both gesture and speech are manifestations of this combined unit of representation. According to another interactionist view—the interface hypothesis—proposed by Kita and Özyürek (2003) with a slightly revised version proposed by Chu and Kita (2016), representational gestures and speech are best characterized as originating from different representations: gestures from imagistic/action representations and language from propositional representations. During the language production process, however, both representations interact at the level of linguistic formulation. Figure 6.1 shows different models proposed for speech and gesture production.

As in the case of production, there is growing evidence that gestures are integrated with speech comprehension. It has been a long-standing finding that addressees pick up information from gestures that accompany speech. Even though most models of speech and gesture have focused on production, recent research has also provided ample evidence that addressees integrate the information coming from both modalities during comprehension, in other words, they are perceived as communicative. Listeners/viewers pay attention to iconic gestures and pick up the information that they encode. For example, Kelly, Barr, Church, and Lynch (1999) showed participants video stimuli where gestures conveyed information additional to that conveyed in speech (producing a gesture pantomiming drinking, while the speech is "I stayed up all night") and asked them to write down what they heard. In addition to the speech they heard, participants included information that was conveyed only in gesture and not in speech (i.e., "I stayed up drinking all night"). In another study, Beattie and Shovelton (1999) showed that listeners/viewers answered questions about the size and relative position of objects in a speaker's message more accurately when gestures were part of the description and conveyed information additional to speech than when they heard speech only.

In a priming study by Kelly, Özyürek, and Maris (2010), participants were presented with action primes (e.g., someone chopping vegetables) followed by targets comprising speech accompanied by gesture. They were asked to press a button if what they heard in speech or saw in gesture depicted the action prime. Participants related primes to targets more quickly and accurately when they contained congruent information (speech: "chop"; gesture: chop) than when they contained incongruent information (speech: "chop"; gesture: twist). Moreover, the degree of incongruence between overlapping speech and gesture affected processing, with fewer errors for weak incongruities (speech: "chop"; gesture: cut) than for strong incongruities (speech: "chop"; gesture: open). This indicates that in comprehension, the semantic relations between the two channels are taken into account, providing evidence against independent processing of the two channels. Furthermore and crucially, this effect was bidirectional and was found to be similar when either speech or gesture targets matched or mismatched the action primes. That is, gesture influences processing of speech and speech influences processing of gesture. Further research has shown that gestures also show semantic priming effects. For example, Yap, So, Yap, and Tan (2010) showed that iconic gestures shown without speech (highly conventionalized gestures such as flapping both hands at the side of the body to mean *bird*) prime the sequentially presented spoken target words.

Finally, evidence for semantic integration between representational gestures and speech is also found in many neurocognitive studies. Several studies have shown that comprehension of iconic gestures activates brain processes known to be involved in semantic processing of speech. First of all, gestures modulate the electrophysiological component N400 (e.g., Özyürek, Willems, Kita, & Hagoort, 2007), which has previously been found to be sensitive to the ease of semantic comprehension of words in relation to a previous context. Second, viewing iconic gestures in the context of speech (matched or mismatched) recruits the left-lateralized frontal–posterior temporal network (left inferior frontal gyrus [LIFG], medial temporal gyrus [MTG], and superior temporal gyrus/sulcus [STG/S]) known to be involved in semantic integration of words in sentences (Hagoort & van Berkum, 2007; Özyürek, 2014; see Emmorey & Özyürek, 2014, for a broader overview). Recent studies by Peeters and colleagues (Peeters, Hagoort, & Özyürek, 2015; Peeters & Özyürek, 2016; Peeters, Snijders, Hagoort, & Özyürek, 2017) have also shown that match/mismatch between an expressed element in speech (e.g., "apple") and pointing to a referent (e.g., an apple) evokes semantic integration, as indexed by modulation of N400, and also recruits LIFG and MTG and STG/S—as found for iconic gestures in the above-mentioned studies.

1.3. Conclusions: Gesture These findings show that gestures are an integral part of language at the level of semantics, syntax, pragmatics, and discourse. Gestures, because of the affordances of the visual modality, can subserve *indicating* and *depicting* aspects of communication—albeit in a different representational format from that found in speech. In visible ways, they allow the grounding of concepts conveyed to the visual here-and-now by the speech component, either by linking speech to objects visibly through pointing, or by reenacting them in a virtual space created among the conversational participants, to convey analog representations of events. In doing so, they play a role in the conceptualization, formulation, and comprehension of utterances. They also recruit language networks in the brain during processing. Thus, although their visual and semiotic properties differ from the linguistic units of spoken language, in relation to the cognitive and neural architecture of language, gestures are an integral part of our language capacity. They are integrated into the language-specific semantic and structural aspects of spoken language and interact with spoken language during the production and processing of utterances. Their role also underscores the claim that language is not a fully modular system at the level of production, comprehension, or in its neural architecture. Many of the features of language—semantic, pragmatic, or even syntactic, can be subserved in orchestration with gestures.

Now we turn to how the visual modality is recruited in sign languages: languages created by Deaf communities, who communicate entirely in the visual modality, and where visible bodily articulators alone express all functions of language and communication.

2. *Visual Modality in Sign Language*

Following the groundbreaking work by linguists and cognitive scientists over the last 50 years, it is now generally recognized that sign languages of Deaf communities, such as ASL (American Sign Language), BSL

(British Sign Language),[5] and Sign Language of the Netherlands are not idiosyncratic compilations of silent gestures/pantomimes used by deaf people but are structured and processed in a similar manner to spoken languages. The one striking difference is that they operate in a wholly nonauditory, visual-spatial modality. In this section we first summarize which aspects of sign language structure and processing are similar to those found in spoken language regardless of the differences in modality. Second, we illustrate which aspects of sign languages reveal modality-specific features and how these compare not only to speech but also to the gestural properties that we see in spoken languages as discussed in section 1. Sign languages thus offer unique insights about the modality-independent versus modality-specific (e.g., iconic, embodied) aspects of our language capacity and its cognitive and neural architecture.

Before we begin it is important to say a few words about the social and linguistic contexts in which sign languages are most likely to emerge. Just as all languages need a community of users, sign languages need a Deaf community, which can only come into existence where deaf people are in contact with one another. Although there are descriptions of deaf people's signing going back hundreds of years, the establishment of schools for deaf children, starting in the late 18th century in Europe, triggered the creation of Deaf communities and sign languages as we know them today. At these schools, communication between children and teachers resulted in the conventionalization of what were previously widely varying *home signs*, gestures used by isolated deaf individuals. In many countries, education for deaf children has only recently begun and this can provide an environment in which new sign languages can emerge. Kegl, Senghas, and Coppola (1999) described how the establishment of the first school for deaf children in Nicaragua in the 1980s led to the beginnings of a national sign language. There are also communities where an unusually high incidence of deafness in a small village results in a sign language used by both deaf and hearing people, even in the absence of schools (Sandler, Meir, Padden, & Aronoff, 2005; see Woll & Ladd, 2010, for a review of a number of these).

Where deaf children are *not* exposed to a sign language, they invariably develop systematic gestural communication within their families. Such communication is called *homesigning* (Coppola & Newport, 2005; Goldin-Meadow, 2003). Homesigning systems do not have a full linguistic structure, unlike sign language. They may have rudimentary features of linguistic structure such as a lexicon, simple morphology, segmentation, and consistent word order. However, there is evidence that homesigners cannot master a sign language fully if they are exposed to one only after childhood (e.g., later than six to seven years of age). In other words, homesigning does not serve as a "first" language when homesigners are later exposed to a conventionalized sign language (Mayberry, 2010).

Although the emergence and sociolinguistic context of sign languages differs from spoken languages, which are—with rare exceptions—transmitted by native speakers and in which the only "new" languages are creoles derived from contact between two different spoken languages, the sign languages used by Deaf communities display many of the linguistic structures we see in spoken languages. It is clear that when deaf individuals are able to communicate with each other, in a very short amount of time the communication system they use shows substantial divergence from the gestures used by speaking people. For example, Senghas, Özyürek, and Kita (2004) investigated co-speech gestures used by Spanish speakers in expressing simultaneously occurring manner and path (for example, *run-downhill*) and compared them to signs used by three cohorts of Nicaraguan signers (cohort 1: deaf homesigners brought together as adults; cohort 2: deaf homesigners brought together as children and exposed to the communication of cohort 1; cohort 3: deaf children exposed to cohorts 1 and 2). While gestures produced by Spanish speakers expressed manner and path elements in one gesture, signer cohorts 2 and 3 started to segment them—akin to the separate verbs for manner and path that we see in some spoken languages—and to combine them sequentially to express the simultaneity of manner and path. A recent study by Özyürek, Furman, and Goldin-Meadow (2014) also observed similar developmental changes in Turkish deaf children's homesigning systems compared to their hearing Turkish caregivers' gestures and the gestures of Turkish-speaking children. Even though Turkish homesigners also started segmenting elements of manner and path into separate units, they were more likely to produce gestures that expressed both elements in one unit, and their patterns resembled those of the first cohort of Nicaraguan signers rather than the second and third cohort signers (Özyürek et al., 2014). These studies indicate that in the context of deafness, visual communication goes beyond the expressive possibilities of the gestures used by speakers. However, as we will discuss, some similarities between the two systems (e.g., the role of iconicity) exist because of the shared affordances of modality between gesture and sign and because sign languages may make use of gestures, as spoken languages do.

FIGURE 6.2 Minimal pairs in BSL.

In section 2.1, a brief description will be provided of both modality-independent and modality-specific characteristics of languages, including phonology, lexicon, and the exploitation of space for grammatical purposes. Although structuring at different linguistic levels is similar across signed and spoken languages, the ways in which these structures can be expressed show modality-specific patterning—this is discussed in sections 2.1.1 and 2.1.2. We will also indicate areas of sign language structure for which there has been a debate about whether they should be analyzed solely in terms of abstract, categorical, grammatical structures or whether their analysis needs to take into account instantiations of gestural form-meaning mappings and/or iconic correspondences. Finally, the processing and neural organization of sign languages will be discussed in relation to those structures that are modality-independent or modality-specific.

2.1. MODALITY-INDEPENDENT ASPECTS OF SIGN LANGUAGES As a result of research undertaken in the past half century on the longer-established sign languages of Europe and North America, it is now recognized that the sign languages used by Deaf communities are complex natural human languages and that they are not derived from the spoken languages of the surrounding hearing communities—even though, due to contact, some features from spoken languages can influence sign language structures.

Meier (2002) listed a number of the noneffects of modality (i.e., the shared properties of spoken and signed languages).

a. Conventional vocabularies: learned pairing of form and meaning
b. Duality of patterning: meaningful units built of meaningless sublexical units, whether orally or manually produced
c. Productivity: new vocabulary may be added to signed and spoken languages
d. Syntactic structure:
    i. Same word classes in spoken and signed languages: nouns, verbs, and adjectives
    ii. Embedding of clauses and recursion
    iii. Trade-offs between word order and morphological complexity in how grammatical relations are marked
e. Acquisition: similar timetables for acquisition of signed and spoken language.

Next we describe some of the features of sign languages that exhibit similar linguistic structures to those found in spoken languages, albeit expressed differently due to different affordances of the visual modality. In many cases, however, especially where visual motivation is evident (i.e., iconicity), these features may also resemble those found in co-speech gesturing.

2.1.1. Phonology Since Stokoe (1960), linguists have seen the phonological structure of signs as consisting of simultaneous combinations of configuration(s) of the hand(s), a location where the sign is articulated, and movement—either a path through signing space or an internal movement of the joints in the hand. Each is understood to be a part of the phonology, because changing one of these parameters can create a minimal pair (see figure 6.2). There have been considerable modifications to Stokoe's framework since 1960, but this model has remained the basic description of sign language phonology.

2.1.2. Morphology Sign language morphology tends to manifest itself in simultaneous combinations of meaningful handshapes, locations, and movements, rather than in affixation. In derivational morphology, for example, handshape can change to

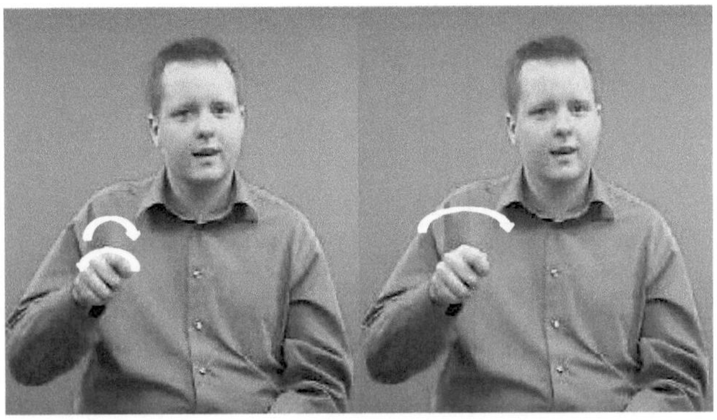

A KEY B LOCK

Figure 6.3 Movement contrast between derivationally related BSL signs KEY (A) and LOCK (B).

reflect numbers. For example in BSL,[6] n-WEEKS, n-O'CLOCK, and n-YEARS-OLD are articulated with conventionalized location and movement, while the handshapes (e.g., of *three* or *five*) incorporated into the time signs indicate the number (e.g., 3-WEEKS, 5-YEARS-OLD).

Signs referring to objects and actions may also differ only in movement, so the verbs LOCK, READ-NEWSPAPER, and EAT are made with long movements, compared to the derivationally related nouns KEY, NEWSPAPER, and FOOD, which have short, repeated movements (figure 6.3).

Other morphological features are also shown by changes in movement and location. Thus, degree is shown through size, speed, onset speed, and length of hold in a movement, with, for example, LUCKY having a smaller, faster, and smoother movement than VERY-LUCKY. Movement changes conveying temporal aspect are frequently visually motivated, so that repeated actions or events are shown through repetition of the sign; duration of an event is paralleled by duration of the sign (signs for shorter events being articulated for less time than signs for longer events) and when an event is interrupted suddenly, the movement of the sign is also interrupted, as shown for example in encoding of telicity (i.e., whether the event expressed is abrupt or continuous) in signs (Strickland et al., 2015). Signs can also change handshape to indicate how a referent is handled. So (I) HAND-OVER-A-FLOWER-TO-EACH-OF-YOU has the same movement as (I) HAND-OVER-AN-ICE-CREAM-TO-EACH-OF-YOU—with the hand moving away from the signer to virtual or real recipients—but with different handshapes that incorporate the shape of a hand as if holding a flower or an ice cream.

Both spoken and signed languages articulate lexical items sequentially. In sign languages, however, the availability of two hands enables the extensive use of simultaneously articulated structures (Vermeerbergen, Leeson, & Crasborn, 2007). Two hands can be used to represent the relative locations and movements of two referents in space and their topic–comment relations. Thus simultaneity is an aspect of sign languages that allows the expression of distinctions found in spoken languages but in a different, visual, format.[7] We will see that this feature also allows the depiction of some iconic structures similar to those found in gestures accompanying spoken languages.

A FILM INDEX FRIEND LIKE-NOT
"(My) friend didn't like that film"

B TABLE BOOK PEN LONG-THIN OBJ-cl
FLAT-OBJ-cl_____

"The pen is next to the book on the table"

Figure 6.4 (A) Example of syntactic space. The referent "film" is located in the upper right of signing space by means of an index, but this does not map onto any real-world location. (B) Example of topographic space (spatialized grammar). In the predicate, the referents "book" and "pen" are replaced with classifiers for "flat object" and "long thin object," respectively, and these handshapes are located adjacent to each other and at the height in signing space of the sign "table."

Sign languages exploit the use of space for grammatical purposes, preferring dimensionality (the analog representation of size and shape) and simultaneity in syntax, while spoken languages prefer linearization and affixation. In earlier literature, on ASL in particular (e.g., Poizner, Klima, & Bellugi, 1987; Padden, 1988), two uses of space for linguistic purposes were contrasted. Topographic space was described as being used to depict spatial relationships and to map referents onto a representation of real space (spatialized grammar), while syntactic space was conceived of as an exploitation of space for purely grammatical purposes, without any mapping to real-world spatial relationships (see figure 6.4 for examples of sentences illustrating these different uses of space). These models were very closely related to the linguistic descriptions used for spoken languages, such as pronouns and agreement in person and number. We will see that recently some of these analyses have been questioned and many researchers now see them as more closely involving iconic gestural features.

2.2. Modality-Specific Aspects of Sign Language: Iconicity and Gestural Elements While significant progress has been made by treating sign languages as having many features in common with those described for spoken languages, the ways in which sign languages differ from spoken languages may have implications for how they are processed and understood by language users.

Perhaps the most obvious of these differences is the way that sign languages exploit the visual modality through iconicity. Iconicity refers to the resemblance between an object or action and the word or sign used to represent that object or action. Early studies emphasized that iconicity might not play a major role in sign language structure or processing. Klima and Bellugi (1979) provided a detailed discussion of what is meant by a sign being *iconic*, pointing out that (a) many signs in ASL (and other sign languages) are noniconic; (b) iconic signs vary from one sign language to another, since different visual motivations for a sign form may be selected (OLD represented by wrinkles in BSL and by a beard in ASL); and (c) they are conventionalized forms, subject to regular processes of phonological change. There have been contrasting findings in relation to the role iconicity plays in sign language processing at the lexical level. Poizner, Bellugi, and Tweney (1981) showed that highly iconic signs were not more easily remembered than signs that are highly opaque; Atkinson, Marshall, Woll, and Thacker (2005) reported that signers with word-finding difficulties following stroke found iconic signs no easier to retrieve than noniconic signs (also see section 2.3 on sign language aphasia and gesture); and Meier, Mauk, Cheek, and Moreland (2008) suggested that iconicity is not a factor in the early sign language acquisition of deaf children.

More recent studies have suggested that iconicity does have a role in the structure of the lexicon and grammar of sign language as well as in processing and learning (Emmorey, 2014; Perniss, Thompson, & Vigliocco, 2010; Strickland et al., 2015; Taub, 2001). For example, Thompson, Vinson, Woll, and Vigliocco (2012) reported that iconic signs are learned earlier than noniconic signs are. Thompson, Vinson, and Vigliocco (2010) also reported effects of iconicity on phonological decision tasks. Studies of ASL at the narrative and discourse level have suggested that in order to understand ASL, the addressee must process *surrogates* (Liddell, 2003) or *depictions* (Dudis, 2004) produced by the signer. Both Liddell and Dudis argued that the signer creates a visual scene and "paints a picture" for the addressee (close to Clark's [2016] notion of *depiction* and *indication* discussed in section 1.1 in relation to gesture), utilizing the visual medium and signing space to convey meaning in ways that are difficult, if not impossible, in spoken languages—except when conveyed through gesture. Thus these authors suggested that sign languages are produced and understood in ways that are very different from spoken languages. However, recent research shows that iconic depiction and indication are not sign language–specific constructions but rather modality-specific forms, as revealed both in signing (ASL) and speaking, when we take into account the gestures speakers use (Quinto-Pozos & Parrill, 2015).

Researchers working on a number of sign languages (Engberg-Pedersen, 1993 [Danish Sign Language]; Liddell, 1990 [ASL]) have argued that the sign language linguistic structures that use sign space to express space and syntactic features are not based on abstract linguistic properties but on inherent analog locative relationships among people, objects, and places. Liddell's later work (2003) has gone further, proposing that verbs such as GIVE are a mix of gestural and linguistic elements and analyzing such signs as being composed of a linguistic part expressed by the handshape and a gestural part linking the referent to a locus.

An increasing number of sign language researchers now assume that mixed forms (i.e., structures involving both linguistic and nonlinguistic components) exist, particularly in classifier constructions. Classifiers are handshapes that provide information about the class by which a noun can be described (see figure 6.4). For example, in the BSL verb VEHICLE-MOVE, the handshape varies according to the class of vehicle (e.g., bicycle, ship, car/bus). Constructions involving classifiers, while originally described in purely linguistic terms as polymorphemic (Supalla, 1986), or semantically multicomponential (Slobin et al., 2003; Morgan & Woll, 2007), have also been seen as blends of nonlinguistic gesture and linguistic structures (Liddell, 2003).

Schembri, Jones, and Burnham (2005) compared the representation of motion events by hearing nonsigners using gesture without accompanying speech and by native signers of three different sign languages. They found that constructions in the three sign languages were strikingly similar, but also that the descriptions of motion events produced by the hearing gesturers showed significant points of correspondence with the signed constructions. In both cases, the location and movement parameters were similar and the handshape component showed the greatest differences. They argue therefore that these data are consistent with the claim that verbs of motion and location in sign languages are blends of gestural and linguistic elements.

Recently, Wilbur (2008) and Strickland et al. (2015) also showed in comparisons of a number of different sign languages that the phonetic realization of the telicity of verbs also contains iconic features (e.g., telic verbs such as *decide*, whose meaning has an intrinsic culmination point, are marked with rapid deceleration to an abrupt stop; whereas atelic verbs such as *ponder*, with no intrinsic culmination, are not). These differences can also be detected by hearing nonsigners and are expressed similarly in gestures.

Thus all this research makes evident the point that sign languages are composed of gestural and linguistic elements just as spoken languages are. However more research is needed to understand the extent to which gestural elements look similar across signed and spoken languages (Özyürek, 2012).

2.3. Processing of Sign Language The main focus in this section is on neuroscience studies of sign language processing. However, it should be noted that substantial numbers of behavioral studies have also explored the cognitive mechanisms underlying sign language perception and production and shown similarities between spoken and sign languages. These include studies of lexical segmentation (Orfanidou, Adam, Morgan, & McQueen, 2010); lexical access—exploring lexicality effects in tasks using signs and nonsigns (Carreiras, Gutiérrez-Sigut, Baquero, & Corina, 2008; Corina & Emmorey 1993); and priming effects at all linguistic levels. Examples of the latter include studies of priming effects related to phonology (Dye and Shih, 2006), morphology (Emmorey, 1991), syntax (Hall, Ferreira, & Mayberry, 2015), and semantics (Bosworth & Emmorey, 2010). There has been relatively little research on production but studies of tip-of-the-finger phenomena (Thompson, Emmorey, & Gollan, 2005) and slips of the hand (Hohenberger, Happ, & Leuninger, 2002) reveal phonological and semantic effects in sign retrieval. For comprehensive reviews of sign language processing research, we refer to Emmorey (2002).

It has been suggested that the right hemisphere (RH) is known to be dominant for a number of visuospatial processing abilities, suggesting that there is a right hemisphere advantage for simultaneous processing (Hellige, 1993). One might therefore infer that spoken language, being more linearized, is left-lateralized, while sign language, which is perceived visually and uses space for grammatical purposes, might be either right-lateralized or show more mixed lateralization. However, case studies of Deaf individuals with acquired brain damage and neuroimaging studies of healthy Deaf subjects have provided evidence for left hemisphere dominance in processing sign languages as well as spoken languages. Deaf signers, like hearing speakers, exhibit language disturbances when left hemisphere (LH) cortical regions are damaged (e.g., Hickok, Love-Geffen, & Klima, 2002; Marshall, Atkinson, Smulovitch, Thacker, & Woll, 2004; Poizner, Klima, & Bellugi 1987; for a review, see MacSweeney, Capek, Campbell, & Woll, 2008). Right hemisphere damage, although it can disrupt visual-spatial abilities (including some involved in sign language processing of spatial language), does not produce sign aphasia (Atkinson et al., 2005).

More evidence for the claim that similar brain structures are involved in signed and spoken languages in healthy subjects comes from a study by MacSweeney et al. (2002), which compared BSL presented to deaf and hearing native signers with audiovisual English presented to hearing monolingual speakers. The perception of BSL and audiovisual English sentences recruited very similar neural systems in native users of those languages. Both languages recruited the perisylvian cortex in the LH. However, there was also RH recruitment by both languages and no differences in the extent of recruitment of the RH by BSL and English (see figure 6.5, columns 1 and 3; see also Bedny & MacSweeney, chapter 37 of this volume). Presumably, this reflects the contribution of both hemispheres to comprehension of the sentences presented, whether the language was spoken or signed. Figure 6.5 also provides evidence that cerebral organization of language in Deaf signers utilizes a left-lateralized frontal/temporal network as in speakers.

At first glance, the robust nature of the left-lateralized sign production system does not seem to be modulated by the iconic forms in sign languages. As we have mentioned, many signs bear some iconic relationship to their real-world referents. For example, a sign may trace the outline of a referent (e.g., HOUSE, which traces the outline of the roof and walls of a house) or may refer to a particular visual characteristic of the referent (e.g., CAT, which traces a cat's whiskers, but means *cat*, rather than whiskers). Despite this, sign-aphasic patients are often unable to produce iconic signs in response to prompts such as "Show me the sign for 'toothbrush,'" although they can produce the same actions elicited as pantomimed gesture—"How do you brush your teeth?" (see Corina et al., 1992; Marshall et al., 2004). That is, they show a dissociation between sign language (impaired) and gesture (unimpaired). Imaging studies, too, suggest that iconicity fails to influence the cortical regions activated in the production of sign language (see Emmorey et al., 2004, and San Jose-Robertson, Corina, Ackerman, Guillemin, & Braun, 2004 for further

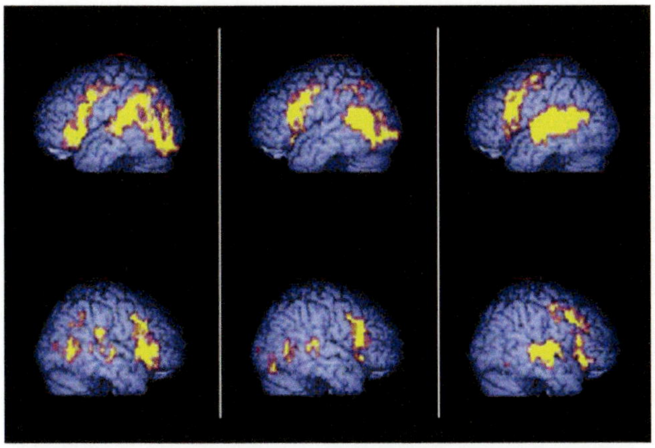

FIGURE 6.5 Color-rendered images of the brain depicting (group) functional MRI activation. Regions activated by BSL perception in Deaf and hearing native signers (first and second columns, respectively) and by audiovisual speech perception in hearing nonsigners (third column). For language in both modalities, and across all three groups, activation is greater in the left than the right hemisphere and perisylvian regions are engaged (from MacSweeney, Woll, Campbell, McGuire, et al. [2002], reprinted with permission).

discussion of the role of iconicity in signed language production).

On the other hand, some modality-specific influences on the brain's processing of sign language have been observed. Neville et al. (1998) suggested that the bilateral activation they observed for ASL could be related to ASL's use of space for linguistic purposes. However, MacSweeney et al.'s (2002) study of BSL sentence processing (figure 6.5), which included sentences with spatial grammar, showed left lateralization comparable to that of spoken languages. These discrepancies underline the need for further studies, using a variety of different signed languages and also a variety of signers with different language experiences and backgrounds. We will see, however, that there is more evidence for the recruitment of the right hemisphere for sign language during processing of spatial language (Emmorey, McCullough, Mehta, Ponto, & Grabowski, 2013) and from lesion studies.

The next question one might ask is the extent to which sign language processing is similar to or different from gesture processing. One way to answer this question has been to compare and contrast the processing of linguistically well-formed material with material that may be superficially similar but which cannot be analyzed linguistically. This type of contrast addresses the question of whether the brain bases for sign language processing are the same as those for the processing of other visual manual actions. In one study, MacSweeney et al. (2004) contrasted BSL utterances with gestures derived from TicTac, the gestural code used by racetrack bookmakers to signal betting odds to each other. The stimuli were modeled by a deaf native signer who constructed "sentences" using hand gestures derived from TicTac codes, adding nonmanual markers (facial gestures) to make these sequences more similar to BSL. Both types of input caused extensive activation throughout the left and right superior temporal lobe when compared to watching the model at rest. That is, much of the upper part of the temporal lobe is involved in attending to gestural displays whether these are linguistically structured or not. However, the brains of the signers who viewed the displays discriminated between the inputs: BSL activated a left-sided region located at the junction of the left posterior superior temporal gyrus and the supramarginal gyrus in the parietal lobe (see figure 6.6) much more than TicTac did. This difference was not due to perceptual differences in the visual quality of the stimuli, because hearing people with no BSL knowledge showed no differences in activation between BSL and TicTac in this region. This region thus appears to be a language-specific region that is not sensitive to the modality of the language it encounters.

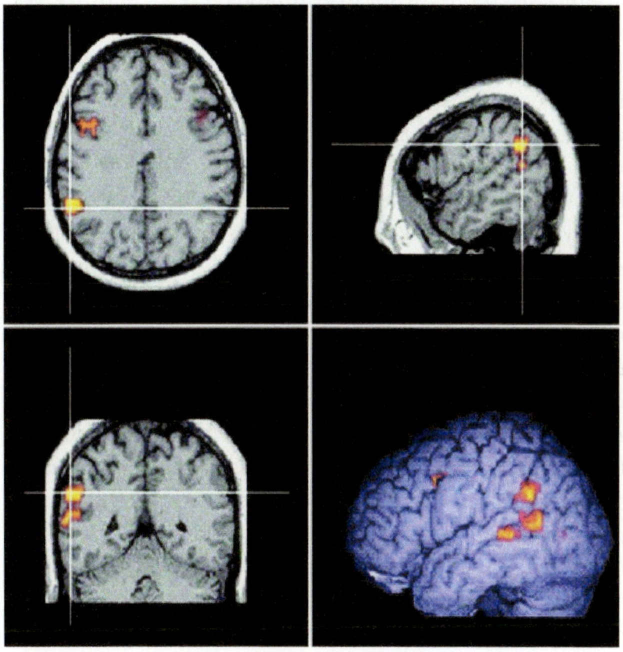

FIGURE 6.6 Colored regions are those recruited to a greater extent by BSL perception than TicTac (nonsense gestures) in Deaf native signers. We interpret these regions as being particularly interested in language processing. Activation up to 5 mm below the cortical surface is displayed. Crosshairs are positioned at Talairach coordinates: $x=-58$, $y=-48$, $z=31$. This is the junction of the inferior parietal lobule and the superior temporal gyrus (from MacSweeney et al. [2004], reprinted with permission).

Even though the comparison of sign and meaningless gesture has activated different cerebral networks, as in MacSweeney et al.'s (2002) study, meaningful gestures (pantomimes) and spoken language are found to recruit overlapping areas. Xu, Gannon, Emmorey, Smith, and Braun (2009) found that comprehending pantomimes (e.g., opening a jar) and their spoken language equivalents both engaged LIFG and left posterior MTG. The authors suggest these are part of a general semantic network for human communication when it comes to comprehension. Emmorey, Xu, Gannon, Goldin-Meadow, and Braun (2010) also found very similar patterns of activation within bilateral posterior temporal cortex when deaf signers passively viewed pantomimed actions and ASL signs, but with evidence for greater activation in the LIFG when viewing ASL signs.

Another domain where we see modality-dependent differences between signed and spoken languages is in the use of spatialized structures to express locative relationships. Liddell (2003) argued that such constructions are partially gestural in nature, and their grammatical status has been debated. Several studies have found differential disruptions in the use and comprehension of sign language sentences that involve spatialized grammar compared to other grammatical constructions. For example, Atkinson et al. (2005) conducted a group study of left and right hemisphere–damaged signers of BSL. They devised tests that included comprehension of single sign and single predicate-verb constructions (e.g., THROW-DART), simple and complex sentences that varied in argument structure and semantic reversibility, locative constructions encoding spatial relationships, constructions involving lexical prepositions, and a final test of classifier placement, orientation, and rotation. Their findings indicated that left hemisphere–damaged (LHD) BSL signers, relative to elderly signing control subjects, exhibited deficits on all comprehension tests, but were better at classifier constructions than lexical prepositions. Right hemisphere–damaged signers (RHD) did not differ from controls on single sign and single predicate-verb construction, or on sentences with a variety of argument structures and semantic reversibility. RHD signers, however, were impaired equally on tests of locative relationships expressed via classifier constructions and prepositions, and on a test of placement, orientation, and rotation of objects. Hickok, Pickell, Klima, and Bellugi (2009) also found the same patterns for classifier production in RHD and LHD patients.

One interpretation of these findings is that the comprehension of spatial language constructions requires not only intact left hemisphere resources, but intact right hemisphere visual-spatial processing mechanisms as well. That is, while both LHD and RHD signers show comprehension deficits, the right hemisphere damaged signers' difficulties stem from more general visual spatial deficits rather than linguistic malfunction per se. The question of whether these visual spatial deficits are nonlinguistic lies at the heart of the debate. In a more recent functional MRI study, Emmorey et al. (2013) also found that viewing both location and motion expressions involving classifier constructions engaged bilateral superior parietal cortex.

Studies of sign language impairments and neural processing thus demonstrate on the one hand, modality-independent aspects of the brain's processing of language; on the other hand, there are also indications that the modality and/or form (i.e., iconic) of the linguistic system may place specific demands on the neural mediation and implementation of language.

2.4. Conclusions: Sign Language Research on sign languages has shown that when the auditory channel is not available, language can exist within the visual modality alone and reveal many of the linguistic structures identified for spoken languages. Unlike spoken languages, however, where new languages are always derived from interactions of existing languages and language users, new sign languages can emerge when deaf people communicate with each other, even if they have no access to previously existing signed or spoken languages, and are then able to transfer that language system to new generations. This is possible because modern humans can scaffold a new sign language on gesture, even when no language is accessible to them. Thus the cognitive architecture of language can be instantiated anew—out of gesture—even in the absence of full conventionalized language input. Thus the human capacity for language structure is not modality specific—to some extent.

However, although signed languages and spoken/written languages share many features in terms of structure, processing, and neural structure, it has become recently more evident that sign languages make use of iconic structures specifically available to the visual modality. Furthermore there is also evidence that modality-specific brain regions might subserve such modality-specific structures.

3. General Conclusions

Use of the visual modality for linguistic expressions and communication is pervasive both in spoken and sign languages and is inevitable when we think of how languages evolved, emerge anew, and are acquired in a face-to-face context. Understanding the role the visual modality plays in language through sign and gesture (in signed or spoken language) is necessary for our understanding of linguistic structure as well as the cognitive and neural architecture of language. By reviewing the role of gesture and sign language in language structure, processing, and neural correlates, we have aimed to offer a joint perspective toward understanding the role that visual modality plays in both spoken and sign language.

The first insight we get when we look at both co-speech gestures and sign languages is that the visual modality can subserve similar linguistic functions and structures to those found in spoken languages—regardless of whether such visual structures accompany spoken language or whether the visual modality takes the whole burden of language and communication. In spoken languages, gestures are semantically, pragmatically, and syntactically, as well as at the discourse level, integrated into the linguistic structures of the spoken languages. In sign languages, visual modality alone can subserve all levels of linguistic structure such as phonology, morphology, syntax, and the lexicon. Thus the visual modality can pattern and function linguistically in similar ways to those we see expressed through speech.

Cognitive and neural processing of visual expressions, as in gesture or sign language, also bears similarities to those of spoken/written structures. Gestures in production and comprehension are influenced by the processing stages of spoken languages (and vice versa). Comprehension of gestures recruits similar brain areas to those used by spoken languages (i.e., the left lateralized frontotemporal network). Similar cognitive and neural processes are involved in processing both sign and spoken languages. There are also similarities between sign and gesture in terms of their cognitive and neural underpinnings (see for reviews Emmorey & Özyürek, 2014; Özyürek, 2014). These findings suggest, first, that gesture should not be excluded from spoken language research, as this misses a great deal of the structures and processes involved in formulating a linguistic message. Second, communication in the visual modality can be supported by cognitive/neural processes that are not specific to any modality.

However the additional insight we get by looking at the visual modality is that both sign language and gesture can also reveal modality-specific characteristics (i.e., those which cannot be expressed through the speech channel) and in similar ways to each other. The visual modality allows visible, analog, gradient, imagistic, and indexical representations to be expressed together with categorical and descriptive ones (in both spoken and sign languages). As such it allows descriptive and categorical aspects of language to be grounded in the here-and-now to convey the Clarkian *depictive* and *indicating* functions (Clark, 2016). This observation indicates that we need to widen our notion of language and its processing possibilities to modality-specific structures visible in both gesture and sign (Goldin-Meadow & Brentari, 2015; Perniss et al., 2010).

Finally, such iconic and analog, gradient representations also recruit modality-specific brain areas—shared by both gesture and sign—outside of what we described as the classical language network—as can be seen in the greater recruitment of parietal areas and right hemisphere in spatial language processing in sign languages. Sign language perception also recruits areas overlapping with those involved in pantomime perception. These show that the cognitive and neural architecture of language is broader than is traditionally assumed and encompasses modality-specific and "embodied" structures and representations—not only the abstract, categorical and arbitrary ones.

All in all, the review provided in this chapter suggests that our understanding of the structures, processes, and neural architecture of language based on data from spoken languages alone needs to be updated if we wish to fully characterize our linguistic capacity and its cognitive and neural architecture. Recent findings have challenged the views that sign language and spoken language are structured and processed identically and that sign languages do not share similarities to gestures or iconic representations. In fact, there is growing research on spoken languages showing how iconicity (motivated form-meaning mappings) may also be a unique characterizing feature of language when non-European languages are considered, such as Japanese or several African languages that contain many more iconic words in their lexicons than European languages (e.g., Dingemanse, Blasi, Lupyan, Christiansen, & Monaghan, 2015). Future research is likely to characterize in more detail the modality-specific and modality-independent aspects of sign language, its cognitive underpinnings and neural correlates. Research on sign languages not of European origin is also needed to generalize the current findings, which are mostly based on ASL and BSL, to other sign languages. Finally we hope that comparisons of spoken and sign languages will also increasingly include audiovisual speech and gesture, rather than only comparing sign languages to heard speech and writing (Goldin-Meadow & Brentari, 2015; Perniss, Özyürek, & Morgan et al., 2015), so that we can more fully understand the modality-specific as well as the modality-independent nature of our language capacity.

NOTES

1. *Deaf* with an uppercase *D* refers to linguistic communities characterized by the use of sign languages. Lowercase *deaf* refers to an individual's audiological status.
2. When talking about types of gestures, it is important to keep in mind that different scholars have proposed different—and sometimes overlapping—categories and semiotic types of gestures used by speakers. Thus the list given here does not include all of the categories proposed so far (see Müller, 2009).
3. Note that gradient expressions can be also expressed and thus depicted using the speech channel, for example, in *loooong* (the extended duration of the vowel meaning *very long* (Okrent, 2002).
4. Even though there is still debate on this topic, gestures are now considered to have double functions: for the self (e.g., to reduce cognitive load, help verbalization, assist lexical access, and form new thought processes) as well as for the addressee. Many studies show that gestures relating to the same event are produced differently (in terms of size and shape or frequency) in relation to changing needs of the addressee (e.g., knowledge, common ground, age, social space) (Alibali, Heath, & Myers, 2001; Campisi & Özyürek, 2013; Özyürek, 2002).
5. ASL and BSL are historically unrelated and mutually unintelligible.
6. All examples are from BSL unless otherwise stated.
7. It should be noted, however, that simultaneity with two hands is an option exercised differently by different sign languages (Perniss, Zwitserlood, & Özyürek, 2015; Saeed, Sutton-Spence and Leeson, 2000).

REFERENCES

Alibali, M. W., Heath, D. C., & Myers, H. J. (2001). Effects of visibility between speaker and listener on gesture production: Some gestures are meant to be seen. *Journal of Memory and Language, 44*, 169–188.

Atkinson, J. A., Marshall, J., Woll, B., & Thacker, A. (2005). Testing comprehension abilities in users of British Sign Language following CVA. *Brain and Language, 94*(2), 233–248.

Bavelas, J. B., Chovil, N., Lawrie, D. A., & Wade, A. (1992). Interactive gestures. *Discourse Processes, 15*, 469–489.

Beattie, G., & Shovelton, H. (1999). Do iconic hand gestures really contribute anything to the semantic information conveyed by speech? An experimental investigation. *Semiotica, 123*, 1–30.

Bernardis, P., & Gentilucci, M. (2006). Speech and gesture share the same communication system. *Neuropsychologica, 44*(2), 178–190.

Bosworth, R. G., & Emmorey, K. (2010). Effects of iconicity and semantic relatedness on lexical access in American Sign Language. *Journal of Experimental Psychology: Learning, Memory, and Cognition, 36*(6), 1573–1581.

Brentari, D. (Ed.) (2010). *Sign languages. Cambridge Language Surveys*. Cambridge: Cambridge University Press.

Campisi, E., & Özyürek, A. (2013). Iconicity as a communicative strategy: Recipient design in multimodal demonstrations for adults and children. *Journal of Pragmatics, 47*, 14–27.

Carreiras, M., Gutiérrez-Sigut, E., Baquero, S., & Corina, D. (2008). Lexical processing in Spanish Sign Language (LSE). *Journal of Memory and Language, 58*, 100–122.

Chu, M., & Kita, S. (2016). Co-thought and co-speech gestures are generated by the same action generation process. *Journal of Experimental Psychology: Learning Memory and Cognition, 42*(2), 257–270.

Chu, M., Meyer, A., Foulkes, L., & Kita, S. (2014). Individual differences in frequency and saliency of speech-accompanying gestures: The role of cognitive abilities and empathy. *Journal of Experimental Psychology: General, 143*(2), 694–709.

Clark, H. H. (1996). *Using language*. Cambridge: Cambridge University Press.

Clark, H. H. (2016). Depicting as a method of communication. *Psychological Review, 123*, 324–347.

Coppola, M., & Newport E. L. (2005). Grammatical subjects in home sign: Abstract linguistic structure in adult primary gesture systems without linguistic input. *Proceedings of the National Academy of Sciences, 102*(52), 19249–19253.

Corina, D. P., & Emmorey, K. (1993). *Lexical priming in American Sign Language*. Paper presented at the meeting of the Linguistic Society of American Conference, Philadelphia, PA.

Corina, D. P., Poizner, H., Bellugi, U., Feinberg, T., Dowd, D., & O'Grady-Batch, L. (1992). Dissociation between linguistic and non-linguistic gestural systems: A case for compositionality. *Brain and Language, 43*, 414–447.

Debreslioska, S., Özyürek, A., Gullberg, M., & Perniss, P. M. (2013). Gestural viewpoint signals referent accessibility. *Discourse Processes, 50*(7), 431–456.

Defina, R. (2016). Do serial verb constructions describe single events? A study of co-speech gestures in Avatime. *Language, 92*(4), 890–910.

de Ruiter, J. (2000). The production of gesture and speech. In D. McNeill (Ed.), *Language and gesture* (pp. 248–311). Cambridge: Cambridge University Press.

de Ruiter, J., Bangerter, A., & Dings, P. (2012). The interplay between gesture and speech in the production of referring expressions: Investigating the tradeoff hypothesis. *Topics in Cognitive Science, 4*(2), 232–248.

Dingemanse, M., Blasi, D. E., Lupyan, G., Christiansen, M. H., & Monaghan, P. (2015). Arbitrariness, iconicity and systematicity in language. *Trends in Cognitive Sciences, 19*(10), 603–615.

Dudis, P. (2004). Body partioning and real space blends. *Cognitive Linguistics, 15*(2), 223–238.

Dye, M. W. G., & Shih, S. (2006). Phonological priming in British Sign Language. In L. M. Goldstein, D. H. Whalen, & C. T. Best (Eds.), *Papers in laboratory of phonology* (Vol. 8, pp. 243–263). Berlin: Mouton de Gruyter.

Emmorey, K. (1991). Repetition priming with aspect and agreement morphology in American Sign Language. *Journal of Psycholinguistic Research, 20*(5), 365–388.

Emmorey, K. (2002). *Language, cognition, and the brain: Insights from sign language research*. Mahwah, NJ: Lawrence Erlbaum.

Emmorey, K. (2014). Iconicity as structure mapping. *Philosophical Transactions of the Royal Society B, 369*(1651), 20130301.

Emmorey, K., Grabowski, T., McCullough, S., Damasio, H., Ponto, L., Hichwa, R., & Bellugi, U. (2004). Motor-iconicity of sign language does not alter the neural systems underlying tool and action naming. *Brain and Language, 89*(1), 27–37.

Emmorey, K., McCullough, S., Mehta, S. H., Ponto, L. B., & Grabowski, T. J. (2013). The biology of linguistic expression impacts neural correlates for spatial language. *Journal of Cognitive Neuroscience, 25*(4), 517–533.

Emmorey, K., & Özyürek, A. (2014). Language in our hands: Neural underpinnings of sign language and co-speech gesture. In M. S. Gazzaniga, & G. R. Mangun (Eds.), *The cognitive neurosciences* (5th ed., pp. 657–666). Cambridge, MA: MIT Press.

Emmorey, K., Xu, J., Gannon, P., Goldin-Meadow, S., & Braun, A. (2010). CNS activation and regional connectivity during pantomime observation: No engagement of the mirror neuron system for deaf signers. *Neuroimage, 49,* 994–1005.

Engberg-Pedersen, E. (1993). *Space in Danish Sign Language: The semantics and morphosyntax of the use of space in a visual language.* Hamburg: Signum.

Floyd, S. (2016). Modally hybrid grammar? Celestial pointing for time-of-day reference in Nheengatú. *Language, 92*(1), 31–64.

Goldin-Meadow, S. (2003). *The resilience of language: What gesture creation in deaf children can tell us about how all children learn language.* New York: Psychology Press.

Goldin-Meadow, S., & Brentari, D. (2015). Gesture, sign and language: The coming of age of sign language and gesture studies. *Behavioral and Brain Sciences, 5,* 1–82.

Gu, Y., Mol, L., Hoetjes, M., & Swerts, M. (2017). Conceptual and lexical effects on gestures: The case of vertical spatial metaphors for time in Chinese. *Language, Cognition and Neuroscience, 32*(8), 1048–1063.

Hall, M. L., Ferreira, V. S., & Mayberry, R. I. (2015). Syntactic priming in American Sign Language. *PLOS ONE, 10*(3), e0119611.

Hagoort, P., & van Berkum, J. (2007). Beyond the sentence given. *Philosophical Transactions of the Royal Society. Series B: Biological Sciences, 362,* 801–811.

Haviland, J. B. (1993). Anchoring, iconicity, and orientation in Guugu Yimidhirr pointing gestures. *Journal of Linguistic Anthropology, 3*(1), 3–45.

Hellige, J. B. (1993). *Hemispheric asymmetry: What's right and what's left.* Cambridge, MA: Harvard University Press.

Hickok, G., Love-Geffen, T., & Klima, E. S. (2002). Role of the left hemisphere in sign language comprehension. *Brain and Language, 82*(2), 167–178.

Hickok, G., Pickell, H., Klima E., & Bellugi, U. (2009). Neural dissociation in the production of lexical versus classifier signs in ASL: Distinct patterns of hemispheric asymmetry. *Neuropsychologia, 47,* 382–387.

Hohenberger, A., Happ, D., & Leuninger, H. (2002). Modality dependent aspects of signed language production: evidence from slips of the hands and their repairs in German Sign Language. In R. P. Meier, K. Cormier, & D. Quinto-Pozos (Eds.), *Modality and structure in signed and spoken languages* (pp. 112–142). Cambridge: Cambridge University Press.

Iverson, J. M., & Goldin-Meadow, S. (1998). Why people gesture when they speak. *Nature, 396,* 228.

Kegl, J., Senghas, A., & Coppola, M. (1999). Creation through contact: Sign language emergence and sign language change in Nicaragua. In M. Degraff, (Ed.), *Language creation and language change* (pp. 179–237). Cambridge, MA: MIT Press.

Kelly, S. D., Barr, D. J., Church, R. B., & Lynch, K. (1999). Offering a hand to pragmatic understanding: The role of speech and gesture in comprehension and memory. *Journal of Memory and Language, 40,* 577–592.

Kelly, S. D., Özyürek, A., & Maris, E. (2010). Two sides of the same coin: Speech and gesture mutually interact to enhance comprehension. *Psychological Science, 21,* 260–267.

Kendon, A. (1986). Some reasons for studying gesture. *Semiotica, 62,* 1–28.

Kendon, A. (2004). *Gesture: Visible action as utterance.* Cambridge: Cambridge University Press.

Kendon, A. (2014). Semiotic diversity in utterance production and the concept of 'language." *Philosophical Transactions of the Royal Society B, 369,* 20130293.

Kita, S. (2009). Cross-cultural variation of speech-accompanying gesture: A review. *Language and Cognitive Processes, 24*(2), 145–167.

Kita, S., & Özyürek, A. (2003). What does cross-linguistic variation in semantic coordination of speech and gesture reveal? Evidence for an interface representation of spatial thinking and speaking. *Journal of Memory and Language, 48*(1), 16–32.

Klima, E., & Bellugi, U. (1979). *The signs of language.* Cambridge, MA: Harvard University Press.

Krauss, R. M., Chen, Y., & Gottesman, R. F. (2000). Lexical gestures and lexical access: A process model. In D. McNeill (Ed.), *Language and gesture* (pp. 261–283). New York: Cambridge University Press.

Krauss, R. M., Dushay, R. A., Chen, Y., & Rauscher, F. (1995). The communicative value of conversational hand gestures. *Journal of Experimental Social Psychology, 31,* 533–552.

Liddell, S. K. (1990). Four functions of a locus: Reexamining the structure of space in ASL. In L. Ceil (Ed.), *Sign language research: Theoretical issues* (pp. 176–198). Washington, DC: Gallaudet University Press.

Liddell, S. K. (2003). *Grammar, gesture and meaning in American Sign Language.* Cambridge: Cambridge University Press.

MacSweeney, M. F., Campbell, R., Woll, B., Giampietro, V., David, A. S., McGuire, P. K., & Brammer, M. J. (2004). Dissociating linguistic and non-linguistic gestural communication in the brain. *Neuroimage, 22*(4), 1605–1618.

MacSweeney, M., Capek, C. M., Campbell, R., & Woll, B. (2008). The signing brain: the neurobiology of sign language. *Trends in Cognitive Science, 12,* 432–440.

MacSweeney, M., Woll, B., Campbell, R., McGuire, P., David, A. S., Williams, S., & Brammer, M. (2002). Neural systems underlying British Sign Language and audio-visual English processing in native users. *Brain, 125,* 1583–1593.

Marshall, J., Atkinson, J., Smulovitch, E., Thacker, A., & Woll, B. (2004). Aphasia in a user of British Sign Language: Dissociation between gesture and sign. *Cognitive Neuropsychology, 21*(5), 537–555.

Mayberry, R. I. (2010). Early language acquisition and adult language ability: What sign language reveals about the critical periods. In M. Marschark & P. E. Spencer (Eds.), *The Oxford handbook of deaf studies, language, and education* (2nd ed., pp. 281–291). Oxford: Oxford University Press.

McNeill, D. (1992). *Hand and mind: What gestures reveal about thought.* Chicago: University of Chicago Press.

McNeill, D. (2005). *Gesture and thought.* Chicago: University of Chicago Press.

McNeill, D., Cassell, J., & Levy, E. (1993). Abstract deixis. *Semiotica, 95,* 5–19.

Meier, R. P. (2002). Why different, why the same? Explaining effects and non-effects of modality upon linguistic structure in sign and speech. In R. P. Meier, K. A. Cormier, & D. G. Quinto-Pozos (Eds.), *Modality and structure in signed and spoken languages* (pp. 1–25). Cambridge: Cambridge University Press.

Meier, R. P., Mauk, C.E., Cheek, A., & Moreland, C. J. (2008). The form of children's early signs: Iconic or motoric determinants? *Language Learning and Development, 4*, 63–98.

Morgan, G., & Woll, B. (2007) Understanding sign language classifiers through a polycomponential approach. *Lingua, 117*(7), 1159–1168.

Morrel-Samuels, P., & Krauss, R. M. (1992). Word familiarity predicts temporal asynchrony of hand gestures and speech. *Journal of Experimental Psychology: Learning, Memory and Cognition, 18*, 615–623.

Müller, C. (2009). Gesture and language. In K. Malmkjaer (Ed.), *The Routledge linguistics encyclopedia* (pp. 214–217). London: Routledge.

Neville, H. J., Bavelier, D., Corina, D., Rauschecker, J., Karni, A., Lalwani, A., ... Turner, R. (1998). Cerebral organization for language in deaf and hearing subjects: Biological constraints and effects of experience. *Proceedings of the National Academy of Sciences USA, 95*, 922–929.

Okrent, A. (2002) A modality-free notion of gesture and how it can help us with the morpheme vs. gesture question in sign language linguistics (or at least give us some criteria to work with). In R. P. Meier, K. A. Cormier, & D. G. Quinto-Pozos (Eds.), *Modality and structure in signed and spoken languages* (pp. 175–198). Cambridge: Cambridge University Press.

Orfanidou, E., Adam, R., Morgan, G., & McQueen, J. M. (2010). Recognition of signed and spoken language: Different sensory inputs, the same segmentation procedure. *Journal of Memory and Language, 62*(3), 272–283.

Özçalışkan, S., Lucero, C., & Goldin-Meadow, S. (2016). Is seeing gesture necessary to gesture like a native speaker? *Psychological Science, 27*(5), 737–747.

Özyürek, A. (2002). Do speakers design their co-speech gestures for their addressees? The effects of addressee location on representational gestures. *Journal of Memory and Language, 46*(4), 688–704.

Özyürek, A. (2012). Gesture. In R. Pfau, M. Steinbach, & B. Woll (Eds.), *Sign language: An international handbook* (pp. 626–646). Berlin: Mouton.

Özyürek, A. (2014). Hearing and seeing meaning in speech and gesture: Insights from brain and behaviour. *Philosophical Transactions of the Royal Society of London. Series B, Biological Sciences, 369*(1651), 20130296.

Özyürek, A., Furman, R., & Goldin-Meadow, S. (2014). On the way to language: Event segmentation in homesign and gesture. *Journal of Child Language, 42*, 64–94.

Özyürek, A., Willems, R. M., Kita, S., & Hagoort, P. (2007). On-line integration of semantic information from speech and gesture: Insights from event-related brain potentials. *Journal of Cognitive Neuroscience, 19*(4), 605–616.

Padden, C. (1988). *Interaction of morphology and syntax in American sign language.* Outstanding Dissertations in Linguistics, Series IV. New York: Garland Press.

Peeters, D., Hagoort, P., & Özyürek, A. (2015). Electrophysiological evidence for the role of shared space in online comprehension of spatial demonstratives. *Cognition, 136*, 64–84.

Peeters, D., & Özyürek, A. (2016). This and that revisited: A social and multimodal approach to spatial demonstratives. *Frontiers in Psychology, 7*, 222.

Peeters, D., Snijders, T. M., Hagoort, P., & Özyürek, A. (2017). Linking language to the visual world: Neural correlates of comprehending verbal reference to objects through pointing and visual cues. *Neuropsychologia, 95*, 21–29.

Perniss, P. M., & Özyürek, A. (2015). Visible cohesion: A comparison of reference tracking in sign, speech, and co-speech gesture. *Topics in Cognitive Science, 7*(1), 36–60.

Perniss, P. M., Özyürek, A., & Morgan, G. (2015). The influence of the visual modality on language structure and conventionalization: Insights from sign language and gesture. *Topics in Cognitive Science, 7*(1), 2–11.

Perniss, P., Thompson, R. L., & Vigliocco, G. (2010). Iconicity as a general property of language: Evidence from spoken and signed languages. *Frontiers in Psychology, 1*, 1–15.

Perniss, P. M., Zwitserlood, I., & Özyürek, A. (2015). Does space structure spatial language? A comparison of spatial expression across sign languages. *Language, 91*(3), 611–641.

Poizner, H., Bellugi, U., & Tweney, R. D. (1981). Processing of formational, semantic, and iconic information in American Sign Language. *Journal of Experimental Psychology: Human Perception and Performance, 7*(5), 1146–1159.

Poizner, H., Klima, E. S., & Bellugi, U. (1987). *What the hands reveal about the brain.* Cambridge, MA: MIT Press.

Quinto-Pozos, D., & Parrill, F. (2015) Signers and co-speech gesturers adopt similar strategies for portraying viewpoint in narratives. *Topics in Cognitive Science, 7*, 12–35.

Saeed, J. I., Sutton-Spence, R., & Leeson, L. (2000). *Constituent order in Irish Sign Language and British Sign Language: A preliminary examination.* Poster presented at the meeting of the 7th International Conference on Theoretical Issues in Sign Language Research (TISLR7), Amsterdam, the Netherlands.

Sandler, W., Meir, I., Padden, C., & Aronoff, M. (2005). The emergence of grammar: Systematic structure in a new language. *Proceedings of the National Academy of Sciences, 102*(7), 2661–2665.

San Jose-Robertson, L., Corina, D. P., Ackerman, D., Guillemin, A., & Braun, A. R. (2004). Neural systems for sign language production: Mechanisms supporting lexical selection, phonological encoding, and articulation. *Human Brain Mapping, 23*, 156–167.

Schembri, A., Jones, C., & Burnham, D. (2005). Comparing action gestures and classifier verbs of motion: Evidence from Australian Sign Language, Taiwan Sign Language, and non-signers' gestures without speech. *Journal of Deaf Studies and Deaf Education, 10*(3), 272–290.

Senghas, A., Kita, S., & Özyürek, A. (2004). Children creating core properties of language: Evidence from an emerging sign language in Nicaragua. *Science, 305*(5691), 1779–1782.

Slobin, D., Hoiting, N., Kuntze, K., Lindert, R., Weinberg, A., Pyers, J., ... Thumann, H., 2003. A cognitive/functional perspective on the acquisition of "classifiers." In K. Emmorey (Ed.), *Perspectives on classifier constructions in sign languages* (pp. 297–310). Mahwah, NJ: Lawrence Erlbaum.

Stokoe, W. (1960). *Sign language structure: An outline of the visual communication systems of the American Deaf.* Silver Spring, MD: Linstok Press.

Strickland, B., Geraci, C., Chemla, E., Schlenker, P., Kelepir, M., & Pfau, R. (2015). *Proceedings of the National Academy of Sciences, 112*(19), 5968–5973.

Supalla, T. (1986). The classifier system in American Sign Language. In C. Craig (Ed.), *Noun classes and categorization* (pp. 181–214). Amsterdam: John Benjamins.

Talmy, L. (1985). Lexicalization patterns: Semantic structure in lexical forms. In T. Shopen (Ed.), *Language typology and syntactic description. Vol. III: Grammatical categories and the lexicon* (pp. 57–149). Cambridge: Cambridge University Press.

Taub, S. F. (2001). *Language from the body: Iconicity and metaphor in American Sign Language.* Cambridge: Cambridge University Press.

Thompson, R. L., Emmorey, K., & Gollan, T. H. (2005). "Tip of the fingers" experiences by deaf signers insights into the organization of a sign-based lexicon. *Psychological Science, 16*(11), 856–860.

Thompson, R. L., Vinson, D. P., & Vigliocco, G. (2010). The link between form and meaning in British sign language: Effects of iconicity on phonological decisions. *Journal of Experimental Psychology: Learning, Memory and Cognition, 36,* 1017–1027.

Thompson, R. L., Vinson, D. P., Woll, B., & Vigliocco, G. (2012). The road to language learning is iconic: Evidence from British Sign Language. *Psychological Science, 23*(12), 1443–1448.

Tversky, B. (2011). Visualizations of thought. *Topics in Cognitive Science, 3,* 499–535.

Vermeerbergen, M., Leeson, L., & Crasborn, O. A. (Eds.). (2007). *Simultaneity in signed languages: Form and function* (Vol. 281). Amsterdam: John Benjamins

Vigliocco, G., Perniss, P., & Vinson, D. (2014). Language as a multimodal phenomenon: Implications for language learning, processing, and evolution. *Philosophical Transactions of the Royal Society B, 369,* 20130292.

Wagner, P., Malisz, Z., & Kopp, S. (2014). Gesture and speech in interaction: An overview. *Speech Communication, 57,* 209–232.

Wilbur, R. B. (2008). Complex predicates involving events, time, and aspect: Is this why sign languages look so similar? In J. Quer (Ed.), *Theoretical issues in sign language research* (pp. 217–250). Hamburg: Signum Press.

Woll, B., & Ladd, P. (2010). Deaf communities. In M. Marschark & P. Spencer (Eds.), *The handbook of deaf studies, language and education* (2nd ed., pp. 159–172). Oxford: Oxford University Press.

Xu, J., Gannon, P., Emmorey, K., Smith, J. F., & Braun A. R. (2009). Symbolic gestures and spoken language are processed by a common neural system. *Proceedings of the National Academy of Sciences, 106*(49), 20664–20669.

Yap, D. F., So, W. C., Yap, M., & Tan, Y. Q. (2011). Iconic gestures prime words. *Cognitive Science, 35,* 171–183.

7 Key Issues and Future Directions: Toward a Comprehensive Cognitive Architecture for Language Use

JAMES M. MCQUEEN AND ANTJE S. MEYER

Language can be used in many different ways. We can, for instance, have a chat with a neighbor, listen to a lecture, write a novel or a manual, cite a poem, or read a scientific text or newspaper. All of these activities involve components of the cognitive architecture that supports linguistic communication. The preceding five chapters in this section discussed important issues that must be addressed in theories of this architecture: the nature of linguistic representations (Jackendoff, chapter 2), the processes involved in understanding spoken language (Dahan & Ferreira, chapter 3) and written language (Andrews & Reichle, chapter 5), those involved in speaking (Roelofs & Ferreira, chapter 4), and the cognitive implications of the multimodal nature of language (Özyürek & Woll, chapter 6). In this chapter, we adopt a broader perspective. We propose a research program to work toward integrative models that account for all of the ways that language is used. Such models should explain speaking, listening, reading, and signing in a coherent and comprehensive way. We discuss which design choices have to be made in generating such models, and how assumptions about the broad architecture of a model and about its components constrain each other.

To start, we categorize language tasks—all the things that people can do with language—by considering the language user's current goal, namely to understand or to produce language, and the modality, namely spoken, written, or signed language. This broad categorization of language tasks and the ways different tasks are related to each other are further discussed in section 1.

We assume that all language tasks involve two types of cognitive components: (i) linguistic and nonlinguistic representations, and (ii) cognitive processes that retrieve or operate on these representations in the order appropriate for the task. Following the principle of parsimony, research should work toward finding the simplest possible characterization of these components. The goal of our proposed research program is thus to describe a single set of representations and processes required for all linguistic tasks and to specify which combination is used and in what way in each task. The set should be as small as possible but sufficient to accomplish all tasks in all languages.

Our working hypothesis implies that there is one language system, consisting of multipurpose linguistic representations (see section 2) and multipurpose procedures to access linguistic representations and perform operations on them (see section 3). Likewise, we propose one domain-general cognitive system (see section 5) that is implicated in linguistic and nonlinguistic tasks. These assumptions do not imply that exactly the same representations or processes are involved in exactly the same way in all tasks; for instance, access to orthographic representations undoubtedly plays a central role in reading but is less important in speaking. Our proposal simply holds that there is one knowledge base and one set of processing mechanisms that are recruited in different ways depending on the specific language task being performed. Some representations or processes may well be unique to a given task.

This characterization of the cognitive architecture for language may suggest that it is a facility that is used over and over again in the same fashion, but in fact a person's language architecture changes with experience across the life span. As discussed in section 4, this plasticity needs to be captured in any theory, as learning and adaptation play crucial roles in many linguistic tasks and establish links between them, with, for instance, learning to comprehend a word being a precursor for someone being able to produce that word.

Our proposal raises three closely related challenges for future research. The first is to specify the minimum representational and processing requirements, that is, to work out which representations and processes are required to perform all language tasks. The second challenge is to delineate which representations and processes are used in each individual task and hence to establish the simplest possible comprehensive model that is consistent with the data pertaining to the task. Third, it needs to be specified how specific processes and representations are combined and ordered in each task.

1. Relationships among Language Tasks

1.1. Producing and Comprehending Language Our proposed research program starts from the assumption that language production and comprehension are different sets of tasks rather than distinct processing systems. Clarifying the relationship between these tasks in terms of shared or unique cognitive processes will be a major step toward developing a comprehensive model of language use.

The tasks of speaking and listening, or producing and understanding sign language, are often said to differ in the direction of the information flow, from concepts to articulation versus from phonetic input to concepts. But comprehending is not speaking (or signing) in reverse. The cognitive challenges arising for speakers and listeners are fundamentally different: Speakers start from a conceptual structure, select and order units and eventually produce overt behavior, which they monitor for appropriateness and correctness (e.g., Levelt, 1989). Listeners do not have to react overtly (and hence probably do not monitor their behavior to the same extent), and they do not have to order units, as the order is provided by the speaker. Their task is to segment the continuous speech stream into words, find the syntactic structure and grasp the speaker's meaning (e.g., Cutler & Clifton, 1999). A main challenge for listeners is to deal with ambiguities arising at all levels of processing (cf. the noisy channel model discussed by Dahan & Ferreira, chapter 3) and to do so at the speed set by the speaker.

Another frequent characterization of the relationship between the tasks of comprehension and production is that they both involve two systems (a production system and a comprehension system). It has long been acknowledged that speakers use a speech production system but also listen to their own overt and inner speech using a speech comprehension system and may use this system to monitor and correct their utterances (e.g. Levelt, 1989; see also Roelofs & Ferreira, chapter 4). In the speech perception literature, it has often been argued that speech recognition depends on the speech production system (Liberman, Cooper, Shankweiler, & Studdert-Kennedy, 1967). In addition, many authors have recently argued for the involvement of the production system in higher-level speech comprehension. Specifically, there is strong evidence for the importance of prediction of upcoming content during reading and listening, and making predictions is often assumed to involve processes that play a central role in production (Pickering & Garrod, 2013). These proposals, that speakers can listen to themselves and that they use their language production system to predict what others might say, fit our experience as speakers and listeners. However, for generating a comprehensive model of language use, they are not particularly helpful since they start from the assumption that there are distinct comprehension and production systems, which nevertheless are both involved in comprehension and production tasks. The proposal that we advocate here does not postulate distinct systems.

We propose that research toward comprehensive models should instead presuppose a single language system. This research then needs to specify which knowledge structures and which processes are implicated in specific production and comprehension tasks and, following on from this, which components of the language system are widely shared between tasks and which are unique to some tasks. This requires conducting research projects that simultaneously consider specific production and comprehension tasks, for instance, word production and word comprehension. In such projects, one might find that some production and comprehension tasks are more closely related to each other than some pairs of tasks within the production family or within the comprehension family.

One important theoretical and empirical challenge in such a research program is to be clear about what it means for a component to be shared across tasks. In a general sense, all language tasks draw on shared conceptual and linguistic knowledge. Though this is rarely discussed, the consensus in the field appears to be that the same conceptual stratum is accessed in production and comprehension. Linguistic knowledge must also be shared since we can only learn to speak a language by hearing it, or to sign a language by seeing it. However, a much debated issue is whether comprehension and production draw on the same representations or whether there are dedicated, but tightly linked representations for the two types of tasks (cf. Meyer & Huettig, 2016). The same-representations view is parsimonious, but it leads to challenging questions about the processes involved in production and comprehension. Most obviously, how do the production deficits arise that are so prominent in children and second-language speakers of a language? And why are estimates of receptive vocabulary usually larger than those of productive vocabulary? The separate-representations view faces a different challenge, namely to explain how task-specific representations are linked such that speakers learn to produce utterances from hearing them.

Studies comparing production and comprehension must zoom in on particular linguistic levels or processing components, such as grammatical, morphological, or phonological encoding of sentences. For each level, we would expect to find the engagement of both shared and unique components. For instance, grammatical encoding processes must occur in production and in comprehension. Perhaps the underlying processes are the same, maybe consisting of combining treelets as proposed by Jackendoff in chapter 2. However, any shared processes must co-occur with unique processes, since grammatical encoding during speaking is driven by the speaker's conceptual representations, which are not usually ambiguous, whereas grammatical encoding during comprehension is driven by the speech input, which is riddled with (at least temporary) ambiguity. Similarly, phonetic and phonological encoding and decoding of utterances may involve shared word-form representations, but phonological encoding during speaking involves serial ordering of segments and preparation of articulatory commands, whereas phonological encoding during listening involves the parsing of a speech signal into words. Ambiguity and ordering requirements may be the main causes of differences in the processes involved in different tasks. Thus, a research program that seeks to establish which representations and processes are used in which tasks may uncover general principles that determine similarities and differences in the processes subserving different tasks.

This discussion has concerned speaking and listening, but similar considerations apply to producing and understanding signed languages and to reading and writing. In all cases, the tasks are different, but are likely to draw on shared as well as unique representations and processes. Functional models of the language system must specify which representations are shared, what it means to share representations, and which processes occur in several tasks or are specific to some of them.

1.2. Using Language in Different Modalities

The modality—spoken, written, or signed language—obviously affects how the linguistic tasks of producing and comprehending language are carried out. Readers only have visual information to rely on; listeners sometimes, for instance, in telephone conversations, have only auditory information to rely on, whereas interlocutors in natural conversations, as Özyürek & Woll (chapter 6) highlight, typically process complex multimodal information streams. Similarly, writers typically use hand movements to produce written output, whereas speakers use their speech motor system and often their hands and body to gesture. The involvement of different sensory and motor systems in different tasks suggests that many of the cognitive components involved in these tasks must be different as well.

In addition to processing differences directly related to the use of the auditory or visual modality, there are processing differences that stem from specific properties of spoken or written language. Many of them are discussed by Dahan and Ferreira (chapter 3) and Andrews and Reichle (chapter 5). For instance, listeners have to segment the speech stream into words and they have to cope with speaker variability as well as disfluencies and errors. The reader's task seems easier, as printed words in many writing systems are separated by blanks, there is limited variability across type fonts (though not in handwriting), and most texts are edited and probably contain fewer errors than spontaneous speech. In contrast, compared to the complex speech input available in everyday conversation, written language is impoverished; there is, for instance, no prosodic information, and there are no gestures and facial expressions clarifying speaker meaning.

In spite of these differences, the knowledge structures and processes involved in using written and spoken language are tightly related. As Andrews and Reichle discuss, children learn to read by linking written words to spoken ones. Moreover, their model of adult reading includes the claim that readers map written forms onto spoken word representations and that higher-level text comprehension processes use largely the same mechanisms as are used in spoken language comprehension. Similarly, the mechanisms involved in conceptual and grammatical encoding are probably shared between speaking and writing.

In short, parallel language tasks carried out in different modalities (e.g., recognizing written and spoken words) undoubtedly involve some shared and some unique cognitive representations and processes. An important theoretical issue is then what it means for representations or processes to be shared. For instance, do speakers retrieve the same word-form representations for speaking and writing and then activate separate output representations, or do they retrieve dedicated modality-specific representations? It needs to be established how tightly linguistic representations are linked to specific input or output modalities (see also section 2.3).

Another important issue is to what extent and how language input or output in different modalities requires the use of distinct domain-general processes. For instance, both reading and listening require rapid incremental processing with an occasional need to revise structures or interpretations generated earlier.

As Andrews and Reichle point out, the mechanisms involved in combining word meanings and generating syntactic structures are likely to be shared across modalities. But the demands on domain-general processes are different. Listeners need to process sentences at the pace set by the speaker and have to base any revisions on working memory representations of earlier input. By contrast, readers can set their own pace for reading, and when reanalysis is necessary they can redirect their gaze to earlier sections of the text. Thus, the attentional demands of reanalyzing spoken and written sentences are different, involving memory retrieval of earlier information or redirecting visual attention to earlier text, respectively. How these differences affect the generation of the grammatical and semantic structure of sentences is largely unknown.

Sign language is a third modality, in addition to spoken and written language. However, the relationship between signing and the other two modalities is different from the relationship between the spoken and written versions of a single oral language (e.g., English) because readers and writers of an oral language can draw on representations that are also implicated in their spoken language, whereas for signers the signed and written languages are different. For instance, users of Dutch Sign Language may be readers of Dutch and English. Thus, the question of whether or not shared representations of specific words or grammatical rules are activated in a person's mind in signing and listening, or signing and reading, does not necessarily arise (mouthing, e.g., concurrent articulation of English words during signing in British Sign Language is an interesting exception). As Özyürek and Woll (chapter 6) demonstrate, however, one can still ask which characteristics of linguistic representations underlying spoken and signed languages are functionally equivalent, which processes occur in both modalities, and which specific constraints arise in each modality.

2. How Is Knowledge about Language Stored in the Mind?

If, as we propose, language production and comprehension are seen as different language tasks (rather than distinct language systems), an important goal for comprehensive models of language is to specify how the tasks are related in terms of the domain-general and linguistic representations and processes they involve. In the simplest model, all language tasks are supported by shared memory representations of language. The single-store assumption is more parsimonious than assuming separate representations for separate tasks. Moreover, the single-store assumption is plausible since, at least for monolingual speakers, there is only one language to represent and there is only one brain and mind to represent it. Whether the assumption is correct is an empirical issue. Future research will also have to adjudicate on several design choices about the representation of linguistic knowledge.

2.1. The Contribution of Linguistic Theory Evidently any psychological theory of language knowledge must be informed by linguistic theory; even the most common descriptive terms, such as words and phrases, are linguistic terms. It is less clear, however, how much linguistic detail needs to be included in cognitive models of the language system. Jackendoff (chapter 2) makes a strong plea for linguistically sophisticated representations: interdisciplinary language science, he argues, would make a fundamental mistake if it ignored the complexity of linguistic structure. In contrast, the other authors in this section present models with relatively simple representations, where broad concepts such as *letters*, *words*, and *sentences* do much of the work.

Psycholinguistic models often fail to consider many aspects of language that quite likely affect speakers' and listeners' behavior. For instance, while there is a substantial body of psycholinguistic work on the representation of the morphological structure of words (for a recent review, see Taft, 2015), there is, as Dahan and Ferreira (chapter 3) point out, still far too little work on prosodic representations. Moreover, psycholinguistic research has largely concentrated on a number of closely related languages. An ambitious goal for a comprehensive model of language use is to account for the ways language tasks are accomplished in all (or at least a variety of structurally different) languages. Sophisticated linguistic descriptions of different languages would be extremely valuable (e.g., by indicating which languages need to be tested) in achieving this goal.

Even though models of language use must include knowledge of linguistic structure, there need not be a one-to-one correspondence between the structures used in accounts of language itself and those used in accounts of language-based communication. This is because the explananda of the two disciplines (language structure and language cognition) are different. Even though much linguistic theorizing is closely tied to claims about cognition and/or is often based on performance data, linguistic theory does not equate with cognitive theory. There can therefore be misalignments between linguistic descriptions and language users' behavior. For instance, while linguistic theory might consider a particular word to be morphologically complex, users may consider it to be morphologically simple (Marslen-Wilson, Tyler, Waksler, & Older, 1994).

process (see Norris, McQueen, & Cutler, 2016, for further discussion).

Many psycholinguistic models take some parts of the system for granted. For example, older models of parsing such as the sausage machine model (Frazier & Fodor, 1978) took strings of words that had already been recognized as input for grammatical processing. Such models thus ignored the possibility that delays in the recognition of the phonological form of words (e.g., due to temporary perceptual ambiguities) could interfere with parsing and meaning construction. In contrast, the constraint-based account proposed by Dahan and Ferreira (chapter 3), in which form, syntax, and meaning jointly constrain comprehension, allows for the possibility of interactions between phonological and grammatical processing. In the future, experimental and computational work aiming at a comprehensive model will need to focus in particular on the interfaces between the main types of processes (such as the interface between form and syntax) to establish whether different stages of processing are indeed separable.

In addition to the question about how many distinct stages of processing may be involved in a given language task, there are two interrelated questions about the way the different stages, if they exist, talk to each other. First, is processing serial or cascaded? Second, is processing unidirectional or bidirectional? At one extreme, if processing goes through fully discrete and serial stages, there is no cascade (i.e., no continuous feedforward flow of information) and no bidirectionality (i.e., no feedback either). At the other extreme, there is cascaded flow of information in both directions. But intermediate accounts, that is, models with cascade but without feedback from one stage of processing to the next, are also possible.

3.2. Cascaded Processing There is consensus that, for both comprehension (Dahan & Ferreira, chapter 3) and production (Roelofs & Ferreira, chapter 4) of spoken language, processing is cascaded. A logically later stage of processing does not need to wait until an earlier stage has been completed; instead, information is passed continuously forward. Cascade of information makes it possible for processing to be, in temporal terms, incremental. For example, in comprehension, information concerning different levels of description (e.g., semantic and phonetic information) appears to be used by the listener as soon as it comes available (see Dahan & Ferreira). Thus, any comprehensive model of the language system is not likely one that assumes that there are only strictly serial stages of processing. However, as Andrews and Reichle (chapter 5) argue with respect to reading, there may, nonetheless, be serial steps in some aspects of language processing.

The consensus on cascaded processing within the speech production and perception literature is a good example of progress; 20 years ago this issue was still open (compare, for instance, cascaded models of speech production such as that of Dell, 1986, with serial models such as that of Levelt, Roelofs, & Meyer, 1999). We suggest that this issue was resolved for several reasons: because the issue was clearly formulated, because the question was simple, and, ultimately, because careful experimentation yielded converging evidence from multiple methods (see Roelofs & Ferreira, chapter 4, for discussion of this evidence for speech production).

3.3. Feedback? In contrast, the question about directionality of information flow has not yet been resolved. Most researchers would agree that extreme modularity (in which flow is strictly unidirectional in all parts of the language system) is untenable. For example, there is evidence of facilitatory effects of lexical neighborhood density in speech production (Vitevitch, 2002) that is difficult to reconcile with purely feedforward processing; in speech recognition, there is evidence that lexical knowledge can be used to retune prelexical processing of speech segments (Norris, McQueen, & Cutler, 2003; Dahan & Ferreira, chapter 3). Nevertheless, there are still many open issues about the limits and the nature of bidirectional processing. Consider, for example, studies on lexical involvement in phonetic categorization (Ganong, 1980). Here, profeedback conclusions (Elman & McClelland, 1988; Gow, Segawa, Ahlfors, & Lin, 2008) contrast starkly with antifeedback conclusions (Kingston, Levy, Rysling, & Staub, 2016; McQueen, Jesse, & Norris, 2009).

Why has this question proved so much harder to resolve than that on cascade? The issue seems clear, and a great deal of research has been done, yet no consensus has been reached. We suggest that the reason why this issue—in the speech perception domain and others (including reading; see Andrews & Reichle, chapter 5)—is still open is that it is not as simple as it appears to be. It is not sufficient to ask only whether or not processing is bidirectional. Instead, researchers need to define the nature of the processing under investigation. If there is feedback from one stage to a preceding stage, what function does it serve? Is it about current linguistic content, for learning or for attentional control, or does it act to bind together representations at different linguistic levels? Apparent contradictions in the literature might be resolved if it can be established that different types of feedback are involved in different experimental tasks and settings.

3.4. Shared Processing Mechanisms across Tasks In addition to describing the broad partitioning of the language system into processing stages and the information flow between them, models of the language system must specify the mechanisms that are involved in each task.

According to our proposal, the same mechanisms are available for all tasks. These would therefore need to be computational primitives that can apply to different types of representations. Various mechanisms that can apply throughout the language system have been proposed in the literature. For instance, all models we know of assume that in comprehension and production, across modalities, lexical items are accessed and selected. These operations are often defined in terms of activation of units and selection among them. Selection may or may not be a competitive process, and it may not involve inhibition. Models concerning the processing of larger units often assume operations that map lexical units onto slots in grammatical frames (Roelofs & Ferreira, chapter 4) or that merge or "clip together" (Jackendoff, chapter 2) successive utterance fragments. To give a final example, many contemporary models stress the importance of predictive processes for all linguistic tasks (Dahan & Ferreira, chapter 3).

An important theoretical and empirical challenge is to determine at which levels of granularity, or in which "vocabulary," processing mechanisms in integrated models of the language system can best be described. Should they be at the level of computational primitives (simple and broadly applicable) or combinations of those primitives (more complex and less broadly applicable)? Another empirical challenge is to demonstrate that the postulated processing mechanisms are indeed sufficient to characterize how people carry out the whole range of linguistic tasks. The history of the field suggests that everyday terms such as *activation*, *selection*, and *inhibition*, which can be transparently implemented in computational models, are the most useful. Parsimony dictates that the processing mechanisms should be the simplest and the most broadly applicable.

Processing mechanisms are likely to differ in their applicability across tasks. For instance, there may be mechanisms that are exclusively involved in segmenting the speech stream and do not play a role in other tasks. Likewise, there may be mechanisms that are of particular importance in processing some languages, but only play a minor role in others (see, e.g., El Aissati, McQueen, & Cutler, 2012, who summarize research on speech segmentation across 10 typologically diverse languages). A key issue to be addressed is which features of the cognitive architecture of the language system are shared across languages and which depend on characteristics of the particular language. In other words, the issue is which, if any, processing universals (rather than linguistic universals, cf. Evans & Levinson, 2009) can be identified.

4. Language Use and Language Learning

In this chapter, we (and our coauthors in this part of the book) focus on language use and do not consider extensively the process of first- or second-language acquisition. However, language use and learning are inextricably bound up with each other. Infants and children learn spoken language in order to be able to use it. Reading is a skill taught and practiced in primary-school classrooms. There is a growing body of evidence that the processing components of the major language tasks are not static but change with language use. Readers, speakers, and listeners are flexible (Dahan & Ferreira, chapter 3). This flexibility also transfers across tasks, such that, for example, learning in speaking can be achieved through listening (Kittredge & Dell, 2016).

While models of reading often include an acquisition component (see Andrews & Reichle, chapter 5), models of speech comprehension and speech production for many years tended to be static. They offered accounts of particular aspects of speech processing but did not capture learning or adaptation in any way. Things are changing, however (see, for instance, Christiansen & Chater, 2016; Dell & Chang, 2014; Kleinschmidt & Jaeger, 2015). A challenge for the development of comprehensive models of the language system will be to further specify the balance that must exist between stability and flexibility. Clearly the processing system adapts in the light of new experience, but it must do so in a way that maintains stable comprehension and production abilities. Another challenge will be to clarify the relative contributions of experience (e.g., hearing a talker speaking a different dialect; cf. Dahan & Ferreira, chapter 3) and maturational change (how the brain and mind change through development across the life span) to these dynamic accounts of language processing. Since learning often involves several language tasks (for instance listening and speaking or listening and reading), work on language learning will contribute to a better understanding of the relationships between these tasks.

As this work advances, it is likely that the traditional boundary between language acquisition and language processing will be further eroded. Ultimately, the goal of language learning is not to acquire linguistic knowledge (e.g., vocabulary, grammatical rules, pronunciation), but to be able to speak and listen and read and write and sign. Second-language learners may sometimes achieve this goal with inadequate or incomplete knowledge but even for first-language learners, the ultimate goal is communication, not the acquisition of knowledge alone. In our proposed approach to the cognitive architecture, learning and adaptation are key processes. From this perspective, however, learning should be seen as supporting language use rather than as passive knowledge acquisition.

5. The Relationship of the Language System to Other Components of the Cognitive System

As we have discussed, language tasks draw on motor and sensory components that are also involved in nonlinguistic tasks. Comprehensive functional models of the language system must spell out how the language system is related to these other components of the cognitive system. Thus, models must specify how linguistic representations are related to sensory and motor representations (see section 2.3) and how the processing requirements of specific tasks (e.g., the need to process auditory or visual information) impinge on linguistic processes.

In addition, language tasks draw on domain-general attention and memory processes. The importance of these systems for language processing has long been recognized. The chapters in the present section discuss the involvement of the domain-general cognitive system in language processing in several ways. For instance, the role of visual attention for language processing is discussed by Roelofs and Ferreira (chapter 4) in their section 4.3 and by Andrews and Reichle (chapter 5) in their section 3.1. Andrews and Reichle highlight the importance of visual attention in linearizing texts, thereby turning the simultaneously present visual information into a temporally ordered sequence of chunks that can be further processed in a working memory buffer and by higher-order processes, in ways very similar to those involved in processing spoken information. The importance of working memory for language processing is also stressed by Roelofs and Ferreira (chapter 4), by Andrews and Reichle (chapter 5), and by Dahan and Ferreira (chapter 3).

A comprehensive model of the language system must indicate how and where specific linguistic processes and domain-general central cognitive processes interact. Currently, the field is fractionated, as researchers have related specific components of the language system (e.g., auditory word processing or comprehension of written sentences) to specific components of attention or memory systems (e.g., inhibitory control or working memory). The components of attention and memory systems are defined in different ways in different studies, and the relationships between these components are far from clear. This state of affairs makes it difficult to determine exactly how attention and memory are involved in different linguistic tasks.

Thus, in addition to design decisions about the language system, decisions are required about the properties of all of the other cognitive systems involved in language tasks. Making such decisions is daunting but not impossible. For instance, Roelofs and collaborators (see references cited in Roelofs & Ferreira, chapter 4, section 4.3) adopted a specific theory of the attentional system (following Posner, 2012, and Miyake et al., 2000) and investigated systematically how different attentional components were involved in different steps in the process of word production. This led to the development of a clearly articulated theory of the involvement of attention in word production. It should be possible to extend this approach to other linguistic tasks. In a comprehensive model of the language system, the involvement of domain-general processes must be clearly specified. In such a model, the smallest possible set of domain-general processes should be implicated, in different combinations in all language tasks.

6. Conclusion

The major challenge for current psycholinguistics is to build a comprehensive theory of all aspects of cognition that supports linguistic behavior. We have taken the stance that there is a single system supporting all language use, rather than separate systems for different language tasks (writing, speaking, signing, and comprehending text, speech, and sign). The theory of this system contains the smallest possible set of cognitive representations and processes that the language user needs to carry out all linguistic tasks and specifies how these representations and processes are combined in each task.

In our proposal, the knowledge base is the same across tasks. It includes a lexicon, with word-form and word-meaning representations, stores containing the sensory and motor components of language, and a grammar (capturing not only syntactic rules but also, e.g., phonological rules). It is unlikely that all components of the knowledge base are used in all tasks. For instance, knowledge about the acoustic properties of a vowel is likely to have little role to play in sign production. Nevertheless, our proposed research strategy is to specify, first, what the minimal set of representations and processes are required for all tasks, and second, which of them are used in which task. As discussed in section 2, it will be necessary to make many other design choices about the knowledge base: to specify how complex linguistic representations are, to delineate which aspects of knowledge are declarative and which are procedural, to clarify the extent to which representations of the meanings of words are or are not distinct from sensory and motor representations of the same concept, and to find the balance in the representation of word form between phonological abstraction and memory for episodic detail.

According to our proposal, the same processes are used across all language tasks and are built from a limited set of basic operations. The set of computational primitives may include retrieval processes (e.g., activation), selection processes (e.g., competition), slot-insertion processes, learning and adaptation processes, and predictive processes. Our proposed research strategy is, once again, to specify the minimal set of processes required for all language tasks and to determine which of them are used in which tasks. It will also be necessary to specify for each how the processes are combined. Are there distinct stages of processing and, if so, how distinct are they, both with respect to the timing of the operation and flow of information?

We have provided an overview of the issues that would need to be addressed in developing a comprehensive model of language use. Many of these issues are quite old. We think that it would be wrong to see this as an indication that understanding of the cognitive architecture of language has failed to advance significantly. Substantial advances have been made and the reason why many of the old questions still need to be answered simply reflects the fact that the questions are hard. Psycholinguists need to keep on chipping away at these hard questions so that our understanding of the cognitive architecture can cumulate.

We have chosen to center our discussion on the issues that will have to be addressed in order to generate a comprehensive model of the language system rather than on the methods that will be used to address them. There are, however, a few methodological points we would like to make. First, developing such a model requires the use of a wide range of research tools including linguistic analysis of language structures, behavioral and neurobiological experiments, and computational modeling. Such work is best done in interdisciplinary teams of scientists. Second, we encourage researchers to use replication and meta-analysis more extensively. This is for a very obvious reason: it needs to be clear what the reproducible core findings are; only then can we build a comprehensive theory of the language system.

Third, though the research goal is to develop the simplest possible model, much can be gained by considering variability between languages, between individuals, and between communicative situations. Considering variation between languages is crucial for discovering processing principles that hold across a wide range of languages and principles that are specific for certain types of languages. Exploiting variability between individuals is essential for correlational studies that relate, for instance, attentional skills to performance in linguistic tasks. Considering variability across communicative situations is crucial for understanding how people adapt to and learn from each other. In order to use and understand variability, new research tools may need to be developed, in particular for studying language use in everyday environments. Ultimately, psycholinguistics should explain not only how people use language in the solitary setting of a traditional laboratory task such as visual lexical decision but also, actually primarily, how they use it in natural settings. While we believe that experimental control is essential, and that research using laboratory tasks has revealed a great deal about normal language processing, techniques that allow for the investigation of language processing in more naturalistic situations such as two-person conversations can offer valuable complementary evidence.

Fourth, we expect that methods will change in the coming decades due to technological advances. Cognitive neuroscience techniques in particular have developed enormously in the last 25 years and will continue to do so. We similarly expect further advances in data collection and analysis that will facilitate much larger-scale investigations than were feasible even a few years ago. The researcher in the 1980s, for example, could barely imagine data collection from thousands of individuals using internet and mobile-device technology. We cannot imagine what will be possible in 2050. What we can say, however, is that whatever the technology used to do so, key questions about the cognitive architecture of language still need to be answered.

We recommend that researchers take a Marrian perspective to answer these questions. The analysis that Marr (1982) offered for vision, in which it could be considered at three levels, the computational, the algorithmic, and the implementational, applies equally well to language processing. The cognitive architecture of language is concerned more with the first two levels (the computational problems that the language user has to solve and the algorithms they use to do so) than with the third level (the neurobiological implementation). To take the Marrian perspective, therefore, is to start at the highest level: what are the computational problems that the listener or reader or signer or speaker has to solve? This question leads naturally to the algorithmic level: what representations and processes have to be in the language system to allow the language user to solve those problems? This perspective is very different from one that might be taken by linguists, in which the nature and structure of language is taken as the starting point rather than the cognition of the language user.

A Marrian perspective is also valuable when considering how the cognitive psychology of language relates to the neurobiology of language. One problem becomes immediately apparent. It is that there are multiple possible implementations for a given algorithm. This can make it very hard to draw strong inferences across levels of analysis, because a given algorithm does not demand a particular implementation, nor the reverse. Nevertheless, there are constraints, and these can work in both directions. That is, cognitive research can lead to proposals for possible implementations, and neurobiological research can indicate which kinds of cognitive algorithms are most plausible.

The Marrian perspective encourages a functional approach to neurobiology, one that goes beyond localization (i.e., questions such as *where in the brain is syntax processed?*) and beyond passive representation (i.e., questions such as *how is syntax represented in the brain?*). The perspective encourages what we believe are more useful questions with respect to the cognitive architecture (i.e., questions such as *how does the brain support grammatical encoding during speech production?*).

The cognitive architecture of language is about cognition and not about either the products of that cognition (overt language) or the neurobiological machinery that supports it. To understand language use, therefore, we have to focus on performance, not on competence. Chomsky (1965) suggested the reverse, but this simply underlines the point we made earlier—that the explananda of cognitive psychology and linguistics are different. The goal of those working on the cognitive architecture of language should be to understand language performance—because performance (much more directly than competence) is an emergent property of the human brain.

All of this is to say that cognitive psycholinguistics is one of the key components of language science. While we predict that there will be more and more biology in language science as the years go by, we also predict that cognitive psycholinguistics will continue to flourish. It will stand alongside linguistics and neurobiology as a key contributor to solving the puzzle that is human language. This means that language science will increasingly need to be done in multidisciplinary teams, with the cognitive architecture—our understanding of the cognitive underpinnings of linguistic communication—playing an essential part.

REFERENCES

Barsalou, L. W. (2008). Grounded cognition. *Annual Review of Psychology, 59*, 617–645.

Chomsky, N. (1965). *Aspects of the theory of syntax*. Cambridge, MA: MIT Press.

Christiansen, M. H., & Chater, N. (2016). The now-or-never bottleneck: A fundamental constraint on language. *Behavioral & Brain Sciences, 39*, e62.

Cutler, A., & Clifton, Jr., C. (1999). Comprehending spoken language: A blueprint of the listener. In C. M. Brown & P. Hagoort (Eds.), *The neurocognition of language* (pp. 123–166). Oxford: Oxford University Press.

Dell, G. S. (1986). A spreading-activation theory of retrieval in sentence production. *Psychological Review, 93*(3), 283–321.

Dell, G. S., & Chang, F. (2014). The P-chain: Relating sentence production and its disorders to comprehension and acquisition. *Philosophical Transactions of the Royal Society B-Biological Sciences, 369*(1634), 20120394.

El Aissati, A., McQueen, J. M., & Cutler, A. (2012). Finding words in a language that allows words without vowels. *Cognition, 124*(1), 79–84.

Elman, J. L., & McClelland, J. L. (1988). Cognitive penetration of the mechanisms of perception: Compensation for coarticulation of lexically restored phonemes. *Journal of Memory and Language, 27*(2), 143–165.

Evans, N., & Levinson, S. C. (2009). The myth of language universals: Language diversity and its importance for cognitive science. *Behavioral and Brain Sciences, 32*(5), 429–492.

Fodor, J. A. (1983). *Modularity of mind: An essay on faculty psychology*. Cambridge, MA: MIT Press.

Frazier, L., & Fodor, J. D. (1978). The sausage machine: A new two-stage parsing model. *Cognition, 6*(4), 291–325.

Ganong, W. F. (1980). Phonetic categorization in auditory word perception. *Journal of Experimental Psychology: Human Perception and Performance, 6*(1), 110–125.

Gow, D. W., Segawa, J. A., Ahlfors, S. P., & Lin, F.-H. (2008). Lexical influences on speech perception: a Granger causality analysis of MEG and EEG source estimates. *NeuroImage, 43*(3), 614–623.

Kingston, J., Levy, J., Rysling, A., & Staub, A. (2016). Eye movement evidence for an immediate Ganong effect. *Journal of Experimental Psychology: Human Perception and Performance, 42*(12), 1969–1988.

Kittredge, A. K., & Dell, G. S. (2016). Learning to speak by listening: Transfer of phonotactics from perception to production [Special issue]. *Journal of Memory and Language, 89*, 8–22.

Kleinschmidt, D. F., & Jaeger, T. F. (2015). Robust speech perception: Recognize the familiar, generalize to the similar, and adapt to the novel. *Psychological Review, 122*(2), 148–203.

Levelt, W. J. M. (1989). *Speaking: From intention to articulation*. Cambridge, MA: MIT Press.

Levelt, W. J. M., Roelofs, A., & Meyer, A. S. (1999). A theory of lexical access in speech production. *Behavioral and Brain Sciences, 22*, 1–38.

Liberman, A. M., Cooper, F. S., Shankweiler, D. P., & Studdert-Kennedy, M. (1967). Perception of the speech code. *Psychological Review, 74*(6): 431–461.

Marr, D. (1982). *Vision*. San Francisco: Freeman.

Marslen-Wilson, W., Tyler, L. K., Waksler, R., & Older, L. (1994). Morphology and meaning in the English mental lexicon. *Psychological Review, 101*(1), 3–33.

McClelland, J. L., & Elman, J. L. (1986). The TRACE model of speech perception. *Cognitive Psychology, 18*(1), 1–86.

McClelland, J. L., & Rumelhart, D. E. (1981). An interactive activation model of context effects in letter perception: Part 1. An account of basic findings. *Psychological Review, 88*(5), 375–407.

McQueen, J. M., Cutler, A., & Norris, D. (2006). Phonological abstraction in the mental lexicon. *Cognitive Science, 30*(6), 1113–1126.

McQueen, J. M., Jesse, A., & Norris, D. (2009). No lexical–prelexical feedback during speech perception or: Is it time to stop playing those Christmas tapes? *Journal of Memory and Language, 61*(1), 1–18.

Meyer, A. S., & Huettig, F. (Eds.). (2016). Speaking and listening: Relationships between language production and comprehension [Special issue]. *Journal of Memory and Language, 89.*

Miyake, A., Friedman, N. P., Emerson, M. J., Witzki, A. H., Howerter, A., & Wager, T. (2000). The unity and diversity of executive functions and their contributions to complex "frontal lobe" tasks: A latent variable analysis. *Cognitive Psychology, 41*(1), 49–100.

Norris, D., & McQueen, J. M. (2008). Shortlist B: A Bayesian model of continuous speech recognition. *Psychological Review, 115*(2), 357–395.

Norris, D., McQueen, J. M., & Cutler, A. (2003). Perceptual learning in speech. *Cognitive Psychology, 47*(2), 204–238.

Norris, D., McQueen, J. M., & Cutler, A. (2016). Prediction, Bayesian inference and feedback in speech recognition. *Language, Cognition and Neuroscience, 31*(1), 4–18.

Page, M. P. A. (2000). Connectionist modelling in psychology: A localist manifesto. *Behavioral and Brain Sciences, 23*(4), 443–467.

Pickering, M. J., & Garrod, S. (2013). An integrated theory of language production and comprehension. *Behavioral and Brain Sciences, 36*(4), 329–347.

Pierrehumbert, J. (2002). Word-specific phonetics. In C. Gussenhoven & N. Warner (Eds.), *Laboratory phonology, 7* (pp. 101–139). Berlin: de Gruyter.

Posner, M. I. (2012). *Attention in a social world.* Oxford: Oxford University Press.

Pulvermüller, F. (2005). Brain mechanisms linking language and action. *Nature Reviews Neuroscience, 6(7),* 576–582.

Seidenberg, M. S., & McClelland, J. L. (1989). A distributed, developmental model of word recognition and naming. *Psychological Review, 96(4),* 523–568.

Taft, M. (2015). The nature of lexical representation in visual word recognition. In A. Pollatsek & R. Treiman (Eds.), *The Oxford handbook on reading* (pp. 99–113). New York: Oxford University Press.

Vitevitch, M. S. (2002). The influence of phonological similarity neighborhoods on speech production. *Journal of Experimental Psychology: Learning, Memory and Cognition, 28(4),* 735–747.

II THE DEVELOPMENT OF LANGUAGE

SECTION EDITORS: CAROLINE F. ROWLAND AND EVAN KIDD

8 Six Questions in Infant Speech and Language Development

ELIZABETH K. JOHNSON AND KATHERINE S. WHITE

Fifty years ago, doing language development research typically meant doing diary studies (i.e., studying orthographic transcriptions of children's language productions, often recorded by the children's own parents; Darwin, 1877; Leopold, 1947). Some elicitation studies also existed (Slobin & Welsh, 1973), as did some studies involving metalinguistic judgments (Gleitman, Gleitman, & Shipley, 1972). But no researcher could have dreamed of the types of tools researchers have at their disposal now. Despite their technological limitations, pioneers in the field managed to characterize many of the key elements of early language acquisition. They showed how quickly children transition from the one-word stage to the multiword stage (Braine, 1976). They documented how variable normal speech-learning trajectories can be (Peters, 1977), as well as how variable language input can be (Snow & Ferguson, 1977). And they provided evidence that children learn grammatical rules, not just patterns of words (Berko, 1958). This early work laid down a solid foundation for the modern-day study of early language acquisition.

Today, researchers interested in early language acquisition have many more tools at their disposal. Behavioral paradigms have been developed to examine how infants perceive speech sounds in the first months of life (Jusczyk, 1997). Eye trackers automatically track the fine-grained time course of children's looking behavior and pupil dilation as the speech signal unfolds (Aslin & McMurray, 2004). Audiovisual equipment and computerized analysis programs have advanced so quickly that it is now relatively easy to build dense corpora of naturalistic parent–child interactions (Roy, Frank, DeCamp, Miller, & Roy, 2015; Weisleder & Fernald, 2013). Brain-imaging equipment allows researchers to safely study children's neural responses to language input (Aslin & Mehler, 2005; Mannel & Friederici, 2008), and ultrasound imaging and heart rate tracking technology allow researchers to observe fetal responses to speech even before birth (Busnel & Granier-Deferre, 1983; DeCasper, Lecanuet, Busnel, Granier-Deferre, & Maugeais, 1994; Kisilevsky et al., 2003).

With all of these additions to their toolbox, today's researchers have been able to ask old questions in new ways, as well as to tackle all sorts of previously unanswerable questions. Research made possible by these new tools has revealed that young infants and toddlers are more language savvy than anyone could have possibly imagined 50 years ago. As we become increasingly aware of how competent children are with language at even the youngest of ages, our theories for how children acquire language have had to undergo dramatic revision.

In this chapter, we will discuss six key questions that have been the focus of intense inquiry over the past decades. All are fundamental questions in the field of early speech and language development that researchers have made great strides toward answering in recent years. But, as is always the case in science, the answers we have today are really only approximations of the truth. As such, our discussion will both outline progress made in addressing these questions in recent decades, as well as highlight how these fundamental questions are still not fully answered (or how the answers we have obtained have simply generated additional questions).

1. How Do Children Acquire the Phoneme Inventory of Their Native Language?

Before the days of infant speech perception labs, the answer to this question appeared to be obvious. The phoneme inventory of a language includes all of those speech sounds (vowels, consonants, and tones) that signal a meaningful contrast between two words. Children were thought to learn the phoneme inventory of their native language by attending to minimal pairs (e.g., *pear* and *bear* in English). That is, when two words entered their lexicon that differed by only a single speech segment, then children had the evidence they needed to postulate that the segments were lexically contrastive in their native language. Prior to acquiring a vocabulary large enough to contain minimal pairs, children were thought to have holistic representations of words that contained little segmental detail. This view was supported by production data showing that children made frequent pronunciation errors (Ferguson & Farwell, 1975; Stoel-Gammon & Cooper, 1984; Vihman, Ferguson, & Elbert, 1986), and comprehension studies suggesting that young children had difficulty discriminating words that differ slightly in their pronunciations (Barton, 1976; Garnica, 1973).

More recently, this view has been turned on its head. Studies with toddlers show that they are, in fact, quite sensitive to even slight mispronunciations of familiar words (Bailey & Plunkett, 2002; Paquette-Smith, Fecher, & Johnson, 2016; Swingley & Aslin, 2002; White & Morgan, 2008), although very young infants may not show this sensitivity for some of their earliest learned words (Bouchon, Floccia, Fux, Adda-Decker, & Nazzi, 2015; Delle Luche, Floccia, Granjon, & Nazzi, 2016). Infant perception studies have also shown that infants are learning about the sound system of their language at an age when they have very few (if any) minimal pair words in their vocabulary (Jusczyk, 1997; Werker & Curtin, 2005). Both behavioral and electrophysiological studies show that, compared to where they start as newborns, 8- to 10-month-old infants do equally well or better at distinguishing between consonants that are meaningfully contrastive in their native language, but are worse at distinguishing between consonants that are *not* contrastive in their native language (Kuhl et al., 2006; Narayan, Werker, & Beddor, 2010; Rivera-Gaxiola, Silva-Pereyra, & Kuhl, 2005; Werker & Tees, 1984). This pattern is explained by the fact that, as a native language category is established, nonnative sounds can be pulled (or assimilated) into that single native category, making them harder to tell apart (Best & McRoberts, 2003). Changes in perception occur even earlier for vowels (Cheour et al., 1998; Kuhl, Williams, Lacerda, Stevens, & Lindblom, 1992; Polka & Werker, 1994). Clearly, given that these changes in perception occur during the first year of life, children are unlikely to be relying on meaningful minimal pairs to work out the phoneme inventory of their native language.

If infants do not depend on minimal pairs to learn about the phoneme inventory of their language, then what other learning strategies might they depend on? One possibility is that infants simply track the frequency distributions of speech sounds (Maye, Werker, & Gerken, 2002). Tracking frequency distributions may be a useful way to learn about the sound categories of a language because of the way these categories are organized. The realization of phonemes is highly variable (more on this later). But if we were to keep track of the different realizations and how often they occur, we would find that the "best" versions of each phoneme occur more frequently, leading to peaks and troughs in the frequency distribution (Lisker & Abramson, 1964). This sort of distributional information could potentially provide young listeners with limited vocabularies a very strong clue to the organization of speech categories.

Of course, no matter how informative distributional cues appear to be in models of language acquisition, they would be useless if infants were not actually able to efficiently track and remember this information. Experimental work has provided some evidence that infants are able to extract this information from their speech input, supporting the notion that phoneme inventories could be bootstrapped from input distributions. We can see this in a couple of ways—first, when we manipulate frequency distributions of sounds in the laboratory, this affects infants' subsequent speech discrimination (Maye et al., 2002); second, in learning their native language, infants learn sound categories that occur more often in their input earlier (Anderson, Morgan, & White, 2003), and distributions that are less clear, or "messier," are harder to learn (Bosch & Sebastián-Gallés, 2003). So it seems that infants likely rely, at least in part, on distributional learning to tune their universal sensitivities to the inventory they hear around them.

However, some distributions are exceptionally difficult to sort out. For example, there is so much variability across individual speakers and phonological contexts in the pronunciation of vowels that if all infants did was track frequency distributions, some vowel categories would appear to significantly overlap with one another (Hillenbrand, Getty, Clark, & Wheeler, 1995). So infants must be using additional information as well. One proposal is that infants are sensitive to not just how often they hear different speech sounds, but also to the contexts (e.g., words) in which these sounds occur (even if they don't know the meanings of these words). This could help them in a number of ways. First, the occurrence of acoustically similar vowels across multiple instantiations of a particular word (e.g., different realizations of the vowel /ɪ/ in multiple tokens of the word *milk*) could help infants group the variable pronunciations of that vowel into a single category (Martin, Peperkamp & Dupoux, 2013). Alternatively (and perhaps in addition), infants may be less likely to mistakenly group two acoustically similar vowels into the same category when they occur in distinct words (e.g., the similar vowels /ɪ/ and /ɛ/ in the words *milk* and *bed*; Feldman, Myers, White, Griffiths, & Morgan, 2013; Feldman, Griffiths, Goldwater, & Morgan, 2013; Thiessen, 2007). Note that these strategies require only that infants track the words in which particular sounds are heard, not that they learn the meanings of these words. The use of lexical information to constrain bottom-up distributional analyses in this way is plausible, because, by the end of the first year, infants have acquired a proto-lexicon of word forms, or at least frequently co-occurring syllables (Halle & de Boysson-Bardies, 1994; Ngon et al., 2013).

Thus, all in all, studies capitalizing on the field's methodological advances have painted an entirely different picture of how children acquire the phoneme inventory of their language. The view that children must learn the meanings of a sizeable number of words before beginning to work out which sounds are contrastive in the native language is no longer thought to be tenable. However, some have questioned whether the distributions of speech sounds are as clear and as easy to learn in natural speech as they are in artificially created speech streams (Martin et al., 2015). And, as we have seen, even if children do learn about the phoneme inventory of their language via distributional statistics, we have very little sense of the full range of information that is available in the input or the statistics that infants have the ability to compute. It is also not clear what exactly infants are learning on the basis of these statistics. Do they truly know the phoneme inventory of their language by 8 to 12 months of age? Or have they just built the foundation on which later phonological learning will depend? How can we tell? One suggestion that the latter might be the case is evidence that children are still refining their perception of phonemes well into childhood (Nittrouer, 2002; Walley, 1993). Moreover, focusing only on statistical distributions tends to overlook some other important differences in the timing and trajectory of learning about different speech segments. For example, the speed with which infants learn about different speech sounds is linked to many other factors besides statistical distributions and frequency, such as perceptual salience (Narayan et al., 2010; Polka, Colantonio, & Sundara, 2001) and perhaps even innate phonological biases (Brown & Matthews, 1997; Fikkert, 2010). Therefore, the argument that children can learn all they need to know about the phoneme inventory via distributional learning in the first 8 to 12 months of life is not an open-and-shut case.

2. *When and How Do Children Find Words in Speech?*

Although we readily perceive spoken language as a string of discrete words, the speech signal does not actually contain any reliable pauses marking word boundaries. Instead, the articulations of words can blend seamlessly with the articulation of neighboring words, leading to ambiguity in where one word ends and the next begins (Cole & Jakimik, 1980). This is known as the word segmentation problem.

Adults solve the word segmentation problem in part by relying on language-specific probabilistic cues to word boundaries. For example, English content words tend to carry word-initial stress (e.g., *baby*, *magazine*, *panther*; Cutler & Carter, 1987; Cutler, 2012), and voiceless plosives are aspirated in word-initial but not word-final position (e.g., the /p/ in *pit* is aspirated whereas the /p/ in *tip* is not). English listeners use these probabilistic patterns to identify word boundaries (e.g., they are biased to perceive stressed syllables as word onsets; Cutler & Norris, 1988). Importantly, however, the probabilistic cues that help you find word boundaries must be learned because they are different in every language (e.g., in Polish stress falls on the penultimate syllable, and in French word-initial plosives are not aspirated).

If adults rely in part on language-specific segmentation cues to solve the word segmentation problem, then how do children learn those cues? One simple assumption could be that parents speak to children in predominantly one-word utterances, essentially feeding children information about word boundaries. However, multiple corpus studies have shown that caregivers typically address children in multiword utterances, with isolated words accounting for as little as 7% of the child's input (Van de Weijer, 1999; see also Brent & Siskind, 2001; Johnson, Lahey, Ernestus, & Cutler, 2013). Another possibility is that children struggle to hear word boundaries for many years until they build up a sizeable lexicon, and then they use top-down lexical knowledge to find word boundaries. But available data does not support this possibility, since both behavioral and electroencephalogram studies have shown that infants both know how words typically sound in the native language and segment words from speech long before their first birthday (e.g., Jusczyk, Houston, & Newsome, 1999; Kooijman, Junge, Johnson, Hagoort, & Cutler, 2013). How is this possible? How do infants learn how to identify word boundaries without first having substantial exposure to words produced in isolation?

One widely discussed explanation for infants' development of word segmentation abilities is that they rely on a domain-general statistical learning mechanism to extract an initial cohort of words from speech. Once an initial cohort of words is identified through such a mechanism, infants could begin bootstrapping language-specific cues to word boundaries. The most popular proposal for such a domain-general statistical learning strategy is the tracking of the transitional probabilities between syllables (Saffran, Aslin, & Newport, 1996). If the transitional probability between two syllables is high, then those syllables are likely to belong to the same word. In contrast, if the transitional probability is low, then those syllables are likely to span a word boundary. Put in more concrete terms, consider the phrase *pretty baby*. The transitional probability between *ba-* and *-by* is likely much higher than the transitional probability between *-tty* and *ba-*. Therefore, the first pair of syllables probably belong to the same word whereas the latter pair of syllables likely span a word boundary.

Infant behavioral studies have provided abundant and compelling support for this hypothesis. In just a few minutes, English-learning infants have been shown to segment words from an artificial language speech stream that contains no cues to word boundaries other than transitional probabilities between syllables (Saffran et al., 1996). This finding has been extended to other language-learning populations (Mersad & Nazzi, 2012), as well as infants as young as five months of age (Johnson & Tyler, 2010). Moreover, infants' ability to track transitional probabilities in an artificial language predicts scores on subsequent tests of child language abilities (Erickson & Thiessen, 2015), and infants show at least some sensitivity to high versus low transitional probabilities in natural speech (Pelucchi, Hay, & Saffran, 2009).

The evidence for infants' use of transitional probabilities to solve the word segmentation problem is so strong that many textbooks simply present this hypothesis as fact. However, just as there are some unresolved issues with how infants might use distributional statistics to learn the phoneme inventory of their native language, there are also some open questions when it comes to how fully children might rely on transitional probabilities to solve the word segmentation problem. One criticism of this hypothesis is that infants' ability to segment an artificial language using transitional probabilities might not scale up to the variability and complexity found in natural language (Johnson, 2012; Yang, 2004). For example, some studies have suggested that infants' ability to track transitional probabilities is very fragile and falls apart when they are presented with a slightly more naturalistic language (e.g., an artificial language containing four words of variable length rather than four words of uniform length; Johnson & Tyler, 2010). Another criticism is that tracking transitional probabilities between syllables might work well for some languages (e.g., English and Italian) but not so well for other languages (e.g., Hungarian and Mandarin). Finally, many have argued that transitional probabilities are not the only language-general cue available to infants. Indeed, some have argued that major prosodic boundaries may provide a far simpler language-general segmentation solution to infants, and there is some evidence that infants rely more on prosodic boundaries than transitional probabilities when they are placed in conflict (Johnson, Seidl, & Tyler, 2014; Shukla, White, & Aslin, 2011). Universal constraints on the form of possible words in a human language may provide another cue (Johnson, Jusczyk, Cutler, & Norris, 2003; Yang, 2004). Clearly, additional work is needed to fully understand precisely how infants solve the word segmentation problem.

3. When and How Do Children Begin Adding Entries to Their Lexicons?

Children typically produce their first word around 12 months of age, but both parents and researchers have long suspected that infants begin understanding words and building up a lexicon well before they begin talking. Initially, all researchers could do to gauge this understanding was ask parents to keep track of when their children showed evidence of understanding words that were spoken in their presence, like *mommy* and *bib*. Based on such parental report surveys, researchers estimated that by 11 months, children understood approximately 50 words on average (Fenson et al., 1994). By the time children reached 18 months, their vocabularies were really taking off, with some estimating that children learn as many as 10 new words a week (Bloom, 2000).

But parental report, on its own, provides only one window onto children's comprehension abilities. In the past 20 years, infant-friendly eye-tracking procedures have become common. These procedures provide researchers with additional insight into infants' early comprehension abilities. In these studies, children are typically presented with images on a screen while a label for one of the objects is produced (e.g., the child will see a picture of a dog and a duck and hear *Look at the dog*). If children understand the word, they will automatically look toward the labeled object. Remarkably, studies using infants' eye movements to assess their comprehension reveal that by 6 months of age, infants already understand the words *mommy* and *daddy* (e.g., they look more to a picture of their mom when they hear *mommy* and a picture of their dad when they hear *daddy*; Tincoff & Jusczyk, 1999), and that 6- to 9-month-olds understand the labels for body parts (e.g., *hand* and *feet*) and many other commonly heard nouns (Bergelson & Swingley, 2012; Tincoff & Jusczyk, 2012). By 6 to 7 months of age, under ideal conditions, children can be taught label-object pairings in the lab (Gogate & Bahrick, 1998; Shukla et al., 2011).

Although the process of word learning might seem intuitively simple (a "teacher" points at an object and provides the label), researchers have long puzzled over its complexity—how do children pick out just the right meanings for words in a world where meanings can be anything? As we gather increasing evidence that comprehension begins earlier than we previously thought, we have had to revise our theories for how children initially learn the meanings of words (see Kucker, chapter 9 of this volume).

The latest research suggests that children rely on different strategies at different ages, and that they also flexibly adjust their referential mapping strategies depending on context. We know that children well under two years of age can use social cues (Baldwin, 1993), perceptual salience (Pruden, Hirsh-Pasek, Golinkoff, & Hennon, 2006), and cross-modal synchrony (Gogate & Bahrick, 1998) to attach labels to objects. By two years of age, children also use various grammatical cues to determine the type of referent (Naigles, 1990; Paquette-Smith & Johnson, 2016; Waxman & Booth, 2001). But at six months of age, infants presumably have very limited (if any) access to some of these types of word-learning strategies. For example, if an infant hasn't yet learned that the ending *-ish* is often found in adjectives (e.g., *sheepish* or *selfish*), then they can't use that ending to help them zoom in on the kinds of meanings that adjectives label. So how does a six-month-old work out the meanings of words?

One language-general word-learning strategy very young infants might rely on is to track information over time. Even if it is impossible to determine what a word means the first time you hear it, perhaps over time you can narrow in on the right meaning. In the case of concrete nouns, one way to do this is to track which object is most reliably present in the environment when a particular word is heard. For example, even though an infant might hear the word *shoe* in the presence of a variety of different objects (and in some cases, even in the absence of a shoe, as in *Where's your shoe?*), over time, shoes will tend to be the objects most reliably present. As we've seen already, infants are remarkably good statistical learners. So it might not be a surprise that by 12 months, infants (in a simplified, but still ambiguous, learning situation) can track the relationships between words and objects over time to figure out which words go with which objects (Smith & Yu, 2008).

Children still need some way to narrow the range of information they are considering. Even if a particular object is most reliably associated with a particular word, how do they know whether the word refers to the whole object, its color, its texture, the fact that it is oriented in a particular way, or any of a variety of other potential meanings? It turns out that infants have certain assumptions about what words should mean, helping them to narrow down these possibilities. For example, one of these assumptions is that words refer to whole objects (believed to be one reason infants have so many nouns in their early vocabulary; Markman, 1990). So infants are not just passively absorbing associations in the world around them—instead, they are making hypotheses about what words mean. Over time, these word-learning strategies build on one another to become more sophisticated. For example, in the process of tracking the relationship between words and objects, infants may discover that an adult's eye gaze frequently signals a reliable association, leading them to the realization that eye gaze is a highly informative cue on its own (Hollich et al., 2000). In other words, infants can use what they know to bootstrap their learning of other regularities (Smith, Jones, Landau, Gershkoff-Stowe, & Samuelson, 2002).

Clearly, the use of new methodologies (e.g., eye tracking) to study early word comprehension has drastically changed our understanding of early word learning. But many questions remain. One important question is how to relate infants' behavior (in this case, word recognition or production) to their underlying understanding. Does the fact that a child looks at the correct picture mean that they have the same understanding of words that an adult has? In particular, do infants understand that words refer, or stand for, their meanings? Or do they instead simply associate labels and objects or events, in the way we might associate the bell of an ice cream truck with a delicious treat? Indeed, some have argued that children's early object–label mappings lack the abstract nature of adults' mappings (see discussion in Sloutsky & Fisher, 2012; Waxman & Gelman, 2009).

Going beyond the meanings of individual words, we can also ask what young word learners know about the semantic relationships among words. In adults, there are extensive networks of connections between words, as demonstrated by semantic priming and eye-tracking studies (Yee & Sedivy, 2006; Moss, Ostrin, Tyler, & Marslen-Wilson, 1995). But thus far no evidence for priming exists in toddlers before the age of 18 months (Arias-Trejo & Plunkett, 2009; Delle Luche, Durrant, Floccia, & Plunkett, 2014; Willits, Wojcik, Seidenberg, & Saffran, 2013), so it is an open question whether words are connected in the same way early in development. Thus, although we now know that children begin building a proto-lexicon in early infancy, we still do not fully understand how they do it or how this early proto-lexicon transforms into the adult lexicon.

4. How Do Children Deal with Variability in the Input?

Most models of language processing assume that words are made up of phonemes. But when speech scientists began looking for cues to phoneme identity in the 1950s, they were startled to discover that the acoustic cues to phonemes are not invariant, but instead differ across contexts (e.g., the acoustics of /d/ when it is followed by /a/ are not the same as when it is followed by /i/). In addition to the adjacent speech context, a variety of other factors also affect the realization of speech sounds, like a speaker's speech rate, gender, and emotion. So how do we know that a /d/ is a /d/? It is difficult enough to imagine how adults solve this problem, let alone how infants do. Over the years, there have been a variety of proposals for how listeners deal with this type of variability, ranging from the presence of individual invariant cues to certain phonetic features (Blumstein & Stevens, 1980), to the use and integration of multiple cues (Toscano & McMurray, 2010), to representations that don't strip off the variability at all (Johnson, 1997; Goldinger, 1998). Although we still don't know exactly how adults solve the lack of invariance problem, we know that they do.

So when do infants solve this problem? When we consider infants' performance in simple speech discrimination tasks involving isolated syllables, the answer appears to be that they solve it quite early. Both behavioral and electrophysiological studies reveal that very young infants are able to recognize that two speech sounds are the same when they differ in pitch or speech rate or are produced by different talkers (Dehaene-Lambertz & Pena, 2001; Eimas & Miller, 1980; Jusczyk, Pisoni & Mullennix, 1992; Polka, Masapollo & Ménard, 2014). For example, six-month-old infants trained to turn their head when they hear a particular vowel will do so even if the pitch of the vowel changes (Kuhl, 1983). Although behavioral measures suggest that very young infants have a somewhat harder time recognizing that a consonant is the same when it occurs in different vowel contexts (Bertoncini, Bijeljac-Babic, Jusczyk, Kennedy, & Mehler, 1988; Eimas, 1999), pupillometry and electrophysiological measures demonstrate that they do this successfully by three to six months (Hochmann & Papeo, 2014; Mersad & Dehaene-Lambertz, 2016).

However, when infants are confronted with the same sorts of variability in the context of word-recognition tasks, they do not always fare as well. For example, even though 7.5-month-olds can recognize that different versions of /pa/ are equivalent, they may not recognize that tokens of a newly trained word that differ in pitch are the same (Houston & Jusczyk, 2000; Singh, White, & Morgan, 2008). Importantly, however, infants don't have this problem if the word is presented in more variable forms during the learning phase (Singh, 2008; Van Heugten & Johnson, 2012) or if the word was familiar to them before arriving in the laboratory (Singh, Nestor, & Bortfeld, 2008).

These successes and failures should be taken as important clues to how early speech perception works. If an infant sometimes struggles to recognize a newly learned word when the pitch changes, what does that mean? And why would infants' ability to cope with these sorts of variability differ depending on context? One possibility that has been advanced is that these sorts of changes can disrupt infants' word recognition because infants aren't quite sure what aspects of a word's pronunciation matter for its identity (Singh, White, & Morgan, 2008). So before they've worked out that pitch alone does not matter in English for distinguishing words, hearing a particular word repeatedly in a given pitch might mislead them into thinking that pitch is relevant (at least for that word). On this view, in the same way that infants track information over time for learning the meanings of words, they may also be trying to determine over time the most reliable acoustic–phonetic properties of the words themselves. Alternatively, it could be that even young infants encode words according to the same lexically contrastive dimensions as adults, but that this ability is fragile, and in difficult tasks, infants fall back on salient features of the signal, such as voice or pitch.

In addition to the above-mentioned types of variation, infants must also cope with speakers who have different language backgrounds (both native speakers of other dialects and nonnative speakers). This type of variation introduces a whole new level of complexity, because the mapping from sounds to words is different across accents. For example, a French-accented /p/ sounds like a /b/ to an English speaker, because of the acoustic cues to /p/ and /b/ in the two languages. This means that once children have figured out how sounds map to words in their own accent, they can't just rigidly apply those mappings to words produced by people with different accents. Moreover, if they are being regularly exposed to more than one variant of their native language(s) in their everyday life, children must work out how the mappings work in each of the different language varieties in their environment (Durrant, Delle Luche, Cattani, & Floccia, 2015; Fennell & Byers-Heinlein, 2014; Floccia, Delle Luche, Durrant, Butler, & Goslin, 2012; Van der Feest & Johnson, 2016).

Infants and children get better at dealing with the variability introduced by accents over development (see

Cristia et al., 2012, for a review), though it's not yet clear how this improvement is linked to other aspects of children's developing phonological and lexical knowledge. For example, when listening to a native accent, 12-month-olds prefer to listen to familiar words over novel words (Hallé & Boysson-Bardies, 1994). But 15-month-olds do not show this preference when words are spoken in an unfamiliar accent (Best, Tyler, Gooding, Orlando, & Quann, 2009), although they do if they are given a suitable chance to adapt to the accent first (van Heugten & Johnson, 2014). By 19 months, toddlers show the preference without the additional exposure. But these listening preference tasks do not necessarily reflect word recognition or comprehension. And studies using an eye-tracking paradigm where children are asked to find a target object on a screen do not always reveal such competencies in toddlers. Although 19- to 20-month-olds recognize words produced in an artificial accent if given previous exposure to the accent (e.g., children exposed to a speaker who pronounces *dog* as *dag* can infer that that same speaker will produce *block* as *black*; White & Aslin, 2011), other studies using similar paradigms have shown that 20-month-olds still have difficulty recognizing words produced in natural unfamiliar accents (e.g., Canadian children have difficulty comprehending Australian-accented words after two minutes of exposure; Van Heugten, Krieger, & Johnson, 2015; see, however, Mulak, Best, Tyler, Kitamura, & Irwin, 2013).

At this point, we still have many more questions than answers about how infants (and adults!) deal with the variability present in speech. We still don't understand how, given an acoustic signal that cues so many things at the same time (e.g., speech sounds, a person's gender, age, identity, and emotion), we can apportion the signal appropriately into the parts that tell us what is being said and the parts that tell us who and how they are saying it. Does the way that this process happens change over development? And how is this linked to the changes we see in children's ability to recognize words across different types of variability, including accent differences? We have a long way to go to understanding these important questions.

5. When Do Children Start Learning about Function Words and Grammar?

Children don't begin speaking in full-fledged sentences. Rather, their first productions are typically one-word utterances such as *mommy* or *doggie*. Other words children tend to find useful, such as *more* or *up*, are also common in children's early productive vocabularies. The likelihood of verbs appearing in children's earliest productions can vary somewhat depending on what language children are learning (Tardif, 1996). But importantly, regardless of what language a child is learning, function words (or morphemes) such as *the* and *a* are often not produced by children until much later in development (Brown, 1973; see, however, Slobin, 1982). Even once children start stringing together more than one word, these early utterances often omit function words and morphemes (e.g., *Mommy sock* rather than *This is mommy's sock*).

Historically, children's omission of function words (and morphemes) was taken to indicate that children did not begin to learn about function words or the grammatical relations they convey until long after they began speaking. That is, even if toddlers possessed some sort of simple pivot grammar (e.g., simple word-ordering regularities), without an understanding of function words (and morphemes), their grammatical competency was a far cry from what they would later possess. One explanation for children's late acquisition of function words was that they are not as perceptually salient as content words. Words such as *the* and *a* are typically unstressed and short in duration. If children couldn't hear them, how could they learn about their use? However, this view was brought into question by studies showing that toddlers understood utterances better when they contained real as opposed to fake function words (e.g., *Give me the duck* vs. *Give me tev duck*; Gerken & McIntosh, 1993). Subsequent studies using both behavioral and electrophysiological measures further showed that even 10.5-month-old infants noticed when function words were omitted or mispronounced in speech (Shafer, Shucard, Shucard, & Gerken, 1998).

Today, it is well accepted that young infants do not simply ignore function words because they lack perceptual saliency. Rather, infants appear to learn about these acoustically distinct and highly frequent lexical items early in development, and they even use them to segment words from speech. For example, 11-month-old English learners segment *brink* from the phrase *the brink* but not from the phrase *kuh brink*, presumably because in the former case they can strip off the known function word *the* to discover the novel word *brink* (Shi, Cutler, Werker & Cruickshank, 2006). French learners may do so even earlier (Shi & Lepage, 2008). Some have suggested that infants' attention to the position of highly frequent function words may actually help them determine the word order of their language (Gervain, Nespor, Mazuka, Horie, & Mehler, 2008). Other studies have shown that 18- to 28-month-olds have already begun learning about the grammatical relationship between function morphemes. For example, English-learning 18-month-olds prefer to listen to passages containing the phrase *is walking* over passages containing the phrase *can walking* (Santelmann & Jusczyk, 1998) and French-learning toddlers know whether the masculine determiner *le* or the feminine determiner *la* should precede familiar words (Van Heugten & Shi, 2009).

Although it is now well accepted that infants pay attention to function words from early in development and that function words (and morphemes) provide infants with a valuable tool for learning new words (and possibly even for learning about a language's typical word order), these findings do not reveal how much infants and toddlers really understand about the grammar of their language. For example, when a child listens longer to a passage that contains real function words rather than fakes ones, what does this really mean? Or when a child listens longer to a passage containing the grammatical phrase *is walking* over the ungrammatical phrase *can walking*, what does this mean? Does it indicate any knowledge of a grammatical rule, or just knowledge of statistical patterns? Certainly we know that children are very sensitive to the frequency of items in their linguistic input (Ambridge, Kidd, Rowland, & Theakston, 2015). So perhaps they just like to listen to things that sound familiar?

There are a few ways to approach this question. One is to argue that grammatical rules are extracted from statistical patterns, and that it is therefore strange to ask whether children's knowledge is grammatical or just statistical. One could point to studies showing that children's statistical learning abilities are correlated with their grammatical competency (Kidd & Arciuli, 2016) as evidence for this view. Another way to approach this question is to examine children's *comprehension* of grammatical structures (which requires a more mature understanding than a simple preference for patterns that are more likely in the native language). A growing number of studies indicate that toddlers extract a fair amount of information from the grammatical structure of utterances. In a now classic study, English-learning two-year-olds used grammatical information presented simultaneously with an action event to determine whether a novel verb was transitive or intransitive (Naigles, 1990). Two-year-olds can do this even if the grammatical information is presented alone (Yuan & Fisher, 2009). German-learning 15-month-olds have been shown to use the presence of a determiner to categorize a novel word as a noun rather than a verb (Höhle, Weissenborn, Kiefer, Schulz, & Schmitz, 2004). Young toddlers have also displayed surprising competency in comprehending questions (Seidl, Hollich, & Jusczyk, 2003). In fact, toddlers not only extract meaning from the grammatical structure of word combinations, but by three years of age, they can use the correspondence between grammatical categories and prosodic position to determine the category of an ambiguous word from the prosody alone (Carvalho, Dautriche, & Christophe, 2016).

In summary, evidence for the very early but gradual emergence of grammatical competency has been slowly accumulating over the years. It appears that children are tuned in to function words and the placement of content words with respect to function words from very early on. Moreover, infants and toddlers appear to use this information to learn new words and bootstrap additional information about their native language structure. However, as in other areas of language, it is unclear to what degree this knowledge resembles the adult's (Tomasello, 2000).

6. Why Is There So Much Individual Variation, and What Does It Mean?

As anyone who has interacted with children knows, there is considerable variability across children in their language abilities (which can cause no end of grief to parents who engage in comparisons). This is particularly obvious in children's productions. Some children talk earlier than others; some children focus for longer on single words, others move more quickly into using longer phrases; some children's speech is relatively clear, others seem more focused on the prosody of their speech, but the details suffer. Although it's harder to observe, there are also differences in children's comprehension. While some of these individual differences can be attributed to factors internal to the child, others are a result of the language-learning environment in which the child finds himself. And sometimes, the two types of factors are difficult to disentangle.

Most of the focus on individual differences in language, at least in infants and young children, has been at the level of vocabulary development. Perhaps this is because this is where the most dramatic (and easily observable) variability is (counting words produced is certainly more straightforward than characterizing development in some other areas of language). But another reason that vocabulary development has been the focus is that it appears to be highly sensitive to children's environment. It has long been known that children who hear more language have larger vocabularies, and that the differences in vocabulary size can be quite dramatic (Hart & Risley, 1995). Children who hear more speech also appear to process speech more efficiently (Hurtado, Marchman, & Fernald, 2008). In recent years, attention has returned to the quality, rather than the quantity, of language input that children receive. In particular, children who hear more child-directed speech tend to have larger vocabularies than children who hear less of that speech style. This difference appears to be in part due to the fact that child-directed speech encourages the child's active engagement in the conversational interaction and other behaviors, such as joint attention (Cartmill et al., 2013; Hirsh-Pasek et al., 2015; Ramirez-Esparza, Garcia-Sierra, & Kuhl, 2014). Other features of child-directed speech, such as increased repetition, might be helpful as well (Newman, Rowe, & Bernstein Ratner, 2016). Hearing more child-directed speech may have facilitative effects on children's efficiency at speech processing (Weisleder & Fernald, 2013). Surprisingly, the amount of child-directed speech is correlated with vocabulary size even in cultures in which the vast majority of children's input is adult-directed speech (Shneidman & Goldin-Meadow, 2012).

Children differ not only in the amount and quality of their input, but also in the type of input they receive. One important difference in the type of input is the number of languages a child is being exposed to. Though changing the number of languages would seem to pose a radically different problem for the language learner, in fact, it leads to less variability in the developmental course than one might think. In speech perception and segmentation tasks, bilingual infants demonstrate some differences from monolinguals, but these tend to disappear when the input is more appropriately tailored to their circumstances or when task demands are altered (Fennell & Byers-Heinlein, 2014; Sebastian-Galles, 2010). For the most part, bilingual infants appear to progress along the same timetable as do monolingual infants (in terms of their language discrimination, their speech categories, their segmentation, their early word processing), though work in this area is really just beginning (for example, we know virtually nothing about grammatical abilities in bilingual infants). However, differences between bilingual and monolingual children may emerge in childhood. For example, bilingual children's vocabulary size in each language is typically smaller than a monolingual's, and the composition of their vocabularies may differ as a result of the contexts in which each language is used. Both vocabulary and grammatical development are affected by the relative amount of exposure bilinguals have to each language, leading to more variability as compared to monolingual development, particularly in a less dominant language (Bedore & Pena, 2008; Hoff et al., 2012; Hoff & Core, 2013). These differences are important to keep in mind for the assessment of potential language delay.

In addition to these environmental factors, there appear to be individual differences in the skills underlying language development that are fairly stable over development (Bornstein & Putnick, 2012; although it is sometimes difficult to disentangle stability in children's language environments and stability in their underlying skills). Recent research has demonstrated considerable continuity between behavior on infant speech perception tasks and later measures of language development. For example, word segmentation abilities in the first year, as assessed by both behavioral and electrophysiological measures, predict word comprehension in the lab at 16 months, productive vocabulary size at the age of two years, and more complex measures of language (including grammar) at four to six years (Junge & Cutler, 2014; Kooijman et al., 2013; Newman, Ratner, Jusczyk, Jusczyk, & Dow, 2006; Singh, Reznick, & Xuehua, 2012). And word processing efficiency at 25 months has been shown to relate to vocabulary size up to eight years (Marchman & Fernald, 2008). It may be somewhat unsurprising that the ability to find words in speech or the ability to process words efficiently is related to later vocabulary. But even more basic speech discrimination in infancy has been shown to predict later language abilities at two years, including not only vocabulary, but also phrase understanding (Tsao, Liu, & Kuhl 2004). Prosodic sensitivity in the first year has been shown to predict vocabulary as well (Cristia & Seidl, 2011). One interesting aspect of these findings is that there appears to be some selectivity in which tasks predict later language, suggesting that those tasks may tap abilities particularly relevant for language, rather than domain-general abilities (although it's not clear whether the absence of some correlations was due to a lack of power). That's not to say that domain-general abilities don't play a role as well. Habituation speed (thought to indicate encoding efficiency), visual recognition memory, and visual sequence learning in infancy have all been associated with vocabulary development (Shafto, Conway, Field, & Houston, 2012).

Despite all of the research that has been done on the role of internal versus environmental factors in language acquisition over the past 50 years, we still do not fully understand how much each of these factors (and the interplay between them) contribute to individual variability during first-language learning. We also do not yet have a satisfactory way to track and study individual differences in young infants (Cristia, Seidl, Junge, Soderstrom, & Hagoort, 2013). Indeed, the group data many researchers depend on most likely mask many crucial subtleties in normal language-learning trajectories. If the field of infant speech perception is to move beyond simply charting trends in language abilities in groups of children, this is an area where researchers need to focus their energy in upcoming years.

Improvements in this area will not only strengthen and broaden our theoretical understanding of normal language development, but they will also open many new possibilities for identifying children at risk early in development.

7. *Where to Go from Here*

Building on the very strong foundation established by the field's pioneers, the study of language development has come a long way in the past 50 years. As the tools available to language researchers have grown increasingly sophisticated and sensitive, we have learned that infants and toddlers have a remarkable toolbox themselves, and are even more linguistically savvy than early researchers could have possibly imagined. This realization, in turn, has pushed our thinking in new directions and forced us to revamp our theories of early childhood language acquisition. In this chapter, we have highlighted some of these recent developments by focusing on six questions key to the field.

One important theme that has emerged is that our old notions of infants progressing along the path to language in a stepwise fashion (from sounds to words to grammar) are not viable. Rather, infants are learning about their language at multiple levels in parallel (e.g., Johnson, 2016). A second theme that has emerged is that our tools, sophisticated as they are, still do not allow us to determine exactly what is in an infant's mind. Yes, infants discriminate sounds and look at pictures that match the words they hear, but what does this mean? A third theme is that, in some cases, it is not clear how the impressive abilities infants demonstrate scale up to the problem of learning language in the real world. Infants can track frequency, transitional probabilities, and the co-occurrence of words and referents across time. But how far do these abilities, demonstrated in simple lab situations, get them in the real world? There are so many other important questions that this review has not even attempted to tackle, such as the relationship between perception and production and the importance of sensitive periods for language development. As our tools continue to develop, we will undoubtedly uncover more answers (and questions). Who knows where we'll be 50 years from now!

REFERENCES

Ambridge, B., Kidd, E., Rowland, C. F., & Theakston, A. L. (2015). The ubiquity of frequency effects in first language acquisition. *Journal of Child Language, 42*, 239–273.

Anderson, J. L., Morgan, J. L., & White, K. S. (2003). A statistical basis for speech sound discrimination. *Language and Speech, 46*, 155–182.

Arias-Trejo, N., & Plunkett, K. (2009). Lexical-semantic priming effects during infancy. *Philosophical Transactions of the Royal Society B, 364*, 3633–3647.

Aslin, R. N., & McMurray, B. (2004). Automated corneal-reflection eye tracking in infancy: Methodological developments and applications to cognition. *Infancy, 6*, 155–163.

Aslin, R. N., & Mehler, J. (2005). Near-infrared spectroscopy (NIRS) for functional studies of brain activity in human infants: Promise, prospects, and challenges. *Journal of Biomedical Optics, 10*, 011009.

Bailey, T. M., & Plunkett, K. (2002). Phonological specificity in early words. *Cognitive Development, 17*, 1265–1282.

Baldwin, D. A. (1993). Early referential understanding: Infants' ability to recognize referential acts for what they are. *Developmental Psychology, 29*, 832–843.

Barton, D. (1976). Phonemic discrimination and the knowledge of words. *Papers and Reports on Child Language Development, 11*, 61–68.

Bedore, L. M., & Pena, E. D. (2008). Assessment of bilingual children for identification of language impairment: Current findings and implications for practice. *International Journal of Bilingual Education and Bilingualism, 11*, 1–29.

Bergelson, E., & Swingley, D. (2012). At 6 to 9 months, human infants know the meanings of many common nouns. *Proceedings of the National Academy of Sciences of the USA, 109*, 3253–3258.

Berko, J. (1958). The child's learning of English morphology. *Word, 14*, 150–177.

Bertoncini, J., Bijeljac-Babic, R., Jusczyk, P. W., Kennedy, L. J., & Mehler, J. (1988). An investigation of young infants' perceptual representations of speech sounds. *Journal of Experimental Psychology: General, 117*, 21–33.

Best, C. T., & McRoberts, G. W. (2003). Infant perception of non-native consonant contrasts that adults assimilate in different ways. *Language and Speech, 46*, 183–216.

Best, C. T., Tyler, M. D., Gooding, T. N., Orlando, C. B., & Quann, C. A. (2009). Development of phonological constancy: Toddlers' perception of native and Jamaican-accented words. *Psychological Science, 20*, 539–542.

Bloom, P. (2000). *How children learn the meanings of words.* Cambridge, MA: MIT Press.

Blumstein, S. E., & Stevens, K. N. (1980). Perceptual invariance and onset spectra for stop consonants in different vowel environments. *Journal of the Acoustical Society of America, 67*, 648–662.

Bornstein, M. H., & Putnick, D. L. (2012). Stability of language in childhood: A multiage, multidomain, multimeasure, and multisource study. *Developmental Psychology, 48*, 477–491.

Bosch, L., & Sebastián-Gallés, N. (2003). Simultaneous bilingualism and the perception of a language-specific vowel contrast in the first year of life. *Language and Speech, 46*, 217–243.

Bouchon, C., Floccia, C., Fux, T., Adda-Decker, M., & Nazzi, T. (2015). Call me Alix, not Elix: Vowels are more important than consonants in own-name recognition at 5 months. *Developmental Science, 18*, 587–598.

Braine, M. D. S. (1976). Children's first word combinations. *Monograph of the Society for Research in Child Development, 41*(1, Serial No. 164).

Brent, M. R., & Siskind, J. M. (2001). The role of exposure to isolated words in early vocabulary development. *Cognition, 81*, B33–B44.

Brown, C., & Matthews, J. (1997). The role of feature geometry in the development of phonemic contrasts. In S. Hannahs & M. Young-Scholten, (Eds.), *Focus on phonological acquisition* (pp. 67–112). Philadelphia, PA: John Benjamins.

Brown, R. (1973). *A first language: The early stages.* Cambridge, MA: Harvard University Press.

Busnel, M. C., & Granier-Deferre, C. (1983). And what of fetal audition? In A. Oliverio & M. Zappella (Eds.), *The behavior of human infants* (pp. 93–126). New York: Plenum Press.

Cartmill, E. A., Armstrong, B. F., Gleitman, L. R., Goldin-Meadow, S., Medina, T. N., & Trueswell, J. D. (2013). Quality of early parent input predicts child vocabulary three years later. *Proceedings of the National Academy of Sciences of the United States of America, 110,* 11278–11283.

Carvalho, A., Dautriche, I., & Christophe, A. (2016). Preschoolers use phrasal prosody online to constrain syntactic analysis. *Developmental Science, 19,* 235–250.

Cheour, M., Ceponiene, R., Lehtokoski, A., Luuk, A., Allik, J., Alho, K., & Näätänen, R. (1998). Development of language-specific phoneme representations in the infant brain. *Nature Neuroscience, 1,* 351–353.

Cole, R. A., & Jakimik, J. A. (1980). A model of speech perception. In R. Cole (Ed.), *Perception and production of fluent speech* (pp. 133–142). Hillsdale, NJ: Erlbaum.

Cristia, A., & Seidl, A. (2011). Sensitivity to prosody at 6 months predicts vocabulary at 24 months. In N. Danis, K. Mesh, & H. Sung (Eds.), *BUCLD 35:* Proceedings of the 35th annual Boston University Conference on Language Development. Somerville, MA: Cascadilla Press.

Cristia, A., Seidl, A., Junge, C., Soderstrom, M., & Hagoort, P. (2013). Predicting individual variation in language from infant speech perception measures. *Child Development, 85,* 1330–1345.

Cristia, A., Seidl, A., Vaughn, C., Schmale, R., Bradlow, A., & Floccia, C. (2012). Linguistic processing of accented speech across the lifespan. *Frontiers in Psychology, 3.*

Cutler, A. (2012). *Native listening: Language experience and the recognition of spoken words.* Cambridge, MA: MIT Press.

Cutler, A., & Carter, D. M. (1987). The predominance of strong initial syllables in the English vocabulary. *Computer Speech and Language, 2,* 133–142.

Cutler, A., & Norris, D. (1988). The role of strong syllables in segmentation for lexical access. *Journal of Experimental Psychology: Human Perception and Performance, 14,* 113–121.

Darwin, C. (1877). A bibliographic sketch of an infant. *Mind, 2,* 285–294.

DeCasper, A. J., Lecanuet, J. P., Busnel, M. C., Granier-Deferre, C., & Maugeais, R. (1994). Fetal reactions to recurrent maternal speech. *Infant Behavior and Development, 17,* 159–164.

Dehaene-Lambertz, G., & Pena, M. (2001). Electrophysiological evidence for automatic phonetic processing in neonates. *Neuroreport, 12,* 3155–3158.

Delle Luche, C., Durrant, S., Floccia, C., & Plunkett, K. (2014). Implicit meaning in 18-month-old toddlers. *Developmental Science, 17,* 948–955.

Delle Luche, C., Floccia, C., Granjon, L., & Nazzi, T. (2016). Infants' first words are not phonetically specified: Own name recognition in British English-learning 5-month-olds. *Infancy, 22*(3), 362–388.

Durrant, S., Delle Luche, C., Cattani, A., & Floccia, C. (2015). Monodialectal and multidialectal infants' representation of familiar words. *Journal of Child Language, 42,* 447–465.

Eimas, P. D. (1999). Segmental and syllabic representations in the perception of speech by young infants. *Journal of the Acoustical Society of America, 105,* 1901–1911.

Eimas, P. D., & Miller, J. L. (1980). Contextual effects in infant speech perception. *Science, 209,* 1140–1141.

Erickson, L. C., & Thiessen, E. D. (2015). Statistical learning of language: Theory, validity, and predictions of a statistical learning account of language acquisition. *Developmental Review, 37,* 66–108.

Feldman, N. H., Griffiths, T. L., Goldwater, S., & Morgan, J. L (2013). A role for the developing lexicon in phonetic category acquisition. *Psychological Review, 120,* 751–778.

Feldman, N. H., Myers, E. B., White, K. S., Griffiths, T. L., & Morgan, J. L. (2013). Word-level information influences phonetic learning in adults and infants. *Cognition, 127,* 427–438.

Fennell, C., & Byers-Heinlein, K. (2014). You sound like Mommy: Bilingual and monolingual infants learn words best from speakers typical of their language environments. *International Journal of Behavioral Development, 38,* 309–316.

Fenson, L., Dale, P. S., Reznick, J. S., Bates, E., Thal, D. J., Pethick, S. J., ... Stiles, J. (1994). Variability in early communicative development. *Monographs of the Society for Research in Child Development,* 59, 1–185.

Ferguson, C. A., & Farwell, C. B. (1975). Words and sounds in early language acquisition. *Language, 51,* 419–439.

Fikkert, P. (2010). Developing representations and the emergence of phonology: Evidence from perception and production. In C. Fougeron, B. Kuehnert, M. Imperio, & N. Vallee (Eds.), *Laboratory phonology 10* (pp. 227–260). Berlin: De Gruyter Mouton.

Floccia, C., Delle Luche, C., Durrant, S., Butler, J., & Goslin, J. (2012). Parent or community: Where do 20-month-olds exposed to two accents acquire their representation of words? *Cognition, 124,* 95–100.

Garnica, O. (1973). The development of phonemic speech perception. In T. E. Moore (Ed.), *Cognitive development and the acquisition of language* (pp. 215–222). New York: Academic Press.

Gerken, L., & McIntosh, B. J. (1993). Interplay of function morphemes and prosody in early language. *Developmental Psychology, 29,* 448–457.

Gervain, J., Nespor, M., Mazuka, R., Horie, R., & Mehler, J. (2008). Bootstrapping word order in prelexical infants: A Japanese-Italian cross-linguistic study. *Cognitive Psychology, 57,* 56–74.

Gleitman, L. R., Gleitman, H., & Shipley, E. F. (1972). The emergence of the child as grammarian. *Cognition, 1,* 137–164.

Gogate, L. J., & Bahrick, L. E. (1998). Intersensory redundancy facilitates learning of arbitrary relations between vowel sounds and objects in seven-month-old infants. *Journal of Experimental Child Psychology, 69,* 133–149.

Goldinger, S. D. (1998). Echoes of echoes? An episodic theory of lexical access. *Psychological Review, 105,* 251–279.

Hallé, P., & de Boysson-Bardies, B. (1994). Emergence of an early lexicon: Infants' recognition of words. *Infant Behavior and Development, 17,* 119–129.

Hart, B., & Risley, R. R. (1995). *Meaningful differences in the everyday experiences of young American children.* Baltimore, MD: Paul H. Brooks.

Hillenbrand, J., Getty, L. A., Clark, M. J., & Wheeler, K. (1995). Acoustic characteristics of American English vowels. *Journal of the Acoustical Society of America, 97,* 3099–3111.

Hirsh-Pasek, K., Adamson, L. B., Bakeman, R., Owen, M. T., Golinkoff, R. M., Pace, A.,... Suma, K. (2015). The contribution of early communication quality to low-income children's language success. *Psychological Science, 26,* 1071–1083.

Hochmann, J.-R., & Papeo, L. (2014). The invariance problem in infancy: A pupillometry study. *Psychological Science, 25,* 2038–2046.

Hoff, E., & Core, C. (2013). Input and language development in bilingually developing children. *Seminars in Speech and Language, 34,* 215–226.

Hoff, E., Core, C., Place, S., Rumiche, R., Señor, M., & Parra, M. (2012). Dual language exposure and early bilingual development. *Journal of Child Language, 39,* 1–27.

Höhle, B., Weissenborn, J., Kiefer, D., Schulz, A., & Schmitz, M. (2004). Functional elements in infants' speech processing: the role of determiners in the syntactic categorization of lexical elements. *Infancy, 5,* 341–353.

Hollich, G. J., Hirsh-Pasek, K., Golinkoff, R. M., Brand, R. J., Brown, E., Chung, H. L., ... Bloom, L. (2000). Breaking the language barrier: An emergentist coalition model for the origins of word learning. *Monographs of the Society for Research in Child Development,* 65, 1–135.

Houston, D. M., & Jusczyk, P. W. (2000). The role of talker-specific information in word segmentation by infants. *Journal of Experimental Psychology: Human Perception and Performance, 26,* 1570–1582.

Hurtado, N., Marchman, V. A., & Fernald, A. (2008). Does input influence uptake? Links between maternal talk, processing speed and vocabulary size in Spanish-learning children. *Developmental Science, 11,* F31–F39.

Johnson, E. K. (2012). Bootstrapping language: Are infant statisticians up to the job? In P. Rebuschat & J. Williams (Eds.), *Statistical learning and language acquisition* (pp. 55–90). Boston: Mouton de Gruyter.

Johnson, E. K. (2016). Constructing a proto-lexicon: an integrative view of infant language development. *Annual Review of Linguistics, 2,* 391–412.

Johnson, E. K., Jusczyk, P. W., Cutler, A., & Norris, D. (2003). Lexical viability constraints on speech segmentation by infants. *Cognitive Psychology, 46,* 65–97.

Johnson, E. K., Lahey, M., Ernestus, M., & Cutler, A. (2013). A multimodal corpus of speech to infant and adult listeners. *Journal of the Acoustical Society of America, 134,* EL534–EL540.

Johnson, E. K., Seidl, A., & Tyler, M. D. (2014). The edge factor in early word segmentation: Utterance-level prosody enables word form extraction by 6-month-olds. *PLOS ONE, 9,* e83546.

Johnson, E. K., & Tyler, M. D. (2010). Testing the limits of statistical learning for word segmentation. *Developmental Science, 13,* 339–345.

Johnson, K. (1997). Speech perception without speaker normalization: An exemplar model. In K. Johnson & J. W. Mullinex (Eds.), *Talker variability in speech perception* (pp. 145–166). New York: Academic Press.

Junge, C., & Cutler, A. (2014). Early word recognition and later language skills. *Brain Sciences, 4,* 532–559.

Jusczyk, P. W. (1997). *The discovery of spoken language.* Cambridge, MA: MIT Press.

Jusczyk, P. W., Houston, D. M., & Newsome, M. (1999). The beginnings of word segmentation in English-learning infants. *Cognitive Psychology, 39,* 159–207.

Jusczyk, P. W., Pisoni, D. B., & Mullennix, J. (1992). Some consequences of stimulus variability on speech processing by 2-month-old infants. *Cognition, 43,* 253–291.

Kidd, E., & Arciuli, J. (2016). Individual differences in statistical learning predict children's comprehension of syntax. *Child Development, 87,* 184–193.

Kisilevsky, B. S., Hains, S. M. J., Lee, K., Xie, X., Huang, H., Ye, H. H., ... Wang, Z. (2003). Effects of experience on fetal voice recognition. *Psychological Science,* 14, 220–224.

Kooijman, V., Junge, C., Johnson, E. K., Hagoort, P., & Cutler, A. (2013). Predictive brain signals of linguistic development. *Frontiers in Psychology, 4,* 1–13.

Kuhl, P. K. (1983). Perception of auditory equivalence classes for speech in early infancy. *Infant Behavior and Development, 6*(2), 263–285.

Kuhl, P. K., Stevens, E., Hayashi, A., Deguchi, T., Kiritani, S., & Iverson, P. (2006). Infants show a facilitation effect for native language phonetic perception between 6 and 12 months. *Developmental Science, 9,* F13–F21.

Kuhl, P. K., Williams, K. A., Lacerda, F., Stevens, K. N., & Lindblom, B. (1992). Linguistic experience alters phonetic perception in infants by six months of age. *Science, 255,* 606–608.

Leopold, W. F. (1947). *Speech development of a bilingual child, 2: Sound learning in the first two years.* Evanston, IL: Northwestern University.

Lisker, L., & Abramson, A. S. (1964). A cross-language study of voicing in initial stops: Acoustical measurements. *Word, 20,* 384–422.

Mannel, C., & Friederici, A. D. (2008). Event-related brain potentials as a window to children's language processing. In I. A. Sekerina, E. M. Fernández, & H. Clahsen (Eds.) *Developmental psycholinguistics: On-line methods in children's language processing* (pp. 29–72). Amsterdam: John Benjamins.

Marchman, V. A., & Fernald, A. (2008). Speed of word recognition and vocabulary knowledge in infancy predict cognitive and language outcomes in later childhood. *Developmental Science, 11,* F9–F16.

Markman, E. M. (1990). Constraints children place on word meanings. *Cognitive Science, 14,* 57–77.

Martin, A., Peperkamp, S., & Dupoux, E. (2013). Learning phonemes with a proto-lexicon. *Cognitive Science, 37,* 103–124.

Martin, A., Schatz, T., Versteegh, M., Miyazawa, K., Mazuka, R., Dupoux, E., & Cristia, (2015). Mothers speak less clearly to infants than to adults: A comprehensive test of the hyperarticulation hypothesis. *Psychological Science, 26,* 341–347.

Maye, J., Werker, J. F., & Gerken, L. (2002). Infant sensitivity to distributional information can affect phonetic discrimination. *Cognition, 82,* B101–B111.

Mersad, K., & Dehaene-Lambertz, G. (2016). Electrophysiological evidence of phonetic normalization across coarticulation in infants. *Developmental Science, 19,* 710–722.

Mersad, K., & Nazzi, T. (2012). When Mommy comes to the rescue of statistics: Infants combine top-down and bottom-up cues to segment speech. *Language Learning and Development, 8,* 303–315.

Moss, H. E., Ostrin, R. K., Tyler, L. K., & Marslen-Wilson, W. D. (1995). Accessing different types of lexical semantic information: Evidence from priming. *Journal of Experimental Psychology: Learning, Memory, and Cognition, 21,* 863–883.

Mulak, K. E., Best, C. T., Tyler, M. D., Kitamura, C., & Irwin, J. R. (2013). Development of phonological constancy: 19-month-olds, but not 15-month-olds, identify words in a non-native regional accent. *Child Development, 84,* 2064–2078.

Naigles, L. (1990). Children use syntax to learn verb meanings. *Journal of Child Language, 17,* 357–374.

Narayan, C., Werker, J. F., & Beddor, P. S. (2010). The interaction between acoustic salience and language experience in developmental speech perception: Evidence from nasal place discrimination. *Developmental Science, 13,* 407–420.

Newman, R., Ratner, N. B., Jusczyk, A. M., Jusczyk, P. W., & Dow, K. A. (2006). Infants' early ability to segment the conversational speech signal predicts later language development: A retrospective analysis. *Developmental Psychology, 42,* 643–655.

Newman, R. S., Rowe, M. L., & Bernstein Ratner, N. (2016). Input and uptake at 7 months predict toddler vocabulary: The role of child-directed speech and infant processing skills in language development. *Journal of Child Language, 43,* 1158–1173.

Ngon, C., Martin, A., Dupoux, E., Cabrol, D., Dutat, M., & Peperkamp, S. (2013). (Non)words, (non)words, (non) words: Evidence for a protolexicon during the first year of life. *Developmental Science, 16,* 24–34.

Nittrouer, S. (2002). Learning to perceive speech: How fricative perception changes, and how it stays the same. *Journal of the Acoustical Society of America, 112,* 711–719.

Paquette-Smith, M., Fecher, N., & Johnson, E. K. (2016). Two-year-olds' sensitivity to subphonemic mismatch during online spoken word recognition. *Attention, Perception, and Psychophysics, 78,* 2329–2340.

Paquette-Smith, M., & Johnson, E. K. (2016). Toddlers' use of grammatical and social cues to learn novel words. *Language Learning and Development, 12,* 328–337.

Pelucchi, B., Hay, J. F., & Saffran, J. R. (2009). Statistical learning in a natural language by 8-month-old infants. *Child Development, 80,* 674–685.

Peters, A. M. (1977). Language learning strategies: Does the whole equal the sum of the parts? *Language, 53,* 560–573.

Polka, L., Colantonio, C., & Sundara, M. (2001). A cross-language comparison of /d/-/ð/ perception: Evidence for a new developmental pattern. *Journal of the Acoustical Society of America, 109,* 2190–2201.

Polka, L., Masapollo, M., & Ménard, L. (2014). Who's talking now? Infants' perception of vowels with infant vocal properties. *Psychological Science, 25,* 1448–1456.

Polka, L., & Werker, J. F. (1994). Developmental changes in perception of nonnative vowel contrasts. *Journal of Experimental Psychology: Human Perception and Performance, 20,* 421–435.

Pruden, S. M., Hirsh-Pasek, K., Golinkoff, R. M., & Hennon, E. A. (2006). The birth of words: Ten-month-olds learn words through perceptual salience. *Child Development, 77,* 266–280.

Ramirez-Esparza, N., Garcia-Sierra, A., & Kuhl, P. K. (2014). Look who's talking: Speech style and social context in language input are linked to concurrent and future speech development. *Developmental Science, 17,* 880–891.

Rivera-Gaxiola, M., Silva-Pereyra, J., & Kuhl, P. K. (2005). Brain potentials to native and non-native speech contrasts in 7- and 11-month-old American infants. *Developmental Science, 8,* 162–172.

Roy, B. C., Frank, M. C., DeCamp, P., Miller, M., & Roy, D. (2015). Predicting the birth of a spoken word. *Proceedings of the National Academy of Sciences of the United States of America, 112,* 12663–12668.

Saffran, J. R., Aslin, R. N., & Newport, E. L. (1996). Statistical learning by 8-month-olds. *Science, 274,* 1926–1928.

Santelmann, L. M., & Jusczyk, P. W. (1998). Sensitivity to discontinuous dependencies in language learners: Evidence for limitations in processing space. *Cognition, 69,* 105–134.

Sebastian-Galles, N. (2010). Bilingual language acquisition: Where does the difference lie? *Human Development, 53,* 245–255.

Seidl, A., Hollich, G., & Jusczyk, P. W. (2003). Early understanding of subject and object wh-questions. *Infancy, 4,* 423–436.

Shafer, V., Shucard, D., Shucard, J., & Gerken, L. (1998). An electrophysiological study of infants' sensitivity to the sound patterns of English speech. *Journal of Speech, Language, and Hearing Research, 41,* 874–886.

Shafto, C. L., Conway, C. M., Field, S. L., & Houston, D. M. (2012). Visual sequence learning in infancy: Domain-general and domain-specific associations with language. *Infancy, 17,* 247–271.

Shi, R., Cutler, A., Werker, J., & Cruickshank, M. (2006). Frequency and form as determinants of functor sensitivity in English-acquiring infants. *Journal of the Acoustical Society of America, 119,* EL61–EL66.

Shi, R., & Lepage, M. (2008). The effect of functional morphemes on word segmentation in preverbal infants. *Developmental Science, 11,* 407–413.

Shneidman, L. A., & Goldin-Meadow, S. (2012). Language input and acquisition in a Mayan village: How important is directed speech? *Developmental Science, 15,* 659–673.

Shukla, M., White, K. S., & Aslin, R. N. (2011). Prosody guides the rapid mapping of auditory word forms onto visual objects in 6-mo-old infants. *Proceedings of the National Academy of Sciences of the United States of America, 108,* 6038–6043.

Singh, L. (2008). Influences of high and low variability on infant word recognition. *Cognition, 106,* 833–870.

Singh, L., Nestor, S. S., & Bortfeld, H. (2008). Overcoming the effects of variation in infant speech segmentation: Influences of word familiarity. *Infancy, 13,* 57–74.

Singh, L., Reznick, J. S., & Xuehua (2012). Infant word segmentation and childhood vocabulary development: A longitudinal analysis. *Developmental Science, 15,* 482–495.

Singh, L., White, K. S., & Morgan, J. L. (2008). Building a word-form lexicon in the face of variable input: Influences of pitch and amplitude on early spoken word recognition. *Language Learning and Development, 4,* 157–178.

Slobin, D. I. (1982). Universal and particular in language acquisition. In L. R. Gleitman & L. R. Wanner (Eds.), *Language acquisition: The state of the art* (pp. 129–170). Cambridge: Cambridge University Press.

Slobin, D. I., & Welsh, C. A. (1973). Elicited imitation as a research tool in developmental psycholinguistics. In C. A. Ferguson & D. I. Slobin (Eds.), *Studies of child language development* (pp. 485–496). New York: Holt, Rinhart & Winston.

Sloutsky, V. M., & Fisher, A. V. (2012). Linguistic labels: Conceptual markers or object features? *Journal of Experimental Child Psychology, 111,* 65–86.

Smith, L. B., Jones, S. S., Landau, B., Gershkoff-Stowe, L., & Samuelson, L. (2002). Object name learning provides on-the-job training for attention. *Psychological Science, 13,* 13–19.

Smith, L. B., & Yu, C. (2008). Infants rapidly learn word-referent mappings via cross-situational statistics. *Cognition, 106,* 1558–1568.

Snow, E. S., & Ferguson, C. A. (Eds.). (1977). *Talking to children.* Cambridge: Cambridge University Press.

Stoel-Gammon, C., & Cooper, J. A. (1984). Patterns of early lexical and phonological development. *Journal of Child Language, 11,* 247–271.

Swingley, D., & Aslin, R. N. (2002). Lexical neighborhoods and the word-form representations of 14-month-olds. *Psychological Science, 13,* 480–484.

Tardif, T. (1996). Nouns are not always learned before verbs: Evidence from Mandarin speakers' early vocabularies. *Developmental Psychology, 32,* 492–504.

Thiessen, E. D. (2007). The effect of distributional information on children's use of phonemic contrasts. *Journal of Memory and Language, 56,* 16–34.

Tincoff, R., & Jusczyk, P. W. (1999). Some beginnings of word comprehension in six-month-olds. *Psychological Science, 10,* 172–175.

Tincoff, R., & Jusczyk, P. W. (2012). Six-month-olds comprehend words that refer to parts of the body. *Infancy, 17,* 432–444.

Tomasello, M. (2000). Do young children have adult syntactic competence? *Cognition, 74,* 209–253.

Toscano, J., & McMurray, B. (2010). Cue integration with categories: Weighting acoustic cues in speech using unsupervised learning and distributional statistics. *Cognitive Science, 34,* 434–464.

Tsao, F.-M., Liu, H.-M., & Kuhl, P. K. (2004). Speech perception in infancy predicts language development in the second year of life: A longitudinal study. *Child Development, 75,* 1067–1084.

Van der Feest, S. V. H., & Johnson, E. K. (2016). Input-driven differences in toddlers' perception of a disappearing phonological contrast. *Language Acquisition, 23,* 89–111.

Van de Weijer, J. (1999). *Language input for word discovery.* Unpublished manuscript, Radboud University, Nijmegen.

Van Heugten, M., & Johnson, E. K. (2012). Infants exposed to fluent natural speech succeed at cross-gender word recognition. *Journal of Speech, Language, and Hearing Research, 55,* 554–560.

Van Heugten, M., & Johnson, E. K. (2014). Learning to contend with accents in infancy: Benefits of brief speaker exposure. *Journal of Experimental Psychology: General, 143,* 340–350.

Van Heugten, M., Krieger, D., & Johnson. E. K. (2015). The developmental trajectory of toddlers' comprehension of unfamiliar regional accents. *Language Learning and Development, 11,* 41–65.

Van Heugten, M., & Shi, R. (2009). French-learning toddlers use gender information on determiners during word recognition. *Developmental Science, 12,* 419–425.

Vihman, M. M., Ferguson, C. A., & Elbert M. (1986). Phonological development from babbling to speech: Common tendencies and individual differences. *Applied Psycholinguistics, 7,* 3–40.

Walley, A. C. (1993). The role of vocabulary development in children's spoken word recognition and segmentation ability. *Developmental Review, 13,* 286–350.

Waxman, S. R., & Booth, A. E. (2001). Seeing pink elephants: Fourteen-month-olds' interpretations of novel nouns and adjectives. *Cognitive Psychology, 43,* 217–242.

Waxman, S. R., & Gelman, S. A. (2009). Early word-learning entails reference, not merely associations. *Trends in Cognitive Sciences, 13,* 258–263.

Weisleder, A., & Fernald, A. (2013). Talking to children matters: Early language experience strengthens processing and builds vocabulary. *Psychological Science, 24,* 2143–2152.

Werker, J. F., & Curtin, S. (2005). PRIMIR: A developmental framework of infant speech processing. *Language Learning and Development, 1,* 197–234.

Werker, J. F., & Tees, R. C. (1984). Cross-language speech perception: Evidence for perceptual reorganization during the first year of life. *Infant Behavior and Development, 7,* 49–63.

White, K. S., & Aslin, R. N. (2011). Adaptation to novel accents by toddlers. *Developmental Science, 14,* 372–384.

White, K. S., & Morgan, J. L. (2008). Sub-segmental detail in early lexical representations. *Journal of Memory and Language, 59,* 114–132.

Willits, J. A., Wojcik, E. H., Seidenberg, M S., & Saffran, J. R. (2013). Toddlers activate lexical semantic knowledge in the absence of visual referents: Evidence from auditory priming. *Infancy, 18,* 1053–1075.

Yang, C. D. (2004). Universal grammar, statistics or both? *Trends in Cognitive Sciences, 8,* 451–456.

Yee, E., & Sedivy, J. C. (2006). Eye movements to pictures reveal transient semantic activation during spoken word recognition. *Journal of Experimental Psychology: Learning, Memory, and Cognition, 32,* 1–14.

Yuan, S. & Fisher, C. (2009). "Really? She blicked the baby?": Two-year-olds learn combinatorial facts about verbs by listening. *Psychological Science, 20,* 619–626.

9 How to Learn a Word: The Dynamic Coupling of Words and Referents in Real and Developmental Time

SARAH C. KUCKER

A single statement can allow humans to communicate. Even a single word can convey a host of information. For example, while making coffee I vaguely wave toward the cabinet and say "Grab a wif." This command involves a host of information for which you, the listener, must decipher. In particular, the key word *wif* must be quickly defined and processed. How do you conclude that I am referring to the coffee mug in the cabinet and rule out the broom in the hall and juice glass on the table? To get your favorite morning drink, you must process *wif* in real time, interpreting it as referring to a white ceramic drinking vessel. But in order to ensure you do not go thirsty tomorrow, you must also learn that *wif* refers to the entire category of liquid-holding ceramic mugs that are similar in shape and material but vary in color and size. Both the real-time in-the-moment processing and the long-term retention and learning are relatively quick, easy processes for adults. But, as adults, we already have a strong handle on language, decades of history in communicating, and are knowledgeable about what makes a word, a word.

Children face similar dilemmas of novel word learning daily, but they have to make sense of the world with significantly less knowledge and with an immature cognitive system. The child's world is a wholly novel, constantly changing influx of sounds and scenes. Given such ambiguity, how do they figure out in mere seconds what *wif* is referring to *and* subsequently add *wif* to their lexicon for future use? Furthermore, in order to become proficient adult communicators in the world and process the hundreds of words they encounter daily, they must learn how to learn from these experiences.

Remarkably, children do appear to learn words rapidly and accurately. And, by many accounts, they do so with relative ease—mapping *and* retaining new words after very few exposures. To the delight and frustration of many researchers, there are dozens of theories and hundreds of studies suggesting different mechanisms and cues that promote this remarkable ability. The current chapter provides a snapshot of how a few prominent mechanisms come together to result in successful word learning. There is a particular focus on empirically based approaches and theories across both real-time word use and long-term retention. Moreover, this chapter aims to highlight what developmental pathways toward word learning best explain language learning from a *child's* view.

1. The Word-Learning Process

1.1. WHAT IS WORD LEARNING? Learning even a single new word is not a trivial task. A word is learned when both the phonological word form and some representation of meaning have been reciprocally linked and added to the lexicon as a single unit. A large portion of research on word learning isolates the elements—examining how children segment the word form from the speech stream, or investigating ways in which children extract meaning. The process of speech perception and segmentation is itself a complicated process, much as Johnson and White describe in chapter 8 of this volume. Semantic–conceptual meaning too is complex. But the acquisition of the link between the word and referent making a holistic word-referent unit is uniquely fundamental to building a vocabulary.

A child's vocabulary consists of hundreds of such word-referent units and changes rapidly. At 8-months, a child comprehends just a few words, but by 24-months, they produce nearly 300 (Fenson et al., 1994). By the time children are in kindergarten, their vocabularies have over 6,000 words. On average, this means children are learning upward of 6 new words each day throughout their first six years of life. These words are encountered in a variety of contexts—words are embedded in sentences, can be part of a command or question, are overheard on the radio and TV, or encountered while reading a book. In the course of a single day, a child hears well over 700 unique words (Fenson et al.,1994; Hart & Risley, 1995) of various kinds—nouns, verbs, adjectives, pronouns, and even nonwords. The language children hear also includes grammar, morphology and syntax, and an overarching structure to acquire at the same time (see chapter 10). This is the complicated context in which word learning is occurring. It is not just vast, but also variable and changing. While the focus of the current chapter is on the process of learning a single noun, the conclusions are widely attributed. Nouns are some of the earliest learned words, and as we'll see, some of the same mechanisms of noun learning can be extended to other categories. And, importantly, the processes involved in noun learning help a child subsequently build a foundation for general vocabulary growth broadly, thereby helping a child learn how to learn.

1.2. The Timescales of Word Learning The moment a child encounters a new word, the path to learning begins. However, processing a new word such as *wif* occurs over at least two levels—the real-time/in-the-moment reaction (i.e., realizing that *wif* is the coffee mug) and the long-term learning (being able to use the word *wif* tomorrow). The real-time in-the-moment process has historically been referred to as *fast mapping* and for years was thought (incorrectly) to capture rapid learning by very young children (Carey & Bartlett, 1978). Since, however, it has become obvious that mapping alone does not guarantee retention (Bion, Borovsky, & Fernald, 2013; Horst & Samuelson, 2008), and a host of work has begun to look further at the process of consolidation and retention (McGregor, 2014; Vlach & Sandhofer, 2012; Wojcik, 2013). Mapping is a real-time process serving communication in-the-moment. Learning, on the other hand, does not need to operate in real time, but is better served through a longer process of strengthening (and weakening) word-referent associations. Moreover, the process of learning can give rise to new processes that support further vocabulary growth (e.g., generalization and category development). In the end, mapping and retaining are likely distinct, but related processes that together adapt and change over development (Kucker, McMurray, & Samuelson, 2015).

Looking at each of these timescales individually (real time, retention, learning to learn) allows for a rich, empirically supported investigation of how different mechanisms and cues may matter for development. However, looking across all timescales can allow for a richer understanding of the developmental process of word learning. If the mechanisms at play during mapping do not matter for retention, it might not be as informative to understanding how a child builds her vocabulary. Likewise, if learning changes over development, how does that inform work on initial mapping? Thus, the work here reviews a few possible theories and mechanisms that contribute to real-time mapping, then examines the role of similar cues for long-term retention, and discusses how these factors promote a cycle of learning. Mechanisms are compared across timescales, with a focus on those with strong empirical support. From this snapshot, the dynamic, multifaceted, and sometimes unpredictable aspect of word learning is apparent.

2. *Real-Time Processing and In-the-Moment Referent Selection*

In 1960, Quine set the framework for researchers for decades to come by outlining a dilemma—while visiting a foreign country, an English-speaking linguist goes on a walk with a local resident who does not speak English. The local resident points to a hillside where there is a small, four-legged, big-eared animal with a white tail hopping along and says "gavagai"; how does that linguist come to learn what *gavagai* means? It could mean *bunny* or *jumping*, but it could also mean *it's a nice day*, *fluffy tail*, *dinner*, or *undetached rabbit parts*. This uncertainty has since been dubbed the problem of referential ambiguity and framed as one of the biggest hurdles facing young children. Like the tourist in Quine's problem, children are thought to sit in a foreign world in which a single new word could refer to a number of possible referents. Seemingly remarkably, children succeed in making sense of this world—when given a novel word, children as young as 10 months overwhelmingly prefer the most novel item as the referent (Mather & Plunkett, 2010). However, Quine's application to child language is likely misguided—adults learning a foreign language are already mature learners with in-the-moment logical processes, strong nonverbal communication skills, and fully functioning long-term memory systems. Children are developing their bodies at the same time as new words and have yet to learn what types of things even get labels. Despite this, some of the oldest, classic theories of mapping arose from Quine's proposal. These classic theories proposed general word constraints and social cues that help children get to the right answer amid all the ambiguity, and they have stuck around since.

As a whole, word-learning constraints propose that the uncertainty inherent in a child's world is reduced by eliminating competitors (e.g., mutual exclusivity, disjunctive syllogism) or highlighting potential referents (e.g., novel-name nameless category). Many of these constraints propose children use high-level cognitive processes to logically deduct likely referents, though at their core they also rely on low-level attentional processes. Importantly, these constraints attempt to understand and explain the same basic mapping behavior—when young children hear a novel label, what subsequently drives them to look at or reach for an unknown item?

The principle of mutual exclusivity rests on the assumption that labels should not overlap in what they refer to. Thus, monolingual children avoid having two labels for a single item (Markman & Wachtel, 1988; Merriman & Bowman, 1989). When they encounter a novel word, mutual exclusivity says that the child rejects already known items as possibilities, leaving only unknown objects. Similarly, if a child encounters a novel word, but only has known referents as options, the whole object assumption maintains that only then will the child forgo mapping to a single item but will instead choose an unknown characteristic or part as the referent. Paired with the taxonomic bias (that children map words to the basic level of a category), it can explain why children can quickly determine that *slinky* means the new silver coiled spring toy and not their shoe, a ball, a specific kind of building block, or a synonym for *toy* (Markman, 1991). Mutual exclusivity has also been implicated in the learning of color terms (Carey & Bartlett, 1978) as well as shape and texture (Heibeck & Markman, 1987). Disjunctive syllogism is a more recent adaptation of this same idea (Halberda, 2003), explaining the cognitive reasoning process of first attending to (and ruling out) known items before increasing focus on novel ones.

Another classic theory, the novel-name nameless category principle, supposes that children are biased to map novel words to unknown, novel objects (Mervis & Bertrand, 1994). The argument is that children direct their attention toward a novel object only after hearing a novel word, suggesting that the novelty of the word interacts with the novelty of the referent to determine behavior. That is, the relative novelty of words and objects is responsible for driving mapping more so than lexical knowledge.

Although each word-learning constraint proposes a different cognitive process for driving "learning," they all attempt to explain the same observable behavior—when given both a known and a novel object along with a novel word, children select (by either looking or reaching for) the unknown, novel item. Likewise, empirical "proof" for these theories uses nearly identical paradigms—multiple known items and a single novel item are presented to a child who is then asked to select a novel item by name (see figure 9.1). This referent selection behavior has been replicated hundreds of times across ages, contexts, and modalities (Bion et al., 2013; Gurteen, Horne, & Erjavec, 2011), as well as with the application of novel facts (Behrend, Scofield, & Kleinknecht, 2001).

The referent selection behavior also changes over development (Bion et al., 2013; Halberda, 2003; Mervis & Bertrand, 1994). Younger children (14- to 18-month-olds) perform much more poorly than older children (30 months). Some suggest this developmental trajectory is due to increases in vocabulary knowledge—after all, only children who had hit the vocabulary spurt demonstrated novel referent selection in Mervis and Bertrand's (1994) study, and there was a direct correlation between productive vocabulary and referent selection performance in Bion et al. Others suggest that developmental changes may be driven by changes in cognitive processing mechanisms (Golinkoff, Mervis, & Hirsh-Pasek, 1994; Nazzi & Bertoncini, 2003). On the other hand, recent theories suggest that development is not a linear trajectory of any single ability but is rather the outcome of the coupling of a dynamic body interacting with multiple mechanisms in a changing world (Spencer et al., 2009).

It is clear that children do certainly operate in-the-moment to select an item in response to a novel word. However, the multiplicity of explanations for the behavior, at best, suggest multiple possible pathways to learning and, at worst, lend doubt to what exactly is driving the behavior. Thus, what follows is a closer analysis of the roles of a few prominent mechanisms—knowledge, novelty, variability, social–pragmatic cues, and ambiguity—in real-time processing.

2.1. Knowledge Mutual exclusivity, the whole object assumption, the taxonomic bias, and disjunctive syllogism all propose to rely heavily on a child's prior knowledge of label–object pairs. Children's referent selection in the classic task is not enough to confirm this fact and differentiate the use of knowledge from the use of novelty. However, follow-up studies have supported the use of vocabulary knowledge in influencing real-time processing of novel words. Referent selection studies often include an array of both known and novel items. Manipulating the known competitor objects such that the child's knowledge could not be brought to bear in the task, changes behavior. For instance, when atypical *known* exemplars are used (e.g., Chihuahua for dog), two- and four-year-olds' ability to select the *novel* target diminishes (Merriman & Schuster, 1991). When known items are well-known (both produced and comprehended), they are less likely to be chosen by two-, three-, and four-year-olds as the referent for a novel word

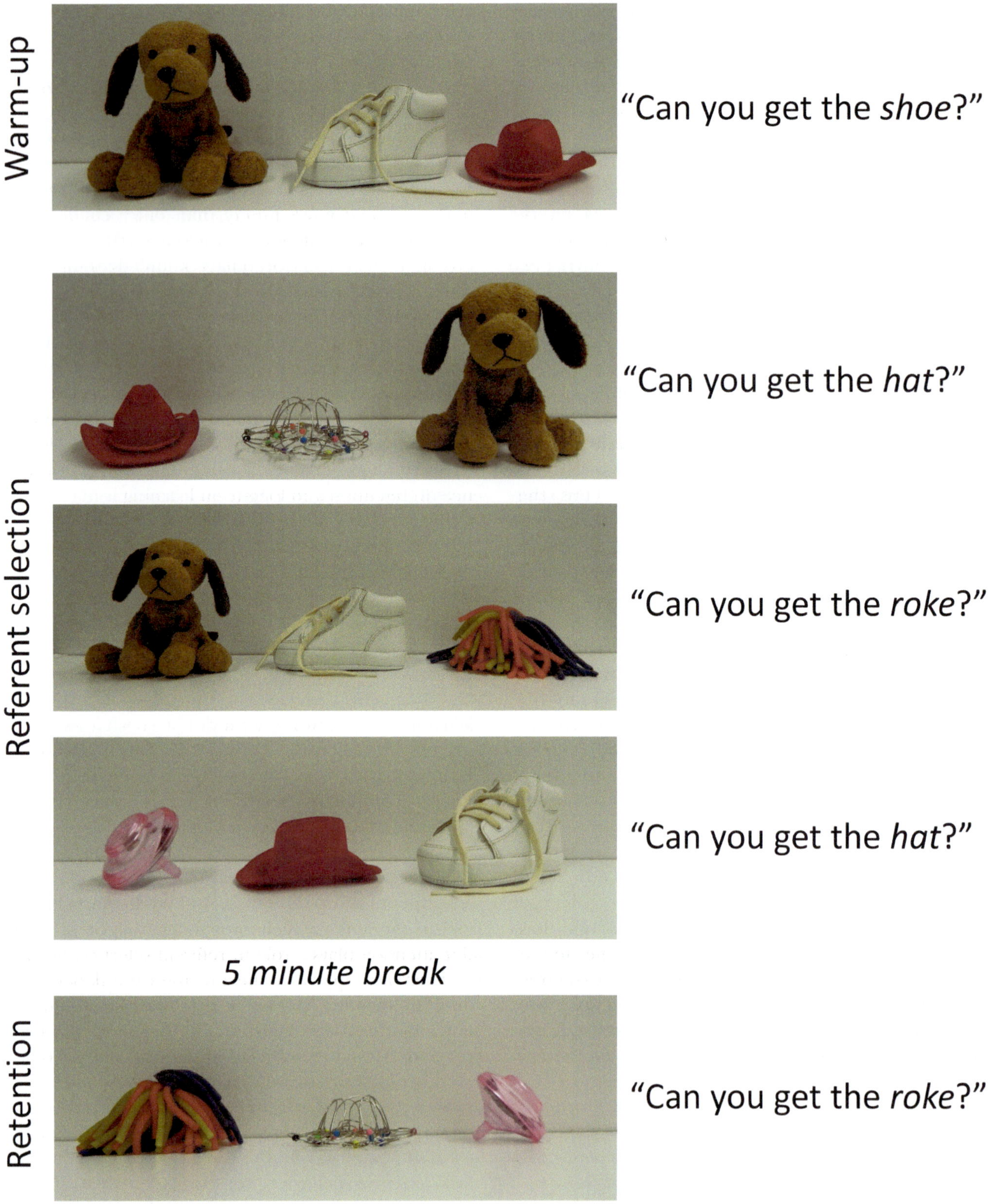

FIGURE 9.1 Traditional referent selection paradigm. The classic arrangement utilized for examining referent selection (fast mapping) abilities of young children through either an eye-tracking looking paradigm or three-dimensional item choice selections.

than are weakly known items (words that are only comprehended; Grassmann, Schulze, & Tomasello, 2015). In another study with 24-month-old children, Borovsky, Ellis, Evans, and Elman (2015) used known items from domains in which the child had robust knowledge and a high number of words (e.g., food words) or from domains in which the child had weak knowledge and a low number of words (e.g., clothing words). Overall, children selected the novel target item at above chance levels for both conditions. However, when eye-movement patterns were examined, children looked quicker and longer to the novel target when distractors were robust, high-knowledge items. In contrast, other similar work with 18- and 24-months-old children and real three-dimensional items found no differences in referent selection between weakly known and well-known trials (Kucker, McMurray, & Samuelson, 2019; Samuelson, Kucker, & Spencer, 2017). Instead, children demonstrated an overriding novelty bias that may have interfered with processing of the known items. Thus, the amount of vocabulary knowledge a child has certainly appears to alter in-the-moment processing, but other factors, such as novelty, may also play a role (see also Mather, 2013).

2.2. NOVELTY More recent work has both confirmed that children's behavior is driven toward novelty and also lends doubt to how much novelty is driven by high-level language-specific processes. One way to extract novelty from knowledge is to eliminate the explicit ability to use vocabulary or labels in the standard task. In a recent study by Horst, Samuelson, Kucker, and McMurray (2011), children were given the classic referent selection task, but critically, none of the items had names. Instead, items could only be differentiated based on familiarity or relative novelty—two items were prefamiliarized before the task, whereas one was "supernovel" or never seen until the trials. They found that at 24 months, children reliably select the *most* novel item from an array as the referent for a novel word (see also Mather & Plunkett, 2012). More recently, Kucker, McMurray, and Samuelson (2018), showed that 18-month-old children are biased to select the novel item regardless of prompt. Here, children were asked to select well-known, easily identifiable items the child previously demonstrated they knew (e.g., dog, shoe), but instead of selecting the target known item, children attended to and selected the novel item. Attention to novelty in the face of current knowledge suggests both that novel word learning (in-the-moment) may be driven by low-level processes and also that at 18 months, current word knowledge (even of well-known words) is weak and can be overridden by other cognitive processes.

2.3. VARIABILITY Variability is a prominent factor in speech perception (Apfelbaum & McMurray, 2011; Rost & McMurray, 2009) and conceptual development (Siegler, 1994), and recent work has begun to suggest it might matter for linking word–referent pairs (Perry, Samuelson, Malloy, & Schiffer, 2010). Variability in much of this work is often defined as the use of changing stimuli or approaches compared to repetitive or static stimuli or approaches. However, variability by definition then does not influence real-time processing because variability is an accumulation across multiple experiences. As we might expect then, much of the work examining variability finds no effect of varying exemplars versus repeated exemplars in referent selection (Axelsson & Horst, 2014; Twomey, Ranson, & Horst, 2014). Although, many of these studies do find differences in retention and long-term learning.

2.4. SOCIAL–PRAGMATIC In addition to knowledge, novelty, and variability, there are also social and pragmatic influences on children's in-the-moment behavior. Joint attention, pointing, and understanding others' intent can help get a child to the "right" answer. Diesendruck and Markson (2001) suggested that three-year-olds avoid lexical overlap much like in mutual exclusivity, but from a pragmatic perspective—in the interest of communicating effectively, a speaker should not intend to label the same thing with different names but would instead give unique names for unique items. In addition, Baldwin (1991) discovered that 16- to 19-month-old infants follow a speaker's nonverbal cues and subsequently map a novel label to an item in line with the speaker's cues. Children here do not map the label to the item in their own focus, suggesting that joint attention plays a role in referent selection behavior. There also appears to be a reciprocal link between children learning from joint attention—as children gain better hand–eye coordination, joint attention abilities appear, and as joint attention appears, so does more hand–eye coordination, both of which predict word learning (Yu & Smith, 2017), though not necessarily for social reasons (Yu & Smith, 2013). Direct pointing to an item while naming it can also be beneficial for helping young children attend to the novel referent, at least initially (Baldwin & Markman, 1989). Akhtar, Carpenter, and Tomasello (1996) also taught children the label for novel objects. This time, there were multiple unknown items that the adult and child interacted with. A new object was added to the mix and without displaying any nonverbal cues, the adult excitedly gave a novel name. The majority of the time, children mapped the novel word to the newest item, which was taken as children using discourse novelty for mapping and understanding the intentions of the adult. Baldwin (1993) similarly showed that children follow an adult's gaze as that adult peers into an opaque container to label an unseen item. Children subsequently chose the item the adult named 73% of the time.

Many of the studies on social-pragmatic cues have been hotly debated for extrapolating social–cognitive processes from simple behavior. For instance, discourse novelty did not alter behavior when relative novelty (not labels) differentiate the items (Horst et al., 2011), lending doubt to the conclusion that social cues are uniquely driving children's behavior. Moreover, in a counterpoint to the study by Baldwin (1993), Samuelson, Smith, Perry, and Spencer (2011) showed that identical behavior could also be achieved through consistent use of physical space to bind words and objects. Here, instead of a hidden item, an adult presented a novel object on a specific spot on the table, removed the novel object, pointed to the (now empty) space, and said a novel word. When asked to retrieve the novel object by name, children reliably chose the item from the same spatial location, suggesting at most, that behavior can be driven by nonobvious properties, and at the very least, that physical space may also drive children's mapping behavior.

2.5. Ambiguity The classic referent selection work based on Quine (1960) all assume an ambiguous learning environment. However, there are also many cases in which ambiguity is artificially eliminated by having only one item present during naming (Dollaghan, 1985; Oviatt, 1980) or when an item is singled out for the child (Markson & Bloom, 1997; Woodward, Markman, & Fitzsimmons, 1994). Ostensive naming studies such as these have nearly perfect real-time mapping simply because children either map correctly or do not map at all (and then are subsequently dropped from the study as nonresponders). In these cases, mapping is not a problem, but such direct naming may not represent the real world. In a potentially damming proposal, Smith and colleagues (Pereira, Smith, & Yu, 2014; Smith, 2013; Yoshida & Smith, 2008; Yurovsky, Smith, & Yu, 2013) proposed that ambiguity is not a problem for in-the-moment behavior in a real-world setting. By their account, the coupling of a child's physical body with the world around them leads to minimal plausible referents for novel words. Put simply, a child's arms are short and thus, when they are holding an item it takes up significantly more of their field of view, thereby leaving them with only one possible referent. Combine this with the fact that mothers tend to synchronously label items when the child is holding it and mapping is easy (Pereira et al., 2014). This is one case where multiple elements come together at precisely the right time in development to aid in mapping.

The obsoleteness of Quine (1960) is further supported by studies showing that children utilize different information than adults. In an eloquent study by Ramscar, Dye, and Klein (2013), two- to three-year-old children, adults, and expert developmental psychologists were given an ambiguous scenario that could be solved in one of two ways. On the first trial, individuals saw two objects (A and B) and heard a single novel word (*dax*). On the second trial, one old object (B) and one newer object (C) were seen along with a newer word (*pid*). Individuals were then given all three items and asked to find the *dax*, the *pid*, and a new word (*wug*). The key question is, which object is the "best" referent for *wug*? One argument is that children could pick object B as the *wug* for logical reasons as it never co-occurred with either of the other words. On the other hand, children might pick object A and C equally as the *wug* because B was the least novel of the items and is therefore somewhat informative of both A and C. Both naïve college students and expert researchers overwhelmingly predicted that children would choose B as the *wug*, consistent with the first hypothesis. Children, however, choose B below chance levels and A and C at chance, consistent with the second hypothesis. That is, an adult's perception of how a child may map new words is not always accurate. A child's view of the world on the other hand, is likely constrained in very different ways through a coupling of the child's own physical development and changing environment (e.g., Smith, 2013). Critically though, such cues may only be relevant so far as ambiguity is a problem. Rather, children are biased toward certain actions, based partially on their history of learning. As children encounter more and more moments of exposure, their knowledge slowly accumulates and words are added to the lexicon long-term.

3. Learning over Time

3.1. The Slow Accumulation of Knowledge Learning is not an instant process, but one that develops over the course of multiple single moments. In what has become a canonical study, Horst and Samuelson (2008) demonstrated that the act of fast mapping does not guarantee retention of the word-referent link. In fact, retention after a single mapping instance is only seen in developmentally older children. Despite the lack of immediate retention, multiple in-the-moment processes accumulate over time to facilitate building strong word-referent links (Kucker et al., 2015). At each moment of exposure, a tiny bit of knowledge is refined. Over multiple exposures, word-referent links are further honed as the new word becomes part of the lexicon. *How* word-referent links are refined differs across theories, however.

The simplest account of knowledge accumulation over time is associative learning. In this case, connections between words and objects are continually built up over exposures, and at the same time, wrong connections are pruned away (McMurray, Horst, & Samuelson, 2012). As evidence grows, associations strengthen, but as counterevidence is presented, the association weakens. Over repetitions, the links between a word and referent get strong enough to support retention and future use.

A variation of the associative account is cross-situational word learning. At any given moment in a day (and even within a single sentence), a child hears multiple words. At the same time, the world around a child houses many objects and potential referents for those words. Cross-situational learning holds that statistics for (and against) word–referent pairs are slowly built (and pruned) as specific words and referents co-occur. Children as young as 12 months evaluate the statistical evidence over the course of multiple trials (Smith & Yu, 2008). Adults too can track multiple co-occurring word–referent pairs over multiple trials (Yu & Smith, 2007). In each of these cases, multiple single words are spoken at the same time as multiple object images are present. Over the course of four or more trials, some words and objects repeat, whereas other pairs vary. On a test of learning for the novel words, children and adults reliably look to and choose the most frequent referent. This finding has been replicated and extended (Dautriche & Chemla, 2014; Fitneva & Christiansen, 2011; Frank, Goodman, & Tenebaum, 2009; Medina, Snedeker, Trueswell, & Gleitman, 2011), but it is most frequently seen in visual looking paradigms after carefully controlled frequent exposure to the words. When objects are not as carefully controlled and some items are more salient and novel than others, learning from cross-situational scenarios drops significantly to chance levels (Smith & Yu, 2013).

On the other hand, Medina, Trueswell, Gleitman, and colleagues (Medina et al., 2011; Trueswell, Medina, Hafri, & Gleitman, 2013) argued for a different explanation of cross-situational behavior. By their account, children's ability to map words and referents over multiple occurrences is not due to complex mathematical calculations and statistical tracking. Rather, upon their first encounter with a word, children make an initial hypothesis about a word's referent. This hypothesis is then confirmed on subsequent exposures and the word is added to the lexicon, or refuted and a new hypothesis is created and subsequently tested. However, much like the classic theories of mapping, the paradigm used here nearly mirrors that in other cross-situational accounts and subsequently, the observable behavior is the same—children and adults correctly identify word–referent pairs after multiple trials.

Regardless of theory, learning appears to occur slowly, over multiple exposures. This is supported empirically by dozens of studies showing that repetition of word-referent links helps retention. In many cases, children do not show evidence of retention after a single mapping (Horst & Samuelson, 2008). However, as exposure and repetition rise, so do the rates of retention. Simply repeating a word while a child is focused on the referent leads to much higher rates of retention (Axelsson, Churchley, & Horst, 2012; Markson & Bloom, 1997; Woodward et al., 1994).

Theories of how knowledge is accumulated vary, ranging from a simple building/pruning of pathways to calculating statistics/co-occurrence, to developing (and testing) a hypothesis. As was the case with mapping, however, many of the theories revolve around the same empirical behavior—more exposure to a new word leads to more retention. Children (and adults) do certainly track word–referent pairs over time, and learning continues past the initial mapping moment. It is possible, however, that what knowledge is gained over repetitions may vary based on the factors present in real time. For instance, processing new words in ambiguous contexts appears to result in more helpful changes in the cognitive architecture at each exposure than new words repeatedly encountered only in direct naming instances. This may have cascading effects for robust retention of the new word.

4. Retention

The canonical mapping studies inspired by Quine (1960) focused on the problem of referential ambiguity and mapping. In so doing, the process of learning and retention was mostly ignored. As we now know, knowledge is accumulated on a slower developmental timescale during which word-referent associations are refined, and learning changes as children develop. Horst and Samuelson (2008) found no retention of new word–referent pairs at 24 months, but using a similar task with 30-month-old infants, Spiegel and Halberda (2011) did find above chance levels of retention. Likewise, Bion et al. (2013) also only found retention at 30 months, but not in the younger 24- or 18-month-old children. Even when items are directly and explicitly named, retention of the new word referent is weak, slowly strengthened over development. However, we know that mapping can have cascading effects on accumulating knowledge and retention. For instance, disambiguation that relies on mutual exclusivity could have different learning outcomes than disambiguation relying only on novelty. Thus, in addition to developmental change, there are many other factors that may influence retention of new word–referent pairs.

4.1. Knowledge Even though the mapping behavior seen in mutual exclusivity and novel-name nameless category work is identical, the differential theories suggest that there may be different pathways by which knowledge can be acquired and retained. A handful of studies have examined referent selection and mapping in contexts that utilize more or less vocabulary knowledge (Borovsky et al., 2015; Kucker et al., 2019; Merriman & Schuster, 1991; Samuelson et al.,2017). However, only Kucker et al. and Samuelson et al. looked at retention of new word–referent pairs mapped in variable knowledge contexts. They compared referent selection and retention of novel words learned in a context with newly learned competitors (e.g., slinky, whisk) compared to well-known competitors (e.g., dog, shoe). Retention varied across both age and condition and, surprisingly, was highest for words learned in a difficult weakly known context. This effect was especially pronounced for the younger 18-month-old children. Thus, the preliminary evidence suggests that knowledge changes retention.

4.2. Novelty Novelty is thought to help children in real time to select the correct referent, but few studies have critically examined whether such an effect extends past in-the-moment processing to also aid in retention. As discussed, 18-month-old children use novelty to find a novel referent, but they also appear biased toward novelty when asked for known items (Kucker et al., 2018) calling into question whether novelty is acting on lexical-processing or low-level perceptual biases. This is further confirmed by work demonstrating that mapping in novelty-driven scenarios does not lead to retention. In a task nearly identical to Horst et al.'s (2011) supernovel study, children mapped novel word–referent pairs based solely on relative novelty, having competitors that had no labels and only varied on visual familiarity. However, neither 18- nor 24-months-old infants retained the word–referent pairs mapped in a novelty-driven context (Samuelson et al., 2017). Computational models simulating this novelty-driven referent selection also predict no retention regardless of development/age (Kucker et al., 2018).

4.3. Variability Variability is generally thought to influence the process of word learning for the better. In one case, Twomey et al. (2014) found that even though two-year-old children do not perform any differently on referent selection for novel exemplars varying in color and material but not shape, compared to a repeated novel exemplar, they did differ significantly on retention; children at two years old retained only novel words mapped across variable examples. On the flip side, other work with three-year-old children showed that variability in the *competitors* hurt retention (Axelsson & Horst, 2014). Variability does certainly influence retention of new word–referent pairs, but how it alters behavior is a result of what varies and at what age.

4.4. Social–Pragmatic Social–pragmatic cues to word learning are perhaps one of the most studied areas in terms of word-referent mapping. As previously discussed, eye gaze, joint attention, pointing, and an understanding of social intent can all get children to the correct referent in-the-moment. However, very few studies examine the long-term consequences of using such cues to map. In fact, Baldwin and Markman (1989) found evidence that pointing does not have long-term benefits. In their study, children as young as 10 months looked longer at an item when it was both labeled and pointed to by an adult. However, labeling only led to initially higher exploration of the item and not retention of the word-referent link. Pointing does not help older 24-month-old children retain novel words either (Axelsson et al., 2012). But on the other hand, Smith and colleagues (Pereira et al., 2014; Smith, 2013; Yoshida & Smith, 2008) work with a head-mounted eye tracker has shown that when the child and parent both focus on the item the child is holding while the parent names it, the child is more likely to retain the word-referent link. Social–pragmatic perspectives argue this is because of the shared social focus, whereas other accounts explain such behavior as limited ambiguity.

4.5. Ambiguity The evidence on the role of ambiguity in long-term learning is mixed. The studies by Smith and colleagues with head-mounted cameras show more learning when mothers name items when the child is holding them (and thus, when there is no ambiguity). This suggests less ambiguity is better for retention. Similarly, if a traditionally ambiguous environment is manipulated to focus 24-month-old children's attention to a single item (thus helping to limit ambiguity), children are more likely to retain the word–referent pair (Axelsson et al., 2012). However, in Axelsson et al.'s case, multiple items were present initially and competitors were subsequently covered or the target was illuminated between mapping and retention. Work by Zosh, Brinster, and Halberda (2013) found more convincing evidence that ambiguity may help retention—when preschool children mapped a novel word–referent pair in the context of both a known and novel item, they were more likely to retain than children who mapped a novel word when only a single novel item was present (see also Coutanche & Thompson-Schill, 2014, for a similar effect with adults).

5. Learning to Learn

In many of these studies with children, a newly learned word refers to a single particular item. However, words generally refer to a whole category of items. For instance, a single word *cup* may indicate my current coffee mug, but it also labels the juice glass and sippy cup. *Cup* can also bring about a conversation about the contents of the cup, the kitchen location, other utensils, morning caffeine (and coffee options), and the coziness and warmth felt in my hands. For some, "cup" may summon an enthusiastic "cheers!" and glass clinking, whereas for others, it's the disposable remnants of a Starbuck's run. All this information is part of word learning, being acquired over development through a host of mechanisms and factors. As children encounter more and more examples of word–referent pairs they not only add words to their vocabulary, but they also learn how to make these generalizations and continue the cycle of mapping and retaining new words. This is perhaps the quintessential goal in studying language development—how do children acquire a whole system that can support the reciprocal, continual cycle of mapping and retaining thousands of word–referent pairs? Various theories have proposed that the mechanisms described in this chapter are key (e.g., Golinkoff et al., 1994; Nazzi & Bertoncini, 2003). However, few have empirically tested how such mechanisms create a change in the system such that either the mechanism itself or the child is refined for future use.

The ways in which a child is trained, or their prior experiences, shape their subsequent behavior (Horst et al., 2009; Samuelson & Horst, 2007). Sustainable cascades like this may be key for understanding children's exponential vocabulary growth during the first few years of life. In one example, children learn what factors or characteristics may be important for the meaning of words. By the time there are between 50 and 150 count nouns in their productive vocabulary (which occurs around 18 months in Western cultures), young children reliably generalize labels to novel items matching in shape (Samuelson, 2002). This shape-bias behavior has been replicated across ages, contexts, and populations (Landau, Smith, & Jones, 1988; Smith, Jones, Landau, Gershkoff-Stowe, & Samuelson, 2002). Consistent across all studies on generalization is a focus on the amount of nouns, specifically shape-based items, in the child's vocabulary. Thus, it appears that more words create more opportunities to extract regularities, such as attention to shape, to aid in the first step of mapping a new word. Such a robust cue is not just learned and strengthened over exposures, but can be carried forward as a cue for mapping at a future point in time.

In another example, Behrend, Scofield, and Kleinknecht (2001) conducted a classic referent selection task with two- to four-year-old children, but instead of a novel label, some trials included the mapping of a novel fact (e.g., "My uncle gave me this"). Although they found no differences in mapping or retention, children only generalized the novel labels, but not novel facts, to new exemplars of the learned items. This suggests that as children develop and acquire words, their ability to generalize new words and perhaps their cycle of learning to learn, narrows in on the features most helpful for language learning.

That generalization component, or the learning to learn ability, is rarely empirically studied. To date, there are no studies examining cascading effects of learning with knowledge or novelty mechanisms. Social–pragmatic studies are limited, but recent work has begun to examine the effect of variability and ambiguity on future learning.

5.1. Variability Variability often appears to have little to no effect on real-time processing, but it does positively influence retention of new words. It also has long-term benefits on shaping a word-learning system. When children map a new word multiple times and each repetition is an exemplar that is constant in shape, but varies in irrelevant dimensions, retention is high (Twomey et al., 2014). Moreover, a recent training study taught 16-month-old children new words with either highly variable stimuli (e.g., *ball* is learned with a football, golf ball, and basketball) or tightly clustered stimuli (e.g., *ball* is learned with a tennis ball and baseball) and then tested both their acquisition of the trained words and overall vocabulary growth outside the lab. Children exposed to variable stimuli not only learned the new items, but also had higher vocabulary growth outside the lab than children who learned with less variable exemplars (Perry et al., 2010). That is, variability helps a child's vocabulary become more flexible in their use of features, but more robust in linking labels with objects and their respective categories. Variability in mapping helps retention, which cascades and helps children learn how to learn.

5.2. Social–Pragmatic Social–pragmatic cues have primarily been studied as a mechanism for real-time mapping. However, children's ability to use social and pragmatic cues increases as they develop and thus, so does their continued language growth. For reasons previously demonstrated, however, it is difficult to identify exactly when social cues are used over cues such as physical space or location.

5.3. Ambiguity Though the evidence for ambiguity's role on mapping and retention are mixed, ambiguity can have long-term learning effects. For instance, a recent model showed a direct correlation between ambiguity during initial vocabulary acquisition and future abilities to learn and map new words—the more ambiguous a model's initial mapping environment, the better the model was at mapping new words in the future (McMurray, Zhao, Kucker, & Samuelson, 2013). Other work with adults in an ambiguous cross-situational paradigm has demonstrated that mapping the wrong word initially benefits long-term retention (Fitneva & Christiansen, 2011). Similar to the process of acquiring knowledge slowly, children also slowly acquire abilities to learn in the future.

Regardless of theoretical approach, children do hear words more than once and learning continues to develop even if they retain that word at first exposure. Even more, children are hearing (and subsequently mapping and learning) more than one word at a time. In fact, children can hear up to 700 words per day and encounter hundreds of objects, people, places, and such. Each word is simultaneously undergoing a process of in-the-moment mapping and longer learning. Words become consolidated into the lexicon and learned not just in and of themselves, but associated with others and incorporated into the entire lexicon of the child, which is itself shifting and growing and changing over time.

6. *Across All Timescales*

Acquiring a vocabulary uses some of the mechanisms and processes described in this chapter, but learning does not advance easily or consistently. From 12 to 18 months, children learn only handful of words, saying an average of 86 words. But from 18 to 24 months, they learn exponentially more, landing at 297 produced words and starting to combine words. From 24 to 30 months, they grow to over 550 words in their vocabulary of a variety of types and have begun to use full sentences. In addition to there being real-time and learning processes operating for word learning, there is also a developmental timescale (McMurray, 2016). The developmental shifts in the process of mapping and retaining new words have been noted for years (Fenson et al., 1994) and dozens of theories have been generated to account for such changes over time. It is debatable, however, whether these rises are due to changes in vocabulary size, growth of knowledge of the world, or physical growth. The evidence suggests that it is a combination—children demonstrate a reliable ability to select the novel item once they hit the vocabulary spurt (Mervis & Bertrand, 1994), children generalize new words to novel exemplars only after accumulating at least 50 nouns in their productive vocabulary (Samuelson, 2002), more visual exploration of an item increases later mapping of a label (Kucker & Samuelson, 2012), and learning from direct naming rises when children's fine motor skills allow them to hold an object steady in their field of view (Smith, 2013). Golinkoff et al. (1994) presented a formal model that proposes that children slowly adapt more refined learning mechanisms as they gain experience in the world (experience defined as both vocabulary growth and general knowledge). More recently, Smith, Spencer, and colleagues (Smith, 2013; Spencer et al., 2009) proposed a dynamic system driven by the confluence of multiple factors within a child. That is, the coupling of a child's own abilities with their physical body and the environment around them creates optimal moments for learning. These moments are unique to a child whose own developmental trajectory builds knowledge and opportunities for the future.

The picture of word learning is complex. Children are encountering hundreds of different words across just as many contexts and are tasked with coming away with robust knowledge for the future. It's a seemingly daunting task, at least to an adult. But, as we have seen, the world from a child's view may look very different and in being so, may also be optimally suited for learning despite the complexity of the task. This chapter has given a snapshot of just a few mechanisms that may play into that process and alter learning, but importantly, they do not all matter at the same times nor to the same extent. Some mechanisms appear helpful for real-time processing, but fail at retention (e.g., novelty), others are mildly helpful across the board (e.g., knowledge), and some are contentious (e.g., ambiguity). Importantly, while children may be using these and similar mechanisms for learning, they are also growing and changing, and development itself will alter the process of learning.

There are multiple interacting pathways to success that create optimal moments for learning over development. Each of those moments cascade and slowly (and surprisingly), a vocabulary arises. The information a child gets at a single moment is highly dependent upon their own current position, attention, and prior knowledge up to that specific moment in time. What they learn is subsequently highly dependent upon the accumulation of individual moments. And those moments do not always add up perfectly! But in the end, knowledge is achieved.

REFERENCES

Akhtar, N., Carpenter, M., & Tomasello, M. (1996). The role of discourse novelty in early word learning. *Child Development, 67*(2), 635.

Apfelbaum, K. S., & McMurray, B. (2011). Using variability to guide dimensional weighting: Associative mechanisms in early word learning. *Cognitive Science, 35*(6), 1105–1138.

Axelsson, E. L., Churchley, K., & Horst, J. S. (2012). The right thing at the right time: Why ostensive naming facilitates word learning. *Frontiers in Psychology, 3*(8), 1–8

Axelsson, E. L., & Horst, J. S. (2014). Contextual repetition facilitates word learning via fast mapping. *Acta Psychologica, 152*, 95–99.

Baldwin, D. A. (1991). Infants' contribution to the achievement of joint reference. *Child Development, 62*(5), 874–890.

Baldwin, D. A. (1993). Early referential understanding: Infants' ability to recognize referential acts for what they are. *Developmental Psychology, 29*(5), 832–843.

Baldwin, D. A., & Markman, E. M. (1989). Establishing word object relations: A first step. *Child Development, 60*, 381–398.

Behrend, D. A., Scofield, J., & Kleinknecht, E. E. (2001). Beyond fast mapping: Young children's extensions of novel words and novel facts. *Developmental Psychology, 37*(5), 698–705.

Bion, R. A. H., Borovsky, A., & Fernald, A. (2013). Fast mapping, slow learning: Disambiguation of novel word-object mappings in relation to vocabulary learning at 18, 24, and 30 months. *Cognition, 126*(1), 39–53.

Borovsky, A., Ellis, E. M., Evans, J. L., & Elman, J. L. (2015). Lexical leverage: category knowledge boosts real-time novel word recognition in 2-year-olds. *Developmental Science, 19*(6), 918–932.

Carey, S., & Bartlett, E. (1978). Acquiring a single new word. *Papers and Reports on Child Language Development, 15*, 17–29.

Coutanche, M. N., & Thompson-Schill, S. L. (2014). Fast mapping rapidly integrates information into existing memory networks. *Journal of Experimental Psychology: General, 143*(6), 2296–2303.

Dautriche, I., & Chemla, E. (2014). Cross situational word learning in the right situations. *Journal of Experimental Psychology: Learning, Memory, and Cognition, 40*(3), 892–903.

Diesendruck, G., & Markson, L. (2001). Children's avoidance of lexical overlap: A pragmatic account. *Developmental Psychology, 37*(5), 630–641.

Dollaghan, C. (1985). Child meets word: Fast mapping in preschool children. *Journal of Speech and Hearing Research, 28*(3), 449–454.

Fenson, L., Philip, P. S., Reznick, J., Bates, E., Thal, D. J., & Pethick, S. J. (1994). Variability in early communicative development. *Monographs of the Society for Research in Child Development, 59*(5), 1–185.

Fitneva, S. A., & Christiansen, M. H. (2011). Looking in the wrong direction correlates with more accurate word learning. *Cognitive Science, 35*(2), 367–380.

Frank, M. C., Goodman, N. D., & Tenebaum, J. B. (2009). Using speakers' referential intentions to model early cross-situational word learning. *Psychological Science, 20*, 257–285.

Golinkoff, R., Mervis, C. B., & Hirsh-Pasek, K. (1994). Early object labels. The case for a developmental lexical principles framework. *Journal of Child Language, 21*(1), 125–155.

Grassmann, S., Schulze, C., & Tomasello, M. (2015). Children's level of word knowledge predicts their exclusion of familiar objects as referents of novel words. *Frontiers in Psychology, 6*(1200), 1–8.

Gurteen, P. M., Horne, P. J., & Erjavec, M. (2011). Rapid word learning in 13- and 17-month-olds in a naturalistic two-word procedure: Looking versus reaching measures. *Journal of Experimental Child Psychology, 109*(2), 201–217.

Halberda, J. (2003). The development of a word-learning strategy. *Cognition, 87*(1), B23–B34.

Hart, B., & Risley, T. R. (1995). *Meaningful differences in the everyday experience of young American children*. Baltimore, MD: Paul H. Brookes.

Heibeck, T. H., & Markman, E. M. (1987). Word learning in children: An examination of fast mapping. *Child Development, 58*, 1021–1034.

Horst, J. S., Ellis, A. E., Samuelson, L. K., Trejo, E., Worzalla, S. L., Peltan, J. R., & Oakes, L. M. (2009). Toddlers can adaptively change how they categorize: Same objects, same session, two different categorical distinctions. *Developmental Science, 12*(1), 96–105.

Horst, J. S., & Samuelson, L. K. (2008). Fast mapping but poor retention by 24-month-old infants. *Infancy, 13*(2), 128–157.

Horst, J. S., Samuelson, L. K., Kucker, S. C., & McMurray, B. (2011). What's new? Children prefer novelty in referent selection. *Cognition, 118*(2), 234–244.

Kucker, S. C., McMurray, B., & Samuelson, L. K. (2015). Slowing down fast mapping: Redefining the dynamics of word learning. *Child Development Perspectives, 9*(2), 74–78.

Kucker, S. C., McMurray, B., & Samuelson, L. K. (2018). Too much of a good thing: How novelty biases and vocabulary influence known and novel referent selection in 18-month-old children and associative learning models. *Cognitive science, 42* (Suppl 2), 463–493.

Kucker, S. C., McMurray, B., & Samuelson, L. K. (2019). *Sometimes it's better to know less: How known words influence referent selection and retention in 18 to 24-month-old children and associative models.* Manuscript under review.

Kucker, S. C., & Samuelson, L. K. (2012). The first slow step: Differential effects of object and word-form familiarization on retention of fast-mapped words. *Infancy, 17*(3), 295–323.

Landau, B., Smith, L. B., & Jones, S. S. (1988). The importance of shape in early lexical learning. *Cognitive Development, 3*(3), 299–321.

Markman, E. M. (1991). Whole object taxonomic and mutual exclusivity. In S. A. Gelman & J. P. Byrnes (Eds.), *Perspectives on language and thought: Interrelations in development* (pp. 72–106). Cambridge: Cambridge University Press.

Markman, E. M., & Wachtel, G. F. (1988). Children's use of mutual exclusivity to constrain the meanings of words. *Cognitive Psychology, 20*(2), 121–157.

Markson, L., & Bloom, P. (1997). Evidence against a dedicated system for word learning in children. *Nature, 385*, 813–815.

Mather, E. (2013). Bootstrapping the early lexicon: How do children use old knowledge to create new meanings? *Frontiers in Psychology, 4*, 96.

Mather, E., & Plunkett, K. (2010). Novel labels support 10-month-olds' attention to novel objects. *Journal of Experimental Child Psychology, 105*(3), 232–242.

Mather, E., & Plunkett, K. (2012). The role of novelty in early word learning. *Cognitive Science, 36*(7), 1157–1177.

McGregor, K. K. (2014). What a difference a day makes: Change in memory for newly learned word forms over 24 hours. *Journal of Speech Language and Hearing Research, 57*(5), 1842–1850.

McMurray, B. (2016). Language at three timescales: The role of real-time processes in language development and evolution. *Topics in Cognitive Science, 8*(2), 393–407.

McMurray, B., Horst, J. S., & Samuelson, L. K. (2012). Word learning emerges from the interaction of online referent selection and slow associative learning. *Psychological Review, 119*(4), 831–877.

McMurray, B., Zhao, L., Kucker, S. C., & Samuelson, L. K. (2013). Pushing the envelope of associative learning: Internal representations and dynamic competition transform association into development. In L. Gogate, & G. Hollich (Eds.), *Theoretical and computational models of word learning: Trends in psychology and artificial intelligence* (pp. 49–80). Hershey, PA: Information Science Reference.

Medina, T. N., Snedeker, J., Trueswell, J. C., & Gleitman, L. R. (2011). How words can and cannot be learned by observation. *Proceedings of the National Academy of Sciences, 108*(22), 9014–9019.

Merriman, W. E., & Bowman, L. (1989). The mutual exclusivity bias in children's word learning. *Monographs of the Society for Research in Child Development, 54*(3/4), 1–129.

Merriman, W. E., & Schuster, J. M. (1991). Young children's disambiguation of object name reference. *Child Development, 62*(6), 1288–1301.

Mervis, C. B., & Bertrand, J. (1994). Acquisition of the novel name-nameless category (N3C) principle. *Child Development, 65*(6), 1646–1662.

Nazzi, T., & Bertoncini, J. (2003). Before and after the vocabulary spurt: Two modes of word acquisition? *Developmental Science, 6*(2), 136–142.

Oviatt, S. L. (1980). The emerging ability to comprehend language: An experimental approach. *Child Development, 51*, 97–106.

Pereira, A. F., Smith, L. B., & Yu, C. (2014). A bottom-up view of toddler word learning. *Psychonomic Bulletin and Review, 21*(1), 178–185.

Perry, L. K., Samuelson, L. K., Malloy, L. M., & Schiffer, R. N. (2010). Learn locally, think globally: Exemplar variability supports higher-order generalization and word learning. *Psychological Science, 21*(12), 1894–1902.

Quine, W. V. O. (1960). *Word and object: An inquiry into the linguistic mechanisms of object reference*. Cambridge, MA: MIT Press.

Ramscar, M., Dye, M., & Klein, J. (2013). Children value informativity over logic in word learning. *Psychological Science, 24*(6), 1017–1023.

Rost, G. C., & McMurray, B. (2009). Speaker variability augments phonological processing in early word learning. *Developmental Science, 12*(2), 339–349.

Samuelson, L. K. (2002). Statistical regularities in vocabulary guide language acquisition in connectionist models and 15-20-month-olds. *Developmental Psychology, 38*(6), 1016–1037.

Samuelson, L. K., & Horst, J. S. (2007). Dynamic noun generalization: Moment-to-moment interactions shape children's naming biases. *Infancy, 11*(1), 97–110.

Samuelson, L. K., Kucker, S. C., & Spencer, J. P. (2017). Moving word learning to a novel space: A dynamic systems view of referent selection and retention. *Cognitive Science, 41*(S1), 52–72.

Samuelson, L. K., Smith, L. B., Perry, L. K., & Spencer, J. P. (2011). Grounding word learning in space. *PLOS ONE, 6*(12), e28095.

Siegler, R.S. (1994). Cognitive variability: A key to understanding cognitive development. *Current Directions in Psychological Science, 3*(1), 1–5.

Smith, L. B. (2013). It's all connected: Pathways in visual object recognition and early noun learning. *American Psychologist, 68*(8), 618–629.

Smith, L. B., Jones, S. S., Landau, B., Gershkoff-Stowe, L., & Samuelson, L. (2002). Object name learning provides on-the-job training for attention. *Psychological Science, 13*(1), 13–19.

Smith, L., & Yu, C. (2008). Infants rapidly learn word-referent mappings via cross-situational statistics. *Cognition, 106*(3), 1558–1568.

Smith, L. B., & Yu, C. (2013). Visual attention is not enough: Individual differences in statistical word-referent learning in infants. *Language Learning and Development, 9*(1), 25–49.

Spencer, J. P., Blumberg, M. S., McMurray, B., Robinson, S. R., Samuelson, L. K., & Tomblin, J. B. (2009). Short arms and talking eggs: Why we should no longer abide the nativist-empiricist debate. *Child Development Perspectives, 3*(2), 79–87.

Spiegel, C., & Halberda, J. (2011). Rapid fast-mapping abilities in 2-year-olds. *Journal of Experimental Child Psychology, 109*(1), 132–140.

Trueswell, J. C., Medina, T. N., Hafri, A., & Gleitman, L. R. (2013). Propose but verify: Fast mapping meets cross-situational word learning. *Cognitive Psychology, 66*(1), 126–156.

Twomey, K. E., Ranson, S. L., & Horst, J. S. (2014). That's more like it: Multiple exemplars facilitate word learning: multiple exemplars facilitate word learning. *Infant and Child Development, 23*(2), 105–122.

Vlach, H. A., & Sandhofer, C. M. (2012). Fast mapping across time: Memory processes support children's retention of learned words. *Frontiers in Psychology, 3*, 46.

Wojcik, E. H. (2013). Remembering new words: Integrating early memory development into word learning. *Frontiers in Psychology, 4,* 151.

Woodward, A. L., Markman, E. M., & Fitzsimmons, C. M. (1994). Rapid word learning in 13- and 18-month-olds. *Developmental Psychology, 30*(4), 553.

Yoshida, H., & Smith, L. B. (2008). What's in view for toddlers? Using a head camera to study visual experience. *Infancy, 13*(3), 229–248.

Yu, C., & Smith, L. B. (2007). Rapid word learning under uncertainty via cross-situational statistics. *Psychological Science, 18*(5), 414–420.

Yu, C., & Smith, L. B. (2013). Joint attention without gaze following: Human infants and their parents coordinate visual attention to objects through eye-hand coordination. *PLOS ONE, 8*(11), e79659.

Yu, C. & Smith, L. B. (2017). Hand–eye coordination predicts joint attention. *Child Development, 88,* 2060–2078.

Yurovsky, D., Smith, L. B., & Yu, C. (2013). Statistical word learning at scale: the baby's view is better. *Developmental Science, 16*(6), 959–966.

Zosh, J. M., Brinster, M., & Halberda, J. (2013). Optimal contrast: Competition between two referents improves word learning. *Applied Developmental Science, 17*(1), 20–28.

10 Insights into Understanding Human Language from Children's Acquisition of Morphology and Syntax: A Historical and Current Perspective on Central Questions in the Field

SHANLEY E. M. ALLEN AND HEIKE BEHRENS

During their first year of life, children are already attending to relationships between words (syntax) and between parts of words (morphology) as they begin to make sense of the speech around them. As they near their second birthday, they move from producing single words in their own speech to putting those words together into sentences. Around their second birthday, children also begin marking words with morphemes including inflections—like the English plural *-s* and past tense *-ed*—that make clear the grammatical relevance of the words and provide information about their grammatical function within the sentence. While understanding and producing the basic ordering of words in simple sentences is mastered relatively early, acquiring the structure of more intricate and complex sentences including passives (*The cake was eaten by John*), relative clauses (*The cake that John ate was delicious*), and questions (*What did John eat?*) occupies children well into their school years. Mastering syntax and morphology constitutes a substantial part of children's language development from shortly after birth into adolescence.

The modern era of research on how children learn morphosyntax dates back to the late 1950s, beginning with longitudinal studies of individual children (e.g., Bloom, 1970; Brown, 1973). Since then, an extensive literature has emerged on the acquisition of numerous morphosyntactic phenomena including questions, negation, causatives, relative clauses, passives, antipassives, datives, word order, argument structure, argument realization, case marking, verbal inflection, tense and aspect marking, and the like. Although much of this research focuses on English-speaking children, studies have also examined the development of morphosyntax in other languages (e.g., Slobin, 1985–1997). Methods have expanded from the early focus on spontaneous speech and diary studies, to experimental methods including elicited production, act-out tasks, and preferential looking, to more recent on-line methods such as eye tracking, electroencephalography, and magnetoencephalography (e.g., Blom & Unsworth, 2010; Crain & Thornton, 1998; Hoff, 2012; McDaniel, McKee, & Smith Cairns, 1996). Thus, we have now amassed a reasonable understanding of the trajectory of development in several languages.

This research is not only interesting in its own right, but has also been at the center of debates about the nature and representation of human language. These debates initially focused on whether nativism or empiricism provides the best explanation of the acquisition of language structure, with roots dating back to the early philosophers. On the nativist side, Plato, most notably in his *Meno* dialogue (~400 B.C.), observed that humans have much more sophisticated knowledge than could be expected on the basis of experience alone and, thus, must be born with at least some innate knowledge (Ionescu, 2007). This is the foundation for the generativist tradition in language acquisition, which holds that syntactic knowledge is innate and represented in a language module in the brain, separate from modules representing other aspects of language and cognition (Chomsky, 1986; Fodor, 1983). The focus is on linguistic competence rather than performance;

children attend to the speech around them only to trigger language-appropriate settings for their already present syntax, and the communicative function of language holds little relevance. On the empiricist side, Locke, in his 1689 work *An Essay Concerning Human Understanding*, argued that all knowledge must instead come from experience gathered from communicative social interactions, which informs language learners about the structural patterns and their function in the target language (Locke & Nidditch, 1975). This is the foundation for the constructivist perspective in language acquisition (also known under labels such as usage-based or emergentist approaches to language development), which holds that no language-specific knowledge is innate, but rather that children come to the language-learning task with cognition-general abilities such as categorization and generalization skills (Tomasello, 2003). Children then use these abilities to analyze patterns in the language spoken around them, gradually moving from item-specific to abstract syntactic knowledge. Development of morphosyntax has played a crucial role in these nativism versus empiricism debates because morphosyntax constitutes knowledge that must be represented in and accessed from the mind, but has a complex enough structure that productive mastery of it cannot not be achieved simply by imitation from the input. Thus, it has been a major source for our understanding of how language is stored and accessed, what aspects of language and cognition are innate versus learned, what processes and mechanisms must underlie learning, and how the brain must be structured so that children get from no language to a solid command of their target language within only five short years.

Over the last decade, the sharp divide between these two perspectives has softened substantially in the light of increasing empirical data showing that neither perspective can be supported in its pure form. Thus, most researchers currently propose that learning language builds on prelinguistic representations and categorization and that children are dependent on learning from the speech input surrounding them for far more morphosyntactic information than simply triggering innate grammatical knowledge. Children are said to have weak abstract representations early on that strengthen considerably over time (Fisher, 2002; Tomasello & Abbot-Smith, 2002). As a result, the focus of research has now largely turned toward the processes and mechanisms of learning, including how children generalize from the input and what language-external and language-internal information children use to learn various aspects of morphosyntax.

In the present chapter, we review the literature on children's development of grammatical morphology and syntax that has been produced over the past almost 60 years. Many reviews of this literature have already appeared, both in textbooks at an introductory level (Berko Gleason & Bernstein Ratner, 2016; Brooks & Kempe, 2012; Clark, 2016, 2017; Gerken, 2009; Hoff, 2014; Kennison, 2014; Rowland, 2014; Saxton, 2010; Tomasello, 2003) or a more advanced level (e.g., Ambridge & Lieven, 2011; Guasti, 2017; O'Grady, 1997), and in handbooks targeted more at researchers in the field (e.g., Bavin & Naigles, 2015; De Villiers & Roeper, 2011; Fletcher & MacWhinney, 1995; Hoff & Shatz, 2007; Lidz, Snyder, & Pater, 2016; MacWhinney & O'Grady, 2015). We do not intend to repeat this material here in detail. Rather, we focus on the ways in which this literature has contributed to our understanding of human language by highlighting the major issues and questions that have driven the field since its inception, as well as the questions that are currently under debate and likely to drive the field for the next several years. In section 1, we offer a brief history of early approaches to morphosyntactic development, which largely focused on determining stages of development and testing generative theories of acquisition. The next three sections each highlight a current topic in the field and the insights it provides into morphosyntactic development. We conclude in section 5 by summarizing the main themes and proposing promising directions for the next stages in research.

1. Early Approaches to the Study of Morphosyntactic Development

1.1. Defining and Characterizing Stages of Development The first focus of early research was documenting the trajectory of development, determining stages in the trajectory, and exploring how best to characterize these stages. It became clear that children largely go through similar stages in a similar trajectory, but that the substantial individual differences in rate of development across children mean that age is not the best way to characterize stages. A pioneering study in the 1950s and 1960s used morphosyntactic indicators to delineate stages (Brown, 1973). Brown and his colleagues recorded longitudinal samples of the speech of three typically developing children—Adam, Eve, and Sarah—from age 1;6 to 4;0. Their development was divided into five stages according to four criteria: the mean length of utterance in morphemes, the longest utterance in morphemes, the order of acquisition of 14 morphemes with grammatical relevance (e.g., preposition *on*, plural *-s*, uncontractible auxiliary), and the order of acquisition of a set of syntactic structures (e.g., adultlike word order in two-word utterances, emergence of coordination). Further work focused in detail on stages within the development of particular structures including negation, questions, coordination, and relative clauses (see review in Zukowski, 2016).

Fairly soon, the focus of investigation shifted from simply characterizing the stages of development to trying to explain why development occurred in that particular order. Researchers sought to correlate stages in morphosyntactic development with metrics of cognitive complexity and/or linguistic complexity. For example, Brown (1973) found that English morphemes encoding only one semantic concept (e.g., plural *-s* encoding number, past *-ed* encoding tense) were acquired earlier than those encoding two concepts (e.g., third singular *-s* encoding both number and tense). Similarly, Klima and Bellugi (1966) found that yes/no questions—that require only one syntactic transformation (subject-auxiliary inversion) according to Chomsky's (1965) transformational grammar—were acquired earlier than wh-questions—that require two syntactic transformations (subject-auxiliary inversion and wh-movement). Although later research revealed that these analyses were too simplistic (e.g., Estigarribia, 2010), they nonetheless constituted an important step in the development of the field.

Overall, this early research was very useful in establishing that children progress in their morphosyntactic development through stages that are reasonably consistent across children and can be clearly characterized. These developmental stages also proved more useful than age for grouping children for research purposes, as well as for assessing children for language delay and deviance. However, this stage-based approach is limited in terms of providing insight about how and why acquisition progresses as it does and how cognitive architecture subsumes and supports language development.

1.2. Crosslinguistic Similarities and Differences Another key line of research beginning around the same time focused on testing whether the trajectory of morphosyntactic development is similar across languages (e.g., Hebrew: Berman, 1980; Hungarian: MacWhinney, 1975). In particular, researchers sought to understand which aspects of development are universal across all children regardless of the language they are learning—perhaps related to the commonalities of cognitive development and world knowledge across all children—and which are language particular in that they are affected by the structure of the particular language a child is learning (Slobin, 1982). For example, several researchers in the 1970s hypothesized that children use agent-action-object-recipient word order early on because this is the order in which an event happens in real life: an agent undertakes an action to convey an object to a recipient (e.g., McNeill, 1975). Slobin pointed out that, while this may seem accurate when only considering data from English-speaking children, it does not hold up in data from other languages. For example, children learning Turkish with default subject (S)–object (O)–verb (V) order typically use SOV from the earliest ages, even though it yields agent-object-action order.

Grammatical inflections provide another example: while they are mastered relatively late in English where the inflectional system is erratic and irregular, they are produced early and relatively error-free in languages where the inflectional system is systematic and regular (e.g., Turkish, see Aksu-Koç & Slobin, 1985; though note that more recent research reveals high error rates for low-frequency inflections in these languages [e.g., Aguado-Orea & Pine, 2015; Rubino & Pine, 1998]). Furthermore, the passive voice, which was assumed to be a late acquisition because of its structural and conceptual complexity on the basis of data from English, is in fact learned and used by age two in languages where it is less structurally complex and more frequent in the input (Allen & Crago, 1996; Demuth, 1989). Overall, crosslinguistic research has been crucial to our increasing understanding of the development of morphosyntax, particularly because the variation in structures across languages has facilitated deeper insight into the role of structure per se on language development (see Slobin, 1985–1997).

1.3. Testing Predictions from Generative Syntax The advent of Chomsky's generative theory of syntax in the 1960s (Chomsky, 1965, 1981, 1986, 1995; Chomsky & Lasnik, 1993) radically changed approaches to the study of language development. Research from this perspective focuses on investigating acquisition data in the light of linguistic theory, actively testing the claims and predictions of generative syntax in order to determine the nature of children's underlying grammatical knowledge. Chomsky's (1981) *principles and parameters theory*, the most influential for language acquisition research in this period, assumes an innate universal grammar (UG) across all human languages represented in a hierarchical treelike structure. Principles are grammatical rules followed by all languages; for example, a verb must have a subject. Parameters are options within the grammar that account for variation between languages; for example, subjects in a given language are either obligatorily present or allowed to be null. Chomsky assumed "poverty of the stimulus"—that the input is too impoverished to support the learning of grammar and thus that grammatical information must be innate (1986)—and modularity—that language acquisition proceeds through a modular "language acquisition device" and involves little interaction with other cognitive abilities (e.g., 1965). Many aspects of this theory have been extensively tested in children's development, focusing on children's grammatical competence rather than their performance. Central questions concern whether and how children's grammatical knowledge is consistent with an innate UG, how children map innate representations to the input, and what mechanisms account for early errors and the transition to adultlike production (Eisenbeiss, 2009; Lidz, 2010; Pinker, 1984).

The main problem for this approach was reconciling the predictions of the theory with the characteristics of children's early language productions. For example, a central tenet of the theory was the *null subject parameter*, which holds that all languages of the world have two possible settings for subjects: (i) subjects may be null, and (ii) subjects may not be null (Rizzi, 1982). According to the theory, English follows the second setting—subjects may not be null. However, contrary to the prediction, English-speaking children until age three or four years typically produce up to 50% null subjects (e.g., *want cookie* vs. *I want a cookie*; see review in Hyams, 2011). If UG is innate, including knowledge of parameters, how can we account for these errors? Hyams posited several theories about how children's null subject errors could be consistent with UG—for example, by following a default Italian-like parameter setting in which null subjects are allowed through licensing by verbal agreement (Hyams, 1986). Ultimately, however, none of these theories fit accurately with the empirical data—for example, Italian-speaking children consistently produce a higher percentage of null subjects than English-speaking children do and in a wider variety of syntactic contexts (Valian, 1991). A second problem was explaining how children switch from an early UG-consistent but incorrect grammar (i.e., null subjects) to the correct grammar for their language (i.e., no null subjects). The main hypothesis was that the initial default parameter setting must be "triggered" to the correct setting by consistent data in the input showing that subjects are obligatory (Wexler & Culicover, 1980), and several plausible triggers were found (e.g., pleonastic subjects as in *it is snowing*). However, the triggering hypothesis predicts instantaneous disappearance of the ungrammatical structure once triggering occurs, and empirical data showed gradual rather than sudden disappearance of null subjects. In addition, research from other theoretical perspectives increasingly revealed that English-speaking children's null subject errors could be explained by other means: strategies to reduce processing load (Bloom, 1990; Valian, 1990), sensitivity to discourse-pragmatics (see review in Allen, Hughes, & Skarabela, 2015), and distributional analysis of the input as shown through computational modeling (Freudenthal, Pine, & Gobet, 2007). Each of these alternative explanations undermined the singular focus on syntactic knowledge within the generative perspective and highlighted the important role of performance in addition to competence.

Despite its problems, research testing Chomsky's syntactic theories advanced the field in several ways. First, it focused the field on explanation rather than description, promoting a scientific, hypothesis-testing approach to studying morphosyntactic development and providing a rich understanding of the intricacies of the acquisition of numerous complex syntactic structures. Second, it led to a richer understanding of syntax in general and to a refining of syntactic principles and parameters. In particular, it was significant in leading to the eventual realization that the concept of parameters in explaining variation in language acquisition and language structure was not defensible when put to the test of real learners and real data. Third, it focused discussion on issues of continuity between child and adult grammars, on mechanisms to facilitate continuity under an assumption of innateness (i.e., triggering, maturation), and on what it means for a grammar to be "learnable." Fourth, it turned attention to children's underlying grammatical competence rather than solely attending to their performance.

However, it became increasingly clear that focusing on syntax in isolation, and on learnability rather than the actual learning trajectory and the processes underlying it, is not sufficient. The generative approach struggled to account for the fact that language is acquired gradually across several years rather than emerging fully adultlike at birth or suddenly changing due to maturation or the triggering of a particular piece of grammatical knowledge. Advances in neuroscience revealed brain structure and function much more in line with a broadly distributed network of activations rather than individual modules subserving individual areas of language (see Skeide, chapter 11 of this volume). It also became clear that factors such as experimental demands, processing constraints, input, and frequency play a significant role in how language is learned and in how children reveal their grammatical knowledge, all of which should have little effect under generative assumptions (Behrens, 2009). Finally, findings from corpus linguistic analyses of adult and child language have shown that many attested structures in language are not derivable from grammatical rules but are more or less idiosyncratic (see surveys in Bybee, 2010; Croft, 2001) and that early child language is much less productive than would be predicted by rule-based accounts (Tomasello, 2003). All of this led to increasing skepticism about the logical necessity of UG, as well as its empirical predictions. Thus, by the mid-1990s, the focus shifted away from investigating morphosyntactic acquisition within the confines of generative theory to explicit debates between generative approaches and the more empirically based information processing and constructivist approaches, which are based on alternative linguistic theories such as construction grammar (e.g., Tomasello, 1998).

1.4. Language Development as Information Processing The information processing approach to language acquisition developed gradually over the 1970s and 1980s, emerging from the rapidly developing field of computer science (Broeder & Murre, 2000; MacWhinney, 1987). The central metaphor is of a child as a computer, calculating probability distributions from the input and adjusting them with increasing experience to gradually reach adultlike competence in language comprehension and production. Research in statistical learning has underlined the validity of such an approach, showing that children are able to generalize from very small amounts of input if that input has the right properties (Gerken & Knight, 2015; see sections 2.2 and 4.1). Central questions include how morphosyntactic information is stored in the brain, how to best characterize mechanisms of language processing within the child, and what the roles of working memory and storage/access capacity are in morphosyntactic production.

The differences between the information processing and generative approaches have been perhaps most clearly articulated through the debate concerning how English past tense morphology is stored and processed (see review in Ambridge & Lieven, 2011, chapter 5.4). Generative researchers posited the dual-route model, in which two different routes for producing past tense verb forms are proposed. In one route, based on symbolic rules, the processor creates a past tense form by identifying an incoming item as a verb root and adding the suffix *-ed*. In the other, based on memory, typically only the exceptions to the rule are stored (e.g., *went, took*), as all other forms can be generated by rule when needed—a computationally more efficient option than storage (Pinker & Ullman, 2002; Yang, 2016). Since rule-based mechanisms are independent from storage, frequency should not play a role in the generation of rule-based forms (i.e., those with *-ed*; Clahsen, 1999). Information processing researchers, in contrast, claimed that all forms can be learned and produced through a single route based on probabilistic learning of forms through the input, as illustrated through a connectionist network (see review in McClelland & Patterson, 2002). Research pertinent to this debate focused on children's use of morphology on novel words (e.g., the *wug* test; Berko, 1958), children's overgeneralizations (e.g., *goed, mouses*), the U-shaped development of irregular forms (Marcus et al., 1992), the fact that the rate of overgeneralization is affected by both frequency and number of phonological neighbors, the correlation between correct production of past tense forms and vocabulary size, and crosslinguistic differences in which and how many form(s) serve as the default (e.g., Marcus, Brinkmann, Clahsen, Wiese, & Pinker, 1995).

This debate served as the debut for serious discussions about whether and how grammatical categories could be learned from the input alone through a connectionist-type network (Hahn & Nakisa, 2000; Rumelhart & McClelland, 1987). It became clear over the years that such a feat was possible and that computational models for how such a network would work in real children produced remarkably similar stages of acquisition and error patterns to those found in actual children (Maslen, Theakston, Lieven, & Tomasello, 2004; Ravid et al., 2008). This debate was key in revolutionizing our thinking about how much syntactic knowledge would have to be innate for children to have targetlike mastery of syntax by age five. The general assumption at present is that, while some rulelike formulation is clearly relevant, there is likely only one single route used in both the acquisition and production of grammatical morphology. The focus currently is on how exactly that system works in terms of both storage of information and access to that information (e.g., Gerken & Knight, 2015; Yang, 2016).

1.5. Constructing Grammar from Language Use The constructivist approach to language acquisition encompasses work from several related perspectives including social-pragmatic, usage-based, emergentist, and functionalist (e.g., Clark, 2016; Lieven & Brandt, 2011; MacWhinney, 1999; O'Grady, 2008; Tomasello, 2003). This approach has been present in the field since its onset but gained considerable traction in the 1990s and 2000s as a result of the increasing prominence of cognitive linguistics, which sees linguistic units as form–function pairings (Langacker, 1987). Within this approach, language is seen as a product of prelinguistic communicative abilities and general cognitive abilities such as categorization and schema formation. The actual utterance with its context-specific meaning is the basic unit for communication and the primary source of information for the language learner. Acquisition proceeds as the learner learns meaningful but potentially unanalyzed chunks from utterances (e.g., *I wanna X*), in order to code their intentions and eventually determine underlying patterns and derive abstract representations of linguistic categories. Central questions include what essential characteristics children attend to in the input, what mechanisms facilitate pattern finding, and what the role of interaction and the social context is in language learning.

The differences between the constructivist and generative approaches have been most clearly laid out in the debate around the learning of argument structure—the specification of the number and types of arguments required for a verb to be well formed (see reviews in Allen, 2015; Ambridge & Lieven, 2011, chapter 6). This work has focused on how children break into the syntactic system, also referred to as the "linking problem" (Ambridge, Pine, & Lieven, 2014), and when and how they develop abstract (rather than item-based) syntactic knowledge. Researchers following the generative approach assume that children have some innate language-specific knowledge about syntactic categories (e.g., noun, verb) and syntactic roles (e.g., subject, object) and then use that information to bootstrap themselves into an abstract representation of argument structure through either semantic bootstrapping (Pinker, 1984, 1989) or syntactic bootstrapping (Gleitman, 1990; Naigles, 1990). As evidence, elicited production studies show that children as young as 2;6 are sensitive to semantic classes, knowledge that can bootstrap them into the syntax of argument structure. In addition, preferential looking studies using video clips reveal that children as young as 1;9 are able to differentiate agent-only actions with novel verbs (e.g., *The duck and the bunny are gorping*) versus agent-patient actions (e.g., *The duck is gorping the bunny*), as well as other argument structures, on the basis of syntax alone (Naigles). However, verb and construction frequency play a clear role in generalization, which would not be predicted by the generative accounts (e.g., Matthews, Lieven, Theakston, & Tomasello, 2007).

Researchers in the constructivist tradition, in contrast, assume that children have no innate knowledge about grammar. Rather, they begin by creating item-specific argument structures on the basis of the input (e.g., *push* requires a *pusher* and a *pushee*) and then use cognition-general innate capacities such as generalization and pattern formation to gradually create abstract syntactic categorizations of the data. Here, numerous spontaneous speech analyses and elicited production studies show that children younger than 3;0 do not easily generalize one argument structure to another possible alternative (e.g., active to passive or vice versa) but that this ability increases with age (e.g., Savage, Lieven, Theakston, & Tomasello, 2003; see also Skeide & Friederici, 2016, for corresponding neurolinguistics evidence). Against this position is evidence suggesting that lack of generalization derives not from lack of abstract knowledge of argument structures but rather from the influence of priming (i.e., persistence in using the construction just heard), appropriate conservatism (i.e., knowledge that not all verbs alternate freely), or high performance demands of the methods used (i.e., producing an explicit response rather than simply showing comprehension) (e.g., Fisher, 2002).

This debate has been significant in fostering open discussion between the different theoretical positions based on empirical data, in gaining much more clarity about the issues at stake in the problem of how children could link the input to innate linguistic knowledge, and in uncovering much important information about how study results are affected by factors other than children's linguistic competence (e.g., memory, task demands). The general assumption now in constructivist theorizing is that even relatively young children (i.e., by 1;7) have weak abstract syntactic representations that can be revealed using methods with low performance demands, and that these representations strengthen with age (e.g., Chan, Meints, Lieven, & Tomasello, 2010; Dittmar, Abbot-Smith, Lieven, & Tomasello, 2011). However, although the constructivist approach accounts well for data from early stages of development, it has focused less on questions of how exactly children transition from lexical schema to abstract representations for more complex syntactic structures, and it has few specific predictions for how development from one stage to another should proceed (e.g., Rowland, Chang, Ambridge, Pine, & Lieven, 2012). Current research focuses on understanding how task demands, individual differences in cognitive abilities, and cross-linguistic structural differences affect children's performance in studies, how these can best be modulated to gain maximum insight into children's linguistic abilities, and how to explain developmental shifts (e.g., Kidd, 2012; Rowland et al., 2012).

In sum, research on children's acquisition of morphosyntax in the first 35 years or so of its modern history has provided many insights. We now know that children's morphosyntax progresses through stages of development that are remarkably similar across children, that language structure has a strong influence on children's developmental trajectory and error patterns, that children possess (weak) abstract grammatical representations as young as we can test them, and that children have relatively complex grammatical knowledge by age three or younger. The fierce debates about the priority of the roles of innate grammatical knowledge versus language input and use have also led to considerable advances. We have discovered that the input is far more central to the learning of morphosyntax than previously acknowledged, although some innate knowledge specific to language may still be necessary as a starting point. And we have found that morphosyntax is not acquired in an isolated way, but rather is acquired in interplay with other domains of language and cognition as well as contextual factors. At this point in the history of the field, attention has turned to several factors in morphosyntactic development that are relevant across generativist and constructivist theories. We highlight three of these in sections 2 to 4; research on the structure and function of the input, a focus on generalization as a key process in language learning, and a deepening in our understanding of the interconnectivity of language and cognition and, thus, the importance of integrating information from multiple linguistic and nonlinguistic domains.

2. *The Abstraction of Linguistic Structures from the Input*

Although there is still a theoretical divide in the field as to the nature and origin of linguistic representations, both formalist (i.e., generative) and functionalist (i.e., information processing, constructivist, emergentist) theories are currently interested in general cognitive processes in the acquisition of morphosyntactic relations and focus on children's pattern recognition and generalization abilities. In this context, the input takes on central importance because learners must extract information from the input in order to build linguistic categories. Thus, considerable research is now focused on understanding whether and how children are sensitive to the information in the input by looking at actual usage data and the affordances of the system with respect to the learnability of linguistic constructions. The main goal is to assess what characteristics of the input are used by children to learn both simple and complex grammar, with or without innate linguistic knowledge (Ibbotson, 2013; Lidz & Gagliardi, 2015). A major focus of constructivist accounts, for example, is the attention to the function of the utterance as a cue to form–function pairings or constructions (Goldberg, 2006), highlighting children's ability to find structural patterns and read intentions in order to identify the function of an utterance in the discourse context. Hence, children's social cognition and inferencing skills (e.g., joint attention, inference based on common ground; Clark, 2015) serve as major sources of information in identifying and parsing linguistic units.

2.1. The Language Learning Environment and the Role of Interaction The input or ambient language is the repository of distributional information and information on form-function relationships, and the area in which children can shape their own language learning environment through active participation in the interaction. Since the input is the primary source of information for the child, differences in the quality and quantity of the input should lead to differences in the learning outcome (Hart & Risley, 1995; Hoff, 2006; Hoff & Core, 2013; Huttenlocher, Vasilyeva, Cymerman, & Levine, 2002; Rowe & Salo, 2014). This includes characteristics of the linguistic richness children experience, both in terms of the total amount of input they hear, and in terms of the lexical and structural variability and complexity of the input. The main finding is that the quantity and quality of the input predicts the children's linguistic productivity in later childhood—not only the richness of their vocabulary, but also their mastery of syntactic complexity and the range of syntactic structures they command. For example, Huttenlocher et al. showed a correlation between parent and teacher input, on the one hand, and children's production and comprehension of complex multiclause sentences, on the other. Interactional style has an effect on language development as well, in terms of whether parents encourage children to elaborate and talk about a wide variety of topics, or whether the interaction is rather prohibitive and consists of warnings and reprimands (Hart & Risley). Also very relevant are pragmatic factors (e.g., the contingency between previous experience and the current situation; see Becker-Bryant, 2015) and the child's role in the interaction (e.g., children prefer child-directed speech over overheard speech; Weisleder & Fernald, 2013). Scaffolding through uptake and recasts further helps children to elaborate their productions, by, for example, learning paradigmatic contrasts in inflectional paradigms through reformulations (Clark, 2007; Clark & de Marneffe, 2012).

2.2. Recognition of Structural Patterns through Statistical Learning One central mechanism that children rely on to extract information from the input is statistical learning. In a seminal study, Saffran, Aslin, and Newport (1996) showed that when confronted with two minutes of continuous input of strings of three-syllable nonce words such as *golabu*, *bidaku*, and *padoti*, children as young as eight months can use simple transitional probabilities between syllables (which are higher within than between words) to segment words out of the input, as well as recognize sequences that had not been part of the input set (e.g., *tiladu*). Initially, these results were taken to demonstrate that children make use of innate linguistic knowledge (Marcus, Vijayan, Bandi Rao, & Vishton, 1999). However, further experiments showed that the same results can be obtained with nonlinguistic stimuli, and in other species. This means that children do not need to access innate linguistic representations, but they can recognize patterns and even "errors" within established patterns based on statistical/distributional information (Hudson Kam & Newport, 2005). Subsequently, children's ability to tolerate variation ("noise") in the data, as well as their ability to extract nonadjacent patterns from such sequences, was tested in order to identify the nature and scope of generalization over distributional evidence (Hudson Kam, 2015; Vuong, Meyer, & Christiansen, 2016). Most recently, Brusini et al. (2017) showed how children can use prosodic and phonological cues to identify word boundaries and function words, which provides them with a skeletal syntactic schema.

While it is well established by now that even infants can track different kinds of regularities in the speech stream, further investigation is needed as to what they generalize over, and in particular when and how they form morphosyntactic categories. Also, the facilitatory and inhibitory effects of correlated cues in naturalistic, more noisy but also more informative environments have to be tested further (see Romberg and Saffran, 2010, 2013; Monaghan, 2017, for reviews on current results and open issues; see also section 4.1 for a further discussion of multiple cue integration).

2.3. Extracting Form-Function Units Children's abilities to find patterns in strings of speech allows them to induce grammatical categories as well as syntactic schemas or skeletons. For example, it has been shown that distributional information such as the identity of a word's immediate left and right neighbor provides information about part-of-speech classification, semantic groups within the noun category, and verb–argument relations (Elman, 2011; Redington, Chater, & Finch, 1998). Likewise, inflectional markers and function words provide cues to part-of-speech classification (Willits, Seidenberg, & Saffran, 2014). Children also pick up nonadjacent constituents with a variable slot in the middle (slot and frame patterns: Braine, 1963; frequent morphological frames: Erkelens, 2009; Mintz, 2003; see sections 4.1 and 5 for more detail). Such frequent frames as *He's X-ing it* provide children with cues to syntactic categories. Childers and Tomasello (2001) demonstrated that children learn novel constructions more easily when the frame is stable (with a pronoun such as *it*) than if it is lexically variable (with a full noun phrase such as *the bathtub*). Importantly, these cues are probabilistic and interactive in that different cues can strengthen similar or same representations, or compete with one another. However, the availability and frequency of frequent frames is language specific (languages with freer word order have lower percentages), and their usefulness seems to be restricted to the early phases of development when children rely more heavily on adjacent, local information to abstract morphosyntactic categories and build syntactic schemas (Stumper, Bannard, Lieven, & Tomasello, 2011).

While the form of the utterance is clearly an important source of data in the input, the function of the utterance also plays a crucial role in extracting useful information. Tomasello (2003) saw intention reading as the second ingredient in language learning: children's pattern-finding skills are determined by the function, not just the distribution of elements in the clause. This is most obvious in the early stages of development when children produce constructional islands in which only a partial analysis of the event is made. For example, a transitive kicking event is reduced to the VO structure *kick X* (see section 3.2). In a series of traceback studies conducted on dense databases, it was shown that children's creative productions (utterances not previously attested in the same form) largely build on previously used blocks or patterns and vary them only slightly, by exchanging some words or connecting established strings (Lieven, Behrens, Speares, & Tomasello, 2003). This tendency is even more pronounced when the traceback is limited to sentence frames with a particular function, such as questions (Dąbrowska & Lieven, 2005). Here, two-year-olds, though not three-year-olds, seem to largely form their creative questions by combining two previously used strings or by slotting a new element into a previously established sentence frame.

It is also important to note that the units children extract and operate with need not correspond to formal constituents. Children also extract functional units or amalgams such as *wanna X*, a process referred to as *chunking* in the second language and corpus linguistic literature (Ellis, 2003; Sinclair, 1991). Crucially, chunks reduce the processing load because they are processed as units (see also Christiansen & Chater, 2016, chapter 5, for chunking in a connectionist modeling of acquisition).

In sum, with a better understanding of the genetic basis of language (Dediu & Christiansen, 2016; Fisher & Vernes, 2015), the versatility of children's learning abilities, and the richness of the input, we now have a very different view of the role of input than that proposed by Chomsky (1986). Learning language from the input is not only possible, but also necessary, especially if the proposed innate syntactic representations do not exist. Research is now focusing on understanding the nuances of how variations in input and access to input affect morphosyntactic development in later stages of development and how individual differences across children affect their use of the input (e.g., Dąbrowska, 2012; Hart & Risley, 1995; Huttenlocher et al., 2002).

3. Generalization Processes in Inductive Models: Deriving Language from Nonlanguage

A second current area of research interest is whether and how children are able to generalize information extracted from the input and transform it into abstract morphological and syntactic knowledge. The focus is largely on inductive models, which highlight how linguistic categories and representations are built on the basis of children's linguistic experience. Generalization is based on analogical reasoning ability and attested learning mechanisms (Behrens, 2017; Gentner & Smith, 2012) and can lead to crosslinguistic and intralinguistic variation. These domain-general learning abilities may be constrained by limited processing abilities that automatically focus children's attention on the relationship between adjacent units. The central question, given the affordances of the linguistic data available to the child from the input, as discussed in section 2, is whether children have appropriate cognitive capacities and processes in place to realistically facilitate abstraction of structure from input data alone.

3.1. Cognitive Underpinnings of Generalization Abilities: Working Memory, Storage, and Categorization Even in the prelinguistic period, children have a number of capacities that are relevant to—but not exclusive to—language learning, such as the pattern recognition and intention reading abilities referred to earlier. However, using these capacities to develop morphosyntactic knowledge requires a significant memory capacity as well as the ability to generalize over stored representations to form abstract categories and to extract rules that operate over these categories. In terms of memory, although children's working memory is limited at this point, that may actually be beneficial for language learning, as proposed by the "starting small" hypothesis (Elman, 1993; Newport, 1990), because it automatically focuses their attention on adjacent elements and the relationship between them (see also the related "short arms" hypothesis discussed by Kucker, chapter 9 of this volume, for word learning). Thus, the growth of the processor, rather than the growth of grammar as in nativist proposals, might allow for the extension of the amount and nature of data that can be taken into account when analyzing the speech stream (or other types of data, for that matter). For example, the computational model of Freudenthal, Pine, and Gobet (2010) showed how crosslinguistic differences in the developmental pattern of subject-verb agreement errors in English, Dutch, Spanish, and French could be explained largely in terms of an interaction between the distributional statistics of the language and the model's utterance final bias, which led it to learn from the ends of utterances first. Again, the processing abilities of the learner and the processability of the input interact: only those linguistic structures can survive that can eventually be learned by the next generation ("learning bottleneck"; Kirby, 2002).

Long-term memory is relevant as well because abstraction over experience requires storage of linguistic experience. Recent research shows that long-term memory has a key role in entrenchment—essentially the increasing memory strength of repeated occurrences—such that highly entrenched units are less likely to be subject to overgeneralization errors (Ambridge, Kidd, Rowland, & Theakston, 2015). These effects go beyond word forms, as storage affects the strength of transitional probabilities. Pattern abstraction can only take place if the units to be compared are stored in memory, and increasing memory strength leads to robustness in representation and automaticity in processing (Diessel, 2016). Arnon and Clark (2011) found that there are fewer overgeneralization errors when the inflected item is part of a well-entrenched string (*brush your teeth*). Likewise, children find it easier to process common multiword sequences than uncommon ones (*sit on a chair* vs. *sit on a rock*; Bannard & Matthews, 2008).

However, more research is needed on the nature of the categorization process and the categories children generalize over. For example, one key question is whether children are guided by the formal syntactic features and categories that are proposed by universalist accounts (Yang, Crain, Berwick, Chomsky, & Bolhuis, 2017) or whether semantic-pragmatic information paves the way for (initial) syntactic role assignment (cf. Chang, Dell, & Bock, 2006; Goldberg, 2004). From a developmental perspective, the question is whether and how these categories change over developmental time. For example, Connor, Fisher, and Roth (2013) proposed a model of syntax acquisition in which lexical knowledge and structural mapping interact over time to allow for increasingly specific syntactic representations and the disambiguation of ambiguous input. From a crosslinguistic perspective, the question is whether the semantic and syntactic categories are universal or language specific. There is no final consensus about whether children's early syntax follows adultlike syntactic principles based on a core set of innate and universal specifications (e.g., Yang et al., 2017) or whether grammatical representations are language specific (Croft, 2001), and possibly even subject to individual variation, as not all learners reach the same end state of linguistic representation (Dąbrowska, 2012).

3.2. Generalization for Simple and Complex Syntax A key finding of usage-based approaches to language acquisition is that children's early representations of simple syntactic structures is "concrete" and item-based (Tomasello, 1992, 2003). Children start out with units of a (partially) unanalyzed nature that do not have adultlike syntactic or semantic representation. Generalization or abstraction over such units, based on commonalities in both form and function, leads to increasingly abstract, productive structures (i.e., children can produce new utterances based on the generalizations they have formed, a process known as schema formation). Adultlike representations develop through growing experience and generalization by means of forming more and more abstract schemas. Limited productivity is attested when certain structures are used only in a small range of possible contexts and/or when children are unable to generalize their knowledge to new items. Such unproductive schemas, which are often referred to as "islands" because they are not yet connected to other schemas, typically develop around pronouns (Lieven, Pine, & Baldwin, 1997; McClure, Pine, & Lieven, 2006; Pine & Lieven, 1993) or verbs (Tomasello, 1992).

While it is widely accepted by now that children's morphosyntactic productivity may be limited in these early phases along the lines suggested by usage-based theorists, there is some disagreement about how limited it is exactly and about when exactly children go beyond the input (see Naigles, Hoff, & Vear, 2009; and the debate between Tomasello, 2000, and Fisher, 2002). Moreover, there is skepticism concerning whether the simple inductive learning mechanisms suffice to learn complex and abstract structures (Lidz & Gagliardi, 2015). However, several studies demonstrate how complex sentences or questions might be gradually learned as abstraction over their formal *and* functional properties. For example, early complex clauses are not only restricted in terms of their morphosyntactic complexity and lexical range, but also in terms of the number and nature of the propositions they encode (Diessel, 2004; Kidd, 2011). Early uses of complement taking verbs such as *I think it's in there* do not show signs of syntactic subordination because the complementizer *that* is typically missing, they are typically restricted to just a few subject-verb combinations and do not instantiate the full morphological paradigm, and they are semantically mono-propositional because they are used like adverbial epistemic markers such as *maybe*, but not as real statements of beliefs (Diessel).

It is also crucial to note that children do not have to start from scratch for every new structure, but instead they draw on their existing knowledge to make inferences to related structures. Elman (2003) demonstrated in a connectionist model how complex structural dependencies can be learned from their component parts, even if the structure itself has never been attested in the input. Abbot-Smith and Behrens (2006) showed how complex past participle constructions (present perfect and passive) are learned with differing ease depending on how well established and semantically contingent their component parts are. If knowledge can be transferred from one construction to the other, very quick learning processes are possible, as the step children have to make is only a small one. That is, development within a usage-based approach is dynamic: initially children need a lot of overtly encoded information that is contingent with a particular situation, but later they generalize over more abstract information, with a wider functional range.

The extension of complexity concerns not only the formal level, but also the conceptual and pragmatic level. Initially, children do not realize all arguments of an event, and they are typically not able to present events from different perspectives by using different verbs with their respective participant marking (Brooks & Tomasello, 1999). The development of further syntactic nuances and of being *linguistically* appropriate with respect to different genres and communicative tasks continues into adulthood (Berman, 2004, 2015; Nippold, 1988, 2006).

In sum, substantial research has shown that children have powerful generalization mechanisms in place and are able to use those to gradually transform chunks and frames extracted from the input into abstract syntactic structures. Current research continues to focus on how those processes work, especially for more complex syntactic structures.

4. *Integration of Multiple Sources of Information*

The third area that we see as a current driving force in understanding the acquisition of morphosyntax is research on how children integrate multiple sources of information in the process of comprehending and producing morphosyntactic structures. This includes information from other areas of language (e.g., phonology, lexicon, semantics, pragmatics), information regarding frequency and distribution of linguistic elements, information from the interactional and discourse context, and information from the physical context and from world knowledge. Although it was commonly assumed under the generative approach that each aspect of language is acquired without reference to the others, it has become clear over the years that a rich network of interacting knowledge is present in our mental representations. Thus, current research focuses on how this interacting knowledge drives the acquisition of morphosyntax, as well as on how children assess, integrate, and update this knowledge in online sentence interpretation and production.

4.1. Integrating Multiple Cues to Learn Syntactic Categories One key research area is how children use multiple cues in the input to identify syntactic categories such as noun and verb. This research has its roots in the statistical learning approach originally used to study word segmentation (Saffran et al., 1996). The first phase of statistical learning research established that children probabilistically track numerous individual cues in the input including those related to prosody, stress, distribution of elements in frequent frames, physical context (e.g., perceptual availability of objects), and social context (e.g., speaker's eye gaze) (see review in Gómez & Gerken, 2000). More recently, research has turned to how children bring together the information from each of these cues, including changes in how the information is integrated throughout development (Romberg & Saffran, 2010).

Research from this perspective on children's identification of syntactic categories capitalizes on the systematicity in mappings between form and meaning, finding that both phonological cues (e.g., consonant voicing, lexical stress, vowel quality) and distributional cues (e.g., co-occurrence relations) reliably signal syntactic categories of words (see reviews in Christiansen & Chater, 2016, chapter 5; Christiansen & Monaghan, 2016). Monaghan, Chater, and Christiansen (2005) used a set of 16 phonological cues, integrated together, to predict whether a given word was a noun (59% accuracy) or a verb (68% accuracy) in a large sample of English child-directed speech. Combining the phonological predictors with distributional cues (co-occurrence relations with the 20 most frequent words in the child-directed speech) improved the accuracy significantly (to 67% for nouns, 71% for verbs). A later study using 53 phonological predictors plus the same distributional cues reached 94% accuracy across English nouns and verbs (Monaghan, Christiansen, & Chater, 2007). Studies of child-directed speech in French and Japanese using the same 53 phonological predictors plus 20 language-appropriate distributional cues reached 91% accuracy for nouns and verbs in French, and 93% in Japanese (Monaghan et al., 2007). Phonologically based form-meaning systematicity is higher for early-learned words than for later-learned words, consistent with the idea that children need more early support in learning morphosyntax from information in the words themselves, before they become more skilled at using sources of information from outside the word (Monaghan, Shillcock, Christiansen, & Kirby, 2014). The phonological cues also seem to be better at predicting verbs than nouns, consistent with the finding that verbs are harder to learn than nouns and thus children need more support for learning verbs (Monaghan et al., 2005).

While it is clear that multiple cues to syntactic categories are present in the input, research on whether and how children actually use this information is still in its infancy. Fitneva, Christiansen, and Monaghan (2009) found that children in second grade (age seven to eight years) can use the 16 phonological cues from Monaghan et al. (2005) to predict whether a novel word identifies an entity (i.e., noun) or an event (i.e., verb), thus illustrating that children are sensitive to the cues in question (although these children are well past the initial stages of acquisition). Although studies with younger children are not yet available, several computational modeling studies have successfully used multiple-cue integration to simulate young children's learning of syntactic categories (e.g., Reali, Christiansen, & Monaghan, 2003).

4.2. Integrating Multiple Cues to Learn Syntactic Roles and Relations Another area where research on information integration has played a central role is children's identification of syntactic roles and relations. Research focuses on what cues children attend to for each role and relation, how children weight and balance information from different cues at different points in development, how cues either compete with or reinforce each other, and how all of these factors differ depending on the particular language a child is learning.

Much of this work is based on Bates and MacWhinney's (1987) *competition model*, in which children are assumed to be simultaneously attending to multiple cues to syntactic relations in the input (e.g., word order, case, agreement, animacy), including weighting their relative importance. Cues in this model differ across languages in reliability (how frequently the cue indicates the target function as opposed to another function), availability (how frequently the cue is present when the target function is present), and cost (how much burden the cue places on working memory). In more than 40 studies using the competition model across 18 languages, results have consistently shown that children weight the relevant cues according to the structure of their language, leading to crosslinguistic differences in the trajectory of learning syntactic roles and relations (MacWhinney, 2012). For example, English-speaking two-year-olds use word order (first noun) to determine which noun in a sentence is the subject or agent, because word order is a highly reliable, available, and low cost cue in English; case is reliable but rarely available, and animacy is available but less reliable. Two-year-olds learning Italian, however, rely more on animacy as a cue to subjecthood/agency because it is a more reliable and valid cue in Italian; word order is less reliable and often not available, while case is reliable but also often not available (Bates et al., 1984). Children pay most attention to highly available cues early in acquisition, but later rely more on highly reliable cues even if they are not highly available (Sokolov, 1988). More recent work in this tradition explores how children acquire different types of sentence frames. For example, two studies on the comprehension of transitive sentences with novel verbs have revealed that German- and Italian-speaking children first acquire those sentence frames where word order cues and case cues collaborate in leading to the correct interpretation (SOV with case-marked pronouns), are later in acquiring sentence frames where only one cue is present (SVO with non-case-marked noun phrases), and are worst at acquiring frames where cues compete with each other (OVS with case-marked pronouns) (Abbot-Smith & Serratrice, 2015; Dittmar, Abbot-Smith, Lieven, & Tomasello, 2008). Overall, this line of work shows that children are sensitive to several cues early on but develop gradually in their ability to deal effectively with conflicting cues or less available cues.

Numerous computational modeling studies have also investigated how children integrate multiple cues to learn syntactic relations, primarily inspired by research on syntactic and semantic bootstrapping (see section 1.5). For example, Connor et al. (2013) used *semantic role labeling* to develop a model that predicts semantic roles and syntactic relations in English child-directed speech. Their system uses a combination of cues deriving from the syntactic bootstrapping tradition including number of nouns in a sentence, position of nouns in the sentence, position of the verb, and animacy, as well as phonological cues that determine which words are nouns and which are verbs (see section 4.1). Alishahi and Stevenson (2010) used hierarchical Bayesian modeling to show how children could learn the semantics–syntax links relevant for syntactic relations on the basis of input data that is already specified for both semantic information (e.g., verb meaning, verb usage) and syntactic information (e.g., which words are nouns vs. verbs, number of arguments, order of arguments). Chang et al. (2006) developed a model more along the lines of semantic bootstrapping, using simple recurrent networks to learn semantics–syntax links in a system where semantic information is prespecified (e.g., semantic roles) but syntactic information is not. This model is based on incremental prediction of upcoming information of the sort typical in sentence processing models (see section 4.4), in which the weighting of each cue changes with exposure to the input. In general, computational models provide a much more finely grained understanding of how multiple cues could work together in determining syntactic roles and relations than we can discover from studies with actual children.

4.3. Integrating Discourse-Pragmatic Cues to Determine Referential Choice A different kind of example of information integration occurs in the research on referential choice in sentence production. This research grows out of the literature on factors influencing information flow in adult discourse, especially related to the form in which referents are realized in spontaneous speech (e.g., Ariel, 1990; Chafe, 1987). Central questions include whether children are sensitive to the same cues as adults, how information from the different cues is integrated, and how use of the cues develops with age.

Like adults, children are strongly influenced by a wide range of discourse-pragmatic accessibility cues in selecting referential forms, including prior mention in discourse, perceptual availability, joint attention, contrast, animacy, and the like (see review in Allen et al., 2015). Studies across some 15 different languages to date reveal that children as young as 18 months are sensitive to these accessibility cues, both individually and in combination, such that they select a more explicit referential form (e.g., lexical noun phrase) for a less accessible referent (e.g., newly introduced in discourse, not perceptually available, inanimate), and a less explicit form (e.g., pronoun, null form) for a more accessible referent (e.g., given in discourse, perceptually available, animate). Hughes and Allen (2015) offered striking evidence of how two- and three-year-old children integrate information from six different cues in their spontaneous productions such that they use incrementally more null and pronominal forms, and fewer lexical forms, as a function of how many cues a given referent is accessible for on a scale of 0 to 6. Interestingly, children seem to be sensitive to discourse-related cues (e.g., immediate prior mention in the conversation) earlier than to perceptually based cues (e.g., listener is looking at the referent) (e.g., Matthews, Lieven, Theakston, & Tomasello, 2006). Despite early sensitivity to discourse-pragmatic cues overall, experimental studies show that children up to at least an age of six years have trouble balancing information from two competing cues in determining referential choice. Furthermore, children show improvement in demanding tasks such as those requiring differentiating two similar referents (e.g., big cup vs. small cup) until at least age nine (e.g., Bahtiyar & Küntay, 2009).

Accessibility cues also affect the choice of syntactic role in which referents are expressed, according to the patterns of *preferred argument structure* (Du Bois, 1987). Newly introduced referents tend to appear as objects of transitive verbs and subjects of intransitive verbs, while the subject of a transitive verb role tends to be reserved for given referents. Children follow preferred argument structure in their spontaneous productions from as early as two years of age learning languages as diverse as Korean (Clancy, 1993, 2003), Inuktitut (Allen & Schröder, 2003), and Hindi (Narasimhan, Budwig, & Murty, 2005). Experimental studies also provide support for a preferred argument structure pattern in older children (Graf, Theakston, Lieven, & Tomasello, 2015; Theakston, 2012). In sum, the research on referential choice in production shows clearly that children integrate multiple cues from discourse-pragmatics in determining aspects of morphosyntax, but take longer to do so in situations of conflict or high communication demand.

4.4. Integrating Multiple Cues during Online Sentence Processing Children's sentence processing is another area in which integration of information from many sources plays a major role, including factors such as knowledge of the argument structure of a verb, frequency of occurrence of a word in a particular construction, prosody, perceptual and referential context, and world knowledge (see reviews in Snedeker, 2013; Snedeker & Huang, 2015; Trueswell & Gleitman, 2007). Key concerns include whether children attend to the same sources of information as adults, whether they process at the same speed as adults, and which features of the system emerge early versus late in development.

One strand of this research has looked at children's interpretation of temporarily ambiguous "garden path" sentences. The pioneering study here presented participants with commands such as *Put the frog on the napkin in the box,* accompanied by a visual display containing one frog sitting on a napkin, one empty napkin, one box, and one distractor item (Trueswell, Sekerina, Hill, & Logrip, 1999). Eye movements to objects in the display were tracked as the sentence unfolded in order to follow participants' incremental interpretation of the sentence. Adults initially understood the first prepositional phrase as the destination for the frog and then revised their interpretation after hearing the second prepositional phrase. Five-year-old children made the same initial interpretation as adults, but they were not able to revise their interpretation as the sentence unfolded. Furthermore, adults could use contextual information (e.g., an additional frog in the display) in their interpretation from the onset of the sentence, but five-year-olds could not. Thus, it appears that children at age five are influenced by lexical biases like adults are, but they are not yet able to integrate information from the discourse context into their sentence interpretation, and they cannot yet use information provided later in the sentence to update an initial interpretation. Several later studies have yielded similar results concerning contextual cues, but they also show that five-year-olds can use prosody and noun plausibility information in their interpretations, albeit at a slower speed than adults can (Snedeker & Huang, 2015). A further line of studies focuses on pronoun interpretation in ambiguous contexts (e.g., *Marion went with Elizabeth to the store, and she bought some candy*) (see review in Sekerina, 2015). These studies show that children aged two to nine years use similar cues as adults do in interpreting the subject pronoun—first-mentioned noun, animacy, gender, and newness among others. However, children take more time than adults do to reach their interpretation, have more trouble than adults do integrating information from competing cues, and tend to focus only on the most reliable cues at younger ages.

Child-adult differences in integrating multiple cues during sentence processing may stem from a variety of sources (see discussion in Snedeker & Huang, 2015). One source is nonlinguistic cognitive abilities; children typically have poorer working memory, less cognitive flexibility, and worse selective attention than adults do, leading to more difficulty integrating several cues simultaneously at a speed fast enough to aid in sentence parsing. A second source appears to be preference for stronger cues, in that children attend to more reliable cues such as lexical verb bias early on and attend only later to less sure cues such as number of nouns in a sentence. Finally, children seem to be much better at "bottom-up" than "top-down" processing. In other words, they can more easily integrate information that appears at a lower level in processing related to syntactic parsing (e.g., prosodic cues, lexical biases such as argument structure), but they have more trouble integrating information that appears at a higher level in the processing hierarchy (e.g., semantic information, referential context).

In sum, several different directions of research reveal that children can effectively integrate information from multiple cues in the input to identify syntactic elements in the input, to gradually uncover the relationship between these elements, and to gradually hone this knowledge throughout development. They attend to similar cues to adults, though they seem to attend first to highly available cues (rather than reliable but less available cues), discourse-based cues (rather than perceptually based cues), strong cues (rather than weak cues), and bottom-up cues (rather than top-down cues). In addition, their integration of these cues is generally slower than that of adults, they have more trouble managing competing cues or integrating cues in situations of high communicative demand, and they have more trouble revising initial incorrect interpretations. Current work continues to focus in more precise detail on which types of information children can integrate in their sentence production and interpretation, the time course of their use of such information, how their abilities develop with age, and why children's processing differs from adult processing. As we learn more, it will be interesting to directly compare cue integration in production versus comprehension since the research just reviewed hints that children may show earlier facility in using multiple cues in production than in comprehension.

5. Outlook

Research on the acquisition of morphology and syntax in children has contributed substantially to our understanding of human language. General cognitive abilities such as pattern recognition and the ability to identify relational analogies across items, as well as the ability to determine statistical information from the input, play a major role in language development. They allow children to generalize from simple patterns to abstract syntax. Language structure also has a strong influence on how children learning different languages attend to the ambient language and what they extract from it most easily. Furthermore, children do not attend to morphology or syntax in isolation, but they are constantly integrating this information with that from other areas of language, pragmatics, and extralinguistic contextual information. All of these abilities are dependent on cognitive underpinnings such as working memory, processing speed, and other executive functions.

Building on this rich foundation, many questions are still open. Research on the nature and influence of the input on morphosyntactic development will help us to understand what exactly in the input is crucial for development and how much input quantity and quality is needed to warrant successful language acquisition. It is important to note that the research so far still focuses on formal features and their distribution as the determinant of morphosyntactic development. Although many current theories emphasize the role of communicative function and intention reading, it is not clear yet how these cues affect children's morphosyntactic parsing. Research on generalization will shed further light on the subtle mechanisms children use to abstract from the input and what that reveals about the roles of memory, storage, and access in morphosyntactic development. Research on information integration will reveal much about children's attention and inhibition mechanisms, the role of top-down versus bottom-up processing, and the timing of which cues come on line earlier versus later. We expect to gain further insight into the exact relationship between the fairly stable developmental processes and stages brought out in the first wave of acquisition research, and the very subtle fine tuning in individual differences, including which differences level out over time and which lead to delays or are indicators of atypical development. We also expect that newer methods including big data analysis and real-time language processing experiments will lead to a deeper understanding of the processes.

REFERENCES

Abbot-Smith, K., & Behrens, H. (2006). How known constructions influence the acquisition of other constructions: The German periphrastic passive and future constructions. *Cognitive Science, 30*, 995–1026.

Abbot-Smith, K., & Serratrice, L. (2015). Word order, referential expression, and case cues to the acquisition of transitive sentences in Italian. *Journal of Child Language, 42*(1), 1–31.

Aguado-Orea, J., & Pine, J. M. (2015). Comparing different models of the development of verb inflection in early child Spanish. *PLOS ONE, 10*(3), e0119613.

Aksu-Koç, A., & Slobin, D. I. (1985). The acquisition of Turkish. In D. I. Slobin (Ed.), *The crosslinguistic study of language acquisition* (Vol. 1, pp. 839–878). Hillsdale, NJ: Erlbaum.

Alishahi, A., & Stevenson, S. (2010). A computational model of learning semantic roles from child-directed language. *Language and Cognitive Processes, 25*(1), 50–93.

Allen, S. E. M. (2015). Verb argument structure. In E. L. Bavin & L. R. Naigles (Eds.), *The Cambridge handbook of child language* (2nd ed., pp. 271–297). Cambridge: Cambridge University Press.

Allen, S. E. M., & Crago, M. B. (1996). Early passive acquisition in Inuktitut. *Journal of Child Language, 23*(1), 129–155.

Allen, S. E. M., Hughes, M. E., & Skarabela, B. (2015). The role of cognitive accessibility in children's referential choice. In L. Serratrice & S. E. M. Allen (Eds.), *The acquisition of reference* (pp. 123–153). Amsterdam: Benjamins.

Allen, S. E. M., & Schröder, H. (2003). Preferred argument structure in early Inuktitut spontaneous speech data. In J. W. Du Bois, L. E. Kumpf, & W. J. Ashby (Eds.), *Preferred argument structure: Grammar as architecture for function* (pp. 301–338). Amsterdam: Benjamins.

Ambridge, B., Kidd, E., Rowland, C. F., & Theakston, A. L. (2015). The ubiquity of frequency effects in first language acquisition. *Journal of Child Language, 42*(2), 239–273.

Ambridge, B., & Lieven, E. V. M. (2011). *Child language acquisition: Contrasting theoretical approaches.* Cambridge: Cambridge University Press.

Ambridge, B., Pine, J. M., & Lieven, E. V. M. (2014). Child language acquisition: Why universal grammar doesn't help. *Language, 90*(3), e53–e90.

Ariel, M. (1990). *Accessing noun-phrase antecedents.* London: Routledge.

Arnon, I., & Clark, E. V. (2011). Why *brush your teeth* is better than *teeth*: Children's word production is facilitated in familiar sentence-frames. *Language Learning and Development, 7*(2), 107–129.

Bahtiyar, S., & Küntay, A. C. (2009). Integration of communicative partner's visual perspective in patterns of referential requests. *Journal of Child Language, 36,* 529–555.

Bannard, C., & Matthews, D. E. (2008). Stored word sequences in language learning: The effect of familiarity on children's repetition of four-word combinations. *Psychological Science, 19,* 241–248.

Bates, E., & MacWhinney, B. (1987). Competition, variation, and language learning. In B. MacWhinney (Ed.), *Mechanisms of language acquisition* (pp. 157–193). Hillsdale, NJ: Erlbaum.

Bates, E., MacWhinney, B., Caselli, C., Devescovi, A., Natale, F., & Venza, V. (1984). A crosslinguistic study of the development of sentence interpretation strategies. *Child Development, 55,* 341–354.

Bavin, E. L., & Naigles, L. R. (Eds.). (2015). *The Cambridge handbook of child language* (2nd ed.). Cambridge: Cambridge University Press.

Becker-Bryant, J. (2015). Pragmatic development. In E. L. Bavin & L. R. Naigles (Eds.), *The Cambridge handbook of child language* (2nd ed., pp. 438–457). Cambridge: Cambridge University Press.

Behrens, H. (2009). Usage-based and emergentist approaches to language acquisition. *Linguistics, 47*(2), 383–411.

Behrens, H. (2017). The role of analogy in language processing and acquisition. In M. Hundt, S. Mollin, & S. Pfenninger (Eds.), *The changing English language: Psycholinguistic perspectives* (pp. 125–239). Cambridge: Cambridge University Press.

Berko, J. (1958). The child's learning of English morphology. *Word, 14,* 150–177.

Berko Gleason, J., & Bernstein Ratner, N. (Eds.). (2016). *The development of language* (9th ed.). Boston: Pearson.

Berman, R. A. (1980). Child language as evidence for grammatical description: Preschoolers' construal of transitivity in the Hebrew verb system. *Linguistics, 18,* 677–701.

Berman, R. A. (Ed.) (2004). *Language development across childhood and adolescence.* Amsterdam: Benjamins.

Berman, R. A. (2015). Language development and use beyond the sentence. In E. L. Bavin, & L. R. Naigles (Eds.), *The Cambridge handbook of child language* (2nd ed., pp. 458–480). Cambridge: Cambridge University Press.

Blom, E., & Unsworth, S. (Eds.). (2010). *Experimental methods in language acquisition research.* Amsterdam: Benjamins.

Bloom, L. (1970). *Language development: Form and function in emerging grammars.* Cambridge, MA: MIT Press.

Bloom, P. (1990). Subjectless sentences in child language. *Linguistic Inquiry, 21,* 491–504.

Braine, M. D. S. (1963). On learning the grammatical order of words. *Psychological Review, 70,* 323–338.

Broeder, P., & Murre, J. (Eds.). (2000). *Models of language acquisition: Inductive and deductive approaches.* Oxford: Oxford University Press.

Brooks, P. J., & Kempe, V. (2012). *Language development.* Chichester, UK: BPS Blackwell.

Brooks, P. J., & Tomasello, M. (1999). Young children learn to produce passives with nonce verbs. *Developmental Psychology, 35,* 29–44.

Brown, R. (1973). *A first language: The early stages.* Cambridge, MA: Harvard University Press.

Brusini, P., Carvalho, A., Dautriche, I., Gutman, A., Cauvet, E., Millotte, S., ... Christophe, A. (2017). Bootstrapping lexical and syntactic acquisition In M. Kail, M. Hickman, & E. Veneziano (Eds.), *Sources of variations in first language acquisition: Languages, contexts, and learners* (pp. 63–80). Amsterdam: Benjamins.

Bybee, J. L. (2010). *Language, usage and cognition.* Cambridge: Cambridge University Press.

Chafe, W. (1987). Cognitive constraints on information flow. In R. Tomlin (Ed.), *Coherence and grounding in discourse* (pp. 21–51). Amsterdam: Benjamins.

Chan, A., Meints, K., Lieven, E., & Tomasello, M. (2010). Young children's comprehension of English SVO word order revisited: Testing the same children in act-out and intermodal preferential looking tasks. *Cognitive Development,* 25, 30–45.

Chang, F., Dell, G. S., & Bock, K. (2006). Becoming syntactic. *Psychological Review, 113*(2), 234–272.

Childers, J. B., & Tomasello, M. (2001). The role of pronouns in young children's acquisition of the English transitive construction. *Child Development, 37,* 739–748.

Chomsky, N. (1965). *Aspects of the theory of syntax.* Cambridge, MA: MIT Press.

Chomsky, N. (1981). *Lectures on government and binding.* Dordrecht, Netherlands: Foris.

Chomsky, N. (1986). *Knowledge of language: Its nature, origin, and use.* New York: Praeger.

Chomsky, N. (1995). *The minimalist program.* Cambridge, MA: MIT Press.

Chomsky, N., & Lasnik, H. (1993). Principles and parameters theory. In J. Jacobs, A. von Stechow, W. Sternefeld, & T. Vennemann (Eds.), *Syntax: An international handbook of contemporary research* (Vol. 1, pp. 506–569). Berlin: Mouton de Gruyter.

Christiansen, M. H., & Chater, N. (2016). *Creating language: Integrating evolution, acquisition, and processing.* Cambridge, MA: MIT Press.

Christiansen, M., & Monaghan, P. (2016). Division of labor in vocabulary structure: Insights from corpus analyses. *Topics in Cognitive Science, 8*(3), 610–624.

Clahsen, H. (1999). Lexical entries and rules of language: A multidisciplinary study of German inflection. *Behavioral and Brain Sciences, 22*, 991–1060.

Clancy, P. M. (1993). Preferred argument structure in Korean acquisition. In E. V. Clark (Ed.), *Proceedings of the 25th Annual Child Language Research Forum* (pp. 307–314). Stanford, CA: CSLI.

Clancy, P. M. (2003). The lexicon in interaction: Developmental origins of preferred argument structure in Korean. In J. W. Du Bois, L. E. Kumpf, & W. J. Ashby (Eds.), *Preferred argument structure: Grammar as architecture for function* (pp. 81–108). Amsterdam: Benjamins.

Clark, E. V. (2007). Children's uptake of new words in conversation. *Language in Society, 36*, 157–182.

Clark, E. V. (2015) Common ground. In B. MacWhinney & W. O'Grady (Eds.), *The handbook of language emergence* (pp. 328–353). London: Wiley-Blackwell.

Clark, E. V. (2016). *First language acquisition* (3rd ed.). Cambridge: Cambridge University Press.

Clark, E. V. (2017). *Language in children.* New York: Routledge.

Clark, E. V., & de Marneffe, M.-C. (2012). Constructing verb paradigms in French: Adult construals and emerging grammatical contrasts. *Morphology, 22*(1), 89–120.

Connor, M., Fisher, C., & Roth, D. (2013). Starting from scratch in semantic role labeling: Early indirect supervision. In T. Poibeau, A. Villavicencio, A. Korhonen, & A. Alishahi (Eds.), *Cognitive aspects of computational language acquisition* (pp. 257–296). Berlin: Springer.

Crain, S., & Thornton, R. (1998). *Investigations in universal grammar: A guide to experiments on the acquisition of syntax and semantics.* Cambridge, MA: MIT Press.

Croft, W. (2001). *Radical construction grammar: Syntactic theory in typological perspective.* Oxford: Oxford University Press.

Dąbrowska, E. (2012). Different speakers, different grammars: Individual differences in native language attainment. *Linguistic Approaches to Bilingualism, 2*(3), 219–253.

Dąbrowska, E., & Lieven, E. (2005). Towards a lexically specific grammar of children's question constructions. *Cognitive Linguistics, 16*(3), 437–474.

Dediu, D., & Christiansen, M. H. (2016). Language evolution: Constraints and opportunities from modern genetics. *Topics in Cognitive Science, 8*(2), 361–370.

Demuth, K. (1989). Maturation and the acquisition of the Sesotho passive. *Language,* 65, 56–80.

De Villiers, J., & Roeper, T. (Eds.). (2011). *Handbook of generative approaches to language acquisition.* Dordrecht, Netherlands: Springer.

Diessel, H. (2004). *The acquisition of complex sentences.* Cambridge: Cambridge University Press.

Diessel, H. (2016). Frequency and lexical specificity in grammar: A critical review. In H. Behrens, & S. Pfänder (Eds.), *Experience counts: Frequency effects in language* (pp. 209–237). Berlin: Mouton de Gruyter.

Dittmar, M., Abbot-Smith, K., Lieven, E., & Tomasello, M. (2008). German children's comprehension of word order and case marking in causative sentences. *Child Development, 79*(4), 1152–1167.

Dittmar, M., Abbot-Smith, K., Lieven, E., & Tomasello, M. (2011). Children aged 2;1-year use transitive syntax to make a semantic-role interpretation in a pointing task. *Journal of Child Language, 38*, 1109–1123.

Du Bois, J. W. (1987). The discourse basis of ergativity. *Language, 63*, 805–855.

Eisenbeiss, S. (2009). Generative approaches to language learning. *Linguistics, 47*, 273–310.

Ellis, N. C. (2003). Constructions, chunking and connectionism: The emergence of second language structure. In C. J. Doughty & M. H. Long (Eds.), *The handbook of second language acquisition* (pp. 63–103). Malden, MA: Blackwell.

Elman, J. L. (1993). Learning and development in neural networks: The importance of starting small. *Cognition, 14*, 179–211.

Elman, J. L. (2003). Generalization from sparse input. In M. Adronis, E. Debenport, A. Pycha, & K. Yoshimura (Eds.), *Proceedings of the 38th Meeting of the Chicago Linguistics Society* (pp. 175–200). Chicago: Chicago Linguistics Society.

Elman, J. L. (2011). Lexical knowledge without a lexicon? *Mental Lexicon, 6*(1), 1–33.

Erkelens, M. (2009). *Learning to categorize verbs and nouns: Studies on Dutch.* Leiden, Netherlands: LOT.

Estigarribia, B. (2010). Facilitation by variation: Right-to-left learning of English yes/no questions. *Cognitive Science, 34*(1), 68–93.

Fisher, C. (2002). The role of abstract syntactic knowledge in language acquisition: A reply to Tomasello (2000). *Cognition, 82*, 259–278.

Fisher, S. E., & Vernes, S. C. (2015). Genetics and the language sciences. *Annual Review of Linguistics, 1*(1), 289–310.

Fitneva, S. A., Christiansen, M. H., & Monaghan, P. (2009). From sound to syntax: Phonological constraints on children's lexical categorization of new words. *Journal of Child Language, 36*(5), 967–997.

Fletcher, P., & MacWhinney, B. (1995). *The handbook of child language.* Cambridge, MA: Blackwell.

Fodor, J. A. (1983). *The modularity of mind.* Cambridge, MA: MIT Press.

Freudenthal, D., Pine, J. M., & Gobet, F. (2007). Understanding the developmental dynamics of subject omission: The role of processing limitations in learning. *Journal of Child Language, 34*(1), 83–110.

Freudenthal, D., Pine, J. M., & Gobet, F. (2010). Explaining quantitative variation in the rate of optional infinitive errors across languages: A comparison of MOSAIC and the variational learning model. *Journal of Child Language, 37*(3), 643–669.

Gentner, D., & Smith, L. L. (2012). Analogical reasoning. In V. S. Ramachandran (Ed.), *Encyclopedia of human behavior* (2nd ed., pp. 130–136). Oxford: Elsevier.

Gerken, L. A. (2009). *Language development.* San Diego, CA: Plural.

Gerken, L., & Knight, S. (2015). Infants generalize from just (the right) four words. *Cognition, 143*, 187–192.

Gleitman, L. (1990). The structural sources of verb meanings. *Language Acquisition, 1*, 3–55.

Goldberg, A. E. (2004). But do we need universal grammar? Comment on Lidz et al. (2003). *Cognition, 94*, 77–84.

Goldberg, A. E. (2006). *Constructions at work: The nature of generalization in language.* Oxford: Oxford University Press.

Gómez, R. L., & Gerken, L. (2000). Infant artificial language learning and language acquisition. *TRENDS in Cognitive Sciences*, *4*(5), 178–186.

Graf, E., Theakston, A., Lieven, E. V. M., & Tomasello, M. (2015). Subject and object omission in children's early transitive constructions: A discourse-pragmatic approach. *Applied Psycholinguistics*, *36*(3), 701–727.

Guasti, M. T. (2017). *Language acquisition: The growth of grammar* (2nd ed.). Cambridge, MA: MIT Press.

Hahn, U., & Nakisa, R. C. (2000). German inflection: Single route or dual route? *Cognitive Psychology*, *41*, 313–360.

Hart, B., & Risley, T. R. (1995). *Meaningful differences in the everyday experience of young American children.* Baltimore, MD: Brookes Publishers.

Hoff, E. (2006). How social contexts support and shape language development. *Developmental Review*, *26*, 55–88.

Hoff, E. (Ed.) (2012). *Research methods in child language: A practical guide.* Oxford: Wiley.

Hoff, E. (2014). *Language development* (5th ed.). Belmont, CA: Cengage Learning.

Hoff, E., & Core, C. (2013). Input and language development in bilingually developing children. *Seminars in Speech and Language*, *34*(4), 215–226.

Hoff, E., & Shatz, M. (Eds.). (2007). *Blackwell handbook of language development.* Malden, MA: Blackwell.

Hudson Kam, C. L. (2015). The impact of conditioning variables on the acquisition of variation in adult and child learners. *Language*, *91*(4), 906–937.

Hudson Kam, C. L., & Newport, E. L. (2005). Regularizing unpredictable variation: The roles of adult and child learners in language formation and change. *Language Learning and Development*, *1*(2), 151–195.

Hughes, M., & Allen, S. E. M. (2015). The incremental effect of discourse-pragmatic sensitivity on referential choice in the acquisition of a first language. *Lingua*, *155*, 43–61.

Huttenlocher, J., Vasilyeva, M., Cymerman, E., & Levine, S. (2002). Language input and child syntax. *Developmental Psychology*, *45*, 337–374.

Hyams, N. (1986). *Language acquisition and the theory of parameters.* Dordrecht, Netherlands: Reidel.

Hyams, N. (2011). Missing subjects in early child language. In J. De Villiers & T. Roeper (Eds.), *Handbook of generative approaches to language acquisition* (pp. 13–52). Dordrecht, Netherlands: Springer.

Ibbotson, P. (2013). The scope of usage-based theory. *Frontiers in Psychology*, *4*, 255.

Ionescu, C. (2007). *Plato's Meno: An interpretation.* Plymouth, MA: Lexington Books.

Kennison, S. M. (2014). *Introduction to language development.* Thousand Oaks, CA: Sage.

Kidd, E. (Ed.) (2011). *The acquisition of relative clauses: Processing, typology, and function.* Amsterdam: John Benjamins.

Kidd, E. (2012). Individual differences in syntactic priming in language acquisition. *Applied Psycholinguistics*, 33, 393–418.

Kirby, S. (2002). Learning bottlenecks and the evolution of recursive syntax. In T. Briscoe (Ed.), *Linguistic evolution through language acquisition* (pp. 173–203). Cambridge: Cambridge University Press.

Klima, E. S., & Bellugi, U. (1966). Syntactic regularities in the speech of children. In J. Lyons, & R. J. Wales (Eds.), *Psycholinguistics papers: The proceedings of the 1966 Edinburgh conference* (pp. 183–208). Edinburgh: Edinburgh University Press.

Langacker, R. W. (1987). *Foundations of cognitive grammar. Vol. 1. Theoretical prerequisites.* Stanford, CA: Stanford University Press.

Lidz, J. (2010). Language learning and language universals. *Biolinguistics*, *4*(2–3), 201–213.

Lidz, J., & Gagliardi, A. (2015). How nature meets nurture: Universal Grammar and statistical learning. *Annual Review of Linguistics*, *1*, 333–353.

Lidz, J., Snyder, W., & Pater, J. (Eds.) (2016). *The Oxford handbook of developmental linguistics.* Oxford: Oxford University Press.

Lieven, E. V. M., Behrens, H., Speares, J., & Tomasello, M. (2003). Early syntactic creativity: a usage-based approach. *Journal of Child Language*, *30*, 33--370.

Lieven, E. V. M., & Brandt, S. (2011). The constructivist approach. *Infancia y Aprendizaje*, *34*(3), 281–296.

Lieven, E. V. M., Pine, J. M., & Baldwin, G. (1997). Lexically-based learning and early grammatical development. *Journal of Child Language*, *24*(1), 187–219.

Locke, J., & Nidditch, P. H. (1975). *The Clarendon edition of the works of John Locke.* Oxford: Clarendon Press.

MacWhinney, B. (1975). Rules, rote, and analogy in morphological formations by Hungarian children. *Journal of Child Language*, *2*, 65–77.

MacWhinney, B. (Ed.). (1987). *Mechanisms of language acquisition.* Hillsdale, NJ: Erlbaum.

MacWhinney, B. (Ed.). (1999). *The emergence of language.* Mahwah, NJ: Erlbaum.

MacWhinney, B. (2012). The logic of the unified model. In S. Gass & A. Mackey (Eds.), *Handbook of second language acquisition* (pp. 211–227). New York: Routledge.

MacWhinney, B., & O'Grady, W. (Eds.). (2015). *The handbook of language emergence.* London: Wiley-Blackwell.

Marcus, G. F., Brinkmann, U., Clahsen, H., Wiese, R., & Pinker, S. (1995). German inflection: The exception that proves the rule. *Cognitive Psychology*, *29*, 186–256.

Marcus, G. F., Pinker, S., Ullman, M., Hollander, M. Rosen, T. J., Xu, F. & Clahsen, H. (1992). Overregularization in language acquisition. *Monographs of the Society for Research in Child Development*, *57*(4), 1–178.

Marcus, G. F., Vijayan, S., Bandi Rao, S., & Vishton, P. M. (1999). Rule learning by seven-month-old infants. *Science*, *283*, 77–80.

Maslen, R. J. C., Theakston, A. L., Lieven, E. V. M., & Tomasello, M. (2004). A dense corpus study of past tense and plural overregularization in English. *Journal of Speech Language and Hearing Research*, *47*(6), 1319–1333.

Matthews, D., Lieven, E., Theakston, A., & Tomasello, M. (2006). The effect of perceptual availability and prior discourse on young children's use of referring expressions. *Applied Psycholinguistics*, *27*(3), 403–422.

Matthews, D., Lieven, E., Theakston, A., & Tomasello, M. (2007). French children's use and correction of weird word orders: A constructivist account. *Journal of Child Language*, *34*(2), 381–409.

McClelland, J., & Patterson, K. (2002). Rules or connections in past tense inflection: What does the evidence rule out? *Trends in Cognitive Sciences*, *6*(11), 465–472.

McClure, C., Pine, J. M., & Lieven, E. V. M. (2006). Investigating the abstractness of children's early knowledge of argument structure. *Journal of Child Language*, *33*(4), 693–720.

McDaniel, D., McKee, C., & Smith Cairns, H. (Eds.). (1996). *Methods for assessing children's syntax.* Cambridge, MA: MIT Press.

McNeill, D. (1975). The place of grammar in a theory of performance. *Annals of the New York Academy of Sciences, 263*(1), 171–177.
Mintz, T. H. (2003). Frequent frames as a cue for grammatical categories in child directed speech. *Cognition, 90*(1), 91–117.
Monaghan, P. (2017). Canalization of language structure from environmental constraints: A computational model of word learning from multiple cues. *Topics in Cognitive Science, 9*(1), 21–34.
Monaghan P., Chater N., & Christiansen M. (2005). The differential role of phonological and distributional cues in grammatical categorization. *Cognition, 96,* 143–182.
Monaghan, P., Christiansen, M. H., & Chater, N. (2007). The phonological-distributional coherence hypothesis: Cross-linguistic evidence in language acquisition. *Cognitive Psychology, 55*(4), 259–305.
Monaghan, P., Shillcock, R. C., Christiansen, M. H., & Kirby, S. (2014). How arbitrary is language? *Philosophical Transactions of the Royal Society B, 369*(1651), 20130299.
Naigles, L. (1990). Children use syntax to learn verb meanings. *Journal of Child Language,* 17, 357–374.
Naigles, L. R., Hoff, E., & Vear, D. (2009). Flexibility in early verb use: Evidence from a multiple-n diary study. *Monographs of the Society for Research in Child Development, 74*(2), 1–112.
Narasimhan, B., Budwig, N., & Murty, L. (2005). Argument realization in Hindi caregiver-child discourse. *Journal of Pragmatics, 37,* 461–495.
Newport, E. L. (1990). Maturational constraints on language learning. *Cognitive Science, 14,* 11–28.
Nippold, M. A. (1988). *Later language development.* San Diego, CA: Singular Publishing Group.
Nippold, M. A. (2006). Language development in school-age children, adolescents, and adults. In K. Brown (Ed.), *Encyclopedia of language and linguistics* (2nd ed., pp. 368–373). Oxford: Elsevier.
O'Grady, W. (1997). *Syntactic development.* Chicago: University of Chicago Press.
O'Grady, W. (2008). The emergentist program. *Lingua, 118,* 447–464.
Pine, J. M., & Lieven, E. V. (1993). Reanalysing rote-learned phrases: Individual differences in the transition to multi-word speech. *Journal of Child Language, 20*(3), 551–571.
Pinker, S. (1984). *Language learnability and language development.* Cambridge, MA: Harvard University Press.
Pinker, S. (1989). *Learnability and cognition: The acquisition of argument structure.* Cambridge, MA: MIT Press.
Pinker, S., & Ullman, M. T. (2002). The past-tense debate: The past and future of the past tense. *Trends in Cognitive Sciences, 6*(11), 456–463.
Ravid, D., Dressler, W. U., Nir-Sagiv, B., Korecky-Kröll, K., Souman, A., Rehfeldt, K., … Gillis, S. (2008). Core morphology in child directed speech: Crosslinguistic corpus analyses of noun plurals. In H. Behrens (Ed.), *Corpora in language acquisition research: Finding structure in data* (pp. 25–60). Amsterdam: Benjamins.
Reali, F., Christiansen, M. H., & Monaghan, P. (2003). Phonological and distributional cues in syntax acquisition: Scaling up the connectionist approach to multiple-cue integration. In R. Alterman & D. Kirsh (Eds.), *Proceedings of the 25th Annual Conference of the Cognitive Science Society* (pp. 970–975). Mahwah, NJ: Erlbaum.
Redington, M., Chater, N., & Finch, S. (1998). Distributional information: A powerful cue for acquiring syntactic categories. *Cognitive Science, 22,* 425–469.
Rizzi, L. (1982). *Issues in Italian syntax.* Dordrecht, Netherlands: Foris.
Romberg, A. R., & Saffran, J. R. (2010). Statistical learning and language acquisition. *Wiley Interdisciplinary Reviews: Cognitive Science, 1*(6), 906–914.
Romberg, A. R., & Saffran, J. R. (2013). All together now: Concurrent learning of multiple structures in an artificial language. *Cognitive Science, 37*(7), 1290–1320.
Rowe, M. L., & Salo, V. C. (2014). Child-directed speech (effects of variation in quality). In P. Brooks & V. Kempe (Eds.), *Encyclopedia of language development* (pp. 78–82). London: Sage.
Rowland, C. F. (2014). *Understanding child language acquisition.* Abingdon, UK: Routledge.
Rowland, C. F., Chang, F., Ambridge, B., Pine, J. M., & Lieven, E. V. M. (2012). The development of abstract syntax: Evidence from structural priming and the lexical boost. *Cognition, 125,* 49–63.
Rubino, R. B., & Pine, J. M. (1998). Subject-verb agreement in Brazilian Portuguese: What low error rates hide. *Journal of Child Language, 25*(1), 35–59.
Rumelhart, D. E., & McClelland, J. L. (1987). Learning the past tense of English verbs: Implicit rules or parallel processing? In B. MacWhinney (Ed.), *Mechanisms of language acquisition* (pp. 195–248). Hillsdale, NJ: Erlbaum.
Saffran, J. R., Aslin, R. N., & Newport, E. L. (1996) Statistical learning by 8-month-old infants. *Science, 274,* 1926–1928.
Savage, C., Lieven, E. V. M., Theakston, A., & Tomasello, M. (2003). Testing the abstractness of children's linguistic representation: lexical and structural priming of syntactic constructions in young children. *Developmental Science, 6*(5), 557–567.
Saxton, M. (2010). *Child language: Acquisition and development.* Thousand Oaks, CA: Sage.
Sekerina, I. A. (2015). Online evidence for children's interpretation of personal pronouns. In L. Serratrice & S. E. M. Allen (Eds.), *The acquisition of reference* (pp. 213–239). Amsterdam: Benjamins.
Sinclair, J. M. (1991). *Corpus, concordance, collocation.* Oxford: Oxford University Press.
Skeide, M. A., & Friederici, A. D. (2016). The ontogeny of the cortical language network. *Nature Reviews Neuroscience, 17*(5), 323–332.
Slobin, D. I. (1982). Universal and particular in the acquisition of language. In E. Wanner & L. R. Gleitman (Eds.), *Language acquisition: The state of the art.* Cambridge: Cambridge University Press.
Slobin, D. I. (1985–1997). *The crosslinguistic study of language acquisition* (Vol. 1–5). Mahwah, NJ: Erlbaum.
Snedeker, J. (2013). Children's sentence processing. In R. P. G. van Gompel (Ed.), *Sentence processing* (pp. 189–220). London: Psychology Press.
Snedeker, J., & Huang, Y. T. (2015). Sentence processing. In E. L. Bavin & L. R. Naigles (Eds.), *The Cambridge handbook of child language* (2nd ed., pp. 409–437). Cambridge: Cambridge University Press.
Sokolov, J. L. (1988). Cue validity in Hebrew sentence comprehension. *Journal of Child Language, 15,* 129–155.
Stumper, B., Bannard, C., Lieven, E. V. M., & Tomasello, M. (2011). "Frequent frames" in German child-directed

speech: A limited cue to grammatical categories. *Cognitive Science, 35*(6), 1190–1205.

Theakston, A. (2012). "The spotty cow tickled the pig with a curly tail": How do sentence position, preferred argument structure, and referential complexity affect children's and adults' choice of referring expression? *Applied Psycholinguistics, 33,* 691–724.

Tomasello, M. (1992). *First verbs: A case study of early grammatical development.* Cambridge: Cambridge University Press.

Tomasello, M. (1998). The return of constructions. *Journal of Child Language, 25,* 431–442.

Tomasello, M. (2000). Do children have adult syntactic competence? *Cognition, 74,* 209–253.

Tomasello, M. (2003). *Constructing a language: A usage-based account of language acquisition.* Cambridge, MA: Harvard University Press.

Tomasello, M., & Abbot-Smith, K. (2002). A tale of two theories: Response to Fisher. *Cognition, 83,* 207–214.

Trueswell, J. C., & Gleitman, L. R. (2007). Learning to parse and its implications for language acquisition. In M. G. Gaskell (Ed.), *The Oxford handbook of psycholinguistics* (pp. 635–655). Oxford: Oxford University Press.

Trueswell, J., Sekerina, I., Hill, N., & Logrip, M. (1999). The kindergarten-path effect: Studying online sentence processing in young children. *Cognition, 73,* 89–134.

Valian, V. (1990). A problem for parameter-setting models of language acquisition. *Cognition, 35*(2), 105–122.

Valian, V. (1991). Syntactic subjects in the early speech of American and Italian children. *Cognition, 40,* 21–81.

Vuong, L. C., Meyer, A. S., & Christiansen, M. H. (2016). Concurrent statistical learning of adjacent and nonadjacent dependencies. *Language Learning, 66*(1), 8–30.

Weisleder, A., & Fernald, A. (2013). Talking to children matters: Early language experience strengthens processing and builds vocabulary. *Psychological Science, 24*(11), 2143–2152.

Wexler, K., and Culicover, P. (1980). *Formal principles of language acquisition.* Cambridge, MA: MIT Press.

Willits, J. A., Seidenberg, M. S., & Saffran, J. R. (2014). Distributional structure in language: Contributions to noun-verb difficulty differences in infant word recognition. *Cognition, 132*(3), 429–436.

Yang, C. (2016). *The price of linguistic productivity: How children learn and break rules of language.* Cambridge, MA: MIT Press.

Yang, C., Crain, S., Berwick, R. C., Chomsky, N., & Bolhuis, J. J. (2017). The growth of language: Universal grammar, experience and principles of computation. *Neuroscience and Biobehavioral Review,* 81, 103–119.

Zukowski, A. (2016). Putting words together: Comprehension and production of morphology and syntax in the preschool years. In J. Berko Gleason & N. Bernstein Ratner (Eds.), *The development of language* (9th ed., pp. 104–136). Boston: Pearson.

11 A Neural Blueprint of Language Acquisition

MICHAEL A. SKEIDE

Understanding language is a sophisticated cognitive ability requiring the brain to integrate multiple types of complex information. In the present chapter, I introduce a neural blueprint of language acquisition, to explain where, when, and how the maturing brain masters the language processing steps that are typical of adult processing. The development of these capacities can be roughly subdivided into two developmental stages: a first stage extending over the first three years of life that mainly involves low-level processes, and a second stage continuing into adolescence that involves high-level processes. *Low-level processing* describes the computation of language representations from the sensory input, whereas *high-level processing* denotes the further analysis of this information under the influence of more complex representations and mechanisms.

1. The Adult Cortical Language Network

A description of the neuroplastic changes of the developing language system requires a description of the neural processes involved in adult language processing. A neurocognitive model of language comprehension provides an inventory of cognitive capacities that have to be acquired by a developing speaker and, moreover, an inventory of the neuronal circuits that have to mature until they can serve their specific functions and reach their full processing efficiency. The present section will focus on areas that have been repeatedly implicated as being functionally specialized for language. An overview of domain-general areas that contribute to language comprehension (e.g., brain areas supporting working memory or cognitive control) will not be considered here (for a review, see Fedorenko & Thompson-Schill, 2014; see also part VI of this volume).

Two comprehensive neurolinguistic models of sentence comprehension covering the entire processing cascade, from perception to interpretation, have been recently proposed: the *dual-stream* model by Friederici (2012) and the *memory, unification, control* model by Hagoort (2013). Both models assume that acoustic, syntactic, and semantic processing levels are fundamental elements of the human language faculty. There is also a general agreement that the incoming speech signal initially undergoes spectrotemporal analyses in the auditory cortex and its vicinity (mid superior temporal gyrus [STG]/superior temporal sulcus [STS]). However, the two models differ in their categorization of the neural implementation of subsequent phonological, syntactic and semantic processes. Moreover, unlike the memory, unification, control model, Friederici's model comprises a suprasegmental acoustic processing level. The dual-stream model will be outlined first before turning to the memory, unification, control model. The dual-stream model will form the conceptual basis for the later perspective on language acquisition.

1.1. Language Comprehension as Dual-Level, Dual-Stream Processing

1.1.1. Low-level processes In Friederici's model, language comprehension is initially driven by low-level processes that are triggered early (with onsets between 20 and 120 ms after an utterance is heard) and completed rapidly (with durations of usually not more than 30–60 ms). At the earliest stage of low-level processing, the acoustic features of phonemes (single sound segments) of each word are extracted and analyzed in the bilateral STS (Hickok & Poeppel, 2007; Overath, McDermott, Zarate, & Poeppel, 2015). If syllables at salient positions within a word (i.e., at the beginning or at the end) provide an unambiguous cue for rapidly reconstructing the word as whole, phonological word forms can be detected within a time window of 20–50 ms (DeWitt & Rauschecker, 2012). Between 40 and 90 ms after utterance onset, phonological word forms undergo morphosyntactic categorization, that is, they are sorted into distinct syntactic groups (such as noun, verb, or preposition) (Herrmann, Maess, Hahne, Schröger, & Friederici, 2011). In parallel, lexical-semantic categorization, which is necessary to determine whether a wordlike acoustic form can be semantically interpreted, occurs between 50 and 80 ms (MacGregor, Pulvermüller, van Casteren, & Shtyrov, 2012) in the left anterior (a) STS and aSTG (MacGregor et al.; Visser, Jefferies, &

Lambon Ralph, 2010). Note that these early decoding and categorization processes run fully automatically and cannot be consciously controlled.

Once an interpretable word form is identified, the corresponding entry in the mental lexicon (carrying semantic and word category information) can be accessed and retrieved between 110 and 170 ms. This process is associated with neural activity in the left aSTG, left aSTS, the left middle temporal gyrus, and several other regions widely distributed over the cortex (Binder, Desai, Graves, & Conant, 2009; Huth, de Heer, Griffiths, Theunissen, & Gallant, 2016; MacGregor et al., 2012; Mitchell et al., 2008). In parallel, as soon as syntactic categories are identified, phrase structures can be built in the left aSTG and aSTS (Brennan et al., 2012), possibly together with the most ventral part of the inferior frontal cortex (Snijders et al., 2009), between 120 and 150 ms after the onset of the utterance (Herrmann et al., 2011).

1.1.2. High-level processes According to Friederici (2012), high-level language comprehension is characterized by processes that occur relatively late (with onsets usually between 200 and 600 ms after the utterance) and proceed relatively slowly (with durations of at least 150 ms) compared with low-level comprehension. Although often running automatically to ensure rapid communication, all of these processes are consciously controllable. Between 200 and 400 ms (Lau, Phillips, & Poeppel, 2008), lexical information is extracted from the left aSTG and aSTS and transferred along the ventrally located inferior fronto-occipital fasciculus (IFOF) (Weiller, Bormann, Saur, Musso, & Rijntjes, 2011) by Brodmann's area 45 (BA45) and BA47 of the left inferior frontal gyrus (IFG) (Binder et al., 2009; Newman, Ikuta, & Burns, 2010; Rodd, Davis, & Johnsrude, 2005), where semantic ambiguity and semantic relations between lexical items are determined in interaction with the inferior parietal cortex (Obleser, Wise, Dresner, & Scott, 2007). At 300 to 500 ms, morphosyntactic information is retrieved from the left aSTG and aSTS by the frontal opercular cortex and BA44, which support the rule-based construction and ordering of complex phrases and sentences (Friederici & Gierhan, 2013). It can be assumed that this information is transferred ventrally along the uncinate fasciculus (Friederici & Gierhan, 2013), but this hypothesis remains to be verified experimentally. BA44, together with the left posterior STG and STS, also supports the construction of a sentence-level argument structure, that is, verb-noun arrays created to produce a meaningful sentence (Dapretto & Bookheimer, 1999; Dick & Tremblay, 2012; Friederici, 2011; Makuuchi, Bahlmann, Anwander, & Friederici, 2009; Meyer, Obleser, Anwander, & Friederici, 2012; Newman et al., 2010). Effective connectivity analyses modeling the causal functional interactions between brain areas during task performance have also revealed that the processing of sentence-level syntax is driven by the left IFG modulating the left posterior superior temporal cortex (den Ouden et al., 2012; Makuuchi & Friederici, 2013). Whether language information stays activated in the temporal cortex under the influence of the inferior frontal cortex remains to be directly tested by future studies (Snijders et al., 2009).

Prosodic information (e.g., intonation, tone, stress, and rhythm information) is processed in a time window of 200 and 500 ms after the utterance onset (Friederici, 2011), predominantly in the right STG and the right IFG (Ethofer et al., 2012; Frühholz, Ceravolo, & Grandjean, 2012; Sammler, Grosbras, Anwander, Bestelmeyer, & Belin, 2015).

After at least 600 ms (Friederici, 2012; Hagoort, 2013), all information is assumed to be integrated into an interpretable conceptual representation in the left posterior STG and STS (Friederici, 2011, 2012), with a potential involvement of the left IFG (Hagoort, 2013; Snijders et al., 2009). To achieve this, syntactic information is exchanged dorsally along the arcuate fasciculus (AF), between BA44 and the posterior (p) STG and STS, and semantic information is exchanged ventrally along the IFOF between BA45 and BA47 and the pSTG (Friederici, 2011, 2012; Newman et al., 2010; Saur et al., 2008; Weiller et al., 2011; Weiller, Musso, Rijntjes, & Saur, 2009; Wilson et al., 2011). Note that the dorsal connection between BA44 and the pSTG and STS that is involved in syntactic processes (Friederici, Bahlmann, Heim, Schubotz, & Anwander, 2006) can be anatomically separated from a second dorsally located fiber bundle that connects the pSTG and STS with the premotor cortex and is involved in the repetition of speech (Saur et al.) (figure 11.1).

1.2. Language Comprehension as Memory, Unification, and Control Processing Cycles Hagoort (2013) assumed that auditory language comprehension is driven by multiple processing cycles that are repeated until an utterance can be interpreted. The starting point of the first stage of the first cycle is the arrival of the speech signal in the left primary auditory cortex. Subsequently, along a dorsal-to-ventral gradient in the left posterior temporal cortex, information about acoustic-phonological properties is retrieved from an area covering the left pSTG/STS, then information about syntactic properties is retrieved from a region spanning the left pSTS to the left posterior middle temporal sulcus. Finally, information about conceptual

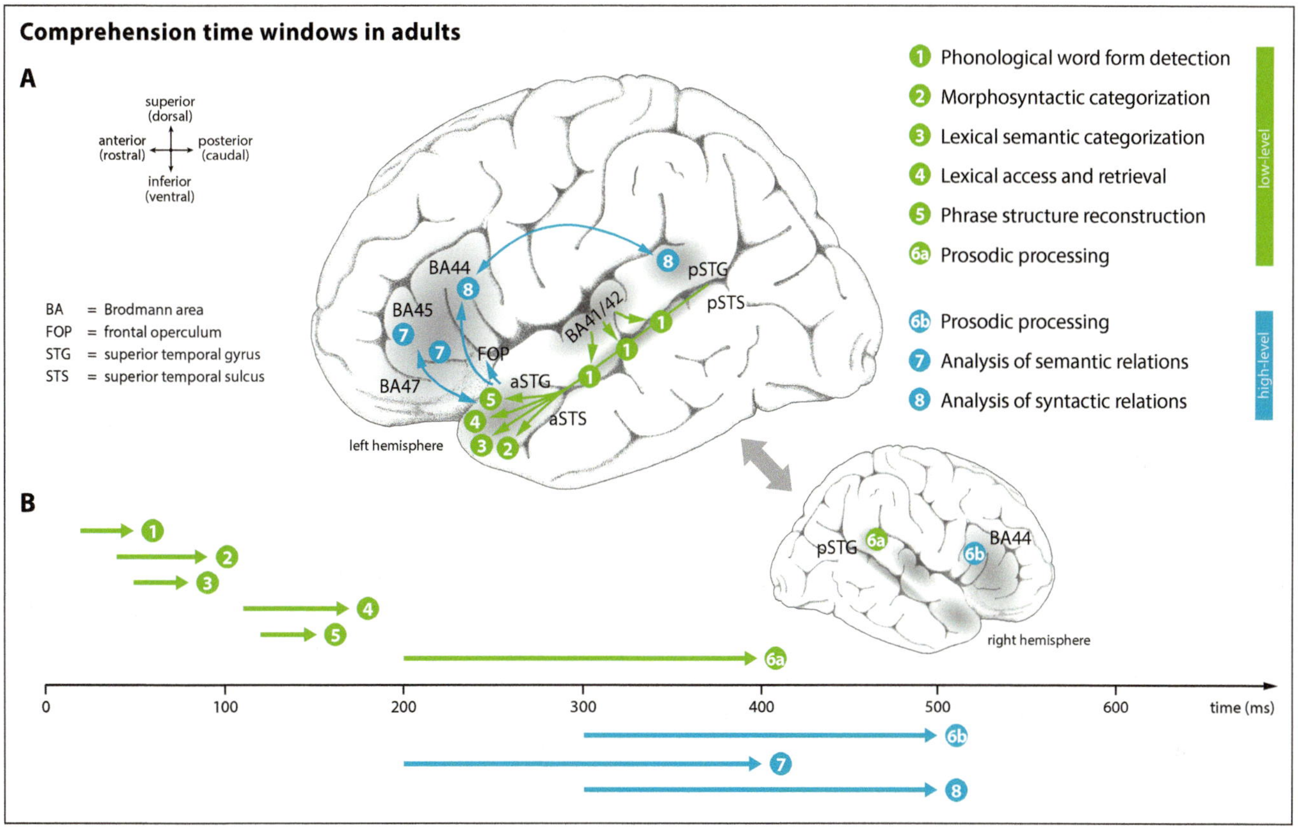

Figure 11.1 The auditory language comprehension network of the adult brain. (A) Gray matter areas in the left perisylvian cortex and their respective contributions to the main stages (1–8) of auditory language comprehension. Arrows indicate the assumed flow of information along the interconnecting white matter fiber tracts. "Low-level" processes are colored in green and "high-level" processes are colored in blue. (B) Typical time windows (in milliseconds) of the processing phases underlying auditory language comprehension. All values are approximate and subject to interindividual variation. (1) Phonological word forms are detected in the left STS within 20–50 ms after the utterance onset by extracting acoustic features of the raw signal delivered by the auditory cortex (DeWitt & Rauschecker, 2012; Hickok & Poeppel, 2007; Overath et al., 2015). (2) Subsequently, phonological word forms are categorized morphosyntactically between 40 and 90 ms (Herrmann et al., 2011) and (3) also categorized semantically between 50 and 80 ms in the left aSTS and aSTG (MacGregor et al., 2012; Visser et al., 2010). (4) Lexical items associated with the phonological word forms are retrieved at around 110–170 ms by the left aSTG and aSTS in an interplay with additional widespread cortical regions (Binder et al., 2009; Huth et al., 2016; MacGregor et al.; Mitchell et al., 2008). (5) In parallel, phrase structures are built at 120–150 ms in the left aSTG and aSTS (Brennan et al., 2012; Herrmann et al.; Snijders et al., 2009). (6a) Phrase-level prosody (e.g., phrasal intonation and word stress) is processed within 200–400 ms after the utterance in the right pSTG (Friederici, 2011). (6b) Sentence-level prosody is processed in a time window of 300–500 ms after the utterance in the right IFG (Ethofer et al., 2012; Frühholz et al., 2012; Sammler et al., 2015). (7) Between 200 and 400 ms, lexical information is transferred from the left aSTG and aSTS along the ventrally located IFOF to BA45 and BA47 of the left IFG, where semantic relations between lexical items are determined (Binder et al.; Lau et al., 2008; Newman et al., 2010; Rodd et al., 2005; Weiller et al., 2011). (8) At 300–500 ms after the utterance, phrase structure information is transmitted ventrally from the left aSTG and aSTS, along the uncinate fasciculus, to the frontal opercular cortex and to BA44 in the left IFG (Friederici & Gierhan, 2013). BA44 is involved not only in ordering phrases but also in building up argument structures together with the left pSTG and pSTS, which are connected to BA44 via the left AF (Dapretto & Bookheimer, 1999; Dick & Tremblay, 2012; Friederici, 2011; Makuuchi et al., 2009; Meyer et al., 2012; Newman et al.). As soon as an utterance has gone through these processing steps, the left pSTG and pSTS, potentially together with the left IFG, can start integrating all information into an interpretable conceptual representation (Friederici, 2011, 2012; Hagoort, 2013; Snijders et al., 2009).

semantic properties of the signal is retrieved from an area extending from the left posterior middle temporal gyrus to the left posterior inferior temporal gyrus. Depending on the type of linguistic information, another dorsal-to-ventral gradient activates in the left inferior parietal cortex, with phonological representations retrieved from the left supramarginal gyrus, syntactic representations retrieved from an area between the left supramarginal gyrus and the left angular gyrus, and semantic representations retrieved from the angular gyrus. These retrieval mechanisms, the core of the so-called memory component, operate sequentially at different but partially overlapping time scales until they are completed at around 250 ms. During the second stage of the first cycle, phonological, syntactic, and semantic information is relayed from the left posterior temporal and inferior parietal cortex to the left IFG over a feedforward pathway that includes the superior longitudinal fasciculus and the AF. The IFG in turn crucially supports the maintenance of the information due to its ability for self-sustained activity. Furthermore, it is involved in the rule-constrained combination of linguistic elements from all three domains along a posterior-dorsal to anterior-ventral gradient, with the posterior-dorsal BA44 and the neighboring ventral BA6 unifying phonological word forms into intonational phrases, the anterior-ventral BA44 and posterior-dorsal BA45 linking together syntactic phrases, and the anterior-ventral BA45 and BA47 integrating lexical items into the discourse context. This so-called unification component selects the optimal candidates out of multiple possible environments and optional alternative links. As soon as the selection procedure is over, the information is sent back from the left IFG to the left posterior temporal cortex along a slow feedback pathway comprising the uncinate fasciculus, the IFOF, and the inferior longitudinal fasciculus. This occurs until the first cycle is completed at around 400 ms. This pathway strengthens the role of the IFG for the maintenance of linguistic information. Usually, full comprehension requires additional recurrent feedforward and feedback signaling cycles between the posterior temporal and the inferior frontal cortex. The higher the competition among suitable candidate combinations, that is, the higher the unification load, the longer it takes to compute the final interpretation of an utterance. Finally, successful comprehension is guided by a so-called control system represented in the anterior cingulate cortex, the dorsolateral prefrontal cortex, and the inferior parietal cortex. The system subserves attentional and pragmatic operations such as attention allocation to a target language during language switching or turn taking in a conversation (see part VI for further details).

2. Methods for Studying the Neural Basis of Language Acquisition

Noninvasive electroencephalography (EEG) is the most frequently employed neurophysiological recording technique in infants and children. Electric currents induced by neuronal activity are recorded directly on the scalp, so that children can move relatively freely during the measurements. Moreover, some paradigms applied in EEG experiments do not even require the participant's attention. These factors make this method especially useful for experiments with newborns and young infants (Benasich et al., 2006; Guttorm et al., 2005). EEG has a high temporal resolution of the order of milliseconds, but its spatial resolution on the scalp is limited to around 5 cm when modeling 64 electrodes (Burle et al., 2015). Additionally, EEG recordings on the scalp have a limited penetrance, making it hard to study subcortical structures. However, successful examinations of childrens' auditory brain stem responses are documented in the literature (Johnson, Nicol, Zecker, & Kraus, 2008).

Depending on ethical regulations, newborns can undergo MRI and children as young as three years can solve simple tasks in the scanner (Skeide, Brauer, & Friederici, 2014). Functional MRI (fMRI) reveals information about the magnitude and the location of neural activity changes in response to external stimulation or intrinsic fluctuations at rest. These neural activity changes are quantified as blood-oxygen-level-dependent signal changes based on the effect of neurovascular coupling. Structural MRI provides morphometric and geometric features of the neural gray and white matter, such as its volume, density, thickness, and surface area. Finally, diffusion-weighted MRI is used to reconstruct the trajectory of white matter pathways interconnecting brain areas. Fractional anisotropy (FA), a measure of tissue diffusivity, can be combined with structural MRI to determine the degree of myelination or the density of fiber tracts, which is in turn related to their maturity. MRI provides a spatial resolution of at least 1 mm, which allows conclusions to be drawn about which brain areas are involved in a particular process. However, the temporal resolution of MRI is limited to about 1 s (Kim, Richter, & Ugurbil, 1997). MRI of healthy young adults has become a routine investigation in neuroscience, but it is difficult to acquire these data in children (Raschle et al., 2012). In order to reduce dropout rates, young participants are prepared in a mock scanner prior to the actual measurement (see https://www.cbs.mpg.de/abteilungen/neuropsychologie/kindersprachlabor). Mock scanner training helps to familiarize the child with the MR procedure and the experimental tasks it has to solve. In practice, the experimenter embeds the training in a game, involving the parents if necessary. Motion sensors that stop the presentation on the screen in case of head movement have proved their worth when instilling in young participants the necessity of lying still in the scanner. Such familiarization should be accompanied by encouraging feedback to create a constructive awareness of the experimental setting. It is important to simulate the actual measurement session as precisely as possible so that the children do not get distracted or scared in the real MR environment. Although training sessions are time-consuming, they markedly reduce dropout rates to less than 40% in three- to four-year-old children and less than 10% in six- to seven-year-old children. An optimal result with respect to data quantity as well as data quality is achieved in a child-friendly, low-noise scanner with a wide tube.

Near-infrared spectroscopy is another method used to measure functional brain activity. The technique is based on 700- to 900-nm wavelength light that penetrates the scalp and is then absorbed differently by cortical areas as a function of their activity-related blood oxygenation. Like EEG, near-infrared spectroscopy allows for a more flexible recording of functional responses than MRI since the registration system is mounted directly on the participant's head. This advantage has made it an important method for language acquisition research in infants and young children. However, the spatial resolution of near infrared spectroscopy is lower than that of fMRI, while its temporal resolution is similarly poor.

3. The Developing Cortical Language Network

3.1. Developing Low-Level Processes of Language Comprehension In this section, we describe the ontological unfolding of low-level language processing in the first three years of life. The skills acquired during this developmental phase enable infants to segment the speech input into phonological word forms. On the basis of this capacity, they can associate phonological word forms with lexical representations that can be semantically categorized. Moreover, the child acquires the ability to sort words into syntactic categories and to group words into phrase structures. A mechanism for integrating syntactic and semantic information to interpret an utterance is established from the second year of life on, at the latest.

3.1.1. Detection of phonological word forms Embryos gather their first experiences with speech in utero (Schnupp, Nelken, & King, 2011). A recent near-infrared spectroscopy study found that preterm infants born between 28 and 32 weeks of gestation are able to detect a deviant phoneme in a sequence of otherwise identical phonemes. These individuals showed hemodynamic mismatch responses in the bilateral posterior superior temporal cortex and the inferior frontal cortex in response to the deviant phoneme (Mahmoudzadeh et al., 2013). Remarkably, it seems that this basic phoneme discrimination capacity is already present when most neurons are still located in the fetal subplate and have not yet migrated into their cortical target layers, as this maturational step is only completed after the 32nd week of gestation. It is unlikely that early phonological skills are solely the result of prenatal environmental experience, since sounds at frequencies above 300 Hz are strongly attenuated in utero. Given this level of acoustic degradation, fetuses in the womb can at best only discriminate between vowels (with lower frequency boundaries at around 200 Hz) but not between more subtly differing consonants, which in speech typically have higher frequencies starting at around 300 Hz (Gomez et al., 2014; Partanen et al., 2013). Further studies suggest that even though preterm infants experience an earlier and richer exposure to speech that is no longer degraded by maternal tissue, speech perception skills in preterm infants are not more advanced than those of age-matched full-term infants (Peña, Pittaluga, & Mehler, 2010; Peña, Werker, & Dehaene-Lambertz, 2012). These findings support the view that the primary, intrauterine speech perception skills are likely to be driven by genetic factors that bias the auditory processing system toward language-specific frequency spectra of the acoustic input. Nevertheless, active exposure to auditory stimuli is crucial for normal language acquisition, as it facilitates the differentiation of the sounds in the child's language environment (Benasich, Choudhury, Realpe-Bonilla, & Roesler, 2014; Kuhl, 2004; Werker & Hensch, 2015; Werker & Tees, 1984).

When listening passively to speech, full-term newborns show strong interhemispheric synchronization of hemodynamic activity in the bilateral posterior STG, but no intrahemispheric synchronization of hemodynamic activity in the left posterior STG and the left IFG, as has been documented in adults (Perani et al., 2011). However, left-hemispheric frontotemporal activity is documented by the age of three months. At this age, infants passively listening to speech show activity bilaterally in the mid and posterior STG during the first 3 s after stimulus onset, and in inferior frontal premotor regions between 7 and 9 s (Dehaene-Lambertz et al., 2006).

These functional findings in newborns (Perani et al., 2011) and three-month-olds (Dehaene-Lambertz et al., 2006) can be reconciled when considering the structural connectivity pattern in adults derived by diffusion-weighted MRI tractography. Specifically, adults exhibit two dorsal fiber tracts that interconnect the posterior temporal cortex: one that projects to the premotor cortex, and one that projects to BA44 of Broca's area (Dubois et al., 2015; Perani et al., 2011). In newborns (Perani et al., 2011) and two- to five-month-old infants (Dubois et al., 2015), the dorsal fiber tract to the premotor cortex is already myelinated, but the one reaching into Broca's area is yet to be myelinated. The pathway that projects to the premotor cortex supports not only the integration of sensory and motor representations during the babbling phase (Dubois et al., 2015; Perani et al., 2011), but the information fed back from the premotor cortex to the temporal cortex may also be crucial for establishing phonemic representations. This is suggested by the observation that six-month-old infants cannot distinguish unfamiliar phonemes from one another when they are prevented from moving their tongue tips to articulate these phonemes (Bruderer, Danielson, Kandhadai, & Werker, 2015). Similar work in adults corroborates the view that this auditory–motor link is not only important for language production but also for language comprehension (Mottonen & Watkins, 2009).

Importantly, when newborns hear a sentence, more activity is observed in the right mid STG than in the left mid STG (Perani et al., 2011). A similar lateralization effect was observed in the temporoparietal cortex of sleeping three-month-old infants (Homae, Watanabe, Nakano, Asakawa, & Taga, 2006). Given that, in adults, the right mid STG is known to support prosodic processing (Ethofer et al., 2012; Frühholz et al., 2012), this finding indicates that newborns not only are sensitive to segmental (phonological) information, but also they are sensitive to suprasegmental (prosodic) information in a sentence (Perani et al., 2011).

The infant's sensitivity to prosody forms the basis for learning how to segment the auditory stream into words, according to the specifics of the target language (Friederici, Friedrich, & Christophe, 2007). Event-related potential (ERP) work reveals that at six months of age, infants only recognize words that have previously been accentuated (that is, marked as acoustically prominent by increasing sound pressure and frequency) but not if they have merely been repeated. It is only at 12 months of age that infants no longer rely on accentuation to detect phonological word forms (Männel & Friederici, 2013). Interestingly, this trajectory coincides with the decrease in the infant's ability to discriminate phonemes that do not belong to the standard inventory of their native target language, as indicated by different patterns of neural oscillations (Bosseler et al., 2013; Gervain, 2015; Kuhl, Ramirez, Bosseler, Lin, & Imada, 2014; Ortiz-Mantilla, Hämäläinen, Musacchia, & Benasich, 2013). The declining flexibility in recognizing phonemes is known as an instance of perceptual narrowing (Pons, Lewkowicz, Soto-Faraco, & Sebastian-Galles, 2009). For example, whereas before the age of six months, English-learning infants can distinguish between not only English phonemes, but also Hindi phonemes, they lose the ability to discriminate Hindi phonemes in the second half year of life (Werker & Tees, 1984). Crucially, losing this ability goes hand-in-hand with an increased sensitivity for acoustic differences between phonemes of the native language (Kuhl et al., 2006).

In the mature brain, suprasegmental prosodic information is processed at the sentence level in the right STG and the right IFG. A similar pattern of dissociable hemodynamic responses in the right hemisphere (relating to prosodic information) and in the left superior temporal cortex (relating to segmental phonological processing) has been observed in four-year-old children (Wartenburger et al., 2007). It is likely, however, that this level of specialization is reached earlier during development. This hypothesis should be tested in follow-up experiments involving younger children.

In sum, language acquisition starts in utero, when the embryo gathers first experiences with acoustic information. Infants learn to discriminate the phonemes of their target language in the first half year of life. From six months of age onward, they are able to use prosodic information to segment speech into phonological word forms. These skills are assumed to be based on the interplay of the bilateral superior temporal auditory system and the inferior frontal motor system.

3.1.2. Lexical access and semantic categorization The refinement of speech segmentation skills in the first half year of life provides the basis for associating phonological word forms with objects in the environment, an ability that has indeed been documented in infants as young as six months (Bergelson & Swingley, 2012). Nine-month-old infants can generalize the meanings of single words into lexical categories according to visual features of the objects they describe, such as their color and shape. At this age, lexical-semantic learning is dependent on the consolidation of recent episodic memory traces during sleep (Friedrich, Wilhelm, Born, &

Friederici, 2015). Support for this finding comes from adult fMRI experiments that suggest that the retrieval of lexical items that were acquired early in life is controlled by a domain-general episodic memory area (the precuneus), whereas the retrieval of later-learned lexical items is controlled by specialized language areas (BA45 and BA47) (Fiebach, Friederici, Müller, von Cramon, & Hernandez, 2003).

An adultlike N400 response that is associated with semantic congruency evaluation can be elicited at ~12 months in early talkers or ~18 months in normal talkers both in the EEG (Friedrich & Friederici, 2010) and the magnetoencephalography (MEG) (Travis et al., 2011) by employing a picture-word priming task. Source localization analyses have revealed that the MEG signal corresponding to the N400 in this task originates mainly from mid and posterior portions of the left temporal cortex (Travis et al., 2011), similar to the pattern found in the adult brain (Lau et al., 2008). Note that a precursor component of the N400 with a later onset at around 600 ms can already be observed at nine and even at six months of age (Friedrich & Friederici, 2011; Junge, Cutler, & Hagoort, 2012). Eighteen-month-old infants seem to learn new words by simultaneously evaluating phonological features and conceptual-semantic and syntactic categories. Indeed, at this age, toddlers looked significantly longer at novel picture–word pairs if the words were from different syntactic and semantic categories to those of a phonologically similar word they had known before (Dautriche, Swingley, & Christophe, 2015).

3.1.3. Syntactic categorization and phrase reconstruction There is behavioral evidence that six-month-old infants are already able to recognize phrases if the speech input carries reliable prosodic cues. They detect noun phrases (for example, *new watches for men*) and verb phrases (for example, *buy the whole supply*) if the syntactic phrase boundary coincides with the prosodic phrase boundary, which in languages like English or German is marked by lengthening of the final syllable and the pause that succeeds it (Soderstrom, Seidl, Nelson, & Jusczyk, 2003). ERP work has shown that, by the age of five months, infants are able to "chunk" speech input into phonological phrases if the pause (a salient acoustic cue) is available, but not if the pause is absent (Männel & Friederici, 2009). An ERP component called the *closure positive shift*, which is associated with the detection of the phonological clause boundary, only emerges in the third year of life and reaches an adultlike appearance at six years of age when children have acquired more advanced syntactic knowledge (Männel & Friederici, 2011).

It is possible that an interhemispheric network comprising the left aSTS and aSTG and the right STG, which are connected by posterior transcallosal fibers, underlies six-months-olds' skills to detect phrase structures using prosodic cues. Low-frequency-fluctuation MRI analysis has revealed interhemispheric functional connectivity between the respective temporal regions in newborns (Perani et al., 2011). Support for this functional role of the corpus callosum in adults comes from a study showing that an incongruence between prosodic intonation and phrase structure, that is, a prosodic boundary occurring before all obligatory syntactic elements had occurred, elicited a mismatch response in healthy adults but not in patients with lesions in the posterior corpus callosum (Sammler, Kotz, Eckstein, Ott, & Friederici, 2010).

Fourteen-month-old toddlers are already able to categorize nouns syntactically based on their knowledge that nouns are preceded by determiners: such toddlers listened significantly shorter to a phrase containing a determiner and a pseudoword than to a phrase containing a pronoun in combination with the same pseudoword (Höhle, Weissenborn, Kiefer, Schulz, & Schmitz, 2004). More advanced syntactic categorization ability can be observed by the age of 18 months: 18-month-olds can already evaluate the compatibility of grammatical gender information between a case-marked article and a noun (van Heugten & Christophe, 2015). For example, French toddlers detected that the case-marked article *la* (which indexes the feminine gender), but not the case-marked article *le* (which indexes the masculine gender), was compatible with the feminine noun *poussette* 'pram'. Two-year-old children showed a P600 EEG response approximately 600 ms after stimulus onset when they encountered two grammatically incompatible phrase types, such as a prepositional and a verb phrase, in German (Oberecker & Friederici, 2006). Thirty-two-month-old children additionally showed an early left anterior negativity signal that originated from left anterior temporal sources (Herrmann et al., 2011), similar to that observed during automatic reconstruction of phrases in adults (Oberecker, Friedrich, & Friederici, 2005).

In children that are younger than three years of age, sentence comprehension is most likely grounded in the temporal cortex and possibly also its ventral connection to the IFG. This connection has been shown not only to support semantic processes at the sentence level (Saur et al., 2008), but also the processing of syntactically simple sentences (Tyler et al., 2011; Wilson et al., 2011) in adults. Nevertheless, very few data directly elucidate this phenomenon. MRI studies involving children below three years of age are needed to explore how the dorsal and ventral connections between the temporal cortex and the IFG contribute to language comprehension in the first three years of life.

3.2. Developing High-Level Processes of Language Comprehension The refinement of basic language skills and increasing working memory resources together drive the emergence of the higher-order semantic and syntactic representations at the sentence level. In the following section, special emphasis is given to studies that have focused on high-level processing of semantic plausibility and phrase order. Other high-level processes are currently only studied in adults (for a review, see Friederici, 2012).

3.2.1. Analysis of semantic relations Neural mechanisms dedicated to extracting the semantic information of a sentence undergo extended refinement in the course of development. ERP effects associated with the evaluation of semantic congruency at the word level have an adultlike appearance from 14 months of age onward and can be considered as precursors for semantic processing at the sentence level (Friedrich & Friederici, 2005a, 2005b). An N400 marking the detection of an incongruence between a verb and a noun (for example, in the sentence *The cat drinks the ball*) can be detected in children before they are two years old, although, compared with the adult N400, this component has a later onset and duration (Friedrich & Friederici, 2005c). In the mature brain, the semantic and syntactic domains are dissociable both in terms of their neuroanatomical localization and in terms of the time windows during which their corresponding functions operate. They only interact at the final stage of sentence comprehension when both types of information have to be integrated into an interpretable conceptual whole in the left posterior STG and STS, and possibly also the left IFG. A recent fMRI study has revealed that this modular architecture of the syntax-semantics interface is implemented in the maturing brain only between seven and nine years of age. In children younger than seven years of age, the hemodynamic activity related to semantics and syntax largely overlaps (Brauer & Friederici, 2007; Nuñez et al., 2011). Moreover, semantic plausibility significantly reduces the hemodynamic resources needed to comprehend syntactically complex sentences and thus seems to provide a cue for sentence comprehension until high-level syntactic computations are fully available in adulthood. This was demonstrated in an experiment that presented either semantically plausible sentences (e.g., "Where is the small beetle, whom the big fox carries?"—in this case, a tall animal was the agent performing a certain action on a small animal) or inversely, semantically implausible sentences (for example, "Where is the big fox, whom the small beetle carries?"). The processing of the noncanonical, but plausible sentences consumed fewer hemodynamic resources than did the noncanonical and implausible sentences. Accordingly, it is likely that young children use their conceptual semantic world knowledge to master syntactic complexity. By contrast, nine- to ten-year-old children who have acquired more advanced syntactic knowledge did not need to constrain their sentence processing by conceptual semantic world knowledge. This was reflected at the neural level where hemodynamic activity underlying semantic processing was anatomically separable from that activity underlying syntactic processing, as in the adult brain (Skeide et al., 2014) (figure 11.2). Follow-up studies are needed to replicate these data to determine whether they can be generalized to other languages.

3.2.2. Analysis of syntactic relations In the third year of language acquisition, children have refined their ability to syntactically categorize words to such a degree that they become able to analyze syntactic relations between subsequent phrases. Moreover, three- to four-year-old children recruit not only the left temporal, but also the left inferior frontal cortex when processing sentences (Skeide et al., 2014; Skeide, Brauer, & Friederici, 2015), suggesting that the basic functional components for high-level syntax processing are established at this age. However, the full neural efficiency for computing syntactic relations between remote phrases in syntactically complex sentences arises gradually and is only established in young adulthood. An ERP study that investigated the use of morphosyntactic information during online sentence comprehension showed that although three-year-olds can detect grammatical case-marking cues, it takes until the age of six before children start to use this information to determine who is doing what to whom in a sentence (Schipke, Friederici, & Oberecker, 2011; Schipke, Knoll, Friederici, & Oberecker, 2012). A combined fMRI–diffusion tensor imaging study of three groups of children aged 3–4, 6–7, or 9–10 years, as well as an adult group, compared hemodynamic activity underlying the processing of well-formed spoken sentences with a canonical word order against sentences with a noncanonical word order. The effect of working memory was removed from the analysis, to allow for a domain-specific investigation of syntactic computation. Children aged three to seven years exhibited no functional selectivity of BA44 or the left posterior STG for syntax. By contrast although children aged nine to ten years, unlike the adults, did not show any such functional selectivity in the left

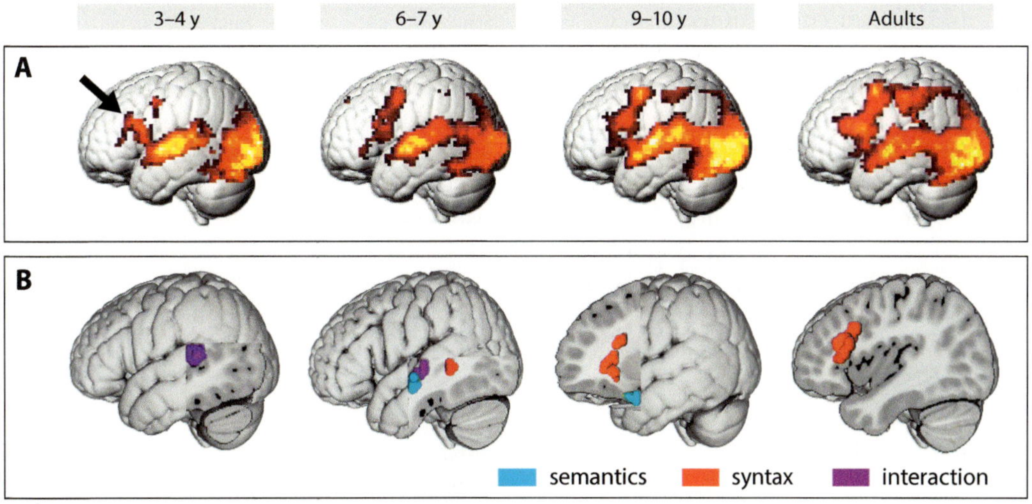

FIGURE 11.2 The functional neural segregation of syntax and semantics in the course of development. (A) Hemodynamic activity in children and adults during a sentence comprehension task (thresholded at $P<.01$; cluster size–and Bonferroni-corrected). The hemodynamic patterns cover similar cortical areas across all age groups. Like older children and adults, three- to four-year-old children recruit not only the left temporal cortex, but also the left inferior frontal cortex when processing sentences (see arrow). Accordingly, it can be assumed that the brain has set the frame for high-level language processing at three years of age. However, the full cortical selectivity for syntactic and semantic information emerges only gradually within this network. The increasing amount of activity in the left inferior parietal cortex could be related to literacy acquisition (Dehaene, Cohen, Morais, & Kolinsky, 2015). (B) The development of cortical selectivity for syntactic and semantic information is shown. Depicted are the results of a fMRI experiment employing a sentence–picture matching task to investigate the processing of syntactic complexity and semantic plausibility information. The effects of syntactic complexity were assessed by comparing responses to sentences with noncanonical word order with sentences with a canonical word order. The effects associated with semantic plausibility were investigated by comparing responses to sentences about conceptually plausible actions of an animal with sentences describing conceptually implausible actions. Depicted are main effects of syntactic complexity (red), semantic implausibility (blue), and the interaction of both factors (purple) ($P<.01$; cluster size- and Bonferroni-corrected). In young children (3–4 and 6–7 years), there is a statistical interaction between the semantics- and syntax-related hemodynamic responses in the mid and posterior portions of the left STG. Only 9- to 10-year-olds show a statistically independent effect of syntax in the left IFG, but, unlike adults, they recruit not only BA44, but also BA45, which might indicate that Broca's area is not yet fully functionally specialized at this age (Skeide et al., 2014).

BA44, they did show some functional selectivity for syntactic complexity in the left posterior STG. Remarkably, although nine- to ten-year-old children understand complex noncanonical sentences with an accuracy above 90%, adults still considerably outperform them with an accuracy of almost 100%, suggesting the left BA44 has a decisive role in reaching full efficiency in processing syntax (Skeide et al., 2015). This view is corroborated by structural morphometric data of five- to eight-year-old children and adults that indicate that increasing syntax skills are inversely correlated with the gray matter volume of the left IFG and the left posterior STG (relative to the total gray matter volume) (Fengler, Meyer, & Friederici, 2015). Accordingly, a maturational mechanism of cortical thinning in these areas, possibly driven by long-lasting synaptic pruning (Petanjek et al., 2011), has been proposed to underlie the sophistication of syntactic abilities. Future studies combining structural and functional imaging techniques are needed to corroborate this notion.

Ordering phrases and building up a sentence-level argument structure requires a common computational effort of the left IFG and the left posterior STG and STS within the dorsal language network (Wilson et al., 2011). A recent study reconstructed the left AF as the dorsal white matter fiber tract that directly connects these areas and also the ventral language pathway along the left IFOF in the four age groups mentioned (Skeide et al., 2015). Subsequently, the individual FA values for each tract of each participant were related to their individual sentence comprehension performance. When controlling for the unspecific effect of age and the domain-general influence of verbal working memory, both the increase in accuracy and the decrease in response times across development were more strongly related to the FA of the AF than to that of the IFOF (figure 11.3). Moreover, the correlational relationship between the FA of the AF and the behavioral response to syntactically more complex noncanonical sentences was markedly stronger than the relationship between the FA of the AF and the behavioral response to

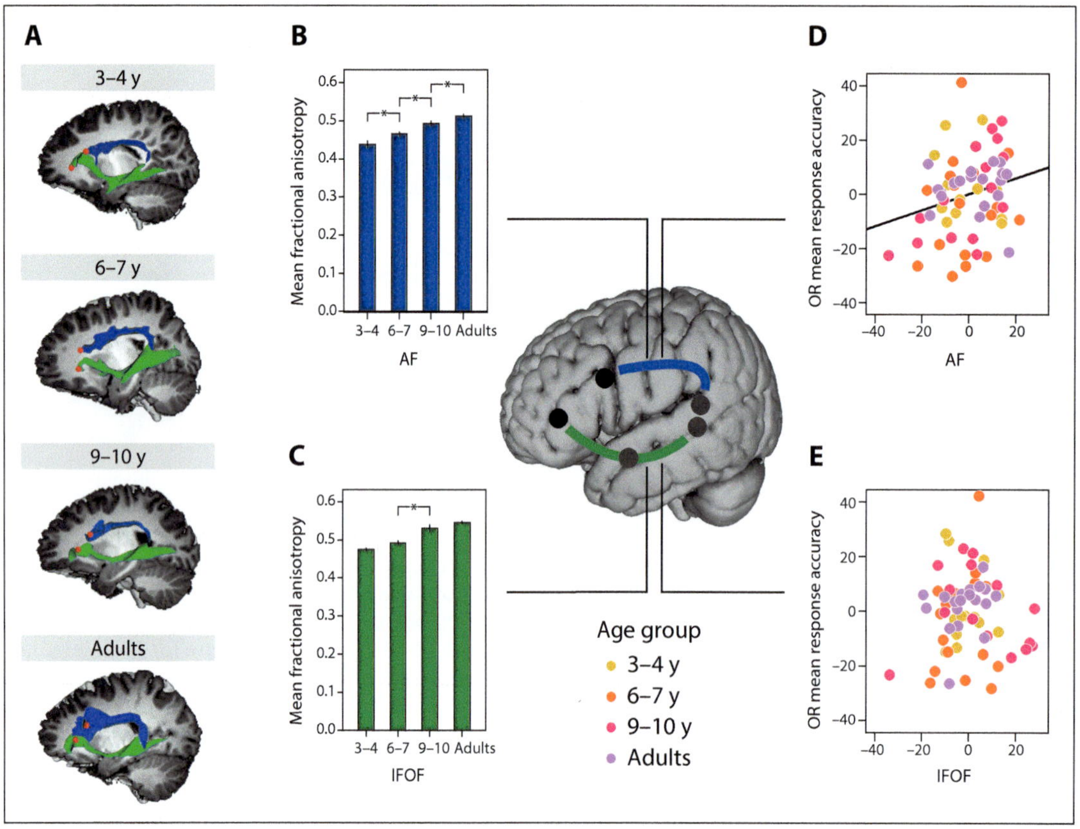

FIGURE 11.3 The ontogenetic emergence of structural networks for complex syntax. Interpreting syntactically complex sentences requires an exchange of syntactic and semantic information between the left IFG and the left STG (Friederici, 2012). In adults, this information transfer depends on a dorsal pathway via the left AF (blue) but not on a ventral pathway via the left IFOF (green) (Wilson et al., 2011). (A) To track the emergence of this dissociation, the corresponding white matter fiber tracts were reconstructed in four age groups and assessed for their FA using diffusion-weighted MRI (B: AF; C: IFOF). The FA of the AF was more strongly correlated with accuracy in comprehending syntactically complex, noncanonical sentences (D) than was the FA of the IFOF (E), whereas no such effect was found for canonical sentences (not shown). The unspecific contributions of age and the domain-general effect of verbal working memory are removed from these associations (Skeide et al., 2015). OR = object relative clauses (syntactically complex sentences).

canonical sentences. Together, these results derived from FA indicate that several other maturational mechanisms, including myelination (Gupta et al., 2012), axon growth (Paus, 2010) and increasing fiber density (Scholz, Klein, Behrens, & Johansen-Berg, 2009), are associated with the developmental refinement of the dorsal syntax network. The healthy maturation of the AF is a necessary precondition for normal language acquisition. Indeed, its disturbance might lead to faulty abstract grammatical representations and might manifest itself as specific language impairment (van der Lely & Pinker, 2014).

To conclude, the functional neural prerequisites for processing sentence-level syntax seem to be present already at three years of age when sentence comprehension is still inefficient. Sentence comprehension abilities improve into adulthood and are related to the functional specialization of the left IFG and the left posterior STG, and, moreover, to the maturation of their structural connection via the AF.

4. Summary

Taken together, at least four weeks before birth the embryo is already equipped with language processing machinery implemented in bilateral superior temporal and inferior frontal cortices, making it possible to analyze basic segmental and suprasegmental features of speech. Through refining their genetically predisposed acoustic skills, infants learn to segment the speech stream into phonological word forms in their first year of life. At the beginning of the second half of the first year of life, infants start associating phonological word forms with objects in their environment. By around 12 months, they not only can access and retrieve items from their lexicon, but they also begin to semantically

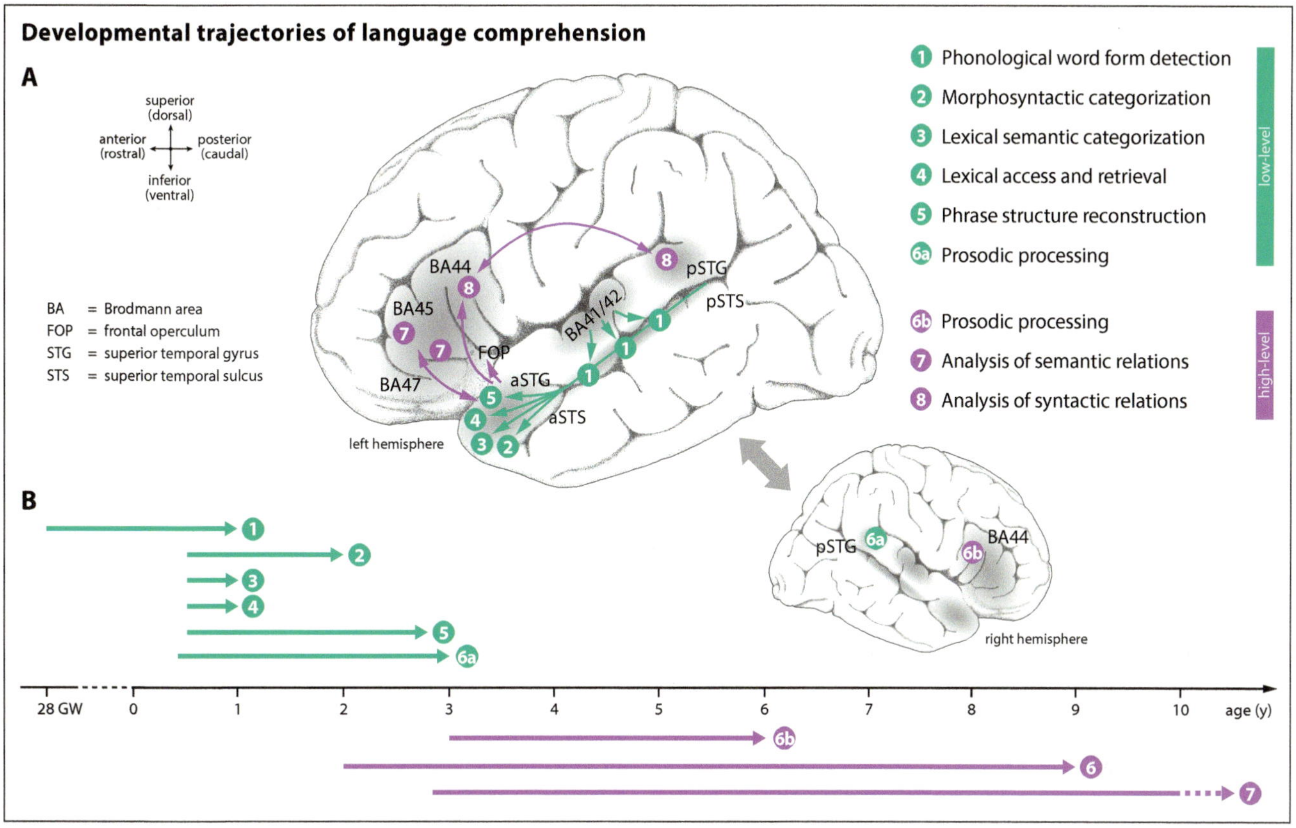

FIGURE 11.4 The developing cortical language comprehension circuit. (A) Neural implementation of the main processing steps (1–8) known from the adult brain. Arrows indicate the assumed flow of information along interconnecting white matter fiber tracts. Low-level processes are colored in turquoise green and high-level processes are colored in purple. (B) A schematic timeline of language acquisition (age in years). The starting points of the lines represent the earliest manifestations and the endpoints of the lines represent the first adultlike appearance of the different neural processing steps. All developmental trajectories are shown at the group level but it is important to note that they underlie interindividual variation. (1) From the 28th gestational week (GW) onward, at the latest, infants are able to extract basic segmental (phonological) and suprasegmental (prosodic) features of speech recruiting the bilateral STS (Mahmoudzadeh et al., 2013). In the first 9 months of life, they have to rely on prosodic cues (for example, accentuation) to detect phonological word forms, but this dependency is no longer present at 12 months of age (Männel & Friederici, 2013). (2, 6a) Prosodic cues enable six-month-old infants to categorize speech inputs according to morphosyntactic criteria, for example, into noun and verb phrases (Soderstrom et al., 2003), but around the second birthday, prosodic information is no longer needed for morphosyntactic categorization (Oberecker & Friederici, 2006). (5) At around 32 months of age, the reconstruction of phrase structures elicits adultlike ERPs marking automatic phrase structure reconstructions that are assumed to have sources in left anterior temporal regions (Oberecker et al., 2005). (8) Sentence-level processing of complex syntax develops gradually until young adulthood and is driven by the increasing selectivity of the left inferior frontal cortex and the maturation of its connection to the posterior superior temporal cortex via the AF (Skeide et al., 2014, 2015). (3) From six months onward, children access and retrieve lexical entries. Around their first birthday, they sort lexical items into conceptual categories and check their semantic compatibility by recruiting central and posterior parts of the left middle and superior temporal cortex (Friedrich et al., 2015; Travis et al., 2011). (4) Two-year-old children can already evaluate the semantic compatibility of elements in a sentence (Friedrich & Friederici, 2010), and this semantic information serves as a cue facilitating the understanding of syntactically complex sentences until around nine years of age (Skeide et al., 2014) (7). (6b) Prosodic processing at the sentence level is documented in three-year-old children, but only at six years of age do children recognize phrases in sentences without having to rely on pause information (Männel, Schipke, & Friederici, 2013).

categorize and compare these items, recruiting central and posterior parts of the left middle and superior temporal cortex to do so. Around their second birthday, children are able to apply their morphosyntactic knowledge to evaluate the grammatical compatibility of adjacent phrase categories in the left anterior superior temporal cortex (figure 11.4).

High-level computation of complex syntax in the left inferior frontal cortex emerges in the fourth year of life and is gradually and slowly refined from this point on. It is only after the age of 10 years that BA44 reaches its full specificity and ultimate efficiency in processing complex syntax. Functional responses to sentence-level semantic information become neuroanatomically separable from functional responses to sentence-level syntactical information around the age of seven years (figure 11.4). These maturational trajectories are assumed to be driven by general learning-related neuroplasticity, such as pruning and myelination of perisylvian nerve cell assemblies and their interconnecting white matter fiber tracts.

EEG studies on infants and toddlers have provided the main data basis for describing the acquisition of low-level language processing capacities and complementary MRI and MEG data are currently rare, particularly for the second and third year of life. Moreover, the exploration of the neurodevelopmental dynamics of higher-order sentence-level processing is still in its beginnings. Although these first results must await replication by follow-up studies, they currently reveal a consistent picture that can be adapted as new data become available.

Several important research questions remain to be answered. For example, subcortical structures, in particular the left caudate nucleus and the thalamus, are known to support the acquisition of a second language, presumably by regulating adaptation to increasing processing demands (Crinion et al., 2006; Friederici, 2006; Wartenburger et al., 2003). However, the role of these structures in the acquisition of the first language is not yet understood. Furthermore, considering the different linguistic levels of language processing, substantial progress has been made in uncovering the neural development of phonology, syntax, and semantics, but very little is known about the neural basis of emerging pragmatic skills.

REFERENCES

Benasich, A. A., Choudhury, N., Friedman, J. T., Realpe-Bonilla, T., Chojnowska, C., & Gou, Z. (2006). The infant as a prelinguistic model for language learning impairments: Predicting from event-related potentials to behavior. *Neuropsychologia, 44*(3), 396–411.

Benasich, A. A., Choudhury, N. A., Realpe-Bonilla, T., & Roesler, C. P. (2014). Plasticity in developing brain: Active auditory exposure impacts prelinguistic acoustic mapping. *Journal of Neuroscience, 34*(40), 13349–13363.

Bergelson, E., & Swingley, D. (2012). At 6–9 months, human infants know the meanings of many common nouns. *Proceedings of the National Academy of Sciences USA, 109*(9), 3253–3258.

Binder, J. R., Desai, R. H., Graves, W. W., & Conant, L. L. (2009). Where is the semantic system? A critical review and meta-analysis of 120 functional neuroimaging studies. *Cerebral Cortex, 19*(12), 2767–2796.

Bosseler, A. N., Taulu, S., Pihko, E., Makela, J. P., Imada, T., Ahonen, A., & Kuhl, P. K. (2013). Theta brain rhythms index perceptual narrowing in infant speech perception. *Frontiers in Psychology, 4*, 690.

Brauer, J., & Friederici, A. D. (2007). Functional neural networks of semantic and syntactic processes in the developing brain. *Journal of Cognitive Neuroscience, 19*(10), 1609–1623.

Brennan, J., Nir, Y., Hasson, U., Malach, R., Heeger, D. J., & Pylkkanen, L. (2012). Syntactic structure building in the anterior temporal lobe during natural story listening. *Journal of Brain and Language, 120*(2), 163–173.

Bruderer, A. G., Danielson, D. K., Kandhadai, P., & Werker, J. F. (2015). Sensorimotor influences on speech perception in infancy. *Proceedings of the National Academy of Sciences USA, 112*(44), 13531–13536.

Burle, B., Spieser, L., Roger, C., Casini, L., Hasbroucq, T., & Vidal, F. (2015). Spatial and temporal resolutions of EEG: Is it really black and white? A scalp current density view. *International Journal of Psychophysiology, 97*(3), 210–220.

Crinion, J., Turner, R., Grogan, A., Hanakawa, T., Noppeney, U., Devlin, J. T., … Price, C. J. (2006). Language control in the bilingual brain. *Science, 312*(5779), 1537–1540.

Dapretto, M., & Bookheimer, S. Y. (1999). Form and content: Dissociating syntax and semantics in sentence comprehension. *Neuron, 24*(2), 427–432.

Dautriche, I., Swingley, D., & Christophe, A. (2015). Learning novel phonological neighbors: Syntactic category matters. *Cognition, 143*, 77–86.

Dehaene, S., Cohen, L., Morais, J., & Kolinsky, R. (2015). Illiterate to literate: Behavioural and cerebral changes induced by reading acquisition. *Nature Review Neuroscience, 16*(4), 234–244.

Dehaene-Lambertz, G., Hertz-Pannier, L., Dubois, J., Meriaux, S., Roche, A., Sigman, M., & Dehaene, S. (2006). Functional organization of perisylvian activation during presentation of sentences in preverbal infants. *Proceedings of the National Academy of Sciences USA, 103*(38), 14240–14245.

den Ouden, D. B., Saur, D., Mader, W., Schelter, B., Lukic, S., Wali, E., … Thompson, C. K. (2012). Network modulation during complex syntactic processing. *Neuroimage, 59*(1), 815–823.

DeWitt, I., & Rauschecker, J. P. (2012). Phoneme and word recognition in the auditory ventral stream. *Proceedings of the National Academy of Sciences USA, 109*(8), E505–E514.

Dick, A. S., & Tremblay, P. (2012). Beyond the arcuate fasciculus: Consensus and controversy in the connectional anatomy of language. *Brain, 135*(12), 3529–3550.

Dubois, J., Poupon, C., Thirion, B., Simonnet, H., Kulikova, S., Leroy, F., … Dehaene-Lambertz, G. (2015). Exploring the early organization and maturation of linguistic pathways in the human infant brain. *Cerebral Cortex, 26*(5), 2283–2298.

Ethofer, T., Bretscher, J., Gschwind, M., Kreifelts, B., Wildgruber, D., & Vuilleumier, P. (2012). Emotional voice areas: Anatomic location, functional properties, and structural connections revealed by combined fMRI/DTI. *Cerebral Cortex, 22*(1), 191–200.

Fedorenko, E., & Thompson-Schill, S. L. (2014). Reworking the language network. *Trends in Cognitive Sciences, 18*(3), 120–126.

Fengler, A., Meyer, L., & Friederici, A. D. (2015). Brain structural correlates of complex sentence comprehension in children. *Developmental Cognitive Neuroscience, 15*, 48–57.

Fiebach, C. J., Friederici, A. D., Müller, K., von Cramon, D. Y., & Hernandez, A. E. (2003). Distinct brain representations for early and late learned words. *Neuroimage, 19*(4), 1627–1637.

Friederici, A. D. (2006). What's in control of language? *Nature Neuroscience, 9*(8), 991–992.

Friederici, A. D. (2011). The brain basis of language processing: From structure to function. *Physiological Reviews, 91*(4), 1357–1392.

Friederici, A. D. (2012). The cortical language circuit: From auditory perception to sentence comprehension. *Trends in Cognitive Sciences, 16*(5), 262–268.

Friederici, A. D., Bahlmann, J., Heim, S., Schubotz, R. I., & Anwander, A. (2006). The brain differentiates human and non-human grammars: Functional localization and structural connectivity. *Proceedings of the National Academy of Sciences USA, 103*(7), 2458–2463.

Friederici, A. D., Friedrich, M., & Christophe, A. (2007). Brain responses in 4-month-old infants are already language specific. *Current Biology, 17*(14), 1208–1211.

Friederici, A. D., & Gierhan, S. M. (2013). The language network. *Current Opinion in Neurobiology, 23*(2), 250–254.

Friedrich, M., & Friederici, A. D. (2005a). Lexical priming and semantic integration reflected in the event-related potential of 14-month-olds. *Neuroreport, 16*(6), 653–656.

Friedrich, M., & Friederici, A. D. (2005b). Phonotactic knowledge and lexical-semantic processing in one-year-olds: Brain responses to words and nonsense words in picture contexts. *Journal of Cognitive Neuroscience, 17*(11), 1785–1802.

Friedrich, M., & Friederici, A. D. (2005c). Semantic sentence processing reflected in the event-related potentials of one- and two-year-old children. *Neuroreport, 16*(16), 1801–1804.

Friedrich, M., & Friederici, A. D. (2010). Maturing brain mechanisms and developing behavioral language skills. *Journal of Brain and Language, 114*(2), 66–71.

Friedrich, M., & Friederici, A. D. (2011). Word learning in 6-month-olds: Fast encoding-weak retention. *Journal of Cognitive Neuroscience, 23*(11), 3228–3240.

Friedrich, M., Wilhelm, I., Born, J., & Friederici, A. D. (2015). Generalization of word meanings during infant sleep. *Nature Communications, 6*, 6004.

Frühholz, S., Ceravolo, L., & Grandjean, D. (2012). Specific brain networks during explicit and implicit decoding of emotional prosody. *Cerebral Cortex, 22*(5), 1107–1117.

Gervain, J. (2015). Plasticity in early language acquisition: The effects of prenatal and early childhood experience. *Current Opinion in Neurobiology, 35*, 13–20.

Gomez, D. M., Berent, I., Benavides-Varela, S., Bion, R. A., Cattarossi, L., Nespor, M., & Mehler, J. (2014). Language universals at birth. *Proceedings of the National Academy of Sciences USA, 111*(16), 5837–5841.

Gupta, N., Henry, R. G., Strober, J., Kang, S. M., Lim, D. A., Bucci, M., ... Rowitch, D. H. (2012). Neural stem cell engraftment and myelination in the human brain. *Science Translational Medicine, 4*(155), 155ra137.

Guttorm, T. K., Leppanen, P. H. T., Poikkeus, A. M., Eklund, K. M., Lyytinen, P., & Lyytinen, H. (2005). Brain event-related potentials (ERPs) measured at birth predict later language development in children with and without familial risk for dyslexia. *Cortex, 41*(3), 291–303.

Hagoort, P. (2013). MUC (memory, unification, control) and beyond. *Frontiers in Psychology, 4*, 416.

Herrmann, B., Maess, B., Hahne, A., Schröger, E., & Friederici, A. D. (2011). Syntactic and auditory spatial processing in the human temporal cortex: An MEG study. *Neuroimage, 57*(2), 624–633.

Hickok, G., & Poeppel, D. (2007). The cortical organization of speech processing. *Nature Review Neuroscience, 8*(5), 393–402.

Höhle, B., Weissenborn, E., Kiefer, D., Schulz, A., & Schmitz, M. (2004). Functional elements in infants' speech processing: The role of determiners in the syntactic categorization of lexical elements. *Infancy, 5*(3), 341–353.

Homae, F., Watanabe, H., Nakano, T., Asakawa, K., & Taga, G. (2006). The right hemisphere of sleeping infant perceives sentential prosody. *Neuroscience Research, 54*(4), 276–280.

Huth, A. G., de Heer, W. A., Griffiths, T. L., Theunissen, F. E., & Gallant, J. L. (2016). Natural speech reveals the semantic maps that tile human cerebral cortex. *Nature, 532*(7600), 453–458.

Johnson, K. L., Nicol, T., Zecker, S. G., & Kraus, N. (2008). Developmental plasticity in the human auditory brainstem. *Journal of Neuroscience, 28*(15), 4000–4007.

Junge, C., Cutler, A., & Hagoort, P. (2012). Electrophysiological evidence of early word learning. *Neuropsychologia, 50*(14), 3702–3712.

Kim, S. G., Richter, W., & Ugurbil, K. (1997). Limitations of temporal resolution in functional MRI. *Magnetic Resonance in Medicine, 37*(4), 631–636.

Kuhl, P. K. (2004). Early language acquisition: Cracking the speech code. *Nature Review Neuroscience, 5*(11), 831–843.

Kuhl, P. K., Ramirez, R. R., Bosseler, A., Lin, J. F., & Imada, T. (2014). Infants' brain responses to speech suggest analysis by synthesis. *Proceedings of the National Academy of Sciences USA, 111*(31), 11238–11245.

Kuhl, P. K., Stevens, E., Hayashi, A., Deguchi, T., Kiritani, S., & Iverson, P. (2006). Infants show a facilitation effect for native language phonetic perception between 6 and 12 months. *Developmental Science, 9*(2), F13–F21.

Lau, E. F., Phillips, C., & Poeppel, D. (2008). A cortical network for semantics: (De)constructing the N400. *Nature Review Neuroscience, 9*(12), 920–933.

MacGregor, L. J., Pulvermüller, F., van Casteren, M., & Shtyrov, Y. (2012). Ultra-rapid access to words in the brain. *Nature Communications, 3*, 711.

Mahmoudzadeh, M., Dehaene-Lambertz, G., Fournier, M., Kongolo, G., Goudjil, S., Dubois, J., ... Wallois, F. (2013). Syllabic discrimination in premature human infants prior to complete formation of cortical layers. *Proceedings of the National Academy of Sciences USA, 110*(12), 4846–4851.

Makuuchi, M., Bahlmann, J., Anwander, A., & Friederici, A. D. (2009). Segregating the core computational faculty of human language from working memory. *Proceedings of the National Academy of Sciences USA, 106*(20), 8362–8367.

Makuuchi, M., & Friederici, A. D. (2013). Hierarchical functional connectivity between the core language system and the working memory system. *Cortex, 49*(9), 2416–2423.

Männel, C., & Friederici, A. D. (2009). Pauses and intonational phrasing: ERP studies in 5-month-old German infants and adults. *Journal of Cognitive Neuroscience, 21*(10), 1988–2006.

Männel, C., & Friederici, A. D. (2011). Intonational phrase structure processing at different stages of syntax acquisition: ERP studies in 2-, 3-, and 6-year-old children. *Developmental Science, 14*(4), 786–798.

Männel, C., & Friederici, A. D. (2013). Accentuate or repeat? Brain signatures of developmental periods in infant word recognition. *Cortex, 49*(10), 2788–2798.

Männel, C., Schipke, C. S., & Friederici, A. D. (2013). The role of pause as a prosodic boundary marker: Language ERP studies in German 3- and 6-year-olds. *Developmental Cognitive Neuroscience, 5*, 86–94.

Meyer, L., Obleser, J., Anwander, A., & Friederici, A. D. (2012). Linking ordering in Broca's area to storage in left temporo-parietal regions: The case of sentence processing. *Neuroimage, 62*(3), 1987–1998.

Mitchell, T. M., Shinkareva, S. V., Carlson, A., Chang, K. M., Malave, V. L., Mason, R. A., & Just, M. A. (2008). Predicting human brain activity associated with the meanings of nouns. *Science, 320*(5880), 1191–1195.

Mottonen, R., & Watkins, K. E. (2009). Motor representations of articulators contribute to categorical perception of speech sounds. *Journal of Neuroscience, 29*(31), 9819–9825.

Newman, S. D., Ikuta, T., & Burns, T., Jr. (2010). The effect of semantic relatedness on syntactic analysis: An fMRI study. *Journal of Brain and Language, 113*(2), 51–58.

Nuñez, S. C., Dapretto, M., Katzir, T., Starr, A., Bramen, J., Kan, E., ... Sowell, E. R. (2011). fMRI of syntactic processing in typically developing children: Structural correlates in the inferior frontal gyrus. *Developmental Cognitive Neuroscience, 1*(3), 313–323.

Oberecker, R., & Friederici, A. D. (2006). Syntactic event-related potential components in 24-month-olds' sentence comprehension. *Neuroreport, 17*(10), 1017–1021.

Oberecker, R., Friedrich, M., & Friederici, A. D. (2005). Neural correlates of syntactic processing in two-year-olds. *Journal of Cognitive Neuroscience, 17*(10), 1667–1678.

Obleser, J., Wise, R. J., Dresner, M. A., & Scott, S. K. (2007). Functional integration across brain regions improves speech perception under adverse listening conditions. *Journal of Neuroscience, 27*(9), 2283–2289.

Ortiz-Mantilla, S., Hämäläinen, J. A., Musacchia, G., & Benasich, A. A. (2013). Enhancement of gamma oscillations indicates preferential processing of native over foreign phonemic contrasts in infants. *Journal of Neuroscience, 33*(48), 18746–18754.

Overath, T., McDermott, J. H., Zarate, J. M., & Poeppel, D. (2015). The cortical analysis of speech-specific temporal structure revealed by responses to sound quilts. *Nature Neuroscience, 18*(6), 903–911.

Partanen, E., Kujala, T., Naatanen, R., Liitola, A., Sambeth, A., & Huotilainen, M. (2013). Learning-induced neural plasticity of speech processing before birth. *Proceedings of the National Academy of Sciences USA, 110*(37), 15145–15150.

Paus, T. (2010). Growth of white matter in the adolescent brain: Myelin or axon? *Brain and Cognition, 72*(1), 26–35.

Peña, M., Pittaluga, E., & Mehler, J. (2010). Language acquisition in premature and full-term infants. *Proceedings of the National Academy of Sciences USA, 107*(8), 3823–3828.

Peña, M., Werker, J. F., & Dehaene-Lambertz, G. (2012). Earlier speech exposure does not accelerate speech acquisition. *Journal of Neuroscience, 32*(33), 11159–11163.

Perani, D., Saccuman, M. C., Scifo, P., Anwander, A., Spada, D., Baldoli, C., ... Friederici, A. D. (2011). Neural language networks at birth. *Proceedings of the National Academy of Sciences USA, 108*(38), 16056–16061.

Petanjek, Z., Judas, M., Simic, G., Rasin, M. R., Uylings, H. B., Rakic, P., & Kostovic, I. (2011). Extraordinary neoteny of synaptic spines in the human prefrontal cortex. *Proceedings of the National Academy of Sciences USA, 108*(32), 13281–13286.

Pons, F., Lewkowicz, D. J., Soto-Faraco, S., & Sebastian-Galles, N. (2009). Narrowing of intersensory speech perception in infancy. *Proceedings of the National Academy of Sciences USA, 106*(26), 10598–10602.

Raschle, N., Zuk, J., Ortiz-Mantilla, S., Sliva, D. D., Franceschi, A., Grant, P. E., ... Gaab, N. (2012). Pediatric neuroimaging in early childhood and infancy: challenges and practical guidelines. *Annals of the New York Academy of Sciences, 1252*, 43–50.

Rodd, J. M., Davis, M. H., & Johnsrude, I. S. (2005). The neural mechanisms of speech comprehension: fMRI studies of semantic ambiguity. *Cerebral Cortex, 15*(8), 1261–1269.

Sammler, D., Grosbras, M. H., Anwander, A., Bestelmeyer, P. E., & Belin, P. (2015). Dorsal and ventral pathways for prosody. *Current Biology, 25*(23), 3079–3085.

Sammler, D., Kotz, S. A., Eckstein, K., Ott, D. V. M., & Friederici, A. D. (2010). Prosody meets syntax: The role of the corpus callosum. *Brain, 133*, 2643–2655.

Saur, D., Kreher, B. W., Schnell, S., Kummerer, D., Kellmeyer, P., Vry, M. S., ... Weiller, C. (2008). Ventral and dorsal pathways for language. *Proceedings of the National Academy of Sciences USA, 105*(46), 18035–18040.

Schipke, C. S., Friederici, A. D., & Oberecker, R. (2011). Brain responses to case-marking violations in German preschool children. *Neuroreport, 22*(16), 850–854.

Schipke, C. S., Knoll, L. J., Friederici, A. D., & Oberecker, R. (2012). Preschool children's interpretation of object-initial sentences: Neural correlates of their behavioral performance. *Developmental Science, 15*(6), 762–774.

Schnupp, J., Nelken, I., & King, A. (2011). *Auditory neuroscience: Making sense of sound.* Cambridge, MA: MIT Press.

Scholz, J., Klein, M. C., Behrens, T. E., & Johansen-Berg, H. (2009). Training induces changes in white-matter architecture. *Nature Neuroscience, 12*(11), 1370–1371.

Skeide, M. A., Brauer, J., & Friederici, A. D. (2014). Syntax gradually segregates from semantics in the developing brain. *Neuroimage, 100*, 106–111.

Skeide, M. A., Brauer, J., & Friederici, A. D. (2015). Brain functional and structural predictors of language performance. *Cerebral Cortex, 26*(5), 2127–2139.

Snijders, T. M., Vosse, T., Kempen, G., van Berkum, J. J., Petersson, K. M., & Hagoort, P. (2009). Retrieval and unification of syntactic structure in sentence comprehension: An FMRI study using word-category ambiguity. *Cerebral Cortex, 19*(7), 1493–1503.

Soderstrom, M., Seidl, A., Nelson, D. G. K., & Jusczyk, P. W. (2003). The prosodic bootstrapping of phrases: Evidence from prelinguistic infants. *Journal of Memory and Language, 49*(2), 249–267.

Travis, K. E., Leonard, M. K., Brown, T. T., Hagler, D. J., Jr., Curran, M., Dale, A. M., … Halgren, E. (2011). Spatiotemporal neural dynamics of word understanding in 12- to 18-month-old-infants. *Cerebral Cortex, 21*(8), 1832–1839.

Tyler, L. K., Marslen-Wilson, W. D., Randall, B., Wright, P., Devereux, B. J., Zhuang, J., … Stamatakis, E. A. (2011). Left inferior frontal cortex and syntax: function, structure and behaviour in patients with left hemisphere damage. *Brain, 134*(2), 415–431.

van der Lely, H. K., & Pinker, S. (2014). The biological basis of language: Insight from developmental grammatical impairments. *Trends in Cognitive Sciences, 18*(11), 586–595.

van Heugten, M., & Christophe, A. (2015). Infants' acquisition of grammatical gender dependencies. *Infancy, 20*(6), 675–683.

Visser, M., Jefferies, E., & Lambon Ralph, M. A. (2010). Semantic processing in the anterior temporal lobes: A meta-analysis of the functional neuroimaging literature. *Journal of Cognitive Neuroscience, 22*(6), 1083–1094.

Wartenburger, I., Heekeren, H. R., Abutalebi, J., Cappa, S. F., Villringer, A., & Perani, D. (2003). Early setting of grammatical processing in the bilingual brain. *Neuron, 37*(1), 159–170.

Wartenburger, I., Steinbrink, J., Telkemeyer, S., Friedrich, M., Friederici, A. D., & Obrig, H. (2007). The processing of prosody: Evidence of interhemispheric specialization at the age of four. *Neuroimage, 34*(1), 416–425.

Weiller, C., Bormann, T., Saur, D., Musso, M., & Rijntjes, M. (2011). How the ventral pathway got lost: And what its recovery might mean. *Journal of Brain and Language, 118*(1), 29–39.

Weiller, C., Musso, M., Rijntjes, M., & Saur, D. (2009). Please don't underestimate the ventral pathway in language. *Trends in Cognitive Sciences, 13*(9), 369–370.

Werker, J. F., & Hensch, T. K. (2015). Critical periods in speech perception: New directions. *Annual Review of Psychology, 66*, 173–196.

Werker, J. F., & Tees, R. C. (1984). Cross-language speech-perception—Evidence for perceptual reorganization during the 1st year of life. *Infant Behavior and Development, 7*(1), 49–63.

Wilson, S. M., Galantucci, S., Tartaglia, M. C., Rising, K., Patterson, D. K., Henry, M. L., … Gorno-Tempini, M. L. (2011). Syntactic processing depends on dorsal language tracts. *Neuron, 72*(2), 397–403.

12 Multilingual Development

VASILIKI CHONDROGIANNI

Bilingual and multilingual children constitute an ever-increasing population of learners across the world. In the United Kingdom alone, the proportion of children whose first language is not English has almost tripled in the last 20 years (NALDIC, 2013). It is estimated that approximately half of the world's population will be functional bilinguals or multilinguals at some stage in their lives (Grosjean, 2010). This increase in the number of bilingual children has led to the evolution of the field from diary or naturalistic data (Ronjat, 1913; Leopold, 1978) toward more systematic and large-scale studies. The profile of the child bilingual leaner studied has also expanded from a focus on simultaneous bilingual (2L1) children (Genesee & Nicoladis, 2007), that is children exposed to both languages from birth (Meisel, 2008) or within the first year of life (De Houwer, 1990), to include sequential bilingual (L2) children. These are children exposed to the second language after properties of their first language (L1) are in place (Unsworth, 2005).

The investigation of bilingual development is important for theoretical and practical reasons. From a theoretical point, bilingual development offers a natural testing ground for acquisition theories that give central stage to external factors in the developmental process, such as input, and that investigate in what way these factors affect various linguistic domains. These theories, usually referred to as *empiricist* theories of language acquisition (Bybee, 2010; Tomasello, 2008), postulate a strong relationship between the different language domains and especially between lexical and grammatical development (Marchman, Martínez-Sussmann, & Dale, 2004). The study of bilingual development is also important for *nativist* accounts of acquisition that assume an innate capacity for language learning and a specified schedule subject to maturational constraints. This innate schedule allows for differential acquisition paths for the various language domains and places less significance on input quantity and quality (Chomsky, 1981, 1995; Valian, 2014). Despite their differences, both theories need to account for the processes and the mechanisms that get the bilingual learner from the initial to the final state of language development (Valian, 2014). Additionally, one would need to explain whether and how factors specific to the child, such as the languages that they speak, children's age of exposure or onset (AoO) to the two languages, as well as factors external to the child, such as input quantity and quality, affect the cognitive architecture and the child's learning mechanism.

From an educational and clinical perspective, a number of studies have shown that L2 children or 2L1 children with unbalanced input in one of their two languages exhibit similar learner profiles to children with language impairment (Kohnert, 2013; Paradis, 2010). This can lead to the misdiagnosis of bilingual children as being language impaired when they are not, or to lack of diagnosis when it is a true case of a language impaired bilingual child (Armon-Lotem, de Jong, & Meir, 2015; Paradis, 2010, for an overview). Understanding how multilingual children acquire their two or more languages is important for tackling these issues and for making sure that they receive necessary and tailored educational and clinical provisions. For that reason, in the present chapter, we examine the factors that affect their performance in the two languages, rather than by continuously comparing them to monolingual children, although this is at times inevitable given the current line of research.

The present chapter reviews state-of-the-art research on bilingual and multilingual development, with a view to highlighting the current issues in the field and identifying the common research thread across the different studies. Broadly, we will cover the following topics: (a) whether the course and rate of bilingual development differ from that of monolingual development, (b) what evidence exists that the two languages of the bilingual individual are differentiated, and (c) how input and age of bilingual exposure interact and affect the degree of crosslinguistic influence. The present chapter is organized as follows. Section 1 examines language development in 2L1 children in the domains of phonology, morphosyntax, and vocabulary and addresses issues of rate and course of acquisition in bilingual children. Section 2 investigates the same issues in L2 children. Section 3 discusses the relationship between different language domains, and section 4 investigates the effects of child-external factors, such as input quantity and quality and socioeconomic status (SES). The focus of this chapter will be more on the linguistic and cognitive aspects of bilingual or multilingual development and less on the social and personality variables that may influence developmental outcomes (for an overview, see Hoff, 2015; de Houwer, 2007). In doing so, we review experimental studies that have used different methodologies and experimental paradigms and that have examined different modalities, from perception and comprehension to production, and we evaluate how these different methodologies may give rise to sometimes contradicting results. We conclude with a general appraisal of the current status of the field and with future directions.

1. Language Development in Simultaneous Bilingual Children

A long-standing debate and the focus of early studies on 2L1 children is whether the two languages of the bilingual individual reflect a single language system characterized by language mixing (Volterra & Taeschner, 1978) or whether they are differentiated and constitute two separate language systems (Genesee, Nicoladis, & Paradis, 1995; de Houwer, 1990). Nowadays, there is accumulating evidence that the two (or more) languages of the multilingual individual are separated but also interact in important ways (Genesee & Nicoladis, 2007; Paradis & Genesee, 1996; Montanari, 2011). Evidence against a unitary system focuses on three main arguments: (i) bilingual children are able to contextually differentiate the use of their two languages, (ii) the order of development reflects the characteristics of the language to be learned, and (iii) language mixing is not a sufficient condition or argument for a single system, because it may merely reflect strategies that bilingual children use in the course of development (Genesee, 1989; Meisel, 1989). Hence, the focus has shifted from whether the two languages of the bilingual child are differentiated to investigating the child-internal and -external factors that trigger and influence the interaction between languages (Byers-Heinlein, 2014; Hoff, 2015).

1.1. Speech Discrimination and Phonological Development in Bilingual Children Bilingual infants are faced with the complex task of perceiving and differentiating between their two phonetic systems and of developing discrete phonological categories that will serve as the basis for word segmentation and learning (Byers-Heinlein, 2014). Even before birth, monolingual and bilingual infants are sensitive to the language spoken by their caregivers in utero (Byers-Heinlein, Burns, & Werker, 2010). After birth, infants as young as four months old can make use of surface cues such as rhythm to discriminate between two languages that are rhythmically different (e.g., differentiating between syllable-timed languages, such as French or Filipino, and stress-timed languages, such as English; Byers-Heinlein & Werker, 2009), and they can achieve this by just watching silent talking faces (Weikum et al., 2007). Bilingual infants are also able to discriminate between the two languages even when they belong to the same rhythmical class (e.g., Catalan and Spanish; Bosch & Sebastián-Gallés, 1997, 2003). This indicates that bilingual children's capacity to discriminate between the two languages is not simply based on the "familiar" versus "different" discrimination, as in the case of monolingual infants, but to subtler sensitivities to rhythmical cues.

Bilingual infants can also form phonetic categories and discriminate between phonetic units belonging to their two languages, although the rate and process of speech discrimination may be different from that of monolingual infants. In a series of studies with Spanish-Catalan bilingual infants, Bosch and Sebastián-Gallés and colleagues have offered evidence of a U-shaped pattern in Catalan-Spanish bilinguals' capacity to discriminate the two mid-front vowels of Catalan (i.e., /e/ vs. /ɛ /; Bosch & Sebastián-Gallés, 2003; Sebastián-Gallés & Bosch, 2009). Both monolingual and bilingual groups discriminated the contrast at the age of four months. However, the 8-month-old bilingual infants had lost the capacity to discriminate this vowel contrast, only to regain it at the age of 12 months. These results have been replicated in a series of studies on vowels and voicing /s-z/ contrasts in both Catalan and Spanish (Bosch & Sebastián-Gallés, 2003; Sebastián-Gallés & Bosch, 2009) and on voice onset time contrasts in English and Spanish by English-Spanish 10- to 12-month-old bilingual infants (Garcia-Sierra et al., 2011). However, this discrimination loss has been challenged in studies examining the perception of consonant contrasts in other language combinations (English-French) and using subtler experimental paradigms, such as eye tracking (Burns, Yoshida, Hill, & Werker 2007; Sundara, Polka, & Molnar, 2008).

Bilingual infants have also been found to retain their "universal listener" capacity for longer than their monolingual counterparts (Kuhl et al., 2006). This is the capacity to recognize both native and non-native sounds, and this is also reflected in their neural responses (Ferjan Ramirez, Ramírez, Clarke, Taulu, & Kuhl, 2017). At the age of 10–12 months, when bilingual infants are still able to make this discrimination, monolingual infants have lost this capacity, indicating a shift from language-general to language-specific discrimination (Kuhl et al.2008; Petitto et al, 2012).

Altogether, the findings from these studies suggest that early sound discrimination in bilingual infants follows the developmental path as that of monolingual children. At the same time, bilingual infants are able to consolidate the frequency and distributional properties of the different sounds in the input across the two sound systems they are exposed to. Furthermore, factors inherent to the bilingual context, such as the proportion of cognate words between languages, processing differences between vowels and consonants, as well as task effects, can give rise to different results across studies (Bosch & Sebastián-Gallés, 2003; Kuhl et al., 2008; Sebastián-Gallés & Bosch, 2009; Sundara et al., 2008; Petitto et al., 2012).

Studies with older 2L1 children have reported an interaction between the two phonological systems of the bilingual individual in the form of crosslinguistic influence that may lead to acceleration, that is to a faster developmental rate in bilinguals versus monolinguals, as well as to deceleration or delay in the acquisition of the target sounds in the two languages (Genesee & Nicoladis, 2007). Evidence for acceleration in the phonological domain has been reported for the acquisition of English consonant clusters by Polish-English 2L1 children (Tamburelli, Sanoudaki, Jones, & Sowinska, 2015), or for Spanish coda consonants by Spanish-German bilingual children in comparison to Spanish monolingual children (Lleó, Kuchenbrandt, Kehoe, & Trujillo, 2003), and even in the context of trilingual acquisition by a Tagalog-Spanish-English child and her acquisition of the target phonetic inventory in relation to her monolingual peers (Montanari, 2011). Conversely, Gildersleeve-Neumann, Kester, Davis, and Peña (2008) and Goldstein and Washington (2001) found evidence for delayed acquisition of English phonology by Spanish-English bilingual three- and four-year-olds. Bilingual children had a lower intelligibility rating and made more overall consonant and vowel errors than their English-speaking monolingual counterparts did. In bilingual development, acceleration and deceleration may occur simultaneously to facilitate acquisition of phonetic inventories across the two languages. In the study by Fabiano-Smith and Goldstein (2010), Spanish-English bilingual children had lower accuracy than did their monolingual counterparts on certain phonological skills (consonants) but not on all. The results from these studies suggest that the two phonological systems interact while remaining discrete and systematic, although not necessarily nativelike (Fabiano-Smith & Barlow, 2010; Fabiano-Smith & Goldstein, 2010).

1.2. Lexical Development in 2L1 Children

Having established the phonetic and phonological categories in their L1, bilingual infants come to the complex task of recognizing a word and learning its meaning. This is a multifaceted task that, among other things, involves segmenting the speech stream into possible words, linking the form of a word to its meaning and representing information about the word in the mental lexicon (Werker & Byers-Heinlein, 2008).

Along with the task of learning two phonological systems, bilingual infants are faced with the task of learning two sets of linguistic labels for a single object or referent. At the age of 14 months, both monolingual and bilingual infants are capable of learning novel labels for objects when the sounds associated with these labels are dissimilar enough, for example, *lif* and *neem*, or of distinguishing between familiar and unfamiliar words (Mills, Coffey-Corina, & Neville, 1997; Conboy & Mills, 2006). However, the two groups differ when acquiring labels that differ minimally, for example, *bih* and *dih*. The associative link between a novel object and its label in a minimal pair context is established in monolingual infants at the age of 17 months, whereas bilingual infants manage to do so at the age of 20 months (Fennell, Byers-Heinlein, & Werker, 2007). Interestingly, bilingual infants are able to discriminate minimal-paired words at 17 months similarly to their monolingual peers when relevant contextual information is provided (Fennell et al., 2007). This suggests that the process of associative learning may be more context-dependent in bilingual infants, reflecting the contextual separation of their two languages (Genesee, 1989).

The way that bilingual infants approach vocabulary learning may also indicate that the constraints that subserve bilingual word learning are either not operative or operate differently in bilingual infants compared to what has been reported for monolingual infants. *Mutual exclusivity* or *disambiguation* is a constraint that limits the number of possible referents a label can have and, hence, aids discrimination between a familiar and a novel label, as the child postulates that new words tend to refer to new referents. This constraint has been found to operate in monolingual lexical development at the age of 17 months (Byers-Heinlein & Werker, 2009; Halberda, 2003). Studies using the looking-while-listening paradigm have found that bilingual and trilingual infants do not make use of mutual exclusivity in the same way as their monolingual peers do. Importantly, the use of this constraint has been shown to be a function of the number of languages spoken by the multilingual individual and the proportion of translational equivalents that the children knew in each language (Byers-Heinlein & Werker, 2009, 2013). However, a closer look at the results reveals that the effects reported for the bilingual group were marginally significant, with small effect sizes (Byers-Heinlein &

Werker, 2009). An additional caveat in the existing mutual exclusivity literature is that studies do not clarify whether the experimental setup evokes a within- or a between-language discrimination. Contextual discrimination may impact on bilingual children's readiness to accept multiple labels for single referents (see Burns et al., 2007, and Sundara et al., 2008, for the issue of context in bilingual phonological discrimination). Given the complexity of the acquisition of vocabulary in monolingual children (Kucker, chapter 9 of this volume) and given the paucity of studies that have examined similar factors in bilingual development, our understanding of how the two languages of the bilingual interact and influence the process and the outcome of lexical development is far from complete.

For theoretical, educational, and clinical purposes, a large body of research in bilingual acquisition has been devoted to whether bilingual children have the same vocabulary knowledge as their monolingual peers do (Oller, Pearson, & Cobo-Lewis, 2007; Pearson, Fernández, Lewedeg, & Oller, 1997). Vocabulary knowledge has been defined in two ways: as (i) vocabulary size and (ii) lexical processing efficiency. In terms of vocabulary size, early studies with bilingual children reported that the vocabulary development in each language of the bilingual child was within the normal range for monolingual children (Pearson, Fernandez, & Oller, 1993). However, these studies were based on small sample sizes that did not lead to statistical differences between the monolingual and the bilingual groups (Hoff et al., 2012). In more recent studies with larger sample sizes of bilingual toddlers younger than three years using caregiver-report questionnaires, such as the MacArthur-Bates Communicative Development Inventories (Fenson et al., 2007), it has been reported that bilingual children have smaller vocabularies than their monolingual peers do when tested in one of their two languages (Hoff et al.; Hurtado, Marchman, & Fernald, 2008; Marchman, Fernald, & Hurtado, 2010; Thordardottir, 2011) and that this difference persists in later childhood (Bialystok et al., 2010; Gross, Buac, & Kaushanskaya, 2015; Eilers, Pearson & Cobo-Lewis, 2006; Umbel, Pearson, Fernández, & Oller, 1992). The gap between monolingual and bilingual children may narrow with age (Bialystok & Feng, 2011; Gathercole & Thomas, 2009; Oller et al., 2007) and may be restricted to specific contexts, for example, academic versus home contexts (Bialystok et al., 2010).

A characteristic of bilingual children's vocabulary is its distributive nature, that is, labels for referents may be distributed across the two or three languages of the bi- or multilingual individual. Differences between monolingual and bilingual vocabularies tend to disappear when bilingual children's vocabularies in both languages are combined (Core, Hoff, Rumiche, & Señor, 2013) or when conceptual scoring is taken into account. Conceptual scoring refers to the total number of concepts for which a child has a word in at least one language, and it has been shown to give a more complete picture of the bilingual child's lexical knowledge, as well as remove any differences between monolingual and bilingual children (Bedore et al., 2012; Kan & Kohnert, 2005; Pearson, Fernández, & Oller, 1993). However, this may depend on the modality being tested with productive vocabulary giving rise to more difficulties than with receptive vocabulary (Gross et al., 2015; Kan & Kohnert, 2005).

Interestingly, vocabulary size has been reported to affect spoken-word recognition in real time. In a series of studies using the MacArthur-Bates Communicative Development Inventories to measure vocabulary size and using subtle eye-tracking techniques to measure processing efficiency, operationalized through the "looking-while-listening" procedure, Fernald and colleagues examined how vocabulary size in Spanish and English monolingual (Fernald, Perfors, & Marchman, 2006; Fernald, Pinto, Swingley, Weinbergy, & McRoberts, 1998; Hurtado et al., 2008) and bilingual children (Marchman et al., 2010) affected the time course of spoken-word recognition from moment to moment. Given that vocabulary size in these studies was found to correlate with input quantity and SES as well, it raises important theoretical implications to which we return in section 3.

1.3. Morphosyntactic Development in 2L1 Children Despite the breadth of studies examining the perception and comprehension of phonological and lexical categories and constraints, research on the morphosyntactic abilities of 2L1 infants and children has been dominated by production studies. Most early naturalistic studies on the morphosyntactic development of preschool 2L1 children have shown that the two languages of the bilingual individual follow the same developmental course as monolingual children, with any differences between the two groups reflecting the dominance of the child's language or exposure patterns (see Genesee & Nicoladis, 2007, and Unsworth, 2013c, for an overview of studies).

Experimental studies with older 2L1 children examining morphological phenomena such as grammatical gender (Gathercole & Thomas, 2009; Unsworth, 2013a; Unsworth et al., 2014) or plural morphology (Thomas, Williams, Jones, Davies, & Binks, 2013) have reported that differences in the rate of acquisition between monolingual and bilingual children reflect language-specific properties. For example, Welsh and Dutch grammatical gender and Welsh noun plurals are notoriously opaque and complex to acquire. Conversely, Greek gender provides plenty of clear morphophonological cues for gender assignment. Greek children acquire gender faster and more accurately than Dutch or Welsh children (Gathercole & Thomas, 2009; Thomas et al., 2013; Unsworth et al., 2014). Importantly, however, 2L1 children did not make different types of errors compared to their monolingual peers (Unsworth, 2013a).

In 2L1 acquisition, phenomena at the interface between syntax and discourse have been susceptible to crosslinguistic influence. Crosslinguistic influence has been argued to result from the structural overlap between the two languages; when language A allows more than one option for a structure, and language B overlaps with one of those options, crosslinguistic influence will occur, and this influence is expected to be unidirectional, from language A to language B but not vice versa (Hulk & Müller, 2000; Müller & Hulk, 2001). The use and distribution of null versus overt pronouns by children speaking non-null subject and null subject language have offered support for crosslinguistic influence. Italian-English bilingual children tend to overextend the use of overt pronouns in English to contexts in Italian where Italian, a null subject language, would not normally use them (Paradis & Navarro, 2003; Serratrice, Sorace, & Paoli, 2004; Sorace, Serratrice, Filiaci, & Baldo, 2009). Crosslinguistic influence seems to be modulated by language dominance (Foroodi-Nejad & Paradis, 2009; Yip & Matthews, 2007) and may take the form of acceleration, as in the case of the acquisition of German articles by Italian-German 2L1 children dominant in Italian (Kupisch, 2007), or as delay, as in the case of Cantonese-English bilingual children's comprehension of object relative clauses (Kidd, Chan, & Chiu, 2015). However, crosslinguistic influence has been shown to be bidirectional (Lillo-Martin, Koulidobrova, de Quadros, & Chen Pichler, 2012) and has been found for narrow syntax (Argyri & Sorace, 2007), syntax-morphology (Foroodi-Nejad & Paradis, 2009), and syntax-semantics/lexicon phenomena (Pérez-Leroux et al., 2009), challenging the initial formulation of Hulk and Müller's theory (Hulk & Müller, 2000; Müller & Hulk, 2001).

2. *Language Development in Sequential Bilingual Children*

L2 children are exposed to the L2 after properties of their L1 are already in place. According to the most prominent current definition, a child L2 learner is defined as one who is exposed to the L2 after the age of three or four years but before the age of seven years (Schwartz, 2004; Unsworth, 2005). The lower cutoff point has been based on specific assumptions about when basic properties of the first language, such as word order and inflectional morphology, are acquired; the upper cutoff point is linked to the notion of "sensitive periods" in L2 acquisition (Johnson & Newport, 1989), according to which there is a decline in neuroplasticity that affects learnability. This decline has been suggested to take effect around the age of seven years, although the exact age may differ for the various language domains (Abrahamsson & Hyltenstam, 2009; Johnson & Newport; Muñoz & Singleton, 2011). Recently, the lower cutoff point has been contested, with studies examining whether AoO effects in childhood emerge earlier than the age of three or four years. If differences can be found between L2 children exposed to the L2 after the age of four, on the one hand, and early L2 children, that is children between the ages of one and three years, on the other, "sensitive periods" to L2 acquisition can be argued to emerge much earlier than previously thought (Meisel, 2008; Unsworth, 2013a). However, addressing AoO effects in childhood is a more complicated matter. Recent studies have shown that AoO effects may be modulated by whether a phenomenon is early acquired in monolingual children. According to Tsimpli (2014), phenomena that are early acquired by monolingual children, such as gender in Greek (Unsworth et al., 2014) or case marking in German (Roesch & Chondrogianni, 2016), are more susceptible to AoO effects in child L2 acquisition, as their associated properties are specified early in L1 children's grammars. Conversely, late-acquired phenomena in L1 acquisition, such as *wh*-questions and passives (Chondrogianni & Marinis, 2011; Roesch & Chondrogianni, 2016), may be subject to L2 exposure effects rather than to AoO effects because they take longer to be fully mastered by L1 children as well. In the following, we review studies investigating how L2 children may differ from their 2L1 peers on phonological, lexical, and morphosyntactic development.

2.1. Acquisition of Phonology in L2 Children Child L2 phonological development can follow similar developmental paths to that of monolingual children. L2 children have been found to make systematic phonological errors, such as cluster reduction, voicing, consonant deletion, or vowel errors also reported for their monolingual peers (Morrow, Goldstein, Gilhool, & Paradis, 2014). This does not mean that the L1 cannot influence the child's L2 phonological development, either in a facilitatory way, when phonemes between the two languages are shared (Goldstein &

Washington, 2001), or in the form of negative transfer. The L1 can interfere with the L2 child's ability to perceive and produce L2 vowels (Darcy & Krüger, 2012) and consonants (McCarthy, Mahon, Rosen, & Evans, 2014), when the L1 and L2 phonological systems differ and when L1 use in the familial or the community context is continuous. In the longitudinal study by McCarthy et al., Sylheti L1–English L2 children failed to felicitously classify English voiced plosives (/b/, /d/, /g/), which do not occur in Sylheti, in a perception and a production task when tested at the nursery. Given that these L2 children were exposed to Sylheti up until nursery entry and to Sylheti-accented variants of English, this non-native input (see also section 3.2.2) interfered with the child's ability to develop target L2 categories. However, L1 influence seemed to be short-lived, as the L2 children did not differ from their monolingual peers a year later.

Child L2 phonological development has been primarily investigated in the context of sensitive periods in L2 acquisition. A number of studies have shown that L2 children may surpass L2 adults and can show nativelike perception and production for L2 contrasts (Baker & Trofimovich, 2005; Flege, Munro, & MacKay, 1995; McCarthy, Evans, & Mahon, 2013; Tsukada et al., 2005). Differences between L2 children and adults have been attributed to differences in neural plasticity, such that L1 phonemic categories become more robust with age, making it harder to acquire new L2 categories in adulthood (Flege, 1995; Iverson et al., 2003). Similarly, L2 children acquire their L2 sounds to a nativelike level, because when they start learning their L2, their L1 categories are less well established than those of adult learners.

2.2. Lexical Development in L2 Children In contrast to 2L1 children, L2 children are cognitively more mature when they come to the task of learning the L2 vocabulary, since much of the conceptual scaffolding has already taken place during the development of their L1. This may result in L2 children accumulating vocabulary faster than their younger bilingual peers, leading to an accelerated rate in vocabulary acquisition (Blom & Bosma, 2016; Goldberg, Paradis, & Crago, 2008). However, learning the label for a referent in an L2 is not an easy task, and studies with L2 children have consistently shown that they underperform on criterion-referenced or vocabulary assessments standardized with monolingual children (Chondrogianni & Marinis, 2011; Engel de Abreu, Baldassi, Puglisi, & Befi-Lopes, 2013; Goldberg, Paradis, & Crago, 2008; Kan & Kohnert, 2005; Sheng, Lu, & Kan, 2011; Umbel et al., 1992); in some cases, they even perform within monolingual language-impaired norms (Paradis, 2010). Vocabulary accumulation in the majority language for L2 learners is a gradual process, even when conceptual vocabulary is considered (Gross et al., 2015). The process of catching up to monolingual norms may take years to complete and could be dependent on input and other contextual factors or schooling, as the comparison between the Spanish-English L2 children in the United States (Miami) (Cobo-Lewis, Pearson, Eilers, & Umbel, 2002) and the L2 English children in Canada (Edmonton) showed (Goldberg, Paradis, & Crago, 2008). Even when matched on SES, the Spanish-English L2 children in Miami who had more frequent daily exposure to L1 Spanish took more years to reach English monolingual norms than their Canadian counterparts, whose exposure was mainly in English and who reached monolingual norms after three years of exposure to the L2; Goldberg, Paradis, & Crago, 2008).

Apart from vocabulary size, L2 children's vocabulary breadth (form-meaning connections) and depth of lexical knowledge (the extent of semantic representations) can differ from that of their monolingual peers. In a large-scale longitudinal study with Urdu/Punjabi L1–Norwegian L2 children tested in kindergarten (mean age: 5;3 years) and in the first year of primary school (mean age: 6;5 years), Karlsen, Halaas Lyster, & Lervåg (2017) reported that the L2 children had weaker skills in L2 vocabulary than their monolingual peers did, but their growth rates did not differ. Interestingly, vocabulary breadth in Urdu/Punjabi in kindergarten predicted the development of L2 vocabulary breadth. This finding indicates an underlying proficiency may facilitate vocabulary learning (Cummins, 1991; and contra Cobo-Lewis et al., 2002; and Hoff et al., 2012) (see also section 3).

2.3. Morphosyntactic Development in L2 Children Child L2 morphosyntactic development has been extensively studied as it has served as the testing ground for both generative and usage-based theories of language acquisition. The generative framework of language acquisition assumes the presence of an innate set of rules (e.g., constituent movement, and categories, such as tense, that can be parameterized across languages; Chomsky & Lasnik, 1993). Much research during the generative era has focused on whether the categories and rules at the initial stages of child L2 development resemble those of child L1 grammars, whether L1 categories influence the initial stages of child L2 acquisition and in what way (Haznedar, 2013; Haznedar & Gavruseva, 2008), what the relationship between overt morphology and underlying grammatical categories is, and whether the child L2 developmental patterns resemble those of L1 children (Unsworth, 2005) or L2 adults (Chondrogianni, 2008). Within this framework, child L2 acquisition was seen as the missing link between child L1 and adult L2 acquisition, as L2 children were assumed to be guided by the same innate mechanisms as L1 children were (Schwartz, 2004). The rational was as follows: if L2 children exhibit a similar acquisition pattern to their L2 adult counterparts, adult L2 acquisition can be assumed to be governed by the same acquisition mechanisms as L1 acquisition is.[1]

At the same time, child L2 acquisition serves as the testing ground for theories of L1 acquisition. If a maturation theory postulates an innate time course of language development (Wexler, 1998), L2 children who are exposed to the L2 after a specific maturation period should not exhibit the same developmental pattern that L1 children do. Thus, different developmental paths in L1 and L2 children should serve as support for L1 maturation theories (Schwartz, 2004). Support for this argument has been provided by child L2 studies examining the acquisition of the auxiliary British English (Paradis & Blom, 2016; Haznedar, 2001; Ionin & Wexler, 2002) and passives (Rothman et al., 2016). Both structures have been argued to be subject to maturational constraints in L1 acquisition. Contrary to L1 children, L2 children show what is called *precocious acquisition* of auxiliary British English; in the study by Rothman et al., L2 children with a later AoO acquired passives faster than L2 children with an earlier AoO, supporting maturational accounts of L1 acquisition.

Recently, usage-based theories of acquisition (Bybee, 2010) have investigated whether low-level, surface morphophonological properties, allomorphic endings, and type and token frequency can influence morphological acquisition in L2 children (Blom & Paradis, 2015; Paradis et al., 2011) in a similar way that has been proposed for L1 children (Plunkett & Marchman, 1993). In a series of studies, Paradis, Blom, and colleagues have shown that L2 children omit more in contexts where regular past tense verbs end in an alveolar /t/ or /d/ than in contexts where they end in another consonant or vowel, and that allomorphic variants of low token frequency, such as the syllabic ending /iz/ for third person—*s*, have lower production rates than consonant clusters in /s/ or /z/ (Blom & Paradis; Blom et al., 2013), similarly to what has been reported in L1 acquisition.

In this section, we will focus on a general overview of findings regarding L1 transfer and developmental patterns found in L2 children.

2.3.1. L1 transfer and effects of the L2 paradigm Despite early studies with bilingual children arguing for a common developmental path between monolingual and bilingual children (Dulay & Burt, 1974), recent empirical and experimental studies have shown that the nature of the L1 can influence both the rate of development as well as the types of errors that children commit. In the domain of morphosyntax, L2 children of preschool and primary school age have been shown to transfer L1 properties at initial stages of acquisition such as word order (Foroodi-Nejad & Paradis, 2009; Haznedar & Schwartz, 1997), as well as to omit elements not present in their L1 such as articles and pronouns (Chondrogianni, 2008; Chondrogianni, Marinis, Edwards, & Blom, 2015; Chondrogianni, Vasic, Marinis, & Blom, 2015; Zdorenko & Paradis, 2008, 2011) or tense morphemes (Blom, Chondrogianni, Marinis, & Vasic, 2015). These transfer effects impact on the time course of morphosyntactic development. L2 children from isolating/noninflecting L1s take longer to acquire the L2 inflectional properties than do children whose L1s mark tense (Blom & Paradis, 2015; Jia & Fuse, 2007; Paradis, Tulpar, & Arppe, 2016) and make more errors of omissions than L2 children from nonisolating / inflecting L1s (Blom et al., 2015; Blom & Paradis, 2015; Chondrogianni & Marinis, 2012; Herschensohn, Stevenson, & Maltmunson, 2005). At the same time, L2 children have been shown to make developmental errors. In a study on Dutch scrambling, Unsworth (2005) showed that English L1–Dutch L2 children exhibited L1 transfer effects by producing sentences in Dutch with an English negation-verb-object word order where scrambling should have occurred (object-negation-verb). At later stages, L2 children made word order errors (negation-object-verb) that have been observed in L1 Dutch-speaking children acquiring scrambling.

Beyond L1 transfer, language-specific factors of the L2 such as the transparency or opacity of the L2 morphophonological paradigm can modulate the course and rate of acquisition. In a study by Unsworth et al. (2014) comparing the production of gender in L2 children learning an opaque morphophonological system such as Dutch and in those learning a transparent system with transparent gender cues such as Greek, L2 Greek children reached higher accuracy than their Dutch age-matched counterparts. Similar, although less pronounced, differences were also observed in a study by Blom et al. (2015) on the production of verbal agreement in Dutch and Greek L2 children of primary school age.

To summarize, studies examining the acquisition of inflectional morphology and syntax in L2 children have shown that transfer is possible at initial stages of acquisition but that this may be short-lived. Errors may be transfer-related (Chondrogianni, 2008) or developmental (Unsworth, 2005) and reflect the nature of the morphophonological paradigm to be acquired (Unsworth et al., 2014; Blom et al., 2015), or they may adhere to general processing constraints (Roesch & Chondrogianni, 2016), an issue that we turn to next.

2.3.2. Comprehension in child L2 acquisition The adoption of more sophisticated and sensitive experimental paradigms in child L2 research, such as online processing techniques and comprehension methods, has allowed researchers to move beyond production studies and investigate the processing mechanisms and constraints that guide L2 children's offline and real-time comprehension and/or processing (see Blom & Unsworth, 2010, for an overview of methods). In a study on the online processing of passives, L2 children exhibited similar processing patterns to their monolingual peers despite having longer latencies (Marinis & Saddy, 2013). In an offline study examining German L2 children's ability to comprehend wh-questions with case-marking cues at sentence-initial or sentence-final positions, Roesch and Chondrogianni (2016) reported that the L2 children behaved similarly to their 2L1 counterparts in that sentence-final cues led to compromised performance (Choi & Trueswell, 2014; Omaki, Davidson-White, Goro, Lidz, & Phillips, 2014).

The use of more subtle techniques has also allowed researchers to investigate long-standing questions within the L2 acquisition literature, especially within the generative framework, such as the relationship between morphology and syntax. Online processing studies targeting morphosyntactic violations in the context of verbal (Blom, de Jong, & Vasic, 2014; Chondrogianni & Marinis, 2012, Herschensohn et al., 2005; Paradis, 2008) and nominal (Vasic, Chondrogianni, Marinis, & Blom, 2012) inflection have found that L2 children are sensitive to these violations despite their variable production of verbal and/or nominal morphology. This finding supports a dissociation between morphology and syntax in the child L2 grammar (Haznedar & Schwartz, 1997) and sheds new light to a long-standing debate.

3. Relationships across Language Domains in Multilingual Development

How do the different domains relate to one another and when can generalizations across domains take place? The question about within-language and cross-language associations across the various language domains has received extensive attention in the literature, primarily due to the theoretical interest of these relationships. Empiricist accounts argue for strong within-language correlations, as the same cognitive learning mechanisms can underpin different domains (e.g., vocabulary and syntax, Conboy & Thal, 2006; Hoff et al., 2012). This relationship is restricted between domains within a particular language but not across the two languages of the bilingual individual (Conboy & Thal, 2006). For nativist accounts, such relationships are less important or not expected (Unsworth, 2013c). Relationships across domains have been traditionally evaluated in two ways: (i) by performing statistical tests, primarily correlations, between different domains both across and within languages, and (ii) by examining the developmental rates between two or more language domains on the basis of naturalistic, experimental, or reference-based data. Differential or asynchronous developmental effects over two or more language domains have been termed "profile effects" (Bohman, Bedore, Peña, Mendez-Perez, & Gillam, 2010; Oller et al., 2007). These effects have been found among the domains of morphosyntax (Chondrogianni & Marinis, 2011; Gathercole, 2006), vocabulary (Cobo-Lewis et al., 2002; Oller et al., 2007), and literacy (Oller et al., 2007). At the same time, significant correlations between vocabulary and grammatical development within languages have been reported in a series of studies (Cobo-Lewis et al., 2002; Conboy & Thal, 2006; Marchman et al., 2004). These results have been taken as evidence that the same mechanisms subserve different areas of language. Importantly, however, these correlations have only been reported within languages but not across, although this issue is subject to much recent research and debate (see Carroll, 2017).

Relationships across domains within and between languages are modulated by the cognitive and conceptual associations implicated within each domain (Kohnert, 2013). Correlations between L1 and L2 vocabulary knowledge have been interpreted as conceptual transfer between the two languages of the bilingual individual, as vocabulary draws on a common underlying proficiency (Cummins, 1991; Karlsen et al., 2017; Melby-Lervåg & Lervåg, 2011).

Working memory is a domain that implicates more general cognitive processes compared to grammar. A number of studies have reported that phonological short-term memory predicts children's performance on vocabulary and grammar in L2 children (Harley & Hart, 1997; Masoura & Gathercole, 1999; Paradis, 2011). However, in most studies, effects of phonological memory on grammar are mediated by L2 vocabulary skill (Engel de Abreu & Gathercole, 2012; French, 2006; Service & Kohonen, 1995), although it seems that phonological working memory remains a significant predictor even when L2 vocabulary skill is controlled for

(French & O'Brien, 2008; Verhagen, Leseman, & Messer, 2015). However, whether relationships between phonological working memory and other domains emerge depends on the design of the nonword repetition task and the language-specific or general properties that have been manipulated (e.g., phonological complexity, phonotactic probability, syllable structure and length, among other things; Chiat, 2015). When the design of the task taps into language-specific factors, strong correlations emerge between exposure and vocabulary measures and performance on nonword repetition (Boerma et al., 2015; Chiat; Gutiérrez-Clellen & Simon-Cereijido, 2010). However, the relationship between phonological working memory and other domains seems to be restricted to each language independently rather than across languages (Messer, Leseman, Boom, & Mayo, 2010; Parra, Hoff, & Core, 2011; but see French & O'Brien, 2008).

These studies highlight the heterogeneity of the findings and the importance of exploring relationships across domains through more robust statistical procedures such as regression analysis, as well as the importance of understanding what the differences in accuracy on the different tasks mean for development.

4. *Child-External Factors in Bilingual Development*

We review three factors that have been shown to influence the language development of simultaneous and sequential bilingual children.

4.1. Input Quantity Input quantity has been operationalized in two ways (Hurtado, Grüter, Marchman, & Fernald, 2014): (i) as absolute amount of L2 exposure, which refers to the frequency with which a particular construction or expression occurs in naturalistic speech (Newport, Gleitman, & Gleitman, 1977), and (ii) as relative amount of exposure, which represents a (detailed) calculation of the child's daily amount of exposure to the two languages on the basis of parental (or teacher) questionnaires/reports (Gutiérrez -Clellen & Kreiter, 2003; Paradis, 2011) concurrently or over time (Unsworth, 2013b). Input quantity has taken central stage in the bilingual language development literature (Gathercole & Thomas, 2009; Paradis et al., 2016; Place & Hoff, 2011, 2016), for both empiricist (Place & Hoff, 2011) and, recently, for nativist (Rankin & Unsworth, 2016) accounts of bilingual language development, and is the object of much recent debate especially in relation to its links with the development of grammar and vocabulary (see Carroll, 2017, and commentaries therein). Given that the amount of input that bilingual children receive in the two languages varies considerably (de Houwer, 2007; Unsworth, 2013b), it is assumed that this variability can account for differences in the rate of acquisition between monolingual and bilingual children, but also within different bilingual populations. To establish the relationship between input quantity and quality and language development, these studies have used both correlations as well as, more recently, linear and nonlinear regressions to establish the precise predictors of language development (e.g., Paradis et al.; Thordardottir & Brandecker, 2013). Children's language abilities are measured either via parental report measures such as the MacArthur-Bates Communicative Development Inventories (Place & Hoff, 2011, 2016) or through standardized or criterion-referenced assessments (Thordardottir & Brandecker, 2013), with fewer studies examining specific morphosyntactic phenomena such as tense morphology (Blom & Paradis, 2013, 2015; Marinis & Chondrogianni, 2010), acquisition of complex structures such as wh-questions (Roesch & Chondrogianni, 2016), or scrambling (Unsworth, 2014). Interestingly, studies with monolingual English- and Spanish-speaking children using subtle processing measures such as eye tracking have shown that variability in language outcomes is linked to early ability in comprehending spoken language in real time (Grüter et al., 2014; Hurtado, Marchman, & Fernald, 2007). This line of research has recently been explored in populations of bilingual children with the same language combination (English and Spanish). In Hurtado et al. (2014), the relative amount of daily exposure to Spanish and English in a group of simultaneous Spanish-English bilingual children observed longitudinally at 30 and 36 months predicted their relative efficiency in real-time language processing in each language. Moreover, early exposure and speed of lexical comprehension predicted later expressive and receptive vocabulary outcomes in Spanish and English.

Input quantity is also linked to language dominance. Bilingual children differ from monolingual children, when tested in their less dominant language, but these differences disappear when tested in their dominant language (Hoff et al., 2012; Paradis et al., 2011; Unworth, 2015). These differences are predominantly found in the domain of vocabulary (Hoff et al., 2012), but they are also found in bilingual children's phonotactic preferences (Sebastián-Gallés & Bosch, 2002) and sensitivity to stress patterns (Abboub, Bijeljac-Babic, Serres, & Nazzi, 2015; Bijeljac-Babic, Serres, Höhle, & Nazzi, 2012).

Studies have also tried to establish how much input is necessary for bilingual children to perform similarly to monolingual children. In a study with English-Spanish toddlers using MacArthur-Bates Communicative

Development Inventories data, Hoff et al. (2012) found that 70% of input in English is enough for them to perform similarly to their monolingual peers. However, Thordardottir (2011) reported that French-English five-year-old children performed within monolingual norms on a standardized vocabulary task after receiving 40% of input or more to the target language; this percentage was higher in the case of expressive vocabulary (60%).

Interestingly and in line with profile effects, input has been shown to influence language domains differentially (Chondrogianni & Marinis, 2011; Paradis, 2011). Thordardottir and Brandecker (2013) found a weaker relationship between amount of exposure and performance on a nonword repetition task, which is regarded as making use of more general cognitive underpinnings of language such as working memory (Gathercole, 2006), compared to sentence repetition, which is a more language-dependent task. However, the degree to which specific morphosyntactic structures are subject to input effects remains controversial (Thordardottir, 2011; Hoff, Welsh, Place, & Ribott, 2014; Unsworth, 2014) and depends on the sensitivity of the measure and the statistical procedure being used to establish that relationship.

Overall, input factors and how they interact with child-internal factors, such as the languages spoken by the bilingual child and the phenomenon to be acquired, merit further investigation using subtler experimental techniques. For example, in the study by Roesch and Chondrogianni (2016) using a picture-selection method, French-German 2L1 children could not implement case-marking cues in the comprehension of German wh-questions as efficiently as their German-speaking L1 peers could, despite the fact that their performance on case was at ceiling. This delay may have resulted from the lack of a corresponding facilitatory structure in their other language (French), or from reduced input in German, which impacted their ability to implement their grammatical knowledge in the context of a complex structure. The nature of the *process* rather than of the *output* is an issue that future research would benefit from addressing further.

4.2. Input Quality Intertwined with input quantity is input quality. Input quality has been defined as input richness and has been measured in different ways: through exposure to different sources, such TV, radio, books, friends/siblings (Jia & Fuse, 2007; Paradis, 2011), and literacy-related activities (Karlsen et al., 2017; Scheele, Leseman, & Mayo, 2010); through the number of different native versus non-native speakers that the child is exposed to (de Houwer, 2009; Fernald, 2006; Place & Hoff, 2011), which may result in higher proportion of code switching (Byers-Heinlein, 2013); or, through the number of speakers of standard or non-standard varieties (Cornips & Hulk, 2006). Bilingual children are more likely to be exposed to non-native speaker input (Fernald, 2006), which may impact on their ability to reach monolingual norms in the domain of vocabulary (Byers-Heinlein, 2013), or to non-standard varieties, which may influence their ability to acquire morphosyntactic phenomena such as gender attribution in the Turkish-Dutch L2 children in the Netherlands (Cornips & Hulk). Recent studies by Byers-Heinlein and colleagues have found that bilingual children prefer native accents to non-native accents, similarly to monolingual children (Souza, Byers-Heinlein, & Poulin Dubois, 2013), and that bilingual infants' exposure to parental input with intrasentential code switching may hinder their ability to develop vocabulary comprehension to monolingual norms at the age of one and a half years and vocabulary production at the age of two years (Byers-Heinlein, 2013). Conversely, variable input may facilitate the formation of more robust phonological representations due to exposure to sufficiently different exemplars (Sebastián-Gallés, & Bosch, 2009). To sum up, these studies suggest that input quality can modulate language development in bilingual children and that the subtleties of input may affect various domains differentially.

4.3. SES Socioeconomic status has emerged as a significant predictor of monolingual (Cartmill et al., 2013; Hoff, 2013) and bilingual children's abilities across a number of studies investigating a range of language domains. SES is most commonly defined on the basis of parental or maternal education (Lawlor et al., 2006) or income (Engel de Abreu et al., 2013). SES may serve as a proxy for a range of other factors, such as parental behavior, which in a recent study was found to be a stronger predictor of children's development than parental educational level (Rindermann & Baumeister, 2015), parents' language use (Hoff & Naigles, 2002), or time spent verbally interacting with parents (Hart & Risley, 1995) or in literacy activities (Jia & Fuse, 2007). SES has been shown to be a significant predictor of bilingual children's vocabulary abilities in terms of vocabulary size (Armon-Lotem, Walters, & Gagarina, 2011; Chondrogianni & Marinis, 2011; Goldberg, Paradis, & Crago, 2008; Hoff et al., 2012), even when conceptual vocabulary is taken into consideration (Mancilla-Martinez & Vagh, 2013), but interestingly, also in terms of processing efficiency (Grüter et al., 2014; Hurtado et al., 2014). In a study by Chondrogianni and Marinis (2011), maternal SES was found to predict L2 children's inability to reach monolingual norms on complex syntax and vocabulary, and similar effects were also reported in Paradis (2011). However, in the study by Bohman et al. (2010), no relationship was found between SES and morphosyntactic development, and the extent to which SES affects children's ability to acquire complex structures remains contested even in the monolingual literature (Vasilyeva, Waterfall, & Huttenlocher, 2008).

5. Conclusions and Future Directions

This review demonstrated that in the last decade, the study of language development in simultaneous and successive bilingual children has made great theoretical and empirical advances, moving away from diary and naturalistic studies to more experimental methods that can help us unravel and better understand the factors that subserve bilingual development. Experimental and empirical research in the last two decades has informed us about how the two languages of the bilingual individual interact, which factors modulate this interaction (e.g., nature of the two languages, age of exposure to the two languages, input quality and quantity, among others), and in what ways bilingual development differs from or is the same as monolingual development. However, the implementation of different experimental paradigms in the bilingual development research, especially in studies with bilingual children of school age (both primary and secondary) is in its infancy and most current studies are primarily offline, output-based measures. These are informative about the outcome of the developmental process but not necessarily about the processes that get the learner from the initial to the final stage of development (Valian, 2014). This is an area that merits further investigation in the field of bilingual research and the field will benefit from future studies exploring these issues.

The study of bilingual development can lead to the refinement of existing theories and concepts in both monolingual and bilingual acquisition. Despite variable input, bilingual children are able to make gains across language domains and to communicatively function using two or more languages. Reassessing the way input has been defined or measured and more robustly investigating and operationalizing the relationship between the different language domains is of seminal theoretical importance for both empiricist and nativist accounts of acquisition. Future studies would benefit from using more fine-grained experimental tools and analyses to establish whether child-internal and child-external factors lead to only quantitative or also to qualitative differences across populations. This investigation could help us understand whether these factors affect bilingual children's cognitive architecture or their capacity to efficiently implement the same learning mechanism across language domains.

NOTE

1. A detailed review of the results and the specific acquisition theories within the generative framework is beyond the scope of the present discussion; the reader is directed to two relatively recent overviews on the topic by Haznedar and Gavruseva (2008) and Lakshmanan (2009), where generative theories of child L2 acquisition are discussed in detail.

REFERENCES

Abboub, N., Bijeljac-Babic, R., Serres, J., & Nazzi, T. (2015). On the importance of being bilingual: Word stress processing in a context of segmental variability. *Journal of Experimental Child Psychology, 132*, 111–120.

Abrahamsson, N., & Hyltenstam, K. (2009). Age of acquisition and nativelikeness in a second language: Listener perception vs. linguistic scrutiny. *Language Learning, 59*, 249–306.

Argyri, E., & Sorace, A. (2007). Crosslinguistic influence and language dominance in older bilingual children. *Bilingualism: Language and Cognition, 10*, 79–99.

Armon-Lotem, S, de Jong, J., & Meir, N. (2015). *Assessing multilingual children: Disentangling bilingualism from language impairment.* London: Multilingual Matters.

Armon-Lotem, S., Walters, J., & Gagarina, N. (2011). The impact of internal and external factors on linguistic performance in the home language and in L2 among Russian-Hebrew and Russian-German preschool children. *Linguistic Approaches to Bilingualism, 1*, 291–317.

Baker, W., & Trofimovich, P. (2005). Interaction of native and second-language vowel system(s) in early and late bilinguals. *Language and Speech, 48*, 1–27.

Bedore, L. M., Peña, E. D., Summers, C. L., Boerger, K. M., Resendiz, M. D., Greene, K., & Gillam, R. B. (2012). The measure matters: Language dominance profiles across measures in Spanish-English bilingual children. *Bilingualism: Language and Cognition, 15*(3),616–629.

Bialystok, E., & Feng, X. (2011). Language proficiency and its implications for monolingual and bilingual children. In A.Y. Durgunoğlu & C. Goldenberg (Eds.), *Language and Literacy Development in Bilingual Settings.* New York: Guilford Publications, Inc.

Bialystok, E., Luk, G., Peets, K. F., & Yang, S. (2010). Receptive vocabulary differences in monolingual and bilingual children. *Bilingualism: Language and Cognition, 13*, 525–531.

Bijeljac-Babic, R., Serres, J., Höhle, B., & Nazzi, T. (2012). Effect of bilingualism on lexical stress pattern discrimination in French-learning infants. *PLOS ONE, 7*, e30843.

Blom, W. B. T., & Bosma, E. (2016). The sooner the better? An investigation into the role of age of onset and its relation with transfer and exposure in bilingual Frisian-Dutch children. *Journal of Child Language, 43*, 581–607.

Blom, W. B. T., Chondrogianni, V., Marinis, T., & Vasic, N. (2015). The acquisition of verbal paradigms in Dutch and Greek L2 children: Cross-linguistic differences and inflectional defaults. *International Journal of Bilingualism, 20*(4), 386–402.

Blom, W. B. T., de Jong, J., Orgassa, A., Baker, A., & Weerman, F. (2013). Verb inflection in monolingual Dutch and sequential bilingual Turkish–Dutch children with and without SLI. *International Journal of Language and Communication Disorders, 48*(4), 382–393.

Blom, W. B. T., de Jong, J., & Vasic, N. (2014). Production and processing of subject-verb agreement in monolingual Dutch children with specific language impairment. *Journal of Speech Language and Hearing Research, 57,* 952–965.

Blom, W. B. T., & Paradis, J. (2013). Past tense production by English second language learners with and without language impairment. *Journal of Speech, Language and Hearing Research, 56,* 281–294.

Blom, W. B. T., & Paradis, J. (2015). Sources of individual differences in the acquisition of tense inflection by English second language learners with and without specific language impairment. *Applied Psycholinguistics, 36,* 953–976.

Blom, W. B. T., & Unsworth, S. (Eds.). (2010). *Experimental methods in language acquisition research* (Vol. 27). Amsterdam: John Benjamins.

Boerma, T., Chiat, S., Leseman, P., Timmermeister, M., Wijnen, F., & Blom, E. (2015). A quasi-universal nonword repetition task as a diagnostic tool for bilingual children learning Dutch as a second language. *Journal of Speech Language and Hearing Research, 58,* 1747–1760.

Bohman, T. M., Bedore, L. M., Pena, E. D., Mendez-Perez, A., & Gillam, R. B. (2010). What you hear and what you say: Language performance in Spanish-English bilinguals. *International Journal of Bilingual Education and Bilingualism, 13,* 325–344.

Bosch, L., & Sebastián-Gallés, N. (1997). Native-language recognition abilities in 4-month-old infants from monolingual and bilingual environments. *Cognition, 65,* 33–69.

Bosch, L., & Sebastián-Gallés, N. (2003). Simultaneous bilingualism and the perception of a language-specific vowel contrast in the first year of life. *Language and Speech, 46,* 217–243.

Burns, T. C., Yoshida, K. A., Hill, K., & Werker, J. F. (2007). The development of phonetic representation in bilingual and monolingual infants. *Applied Psycholinguistics, 28,* 455–474.

Bybee, J. L. (2010). *Language, usage and cognition.* Cambridge: Cambridge University Press.

Byers-Heinlein, K. (2013). Parental language mixing: Its measurement and the relation of mixed input to young bilingual children's vocabulary size. *Bilingualism: Language and Cognition, 16,* 32–48.

Byers-Heinlein, K. (2014). Languages as categories: Reframing the "one language or two" question in early bilingual development. *Language Learning, 64*(2), 184–201.

Byers-Heinlein, K., Burns, T. F., & Werker, J. F. (2010). The roots of bilingualism in newborns. *Psychological Science, 21*(3), 343–348.

Byers-Heinlein, K., & Werker, J. F. (2009). Monolingual, bilingual, trilingual: Infants' language experience influences the development of a word learning heuristic. *Developmental Science, 12,* 815–823.

Byers-Heinlein, K., & Werker, J. F. (2013). Lexicon structure and the disambiguation of novel words: Evidence from bilingual infants. *Cognition, 128,* 407–416.

Carroll, S. E. (2017). Experience and input in bilingual development. *Bilingualism: Language and Cognition, 20*(1), 3–16

Cartmill, E. A., Armstrong, B. F., Gleitman, L. R., Goldin-Meadow, S., Medina, T. N., & Trueswell, J. C. (2013). Quality of early parent input predicts child vocabulary 3 years later. *Proceedings of the National Academy of Science (PNAS), 10,* 11278–11283.

Chiat, S. (2015). Non-word repetition. In S. Armon-Lotem, J. de Jong, & N. Meir (Eds.), *Methods for assessing multilingual children: Disentangling bilingualism from language impairment* (pp. 125–150). Bristol, UK: Multilingual Matters.

Choi, Y., & Trueswell, J. C. (2014). Children's (in)ability to recover from garden paths in a verb-final language: Evidence for developing control in sentence processing. *Journal of Experimental Child Psychology, 106,* 41–61.

Chomsky, N. (1981). *Lectures on government and binding.* Dordrecht, Netherlands: Foris.

Chomsky, N. (1995). *The minimalist program.* Cambridge, MA: MIT Press.

Chomsky, N., & Lasnik, H. (1993). The theory of principles and parameters. *Syntax: An International Handbook of Contemporary Research, 1,* 506–569.

Chondrogianni, V. (2008). Comparing child and adult L2 acquisition of the Greek DP: Effects of age and construction. In B. Haznedar & E. Gavruseva (Eds.), *Current trends in child second language acquisition: A generative approach.* (pp. 97–144). Amsterdam: John Benjamins.

Chondrogianni, V., & Marinis, T. (2011). Differential effects of internal and external factors on the development of vocabulary, tense morphology and morpho-syntax in successive bilingual children. *Linguistic Approaches to Bilingualism, 1,* 318–342.

Chondrogianni, V., & Marinis, T. (2012). Production and processing asymmetries in the acquisition of tense morphology by sequential bilingual children. *Bilingualism: Language and Cognition, 15*(1), 5–21.

Chondrogianni, V., Marinis, T., Edwards, S., & Blom, W. B. T. (2015). Production and on-line comprehension of definite articles and clitic pronouns by Greek sequential bilingual children and monolingual children with specific language impairment. *Applied Psycholinguistics, 36,* 1155–1191.

Chondrogianni, V., Vasic, N., Marinis, T., & Blom, W. B. T. (2015). Production and on-line comprehension of definiteness in English and Dutch by monolingual and sequential bilingual children. *Second Language Research, 31,* 309–341.

Cobo-Lewis, A. B., Pearson, B. Z., Eilers, R. E., & Umbel, V. C. (2002). Effects of bilingualism and bilingual education on oral and written Spanish skills: A multifactor study of standardized test outcomes. In D. K. Oller & R. E. Eilers (Eds.), *Language and literacy in bilingual children* (pp. 98–117). Clevedon, UK: Multilingual Matters.

Conboy, B. T., & Mills, D. L. (2006). Two languages, one developing brain: Event-related potentials to words in bilingual toddlers. *Developmental Science, 9,* F1–F12.

Conboy, B. T., & Thal, D. J. (2006). Ties between the lexicon and grammar: Cross-sectional and longitudinal studies of bilingual toddlers. *Child Development, 77,* 712–735.

Core, C., Hoff, E., Rumiche, R., & Señor, M. (2013). Total and conceptual vocabulary in Spanish-English bilinguals from

22 to 30 months: Implications for assessment. *Journal of Speech, Language and Hearing Research, 55,* 1637–1649.

Cornips, L., & Hulk, A. (2006). External and internal factors in bilingual and bidialectal language development: Grammatical gender of the Dutch definite determiner. In L. Claire, L. White, & C. Jourdan (Eds.), *L2 acquisition and creole genesis: Dialogues* (pp. 355–378). Amsterdam: John Benjamins.

Cummins, J. (1991). Interdependence of first and second language proficiency in bilingual children. In B. Ellen (Ed.), *Language processing in bilingual children* (pp. 70–89). Cambridge: Cambridge University Press.

Darcy, I., & Krüger, F. (2012). Vowel perception and production in Turkish children acquiring L2 German. *Journal of Phonetics, 40,* 568–581.

De Houwer, A. (1990). *The Acquisition of Two Languages from Birth: A Case Study* (Cambridge Studies in Linguistics). Cambridge: Cambridge University Press.

De Houwer, A. (2007). Parental language input patterns and children's bilingual use. *Applied Psycholinguistics, 28,* 411–424.

De Houwer, A. (2009). *Bilingual first language acquisition.* London: Multilingual Matters.

Dulay, H., & Burt, M. (1974). Natural sequences in child second language acquisition. *Language Learning, 24,* 37–53.

Eilers, R., Pearson, B. & Cobo-Lewis, A. (2006). Social factors in bilingual development: The Miami experience. In P. McCardle & E. Hoff (Eds.), *Childhood bilingualism: research on infancy through school age* (pp. 68–90). Bristol, Blue Ridge Summit: Multilingual Matters.

Engel de Abreu, P. M. J., Baldassi, M., Puglisi, M. L., & Befi-Lopes, D. M. (2013). Crosslinguistic and cross-cultural effects on verbal working memory and vocabulary: Testing language minority children with an immigrant background. *Journal of Speech, Language and Hearing Research, 56,* 630–642.

Engel de Abreu, P. M. J., & Gathercole, S. E. (2012). Executive and phonological processes in second-language acquisition. *Journal of Educational Psychology, 104*(4), 974–986.

Fabiano-Smith, L., & Barlow, J. A. (2010). Interaction in bilingual phonological acquisition: Evidence from phonetic inventories. *International Journal of Bilingual Education and Bilingualism, 13,* 81–97.

Fabiano-Smith, L., & Goldstein, B. A. (2010). Phonological acquisition in bilingual Spanish-English speaking children. *Journal of Speech, Language, and Hearing Research, 53,* 160–178.

Fennell, C. T., Byers-Heinlein, K., & Werker, J. F. (2007). Using speech sounds to guide word learning: The case of bilingual infants. *Child Development, 78,* 1510–1525.

Fenson, L., Marchman, V. A., Thal, D., Dale, P. S., Reznick, J. S., & Bates, E. (2007). *MacArthur-Bates Communicative Development Inventories: User's guide and technical manual 2.* Baltimore, MD: Brookes Publishing.

Ferjan Ramírez, N., Ramírez, R. R., Clarke, M., Taulu, S., & Kuhl, P. K. (2017). Speech discrimination in 11 month old bilingual and monolingual infants: A magnetoencephalography study. *Developmental Science, 20,* e12427.

Fernald, A. (2006). When infants hear two languages: Interpreting research on early speech perception by bilingual children. In P. D. McCardle & E. Hoff (Eds.), *Childhood bilingualism: Research on infancy through school age* (pp. 19–29). Clevedon, UK: Multilingual Matters.

Fernald, A., Perfors, A., & Marchman, V. A. (2006). Picking up speed in understanding: Speech processing efficiency and vocabulary growth across the 2nd year. *Developmental Psychology, 42,* 98–116.

Fernald, A., Pinto, J. P., Swingley, D., Weinbergy, A., & McRoberts, G. W. (1998). Rapid gains in speed of verbal processing by infants in the 2nd year. *Psychological Science, 9,* 228–231.

Flege, J. E. (1995). Second language speech learning: Theory, findings, and problems. *Speech perception and linguistic experience: Issues in cross-language research, 92,* 233–277.

Flege, J. E., Munro, M. J., & MacKay, I. R. (1995). Effects of age of second-language learning on the production of English consonants. *Speech Communication, 16*(1), 1–26.

Foroodi-Nejad, F., & Paradis, J. (2009). Crosslinguistic transfer in the acquisition of compound nouns in Persian-English bilinguals. *Bilingualism: Language and Cognition, 12,* 411–427.

French, L. M. (2006). *Phonological working memory and second language acquisition: A developmental study of Francophone children learning English in Quebec.* Lewiston, NY: Edwin Mellen Press.

French, L. M., & O'Brien, I. (2008). Phonological memory and children's second language grammar learning. *Applied Psycholinguistics, 29,* 463–487.

Garcia-Sierra, A., Rivera-Gaxiola, M., Percaccio, C. R., Conboy, B. T., Romo, H., Klarman, L., & Kuhl, P. K. (2011). Bilingual language learning: An ERP study relating early brain responses to speech, language input, and later word production. *Journal of Phonetics, 39,* 546–557.

Gathercole, S. E. (2006). Non-word repetition and word learning: The nature of the relationship. *Applied Psycholinguistics, 27,* 513–543.

Gathercole, V. C. M., & Thomas, E. M. (2009). Bilingual first-language development: Dominant language takeover, threatened minority language take-up. *Bilingualism: Language and Cognition, 12,* 213–237.

Genesee, F. (1989). Early bilingual development: One language or two? *Journal of Child Language, 16,* 161–179.

Genesee, F., & Nicoladis, E. (2007). Bilingual first language acquisition. In: E. Hoff & M. Shatz (Eds.), *Blackwell handbook of language development.* Malden, MA: Blackwell.

Genesee, F., Nicoladis, E., & Paradis, J. (1995). Language differentiation in early bilingual development. *Journal of Child Language, 22,* 611–631.

Gildersleeve-Neumann, C. E., Kester, E. S., Davis, B. L., & Pena, E. D. (2008). English speech sound development in preschool-aged children from bilingual English-Spanish environments. *Language, Speech, and Hearing Services in Schools, 39,* 314–328.

Goldberg, H., Paradis, J., & Crago, M. (2008). Lexical acquisition over time in minority L1 children learning English as L2. *Applied Psycholinguistics, 29,* 41–65.

Goldstein, B., & Washington, P. S. (2001). An initial investigation of phonological patterns in typically developing 4-year-old Spanish-English bilingual children. *Language, Speech, and Hearing Services in Schools, 32,* 153–164.

Grosjean, F. (2010). *Bilingual: Life and reality.* Cambridge, MA: Harvard University Press.

Gross, M., Buac, M., & Kaushanskaya, M. (2015). Conceptual scoring of receptive and expressive vocabulary measures in simultaneous and sequential bilingual children. *American Journal of Speech and Language Pathology, 23,* 574–586.

Grüter, T., Hurtado, N., Marchman, V. A., & Fernald, A. (2014). Language exposure and online processing efficiency in bilingual development: Relative versus absolute measures. In T. Grüter & J. Paradis (Eds.), *Input and experience in bilingual development* (pp. 15–36). Amsterdam: John Benjamins.

Gutiérrez-Clellen, V. F., & Kreiter, J. (2003). Understanding child bilingual acquisition using parent and teacher reports. *Applied Psycholinguistics, 24,* 267–288.

Gutiérrez-Clellen, V. F., & Simon-Cereijido, G. (2010). Using nonword repetition tasks for the identification of language impairment in Spanish-English-speaking children: Does the language of assessment matter? *Learning Disabilities Research and Practice, 25,* 48–58.

Halberda, J. (2003). The development of a word-learning strategy. *Cognition, 87,* B23–B34.

Harley, B., & Hart, D. (1997). Language aptitude and second language proficiency in classroom learners of different starting ages. *Studies in Second Language Acquisition, 19*(3), 379–400.

Hart, B., & Risley, T. R. (1995). *Meaningful differences in the everyday experience of young American children.* Baltimore, MD: P. H. Brookes.

Haznedar, B. (2001). The acquisition of the IP system in child L2 English. *Studies in Second Language Acquisition, 23*(1), 1–39.

Haznedar, B. (2013). Child second language acquisition from a generative perspective. *Linguistic Approaches to Bilingualism, 3*(1), 26–47.

Haznedar, B., & Gavruseva, E. (2008). Recent perspectives in child second language acquisition. In B. Haznedar & E. Gavruseva (Eds.), *Current trends in child second language acquisition* (pp. 3–15). Amsterdam: John Benjamins.

Haznedar, B., & Schwartz, B. (1997). *Are there optional infinitives in child L2 acquisition?* Paper presented at the meeting of the Boston University Conference on Language Development, Boston, MA.

Herschensohn, J., Stevenson, J., & Waltmunson, J. (2005). Children's acquisition of L2 Spanish morphosyntax in an immersion setting. *IRAL, 43,* 193–217.

Hoff, E. (2013). Interpreting the early language trajectories of children from low SES and language minority homes: Implications for closing achievement gaps. *Developmental Psychology, 49,* 4–14.

Hoff, E. (2015). Language development in bilingual children. In E. Bavin & L. Naigles (Eds.), *The Cambridge handbook of child language* (pp. 483–503). Cambridge: Cambridge University Press.

Hoff, E., Core, C., Place, S., Rumiche, R., Señor, M., & Parra, M. (2012). Dual language exposure and early bilingual development. *Journal of Child Language, 39,* 1–27.

Hoff, E., & Naigles, L. (2002). How children use input to acquire a lexicon. *Child Development, 73,* 418–433.

Hoff, E., Welsh, S., Place, S., & Ribot, K. M. (2014). Properties of dual language input that shape development and properties of environments that shape dual language input. In T. Grüter & J. Paradis (Eds.), *Input and experience in bilingual development* (pp. 119–140). Amsterdam: John Benjamins.

Hulk, A., & Müller, N. (2000). Bilingual first language acquisition at the interface between syntax and pragmatics. *Bilingualism: Language and Cognition, 3*(3), 227–244.

Hurtado, N., Grüter, T., Marchman, V. A., & Fernald, A. (2014). Relative language exposure, processing efficiency and vocabulary in Spanish-English bilingual toddlers. *Bilingualism: Language and Cognition, 17,* 189–202.

Hurtado, N., Marchman, V. A., & Fernald, A. (2007). Spoken word recognition by Latino children learning Spanish as their first language. *Journal of Child Language, 34*(2), 227–249.

Hurtado, N., Marchman, V. A., & Fernald, A. (2008). Does input influence uptake? Links between maternal talk, processing speed and vocabulary size in Spanish-learning children. *Developmental Science, 11,* F31–F39.

Ionin, T., & Wexler, K. (2002). Why is "is" easier than "-s"?: Acquisition of tense/agreement morphology by child second language learners of English. *Second Language Research, 18*(2), 95–136.

Iverson, P., Kuhl, P. K., Akahane-Yamada, R., Diesch, E., Tohkura, Y. I., Kettermann, A., & Siebert, C. (2003). A perceptual interference account of acquisition difficulties for non-native phonemes. *Cognition, 87*(1), B47–B57.

Jia, G., & Fuse, A. (2007). Acquisition of English grammatical morphology by native Mandarin-speaking children and adolescents: Age-related differences. *Journal of Speech, Language and Hearing Research, 50,* 1280–1299.

Johnson, J. S., & Newport, E. L. (1989). Critical period effects in second language learning: The influence of maturational state on the acquisition of ESL. *Cognitive Psychology, 21,* 60–99.

Kan, P. F., & Kohnert, K. (2005). Preschoolers learning Hmong and English: Lexical-semantic skills in L1 and L2. *Journal of Speech Language and Hearing Research, 48,* 372–383.

Karlsen, J., Halaas Lyster, S.-A., & Lervåg, A. (2017). Vocabulary development in Norwegian L1 and L2 learners in the kindergarten-school transition. *Journal of Child Language, 44,* 402–426.

Kidd, E., Chan, A., & Chiu, J. (2015). Cross-linguistic influence in simultaneous Cantonese-English bilingual children's comprehension of relative clauses. *Bilingualism: Language and Cognition, 18,* 438–452.

Kohnert, K. (2013). *Language disorders in bilingual children and adults.* San Diego, CA: Plural Publishing.

Kuhl, P. K., Conboy, B. T., Coffey-Corina, S., Padden, D., Rivera-Gaxiola, M., & Nelson, T. (2008). Phonetic learning as a pathway to language: New data and native language magnet theory expanded (NLM-e). *Philosophical Transactions of the Royal Society of London B: Biological Sciences, 363,* 979–1000.

Kuhl, P. K., Stevens, E., Hayashi, A., Deguchi, T., Kiritani, S., & Iverson, P. (2006). Infants show a facilitation effect for native language phonetic perception between 6 and 12 months. *Developmental Science, 9,* F13–F21.

Kupisch, T. (2007). Determiners in bilingual German–Italian children: What they tell us about the relation between language influence and language dominance. *Bilingualism: Language and Cognition, 10,* 57–78.

Lakshmanan, U. (2009). Child second language acquisition. In W. Ritchie & T. K. Bhatia (Eds.), *The new handbook of second language acquisition* (pp. 377–399). Sheffield, UK: Emerald.

Lawlor, D. A., Najman, J. M., Batty, G. D., O'Callaghan, M. J., Williams, G. M., & Bor, W. (2006). Early life predictors of childhood intelligence: Findings from the Mater-University study of pregnancy and its outcomes. *Pediatric and Perinatal Epidemiology, 20,* 148–162.

Leopold, W. (1978). A child's learning of two languages. In E. Hatch (Ed.), *Second language acquisition: A book of readings*. Rowley, MA: Newbury House.

Lillo-Martin, D., Koulidobrova, H., de Quadros, R. M., & Chen Pichler, D. (2012). Bilingual language synthesis: Evidence from WH-questions in bimodal bilinguals. In A. K. Biller, E. Y. Chung, & A. E. Kimball (Eds.), *Proceedings of the 36th Annual Boston University Conference on Language Development* (pp. 302–314). Somerville, MA: Cascadilla Press.

Lleó, C., Kuchenbrandt, I., Kehoe, M., & Trujillo, C. (2003). Syllable final consonants in Spanish and German monolingual and bilingual acquisition. In N. Müller (Ed.), *(In)vulnerable domain in multilingualism* (pp. 191–220). Amsterdam: John Benjamins.

Mancilla-Martinez, J., & Vagh, S. B. (2013). Growth in toddlers' Spanish, English, and conceptual vocabulary knowledge. *Early Childhood Research Quarterly, 28*, 555–567.

Marchman, V. A., Fernald, A., & Hurtado, N. (2010). How vocabulary size in two languages relates to efficiency in spoken word recognition by young Spanish-English bilinguals. *Journal of Child Language, 37*, 817–840.

Marchman, V. A., Martínez-Sussmann, C., & Dale, P. S. (2004). The language-specific nature of grammatical development: Evidence from bilingual language learners. *Developmental Science, 7*, 212–224.

Marinis, T., & Chondrogianni, V. (2010). Production of tense marking in successive bilingual children: When do they converge with their monolingual peers? *International Journal of Speech-Language Pathology, 12*, 19–28.

Marinis, T., & Saddy, D. (2013). Parsing the passive: Comparing children with specific language impairment to sequential bilingual children. *Language Acquisition, 20*, 155–179.

Masoura, E. V., & Gathercole, S. E. (1999). Phonological short-term memory and foreign language learning. *International Journal of Psychology, 34*, 383–388.

McCarthy, K. M., Evans, B. G., & Mahon, M. (2013). Acquiring a second language in an immigrant community: The production of Sylheti and English stops and vowels by London Bengali speakers. *Journal of Phonetics, 41*, 344–358.

McCarthy, K. M., Mahon, M., Rosen, S., & Evans, B. G. (2014). Speech perception and production by sequential bilingual children: A longitudinal study of voice onset time acquisition. *Child Development, 85*, 1965–1980.

Meisel, J. M. (1989). Early differentiation of languages in bilingual children. In K. Hyltenstam & L. Obler (Eds.), *Bilingualism across the lifespan: Aspects of acquisition, maturity and loss* (pp. 13–40). Cambridge: Cambridge University Press.

Meisel, J. M. (2008). Child second language acquisition or successive first language acquisition? In B. Haznedar & E. Gavruseva (Eds.), *Current trends in child second language acquisition* (pp. 55–80). Amsterdam: John Benjamins.

Melby-Lervåg, M., & Lervåg, A. (2011). Cross-linguistic transfer of oral language, decoding, phonological awareness and reading comprehension: A meta-analysis of the correlation evidence. *Journal of Research in Reading, 34*, 114–135.

Messer, M. H., Leseman, P. P. M., Boom, J., & Mayo, A. Y. (2010). Phonotactic probability effect in nonword recall and its relationship with vocabulary in monolingual and bilingual preschoolers. *Journal of Experimental Child Psychology, 105*, 306–323.

Mills, D. L., Coffey-Corina, S., & Neville, H. J. (1997). Language comprehension and cerebral specialization from 13 to 20 months. *Developmental Neuropsychology, 13*, 397–445.

Montanari, S. (2011). Phonological differentiation before age two in a Tagalog-Spanish-English trilingual child. *International Journal of Multilingualism, 8*, 5–21.

Morrow, A., Goldstein, B. A., Gilhool, A., & Paradis, J. (2014). Phonological skills in English language learners. *Language, Speech, and Hearing Services in Schools, 45*(1), 26–39.

Müller, N., & Hulk, A. (2001). Crosslinguistic influence in bilingual language acquisition: Italian and French as recipient languages. *Bilingualism: Language and Cognition, 4*, 1–21.

Muñoz, C., & Singleton, D. (2011). A critical review of age-related research on L2 ultimate attainment. *Language Teaching, 44*, 1–35.

National Association for Language Development in the Curriculum. (2013). *EAL pupils 1997–2013*. London: NALDIC.

Newport, E., Gleitman, H., & Gleitman, L. (1977). Mother, I'd rather do it myself: Some effects and non-effects of maternal speech style. In C. E. Snow & C. Ferguson (Eds.), *Talking to children* (pp. 109–149). Cambridge: Cambridge University Press.

Oller, D. K., Pearson, B. Z., & Cobo-Lewis, A. B. (2007). Profile effects in early bilingual language and literacy. *Applied Psycholinguistics, 28*, 191–230.

Omaki, A., Davidson-White, I., Goro, T., Lidz, J., & Phillips, C. (2014). No fear of commitment: Children's incremental interpretation in English and Japanese wh-questions. *Language Learning and Development, 10*, 206–233.

Paradis, J. (2008). Tense as a clinical marker in English L2 acquisition with language delay/impairment. In E. Gavruseva & B. Haznedar (Eds.), *Current trends in child second language acquisition: A generative perspective* (pp. 337–356). Amsterdam: John Benjamins.

Paradis, J. (2010). The interface between bilingual development and specific language impairment: Keynote article for special issue with peer commentaries. *Applied Psycholinguistics, 31*, 3–28.

Paradis, J. (2011). Individual differences in child English second language acquisition: Comparing child-internal and child-external factors. *Linguistic Approaches to Bilingualism, 1*, 213–237.

Paradis, J., & Blom, W. B. T. (2016). Do early successive bilinguals show the English L2 pattern of precocious BE acquisition? *Bilingualism: Language and Cognition, 3*, 630–635.

Paradis, J., & Genesee, F. (1996). Syntactic acquisition in bilingual children: Autonomous or independent? *Studies in Second Language Acquisition, 18*, 1–25.

Paradis, J., & Navarro, S. (2003). Subject realization and crosslinguistic interference in the bilingual acquisition of Spanish and English: What is the role of input? *Journal of Child Language, 30*, 371–393.

Paradis, J., Nicoladis, E., Crago, M., & Genesee, F. (2011). Bilingual children's acquisition of the past tense: A usage-based approach. *Journal of Child Language, 38*(3), 554–578.

Paradis, J., Tulpar, Y., & Arppe, A. (2016). Chinese L1 children's English L2 verb morphology over time: Individual variation in long-term outcomes. *Journal of Child Language, 43*, 553–580.

Parra, M., Hoff, E., & Core, C. (2011). Relations among language exposure, phonological memory, and language development in Spanish-English bilingually developing

2-year-olds. *Journal of Experimental Child Psychology, 108,* 113–125.

Pearson, B. Z., Fernández, S. C., Lewedeg, V., & Oller, D. K. (1997). The relation of input factors to lexical learning by bilingual infants. *Applied Psycholinguistics, 18,* 41–58.

Pearson, B. Z., Fernández, S., & Oller, D. K. (1993). Lexical development in bilingual infants and toddlers: Comparison to monolingual norms. *Language Learning, 43,* 93–120.

Pérez-Leroux, A. T., Pirvulescu, M., & Roberge, Y. (2009). Bilingualism as a window into the language faculty: The acquisition of objects in speaking-speaking children in bilingual and monolingual contexts. *Bilingualism: Language and Cognition, 12,* 97–112.

Petitto, L. A., Berens, M. S., Kovelman, I., Dubins, M. H., Jasinska, K., & Shalinsky, M. (2012). The "perceptual wedge hypothesis" as the basis for bilingual babies' phonetic processing advantage: New insights from fNIRS brain imaging. *Brain and Language, 121,* 130–143.

Place, S., & Hoff, E. (2011). Properties of dual language exposure that influence two-year-olds' bilingual proficiency. *Child Development, 82,* 1834–1849.

Place, S., & Hoff, E. (2016). Effects and non-effects of input in bilingual environments on dual language skills in 2 ½- year-olds. *Bilingualism: Language and Cognition, 19,* 1023–1041.

Plunkett, K., & Marchman, V. (1993). From rote learning to system building: Acquiring verb morphology in children and connectionist nets. *Cognition, 48,* 21–69.

Rankin, T., & Unsworth, S. (2016). Beyond poverty: Engaging with input in generative SLA. *Second Language Research, 32,* 563–572.

Rindermann, H., & Baumeister, A. E. E. (2015). Parent's SES vs. parental behavior and children's development: A reanalysis of the Hart and Risley study. *Learning and Individual Differences, 37,* 133–138.

Roesch, A. D., & Chondrogianni, V. (2016). "Which mouse kissed the frog?" Effects of age of onset, length of exposure, and knowledge of case marking on the comprehension of wh-questions in German-speaking simultaneous and early sequential bilingual children. *Journal of Child Language, 43*(3), 635–661.

Ronjat, J. (1913). *Le developement du langage observe chez un enfant bilingue.* Paris: Champion.

Rothman, J., Long, D. Iverson, M., Judy T., Lingwall, A., & Chakravarty, T. (2016). Older age of onset in child L2 acquisition can be facilitative: Evidence from the acquisition of English passives by Spanish natives. *Journal of Child Language, 43*(3), 662–686.

Scheele, A. F., Leseman, P. P. M., & Mayo, A. Y. (2010). The home language environment of monolingual and bilingual children and their language proficiency. *Applied Psycholinguistics, 31,* 117–140.

Schwartz, B. (2004). Why child L2 acquisition? In J. Van Kampen & S. Baauw (Eds.), *Proceedings of generative approaches to language acquisition* (Vol. 3, pp. 47–66). Utrecht, Netherlands: LOT Occasional Series.

Sebastián-Gallés, N., & Bosch, L. (2002). Building phonotactic knowledge in bilinguals: Role of early exposure. *Journal of Experimental Psychology: Human Perception and Performance, 28,* 974–989.

Sebastián-Gallés, N., & Bosch, L. (2009). Developmental shift in the discrimination of vowel contrasts in bilingual infants: Is the distributional account all there is to it? *Developmental Science, 12,* 874–887.

Sebastián-Gallés, N., Vera-Constán, F., Larsson, J. P., Costa, A., & Deco, G. (2009). Lexical plasticity in early bilinguals does not alter phoneme categories: II. Experimental evidence. *Journal of Cognitive Neuroscience, 21,* 2343--2357.

Serratrice, L., Sorace, A., & Paoli, S. (2004). Subjects and objects in Italian-English bilingual and monolingual acquisition. *Bilingualism: Language and Cognition, 7,* 183–206.

Service, E., & Kohonen, V. (1995). Is the relation between phonological memory and foreign language learning accounted for by vocabulary acquisition? *Applied Psycholinguistics, 16,* 155–172.

Sheng, L., Lu, Y., & Kan, P. F. (2011). Lexical development in Mandarin-English bilingual children. *Bilingualism: Language and Cognition, 14,* 579–587.

Sorace, A., Serratrice, L., Filiaci, F., & Baldo, M. (2009). Discourse conditions on subject pronoun realization: Testing the linguistic intuitions of older bilingual children. *Lingua, 119*(3), 460–477.

Souza, A. L., Byers-Heinlein, K., & Poulin Dubois, D. (2013) Bilingual and monolingual children prefer native-accented speakers. *Frontiers in Psychology, 4,* 953.

Sundara, M., Polka, L., & Molnar, M. (2008). Development of coronal stop perception: Bilingual infants keep pace with their monolingual peers. *Cognition, 108,* 232–242.

Tamburelli, M., Sanoudaki, E., Jones, G., & Sowinska, M. (2015). Acceleration in the bilingual acquisition of phonological structure: Evidence from Polish-English bilingual children. *Bilingualism: Language and Cognition, 18,* 713–725.

Thomas, E. M., Williams, N., Jones, L. A., Davies, S., & Binks, H. (2013). Acquiring complex structures under minority language conditions: Bilingual acquisition of plural morphology in Welsh. *Bilingualism: Language and Cognition, 17,* 478–494.

Thordardottir, E. (2011). The relationship between bilingual exposure and vocabulary development. *International Journal of Bilingualism, 15,* 426–445.

Thordardottir, E., & Brandecker, M. (2013). The effect of bilingual exposure versus language impairment on nonword repetition and sentence imitation scores. *Journal of Communication Disorders, 46,* 1–16.

Tomasello, M. (2008). *Origins of human communication.* Cambridge, MA: MIT Press.

Tsimpli, I.-M. (2014). Early, late or very late? Timing acquisition and bilingualism. *Linguistic Approaches to Bilingualism, 4,* 283–313.

Tsukada, K., Birdsong, D., Bialystok, E., Mack, M., Sung, H., & Flege, J. (2005). A developmental study of English vowel production and perception by native Korean adults and children. *Journal of Phonetics, 33,* 263–290.

Umbel, V. M., Pearson, B. Z., Fernández, M. C., & Oller, D. K. (1992). Measuring bilingual children's receptive vocabularies. *Child Development, 63,* 1012–1020.

Unsworth, S. (2005). Child L2, adult L2, child L1: Differences and similarities. A study on the acquisition of direct object scrambling in Dutch (Unpublished doctoral dissertation). Utrecht University, Utrecht, Netherlands.

Unsworth, S. (2013a). Assessing age of onset effects in (early) child L2 acquisition. *Language Acquisition, 20,* 74–92.

Unsworth, S. (2013b). Assessing the role of current and cumulative exposure in simultaneous bilingual acquisition: The case of Dutch gender. *Bilingualism: Language and Cognition, 16,* 86–110.

Unsworth, S. (2013c). Current issues in multilingual first language acquisition. *Annual Review of Applied Linguistics, 33,* 21–50.

Unsworth, S. (2014). Comparing the role of input in bilingual acquisition across domains. In T. Grüter & J. Paradis (Eds.), *Input and experience in bilingual development.* Amsterdam: John Benjamins.

Unsworth, S. (2015). Amount of exposure as a proxy for dominance in bilingual language acquisition. In C. Silva-Corvalan & J. Treffers-Daller (Eds.), *Language dominance in bilinguals: Issues of measurement and operationalization* (pp. 165–173). Cambridge: Cambridge University Press.

Unsworth, S., Argyri, F., Cornips, L., Hulk, A., Sorace, A., & Tsimpli, I. (2014). On the role of age of onset and input in early child bilingualism in Greek and Dutch. *Applied Psycholinguistics, 35,* 765–805.

Valian, V. (2014). Arguing about innateness. *Journal of Child Language, 41,* 78–92.

Vasic, N., Chondrogianni, V., Marinis, T., & Blom, W. B. T. (2012). Processing of gender in Turkish-Dutch and Turkish-Greek child L2 learners. In A. K. Biller, E. Y. Chung, & A. E. Kimball (Eds.), *Proceedings of the 36th Annual Boston University Conference on Language Development* (pp. 646–659). Somerville, MA: Cascadilla Press.

Vasilyeva, M., Waterfall, H., & Huttenlocher, J. (2008). Emergence of syntax: Commonalities and differences across children. *Developmental Science, 11,* 84–97.

Verhagen, J., Leseman, P., & Messer, M. (2015). Phonological memory and the acquisition of grammar in child L2 learners. *Language Learning, 65*(2), 417–448.

Volterra, V., & Taeschner, T. (1978). The acquisition and development of language by bilingual children. *Journal of Child Language, 5,* 311–326.

Weikum, W. M., Vouloumanos, A., Navarra, J., Soto-Faraco, S., Sebastián-Gallés, N., & Werker, J. F. (2007). Visual language discrimination in infancy. *Science, 316*(5828), 1159.

Werker, J. F., & Byers-Heinlein, K. (2008). Bilingualism in infancy: First steps in perception and comprehension. *Trends in Cognitive Sciences, 12,* 144–151.

Wexler, K. (1998). Very early parameter setting and the unique checking constraint: A new explanation of the optional infinitive stage. *Lingua, 106,* 23–79.

Yip, V., & Matthews, S. (2007). *The bilingual child: Early development and language contact.* Cambridge: Cambridge University Press.

Zdorenko, T., & Paradis, J. (2008). The acquisition of articles in child second language English: Fluctuation, transfer or both? *Second Language Research, 24,* 227–250.

Zdorenko, T., & Paradis, J. (2011). Articles in child L2 English: When L1 and L2 acquisition meet at the interface. *First Language, 21,* 1–25.

13 Key Issues and Future Directions: How Do Children Acquire Language?

CAROLINE F. ROWLAND AND EVAN KIDD

In part II, the authors have provided a clear overview of some of the key findings in language acquisition research, and the theories that have been proposed to explain them. The section covers a wide range of topics. It starts with how children learn to interpret and reproduce speech (Johnson & White, chapter 8) and then explores how they learn to map words to meanings (Kucker, chapter 9) and put words and morphemes together into grammatical sentences (Allen & Behrens, chapter 10). Skeide (chapter 11) then outlines how all these learning tasks might be mapped to structural networks in the developing brain, and Chondrogianni (chapter 12) discusses the questions that arise when we consider how children learn more than one language at a time. From this overview, it is immediately apparent that research in language acquisition covers many topics and that it crosses the disciplinary boundaries between developmental psychology, linguistics, and neuroscience. Despite this diversity, some common themes emerge, which we summarize here.

The clearest theme that emerges is the complexity of the task facing language learning children. They have to acquire a language-specific phonemic inventory, segment a continuous speech stream to identify words (Johnson & White), and map words onto meanings (Kucker). They must then extract abstract, hierarchical, grammatical patterns from a linear input that provides only indirect evidence for grammatical rules and learn a complex set of discourse-pragmatic cues that govern referential choice (Allen & Behrens). Yet almost all children achieve this in early childhood, despite extremely variable input from child to child, and multilingual children achieve it for two or more languages simultaneously (Chondrogianni).

Each new language advance builds on a basic foundation of previous learning. For example, once infants learn some language-specific properties of speech (e.g., phonemes, dominant stress patterns), they can begin to segment their input (e.g., via attention to distributional information, see Johnson & White). Successful segmentation of the input allows infants to recognize, and store, the phonetic detail of words as independent units, which enables them to start to map words onto meaning (Kucker). Mapping words onto meaning is, in turn, the first step toward the discovery of morphology and syntax (Allen & Behrens). However, this does not mean that language acquisition is necessarily stagelike because it is not essential to complete the acquisition process for a precursor skill before progressing to the next learning task. Instead, it seems that "infants are learning about their language at multiple levels in parallel" (Johnson & White, p. 108) and that different subsystems interact throughout the acquisition process. For example, English-speaking children probably need only acquire a few content words before they learn the dominant trochaic stress pattern (Cutler & Carter, 1987) that they can then start to use as a cue for segmenting further words from the speech stream. Such interactions appear to be present across the whole system. For instance, Skeide presents neuroimaging data to suggest that the hemodynamic activity supporting semantic and syntactic processing are largely overlapping before the age of seven years (Brauer & Friederici, 2007; Nunez et al., 2011). Additionally, Chondrogianni discusses how bilingual children often experience crosslinguistic transfer when acquiring aspects of syntax that are discourse-conditioned (e.g., subject expression).

A second theme that emerges concerns the nature of the learning mechanisms. Given the complex and dynamic nature of the system, there must be multiple, powerful learning mechanisms in the developing brain for language acquisition to occur. We say mechanisms (plural) rather than mechanism (singular), because learning involves solving a number of different tasks and thus is bound to require more than one type of learning mechanism. However, as we have seen in the chapters in this part, there are still fierce debates about what these mechanisms are. We suggest that at least three different types of mechanism will turn out to be essential.

First, children must be able to process language online; what Kucker calls "the real-time/in the moment reaction" (p. 114) (i.e., *realizing that wif is the coffee mug*). We can only acquire a language if we are able to process incoming input, and identify, at some level, the meaning that the speaker is trying to convey. Thus, we will need a mapping mechanism that integrates cues from the stimuli ("in the moment") to map the speech signal to the correct meaning. Discussions of this online mapping process are most prevalent in the word-learning literature, for example, to integrate speech, visual, and sociopragmatic cues to map words to correct referents (Kucker). However, online processing is equally important to other language-learning tasks, such as identifying when a minimal pair of phonemes marks a semantic distinction (for instance, in the Italian *capello* 'hair' vs. *cappello* 'hat'), identifying word-order constraints on meaning (e.g., that the sentence *the girl kissed the boy* refers to the action that a girl is bestowing on a boy), or integrating discourse level cues to interpret morphosyntactic cues. Thus, the ability to process language rapidly is an essential component of the acquisition process, even if children sometimes do it imperfectly (e.g., five-year-olds tend to misinterpret temporarily ambiguous garden path sentences such as *Put the frog on the napkin in the box*; Trueswell, Sekerina, Hill, & Logrip, 1999; see Allen & Behrens, section 4.4).

However, online processing is not enough. A second mechanism is required: children need to be able to store particular instances of language use in long-term memory for later retrieval. As Kucker states, this may well involve separable mechanisms: "Mapping is a real-time process serving communication in-the-moment. Learning, on the other hand, ... is better served through a longer process of strengthening (and weakening) word-referent associations" (p. 114). No doubt the long-term storage of information aids rapid online processing; for instance, having more robust phonemic knowledge allows better recognition of allophonic variants of phonemes and better segmentation of the input (Johnson & White), and knowledge of frequent frames may enable the identification of syntactic categories (Allen & Behrens).

Third, children need to be able to consolidate the information they have retained and generalize over particular instances of knowledge to build categories. For example, they must learn to generalize the word *dog* to all types of dogs (big dogs, small dogs, dogs in books, dogs on the TV) while excluding all other animals from the dog category (Kucker). They must learn to categorize words into grammatical categories (nouns, verbs, adjectives, etc.) and to generalize across particular instances of a category (e.g., nouns) to extract morphological markers (e.g., the English plural morpheme *-s*; Allen & Behrens). Some of these tasks may require a mechanism that accumulates information slowly over time (e.g., via associative learning processes; McMurray, Horst, & Samuelson, 2012), whereas others may rely on an error-based, prediction mechanism (Chang, Dell, & Bock, 2006) or hypothesis-driven learning (Medina, Snedeker, Trueswell, & Gleitman, 2011). Some of these mechanisms, such as those that are sensitive to prosodic information (Perani et al., 2011), may rely solely on low level, automatic processing whereas others, such as the emergence of semantic and syntactic representations at the sentence level, may need higher level processing involving further analysis by more complex mechanisms (Skeide). Some processes may even rely on memory consolidation during sleep. For example, in nine-month-old infants, lexical-semantic learning seems to be dependent on the consolidation of recent episodic memory traces during sleep (Friedrich, Wilhelm, Born, & Friederici, 2015; see Skeide). All these mechanisms will need to be capable of learning two or more languages simultaneously; for instance, to learn to use cues such as rhythm (stress/syllable timed languages) to distinguish between languages or to learn to discriminate between phonetic units belonging to the two languages (Chondrogianni).

Another clear point of commonality across the chapters is the fact that the learning mechanisms must be constrained or biased to use the input in certain ways. As Saffran has stated, "learners must be able to detect input statistics that are pertinent to linguistic structure amid all the irrelevant information in the input" (2002: 173), and to do this, the mechanisms must be biased to pay attention to some regularities in the input and not others. The debate about what kinds of constraints must be built into our learning mechanisms has been one of the most fiercely argued in the history of language acquisition research. It can be tracked back nearly 60 years to Chomksy's 1959 review of Skinner's book *Verbal Behavior*, which argued that behaviorist learning processes were incapable of learning a higher order cognitive function such as language. Since then, we have engaged in a long, productive debate about whether low-level domain-general cognitive constraints are sufficient for learning or whether innate knowledge of complex, abstract linguistic principles is required (see Allen & Behrens). However, this literature has, perhaps, focused too narrowly on these two, polar opposites. Instead, it is likely that the learning mechanism is constrained on a number of different levels.

Candidate low-level constraints include physical constraints (the binding of an object and word in physical space; see Kucker), as well as statistical biases; for example, frequent (Johnson & White), reliable (Allen & Behrens), or perceptually salient (Johnson & White) statistical cues may be weighted more highly (Bates & MacWhinney, 1987). There may also be higher-order constraints such as sociopragmatic cues (Kucker), as well as language-specific cues such as auditory biases (Skeide), phonological constraints (Johnson & White), constraints on the form of possible words (Johnson & White), or word-meaning mapping biases such as mutual exclusivity (Kucker). Sensitivity to some of these constraints may be built into the learning architecture by the genetic code, but others, such as mutual exclusivity, are likely to be a product of learning themselves (Halberda, 2003). Some tasks may need more powerful innate linguistic knowledge to constrain learning (e.g., language-specific knowledge about syntactic categories and roles; Allen & Behrens), although this is still a highly controversial proposal (see Ambridge, Pine, & Lieven, 2014, including commentaries).

To date, the field has also not paid enough attention to the fact that children's learning strategies are highly likely to change with age. For example, substantial evidence now suggests that referential mapping strategies change with development, with two-year-olds relying more on social cues, perceptual salience, and cross-modal synchrony and older children relying more on grammatical cues (Johnson & White; Kucker). Similarly, children can detect grammatical case marking cues by three years of age but may not be able to use this information to determine who is doing what to whom in a sentence until six years of age (Schipke, Friederici, & Oberecker, 2011; see Skeide). Some of these changes may be driven by the child's increasing knowledge of the world and of their language. For example, it is likely that the shape bias, which biases children to extend novel noun labels to similarly shaped objects, is learned through experience of how object labels tend to be extended (Kucker). Similarly, two-year-olds' ability to use abstract syntactic relations to comprehend novel transitive sentences seems to rely on the size of their vocabulary (Matthews, Lieven, Theakston, & Tomasello, 2007; see Allen & Behrens). However, others may wait upon the maturation of the necessary connections in the brain (Skeide). It is also highly likely that the developmental pattern is different in critical ways in multilingual children. For example, Chondrogianni summarizes work showing that bilingual children do not seem to develop a mutual exclusivity bias, or at least develop a much more context-dependent one, and that in second language (L2) learners, there may be delayed or different patterns of acquisition dependent on the characteristics of the L1 and its similarity to their L2. Much more work needs to be done to determine how the learning process itself changes throughout development, particularly in the field of developmental neuroscience, where we need to understand more about how the neural circuits change with age and with the accumulation of new knowledge. What is clear, however, is that only a developmentally dynamic model of language acquisition will be able to adequately capture the learning process.

Another theme we extract from this part concerns methodology. In the last few decades, language acquisition research has been revolutionized by methodological advances that have opened up new avenues of research, as well as making old methods much easier to implement. Noninvasive electroencephalography and eye tracking provide implicit measures of online language processing, enabling us to develop hypotheses about language acquisition in prelinguistic infants. Headturn preference and habituation methodologies allow us to determine what one-month-, -week- or even -day-old infants are able to discriminate (individual difference notwithstanding; e.g., Junge, Kooijman, Hagoort, & Cutler, 2012). New recording technology makes it easier for us to track and analyze how children use language in their natural environment and enables us to perform detailed analyses of the linguistic environment in which they are immersed. Even the more traditional methods (e.g., elicitation, forced-choice pointing) are easier to implement, and the data easier to analyze, with new presentation software and statistical programs. More advances are on the horizon, including techniques to make functional MRI and magnetoencephalography scanning easier with young children, open source tools for (semi)automatic transcription and coding of naturalistic data, and virtual reality labs in which we can create controlled three-dimensional learning environments for our experiments.

However, new methods bring new challenges as well as opportunities. First, there is the problem of integrating results from different methodologies. For example, what are we to make of the fact that six-month-olds look longer at a picture of their mom when they hear *mommy* and look longer at a picture of their dad when they hear *daddy* in a classic preferential looking design (Tincoff & Jusczyk, 1999; see Johnson & White)? Does this mean that these children have the same referential understanding of the words mommy and daddy as adults, or, as Johnson and White speculate, do they "simply associate labels and objects or events, in the way we might associate the bell of an ice cream truck with a delicious treat" (p. 103)? If the latter, how does this early type of understanding develop into the adultlike concept? Similar issues arise when we consider artificial language learning studies, where infants are trained on novel languages in the lab. For example, infants can learn to recognize word boundaries using transitional probabilities in artificial language experiments in which each "word" is three syllables long but cannot do it when the words are of a more naturalistic variable length (Johnson & Tyler, 2010; see Johnson & White). There are big questions still to be asked about how children use cues to language in the real, messy, complex world to develop adultlike linguistic representations.

Finally, we suggest that there needs to be more research on individual differences in future. Many of the findings cited in this part are based on between-subjects group-based experiments, which mirror the focus of research in language acquisition in general. However, there are large and stable individual differences both in children's environments, and in the sequence and speed of children's acquisition, which will need to be explained by any complete theory of language development (Kidd, Donnelly, & Christiansen, 2018). The quality and quantity of the language in children's environment clearly plays a role, both in monolingual and multilingual acquisition. However, other factors may be equally important, such as the speed with which children process linguistic information (Marchman & Fernald, 2008). Unfortunately, however, most of our current methodological designs simply measure group effects, which means they are often not sensitive enough to capture individual variation. Adapting these measures will require some thought; we will almost certainly need to test larger sample sizes, we will need to design our stimuli so as to maximize individual variation across children, and we will need to implement test-retest procedures to determine how much of the variation is meaningful and how much due to factors beyond our control, such as the infant's willingness to cooperate at the time of testing. This may not be a simple task.

In this chapter we have summarized some of the common themes that emerged in part II and highlighted what we think are the most promising directions for future research. The discussion demonstrates the complexity of the learning task, which requires that children master a number of different skills. The development of such skills relies on a strong basic foundation from previous learning, but this does not mean that acquisition is necessarily stagelike; instead, children are tackling many of the different tasks involved in parallel. The complexity of the learning task also requires that we posit multiple powerful learning mechanisms in the child's developing brain. Unfortunately, we do not yet know enough about what these mechanisms are or what constraints are built in that enable them to extract the relevant information from the input. We also do not know how children's learning strategies change over development, as the accumulation of knowledge, and the maturation of the brain, changes the nature of the learning task. In addition, although we already know a substantial amount about how different regions of the brain might support language learning, we need more work on how the brain builds the functional adult cortical language network. Finally, we have discussed the methodological advances that have made many of these insights possible and summarized some of the benefits, but also the challenges, that these advances bring to the study of language acquisition. In future, we will need to work to identify what the results from such studies tell us about the nature of children's developing linguistic representations, and we will need to learn how to adapt new methodologies to enable us to explore individual differences in acquisition.

Acknowledgments

This work was supported by the International Centre for Language and Communicative Development (LuCiD) at the University of Liverpool, funded by the Economic and Social Research Council (UK) (ES/L008955/1), and the Australian Research Council Centre of Excellence for the Dynamics of Language (CE140100041).

REFERENCES

Ambridge, B., Pine, J. M., & Lieven, E. V. M. (2014). Child language acquisition: Why universal grammar doesn't help. *Language, 90*, e53–e90.

Bates, E., & MacWhinney, B. (1987). Competition, variation, and language learning. In B. MacWhinney (Ed.), *Mechanisms of language acquisition* (pp. 157–193). Hillsdale, NJ: Erlbaum.

Brauer, J., & Friederici, A. D. (2007). Functional neural networks of semantic and syntactic processes in the developing brain. *Journal of Cognitive Neuroscience, 19*(10), 1609–1623.

Chang, F., Dell, G. S., & Bock, K. (2006). Becoming syntactic. *Psychological Review, 113*(2), 234–272.

Chomsky, N. (1959). A review of B. F. Skinner's *Verbal Behavior. Language, 35*(1), 26–58.

Cutler, A., & Carter, D. M. (1987). The predominance of strong initial syllables in the English vocabulary. *Computer Speech and Language, 2*, 133–142.

Friedrich, M., Wilhelm, I., Born, J., & Friederici, A. D. (2015). Generalization of word meanings during infant sleep. *Nature Communications, 6*, 6004.

Halberda, J. (2003). The development of a word-learning strategy. *Cognition, 87*, B23–B34.

Johnson, E. K., & Tyler, M. D. (2010). Testing the limits of statistical learning for word segmentation. *Developmental Science, 13*, 339–345.

Junge, C., Kooijman, V., Hagoort, P., & Cutler, A. (2012). Rapid recognition at 10 months as a predictor of language development. *Developmental Science, 15*, 463–473.

Kidd, E., Donnelly, S., & Christiansen, M. H. (2018). Individual differences in language acquisition and processing. *Trends in Cognitive Sciences, 22*(2), 154–169.

Marchman, V. A., & Fernald, A. (2008). Speed of word recognition and vocabulary knowledge in infancy predict cognitive and language outcomes in later childhood. *Developmental Science, 11*, F9–F16.

Matthews, D., Lieven, E., Theakston, A., & Tomasello, M. (2007). French children's use and correction of weird word orders: A constructivist account. *Journal of Child Language, 34*(2), 381–409.

McMurray, B., Horst, J. S., & Samuelson, L. K. (2012). Word learning emerges from the interaction of online referent selection and slow associative learning. *Psychological Review, 119*(4), 831–877.

Medina, T. N., Snedeker, J., Trueswell, J. C., & Gleitman, L. R. (2011). How words can and cannot be learned by observation. *Proceedings of the National Academy of Sciences, 108*(22), 9014–9019.

Nunez, S. C., Dapretto, M., Katzir, T., Starr, A., Bramen, J., Kan, E., ... Sowell, E. R. (2011). fMRI of syntactic processing in typically developing children: Structural correlates in the inferior frontal gyrus. *Developmental Cognitive Neuroscience, 1*(3), 313–323.

Perani, D., Saccuman, M. C., Scifo, P., Anwander, A., Spada, D., Baldoli, C., ... Friederici, A. D. (2011). Neural language networks at birth. *Proceedings of the National Academy of Sciences, 108*(38), 16056–16061.

Saffran J. R. (2002). Constraints on statistical language learning. *Journal of Memory and Language, 47*, 172–196.

Schipke, C. S., Friederici, A. D., & Oberecker, R. (2011). Brain responses to case-marking violations in German preschool children. *Neuroreport, 22*(16), 850–854.

Tincoff, R., & Jusczyk, P. W. (1999). Some beginnings of word comprehension in six-month-olds. *Psychological Science, 10*, 172–175.

Trueswell, J., Sekerina, I., Hill, N., & Logrip, M. (1999). The kindergarten-path effect: Studying online sentence processing in young children. *Cognition, 73*, 89–134.

III COMMUNICATION WITH AND BEFORE LANGUAGE

SECTION EDITORS: STEPHEN C. LEVINSON AND IVAN TONI

14 Interactional Foundations of Language: The Interaction Engine Hypothesis

STEPHEN C. LEVINSON

1. Communication without Language

This chapter argues that humans are endowed with a special interactive ability. It is this ability that underlies most of our species' great achievements: our language, our ability to cooperate, and hence our capacity as a culture-bearing species. In particular, I argue that it is not language that makes human communication possible, but rather a special underlying communicative ability that makes language possible. Despite the fact that this view runs against the orthodoxy, there are plenty of reasons to believe in its veracity.

For a start, there are humans born who are cut off from the normal means to learn language—they are profoundly deaf, and unless they have the luck to be born in a culture where there is a developed sign language, or where cochlear implants are mandatory, they are forced to invent from scratch some kind of sign language to communicate with those who look after them (Goldin-Meadow, 2003). Such "homesign" systems have been described from around the world (Van Deusen-Phillips, Goldin-Meadow, & Miller, 2001), and have certain features in common: lexical signs and simple sentence structures tend to emerge, despite great difference in the kind of feedback such children get. Such homesign systems have formed the basis out of which more developed sign languages have sometimes emerged (Senghas, Kita, & Özyürek, 2004).

That we have the means to communicate effectively without language is also clear from encounters with other cultures—every early explorer (and even adventurous tourist) has had some experience of this. In 1930, the Leahy brothers penetrated the central highlands of New Guinea, at that time thought to be uninhabited—their first encounters with the thriving indigenous population and their gestural communications were captured on film (see *First Contact* [Connolly & Anderson, 1987], a film incorporating the early footage) and demonstrate the effectiveness of gesture and pantomime under potentially hostile conditions.

Additional evidence comes from interactions with prelinguistic children; for example, one-year-olds can communicate about absent entities or indicate a searched-for item, by pointing to the item or the place where it was (Liszkowski, 2013). Reciprocal interactions develop with infants from at least three months after birth, and by nine months, there is little doubt that a dawning understanding of communicative intentions is developing. This is discussed at length in section 4.

Various language impairments are also revealing. Severely aphasic patients can communicate remarkably well in their home environments with limited resources—with little more than assent and dissent, complex communication can be sustained (Goodwin, 1995). Children with severe speech impediments, such as some Down's syndrome individuals, may be fluid communicators, while others with intact language, such as high-functioning autistic children, may have intact language but lack interactive intuitions and motivations and so be unable to use it effectively (a double dissociation we return to in the conclusions). Understanding precisely what such autistic individuals may lack in the way of interactional abilities may ultimately offer the key to the topic of this chapter (Frith, 2003).

Finally there is experimental evidence about our abilities to communicate without language, or even without any communicative conventions. Galantucci (2005) showed that when players of a coordination game, separated in space, are deprived of an easy means of linguistic or symbolic communication, they can invent an arbitrary symbol system and coordinate on its use. De Ruiter et al. (2010) used a similar method to show that coordination and communication was possible even when participants were deprived of the chance to build conventional symbols by rapid changes in the stimuli. The team went on to show that the sender and receiver during such exchanges activated overlapping areas of the brain, especially the right posterior superior temporal sulcus, a region previously associated with the attribution of intention (Noordzij et al., 2009).

The idea that communication does not rely on conventional symbols, but rather that such symbols rely on an underlying communicative ability, can be traced to the philosopher H. P. Grice (1957, 1989). His analysis suggested that all meaning is ultimately related to intention recognition:

S $means_{NN}$ z by U iff:
1. S intends U to cause z in H
2. S intends 1 to be achieved by H recognizing 1.

Here $meaning_{NN}$ stands for *nonnatural or intentional meaning*, sometimes talked about in terms of M-intentions (meaning intentions). So if I want to silently communicate to you, across the table in the seminar room, that you have a cappuccino moustache, I could vigorously wipe my upper lip: my intention is to cause a belief in you (that you have some foodstuff on your upper lip) by means of your recognizing that intention, and so coming to believe that you have some foodstuff on your upper lip! Why else would I be rubbing my lip so vigorously while gazing at you? Grice, and subsequent commentators, have argued that this special reflexive intention recognition could underlie complex symbol systems: once you have solved one of these joint intentional puzzles using some action, the next time we use the same action you'll leap to the same conclusion. Schelling (1960) showed that tacit communication, even without any overt signal, allows people to coordinate actions if they share the goal—each will think what the other might be thinking (these are games of pure coordination); furthermore, repetitive games give rise to salient solutions using earlier precedent. Lewis (1969) built on this and argued that all conventions have the character of arbitrary solutions on which participants jointly coordinate and that language is a huge edifice of such conventions. These authors thus provide us with an ultimate reduction of meaning to intention recognition, together with ways in which such complex inferences can be short-circuited by building up conventions in a community (we return to the psychological reality that may underlie this philosophical reduction in section 4).

We thus have both the demonstration of the ability to communicate complex propositions without language (homesign, first contact situations, and experimental semiotics and pragmatics) and some theoretical reconstruction of how it is possible. It is tempting to think that this communicative ability is in abeyance once we have complex linguistic communication at our disposal, but nothing could be further from the truth. Even the simplest sentence requires inferential resolution: If I say *The cat is on the mat*, already the definite articles constitute a coordination problem—which cat, which mat—and what sense of *on* is intended (is the cat a picture woven into the mat?)? But in interactive language use, the inferential requirements are compounded: conversation consists of exchanges of *speech acts*, sequences of actions such as greetings, offers, acceptances, requests, compliances, assessments or appreciations, and the like, typically coming in pairs of initiating action and response. But speech acts are often not directly signaled by the form of the utterance—there's no one-to-one mapping between form and function (Levinson, 2013): if I say *What are you doing tonight?* it might be a prelude to an invitation, proposal, or request rather than a question about your domestic routines—language use would be impossible without the deployment of the same inferential abilities that we see involved in radical tacit communication (Levinson, 2000; Sperber & Wilson, 1986).

2. The Puzzle of Language Diversity

There are some 7,000 languages today, grouped into some 430 families or highest level clades that can be reconstructed (Hammarström, Forkel, Haspelmath, & Bank, 2016). Languages of different families tend to show little similarity in their sound systems (with inventories of phonemes from 11 to 140), their word formation (little morphology or so elaborate that it more or less exhausts the syntax), their syntax (all manner of word orders, incomparable constructions, and such) or their semantics (absence of conditionals, proper color words, varied segmentation of the body into body parts, and such). Beneath the diversity, some structural parallels can be noted, but the lowest common denominators are few and far between (Evans & Levinson, 2009). In a study attempting to map the way languages have colonized the possible "design space," we found that as far as the mapping of semantic and functional distinctions into morphosyntax is concerned, unrelated languages are widely dispersed in their solutions (Levinson, Hammarström, & Roberts, 2018).

How then to account for the fact that we are the only language-bearing mammal? If language structure was built into the genes, this ought to be clearly revealed by crosslinguistic comparison. If instead there is just some special abstract facility for building complex syntax, as Berwick and Chomsky (2016) assumed, then that itself won't account for our abilities to learn such diverse languages. The more special, restricted, and uniform any such endowment is, the less it can account for language diversity. An alternative line of thought would emphasize the constraints from general cognition (perception and memory, for example; see Christiansen & Chater,

2008; Christiansen & Chater, 2016) that will limit the possible target languages, but such constraints are unlikely to differ greatly from those in our nearest primate cousins, and thus will probably not account for our unique language abilities. For that, what we crucially need are learning mechanisms. One such mechanism is our capacity for vocal learning, that is, the ability to hear a sound and mimic it in production. To date, there is not much evidence of this ability elsewhere in the primate order (as opposed, for example, to the passerines). But vocal learning only primes the pump for learning sound systems, crucial though that is.

It is here that the need becomes evident for some mechanism that can motivate and enable the child to learn the structure of the native language it must acquire. We suggest here that the child is endowed both with strong instincts to communicate and with instinctual understanding of the interactional system that humans use to communicate (Lee, Mikesell, Joaquin, Mates, & Schumann, 2009; Levinson, 2006). Using this understanding, the infant can bootstrap itself into the world of intersubjectivity (see section 4). On this account, languages are free to vary under drift and cultural selection because human language capacities lie very largely in an invariant, underlying, communicative infrastructure rather than in a universal "blueprint" for language structure.

3. Universals of Interactive Language Use

If there is an instinctual basis for patterns of interactive language use, we would expect strongly universal patterns of use to be evident. Here we make the case that these can be clearly discerned.

The conversational deployment of language is its core use—face-to-face interaction is the central niche in which it is learned, in which it evolved, and the home for the bulk of all language use. We each output on average a couple of hours a day, over 15,000 words, but distributed into small bursts of communication (on average around two seconds long), which alternate with the output of others. Any Martian ethologist would find this pattern of distributed recurrent short bursts of signaling one of the most notable features of human communication, observing that the groups involved can be of different sizes, that participants tend to face or orient to one another, and that the current speaker alone tends to produce concurrent hand movements, facial gestures, and such. Research on language use in conversation has mostly been undertaken by the nonexperimental sciences, partly because the phenomena are hard to capture in a laboratory setting, and partly because the study of language has often been construed more abstractly, outside its original and core functional niche. Extensive corpus work has been done, however, especially by the conversational analysts (see, e.g., Schegloff, 2007), but also by corpus phoneticians (Heldner & Edlund, 2010; Levinson & Torreira, 2015) and students of gesture and situated communication (Seyfeddinipur & Gullberg, 2014).

Although there is substantial cultural patterning of conversational behavior, its fundamental organization has recently been shown to be strongly universal. The alternating turn-taking, for example, has distinctive properties recurrent across all languages that have been examined: the timing is precise, with gaps of ~200 ms, overlapping speaking tends to be minimal and brief (~5% of the speech stream), and the exchange of roles seems organized by the same principles outlined by Sacks, Schegloff, and Jefferson (1974). A speaker gets a minimal clauselike unit as an initial turn, which can be extended by tacit or explicit agreement to another such unit, but after each such unit, speaker transition may occur (unless a speaker is specifically selected, the first starter becomes the next speaker). A study of 10 languages from five continents showed only minor differences in timing (Stivers et al., 2009).

The turn-taking system with its rapid transitions has interesting implications for language processing: the typical modal 200-ms gap is much tighter than the latency of speech planning and encoding (600 ms for a single word, 1000 ms or more for a clause), implying that next speakers must plan their utterances well before the incoming speech has ceased. This would require predicting both the content and the timing of the incoming turn, so that the relevant response can be planned to come in on time. We believe, on the basis of a series of experiments, that the recipient of an incoming turn tries early to predict the speech act (whether the utterance is a question or offer, for example) and begins planning the appropriate response as early as possible, withholding the response (if necessary) until turn-end signals are detected in the incoming turn, as sketched in figure 14.1 (see Bögels, Magyari, & Levinson, 2015; Levinson, 2016; Levinson & Torreira, 2015). This would account for the relatively stable response time, with a mode of 200 ms, close to the limit of human responses to simple "go" signals. Bearing in mind that response times for all tasks increase logarithmically as the number of choices increase (Hick's Law), and that a speaker selects from a vocabulary of at least 20,000 words, it is clear that this is a remarkable feat. The system of short response times of around 200 ms thus puts very considerable pressure on the processing systems for language.

It is worth asking why this system is the way it is. The possibility arises that this system arose prior to the

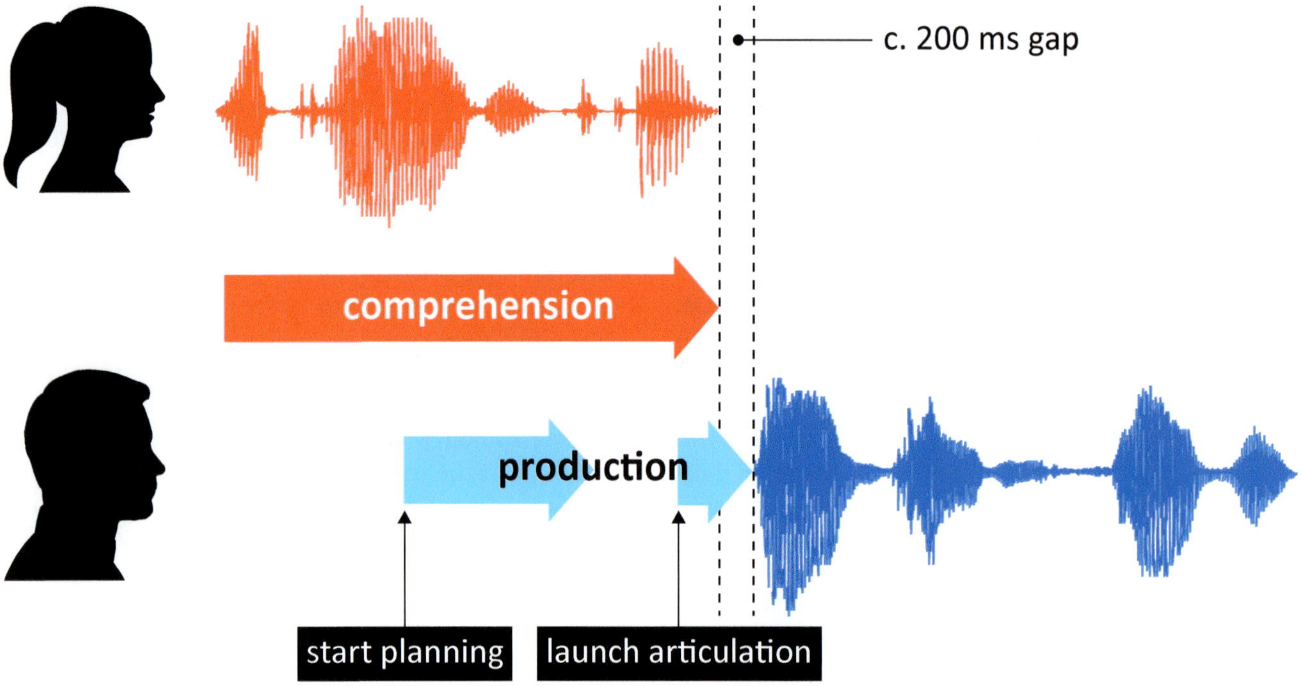

FIGURE 14.1 Overlapping comprehension and production in conversation (after Levinson, 2016).

evolution of complex language, and that the linguistic system has itself had to adjust to it. Infants today have to learn to squeeze ever increasingly complex encodings into these same short two-second bursts with quick response times, and struggle with it throughout the childhood years (see section 4). Perhaps the same slow development took place in phylogenetic time over the million years leading up to complex language (Dediu & Levinson, 2012).

In any case, the turn-taking system is a highly specific part of the ethology of human communication, which requires some account. One may ask what hurries participants on? One factor is that given the norm of a quick response time, delayed responses come to have a specific semiotics, signaling difficulties in responding in the desirable direction. For example, after a request, a delay is a likely harbinger of a refusal (Kendrick & Torreira, 2015; Pomerantz & Heritage, 2013). To avoid such suggestions interlocutors have to keep the pace up.

Turn-taking is just one layer of universal organization characteristic of interactive language use. Another such system is *repair*—a system for correction, clarification, and the maintenance of intersubjectivity. Schegloff, Jefferson, and Sacks (1977) outlined an ordered system of priorities, where provision is first provided for self-repair mid-turn, then at turn-transition, and then by other-initiation in next turn. Considerable crosslinguistic work has been done on the system of other-initiated repair (Enfield et al., 2013), and across languages it occurs at a rate of 1.4 times per minute (Dingemanse et al., 2015). The rate gives one some idea of its functional importance: without this means of correcting intersubjective understandings, our communications would rapidly derail.

In a study of a dozen languages from five continents, it was found that there are remarkable parallels in the shape and format of these initiators of repair, with most languages having a schwa-based particle (cf. English *huh?*) where the whole prior turn is within the scope of doubt (e.g., it was not heard or understood; Dingemanse, Torreira, & Enfield, 2013) and then a series of ways of targeting precisely items of lesser scope (Dingemanse et al., 2015). Even the bodily behavior of repair initiators is similar across cultures, often involving a freezing of kinesic movement (Floyd, Manrique, Rossi, & Torreira, 2016). Without a system of this kind, our understandings in interaction would soon go their separate ways.

Another area where crosslinguistic work shows remarkable parallels is in the higher order organization of turns into sequences. As we noted, conversations consist of exchanges of *actions* or speech acts, where these are very variably coded in linguistic form. A basic sequence is composed of an initiating action, followed by its response, such as questions and their answers, offers and acceptances, requests and compliance. Such "adjacency pairs" form a core on which elaborate extensions of structure can be built (Schegloff, 2007). There can be presequences such as the following preoffer sequence (p. 30):

A: Whatcha doin'? ← *Pre-offer* } *Pre-sequence*
B: Not much. ← *Go ahead*

A: Y'wanna drink? ← *Offer/Proposal* } *Base adjacency pair*
B: Okay. ← *Acceptance*

There can be insert sequences such as the following other-initiated repair (OIR) sequence (p. 102):

A: Have you ever tried a clinic? *Question (base first pair-part)*
B: <u>What</u>? } *Insert sequence (OIR)*
A: Have you ever tried a clinic?
B: ((sigh)) No, I don't want to go to a clinic *Answer (base second pair-part)*

And there can be post-adjacency-pair expansions, as in the following example after a repair insert (from p. 152 with many simplifications):

A: You sound HA:PPY hh ← *base first pair-part*
B: u- I sound ha:p[py?] } *Insert sequence (OIR)*
A: [Yee]uh
(0.3 sec.)
B: No ← *base second pair-part*
A: Nno:? } *Post expansion*
B: No

As even these simplified examples make clear, the structures that can be built with these building blocks can be quite complex. In fact, using insert sequences, far deeper center-embedding is found in conversational structure than can be found in language syntax, as illustrated next (Levinson, 2013, 155, ex. 14, after Merritt 1976):

(14) Merritt (1976)

C: Do you have master carbons?
(Q(prerequest: 0)
S: (pause) Yes, I think we do
(A:0)
Wat kind do you want?
(Q:1)
C: How many kinds do you have?
(Q:2)
S: Well, there are carbons for gelatin
(A:2)
duplicators, and carbons for spirits
C: Well, I'll take the carbons for spirits, please
(A:1)
S: ((goes to get))
(Action: 0)

A recent study (Kendrick et al., 2014) showed that virtually all these complex patterns of usage can be found in a dozen different languages of 11 distinct unrelated language families. So once again, these patterns seem to have a "natural" origin.

We have here reviewed three dimensions of conversational organization—turn-taking, repair, sequence organization—that have been studied in careful, controlled comparisons across 10 or more languages around the world. Another domain for which we have much comparative material is the greeting behavior that initiates verbal interaction (see Duranti, 1997). We can anticipate that many other features of the conversational niche will turn out to be equally pan-human, for example the use of laughter, gaze, and gesture, which are known to be culturally curtailed in variable ways (see, e.g., Rossano, Brown, & Levinson, 2009), but which nevertheless can be found in probably all societies (see, e.g., Sidnell, 2009). This universality stands in contrast to the extraordinary variability of human language itself across social groups, as mentioned at the outset.

What does the strong universality betoken? It could perhaps be simply a question of optimal design for the niche—a set of recurrent best solutions to recurrent problems in communication (Schegloff, 2006). But more than that seems involved. Let us consider turn-taking in a bit more detail. Would functional considerations inevitably lead communicators to take turns at speaking? A number of observations argue against a purely functional explanation, an argument for parallel cultural evolution or emergence. For example, could turn-taking be basically motivated by the difficulty of listening and speaking at the same time? Miller (1947) showed that one voice is in fact poor masking for another—only when there are many background voices is intelligibility deeply compromised. Perhaps a more persuasive point is that turn-taking in the sign languages of the deaf appears to be identical to that of spoken languages (de Vos, Torreira, & Levinson, 2015): here there are two distinct channels, vision and action, and hence no channel masking occurs, but nevertheless normal conversational turn-taking occurs. Another functional motivation might be what Sacks et al. (1974) called the "proof procedure" offered by the turn-taking system: if the speaker issues only a short utterance, he can find out rapidly from the response whether he was immediately understood. That might motivate rapid turn-taking as a rapid system of checks. However, that predicts a counterfactual, namely that people would respond as soon as they can predict what speech act is intended. We know that speech act interpretation is remarkably early, sometimes if context permits as early as the first word or two of the incoming turn (Gisladottir, Chwilla, & Levinson, 2015), yet responders clearly aim to delay their turn until the incoming turn is completed. In fact, overlap occurs in only 5% of the speech stream (Levinson & Torreira, 2015). A final argument against a functional account of the universal patterns is that other turn-taking systems are not only conceivable, but actual—they occur in special institutional settings, such as classrooms, law courts, or parliaments. These in contrast to conversational turn-taking vary widely across cultures and indeed institutions, and appear functional within those contexts. So the universality of the conversational mode of turn-taking requires an independent explanation.

In the following sections, we produce some prima facie evidence for a much deeper explanation of these universal patterns, tracing their origin in ontogenesis and phylogenesis.

4. Ontogenesis of Interactive Skills

The phenomenon of *protoconversation* first came under scientific scrutiny in the 1970s (Bates, Camaioni, & Volterra, 1975; Bateson, 1971; Bruner, 1975; Trevarthen, 1977): infants begin early to show contingent responses to a caretaker's communicative displays. Let us first consider imitation. The ability to imitate has been thought to be instinctual, on the basis of manifestation in the first hours after birth (Meltzoff & Moore, 1977), but recent studies challenge this (Oostenbroek et al., 2016): it is at least an early-learned and complex mapping from visual image to motor output. By one month infants are cooing, and by three months infants can do vocal imitation of vowels, although the output more closely matches adult categories by five months (Kuhl & Meltzoff, 1996). This vocal imitation (apparently unique in the primates) must have an instinctual basis, but the match to target requires maturation of the vocal tract and learning of the local target categories.

Early proto-conversation involves more than the contingent matching of signals, it also requires control of timing—the turn-taking of conversation requires coordination between two or more parties, the one taking the floor, the other desisting. A series of recent studies has explored the development of this coordination from three months to three-and-a-half years and beyond. Three-month-old infants respond with the same latency as their mothers (median ~550 ms), but interestingly this response time gets greater with age up until nine months (1100 ms), when it decreases again (Hilbrink, Gattis, & Levinson, 2015); however, over the same time interval, the amount of overlapping turns produced by the infant diminish from roughly 40% to ~25%, but overlaps remain short in duration (between 600 and 500 ms). We interpret this as an instinctual tendency for response, but with temporal control developing over time.

From 18 months old, when language production begins in earnest, children respond to questions with latencies of about 870 ms, with this decreasing over the next two years to about 530 ms (Casillas, Bobb, & Clark, 2016). This is slow by adult standards and remains slow through to middle childhood as the complexity of the language increases both in input and output (Stivers, Sidnell, & Bergen, 2018). The delay in response is fairly clearly due to output difficulties, since children from two-and-a-half years of age already use linguistic cues to anticipate the ends of turns like adults do (Lammertink, Casillas, Benders, Post, & Fikkert, 2015).

Part of early interactional abilities involves the understanding of another universal interactional system, the repair system. From their first words, children perform self-repair quickly, thus forestalling other-initiated repair, and by 2;0 years old children respond to other-initiated repairs or requests for clarification about 75% of the time, mostly with helpful reformulations rather than repeats (Casillas, 2014). Their repair system appears to have all the hallmarks of the adult system remarkably early in development.

Turn-taking and repair have to do with the "mechanics" of interactive communication, but what about their understanding of the intentional background to communication? We earlier pointed to the universality of conversational sequences involving paired action types, such as questions and answers. Infants start producing "protoimperatives" (reaching for something while looking at their caretakers) and "protodeclaratives" (showing and giving of objects) by about 10 months old (Bates et al., 1975; Casillas & Hilbrink, 2018), and then pointing by 12 months (Liszkowski, Brown, Callaghan, Takada, & de Vos, 2012). These simple gestures are interpreted by caregivers as requests and declarations of interest, but what the infant actually has in mind is harder to determine. This brings us back to Grice's intentional analysis of meaning: The speaker or actor intends to cause an effect in the recipient just by getting the recipient to recognize that intention. This requires understanding and reconstructing other's mental states, but it has even been doubted that infants have any clear understanding of other's minds—children, after all, fail classic theory of mind tasks right until age four years (Baron-Cohen, Leslie, & Frith, 1985; but see Onishi & Baillargeon, 2005, who actually show false-belief understanding from 15 months). On the other hand, by 9 months infants begin to sustain joint attention and follow gaze, and by 12 months they understand and produce referential pointing and helpfully point to displaced objects to help adults (Liszkowski, Carpenter, Striano, & Tomasello, 2006), showing understanding of other's mental goals. The different results can be reconciled by noting that there are significant differences between explicit mind reading (required by the classical tasks) and implicit mind reading, which is indubitably early (Frith & Heyes, 2014).

If one reconsiders the Gricean analysis of meaning:

S $means_{NN}$ z by U iff:
1. S intends U to cause z in H
2. S intends (1) to be achieved by H recognizing (1)

there is little doubt that by the first year of life, infants grasp the first clause—they understand that communicating involves causing effects in other minds. What is less clear is whether they grasp the second clause, that communication involves the intention to make this mind-changing intention overt. Tomasello, Carpenter, and Liszkowski (2007) argued that the child does grasp this in the second year, at which time he or she is able to discount accidental behaviors and correct misunderstandings even when, for example, a request has been satisfied. Csibra (2010) suggested that human infants may be born with an innate sensitivity to an inflection as it were (motherese, gaze, and such) that marks an action as communicative, so guaranteeing the ability to recognize the second Gricean clause.

But regardless of the age at which this higher-level intention is recognized, it is important to see that there is a simple bridge to this higher-level intention recognition provided by the ritualization of behavior—where Y is an initial move followed by X, producing Y may elicit X. The infant may reach toward an object, and have it provided, without having a clear understanding of request behavior at all: it is just an observed contingency. Success at that level will lead to repetition, by which time reaching is a kind of tool that works by manipulating the recipient (Grice's condition 1). This level may be achieved not only by infants in the last months of the first year, but also by infant apes (see section 5). The next level is for the child to realize that the tool only works when she has the recipient's attention and willingness and the tool itself manifests its intention—the recipient has to discern the desire for recognition, and now we have the second clause or something close to it.

The early availability and universality of the interaction system predicts that children growing up in a linguistic vacuum would nevertheless learn to communicate, and to do so in remarkably similar ways—a prediction borne out by the phenomenon of homesign, where deaf children construct gestural signs in the absence of a cultural tradition (Goldin-Meadow, 2003).

To sum up this section, it is clear that the human infant is inducted into the interactive world remarkably early. The comparison to other primates (see section 5) suggests that the sheer rapidity, automaticity, and almost exceptionless success of this induction has an innate basis, along the lines of Darwin's (1871, pp. 55–56) "instinctive tendency to acquire an art." Within three months of birth, the infant is already an interactive being, acquiring communicative abilities proper within or soon after the first year of life. All this happens long before the child is producing language, which argues that it is not language that induces communicative and interactive abilities, but that these interactive instincts are the engine that facilitates the learning of language.

5. The Phylogenetic Background

In trying to understand the evolution of our interactive abilities, we turn inevitably to our nearest cousins, the other primates. For various reasons, the record is patchy—there has been a tendency for example to observe "focal animals," so yielding ethograms for individual animals rather than interacting pairs. Nevertheless there are significant traces of the origins of our interactive abilities in the primate record.

Vocal turn-taking, for example, has been reported patchily across all the major clades of the primate order (see figure 14.2). Thus we have accounts of turn-taking from among the lemurs, such as *Lepilemur edwardsi* (Mendez-Cardenas & Zimmermann, 2009); from New World monkeys, such as the common marmoset, *Callithrix jacchus* (Chow, Mitchell, & Miller, 2015; Takahashi, Narayanan, & Ghazanfar, 2013), the pygmy marmoset, *Cebuella pygmaea* (Snowdon & Cleveland, 1984), the coppery titi, *Callicebus cupreus* (Müller & Anzenberger, 2002), and squirrel monkeys of the *Saimiri* genus (Symmes & Biben, 1988); from the Old World monkeys such as Campbell's monkey, *Cercopithecus campbelli* (Lemasson et al., 2011); and finally from the lesser apes, the siamangs, *Hylobates syndactylus* (Geissmann & Orgeldinger, 2000; Haimoff, 1981). Homoplasy—parallel evolution—cannot be ruled out, but the situation may be just as in the birds where vocal learning is discontinuously represented, suggesting easy gain, loss, and regain (Jarvis et al., 2014).

Particularly interesting are various reports on the learning of this turn-taking system by youngsters—just like human infants, juvenile Campbell's monkeys have the instinct to reciprocate, but the timing and avoidance of overlap have to be learned (Lemasson et al., 2011). Marmoset mothers reinforce the learning by

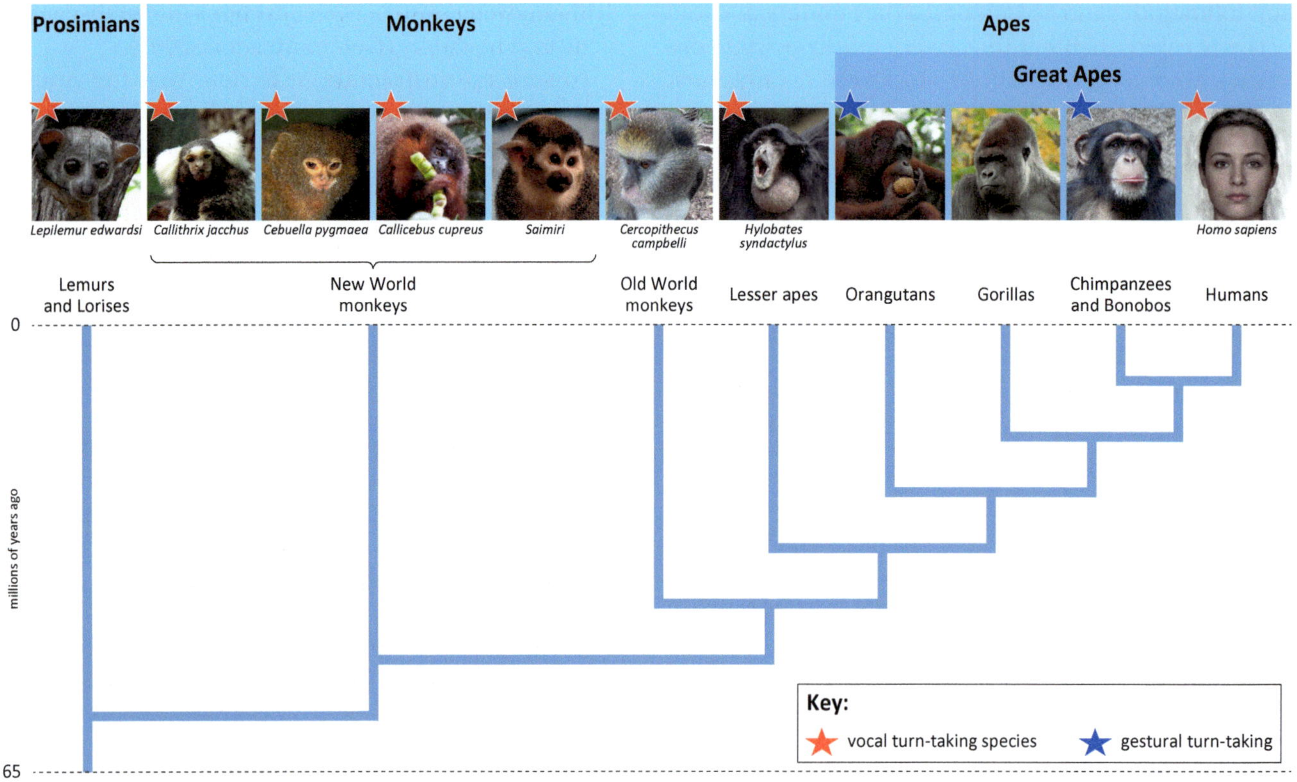

FIGURE 14.2 Vocal turn-taking represented across the primate order (from Levinson, 2016).

penalizing overlaps by the infant with nonresponse, and interrupting their infants when the response is of the wrong type (Chow et al., 2015; Takahashi, Fenley, & Ghazanfar, 2016). Vocal turn-taking seems, on the current evidence, to be particularly associated with pair-bonding species.

However, the great apes, apart from us, stand apart; their vocalizations have been described as involuntary and show little evidence of fast turn-taking (Call & Tomasello, 2007). But they use gesture in a flexible way to negotiate leaving together, food exchange, and the like (Pika, Liebal, Call, & Tomasello, 2005). The shape of the gestures is plausibly derived from ritualization of instrumental actions (Halina, Rossano, & Tomasello, 2013). Gestural turn-taking in bonobos has in fact a similar rapidity to human vocal alternation (Rossano, 2013). But the most interesting findings emerging from work on the great apes' gestural communication concerns parallels to the action sequences we reviewed in section 3. Orangutans, for instance, request and offer food using gestures of various kinds (approaching for mouth-to-mouth transfer, open-hand begging, apparent reaching), with success between a quarter and half the time (Rossano & Liebal, 2014). Requests may often be repeated, pursuing the goal unless rejected, with gaze maintained. A study of joint travel initiations among bonobos and their infants showed similar joint attention and sequences of eliciting and responding actions (Hutchins & Johnson, 2009; Rossano, 2013). In a larger scale comparison of bonobo and chimpanzee travel initiations, Fröhlich et al. (2016) showed that these sequences are systematic and show differences across the two closely related species, with bonobos showing more humanlike gaze engagement and speed of response.

All in all, there is increasing evidence of precursors to human interactional behavior among our nearest phylogenetic relatives.

6. *The Interaction Engine Hypothesis*

This chapter has brought together some systematic arguments for positing a distinct interactional ability that underlies our communicative behavior and language in particular. Communicative abilities can be dissociated from language: they are antecedent in ontogeny and phylogeny, and they permit humans to communicate effectively without language where deprived of it. These abilities are in their crucial properties universal—relatively invariant—across cultural groups, in strong contrast to the cultural specificity of linguistic coding. Furthermore, it is this interactional capacity that underlies the very possibility of language: languages are learned in the interactional niche, within the scaffolding that this provides, with joint attention to the referents being named, with exercise of other-initiated repair, and with the whole functional framework that sequences of actions (speech acts) provide. It is this interactional ability that I have dubbed "the interaction engine."

The objection may be made that this interactional ability is not one thing, but rather an assemblage of various talents and proclivities, with different phylogenetic origins and different ontogenetic patterns of development. All that is indubitable—the interaction engine is not a Fodorean module, or a Chomskyan single miraculous mutation, or a psychological faculty. It is rather a loose assemblage of various abilities, instincts, and motivations that work together to make possible the miracle of human communication. It is the bric-a-brac of useful oddments assembled through a long phylogeny, part of which may be reconstructed by comparison across species and cultures and evolutionary forbears. The sequence may have gone as follows (Levinson & Holler, 2014), perhaps cashing out Darwin's (1871, p. 106) observation that "the half-art and half-instinct of language still bears the stamp of its gradual evolution."

a. The fact that vocal turn-taking behavior is found scattered across all the major clades of the primates suggests that this may have been the foundational element in this accumulation. As we have seen it puts tremendous pressure on language processing (see also Levinson, 2016), and children are not able to match adult standards until well into middle childhood. The suggestion is that we have inherited a specific tempo, with the 1–2 s bursts separated by 2–300 ms, which was an ecology that language grew into and had to adapt to—early vocal turn-taking may have had the simplest functions, like the signals of proximity and well-being in marmosets (Takahashi et al., 2013). The complexity of linguistic signals reflects this compression into preexisting slots.
b. In the Hominidae, gesture has played an important role in communication (Call & Tomasello, 2007), implicating gaze, and humans show anatomical adaptation to the importance of the visibility of gaze direction in the white sclera of our eyes (Kobayashi & Kohshima, 2001) and their horizontal elongation (Mayhew & Gómez, 2015), which may be one of the earliest physiological adaptations for human communication. Gaze is important for establishing mutual attention, but gaze and gesture also directly afford spatial communication, suggesting that this may have played a crucial role in early hominin communication (Levinson & Holler, 2014). The face-to-face pose that would make this useful is the basis for what Goffman (1961, p. 18) called the "eye-to-eye ecological huddle" characteristic of human communication. Sustained joint attention makes possible the joint action that provides the working proof of other minds and the possibility of working with them.
c. Simple contingent action sequencing is found in gestural communication in the great apes (Rossano & Liebal, 2014), showing that for some actions such as requesting and offering, more complex communication media are not required (often gestures alone suffice in human interaction; Rossi, 2014). This precursor for human language seems restricted to certain social relations, mother-infant in particular, so one of the human tricks has been to generalize maternal empathy across our social relationships.
d. Vocal communication must slowly have developed in complexity in this interactional niche formed by turn-taking, gesture, and gaze. *Homo erectus* seems to have lacked voluntary breath control, but by the time of the common ancestor to Neanderthals and modern humans (well over half a million years ago) this special rewiring of the peripheral nervous system for elaborate speech was in place (Dediu & Levinson, 2013), responding no doubt to the adaptive pressures for expressive vocal language. The antecedent fast-paced turn-taking system explains the highly compressed nature of conversational speech.

Because this suite of capacities, the interaction engine, has probably been acquired en route through hominin evolution does not entail that it does not act as a package—even bipedalism required multiple adaptations over a long period, from adjustments to pelvis, vertebral column, and its articulation to the skull, not to mention balance and respiration. Some evidence for the package comes from *autism* (Frith, 2003; Frith & Hill, 2003), where the classical syndrome exhibits gaze aversion, difficulties with face recognition and emotion displays, and difficulties with joint attention and theories of mind tasks. Combined with this is lack of motivation for communication, problems with understanding gestures, and where, as often, language is well-formed, remarkable difficulties in the pragmatics of usage: there is no search for the underlying action or intent. Turn-taking involves the switching of deictic roles, so what was *you* becomes *I* and what was *there* becomes *here*, but these switches are difficult for autistic children, who often echo utterances without deictic adjustment.

Autistic viewers of conversation fail, unlike normal children from two years old and up (Lammertink et al., 2015), to anticipate speaker switches by switching gaze to the recipient or to the direction of pointing gestures (Klin, Jones, Schultz, Volkmar, & Cohen, 2002). Interestingly, these deficits cannot be attributed directly to low-level physiological causes such as failure of detection of looking direction—they seem to depend crucially on failures to understand joint attention and "theory of mind" (Baron-Cohen, 1995). Comparisons with other clinical populations are also revealing: Down's syndrome children, for instance, have full social and communicative competence despite their mental retardation (Baron-Cohen et al., 1985). A further recent development is the demonstration that genetic factors predisposing to autism also correlate with lesser interpersonal and communication skills in the general population (Robinson et al., 2016).

To conclude, the hypothesis is that the signal achievements of our species—language and cultural accumulation—rely crucially on a relatively neglected substrate of human interactional abilities, which seem largely instinctual in character, and which together have formed the crucible for human development both over the individual lifetime and the lifetime of the species.

Acknowledgments

This report builds on extensive work on interaction structure conducted in my department from 2005: I am grateful to all my colleagues, and the Max Planck Gesellschaft and the European Research Council for generous funding. This chapter benefitted greatly from comments by Susan Goldin-Meadow on a draft.

REFERENCES

Baron-Cohen, S. (1995). *Mindblindness and the language of the eyes: An essay in evolutionary psychology.* Cambridge, MA: MIT Press.

Baron-Cohen, S., Leslie, A. M., & Frith, U. (1985). Does the autistic child have a 'theory of mind'? *Cognition, 21*(1), 37–46.

Bates, E., Camaioni, L., & Volterra, V. (1975). Performatives prior to speech. *Merrill-Palmer Quarterly, 21,* 205–226.

Bateson, M. C. (1971). The interpersonal context of infant vocalization. *Quarterly Progress Report of the Research Laboratory of Electronics, 100,* 170–176.

Berwick, R. C., & Chomsky, N. (2016). *Why only us: Language and evolution.* Cambridge, MA: MIT Press.

Bögels, S., Magyari, L., & Levinson, S. C. (2015). Neural signatures of response planning occur midway through an incoming question in conversation. *Scientific Reports, 5,* 12881.

Bruner, J. S. (1975). The ontogenesis of speech acts. *Journal of Child Language, 2,* 1–19.

Call, J., & Tomasello, M. (Eds.) (2007). *The gestural communication of apes and monkeys.* Mahwah, NJ: Lawrence Erlbaum.

Casillas, M. (2014). Turn-taking. In D. Matthews (Ed.), *Pragmatic development in first language acquisition* (pp. 53–70). Amsterdam: Benjamins.

Casillas, M., Bobb, S. C., & Clark, E. V. (2016). Turn-taking, timing, and planning in early language acquisition. *Journal of Child Language, 43*(6), 1310–1337.

Casillas, M., & Hilbrink, E. (2018). *Communicative act development.* Manuscript in preparation.

Chow, C. P., Mitchell, J. F., & Miller, C. T. (2015). Vocal turn-taking in a non-human primate is learned during ontogeny. *Proceedings of the Royal Society B, 282*(1807), 20150069.

Christiansen, M. H., & Chater, N. (2008). Language as shaped by the brain. *Behavioral and Brain Sciences, 31,* 489–558.

Christiansen, M. H., & Chater, N. (2016). The Now-or-Never bottleneck: A fundamental constraint on language. *Behavioral and Brain Sciences, 39,* e62.

Connolly, B., & Anderson, R. (1987). *First contact.* New York: Film Makers Library.

Csibra, G. (2010). Recognizing communicative intentions in infancy. *Mind and Language, 25,* 141–168.

Darwin, C. (1871). *The descent of man.* London: John Murray.

Dediu, D., & Levinson, S. C. (2012). Abstract profiles of structural stability point to universal tendencies, family-specific factors, and ancient connections between languages. *PLOS ONE, 7*(9), e45198.

Dediu, D., & Levinson, S. C. (2013). On the antiquity of language: The reinterpretation of Neanderthal linguistic capacities and its consequences. *Frontiers in Language Sciences, 4,* 397.

De Ruiter, J. P. A., Noordzij, M. L., Newman-Norlund, S., Hagoort, P., Levinson, S. C., & Toni, I. (2010). Exploring the cognitive infrastructure of communication. *Interaction Studies, 11,* 51–77.

de Vos, C., Torreira, F., & Levinson, S. C. (2015). Turn-timing in signed conversations: Coordinating stroke-to-stroke turn boundaries [Special issue]. *Frontiers in Psychology, 6*(268) (on turn-taking in human communication).

Dingemanse, M., Roberts, S. G., Baranova, J., Blythe, J., Drew, P., Floyd, S., ... Enfield, N. J. (2015). Universal principles in the repair of communication problems. *PLOS ONE, 10*(9), e0136100.

Dingemanse, M., Torreira, F., & Enfield, N. J. (2013). Is "Huh?" a universal word? Conversational infrastructure and the convergent evolution of linguistic items. *PLOS ONE, 8*(11), e78273.

Duranti, A. (1997). Universal and culture-specific properties of greetings. *Journal of Linguistic Anthropology, 7*(1), 63–97.

Enfield, N. J., Dingemanse, M., Baranova, J., Blythe, J., Brown, P., Dirksmeyer, T., ... Torreira, F. (2013). Huh? What?: A first survey in 21 languages. In M. Hayashi, G. Raymond, & J. Sidnell (Eds.), *Conversational repair and human understanding* (pp. 343–380). New York: Cambridge University Press.

Evans, N., & Levinson, S. C. (2009). The myth of language universals: Language diversity and its importance for cognitive Science. *Behavioral and Brain Sciences, 32*(5), 429–492.

Floyd, S., Manrique, E., Rossi, G., & Torreira, F. (2016). Timing of visual bodily behavior in repair sequences: Evidence from three languages. *Discourse Processes, 53*(3), 175–204.

Frith, C., & Heyes, C. (2014). The cultural evolution of mind reading. *Science, 344*, 6190.

Frith, U. E. (2003). *Autism: Explaining the enigma* (2nd ed.). Oxford: Blackwell.

Frith, U. E., & Hill, E. E. (2003). *Autism: Mind and brain.* Oxford: Oxford University Press.

Fröhlich, M., Kuchenbuch, P., Müller, G., Fruth, B., Furuichi, T., Wittig, R. M., & Pika, S. (2016). Unpeeling the layers of language: Bonobos and chimpanzees engage in cooperative turn-taking sequences. *Scientific Reports, 6*, 25887.

Galantucci, B. (2005). An experimental study of the emergence of human communication systems. *Cognitive Science, 25*, 737–767.

Geissmann, T., & Orgeldinger, M. (2000). The relationship between duet songs and pair bonds in siamangs. *Hylobates syndactylus. Animal Behaviour, 60*, 805–809.

Gisladottir, R. S., Chwilla, D. J., & Levinson, S. C. (2015). Conversation electrified: ERP correlates of speech act recognition. *PLOS ONE, 10*(3), *e0120068.*

Goffman, E. (1961). *Encounters: Two studies in the sociology of interaction.* New York: Bobbs-Merrill.

Goodwin, C. (1995). Co-constructing meaning in conversations with an aphasic man. *Research on Language and Social Interaction, 28*(3), 233–260.

Goldin-Meadow, S. (2003). *The resilience of language: What gesture creation in deaf children can tell us about how all children learn language.* Oxford: Psychology Press.

Grice, H. P. (1957). Meaning. *Philosophical Psychology, 67*, 377–388.

Grice, H. P. (1989). *Studies in the ways of words.* Cambridge, MA: Harvard University Press.

Haimoff, E. H. (1981). Video analysis of siamang *(Hylobates syndactylus)* songs. *Behaviour, 76*, 128–151.

Halina, M., Rossano, F., & Tomasello, M. (2013). The ontogenetic ritualization of bonobo gestures. *Animal Cognition, 16*(4), 653–666.

Hammarström, H., Forkel, R., Haspelmath, M., & Bank, S. (2016). *Glottolog 2.7.* Jena, Germany: Max Planck Institute for the Science of Human History. Retrieved from http://glottolog.org.

Heldner, M., & Edlund, J. (2010). Pauses, gaps and overlaps in conversations. *Journal of Phonetics, 38*, 555–568.

Hilbrink, E., Gattis, M., & Levinson, S. C. (2015). Early developmental changes in the timing of turn-taking: A longitudinal study of mother-infant interaction. *Frontiers in Psychology, 6*, 1492.

Hutchins, E., & Johnson, C. 2009. Modeling the emergence of language as an embodied collective cognitive activity. *Topics in Cognitive Science, 1*(3), 523–546

Jarvis, E. D., Mirarab, S., Aberer, A. J., Li, B., Houde, P., Li, C., ... Suh, A. (2014). Whole-genome analyses resolve early branches in the tree of life of modern birds. *Science, 346*(6215), 1320–1331.

Kendrick, K. H., Brown, P., Dingemanse, M., Floyd, S., Gipper, S., Hayano, K., ... Levinson, S. C. (2014). *Sequence Organization: A Universal Infrastructure for Action.* Paper presented at the 4th International Conference on Conversation Analysis, University of California at Los Angeles.

Kendrick, K. H., & Torreira, F. (2015). The timing and construction of preference: A quantitative study. *Discourse Processes, 52*(4), 255–289.

Klin, A., Jones, W., Schultz, R., Volkmar, F., & Cohen, D. (2002). Visual fixation patterns during viewing of naturalistic social situations as predictors of social competence in individuals with autism. *Archives of General Psychiatry, 59*(9), 809–816.

Kobayashi, H., & Kohshima, S. (2001). Unique morphology of the human eye and its adaptive meaning: Comparative studies on external morphology of the primate eye. *Journal of Human Evolution, 40*(5), 419–435.

Kuhl, P. K., & Meltzoff, A. N. (1996). Infant vocalizations in response to speech: Vocal imitation and developmental change. *Journal of the Acoustical Society of America, 100*(4), 2425–2438.

Lammertink, I., Casillas, M., Benders, T., Post, B., & Fikkert, P. (2015). Dutch and English toddlers' use of linguistic cues in predicting upcoming turn transitions. *Frontiers in Psychology, 6*, 495.

Lee, N., Mikesell, L., Joaquin, A., Mates, A., & Schumann, J. (2009). *The interactional instinct: The evolution and acquisition of language.* Oxford: Oxford University Press.

Lemasson, A., Glas, L., Barbu, S., Lacroix, A., Guilloux, M., Remeuf, K., & Koda, H. (2011). Youngsters do not pay attention to conversational rules: Is this so for nonhuman primates? *Scientific Reports, 1*, 22.

Levinson, S. C. (2000). *Presumptive meanings: The theory of generalized conversational implicature.* Cambridge, MA: MIT Press.

Levinson, S. C. (2006). On the human 'interaction engine'. In N. J. Enfield, & S. C. Levinson (Eds.), *Roots of human sociality: Culture, cognition and interaction* (pp. 39–69). Oxford: Berg.

Levinson, S. C. (2013). Recursion in pragmatics. *Language, 89*, 149–162.

Levinson, S. C. (2016). Turn-taking in human communication, origins, and implications for language processing. *Trends in Cognitive Sciences, 20*(1), 6–14.

Levinson, S. C., Hammarström, H., & Roberts, S. (2018). *How is the design space for language actually populated?* Manuscript in preparation.

Levinson, S. C., & Holler, J. (2014). The origin of human multi-modal communication. *Philosophical Transactions of the Royal Society B, 369*, 20130302.

Levinson, S. C., & Torreira, F. (2015). Timing in turn-taking and its implications for processing models of language. *Frontiers in Psychology, 7*, 731.

Lewis, D. (1969). *Convention: A philosophical study.* Cambridge, MA: Harvard University Press.

Liszkowski, U. (2013). Using theory of mind. *Child Development Perspectives, 7*(2), 104–109.

Liszkowski, U., Brown, P., Callaghan, T., Takada, A., & De Vos, C. (2012). A prelinguistic gestural universal of human communication. *Cognitive Science, 36*, 698–713.

Liszkowski, U., Carpenter, M., Striano, T., & Tomasello, M. (2006). 12- and 18-month-olds point to provide information for others. *Journal of Cognition and Development, 7*(2), 173–187.

Mayhew, J. A., & Gómez, J. C. (2015). Gorillas with white sclera: A naturally occurring variation in a morphological trait linked to social cognitive functions. *American Journal of Primatology, 77*(8), 869–877.

Meltzoff, A., & Moore, M. K. (1977). Imitation of facial and manual gestures of human neonates. *Science, 198*, 75–78.

Mendez-Cardenas, M. G., & Zimmermann, E. (2009). Duetting: A mechanism to strengthen pair bonds in a dispersed pair-living primate *(Lepilemur edwardsi). American Journal of Physiological Anthropology, 139*, 523–532.

Merritt, M. W. (1976). *Resources for saying in service encounters.* (Unpublished doctoral dissertation). University of Pennsylvania, Philadelphia.

Miller, G. A. (1947). The masking of speech. *Psychological Bulletin, 44,* 105–129.

Müller, A. E., & Anzenberger, G. (2002). Duetting in the Titi monkey *Callicebus cupreus*: Structure, pair specificity and development of duets. *Folia Primatologica, 73,* 104–115.

Noordzij, M. L., Newman-Norlund, S. E., de Ruiter, J. P., Hagoort, P., Levinson, S. C., & Toni, I. (2009). Brain mechanisms underlying human communication. *Frontiers in Human Neuroscience, 3,* 14.

Onishi, K. H., & Baillargeon, R. (2005). Do 15-month-old infants understand false beliefs? *Science, 308*(5719), 255–258.

Oostenbroek, J. M., Suddendorf, T., Nielsen, M., Redshaw, J., Kennedy-Costantini, S., Davis, J., … Slaughter, V. (2016). Comprehensive longitudinal study challenges the existence of neonatal imitation in humans. *Current Biology, 26*(10), R412–R414.

Pika, S., Liebal, K., Call, J., & Tomasello, M. (2005). The gestural communication of apes. *Gesture, 5*(1/2), 41–56.

Pomerantz, A., & Heritage, J. (2013). Preference. In T. Stivers & J. Sidnell (Eds.), *Handbook of Conversation Analysis* (pp. 210–228). Chichester, UK: Wiley-Blackwell.

Robinson, E. B., St. Pourcain, B., Anttila, V., Kosmicki, J. A., Bulik-Sullivan, B., Grove, J., … & Martin, J. (2016). Genetic risk for autism spectrum disorders and neuropsychiatric variation in the general population. *Nature Genetics, 48*(5), 552–555.

Rossano, F. (2013). Sequence organization and timing of bonobo mother-infant interactions. *Interaction Studies, 14,* 160–189.

Rossano, F., Brown, P., & Levinson, S. C. (2009). Gaze, questioning and culture. In J. Sidnell (Ed.), *Conversation analysis: Comparative perspectives* (pp. 187–249). Cambridge: Cambridge University Press.

Rossano, F., & Liebal, K. (2014). "Requests" and "offers" in orangutans and humans. In P. Drew & E. Couper-Kuhlen (Eds.), *Requesting in social interaction* (pp. 333–362). Amsterdam: John Benjamins.

Rossi, G. (2014). When do people not use language to make requests? In P. Drew & E. Couper-Kuhlen (Eds.), *Requesting in social interaction* (pp. 303–334). Amsterdam: John Benjamins.

Sacks, H., Schegloff, E., & Jefferson, G. (1974). A simplest systematics for the organization of turn-taking in conversation. *Language, 50*(4), 696–735.

Schegloff, E. (2006). Interaction: The infrastructure for social institutions, the natural ecological niche for language, and the arena in which culture is enacted. In N. J. Enfield & S. C. Levinson (Eds.), *Roots of human sociality: Culture, cognition and interaction* (pp. 123–148). Oxford: Berg.

Schegloff, E. (2007). *Sequence organization in interaction: A primer for conversation analysis: Sequence organization.* Cambridge: Cambridge University Press.

Schegloff, E., Jefferson, G., & Sacks, H. (1977). The preference for self-correction in the organization of repair in conversation. *Language, 53*(2), 361–382.

Schelling, T. (1960). *The strategy of conflict.* Cambridge, MA: MIT Press.

Senghas, A., Kita, S., & Özyürek, A. (2004). Children creating core properties of language: Evidence from an emerging sign language in Nicaragua. *Science, 305*(5691), 1779–1782.

Seyfeddinipur, M., & Gullberg, M. (Eds.) (2014). *From gesture in conversation to visible action as utterance: Essays in honor of Adam Kendon.* Amsterdam: John Benjamins.

Sidnell, J. (2009). *Conversation analysis: Comparative perspectives.* Cambridge: Cambridge University Press.

Snowdon, C. T., & Cleveland, J. (1984). "Conversations" among pygmy marmosets. *American Journal of Primatology, 7,* 15–20.

Sperber, D., & Wilson, D. (1986). *Relevance.* Oxford: Basil Blackwell.

Stivers, T., Enfield, N. J., Brown, P., Englert, C., Hayashi, M., Heinemann, T., … & Levinson, S. C. (2009). Universals and cultural variation in turn-taking in conversation. *Proceedings of the National Academy of Sciences, 106*(26), 10587–10592.

Stivers, T., Sidnell, J., & Bergen, C. (2018). Children's responses to questions in peer interaction: A window into the ontogenesis of interactional competence. *Journal of Pragmatics, 124,* 14–30.

Symmes, D., & Biben, M. (1988). Conversational vocal exchanges in squirrel monkeys. In D. Todt, P. Goedeking, & D. Symmes (Eds.), *Primate vocal communication* (pp. 123–132). Berlin: Springer.

Takahashi, D. Y., Fenley, A. R., & Ghazanfar, A. A. (2016). Early development of turn-taking with parents shapes vocal acoustics in infant marmoset monkeys. *Philosophical Transactions of the Royal Society B, 371,* 20150370.

Takahashi, D. Y., Narayanan, D. Z., & Ghazanfar, A. A. (2013). Coupled oscillator dynamics of vocal turn-taking in monkeys. *Current Biology, 23,* 2162–2168.

Tomasello, M., Carpenter, M., & Liszkowski, U. (2007). A new look at infant pointing. *Child Development, 78*(3), 705–722.

Trevarthen, C. (1977). Descriptive analyses of infant communicative behavior. In H. R. Schaffer (Ed.), *Studies in mother-infant interaction* (pp. 227–270). London: Academic Press.

Van Deusen-Phillips, S. B., Goldin-Meadow, S., & Miller, J. P. (2001). Enacting stories, seeing worlds: Similarities and differences in the cross-cultural narrative development of linguistically isolated deaf children. *Human Development, 44,* 311–336.

15 The Structure and Timing of Human versus Primate Social Interaction

FEDERICO ROSSANO

Humans' most powerful tool is arguably language. It facilitates coordination and cooperation, creates commitments and social institutions, and is a key resource for high-fidelity cultural transmission. Furthermore, language allows humans to exchange information, share emotions and beliefs, and communicate about absent entities and about past and future events. Scholars have speculated about the origins of this impressive faculty for centuries. The impossibility of providing a definitive answer to this question has famously led to the ban on any further publication on the origin of language by the Societé de Linguistique de Paris in 1866. Through technological innovation, better evolutionary models, and the development of research on the communicative behavior of multiple animal species, there has been a recent resurgence of research on language evolution. At different times, continuity or discontinuity have been proposed between human language and animal communication. The communicative abilities of birds and nonhuman primates have been the natural source of comparison to human language, claiming similarities between the systems because of convergent evolution (for birds) or common descent (for primates). What has been mostly neglected, however, is how language builds on top of an infrastructure for social interaction and the cognitive ability to cooperate and/or manipulate others that had to be in place before language evolution (see Bruner, 1975). The general agreement at the time of this writing is that nonhuman primates' (henceforth, primates) vocalizations are innate and mostly inflexible (primates can partly control intensity and when to produce vocalizations but cannot learn new ones, see Seyfarth & Cheney, 2003; Zuberbühler, 2000b) while their gestural communication is flexible and under full intentional control and at least partially learned (Call & Tomasello, 2007; Halina, Rossano, & Tomasello, 2013; Schneider, Call, & Liebal, 2012). In this chapter, I will outline what we know about the structure of primate communication (with a particular emphasis on gestural communication in great apes) and explicate how the traditional way of including primate communication under "animal communication" as separate from "human communication" fails to take into account something already outlined by Darwin in *The Descent of Man* (1871): the key importance of cognition for language evolution.

1. Animal Communication, Information versus Manipulation and the Need for a Social Action Perspective

The study of animal communication is divided between two main perspectives: the Manipulation and the Information perspectives. The former mostly focuses on how evolution through natural selection has selected signals that are designed to influence others and believes that communication is fundamentally competitive in nature (Dawkins & Krebs, 1978; Owings & Morton, 1997; Owren, Rendall, & Ryan, 2010). Scholars that adopt this perspective mostly reject the idea that signals convey information. They also reject the communication model by Shannon and Weaver (1949) because they believe that relying on human communication as a model for animal communication is not only an anthropomorphic sin but also an evolutionary nonsense because of its backward direction (like using bipedalism to explain quadrupedalism, see Owren et al., 2010).

In contrast, the Information perspective heavily relies on Shannon and Weaver's model of communication and on Wiener's (1948) cybernetic theory of communication and feedback control. The idea is that communication consists in encoding and decoding information via the production and perception of signals. To that degree, it is possible to map the meaning of signals and a signal could be assessed in terms of its *honesty* (Zahavi, 1975), that is, in terms of how it reflects specific characteristics of the signaler or its intentions. Signals encode information and the main task of the perceiver is to decode the signal. The Information perspective took hold with the advent of the cognitive revolution and the use of computational metaphors to describe animal communication. Smith (1977) was one of the first to famously suggest that a one-to-one signal mapping was too simplistic and rather perceivers must invest effort to interpret the signals they are exposed to. Around the same time, the discovery that vervet monkey alarm calls might have clear semantic meanings (Seyfarth, Cheney, & Marler, 1980) provided further strength to a perspective aimed at unpacking the meaning of animal signals and that believed in the possibility of developing species-typical lexicons, especially for primates (Seyfarth & Cheney, 2003; Seyfarth et al. 2010; Hobaiter & Byrne, 2014; Zuberbühler, 2000a, 2000b). Initially, scholars adopting this perspective would think of vocalizations (especially alarm calls) as referential (e.g., Seyfarth et al., 1980; Zuberbühler, 2000a). More recently, however, there has been a tendency to refer to vocalization as *functionally referential* (see, e.g., Evans, Evans, & Marler, 1993; Slocombe & Zuberbühler, 2005), which means that signals do not reflect signalers' communicative intentions but rather are provoked by external stimuli and convey information about those stimuli that can be decoded by receivers, functioning like words (see, however, Rendall, Owren, & Ryan, 2009; Wheeler & Fisher, 2012, for criticisms of this concept). Notwithstanding the level of referentiality of animal signals, the logic of this approach invokes that a certain amount of cognition is necessary for the decoding of communicative signals. Most importantly, this approach claims that the transfer of information between animals is a key process for animal communication.

Ethologists have conceptualized animal communication as distinct from the referential approach in a further way that can be labeled the Motivational perspective (Marler, Evans, & Hauser, 1992). The idea originated in Darwin's *Expression of Emotions in Animals and Man* (1872) and can be summarized as the notion that animal signals are fundamentally externalizations of internal emotional states and their "meaning" can be interpreted by combining the environmental context in which it occurred and the motivational state of the signaler. In other words, signals only make sense in context and necessarily reflect signalers' internal states, contrary to how humans can use words, for example.

In general, ethologists and behavioral ecologists have long tried to avoid the use of cognitive terminology to make sense of animal communication, basically avoiding terms such as *intentionality* or *theory of mind.* This has led to a fascinating side effect: when ethologists talk about animal communication, there are really two types: human communication versus all other animal species. In other words, while they would agree that humans are different from most other animals, they tend to use a single model to refer to all other animal species. Thus, the communicative behaviors of bees, moths, ants, wolf spiders, crayfish, and chimpanzees can supposedly all be investigated using the same acognitive framework (for an overview on animal signals, see, e.g., Maynard-Smith & Harper, 2003; and for specific claims about the communicative behaviors of each of the species mentioned, see, e.g., Acquistapace, Aquiloni, Hazlett, & Gherardi, 2002; Higham & Hebets, 2013; Partan & Marler, 1999, 2005; Rendall et al. 2009; Uetz, Roberts, & Taylor, 2009; Von Frisch, 1955).

While the idea of a single framework for all animal species but humans might on the surface seem appealing, it blinds researchers to a very simple fact: the cognitive abilities and social life of a chimpanzee do not resemble those of moths, ants, spiders, and other animals. Akin to biologists critiquing psychologists for using human models of communication to study animal communication, psychologists should feel compelled to point out that what biologists are doing is comparable to using the same concepts to describe a tricycle and a space shuttle.

For example, we know that great apes recognize each other's goals by inferring them through others' behaviors. Great apes know what other individuals can see from their position (Hare, Call, Agnetta, & Tomasello, 2000) and know what others know (Hare, Call, & Tomasello, 2001). Furthermore, a recent study has shown that great apes can also pass the implicit version of the false-belief task (Krupenye, Kano, Hirata, Call, & Tomasello, 2016). In other words, great apes exhibit all the key skills related to what has been called *Theory of Mind* (Premack & Woodruff, 1978; see Call & Tomasello, 2008, for a review). However, this should not conflate great apes' cognitive and behavioral abilities with those of humans. For example, great apes are mostly competitive, as opposed to cooperative (Hare & Tomasello, 2004; Tomasello, 2009), and they do not seem to engage in shared intentionality as humans do (Tomasello, Carpenter, Call, Behne, & Moll, 2005). One of the consequences of these differences is that great apes do not use pointing gestures as a signal for one another while all humans do (see, e.g., Tomasello, 2006).

Given our knowledge of the cognitive abilities of great apes and given the predominant communicative models for animal communication, it is time for a change of perspective in the study of primate communication. I am not suggesting adopting a different perspective on communication for the entire animal kingdom. Rather, I believe that the perspective that will follow can certainly be applied to nonhuman great apes and probably many other social animals that exhibit the necessary cognitive abilities to make it realistically feasible. The suggestion here is to introduce in the study of primate communication the notion of *social action* put forward by the sociologist Max Weber in his work *Economy and Society* (1978). According to Weber's definition:

> We shall speak of "action" insofar as the acting individual attaches a subjective meaning to his behavior—be it overt or covert, omission or acquiescence. Action is "social" insofar as its subjective meaning takes account of the behavior of others and is thereby oriented in its course. (p. 4)

He then added:

> Social action, which includes both failure to act and passive acquiescence, may be oriented to the past, present, or expected future behavior of others... not every kind of action, even of overt action, is "social" in the sense of the present discussion. Overt action is non-social if it is oriented solely to the behavior of inanimate objects.... Not every type of contact of human beings has a social character; this is rather confined to cases where the actor's behavior is meaningfully oriented to that of others. (p. 22)

Weber mentioned that it is possible for actions to be similar and rather uniform in their production across individuals because they are the best solutions to the problem that the individual is facing. In other ways, order can develop not because of norms but because individuals with similar cognitive endowments and needs are confronted with similar contextual factors. Moreover, Weber reminded us that just because multiple individuals perform similar behaviors reliably at the same time of the day and in the same location, it does not imply that there is a norm for it and that failure to engage in that social action would lead to sanctioning. For example, people tend to have breakfast in the morning, yet eating breakfast in the morning is not a social norm but rather a custom, since failure to have breakfast will not be met with sanctioning (unless the subject is a child interacting with a parent).

In sum, I suggest we move beyond the notion of signals and their use to either transfer information or manipulate others. Rather, we should focus on social action, that is, on what participants in interaction are trying to do and on the fact that their behavior is rationally purposeful and is calibrated on what they know about their interlocutors and what they see them do and understand. Participants in interaction monitor and interpret each other's behavior. Lack of response or inaction is a social action in and of itself in the context of social interaction. Moreover, using the terminological distinction further provided by Schutz (1967), social action can have a *because-motive* (i.e., being motivated by past experience) or an *in-order-to motive* (i.e., being motivated by an expected future outcome). Finally, a proper understanding of social action requires both an understanding of an individual's motives to produce such action and evidence that what we are investigating is a reliable and systematic behavioral practice produced by the individuals under investigation.

2. Social Action in Humans

Fifty years ago, sociologist Harvey Sacks provided a characterization of social behavior in social interaction that allows us to further expand the notion of social action previously introduced: the notion that communicative behavior is calibrated on the recipients and it is publicly recognizable. Sacks claimed: "For Members, activities are observables. They see activities. They see persons doing intimacy, they see persons lying, etc.... And that poses for us this task: Finding how is it that people can produce sets of actions that provide that others can see such things" (1992, p. 119). In other words, how do human beings decide in interaction what to say, how to do it and when it should be done? And how can they do it so rapidly?

Conversation analysts have labeled this problem (i.e., producing and comprehending communicative moves) the *action formation* (from the speaker's point of view), and the *action recognition/ascription* problem (from the recipient's point of view). It concerns the resources that participants deploy in social interaction to make their actions (e.g., requests, offers, invitations, complaints, sanctions, compliments) recognizable and intelligible to coparticipants so that they can obtain what they wish (Schegloff, 2007; Levinson, 2013). Conversation analysts ask, "How is intelligible social action possible?" (see Heritage, 2004). The intelligibility of social action allows for some level of mutual understanding, which provides for the successful engagement of cooperative interactions, and thus, makes social life possible. Before unpacking the relevance of this question for this chapter, however, three possible misunderstandings should be clarified.

First, *action recognition* does not necessarily mean *intention recognition*. It would be misleading to assume that participants always aim at getting their intentions recognized by their recipients. Deception is a currency often deployed in human interaction. Yet there is a public, shared dimension to human behavior and human language that can be observed, recognized, and reproduced (see, e.g., Goodwin, 2000), independently of the actual real interests, desires, and concerns of the participants who deploy them. Ambiguity in the signal can be strategically deployed to minimize the social accountability of our behavior through deniability (e.g., in light of a potential escalating conflict, it allows us to claim/pretend that we were simply asking about something while we were actually criticizing).

Second, *mutual understanding*, at least in terms of what interactants do with action formation and ascription, entails inferential processes that are necessarily cognitive. Any effort to recipient/audience design (Clark & Murphy, 1982; Sacks & Schegloff, 1979) an utterance or a body movement shows how much we rely on our shared history with the person addressed, on our common ground with them (Clark, Schreuder, & Buttrick, 1983; Clark & Brennan, 1991). These cognitive processes rely on memory, perception, attention, and some level of Theory of Mind (see, e.g., Tomasello, 2010) but also on our capacity to interpret each other's behavior as intentionally driven and our capacity to project what could happen next more than one step ahead, as in a chess game. In interaction, individuals have to relate with an ever-changing environment, where every new utterance or social action by a participant is at the same time "context shaped" and "context renewing" (Heritage, 1984). The environment that human beings had to evolutionarily adapt to was not only the one physically surrounding them but also the contingencies of the social one (see, e.g., Tomasello, 2010).

Third, engaging in a conversation is a cooperative act (see, e.g., Goodwin, 2011; Tomasello, 2010). The importance of cooperation for the success of social interaction has a long scholarly tradition (see, e.g., Grice, 1975; Bruner, 1975; Sperber & Wilson, 1986; Tomasello, 2010). The low price that we place to most of our talk during ordinary conversation makes us underestimate how cooperative human communication is. Accordingly, one might wonder how social interactions in species (e.g., great apes) in which individuals tend to be more competitive with each other would differ from human social interactions (see Hare & Tomasello, 2004). This chapter will show how some of our closest evolutionary relatives (nonhuman great apes) act in specific kinds of social interaction and will try to unpack some similarities between human social interaction and primate social interaction.

3. Human Social Interaction

One of the most intriguing computational mysteries of human social interaction concerns how people design a first communicative move (e.g., requesting) in order to elicit a specific and appropriate response (e.g., granting a request) by the addressed recipient. Many studies of human social interaction have shown that people are very apt at calibrating their utterances to their specific addressees. However, the computational mystery does not only concern our capacity to initiate a course of action via turns-at-talk (Sacks, Schegloff, & Jefferson, 1974) and embodied behavior (e.g., Goodwin, 1981, 2000), but also our capacity to interpret these constellations of behaviors when they occur and to produce appropriate responses at a speed that is absolutely remarkable. In face-to-face interaction, the average time it takes to transition between a question and a response is around 0.2 s (Stivers et al., 2009; see also De Ruiter, Mitterer, & Enfield, 2006; Levinson, 2016). Even when responses are delayed, most are not delayed by more than one second (Jefferson, 1988; Rossano, 2012), unless other contingencies are at play (e.g., the person is engaged in another task simultaneously). Human beings are particularly cooperative in that almost 90% of questions in ordinary conversation obtain some form of response (Rossano, 2012; Stivers et al., 2009).

Recently, psycholinguists have paid special attention to the timing between utterances (turn-taking) (see, e.g., Christiansen & Chater, 2016; Garrod & Pickering, 2004; Levinson, 2016), highlighting the fact that our capacity to produce such fast-paced interaction challenges, for example, the traditional Shannon and Weaver communication model (1949). We now know that participants need to start planning their responsive turn *before* the previous turn is actually complete (see Levinson, 2016). In other words, the inferential processes at work in social interaction must be extensive and an encoding-decoding model does not quite capture how meaning is interpreted and monitored online. On the other hand, researchers interested in animal communication have begun making claims about the similarity between human turn-taking and turn-taking in the vocalizations of primates (Chow, Mitchell, & Miller, 2015) and different species of birds (see, e.g., Henry, Craig, Lemasson, & Hausberger, 2015). While this research brings forward important information about the possible precursors to human turn-taking in conversation, it is important to emphasize the difference between three types of turn-taking (see figure 15.1).

The first type, *mimetic*, can be found in multiple species and basically occurs whenever two animals interacting produce the same vocalization (e.g., a pant-hoot for chimpanzees) but minimize overlap. This type of turn-taking does not require any special inferential process and can be elicited (primed) by the production of the first vocalization.

The second type, *alternating*, can also be found in several species but especially in birds. It occurs whenever

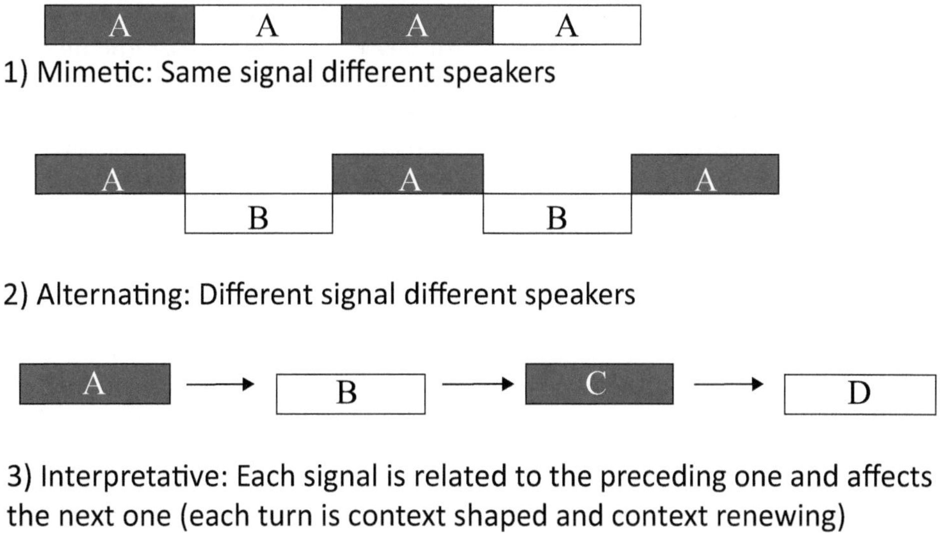

FIGURE 15.1 Different types of turn-taking in the animal kingdom.

animals take turns in producing their own song, for example. This is cognitively more complex in that it requires the production of a different vocalization than the one perceived and still requires the capacity to minimize overlap and produce (mostly) nonoverlapping vocalizations.

The third type, *interpretative*, has been clearly documented in humans (Sacks et al., 1974) and nonhuman great apes (see, e.g., Rossano, 2013; Rossano & Liebal, 2014). For example, this occurs when animal 1 produces a signal (turn A) that is not responded adequately by animal 2 (turn B) and consequentially leads to a pursuit/repair of the initial signal by animal 1 (turn C). This usually results in animal 2 next providing a different response (turn D). In other words, multiple signals are exchanged in pursuit of intersubjectivity, resulting in expansions of a basic two-turn exchange. In this kind of turn-taking system, individuals must deal with an ever-changing environment in which every new utterance (or social action) is simultaneously "context shaped" and "context renewing" (Heritage, 1984). Or to paraphrase Schutz (1967), it has simultaneously a "because-motive" and an "in-order-to motive."

I believe that there is no current comparison nor evolutionary theory accounting for the transition from one type of turn-taking system to another. Similarly, we do not know how widespread each type is in the animal kingdom. If we adopt a cognitive perspective in our approach to animal communication, producing an interpretive type of turn-taking is significantly more cognitively demanding than the other two and will be most likely rare. It should also be noted that, as of now, researchers are mostly focusing on either vocal or visual modality (gestural/facial expression), and only minimal work has been done in terms of combining these modalities into a multimodal approach to the study of animal turn-taking abilities (see Partan & Marler, 1999; Slocombe, Waller, & Liebal, 2011, for work on multimodality in animal communication in general).

Moreover, turns like the ones in the interpretive turn-taking model are produced sequentially and their sequentiality is meaningful. In other words, to understand what an animal is trying to do in terms of social action, it is not sufficient to identify the *composition* of the signal but also its *position* in a sequence of signals within an interactional bout (Schegloff, 2007).

4. The Sequential Organization of Action

For the last 50 years, conversation analysts have investigated human social interaction as "the primordial site of human sociality" (Schegloff, 1996, p. 4; 2006, p. 70) and a major contribution to that endeavor has been outlining the sequential organization of social action via turns at talk. The *adjacency pair* is the most fundamental structure in this organization; one that can be recursively reproduced and expanded in conversation (Schegloff, 1968; Schegloff, 2007; Schegloff & Sacks, 1973; Sacks, 1992). In its minimal, unexpanded form, an adjacency pair has the following features:

a. It is composed of two turns/utterances/social actions
b. by different participants
c. that are adjacently placed
d. and are relatively ordered into *first pair parts* (FPPs) and *second pair parts* (SPPs). FPPs are actions in first position that initiate some exchange (e.g., questions, requests, offers) and SPPs are in second

position and are responsive to FPPs (e.g., answers, grants, acceptances).

e. These components are pair-type related; that is, not every SPP can follow any FPP, so that, for example, a greeting makes relevant a reciprocal greeting, a question an answer, and so on (Schegloff, 2007, p. 13).

The same utterance produced in first or second position can implement very different kinds of actions. For example, *it's raining!* in first position could just be a noticing while *it's raining!* in second position to an invitation to go for a walk would count as a rejection of the invitation. Accordingly, the sequential position of any action in an adjacency pair plays an integral role in both action recognition and ascription.

Conversation analysts have interpreted the timing of a responsive action as not neutral. Preferred[1] responses are produced without delay (Heritage, 1984; Levinson, 1983; Schegloff, 1988); that is, usually within 0.2 s of the completion of FPP (Levinson, 2016; Stivers et al., 2009). Dispreferred responses are usually marked by delay, hesitations, accounts, and such. The absence of response, or delays longer than 0.2 s, can indicate a recipient's trouble in understanding a speaker's FPP (Pomerantz, 1984), and one possible remedy for such trouble is for the speaker to pursue a response by producing another FPP (e.g., Davidson, 1984; Pomerantz, 1984; Rossano, 2012). According to the rule of *standard maximum silence* proposed by Jefferson (1989), participants in interaction tend to provide recipients up to one second to respond before pursuing it further or moving on. Sometime, delayed responses can take longer than one second, especially in face-to-face interaction, since participants can perceive via embodied behavior whether a recipient is going to respond or not and which factors might be causing further delays (Rossano, 2012). If participants do not respond to a first pursuit, speakers of the FPP can pursue a line of action further by producing a further pursuing utterance. In ordinary conversation, speakers of an FPP appear to wait the same amount of time for a recipient to respond between an initial FPP, a first or even second pursuit (Rossano, 2012).

Regarding a recipient's responsive action, it is possible that a recipient produces an inapposite SPP, which can display a problematic understanding of the FPP. Speaker A can correct herself and produce what has been called a *third position repair* (Schegloff, 1992) that usually takes the form of *I did not mean X, I meant Y.* Schegloff asserted that this type of repair is "the last structurally provided defense of intersubjectivity in conversation" (p. 1295).

Although the majority of these findings may appear to relate specifically to dyadic interaction, they apply to multiparty interactions as well. In a situation where many individuals share the same space and, thus, can perceive the same communicative behavior from an individual, parties frequently establish and negotiate *participation frameworks* (Goffman, 1981; Goodwin & Goodwin, 2004). Participants in an interaction tend to indicate who is part of the conversation and who is not (and who just happens to be a bystander), who is being addressed and who should respond. When one party looks at another, she not only monitors the other participants' attention but also ensures that her recipient(s) know(s) that he is being addressed. Establishing a participation framework, then, is often the first step for participants to achieve sequentially organized social interaction.

Recent work by Rossano and colleagues showed preliminary evidence that great apes can produce adjacency pair structures (Rossano, 2013; Rossano & Liebal, 2014) and can pursue responses when they do not obtain the response they seemed to be looking for (Liebal, Call & Tomasello, 2004; Hobaiter & Byrne, 2011). In what follows, I provide further evidence for the existence of adjacency pair structures, pursuit practices, and the first evidence of their ability to engage in third position repairs. Furthermore, I will emphasize the similarity between the timing of sequential behavior in great apes and the timing of human conversation.

5. Adjacency Pairs

In what follows, I present examples of behavioral data collected in zoos over the last few years. Most of the interactions concern mother-offspring dyads since they are likely to constitute the most cooperative type of relationship in primates. I specifically choose to show data from bonobos (the primate most closely related to humans), gorillas, and orangutans to convey that contrary to the paucity of data on great apes that are not chimpanzees, the claims here reported are likely to apply to all great ape species (at least within a captive environment).

In the first four examples (figures 15.2–15.5), I show how great apes can engage in basic adjacency pairs while interacting spontaneously with conspecifics. Notably, these adjacency pair sequences are carried out with a similar orientation to the precision timing of preferred responses to FPPs reported in naturally occurring human conversation.

In the first example (figure 15.2), K initiates a request to be carried by grabbing mother's side. In frame A

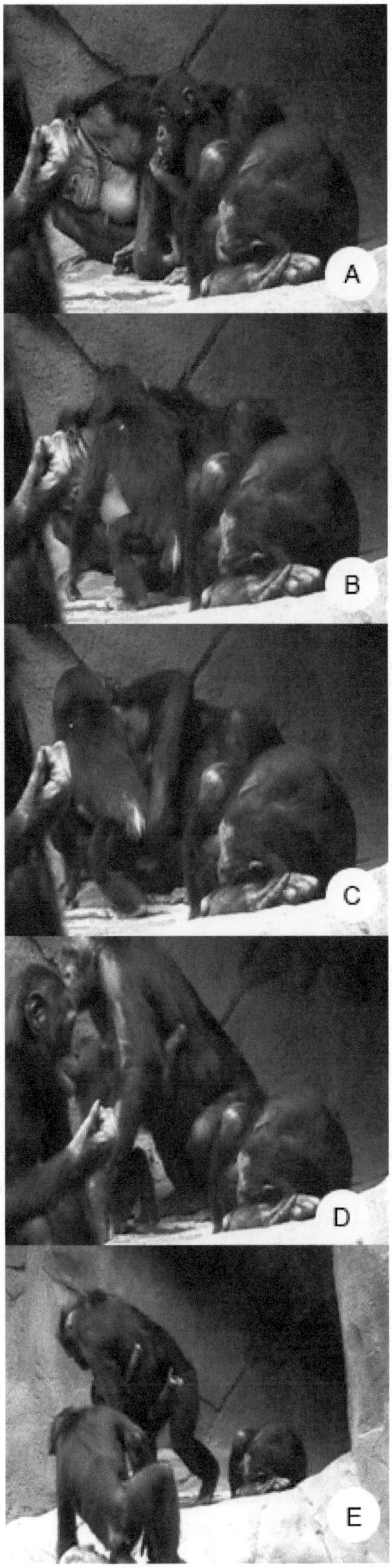

FIGURE 15.2 Bonobo, grab side, infant K.

(figure 15.2), K is initially standing close to her mother, who is lying down. There is another adult individual lying down to the infant's right side and another standing close by. This configuration of individuals in this environment is rather common and reminds us that, in nonexperimental settings, dyadic exchanges do not take place because there are only two individuals in the environment, but because such exchanges are an achievement accomplished by participants. In frame B, K turns her body around to face mother and looks at her. K's movement establishes a specific participation framework (Goffman, 1981; Goodwin & Goodwin, 2004) in which mother becomes the addressed recipient for K's incipient action, and the other two individuals nearby become bystanders. In doing so, K not only can monitor mother's attention, but she also selects mother as the individual who will be the target of her next move.

After positioning herself in front of mother, K produces an FPP: K grabs mother's side with her right hand and slightly pulls (frame C). Mother gets up and begins the carry. In this first example, then, we can see the following structure:

1. Infant moves to initiate a participation framework that selects mother as addressee and facilitates K's monitoring of mother's attention, K's performance of the FPP, and mother's response (frame B);
2. Infant produces FPP: grabs mother's side (frame C); and,
3. Mother produces SPP: immediately recognizes infant's request and satisfies it (frames D and E).

The timing passing between frames C and D is remarkably short. As soon as the infant places her hand on mother's body, slightly lifting her arm, mother gets up. The delay in responding is less than 0.2 s.

In example 2 (see figure 15.3), infant F elicits a carry by reaching towards mother, while hanging from a mesh. Mother begins responding before F's reaching gesture is fully articulated. In frame A in figure 15.3, infant F is hanging on the mesh looking downward while mother monitors her from behind. F suddenly turns her head toward mother and stretches her left hand toward mother while looking toward her (therefore selecting her as the addressed recipient, both through gaze and the directionality of the arm reach). Before F even completes the stroke of her gesture, however, mother begins moving her right arm toward her (figure 15.3B). By the time F reaches completely toward mother, mother places her right hand under F's armpit to grab her (frame C). Mother then pulls F toward her venter so that she can grab onto her and initiate the carry (frame D). In example 2, we see that mothers can not only recognize the infant's gesture before its

FIGURE 15.3 Bonobo, reach, infant F.

articulation is brought to completion but also anticipate the production of the response the infant is in the process of soliciting. In doing so, mother displays the capacity to project what the infant is trying to do before the signal is fully articulated, displaying skills (action projection and action recognition) that we know are reliably deployed by humans when they talk to each other.

6. Delay and Pursuit

One advantage of the production of visible behavior over vocalizations is that some body postures and the stroke of many gestures can be sustained in cases when an addressed recipient may be momentarily distracted or hesitant. Example 3 in figure 15.4 is a case in point. In this example, the adult male orangutan offers food to an adult female and her juvenile daughter, and they hesitate to respond. In figure 15.4A, we can see the adult male sitting against a wall with a bag of food in front of him and a branch in his hand. The male is looking toward the other three animals in front of him. Specifically, an adult female D and her juvenile daughter T are staring at the male while standing on four limbs and leaning toward the male while another juvenile male is sitting behind D and T. The male and the two females engage in mutual gaze for about one second when the male then turns the branch that he was chewing toward the females (frame B) who then move a little closer to him. The male is looking at the females for about 0.8 s while still holding the branch toward them until the adult female D hesitantly reaches toward the branch (frame C) and takes the branch away from the male's hand (frame D). Once the food has reached the female and the offer sequence has therefore been completed, the male turns his gaze away from the females and returns toward the bag with food.

In this example, we can see that the delay in responding lasts less than a second, like in humans, yet it is not clear whether it has to do with resisting accepting food or rather because of the delicacy of the situation. Indeed, a too quick attempt at grabbing the branch might be met with resistance (see Rossano & Liebal, 2014; Liebal & Rossano, 2017). In this case, the adult male persists in holding a branch toward the females while looking toward them as well. This configuration

FIGURE 15.4 Orangutan, offer, adult male B.

FIGURE 15.5 Orangutan, request rejected, adult female D.

constitutes a food offer in terms of social action, as described by Rossano and Liebal in analogous data.

However, an adjacency pair sequence does not always achieve the preferred response that the sequence initiator attempts to elicit. In the example in figure 15.5, we can see a request of food that is met with rejection rather than acceptance. Adult male B is sitting with a bag of food with three adult females and the babies of two of them seated in front of him. All the females are looking at the bag (frame A). Adult female P turns toward her left, away from the male, while adult female D produces a begging gesture toward the male, while looking at him (frame B). As soon as the begging gesture reaches its stroke, the male quickly pushes D's hand away (frame C). The male reengages with the food bag by licking the paper bag that retains traces of honey and D looks away from him, toward her right.

In this example, we can see how an individual can reject a request by pushing away the begging hand, leading to the request being abandoned. The requester does not pursue it any further. By deploying a begging gesture, D is recognizably not attempting to take/steal the food from the male. By pushing D's hand away the male conveys that he is not willing to share more food with her.

7. Pursuing Responses

The previous examples have shown social actions that were successfully (figures 15.2–15.4) or unsuccessfully (figure 15.5) accomplished after the signaler had clearly selected via gaze one individual as the addressed recipient of their communications. The timing between the initiating action (FPP) and the responsive behavior (SPP) has been shown to be remarkably fast, comparable to turn-taking in adult humans. If this is the timing of a "preferred" response, then we can ask: what happens when the recipient does not produce the expected or preferred response? Just because an individual does not respond immediately, it does not mean that one never will. So how long would apes wait for an appropriate response and what can they do if the response is not immediately forthcoming or is not the one they were aiming for?

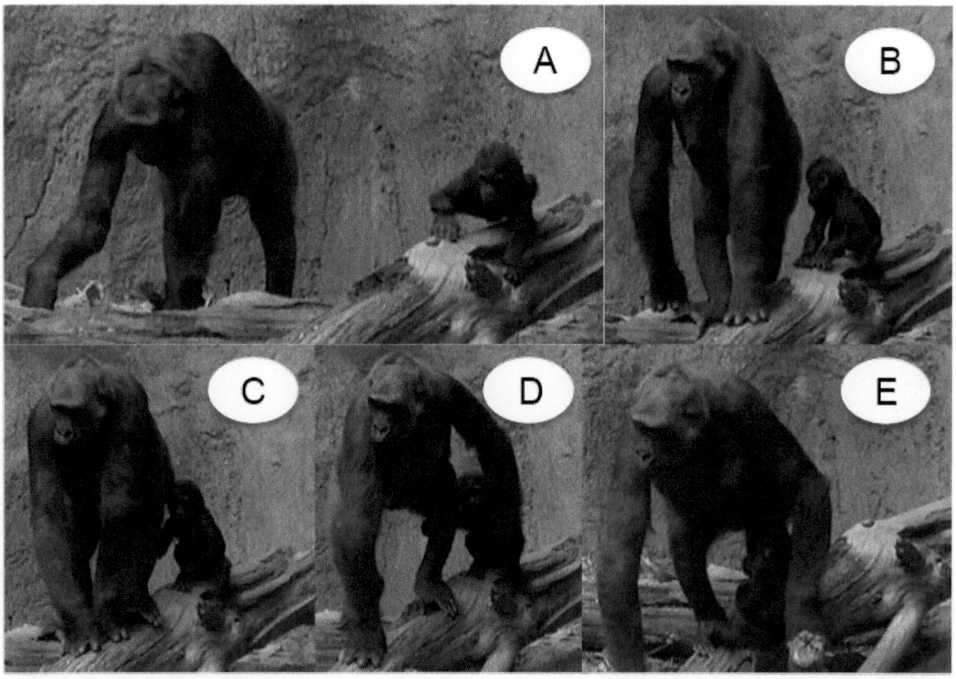

FIGURE 15.6 Gorilla, leg carry, adult female K.

In the following examples, I show pursuits in gorillas and bonobos and a possible account for the signals that they decide to deploy. The example in figure 15.6 shows a gorilla mother (K) attempting to get her infant (D) to grab her leg so that she can get carried around the enclosure, yet the initial move fails to elicit the expected response immediately. In frame 15.6A, infant D is slowly moving on a tree trunk while her mother K is picking some food from the ground, not looking at her. K moves in front of D by placing her left leg just in front of D and stops moving. This body configuration corresponds to an invitation to grab her leg to begin a leg carry. D first looks at her mother's leg (frame B) and touches it with one hand (frame C) but clearly does not grab onto it. One second after D has touched K's leg, K taps onto D's leg without even turning towards her (frame D). As soon as K taps on D's leg, D places both arms around K's leg, K begins moving and carrying D around the enclosure on her left leg (frame E).

In this example, we can see that a delay in responding leads to a pursuit performed with a different gesture. D had not completely misunderstood the initial signals since she had moved toward mother's leg and was touching it. Rather D had not proceeded with grabbing the leg so that she could be carried around. The tap on the leg works as a pursuit in that following D's failure to grab onto K's leg, it invites the infant to get on her leg so that K can begin the leg carry.

The example in figure 15.7 is an instance of a bonobo infant's (K) pursuit using a different gesture, which also occurs following an ambiguous action by mother. Mother moves in such a way as to, at least initially, suggest that she will satisfy the infant's request.

In frame 15.7A, K positions herself in front of mother then produces the grab side gesture to signal her "request" (frame B). Mother lifts herself up, but rather than move to engage in a carry, she instead starts scratching her side (frame C), precisely where K had grabbed and pulled her. Seeing that mother has performed a different responsive action, K produces another request to be carried, this time in a different way—she touches mother's side slightly lower down and then removes her hand (frame D). Between K's lowering her hand and mother gently tapping on K's back and engaging in the carry (frames F and G), 1.2 s pass. This is once again akin to the standard maximum silence of ordinary conversation in humans (Jefferson, 1988), and indeed, what happens within this second is K's success in pursuing a response from mother. Finally, note that the grab side gesture is produced when mother is lying down (see also figure 15.2), while the touch side is produced when mother is sitting up, indicating one possible difference in terms of contingencies affecting the implementation of these two tactile gestures.

In the next example in figure 15.8, mother is lying down. K is moving toward the camera and therefore is in front of mother, who is monitoring K from behind

Figure 15.7 Bonobo, grab plus touch side 1 pursuit, infant K.

(frame A). K then turns and moves back closer toward mother, where she stops and looks at her face (frame B). While looking at mother, K produces the first "request" by producing a grab side gesture (frame C), but mother does not move (frame D). This FPP fails to obtain an immediate response. K waits for a response for one second and then pursues uptake from mother using another grab side gesture (frame E), but again, mother does not respond (frame F). This time, K waits 0.9 s for a response, looks at mother, and then redoes the same gesture a third time (grab side) as she continues to look at mother (frame G). Once again, mother does not move (frame H).

After three clear attempts by K to request a carry from mother, K waits another 1.5 s before moving out in front of mother and producing a spin gesture (an idiosyncratic gesture that this infant produces to request mother to carry her, see Halina, Rossano, & Tomasello, 2013; Rossano, 2018). In this example, the spin gesture entails turning almost 360° rather quickly with K ending up facing mother again (frame I). Unfortunately for K, even this time, mother does not respond (frame J).

Following the spin gesture, K waits another 0.9 s before moving back toward mother and touching her leg (frame K), which is not a gesture signaling a request to be carried, and indeed mother does not respond to it as such. Although mother rises and puts her hand on K's back, her body posture indicates that she is not going to proceed with a carry. Indeed, she remains in a different configuration than what is required to carry the infant; that is, she holds herself up on her right arm and clearly remains seated on the ground (frame L), as opposed to leaning forward, lifting her knees, and rising off the ground (see figures 15.2 and 15.7). Once mother has remained in this configuration for one second, K gives up, and moves away from mother (frame M).

It is important to point out that in this example (as in figures 15.2 and 15.7), the infant uses the tactile grab side gesture when mother is lying down. In the absence of a response from mother who continues lying down, the infant pursues a response using the same tactile gesture two more times. Only after the grab side gesture has failed three times, K produces a less equivocal gesture for requesting a carry, the spin. In other words, K selects the gesture that is least likely to be misunderstood (a touch or a grab, in particular in the mother's breast area, can at times successfully elicit nursing).

Concerning the timing of each pursuit, K waits approximately one second each time before pursuing a response, although she waits somewhat longer (1.5 s) after having failed three times to elicit a response from mother using the grab side gesture and before implementing the spin gesture. The amount of time K waits before producing the spin gesture may have to do with the shifting from one gestural type to another (i.e., from a tactile gesture to a visible one). The examples in figures 15.7 and 15.8 demonstrate how K displays a sensitivity to a one-second boundary for response time. Even in the case of pursuing more than once, K appears to provide mother with approximately the same amount of time to respond, as has been claimed for ordinary conversation among humans (Rossano, 2012).

8. Third Position Repair

In the example of figure 15.9, infant K first requests to be carried and then produces a third position repair following a response from mother that suggests a possible misunderstanding between the two. There is another adult individual sitting nearby mother and infant, so K first positions herself in front of mother and looks at her. After selecting mother as her addressed recipient, K produces the touch side gesture (frame A) and raise arm (frame B) without a break in the two movements. Thus, she combines two gestures associated with carry (demonstrating combinatorial abilities), but because they are both produced with her right arm, she combines them sequentially; that is, she produces them one after the other. After holding her arm up for approximately 0.5 s, she initiates a pursuit by moving her right arm toward mother's side (as in frame A; see frame C). However, before K reaches mother's body, mother begins moving her right arm toward K. Instead of producing the expected SPP (i.e., getting up to begin a ventral carry), mother moves closer to K but remains on the ground and pushes K closer to her with her other hand/arm (frame D). K recognizes that the requested action is not underway and moves out from under mother's arm and again positions herself in front of mother. Recall that in the example of figure 15.8, when mother lifted her back up but did not get up and placed her hand behind K's back, K waited one second before moving away, as her rising might have been preliminary to a carry. By contrast, in this example, K moves out of mother's grasp immediately, thereby suggesting that she interprets mother's behavior as not preliminary to a carry (mother is not lifting herself up in this case) but indeed as a different action altogether. At this point, K again faces mother and pursues the request for carry again, this time producing a 360° spin (frame E). Following this gesture, K waits approximately one second for mother to respond, while continuing to look at her. Mother then leans forward toward K, raises herself off the ground while placing her right hand on K's back

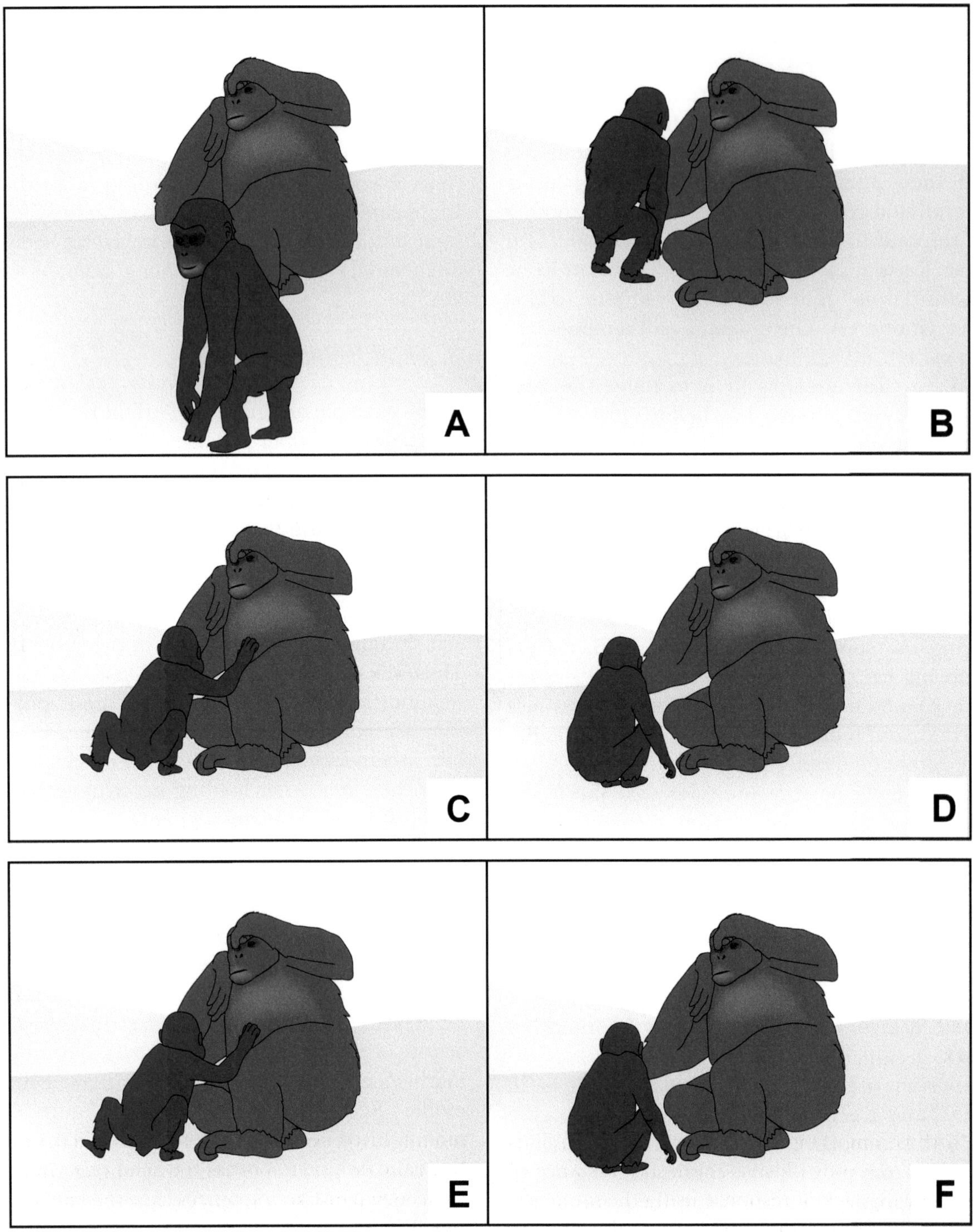

FIGURE 15.8 Infant K. (A, B) Approach and look at mother's face. (C, D) Grab side and wait. (E, F) Second grab side and wait. (G, H) Third grab side and wait. (I, J) Spin and wait. (K–M) Touch, mother gets up but no carry.

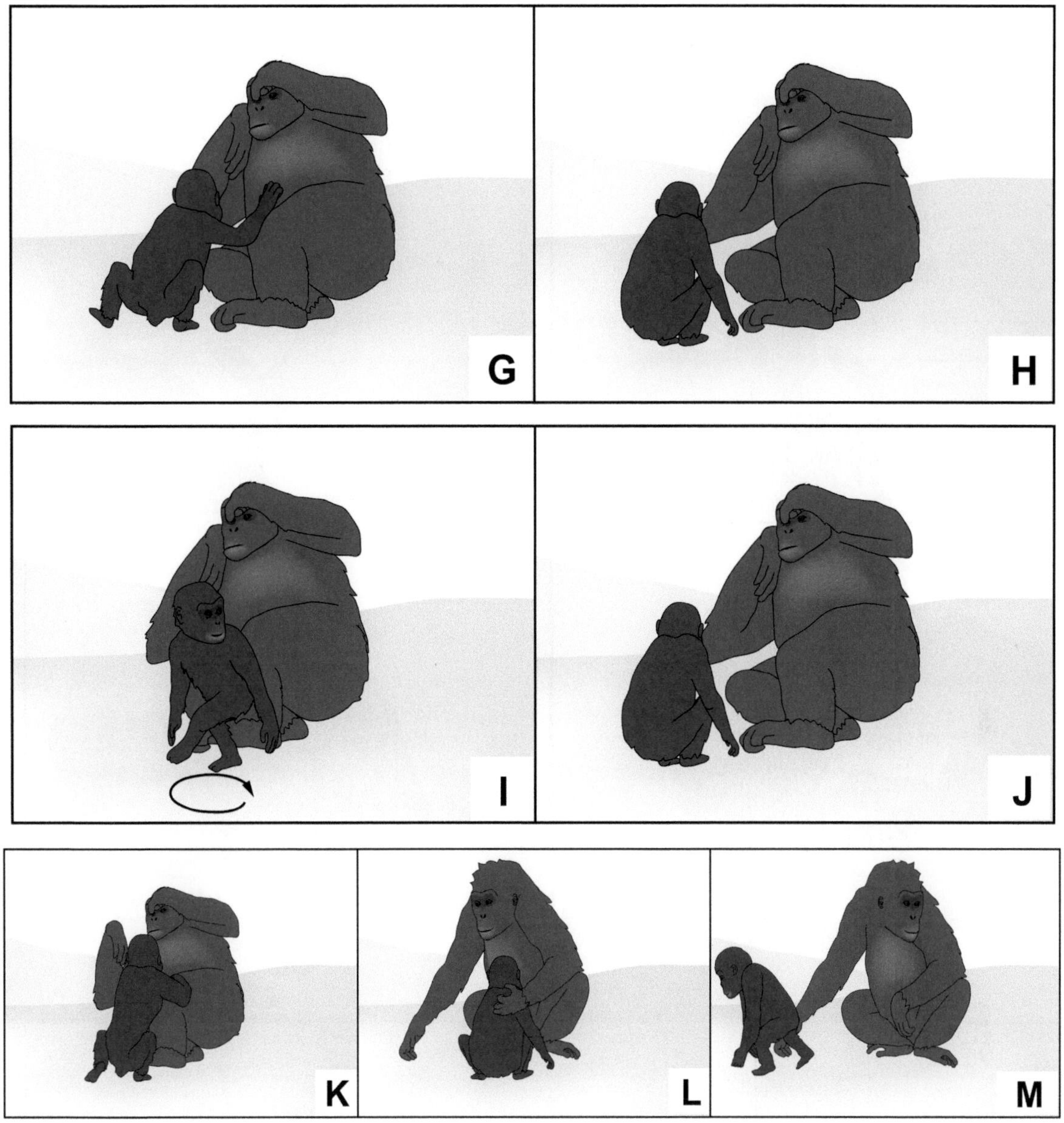

Figure 15.8 (continued)

(frame F), thus, engaging in a carry (frame G). This example, then, not only illustrates another instance of a pursuit following lack of response using the same gestural form (as in figures 15.8 and 15.9), it also demonstrates that infant bonobos are capable of recognizing a misunderstanding in the interpretation of a sequence initiating action, and thus, their capacity to repair such misunderstanding.

9. Discussion

The data shown in this chapter shows that nonhuman great apes (in particular mother-infant dyads) structure their communicative signals in a manner that strongly resembles the structure of human social interaction. Not only do great apes regularly engage in successful adjacency pair-like sequences, but they also do so at a speed remarkably similar to the timing of human ordinary conversation. I have also shown in the example of figure 15.9 that a structure akin to third position repair occurs in great apes' interactions in situations where mother's response displays a different understanding of what the infant was requesting. Moreover, I have shown how infant bonobos pursue a response from their mothers when mothers do not provide one, by either repeating the same gesture or deploying a new one with the same goal in mind. Repeating the same gesture or using a new one is not a random decision. If mother does not

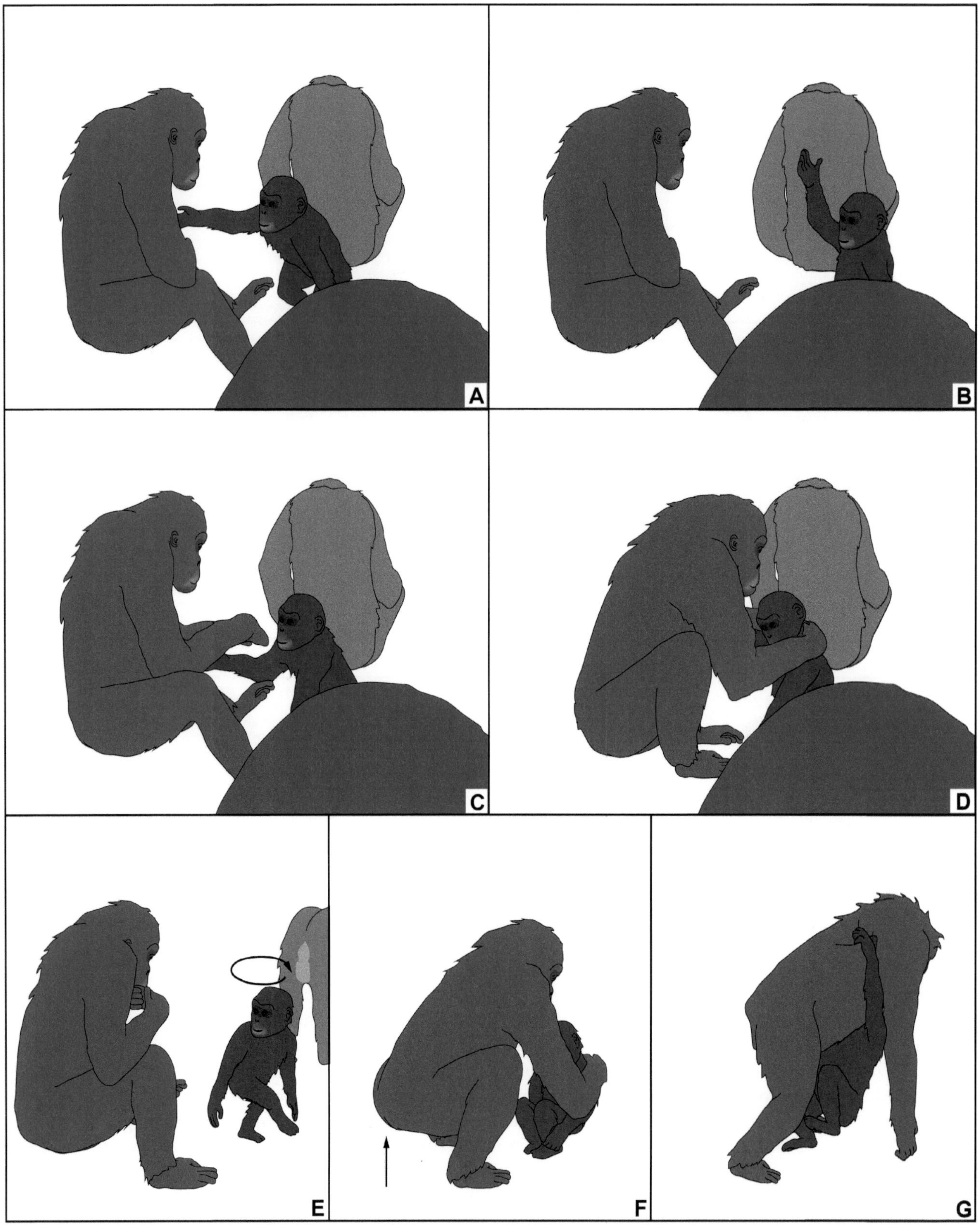

FIGURE 15.9 Bonobo, pursuit plus third position repair, infant K.

react at all, infants tend to produce the same gesture, whereas if mother responds but not in a way that addresses the infant's sequence-initiating action, then they will produce a different gesture. However, if the same gesture fails to elicit a response multiple times, I have shown that an infant may resort to producing a different gesture that is more uniquely associated with the action being solicited (i.e., a carry) than the previous one. I have suggested that one reason for doing so may be to minimize the chances of misunderstanding concerning what the infant is requesting.

This ordering of gestures raises the question of whether there is a possible ranking to them. That is, certain forms may be more effective than others (and their efficacy may change over time), they may convey different interactional nuances, or their deployment may have to do with their ontogenetic origins (e.g., the last gesture to develop in an infant's repertoire may be the first one to be deployed). In chimpanzees, for example, tactile gestures often precede visual gestures when combined in a sequence (Liebal, Call, & Tomasello, 2004). While this often appears to be the case in the bonobo interactions shown here, there are cases in this collection in which a visual gesture occurs first and they can be accounted for. In these cases, initiating a request to be carried with a tactile gesture is difficult since the infant is not close enough to be able to touch mother. Previous literature did not indicate whether there might be specific criteria for deploying one tactile gesture over another. I have shown that K, for example, tends to use the *grab* gesture when mother is lying down, while she uses the *touch* gesture when mother is sitting up. Both gestures can lead to a carry, but the choice between using one over the other does not appear to be random.

With respect to the timing of these sequences, it appears that a great ape is willing to wait for the recipient to respond approximately the same amount of time previously documented in human ordinary interaction (within the range of one second). This potential sensitivity to a one-second boundary for response time, together with the timing it usually takes for a successful carry to occur once it has been solicited (0.2 s or less), suggests that it is not just the *structure* of human social interaction but also its *tempo* that might have a more ancient evolutionary origin than we think. And both may have little or nothing to do with language per se.

Finally, I have tried to outline how the field of animal communication and more specifically primate communication would benefit from adopting an interactional perspective on primate communication. In particular, Weber's notion of social action, especially when combined with a conversation analytic perspective on animal behavior and on turn-taking in other fields, allows the field investigating animal communication to see the interactional dynamics of meaning making in primates and their capacity to conduct a subjective existence, at least to some degree.

Overall, this chapter has outlined a framework that allows for a more fine-grained analysis of primate social interactions; one that has shown that the capacity to engage in fast-paced, repeated, cooperative adjacency-pair structures, to pursue responses flexibly by selecting among different gestures, and to repair misunderstandings concerning the interpretation of those gestures, may not be uniquely human skills. Ultimately, once we systematically identify the interactional skills we share with other primates, we are better equipped to understand the distinctive role language actually played in human evolution and what made it possible.

NOTE

1. Here I am adopting the notion of preference of social actions outlined in Schegloff (2007, pp. 58–96). The terms *preferred* and *dispreferred* refer to a structural relationship rather than a psychological one. In particular, some responsive actions align with the action produced in first position and allow for the interactional project to progress (e.g., accepting an offer, agreeing with a proposition), while others do not align and in some sense halt or disallow the project's forward trajectory (e.g., rejecting an offer, disagreeing with a proposition).

REFERENCES

Acquistapace, P., Aquiloni, L., Hazlett, B. A., & Gherardi, F. (2002). Multimodal communication in crayfish: Sex recognition during mate search by male *Austropotamobius pallipes. Canadian Journal of Zoology, 80*(11), 2041–2045.

Bruner, J. S. (1975). The ontogenesis of speech acts. *Journal of Child Language, 2*(1), 1–19.

Call, J., & Tomasello, M. (Eds.). (2007). *The gestural communication of apes and monkeys.* New York: Taylor & Francis Group/Lawrence Erlbaum Associates.

Call, J., & Tomasello, M. (2008). Does the chimpanzee have a theory of mind? 30 years later. *Trends in Cognitive Science, 12,* 187–192.

Chow, C. P., Mitchell, J. F., & Miller, C. T. (2015). Vocal turn-taking in a non-human primate is learned during ontogeny. *Proceedings of the Royal Society of London B, 282*(1807), 20150069.

Christiansen, M. H., & Chater, N. (2016). The Now-or-Never bottleneck: A fundamental constraint on language. *Behavioral and Brain Sciences, 39,* e62.

Clark, H. H., & Brennan, S. E. (1991). Grounding in communication. *Perspectives on Socially Shared Cognition, 13,* 127–149.

Clark, H. H., & Murphy, G. L. (1982). Audience design in meaning and reference. *Advances in Psychology, 9,* 287–299.

Clark, H. H., Schreuder, R., & Buttrick, S. (1983). Common ground at the understanding of demonstrative reference.

Journal of Verbal Learning and Verbal Behavior, 22(2), 245–258.
Darwin, C. (1871). *The descent of man, and selection in relation to sex* (Vol. 2). New York: D. Appleton.
Darwin, C. (1872). *The expression of emotion in animals and man.* London: Methuen.
Davidson, J. (1984). Subsequent versions of invitations, offers, requests, and proposals dealing with potential or actual rejection. In J. M. Atkinson & J. Heritage (Eds.), *Structures of social action* (pp. 102–128). Cambridge: Cambridge University Press.
Dawkins, R., & Krebs, J. R. (1978). Animal signals: Information or manipulation. *Behavioural Ecology: An Evolutionary Approach, 2*, 282–309.
De Ruiter, J. P., Mitterer, H., & Enfield, N. J. (2006). Projecting the end of a speaker's turn: A cognitive cornerstone of conversation. *Language, 82*(3), 504–524.
Evans, C. S., Evans, L., & Marler, P. (1993). On the meaning of alarm calls: Functional reference in an avian vocal system. *Animal Behaviour, 46*(1), 23–38.
Garrod, S., & Pickering, M. J. (2004). Why is conversation so easy? *Trends in Cognitive Sciences, 8*(1), 8–11.
Goffman, E. (1981). Footing. In E. Goffman (Ed.), *Forms of talk* (pp. 124–159). Philadelphia: University of Pennsylvania Press.
Goodwin, C. (1981). *Conversational organization: Interaction between speakers and hearers.* New York: Academic Press.
Goodwin, C. (2000). Action and embodiment within situated human interaction. *Journal of Pragmatics, 32*(10), 1489–1522.
Goodwin, C. (2011). Building action in public environments with diverse semiotic resources. *Versus, 112–113*, 169–182.
Goodwin, C., & Goodwin, M. H. (2004). Participation. In A. Duranti (Ed.), *A companion to linguistic anthropology* (pp. 222–244). Oxford: Basil Blackwell.
Grice, H. P. (1975). Logic and conversation. In P. Cole & J. L. Morgan (Eds.), *Syntax and semantics. Vol. 3. Speech Acts* (pp. 225–242). New York: Seminar Press.
Halina, M., Rossano, F., & Tomasello, M. (2013). The ontogenetic ritualization of bonobo gestures. *Animal Cognition, 16*(4), 653–666.
Hare, B., Call, J., Agnetta, B., & Tomasello, M. (2000). Chimpanzees know what conspecifics do and do not see. *Animal Behaviour, 59*(4), 771–785.
Hare, B., Call, J., & Tomasello, M. (2001). Do chimpanzees know what conspecifics know? *Animal Behaviour, 61*(1), 139–151.
Hare, B., & Tomasello, M. (2004). Chimpanzees are more skillful in competitive than in cooperative cognitive tasks. *Animal Behaviour, 68*(3), 571–581.
Henry, L., Craig, A. J., Lemasson, A., & Hausberger, M. (2015). Social coordination in animal vocal interactions: Is there any evidence of turn-taking? The starling as an animal model. *Frontiers in Psychology, 6*, 1416.
Heritage, J. (1984). *Garfinkel and ethnomethodology.* Cambridge: Polity Press.
Heritage, J. (2004). Conversation analysis and institutional talk: Analysing data. In D. Silverman (Ed.), *Qualitative research: Theory, method and practice* (pp. 222–245). London: Sage Publications.
Higham, J. P., & Hebets, E. A. (2013). An introduction to multimodal communication. *Behavioral Ecology and Sociobiology, 67*(9), 1381–1388.
Hobaiter, C., & Byrne, R. W. (2011). Serial gesturing by wild chimpanzees: Its nature and function for communication. *Animal Cognition, 14*(6), 827–838.
Hobaiter, C., & Byrne, R. W. (2014). The meanings of chimpanzee gestures. *Current Biology, 24*(14), 1596–1600.
Jefferson, G. (1988). Notes on a possible metric which provides for a "standard maximum" silence of approximately one second in conversation. In D. Roger & P. Bull (Eds.), *Conversation: An interdisciplinary perspective* (pp. 166–196). Clevedon, UK: Multilingual Matters.
Jefferson, G. (1989). *Preliminary notes on a possible metric which provides for a "standard maximum" silence of approximately one second in conversation.* Clevedon, UK: Multilingual Matters.
Krupenye C., Kano, F., Hirata, S., Call, J., & Tomasello M. (2016). Great apes anticipate that other individuals will act according to false beliefs. *Science, 354*(6308), 110–114.
Levinson, S. C. (1983). Conversational structure. In S. C. Levinson (Ed.), *Pragmatics* (pp. 284–370). Cambridge: Cambridge University Press.
Levinson, S. C. (2013). Action formation and ascription. In J. Sidnell & T. Stivers (Eds.), *Handbook of conversation analysis* (pp. 103–130). Malden, MA: Wiley-Blackwell.
Levinson, S. C. (2016). Turn-taking in human communication: Origins and implications for language processing. *Trends in Cognitive Sciences, 20*(1), 6–14.
Liebal, K., Call, J., & Tomasello, M. (2004). Use of gesture sequences in chimpanzees. *American Journal of Primatology, 64*, 377–396.
Liebal, K., & Rossano, F. (2017). The give and take of food sharing in Sumatran orang-utans (Pongo abelii) and chimpanzees (Pan troglodytes). *Animal Behaviour, 133*, 91–100.
Marler, P., Evans, C. S., & Hauser, M. D. (1992). Animal signals: Motivational, referential, or both. In H. Papoušek, U. Jürgens, & M. Papoušek (Eds.), *Studies in emotion and social interaction. Nonverbal vocal communication: Comparative and developmental approaches* (pp. 66–86). New York: Cambridge University Press.
Maynard-Smith, J., & Harper, D. (2003). *Animal signals.* Oxford: Oxford University Press.
Owings, D. H., & Morton, E. S. (1997). The role of information in communication: An assessment/management approach. In D. H. Owings., M. D. Beecher, & N. S. Thompson (Eds.), *Perspectives in Ethology. Vol. 12. Communication* (pp. 359–390). Boston: Springer.
Owren, M. J., Rendall, D., & Ryan, M. J. (2010). Redefining animal signaling: Influence versus information in communication. *Biology and Philosophy, 25*(5), 755–780.
Partan, S., & Marler, P. (1999). Communication goes multimodal. *Science, 283*(5406), 1272–1273.
Partan, S. R., & Marler, P. (2005). Issues in the classification of multimodal communication signals. *American Naturalist, 166*(2), 231–245.
Pomerantz, A. (1984). Pursuing a response. In J. M. Atkinson & J. Heritage (Eds.), *Structures of social action* (pp. 152–164). Cambridge: Cambridge University Press.
Premack, D., & Woodruff, G. (1978). Does the chimpanzee have a theory of mind? *Behavioural and Brain Sciences, 1*, 515–526.
Rendall, D., Owren, M. J., & Ryan, M. J. (2009). What do animal signals mean? *Animal Behaviour, 78*(2), 233–240.
Rossano, F. (2012). *Gaze behavior in face-to-face interaction* (Doctoral dissertation). MPI Series in Psycholinguistics 71.

Retrieved from https://repository.ubn.ru.nl/handle/2066/99151.

Rossano, F. (2013). Sequence organization and timing of bonobo mother-infant interactions. *Interaction Studies, 14*(2), 160–189.

Rossano, F. (2018). All the change we cannot see: The methodological problem of studying the development of primate communicative behavior. Manuscript under review.

Rossano, F., & Liebal, K. (2014). "Requests" and "offers" in orangutans and human infants. In P. Drew & E. Couper-Kuhlen (Eds.), *Requesting in Social Interaction* (vol. 26, pp. 335–364). *Studies in Language and Social Interaction Series.* Amsterdam: John Benjamins.

Sacks, H. (1992). Adjacency pairs: Scope of operation. In G. Jefferson (Ed.), *Lectures on conversation* (Vol. 2, pp. 521–532). Oxford: Blackwell.

Sacks, H., & Schegloff, E. A. (1979). Two preferences in the organization of reference to persons in conversation and their interaction. In G. Psathas (Ed.), *Everyday language: Studies in ethnomethodology* (pp. 15–21). New York: Irvington.

Sacks, H., Schegloff, E. A., & Jefferson, G. (1974). A simplest systematics for the organization of turn-taking for conversation. *Language, 50*, 696–735.

Schegloff, E. A. (1968). Sequencing in conversational openings. *American Anthropologist, 70*(6), 1075–1095.

Schegloff, E. A. (1988). On an actual virtual servo-mechanism for guessing bad news: A single case conjecture. *Social Problems, 35*(4), 442–457.

Schegloff, E. A. (1992). Repair after next turn: The last structurally provided for place for the defense of intersubjectivity in conversation. *American Journal of Sociology, 95*(5), 1295–1345.

Schegloff, E. A. (1996). Issues of relevance for discourse analysis: Contingency in action, interaction and co-participant context. In E. H. Hovy & D. Scott (Eds.), *Interdisciplinary perspectives on discourse* (pp. 3–35). Heidelberg: Springer Verlag.

Schegloff, E. A. (2006). Interaction: The infrastructure for social institutions, the natural ecological niche for language, and the arena in which culture is enacted. In N. J. Enfield & S. C. Levinson (Eds.), *Roots of human sociality: Culture, cognition and interaction* (pp. 70–96). Oxford: Berger.

Schegloff, E. A. (2007). *Sequence organization in interaction: A primer in conversation analysis.* Cambridge: Cambridge University Press.

Schegloff, E. A., & Sacks, H. (1973). Opening up closings. *Semiotica, 8*, 289–327.

Schneider, C., Call, J., & Liebal, K. (2012). What role do mothers play in the gestural acquisition of bonobos (*Pan paniscus*) and chimpanzees (*Pan troglodytes*)? *International Journal of Primatology, 33*(1), 246–262.

Schutz, A. (1967). *The phenomenology of the social world.* Evanston, IL: Northwestern University Press.

Seyfarth, R. M., & Cheney, D. L. (2003). Signalers and receivers in animal communication. *Annual Review of Psychology, 54*(1), 145–173.

Seyfarth, R. M., Cheney, D. L., Bergman, T., Fischer, J., Zuberbühler, K., & Hammerschmidt, K. (2010). The central importance of information in studies of animal communication. *Animal Behaviour, 80*(1), 3–8.

Seyfarth, R. M., Cheney, D. L., & Marler, P. (1980). Vervet monkey alarm calls: Semantic communication in a free-ranging primate. *Animal Behaviour, 28*(4), 1070–1094.

Shannon, C. E., & Weaver, W. (1949). *The mathematical theory of communication.* Urbana, IL: University of Illinois Press.

Slocombe, K. E., Waller, B. M., & Liebal, K. (2011). The language void: The need for multimodality in primate communication research. *Animal Behaviour, 81*(5), 919–924.

Slocombe, K. E., & Zuberbühler, K. (2005). Functionally referential communication in a chimpanzee. *Current Biology, 15*(19), 1779–1784.

Smith, W. J. (1977). *The behavior of communicating.* Cambridge, MA: Harvard University Press.

Sperber, D., & Wilson, D. (1986). *Relevance: Communication and cognition.* Cambridge, MA: Harvard University Press.

Stivers, T., Enfield, N. J., Brown, P., Englert, C., Hayashi, M., Heinemann, T., … Levinson, S. C. (2009). Universals and cultural variation in turn-taking in conversation. *Proceedings of the National Academy of Sciences, 106*(26), 10587–10592.

Tomasello, M. (2006). Why don't apes point? In N. J. Enfield & S. C. Levinson (Eds.), *Roots of human sociality: Culture, cognition and interaction* (pp. 506–524). London: Berg.

Tomasello, M. (2009). *Why we cooperate.* Cambridge, MA: MIT Press.

Tomasello, M. (2010). *Origins of human communication.* Cambridge, MA: MIT Press.

Tomasello, M., Carpenter, M., Call, J., Behne, T., & Moll, H. (2005). Understanding and sharing intentions: The origins of cultural cognition. *Behavioral and Brain Sciences, 28*(5), 673–727.

Uetz, G. W., Roberts, J. A., & Taylor, P. W. (2009). Multimodal communication and mate choice in wolf spiders: Female response to multimodal versus unimodal signals. *Animal Behaviour, 78*(2), 299–305.

Von Frisch, K. (1955). *The dancing bees.* New York: Harcourt, Brace.

Weber, M. (1978). *Economy and society: An outline of interpretive sociology.* Berkeley: University of California Press.

Wheeler, B. C., & Fischer, J. (2012). Functionally referential signals: A promising paradigm whose time has passed. *Evolutionary Anthropology: Issues, News, and Reviews, 21*(5), 195–205.

Wiener, N. (1948). *Cybernetics: Control and communication in the animal and the machine.* New York: Wiley.

Zahavi, A. (1975). Mate selection: A selection for a handicap. *Journal of Theoretical Biology, 53*(1), 205–214.

Zuberbühler, K. (2000a). Interspecies semantic communication in two forest primates. *Proceedings of the Royal Society of London B: Biological Sciences, 267*(1444), 713–718.

Zuberbühler, K. (2000b). Referential labelling in Diana monkeys. *Animal Behaviour, 59*(5), 917–927.

16 The Resilience of Language: Homesign

SUSAN GOLDIN-MEADOW

Human language has evolved to be learnable, else it could not be passed down from generation to generation. We can therefore look to language to discover structures that are compatible with human cognition. But many years have passed since the first spoken language was invented, and many changes have taken place. Thus, although language as we currently know it is necessarily compatible with human learning, it may reflect pressures other than human cognition. One way to discover the properties of language that reflect human predispositions to structure communication is to examine a language in its infancy. We cannot study a spoken language at its beginnings simply because new spoken languages evolve out of old spoken languages and thus do not provide an opportunity to examine the linguistic structures that humans introduce into language de novo—that is, the linguistic structures so fundamental to human language, so resilient, that they can be reinvented by a human child without guidance from a language model. We turn to language in the manual modality, to sign language, to study the resilient properties of language.

Sign languages are manual communication systems created within Deaf communities (sometimes in conjunction with hearing members of the community, e.g., Nyst, 2012) that are handed down from generation to generation. Sign languages are autonomous linguistic systems not based on the spoken languages of the surrounding hearing community (Bellugi & Studdert-Kennedy, 1980; Klima & Bellugi, 1979). Like spoken languages, they are, however, structured at syntactic (Liddell, 1980; Lillo-Martin, 1991; Neidle, Kegl, MacLaughlin, Bahan, & Lee, 2000), morphological (Aronoff, Meir, Padden, & Sandler, 2003; Emmorey, 2003; Fischer, 1973; Schick, 1990; Supalla, 1986), and phonological (Brentari, 1998; Sandler, 1989) levels. If exposed to a sign language by their deaf parents, deaf children will acquire that language as naturally as hearing children acquire spoken language from their hearing parents (Newport & Meier, 1985; Lillo-Martin, 1999). But the genetics of deafness are complicated and most deaf children are born to hearing (rather than deaf) parents, who are not likely to know a sign language. We might expect that deaf children who are not exposed to a sign language by their hearing parents, and whose hearing losses are so severe that they cannot learn the spoken language that surrounds them (even with amplification and oral education), would be unable to communicate. But deaf children in these circumstances do communicate with the hearing individuals in their worlds and use gestures, called *homesigns*, to do so (Lenneberg, 1964; Moores, 1974; Tervoort, 1961). Homesign has been shown to have many, although not all, of the properties of natural language (Goldin-Meadow, 2003a) and thus is a linguistic system developed by an individual without benefit of a conventional language model—a language developed de novo.

The first section of this chapter reviews the properties of language that have thus far been identified in homesign—the resilient properties of language. But, as noted earlier, homesign does not contain all of the properties of natural language. The second section follows homesign as it grows from isolated gesture systems used by deaf individuals, each in a hearing household, to a shared system used by a community of signers, and thus examines the conditions needed for homesign to become a full-fledged language. Because homesigners use their communication systems with hearing (rather than deaf) communication partners, it is possible that the linguistic structures found in homesign are introduced into the system not by the deaf homesigner, but by the signer's hearing partners. The third section thus examines the input to homesign provided by *co-speech gestures* produced by hearing communication partners. The final section takes an experimental approach to the question of language creation and explores manual communication constructed on-the-spot by hearing individuals asked to describe events using only their hands without speech. These *silent gestures*, as they are known, are structured differently from co-speech gestures. They contain a number of properties found in natural language and, interestingly, do not necessarily mirror the spoken language of the hearing individual who is producing them. They are thus a type of manual communication that can be compared with emerging and established sign languages.

1. Manual Communication Developed without a Language Model: The Linguistic Properties of Homesign

Homesigners use their gestures for many of the functions to which conventional languages are put. They use homesigns not only to get others to do things for them (i.e., to make requests), but also to share ideas and request information (i.e., to make comments and ask questions). Homesigners even use their gestures to serve some of the more sophisticated functions of language—to tell stories (Van Deusen-Phillips, Goldin-Meadow, & Miller, 2001), to comment on their own and others' gestures, and to talk to themselves (Goldin-Meadow, 1993). In this sense, homesigners' gestures are qualitatively different from the gestures produced by language-trained apes, who use whatever language they are able to learn to change peoples' behavior, not to change their ideas (see, for example, Greenfield & Savage-Rumbaugh, 1991). Homesigners' gestures serve the functions of language.

Homesigners' gestures also take on the forms of language. They are characterized by linguistic properties despite the fact that the homesigner does not have a usable model of a conventional language to guide his or her gesture creation (Goldin-Meadow, 2003a). I describe the properties of language that have been identified in child homesign in the following sections.

1.1. Lexicon Homesigners use both pointing gestures and iconic gestures to communicate. Their gestures, rather than being mimelike displays, are discrete units, each of which conveys a particular meaning. Moreover, the gestures are not tied to a specific situation—a *twist* gesture, for instance, can be used to request someone to twist open a jar, to indicate that a jar has been twisted open, to comment that a jar cannot be twisted open, or to tell a story about twisting open a jar that is not present in the room. In other words, the homesigner's gestures can be used in any context and can even be used to talk about events and objects that are not present (Morford & Goldin-Meadow, 1997). In this sense, the gestures warrant the label *sign*.

Child homesigners use their pointing gestures to refer to the same range of objects that young hearing children refer to using, first, pointing gestures and, later, words—and in the same distribution (Feldman, Goldin-Meadow, & Gleitman, 1978). Both groups of children refer most often to inanimate objects, followed by people and animals. They also both refer to body parts, food, clothing, vehicles, furniture, and places, but less frequently.

Child homesigners use iconic gestures more frequently than most hearing children learning spoken language. Their iconic gestures function like nouns, verbs, and adjectives in conventional languages (Goldin-Meadow, Butcher, Mylander, & Dodge, 1994), although there are fundamental differences between iconic gestures and words. The form of an iconic gesture captures an aspect of its referent; the form of a word does not. Interestingly, although iconicity is present in many of the signs of American Sign Language (ASL), deaf children learning ASL do not seem to notice. Most of their early signs are either not iconic (Bonvillian, Orlansky, & Novack, 1983) or, if iconic from an adult's point of view, not recognized as iconic by the child (Schlesinger, 1978; but see Thompson, Vinson, & Vigliocco, 2009, 2010, for evidence that iconicity does play a small role in the acquisition of sign). In contrast, deaf individuals inventing their own homesigns are forced by their social situation to create gestures that not only begin transparently but remain so. If they did not, no one in their worlds would be able to take any meaning from the gestures they create. Homesigns therefore have an iconic base.

Despite the fact that the gestures in a homesign system need to be iconic to be understood, they form a stable lexicon. Homesigners could create each gesture anew every time they use it, as hearing speakers seem to do with their gestures (McNeill, 1992). If so, we might still expect some consistency in the forms the gestures take simply because the gestures are iconic and iconicity constrains the set of forms that can be used to convey a meaning. However, we might also expect a great deal of variability around a prototypical form—variability that would crop up simply because each situation is a little different, and a gesture created specifically for that situation is likely to reflect that difference. In fact, it turns out that there is relatively little variability in the set of forms a child homesigner uses to convey a particular meaning. The child tends to use the same form, say, two fists breaking apart in a short arc to mean *break*, every single time that child gestures about breaking, no matter whether it is a cup breaking, or a piece of chalk breaking, or a car breaking (Goldin-Meadow et al., 1994). Thus, the homesigner's gestures adhere to standards of form, just as a hearing child's words or a deaf child's signs do (Singleton, Morford, & Goldin-Meadow, 1993). The difference is that the homesigner's standards are idiosyncratic to the creator rather than shared by a community of language users.

1.2. Morphology Languages, both signed and spoken, build up words combinatorially from a repertoire of a few dozen smaller meaningless units. We do not yet know whether homesign has phonological structure (but see Brentari, Coppola, Mazzoni, & Goldin-Meadow, 2012). However, there is evidence that homesigns are composed of parts, each of which is associated with a particular meaning; that is, they have morphological structure (Goldin-Meadow, Mylander, & Butcher, 1995; Goldin-Meadow, Mylander, & Franklin, 2007). Homesigners could have faithfully reproduced in their gestures the actions that they actually perform. They could have, for example, created gestures that capture the difference between holding a balloon string and holding an umbrella. But they do not. Instead, the children's gestures are composed of a limited set of handshape forms, each standing for a class of objects, and a limited set of motion forms, each standing for a class of actions. These handshape and motion components combine freely to create gestures, and the meanings of these gestures are predictable from the meanings of their component parts. For example, a hand shaped like an "O" with the fingers touching the thumb, that is, an *OTouch* handshape form, combined with a *Revolve* motion form means "rotate an object <2 inches wide around an axis," a meaning that can be transparently derived from the meanings of its two component parts (*OTouch* = handle an object <2 inches wide; *Revolve* = rotate around an axis).

Importantly in terms of arguing that a morphological system underlies a child's homesigns, the vast majority of gestures that each deaf child produces conforms to the morphological description for that child; and this morphological description can be used to predict the forms and meanings of the new gestures that the child produces. Thus, homesign exhibits a simple morphology, one that is akin to the morphologies found in conventional sign languages. Interestingly, it is much more difficult to impose a coherent morphological description that can account for the gestures that hearing speakers produce (Goldin-Meadow et al., 1995, 2007; see section 3 for additional discussion), suggesting that morphological structure is not an inevitable outgrowth of the manual modality but is instead a linguistic characteristic that homesigners impose on their communication systems.

1.3. Sentence Structure Child homesigners frequently combine their gestures with other gestures, unlike hearing children who rarely produce gesture + gesture combinations (Goldin-Meadow & Morford, 1985). We consider a string of gestures to be a single unit if the child does not pause or relax his hand between gestures (Goldin-Meadow & Mylander, 1984). For example, a homesigner combined a point at a toy grape with an EAT gesture to comment on the fact that grapes can be eaten, and at another time combined the EAT gesture with a point at a visitor to invite her to lunch with the family. The same homesigner combined all three gestures into a single sentence to offer the experimenter a snack. Interestingly, child homesign sentences convey the same meanings that young children learning conventional languages, signed (Newport & Meier, 1985) or spoken (Brown, 1973), typically convey with their sentences (Goldin-Meadow & Mylander, 1984).

In addition, homesign sentences display linguistic structure. For example, the child homesigner's gesture sentences are organized around predicate frames (e.g., *x* dances; *x* moves to *y*; *x* hits *y*; *x* gives *y* to *z*) and thus are structured at an underlying level (Goldin-Meadow, 1985). Moreover, the gesture sentences often contain more than one predicate frame or proposition and, in this sense, constitute complex sentences (e.g., point at drum–BEAT–point at straw–SIP, all produced within the bounds of a single gesture sentence to describe a scene in which a solder is beating a drum and a cowboy is sipping a straw; Goldin-Meadow, 1982).

Child homesigners' gesture sentences also exhibit three devices for marking who does what to whom. Homesigners indicate the thematic role of a referent by preferentially producing (as opposed to omitting) gestures for referents playing particular roles. Homesigners in both the United States and China are more likely to produce a gesture for the patient (e.g., the eaten cheese in a sentence about eating) than to produce a gesture for the actor (e.g., the eating mouse; Goldin-Meadow & Mylander, 1998).

Child homesigners' second device for indicating thematic roles is to place gestures for objects playing particular roles in set positions in a sentence. In other words, they use linear position to indicate who does what to whom (Feldman et al., 1978). Surprisingly, homesigners in the United States and China use the same particular linear orders in their gesture sentences despite the fact that each child is developing his or her system alone without contact with other deaf children and in different cultures (Goldin-Meadow & Mylander, 1998). Homesigners tend to produce gestures for patients in the first position of their sentences, before gestures for verbs (point at cheese–EAT) and before gestures for endpoints of a transferring action (point at cheese–point at table). They also produce gestures for verbs before gestures for endpoints (GIVE–point at table). In addition, they produce gestures for intransitive actors before gestures for verbs (point at mouse–RUN).

Child homesigners' third device for indicating thematic roles is to displace verb gestures toward objects playing particular roles, as opposed to producing them in neutral space (at chest level). These displacements are reminiscent of *inflections* in conventional sign languages (Padden, 1983; 1990). Homesigners tend to displace their gestures toward objects that are acted upon and thus use their inflections to signal patients. For example, displacing a BLOW gesture toward a bubble jar signals that the jar (or one like it) is the object to be acted upon (Goldin-Meadow et al., 1994). Thus, homesign sentences adhere to simple syntactic patterns marking who does what to whom.

1.4. Grammatical Categories There is evidence that child homesign has at least one grammatical category—a distinction between nouns and verbs. Goldin-Meadow et al. (1994) studied a homesigner beginning at age 2;10 and found that, initially, the child distinguished nouns from verbs by using nonoverlapping gesture inventories for the two roles—one set of gestures functioned as nouns, another as verbs. In addition, Hunsicker and Goldin-Meadow (2013) found that, early in development, this same child used one type of handshape in his nouns (*object* handshapes mirroring a perceptual aspect of the object) and a different type of handshape in his verbs (*handling* handshapes mirroring the hand as it manipulates the object).

However, after the age of 3;3, this child began to use the same gesture stems in both grammatical roles and, at that point, began to distinguish nouns from verbs using both morphological and syntactic devices. For example, if the child used TWIST as a verb, that gesture would likely be produced near the jar to be twisted open (i.e., it would be inflected); it would not be abbreviated (it would be produced with several twists rather than one); and it would be produced *after* a pointing gesture at the jar (that–TWIST). In contrast, if the child used that same stem TWIST as a noun to mean *jar*, the gesture would likely be produced in neutral position near the chest (i.e., it would not be inflected); it would be abbreviated (produced with one twist rather than several); and it would occur *before* the pointing gesture at the jar (TWIST–that). Thus, the child distinguished nouns from verbs morphologically (nouns are abbreviated, verbs inflected) and syntactically (nouns occur in initial position of a two-gesture sentence, verbs in second position after the patient). Interestingly, the homesigner's adjectives sit somewhere in between—as they often do in natural languages (Thompson, 1988). The homesigner marked adjectives like nouns morphologically (BROKEN is abbreviated but not inflected) and like verbs syntactically (BROKEN is produced in the second position of a two-gesture sentence).

1.5. Sentential Markers for Negations, Questions, Past, and Future Child homesigners' gesture sentences contain at least two forms of sentence modification—negation and questions (Franklin, Giannakidou, & Goldin-Meadow, 2011). Young homesigners express two types of negative meanings: *rejection* (e.g., when offered a toy, the homesigner shakes his head, indicating that he does not want the object) and *denial* (e.g., the homesigner points to his chest and then gestures SCHOOL [holding his hands as though praying because he attended a Catholic school] while shaking his head, to indicate that he is not at school). In addition, they express three types of questions: *where* (e.g., the homesigner produces a two-handed flip when searching for a piece of a toy), *what* (e.g., the homesigner produces the flip when trying to figure out which object his mother wants), and *why* (e.g., the homesigner produces the flip when trying to figure out why a toy fell). As these examples suggest, different forms are used to convey these two different meanings—the side-to-side headshake for negative meanings, the manual flip for question meanings. These gestures are co-opted from the hearing speakers' gestures that surround the homesigners. But the homesigners use these gestures as sentence modulators and produce them in systematic positions in their gesture sentences: headshakes appear at the beginning of sentences, flips at the end.

Child homesigners also use particular gestures to make reference to the past and future (Morford & Goldin-Meadow, 1997). For example, one homesigner produced a gesture, not observed in the gestures of his hearing parents, to refer to both remote future and past events—needing to repair a toy (future) and having visited Santa (past). The gesture is made by holding the hand vertically near the chest, palm out, and making an arcing motion away from the body. Another child homesigner invented his own gesture to refer only to past events. In addition to these two novel gestures, homesigners have been found to modify a conventional gesture to use as a future marker. The gesture, formed by holding up the index finger, is typically used to request a brief delay or time-out and is glossed as "wait one minute." Homesigners used the form for its conventional meaning, but they also use it to identify their intentions, that is, to signal the immediate future. For example, one child homesigner produced the gesture and then pointed at the toy bag to indicate that he was going to get a toy from the bag. Hearing speakers use the WAIT gesture to get someone's attention, never to refer to the immediate future. The form of the homesigners' gesture is borrowed from hearing speakers' gestures but it takes on a meaning of its own.

1.6. Hierarchical Structure As noted earlier, homesigners refer to entities either by *pointing* at the entity, or by producing an *iconic* gesture evoking some aspect of the entity. There is evidence that these two types of nominal gestures can be combined to form a larger unit akin to a noun phrase (Hunsicker & Goldin-Meadow, 2012). For example, initially, a child homesigner might point at a penny and then at himself (*point at penny*–point at self), or produce an iconic gesture for the penny and then point at himself (*PENNY*–point at self), to ask someone to give him a penny. But later in development, the homesigner produced sentences containing both nominal gestures—the iconic gesture along with the point at the penny (*PENNY–point at penny*–point at self). The iconic + point combination thus occupied the patient slot in the sentence and functioned as a single unit. Indeed, iconic + point combinations of this sort serve the same semantic and syntactic functions as pointing gestures and iconic gestures do when used on their own to refer to objects. In other words, the iconic + point combination substitutes for the smaller units and, in this way, functions as a noun phrase, which is embedded within a larger sentence ([PENNY–point at penny]–point at self) and thus is an instance of hierarchical structure (Hunsicker & Goldin-Meadow, 2012).

1.7. Marking Causal Structure Languages of the world display a variety of strategies for expressing causation. Despite this variety, children across the globe all tend to go through a period of overgeneralization during which they inappropriately extend the causative structure in their language (e.g., English, Bowerman, 1974; Portuguese, Figueira, 1984; Japanese, Murasugi, Hashimoto, & Fuji, 2007; Taiwan Southern Min, Lin & Tsay, 2008; Hebrew, Berman, 1993; Inuktitut, Allen, 1998). Broadly speaking, children's first verbs are used appropriately but show little evidence of internal causal structure. Children demonstrate that they have inferred a productive linguistic rule for expressing causation when they begin to produce causative overgeneralizations, for example, saying "I'm singing him" to mean *I'm making this toy sing* (Bowerman, 1974). Over time, they learn the constraints on when to apply these productive rules in the language they are learning (e.g., English allows *I broke the mug* but not *I disappeared the mug*). Acquiring causal language thus involves analyzing linguistic forms in terms of causal structure, as well as mastering the nuances of the particular causative devices in the native language.

Rissman and Goldin-Meadow (2017) asked whether a child homesigner could create linguistic devices for encoding causation and, if so, whether the emergence of this causal language would follow a trajectory similar to the one observed for children learning language from linguistic input. They found a striking U-shaped curve, where use of *handling* handshape in causative contexts was initially high (parallel to the stage when children's causal verbs are correct on the surface, but unanalyzed); followed by a dip during which the homesigner used many *object* handshapes to describe causal events; followed finally by a rebound in use of *handling* handshapes to the initial level. Importantly, the dip occurred close to the age at which the homesigner started producing verbal compounds, which are evidence for linguistic event decomposition. The homesigner had thus invented causal language that has both internal structure (the *handling* handshape marks CAUSE) and systematicity (the *handling* handshape near-obligatorily marks the causative component of an event), but only after a significant delay reflecting a reorganization of the system. This reorganization cannot be attributed to input from a conventional language model as it often is in children learning language from their parents, as the homesigner had no model. These results not only suggest that the ability to encode causation linguistically can emerge in the absence of linguistic input, but also that exposure to linguistic input is not the only factor guiding children from one stage to the next in their production of causal language.

1.8. Prosody Applebaum, Coppola, and Goldin-Meadow (2014) examined prosody in a child homesigner and found that the child used features known to mark phrase and utterance boundaries in established sign languages (Nespor & Sandler, 1999) to consistently mark the ends of his gesture sentences (phrasing). Interestingly, however, he did not use these features to mark phrase- or sentence-internal boundaries. Prosody can thus be found in homesign but its use is limited. In the future, it will be important to compare the homesigner's use of prosody with the same features in deaf children's signs (e.g., Brentari, Falk, & Wolford, 2015) to determine whether the limitations in the homesigner's prosodic structure stem from the fact that he is creating his system, or from the fact that he is a child.

2. *From Homesign to Sign Language: Watching Language Grow in Nicaragua*

Although homesigns are structured, they do not display all of the properties found in natural languages, perhaps because each homesigner is developing a gesture system on his or her own. What would happen if individual homesigners were brought together into a community? Just such a situation arose in Nicaragua when a group of homesigners came together for the first time. Over time, a sign language with much of the grammatical complexity of established sign languages evolved and Nicaraguan Sign Language (NSL), as this newly emergent language is called, was born.

Prior to 1977, deaf individuals in Nicaragua had few opportunities to interact with one another. Between 1977 and 1983, two educational programs were established that served more than 400 deaf students; although instruction was not in sign, students were able to freely communicate with each other using gestures. The deaf children and adolescents in these programs had not previously been exposed to a spoken, written, or signed language. These individuals were like the homesigning children described earlier. But when brought together for the first time, these homesigners had the opportunity to see *others* use homesign. In other words, for the first time, they not only were producers, but were also receivers, of homesign. This first cohort of NSL thus allows us to ask whether having a community, in particular, being both a receiver and a producer of a language, changes the structure of that language.

NSL gives us the opportunity to ask one additional question about language emergence. Many deaf individuals in Nicaragua are now exposed early in life to NSL—these signers thus *learn* their language from others. These subsequent cohorts of NSL thus allow us to ask whether transmitting a language to a new set of users changes the structure of that language. By comparing current day homesigners in Nicaragua with groups whose circumstances have allowed them to go beyond homesign, we can begin to develop hypotheses about which properties of language are "resilient" (appear in all systems) and which are "fragile" (appear only in select conditions), and which conditions foster the development of these relatively fragile properties. Unlike the United States, where deaf individuals rarely remain homesigners after childhood (they learn a conventional sign language or receive cochlear implants and focus on spoken language), homesigners in Nicaragua often continue to use homesign into adulthood. Because adult homesigners are better able to respond to experimental tasks (e.g., that elicit specific types of structures) than child homesigners are, researchers can conduct analyses on structures not previously reported in studies of child homesign (which, for the most part, relied on spontaneous production of homesign in naturalistic interactions). The next sections describe these analyses.

2.1. LEXICON: POINTS TAKE ON NEW GRAMMATICAL FUNCTIONS As noted earlier, homesigners, without the benefit of a language model or a community, effortlessly use points for all language functions. Coppola and Senghas (2010) asked whether these functions have systematically different forms and, if so, whether the differences appear in homesign, or whether they require a community of users or a language model to emerge. Coppola and Senghas studied pointing gestures in adult homesigners in Nicaragua, the first cohort of NSL signers who co-created the language, and subsequent cohorts of NSL signers who learned the system from preceding cohorts. All participants saw the same cartoon and were asked to retell the cartoon to a communication partner (for signers, other members of their cohort; for homesigners, a hearing individual with whom the homesigner communicated regularly[1]). Coppola and Senghas analyzed all of the deictic points at empty spaces produced by the groups and classified them according to whether they referred to a location (*locative* points, *he climbed <u>up to the house</u>*, while pointing upward) or to a character (*nominal* points, *<u>he</u> was frightened*, while pointing to a location in neutral space). They found that the two types of points differed in form. Nominal points were articulated quickly, in the area directly in front of a signer's torso, with little or no path movement; in addition, signers rarely followed the direction of a nominal point with their gaze. In contrast, locative points were articulated more slowly and followed larger paths; signers often followed the direction of a locative point with their gaze.

All groups of signers produced locative points and did so at about the same rate. But nominal points differed as a function of group. Homesigners and cohort 1 NSL signers rarely used nominal points. This type of point appeared with some frequency only in cohorts that had learned the system from preceding cohorts, increasing steadily with each subsequent cohort. The fact that homesigners and cohort 1 signers both rarely used nominal points suggests that being a producer *and* a receiver of a language system is not sufficient for this particular form to emerge in the system. The form seems to become a stable part of the language only when the language is transmitted to the next cohort; that is, only when it is learned from a language model.

2.2. Morphology

2.2.1. Spatial modulation As noted earlier, many established sign languages use loci in space, and movement toward and away from these loci, to indicate shared reference or co-reference (Abner, 2012; Fenlon, Schembri, & Cormier, 2018; Liddell, 2003; Meir, 2002; Padden, 1983; Schlenker, 2018). These spatial modulations, as they are known, are used to situate arguments in signing space; the marked spaces can then be indicated later to refer back to the arguments. For example, a signer might sign PAY toward her right side and then subsequently sign GIVE toward the same location, indicating that the same person was both paid and given to.

Senghas and Coppola (2001) found that cohort 1 signers differed from cohort 2 signers in how they used space to express co-reference. Controlling for number of years of sign language exposure, Senghas and Coppola found that early-exposed cohort 2 signers (who were exposed to the sign language before the age of 10 years) produced significantly more spatial modulations than early-exposed cohort 1 signers (who were brought together with other homesigners before the age of 10). Moreover, cohort 2 signers were also more likely than cohort 1 signers to reuse their spatial modulations to track reference across utterances; in other words, to organize their discourse.

Spatial modulations thus become a differentiated and multipurpose set of structures only in later NSL cohorts, *after* the language has been transmitted to, and learned by, a new set of signers. Additional studies on Nicaraguan homesigners using these same materials are needed to determine the impact that sharing a language with a community (i.e., being both a producer *and* receiver of the language) has on how spatial modulations are used to indicate arguments.

2.2.2. Morphosyntax A distinction central to all natural languages is agency. For example, the sentence *Mary closed the door* contains an agent (Mary), but the sentence *The door closed* does not. Several established sign languages vary handshape type in the verb to capture this opposition (Benedicto & Brentari, 2004; Schick, 1987). In complex verbs known as classifier predicates, handshapes representing properties of the object (*object* handshapes) are used in the predicate in sentences with a nonagentive subject (*The door closed* where *closed* is produced with a flat palm handshape). In contrast, handshapes representing how objects are handled (*handling* handshapes) are used in the predicate in sentences with an agentive subject (*Mary closed the door* where *closed* is produced with a fist handshape).

In a study targeting predicates of this type, Goldin-Meadow, Brentari, Coppola, Horton, and Senghas (2015) showed adult Nicaraguan homesigners, cohort 1 NSL signers, and cohort 2 NSL signers events that varied in whether an agent was present (e.g., a person taking a pen off a table) or not (e.g., a pen sitting on a table). They found that all groups used more *object* handshapes when describing no-agent events and more *handling* handshapes when describing agent events. Handshape type in the predicate thus varies as a function of grammatical context, and the variation is systematic. There is thus *morphosyntactic variation* as a function of agency, even in adult homesigners.

2.2.3. Morphophonology Brentari and Eccarius (2010) examined established sign languages to determine whether *object* and *handling* handshapes, which are morphological categories, also exhibit systematic phonological differences, in particular, differences in finger complexity. They grouped handshapes into low, medium, and high complexity forms on the basis of selected fingers (which fingers are moved and contact the body) and joint involvement (whether the knuckles and/or finger joints are flexed). They found, in three unrelated sign languages, that *object* and *handling* morphological categories differed in both phonological parameters—*object* handshapes were high in selected finger complexity and low in joint complexity; conversely, *handling* handshapes were low in selected finger complexity and high in joint complexity.

Brentari et al. (2012) asked whether four adult homesigners in Nicaragua displayed the selected finger complexity patterns found in established sign languages. They found that three of the four homesigners displayed the pattern—high complexity in selected fingers in *object* handshapes, and low complexity in selected fingers in *handling* handshapes. Interestingly, hearing speakers in the United States and Italy, asked to describe the same scenes using gesture without speech (i.e., in silent gesture) did not display this pattern, suggesting that the pattern is not inevitable in the manual modality—it is a pattern that grows out of linguistic structure, not the hand (see section 4 for additional discussion of this point). This pattern has predictive power—these form-form pairings based on an internally consistent, albeit simple, phonology provide the signer with a systematic way to add new forms, which should conform to this pattern. Additional studies on these patterns in the successive cohorts of NSL are needed to determine the impact that transmitting the sign language to a new set of learners has on these morphophonological patterns.

2.3. Grammatical Categories

2.3.1. Distinguishing between Nouns and Verbs The form of a nominal lexical item does not vary with syntactic context in established sign languages; for example, the handshape used to refer to a pen when it is placed on the table, an agent context, is the same as the handshape used when the pen is lying on the table, a no-agent context. In contrast, the form of a classifier predicate *does* vary with these syntactic contexts; for example, the handshape used to describe placing the pen on the table, an agent context, differs systematically from the handshape used to describe the pen lying on the table, a no-agent context. Using an experimental elicitation paradigm, Goldin-Meadow et al. (2015) found that homesigners, as well as cohort 1 and cohort 2 NSL signers, all followed this pattern, thus displaying a distinction between nominal lexical items and predicate classifier constructions that is fundamental to established sign languages. Nominals and predicates are different types of word classes, even in homesign, suggesting that these grammatical categories are available to individuals without either a linguistic community or a language model.

Although a distinction between nouns and verbs has been found in all established sign languages studied thus far, the particular devices used to mark the distinction have also been found to vary crosslinguistically (in ASL, Supalla & Newport, 1978; Russian Sign Language, Kimmelman, 2009; Australian Sign Language, Johnston, 2001; Jordanian Sign Language, Hendriks, 2008; Turkish Sign Language, Kubus, 2008; the Sign Language of the Netherlands, Schreurs, 2006; Italian Sign Language, Pizzuto & Corraza, 1996; Austrian Sign Language, Hunger, 2006; Quebecois Sign Language, Voghel, 2005; see Abner, Flaherty, Stangl, Brentari, & Goldin-Meadow, 2019, for a review). We might then expect the various language users of NSL to differ in the particular devices they use to mark the noun-verb distinction. Abner et al. (2019) used an experimental paradigm to elicit noun-verb pairs in homesigners and successive cohorts of NSL signers. They showed participants vignettes of an object used in an atypical way (e.g., digging with a camera, a vignette designed to elicit a label for "camera"), and the same object used in a typical way (e.g., shooting with the camera, a vignette designed to elicit a label for "shoot"). They discovered that all of the groups (homesigners and each of the successive cohorts of NSL) used two devices found in some established sign languages to mark the noun-verb distinction—sign distribution (i.e., placing verbs in final position) and movement size (i.e., making the movement in the verb with a proximal joint, which results in bigger movements).

However, Abner et al. (2019) also found an interesting progression in a third device, movement repetition (repeating the movement in the noun sign, but not the verb sign), a device also found in some established sign languages. Movement repetition marked the noun-verb distinction in homesign, but *only* in an iconic context (on noun-verb pairs whose associated action was iterable, e.g., on the noun associated with chopping with a machete, a repeatable action, and not on the noun associated with putting on a ring, a one-time action). Movement repetition was not used contrastively at all in cohort 1. In contrast, the later NSL cohorts, each of whom had learned NSL from the preceding cohort, used movement repetition to mark the noun-verb distinction in *both* iconic and noniconic contexts (on noun-verb pairs whose associated action was either iterable or noniterable). The findings suggest that transmitting a communication system from one generation to the next may be essential for an iconic noun-verb marking device to go beyond its iconic routes.

2.3.2. The Notion of Subject Another grammatical category that has been studied in Nicaraguan homesign is the notion of subject. Grammatical subjects do not have a simple semantic correlate. No fixed criteria exist to categorically identify a noun phrase as a subject, but a set of common, multidimensional criteria can be applied across languages (Keenan, 1976). A hallmark of subject noun phrases crosslinguistically is the range of semantic roles they display. The subject of a sentence will likely be an agent (one who performs an action), but many other semantic roles can be the subject. For example, the theme or patient can be a subject (*The door* opened), as can an instrument (*The key* opened the door) or instigator (*The wind* opened the door).

Coppola and Newport (2005) studied three adult homesigners in Nicaragua and found that the primary argument of the utterance tended to be placed in clause-initial position, regardless of the semantic role of the argument. To confirm that this argument was truly a subject and not a discourse topic, Coppola and Newport (2005) varied the contexts so that sometimes the argument was new information and sometimes it was old information. The pattern was unchanged. These findings indicate that the construct grammatical *subject* is available to individuals without either a linguistic community or a language model.

3. Co-speech Gesture: Does It Influence the Structure Found in Homesign?

Homesigners, by definition, are not exposed to a conventional sign language and thus could not have fashioned their communication systems after such a model. They are, however, exposed to the gestures that hearing speakers produce when they talk. Although these co-speech gestures are not typically characterized by linguistic properties (McNeill, 1992), it is possible that hearing individuals alter their gestures when communicating with homesigners, introducing linguistic properties into those gestures. If so, hearing speakers' gestures could serve as a model for the structure in homesign.

To examine this possibility in the U.S. homesigners they studied, Goldin-Meadow and Mylander (1983, 1984) analyzed the gestures that the primary caregiver (the mother in every case) of six homesigners produced when talking to their deaf children. They attempted to look at the gestures through the eyes of a child who cannot hear and thus turned off the sound and coded the mothers' gestures as though they had been produced without speech. Not surprisingly, all six mothers used both pointing and iconic gestures when talking to their children. Moreover, the mothers used pointing and iconic gestures in roughly the same distribution as their homesigning children did. However, the mothers' gestures differed from their children's along a number of important dimensions.

First, mothers produced fewer different *types* of iconic gestures than the homesigners did, and they also used only a small subset of the *particular* iconic gestures that the homesigners used (Goldin-Meadow & Mylander, 1983, 1984).

Second, mothers produced very few strings containing more than one gesture. Like most English-speakers (McNeill, 1992), the mothers tended to produce one gesture per spoken clause, and they rarely combined multiple gestures into a single, motorically uninterrupted unit. Moreover, the very few multigesture strings that the mothers did produce did *not* exhibit the same structural regularities as their children's homesigns did (Goldin-Meadow & Mylander, 1983, 1984). The mothers thus did not appear to have structured their gestures at the sentence level.

Nor did the mothers structure their gestures at the word level. Each mother used her gestures in a more restricted way than her homesigning child did, omitting many of the handshape and motion morphemes that the homesigner produced (or using the ones she did produce more narrowly than the homesigner), and omitting completely a very large number of the handshape-motion combinations that the homesigner produced. Indeed, there was no evidence at all that the mothers' gestures could be broken into meaningful and consistent parts (Goldin-Meadow et al., 1995).

Finally, a mother's iconic gestures were not stable in form and meaning over time, whereas her child's homesigns were. Moreover, the mother did not distinguish between gestures serving a noun role and gestures serving a verb role; the homesigner did (Goldin-Meadow et al., 1994).

Thus, at least in the United States, co-speech gesture is structured differently from homesign and therefore cannot serve as a model for the linguistic structure found in homesign. But there is reason to believe that the gestures hearing speakers produce in Nicaragua could be different. First, Nicaragua is a gesture-rich culture. The hearing culture has a large repertoire of gestural emblems (Coppola, 2015) and hearing individuals have been observed to gesture even when not talking. In addition, the oralist tradition in deaf education (focusing on vocal communication to the exclusion of all manual communication) is not strong in Nicaragua (Polich, 2005), leaving room for hand gestures to be used when interacting with deaf individuals. The gestural raw materials available to Nicaraguan homesigners may therefore be more structured than those found in other cultures. If so, having access to these rich raw materials might have led to more complex homesign systems, which, in turn, could have served as the ideal building blocks for NSL.

To explore this possibility, Flaherty, Hunsicker, and Goldin-Meadow (2019) observed 7 Nicaraguan deaf homesigning children during naturalistic play at home with their hearing family members and compared them to four previously described homesigners in the United States. As expected, there were cross-cultural differences between Nicaragua and the United States in the gestures that hearing family members produced when interacting with homesigners. But, interestingly, there was no evidence that these differences had an effect on homesign. On no measure did Nicaraguan homesigners display more complexity than U.S. homesigners did, despite the fact that Nicaraguan homesigners received more complex gestural input than U.S. homesigners did. Furthermore, as a group, homesigners in both cultures produced more complex constructions (more complex sentences and more complex noun phrases) than their hearing family members did. Importantly, there was no correlation between individual homesigners' complexity scores and their hearing family members' gestural complexity scores.

Do homesigners learn to structure their homesign systems from hearing speakers' co-speech gestures? Probably not, even in Nicaragua—although it may be necessary for homesigners to see hearing people gesturing in communicative situations in order to get the idea that gesture can be appropriated for the purposes of communication. But in terms of how homesigners *structure* their communication, there is no evidence that this structure comes from hearing speakers' co-speech gestures. The hearing speakers' gestures do not have structure when looked at with tools used to describe homesign (although they are structured when looked at with tools used to describe co-speech gesture, that is, when they are described in relation to speech, which homesigners cannot hear; see, for example, Goldin-Meadow, 2003b).

4. Silent Gesture: An Experimental Approach to Language Creation

It is difficult to create spoken forms that can be understood on the spot. But the manual modality, in large part because of its iconic potential, allows us to construct forms that are immediately understood, thus requiring essentially no time to develop. We can, as a result, ask hearing individuals to use gesture without speech to describe events, to create *silent gestures* that are formed in the moment. Silent gestures have been found to be structured differently from co-speech gestures and, in fact, have linguistic properties not found in co-speech gestures (Goldin-Meadow, McNeill, & Singleton, 1996). Silent gestures have been mined primarily for word order in speakers of a wide variety of languages—English, Chinese, Turkish, Spanish, Italian, Hebrew, Japanese, and Korean (Gibson et al., 2013; Gershkoff-Stowe & Goldin-Meadow, 2002; Goldin-Meadow, So, Özyürek, & Mylander, 2008; Hall, Ferreira, & Mayberry, 2013; Langus & Nespor, 2010; Meir, Lifshitz, Ilkbasaran, & Padden, 2010). Despite the fact that the canonical word orders for simple transitive sentences in the languages spoken in these countries differ, the gesture order used by the silent gesturers to describe a prototypical event encoded in a transitive sentence (i.e., an animate acting on an inanimate) is identical in all countries. Silent gestures thus do not appear to be a mere translation into gesture of the language that the gesturer routinely speaks, but reflect the construction of new forms on the spot (see also Özçalışkan, Lucero, & Goldin-Meadow, 2016).

Thus, when asked to use gesture on its own, silent gesturers do not use gesture as they typically do when they speak, but rather transform their gestures into a system that has some linguistic structure. Importantly, however, silent gesture lacks many of the linguistic properties found in established sign languages. Indeed, it even lacks a number of properties that are found in homesign (see Goldin-Meadow, 2015), despite the fact that silent gesturers are all fluent speakers of a conventional language, whereas homesigners do not have access to a usable model for language. But homesigners have been using their gestures as a primary communication system for many years, and silent gesturers create their gestures on the spot. Silent gestures may reflect heuristics for communicating information in a structured manner but, because they are developed on the fly, they do not function as part of a *linguistic system*. As a result, we may be able to use comparisons with silent gesture to determine whether a property found in homesign is part of a linguistic system. A property found in homesign but not in silent gesture may be a good candidate for a systemic linguistic property, as opposed to a one-off heuristic for communicating information.

5. Conclusion

The manual modality is an ideal medium within which to explore the properties of language that reflect human predispositions to structure communication simply because (unlike current-day spoken languages) language can be developed de novo in the manual modality. We find that children whose hearing losses prevent them from acquiring spoken language, and whose hearing parents have not exposed them to a conventional sign language, nevertheless communicate and turn to the manual modality to do so. The *homesigns* that these individuals generate have many of the structural properties found in language—it is these properties, which have been detailed in this chapter, that humans themselves bring to language, the resilient properties of language.

But, homesign does not contain all of the properties found in natural language. The chapter capitalizes on a situation in Nicaragua that brought homesigners together for the first time, leading to the creation of a new sign language, NSL. We were thus able to follow homesign as it evolves from isolated gesture systems used by deaf individuals to a shared system used by a community of signers and, in the process, explore the conditions needed for homesign to become a full-fledged language. We have also been able to explore potential precursors to homesign—the spontaneous gestures that hearing individuals produce when they talk. We find, however, that the linguistic structures that characterize homesign cannot be traced back to the signer's hearing communication partner's co-speech gestures, but are instead the homesigner's own invention.

The chapter ends by exploring an experimental approach to language creation—asking hearing individuals to describe events using their hands and not their mouths. These silent gestures, which are created on-the-spot, take on some, but not all, of the properties of natural language. Importantly, silent gestures do not even contain all of the linguistic properties found in homesign. Silent gesture thus provides a backdrop for assessing which properties of homesign are likely to be part of a rich, linguistic system developed over a period of years to solve the problem of human communication.

Acknowledgments

This research was supported by grant no. R01 DC00491 from the National Institute on Deafness and Other Communication Disorders and grant no. BCS 1654154 from the National Science Foundation. Thanks to my many collaborators for their invaluable help in uncovering the structure of homesign and to the children and their families for welcoming us into their homes.

NOTE

1. Carrigan and Coppola (2017) found that the hearing communication partners who interact routinely with adult homesigners in Nicaragua do *not* understand the homesigners' gestures well—in fact, ASL signers, who did not know the Nicaraguan homesigners at all, gleaned more information from the homesigners' gestures than did the homesigners' long-term communication partners.

REFERENCES

Abner, N. (2012). There once was a verb: The predicative core of possessive and nominalization structures in American Sign Language. (Unpublished doctoral dissertation) . https://linguistics.ucla.edu/research/phd-dissertations/

Abner, N., Flaherty, M., Stangl, K., Brentari, D., & Goldin-Meadow, S. (2019). The noun-verb distinction in established and emergent languages. *Language*, in press.

Allen, S. (1998). Categories within the verb category: Learning the causative in Inuktitut. *Linguistics, 36*(4), 633–677.

Applebaum, L, Coppola, M., & Goldin-Meadow, S. (2014). Prosody in a communication system developed without a language model. *Sign Language and Linguistics, 17*(2), 181–212.

Aronoff, M., Meir, I., Padden, C., & Sandler, W. (2003). Classifier constructions and morphology in two sign languages. In K. Emmorey (Ed.), *Perspectives on classifier constructions in sign languages* (pp. 53–84). Mahwah, NJ: Erlbaum.

Bellugi, U., & Studdert-Kennedy, M. (Eds.) (1980). *Signed and spoken language: Biological constraints on linguistic form.* Deerfield Beach, FL: Verlag Chemie.

Benedicto E., & Brentari, D. (2004). Where did all the arguments go?: Argument-changing properties of classifiers in ASL. *Natural Language and Linguistic Theory, 22*, 1–68.

Berman, R. A. (1993). Marking of verb transitivity by Hebrew-speaking children. *Journal of Child Language, 20*(3), 641–669.

Bonvillian, J. D., Orlansky, M. D., & Novack, L. L. (1983). Developmental milestones: Sign language acquisition and motor development. *Child Development, 54*, 1435–1445.

Bowerman, M. (1974). Learning the structure of causative verbs: A study in the relationship of cognitive, semantic and syntactic development. *Papers and Reports on Child Language Development, 8*, 142–178.

Brentari, D. (1998). *A prosodic model of sign language phonology.* Cambridge, MA: MIT Press.

Brentari, D., Coppola, M., Mazzoni, L., & Goldin-Meadow, S. (2012). When does a system become phonological? Handshape production in gesturers, signers, and homesigners. *Natural Language and Linguistic Theory, 30*, 1–31.

Brentari, D., & Eccarius, P. (2010). Handshape contrast in sign languages. In D. Brentari (Ed.), *Sign languages* (pp. 284–311). Cambridge: Cambridge University Press.

Brentari, D., Falk, J., & Wolford, G. (2015). The acquisition of prosody in American Sign Language. *Language, 91*(3), e144–e168.

Brown, R., (1973). *A first language.* Cambridge, MA: Harvard University Press.

Carrigan, E. M., & Coppola, M. (2017). Successful communication does not drive language development: Evidence from adult homesign. *Cognition, 158*, 10–27.

Coppola, M. (2015). *Where do words come from? Social interactions and conventionalization of the lexicon in an emerging sign language in Nicaragua.* Paper presented at the First International Workshop on Emerging Sign Languages in the Americas, Mexico City.

Coppola, M., & Newport, E. L. (2005). Grammatical "subjects" in home sign: Abstract linguistic structure in adult primary gesture systems without linguistic input. *Proceedings of the National Academy of Sciences, 102*, 19249–19253.

Coppola, M., & Senghas, A. (2010). Deixis in an emerging sign language. In D. Brentari (Ed.), *Sign languages: A Cambridge language survey* (pp. 543–569). Cambridge: Cambridge University Press.

Emmorey, K. (Ed.). (2003). *Perspectives on classifier constructions in sign language.* Mahwah, NJ: Erlbaum.

Feldman, H., Goldin-Meadow, S., & Gleitman, L. R. (1978). Beyond Herodotus: The creation of language by linguistically deprived deaf children. In A. Lock (Ed.), *Action, symbol, and gesture: The emergence of language* (pp. 351–414). New York: Academic Press.

Fenlon, J., Schembri, A., & Cormier, K. (2018). Modification of indicating verbs in British Sign Language: A corpus-based study. *Language 94*(1), 84–118.

Figueira, R. A. (1984). On the development of the expression of causativity: a syntactic hypothesis. *Journal of Child language, 11*(1), 109-127.

Fischer, S. (1973). Two processes of reduplication in American Sign Language. *Foundations of Language, 9*, 469-480.

Flaherty, M., Hunsicker, D., & Goldin-Meadow, S. (2019). *Structural biases that children bring to language learning: A cross-cultural look at gestural input to homesign.* Revised manuscript, under review.

Franklin, A., Giannakidou, A., & Goldin-Meadow, S. (2011). Negation, questions, and structure building in a homesign system. *Cognition, 118*, 398–416.

Gershkoff-Stowe, L., & Goldin-Meadow, S. (2002). Is there a natural order for expressing semantic relations? *Cognitive Psychology, 45*(3), 375–412.

Gibson, E., Piantadosi, S. T., Brink, K., Bergen, L., Lim, E., & Saxe, R. (2013). A noisy-channel account of crosslinguistic word order variation. *Psychological Science, 24*(7), 1079–1088.

Goldin-Meadow, S. (1982). The resilience of recursion: A study of a communication system developed without a conventional language model. In E. Wanner & L. R. Gleitman

(Eds.), *Language acquisition: The state of the art* (pp. 51–77). New York: Cambridge University Press.

Goldin-Meadow, S. (1985). Language development under atypical learning conditions: Replication and implications of a study of deaf children of hearing parents. In K. Nelson (Ed.), *Children's language* (Vol. 5, pp. 197–245). Hillsdale, NJ: Lawrence Erlbaum and Associates.

Goldin-Meadow, S. (1993). When does gesture become language? A study of gesture used as a primary communication system by deaf children of hearing parents. In K. R. Gibson & T. Ingold (Eds.), *Tools, language and cognition in human evolution* (pp. 63–85). New York: Cambridge University Press.

Goldin-Meadow, S. (2003a). *The resilience of language.* Philadelphia: Taylor and Francis.

Goldin-Meadow, S. (2003b). *Hearing gesture: How our hands help us think.* Cambridge: Harvard University Press.

Goldin-Meadow, S. (2015). The impact of time on predicate forms in the manual modality: Signers, homesigners, and silent gesturers. *TopiCS, 7,* 169–184.

Goldin-Meadow, S., Brentari, D., Coppola, M., Horton, L., & Senghas, A. (2015). Watching language grow in the manual modality: Nominals, predicates, and handshapes. *Cognition, 135,* 381–395.

Goldin-Meadow, S., Butcher, C., Mylander, C., & Dodge, M. (1994). Nouns and verbs in a self-styled gesture system: What's in a name? *Cognitive Psychology, 27,* 259–319.

Goldin-Meadow, S., McNeill, D., & Singleton, J. (1996). Silence is liberating: Removing the handcuffs on grammatical expression in the manual modality. *Psychological Review, 103*(1), 34–55.

Goldin-Meadow, S., & Morford, M. (1985). Gesture in early child language: Studies of deaf and hearing children. *Merrill-Palmer Quarterly, 31,* 145–176.

Goldin-Meadow, S., & Mylander, C. (1983). Gestural communication in deaf children: The non-effects of parental input on language development. *Science, 221,* 372–374.

Goldin-Meadow, S., & Mylander, C. (1984). Gestural communication in deaf children: The effects and non-effects of parental input on early language development. *Monographs of the Society for Research in Child Development, 49,* 1–121.

Goldin-Meadow, S., & Mylander, C. (1998). Spontaneous sign systems created by deaf children in two cultures. *Nature, 91,* 279–281.

Goldin-Meadow, S., Mylander, C., & Butcher, C. (1995). The resilience of combinatorial structure at the word level: Morphology in self-styled gesture systems. *Cognition, 56,* 195–262.

Goldin-Meadow, S., Mylander, C., & Franklin, A. (2007). How children make language out of gesture: Morphological structure in gesture systems developed by American and Chinese deaf children. *Cognitive Psychology, 55,* 87–135.

Goldin-Meadow, S., So, W.-C., Özyürek, A., & Mylander, C. (2008). The natural order of events: How speakers of different languages represent events nonverbally. *Proceedings of the National Academy of Science, 105,* 9163–9168.

Greenfield, P. M., & Savage-Rumbaugh, E. S. (1991). Imitation, grammatical development, and the invention of protogrammar by an ape. In N. A. Krasnegor, D. M. Rumbaugh, R. L. Schiefelbusch, & M. Studdert-Kennedy (Eds.), *Biological and behavioral determinants of language development* (pp. 235–262). Hillsdale, NJ: Earlbaum Associates.

Hall, M. L., Ferreira, V. S., & Mayberry, R. I. (2013). Investigating constituent order change with elicited pantomime: A functional account of SVO emergence, *Cognitive Science, 38,* 943–972.

Hendriks, B. (2008). *Jordanian Sign Language: Aspects of grammar from a cross-linguistic perspective* (LOT nr. 193). Utrecht, Netherlands: Landelijke Onderzoekschool Taalwetenschap.

Hunger, B. (2006). Noun/verb pairs in Austrian Sign Language (OGS). *Sign Language and Linguistics, 9*(1–2), 1–2.

Hunsicker, D., & Goldin-Meadow, S. (2012). Hierarchical structure in a self-created communication system: Building nominal constituents in homesign. *Language, 88,* 732–763.

Hunsicker, D., & Goldin-Meadow, S. (2013). How handshape type can distinguish between nouns and verbs in homesign. *Gesture, 13,* 354–376.

Johnston, T. (2001). Nouns and verbs in Australian Sign Language: An open and shut case? *Journal of Deaf Studies and Deaf Education, 6*(4), 235–257.

Keenan, E. (1976). Towards a universal definition of subject. In C. N. Li (Ed.), *Subject and topic.* New York: Academic Press.

Kimmelman, V. (2009). Parts of speech in Russian Sign Language: The role of iconicity and economy. *Sign Language and Linguistics, 12*(2), 161–186.

Klima, E., & Bellugi, U. (1979). *The signs of language.* Cambridge, MA: Harvard University Press.

Kubus, O. (2008). *An analysis of Turkish Sign Language (TID) phonology and morphology* (Unpublished master's thesis). Middle East Technical University, Turkey.

Langus, A., & Nespor, M. (2010). Cognitive systems struggling for word order. *Cognitive Psychology, 60,* 291–318.

Lenneberg, E. H. (1964). Capacity for language acquisition. In J. A. Fodor & J. J. Katz (Eds.), *The structure of language: Readings in the philosophy of language* (pp. 579–603). Upper Saddle River, NJ: Prentice Hall.

Liddell, S. (1980). *American Sign Language syntax.* The Hague, the Netherlands: Mouton.

Liddell, S. (2003). *Grammar, gesture, and meaning in American Sign Language.* Cambridge: Cambridge University Press.

Lillo-Martin, D. (1991). *Universal grammar and American Sign Language: Setting the null argument parameters.* Dordrecht, the Netherlands: Kluwer.

Lillo-Martin, D. (1999). Modality effects and modularity in language acquisition: The acquisition of American Sign Language. In W. C. Ritchie & T. K. Bhatia (Eds.), *The handbook of child language acquisition* (pp. 531–567). New York: Academic Press.

Lin, H.-L., & Tsay, J, S. (2008). Acquiring causatives in Taiwan Southern Min. *Journal of Child Language, 35*(2), 467–487.

McNeill, D. (1992). *Hand and mind: What gestures reveal about thought.* Chicago: University of Chicago Press.

Meir, I. (2002). Across-modality perspective on verb agreement. *Natural Language and Linguistic Theory, 20,* 413–450.

Meir, I., Lifshitz, A., Ilkbasaran, D., & Padden, C. (2010). The interaction of animacy and word order in human languages: A study of strategies in a novel communication task. In A. D. M. Smith, M. Schouwstra, B. de Boer, & K. Smith (Eds.), *Proceedings of the Eighth Evolution of Language Conference* (pp. 455–456). Singapore: World Scientific Publishing Co.

Moores, D. F. (1974). Nonvocal systems of verbal behavior. In R. L. Schiefelbusch & L. L. Lloyd (Eds.), *Language perspectives: Acquisition, retardation, and intervention.* Baltimore: University Park Press.

Morford, J. P., & Goldin-Meadow, S. (1997). From here to there and now to then: The development of displaced reference in homesign and English. *Child Development, 68,* 420–435.

Murasugi, K., Hashimoto, T., & Fuji, C. (2007). VP-shell analysis for the acquisition of Japanese intransitive verbs, transitive verbs, and causatives. *Linguistics, 45,* 615.

Neidle, C., Kegl, J., MacLaughlin, D., Bahan, B., & Lee, R. G. (2000). *The syntax of American Sign Language.* Cambridge, MA: MIT Press.

Nespor, M., & Sandler, W. (1999). Prosody in Israeli Sign Language. *Language and Speech, 42,* 143–176.

Newport, E. L., & Meier, R. P. (1985). The acquisition of American Sign Language. In D. I. Slobin (Ed.), *The crosslinguistic study of language acquisition, Vol. 1. The data* (pp. 881–938). Hillsdale, NJ: Erlbaum.

Nyst, V. A. S. (2012). Shared sign languages. In M. Pfau, M. Steinbach, & B. Woll (Eds.), *Sign language* (pp. 552–574). Berlin: Mouton de Gruyter.

Özçalışkan, S., Lucero, C., & Goldin-Meadow, S. (2016). Does language shape silent gesture? *Cognition, 148,* 10–18.

Padden, C. (1983). Interaction of morphology and syntax in American Sign Language (Unpublished doctoral dissertation). University of California, San Diego.

Padden, C. (1990). The relation between space and grammar in ASL verb morphology. In C. Lucas (Ed.), *Sign language research: Theoretical issues* (pp. 118–132), Washington, DC: Gallaudet University Press.

Pizzuto, E., & Corazza, S. (1996). Noun morphology in Italian Sign Language (LIS). *Lingua, 98*(1), 169–196.

Polich, L. (2005). *The emergence of the deaf community in Nicaragua.* Washington, DC: Gallaudet University.

Rissman, L., & Goldin-Meadow, S. (2017). The development of causal structure without a language model. *Language Learning and Development, 13*(3), 286–299.

Sandler, W. (1989). *Phonological representation of the sign.* Dordrecht, the Netherlands: Foris.

Schick, B. (1987). *The acquisition of classifier predicates in American Sign Language* (Order No. 8807670). Available from ProQuest Dissertations & Theses Global (303485642). Retrieved from https://search.proquest.com/docview/303485642?accountid=105521.

Schick, B. (1990). Classifier predicates in American Sign Language. *International Journal of Sign Linguistics, 1*(1), 15–40.

Schlenker, P. (2018). Visible meaning: Sign language and the foundations of semantics. *Theoretical Linguistics, 44,* 123–208.

Schlesinger, H. (1978). The acquisition of bimodal language. In I. Schlesinger & L. Namir (Eds.), *Sign language of the deaf: Psychological, linguistic, and sociological perspectives* (pp. 57–93). New York: Academic Press.

Schreurs, L. (2006). *The distinction between formally and semantically related noun-verb pairs in Sign Language of the Netherlands (NGT)* (Unpublished master's thesis). University of Amsterdam.

Senghas, A., & Coppola, M. (2001). Children creating language: How Nicaraguan Sign Language acquired a spatial grammar. *Psychological Science, 12,* 323–328.

Singleton, J. L., Morford, J. P., & Goldin-Meadow, S. (1993). Once is not enough: Standards of well-formedness in manual communication created over three different timespans, *Language, 69,* 683–715.

Supalla, T. (1986). *The classifier system in American Sign Language.* In C. Craig (Ed.), *Noun classes and categorization* (pp. 181–215). Philadelphia: John Benjamins.

Supalla, T., & Newport, E. L. (1978). How many seats in a chair? The derivation of nouns and verbs in American Sign Language. In P. Siple (Ed.), *Understanding language through sign language research* (pp. 91–132). New York: Academic Press.

Tervoort, B. T. (1961). Esoteric symbolism in the communication behavior of young deaf children. *American Annals of the Deaf, 106,* 436–480.

Thompson, R. L., Vinson, D. P., & Vigliocco, G. (2009). The link between form and meaning in American sign language: Lexical processing effects. *Journal of Experimental Psychology: Language, Memory, and Cognition, 35,* 550–557.

Thompson, R. L., Vinson, D. P., & Vigliocco, G. (2010). The link between form and meaning in British sign language: Lexical processing effects in a phonological decision task. *Learning, Memory, and Cognition, 36,* 1017–1027.

Thompson, S. A. (1988). A discourse approach to the crosslinguistic category "adjective". In John A. Hawkins (Ed.), *Explaining language universals* (pp. 167–185). Cambridge, MA: Basil Blackwell.

Van Deusen-Phillips, S. B., Goldin-Meadow, S., & Miller, P. (2001). Enacting stories, seeing worlds: Similarities and differences in the cross-cultural narrative development of linguistically isolated deaf children. *Human Development, 44,* 311–336.

Voghel, A. (2005). Phonologically identical noun-verb pairs in Quebec Sign Language (LSQ): Form and context. *Toronto Working Papers in Linguistics, 25,* 68–75.

17 Depicting in Communication

HERBERT H. CLARK

For most people—scholars and dabblers alike—human communication is language pure and simple. It is English or French or Japanese as we find it in conversation, on television, in books and newspapers. But when people are face-to-face, they also use methods that are *not* represented in words, phrases, and sentences. They gesture with their hands. They nod. They grimace. They show things to each other. Actions like these, typically combined with language, are ubiquitous and yet neglected by both scholars and dabblers. This chapter is about a feature of many of these actions: depicting.

Imagine two university students, Ann and Burton, assembling a TV stand with only a picture of the final product as a guide:

(1) Ann Okay, so let's put the bottom on first.
Burton We'll just- Okay, so, (*holding up a piece of wood*) this is the bottom.
Ann That's the, yeah that's the lower part.[1]

When Ann tells Burton, "*this* is the bottom," she holds up a piece of wood for Burton, and he agrees, "yeah that's the lower part." Ann and Burton rely not only on *describing things* (with their choice of words), but on *indicating things* (by holding up a piece of wood, and by nodding at it).

In conversation, people also rely on a third method of communication—*depicting* (Clark, 2016). A bit later Ann and Burton have this exchange:

(2) Ann Does (*holding a piece of wood*) this go the other way. Oh, we can turn it (*turning the piece 90°*) *this way*, yeah.
Burton Yeah, turn it *that way*.

As Ann says "this way," she reorients a wooden piece to *depict* the way she thinks it ought to go, and when Burton says "that way," he refers to the same depiction. Face-to-face, people depict with their hands (so-called iconic gestures; McNeill, 1992), voices (as in quotations), faces, and entire bodies (as in full-scale demonstrations). Even on the telephone, they depict with their voices. And the more lively the talk, the more common the depictions.

Many depictions are created de novo, as in (2), but others are made with prefabricated parts. A witness at an American trial, for example, could depict the snort of a pig in two ways:

(3) And then I heard "(*produces a spontaneous vocal imitation of a pig snorting*)."
(4) And then I heard "oink-oink."

In (3) the witness created a depiction de novo by mimicking a pig snorting. She could have used the same imitation in speaking French. This is a *free* depiction. In (4) she built her depiction on *oink-oink*, an onomatopoeic word that is conventionally used for depicting pig snorts in English. She would have used *groin-groin* in French and *bu-bu* in Japanese. Words like these belong to codes—systems of conventional signs such as English, French, or Japanese. I will call depictions like the one in (4) *codified* depictions.

This chapter is about depicting as a method of communication. I begin with what it means to depict something. I then turn to free depictions and finish up with codified depictions.

1. Icons and Iconicity

Depictions in language use are traditionally described as *iconic*—as possessing *iconicity*. The modern concept of icon was developed by the philosopher Charles Sanders Peirce (1839–1914) in his theory of signs (Peirce, 1932, 1974). For Peirce, a *sign* is a physical manifestation that signifies an *object* via an interpretation. Hans Holbein's portrait of Henry VIII, for example, is a sign (the physical painting) that signifies its object (the historical Henry VIII) via an interpretation (the painting's resemblance to the historical Henry VIII). Signs, in turn, come in three main types—symbols, indexes, and icons.

1.1. Symbols, Indexes, and Icons *Symbol.* A symbol is a sign that signifies its object *by rule*. The word *dog* is a symbol because it signifies a particular species of animal by a rule of English; the same species would be signified in French by *chien*, in Dutch by *hond*, and in Spanish by *perro*. In certain cultures, crossing one's fingers signifies good luck by rule, so it is also a symbol. Gestures that signify by rule are called *emblems* (Ekman & Friesen, 1969; Kendon, 2004; McNeill, 1992).

Most symbols belong to systems of conventions, codes that are shared by cultural communities. English, French, and Japanese are language codes, but so are the specialized jargons used by lawyers, obstetricians, and gold miners (Clark, 1998). Emblems also belong to codes. There are systematic conventions for the emblems used by the French (Calbris, 2011), by American football referees, and by traders on the Chicago Mercantile Exchange.

Index. An index is a sign that signifies its object by means of a *physical connection to the object.* A roadside arrow (labeled "San Francisco") signifies its object (the direction to San Francisco) by a vector that physically connects the arrow to the direction. Weathervanes index the direction of the wind, and smoke indexes the fire that produced it.

Icon. An icon is different again. It signifies its object via a *perceptual resemblance* to the object. The resemblance may be visual, as in Holbein's painting of Henry VIII, or auditory, as in a mechanical duck call, or a combination of senses.

Many signs, Peirce noted, are *mixed* signs—combinations of two or three of these types. When Ann tells Burton "I'll hold this," her utterance of "I" counts as both a symbol, because it signifies "the person speaking now" by a rule of English, and as an index, because it locates Ann as the source of the speech. The same holds for *you, here, now, this way,* and other such expressions. As for "oink-oink" in (4), it is a symbol of English, but it is *felt* to be an icon with an auditory resemblance to a pig snort. So it, too, is a mixed sign.

1.2. Images, Diagrams, and Metaphors Icons, according to Peirce, divide further into *images, diagrams,* and *metaphors.* The most basic are imagistic icons, such as Holbein's portrait of Henry VIII. By "image," Peirce apparently meant a *physical analog* based on any of the senses—vision, audition, taste, smell, touch, proprioception.

Diagrams, in contrast, are "not pure icons." They "represent the relations of the parts of one thing by analogous relations in their own parts... [M]any diagrams resemble their objects not at all in looks; it is only in respect to the relations of their parts that their likeness consists" (Peirce, 1902/1955, pp. 105–107; see Waugh, 1994). A map, for example, is a diagram of a territory. It can range from detailed (such as a topographic map of the Grand Canyon) to schematic (such as the map of the London Underground). Algebra equations, line graphs, and chess boards are all diagrams.

Metaphors are even more abstract. The verb *plow* in "the canoe *plowed* through the lake" signifies an action (what the canoe did in moving through the lake) by naming an action in an analogous scene (what a plow would do in moving through a field). We understand the metaphor by interpreting the symbols *canoe, plow, through,* and *lake,* imagining the two scenes, and adjusting the lake scene to match the field scene in the appropriate way.

As interesting as diagrams and metaphors are, I will focus on imagistic icons. Understanding them will be hard enough.

1.3. Describing, Indicating, and Depicting Although Peirce spoke of *signs,* he said little about *signals*—the use of signs in communicative acts. In the spirit of Peirce, a signal would be the display of one or more signs for others to interpret. In (1), when Ann tells Burton, "Yeah that's the lower part," she *displays* a complex sign that contains *yeah, lower,* and *part* as pure symbols, and *that, is,* and *the* as combined symbols and indexes.

The idea, then, is that there are three basic *methods* of communication, one for symbols, one for indexes, and one for icons (Clark, 1996, 2016):

a. *Describing* is the communicative display of symbols.
b. *Indicating* is the communicative display of indexes.
c. *Depicting* is the communicative display of icons.

People typically use the three methods in combination—in *composite* signals. In (1), when Ann says, "Yeah that's the lower part," she is using a composite of describing and indicating. And in (2), when she says, "Oh, we can turn it (*turning the piece 90°*) this way, yeah," she is using a composite of describing, indicating, *and* depicting.

2. Using Depictions

Depictions fall into two broad categories—*exhibited* and *performed.* Exhibited depictions are artifacts that people exhibit to others. These include paintings, sketches, blueprints, photographs, statues, movies, animated cartoons, TV sitcoms, and radio plays. Performed depictions, in contrast, are actions that people perform for others in person—with their hands, arms, head, face, eyes, voice, and body, with or without physical props. These include depictive manual gestures, depictive facial expressions, quotations of all kinds, full-scale demonstrations, and make-believe play (Clark, 2016).

What is the difference? Exhibited depictions are created at one place and time and exhibited at another. The exhibitors may be different from the creators and have different goals.[2] Performed depictions, in contrast, are created *and* displayed by a single person, sometimes in collaboration with others, at a *single* place and time in a *single* set of actions with a *single* set of goals. It is performed depictions that are integral to language use.

2.1. Varieties of Depicting Performed depictions come in many shapes, sizes, and uses. As illustration, consider these examples:

(5) [The speaker had already described a "whirligig" (a stick with a propeller on top) and how to launch it by twirling the stick between the palms. Now he was describing how he caught it. He did all this without a whirligig.] And you just (*using only his fingers, he depicts a very tiny twirling action, followed smoothly by a life-sized enactment of catching the whirligig*) twirl it and catch it, that's the idea. (Gerwing & Bavelas, 2004, p. 175; see Bavelas & Gerwing, 2007)

(6) [The speaker was narrating a story about being a passenger in a friend's truck.] He drove, he drove too close and we had one front wheel off the cliff (*displays fear by raising her eyebrows and widening her eyes*) my side, and the truck was going over the edge. (Bavelas & Gerwing, 2007)

(7) [A caller was speaking to the hosts of the radio program *Car Talk*.] My son and I noticed that the car has a-uh, a worry noise, kind of a "(*low growling noise*)" like when you used to put a baseball card in a-uh, the spokes of your bike. (Clark, 2016, p. 325)

(8) [MB was discussing a piano piece as played by Bela Bartok.] He doesn't play (*performing four measures on the piano incorrectly, while singing softly*), but rather he plays—and he does it better than I do—(*performing the same four measures correctly, while singing softly*). (Ibid.)

As these examples suggest, depictions vary in the body parts used—hands, face, head, voice, or entire body. They vary in how they are created: either by *enacting* what someone did (catch a whirligig, show fear, lift someone's chin, or play the piano), or by *realizing a prop* for an object or event (the car noise). They vary in whether they include *physical* props (the piano), *imaginary* props (the whirligig), or no props at all. And they vary in the space they are created in. Examples (5), (6), and (8) are performed in a three-dimensional (3D) space, and example (7) in a 0D space. The challenge is to account for these and other variations.

Every performed depiction has both a *content* and a *function*. In (5) through (8), the depictions range in content from a person playing with a whirligig to Bartok playing the piano. They also range in function—in how they are used.

2.2. Discourse Functions In everyday discourse, depictions are found *embedded* as syntactic elements of linguistic utterances, *indexed* by phrases within linguistic utterances, *adjoined* to words or phrases in linguistic utterances, and used as separate, *independent* contributions to the discourse.

Embedded depictions. An embedded depiction is one that serves *in place of* a conventional noun, verb, adjective, noun phrase, and so on, in a spoken utterance. In (7), the caller depicted a car noise in place of a noun (N) in "the car has a worry noise, kind of a N." And in (8), MB depicted two ways of playing in place of the two noun phrases (NP) in "he doesn't play NP but rather he plays NP."[3]

Indexed depictions. An indexed depiction is one that is required by an expression such as *like this* or *this long*. When I tell a friend, "I caught a fish (*holding my hands 45 cm apart*) this long," I use "this long" to *index* my gesture. In (2), Ann used "this way" to index her depiction of the way the block should be turned.

Adjunct depictions. An adjunct depiction is one that is timed to accompany a word, phrase, or other stretch of speech as a *nonrestrictive modifier*. The stretch of speech it modifies—usually a word, phrase, or clause—is called its *affiliate*. In (3), the speaker depicted two scenes, first the twirling of the whirligig and then himself catching it. He timed the twirling motion to overlap "twirl it" and the catching action to overlap "catch it."

Embedded and indexed depictions are mandatory, but adjunct depictions are not. The sentence "the car has a worry noise, kind of a '(*low growling noise*)'" would have been incomplete without the low growling, and "I caught a fish (*holding my hands 45 cm apart*) this long" would have been incomplete without the gesture. And yet the sentence "you just (*gestures depicting twirling and then catching*) twirl it and catch it" would have made sense without the adjunct depictions. It isn't that adjunct depictions aren't informative. Like all nonrestrictive modifiers, they contribute content to the ongoing discourse.[4]

Independent depictions. An independent depiction is one that makes a separate contribution to the discourse. Here is an example:

(9) [Morgan was answering an interviewer's question about how she happened to audition for a role on MTV.] My friend Gary really talked me into it. "Come on. We gotta audition together. We gotta be on it together." And I was kind of like, "Whatever," you know. And um ... (Clark, 2016, p. 326)

When Morgan said, "come on ... together," her depiction was a separate contribution to her anecdote. Quotations are often embedded in expressions like "he said NP" or "he was like NP." But Morgan produced "come on…together" without embedding, indexing, or adjoining it to another utterance. In a collection of oral narratives studied by Tannen (1986), 26% of the quotations were independent depictions.

The same depiction can be used for any of these functions—if the conditions are right. Here are four demonstrations of piano playing from a lecture by a well-known pianist I am calling MB, one for each of the four functions:

(10) *Embedded.* [MB was discussing a piano piece played by Bela Bartok.] He doesn't play (*demonstrates four measures on the piano while singing*), but rather he plays—and he does it better than I do—(*demonstrates the same four measures while singing, but differently*).

(11) *Indexed.* Now what would that say about a piece like this: (*MB plays four measures of a sonata*).

(12) *Adjunct.* All of us learned (*playing a scale downward on the piano*) to play scales that we must go over, and it must be inaudible.

(13) *Independent.* For example, (*MB plays two measures of Mozart on the piano*).

Here are iconic gestures used with the same four functions:

(14) *Embedded.* [MB was lecturing about piano playing] It was very easy to tell the difference because the (*cringing his face, arms, shoulder briefly*) was gone out of it, you know.

(15) *Indexed.* [MB was speaking about reading a phrase in a score.] He was teaching a master class (*drawing an upward arc in the air with R index finger*) and here was one bow by Mozart, (*bowing with his R arm back and forth on an imaginary violin*) and the violinist was going back and forth.

(16) *Adjunct.* [MB was describing how to perform music.] (*sliding curled L hand up and then down the neck of an invisible violin*). We can slide (*pauses L hand and raises R hand to touch throat*) on the voice (*continues with L hand*) or violin. We (*slides L hand on neck of violin in a rapid vibrato*) can vibrate. That's all we've got.

(17) *Independent.* [Speaker was narrating *Little Red Riding Hood*]. Then the wolf walked up to grandmother's house. [*Knocks three times on an imaginary door.*]

The function of a depiction (how it is used) is largely independent of its content (what it depicts and how). Largely, but not entirely. I cannot overlap a spoken quotation, such as "How stupid," with the spoken phrase I want it to modify. Nor can I depict a raised voice without an utterance to depict it in.

3. Content of Depictions

Depictions, in Peirce's account, are signs that signify objects by perceptual resemblance to those objects. For Peirce, a sign and the object it signifies are distinct entities. The artist René Magritte illustrated this principle in his painting of a briarwood pipe over the words *Ceci n'est pas une pipe* 'This is not a pipe'. His painting (a sign) was a 2D canvas surface covered with paint, yet what it depicted (its object) was a 3D pipe made of briarwood. The *pas-une-pipe principle*, as I will call it, is this: A depiction is *not* what it depicts. Alfred Korzybski (1948) put it another way: "A map is not the territory it represents." This principle applies not only to exhibited depictions—paintings and maps—but to performed depictions.

3.1. Depictions as Physical Analogs On late night television, the actress Chloe Sevigny was telling TV host Conan O'Brian how a cinematographer had asked her to hold her chin up during a scene he was shooting. Here is part of her narrative (Clark, 2016, p. 332):

(18) Before every take, the (*directing her hand, palm down, sagittally toward her throat, patting her chin from underneath, and lifting her chin with each pat*) cinematographer would be like, "Lift your chin." So when I watched the movie, I'm…

In the middle of her utterance, Sevigny depicted the cinematographer putting his hand under her chin and saying "Lift your chin." To do that, she dramatized, or staged, a scene in the here-and-now that her audience was to use in imagining a scene from her past. The first, or *proximal*, scene was the depiction (what Sevigny displayed to her audience). The second, or *distal*, scene was what it depicted (what the cinematographer had done). True to Magritte, the depiction was not what it depicted.

For someone's action to be a depiction, it must be construed as a *physical analog* of what it depicts (for a model of analogy, see Gentner, 1983). The depiction must have an analog mapping into the distal scene (the arrow means "is a physical analog of"):

proximal scene	→	distal scene
(the depiction)		(what is depicted)

Table 17.1 lists the proximal and distal scenes for examples (5) through (9) and (18). The whirligig scenes were visual analogs; the car noise was an auditory analog; and the rest were combined visual and auditory analogs. As in all analogies, the mapping from proximal to distal scenes is selective and incomplete—often radically so (see Gentner; Holyoak & Thagard, 1989).

TABLE 17.1

Proximal and distal scenes for seven depictions

	Example	Proximal scene (depiction)	Distal scene (what is depicted)
(5)	Whirligig	Speaker twirling his finger	Moving propeller of whirligig
(5)	Whirligig	Speaker catching invisible object	Person catching actual whirligig
(6)	Fear	Speaker raising eyebrows and widening eyes	Person raising eyebrows and widening eyes in actual fear
(7)	Car noise	Speaker producing low growling noise	Actual sound of speaker's car
(8)	Playing piano	MB playing four measures on piano	Bartok playing four measures on piano
(9)	Quotation	Morgan uttering "come on ... together"	Gary uttering "come on ... together"
(18)	Lifting chin	Sevigny putting her own hand under chin and uttering "lift your chin"	Cinematographer putting *his* own hand under S's chin and saying "lift your chin"

3.2. DEPICTIONS AS STAGED SCENES How do people create these scenes, and why? Sevigny's actions in (18) exemplify six properties that are general to performed depictions (see Clark, 2016, for details):

a. *An act of depicting entails both a proximal and a distal scene.* Sevigny created a proximal scene in the here-and-now. In doing so, she signified a distal scene, what the cinematographer did.
b. *Depictions are physical analogs of what they depict.* Sevigny intended the proximal scene, her speech and gestures, to be a physical analog of the distal scene, the cinematographer getting her to lift her chin.
c. *Depictions are a type of joint pretense.* When Sevigny uttered "Lift your chin" and pushed up her chin, she wasn't "literally or really or actually" doing these things, as Goffman (1974, p. 47) put it. She was only *making as if* she was doing them. Making as if is pretense. Sevigny intended her audience to recognize the pretense and establish it as joint pretense.
d. *Depictions are staged scenes.* Sevigny created the proximal scene by choosing a space, or *stage*, in front of her audience and, in that space, *enacting* a cinematographer trying to get her to lift her chin. As we will see, scenes can be staged in many ways.
e. *Depictions serve specific functions in discourse.* Sevigny intended her depiction to function as the NP in "the cinematographer would be like NP."
f. *In depictions, the intended recipients participate in the staged scenes as spectators.* Sevigny tacitly treated her audience as silent coparticipants in the scene she staged. They were to play the role of spectators—covert observers of what the cinematographer asked her to do.

If these are properties of staged scenes in general, they should also be properties of the theater, and they are (see Clark, 2016, pp. 329–330). Shakespeare's *Hamlet*, for example, is really a depiction of fictional events in Elsinore Castle in 11th-century Denmark. The events created on stage are proximal scenes, and they are physical analogs of distal scenes surrounding Hamlet in Elsinore Castle. Suppose we see the actor Kenneth Branagh, as Hamlet, standing opposite the actress Kate Winslet, as Ophelia, reciting the line, "Get thee to a nunnery." We recognize that Branagh isn't "actually or really or literally" telling Winslet to go to a nunnery and that she knows this. We recognize that the two of them are engaged *with us* in a joint pretense: (a) that Hamlet is speaking to Ophelia and (b) that we are covert observers of Hamlet and Ophelia's conversation.

3.3. ACTOR VERSUS PROP DEPICTIONS In the theater, the people who execute scenes are either *actors* or *prop managers*. An actor is a person *on stage* who *enacts a character* in the distal scene. A prop manager is a person *off stage* who *realizes props* that *are* on stage and that map into objects, states, or events in the distal scene. In *Hamlet*, Branagh appears on stage and enacts Hamlet saying "Get thee to a nunnery." Another member of the theater company, a prop manager, remains off stage and creates props such as the howling of the wind. Branagh's depiction is an *actor depiction*, and the prop manager's is a *prop depiction*.

The division between actor and prop depictions is fundamental to performed depictions. The split has long been recognized in iconic gestures (Kendon, 2004, chapter 6) and in sign languages (e.g., Emmorey, 2002). When the speaker in (5) twirled his index finger, he depicted the whirligig's motion *from offstage*—as if he were outside the scene. This was a prop depiction. Immediately afterward, he became an actor *on stage* and enacted a person catching the whirligig as it came down. This was an actor depiction.

Actor and prop depictions are found in performed depictions of all types. The piano demonstrations in (10) through (13) were actor depictions, whereas the simulated car noise in (7) was a prop depiction. A producer can even take two roles at once, creating a *hybrid* depiction (Clark, 2016, pp. 331–332). In make-believe play, one little girl propped a doll up while speaking in the doll's voice, "(*whining in a petulant voice*) I want my teddy bear." The child was off stage in handling the doll (a prop depiction), but on stage in enacting its speech (an actor depiction).

3.4. Spatialization of Scenes Depictions are performed for others in space. Producers must choose for each depiction where to place it, how to orient it, and how large to make it. They have to *spatialize* each depiction. The general rule is this: In creating a depiction, select a spatialization that best suits your purpose and is optimal for recipients. Spatializing a depiction entails five parameters: place, orientation, scale, granularity, and dimensionality.

Placement. People normally try to place depictions in locations that are optimal for their addressees to see, hear, and interpret. In telling stories (as in (5)), narrators place most of their gestures in a *gesture space*—an imaginary 3D bubble in front of the body (McNeill, 1992). This space is optimal not only for performing gestures but for displaying them to their interlocutors (Gullberg & Holmqvist, 1999). When speaking to two addressees in alternation, people will direct their gestures toward the current addressee (Özyürek, 2002). When tennis coaches demonstrate a serve, they stand in front of their students, and in (6), MB demonstrated Bartok's playing on a piano he had placed in front of his audience.

Orientation. People often have a choice in the orientation of a depiction. Here is another depiction from Sevigny:

(19) [Chloe Sevigny was speaking to Conan O'Brian.] (*facing O'Brien*) When I was little, I used to smack my lips when I ate, you know (*turning 90° to her right, smacks her lips three times*). (*turning back to O'Brien*) [laughs].

Sevigny turned away from O'Brien as she smacked her lips and then turned back. Why? One reason was to perform the scene in a marked space, or stage, to make it clear she was *not* smacking her lips at O'Brien. *Orientation shifts* are common in quotations, where speakers distinguish their own speech from the speech they are quoting (Clark, 2016; Emmorey, 2002; Sidnell, 2006).

Scale and granularity. In many depictions, people also have a choice of scale and granularity. Consider, again, the two whirligig gestures in (3). Earlier in the narrative, the speaker had depicted the twirling of the whirligig several times. His first depiction was large and precise, but by the time he said, "And you just twirl it and catch it," he had down-sized the depiction to "a very tiny twirling action." As Gerwing and Bavelas (2004) showed, people reduce the size and detail of a repeated depiction—when it is no longer new information—just as they would reduce a repeated reference (from "a very large dog" to "the dog" to "it"). In contrast, when the speaker depicted a person *catching* the whirligig, he used an expansive, detailed gesture because this was information he was depicting for the first time.

Ordinarily, people decide on the placement, orientation, scale, and granularity of a depiction as a package, as here:

(20) [MB was lecturing to an audience about musical notation] It means simply (*tracing a 30 cm long sideways V in the air with R thumb and forefinger near his R shoulder*) what we would call the tapering.

MB depicted a diminuendo sign as an adjunct to "what we would call the tapering." In a musical score, the diminuendo sign would be about 3 cm long on a horizontal surface. In his lecture, MB made it 30 cm long, oriented vertically, and high above his shoulder. He did all this to make it easy to see and interpret.

Dimensionality. The space occupied by a depiction can be 0D, 1D, 2D, or 3D. Most of the depictions cited so far have been in 3D—the whirligig, the facial gesture of fear, the cinematographer, and the pianist playing. But the diminuendo sign in (20) was depicted on a 2D surface, and the gesture with "I caught a fish this long" would have been in 1D. The depiction of the car noise in (7) was effectively in 0D because it was produced by the voice at a single point in space. Producers must take these dimensions into account in spatializing a depiction. For 1D, 2D, or 3D depictions, they must select a place, orientation, scale, and granularity. For 0D depictions, they must select a place, scale, and granularity.

3.5. Temporal Placement Depictions are also performed for others in time. On the one hand, the overall timing of a depiction—its *temporal placement*—is used to mark its function in discourse. In (8), the two demonstrations of piano playing were timed to replace the NPs in "he doesn't play NP, but rather he plays—and he does it better than I do—NP." On the other hand, the *internal timing* of a depiction is part of the content. In (8), MB played the musical excerpts at an appropriate tempo.

Temporal placement is used for marking the function of a depiction via three main techniques:

a. *Slotting.* Slotting is used for marking the role of embedded depictions. The engine noise in (7) was performed in a slot for a noun. The two piano demonstrations in (8) were performed in slots for NPs. And in (14), the speaker's cringe was performed in a slot for the N in "the N was gone out of it." Slotting is also used for marking the role of independent depictions. In (9), "come on ... together" was performed in a slot for the next contribution in Morgan's narrative. In all these examples, the depictions were slotted into the utterances without breaks or overlap.
b. *Overlap.* Temporal overlap is used for marking the affiliates of adjunct depictions. In (5), the depiction of twirling was timed to overlap "twirl it," and the depiction of catching the whirligig, to overlap "catch it." And in (20), the depiction of the diminuendo sign was timed to overlap its affiliated phrase "what we call the tapering." Speakers generally perform this overlap with remarkable precision (Kendon, 2004).
c. *Fusion.* Certain depictions can only be performed in fusion with parts of an utterance. Depicting loud and soft voices is an example:

(21) [MB was speaking to an audience of 40 people.] First of all, I'm using right now a medium-sized voice, because I'm talking to you over there. If this were our large Bailey (*raising voice*) Concert Hall, I would probably be speaking like this. (*returning to "medium-sized" voice*) And if we were sitting in my living room, (*lowering voice*) I'd probably be speaking like this.

MB depicted a medium-sized, a loud, and a soft voice by fusing his normal, raised, and lowered voice with different parts of his utterance. He indexed each depiction with "right now" or "like this" as he was performing them.

Fusing is also used in adjunct depictions. One way to depict speed, for example, is to accelerate the pronunciation of a word or phrase, as here:

(22) [A woman was describing some gunmen she witnessed to a TV reporter.] I come out here to the steps [pause] (*accelerated pronunciation, hand moving quickly to her right*) take off down that way ... and then I see some people (*accelerated pronunciation, hand moving quickly to her left*) running that way.

With the acceleration of "take off," the woman created an adjunct depiction fused with "take off." At the same time, she accelerated her hand to the left to depict speed in that direction. As a result, she modified "take off" with two adjunct depictions at once—her accelerated pronunciation and her accelerated gesture. She modified "running" with two similar depictions. Other fused depictions will come up in examples (32) and (33).

3.6. Internal Timing The time course of actions *within* a depiction, its internal timing, usually matches the scene it depicts. In (6), Sevigny said "lift your chin" at the speed the cinematographer would have said it. And yet the time course of a depiction can be modified for special purposes (Clark, 2016, pp. 339–340):

a. *Holds.* People often hold, or pause, a depiction at particular junctures. A tennis coach demonstrating a serve might pause briefly at the top of the serve to focus on that feature. A piano teacher might pause briefly in the middle of a melodic phrase to illustrate a break in the phrasing.
b. *Slow motion.* A tennis coach might perform a serve in slow motion, and a piano teacher might play a phrase at half tempo, both to draw attention to features being depicted.
c. *Condensation.* A piano teacher might speed through a part of a demonstration that is of little interest. A ballet teacher might depict a spin with a gesture many times faster than normal (Kirsh, 2011).
d. *Truncation.* People can also truncate depictions prematurely. One story teller reported, "I heard a man in the next room praying 'Our father who art,'" cutting off the prayer once it was recognizable as the Lord's Prayer.

4. Codified Depictions

The focus so far has been on free depictions. But what about codified depictions? The most familiar of these are based on *ideophones*, onomatopoeic words that belong to particular languages. *Oink-oink*, *cock-a-doodle-do*, and *bang* are ideophones in English; *guoin-guoin*, *cocorico*, and *pan* are the equivalents in French. People who don't know English or French can only guess at what these words signify. And people's guesses are, in general, terrible. In a study of five categories of ideophones, people were only 5% above chance for four of the categories and only 15% above chance for the fifth (Dingemanse, Schuerman, Reinisch, Tufvesson, & Mitterer, 2016).

But why? If the witness in (4) had told the American jury, "And then I heard 'guoin-guoin'" why should they

Table 17.2
Aspects of Michelangelo's David

Aspect	Type	Value
Shape	Depictive	Standing-male-nude-with-slingshot
Size	Supportive	5.17 m tall
Material	Supportive	Carrara marble
Surface color	Supportive	Off-white
Animation	Supportive	Inanimate
Plinth	External	3 m cube
Surface damage	External	Pitting, cracks, repairs

be baffled? If the French hear *guoin-guoin* as a pig snort, why don't English speakers hear it that way too? For an answer, we need to look inside depictions.

4.1. Aspects of Depictions Most depictions can be partitioned—though not always neatly—into *aspects*, features that apply to a depiction as a whole (Clark & Gerrig, 1990). Michelangelo's statue of David (in the Galleria dell'Accademia in Florence) might be partitioned into the aspects listed in table 17.2. The aspect "shape" has the value "standing-male-nude-with-slingshot," which fits the biblical David, so it is a *depictive* aspect. The aspects "size," "material," "surface color," and "animation," in contrast, have values that do *not* fit the biblical David, so they are *not* depictive. Still, every statue has *some* size, *some* material, *some* surface color, and *some* animation, so even if these aren't depictive, they are necessary. They are *supportive* aspects. And, finally, the aspects "plinth" and "surface damage" are neither depictive nor supportive. They are *external* aspects.

Depictions, in this analysis, partition into at least three types of aspects, as illustrated again for Michelangelo's *David*:

a. *Depictive aspects* (shape)
b. *Supportive aspects* (size, material, surface color, animation)
c. *External aspects* (plinth, surface damage)

People viewing *David* would be wrong to believe Michelangelo thought the biblical David was 5.17 m tall, made of Carrara marble, immobile, and pitted. Performed depictions are no different. In (18), the diminuendo sign was depicted by MB tracing a 30 cm long sideways V in the air. "Shape" was depictive, while "size," "orientation," and "placement" were merely supportive. MB's audience would be wrong to think he thought diminuendo signs were 30 cm long, in a vertical plane, and in the air.

4.2. Aspects of Codified Depictions When ideophones are used in codified depictions, they always include "language," "phonological form," and "conventional meaning" as *supportive* aspects. The word use of *thump* in Lewis Carroll's "My heart went 'thump'" might be partitioned as shown in table 17.3. The first three aspects are depictive, because they match the low muffled sound of a heartbeat, and the next three are supportive—necessary but not depictive. In German, Danish, and Turkish, the counterparts for *thump* are *bumm*, *dunk*, and *güm*. All four of these words have the same depictive aspects. Where they differ is in the supportive aspects "language" and "phonological form."

Codified depictions can also be based on emblems (conventional gestures). In many cultures, for example, the use of fingers for depicting number is governed by special codes. In my own culture, two is depicted with the index and middle fingers, not with the thumb and index finger, or with the ring finger and pinky, as in some cultures. Also in my culture, a "hand phone" is conventionally formed from a fist and extended thumb and pinky. This, too, is part of a larger code.

4.3. Uses of Ideophones Ideophones are word *types*, but codified depictions are created from *tokens* of these types. A token of *bang*, for example, can be used unadorned to depict a single gunshot, as "I heard a 'bang.'" A token of *bang* can also be lengthened to depict a long gunshot ("I heard a 'ba-a-ang'") or boomed out to depict a loud gunshot ("I heard a 'bang!!!'"). And multiple tokens of *bang* can be used to depict multiple gunshots:

(23) [A woman, the same as in 22, was telling the TV reporter about a shooting.] Then I hear this "(*in a louder voice with deliberate timing*) bang bang (0.5 s) bang bang bang" you know.

Here the woman used five tokens of *bang* to depict a detailed auditory scene. There were not only five

Table 17.3
Aspects of the use of thump *in "my heart went 'thump'"*

Aspect	Type	Value
Unit	Depictive	Closed syllable
Unit's vowel	Depictive	Low vowel
Unit's final cluster	Depictive	Nasal
Language	Supportive	English
Phonological form	Supportive	/θʌmp/
Conventional meaning	Supportive	"A dead, dull, somewhat hard sound"

gunshots, but they were loud, fast, and in a two-plus-three rhythm. Codified depictions, in short, are staged scenes that use *tokens* of ideophones or emblems as elements (see Dingemanse, 2013; Dingemanse & Akita, 2017).

Ideophones, viewed as word types, lie on a continuum from well entrenched to novel,[5] as in this rough classification:

Primary ideophones. Word types such as *oink-oink, cock-a-doodle-do,* and *bang* are primarily vehicles for depicting pig snorts, rooster crows, and gunshots. A token of *oink* invariably carries the meaning "pig snort."

Secondary ideophones. Other word types *can* be used as ideophones even though that is not their primary use. Here are two genuine examples:

(24) And my nose went "sniff, sniff, sniff."

(25) Once we got to Evelyn's house, I knocked on the door. *Knock! Knock! Knock!* And just as I was hoping, ...

In (24), the speaker used three tokens of *sniff* to depict three acts of sniffing. Although *sniff* is primarily a verb, the speaker treated it as an ideophone with the supportive aspect "act of drawing in air audibly through the nose" and used three tokens to depict three such acts.

Novel ideophones. Ideophones can also be invented for the nonce, as in these genuine examples:

(26) The pounding of the cylinders increased: *ta-pocket-pocketa-pocketa-pocketa-pocketa.* [James Thurber, *The Secret Life of Walter Mitty*]

(27) I took my broom and went *sweep, sweep, sweep* up the road.

(28) Then she walked off the lawn to the meadow whose corner to the right I can just see. *Trail trail,* went her long dress over the sopping grass. [E. M. Forster, *Howard's End*]

Thurber invented *pocketa* in (26) as imitative of the "pounding of the cylinders" and used five tokens of it to depict five events of cylinder-pounding. We infer the supportive aspect "unit's meaning" from the surrounding narrative. In languages with a wealth of ideophones, it is common to invent new ideophones by elaborating on old ones (Dingemanse, 2014).

Sweep and *trail* are ideophones created from verbs. In (27), the speaker produced *sweep* three times to depict three acts of sweeping; he based the ideophone on the verb *sweep,* which gave it the meaning "act of swinging a broom." The narrator in (28) did much the same. He depicted two events of a dress trailing in the grass, using the verb *trail* as the source for the meaning of the ideophone. Without these meanings, there is little to guide us in imagining the two distal scenes.

4.4. Functions of Codified Depictions Codified depictions should be available for the same functions as free depictions, and they are.

Embedded depictions. In English, tokens of *oink* can be embedded as nouns ("I heard an 'oink' outside"), as verbs ("The beast oinked at me"), as adverbs ("The beast went 'oink oink oink oink oink'"), and as other constituents. Codified depictions have much the same range as quotations (e.g., Clark & Gerrig, 1990; Pascual, 2006).

Indexed depictions. Codified depictions are readily indexed by expressions such as *like this* or *here.* One can say "I heard a beast go like this: 'Oink oink—oink oink,'" or "Here is what I heard outside my window: 'Cock-a-doodle-do.'"

Adjunct depictions. For an adjunct depiction to overlap a spoken word or phrase, it cannot itself be a spoken word or phrase, but it *can* be a conventional gesture. In a noisy bar, I can tell the bartender, "I'd like (*holding up two fingers*) two glasses of Guinness." And a friend might tell me "I tried to (*placing a hand phone next to her ear*) call you last night."

Independent depictions. When asked "What did you hear?" I can answer "Oink oink" as a separate, independent contribution. And when asked "How many beers would you like?" I can hold up two fingers as an independent contribution.

4.5. Codified Depictions across Languages Languages vary enormously in the prevalence of ideophones. Languages like English, French, and Dutch have only small desultory sets of them, and they are used only occasionally. But languages like Japanese, Korean, and Siwu (spoken in Ghana) have large, highly structured sets of ideophones, and they are an important part of everyday talk (Dingemanse, 2013). For scholars of English, French, and Dutch, it is tempting to pass off ideophones as linguistic oddities. For scholars of Japanese, Korean, and Siwu, that would be impossible.

According to Dingemanse (2013), ideophones contrast with other words in several ways:

> Across languages, ideophones tend to be relatively syntactically independent, delivered with prosodic foregrounding that sets them off from the rest of the utterance. They are often introduced by quotative markers or "say" or "do" verbs, emphasizing their performative nature. All of these features work together to mark ideophones as depictions, much as the frame around a painting tells us to interpret it as a painting and not as the wallpaper (p. 144).

But how do ideophones depict, and how are they used in communication? For answers, let us look at a language that is rich in ideophones.

4.6. Japanese Mimetics Japanese has a class of over 1,600 ideophones called *mimetics*. Here are three examples with approximate English translations (from Hamano, 1998):

(29) I did *pin-pin* ("I struck the guitar string twice") (p. 19)

(30) He does *puri-puri*, doesn't he? ("He is irritable, isn't he?") (p. 114)

(31) So, Gon *kossori* ("stealthily") entered the house through the back door (p. 12)

As linguistic forms, mimetics are usually easy to identify. Almost all of them are either reduplications of a one- or two-syllable stem, such as *pin-pin* and *puri-puri*, or one- or two-syllable stems with the suffix *-ri*, as in *kossori* (Dingemanse & Akita, 2017). As for their functions, mimetics are used as adverbs, as noun modifiers, or as predicates (Shibatani, 1990, p. 154). In (31), *kossori* is an adverb, and in (29) and (30), *pin-pin* and *puri-puri* are predicates.

Mimetics are, indeed, an integral part of Japanese syntax. A good example are the expressions for laughing and crying. Japanese has one nonmimetic word for laughing (*warau*) and another for crying (*naku*), but none for chuckling, giggling, sobbing, whimpering, and the like. Japanese expresses these notions instead with *warau* or *naku* plus a mimetic, as in these examples (Shibatani, 1990, p. 155):

Laughing	'laugh' *ha-ha-ha* to warau; 'smile' *niko-niko* to warau; 'chuckle' *kutu-kutu* to warau; 'haw-haw' *wa-ha-ha* to warau; 'giggle' *gera-gera* to warau
Crying	'cry' *waa-waa* naku; 'weep' *meso-meso* naku; 'sob' *kusun-kusun* naku; 'blubber' *oi-oi* naku; 'whimper' *siku-siku* naku

To giggle, for example, is literally 'to laugh uttering *gera-gera*,' and to blubber is 'to cry *oi-oi*.' So Japanese has as many expressions for laughing and crying as English, but the Japanese expressions may seem more vivid because they depict as well as describe.

4.7. Composite Depictions Japanese offers good evidence that codified depictions are created from tokens of ideophones (Dingemanse & Akita, 2017). Consider these examples of people talking to an interviewer about the 2011 tsunami (with Dingemanse & Akita's glosses in parentheses):

(32) Things like debris moved *ju:k:uri* ('sloowly')

(33) I heard the sound like *zabu:n-zabu:n* ('splaash-splaash').

(34) The tsunami I had in mind was something like *cita-cita-cita-cita* ('lap lap lap lap').

As word types, *jukuri* depicts slowness, *zabun-zabun* depicts the sound of splashing, and *cita-cita* depicts the sound of water lapping. The token of *jukuri* in (32), however, was lengthened to depict "very slowly"; the token of *zabun-zabun* in (33) was lengthened to depict a *great deal* of splashing; and the token of *cita-cita* in (34) was reduplicated to depict a *great amount* of water lapping. The speakers modified each token appropriately for the scene they were staging.

Japanese also offers good evidence that codified depictions are spatialized—distinguished from the surroundings by their placement, orientation, scale, granularity, and dimensionality. Here is a depiction that incorporates a token of *zoro-zoro*, a mimetic that can be glossed "one after another in line" (p. 16):

(35) [A man was telling an interviewer about people rushing by him.] Yeah, everyone ran (*both arms swinging in front of him R to L*) zoro-zoro ('one after another in line') through (*swinging both arms R to L, but more briefly*) here.

In the slot for a modifier of *ran*, the speaker staged a life-sized scene in which people were running in front of him *zoro-zoro* ('one after another in line') from right to left. He created the scene by swinging his arms, turning his head and torso, following the imagined people with his eyes, and, simultaneously, uttering a token of *zoro-zoro*. As Dingemanse (2013) showed, the gestures used with ideophones, as in (35), are *not* part of the ideophones as word types. They are free depictions created de novo. And they are common. In one Japanese sample (Dingemanse & Akita), 85% of the depictions with mimetics included gestures. In another sample (Kita, 1997), the proportion was 94%.

4.8. Sensory-based Scenes Ideophones, Dingemanse (2012, p. 663) argued, depict *sensory-based scenes*. Across languages, these scenes fall into five categories:

a. *Sounds*. In (29), *pin-pin* depicts the pinging of a guitar string.
b. *Movements*. In (31), *kossori* depicts a manner of entering a house. *Kuru-kuru* depicts something turning.
c. *Visual patterns*. *Hira-hira* depicts something fluttering.
d. *Other sensory modalities*. *Zara-zara* depicts roughness in texture.
e. *Inner feelings and cognitive states*. In (30), *puri-puri* represents a person's irritability.

These categories form a hierarchy. If a language has ideophones in category *b*, it will also have them in category *a*, but not necessarily vice versa. And so on down the list. So there should be no language with ideophones in *d* and *e*, but not in *a*, *b*, and *c*. In English, most ideophones are in category *a*. In Japanese, the 1,600 mimetics span all five categories.

It is easy to see how mimetics might be used to depict sounds. Japanese *pin-pin*, like English *ping* and *plunk*, bears an auditory resemblance to a plucked guitar. But what about the other four categories? How does *hira-hira* depict fluttering, *zara-zara* rough texture, and *puri-puri* irritability?

For Japanese mimetics, most answers have focused on sound symbolism and reduplication (Hamano, 1998; Shibatani, 1990). For example, the voiced consonants *g*, *d*, and *z*, "relate to heavier or louder sounds, or stronger, bigger, rougher actions or states" in comparison to their voiceless counterparts *k*, *t*, and *s* (Shibatani, p. 156). So *gasa-gasa* depicts a heavy rustling sound and *kasa-kasa* a light rustling sound. *Don-don* depicts a banging sound and *ton-ton* a light knocking sound. *Zara-zara* depicts a rough texture and *sara-sara*, a light texture. And "reduplication, as in many languages, symbolizes repetitive sounds or movements, e.g., *goro-goro* 'rumbling,' *kuru-kuru* 'turning'." (p. 155).

Still, sound symbolism and reduplication go only so far. They offer far too few contrasts mathematically to distinguish all 1,600 mimetics. If so, we must look for other accounts.

5. *Interpretive Frameworks*

Free depictions are always interpreted with respect to *interpretive frameworks* (Clark, 2016, p. 328). They have to be. When I make a small circular hand gesture for you, I might be depicting a bike wheel, a hydrogen atom, a ship's porthole, or countless other scenes. To know which, you must infer the framework I intended. Am I talking about bikes, atoms, cruise ships, or what? Typically, I will set the framework in advance, as with "My bike was lying there, but (*gesturing*) the front wheel was missing" or "In the hydrogen atom, there is (*gesturing*) only one electron."

Unlike free depictions, codified depictions come with interpretive frameworks that are prefabricated—at least in part. Consider a story by 12-year-old John L. about the "baddest girl in the neighborhood," and "if you didn't bring her candy to school, she would punch you in the mouth" (Labov, 1972, pp. 358–359):

(36) So I go to school, and this girl says, "Where's the candy?"
I said "I don't have it."
She say "Powww!"

In the third quotation, John depicted what the girl did—punch him in the mouth—by producing a loud, lengthened token of the ideophone *pow*.[6] He relied on the conventional meaning of *pow* ("the sound of a blow, punch, shot, etc.") to help his interlocutor establish the interpretive framework for his depiction. He would have failed if he had used French "pafff!" or Japanese "bannn!" instead.

Ideophones are paradoxical. The word *pow* has a phonetic form, */pau/*, that is construed as a depiction of a fist punch, and that makes *pow* an icon. But *pow* also has a conventional meaning ("hitting sound") that is specific to English. In Dutch, the phonetic form */pau/* is paired with the conventional meaning "peacock," and in French, the conventional meaning "hitting sound" is paired with the form /paf/. That makes *pow* a symbol as well as an icon. It is this duality that distinguishes ideophones from other words in the lexicon.

Variations in ideophones across languages are to be expected. The lexicon of a language is a system. Whenever a new ideophone is added to a lexicon (such as *tweet* in the 1500s or *oink* in the 1900s in English), its form and meaning are shaped by the words already present, and they constrain the form and meaning of the words added later. As Bolinger (1949, 1950) argued: "Whatever [a word's] origin, be it as pictorial as an imitative word or as abstract as the number *ten*, once part of the individual's equipment it can no longer be arbitrary, and cannot 'just as well' be something else" (1949, p. 53). Japanese mimetics have surely evolved under similar constraints.

Codified depictions, then, are a species of performed depiction. Kunene (1965, p. 32, cited by Dingemanse, 2013, p. 153) said as much:

> The most remarkable thing about the ideophone is that it constitutes a dramatic presentation of past and future events and states. By means of it, the speaker conjures up before the eyes of his listeners events which are not actually happening, or states which are not existing, at the time of speaking.

In Kunene's view, the speaker uses an ideophone to *stage a proximal scene* ("a dramatic presentation") that *maps into a distal scene* ("of past or future events or states"). Why? To get his or her listeners *to imagine* ("to conjure up before their eyes") *the distal scene* ("the events that are not actually happening, and the states that do not actually exist"). To use an ideophone is to stage a scene, and that puts codified depictions on a par with free depictions.

6. *Conclusion*

Free depictions are ubiquitous in language use. They range from brief scenes, such as the auditory depiction in "the car has kind of a (*low growling noise*)" to life-sized enactments, such as Sevigny's depiction of the cinematographer getting her to lift her chin. They can be embedded in, indexed by, or adjoined to other utterances, or used as independent acts of communication. Regardless of their function, they add to the ongoing discourse.

So do codified depictions. Most of these are created with ideophones (such as *pow* for the sound of a punch), but others are formed around emblems (such as two fingers for the number 2). Ideophones as a class may seem peripheral in English or French, but they are essential to the grammars of Japanese and Korean. The conclusion is clear. Depicting is present not only in language *use*, but in language *structure*.

NOTES

1. In the examples cited in this paper, the speech is in roman print, a gesture or other depiction is described in italics within parentheses, and the speech it overlaps is underlined.
2. So-called live recordings (as of TV talk shows and pop concerts) are artifacts that are based on live performances, but are exhibited by others.
3. If *play* is interpreted as an intransitive verb, these two depictions would be adverbs (ADV). Compare the quotations in "he said 'Hi'" and "he went 'Hi.'" The first is an NP and the second an ADV (Clark & Gerrig, 1990).
4. In the gesture literature, all gestural depictions tend to be classified as "co-speech gestures" even though the term is only accurate for adjunct depictions.
5. Denominal verbs lie on a similar continuum, from *table* and *boycott*, which are well entrenched, to *porch* and *houdini*, which are novel (Clark & Clark, 1979).
6. John may well have produced a hitting gesture as well, but the transcript didn't include gestures.

REFERENCES

Bavelas, J. B., & Gerwing, J. (2007). Conversational hand gestures and facial displays in face-to-face dialogue. In K. Fiedler (Ed.), *Social communication* (pp. 283–308). New York: Psychology Press.

Bolinger, D. L. (1949). The sign is not arbitrary. *Boletín del Instituto Caro y Cuervo, 5*, 56–62.

Bolinger, D. L. (1950). Rime, assonance, and morpheme analysis. *Word, 6*(2), 117–136.

Calbris, G. (2011). *Elements of meaning in gesture.* Amsterdam: John Benjamins Publishing.

Clark, E. V., & Clark, H. H. (1979). When nouns surface as verbs. *Language* (Vol. 55), 767–811.

Clark, H. H. (1996). *Using language.* Cambridge: Cambridge University Press.

Clark, H. H. (1998). Communal lexicons. In K. Malmkjaer & J. Williams (Eds.), *Context in language learning and language understanding* (pp. 63–87). Cambridge: Cambridge University Press.

Clark, H. H. (2016). Depicting as a method of communication. *Psychological Review, 123*(3), 324–347.

Clark, H. H., & Gerrig, R. J. (1990). Quotations as demonstrations. *Language*, (Vol. 66), 764–805.

Dingemanse, M. (2012). Advances in the cross-linguistic study of ideophones. *Language and Linguistics Compass, 6*(10), 654–672.

Dingemanse, M. (2013). Ideophones and gesture in everyday speech. *Gesture, 13*(2), 143–165.

Dingemanse, M. (2014). Making new ideophones in Siwu: Creative depiction in conversation. *Pragmatics and Society, 5*(3), 384–405.

Dingemanse, M., & Akita, K. (2017). An inverse relation between expressiveness and grammatical integration: On the morphosyntactic typology of ideophones, with special reference to Japanese. *Journal of Linguistics, 53*(3), 501–532.

Dingemanse, M., Schuerman, W., Reinisch, E., Tufvesson, S., & Mitterer, H. (2016). What sound symbolism can and cannot do: Testing the iconicity of ideophones from five languages. *Language, 92*(2), e67–e83.

Ekman, P., & Friesen, W. V. (1969). The repertoire of nonverbal behaviour. *Semiotica, 1*, 49–98.

Emmorey, K. (2002). *Language, cognition, and the brain: Insights from sign language research.* Mahwah, NJ: Lawrence Erlbaum.

Gentner, D. (1983). Structure-mapping: A theoretical framework for analogy. *Cognitive Science, 7*(2), 155–170.

Gerwing, J., & Bavelas, J. B. (2004). Linguistic influences on gesture's form. *Gesture, 4*(2), 157–195.

Goffman, E. (1974). *Frame analysis: An essay on the organization of experience.* Cambridge, MA: Harvard University Press.

Gullberg, M., & Holmqvist, K. (1999). Keeping an eye on gestures: Visual perception of gestures in face-to-face communication. *Pragmatics and Cognition, 7*(1), 35–63.

Hamano, S. (1998). *The sound-symbolic system of Japanese.* Stanford, CA: CSLI Publications.

Holyoak, K. J., & Thagard, P. (1989). Analogical mapping by constraint satisfaction. *Cognitive Science, 13*(3), 295–355.

Kendon, A. (2004). *Gesture: Visible action as utterance.* Cambridge: Cambridge University Press.

Kirsh, D. (2011). How marking in dance constitutes thinking with the body. *Versus: Quaderni di Studi Semiotici, 113–115*, 179–210.

Kita, S. (1997). Two-dimensional semantic analysis of Japanese mimetics. *Linguistics, 35*, 379–415.

Korzybski, A. (1948). *Science and sanity: An introduction to non-Aristotelian systems and general semantics.* Lakeville, CT: International Non-Aristotelian Library Publishing.

Kunene, D. P. (1965). The ideophone in southern Sotho. *Journal of African Languages, 4*, 19–39.

Labov, W. (1972). *Language in the inner city: Studies in the Black English vernacular.* Philadelphia: University of Pennsylvania Press.

McNeill, D. (1992). *Hand and mind: What gestures reveal about thought.* Chicago: University of Chicago Press.

Özyürek, A. (2002). Do speakers design their cospeech gestures for their addressees? The effects of addressee location on representational gestures. *Journal of Memory and Language, 46*(4), 688–704.

Pascual, E. (2006). Fictive interaction within the sentence: A communicative type of fictivity in grammar. *Cognitive Linguistics, 17*(2), 245–267.

Peirce, C. S. (1932). The icon, index, and symbol. In C. Hartshorne & P. Weiss (Eds.), *Collected papers of Charles Sanders*

Peirce (Vol. 2, pp. 156–173). Cambridge, MA: Harvard University Press.

Peirce, C. S. (1955). Logic as semiotic: The theory of signs. In J. Buchler (Ed.), *Philosophical writings of Peirce* (pp. 98–119). New York: Dover.

Peirce, C. S. (1974). *Collected papers of Charles Sanders Peirce* (Vol. 3). Cambridge, MA: Harvard University Press.

Shibatani, M. (1990). *The languages of Japan*. Cambridge: Cambridge University Press.

Sidnell, J. (2006). Coordinating gesture, talk, and gaze in reenactments. *Research on Language and Social Interaction, 39*(4), 377–409.

Tannen, D. (1986). Introducing constructed dialogue in Greek and American conversational and literary narrative. In F. Coulmas (Ed.), *Direct and indirect speech* (pp. 311–332). Berlin: Mouton.

Waugh, L. R. (1994). Degrees of iconicity in the lexicon. *Journal of Pragmatics, 22*(1), 55–70.

18 Conceptual Alignment as a Neurocognitive Mechanism for Human Communicative Interactions

IVAN TONI AND ARJEN STOLK

1. Experimental Semiotics

In the past few years there has been renewed interest in using empirical investigations to understand the principles and the mechanisms that allow humans to communicate (de Ruiter et al., 2010; Fusaroli & Tylén, 2016; Galantucci, 2005; Selten & Warglien, 2007; Scott-Phillips, Kirby, & Ritchie, 2009). The theoretical framework of this research field, loosely grouped under the label of experimental semiotics (Galantucci & Garrod, 2012), is broadly grounded in the view of "language use as joint action" (Clark, 1996). The novelty of this research field consists in using experimentally controlled approaches, in a genuinely interactive context, to capture generative elements of human communication. Those generative elements are at the root of our ability to share thoughts with other people during our daily social interactions (Levinson, 2006). Yet, paradoxically, interactive contexts as well as generative communicative elements have often remained outside the focus of experimentally oriented linguistic studies. For instance, traditional psycholinguistics has largely focused on encoding–decoding of linguistic material as implemented by individual agents, away from the context of interaction with other interlocutors. Generative linguistics has also been largely uninterested in the actual use of language, claiming that internal structural dependencies cannot be understood through contingent externalizations (Everaert, Huybregts, Chomsky, Berwick, & Bolhuis, 2015). Experimental semiotics takes interactive contexts and generative elements seriously, while deepening our understanding of human communication beyond purely linguistic means. There is a theoretical reason supporting this approach. Language is but one of several ways in which human communicative acts can be manifested. Understanding the conditions necessary for the communicative use of language avoids the parochial yet widespread stance of considering only language-based communication as worthwhile. There is also an empirical reason for pursuing this approach to study communication. In humans, it is hard to gain experimental access to core generative processes supporting our communicative abilities, given the presence of a large set of preestablished symbols (e.g., our daily spoken idiom). The presence of those preestablished conventional symbols makes it hard to distinguish the generation of new communicative elements from the retrieval of previous communicative elements. Starting from the premise of those two reasons, experimental semiotics has proven effective in bringing generative elements of human communication under the lens of quantitative and reproducible experimental procedures, without sacrificing truly interactive aspects. The general approach is to examine communicative behaviors evoked by procedurally well-defined problems where conventional communicative tools, such as spoken language, are not available. This approach allows for quantifying the generation of novel communicative solutions and understanding the conditions under which communication arises, both within as well as across iterated communicative turns (Galantucci & Garrod, 2011).

Besides broadening the study of human communication beyond language, this new wave of empirical work on human communication has been characterized by using experimental situations that capture its inherent interpersonal nature. This is an important issue to consider when interpreting the scope and the relevance of findings from this field of inquiry. To date, the bulk of research in cognitive neuroscience and language sciences has usually neglected language use and rarely paid attention to the neurocognitive consequences of having agents producing and comprehending communicative behaviors embedded in a genuinely interactive setting. In fact, laboratory-based studies of human communication have rarely been approached in terms of mutual understanding (Hari, Henriksson, Malinen, & Parkkonen, 2015). The issue of the triadic representation of a referent, a central issue in human communication, provides a case in point (Grice, 1957; Tomasello, 2008). It has long been acknowledged that humans regulate most communicative interactions triadically, that is, by directing the mental states of an addressee to a mutually known referent (Tomasello, Carpenter, Call, Behne, & Moll, 2005). Put differently, most human communicative behaviors do not rely on stereotyped signal–response associations as those often used across several animal taxa (Dyer, 2002; Mather, 2004; Seyfarth, Cheney, & Marler, 1980). In fact, most human communicative behaviors are structurally ambiguous, that is, the relation between behavior and informative intentions is many-to-many (Levinson, 2006). One way in which that myriad behavior–meaning mappings can be disambiguated relies on constraints imposed on the interlocutors by what they commonly presume is known and believed by the other under the current circumstances (common ground; Brennan & Clark, 1996). Studying individual agents producing scripted utterances or processing isolated sentences is hardly conducive to capture those constraints, that is, the propositions jointly taken for granted as well as the likely joint goals, norms, and affordances embedded in a communicative event. The challenge is to understand how interlocutors define those constraints precisely enough to select and understand their communicative behaviors, despite differences in their private perspectives and background knowledge of the situation at hand.

The emphasis of experimental semiotics on respecting the interpersonal and generative nature of human communication has led to several studies that have captured fundamental features of the problems solved during human communication (de Ruiter et al., 2010; Fay, Arbib, & Garrod, 2013; Fusaroli & Tylén, 2016; Galantucci, 2005; Garrod, Fay, Lee, Oberlander, & MacLeod, 2007; Healey, Swoboda, Umata, & King, 2007; Selten & Warglien, 2007; Scott-Phillips et al., 2009; Roberts, Lewandowski, & Galantucci, 2015). These features will be the focus of the next section. Here we would like to emphasize that the approach used in experimental semiotics is clearly different from other recent attempts to characterize human social cognition, largely based on either passive observation of human behaviors (Rizzolatti & Sinigaglia, 2016), dynamic interactions between multiple agents lacking a referent (Noy, Dekel, & Alon, 2011; Sebanz, Bekkering, & Knoblich, 2006), or strictly choreographed interactions where participants need to retrieve extensively trained communicative behaviors, rather than generating new ones (Silbert, Honey, Simony, Poeppel, & Hasson, 2014). Experimental semiotics has also been instrumental in focusing the research community toward an important and defining aspect of human communication, often missed by earlier work on action observation and joint action. Namely, experimental semiotics has not sacrificed the open-ended nature of human communication for the sake of experimental control. Many studies on action observation, and even some studies on the emergence of novel communicative systems (Selten & Warglien, 2007) have relied on experimentally constrained sets of stimulus–meaning mappings. This approach removes one of the most fundamental and characteristic features of the human communicative faculty, namely its ability to disambiguate a referent from a potentially infinite set of referents. Other studies have also inadvertently simplified the communicative problem by removing the need (for both communicator and addressee) to disambiguate communicative behaviors from instrumental behaviors (i.e., the problem of "signaling signalhood"; Scott-Phillips et al., 2009). This situation arises when experimenters prespecify the signals available to participants (Kirby, Griffiths, & Smith, 2014; Selten & Warglien; Smith, 2004). Experimental semiotics has also been careful to avoid confounding the generation of communicative behaviors with collateral by-products of a rigid experimental setting. An example of those experimental by-products are the cognitive control demands often imposed by apparently trivial task instructions. For instance, instead of verbally imposing a task constraint that needs to be remembered and continuously controlled (e.g., "do not speak"), experimental semiotics studies would rather nudge participants toward generating novel communicative signals by removing the possibility of using speech through computer-mediated interactions (de Ruiter et al., 2010; Galantucci, 2005). Similarly, instructing participants through task contingencies, rather than verbal instructions, is functionally closer to the natural ethology of human communication, promoting the spontaneous deployment of pragmatic inferences and shared goals (Oosterwijk et al., 2017).

2. *Cognitive Mechanisms of Conceptual Alignment*

Experimental semiotics uses the generation of novel shared symbols as a privileged window into the mechanisms supporting the negotiation of meaning across interacting agents, contingent on the interaction dynamics. Given that novel symbols lack a preexisting shared representation, jointly establishing their meaning relies on converging on a common ground of knowledge and beliefs across communicators, more so than the meaning of already known words and gestures. Accordingly, the generation of novel shared symbols requires a mechanism that allows us to rapidly converge on a shared meaning, constraining a potentially infinite cognitive search space of mappings between symbols and their possible interpretations. In order to alter an interlocutor's mental state according to the content of the communicative behavior, that mechanism should involve predictions based on presumed knowledge and beliefs of the interlocutor, conceptual knowledge that needs to be continuously updated and sharpened according to the shared history of the interaction (Brennan, Galati, & Kuhlen, 2010; Clark, 1996). For instance, we can direct the attention of an addressee to a remote referent by means of a proximate sign, as when a diner points to an empty glass to request a new bottle of wine from the waiter. The motoric simplicity of the pointing movement might easily lead one to underestimate the computational complexity of selecting that particular behavioral vehicle for conveying the intended meaning. Part of the complexity arises from the fact that, when the diner points to her empty glass, she uses a considerable scaffold of conceptual information to minimize the referential ambiguity of the movement, and she expects the waiter to use a similar scaffold to interpret the movement. The diner might use knowledge that wine can be ordered from waiters, but not from cleaners; knowledge that waiters know that customers may like a refill when their glass is empty; knowledge that the waiter saw the pointing being directed toward a red-wine glass, and that the waiter can differentiate red- from white-wine glasses; and so on. That knowledge is not immediately available in the environment. Furthermore, it is not obvious how a communicator can rapidly navigate through her general world knowledge to converge on elements that are relevant for that particular communicative behavior and to do that on first encounter with that unique set of circumstances (see also de Ruiter et al., 2010; Galantucci, 2005). This core phenomenon of human communication cannot be easily accommodated by traditional reinforcement learning theories (Sutton & Barto, 1998), game theories (Pinker, Nowak, & Lee, 2008), fast and frugal heuristics (Gigerenzer, 2008), or Bayesian models (Tenenbaum, Griffiths, & Kemp, 2006). Those models, theories, and heuristics require either a priori internal models of all possible novel signals (i.e., the models have meanings of symbols built in as conventions) or an unrealistically large number of training trials (Steels, 2003). We have recently suggested a possible computational path that does not seem to suffer from these problems: structure mapping theory, or analogical reasoning (Blokpoel, 2015; Gentner, 1983). In analogical reasoning, one uses representations of the relational structure of concepts to find analogical matches between different concepts (e.g., *the atom is like a solar system, with electrons circling the nucleus in much the same way as planets circle the sun*). Based on such matches, one can then transfer knowledge from a base concept to a target, generating new concepts (e.g., *perhaps the revolving of electrons is caused by the attraction of the nucleus like the revolving of the planets is caused by attraction of the sun*). This kind of reasoning might have the combination of generative power and conceptual sensitivity adequate to support the computational requirements for fast, even one-trial, convergence on a communicative behavior that can plausibly convey an intended content to an addressee. However, even assuming that this (untested) suggestion is viable, it remains unclear how a communicator can finesse those inferences over the presumed knowledge landscape of the waiter (Blokpoel et al., 2011), as it often happens when communicative signals are adjusted to the presumed characteristics of an addressee (*audience design*; Clark & Carlson, 1982; Clark & Murphy, 1982; Newman-Norlund et al., 2009). We have recently suggested that a further set of constraints used by interlocutors to rapidly infer the meaning of communicative signals might be found by embedding those symbols in a conceptual space whose activation predates in time the processing of the signals themselves and that consider multiple counterfactual communicative scenarios (Stolk, Noordzij, Verhagen, et al., 2014; Stolk, Verhagen, & Toni, 2016). The suggestion is that, even during a simple conversation, we continuously update and sharpen our (conceptual) priors according to the recent history of the communicative interaction, generating possible-world scenarios based on the presumed conceptual space shared with the interlocutor. In fact, several possible-world contexts might need to be continuously assessed to achieve the flexibility characteristic of human interactions (e.g., in the pointing example, when the waiter does not take the order, we might rapidly adjust to the possibility that the restaurant is closing, or that the presumed waiter is in fact a customer). This suggestion implies that alternative possible-world scenarios, including those not currently guiding the interpretation of the communicative behavior, are concurrently evaluated. Besides monitoring the effectiveness of the conceptual framework currently guiding behavior, evidence in favor of several counterfactual conceptual frameworks might need to be concurrently evaluated to enable an individual to adaptively and rapidly adjust to the fleeting communicative setting. The effectiveness of the ongoing conceptual framework is evaluated based on its reliability, defined as the likelihood of actual action outcomes given the current situation. The reliability of the ongoing framework thus indicates whether the situation has changed and the current conceptual framework should be adjusted accordingly (Donoso, Collins, & Koechlin, 2014; Koechlin, 2016). To optimally infer when to switch, and to which alternative regulatory strategy, the reliability of multiple previously used strategies should be concurrently monitored in an inferential buffer (Donoso et al., 2014). As the interaction unfolds, at each utterance, communicators continuously update their conceptual spaces, probe counterfactual spaces, and possibly generate additional conceptual spaces for interpreting the existing set of communicative behaviors. This conceptual alignment process builds on semantic structures operating over multiple timescales, from the fleeting idiosyncrasies of an ongoing communicative interaction to the long-term verbal and nonverbal semantic regularities acquired throughout life (Bozeat, Lambon Ralph, Patterson, Garrard, & Hodges, 2000; Regier & Kay, 2009).

3. Neural Mechanisms of Conceptual Alignment

Building on the approach detailed in the previous sections, experimental semiotics has made it possible to empirically interrogate, at the implementation level, long-standing theoretical issues (Clark, 1996; Grice, 1957; Levinson, 2006; Wilson & Sperber, 2004) and to provide new insight into the neural mechanisms of human communication. Recent reviews have summarized several important findings (Hasson & Frith, 2016; Stolk et al., 2016). Here we will focus on recent work particularly relevant to the issue of conceptual alignment, providing empirical evidence for three major insights. First, it has been shown that both production and comprehension of novel communicative behaviors is supported by a right-lateralized frontotemporal network (Noordzij et al., 2009; Stolk et al., 2013), known to be necessary for implementing the flexible conceptual processes supporting pragmatic and mental-state inferences, embedding utterances into their conversational context (Beeman, 1993; Bašnáková, Weber, Petersson, van Berkum, & Hagoort, 2013; Han et al., 2013; Sabbagh, 1999; Stolk, D'Imperio, di Pellegrino, & Toni, 2015). Those observations converge in linking a right-lateralized frontotemporal network to the processing of conceptual knowledge attributed to other human agents, dynamically adjusting that knowledge to the current situation. Second, neural activity in this right-lateralized frontotemporal network predates in time the production and comprehension of the communicative behaviors (Stolk et al., 2013; Stolk, Noordzij, Verhagen, et al., 2014). This observation fits with the notion that communicative behaviors are disambiguated in relation to a dynamic conceptual space defined by the ongoing communicative interaction. Interference studies of anterior frontal and posterior temporal nodes of that right-lateralized frontotemporal network confirm and qualify the contribution of those cortical regions. Ventromedial prefrontal lesion patients, known for their social conduct deficits (Beer, John, Scabini, & Knight, 2006), were found lacking the specific ability to constrain their ongoing communication with knowledge implied by the circumstances of an interaction, including the presumed abilities of an interlocutor (Stolk et al., 2015). Transient disruption of neural function in the right posterior superior temporal sulcus diminishes the ability to constrain the interpretation of communicative behaviors using knowledge abstracted from the recent communicative history (Stolk, Noordzij, Volman, et al., 2014). These causal observations emphasize the constraining role of contingently updated conceptual knowledge in social interaction. Third, neural activity in the same right-lateralized frontotemporal network is also modulated by experimental manipulations of the common ground shared by interlocutors.

This evidence has been gathered thanks to the novel and controlled environments used in experimental semiotics. Those environments offer the possibility of having interlocutors encounter situations for which they have previously converged on a shared communicative solution and contrast those situations with others where a shared communicative solution has not been developed yet (Stolk, Noordzij, Verhagen, et al., 2014). Communicating in those circumstances had behavioral and neural consequences. Behaviorally, interlocutors understood one another better, as indexed by the number of communicative problems jointly completed. Neurally, the same frontotemporal network identified earlier was more involved, both during production and comprehension of a communicative behavior. Given the ability of these experimental settings to minimize access to existing communicative conventions, it becomes possible to understand how and when interlocutors define the individual constraints that disambiguate each other's communicative behaviors. One subregion of that network, the superior temporal gyrus, showed neural dynamics matched to the behavioral dynamics of mutual understanding. A neural dynamics coupled across interlocutors was selectively evoked during communicative interactions in which the signal–meaning mappings were still open for negotiation. Crucially, the neurodynamic coupling between interlocutors occurred over temporal scales independent of the occurrence of the communicative behaviors themselves. These observations define a neural basis for the notion that we continuously update and sharpen our conceptual spaces according to the recent history of the communicative interaction, in line with the work we have described. Moreover, the simultaneous neural changes in superior temporal gyrus during meaning negotiation across interlocutors suggest that we mutually coordinate and adjust our individual knowledge spaces.

4. *Outstanding Issues*

To date, most research efforts have been directed at characterizing the cognitive complexity of human communicative abilities (Clark, 1996; Tomasello, 2008; Wilson & Sperber, 2004). Yet, any communicative system, irrespectively of its complexity, would be useless without a motivational system that makes an organism devote cognitive resources toward producing and interpreting communicative behaviors. What are the mechanisms that make us invest resources into changing the mental state of another person and liberally share goals with other agents, an apparently insurmountable obstacle for several other primates (Herrmann, Call, Hernandez-Lloreda, Hare, & Tomasello, 2007; Melis, Hare, & Tomasello, 2006)? Can the cognitive exploration of conceptual spaces relevant to the current communicative situation be controlled by relatively simple hormonal factors, as shown to be the case for the cognitive processes supporting economic transactions (Kosfeld, Heinrichs, Zak, Fischbacher, & Fehr, 2005)? We have recently started to explore whether oxytocin, a neuropeptide known to influence how humans share material resources (Meyer-Lindenberg, Domes, Kirsch, & Heinrichs, 2011; Shamay-Tsoory & Abu-Akel, 2016), is also involved in modulating how we share knowledge during a communicative exchange (de Boer et al., 2017), and how those modulations are neuronally implemented (Stolk, Kokal, de Boer, Oostenveld, & Toni, 2018). We found that oxytocin administration (intranasally, placebo-controlled) drives communicators to generate signals of higher referential quality, that is, signals that disambiguate more communicative problems, and to more rapidly adjust those communicative signals to what the addressee understands. The combined effects of oxytocin on referential quality and audience design fit with the notion that oxytocin administration leads communicators to explore more pervasively behaviors that can convey their intention and to consider more deeply alternative models of their addressees. These findings suggest that, besides affecting prosocial drive and salience of social cues (Shamay-Tsoory & Abu-Akel, 2016; Zak, Stanton, & Ahmadi, 2007), oxytocin influences how we share knowledge by promoting cognitive exploration (Ye, Stolk, Toni, & Hagoort, 2016). We also found that those modulatory effects of oxytocin on human communicative behavior have a neural counterpart in the same right-lateralized frontotemporal network identified earlier as critical for processing conceptual knowledge and adjusting it to the current communicative context (Stolk, Kokal, et al., 2018). More precisely, individuals with stronger activity in this network adjusted more readily their communicative behaviors to what their interlocutor actually understood during their interactions and were less biased by their prior stereotypical beliefs about the abilities of their interlocutor. This finding points to a neurophysiological mechanism through which oxytocin influences how we keep our thoughts aligned to the current situation and to each other, driving us away from perseverating in exploiting the same semantic structures. This research line illustrates how relatively simple motivational factors, as those mobilized and modulated by hormones, might have substantial cognitive consequences, possibly by boosting preexisting social tendencies through a reduction in social anxiety (Chang, Barter, Ebitz, Watson, & Platt, 2012; Radke et al., 2015). The crucial effect is that reduced social anxiety can then release the expression of cognitive competences that would be otherwise inhibited by competitive social dynamics (Burkart, Hrdy, & van Schaik, 2009; Hare, 2017; Melis et al., 2006), driving individuals to take more risky cognitive foraging decisions by enhancing their exploratory tendencies.

The neurocognitive mechanisms underlying communicative alterations is another research field that, in our opinion, will provide important insights when addressed through the approach and tools of experimental semiotics. For instance, there is evidence that neural degeneration of frontotemporal structures interferes with the ability to hold coherent conversations (Mates, Mikesell, & Smith, 2010) and to exhibit socially appropriate behavior (Josephs et al., 2009), which is possibly a reflection of an inability to cope with the cognitive alignment demands emphasized earlier. Ongoing behavioral work using tools from experimental semiotics might be able to precisely quantify communicative alterations in patients with frontotemporal dementia. By exploiting the natural variability in the location and extent of damaged brain tissue in this pathology, it might become possible to triangulate the relation between brain tissue loss and deficits in cognitive alignment, as the pathology progresses. Ultimately, it will be important to implement systematic comparisons between patient groups with communicative alterations. This pathophysiological comparative approach might isolate cognitive phenotypes that unify apparently disparate disorders (e.g., schizophrenia, autism spectrum disorder, frontotemporal dementia). For instance, it might emerge that patients with either one of those disorders have problems in creating and maintaining a shared conceptual space with an interlocutor. Preliminary evidence from our group suggests that this alteration is present in people with high-functioning autism spectrum disorder (ASD), despite relatively intact abilities and motivation to generate otherwise intelligible solutions to novel communicative challenges (Wadge, Brewer, Bird, Toni, & Stolk, 2019). By the same token, the comparative approach might pinpoint cognitive phenotypes exclusive to either one of those disorders. For instance, it might become possible to delineate the biological boundaries of the communication impairments that define one of the subdimensions of ASD (Bishop, Havdahl, Huerta, & Lord, 2016; Pina-Camacho et al., 2012). Given the aberrant functional and structural connectivity found in ASD throughout the frontotemporal network (Pina-Camacho et al., 2012; Voineagu et al., 2011), it might emerge that ASD patients have specific impairments in parsing apparently continuous behaviors into communicative and instrumental elements, preventing them from focusing on the communicatively relevant elements of a social interaction (e.g., in the diner example, the customer might communicatively point with the right hand while adjusting her hair with the left hand).

Ultimately, we believe that neurological patients will be crucial for understanding the neural mechanisms implementing conceptual alignment. Namely, patients implanted with high-density multielectrode arrays for the treatment of drug-resistant epilepsy (electrocorticography) will provide fundamental insights into the neuronal mechanisms supporting cognitive alignment. The main reason is that, despite the procedural control afforded by experimental semiotics protocols, it remains extremely hard to reliably capture neuronal traces associated with transient and abstract events such as conceptual alignment across interlocutors. Those events, being contingent on the shared history of interaction during a communicative exchange, cannot be simply averaged across multiple repeated instances, as is standard and necessary practice in most of the cognitive neurosciences. Electrocorticography can record neuronal populations with high signal-to-noise, opening up the possibility of investigating trial-unique events such as history-dependent episodes of conceptual alignment, analogous to how humans establish mutual understanding during social interaction.

5. Conclusion

We have outlined how experimental semiotics has opened the way for approaching human communication in terms of mutual understanding and offered a theoretically and empirically grounded framework that provides a neurocognitive account of our ability to disambiguate everyday communicative behaviors. In contrast to existing accounts of our communicative abilities that have focused on the information content of the symbols we use to understand each other during communicative interactions, the conceptual alignment framework emphasizes the dynamic conceptual space human communicators mutually coordinate, informed by what they commonly presume is known and believed by the other under the current circumstances. The empirical evidence suggests we continuously update and sharpen our conceptual spaces as interaction unfolds, providing an elusive yet critical backdrop for disambiguating a referent from a potentially infinite set of referents and for generating even completely novel symbols.

REFERENCES

Bašnáková, J., Weber, K., Petersson, K. M., van Berkum, J., & Hagoort, P. (2013). Beyond the language given: The neural correlates of inferring speaker meaning. *Cerebral Cortex, 24*(10), 2572–2578.

Beeman, M. (1993). Semantic processing in the right hemisphere may contribute to drawing inferences from discourse. *Brain and Language, 44*(1), 80–120.

Beer, J. S., John, O. P., Scabini, D., & Knight, R. T. (2006). Orbitofrontal cortex and social behavior: Integrating self-monitoring and emotion-cognition interactions. *Journal of Cognitive Neuroscience, 18*(6), 871–879.

Bishop, S. L., Havdahl, K. A., Huerta, M., & Lord, C. (2016). Subdimensions of social-communication impairment in autism spectrum disorder. *Journal of Child Psychology and Psychiatry, 57*(8), 909–916.

Blokpoel, M. (2015). *Understanding understanding: A computational-level perspective* (Doctoral dissertation). Donders Graduate School for Cognitive Neuroscience Series 195. Retrieved from repository of the Radboud Universiteit, Nijmegen http://repository.ubn.ru.nl/handle/2066/144897.

Blokpoel, M., Kwisthout, J. H. P., Wareham, H. T., Haselager, W. F. G., Toni, I., & van Rooij, I. J. E. I. (2011). The computational costs of recipient design and intention recognition in communication. *Proceedings of the 33rd Annual Meeting of the Cognitive Science Society* (pp. 465–470). Austin, TX: Cognitive Science Society.

Bozeat, S., Lambon Ralph, M. A., Patterson, K., Garrard, P., & Hodges, J. R. (2000). Non-verbal semantic impairment in semantic dementia. *Neuropsychologia, 38*(9), 1207–1215.

Brennan, S. E., & Clark, H. H. (1996). Conceptual pacts and lexical choice in conversation. *Journal of Experimental Psychology: Learning, Memory, and Cognition, 22*(6), 1482.

Brennan, S. E., Galati, A., & Kuhlen, A. K. (2010). Two minds, one dialog: Coordinating speaking and understanding. *Psychology of Learning and Motivation, 53*, 301–344.
Burkart, J. M., Hrdy, S. B., & Van Schaik, C. P. (2009). Cooperative breeding and human cognitive evolution. *Evolutionary Anthropology: Issues, News, and Reviews, 18*(5), 175–186.
Chang, S. W. C., Barter, J. W., Ebitz, R. B., Watson, K. K., & Platt, M. L. (2012). Inhaled oxytocin amplifies both vicarious reinforcement and self reinforcement in rhesus macaques *(Macaca mulatta)*. *Proceedings of the National Academy of Sciences, 109*(3), 959–964.
Clark, H. H. (1996). *Using language*. Cambridge: Cambridge University Press.
Clark, H. H., & Carlson, T. B. (1982). Hearers and speech acts. *Language, 58*(2), 332–373.
Clark, H. H., & Murphy, G. L. (1982). Audience design in meaning and reference. *Advances in Psychology, 9*, 287–299.
de Boer, M., Kokal, I., Blokpoel, M., Liu, R., Stolk, A., Roelofs, K., ... Toni, I. (2017). Oxytocin modulates human communication by enhancing cognitive exploration. *Psychoneuroendocrinology, 86*(supplement C), 64–72.
de Ruiter, J. P., Noordzij, M. L., Newman-Norlund, S., Newman-Norlund, R., Hagoort, P., Levinson, S. C., & Toni, I. (2010). Exploring the cognitive infrastructure of communication. *Interaction Studies, 11*(1), 51–77.
Donoso, M., Collins, A. G., & Koechlin, E. (2014). Foundations of human reasoning in the prefrontal cortex. *Science, 344*(6191), 1481–1486.
Dyer, F. C. (2002). The biology of the dance language. *Annual Review of Entomology, 47*(1), 917–949.
Everaert, M. B. H., Huybregts, M. A. C., Chomsky, N., Berwick, R. C., & Bolhuis, J. J. (2015). Structures, not strings: Linguistics as part of the cognitive sciences. *Trends in Cognitive Sciences, 19*(12), 729–743.
Fay, N., Arbib, M., & Garrod, S. (2013). How to bootstrap a human communication system. *Cognitive Science, 37*(7), 1356–1367.
Fusaroli, R., & Tylén, K. (2016). Investigating conversational dynamics: Interactive alignment, interpersonal synergy, and collective task performance. *Cognitive Science, 40*(1), 145–171.
Galantucci, B. (2005). An experimental study of the emergence of human communication systems. *Cognitive Science, 29*(5), 737–767.
Galantucci, B., & Garrod, S. (2011). Experimental semiotics: A review. *Frontiers in Human Neuroscience, 168*(6), 161.
Galantucci, B., & Garrod, S. (2012). *Experimental semiotics: Studies on the emergence and evolution of human communication*. Philadelphia: John Benjamins.
Garrod, S., Fay, N., Lee, J., Oberlander, J., & MacLeod, T. (2007). Foundations of representation: Where might graphical symbol systems come from? *Cognitive Science, 31*(6), 961–987.
Gentner, D. (1983). Structure-mapping: A theoretical framework for analogy. *Cognitive Science, 7*, 155–170.
Gigerenzer, G. (2008). Why heuristics work. *Perspectives on Psychological Science, 3*(1), 20–29.
Grice, H. P. (1957). Meaning. *Philosophical Review, 66*, 377–388.
Han, Z., Bi, Y., Chen, J., Chen, Q., He, Y., & Caramazza, A. (2013). Distinct regions of right temporal cortex are associated with biological and human-agent motion: Functional magnetic resonance imaging and neuropsychological evidence. *Journal of Neuroscience, 33*(39), 15442–15453.
Hare, B. (2017). Survival of the friendliest: *Homo sapiens* evolved via selection for prosociality. *Annual Review of Psychology, 68*, 155–186.
Hari, R., Henriksson, L., Malinen, S., & Parkkonen, L. (2015). Centrality of social interaction in human brain function. *Neuron, 88*(1), 181–193.
Hasson, U., & Frith, C. D. (2016). Mirroring and beyond: Coupled dynamics as a generalized framework for modelling social interactions. *Philosophical Transactions of the Royal Society B, 371*(1693), 20150366.
Healey, P. G., Swoboda, N., Umata, I., & King, J. (2007). Graphical language games: Interactional constraints on representational form. *Cognitive Science, 31*(2), 285–309.
Herrmann, E., Call, J., Hernandez-Lloreda, M. V., Hare, B., & Tomasello, M. (2007). Humans have evolved specialized skills of social cognition: The cultural intelligence hypothesis. *Science, 317*(5843), 1360–1366.
Josephs, K. A., Whitwell, J. L., Knopman, D. S., Boeve, B. F., Vemuri, P., Senjem, M. L., ... Jack, C. R. (2009). Two distinct subtypes of right temporal variant frontotemporal dementia. *Neurology, 73*(18), 1443–1450.
Kirby, S., Griffiths, T., & Smith, K. (2014). Iterated learning and the evolution of language. *Current Opinion in Neurobiology, 28*, 108–114.
Koechlin, E. (2016). Prefrontal executive function and adaptive behavior in complex environments. *Current Opinion in Neurobiology, 37*, 1–6.
Kosfeld, M., Heinrichs, M., Zak, P. J., Fischbacher, U., & Fehr, E. (2005). Oxytocin increases trust in humans. *Nature, 435*, 673–676.
Levinson, S. C. (2006). On the human "interaction engine." In N. J. Enfield, S. C. Levinson, & L. C. Aiello (Eds.), *Roots of human sociality: Culture, cognition and interaction* (pp. 39–69). Duck, NC: Berg.
Mates, A. W., Mikesell, L., & Smith, M. S. (Eds.) (2010). *Language, interaction and frontotemporal dementia: Reverse engineering the social mind*. London: Equinox. Retrieved from http://scholar.google.com/scholar?cluster=8930796620766635945&hl=en&oi=scholarr.
Mather, J. A. (2004). Cephalopod skin displays: From concealment to communication. In D. K. Oller & U. Griebel (Eds.), *Evolution of communication systems* (pp. 193–214). Cambridge, MA: MIT Press.
Melis, A. P., Hare, B., & Tomasello, M. (2006). Engineering cooperation in chimpanzees: Tolerance constraints on cooperation. *Animal Behaviour, 72*, 275–286.
Meyer-Lindenberg, A., Domes, G., Kirsch, P., & Heinrichs, M. (2011). Oxytocin and vasopressin in the human brain: Social neuropeptides for translational medicine. *Nature Reviews Neuroscience, 12*(9), 524–538.
Newman-Norlund, S. E., Noordzij, M. L., Newman-Norlund, R. D., Volman, I. A. C., Ruiter, J. P. de, Hagoort, P., & Toni, I. (2009). Recipient design in tacit communication. *Cognition, 111*(1), 46–54.
Noordzij, M. L., Newman-Norlund, S. E., de Ruiter, J. P., Hagoort, P., Levinson, S. C., & Toni, I. (2009). Brain mechanisms underlying human communication. *Frontiers in Human Neuroscience, 3*, article 14.
Noy, L., Dekel, E., & Alon, U. (2011). The mirror game as a paradigm for studying the dynamics of two people

improvising motion together. *Proceedings of the National Academy of Sciences, 108*(52), 20947–20952.

Oosterwijk, A. M., de Boer, M., Stolk, A., Hartmann, F., Toni, I., & Verhagen, L. (2017). Communicative knowledge pervasively influences sensorimotor computations. *Scientific Reports, 7*(1), 4268.

Pina-Camacho, L., Villero, S., Fraguas, D., Boada, L., Janssen, J., Navas-Sánchez, F. J., ... Parellada, M. (2012). Autism spectrum disorder: Does neuroimaging support the DSM-5 proposal for a symptom dyad? A systematic review of functional magnetic resonance imaging and diffusion tensor imaging studies. *Journal of Autism and Developmental Disorders, 42*(7), 1326–1341.

Pinker, S., Nowak, M. A., & Lee, J. J. (2008). The logic of indirect speech. *Proceedings of the National Academy of Sciences, 105*(3), 833–838.

Radke, S., Volman, I., Mehta, P., van Son, V., Enter, D., Sanfey, A., ... Roelofs, K. (2015). Testosterone biases the amygdala toward social threat approach. *Science Advances, 1*(5), e1400074.

Regier, T., & Kay, P. (2009). Language, thought, and color: Whorf was half right. *Trends in Cognitive Sciences, 13*(10), 439–446.

Rizzolatti, G., & Sinigaglia, C. (2016). The mirror mechanism: A basic principle of brain function. *Nature Reviews: Neuroscience, 17*(12), 757–765.

Roberts, G., Lewandowski, J., & Galantucci, B. (2015). How communication changes when we cannot mime the world: Experimental evidence for the effect of iconicity on combinatoriality. *Cognition, 141*, 52–66.

Sabbagh, M. A. (1999). Communicative intentions and language. *Brain and Language, 70*, 29–69.

Scott-Phillips, T. C., Kirby, S., & Ritchie, G. R. S. (2009). Signalling signalhood and the emergence of communication. *Cognition, 113*(2), 226–233.

Sebanz, N., Bekkering, H., & Knoblich, G. (2006). Joint action: Bodies and minds moving together. *Trends in Cognitive Sciences, 10*, 70–76.

Selten, R., & Warglien, M. (2007). The emergence of simple languages in an experimental coordination game. *Proceedings of the National Academy of Sciences, 104*, 7361–7366.

Seyfarth, R. M., Cheney, D. L., & Marler, P. (1980). Monkey responses to three different alarm calls: Evidence of predator classification and semantic communication. *Science, 210*, 801–803.

Shamay-Tsoory, S. G., & Abu-Akel, A. (2016). The social salience hypothesis of oxytocin. *Biological Psychiatry, 79*(3), 194–202.

Silbert, L. J., Honey, C. J., Simony, E., Poeppel, D., & Hasson, U. (2014). Coupled neural systems underlie the production and comprehension of naturalistic narrative speech. *Proceedings of the National Academy of Sciences, 111*(43), E4687–E4696.

Smith, K. (2004). The evolution of vocabulary. *Journal of Theoretical Biology, 228*(1), 127–142.

Steels, L. (2003). Evolving grounded communication for robots. *Trends in Cognitive Sciences, 7*, 308–312.

Stolk, A., D'Imperio, D., di Pellegrino, G., & Toni, I. (2015). Altered communicative decisions following ventromedial prefrontal lesions. *Current Biology, 25*(11), 1469–1474.

Stolk, A., Kokal, I., de Boer, M., Oostenveld, R., & Toni, I. (2018). *Oxytocinergic modulation of human adaptive communication and broadband neuronal dynamics.* Manuscript in preparation.

Stolk, A., Noordzij, M. L., Verhagen, L., Volman, I., Schoffelen, J.-M., Oostenveld, R., ... Toni, I. (2014). Cerebral coherence between communicators marks the emergence of meaning. *Proceedings of the National Academy of Sciences, 111*(51), 18183–18188.

Stolk, A., Noordzij, M. L., Volman, I., Verhagen, L., Overeem, S., van Elswijk, G., ... Toni, I. (2014). Understanding communicative actions: A repetitive TMS study. *Cortex, 51*(supplement C), 25–34.

Stolk, A., Verhagen, L., Schoffelen, J.-M., Oostenveld, R., Blokpoel, M., Hagoort, P., ... Toni, I. (2013). Neural mechanisms of communicative innovation. *Proceedings of the National Academy of Sciences, 110*(36), 14574–14579.

Stolk, A., Verhagen, L., & Toni, I. (2016). Conceptual alignment: How brains achieve mutual understanding. *Trends in Cognitive Sciences, 20*(3), 180–191.

Sutton, R. S., & Barto, A. G. (1998). *Reinforcement learning an introduction.* Cambridge, MA: MIT Press.

Tenenbaum, J. B., Griffiths, T. L., & Kemp, C. (2006). Theory-based Bayesian models of inductive learning and reasoning. *Trends in Cognitive Sciences, 10*(7), 309–318.

Tomasello, M. (2008). *Origins of human communication.* Cambridge, MA: MIT Press.

Tomasello, M., Carpenter, M., Call, J., Behne, T., & Moll, H. (2005). Understanding and sharing intentions: The origins of cultural cognition. *Behavioral and Brain Sciences, 28*(5), 675–691.

Voineagu, I., Wang, X., Johnston, P., Lowe, J. K., Tian, Y., Horvath, S., ... Geschwind, D. H. (2011). Transcriptomic analysis of autistic brain reveals convergent molecular pathology. *Nature, 474*(7351), 380–384.

Wadge, H., Brewer, R., Bird, G., Toni, I., & Stolk, A. (2019). Communicative misalignment in autism spectrum disorder. *Cortex*. doi:10.1016/j.cortex.2019.01.003

Wilson, D., & Sperber, D. (2004). Relevance theory. L. R. Horn & G. Ward (Eds.), *The handbook of pragmatics* (pp. 607–632). Oxford: Blackwell.

Ye, Z., Stolk, A., Toni, I., & Hagoort, P. (2016). Oxytocin modulates semantic integration in speech comprehension. *Journal of Cognitive Neuroscience, 29*(2), 267–276.

Zak, P. J., Stanton, A. A., & Ahmadi, S. (2007). Oxytocin increases generosity in humans. *PLOS ONE, 2*(11), e1128.

19 Key Issues and Future Directions: Interactional Foundations for Language

STEPHEN C. LEVINSON AND IVAN TONI

1. The Design Features of Interactive Abilities That Lie behind Successful Language Use

The faculty for language is so bound up with the success of our species that without it we would still probably be a middle-sized inarticulate ape in Africa. Certainly, there would be no large-scale polities, no elaborated culture and technology, and no science. It may then seem churlish to question its preeminence, its ability to single-handedly transform us into cultural animals with a collective consciousness. But that is what this part of this volume sets out to do.

Consider for a moment a list of the failings of language (where "language" is understood just as the meaning of the words expressed), many of which have been usefully exposed by attempts to make machines comprehend natural languages (Jurafsky, 2003):

a. Natural language vocabularies are limited, hence most words are semantically generally over many meanings (consider *Aunt* which might denote one's father's sister, or mother's sister, or father's brother's wife, or mother's brother's wife, not to mention the many possible denotations of *great aunt*), or just plain vague (How big is a *heap*? When does a *mist* become *fog*, or a *bush* become a *tree?*).
b. Virtually all natural language sentences are multiply ambiguous, as in *He saw her duck* or *Visiting academics can be tiresome.*
c. Virtually all utterances require a good dose of common-sense understanding under interpretive rules of thumb: *It's not impossible he'll still come* suggests it is unlikely (Levinson, 2000), *The room was huge and the speaker couldn't be heard* suggests that the speaker was in that room and the size of it was the reason he couldn't be heard (Clark, 1977).
d. Language has distinct limits: it seems ill equipped to express shapes, precise hues, faces, or emotions (Levinson & Majid, 2014).
e. The main point of an utterance, the speech act expressed (e.g., advising, offering, requesting), is rarely encoded directly, but rather contextually inferred (Levinson, 2013).

The list could be multiplied, but the point is clear: language only works because it is supplemented by a great deal of further information, including conventions or heuristics under which it is used. One of the enduring insights of pragmatics—the study of language usage—is that language presupposes, indeed rides on, a vast infrastructure of human capacities for communication, partially independent of language. This is not simply the insight that there are forms of human communication independent of language—pictorial signs, gesture and "body language," music, and so forth—but rather the insight that language could not possibly work without a prior framework of mutual assumptions, interpretative heuristics, and interactional norms. Most observers and practitioners of the language sciences hugely overrate the independent efficacy of language alone—it is a seemingly miraculous ability, but the miracle lies very largely in these unseen foundations that allow it to operate.

This section of the book seeks to throw some light on these underlying, unseen foundations on which our linguistic abilities rest. These foundations play a crucial role in nearly every current use of language, but also in the creative abilities and the cooperative assumptions under which new languages have arisen, and they throw a great deal of light on the great unsolved mystery of language, namely how it ever evolved in the first place. But what are these foundations exactly?

The chapters gathered here agree that the focus should be on interactive uses of language, for conversation and task-oriented interactive language use forms the great bulk of ordinary language usage and the context in which language is learned by children and almost certainly arose in evolution. So what are the hidden background properties that make language work so well in this interactional niche? One way to think about this is to entertain the following thought experiment: What would we need to build into a robot, beyond and above linguistic abilities themselves, for such a machine to be able to use language in a humanoid way? Table 19.1 provides a provisional and partial list (drawn partly from Levinson, 2019). The list starts off by pointing out that natural interactive language usually comes in a multimodal format: there is first of all the prosody (the duration, pitch, amplitude not intrinsically specified by the linguistic system) of the vocal signal; then there are the gestures of the hands, the raised eyebrows and furrowed brow of the face, the nods and even the blinks (Hömke, Holler, & Levinson, 2017), that are all packaged up to construct a single complex multimodal message (see Clark, chapter 17 of this volume, for illustrations of the expressive power of multimodal communication). In some cases, the entire burden of communication is shifted off the voice to the hands (see Goldin-Meadow, chapter 16).

The list continues with aspects of the context of verbal interaction, which canonically takes place under the umbrella of cooperative assumptions, with far-reaching inferential consequences (Grice, 1989). Interactive language usage normally takes place in the context of, and indeed constitutes, a form of joint action: participants have some shared goals, and to that end they each contribute to them. Next comes the physical context in its epistemic wrapper as it were: the mutual knowledge of what is salient in the environment, affording, for example, pointing to it, and more broadly the shared knowledge that each participant will assume the other shares. Arguably, the whole point of communication is primarily to convert a knowledge asymmetry into symmetrical, matched epistemic states (see Toni & Stolk, chapter 18). Then comes the discourse context—the parsed sequential environment in which an individual utterance is enacted. For interactive language use, the alternating turns at talk provide a crucial environment for response interpretation—whether some utterance is an answer,

TABLE 19.1
Some ingredients, beyond language, for a language-using robot (after Levinson, 2019)

MEDIA	
(Function: communication)	Multimodal signals Language
COOPERATIVE UMBRELLA	Joint action: shared goals with distributed sub-goals; Additional motivations, e.g. politeness, signaling of appropriate affect
EPISTEMIC CONTEXT	Salient properties of physical context; presumed 'common ground' and background knowledge
DISCOURSE CONTEXT, ACTION SEQUENCES	
(a) Alternating turns (Functions include legibility, opportunities for repair)	Speech act mapped to language
(b) Action sequences (Functions include structuring exchange)	Adjacency pairs (e.g. Q-A) Complex sequences: e.g. Insert pairs (e.g. Q-Q-A-A)
(c) Simultaneous actions (Function: coordination, ritual)	e.g. shaking hands, laughing
META-COMMUNICATION	
(Function: check communication)	Repair
(Function: confirm message receipt)	Feedback tokens (e.g. *uhuh*)
TIMING	
(Function: indicates state of processing is 'on time')	Turn-taking timing
(Function: 'clock speed' check)	Synchronicity (as in shaking hands)
(Function: 'message received now')	Timing of feedback
LEGABILITY	
(a) of attention (Function: indicates current focus of processing)	e.g. gaze readability
(b) of intention (Function: aid predictive processing)	e.g. of gesture signal vs. instrumental action

an assertion, a correction, and so on, will depend on the prior utterance. This involves a complex mapping of utterances onto speech acts, for which we currently lack deep understanding (Levinson, 2013). Then there is a sort of "grammar" of action sequences, which can be quite elaborate (see Levinson, chapter 14). The whole system only works because there are ways of invoking "repair" when a prior utterance cannot be construed—this mechanism of repair is fundamentally metacommunicative, as is the system of back-channel signals that acknowledge receipt without intruding on the speaker's turn. All of this is conducted under tight timing constraints, for example, turns at talking are usually separated by no more than 200 ms, literally near the duration of a blink of the eye, and delays of 600 ms or more in response begin to be interpreted as a hearer's problem (see Bögels & Levinson, 2017). Finally, and especially problematic for a robot, all these signals have to be legible—thus if my gaze is directed at the bottle when I say "this," the robot would have to be able to "read" the direction of my gaze. Most complex of all is that understanding an utterance may require "mind reading"—the ability to recover not only the surface content of an utterance, but the underlying role it is meant to play in the joint action in which we are engaged.

Summarizing, we can say that building an interactive robot with these properties would be extremely challenging (see Gluck and Laird, 2019), not least because for many of these properties we have only the haziest idea about the underlying cognitive mechanisms that instantiate them. We return to the future challenges in section 3.

2. The Chapters in This Part

The individual chapters in this part can be read as contributing to the overall picture of this large, implicit basis for human communication. Clark points to a fundamental gap in our understanding of human communication. We tend to think about communication as the transfer of propositional information—a view that founders on many rocks in fact, including the fact that what is traded in interactional exchange is speech acts, not propositions. Clark notes that many utterances *depict*, that is, convey an iconic representation of a described situation, as when a gesture indicates the manner in which an action was done, or a facial expression depicts the horror of a witnessed scene. In an elaborate taxonomy of such depictions, he notes that sometimes they fill a slot in a sentence frame, as in *Suddenly the car went grrrrrrunk!* where the acoustic depiction fills an adverbial slot. We have no adequate theory about how such depictions, acoustic or gestural, can be unified with the rest of the message—depictions are analog, "pictorial" elements that can nevertheless contribute to the content of utterances in crucial ways.

Goldin-Meadow looks at the "big-bang" of language origins, as discernable in the microcontexts of a deaf child shielded from institutional help, growing up with hearing parents who do not command sign language. Under these circumstances, infants actually create a mini sign language, one that may be understood by the parents but not symmetrically mastered by them. Known as *homesign*, these systems are idiosyncratic, and yet they show some striking similarities in design and usage. They have a strong basis in the iconicity and enactment that Clark has also noted as characteristic of natural speech. These systems over time come to exhibit grammatical categories such as noun versus verb, hierarchical structure, sentence modifiers such as negation and interrogation, and other properties typical of language. Since the parental input, typically speech and gesture, has none of these properties discernible to the deaf child, the child appears to create them de novo—there is some interaction between human nature and the communicative situation that builds these mini systems. Without cultural elaboration they remain highly limited, however, but the transformation of such systems into much richer communication systems can be witnessed when multiple homesigners are brought together, as in the birth of the conventional Nicaraguan sign language.

Homesigners and their caretakers succeed in understanding one another to the extent that they do because they build on a fundamental incremental mechanism, the generation of "common ground," things we know each other knows. Toni and Stolk sketch recent developments in our understanding of this "conceptual alignment," the sharing of intentions, mental models, and interaction history: having once used a phrase like *the squiggle* to successfully refer, we know we can continue to do so. They point to the new field of experimental semiotics, where participants have to create successful signals de novo, just like the homesigners, but in controlled experimental circumstances that give insight into the underlying reasoning and its neurocognitive implementation. An interesting new development is the demonstration that oxytocin, a relatively simple neuropeptide, can induce improved performance in these communicative tasks. This observation suggests that relatively simple motivational factors, such as those mobilized and modulated by hormones, might have substantial cognitive consequences (Theofanopoulou, Boeckx, & Jarvis, 2017).

Levinson sketches a great slew of interactional assumptions and behaviors that we bring to verbal exchanges, including the motivational system that promotes intense and prolonged communication. He argues that these are deeply embedded in human nature, and are thus strongly universal across all cultures and languages. They consist of elements such as the exchange of short bursts of nonoverlapping speech with rapid transitions of speakers, of strikingly similar action types and sequences, and even the rates and types of interactive repair. He argues that these have a prelinguistic origin, since they can be partially seen very early on in infancy, and even in part across other primate species. It is this strong preexisting background of interactional skills and expectations, and indeed motivations, that makes it possible for infants to bootstrap themselves up into the local language with all of its conventions. This "interaction engine" points to a likely deep phylogenetic background, and hints that there may be strong continuities with other species underlying the perceived only-human Rubicon of linguistic abilities.

Rossano (chapter 15) provides much further evidence for parallels in interactional patterns between humans and the great apes, our nearest cousins. He argues that studies of animal communication have been hampered by an information theoretic framework; instead he argues we should adopt the same kind of framework that conversation analysts (like Schegloff, 2007) have applied to human interaction, namely thinking about communication as the exchange of social actions (cf. speech acts like requesting, offering, refusing, and such). He argues that by adopting this perspective, one sees strong commonalities across human and ape interactions, down to the very kinds of social action and their temporal properties. The interactional perspective thus serves to bridge the Rubicon between human and animal communication systems, and in so doing points to the evolutionary origins of our communicational prowess.

3. Future Challenges

As we said in the introduction, for most of the properties that characterize efficient language use we have little understanding of the underlying cognitive mechanisms that drive them. Nor do we understand the origins of the often quite striking uniformities across languages and cultures—native ethology, rational agency, or emergent adaptation of tools for the communicative job. Grice (1989) outlined some principles that he felt followed from the cooperative nature of communicative exchange (see Levinson, 2000, for an account in terms of simple heuristics), which are responsible for a wide range of inferences beyond what is actually said. But how many such principles are there? And, again, where do they come from? The elephant in the room here is perhaps the nature of communicative intention recognition. Philosophical reconstructions of this process seem psychologically unrealistic (e.g., threatening an infinite regress of my thinking about what you will think my intent is, based on your thinking about my thinking about your thinking…), even though some mirroring of design processes seems to show up in neuroimaging studies (Stolk et al., 2013). Computational models seem to demonstrate the intractability of intention reconstruction outside a narrow range of goals (Blokpoel, 2015). It will be easier to make progress on some of the more superficial features in table 19.1, for example, it is currently unknown how multimodal signals are composed into a single message—we know that the bits that go together often do not precisely align in time, so we have here a substantial "binding" problem both from a comprehension perspective (Habets, Kita, Shao, Özyürek, & Hagoort, 2011; Holler et al., 2015) and, even more challenging, a production perspective (see Chu & Hagoort, 2014). We have recently made good progress on some of the timing questions (see, e.g., Levinson, 2016), the meta-communicative systems (see, e.g., Dingemanse et al., 2015, Hömke et al., 2017), and the action-sequencing systems (Schegloff, 2007; Kendrick et al., 2014), although only in the first do we have a sketch of the cognitive processes involved.

Future work could be usefully directed at the following targets:

- Developing tasks and experimental models that can illuminate the cognitive processing behind the recognition of intentions (cf. de Ruiter et al., 2010) and the creative creation of one-off signals and depictions.
- Understanding the *binding problem* in multimodal signals—determining what bits in the parallel streams of signals belong together, how participants know, and how communicators orchestrate these multiple channels.
- Building processing models for multimodal processing, common ground updating, signal disambiguation that can make predictions in the neurocognitive domain.
- Determining how context is brought to bear so rapidly to disambiguate and resolve reference, grammatical ambiguity, speech act assignment, and the like.
- Determining how much of the "interaction engine" is native endowment, how much is learned, and how much is naturally emergent in the context of interactive communication.
- Examining how best can we pursue the leads that seem to point to deep underlying homologies

across at least some of our nearest cousins, the other primate species.

- Using human neuroscience methods with high signal-to-noise (e.g., electrocorticography), it might become possible to reliably capture neuronal traces supporting conceptual alignment across interlocutors, that is, transient and nonstationary events contingent on the shared history of interaction during a communicative exchange.

REFERENCES

Blokpoel, M. (2015). *Understanding understanding: A computational-level perspective* (Doctoral dissertation), Donders Graduate School for Cognitive Neuroscience Series 195. Retrieved from Radboud Repository of the Radboud University Nijmegen, http://hdl.handle.net/2066/144897.

Bögels, S., & Levinson, S. C. (2017). The brain behind the response: Insights into turn-taking in conversation from neuroimaging. *Research on Language and Social Interaction, 50*(1), 71–89.

Chu, M., & Hagoort, P. (2014). Synchronization of speech and gesture: Evidence for interaction in action. *Journal of Experimental Psychology: General, 143*(4), 1726–1741.

Clark, H. H. (1977). Inferences in comprehension. In D. LaBerge & S. J. Samuels (Eds.), *Basic processes in reading: Perception and comprehension* (pp. 243–263). Hillsdale, NJ: Lawrence Erlbaum.

De Ruiter, J. P., Noordzij, M. L., Newman-Norlund, S., Newman-Norlund, R., Hagoort, P., Levinson, S. C., & Toni, I. (2010). Exploring the cognitive infrastructure of communication. *Interaction Studies, 11*(1), 51–77.

Dingemanse, M., Roberts, S. G., Baranova, J., Blythe, J., Drew, P., Floyd, S., Gisladottir, R. S., & Enfield, N. J. (2015). Universal principles in the repair of communication problems. *PLOS ONE, 10*(9), e0136100.

Gluck, K. A. & Laird, J. E. (Eds.). (2019). *Interactive Task Learning: Humans, Robots, and Agents Acquiring New Tasks through Natural Interactions.* Cambridge, MA: MIT Press.

Grice, H. P. (1989). *Studies in the way of words.* Cambridge, MA: Harvard University Press.

Habets, B., Kita, S., Shao, Z., Özyürek, A., & Hagoort, P. (2011). The role of synchrony and ambiguity in speech–gesture integration during comprehension. *Journal of Cognitive Neuroscience, 23,* 1845–1854.

Holler, J., Kokal, I., Toni, I., Hagoort, P., Kelly, S. D., & Ozyurek, A. (2015). Eye'm talking to you: Speakers' gaze direction modulates co-speech gesture processing in the right MTG. *Social Cognitive and Affective Neuroscience, 10,* 255–261.

Hömke, P., Holler, J., & Levinson, S. C. (2017). Eye blinking as addressee feedback in face-to-face conversation. *Research on Language and Social Interaction, 50*(1), 54–70.

Jurafsky, D. (2003). Pragmatics and computational linguistics. In L. Horn & G. Ward (Eds.), *Handbook of pragmatics* (pp. 578–604). Oxford: Blackwell.

Kendrick, K. H., Brown, P., Dingemanse, M., Floyd, S., Gipper, S., Hayano, K., & Levinson, S. C. (2014). Sequence organization: A universal infrastructure for action. Paper presented at the *4th International Conference on Conversation Analysis,* University of California at Los Angeles.

Levinson, S. C. (2000). *Presumptive meanings.* Cambridge, MA: MIT Press.

Levinson, S. C. (2013). Action formation and ascription. In T. Stivers & J. Sidnell (Eds.), *The handbook of conversation analysis* (pp. 103–130). Chichester, UK: Wiley-Blackwell.

Levinson, S. C. (2016). Turn-taking in human communication, origins, and implications for language processing. *Trends in Cognitive Sciences, 20*(1), 6–14.

Levinson, S. C. (2019). Natural forms of purposeful interaction among humans: What makes interaction effective? In K. A. Gluck & J. E. Laird (Eds.), *Interactive Task Learning: Humans, Robots, and Agents Acquiring New Tasks through Natural Interactions.* Cambridge, MA: MIT Press.

Levinson, S. C., & Majid, A. (2014). Differential ineffability and the senses. *Mind and Language, 29*(4), 407–427.

Schegloff, E. A. (2007). *Sequence organization in interaction.* Cambridge: Cambridge University Press.

Stolk, A., Verhagen, L., Schoffelen, J.-M., Oostenveld, R., Blokpoel, M., Hagoort, P., … & Toni, I. (2013). Neural mechanisms of communicative innovation. *PNAS, 110*(36), 14574–14579.

Theofanopoulou, C., Boeckx, C., & Jarvis, E. D. (2017). A hypothesis on a role of oxytocin in the social mechanisms of speech and vocal learning. *Proceedings of the Royal Society B, 284*(1861), 20170988. doi:10.1098/rspb.2017.0988.

IV MODELING LANGUAGE

SECTION EDITOR: WILLEM ZUIDEMA

20 Speech Perception and Production

BART DE BOER

Computer models have played an important role in the study of speech as long as computers have been available to researchers and engineers (de Boer & Fitch, 2010; Schroeder, 1993). For instance, the first digital computers in Sweden (called BARK and BESK), along with one of their first analog computers (LEA) were used by Fant (1960) for research into speech production. A lot of this early work had a very practical motivation: much of the research on speech production and perception had the ultimate aim to put as many simultaneous telephone calls on a transatlantic cable as possible. However, a lot of this early work turned out to be extremely helpful in better understanding how humans produce and perceive speech. Therefore, computer models started to be used by phoneticians with less application-oriented motivations as well. This eventually led to a split in the field: engineers interested in pursuing practical applications of computers in producing and processing speech on the one hand, and phoneticians using computer models to better understand how humans produce and process speech on the other. This chapter will focus on the latter kind of research.

This chapter first gives an overview of engineering approaches to put the modeling research in context. Different aspects of research models for speech will be discussed in section 2, and the ways such models can interact with experiment and observation are discussed in section 3. The chapter ends with two case studies to illustrate the concepts that are discussed.

1. Engineering-Based Methods

Engineering-based methods for speech recognition and speech synthesis have certainly been successful, as witnessed by readily available applications in smartphones and other portable computer devices. However, the methods used are aimed at performance, not at achieving cognitive realism. For instance, in computer speech synthesis, there are several techniques that can be used to generate speech. One of these is articulatory synthesis (Taylor, 2009, chapter 13), which models the physical process of speech production directly and is therefore also very useful to understand the process of speech production. Although this method was an important focus of early work because it has very low memory requirements (which was important for early computers) it is never used in practical applications anymore, because it is very hard to achieve acceptably realistic speech quality with it. In practice, a different technique is used nowadays: concatenation synthesis (Taylor, 2009, chapters 14–16). In this technique, stretches of speech are selected from a large database of example speech and concatenated (with slight modifications of pitch, volume, and duration) in such a way as to approximate a target utterance as closely as possible with as little modification as possible. Results get better as the database gets bigger. Obviously, this technique has huge memory requirements, but this is not an issue for modern computers anymore. Unfortunately for phoneticians, concatenation synthesis is not very informative about how human speech works.

A similar situation holds for speech processing: here too, engineers use techniques because they work well and because they are efficiently implementable, not because they are cognitively plausible or informative about how the human brain may work. A case in point is the use of hidden Markov models (Rabiner, 1989) in combination with mel-weighted cepstral coefficients (e.g., Huang, Acero, & Hon, 2001, section 6.5.2). Hidden Markov models are statistical formalisms that can describe (temporal) sequences with random variation. They can calculate probabilities of an observed sequence efficiently, and they can also be trained efficiently on the basis of a limited number of examples. To a certain extent, speech can be considered as a sequence of acoustic features with random variation, but the statistical properties of human speech are rather different from the sequences generated by hidden Markov models. For instance, hidden Markov models do not model the duration of human speech sounds very well, and research is ongoing to find models that model duration better (see Chiappa, 2014; Yu, 2010, and references therein). Even more problematic is the inherent division between recognition of temporal features and spectral features. In a standard speech recognition system, the signal is split up in very short (i.e., 5–50 ms) frames that are then analyzed using some kind of spectral features, usually a variant of mel-weighted cepstral coefficients (Huang et al., 2001, section 6.5.2) or perceptual linear predictive analysis coefficients (Huang et al., 2001, section 6.5.3). These provide information about short-term spectral features. The hidden Markov model then tries to detect temporal sequences. However, this division of labor is not very logical, given that in human speech many interesting events, such as intonation patterns, formant transitions and stops have inherently both spectral and temporal features. The trouble that traditional methods have for dealing with such events has been identified as an important problem in speech recognition (Hinton et al., 2012), and Hinton et al. therefore suggested that the success of deep learning methods (trainable neural network models with a large number of interconnected layers, i.e., there are long "deep" paths between inputs and outputs, also see section 4.2 on neural models) is partly due to their better ability to deal with them. Moreover there is active research into finding methods that can deal with inherently spectrotemporal features better than traditional methods can (e.g., Ezzat, Bouvrie, & Poggio, 2007).

The fact that speech recognition is not entirely cognitively plausible does not necessarily mean that it cannot be to some extent informative about how humans may process speech. After all, through implementing the task of speech recognition in practically working systems, one has to tackle all the problems that are involved in dealing with continuous, noisy, variable speech. Many of these problems are avoided in more highly idealized linguistic models of speech processing (for instance, some of the neural models discussed in section 4.2). It is interesting that statistical methods are as successful as they are, given that they are not traditionally considered by linguists (but they are gaining in popularity, see section 2). Perhaps statistical pattern recognition is more important in speech recognition than is assumed in phonetics and phonology, analogously to the way it turns out that statistical sequence recognition can deal better with the complexity of syntax than was thought (Frank, Bod, & Christiansen, 2012). Also, deep learning methods are intriguing in that they achieve high recognition success using computational methods that are similar to the structure of the human cortex: both consist of many layers that pass information to each other. Similarly, it may be that human speech production uses concatenation-like mechanisms to some extent for elements of speech that are often used, rather than building every utterance from scratch on the basis of phonological building blocks.

Conversely, engineering approaches may need to pay more attention to models of human cognition, as it appears that engineering approaches still need much more training data than humans do to reach the same level of performance (Moore, 2003) and are much less robust to increasing noise levels than human speech recognition is (Hannun et al., 2014 do not show much better performance than that reported in Lippmann, 1997).

Although there is therefore undoubtedly much to be gained from a closer interaction between engineering and linguistic work, the remainder of this chapter will focus on computer models that have been built specifically for investigating how humans process speech, rather than for solving practical problems.

2. Research Models

To understand the range of different models that exist, it is useful to define a few dimensions with which to classify them. An obvious one is the aspect of speech to which the models are applied, for instance, production, perception, processing, learning, universals, sound change, and evolution of speech. Computer models have been applied to all these aspects, and some examples will be reviewed and discussed in somewhat more detail in the case studies. However, it is useful to first introduce two somewhat more fundamental classifications of models. The first is the classification of computer models into explanatory and predictive ones (introduced in the context of agent-based modeling by Gilbert & Troitzsch, 2005). Explanatory models operate at a relatively high level of abstraction and are intended to obtain qualitative insights into a given phenomenon. Predictive models are intended to actually reproduce and possibly predict a given phenomenon and are often built for particular practical purposes. An example of an explanatory model outside the domain of language could be a simple model of traffic flow in order to understand how traffic jams can form on an otherwise unobstructed highway. A predictive model in this domain would be a model to predict the traffic density in a particular city at a particular time for a navigation and route-planning application. In the domain of speech, the two-mass model of vocal fold vibration (Ishizaka & Flanagan, 1972) is an example of an explanatory model, while a model to help surgeons predict the effects of laryngeal surgery (an ongoing project at the Dutch cancer institute, NKI) would be an example of a predictive model. The distinction is not absolute, however. In the study of speech and in linguistics in general, many models are somewhere in between the two extremes: they are not entirely abstract because one wants to be able to verify the predictions of the model with real data, but on the other hand one does not require of most models that they predict real linguistic behavior 100% accurately. The precise trade-off depends on how general one wants the conclusions of a model to be and on the level of abstraction of the research. In some cases, models can in fact be very abstract but still interesting.

A related classification is whether a model is analytical, numerical, or a direct simulation. In an analytical model, a closed-form equation is available that describes the model's behavior completely. Although closed-form solutions can provide a lot of insight into a phenomenon, they can usually only be obtained for the simplest possible systems. Because closed-form solutions are considered something of a holy grail in certain fields, sometimes models are simplified so much that essential complexity is removed. Although such models may be of great interest from a mathematical point of view, they are less so from a linguistic point of view (e.g., Patriarca & Leppänen, 2004). The next level of complexity is that of numerical models. Here a complete mathematical description of the model's behavior can be derived, but this description cannot be solved analytically. In this case, it is often possible to use numerical methods to approximate a solution as closely as desired. Although such numerical approximations do not provide the precise insights that an analytical solution does, often much less simplification is necessary and therefore numerical models are applicable to a much larger class of problems. Finally, the most realistic models simulate a phenomenon directly or with very little simplification. Often such models explicitly model random fluctuation and are therefore difficult if not impossible to solve numerically (numerical solvers work on the assumption that there is a well-defined solution and this is not the case if there is random fluctuation). Although such models are the most realistic class of models, they are like real systems in the sense that one needs to do a series of experiments with them (with different parameter settings and different random values) in order to get insight into how they work.

An example for speech could be models of the vocal tract. A strongly simplified model of the vocal tract consisting of two tubes of equal length can be solved analytically for the frequencies of its resonances using equation 3.84 from Flanagan (1965) resulting in:

$$F = \frac{c}{l}\left(n \times \pi \pm \tan^{-1}\sqrt{A_{\text{front}}/A_{\text{back}}}\right)$$

where F is the resonance frequency, c is the speed of sound, l is the length of the tubes, n an integer number, and the A's are the areas of the tubes. This can be calculated on a pocket calculator and would take a few dozen nanoseconds to calculate on a modern desktop computer. Such a model is useful if one is interested in the qualitative relation between articulatory parameters and resonance frequencies. When one wants to get a better approximation of a speech sound, taking into account radiation and losses, one needs to use a more complete model, along the lines of the more detailed model by Flanagan or the model described by Maeda (1982). These models depend on approximating the vocal tract by a series of tubes and can calculate their response to a sinusoidal signal of a given frequency. Calculating their response requires calculating the response at a range of frequencies of interest, and this would take probably around a microsecond per frequency, so using a good peak-finding algorithm, finding say the first four formants would take on the order of a hundred microseconds on a modern desktop computer. When one wants to investigate the interaction of vocal fold vibration and the vocal tract and see whether this can cause chaotic behavior of the vocal folds (e.g., de Boer, 2012), one needs to build a full simulation and run it multiple times because chaotic systems are by definition sensitive to initial conditions. Each individual run of such a model takes on the order of a second to run. Thus, this classification not only has implications for what kinds of conclusions one can draw, but also how much computation time is required.

Models intended to investigate different aspects of speech tend to fall into different categories. Models for investigating production tend to be numerical models or direct simulations and closer to being predictive than to being explanatory (see examples in the case studies). Models of perception tend to be numerical and tend to be predictive (e.g., Bladon & Lindblom, 1981; Mantakas, Schwartz, & Escudier, 1986; Martínez et al., 2004; Smith & Lewicki, 2006).

Models of processing and learning tend to be explanatory and direct simulations, sometimes based on generic machine learning techniques (de Boer & Kuhl, 2003; Vallabha, McClelland, Pons, Werker, & Amano, 2007) and sometimes based on specific linguistic models (e.g., Boersma, Escudero, & Hayes, 2003, used bidirectional optimality theory). Recently, statistical models of speech acquisition have also been gaining in popularity, probably due to a better understanding of how (Bayesian) statistical learning works, because of the success of these methods in other areas of language, and because there appears to be a trend toward more general learning mechanisms to explain linguistic behavior. Usually these models are based on pre-extracted and somewhat idealized features, such as a single formant with Gaussian distribution (McMurray, Aslin, & Toscano, 2009) or bivariate Gaussians based on two formants (Dillon, Dunbar, & Idsardi, 2013; Feldman, Griffiths, Goldwater, & Morgan, 2013). Another type of learning models, specifically connectionist models will be reviewed in the second case study.

There are many models focusing on the problem of speech segmentation, but many of these tend to be from the engineering tradition and cognitive realism does not tend to be a criterion (see Versteegh et al., 2015, for some recent references) although some models do address cognitive questions (e.g., Räsänen, 2011; Rasilo, Räsänen, & Laine, 2013).

Models of universals (Liljencrants & Lindblom, 1972; Lindblom, MacNeilage, & Studdert-Kennedy, 1983) are related to those of sound change and evolution. Models of change and evolution tend to be direct simulations, often agent-based models (Baronchelli, Chater, Pastor-Satorras, & Christiansen, 2012; de Boer, 2000; Oudeyer, 2005) but numerical (Wang & Minett, 2005) and analytic models (Abrams & Strogatz, 2003; Griffiths & Kalish, 2007; Kirby, Dowman, & Griffiths, 2007) are used as well. Some of these models, such as Oudeyer's use neurally inspired methods, but most tend to be less neurally plausible (for reviews, see de Boer, 2002; de Boer & Fitch, 2010; Gong, Shuai, & Zhang, 2014).

3. Interaction of Models, Analysis, and Experiments

The more abstract a computer model is, the more indirect its relation to real data can be. This can lead to a shift in focus from studying the underlying questions to studying the properties of the models themselves. Especially in areas where data is scarce and models are highly abstract, finding the proper balance between models and data can be a challenge in itself (see Vogt & de Boer, 2010, and other papers in the special issue). Although for speech this is often less problematic than for other aspects of language (because it is a physical signal), it is nevertheless useful to consider how models, analysis, and experiments interact.

First of all, what are the reasons for building explicit computer models or mathematical models? Running explicit models is important when the complexity of the problem is such that it is impossible to know what the theory would predict using ordinary pen-and-paper analysis. Problems can become complex either because the inherent complexity of the system (such as the acoustics of a vocal tract or the complexity of a language-learning brain) or when there are lots of data involved (such as in learning the speech sounds of a language). In fact, there are cases when even relatively simple systems can show very complex behavior—this is often the case in chaotic systems, for instance in certain situations, vocal fold vibration may be chaotic (de Boer, 2012; Fitch, Neubauer, & Herzel, 2002). A somewhat different reason for using models was proposed by Langton (1989). He proposed that as an alternative to the analytic method of doing science, where one first observes a system and then tries to construct theories of it, one could take a synthetic approach where one tries to find a model that shows the same behavior as a given system in the real world. Because he introduced this in the context of models for understanding biology (or rather "life as it could be") the approach was dubbed artificial life. Especially within the area of language evolution, many early models were built using the artificial life approach (Brighton & Kirby, 2001; de Boer, 1997; de Boer & Vogt, 1999; Kirby, 2002; Smith, Kirby, & Brighton, 2003; Steels, 1995). One reason that this approach has become less popular recently is that it is particularly prone to becoming focused on the models themselves rather than on the phenomena they are supposed to explain (Smith, 1995).

The relation with real data is therefore something that should be kept in mind all the time when working with models. For models of speech production, the relation with experimental and observational data is relatively unproblematic: one can just investigate how well the model approximates human speech. For speech perception, one can also directly compare the behavior of the model with human behavior by investigating whether the model confuses speech sounds in the same way as humans do, and whether it degrades in the same way with increasing noise, speaker variation, and accent. Perhaps surprisingly, such comparisons are very rarely done (Lippmann, 1997; Moore, 2003). For models of learning and representation, it is already more complex to evaluate the correspondence of the model and the data. This is because there are many different models that could explain the data. One way to address this issue is to see what other phenomena (apart from the phenomena that were used to build and tune the model) the model explains. Another (related) approach is to study the behavior of the model and to derive predictions from it that can be tested empirically.

The relation between models and data is most problematic in models of language evolution. Models of language change can be used successfully to investigate specific cases (such as the sound system of New Zealand English: Baxter, Blythe, Croft, & McKane, 2009) but the above-mentioned problem of knowing whether one has the right model also applies in this case, and there is very little detailed data about sound change, while doing experiments is obviously problematic. This argument applies even more strongly to models of language evolution and language emergence, and this is one of the reasons why recent work has focused on developing methods for doing speech evolution experiments in the laboratory (Little, Eryılmaz, & de Boer, 2015; van der Ham & de Boer, 2015; Verhoef, 2012; Verhoef, Kirby, & de Boer, 2014). These experiments fit in the tradition of experiments discussed in chapter 18 of this volume by Toni and Stolk. In the case studies, the relation between models and data will be discussed in more detail.

4. Case Studies

In order to make the above-mentioned points more concrete, two case studies will be presented in some detail. The first focuses on the evolution of the vocal tract and is mostly concerned with numerical models. Some of these models are explanatory, others are closer to predictive models. The second focuses on connectionist and neural models of speech. The models in this section will mostly be direct simulations and mostly about speech processing. These models tend to stay relatively close to real human behavior and therefore aim (or pretend) to be predictive models, although in practice many abstractions need to be made and therefore the models fall somewhere between explanatory and predictive.

4.1. Evolution of the Vocal Tract The evolution of the vocal tract and its role in the evolution of language was one of the first applications of computer models to linguistics questions that were not directly about practical issues. Lieberman, Klatt, and Wilson (1969) first modeled the vocal abilities of extant primate species and compared these to those of humans. Lieberman and colleagues then extended this work to modeling the reconstructed vocal abilities of Neanderthals (Lieberman & Crelin, 1971; Lieberman, Crelin, & Klatt, 1972). In both cases, they concluded that the modern human vocal tract can produce a substantially larger range of speech sounds than the other primate's vocal tracts or that of Neanderthals.

The reason for using a model was twofold: First, Lieberman et al. (1972) were interested in the *potential* of the different vocal tracts. It is likely that in the *actual* vocalizations of extant primates, not all possible vocal tract configurations are exploited, and therefore that just investigating observed vocalizations severely underestimates the potential of primates' vocal tracts. Second, vocal tracts consist of soft tissue, and therefore do not fossilize well. For this reason, there is no fossil evidence for the precise configuration of the Neanderthal vocal tract (nor for that of any other ancestral hominid for that matter). Therefore, one needs to have an explicit model of primate and Neanderthal vocal tracts in order to investigate their articulatory abilities. The reason for using a computer model was that the calculations involved in going from an articulation (i.e., a given position of the jaw, tongue, lips, pharynx, velum) to a sound are very complex, but that at the time of writing, accurate computer models for this task were becoming available.

Lieberman et al.'s (1972) procedure was as follows: for living primates and humans, plaster or silicone casts were made of different vocal tract configurations. These were then sliced into slices of 0.5 cm thickness. The areas of these slices were measured and these were used to create a so-called *lossless tube model* of the vocal tract. For physical reasons, the acoustic properties of any vocal tract configuration for the frequencies in the range of interest for speech (100–3,000 Hz) can be approximated by a concatenation of simple cylindrical tubes (simplifying away from the complex three-dimensional shape of a real vocal tract) as long as the cross-sectional area at each point in this tube is the same as the area in the corresponding position of the real vocal tract. If one is only interested in the formant frequencies (i.e., the resonance frequencies) of the vocal tract rather than in a realistic-sounding synthesis, one also does not need to model the damping effects of the soft tissues of the vocal tract nor the effects of sound radiation at the lips. Although it involves a considerable amount of calculation, the formant frequencies can be calculated accurately from the lossless tube model.

Although Lieberman et al.'s (1972) work was groundbreaking, it attracted considerable critique. Paleontologists soon noted that the Neanderthal reconstruction that was used was inaccurate, leading to an overestimate of the required size of the tongue, and therefore an erroneously high larynx position (Schepartz, 1993). From the modeling perspective, the work was problematic because it used a very limited number of hand-picked articulations. It is unlikely that these few articulations really sample the extreme possibilities of the different vocal tracts. A solution to this problem is to extend the model and replace the hand-derived areas by ones calculated through an articulatory model. Articulatory models take as input the position of the different articulators (jaw, tongue, lips, larynx, velum) and calculate the areas of the tubes necessary to approximate that articulation. From this, the formants can then be calculated with the same method that Lieberman et al. (1972) used. Although articulatory models were available around the time of Lieberman's research (Henke, 1966; Mermelstein, 1973), it was only much later that articulatory synthesis and systematic exploration of the possible articulation space was performed with reconstructed Neanderthal vocal tracts (Boë, Heim, Honda, & Maeda, 2002).

The computer model that Boë et al. used was based on Maeda's (1989) articulatory model, modified to model vocal tracts with different sizes and different ratios of horizontal (oral) and vertical (pharyngeal) lengths (Boë & Maeda, 1998) so that different positions of the larynx could be modeled. They also used a more modern reconstruction of the Neanderthal vocal tract. They explored the possible range of articulations by randomly generating 10,000 different articulations and measuring what area of the space of the first and second formant (generally considered representative of the articulatory abilities of a given vocal tract) these cover. Their findings were that the Neanderthal vowel space is essentially the same size as that of modern humans.

Two potential problems with their model were noted by de Boer and Fitch (2010). The first problem has to do with the nature of the Maeda articulatory synthesizer: it is based on a factor analysis of 1,000 different articulations of a modern human vocal tract. These factors can be identified with different articulatory movements (jaw, tongue body front-back, tongue body high-low, tongue tip, and lip movements). However, models based on factor analysis can in principle generate unrealistic articulations even if one only uses realistic articulator values: it is possible that certain combinations result in impossible configurations. Even if this problem is probably not very important for the Maeda model, the second problem is more important. Because of the way that the original model is modified to correspond to a Neanderthal vocal tract (through scaling of the vertical part of the model), it does not take into account the constraints on muscle action that are different for the flat ancestral primate tongue and the round modern human tongue. The flat ancestral tongue is not as capable as the modern human tongue to make front-back displacements of the tongue body. This constraint is probably not captured properly by the modified Maeda model. Both problems lead to overestimation of the acoustic space, and therefore Boë et al.'s (2002) estimate of the reconstructed Neanderthal vowel space is probably somewhat too large. More recent combined modeling and experimental work by Fitch, de Boer, Mathur, and Ghazanfar (2016) based on X-ray images of a macaque vocal tract (presumed to have comparable anatomy to our latest common ancestor) has shown that the usable acoustic space is enough for understandable speech (even though it is somewhat smaller than that of modern humans).

The reconstruction of ancestral vocal tracts will remain speculative until fossils of soft tissue become available (which is extremely unlikely) or until ancient DNA analysis will be able to tell us more about ancestral vocal tracts. However, there is another way that computer models can be used to investigate the evolution of the vocal tract and that is by investigating whether it has undergone selective pressure related to vocalization. One way that selective pressure can be detected is by showing that a certain parameter (e.g., larynx position) is optimal for a certain function (e.g., producing as large a range of vowels as possible). This argument has been proposed by Parker and Smith (1990) who dubbed it optimality theory (not to be confused by the phonological theory of the same name). Optimality in this context means that any change of the parameter will result in a less functional system. For the vocal tract, this translates to any change in (modern human) larynx position resulting in a smaller vowel space.

This matter was investigated by de Boer (2010) who found, using a strongly simplified articulatory model, based on Mermelstein's (1973) model and the vowel space exploration method of Boë et al. (2002) outlined, that the position of the larynx that maximizes the vowel space (given all other articulatory constraints of the human vocal tract) is the one found in modern human females. The suboptimal lower larynx position of human males had already been proposed as the result of evolutionary pressure to exaggerate size (Fitch & Reby, 2001). However, Badin, Boë, Sawallis, and Schwartz (2014) showed that a weakness of de Boer's (2010) model is that it does not properly model the lips (due to too much simplification). When lips are taken into account, the position of the larynx becomes less important: manipulation of the lips can compensate for the reduced freedom of motion of the tongue due to the suboptimal larynx position. This indicates that perhaps the larynx position is not optimized for speech, but it remains to be investigated whether the degree of lip control that Badin et al. assume is realistic for human and other primate vocal tracts. An interesting possibility that this work raises is that the human lips may have evolved for speech as well.

4.2. Neural and Connectionist Models for Speech There are many models of speech recognition and acquisition that are explicitly or implicitly inspired by the way the brain works. Such models tend to consist of many units, for instance, representing features, phonemes, or words, that are connected in a network through which they can activate or inhibit each other. Such models are therefore called connectionist models. Although most of these models (reviewed in, e.g., Luce & McLennan, 2005; Magnuson, Mirman, & Harris, 2012) are directly inspired by the fact that the human brain seems to be connectionist in nature, there have been models that arrived at a connectionist architecture through trying to model human *behavior* as precisely as possible without being explicitly inspired by the brain (e.g., Klatt, 1979). Nevertheless, even the models that are inspired by the architecture of the brain are usually not organized in a way that explicitly models brain regions (although there are exceptions such as the DIVA model; Guenther, 2006). Instead the models tend to be organized around levels of linguistic representation (features, phonemes, and words, for instance) or functional modules (acoustics, memory, and articulation, for instance).

Perhaps the most influential connectionist model of speech recognition is TRACE. It was originally published in two versions: TRACE I which could work with real speech input (Elman & McClelland, 1986) and TRACE II which worked with more abstract, but more easily manipulated inputs (McClelland & Elman, 1986). TRACE has nodes representing features, phonemes, and words. In TRACE I, activation of features is calculated directly from a speech signal. Features have a continuous level of activation, and tend therefore (in the ideal case) to become gradually more active over time, stay active for a while and then decline gradually over time. In TRACE II, this process is modeled without real speech input by representing speech input as "scores" of feature activations over time. Phonemes are represented as templates of subsequent feature activations. To handle occurrences at different positions (shifts) in time, phoneme representations are reduplicated. Similarly, words are represented as reduplicated sequences of phoneme activations. Whereas features only activate phonemes, phonemes and words activate each other; hence, whenever a word becomes more likely, its constituent phonemes also become more likely. At the same time, phonemes inhibit each other, as do words. This helps the model to only recognize a single word at a time. Neither version of TRACE learns. The connections of the model are hand-wired.

Although TRACE has been quite successful in replicating human word recognition behavior, many alternative models of spoken word recognition have been proposed, often as a modification of or a reaction to TRACE. Shortlist (Norris, 1994) reduces the unrealistic reduplication of phoneme and word nodes in TRACE by establishing shortlists of candidate words. TISK (Hannagan, Magnuson, & Grainger, 2013) likewise reduces reduplication by introducing nodes that provide time-invariant representations of diphones and words, based on techniques that were developed in visual object recognition (which has to deal with a similar what-versus-where problem). Both models use the artificial input features of TRACE II and therefore have not been demonstrated to be able to deal with real speech input. Likewise, as in TRACE, the connection strengths are hand-wired.

Two other models that use hand-wired connections but that stem from a very different connectionist tradition are Grossberg's ARTPHONE (Grossberg, Boardman, & Cohen, 1997) and ARTWORD (Grossberg & Myers, 2000) models. These models are based on his adaptive resonance theory (which in principle is capable of learning, but this is not used in these models). This is a neural network paradigm that works in continuous time (as opposed to the TRACE-like models in which time is split up into discrete steps) so that it can model perception of time series and effects of timing very well. The models use similar levels of representation as TRACE, such as features, phonemes, words, and similar mechanisms, such as activation between layers and lateral inhibition within layers. However, because the models work in continuous rather than discretized time, they do not need the reduplication of the TRACE model.

Learning models for spoken word recognition in the cognitive science literature are generally based on some variant of recurrent neural networks: multilayer neural networks in which one of the hidden layers or the output layer at time $t-1$ is copied back to the input layer at time t. As this adds a representation of past behavior to the network, recurrent neural networks can deal with time series, and hence with speech. Such models can be trained using backpropagation (e.g., Hertz, Palmer, & Krogh, 1991).

The distributed cohort model (Gaskell & Marslen-Wilson, 1997) is such a model, which is trained to recognize words (and semantic features) on the basis of an input stream of phonetic features. Even though the model is quite simple in its architecture, it is capable of reproducing aspects of human spoken-word recognition behavior. Another, much more complex model that is also based on recurrent neural networks was proposed by Plaut and Kello (1999). This model not only learns to recognize words on the basis of streams of phonetic features derived from strings of input features, but it also learns how to reproduce this speech using a simplified articulatory model. Both models do work on the level of idealized acoustic/phonetic features and do not work with real speech.

A neural model that does work with real speech is the DIVA model proposed by Guenther and colleagues (Guenther, 2006; Guenther, Ghosh, & Tourville, 2006). This model is not only equipped with a real articulatory synthesizer and a model of sound perception, but it also tries to map the modules in its architecture directly to the corresponding brain regions in humans. Moreover, the model does not make use of backpropagation as a learning mechanism, but only updates connections using local information. Although the actual method used is not inspired by neural learning mechanisms, it is argued that it is in fact equivalent to neurologically plausible learning methods (Guenther et al., 2006, appendix B). The model is used to predict brain activity as measured by functional MRI as well as to predict the effects of brain lesions by selectively disabling parts of the model.

There are two more classes of neural networks that, although in principle, are neurologically plausible and powerful do not appear to be used much (if at all) in psycholinguistic research related to speech. These are spiking neural networks and deep neural networks. Spiking neural networks (Näger, Storck, & Deco, 2002) are networks that model the temporal behavior of neurons directly. Instead of slicing up time into discrete intervals and modeling activation of neurons as a simple number, spiking neural networks model the actual spiking behavior of real neurons: activity of a neuron is modeled by the frequency with which they emit spikes of activity. This means that spiking neural models require much more computation to simulate. Moreover, it is still not entirely understood how such models should learn in a neurologically plausible way. Nevertheless, given that spiking neural models are the most realistic neural models and that they can model temporal phenomena precisely, they would appear a potentially promising way to model human speech perception—especially those aspects where timing is crucial.

Deep neural networks make up a diverse class of network architectures that have in common that there can be many layers of neurons between input and output; in the connectionist architectures discussed, there are usually no more than one or two layers of neurons between input and output. Deep neural networks are in a sense more realistic than the traditional models, because it is known that in the brain there are also many layers of neurons between inputs and outputs, while the cerebral cortex also consists of multiple layers. Developments in algorithms for training neural networks, as well as recent increases in computing power and availability of data, have made using deep neural networks feasible. However, it seems deep networks have not been used for psycholinguistic research; a recent review (Joanisse & McClelland, 2015) did not contain references to psycholinguistic deep learning models. Although this is an echo of the divide between engineering approaches and psycholinguistic approaches in computer modeling mentioned, it is perhaps less surprising than it seems, as in general deep learning networks need large amounts of diverse training data, whereas psycholinguistic research is usually based on small, selected data sets. However, given the availability of freely usable and easy-to-use software packages for deep learning (e.g., TensorFlow) and the potential realism of deep neural models, it is only a matter of time before deep neural networks will be used for psycholinguistic models.

5. *Conclusion*

Speech is an aspect of language that lends itself very well to computer modeling and because of the complexity of the signal and the neural processes involved, computer models can make a useful contribution to the study of speech. There are many ways in which computer models can be brought to bear on the study of speech from very abstract models to very concrete ones, and from the models of speech production on a very individual level to models of the evolution of speech at a population level. Interestingly, there is deep divide in computer modeling of speech between engineering approaches and psycholinguistic approaches that has existed for over 40 years of research (and still persists). There are however a number of techniques from engineering that may be promising for psycholinguistic research: statistical techniques, articulatory models, and deep neural networks to name a few. It is hoped that this chapter can make a small contribution toward incorporating these techniques into psycholinguistic research.

REFERENCES

Abrams, D. M., & Strogatz, S. H. (2003). Modelling the dynamics of language death. *Nature, 424*, 900.

Badin, P., Boë, L.-J., Sawallis, T. R., & Schwartz, J.-L. (2014). Keep the lips to free the larynx: Comments on de Boer's articulatory model (2010). *Journal of Phonetics, 46*, 161–167.

Baronchelli, A., Chater, N., Pastor-Satorras, R., & Christiansen, M. H. (2012). The biological origin of linguistic diversity. *PLOS ONE, 7*(10), e48029.

Baxter, G. J., Blythe, R. A., Croft, W., & McKane, A. J. (2009). Modeling language change: An evaluation of Trudgill's theory of the emergence of New Zealand English. *Language Variation and Change, 21*(2), 257–296.

Bladon, R. A. W., & Lindblom, B. (1981). Modeling the judgement of vowel quality differences. *Journal of the Acoustical Society of America, 69*(5), 1414–1422.

Boë, L.-J., Heim, J.-L., Honda, K., & Maeda, S. (2002). The potential Neanderthal vowel space was as large as that of modern humans. *Journal of Phonetics, 30*(3), 465–484.

Boë, L.-J., & Maeda, S. (1998). Modélisation de la croissance du conduit vocal. *Journée d'Études Linguistiques: La voyelle dans tous ses états* (pp. 98–105). Nantes: Université de Nantes.

Boersma, P., Escudero, P., & Hayes, R. (2003). Learning abstract phonological from auditory phonetic categories: An integrated model for the acquisition of language-specific sound categories. In M.J. Solé, D. Recansens, & J. Romero (Eds.), *Proceedings of the 15th International Congress of Phonetic Sciences* (pp. 1013–1016). Universitat Autonoma de Barcelona, Barcelona.

Brighton, H., & Kirby, S. (2001). The survival of the smallest: Stability conditions for the cultural evolution of compositional language. In J. Kelemen & P. Sosik (Eds.), *Advances in artificial life: Proceedings of the 6th European Conference in Artificial Life* (pp. 592–601). Berlin: Springer Verlag.

Chiappa, S. (2014). Explicit-duration Markov switching models. *Foundations and Trends in Machine Learning, 7*(6), 803–886.

de Boer, B. (1997). Generating vowel systems in a population of agents. In P. Husbands & I. Harvey (Eds.), *Fourth European Conference on Artificial Life* (pp. 503–510). Cambridge, MA: MIT Press.

de Boer, B. (2000). Self organization in vowel systems. *Journal of Phonetics, 28*(4), 441–465.

de Boer, B. (2002). Evolving sound systems. In A. Cangelosi & D. Parisi (Eds.), *Simulating the evolution of language* (pp. 79–97). Berlin: Springer Verlag.

de Boer, B. (2010). Investigating the acoustic effect of the descended larynx with articulatory models. *Journal of Phonetics, 38*(4), 679–686.

de Boer, B. (2012). Air sacs and vocal fold vibration: Implications for evolution of speech. *Theoria et Historia Scientiarum, 9*, 13–28.

de Boer, B., & Fitch, W. T. (2010). Computer models of vocal tract evolution: An overview and critique. *Adaptive Behavior, 18*(1), 36–47.

de Boer, B., & Kuhl, P. K. (2003). Investigating the role of infant-directed speech with a computer model. *Acoustics Research Letters Online, 4*(4), 129–134.

de Boer, B., & Vogt, P. (1999). Emergence of speech sounds in changing populations. In D. Floreano, J.-D. Nicoud, & F. Mondada (Eds.), *Advances in artificial life, Lecture notes in artificial intelligence 1674* (pp. 664–673). Berlin: Springer Verlag.

Dillon, B., Dunbar, E., & Idsardi, W. (2013). A single-stage approach to learning phonological categories: Insights from Inuktitut. *Cognitive Science, 37*(2), 344–377.

Elman, J. L., & McClelland, J. L. (1986). Exploiting lawful variability in the speech wave. In J. S. Perkell & D. H. Klatt (Eds.), *Invariance and variability in speech processes* (pp. 360–380). Hillsdale, NJ: Erlbaum.

Ezzat, T., Bouvrie, J. V., & Poggio, T. (2007). *Spectro-temporal analysis of speech using 2-d Gabor filters.* Paper presented at the 8th Annual Conference of the International Speech Communication Association, Antwerp, Belgium.

Fant, G. (1960). *Acoustic theory of speech production.* The Hague, the Netherlands: Mouton.

Feldman, N. H., Griffiths, T. L., Goldwater, S., & Morgan, J. L. (2013). A role for the developing lexicon in phonetic category acquisition. *Psychological Review, 120*(4), 751–778.

Fitch, W. T., de Boer, B., Mathur, N., & Ghazanfar, A. A. (2016). Monkey vocal tracts are speech-ready. *Science Advances, 2*(12), e1600723. doi:10.1126/sciadv.1600723.

Fitch, W. T., Neubauer, J., & Herzel, H. (2002). Calls out of chaos: The adaptive significance of nonlinear phenomena in mammalian vocal production. *Animal Behaviour, 63*(3), 407–418.

Fitch, W. T., & Reby, D. (2001). The descended larynx is not uniquely human. *Proceedings of the Royal Society of London Series B—Biological Sciences, 268*, 1669–1675.

Flanagan, J. L. (1965). *Speech analysis, synthesis and perception.* Berlin: Springer.

Frank, S., Bod, R., & Christiansen, M. H. (2012). How hierarchical is language use? *Proceedings of the Royal Society B, 279*(1747), 4522–4531.

Gaskell, M. G., & Marslen-Wilson, W. D. (1997). Integrating form and meaning: A distributed model of speech perception. *Language and Cognitive Processes, 12*(5–6), 613–656.

Gilbert, N., & Troitzsch, K. G. (2005). *Simulation for the social scientist* (2nd ed.). Maidenhead, UK: Open University Press.

Gong, T., Shuai, L., & Zhang, M. (2014). Modelling language evolution: Examples and predictions. *Physics of Life Reviews, 11*(2), 280–302.

Griffiths, T. L., & Kalish, M. L. (2007). Language evolution by iterated learning with Bayesian agents. *Cognitive Science, 31*(3), 441–480.

Grossberg, S., Boardman, I., & Cohen, M. (1997). Neural dynamics of variable-rate speech categorization. *Journal of Experimental Psychology: Human Perception and Performance, 23*(2), 481.

Grossberg, S., & Myers, C. W. (2000). The resonant dynamics of speech perception: Interword integration and duration-dependent backward effects. *Psychological Review, 107*(4), 735.

Guenther, F. H. (2006). Cortical interactions underlying the production of speech sounds. *Journal of Communication Disorders, 39*, 350–365.

Guenther, F. H., Ghosh, S. S., & Tourville, J. A. (2006). Neural modeling and imaging of the cortical interactions underlying syllable production. *Brain and Language, 96*, 280–301.

Hannagan, T., Magnuson, J. S., & Grainger, J. (2013). Spoken word recognition without a TRACE. *Frontiers in Psychology, 4*, 563.

Hannun, A., Case, C., Casper, J., Catanzaro, B., Diamos, G., Elsen, E., ... Ng, A. Y. (2014). Deep speech: Scaling up end-to-end speech recognition. *ArXiv Preprint.* Advanced online publication. arXiv:1412.5567v2.

Henke, W. L. (1966). *Dynamic articulation model of speech production using computer simulation.* Cambridge, MA: MIT Press.

Hertz, J. A., Palmer, R. G., & Krogh, A. S. (1991). *Introduction to the theory of neural computation.* Redwood City, CA: Addison-Wesley.

Hinton, G., Deng, L., Yu, D., Dahl, G. E., Abdel-rahman, M., Jaitly, N., ... Kingsbury, B. (2012). Deep neural networks for acoustic modeling in speech recognition. *IEEE Signal Processing Magazine, 29*(6), 82–97.

Huang, X., Acero, A., & Hon, H.-W. (2001). *Spoken language processing.* Upper Saddle River, NJ: Prentice Hall PTR.

Ishizaka, K., & Flanagan, J. L. (1972). Synthesis of voiced sounds from a two-mass model of the vocal cords. *Bell System Technical Journal, 51*(6), 1233–1268.

Joanisse, M. F., & McClelland, J. L. (2015). Connectionist perspectives on language learning, representation and

processing. *Wiley Interdisciplinary Reviews: Cognitive Science, 6*(3), 235–247.
Kirby, S. (2002). Natural language from artificial life. *Artificial Life, 8*(2), 185–215.
Kirby, S., Dowman, M., & Griffiths, T. L. (2007). Innateness and culture in the evolution of language. *Proceedings of the National Academy of Sciences, 104*(12), 5241–5245.
Klatt, D. H. (1979). Speech perception: A model of acoustic-phonetic analysis and lexical access. *Journal of Phonetics, 7*, 279–312.
Langton, C. G. (1989). *Artificial life.* Reading, MA: Addison Wesley.
Lieberman, P. H., & Crelin, E. S. (1971). On the speech of Neanderthal man. *Linguistic Inquiry, 2*, 203–222.
Lieberman, P. H., Crelin, E. S., & Klatt, D. H. (1972). Phonetic ability and related anatomy of the newborn and adult human, Neanderthal man, and the chimpanzee. *American Anthropologist, 74*, 287–307.
Lieberman, P. H., Klatt, D. H., & Wilson, W. H. (1969). Vocal tract limitations on the vowel repertoires of rhesus monkey and other nonhuman primates. *Science, 164*, 1185–1187.
Liljencrants, J., & Lindblom, B. (1972). Numerical simulations of vowel quality systems. *Language, 48*, 839–862.
Lindblom, B., MacNeilage, P. F., & Studdert-Kennedy, M. (1983). Self-organizing processes and the explanation of language universals. *Linguistics, 21*, 181–203.
Lippmann, R. P. (1997). Speech recognition by machines and humans. *Speech Communication, 22*(1), 1–15.
Little, H., Eryılmaz, K., & de Boer, B. (2015). Linguistic modality affects the creation of structure and iconicity in signals. In D. C. Noelle, R. Dale, A. S. Warlaumont, J. Yoshimi, T. Matlock, C. D. Jennigs, & P. P. Maglio (Eds.), *Proceedings of The 37th Annual Cognitive Science Society Meeting.* Pasadena, CA: Cognitive Science Society.
Luce, P. A., & McLennan, C. T. (2005). Spoken word recognition: The challenge of variation. In D. B. Pisoni & R. E. Remez (Eds.), *Handbook of speech perception* (pp. 591–609). Malden, MA: Blackwell.
Maeda, S. (1982). A digital simulation method of the vocal-tract system. *Speech Communication, 1*(3–4), 199–229.
Maeda, S. (1989). Compensatory articulation during speech: Evidence from the analysis and synthesis of vocal tract shapes using an articulatory model. In W. J. Hardcastle & A. Marchal (Eds.), *Speech production and speech modelling* (pp. 131–149). Dordrecht, the Netherlands: Kluwer Academic.
Magnuson, J. S., Mirman, D., & Harris, H. D. (2012). Computational models of spoken word recognition. In M. Spivey, K. McRae, & M. Joanisse (Eds.), *The Cambridge handbook of psycholinguistics* (pp. 76–103). Cambridge: Cambridge University Press.
Mantakas, M., Schwartz, J. L., & Escudier, P. (1986). Modèle de prédiction du 'deuxiéme formant effectif'F2': Application à l'étude de la labialité des voyelles avant du français. *Proceedings of the 15th journées d'étude sur la parole. Société Française d'Acoustique*, 157–161.
Martínez, I., Rosa, M., Arsuaga, J.-L., Jarabo, P., Quam, R., Lorenzo, C., … Carbonell, E. (2004). Auditory capacities in Middle Pleistocene humans from the Sierra de Atapuerca in Spain. *Proceedings of the National Academy of Sciences, 101*(27), 9976–9981.
McClelland, J. L., & Elman, J. L. (1986). The TRACE model of speech perception. *Cognitive Psychology, 18*(1), 1–86.
McMurray, B., Aslin, R. N., & Toscano, J. C. (2009). Statistical learning of phonetic categories: Insights from a computational approach. *Developmental Science, 12*(3), 369–378. doi:10.1111/j.1467-7687.2009.00822.x.
Mermelstein, P. (1973). Articulatory model for the study of speech production. *Journal of the Acoustical Society of America, 53*(4), 1070–1082.
Moore, R. K. (2003). A comparison of the data requirements of automatic speech recognition systems and human listeners. In Proceedings of *8th European Conference on Speech Communication and Technology* (pp. 2582–2584). Geneva: ISCA.
Näger, C., Storck, J., & Deco, G. (2002). Speech recognition with spiking neurons and dynamic synapses: A model motivated by the human auditory pathway. *Neurocomputing, 44*, 937–942.
Norris, D. (1994). Shortlist: A connectionist model of continuous speech recognition. *Cognition, 52*(3), 189–234.
Oudeyer, P.-Y. (2005). The self-organization of speech sounds. *Journal of Theoretical Biology, 233*(3), 435–449.
Parker, G. A., & Smith, J. M. (1990). Optimality theory in evolutionary biology. *Nature, 348*, 27–33.
Patriarca, M., & Leppänen, T. (2004). Modeling language competition. *Physica A: Statistical Mechanics and Its Applications, 338*(1), 296–299.
Plaut, D. C., & Kello, C. T. (1999). The emergence of phonology from the interplay of speech comprehension and production: A distributed connectionist approach. In B. MacWhinney (Ed.), *The emergence of language* (pp. 381–415). Mahwah, NJ: Lawrence Erlbaum.
Rabiner, L. R. (1989). A tutorial on hidden Markov models and selected in speech recognition. *Proceedings of the IEEE, 77*(2), 257–286.
Räsänen, O. J. (2011). A computational model of word segmentation from continuous speech using transitional probabilities of atomic acoustic events. *Cognition, 120*(2), 149–176.
Rasilo, H., Räsänen, O. J., & Laine, U. (2013). Feedback and imitation by a caregiver guides a virtual infant to learn native phonemes and the skill of speech inversion. *Speech Communication, 55*, 909–931.
Schepartz, L. A. (1993). Language and modern human origins. *Yearbook of Physical Anthropology, 36*, 91–126.
Schroeder, M. R. (1993). A brief history of synthetic speech. *Speech Communication, 13*(1–2), 231–237.
Smith, E. C., & Lewicki, M. S. (2006). Efficient auditory coding. *Nature, 439*(7079), 978–982.
Smith, J. M. (1995). Life at the edge of chaos? *New York Review of Books, 42*(2), 28–30.
Smith, K., Kirby, S., & Brighton, H. (2003). Iterated learning: A framework for the emergence of language. *Artificial Life, 9*(4), 371–386.
Steels, L. (1995). A self-organizing spatial vocabulary. *Artificial Life, 2*(3), 319–332.
Taylor, P. (2009). *Text-to-speech synthesis.* Cambridge: Cambridge University Press.
Vallabha, G. K., McClelland, J. L., Pons, F., Werker, J. F., & Amano, S. (2007). Unsupervised learning of vowel categories from infant-directed speech. *Proceedings of the National Academy of Sciences, 104*(33), 13273–13278.
van der Ham, S., & de Boer, B. (2015). *Statistical learning of categories is not specialized for speech.* Paper presented at the Interdisciplinary Advances in Statistical Learning, San Sebastian, Spain.

Verhoef, T. (2012). The origins of duality of patterning in artificial whistled languages. *Language and Cognition, 4*(4), 357–380.

Verhoef, T., Kirby, S., & de Boer, B. (2014). Emergence of combinatorial structure and economy through iterated learning. *Journal of Phonetics, 43*(C), 57–68.

Versteegh, M., Thiolliere, R., Schatz, T., Cao, X. N., Anguera, X., Jansen, A., & Dupoux, E. (2015). *The zero resource speech challenge 2015.* Paper presented at the 14th Annual Conference of the International Speech Communication Association, Dresden, Germany.

Vogt, P., & de Boer, B. (2010). Language evolution: Computer models for empirical data. *Adaptive Behavior, 18*(1), 5–11.

Wang, W. S.-Y., & Minett, J. W. (2005). The invasion of language: Emergence, change and death. *Trends in Ecology and Evolution, 20*(5), 263–269.

Yu, S.-Z. (2010). Hidden semi-Markov models. *Artificial Intelligence, 174*(2), 215–243.

21 Neural Network Models of Language Acquisition and Processing

STEFAN L. FRANK, PADRAIC MONAGHAN, AND CHARA TSOUKALA

1. Neural Networks in Cognitive Science

Artificial neural network models (also known as *Parallel Distributed Processing* or *Connectionist* models) have been highly influential in cognitive science since the mid-1980s. The original inspiration for these systems comes from information processing in the brain, which emerges from a large number of (nearly) identical, simple processing units (neurons) that are interconnected into a network. Each unit receives activation from other units or by stimulation from the external world and generates an output activation that is a function of the total input activation received. The unit then feeds the output activation onward to the units to which it is connected. Information processing is thus implemented in terms of activation flowing through this network.

Each connection between two units has a weight that determines how strongly the first unit affects the second. These weights can be adapted, which constitutes learning, or "training" as it is commonly known in the neural network literature. Algorithms for network training can be roughly divided into *supervised* and *unsupervised* methods. Supervised training is applied when a specific and known input-to-output mapping is required (e.g., learning to transform orthographic to phonological representations). To accomplish this, the network is provided with a representative set of "training examples" of inputs and the corresponding target outputs. It then processes each example and the difference between the networks' actual output and the target output leads to an update of the connection weights such that, next time, the output error will be smaller. By far the best known and most used method for supervised training is the Backpropagation algorithm (Rumelhart, Hinton, & Williams, 1986) that makes the network's output activations for the training examples gradually converge toward the target outputs. Unsupervised training, in contrast, makes the network adapt to (aspects of) the statistical structure of input examples without mapping to target outputs (e.g., discovery of regularities in the phonological structure of language). These networks are well-suited to uncovering statistical structure present in the environment without requiring that the modeler be aware what the structure is. One well-known example of an unsupervised training method is the learning rule proposed by Hebb (1949): strengthen connections between units that are simultaneously active and weaken the connections between two units if only one is active.

In spite of the superficial similarities between artificial and biological neural networks (i.e., interconnectivity and stimulation passing between neurons to determine their activation, and learning by adaptation of connection strengths), these cognitive models are not usually claimed to simulate processing at the level of biological neurons. Rather, neural network models form a description at Marr's (1982) algorithmic level, that is, they specify cognitive representations and operations while ignoring the biological implementation.

Neural networks underwent a surge of popularity in the 1990s, but from the early 21st century, they were somewhat overshadowed by symbolic probabilistic models. However, neural networks have enjoyed a recent revival partly due to the success of *deep learning* models, which display state-of-the-art performance on a wide range of artificial intelligence tasks (LeCun, Bengio, & Hinton, 2015). For the most part, the field of cognitive modeling is still to catch up with these novel developments. Consequently, the currently most influential connectionist cognitive models are of the more traditional variety. We return to this issue in section 5.

1.1. Feedforward and Recurrent Networks Connectionist models are not amorphous networks in which everything is connected to everything else. Rather, a particular structure is imposed, for example, by grouping units into a number of layers and allowing activation to flow only from each layer to the next. The first layer receives inputs from the environment, the final layer produces the corresponding output, and any intermediate layer is known as *hidden*. Although this so-called *feedforward* architecture can (at least in theory) approximate any computable input-to-output function, it is unable to handle input that comes in over time. This is because the network has no working memory: each input is immediately overwritten by the next. Hence, the feedforward network is not the most appropriate model for simulating language processing, which is a fundamentally temporal phenomenon.

Elman (1990), in his seminal paper "Finding Structure in Time," proposed a solution: include a set of *recurrent* connections with trainable weights that link each unit of the single hidden layer to all hidden-layer units. Consequently, the hidden layer receives both the current environmental input and its own previous activation state, which, in turn, depends on the state before that, and so on. In this manner, the model is equipped with a working memory and can therefore encode sequential information, or "structure in time," making it well suited to processing language as it unfolds over time. This particular architecture became known as the Simple Recurrent Network (SRN) but forms part of a larger class of Recurrent Neural Networks (RNNs) that have connections through which (part of) the network's current activation feeds back to the network itself.

1.2. Neural Network Models and Linguistic Theory Connectionist models of language acquisition and processing offer a view of the human language system that is very different from traditional, symbolic models in cognition. For one, neural networks do not distinguish competence (i.e., language knowledge) from performance (i.e., language behavior). Instead, knowledge becomes instantiated in network connection weights in order for the network to display particular performance. In a sense, it forms procedural rather than declarative knowledge: it is know-*how*, not know-*that*. Hence, there is no way for the network to assess its own knowledge. As Clark and Karmiloff-Smith (1993, p. 495) put it: "it is knowledge in the system, but it is not yet knowledge to the system."

Second, language researchers from the nativist tradition have famously argued that infants must possess innate, language-specific knowledge or learning mechanisms, because otherwise language acquisition in the absence of negative evidence would be impossible (e.g., Chomsky, 1965; Gold, 1967; Pinker, 1989; among many others). In contrast, empiricists claim that language acquisition requires only domain-general mechanisms. Connectionism falls squarely into the empiricist camp because the representations and learning mechanisms built into neural networks are not specific to language and the networks receive no negative evidence during training. Hence, successful neural network learning of (relevant aspects of) syntax would undermine the nativist position.

A third major difference with traditional linguistic thinking is that neural networks do not represent discrete categories (be it phonemes, words, parts of speech, or any other category) unless these are explicitly assigned to the network's units a priori. However, in most (and, arguably, the most insightful) models, representations are learned in the hidden layer(s) rather than assigned, which results in "fuzzy" categories. To the extent these learned representations explain (psycho) linguistic phenomena, neural network models reject a causal role for explicit representation of typical linguistics constructs. The alternative proposed by connectionism is that apparently symbolic behavior emerges from statistical regularities present in the language and in the mapping between form and meaning. By capturing these regularities, neural networks are able to account for a range of phenomena in human language acquisition, comprehension, and production.

Neural network models and their ability to discover structure in the environment have enabled researchers to test the mechanisms required for many aspects of language behavior. For instance, such models have been applied to determine whether language structures need to be prespecified before exposure to language, or whether general purpose statistical processes, embedded in neural networks, are sufficient for reflecting human performance. Further, as language corpora have become more and more representative of language-learning environments, neural network models have provided insight into the sources of information present within the communicative environment that can contribute to learning and processing. In the following sections, we review key areas of language behavior where neural networks have been applied to explore environmental features that are useful for language learning, comprehension, and production.

2. *Word Learning*

In order to learn a spoken word, the listener has to be able to identify its phonological form, isolate it from continuous speech, and relate the word's form to its meaning in the environment. Each of these tasks is hugely challenging for the learner. Neural network models have been used to illustrate the inherent difficulties of these tasks, as well as to test potential solutions that the learner may bring to the situation to support acquisition, and subsequent online processing of words in speech.

2.1. Identifying Words from Continuous Speech Neural network models have been useful in uncovering the statistical properties of speech input that can contribute to speech segmentation. Elman (1990) investigated how statistical transitional information from sequences of phonemes that are present in child-directed speech could provide clues to word boundaries. His SRN model had as input a sequence of phonemes from an artificial corpus designed to mimic child-directed speech, and was trained to predict the next phoneme in the sequence. This SRN demonstrated that transitions between phonemes within words tend to be much more predictable than transitions across word boundaries. Thus, the model was able to detect the unpredictable nature of phoneme transitions at word boundaries and predicted future behavioral studies on infants' sensitivity to transitional probabilities between syllables in detecting potential words in continuous speech (Saffran, Aslin, & Newport, 1996). Cairns, Shillcock, Chater, and Levy (1997) demonstrated that a similar model was able to scale up to detect word boundaries, as well as to learn phonotactic regularities of syllable structure within words, from a transcribed corpus of natural language child-directed speech. These neural network models of segmentation also indicated limitations of characterizing the environment merely in terms of transitions between phonemes in speech, providing further insight into the process of early speech segmentation in language acquisition. One limitation was that transitional probabilities alone provide a glass ceiling for overall accuracy of segmentation. Gambell and Yang (2003) showed that transitional probabilities between syllables in child-directed speech, though able to account for segmentation at a rate significantly greater than chance, still missed very many word boundaries. So, some aspect of processing or of the environment was lacking from this modeling approach. To address this, Christiansen, Allen, and Seidenberg (1998) employed an SRN with natural language input, adding additional information about utterance boundaries and stress position within words. They found that these combined cues outperformed a model that used only one of these individual cues, indicating that a language-learning system benefits from sensitivity to combined cues for segmentation that are present in the speech environment. EEG studies of speech segmentation have shown relations between these proposed prosodic and statistical cues as predicted by computational models, demonstrating human sensitivity to these cues consistent with neural processing of sequence prediction (e.g., Cunillera, Toro, Sebastián-Gallés, & Rodríguez-Fornells, 2006).

These models have been extremely successful in illuminating the information that is present in the speech environment and that is also extractable using simple statistical processes that are potentially similar to those employed by the human language learner. But the models are also chronically limited by the fact that they do not construct a vocabulary. This means that they are unable to process ambiguities in speech, limited as they are by very local statistical structures. A classic example is the pair "recognize speech" and "wreck a nice beach" (Fosler-Lussier, Amdal, & Kuo, 2005), which would be indistinguishable in these simple, local-information approaches. A model with a vocabulary, with expectations (or priors) about individual and combined probabilities of combinations of words, would be able to more accurately segment continuous speech (Kamper, Wang, & Livescu, 2016; Monaghan & Christiansen, 2010; Philips & Pearl, 2015).

The TRACE model (McClelland & Elman, 1986) of auditory word recognition provided such a link between top-down lexical information with bottom-up statistical information about relations between acoustic features of speech and phonemes (Allopenna, Magnuson, & Tanenhaus, 1998). TRACE has recently been adapted to incorporate additional information from the visual environment to simulate multiple cue combination in speech processing (Smith, Monaghan, & Huettig, 2017). However, the TRACE model was not designed to process sequences of words embedded within continuous speech. To address this requirement to acquire lexical items online and simultaneously apply them to segment speech, French, Addyman, and Mareschal (2011) constructed a neural network model (TRACX) that learned to construct and identify chunked sequences of phonemes from continuous speech, using backward transitional probabilities to initially detect the word candidates from the language. These were then used for recognition of sequences of encountered speech, rather than, as with the SRN models, for prediction of the next phoneme in the sequence. The model was not only shown to outperform SRN models, but it also provided a framework for how a single statistical processing system can both generate and utilize word candidates from continuous speech.

These modeling approaches have thus shown the value of individual and combined cues present in natural language contributing to speech segmentation, turning a notoriously difficult problem into a task substantially more tractable. Furthermore, they highlight the kinds of computations that the learner must develop—and the processing interface between online, speech processing and the statistical probabilities that have to be encoded within a growing vocabulary, to move toward a highly accurate and fast speech segmentation system. Though these approaches have tended to take an abstracted input, ignoring phonetic variability and noise in the acoustic signal, adding such information provides challenges but also opportunities for additional cues to be discovered and derived from the input. Bayesian approaches to discovery of word-level acoustic forms using unsupervised training demonstrate the possibility of this in artificial computational systems (Kamper, Jansen, & Goldwater, 2016). One such example is increased word-final phoneme duration that is a further useful, and useable, cue for speech segmentation (e.g., Scharenborg, 2010) and that could be incorporated into a neural model of speech processing.

2.2. Mapping Words to Meaning A second task for word learning in language acquisition is to determine the referent of a word once it has been isolated from continuous speech. This task is famously difficult to accomplish, because an utterance tends to contain several words, only one of which is referring to a particular aspect of the environment, and because the environment contains an infinite number of potential referents for each word (Quine, 1960)—the word *pig* could refer to the small cuddly toy that a parent is waving in a child's face, but it could also refer to the action of waving, to the softness of the material of the toy, the color, its general shape, or to the communicative act itself. Hence, the task can be conceived as, at the very least, a many-to-many mapping problem.

Similar to the models of speech segmentation, there are numerous models of word-meaning acquisition that provide abstract characterizations of the task, with the intention to highlight the environmental properties that enable this mapping task to be accomplished. These models explore the information present around the child acquiring her language in resolving this link between word and world. More recently, the models have become more concretely related to neural processing in order to incorporate known perceptual and attentional constraints that, paradoxically perhaps, demonstrate even greater tractability of the problem (Regier, 2003; Yurovsky, Smith, & Yu, 2013).

Associations between occurrences of words and occurrences of their intended referent have been proposed as a mechanism driving the formation of word-meaning mappings (Plunkett, 1997). Yu and Ballard (2007) implemented a model of associations between words and objects that was exposed to small corpora of child-directed speech, where objects in the environment had also been encoded in the corpus transcription. When a particular word was spoken in the presence of a particular object, the strength of association between the word and the object was increased. They found that the strongest associations were often between words and the target object, such as the words *oinko* and *pig* being strongly associated with a pig soft toy in the child's environment.

Yu and Ballard's model was further tested by adding in additional cues that constrained either the referring word or the referent in the environment. They included prosodic information about words that were emphasized by being prosodically distinct from the rest of the utterance and increased the possibility of strengthening associations for those words. They also included information about objects in the environment for which the speaker and the child held joint attention during the utterance, again increasing the association strength for the attended object. Each of these information cues improved the model's performance further. The model was able to exploit cross-situational correspondences between words and objects in the environment, such that from a single exposure to a word and an object, the relation between the word and referent is not evident (because of the presence of other words and environmental stimuli at the same time), but over multiple instances the likelihood of the object being named when present increases. Such cross-situational information was shown to be used by infants in acquiring word-meaning mappings (Smith & Yu, 2008), supporting the computational models suggesting that associationism may underlie the acquisition of word meanings.

This associationist modeling approach, however, has been criticized by researchers who propose instead that children do not learn by association between words and environmental properties, but rather that children propose and self-test hypotheses for a possible word-meaning link. Yu and Smith (2012) compared both of these computational approaches and found that an associationist model provided an adequate fit to the data and that even apparent complexities that seem to suggest hypothesis testing (e.g., children applying strategies such as mutual exclusivity: using a novel word for an unknown object appearing among known objects) can be accounted for within an associationist model. Similarly, apparent developmental differences between children and adults in terms of whether they employ an associative approach or a strategic, referential approach can also be accounted for by a model with a single associative mechanism (Fazly, Alishahi, & Stevenson, 2010).

However, other computational models have suggested that combinations of associative and strategic word-learning mechanisms are required to adequately describe performance. McMurray, Horst, and Samuelson (2012) proposed a hybrid network model with two components. The first was an associative network that learned mappings between words in an utterance and objects in the environment with a very slow learning rate and was able to acquire word-object mappings from cross-situational statistics. The second component was a quick-responding referent selection mechanism that was able to strategically select a potential referent for a word, under circumstances where children can demonstrate fast mapping between words and potential referents, such as conditions of mutual exclusivity. The model predicted that word learning would be more robust from multiple presentations of a word, but less stable when learning is inaugurated under conditions of fast mapping. Twomey, Ranson, and Horst (2014) subsequently confirmed these predictions of the model in children learning novel words: fast mapping and associative learning resulted in immediate word-object mappings, but only associative learning resulted in long-term learning.

Additional attentional and perceptual constraints have recently been included in these models of associative word learning to align the referent selection mechanism with the cognitive processes known to impact on visual object processing. Samuelson, Kucker, and Spencer (2017) showed how this can further improve performance on word learning in a manner consistent with approaches that closely describe the actual multimodal input that children experience (Yurovsky et al., 2013). Furthermore, such associative models are able to generalize properties of the environment to learn category-level information, accounting for children's distinctions in learning names for shapes and names for materials (Colunga & Smith, 2005) and learning individual and category labels (Mayor & Plunkett, 2010). Such approaches provide a valuable bridge between a long tradition of neural network modeling in perception and attention (e.g., Mozer, 1991), embedding our understanding of language-learning models within broader, domain-general processes.

3. Syntactic Development

3.1. Lexical Category Learning How children acquire category knowledge about words has been a hotly debated issue, at the heart of the nature-nurture debate in language acquisition. Nativists assume that the lexical categories, or semantic features to which these categories relate, are innately specified in the learner, and individual words' membership to these categories is then determined from exposure to the language. Alternatively, empiricist approaches propose that the categories themselves are acquired as a result of experience, without the need for innately specified structure in advance of exposure to the language. Neural network modeling approaches have been key to determining the extent to which such innate knowledge is required in advance to simulate the acquisition of lexical categories within the language, and, relatedly, these approaches have explored the type of computational mechanisms that might be able to drive the discovery of these categories.

Elman (1990) applied the same SRN approach as used for the speech segmentation task to prediction of words in utterances, again using a small artificial corpus. The model's input was a sequence of words in the utterance and the model had to predict (i.e., generate as output) the next word in the sequence. Utterances were constructed from one of four sentence frames, and there were 29 different words that were either animate or inanimate nouns or transitive or intransitive verbs. After training, the hidden unit states of the model were investigated to determine whether it had learned to reflect the lexical categories in the language in its internal structure, resulting only from the statistical sequential information from the sentences. A hierarchical cluster analysis revealed distinctions in the model's representations of nouns versus verbs. Furthermore, clusters within the noun category were also evident, with animate and inanimate nouns forming distinct groups. More recently, Mikolov, Chen, Corrado, and Dean (2013) showed that recurrent and feedforward networks can be trained on very large natural language corpora and that the resulting connection weights from input units display semantic/syntactic word clustering, similar to what was demonstrated by Elman but on a much more realistic scale. Thus, apparent semantic features that relate to syntactic properties (such as nouns versus verbs, as well as features within a syntactic category, such as animacy) were detectable by applying generic distributional statistical learning to sequences of words.

The extent to which general purpose learning of lexical categories scaled up to a more realistic natural language environment was tested computationally by Redington, Chater, and Finch (1998). Using a large corpus of English child-directed speech, they measured co-occurrences between pairs of words that were either adjacent or separated by one word. Cluster analysis was then applied to these counts of associations, such that words with similar patterns of co-occurrences should be clustered together. The results were spectacular, demonstrating for the first time, with large-scale corpora, that lexical categories were extractable with accuracy using simple statistical processes applied to natural language exposure (see Kiss, 1973, for a smaller-scale version, and Schütze, 1993, for large-scale use of distributional information for lexical clustering in a language engineering context). As with Elman's computational investigation, the clusters in the co-occurrence model respected broad syntactic category distinctions, such as between nouns and verbs, but they also reflected nuanced distinctions within those categories, such as fruits and animals forming individual clusters. Hence, the syntactic categories were indeed just one level of a hierarchy of lexical categories, which is consistent with constructionist approaches to describing grammatical structure (Goldberg, 2009).

The Redington et al. approach utilized generic, simple statistical processing mechanisms that are consistent with the associative learning approaches that have been tested for effectiveness for speech segmentation (see section 2.1) and word-meaning (section 2.2) mappings. However, the extent to which these computations are tractable to a small child acquiring their language remained an open question. Redington et al. counted co-occurrences between each word in the language and the 150 most frequently occurring words in the same corpus, but in four different positions (two words before, one word before, one word after, two words after). Thus, for 1,000 target words, the system is required to store 600,000 separate counts. This is likely to exceed somewhat the child's working memory capacity. Mintz (2003) proposed an alternative method by which co-occurrence information could give rise to syntactic categories based on input alone that was closer to respecting known limitations about working memory. He identified the 35 most frequent pairs of words separated by one other word (e.g., "put __ in") and determined the set of words that occurred within the pair (e.g., *it, them, him, things, teddy, dolly,* ...). Then, the extent to which words occurring within these frames were of the same syntactic category was measured. The results were highly accurate, with more than 90% of words belonging to the same category within each frame. Hence, using a very simple statistical mechanism, based on slightly more specific contextual information than was used by Redington et al., accurate lexical categorization could proceed.

Though accurate, the frequent frames approach was unable to classify the majority of words in the language, as most words did not occur within these frequent frames. St. Clair, Monaghan, and Christiansen (2010) applied a neural network to test the optimal statistical information available from child-directed speech in order to generate knowledge about syntactic categories. Specifically, they tested whether sparse, but accurate, frequent frames or whether less specific information, broader in coverage, about just the preceding word and the following word resulted in more effective classification of child-directed speech into syntactic categories. St. Clair et al.'s (2010) model was a feedforward network, with distributional information from the corpora of child-directed speech as input. The model was required to produce at the output the syntactic category of each word. The distributional information presented as input represented either the specific frames of Mintz's (2003) approach or the preceding and the succeeding word. The modeling demonstrated that training on Mintz's specific frames resulted in substantially lower accuracy of syntactic category detection than training with separate preceding and succeeding contextual information. This model indicated that distributional information was sufficient to generate hypotheses about the category of words in child-directed speech and furthermore that this distributional information was most effective when processed in terms of very local, flexible contextual information about only the preceding and the succeeding word.

Neural network models of lexical categorization have largely followed the empiricist approach of determining how general purpose statistical mechanisms can apply to the case of language acquisition. However, this work has proceeded somewhat independently of advances in neural network studies of perceptual and attentional development. Exceptions, such as Samuelson et al. (2017), show the benefit of combining these approaches for word learning. For studies of categorization, there are numerous related neural networks of perceptual category learning that are consistent with domain-general accounts of language learning that demonstrate more closely how computational modeling can relate to neural processing in the brain (see Schultz, 2012, for a review). For instance, young infants are able to categorize stimuli into separate categories, based on distinctions in the visual form of those stimuli (Younger, 1985). Westermann and Mareschal (2012) tested these behaviors using two computational models. The first model simulated perceptual processing of stimuli from different categories using a model with changing receptive field sizes processing the visual input. Visual development has been proposed to adjust from larger to smaller receptive field sizes during infancy (Spencer, Simmering, Schutte, & Schöner, 2007). Implementing this property into the hidden layer of an autoassociator model, which learned to reproduce a visual input at output, was able to simulate visual category processing development in children's early years.

An additional simulation by Westermann and Mareschal (2012) investigated the processing of different memory systems during learning of visual categories. In this model, two auto-associator neural network models mapped the same visual input onto representations of the visual form at two output layers, via two hidden layers. These networks were yoked by interconnecting the networks' hidden layers. One hidden layer was trained with a fast learning rate, representing hippocampal learning, whereas the other was trained with a slow learning rate, representing cortical learning (McClelland, McNaughton, & O'Reilly, 1995). After training on a set of perceptual category stimuli, the model was presented repeatedly with a test trial, to simulate infant testing in habituation studies. How long the model took to adapt to the stimulus was taken as a measure of the model's processing of the category stimuli. The model was able to simulate developmental shifts in category learning, and, importantly, the role of hippocampal and cortical systems could be appraised separately for the extent to which recent and long-term memory of stimuli accounted for the effects.

Such closer links between models of perceptual system performance and models that have previously investigated more abstract treatments of linguistic input, are areas where cognitive models of language could be further informed by neural processing. As we shall see in the next sections, neural network models of syntactic processing have made substantial strides in drawing closer together neural processing of the brain's language neural network with computational approaches to studying language.

3.2. Learning Syntactic Structure Most, if not all, connectionist models of the acquisition of syntactic structure have been based on models of sentence comprehension or production, which will be discussed in section 4. Here, we focus on the application of RNN sentence processing models to account for aspects of acquisition.

Most often, neural networks that learn sentence processing are trained to generate as output a prediction of the next word at each point of the utterance. Elman (1993) demonstrated that such a next-word prediction SRN can learn a semirealistic, miniature language with multiple embedded structures. He claimed that the network, somewhat like a child, needs to "start small," that is, it has to be trained on simple sentences first or must start out with limited short-term memory capacity. However, Rohde and Plaut (1999) later showed that the inclusion of semantic (in addition to syntactic) dependencies in the language allows an SRN to learn the language without starting small. These results suggest that a notion of embedded syntactic structure can be learned by systems without hierarchical structure built in, that is, it does not need to be an innate capacity.

Nevertheless, the acquisition of relative (embedded) structures is challenging for children. Diessel and Tomasello (2005) asked English and German four-year-olds to repeat six types of embedded clauses. Even though English and German have very different syntactic structures, both groups of children displayed the same ordering range of production difficulties, with intransitive subject relatives ("That's the woman who played the piano this morning") being the easiest and genitive relatives ("That's the girl whose dog was chasing a cat yesterday") the most difficult to repeat. Fitz and Chang (2008; see also Fitz, Chang, & Christiansen, 2011) simulated this task in a sophisticated RNN model: The *Dual-path model of sentence production* (discussed in more detail in section 4.2), which assumes separate processing streams for syntactic and semantic information. The model was able to produce embedded structures with more than 90% accuracy and to approximate most of Diessel and Tomasello's experimental results. Furthermore, Fitz and Chang found that how well and fast a structure is acquired depends on its frequency and its similarities to previously learned structures.

Naturally, children who are acquiring their first language are exposed to more linguistic cues than merely the word-order patterns available to standard SRNs. Reali, Christiansen, and Monaghan (2003) included additional lexical cues during SRN training so that the network also had access to information about word length, phonology, and prosody. Another noticeable thing about this simulation is that the model was not trained on a toy language but on over 10,000 utterances taken from a corpus of child-directed speech. However, the words were replaced by their lexical categories to reduce the required size of the network. Christiansen, Dale, and Reali (2010) took this realistic approach one step further by representing each of the 1,371 word types in the corpus by a unique vector encoding 16 realistic phonetic cues. RNNs were then trained on sequences of these vectors, corresponding to the child-directed utterances, where the network's task was to predict the upcoming lexical category at each point. As a control condition, other RNNs were trained with a random assignment of the phonetic vectors to words. The networks with the nonrandom assignment of vectors showed improved generalization, suggesting that infants can indeed make use of phonetic cues to learn grammar. An analysis of the networks' hidden representations revealed that the use of phonetic cues improves the representation of abstract lexical categories (at least for nouns and verbs). Similarly, natural language processing (NLP) systems, whose primary aim is not to reflect cognitive performance but rather achieve highest possible accuracy, have utilized "character language models," whereby individual letters rather than words are provided as input to the model, and the model must discover the categories of the words to which the letters belong. Such models can outperform models that take as input word-level information (e.g., Kim, Jernite, Sontag, & Rush, 2016).

The Reali and Christiansen studies (Christiansen et al., 2010; Reali et al., 2003) demonstrate neural networks' ability to learn about the general word-order patterns in child-directed speech. However, the argument for innate language-specific knowledge often centers on particular constructions that children are claimed to learn without sufficient input and without producing errors. One hallmark example is the formation of polar interrogatives: turning *the man who is eating is hungry* into a yes/no question requires fronting the correct auxiliary verb (i.e., *is the man who is eating hungry?* and not *is the man who eating is hungry?*). Reali and Christiansen (2005) demonstrated that the Reali et al. model is in fact able to perform correct auxiliary fronting, which, therefore, may not require innate, language-specific knowledge.

The nativism versus empiricism debate has also been tackled by the Dual-path model of sentence production. Chang, Dell, And Bock (2006) used the model to investigate the syntax learning mechanism early in development. Elicited production tasks (e.g., Tomasello, 2000) tend to support empiricism (where abstractions are learned through experience only, so there is no such thing as innate language competence), whereas some preferential-looking studies (e.g., Naigles, 1990) support nativism. The model simulated both tasks and was evaluated against empirical data (i.e., the patterns found in Tomasello). Using the same model and input it was shown that preferential-looking preceded production, which led to the conclusion that children learn to abstract syntactic features through experience but that the abstraction mechanism is possible in part because of a preexisting separation in the brain between "neurons that learn sequences and neurons that encode concepts" (Chang et al., p. 264), which correspond to the model's syntactic and semantic processing streams, respectively.

Moving beyond the nativism-empiricism debate, connectionist models have provided explanations for several phenomena in first language acquisition. For example, children tend to make overgeneralization errors in the use of locative verbs (e.g., *fill, pour, spray*) where they seem to understand the verbs' meaning and the sentence structures, but sometimes produce ungrammaticalities such as *I filled water into the glass* instead of *I filled the glass with water.* Twomey, Chang, and Ambridge (2014) investigated whether the Dual-path model shows the same behavior when it learns syntactic structures. Indeed, the model showed an initial preference for producing first the theme (*water*) and then the location (*glass*), which leads to overgeneralization errors on verbs that have a location-first bias, such as *fill.* The output of the model was consistent with a corpus analysis on child-directed speech as children seemed to master location-first structures faster than theme-first structures. Over development, the bias toward this structure was reduced and gradually disappeared.

One influential view of language development is that knowledge of syntax helps children learn semantics, a phenomenon known as syntactic bootstrapping (Gleitman, 1990). Conversely, semantic bootstrapping (Pinker, 1984) occurs when knowledge of word meaning supports the acquisition of syntax. In both cases, correlations between syntax and semantics need to be discovered, which is precisely what RNNs do when they learn to map sentence forms to semantic representations. Indeed, such models display syntactic and semantic bootstrapping as an emergent side effect of learning (Desai, 2002, 2007; Frank & Vigliocco, 2011).

4. Sentence Processing

4.1. Sentence Comprehension With only very few exceptions (e.g., Sturt, Costa, Lombardo, & Frasconi, 2003), neural network models of sentence comprehension apply the RNN architecture. If a network is trained on next-word prediction (the standard task in SRNs), it extracts statistical knowledge of the language's word-order patterns (i.e., a form of syntactic knowledge) without explicitly implementing traditional syntactic structures. Early SRN studies (Elman, 1991; Rodriguez, Wiles, & Elman, 1999; Servan-Schreiber, Cleeremans, & McClelland, 1991) aimed to show that these models can nevertheless represent hierarchical syntactic structure implicitly in the organization of their internal state space. It wasn't until several years later that SRNs (and RNNs in general) were used to explain particular phenomena in human sentence processing.

One aspect of sentence comprehension where RNN models have been particularly influential is in simulations of embedded structure processing. Here, a direct advantage of connectionist over traditional, grammar-based accounts is that the former provide an explanation for the difficulty of processing multiple embeddings (as in: *the cook*$_i$ *who the thief*$_j$ *who the wife*$_k$ *loved*$_k$ *robbed*$_j$ *served*$_i$ *food*). In a recurrent network, each next level of embedding needs to be represented in the same network units, resulting in interference that grows stronger as the number of embedded phrases increases. This leads to increased next-word prediction error, which is generally taken to be indicative of processing difficulty in the network (Christiansen & Chater, 1999). Hence, increased processing difficulty for deeper embedding is inherent to RNNs, as it is to humans. In addition, Christiansen and MacDonald (2009) demonstrated that RNNs, like humans, suffer more difficulty with embedded structures (e.g., *the cook*$_i$ *who the thief*$_j$ *saw*$_j$ *serving*$_i$ *food*…) than with cross-dependent recursive structures (which exist in Dutch and Swiss German and correspond to *the cook*$_i$ *who the thief*$_j$ *food serving*$_i$ *saw*$_j$…"), while right-branching structures (*the thief*$_i$ *saw*$_i$ *the cook*$_j$ *serving*$_j$ *food*) are the easiest. They also showed that RNNs experience "grammaticality illusion": the phenomenon that removing a required verb from English sentences with double-embedded relative clauses (e.g., *the cook*$_i$ *who the thief*$_j$ *who the wife*$_k$ *loved*$_k$ *served*$_i$ *food*) makes these (now ungrammatical) sentences appear more acceptable to human readers, who also read them faster. Interestingly, the reversed effect occurs in German and Dutch (i.e., the correct double-embedded structures are read faster than those with a missing verb), a phenomenon that has also been demonstrated in RNNs (Engelmann & Vasishth, 2009; Frank, Trompenaars, & Vasishth, 2016).

English object-relative clauses (e.g., *the cook who the thief saw*…) are more difficult to process than subject-relative clauses (*the cook who saw the thief*…), albeit less so for people with higher exposure to object-relative structures. RNNs behave similarly (MacDonald & Christiansen, 2002). The situation is more complex in Chinese, where the reversed pattern holds when the relative clause modifies the sentence's grammatical subject but not when it modifies the object. Hsiao and MacDonald (2013) replicate this pattern, as well as interactions with noun animacy, in an RNN simulation.

One criticism that can be raised against all simulations referred to thus far (with the exception of Frank et al., 2016) is that the models are trained and tested on hand-crafted, miniature languages. Hence, they have no knowledge of the true language and are unable to simulate the processing of actual experimental stimuli. Comparisons between model simulations and human behavior will therefore have to remain qualitative rather than quantitative. However, recent technical developments make it more feasible to train RNNs on large-scale realistic corpora and to evaluate them on the same items used in experiments. The size of next-word prediction errors by these more realistic RNNs explain general patterns of reading times (Frank, 2013; Hahn & Keller, 2016; Monsalve, Frank, & Vigliocco, 2012) and brain activity as measured by EEG (Frank, Otten, Galli, & Vigliocco, 2015) or magnetoencephalography (Wehbe, Vaswani, Knight, & Mitchell, 2014) from participants reading naturalistic materials. When RNNs are directly compared to symbolic grammars, the RNNs often fit the human data better and the grammars do not account for additional variance (Frank & Bod, 2011; Frank et al., 2015; but see Fossum & Levy, 2012, for conflicting results), suggesting that RNNs form more adequate models of cognitive processing difficulty than traditional grammars do.

The ability of RNN prediction error to account for these general reading-time and brain-activity patterns may seem to suggest that many psycholinguistic phenomena can be simulated in this manner. However, this approach has not been very successful in explaining much beyond embedded clause processing. For example, Tabor, Juliano, and Tanenhaus (1997; Tabor & Tanenhaus, 1999) simulated garden path effects (Frazier & Rayner, 1982) in an RNN, but, rather than using prediction error as the relevant measure, they included an additional mechanism that steers the network's internal state toward one of several state-space clusters that correspond to the possible structures of the syntactic ambiguity. The time required to reach a cluster was taken as a predictor of reading time. These simulation results suggest that garden path effects do not rely (solely) on prediction but also require sentence interpretation, something that is (arguably) not reflected in RNN prediction error. Indeed, the equivalent of prediction error in incremental symbolic probabilistic parsers, which do generate syntactic interpretations, has successfully been used to account for human garden path effects (Brouwer, Fitz, & Hoeks, 2010; Hale, 2001; Levy, 2008). Again, this suggests that next-word prediction does not suffice to simulate the garden path phenomena.

However, sentence-processing RNNs are not restricted to next-word prediction. An alternative is to train these networks to transform input sentences into representations of their meaning. As was the case for next-word prediction models, early form-to-meaning RNNs mostly formed proofs of concept that connectionist models can simulate the incremental mapping from sentence input to semantic output and were much less concerned with explaining human performance (McClelland, St. John, & Taraban, 1989; Miikkulainen, 1996; Miikkulainen & Dyer, 1991; St. John & McClelland, 1990). More recent RNN models of semantic interpretation have been used to explain findings from EEG studies.

Hinaut and Dominey (2013) equated the recurrent part of their model with the human brain's frontal cortical network, while the semantic output units are claimed to correspond to the striatum. The amount of activation change in the simulated striatum (i.e., the amount of semantic reinterpretation required to integrate the current word) is taken to simulate the size of the P600 event-related potential (ERP) component. In this manner, the model reproduces the finding that object-relative clauses result in greater P600 than subject-relative clauses, at the point where the relative clause type is disambiguated. According to the model, this effect is caused not by structural differences between the two sentence types but simply by the fact that object relatives are less frequent and, consequently, less preferred as an initial interpretation.

All models discussed so far receive only linguistic input for training and evaluation. However, as observed in models of word learning and comprehension (e.g., Samuelson et al., 2017; Smith et al., 2017), real-life language use rarely takes place without nonlinguistic context. The CIANET model (Mayberry, Crocker, & Knoeferle, 2009) simulates the comprehension of a sentence in visual context. It receives as input not only a sentence but also a simplified representation of a visual scene and outputs at each point of the sentence both a semantic interpretation and a simulated eye gaze toward one of two relevant parts of the visual scene. This model managed to capture effects of word order, of stereotypicality of the described action, and of conflict between the stereotypical actions of depicted agents and the action described in the sentence. Crocker, Knoeferle, and Mayberry (2010) applied the same model to account for ERP effects. They took the amount of change in recurrent layer activation when processing a word as an index of Left Anterior Negativity and P600 size. In this manner, the model could account for effects of word order, disambiguation, and the presence of a visual scene. Interestingly, different recurrent layer units turned out to be responsible for the simulated Left Anterior Negativity versus P600 effects, indicating that the network functionally separates the two, as does the brain.

Rabovsky, Hansen, and McClelland (2018) took the amount of activation change in hidden units of the McClelland et al. (1989) model as an index of N400 size, and thereby explained several well-known findings. For one, the simulated N400 was stronger for less expected words and for words earlier in the sentence. Also, semantically incongruent words resulted in stronger N400, but much less so if they were from the same semantic category as the expected, congruent word. Finally, the model accounted for the so-called *semantic illusion*, where in a sentence like *for breakfast, the eggs eat* the strong association between *breakfast, eggs* and *eat* results in a relatively weak N400 in spite of the semantic incongruency.

Note that Rabovsky et al. (2018) related hidden-unit update to N400 size, whereas Crocker et al. (2010) took basically the same quantity to reflect the P600. A recent RNN model of semantic illusions by Brouwer, Crocker, Venhuizen, and Hoeks (2017) has multiple hidden layers, where one is devoted to lexical retrieval and another to the integration of concept into a semantic representation of the sentence. The N400 is hypothesized to index lexical retrieval difficulty (modeled by the amount of activation change in the lexical retrieval layer) and the P600 corresponds to the update of the sentence interpretation (reflected in the amount of activation change in the integration layer). Indeed, this division of labor accounted for many (if not all) ERP effects observed in the comprehension of semantic illusion sentences.

4.2. Sentence Production There has been substantially less work on models of sentence production than sentence comprehension. It may seem straightforward to construct a production model by running a sentence comprehension model backward, and this is indeed how two recent connectionist models of production were developed (Calvillo, Brouwer, & Crocker, 2016; Hinaut et al., 2015). However, the most successful and empirically validated sentence production models were specifically designed to simulate production.

The first neural network model of sentence production, the so-called *structural priming* model (Dell, Chang, & Griffin, 1999; see also Chang, Dell, Bock, & Griffin, 2000), was developed to simulate syntactic priming: the tendency of the speakers to repeat the structure of recently spoken or heard sentences (Bock, 1986; Bock & Loebell, 1990). The model assumes a close link between sentence comprehension and production; comprehension of what has been said or heard so far influences the production of a sentence. The model encodes the intended meaning (or "message") by units that represent role-concept pairs (e.g., *agent-CHILD* or *patient-MAN*) that form input to the hidden layer during production of the whole sentence. The output layer units represent words, where the most active unit is taken to be the produced word and is fed back into the network, which thereby receives information about what has been produced so far. This model was able to successfully account for several structural priming phenomena, for instance, if *Boys chase dogs* was used as prime (i.e., active rather than passive voice), the model would produce the message *agent-GIRL;*

action-FEED; patient-CAT as *girl feeds cat* instead of *cat is fed by girl* (Dell et al., 1999). However, it failed to show priming between transitive locatives (*Boys chase dogs near car*) and prepositional datives (*Boys give dog to girl*), which is empirically shown by Bock and Loebell (1990). Another limitation of the model was that, because each concept-role pair is represented in a single unit, the agent-concept MAN is different from the patient-concept MAN. Consequently, the model is unable to generalize its ability to produce *the man is chasing a dog* to the ability to produce *the dog is chasing a man.* This violates the property of systematicity, which Fodor and Pylyshyn (1988) argued is a fundamental feature of human cognition that neural networks do not possess.

Chang (2002) proposed and compared two neural network models of sentence production, Prod-SRN and Dual-path. Prod-SRN is a simple extension of the structural priming model, tested on a more advanced morphology and closer to a typical SRN, but it is still lacking systematicity. Dual-path, which is still the most influential neural network model of sentence production, was the first to overcome the limitation of generalization. It does so by creating temporary bindings between a layer for roles and a layer for concepts, so that there is only one unit for MAN, irrespective of its semantic role. These bindings, along with the event semantics (information about tense and aspect, e.g., PRESENT SIMPLE), form the model's *semantic stream.* The model has a second stream (hence its name), the *syntactic stream,* which is an SRN that allows the model to learn syntactic categories. This way, the model was not only able to generalize words to new positions, but also to generalize a noun as a verb; this is something that speakers usually do with proper nouns, for example, *Skype* becomes *skyping.* Chang (2002) compared Prod-SRN to the Dual-path model, and the latter was able to generalize 82% of the time, whereas Prod-SRN only reached 6%. The models were also tested on unseen adjective-noun pairs and identity construction (e.g., *a cat is a cat*); Dual-path outperformed Prod-SRN in all tests. The model also expanded on Gordon and Dell's (2003) simple model of aphasic production, offering a natural explanation of two different types of aphasia, agrammatism and anomia (see also Dell & Chang, 2014).

Chang et al. (2006) applied the Dual-path model to a wider range of structural priming phenomena. The model displayed similar priming whether the prime had been produced or only comprehended. It was also able to account for long-term priming, as the extent of structural priming was not dependent on the number of fillers between the sentences. Furthermore, Chang et al. showed that the strong but short-lived tendency to repeat previously heard or said words (the so-called lexical boost) is due to a different mechanism from structural priming. This prediction was confirmed experimentally two years later (Hartsuiker, Bernolet, Schoonbaert, Speybroeck, & Vanderelst, 2008).

To test whether the model could handle a language that is typologically very different from English, Chang (2009) tested Dual-path on Japanese. Despite the different word orders between these two languages, the model was able to exhibit similar levels of grammaticality (93% for English and 95% for Japanese). Furthermore, the model was able to explain differences in production preferences between speakers of these two languages. For instance, in English, long phrases are usually placed after short ones (e.g., *The woman sent a book to the man that she met while traveling* is preferred over *The woman sent the man that she met while traveling a book*). This phenomenon is called heavy noun phrase shift (Ross, 1967), and English exhibits a short-before-long bias whereas Japanese has the opposite. The model was able to account for this crosslinguistic difference. Chang showed that the phenomenon is caused by a difference in the relative importance of the meaning in the positions ("choice points") where the word orders differed. For English, the choice point was right after the verb, whereas in Japanese the choice point is at the beginning of the sentence as verbs tend to occur at the end of the sentence.

The Dual-path model was further able to account for crosslinguistic differences in lexical/conceptual accessibility between English and Japanese (Chang 2009). English speakers tend to prefer using animate elements early in the sentence (McDonald, Bock, & Kelly, 1993), which can lead to the usage of less common structures such as passives (e.g., *the man was almost hit by a car*). At the same time, they do not have animacy preference in conjunctions: speakers of English find *the man and the car* as acceptable as *the car and the man.* Therefore, it was hypothesized that animacy can influence the functional level but not the positional level from Garrett's (1988) theory of sentence production. However, if this were the case, animacy wouldn't affect word order in Japanese as this language uses case markers to indicate roles and repositioning words doesn't affect the meaning (e.g., in the passive sentence, *man* would be marked as the subject that receives the action *hit,* regardless of word order). Nevertheless, it has been shown that animacy affects word position (Branigan, Pickering, & Tanaka, 2008). Using the Dual-path model, Chang noticed that these preferences were related to the frequency of the input; by giving it sentences where animate words were used early in the sentence, the model learned stronger connections between animate concepts and words than for inanimate ones.

One criticism against models of sentence production is that all current models use a miniature (artificial) language instead of natural stimuli. Another limitation of the current models of sentence production is that none of them consider the phonological level of sentence production (Garrett, 1988). Rather, they all focus on message planning (conceptualization) and sentence formation. Furthermore, even though in most models the hypothesis that comprehension influences production is supported, the exact connection between these has yet to be established.

5. Conclusion

Connectionist models have been instrumental in explaining a range of human language behaviors, from word segmentation and word-meaning mapping to sentence processing and syntactic development. These models' successes have demonstrated the richness of the environment for language learning, contributing constructively to debates over empiricist versus nativist positions of language acquisition. They have also been useful in determining the knowledge required by an information processing system for simulating human behavior, addressing questions of the extent to which language processing is hierarchical or sequential, the interface between syntax and semantics, and the role of prediction and richer interpretation in sentence processing.

In the machine learning and NLP literature, deep learning neural networks currently outperform other systems on a range of NLP tasks. For example, very large corpora of natural texts can now be used to train RNNs on next-word prediction (Mikolov, Deoras, Povey, Burget, & Černocký, 2011) and to let feedforward networks extract lexical semantics from distributional patterns (Mikolov et al., 2013). RNNs have been applied to select which image is described by a sentence (Chrupała, Kádár, & Alishahi, 2015) and, conversely, for generating sentences that describe a given image (Chen & Zitnick, 2015), suggesting that the problem of representing semantics in production models (see section 4.2) can be solved by using images as meaning representations instead of pairs of semantic roles and fillers.

With a few exceptions (e.g., Hahn & Keller, 2016; Kamper et al., 2016), such NLP models are not generally intended to simulate human cognitive processing and, consequently, their evaluation against human data is still in its infancy. As a case in point, the RNN model by Mikolov et al. (2011) shows excellent next-word prediction performance on natural language, but unlike the models by Elman (1993) and Rohde and Plaut (1999) discussed in section 3.2., its behavior has not been related to human performance or acquisition, for example, regarding embedded clause processing. Nevertheless, the remarkable performance of current neural networks may well suggest that they embody relevant aspects of the human language system, affording great potential for further applications in the psycholinguistic community.

REFERENCES

Allopenna, P. D., Magnuson, J. S., & Tanenhaus, M. K. (1998). Tracking the time course of spoken word recognition using eye movements: Evidence for continuous mapping models. *Journal of Memory and Language, 38,* 419–439.

Bock, J. K. (1986). Meaning, sound, and syntax: Lexical priming in sentence production. *Journal of Experimental Psychology: Learning, Memory, and Cognition, 12,* 575–586.

Bock, K., & Loebell, H. (1990). Framing sentences. *Cognition, 35,* 1–39.

Branigan, H. P., Pickering, M. J., & Tanaka, M. (2008). Contributions of animacy to grammatical function assignment and word order during production. *Lingua, 118,* 172–189.

Brouwer, H., Crocker, M. W., Venhuizen, N. J., & Hoeks, J. C. J. (2017). A neurocomputational model of the N400 and the P600 in language processing. *Cognitive Science, 41*(S6), 1318–1352.

Brouwer, H., Fitz, H., & Hoeks, J. (2010). Modeling the noun phrase versus sentence coordination ambiguity in Dutch: Evidence from surprisal theory. In Hale, J.T. (Ed.), *Proceedings of the 2010 Workshop on Cognitive Modeling and Computational Linguistics* (pp. 72–80). Stroudsburg, PA: Association for Computational Linguistics.

Cairns, P., Shillcock, R., Chater, N., & Levy, J. (1997). Bootstrapping word boundaries: A bottom-up corpus-based approach to speech segmentation. *Cognitive Psychology, 33,* 111–153.

Calvillo, J., Brouwer, H., & Crocker, M. W. (2016). Connectionist semantic systematicity in language production. In Papafragou, A., Grodner, D., Mirman, D., & Trueswell, J.C. (Eds.), *Proceedings of the 38th Annual Conference of the Cognitive Science Society* (pp. 2555–2560). Austin, TX: Cognitive Science Society.

Chang, F. (2002). Symbolically speaking: A connectionist model of sentence production. *Cognitive Science, 26,* 609–651.

Chang, F. (2009). Learning to order words: A connectionist model of heavy NP shift and accessibility effects in Japanese and English. *Journal of Memory and Language, 61,* 374–397.

Chang, F., Dell, G. S., & Bock, K. (2006). Becoming syntactic. *Psychological Review, 113,* 234–272.

Chang, F., Dell, G. S., Bock, K., & Griffin, Z. M. (2000). Structural priming as implicit learning: A comparison of models of sentence production. *Journal of Psycholinguistic Research, 29,* 217–230.

Chen, X., & Zitnick, C. L. (2015). Mind's eye: A recurrent visual representation for image caption generation. In *IEEE Conference on Computer Vision and Pattern Recognition* (pp. 2422–2431). Boston, MA: IEEE.

Chomsky, N. (1965). *Aspects of the theory of syntax.* Cambridge, MA: MIT Press.

Christiansen, M. H., Allen, J., & Seidenberg, M. S. (1998). Learning to segment speech using multiple cues: A connectionist model. *Language and Cognitive Processes, 13,* 221–268.
Christiansen, M. H., & Chater, N. (1999). Toward a connectionist model of recursion in human linguistic performance. *Cognitive Science, 23,* 157–205.
Christiansen, M. H., Dale, R., & Reali, F. (2010). Connectionist explorations of multiple-cue integration in syntax acquisition. In S. P. Johnson (Ed.), *Neoconstructivism: The new science of cognitive development* (pp. 87–108). New York: Oxford University Press.
Christiansen, M. H., & MacDonald, M. C. (2009). A usage-based approach to recursion in sentence processing. *Language Learning, 59,* 126–161.
Chrupała, G., Kádár, À., & Alishahi, A. (2015). Learning language through pictures. In Zong, C., & Strube, M. (Eds.), *Proceedings of the 53rd Annual Meeting of the Association for Computational Linguistics* (pp. 112–118). Stroudsburg, PA: Association for Computational Linguistics.
Clark, A., & Karmiloff-Smith, A. (1993). The cognizer's innards: A psychological and philosophical perspective on the development of thought. *Mind and Language, 8,* 487–519.
Colunga, E., & Smith, L. B. (2005). From the lexicon to expectations about kinds: A role for associative learning. *Psychological Review, 112,* 347–382.
Crocker, M. W., Knoeferle, P., & Mayberry, M. R. (2010). Situated sentence processing: The coordinated interplay account and a neurobehavioral model. *Brain and Language, 112,* 189–201.
Cunillera, T., Toro, J. M., Sebastián-Gallés, N., & Rodríguez-Fornells, A. (2006). The effects of stress and statistical cues on continuous speech segmentation: An event-related brain potential study. *Brain Research, 1123,* 168–178.
Dell, G. S., & Chang, F. (2014). Relating sentence production and its disorders to comprehension and acquisition. *Philosophical Transactions of the Royal Society B, 369,* 20120394.
Dell, G. S., Chang, F., & Griffin, Z. M. (1999). Connectionist models of language production: Lexical access and grammatical encoding. *Cognitive Science, 23,* 517–542.
Desai, R. (2002). Bootstrapping in miniature language acquisition. *Cognitive Systems Research, 3,* 15–23.
Desai, R. (2007). A model of frame and verb compliance in language acquisition. *Neurocomputing, 70,* 2273–2287.
Diessel, H., & Tomasello, M. (2005). A new look at the acquisition of relative clauses. *Language, 81,* 882–906.
Elman, J. L. (1990). Finding structure in time. *Cognitive Science, 14,* 179–211.
Elman, J. L. (1991). Distributed representations, simple recurrent networks, and grammatical structure. *Machine Learning, 7,* 195–225.
Elman, J. L. (1993). Learning and development in neural networks: The importance of starting small. *Cognition, 48,* 71–99.
Engelmann, F., & Vasishth, S. (2009). Processing grammatical and ungrammatical center embeddings in English and German: A computational model. In A. Howes, D. Peebles, & R. Cooper (Eds.), *Proceedings of 9th International Conference on Cognitive Modeling* (pp. 240–245). Manchester, UK.
Fazly, A., Alishahi, A., & Stevenson, S. (2010). A probabilistic computational model of cross-situational word learning. *Cognitive Science, 34,* 1017–1063.
Fitz, H., & Chang, F. (2008). The role of the input in a connectionist model of the accessibility hierarchy in development. In Chan, H., Jacob, H., & Kapia, E. (Eds.), *32nd Annual Boston University Conference on Language Development* (pp. 120–131). Somerville, MA: Cascadilla Press.
Fitz, H., Chang, F., & Christiansen, M. H. (2011). A connectionist account of the acquisition and processing of relative clauses. In E. Kidd (Ed.), *The acquisition of relative clauses: Processing, typology and function* (pp. 39–60). Amsterdam: John Benjamins.
Fodor, J. A., & Pylyshyn, Z. W. (1988). Connectionism and cognitive architecture: A critical analysis. *Cognition, 28,* 3–71.
Fosler-Lussier, E., Amdal, I., & Kuo, H. K. J. (2005). A framework for predicting speech recognition errors. *Speech Communication, 46,* 153–170.
Fossum, V., & Levy, R. (2012). Sequential vs. hierarchical syntactic models of human incremental sentence processing. In Reitter, D., & Levy, R. (Eds.), *Proceedings of the 3rd Workshop on Cognitive Modeling and Computational Linguistics* (pp. 61–69). Stroudsburg, PA: Association for Computational Linguistics.
Frank, S. L. (2013). Uncertainty reduction as a measure of cognitive effort in sentence comprehension. *Topics in Cognitive Science, 5,* 475–494.
Frank, S. L., & Bod, R. (2011). Insensitivity of the human sentence-processing system to hierarchical structure. *Psychological Science, 22,* 829–834.
Frank, S. L., Otten, L. J., Galli, G., & Vigliocco, G. (2015). The ERP response to the amount of information conveyed by words in sentences. *Brain and Language, 140,* 1–11.
Frank, S. L., Trompenaars, T., & Vasishth, S. (2016). Cross-linguistic differences in processing double-embedded relative clauses: Working-memory constraints or language statistics? *Cognitive Science, 40,* 554–578.
Frank, S. L., & Vigliocco, G. (2011). Sentence comprehension as mental simulation: An information-theoretic perspective. *Information, 2,* 672–696.
Frazier, L., & Rayner, K. (1982). Making and correcting errors during sentence comprehension: Eye-movements in the analysis of structurally ambiguous sentences. *Cognitive Psychology, 14,* 178–210.
French, R. M., Addyman, C., & Mareschal, D. (2011). TRACX: A recognition-based connectionist framework for sequence segmentation and chunk extraction. *Psychological Review, 118*(4), 614–636. doi:10.1037/A0025255.
Gambell, T., & Yang, C. D. (2003). Scope and limits of statistical learning in word segmentation. In Moulton, K., & Wolf, M. (Eds.), *Proceedings of the 34th Northeastern Linguistic Society Meeting* (pp. 29–30). New York: Stony Brooks University.
Garrett, M. F. (1988). Processes in language production. In F. J. Newmeyer (Ed.), *Linguistics: The Cambridge survey: III Language: Psychological and biological aspects* (pp. 69–96). Cambridge: Cambridge University Press.
Gleitman, L. (1990). The structural sources of verb meanings. *Language Acquisition, 1,* 3–55.
Gold, E. M. (1967). Language identification in the limit. *Information and Control, 16,* 447–474.
Goldberg, A. E. (2009). The nature of generalization in language. *Cognitive Linguistics, 20,* 93–127.
Gordon, J. K., & Dell, G. S. (2003). Learning to divide the labor: An account of deficits in light and heavy verb production. *Cognitive Science, 27,* 1–40.
Hahn, M., & Keller, F. (2016). Modeling human reading with neural attention. In Su, J., Duh, K., & Carreras, X. (Eds.),

Proceedings of the 2016 Conference on Empirical Methods in Natural Language Processing (pp. 85–95). Stroudsburg, PA: Association for Computational Linguistics.

Hale, J. T. (2001). A probabilistic Earley parser as a psycholinguistic model. In *Proceedings of the Second Meeting of the North American Chapter of the Association for Computational Linguistics* (Vol. 2, pp. 159–166). Stroudsburg, PA: Association for Computational Linguistics.

Hartsuiker, R. J., Bernolet, S., Schoonbaert, S., Speybroeck, S., & Vanderelst, D. (2008). Syntactic priming persists while the lexical boost decays: Evidence from written and spoken dialogue. *Journal of Memory and Language, 58*, 214–238.

Hebb, D. O. (1949). *The organization of behavior.* New York: Wiley & Sons.

Hinaut, X., & Dominey, P. F. (2013). Real-time parallel processing of grammatical structure in the fronto-striatal system: A recurrent network simulation study using reservoir computing. *PLoS ONE, 8*, e52946.

Hinaut, X., Lance, F., Droin, C., Petit, M., Pointeau, G., & Dominey, P. F. (2015). Corticostriatal response selection in sentence production: Insights from neural network simulation with reservoir computing. *Brain and Language, 150*, 54–68.

Hsiao, Y., & MacDonald, M. C. (2013). Experience and generalization in a connectionist model of Mandarin Chinese relative clause processing. *Frontiers in Psychology, 4*, 767.

Kamper, H., Jansen, A., & Goldwater, S. (2016). Unsupervised word segmentation and lexicon discovery using acoustic word embeddings. *IEEE/ACM Transactions on Audio, Speech and Language Processing (TASLP), 24*(4), 669–679.

Kamper, H., Wang, W., & Livescu, K. (2016). Deep convolutional acoustic word embeddings using word-pair side information. In *IEEE International Conference on Acoustics, Speech and Signal Processing (ICASSP)* (pp. 4950–4954). Shanghai, China: IEEE Signal Processing Society.

Kim, Y., Jernite, Y., Sontag, D., & Rush, A. M. (2016). Character-aware neural language models. In Schuurmans, D. & Wellman, M. (Eds.), *Proceedings of the 30th AAAI Conference on Artificial Intelligence* (pp. 2741–2749). Phoenix, AZ: AAAI Press.

Kiss, G. R. (1973). Grammatical word classes: A learning process and its simulation. *Psychology of Learning and Motivation, 7*, 1–41.

LeCun, Y., Bengio, Y., & Hinton, G. (2015). Deep learning. *Nature, 521*, 436–444.

Levy, R. (2008). Expectation-based syntactic comprehension. *Cognition, 106*, 1126–1177.

MacDonald, M. C., & Christiansen, M. H. (2002). Reassessing working memory: A comment on Just & Carpenter (1992) and Waters & Caplan (1996). *Psychological Review, 109*, 35–54.

Marr, D. (1982). *Vision.* San Francisco: W. H. Freeman and Company.

Mayberry, M. R., Crocker, M. W., & Knoeferle, P. (2009). Learning to attend: A connectionist model of situated language comprehension. *Cognitive Science, 33*, 449–496.

Mayor, J., & Plunkett, K. (2010). A neurocomputational account of taxonomic responding and fast mapping in early word learning. *Psychological Review, 117*, 1–31.

McClelland, J. L., & Elman, J. L. (1986). The TRACE model of speech perception. *Cognitive Psychology, 18*, 1–86.

McClelland, J. L., McNaughton, B. L., & O'Reilly, R. C. (1995). Why there are complementary learning systems in the hippocampus and neocortex: Insights from the successes and failures of connectionist models of learning and memory. *Psychological Review, 102*, 419–457.

McClelland, J. L., St. John, M. F., & Taraban, R. (1989). Sentence comprehension: A parallel distributed processing approach. *Language and Cognitive Processes, 4*, 287–335.

McDonald, J. L., Bock, K., & Kelly, M. H. (1993). Word and world order: Semantic, phonological, and metrical determinants of serial position. *Cognitive Psychology, 25*, 188–230.

McMurray, B., Horst, J. S., & Samuelson, L. K. (2012). Word learning emerges from the interaction of online referent selection and slow associative learning. *Psychological Review, 119*, 831–877.

Miikkulainen, R. (1996). Subsymbolic case-role analysis of sentences with embedded clauses. *Cognitive Science, 20*, 47–73.

Miikkulainen, R., & Dyer, M. G. (1991). Natural language processing with modular PDP networks and distributed lexicon. *Cognitive Science, 15*, 343–399.

Mikolov, T., Chen, K., Corrado, C., & Dean, J. (2013). Efficient estimation of word representations in vector space. In *Proceedings of the ICLR Workshop.*

Mikolov, T., Deoras, A., Povey, D., Burget, L., & Černocký, J. (2011). Strategies for training large scale neural network language models. In *IEEE Workshop on Automatic Speech Recognition and Understanding* (pp. 196–201). Waikoloa, HI: IEEE.

Mintz, T. (2003). Frequent frames as a cue for grammatical categories in child directed speech. *Cognition, 90*, 91–117.

Monaghan, P., & Christiansen, M. H. (2010). Words in puddles of sound: Modelling psycholinguistic effects in speech segmentation. *Journal of Child Language, 37*, 545–564.

Monsalve, I. F., Frank, S. L., & Vigliocco, G. (2012). Lexical surprisal as a general predictor of reading time. In W. Daelemans (Ed.), *Proceedings of the 13th Conference of the European Chapter of the Association for Computational Linguistics* (pp. 398–408). Stroudsburg, PA: Association for Computational Linguistics.

Mozer, M. C. (1991). *The perception of multiple objects: A connectionist approach.* Cambridge, MA: MIT Press.

Naigles, L. (1990). Children use syntax to learn verb meanings. *Journal of Child Language, 17*, 357–374.

Philips, L., & Pearl, L. (2015). The utility of cognitive plausibility in language acquisition modeling: Evidence from word segmentation. *Cognitive Science, 39*, 1824–1854.

Pinker, S. (1984). *Language learnability and language development.* Cambridge, MA: Harvard University Press.

Pinker, S. (1989). *Learnability and cognition: The acquisition of argument structure.* Cambridge, MA: MIT Press.

Plunkett, K. (1997). Theories of early language acquisition. *Trends in Cognitive Sciences, 1*, 146–153.

Quine, W. (1960). *Word and object.* Cambridge, MA: MIT Press.

Rabovsky, M., Hansen, S., & McClelland, J. L. (2018). Modelling the N400 brain potential as change in a probabilistic representation of meaning. *Nature Human Behavior, 2*, 693–705.

Reali, F., & Christiansen, M. H. (2005). Uncovering the richness of the stimulus: Structural dependence and indirect statistical evidence. *Cognitive Science, 29*, 1007–1028.

Reali, F., Christiansen, M. H., & Monaghan, P. (2003). Phonological and distributional cues in syntax acquisition: Scaling up the connectionist approach to multiple-cue Integration. In R. Alterman & D. Kirsh (Eds.), *Proceedings of the 25th Annual Conference of the Cognitive Science Society* (pp. 970–975). Mahwah, NJ: Cognitive Science Society.

Redington, M., Chater, N., & Finch, S. (1998). Distributional information: A powerful cue for acquiring syntactic structures. *Cognitive Science, 22,* 425–469.

Regier, T. (2003). Emergent constraints on word-learning: A computational review. *Trends in Cognitive Sciences, 7,* 263–268.

Rodriguez, P., Wiles, J., & Elman, J. L. (1999). A recurrent neural network that learns to count. *Connection Science, 11,* 5–40.

Rohde, D. L. T., & Plaut, D. C. (1999). Language acquisition in the absence of explicit negative evidence: How important is starting small? *Cognition, 72,* 67–109.

Ross, J. R. (1967). *Constraints on variables in syntax.* Harvard, MA: MIT Press.

Rumelhart, D. E., Hinton, G. E., & Williams, R. J. (1986). Learning internal representations by error propagation. In D. E. Rumelhart, J. L. McClelland, & The PDP Research Group (Eds.), *Parallel distributed processing: Explorations in the microstructure of cognition.* Harvard, MA: MIT Press.

Saffran, J. R., Aslin, R. N., & Newport, E. L. (1996). Statistical learning by 8-month-old infants. *Science, 274,* 1926–1928.

Samuelson, L. K., Kucker, S. C., & Spencer, J. P. (2017). Moving word learning to a novel space: A dynamic systems view of referent selection and retention. *Cognitive Science, 41,* 52–72.

Scharenborg, O. (2010). Modeling the use of durational information in human spoken-word recognition. *Journal of the Acoustical Society of America, 127,* 3758–3770.

Schultz, T. R. (2012). A constructive neural-network approach to modeling psychological development. *Cognitive Development, 27,* 383–400.

Schütze, H. (1993). Part-of-speech induction from scratch. In Schubert, L. (Ed.), *Proceedings of the 31st Annual Meeting of the Association for Computational Linguistics* (pp. 251–258). Stroudsburg, PA: Association for Computational Linguistics.

Servan-Schreiber, D., Cleeremans, A., & McClelland, J. L. (1991). Graded state machines: The representation of temporal contingencies in simple recurrent networks. *Machine Learning, 7,* 161–193.

Smith, A. C., Monaghan, P., & Huettig, F. (2017). The multimodal nature of spoken word processing in the visual world: Testing the predictions of a multimodal parallel integration model. *Journal of Memory and Language, 93,* 276–303.

Smith, L., & Yu, C. (2008). Infants rapidly learn word-referent mappings via cross-situational statistics. *Cognition, 106,* 1558–1568.

Spencer, J. P., Simmering, V. R., Schutte, A. R., & Schöner, G. (2007). What does theoretical neuroscience have to offer the study of behavioral development? Insights from a dynamic field theory of spatial cognition. In J. M. Plumert & J. P. Spencer (Eds.), *The emerging spatial mind* (pp. 320–361). New York: Oxford University Press.

St. Clair, M. C., Monaghan, P., & Christiansen, M. H. (2010). Learning grammatical categories from distributional cues: Flexible frames for language acquisition. *Cognition, 116,* 341–360.

St. John, M. F., & McClelland, J. L. (1990). Learning and applying contextual constraints in sentence comprehension. *Artificial Intelligence, 46,* 217–257.

Sturt, P., Costa, F., Lombardo, V., & Frasconi, P. (2003). Learning first-pass structural attachment preferences with dynamic grammars and recursive neural networks. *Cognition, 88,* 133–169.

Tabor, W., Juliano, C., & Tanenhaus, M. K. (1997). Parsing in a dynamical system: An attractor-based account of the interaction of lexical and structural constraints in sentence processing. *Language and Cognitive Processes, 12,* 211–271.

Tabor, W., & Tanenhaus, M. K. (1999). Dynamical models of sentence processing. *Cognitive Science, 23,* 491–515.

Tomasello, M. (2000). Do young children have adult syntactic competence? *Cognition, 74,* 209–253.

Twomey, K. E., Chang, F., & Ambridge, B. (2014). Do as I say, not as I do: A lexical distributional account of English locative verb class acquisition. *Cognitive Psychology, 73,* 41–71.

Twomey, K. E., Ranson, S. L., & Horst, J. S. (2014). That's more like it: Multiple exemplars facilitate word learning. *Infant and Child Development, 23,* 105–122.

Wehbe, L., Vaswani, A., Knight, K., & Mitchell, T. (2014). Aligning context-based statistical models of language with brain activity during reading. In Moschitti, A., Pang, B., & Daelemans, W. (Eds.), *Proceedings of the 2014 Conference on Empirical Methods in Natural Language Processing* (pp. 233–243). Stroudsburg, PA: Association for Computational Linguistics.

Westermann, G., & Mareschal, D. (2012). Mechanisms of developmental change in infant categorization. *Cognitive Development, 27,* 367–382.

Younger, B. A. (1985). The segregation of items into categories by ten-month-old infants. *Child Development, 56*(6), 1574–1583.

Yu, C., & Ballard, D. H. (2007). A unified model of early word learning: Integrating statistical and social cues. *Neurocomputing, 70,* 2149–2165.

Yu, C., & Smith, L. B. (2012). Modeling cross-situational word-referent learning: Prior questions. *Psychological Review, 119,* 21–39.

Yurovsky, D., Smith, L. B., & Yu, C. (2013). Statistical word learning at scale: The baby's view is better. *Developmental Science, 16,* 959–966.

22 Cognitive Models of Syntax and Sentence Processing

VERA DEMBERG AND FRANK KELLER

1. Introduction

1.1. Motivation Cognitive models of sentence processing implement cognitive processing theories. They should hence display the same processing difficulties as humans and be able to predict processing ease where humans experience processing ease. This is usually not the case for general computational natural language processing models, which may behave very differently from humans in general in terms of when processing difficulties occur. For instance, a parser may not show difficulties for sentences where humans are garden-pathed, or may correctly process center embedding constructions that humans find difficult, while having a lot of trouble in recovering the intended structure of sentences containing minor ungrammaticalities (which is usually not difficult for humans), or having difficulty in dealing with ambiguities that pose few problems for humans.

Computational cognitive models are useful for cognitive science because they spell out all necessary assumptions of a theory explicitly and can show effects of complex interactions of different aspects of a processing theory. Cognitive models can also be a tool for identifying what additional assumptions would have to be made to implement a theory. An implemented model makes a theory testable empirically, including not only statements about which construction should be easier or more difficult than which other construction, but also providing an estimate for the *size* of an effect. It can also generate quantitative predictions for so-far untested constructions, which can in turn be tested empirically to provide additional support for the model or to contradict the model.

Cognitive models can hence give us insight into cognition, by attempting to give an explanation for the phenomena that we observe, or by at least providing an elegant description of these phenomena. They also have potential application in broad areas of natural language processing, if systems could get closer to human levels of robustness in dealing with noise and ungrammatical input. Further applications for which cognitive modeling is highly relevant include readability assessment and dialog systems: better natural language-based human-computer interaction could be achieved if dialog systems can accurately model human processing and can thus adapt the utterances generated by the system more optimally to human comprehension.

A computational cognitive model typically consists of an architecture that specifies the representations used by the computational model (see section 2.1), a processing algorithm (see section 2.2), and a linking theory (see section 4). Applied to models of human sentence processing, the linking theory specifies how the parsing operations of the algorithm are connected to processing difficulty effects such as longer reading times, certain types of event-related potential (ERP) effects, or changes in acceptability.

In this chapter, we will focus on mostly syntactic aspects of sentence processing, an area that has received a lot of attention in the human language processing community. Note that a comprehensive answer to our modeling challenge would also include phonological and morphological processing, semantic inference, discourse processing, and other nonsyntactic aspects of language processing. Furthermore, established results regarding the interface between language processing and nonlinguistic cognition (e.g., the sensorimotor system) should ultimately be accounted for in a model of human language cognition.

1.2. Key Properties The central tenet of computational modeling in psycholinguistics is to capture key properties of human language processing, as established by psycholinguistic experimentation. Striking properties of the human language processor are its *efficiency and robustness*. For the vast majority of sentences, it will effortlessly and rapidly deliver the correct analysis, even in the face of noise and ungrammaticalities. There is considerable experimental evidence that shallow processing strategies are used to achieve this. The processor also achieves *broad coverage*: it can deal with a wide variety of syntactic constructions and is not restricted by the domain, register, or modality of the input.

Evidence from psycholinguistic research shows that human language comprehension is *incremental.* Comprehenders do not wait until the end of the sentence before they build a syntactic representation for the sentence; rather, they construct a sequence of partial representations for sentence prefixes. (A sentence prefix is the subsequence of a sentence from the first word to the word currently being processed.) Experimental results indicate that each new word that is read or heard triggers an update of the representation constructed so far (Konieczny, 2000; Tanenhaus, Spivey-Knowlton, Eberhard, & Sedivy, 1995). There is also evidence for connectedness in human language processing (Sturt & Lombardo, 2005). *Connectedness* means that all input words are attached to the same syntactic structure (though connected structures can be constructed in parallel); comprehenders build no unconnected tree fragments, even for the sentence prefixes that arise during incremental processing.

Furthermore, the processor is able to make *predictions* about upcoming material on the basis of sentence prefixes. Evidence for prediction comes, for example, from the finding that people are able to anticipate the argument of a verb (increased fixations on the argument in a visual world paradigm),[1] even before this argument is heard (Kamide, Scheepers, & Altmann, 2003). Additional evidence for prediction is provided by experiments in which N400 effects are observed when the form of the determiner does not match the anticipated noun (Delong, Urbach, & Kutas, 2005) or by the processing of *either… or* constructions, where the word *or* and a following noun phrase are read faster in contexts that include *either* (Staub & Clifton, 2006). This effect can be explained if we assume that *or* and the second conjunct are predicted when the parser reaches *either.*

Other key properties of human language processing are the facts that it operates with limited memory and that structures in memory are subject to decay and interference. In particular, the processor is known to incur a distance-based *memory cost*: combining the head of a phrase with its syntactic dependents is more difficult the more dependents have to be integrated and the further away they are. This integration process is also subject to interference from similar items that have to be held in memory at the same time.

2. Formalisms and Algorithms

2.1. Formalisms

2.1.1. Context-free grammar In order to build a model of human sentence processing, we have to make assumptions about how sentence structure is formalized and processed. Oftentimes, the psycholinguistic literature has couched these assumptions in the form of a context-free grammar (CFG). CFGs model phrase structure, that is, the hierarchical relationships that hold between the different elements of a sentence. A CFG typically decomposes a sentence into phrasal categories (e.g., S for sentence, NP for noun phrase, VP for verb phrase) and lexical categories (parts of speech) such as determiner (Det), noun (N), and verb (V). These categories are then combined into CFG rules, such as the following:

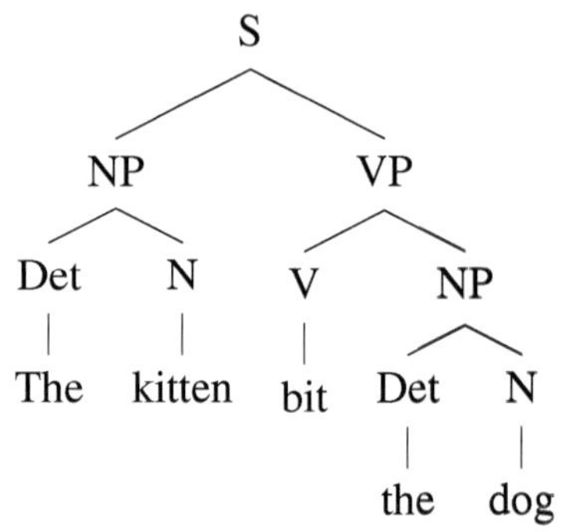

FIGURE 22.1 Example of a syntax tree.

(1)	S	→	NP VP NP	Det	→	The
	NP	→	Det N	N	→	Kitten
	VP	→	V NP	N	→	Dog
				V	→	Bit

A *derivation* is the sequence that results from applying a sequence of context-free rules, starting from a start symbol (here S). As an example, consider the derivation for the sentence *the kitten bit the dog* relative to the grammar in (1):

(2) S ⇒ NP VP ⇒ NP V NP ⇒ NP V Det N ⇒ NP bit Det N ⇒ NP bit Det dog ⇒ NP bit the dog ⇒ Det N bit the dog ⇒ the N bit the dog ⇒ the kitten bit the dog

A derivation assigns a *syntax tree* to a sentence. The syntax tree notates which rules have been applied in the derivation (but not the order in which they have been applied). An example is figure 22.1, which gives the syntax tree for the derivation in (2).

Crucially, a sentence can have multiple syntax trees, each of which can have a different interpretation, a phenomenon called *syntactic ambiguity.* The following sentences are syntactically ambiguous:[2]

(3) a. She sat on the chair covered in dust.
b. I put the book on the table in the kitchen.
c. Milk drinkers are turning to powder.
d. Old school pillars are replaced by alumni.

In all of these cases, the whole sentence is ambiguous, which is referred to as *global ambiguity.*

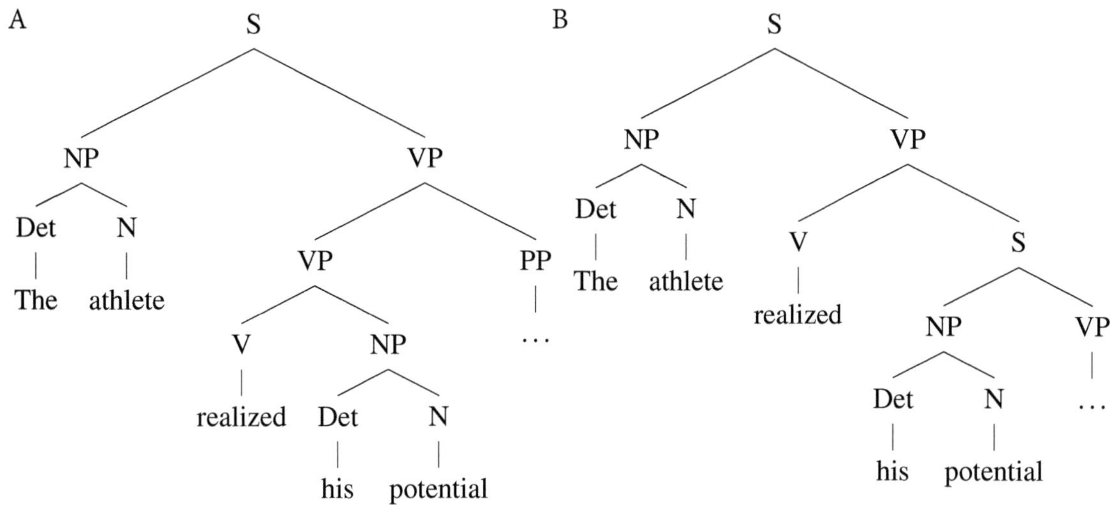

FIGURE 22.2 Example of local ambiguity.

As the human sentence processor builds the syntax tree for the input incrementally (see s 1.2), it can encounter an ambiguity that disappears later when more of the input is read. This is referred to as *local ambiguity*. An example of a local ambiguity is:

(4) The athlete realized his potential…
  a. at the competition.
  b. could make him a world-class sprinter.

The local syntactic trees for (4a) and (4b) are given in figure 22.2.

Both trees are compatible with the input up until *potential*: only the next word disambiguates. However, the processor commits to a single (wrong) tree early on and trips up when later input is inconsistent with that tree. Presumably, the processor now has to compute a new tree that is consistent with the input, which results in longer reading times, reverse eye movements, lower comprehension accuracies, and so on (we will come back to this in section 3.1).

2.1.2. Tree-adjoining grammar Tree-adjoining grammar (TAG; Joshi, Levy, & Takahashi, 1975) is an extension of CFG that assumes that a grammar consists of lexicalized tree fragments rather than context-free rules. TAG assumes two operations that can combine elementary trees: substitution combines an initial tree with a partial tree (this happens at substitution nodes, marked with "↓"); adjunction combines an auxiliary tree with a partial tree (this happens at adjunction nodes, marked with "*"). A TAG derivation consists of a sequence of substitution and adjunction operations, applied to the initial and auxiliary trees associated with the words in the sentence.

Standard TAG derivations are not incremental; that is, words are not processed in the left-to-right order of the input. As we saw in section 1.2, incrementality is a key property of the human sentence processor, which means that TAG needs to be extended in order to be utilized in a model of human sentence processing. Such an extension has been proposed in the form of Psycholinguistically Motivated Tree-adjoining Grammar (PLTAG; Demberg & Keller, 2008b). PLTAG parsing proceeds strictly incrementally; all words in the input are always connected under a single syntactic node. This strict incrementality is made possible through a *prediction* mechanism that allows the parser to insert syntactic structure into the unfolding representation before its syntactic head has been encountered in the input. For example, when observing the NP *John* at the beginning of a sentence, the parser would predict that this NP might be the subject of a sentence and would predict that a VP will be encountered. (In standard TAG by contrast, the VP and role of the NP in the sentence would only be constructed once the verb is actually encountered in the input.) Figure 22.3 illustrates how this prediction mechanism allows PLTAG to connect the words *Peter* and *often* before the verb *sleeps* is encountered. The prediction mechanism makes use of so-called prediction trees, a new type of elementary tree that carries indices on each node (which mark their status as predicted and not yet grounded in input), as shown in figure 22.3D. Note that prediction trees (as opposed to standard TAG trees) need not contain a lexical anchor.

The prediction mechanism is complemented by a *verification* mechanism that matches predictions against subsequent input, and when matching removes the prediction markers to indicate that the previously predicted structure is now grounded in actually observed material (the structure that comes with the verb *sleeps* in our example).

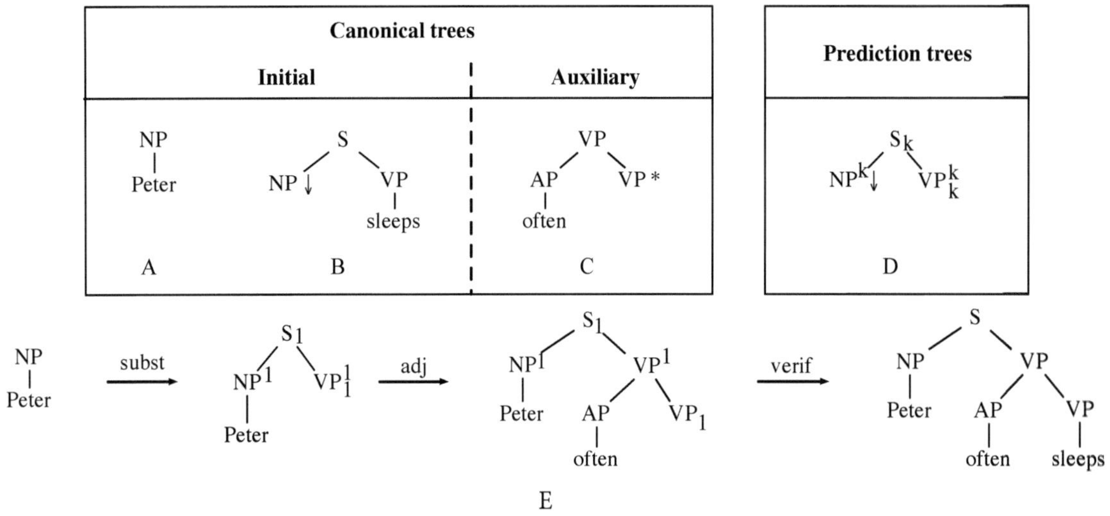

FIGURE 22.3 (A–D) An example of a PLTAG grammar, containing initial and auxiliary trees (shared with standard TAG), and an example of a prediction tree (which is specific to PLTAG) are shown. (E) An incremental derivation in PLTAG using the trees (A) to (D) is shown. From Demberg et al. (2013).

PLTAG is a fully implemented model that consists of an algorithm for extracting a PLTAG grammar from an annotated corpus, a parsing algorithm that processes PLTAG structures incrementally, and a probability model over PLTAG derivations. We will return to this in section 6.3.

2.1.3. Dependency grammar Traditional CFGs capture the *configurational structure* of sentences; that is, they describe which phrases a sentences is composed of and how these phrases are ordered relative to each other, both sequentially and hierarchically. However, this view of syntax has been challenged, as it makes it difficult to analyze languages that allow discontinuous constituents or have free or flexible word order.

Dependency grammar is one of the answers to this challenge. It is a formalism that starts from the assumption that the grammatical relations between words, rather than their configurations, are key to sentence structure. For instance, the important thing about verbs such as *like* is that they license two NPs: an agent, found in subject position or with nominative inflection, and a patient, found in object position or with accusative inflection. Which arguments are licensed, and which roles they play, depends on the verb. The configuration of these arguments is secondary.

Dependency grammar therefore models the grammatical relations that hold between words; the claim is that these can be identified even in nonconfigurational languages. A dependency structure consists of *dependency relations*, which are binary and asymmetric. A relation is a tuple of a *head* h, a dependent d, and a *label*, which identifies the relation between h and d. An example of a dependency structure can be found in figure 22.4.

Formally, a dependency structure of a sentence is a graph with the words of the sentence as its nodes, linked by directed, labeled edges, with the following properties: (a) connectedness: every node is related to at least one other node, and (through transitivity) to ROOT; (b) single headedness: every node (except root) has exactly one incoming edge (from its head); (c) acyclicity: the graph cannot contain cycles of directed edges.

These conditions ensure that the dependency structure is a tree. We distinguish projective and nonprojective dependency trees: A dependency tree is *projective* with respect to a particular linear order of its nodes if, for all edges $h \rightarrow d$ and nodes w, w occurs between h and d in linear order only if w is dominated by h (i.e., in the transitive closure of the arc relation). A dependency tree is *nonprojective* if w can occur between h and d in linear order without being dominated by h. This is illustrated in figure 22.5.

Dependency grammar is the formalism that underlies Dependency Locality Theory (DLT), one of the key approaches to psycholinguistic modeling, which will be discussed in section 6.1.

2.2. PARSING ALGORITHMS A parsing algorithm takes a grammar and a sentence and derives a syntax tree for the sentence. A range of parsing algorithms have been proposed in the literature (for an overview, see Jurafsky & Martin, 2009). The most basic distinction is between top-down and bottom-up parsing algorithms. A top-down parser starts with the start symbol of the

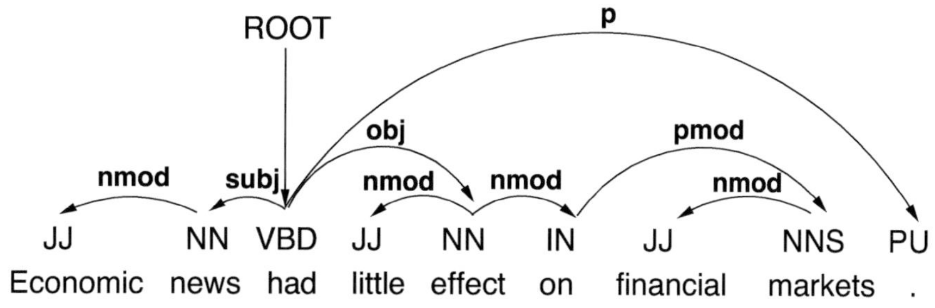

FIGURE 22.4 An example of a dependency graph. The arc labels are grammatical relations (subject, object, predicate, modifier, and combinations of these); the targets of the arrows are part-of-speech labels (JJ for adjective, NN for noun, VBD for verb, IN for preposition, NNS for singular noun, PU for punctuation).

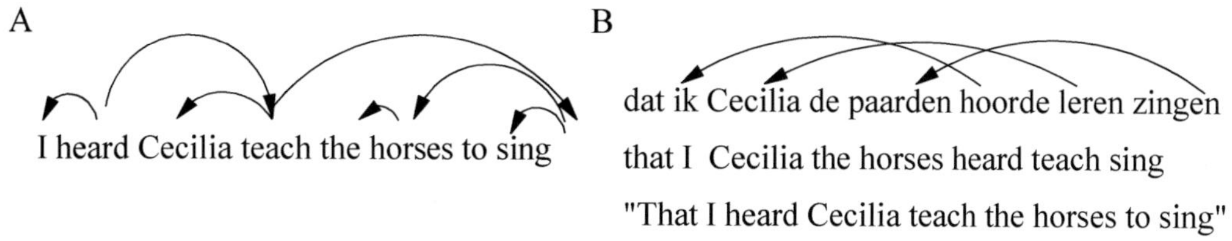

FIGURE 22.5 An example of projective (A) and nonprojective (B) dependency trees.

grammar and expands each phrasal category until it reaches the leaf nodes (the words of the sentence). We already encountered an example of a top-down derivation in (2) for the sentence *the kitten bit the dog* with respect to the grammar in (1). A bottom-up parser starts with the leaf nodes and replaces each of them with phrasal categories until it reaches the start symbol. The bottom-up derivation for the same sentence and grammar is as follows:

(5) the kitten bit the dog ⇒ Det kitten bit the dog ⇒ Det N bit the dog ⇒ Det N V the dog ⇒ Det N V Det dog ⇒ Det N V Det N ⇒ NP V Det N ⇒ NP V NP ⇒ NP VP ⇒ S

As we saw in section 2.1.1, a derivation for a sentence using a CFG can be ambiguous; that is, it can assign more than one syntax tree to a given sentence, either locally or globally. Ambiguity occurs for every realistically large grammar, which typically has thousands of rules. This means that naive top-down and bottom-up parsers have a large search space to explore: for example, in (2), there could be many alternative ways of expanding S or NP; in (5), there could be multiple ways of expanding *bit* or *dog*, each allowing the parser to derive multiple phrasal categories.

The standard solution to this problem is to employ a *chart*, a data structure that stores subtrees for each phrase the parser has discovered. These subtrees can then be used in subsequent analyses that require this phrase, without having to be constructed again, making parsing much more efficient. Standard examples of chart parsing algorithms are the CKY algorithm and the Earley algorithm (Jurafsky & Martin, 2009). Of particular interest from a psycholinguistic point of view is left-corner parsing, a chart-parsing strategy that combines top-down and bottom-up information and has been argued to be cognitively plausible based on its memory requirements (Abney & Johnson, 1991). While we cannot give a formal definition of the left-corner algorithm here, figure 22.6 shows an example of a left-corner derivation (again for *the kitten bit the dog* with respect to (1)).

The key idea is that the parser starts from the bottom up with *the*, but then makes a top-down prediction of a rule whose left corner is Det, that is, the rule NP → Det N. Next, it completes the NP from the bottom up (by parsing *kitten*), which allows it to make another top-down prediction, as NP is the left corner of the rule S → NP VP. Top-down and bottom-up predictions alternate until the input is consumed.

All the parsing algorithms we discussed so far are designed for CFG, but they can be adapted for TAG (Joshi & Schabes, 1997). For dependency grammars, a different set of parsing algorithms is typically employed, including transition-based parsing (Nivre, 2003) and the maximum spanning tree algorithm (McDonald, Pereira, Ribarov, & Hajič, 2005). An early implemented broad-coverage incremental probabilistic processing model was suggested by Crocker and Brants (2000). Their model did not claim cognitive plausibility based on the architecture (cascaded Markov models) but through its broad-coverage incremental parsing process.

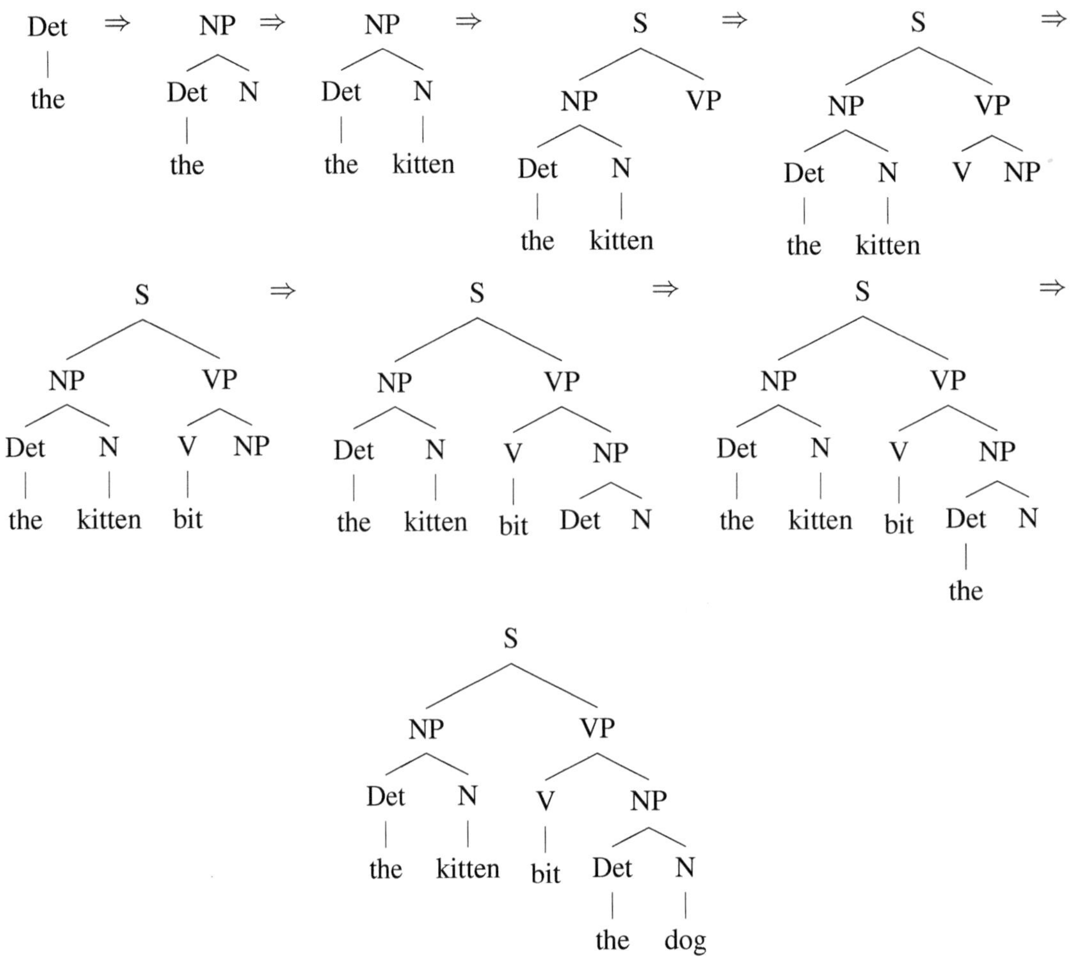

FIGURE 22.6 Example of a left-corner derivation.

3. Specific Constructions

3.1. GARDEN PATHS Garden path sentences are one of the classic human language processing phenomena. A *garden path effect* refers to a situation where a language processing difficulty becomes so severe that the reader becomes aware of the difficulty of the sentence. The reader is initially "lured" into a very probable interpretation, which later turns out to be incompatible with the end of the sentence. The difficulty of reanalyzing the sentences can be so strong that correct analysis cannot be recovered, and the sentence is judged to be ungrammatical.

The most famous example is probably the sentence of Bever (1970) shown in (6):

(6) The horse raced past the barn fell.

The reader initially analyzes *raced* as a simple past form and hence the main verb of the sentence. However, this is incompatible with *fell*. For the correct interpretation of the sentence, *raced past the barn* must be analyzed as a reduced relative clause, and *raced* hence as a past participle.

Garden path effects are usually not only caused by difficult syntactic constructions, but are also dependent on semantics; that is, they often consist of syntactically slightly difficult structures that in addition are made implausible by the semantics of the sentence.

There is a qualitative difference between a minor processing difficulty that people often do not become aware of and complete processing breakdown, where some people are not able to recover the correct analysis at all. Parallel models explain this difference by assuming that the processor only focuses on a finite number of most probable analyses and that the correct analysis has fallen out of the reader's search beam[3] because it was highly unlikely compared to alternative analyses. Hence, the sentence must be reanalyzed from scratch, which is only successful if enough memory resources can be made available for the larger beam needed to process the sentence (Altmann & Steedman, 1988; Gibson, 1991).

3.2. Center Embedding A well-known construction is *center embedding*, where several (of the same type of) structures nested inside one another make sentences increasingly difficult to process (e.g., Chomsky, 1957; Miller & Isard, 1964). Consider the sentences from Gibson (1998) in example (7):

(7) a. The intern [who the nurse supervised] had bothered the administrator [who lost the medical reports].
   b. The administrator [who the intern [who the nurse supervised] had bothered] lost the medical reports.

Sentence (7b) has been shown to be considerably more difficult to process than (7a) based on complexity ratings. The difficulty with such sentences is usually modeled via memory access, that is, as arising from having to keep a large number of similar open structures accessible in memory (Gibson; Warren & Gibson, 2002), or more specific constraints on the number of clauses that can be parsed at the same time (Kimball, 1973; Lewis, 1996; Stabler, 1994).

3.3. Locality and Antilocality The term *locality effects* is used to refer to increased processing difficulty (measured typically in terms of increased reading times) for the second element of a long-distance dependency. Longer processing times are usually observed for longer distance dependencies and are essentially a more general notion of center embedding. Locality effects are one possible explanation for the subject relative clause versus object relative clause asymmetry (i.e., the fact that object relative clauses are harder to process than subject relative clauses are). While subject and object relative clauses have the same number of embeddings, an object relative clause has longer dependencies.

Antilocality effects have been shown for a number of languages, including German (Konieczny, 2000; Konieczny & Döring, 2003) and Hindi (Vasishth & Lewis, 2006). Antilocality effects refer to the finding that reading times can be shorter at the head when intervening material is inserted between the head and its arguments. Examples of antilocality effects in German include results presented by Konieczny and Konieczny and Döring. The experiment reported by Konieczny found that the verb in verb-final constructions in German is read faster when more material (just one argument vs. an additional prepositional phrase vs. a longer, modified prepositional phrase) occurred before the verb. This finding is contrary to the locality effect found in English center embedding and the subject/object relative clause asymmetry. A similar experiment was conducted by Konieczny and Döring, who also controlled for length of the intervening material between conditions. An example of their materials is shown in (8).

(8) a. Die Einsicht, dass [NP-NOM der Freund] [NP-DAT dem Kunden] [NP-ACC das Auto aus Plastik] verkaufte ...
   the insight, that the friend the client the car of plastic sold, ...
   "The insight that the friend sold the client the plastic car ..."
   b. Die Einsicht, dass [NP-NOM der Freund [NP-GEN des Kunden]] [NP-ACC das Auto aus Plastik] verkaufte, ...
   the insight, that the friend of the client the car of plastic sold, ...
   "The insight that the friend of the client sold the plastic car ..."

In materials following the example in (8), reading times on the verb *verkaufte* are shorter in (8a) than in (8b), even though the length of interfering elements is exactly identical. Such antilocality effects can be accounted for in terms of entropy or surprisal (see section 4).

3.4. Local Coherence It has been observed that processing difficulty can sometimes occur in sentences that are neither ambiguous nor particularly complex (Tabor, Galantucci, & Richardson, 2004). An example of this is the sentence in (9a), which has the same syntactic complexity as (9c).

(9) a. The coach smiled at the player tossed a frisbee by ...
   b. The coach smiled at the player who was tossed a frisbee by ...
   c. The coach smiled at the player thrown a frisbee by ...
   d. The coach smiled at the player who was thrown a frisbee by ...

The important difference between the sentences is that in (9a) the word sequence *the player tossed a frisbee* is a coherent string of words where *the player* would be the subject of a main verb *tossed*, while *the player thrown a frisbee* cannot be interpreted as such. While (9a) can be expected to be the most difficult sentence among the sentences in (9), because reduced relative clauses are more difficult than nonreduced relative clauses are, and ambiguous verb forms are more difficult than unambiguous verb forms are, the observed difficulty effect was stronger than would be expected by adding the verb ambiguity and reduced relative clause effects.

A common explanation for the effect is that the locally coherent interpretation of *the player tossed a frisbee* interferes with the globally coherent analysis of the sentence and has therefore been argued to provide evidence against a view of strictly incremental processing, as the locally coherent analysis should not be calculated in the first place, because *the player* already has a different function in the sentence and cannot possibly be the subject of *tossed*, and *tossed* cannot be the main verb of the sentence, as there is already a main verb, *smiled*. Alternatively, this type of effect could be explained by a noisy channel model, where readers maintain uncertainty about what they have processed so far (under this account, readers take into consideration that they may have misheard or misread or that the speaker may have mispronounced/misspelled). The reader would then generate hypotheses compatible with plausible input that differs from the actually perceived input (Levy, Bicknell, Slattery, & Rayner, 2009). This kind of explanation would still be compatible with rational incremental language processing.

3.5. Competition *Competition effects* are used to refer to the observation that ambiguous words or structures cause higher processing difficulty than unambiguous words or structures. Such effects were initially observed for lexical ambiguities, in experiments that compared reading times on words that were unambiguous, ambiguous with one predominant meaning, or ambiguous with two meanings being similarly prominent. Experimental results showed that fixation times on words with two similarly prevalent meanings were longer than on the other two classes (Duffy, Morris, & Rayner, 1988), which was then explained as both analyses "competing" with one another and hence taking longer to be processed. MacDonald, Pearlmutter, and Seidenberg (1994) suggested that syntactic ambiguities might show similar effects, with supporting evidence provided by experiments on prepositional phrase attachment ambiguities (Spivey-Knowlton & Sedivy, 1995).

3.6. Facilitating Ambiguity In the so-called *facilitating ambiguity* effect (Traxler, Pickering, & Clifton, 1998), reading times can be faster (as opposed to slower, which would be in line with competition effects) under some circumstances in an ambiguous region compared to an unambiguous region. Consider example (10): the reflexive pronoun *herself/himself* in (10a) and (10b) is unambiguous in that it can only refer to the daughter/colonel, respectively. In sentence (10c), however, *himself* is ambiguous as to whether it refers to the son or the colonel.

(10) a. The daughter$_i$ of the colonel$_j$ who shot herself$_{i/*j}$ on the balcony had been very depressed.
b. The daughter$_i$ of the colonel$_j$ who shot himself$_{*i/j}$ on the balcony had been very depressed.
c. The son$_i$ of the colonel$_j$ who shot himself$_{i/j}$ on the balcony had been very depressed.

Reading times were found to be faster on the *himself/herself* and immediately following region in the ambiguous case (10c). This finding is difficult to explain for models that assume competition. Models that can account for facilitating ambiguity effects model the effect by avoiding fully disambiguating these sentences until necessary.

3.7. Commitment and Digging-in Digging-in effects refer to the finding that a wrong syntactic analysis becomes harder and harder to reanalyze the longer the ambiguous region is. As an example, consider the sentences in (11). Sentences (11a) and (11b) are initially ambiguous at the NP *the book* with respect to whether the NP is an argument of the verb *write* or the subject of the main phrase, while sentences (11c) and (11d) are not (because the verb already has an argument, *the essay*). Subjects initially interpret *the book* as an object of *write* because it is a semantically very likely object of *write* and because *write* is more often seen as a transitive verb than as an intransitive one.

(11) a. As the author wrote the book grew. (ambiguous, short)
b. As the author wrote the book describing Babylon grew. (ambiguous, long)
c. As the author wrote the essay the book grew. (unambiguous, short)
d. As the author wrote the essay the book describing Babylon grew. (unambiguous, long)

It has been shown using acceptability judgments (Ferreira & Henderson, 1991) and reading times (Tabor & Hutchins, 2004) that (11b) is much more difficult than (11a) and the control condition (11d). One would expect (11b) to be the most difficult condition anyway, as it is more complex than (11b) and more ambiguous than (11d). However, Tabor and Hutchins found a difficulty effect on the last word of (11b) (an interaction between length and ambiguity before encountering *grew*) that goes beyond the main effects of ambiguity and complexity. This effect may be related to the even stronger garden path effects where recovery of the correct analysis may fail completely, or to effects related to the strength of context constraint.

Context constraint has been shown to facilitate processing of an expected word. However, there are also effects indicating that there may be additional effects related to the overall level of constraint of a context on processing. Federmeier, Wlotko, De Ochoa-Dewald, and Kutas (2007) showed in an ERP study that while the N400 related to predictability is the same in weakly and strongly constraining contexts, strongly constraining contexts show a larger late positivity.

4. *Linking Theories*

We have now seen a range of different grammar formalisms that can describe syntactic structure, as well as different parsing algorithms that describe how to use a grammar to construct the syntactic representation for linguistic input. The function of a so-called *linking theory* or *linking hypothesis* is to suggest what aspects of the parsing process are effortful and hence lead to observable processing difficulty. The linking theory, together with a grammar and a parsing algorithm, constitutes a *processing model* (see section 6), which we can then test as to whether it can account for the data, that is, both broad-coverage processing and experimental data as outlined in section 3.

4.1. Early Approaches Early attempts at accounting for the occurrence of processing difficulty (with a specific focus on garden path sentences) suggested that language processing may be subject to a range of constraints or preferences that are used to rank alternative parses for a sentence, such that some structures would be predicted to be preferred over others. Processing difficulty would be related to cases where a dispreferred analysis would turn out to be the only viable one. Very strongly dispreferred structures are pruned, which provides a mechanism to account for garden path sentences (cases of extreme processing difficulty). The most important constraints were based on locality preferences (e.g., Right Association, Kimball, 1973; Local Association, Fodor & Frazier, 1978; Late Closure, Frazier, 1979; Final Arguments, Ford, Bresnan, & Kaplan, 1982; Attach Low and Parallel, Hobbs & Bear, 1990; and the Recency Preference, Gibson, 1991). These hypotheses were only descriptive and not very generalizable and have since been replaced by linking hypotheses that can apply to a larger range of structures and phenomena.

4.2. Memory Decay and Interference Based on the observed processing difficulty that humans encounter in sentences with center embeddings or long dependencies (locality effects), several linking theories have been suggested that attribute processing difficulty to retrieving material that has already been processed from memory to integrate it with new upcoming information. Variants of these ideas also take into account whether there are strong competitors, that is, the degree to which other entities in memory may lead to interference effects when retrieving the relevant entities. The most well-known linking theory that describes this aspect is the integration cost component of DLT (Gibson, 1998, 2000). Processing difficulty is suggested to be related to the distance between a dependent and its head: in order to semantically integrate the two concepts, the dependent needs to be retrieved from memory. Different ways of quantifying the distance have been suggested in the literature and include counting the number of new discourse referents that have been encoded since processing the dependent (Gibson, 1998, pp. 12–13; Warren & Gibson, 2002), simply counting intervening words (Demberg, Keller, & Koller, 2013; Temperley, 2007), or counting the number of intervening syntactic heads (Alexopoulou & Keller, 2007). Note that processing difficulty estimates for locality costs depend on the grammar formalism used to describe the linguistic structure.

4.3. Short-term Memory Load A related idea is that processing may slow down under high memory load—if there are many open dependencies, these may have to be maintained in working memory, and the process of maintaining entities in working memory may slow down processing overall. This idea is formalized as part of the storage cost component of DLT (Gibson, 1998, 2000). Earlier proposals suggested that processing difficulty may be related to the depth of embedding of the current word, and this idea has also been taken up and tested in later proposals (e.g., Pynte, New, & Kennedy, 2008; Schuler, AbdelRahman, Miller, & Schwartz, 2010), where processing difficulty is hypothesized to occur when a new dependency needs to be added to the memory or a stored dependent needs to be retrieved from memory for integration. In DLT (see section 6.1), a combination of the costs of integration and storage is used to also explain how ambiguous input is resolved (preference for structures with lower memory requirements). Storage cost depends on the grammar formalism over which it operates, as different grammar formalisms may suggest flatter versus deeper structures or may posit a different number of open dependencies (e.g., differences in subcategorization frames affect the number of expected open dependencies) for a given linguistic input.

4.4. Frequency-Based Linking Frequencies have early on been found to significantly correlate with processing difficulty and reading times (slower reading on infrequent words). Frequency-based linking hypotheses hold that frequency effects may not only occur for lexical access but also for syntactic structures (more frequent structures are processed faster than infrequent ones). Underlyingly, this could for instance be due to faster memory retrieval for words and structures that are used more often and may thus be easier to activate, or to the retrieval of larger structures, so that less online integration is necessary. Jurafsky (1996) first proposed to use probabilistic context-free grammars (PCFGs[4]) to estimate probabilities of alternative analyses and used the probabilities to explain garden path sentences: Only the most probable parses (according to the PCFG) would be kept in memory. The improbable ones were pruned using beam search, which discards highly improbable analyses. For interpretations that were pruned, the parser would have to backtrack, which explains the processing difficulty for garden path sentences.

Frequency estimates for linguistic structures obviously very strongly depend on the form and size of linguistic structures that are assumed by the grammar formalism. Furthermore, frequency estimates also depend on the corpus from which they are estimated.

4.5. Conflict A question that arises in the construction of probability-based models is whether probabilities from different sources (n-grams, parsing, valency) are needed, and how to combine them. Narayanan and Jurafsky (2002) used a belief network with probabilities from a range of different sources. A belief network is more powerful than a PCFG, but it is harder to justify the contributing factors and their relationship to one another.

Padó, Crocker, and Keller (2009) suggested a linking hypothesis according to which processing difficulty ensues if probabilities from two different processing components (e.g., syntactic structure and semantics) do not agree with each other, that is, when they bias the interpretation toward different analyses. This linking hypothesis potentially has some points of overlap with the competition models, except that for the conflict hypothesis, the source of the competition between alternative analyses stems from potentially different processes within comprehension (syntax vs. semantics). This linking hypothesis hence requires an architecture in which syntax and semantics (or whatever domains from which costly conflicts can arise) are different processes.

4.6. Information Content and Surprisal A measure that has received a lot of attention in the sentence processing literature is surprisal (Hale, 2001; Levy, 2008a). Surprisal quantifies the amount of new information conveyed by a word in context, and the surprisal at word w_i is formalized as $-\log P(w_i|w_1 \ldots w_{i-1})$, that is, as the probability of observing word w_i given that words $w_1 \ldots w_{i-1}$ have already been processed. The linking hypothesis for surprisal is hence that processing difficulty is proportional to the amount of new information that needs to be processed. Surprisal as such is agnostic with respect to the grammar formalism that is used as a processing mechanism to derive the predictability estimates, but just as for frequencies, actual surprisal values depend both on the grammar formalism and the data used to estimate the model. Using surprisal as a linking hypothesis also puts constraints on the processing algorithm, which needs to be incremental to properly estimate the probability of a word given previous content. An extension of the concept of surprisal is the *noisy channel hypothesis* (Levy, 2008b), which proposes that language comprehenders always maintain some uncertainty about the linguistic input. This overall uncertainty about the input may be due to uncertainty in visual perception during reading, uncertainty about auditory processing due to noise, or uncertainty due to cognitive processes (e.g., imperfect encoding or short-term memory). With a noisy channel extension to the model, the identity of word w_i and earlier words that the model conditions on are not certain, but can themselves be described using a probability distribution. However, no broad-coverage noisy channel model exists, as it is unclear to date how to adequately estimate the probabilities of the input observations.

4.7. Uncertainty and Competing Analyses Another family of linking hypotheses is related to the idea that ambiguity in the linguistic input may cause processing difficulty: The competition hypothesis suggests that it may be difficult if two or more alternative linguistic analyses have a similar likelihood. These analyses may then compete against one another until one of the alternatives reaches criterion. The same basic concept is also reflected in using *entropy* as a measure for processing difficulty: processing difficulty at a word is suggested to be correlated to the amount of entropy at that word. There are different notions, however, regarding what entropy is meant: entropy with respect to alternative analyses up to the current word (this is most similar to the competition models), entropy about the upcoming next word (Roark, Bachrach, Cardenas, & Pallier, 2009; Willems, Frank,

Nijhof, Hagoort, & Van den Bosch, 2016), or entropy about the rest of the sentence (Linzen & Jaeger, 2016).

4.8. Entropy Reduction The entropy reduction hypothesis (Hale, 2003, 2006) suggests that processing difficulty may be related to changing from a state of high uncertainty about the rest of the sentence to a state of lower uncertainty about the rest of the sentence during parsing. There are, however, some conceptual difficulties with this notion (e.g., uncertainty can also increase when processing a sentence, which is not related to any cost or benefit), as discussed by Linzen (2015) and Linzen and Jaeger (2016). Additionally, there is the practical difficulty that entropy with respect to the complete rest of the sentence is extremely difficult to estimate. A number of authors have empirically evaluated the entropy reduction hypothesis (Frank, 2013; Linzen & Jaeger; Roark et al., 2009); note, however, that these implementations are based on different parsers and use different amounts of look-ahead (entropy reduction with respect to the next word, next four words, or whole sentence). Linzen and Jaeger showed that the amount of look-ahead massively affects entropy reduction estimates.

5. Data Sets and Evaluation

5.1. Test Sets A key challenge that needs to be addressed in order to develop cognitively plausible models of human language processing concerns test data and model evaluation. Here, the state of the art in psycholinguistic modeling lags significantly behind standards in computational linguistics. Most of the models that will be discussed in section 6 have not been evaluated rigorously. The authors typically describe their performance on a small set of hand-picked examples; often, no attempt is made to test on a range of items from the experimental literature and to evaluate directly against experimentally obtained measures of human language processing (e.g., reading times). This makes it very hard to obtain a realistic estimate of how well the models achieve their aim of capturing human language processing.

More recent work in psycholinguistic modeling has started to use standardized test sets for model evaluation, similar to what is commonplace for tasks in computational linguistics. In computational linguistics, parsers are evaluated against the Penn Treebank, word-sense disambiguation systems against the SemEval data sets, co-reference systems against the Tipster corpus or the Automatic Content Extraction corpus, and so on. In psycholinguistic modeling, evaluation is more complicated: two types of test data are required. The first type of test data consists of a collection of representative experimental results. Such data sets should contain the actual experimental materials (sentences or discourse fragments) used in the experiments, together with the experimental measurements obtained (e.g., reading times, eye-movement records, ERP data). The experiments included in this test set should be chosen to cover a wide range of experimental phenomena, for example, garden paths, syntactic complexity, and memory effects (see section 3). Such test sets enable the standardized evaluation of psycholinguistic models by comparing the model predictions (e.g., rankings, surprisal values, memory costs) against experimental measures on a large set of items. This way both the coverage of a model (how many phenomena it can account for) and its accuracy (how well it fits the experimental data) can be assessed.

Experimental test sets should be complemented by test sets based on corpus data. In order to assess the efficiency, robustness, and broad coverage of a model, a corpus of unrestricted, naturally occurring text is required. The use of contextualized language data makes it possible to assess not only syntactic models, but also models that capture discourse effects. These corpora need to be annotated with experimental measures, for example, eye-tracking or reading time data. Some relevant corpora have already been constructed (see the overview in table 22.1) and various authors have used them for model evaluation (Demberg & Keller, 2008a; Ferrara Boston, Hale, Kliegl, Patil, & Vasishth, 2008; Frank, 2009; Mitchell, Lapata, Demberg, & Keller, 2010; Patil, Vasishth, & Kliegl, 2009; Pynte et al., 2008; Roark et al., 2009).

5.2. Behavioral and Neural Data As outlined in section 5.1, a number of authors have evaluated psycholinguistic models against eye-tracking or reading time corpora. Part of the data and evaluation challenge is to extend this evaluation to neural data as provided by ERP or brain imaging studies (e.g., using functional MRI). Neural data sets are considerably more complex than behavioral ones, and modeling them is an important new task that the community is only beginning to address. Some recent work has evaluated models of word semantics against ERP (Frank, Otten, Galli, & Vigliocco, 2015; Murphy, Baroni, & Poesio, 2009) or functional MRI or magnetoencephalography data (Hale, Lutz, Luh, & Brennan, 2015; Mitchell et al., 2008; van Schijndel, Murphy, & Schuler, 2015; Wehbe, Ashish, Knight, & Mitchell, 2015; Willems et al., 2016). This is a very promising direction, and the challenge is to

TABLE 22.1

Test corpora that have been used for psycholinguistic modeling of sentence processing

Corpus	Language	Words	Participants	Method	Annotation	Reference
Dundee corpus	English, French	50,000	10	Eye tracking	PoS, dep	Kennedy and Pynte (2005); Barrett, Agic, and Sogaard (2015)
Potsdam corpus	German	1,138	222	Eye tracking	—	Kliegl, Nuthmann, and Engbert (2006)
MIT corpus	English	3,534	23	Self-paced reading	—	Bachrach (2008)
University College London corpus	English	4,946	117	Self-paced reading	PoS	Frank, Monsalve, Thompson, and Vigliocco (2013)
		1,931	43	Eye tracking, ERP		

Note that the Potsdam Corpus and the UCL Corpus consist of isolated sentences, rather than of continuous text. Words: number of word tokens in the corpus; Part: number of participants; PoS: parts of speech; dep: dependency parses.

extend this approach to the sentence and discourse level (see Bachrach, 2008). Again, it will be necessary to develop standardized test sets of both experimental data and corpus data.

5.3. EVALUATION MEASURES We also anticipate that the availability of new test data sets will facilitate the development of new evaluation measures that specifically test the validity of psycholinguistic models. Established evaluation measures from computational linguistics, such as Parseval, are of limited use, as they can only test the linguistic, not the behavioral or neural, predictions of a model.

So far, many authors have relied on qualitative evaluation: if a model predicts a difference in (for instance) reading time between two types of sentences where such a difference was also found experimentally, then that counts as a successful test. In most cases, no quantitative evaluation is performed, as this would require modeling the reading times for individual items and individual participants. Suitable procedures for performing such tests do not currently exist, though linear mixed-effects models (Baayen, Davidson, & Bates, 2008) provide a way of dealing with item and participant variation.

Further issues arise from the facts that we often want to compare model fit for multiple experiments (ideally without reparametrizing the models) and that various mutually dependent measures are used for evaluation, for example, processing effort at the sentence, word, and character level. An important open challenge in this area is to develop evaluation measures and associated statistical procedures that can deal with these problems.

6. Models

The earliest approaches were *ranking-based models*, which make psycholinguistic predictions based on the ranking of the syntactic analyses produced by a probabilistic parser. Jurafsky (1996) assumed that processing difficulty is triggered if the correct analysis falls below a certain probability threshold (i.e., is pruned by the parser). Similarly, Crocker and Brants (2000) assumed that processing difficulty ensues if the highest-ranked analysis changes from one word to the next. Both approaches have been shown to successfully model garden path effects. Being based on probabilistic parsing techniques, ranking-based models generally achieve a broad coverage, but their efficiency and robustness have not been evaluated. Also, they are not designed to capture syntactic prediction or memory effects (other than search with a narrow beam in Brants & Crocker, 2000).

The ranking-based approach has been generalized by *surprisal models*, which predict processing difficulty based on the change in the probability distribution over possible analyses from one word to the next (Demberg & Keller, 2008a; Ferrara Boston et al., 2008; Hale, 2001; Levy, 2008a; Roark et al., 2009). These models have been successful in accounting for a range of experimental data, and they achieve broad coverage. They also instantiate a limited form of prediction, namely, they build up expectations about the next word in the input. On the other hand, the efficiency and robustness of these models has largely not been evaluated, and memory costs are not modeled (again except for restrictions in beam size).

Prediction-based models explicitly predict syntactic structure for upcoming words (Demberg & Keller,

2008b, 2009), thus accounting for experimental results on predictive language processing. This approach implements a strict form of incrementality by building fully connected trees. Memory costs are modeled directly as a distance-based penalty that is incurred when a prediction has to be verified later in the sentence.

The *stack-based model* (Schuler et al., 2010) imposes explicit, cognitively motivated memory constraints on the parser, in effect limiting the stack size available to the parser. This delivers robustness, efficiency, and broad coverage, but it does not model syntactic *prediction*. A series of psycholinguistic evaluation studies that build on this model have recently been carried out (van Schijndel, Exley, & Schuler, 2013; van Schijndel et al., 2015).

6.1. Dependency Locality Theory DLT, suggested by Gibson (1998), operates on top of a dependency analysis. Processing difficulty according to DLT is related to the use of the computational resources consumed by the processor. Two distinct cost components can be distinguished: (i) integration cost associated with encoding new referents and with integrating referents into the structures already built at a given stage in the computation, and (ii) storage cost involved in the storage of parts of the input that may be used in parsing later parts of an input.

Integration costs and storage cost interact through the concepts of energy units, memory units, and time units. According to the model, there is only a limited number of energy units available at each point in time, so working memory resources can be used up by having to remember many dependencies (thus using up lots of memory units), in which case there will be less resources for actual integrations (as measured using integration cost), in turn causing them to take more time. The relationship among energy units, memory units, and time units was formalized as $EU = MU \times TU$. In the case of ambiguity, analyses that require fewer energy units are preferred.

DLT has been shown to account for a range of linguistic effects including the subject/object relative clause asymmetry, difficulty of center embeddings, garden paths, filler-gap dependencies, heavy NP shift, and extraposition (Gibson, 1998, 2000; Warren & Gibson, 2002; Roark et al., 2009). Broad coverage implementations exist only of the DLT integration cost component (which Gibson, 1998, suggested is the more important component), based on the MiniPar parser (Demberg & Keller, 2008a). A later reimplementation calculates integration cost on top of Stanford dependencies, which achieves better performance (see Demberg, 2013). The broad coverage evaluations showed that there is scant evidence for an effect of integration cost on reading times in general texts (see Demberg & Keller, 2008a, for a detailed analysis, as well as confirmatory evidence from van Schijndel & Schuler, 2013).

6.2. Information-Theoretic Models Information-theoretic measures (surprisal, entropy, entropy reduction; see section 4.6) can be calculated based on incremental generative language models. A first implemented (toy) model, based on a PCFG, was suggested by Hale (2001). Surprisal as an information-theoretic measure is in principle agnostic with respect to the underlying way of estimating information content (i.e., based on word sequences, syntactic structures, semantic expectancies; see also Levy, 2008a); various alternative model implementations have been proposed, including top-down PCFG parsing (Roark, 2001; Roark et al., 2009), dependency-parser–based surprisal model (Ferrara Boston et al., 2008), TAG-based (Demberg et al., 2013), generalized categorial grammars (van Schijndel & Schuler, 2013), as well as a sequence model working on part-of-speech tag sequences (Frank & Bod, 2011) (see also chapter 21 by Frank, Monaghan, and Tsoukala in this volume, who introduce this and related models in more detail). Most current models include representations of hierarchical syntactic structure for surprisal prediction, as this has been shown to more accurately fit human data (for more detail on a controversy of the empirical usefulness of representing hierarchical structure in information theoretic models, see Frank & Bod, 2011, and Fossum & Levy, 2012).

There is no specific reason why surprisal estimates should only model lexical information and syntax. Suggestions for how to include semantic information into surprisal models have been proposed (Blacoe & Lapata, 2012; Mitchell et al., 2010; Sayeed, Fischer, & Demberg, 2015).

To determine surprisal accurately, a full parallel search would be necessary. Due to performance reasons, however, all widely used parsers use beam search; that is, they do not compute all possible analyses to determine surprisal. Some of these models (Roark et al., 2009; Frank & Bod, 2011) have also been used to calculate related information-theoretic measures such as entropy and entropy reduction. A particular challenge for good estimates of entropy or entropy reduction is that one would need to estimate the entropy of the whole rest of the sentence; that is, we would need to calculate a probability distribution over the unseen rest of the sentence at each word. As this is very hard and computation-intensive to estimate, most implementations only regard a limited context of the next word or next few words; see also Linzen and Jaeger (2016).

Surprisal models have been shown to explain a wide range of psycholinguistic phenomena (see Levy, 2008a, for a more detailed discussion), and furthermore perform well on broad-coverage modeling on eye-tracking corpora (Demberg & Keller, 2008a; Ferrara Boston et al., 2008; Roark et al., 2009; van Schijndel & Schuler, 2013). Frank et al. (2015) furthermore showed that surprisal estimates are predictive of the size of the ERP component N400.

6.3. PSYCHOLINGUISTICALLY MOTIVATED TREE-ADJOINING GRAMMAR PLTAG is a computational model that explicitly accounts for prediction effects in human parsing. It is psycholinguistically motivated in that it simulates strict incrementality and incorporates an explicit prediction and verification process, as well as memory decay (see section 1.2 for a discussion of these key properties of human sentence processing). The PLTAG model consists of a grammar formalism, also called PLTAG (Demberg & Keller, 2008b; see section 2.1.2), together with a parser and a probability model for this formalism (Demberg et al., 2013), and a linking theory that relates PLTAG to behavioral data (Demberg & Keller, 2009).

PLTAG has two mechanisms that account for processing difficulty. First, surprisal is used to quantify difficulty in terms of updates to the parser's probability distribution over possible analyses as the sentence unfolds. Surprisal can be calculated straightforwardly from the probabilities of the (always fully connected) prefix trees of a sentence (a prefix tree is a partial tree that spans a prefix of a sentence). Second, difficulty can arise at verification time, when previously predicted structures are checked against what is actually encountered in the input. By hypothesis, the amount of difficulty generated by verification depends on two things: (i) how difficult the prediction was: the less probable the predicted structure, the higher the verification difficulty; and (ii) how recently the prediction was made: the more the prediction has decayed, the higher the verification difficulty.

To model memory decay, the processor needs to keep track of when each syntactic node was predicted. Technically, this is realized by assuming that a predicted node carries a timestamp that is initialized when the node is committed to memory. Based on the time stamp, the duration between initial prediction and verification can be quantified. The verification process thus attributes difficulty to the memory retrieval process that happens when newly encountered evidence is matched against a structure whose representation has decayed since it was last activated.

It is instructive to compare PLTAG verification cost to DLT integration cost. While both are distance-based costs, they differ in how distance is calculated. In DLT, distance is a linear function of the number of discourse referents between an argument and its head; verification cost, on the other hand, is computed using an exponential decay function based on the number of words intervening between the point at which the prediction was made and the point at which it is verified. Furthermore, verification cost not only depends on distance, but also on the difficulty of the prediction. This is modeled in terms of the probability of the predicted structure: the lower the probability of the predicted structure, the higher the cost of verifying it (see Demberg et al., 2013, for a formalization of this intuition).

PLTAG and DLT therefore differ in their assumptions about what causes processing difficulty: DLT integration cost attributes difficulty to integrating the head and its dependents. In PLTAG, the relevant syntactic structure has already been predicted at the point of verification; hence, verification cost is not about integrating structure (which we assume to be accounted for by surprisal), but about retrieving previously predicted structure and making sure that only analyses that are consistent with that predicted structure are built.

From the perspective of the comprehension system, the purpose of the verification operation is to check whether the predicted input matches the actually observed input and hence to ensure the consistency of the overall analysis. Verification is an additional operation that a nonpredictive parser does not require; it therefore makes sense that verification carries a cost and that this cost depends on the difficulty of the memory access required. Note that this cost can be safely assumed to be lower than the cost of accidentally constructing inconsistent analyses based on unobserved input, which are then never verified (perhaps until the end of the sentence). It turns out that DLT integration cost and PLTAG verification cost often assign difficulty at the same points in a sentence: in PLTAG, a head is predicted when its dependent (or a dependent of its dependent, as needed for maintaining connectedness) is encountered. This prediction is then verified when the argument is encountered, which is exactly the point at which the integration is assumed to take place in DLT.

PLTAG hence provides a way of combining surprisal-based processing cost with distance-based processing cost. It assumes that the overall processing cost of a word is the sum of the surprisal and the verification cost incurred at that word. Given that surprisal-based and distance-based costs have been argued to be complementary (Demberg & Keller, 2008a; Staub, 2010), PLTAG can be expected to provide a more complete account than either surprisal or DLT; that is, it should be able to capture a wider range of experimental results. At the same time, PLTAG provides a unified theoretical framework that explains both types of processing cost: verification cost is a natural consequence of assuming a prediction-based parser, and a prediction-based parser is in turn a consequence of assuming that human sentence processing is incremental and builds fully connected structures. Furthermore, PLTAG allows the two types of cost to be computed over the same representations (TAG structures): surprisal is associated with the forward-looking, predictive aspect of TAG parsing, while verification is the backward-looking, integrative aspect of TAG parsing. An alternative would be an additive combination of surprisal and DLT (for example); however, such an approach fails to provide a single theoretical framework (it has no way of explaining why the two types of cost arise and how they are related) and requires two separate representations (a phrase structure representation for surprisal cost and a dependency representation for DLT integration cost).

There is also independent experimental evidence that points toward an integrated model of surprisal-based and distance-based processing cost. For instance, Vasishth and Drenhaus (2011), using a number of experimental paradigms (self-paced reading, eye-tracking, and ERPs), investigate relative clause processing in German and find standard distance-based effects at the relative clause verb (as predicted by DLT). However, they also report a surprisal-based effect in first-pass regression probabilities in their eye-tracking study (increased distance between the verb and its argument, i.e., higher predictability, corresponds to fewer regressions), concluding that a complete theory needs to integrate both types of effects. This is corroborated by recent findings by Levy and Keller (2013), who report eye-tracking data for ditransitive sentences in German that indicate not only a decrease in processing difficulty at the verb with an increasing number of verb arguments (as predicted by surprisal), but also an increase in processing difficulty with an increasing distance between the arguments and the verb (as predicted by DLT integration cost). These results strongly suggest that a complete theory needs to combine both types of mechanisms.

6.4. Similarity-Based Interference Lewis and Vasishth (2005) proposed a model of sentence processing that uses the cognitive architecture Adaptive Control of Thought—Rational (Anderson, 1996). ACT-R implements cognitively plausible mechanisms, such as a working memory architecture. ACT-R has also been used to model a large range of other cognitive processes. The *memory and activation model* attempts to explain processing phenomena through memory-retrieval effects. The underlying mechanisms of memory retrieval are rehearsal, spreading activation, and decay. The model grammar is a PCFG, and it uses left-corner parsing as a processing algorithm to determine top-down predictions about what types of words or structures are needed to build a sentence, simultaneously with bottom-up evidence for what words are encountered in the input. When a word is retrieved from memory, its activation is boosted (this explains, e.g., lexical frequency effects: items that are retrieved very often have higher activation). The model also implements a steady activation decay according to the power law of forgetting, which is applied to all of the items in memory.

The model accounts for locality effects (like the subject/object relative clause asymmetry and center embedding) through decay and resulting lower activation of words that need to be retrieved for integration after seeing a lot of intervening material. It can also account for some antilocality effects through activation of the head through intervening arguments. Furthermore, the theory can explain interference effects (retrieval is hindered by activation of similar items) and storage load effects (if more items need to be stored, there are also more interference effects at retrieval).

6.5. Other Models Another incremental broad-coverage model is the hierarchical sequential model using categorial grammar, also known as the ModelBlocks parser (Wu, Bachrach, Cardenas, & Schuler, 2010; van Schijndel & Schuler, 2013; van Schijndel et al., 2013). A difficulty for incremental parsing is that linguistic structure (in English) is predominantly right-branching. With a right-branching structure, one cannot build a fully connected structure for partial sentences read from the left. The ModelBlocks parser deals with this challenge by performing a right-corner transform; that is, it rewrites syntax trees such that all right-branching structures are turned into left-branching structures. Parsing with this model is then a hierarchical sequence process that keeps track of the parsing states and can increase or decrease the size of the stack (the size of the stack reflects the number of open dependencies). Processing difficulty from this model is related to whether stack depth is increased or decreased, with the empirical finding that increasing stack depth leads to increased difficulty and decreased stack depth leads to a facilitation, which is the opposite

Table 22.2

Key properties of human language processing and their instantiation in various models of sentence processing

Property	Evidence	Model			
		Rank	Surp	Pred	Stack
Efficiency and robustness	Ferreira, Christianson, and Hollingworth (2001); Sanford and Sturt (2002)	–	–	–	+
Broad coverage	Crocker and Brants (2000)	+	+	–	+
Incrementality and connectedness	Tanenhaus et al. (1995); Konieczny (2000); Sturt and Lombardo (2005)	+	+	+	+
Prediction	Kamide, Altmann, and Haywood (2003); Delong et al. (2005); Staub and Clifton (2006)	–	±	+	–
Memory cost	Gibson (1998); Vasishth and Lewis (2006)	–	–	+	+

See section 6 for details. Pred = prediction-based models; rank = ranking-based models; stack = stack-based models; surp = surprisal-based models.

of the prediction in locality theory (Wu et al.; van Schijndel & Schuler).

A linking theory has also been proposed for top-down parsing with *minimalist grammars* (Kobele, Gerth, & Hale, 2013): the amount of time (i.e., the number of steps during parsing) that a derivation tree node spends on the parser's stack before being popped is related to increased processing difficulty in the form of higher memory demands.

More details on different incarnations of *competition models* are provided by McRae, Spivey-Knowlton, and Tanenhaus (1998); Tabor, Juliano, and Tanenhaus (1997); and Tabor and Tanenhaus (2001), but see van Gompel, Pickering, Pearson, and Liversedge (2005) who provided an overview of competition models and argue against them. Recently, Lau, Clark, and Lappin (2017) proposed a bottom-up probabilistic parser as a model of gradient grammaticality judgments.

6.6. Summary To summarize, the challenge is to develop a computational model that captures the key properties of human language processing (see section 1.2). Table 22.2 lists these key properties, along with some representative references to experimental work demonstrating these key properties. The table also gives a qualitative evaluation of four main model types (ranking, surprisal, prediction, and stack-based models) with respect to the properties.

7. Discussion

In this chapter, we discussed the state of the art in computational modeling of human sentence processing. Developing computational models is of scientific importance in so far as models are implemented theories: models of language processing allow us to test scientific hypotheses about the cognitive processes that underpin language processing. We argued that this type of precise, formalized hypothesis testing is only possible if standardized data sets and uniform evaluation procedures are available. Ultimately, this approach enables qualitative and quantitative comparisons between theories, and thus enhances our understanding of a key aspect of human cognition, language processing.

In this chapter, we have focused on syntactic processing. However, there is strong evidence that human language processing is driven by an interaction of syntactic, semantic, and discourse processes. There is considerable experimental evidence that verb senses, selectional restrictions, and thematic roles interact with syntactic ambiguity resolution. Another large body of research has elucidated the interaction of discourse processing and syntactic processing. A key challenge in computational psycholinguistics is therefore the development of language processing models that combine syntactic processing with semantic and discourse processing. So far, this challenge is largely unmet: there are some examples of models that integrate semantic processes such as thematic role assignment into a parsing model (Narayanan & Jurafsky, 2002; Padó et al., 2009; Konstas & Keller, 2015), but models that combine syntactic and discourse are virtually nonexistent.

A second challenge for the field is to bring together modeling work in language processing and language development. The human language processor is the product of an acquisition process that does not require explicit training and has access to only limited data: children aged 12 to 36 months are exposed to between 10 and 35 million words of input (Hart & Risley, 1995).

The challenge therefore is to develop a model of language acquisition that works with such small data sets, while also giving rise to a language processor that meets the key criteria in table 22.2.

Finally, it remains to note that there is also an applied side to the modeling enterprise. Once computational models of human language processing are available, they can be used to predict the difficulty that humans experience when processing text or speech. This is useful for a number of applications: for instance, natural language generation would benefit from being able to assess whether machine-generated text or speech is easy to process. For text simplification (e.g., for children or impaired readers), such a model is even more essential. It could also be used to assess the readability of text, which is of interest in educational applications (e.g., essay scoring). In machine translation, evaluating the fluency of system output is crucial, and a model that predicts processing difficulty could be used for this, or to guide the choice between alternative translations, and maybe even to inform human postediting.

NOTES

1. In the visual world paradigm, participants see a visual scene while hearing a sentence. Their eye movements on the scene in reaction to the sentence are recorded, providing information about the current state of their sentence processing.
2. Examples are from http://www.fun-with-words.com/ambiguous_garden_path.html.
3. The analyses in the search beam are the ones that the processor is actively pursuing. Typically a probability threshold is used to determine which analyses are removed from the beam (pruned).
4. A PCFG is a context-free grammar in which each grammar rule is associated with a probability. The overall probability of a parse tree is defined as the product of the probabilities of all grammar rules that are applied when generating the tree.

REFERENCES

Abney, S. P., & Johnson, M. (1991). Memory requirements and local ambiguities of parsing strategies. *Journal of Psycholinguistic Research, 20*(3), 233–250.

Alexopoulou, T., & Keller, F. (2007). Locality, cyclicity and resumption: At the interface between the grammar and the human sentence processor. *Language, 83*(1), 110–160.

Altmann, G. T. M., & Steedman, M. J. (1988). Interaction with context during human sentence processing. *Cognition, 30*(3), 191–238.

Anderson, J. R. (1996). ACT: A simple theory of complex cognition. *American Psychologist, 51*(4), 355.

Baayen, R. H., Davidson, D. J., & Bates, D. M. (2008). Mixed-effects modeling with crossed random effects for subjects and items. *Journal of Memory and Language, 59*(4), 390–412.

Bachrach, A. (2008). Imaging neural correlates of syntactic complexity in a naturalistic context (Unpublished doctoral dissertation). Massachusetts Institute of Technology, Cambridge.

Barrett, M. J., Agic, Z., & Sogaard, A. (2015). The Dundee treebank. In M. Dickinson, E. Hinrichs, A. Patejuk, & A. Przepiórkowski (Eds.), *Proceedings of the 14th International Workshop on Treebanks and Linguistic Theories* (pp. 242–248). Warsaw, Poland.

Bever, T. G. (1970). The cognitive basis for linguistic structures. In J. Hayes (Ed.), *Cognition and the development of language* (pp. 279–362). New York: Wiley.

Blacoe, W., & Lapata, M. (2012). A comparison of vector-based representations for semantic composition. In J. Tsujii, J. Henderson, & M. Pasca (Eds.), *Proceedings of the 2012 Joint Conference on Empirical Methods in Natural Language Processing and Computational Natural Language Learning* (pp. 546–556). Stroudsburg, PA: Association for Computational Linguistics.

Brants, T., & Crocker, M. W. (2000). Probabilistic parsing and psychological plausibility. In *Proceedings of the 18th International Conference on Computational Linguistics* (pp. 111–117). Saarbrücken, Germany.

Chomsky, N. (1957). *Syntactic structures.* The Hague, the Netherlands: Mouton.

Crocker, M. W., & Brants, T. (2000). Wide-coverage probabilistic sentence processing. *Journal of Psycholinguistic Research, 29*(6), 647–669.

Delong, K. A., Urbach, T. P., & Kutas, M. (2005). Probabilistic word pre-activation during language comprehension inferred from electrical brain activity. *Nature Neuroscience, 8*, 1117–1121.

Demberg, V. (2013). Integration costs on auxiliaries? A self-paced reading study using WebExp. In M. Knauff, M. Pauen, N. Sebanz, & I. Wachsmuth (Eds.), *Proceedings of the 35th Annual Conference of the Cognitive Science Society* (pp. 2160–2165). Berlin: Cognitive Science Society.

Demberg, V., & Keller, F. (2008a). Data from eye-tracking corpora as evidence for theories of syntactic processing complexity. *Cognition, 101*(2), 193–210.

Demberg, V., & Keller, F. (2008b). A psycholinguistically motivated version of TAG. In C. Gardent & A. Sarkar (Eds.), *Proceedings of the 9th International Workshop on Tree Adjoining Grammars and Related Formalisms* (pp. 25–32). Stroudsburg, PA: Association for Computational Linguistics.

Demberg, V., & Keller, F. (2009). A computational model of prediction in human parsing: Unifying locality and surprisal effects. In N. Taatgen & H. van Rijn (Eds.), *Proceedings of the 31st Annual Conference of the Cognitive Science Society* (pp. 1888–1893). Amsterdam, the Netherlands: Cognitive Science Society.

Demberg, V., Keller, F., & Koller, A. (2013). Incremental, predictive parsing with psycholinguistically motivated tree-adjoining grammar. *Computational Lingusitics, 39*(4), 1025–1066.

Duffy, S. A., Morris, R. K., & Rayner, K. (1988). Lexical ambiguity and fixation times in reading. *Journal of Memory and Language, 27*, 429–446.

Federmeier, K. D., Wlotko, E. W., De Ochoa-Dewald, E., & Kutas, M. (2007). Multiple effects of sentential constraint on word processing. *Brain Research, 1146*, 75–84.

Ferrara Boston, M., Hale, J., Kliegl, R., Patil, U., & Vasishth, S. (2008). Parsing costs as predictors of reading difficulty: An evaluation using the Potsdam Sentence Corpus. *Journal of Eye Movement Research, 2*(1), 1–12.

Ferreira, F., Christianson, K., & Hollingworth, A. (2001). Misinterpretations of garden-path sentences: Implications for models of sentence processing and reanalysis. *Journal of Psycholinguistic Research, 30*(1), 3–20.

Ferreira, F., & Henderson, J. (1991). Recovery from misanalyses of garden-path sentences. *Journal of Memory and Language, 30*(6), 725–745.

Fodor, J. D., & Frazier, L. (1978). The sausage machine: A new two-stage parsing model. *Cognition, 6*, 291–325.

Ford, M., Bresnan, J., & Kaplan, R. M. (1982). A competence-based theory of syntactic closure. In J. Bresnan (Ed.), *The mental representation of grammatical relations* (pp. 727–796). Cambridge, MA: MIT Press.

Fossum, V., & Levy, R. (2012). Sequential vs. hierarchical syntactic models of human incremental sentence processing. In D. Reitter & R. Levy (Eds.), *Proceedings of the 3rd Workshop on Cognitive Modeling and Computational Linguistics* (pp. 61–69). Stroudsburg, PA: Association for Computational Linguistics.

Frank, S. L. (2009). Surprisal-based comparison between a symbolic and a connectionist model of sentence processing. In N. Taatgen & H. van Rijn (Eds.), *Proceedings of the 31st Annual Conference of the Cognitive Science Society* (pp. 1139–1144). Amsterdam: Cognitive Science Society.

Frank, S. L. (2013). Uncertainty reduction as a measure of cognitive load in sentence comprehension. *Topics in Cognitive Science, 5*, 475–494.

Frank, S. L., & Bod, R. (2011). Insensitivity of the human sentence-processing system to hierarchical structure. *Psychological Science, 22*(6), 829–834.

Frank, S. L., Monsalve, I., Thompson, R., & Vigliocco, G. (2013). Reading-time data for evaluating broad-coverage models of English sentence processing. *Behavior Research Methods, 45*, 1182–1190.

Frank, S. L., Otten, L., Galli, G., & Vigliocco, G. (2015). The ERP response to the amount of information conveyed by words in sentences. *Brain and Language, 40*, 1–11.

Frazier, L. (1979). On comprehending sentences: Syntactic parsing strategies (Unpublished doctoral dissertation). University of Connecticut, Storrs.

Gibson, E. (1991). A computational theory of human linguistic processing: Memory limitations and processing breakdown (Unpublished doctoral dissertation). Carnegie Mellon University, Pittsburgh.

Gibson, E. (1998). Linguistic complexity: Locality of syntactic dependencies. *Cognition, 68*, 1–76.

Gibson, E. (2000). Dependency locality theory: A distance-based theory of linguistic complexity. In A. Marantz, Y. Miyashita, & W. O'Neil (Eds.), *Image, Language, Brain: Papers from the First Mind Articulation Project Symposium* (pp. 95–126). Cambridge, MA: MIT Press.

Hale, J. (2001). A probabilistic Earley parser as a psycholinguistic model. In *Proceedings of the 2nd Conference of the North American Chapter of the Association for Computational Linguistics* (Vol. 2, pp. 159–166). Stroudsburg, PA: Association for Computational Linguistics.

Hale, J. (2003). The information conveyed by words. *Journal of Psycholinguistic Research, 32*, 101–122.

Hale, J. (2006). Uncertainty about the rest of the sentence. *Cognitive Science, 30*(4), 609-642.

Hale, J., Lutz, D. E., Luh, W.-M., & Brennan, J. R. (2015). Modeling fMRI time courses with linguistic structure at various grain sizes. In T. O'Donnell & M. van Schijndel (Eds.), *Proceedings of the 6th Workshop on Cognitive Modeling and Computational Linguistics* (pp. 89–97). Stroudsburg, PA: Association for Computational Linguistics.

Hart, B., & Risley, T. R. (1995). *Meaningful differences in the everyday experience of young American children.* Baltimore, MD: Paul H. Brookes.

Hobbs, J. R., & Bear, J. (1990). Two principles of parse preference. In H. Karlgren (Ed.), *Proceedings of the 13th Conference on Computational Linguistics* (pp. 162–167). Stroudsburg, PA: Association for Computational Linguistics.

Joshi, A., Levy, L., & Takahashi, M. (1975). Tree adjunct grammars. *Journal of the Computer and System Sciences, 10*(1), 136–163.

Joshi, A., & Schabes, Y. (1997). Tree-adjoining grammars. In G. Rozenberg & A. Salomaa (Eds.), *Handbook of formal languages* (Vol. 3., pp. 69–123). Berlin: Springer.

Jurafsky, D. (1996). A probabilistic model of lexical and syntactic access and disambiguation. *Cognitive Science, 20*(2), 137–194.

Jurafsky, D., & Martin, J. H. (2009). *Speech and language processing: An introduction to natural language processing, computational linguistics and speech recognition* (2nd ed.). Upper Saddle River, NJ: Pearson Education.

Kamide, Y., Altmann, G. T. M., & Haywood, S. L. (2003). The time-course of prediction in incremental sentence processing: Evidence from anticipatory eye movements. *Journal of Memory and Language, 49*, 133–156.

Kamide, Y., Scheepers, C., & Altmann, G. T. M. (2003). Integration of syntactic and semantic information in predictive processing: Cross-linguistic evidence from German and English. *Journal of Psycholinguistic Research, 32*, 37–55.

Kennedy, A., & Pynte, J. (2005). Parafoveal-on-foveal effects in normal reading. *Vision Research, 45*, 153–168.

Kimball, J. (1973). Seven principles of surface structure parsing in natural language. *Cognition, 2*(1), 15–47.

Kliegl, R., Nuthmann, A., & Engbert, R. (2006). Tracking the mind during reading: The influence of past, present, and future words on fixation durations. *Journal of Experimental Psychology: General, 135*(1), 12–35.

Kobele, G. M., Gerth, S., & Hale, J. (2013). Memory resource allocation in top-down minimalist parsing. In G. Morrill & M.-J. Nederhof (Eds.), *Formal Grammar. FG 2013, FG 2012. Lecture Notes in Computer Science, vol 8036* (pp. 32–51). Springer, Berlin.

Konieczny, L. (2000). Locality and parsing complexity. *Journal of Psycholinguistic Research, 29*(6), 627–645.

Konieczny, L., & Döring, P. (2003). Anticipation of clause-final heads: Evidence from eye-tracking and SRNs. In P. P. Slezak (Ed.), *Proceedings of the Joint International Conference on Cognitive Science, ICCS/ASCS, Sydney* (pp. 13–17).

Konstas, I., & Keller, F. (2015). Semantic role labeling improves incremental parsing. In C. Zong & M. Strube (Eds.), *Proceedings of the 53rd Annual Meeting of the Association for Computational Linguistics* (pp. 1191–1201). Stroudsburg, PA: Association for Computational Linguistics.

Lau, J. H., Clark, A., & Lappin, S. (2017). Grammaticality, acceptability, and probability: A probabilistic view of linguistic knowledge. *Cognitive Science, 41*(5), 1202–1241.

Levy, R. (2008a). Expectation-based syntactic comprehension. *Cognition, 106*(3), 1126–1177.

Levy, R. (2008b). A noisy-channel model of rational human sentence comprehension under uncertain input. In M. Lapata & H. Tou Ng (Eds.), *Proceedings of the Conference on Empirical Methods in Natural Language Processing* (pp. 234–243). Stroudsburg, PA: Association for Computational Linguistics.

Levy, R., Bicknell, K., Slattery, T., & Rayner, K. (2009). Eye movement evidence that readers maintain and act on uncertainty about past linguistic input. *Proceedings of the National Academy of Sciences, 106*(50), 21086–21090.

Levy, R., & Keller, F. (2013). Expectation and locality effects in German verb-final structures. *Journal of Memory and Language, 68*(2), 199–222.

Lewis, R. L. (1996). A theory of grammatical but unacceptable embeddings. *Journal of Psycholinguistic Research, 25*(93), 116.

Lewis, R. L., & Vasishth, S. (2005). An activation-based model of sentence processing as skilled memory retrieval. *Cognitive Science, 29*, 1–45.

Linzen, T. (2015). Probabilistic linguistic representations: Between learning and processing (Unpublished doctoral dissertation). New York University.

Linzen, T., & Jaeger, T. F. (2016). Uncertainty and expectation in sentence processing: Evidence from subcategorization distributions. *Cognitive Science, 40*(6), 1382–1411.

MacDonald, M. C., Pearlmutter, N. J., & Seidenberg, M. S. (1994). Lexical nature of syntactic ambiguity resolution. *Psychological Review, 101*, 676–703.

McDonald, R., Pereira, F., Ribarov, K., & Hajič, J. (2005). Non-projective dependency parsing using spanning tree algorithms. In *Proceedings of the Human Language Technology Conference and the Conference on Empirical Methods in Natural Language Processing* (pp. 523–530). Stroudsburg (PA): Association for Computational Linguistics.

McRae, K., Spivey-Knowlton, M. J., & Tanenhaus, M. K. (1998). Modeling the influence of thematic fit (and other constraints) in on-line sentence comprehension. *Journal of Memory and Language, 38*(3), 283–312.

Miller, G., & Isard, S. (1964). Free recall of self-embedded English sentences. *Information and Control, 7*(3), 292–303.

Mitchell, J., Lapata, M., Demberg, V., & Keller, F. (2010). Syntactic and semantic factors in processing difficulty: An integrated measure. In J. Hajic, S. Carberry, S. Clark & J. Nivre (Eds.), *Proceedings of the 48th Annual Meeting of the Association for Computational Linguistics* (pp. 196–206). Stroudsburg, PA: Association for Computational Linguistics.

Mitchell, T. M., Shinkareva, S. V., Carlson, A., Chang, K.-M., Malave, V. L., Mason, R. A., & Just, M. A. (2008). Predicting human brain activity associated with the meanings of nouns. *Science, 320*(5880), 1191–1195.

Murphy, B., Baroni, M., & Poesio, M. (2009). EEG responds to conceptual stimuli and corpus semantics. In P. Koehn & R. Mihalcea (Eds.), *Proceedings of the Conference on Empirical Methods in Natural Language Processing* (pp. 619–627). Stroudsburg, PA: Association for Computational Linguistics.

Narayanan, S., & Jurafsky, D. (2002). A Bayesian model predicts human parse preference and reading time in sentence processing. In T. G. Dietterich, S. Becker, & Z. Ghahramani (Eds.), *Advances in Neural Information Processing Systems 14* (pp. 59–65). Cambridge, MA: MIT Press.

Nivre, J. (2003). An efficient algorithm for projective dependency parsing. In *Proceedings of the International Workshop on Parsing Technologies* (pp. 149–160). Stroudsburg, PA: Association for Computational Linguistics.

Padó, U., Crocker, M. W., & Keller, F. (2009). A probabilistic model of semantic plausibility in sentence processing. *Cognitive Science, 33*(5), 794–838.

Patil, U., Vasishth, S., & Kliegl, R. (2009). Compound effect of probabilistic disambiguation and memory retrievals on sentence processing: Evidence from an eye-tracking corpus. In A. Howes, D. Peebles, & R. Cooper (Eds.), *Proceedings of 9th International Conference on Cognitive Modeling.* Manchester, England.

Pynte, J., New, B., & Kennedy, A. (2008). On-line contextual influences during reading normal text: A multiple-regression analysis. *Vision Research, 48*(21), 2172–2183.

Roark, B. (2001). Probabilistic top-down parsing and language modeling. *Computational Linguistics, 27*(2), 249–276.

Roark, B., Bachrach, A., Cardenas, C., & Pallier, C. (2009). Deriving lexical and syntactic expectation-based measures for psycholinguistic modeling via incremental top-down parsing. In P. Koehn & R. Mihalcea (Eds.), *Proceedings of the Conference on Empirical Methods in Natural Language Processing* (pp. 324–333). Stroudsburg, PA: Association for Computational Linguistics.

Sanford, A. J., & Sturt, P. (2002). Depth of processing in language comprehension: Not noticing the evidence. *Trends in Cognitive Sciences, 6*, 382–386.

Sayeed, A., Fischer, S., & Demberg, V. (2015). Vector-space calculation of semantic surprisal for predicting word pronunciation duration. In C. Zong & M. Strube (Eds.), *Proceedings of the 53rd Annual Meeting of the Association for Computational Linguistics and the 7th International Joint Conference on Natural Language Processing* (Vol. 1, pp. 763–773). Stroudsburg, PA: Association for Computational Linguistics.

Schuler, W., AbdelRahman, S., Miller, T., & Schwartz, L. (2010). Broad-coverage parsing using human-like memory constraints. *Computational Linguistics, 26*(1), 1–30.

Spivey-Knowlton, M., & Sedivy, J. (1995). Resolving attachment ambiguities with multiple constraints. *Cognition, 55*, 227–267.

Stabler, E. P. (1994). The finite connectivity of linguistic structure. In C. Clifton, L. Frazier, & K. Rayner (Eds.), *Perspectives on Sentence Processing* (pp. 303–336). Hillsdale, NJ: Erlbaum.

Staub, A. (2010). Eye movements and processing difficulty in object relative clauses. *Cognition, 116*, 71–86.

Staub, A., & Clifton, C. (2006). Syntactic prediction in language comprehension: Evidence from either ... or. *Journal of Experimental Psychology: Learning, Memory, and Cognition, 32*, 425–436.

Sturt, P., & Lombardo, V. (2005). Processing coordinated structures: Incrementality and connectedness. *Cognitive Science, 29*(2), 291–305.

Tabor, W., Galantucci, B., & Richardson, D. (2004). Effects of merely local syntactic coherence on sentence processing. *Journal of Memory and Language, 50*(4), 355–370.

Tabor, W., & Hutchins, S. (2004). Evidence for self-organized sentence processing: Digging-in effects. *Journal of Experimental Psychology-Learning Memory and Cognition, 30*(2), 431–450.

Tabor, W., Juliano, C., & Tanenhaus, M. K. (1997). Parsing in a dynamical system: An attractor-based account of the interaction of lexical and structural constraints in sentence processing. *Language and Cognitive Processes, 12*(2–3), 211–271.

Tabor, W., & Tanenhaus, M. K. (2001). Dynamical systems for sentence processing. In M. Christiansen & N. Chater (Eds.), *Connectionist psycholinguistics* (pp. 177–211). Westport, CT: Ablex Publishing.

Tanenhaus, M. K., Spivey-Knowlton, M. J., Eberhard, K. M., & Sedivy, J. C. (1995). Integration of visual and linguistic information in spoken language comprehension. *Science, 268*, 1632–1634.

Temperley, D. (2007). Minimization of dependency length in written English. *Cognition, 105*(2), 300–333.

Traxler, M. J., Pickering, M. J., & Clifton, C. (1998). Adjunct attachment is not a form of lexical ambiguity resolution. *Journal of Memory and Language, 39*, 558–592.

van Gompel, R. P., Pickering, M. J., Pearson, J., & Liversedge, S. P. (2005). Evidence against competition during syntactic ambiguity resolution. *Journal of Memory and Language, 52*, 284–307.

van Schijndel, M., Exley, A., & Schuler, W. (2013). A model of language processing as hierarchic sequential prediction. *Topics in Cognitive Science, 5*, 522–540.

van Schijndel, M., Murphy, B., & Schuler, W. (2015). Evidence of syntactic working memory usage in MEG data. In T. O'Donnell & M.van Schijndel (Eds.), *Proceedings of the 6th Workshop on Cognitive Modeling and Computational Linguistics* (pp. 79–88). Stroudsburg, PA: Association for Computational Linguistics.

van Schijndel, M., & Schuler, W. (2013). An analysis of frequency- and memory-based processing costs. In L. Vanderwende, H. Daumé III, & K. Kirchhoff (Eds.), *Proceedings of the 2013 Conference of the North American Chapter of the Association for Computational Linguistics: Human Language Technologies* (pp. 95–105). Stroudsburg, PA: Association for Computational Linguistics.

Vasishth, S., & Drenhaus, H. (2011). Locality in German. *Dialogue and Discourse, 1*, 59–82.

Vasishth, S., & Lewis, R. L. (2006). Argument-head distance and processing complexity: Explaining both locality and antilocality effects. *Language, 82*(4), 767–794.

Warren, T., & Gibson, E. (2002). The influence of referential processing on sentence complexity. *Cognition, 85*(1), 79–112.

Wehbe, L., Ashish, V., Knight, K., & Mitchell, T. (2015). Aligning context-based statistical models of language with brain activity during reading. In L. Màrquez, C. Callison-Burch, & J. Su (Eds.), *Proceedings of the Conference on Empirical Methods in Natural Language Processing.* Stroudsburg, PA: Association for Computational Linguistics.

Willems, R., Frank, S., Nijhof, A., Hagoort, P., & Van den Bosch, A. (2016). Prediction during natural language comprehension. *Cerebral Cortex, 26*(6), 2506–2516.

Wu, S., Bachrach, A., Cardenas, C., & Schuler, W. (2010). Complexity metrics in an incremental right-corner parser. In J. Hajič, S. Carberry, S. Clark, & J. Nivre (Eds.), *Proceedings of the 48th Annual Meeting of the Association for Computational Linguistics* (pp. 1189–1198). Stroudsburg, PA: Association for Computational Linguistics.

23 Vector-Based and Neural Models of Semantics

WILLEM ZUIDEMA AND PHONG LE

There is a rich tradition in modeling the meanings of words, phrases, sentences, and discourse, going back, in some cases, thousands of years. For most of that long history, the dominant approach has been an approach based on some form of symbolic logic. This approach has been very successful, in many respects, but it has proven difficult to directly relate such models of semantics to research in (experimental) psychology and cognitive neuroscience. In this chapter, we discuss an alternative approach to modeling meaning based on numerical vectors, using tools from some other branches of mathematics: linear algebra and differential calculus. These models are very attractive from the neural perspective, because numerical vectors are the natural representation for patterns of neural activity, as well as from the learning perspective, because such vectors are compatible with artificial neural network models, and, if certain conditions are met, rich sets of tools for optimization (learning) developed in this paradigm become available. We will get back to these attractive properties in section 1.4.

But are numerical vectors not too weak a representation to account for the intricacies of natural language semantics? Hasn't the rich tradition in lexical and compositional semantics revealed the need to have nontrivial, structured, symbolic models? The goal of this chapter is to show that we can have our cake and eat it: we can use vectorial representations that are compatible with neuroscience and that are optimized using techniques from the neural network toolbox, while still being able to account for nontrivial semantic phenomena. Vector-space models therefore form a class that is a key intermediary between neural and behavioral data and the rich modeling traditions in semantics (and are, we argue, complementary to symbolic approaches, rather than replacements for them).

1. Words: Vector-Space Models of Lexical Semantics

Many approaches that tackle the problem of how to represent word meaning by vectors, are based on the insight of Firth: "You shall know a word by the company it keeps" (Firth, 1957). It means that the meaning of a word is determined by (or at least is correlated with) the words that surround it, and word meanings can thus be computed by analyzing the *distribution* of words, their occurrence and co-occurrence frequencies, in large text corpora. We start our discussion of vectorial models of semantics by looking in some more detail at *distributional* methods; in section 1.3 we also briefly mention nondistributional ways to determine vectorial meaning representations.

1.1. An Example To illustrate the key ideas behind distributional semantics, consider the following example (used in slides[1] of S. Evert). Imagine that we are reading an English book and encounter the strange word *bardiwac* in some sentences:

- He handed her her glass of *bardiwac.*
- Beef dishes are made to complement the *bardiwac.*
- Nigel staggered to his feet, face flushed from too much *bardiwac.*
- Malbec, one of the lesser-known *bardiwac* grapes, responds well to Australia's sunshine.
- I dined off bread and cheese and this excellent *bardiwac.*
- The drinks were delicious: blood-red *bardiwac* as well as light, sweet Rhenish.

Although we have no idea about the meaning of the word *bardiwac,* we can guess it using contexts. From the first sentence, we can guess that *bardiwac* is a kind of (drinkable) liquid. The second sentence suggests that it is drunk with beef dishes. With the third sentence, it seems that *bardiwac* can make us drunk. And so on. So finally, we can guess (with a very high confidence) that *bardiwac* is a kind of red strong wine.

Because we are human beings, we can guess the meaning of an unknown word easily. How can we simulate this process on a computer? The intuitions behind distributional semantics suggest a computational model can be based on counts of co-occurrences of words. In the following, we will work out a simple formalization of this intuition (before mentioning a number of recent

edge of the sword each of this
Lets fall his sword before your Highness'
it me; my sword shall soon dispatch
Sooner this sword shall plough thy
claps me his sword upon the table
I see a sword out, my finger
What, the sword and the word!
Even with the sword that kill'd thee.
If to my sword his fate be
can shake my sword or hear the
He who the sword of heaven will
She takes a sword and runs at
I and my sword will earn our
It eats the sword it fights with.
But for thy sword and fortune, trod
For what the sword cuts down or

u	$\#(u\|w)$	$P(u\|w)$	$P(u\|w)/P(u)$
claps	1	0.002	1.333
cuts	1	0.002	1.333
dispatch	1	0.002	1.333
drink	0	0.0	0.0
down	1	0.002	1.333
earn	1	0.002	1.333
eats	1	0.002	1.333
fall	1	0.002	1.333
fate	1	0.002	1.333
fights	1	0.002	1.333
finger	1	0.002	1.333
fortune	1	0.002	1.333
hear	1	0.002	1.333
heaven	1	0.002	1.333
Highness	1	0.002	2.666
kill'd	1	0.002	1.333
plough	1	0.002	1.333
refuse	0	0.0	0.0
see	1	0.002	1.333
shake	1	0.002	1.333
shall	2	0.004	0.987
table	1	0.002	1.333
takes	1	0.002	1.333
trod	1	0.002	1.333

$$\begin{pmatrix} 1.8 \\ 1.3 \\ 0.8 \\ 0.9 \\ 0.2 \\ 2.8 \\ 2.0 \\ 0.1 \\ 1.7 \\ 1.8 \\ 2.5 \\ 0.5 \\ 0.7 \end{pmatrix}$$

FIGURE 23.1 (Left) An example toy corpus (a random selection of word sequences containing the word *sword* in the Shakespeare texts on Project Gutenberg, with a three-word context window). (Right) Co-occurrence frequencies with *sword* of various other words u in a window of size $k=3$, conditional probabilities, hypothetical PMI values (illustrating that a high-frequency function word such as *shall* is less informative than a low-frequency word such as *Highness*) and the resulting hypothetical word embedding vector (after dimensionality reduction).

variants, including neural network approaches that *predict* rather than *count*).

For simplicity, we define a *context* of a target word in a sentence as the set of all words near the target word, that is, within a *window* of k words before or after the target word (if $k=\infty$, the context is the entire sentence.) Let V be a vocabulary. A very simple method is to count how many times a word $u \in$ V co-occurs with the target word $w \in$ V in the k-sized window. This is equivalent to estimating the conditional distribution $Pr(U=u|W=w)$ using maximum likelihood. An illustration is shown in figure 23.1. First, we collect sentences containing the target word w. We then for each word u count how many times u co-occurs with w within a defined window size k. Finally, we extract a vector containing all of those counts, which is used to represent the target word w. We estimate the conditional distribution simply with:

$$P(U=u|W=w) = \frac{\# u \text{ and } w \text{ co-occur}}{\# w} \tag{1}$$

However, in order to come up with useful vectors, there are some details that should be considered. First, which u's should be counted. It is clear that function words (e.g., *a, an, how, what*), which are extremely frequent, carry little information about the meanings of target words. By contrast, rare words tend to contribute more important information. We thus need one more step called a *weighting scheme*. One of popular schemes is weighting with pointwise mutual information (PMI), which corresponds to weighting with the inverse of the probability that u appears in the corpus, $P(U=u)$. Intuitively, this means that rare words get higher weights than frequent words do. The embedding $\vec{e}(w)$ of word w (before dimensionality reduction) is then given as:

$$\vec{e}(w)[u] = PMI(u,w) = \log\left(\frac{(P(U)=u \mid W=w)}{P(U=u)}\right) \tag{2}$$

In practice, the resulting vectors are high dimensional (easily many thousands of dimensions or more). Therefore, a dimensionality reduction method such as principal component analysis or non-negative matrix factorization is often employed as a final step to compute reduced embeddings $\vec{e'}$ (see Dinu & Lapata, 2010, for an example).

The model we just derived is a simple instance of a *vector-space model*, where every word is assigned a vector (based on co-occurrence statistics, or *distributional information*), and the set of vectors reflects similarities and differences in the contexts in which words find themselves. In distributional semantics this distributional similarity is, in turn, taken to correlate with semantic similarity (the *distributional hypothesis*).

1.2. A VARIETY OF METHODS IN DISTRIBUTIONAL SEMANTICS Many variants of distributional semantics models exist (see Lowe, 2001, for an overview of the relevant dimensions at which they differ, and Erk, 2012, for a review of the state of the art a decade later). In cognitive science, the still most well-known variant is Latent Semantic Analysis (Landauer & Dumais, 1997), which differs from the model sketched in section 1.1 in that it counts how often a word is used in each of very many documents (i.e., a word–document matrix rather than a word–word matrix is computed; additionally tf/idf rather than PMI is the weight function, and singular value decomposition is used for dimensionality reduction).

Depending on how distributional context is defined (e.g., sentence context or document context) and on how different aspects of the context are weighted (e.g., with distance, or with position to the left or right of the target word), the configurations differ (as do their applicabilities in specific tasks).

In recent years, a class of alternative models has become popular where the key idea is that embeddings are optimized such that they allow prediction of missing or next words (e.g., Collobert et al., 2011, Mikolov, Sutskever, Chen, Corrado, & Dean, 2013). This approach can be illustrated with "filling in blanks" questions in Test of English as a Foreign Language and International English Language Testing System tests, as in the following example:

_________ is the study of numbers, equations, functions, and geometric shapes and their relationships.

a. Physics
b. Mathematics
c. Geography
d. Theology

To arrive at the correct answer in such a task (here: "mathematics"), a model needs (a) to summarize the relevant aspects of meaning of all the words in the given sentence and the way they are put together, (b) have a representation for each of the candidate answers, and (c) have a way to evaluate how well (b) matches (a). Success at this task thus implies knowledge of the meaning of all the words; in a vector-space model optimized for this task, vectors for words should thus reflect their meanings.

More formally: let us assume that each word $v \in \mathrm{V}$ is represented by a vector $\mathbf{v} \in \mathrm{R}^d$ and there is a mechanism for computing the probability $P(W=w|U=(u_1, u_2, \ldots, u_l))$ of the event that a target word w appears in a context $(u_1, u_2, \ldots, u_l)$. We are going to find a vector $\mathbf{v}$ for each word v such that those probabilities are as high as possible for each w and its context $(u_1, u_2, \ldots, u_l)$.

Bengio, Ducharme, Vincent, and Janvin (2003) provided, to our knowledge, the very first approach in this class. Mikolov, Karafiát, Burget, Černocký, and Khudanpur (2010) employed (simple) recurrent network networks. The very popular *word2vec* models, introduced by Mikolov, Chen, Corrado, & Dean (2013), are optimized to predict the target word given the context word embeddings (the continuous bag of words model), or vice versa, to predict context words given the target word embedding (the skipgram model). Pennington, Socher, and Manning (2014) presented a method called *GloVe* and showed that it performs on par with the word2vec method on many tasks. Interestingly, Levy and Goldberg (2014) argued that prediction and count-based approaches might, mathematically, not be so different after all. They derive a count-based objective function that the word2vec-skipgram method is approximating.

1.3. Successes: Proximity and Cognitive Semantic Similarity The most natural property of continuous vector spaces is that they immediately let us think about sets of words in terms of their mutual distances: for every two points in a vector space, we can compute the distance, and, inversely, proximity, using standard distance metrics from algebra and geometry. In vector-space models of semantics, we want *proximity* in that space to reflect semantic *similarity*. A popular distance metric is cosine, which equals 1 if two vectors are pointing in exactly the same direction and 0 if the two vectors are orthogonal. The cosine of two vectors w and v can be computed as the sum of products of the corresponding elements scaled by the vector lengths:

$$\cos(\vec{w}, \vec{v}) = \frac{\sum_{i=1}^{n} w_i v_i}{\|\vec{w}\| \cdot \|\vec{v}\|}$$

With a method for computing embeddings (say: word2vec–continuous bag of words), a corpus (say: English Wikipedia) and a distance metric (say: cosine) in place, we can start to investigate whether the proximity of the resulting vectors indeed corresponds to semantic similarity. In figure 23.2, we give a few examples of the nearest neighbors in such a word vector model of an arbitrary noun, verb, and adjective. These tables illustrate that words find themselves in the neighborhood of other words with, intuitively, strong semantic similarity.

And indeed, a large literature, from the 1990s onward, has demonstrated that word vectors from distributional semantics can (with the right setting of the parameters) strongly correlate with human similarity and typicality judgments and can be strongly predictive of semantic priming effects. Recent large-scale comparisons (Baroni & Lenci, 2010, Pereira, Gerhsman, Ritter, & Botvinick, 2016) confirm this point and moreover show that the GloVe and word2vec methods now so popular in natural language processing (NLP), correlate even better with human data of various sorts than classical methods such as Latent Semantic Analysis. More recently,

dizziness	0.85	write	0.77	blue	0.76
nausea	0.81	learn	0.47	white	0.70
drowsiness	0.81	literacy	0.45	yellow	0.69
convulsions	0.78	print	0.40	red	0.65
vomiting	0.78	writing	0.39	purple	0.62
spasms	0.77	text	0.39	orange	0.60
constipation	0.77	edit	0.38	pink	0.56
migraine	0.77	how	0.38	magenta	0.54
cramps	0.77	readers	0.37	dark	0.52
duodenal	0.76	find	0.37	multicoloured	0.52
A NNs of "headache"		B NNs of "read"		C NNs of "green"	

FIGURE 23.2 Nearest neighbors (NNs) of three words with cosine similarities, using embeddings obtained by applying word2vec on the English Wikipedia corpus. This figure was created using data from the "Word2vec playground" (http://deeplearner.fz-qqq.net/) in June 1997 (site no longer available; a similar service is available at, e.g., https://rare-technologies.com/word2vec-tutorial/).

they have also been successfully employed in other areas of linguistics, including historical linguistics (e.g., Fremann & Lapata, 2016; Hamilton, Leskovec, & Jurafsky, 2016).

Mitchell et al. (2008) pioneered the use of distributional methods in analyzing neuroimaging data and showed that they can be used to predict, with some success, BOLD-responses of subjects exposed to word-picture pairs of concrete nouns such as *celery* or *airplane.* In the original study, a very simple distributional model was used, where the nouns used as stimuli are represented with a vector recording co-occurrence frequencies with 25 hand-picked verbs (such as *swallow* and *fly*). Later work (Murphy, Talukdar, & Mitchell, 2012) showed that more advanced techniques give even better results, with the best results for a model that makes use of the grammatical structure of sentences in which words find themselves (modeled using labeled dependency parses). These studies use a leave-2-out paradigm, where a model is evaluated on its ability to decide, based on the functional MRI (fMRI) data, which of two words were presented to the subject, given labeled fMRI data for all remaining words. The baseline performance for random guessing is 50%; the best models score around 83%.

Most successful demonstrations of vector-space models are thus based on distributional semantics, using language-internal statistics about words co-occurring with other words. However, it is clear that the meaning of words must also somehow be grounded in the senses and the world. Vector-space models are not restricted to distributional information. Vectors can also encode other types of information and are compatible with approaches based on embodiment and/or conceptual spaces, although it has been difficult to define this type of information at similar scales and levels of precision as the automatic, corpus-based methods for distributional semantics have made possible. Recent progress in vectorial models of vision and object recognition, however, provides a possible resource to fill this gap (see Baroni, 2016, for a recent review of the relevant literature in this area).

1.4. THEORETICAL ADVANTAGES OF VECTORIAL REPRESENTATIONS Vector representations of meaning are not only successfully used in practice for a variety of tasks, they are also attractive from a theoretical perspective as an intermediary among linguistic, cognitive science, and neuroscientific notions of meaning. This is first and foremost because numerical vectors are widely used to describe the activity of a layer of neurons, where the value of an element in the vector is representing the firing rate of a single neuron. But vector representations are not limited to such one-to-one mappings and to rate-coding. We can also think of an element in the vector as representing one of the other parameters of a neuron besides firing rate: oscillation phase, depletion of chemicals, changes in synaptic strength, and so on. Each neuron is then represented by multiple elements in the vector. Alternatively, one element might represent the average activity of a (large) group of neurons (e.g., a voxel in fMRI data) or of some other aggregate variable representing (part of) the information a group of neurons codes for. Interestingly, choosing between different neural interpretations is not always necessary (although it becomes very relevant when relating vectors to empirical neural data or trying to use neurobiological findings to constrain the model space; for example, when selecting an appropriate similarity metric or learning procedure).

For learning, numerical vectors are an attractive representation as they allow optimization using standard tools from machine learning. And, indeed, in recent *deep learning* work, many successful applications

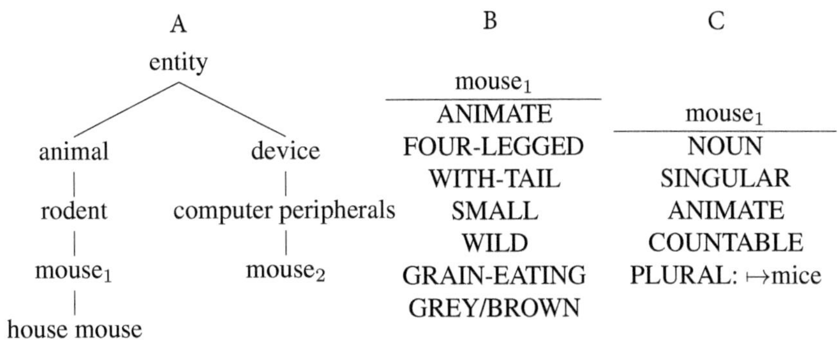

FIGURE 23.3 Examples of the typical information included in symbolic lexicons about words: (A) words might be polysemous, motivating multiple entries in a lexicon. Each entry can be placed in an ontology, detailing superset and subset or other relations it has with other entries. (B) The meaning of words can often be broken down into smaller meaning components/semantic primitives, allowing similarity relations between meanings to be made explicit. (C) The lexical entry often also contains syntactic information about the word form, regulating its behavior in when combined with other words to form sentences.

of this toolbox to questions relating to natural language meaning are reported, often exploiting the huge amount of texts available on the Internet.

1.5. LIMITATIONS OF SIMILARITY MODELS So far we have discussed successes in distributional semantics based on measuring proximity between word vectors, mostly treating semantic similarity as a scalar property, that is, as something that can be described with a single number. However, semantics, even at the word level, involves much more than similarity. Symbolic models of lexical semantics also try to account for phenomena such as the fact that one word might have multiple meanings (polysemy) and that the meanings of words have systematic *relations* with each other such as hypernymy (subset-superset relations, such as the relation between *animal* and *dog*), antonymy (being each other's opposite, such as *large* vs. *small*), and differing in degrees (such as in comparatives, *small, smaller, smallest*). Moreover, symbolic models of the lexicon also carry information about the role that a word plays in a sentence, such as its syntactic category (part-of-speech tag) and morphological information. Figure 23.3 illustrates some of the information about a word like *mouse* that a symbolic lexicon would specify.

1.6. LEXICAL RELATIONS IN VECTOR-SPACE MODELS Much current work is devoted to working out how such phenomena can be accounted for in vector-space models. An important reason for optimism in the field about the ability of vector-space models to account for such aspects of meaning, is the discovery of Mikolov and colleagues (Mikolov, Yih, & Zweig, 2013) that many of the mentioned relations between words, including the morphosyntactic relations between words, seem to be reflected by geometric relations (mostly linear shifts) in the vector space. Mikolov, Yih, and Zweig's by now most famous example is the demonstration that the linear shift in vector space to go from *king* to *queen* is very similar to the linear shift to go from *man* to *woman*, such that we can predict the vector for *queen* using addition and subtraction:

$$\overrightarrow{king} - \overrightarrow{man} + \overrightarrow{woman} = \overrightarrow{queen}$$

In figures 23.4 and 23.5, we show a number of similar predictions, using again the word2vec model. These examples try to capture several part-whole relations, the comparative transform, the past tense and present participle morphology, antonymy relations, and the regular plural morphology in English. These examples do not prove that the mentioned relations are consistently corresponding to linear shifts in vector space, but they do illustrate that the assignment of vectors to words is quite structured and that the high-dimensional vector space allows for many relations to be captured in this way.

Much more sophisticated analysis of these relation can be found in, for instance, Henderson and Popa (2016), who showed that word2vec is closely related to models that optimize for lexical entailment and Cotterell, Schütze, and Eisner (2016), who extended distributional methods to account for word morphology.

2. Sentences: Vector-Space Models of Compositional Semantics

Perhaps the biggest challenge for vector-space models of semantics is how to compute meaning representation for sentences and multiword phrases. In theory, the distributional statistics used for computing word representations can be applied to larger items such as phrases

berlin	0.64	bologna	0.56	legs	0.53	inches	0.56
munich	0.59	turin	0.55	toes	0.53	meters	0.55
leipzig	0.58	venice	0.53	feet	0.52	tall	0.53
vienna	0.55	milan	0.52	instep	0.52	weighs	0.53
dresden	0.53	florence	0.51	waist	0.51	weigh	0.52
holtenau	0.51	vienna	0.51	heel	0.51	backboard	0.50
frankfurt	0.50	nabucco	0.51	soles	0.50	metres	0.50
hamburg	0.49	palermo	0.50	shoulder	0.49	octaves	0.50
bonn	0.49	fossa	0.49	neck	0.49	centimetres	0.49
rheinische	0.48	teatro	0.48	shoulders	0.49	millimeter	0.49
A France:Paris::Germany:?		B France:Paris::Italy:?		C hand:fingers::foot:?		D hands:fingers::feet:?	

FIGURE 23.4 Nearest neighbors from the predicted vector in four analogy tasks. In the analogy *A is to B as C is to D*, a simple prediction for the vector for *D* is given by $C+B-A=-(A-B)+C$. The used embeddings are obtained by applying word2vec on the English Wikipedia corpus. This figure was created using data from the "Word2vec playground" (http://deeplearner.fz-qqq.net/) in June 2017 (site no longer available; a similar service is available at, e.g., https://rare-technologies.com/word2vec-tutorial/).

nicer	0.69	talk	0.74	talk	0.77
nice	0.67	talked	0.71	talking	0.76
happier	0.61	talking	0.63	talked	0.61
better	0.58	spoke	0.60	discussing	0.59
smarter	0.55	spoken	0.50	chatting	0.55
easier	0.53	discussing	0.48	telling	0.46
harder	0.52	approached	0.47	spoke	0.45
prettier	0.52	laughed	0.46	joking	0.45
bigger	0.51	speak	0.46	chatter	0.45
heck	0.50	chatting	0.44	spoken	0.44
A smart:smarter::nice:?		B walk:walked::talk:?		C walk:walking::talk:?	

good	0.61	good	0.85	headache	0.80
black	0.52	bad	0.62	headaches	0.74
great	0.48	great	0.62	nightmares	0.47
decent	0.47	decent	0.58	problems	0.42
bad	0.47	terrific	0.55	nightmare	0.41
terrific	0.44	tough	0.53	pain	0.38
poor	0.43	nice	0.52	anxiety	0.37
nice	0.43	terrible	0.52	distraction	0.35
terrible	0.43	excellent	0.52	symptoms	0.35
excellent	0.43	fantastic	0.51	cough	0.35
D green:black::good:?		E white:black::good:?		F message:messages::headache:?	

FIGURE 23.5 Nearest neighbors from the predicted vector in four analogy tasks. In the analogy *A is to B as C is to D*, a simple prediction for the vector for *D* is given by $C+B-A=-(A-B)+C$. Retrieved from Liu (2016).

and sentences as well. However, in practice, even in very large corpora, any particular sentence is extremely unlikely to occur multiple times. For words, phrases, or sentences occurring only once, we only have one chance to collect statistics on the items they co-occur with or that they can be predicted from, which is, in general, not enough to build useful semantic representations. Lack of data is therefore a serious problem for approaches that simply transfer the word-based techniques to multiword phrases and sentences.

The solution that formal semantics research found is to rely on the principle of compositionality, which says that "The meaning of a whole is a function of the meanings of the parts and of the way they are syntactically combined" (Partee, 1995). With the lambda calculus, composition turns out to be very elegant: lambda function application and reduction. For instance,

Lola:	*lola*
runs:	$\lambda x.run(x)$
Lola runs:	$(\lambda x.run(x))(lola) = run(lola)$

Many variants of formal semantics have been worked out, generally relying on a hierarchical, syntactic analysis of a sentence and an expressive logic to describe sentence meaning (including devices to represent temporal information and to represent thoughts or statements about thoughts or statements). In this approach, words get assigned potentially rather complex lambda expressions that are needed to compute meaning representations of sentences from those of words. For instance, a ditransitive verb *give* will get assigned an expression that regulates that the three arguments, giver, givee, and gift, will be labeled with the right variables, as well as with the information that the verb is in present tense. A word like *some* (in a sentence like *Some mammals swim*) will be assigned an expression that regulates that the sentence meaning will contain an existential quantifier (∃) and that the properties from its first argument (mammals) and from the second argument (swim) end up in the right place in that expression. On the other hand, the lexical meaning of content words typically remains unanalyzed, that is, it is assumed to be atomic. For instance, the word *mouse* just means MOUSE.

Vectorial lexical semantics, as reviewed in section 1, holds the promise of providing a more satisfactory account of the lexical semantics, but how to combine the strength of distributional and formal semantics is far from trivial. But since 2008 a number of promising approaches have been worked out to also compute useful vector representations for sentences. We will briefly review four approaches, of which the first three, like formal semantics take the *hierarchical* structure of a sentence as given. The fourth approach, in contrast, treats sentences as *sequences* of words.

2.1. Hierarchical Approaches We take as given that $\mathbf{x}$ and $\mathbf{y}$ are vectors representing two items x and y, respectively, and that $\mathbf{x}$ and $\mathbf{y}$ are to be combined according to the assumed hierarchical analysis of the sentence. The central task of vector-space models for phrases and sentences is then to find a composition function f so that the output is a vector representing constituent $p = xy$:

$$\mathbf{p} = f(\mathbf{x}, \mathbf{y}, R, K) \tag{3}$$

where R is the syntactic rule that combines x and y, and K is background knowledge (Mitchell & Lapata, 2008).

Computing a vector for a sentence is thus, like in formal semantics, carried out in a bottom-up manner on a given parse tree. An illustration is given in figure 23.6. Assuming that there are vectors representing words, we first compute a vector for the verb phrase:

$$\overrightarrow{like\ it} = f(\overrightarrow{like}, \vec{it}, VP \rightarrow ADJ\ PRP)$$

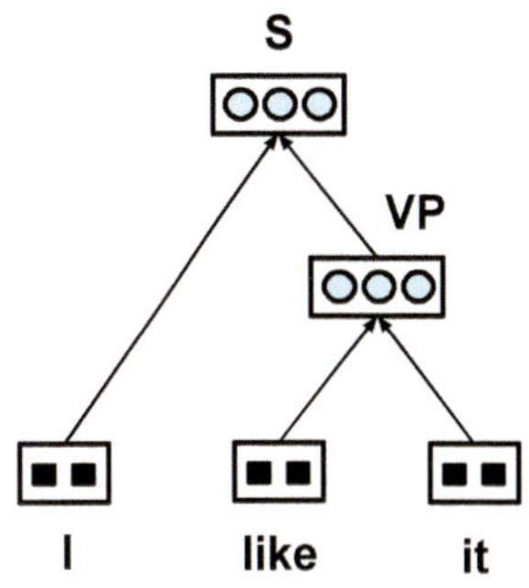

Figure 23.6 Compute sentence vector in a bottom-up manner. First, we compute a vector for the verb phrase (VP) and then compute a vector for the subject (S).

and then go higher up in the hierarchy to compute a vector for the whole sentence

$$\overrightarrow{I\ like\ it} = f(\vec{I}, \overrightarrow{like\ it}, S \rightarrow NP\ VP)$$

Approaches in this class mainly differ by the composition functions they make use of. Next, we will discuss three approaches.

2.1.1. Approach 1: Additive and multiplicative composition The simplest approach is to use vector element-wise operations such as addition and multiplication (Mitchell & Lapata, 2008)

- Additive: $\mathbf{p} = \mathbf{x} + \mathbf{y}$
- Multiplicative: $\mathbf{p} = \mathbf{x} \odot \mathbf{y}$ (i.e., $p_i = x_i \times y_i$)

where $p = xy$ is formed by combined x and y.

Despite its simplicity, this approach works surprisingly well in composing noun-noun and adjective-noun phrases (Blacoe & Lapata, 2012). However, from the theoretical point of view, this approach is unsatisfactory. Because those operations are commutative, word order and hierarchical structure of a sentence become irrelevant, that is,

$$\overrightarrow{man\ bit\ dog} = \overrightarrow{dog\ bit\ man}$$

A more sophisticated way is to linearly combine those two operations:

$$\mathbf{p} = \alpha\mathbf{x} + \beta\mathbf{y} + \gamma\mathbf{x}_i \odot \mathbf{y}$$

2.1.2. Approach 2: Tensor-product–based composition The idea is to resemble composition in formal semantics. In formal semantics, words such as adjectives and verbs are *functors* that are to modify *arguments* representing other words such as nouns, and composition is simply function application. Based on this, with the slogan "composition is largely a matter of *function application*" Baroni, Bernardi, and Zamparelli (2013) proposed using tensors[2] to represent words:

- first-order tensors (i.e., vectors $\in R^n$) are to represent nouns,
- second-order tensors (i.e., matrices $\in R^{n \times m}$) are to represent adjectives, intransitive verbs,
- third-order tensors (i.e., $\in R^{n \times m \times k}$) are for transitive verbs, and so on.

Composition is carried out through tensor production. For example, composing an adjective and a noun is a matrix-vector multiplication like:

$$\overrightarrow{old\ dog} = f_{old}(\overrightarrow{dog}) = \mathbf{OLD} \times \overrightarrow{dog} \qquad (4)$$

Training this model corresponds to finding proper tensors for words. Given the fact that approaches for word vectors are not applicable to second- and higher-order tensors, Baroni et al. (2013) and Grefenstette, Dinu, Zhang, Sadrzadeh, and Baroni (2013) employed linear regression with co-occurrence counts as training signals. First, co-occurrence count is used to find vectors for nouns like *dog*, vectors for noun phrases *old X* such as *old dog, old cat*. From those vectors representing *old X* and *X*, linear regression is used to estimate a matrix for *old* (see figure 23.7). Higher-order tensors are estimated with the same mechanism but with a technique called multistep linear regression.

From a theoretical point of view, the function-application–based approach is a beautiful model, which could be considered as inheriting the philosophy of formal semantics. However, learning tensors for lexical items is its most serious weakness. This is because the model requires a very large number of parameters (a single transitive verb needs n^3, i.e., $1M$ if $n = 100$,

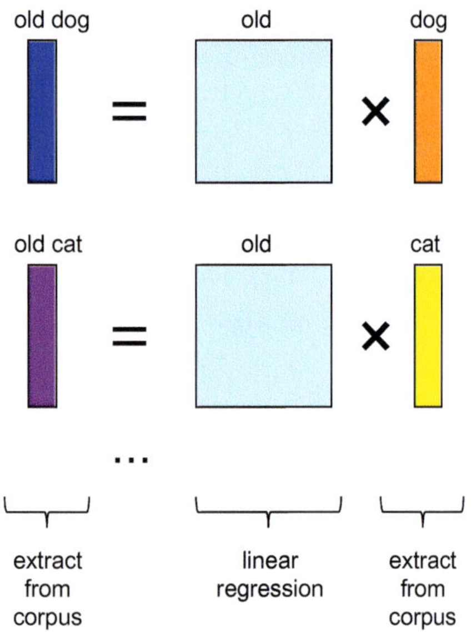

FIGURE 23.7 Learning a matrix for an adjective using linear regression.

parameters). Besides, we need to collect context distributions of phrases like *old dog* as in the preceding example. This becomes problematic when phrases are rare, and phrases tend to get very rare rapidly if they are composed of more than two words.

The most successful application of this idea, according to Baroni, is to composition in morphology (Marelli & Baroni, 2015). Marelli and Baroni found, for instance, that the composition of representations for *column* and *–ist* predicts that it's similar to *journalist* and thus correctly picks out the relevant sense (as *column* is similar to *pillar*).

2.1.3. Approach 3: Recursive neural networks Because two-layer feedforward neural networks are universal approximators (Cybenko, 1989), they can theoretically approximate *true* composition functions (if they exist and are measurable). Therefore, it is natural to think of using a feedforward neural net as a composition function. This leads to a model called Recursive neural network (RxNN). This idea was first proposed by Pollack (1990) and Goller and Küchler (1996). It then became popular in NLP thanks to the work series of Socher and colleagues (Socher, Manning, & Ng, 2010; Socher, Pennington, Huang, Ng, & Manning, 2011; Socher, Huval, Manning, & Ng, 2012).

A composition function for this model is a network consisting of two layers (figure 23.8A): an input layer with $2n$ nodes for pairs of two children and an output layer with n nodes for parents. Technically speaking, given a production $p \rightarrow x\,y$, the computation is given by

$$\mathbf{p} = a(\mathbf{W}_1\mathbf{x} + \mathbf{W}_2\mathbf{y} + \mathbf{b}) \qquad (5)$$

where $\mathbf{x}, \mathbf{y}, \mathbf{p} \in R^n$ are vectors representing x, y, p; $\mathbf{W}_1, \mathbf{W}_2 \in R^{n \times n}$, $\mathbf{b} \in R^n$ are weight matrices; and bias vector a is an activation function (e.g., *tanh*). The RxNN model differs from approaches 1 and 2 in that it is nonlinear (if the activation function a in equation 5 is nonlinear, e.g., *sigmoid, tanh*).

Although it is not clear whether the true composition functions are linear or nonlinear, the nonlinearity obviously makes the RxNN model more expressive and a wide range of choices for a make it more flexible. Although the RxNN model is simple, its elegance in terms of modeling and training makes it popular in the NLP community. A bundle of its extensions have been proposed, mainly along two directions: (i) proposing more expressive composition functions (e.g., Le & Zuidema, 2015a; Socher et al., 2012; Socher et al., 2013; Tai, Socher, & Manning, 2015) and (ii) extending the topology (Irsoy & Cardie, 2013; Le & Zuidema, 2014, 2015b).

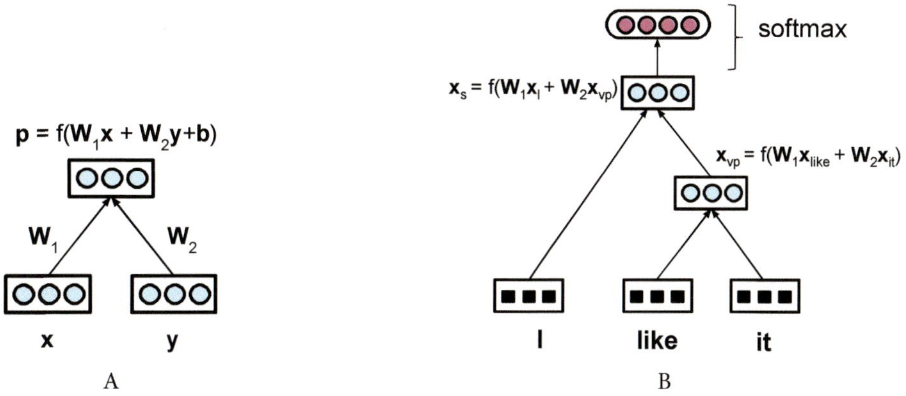

FIGURE 23.8 (A) A feedforward neural network as a composition function. (B) RxNN for sentence classification.

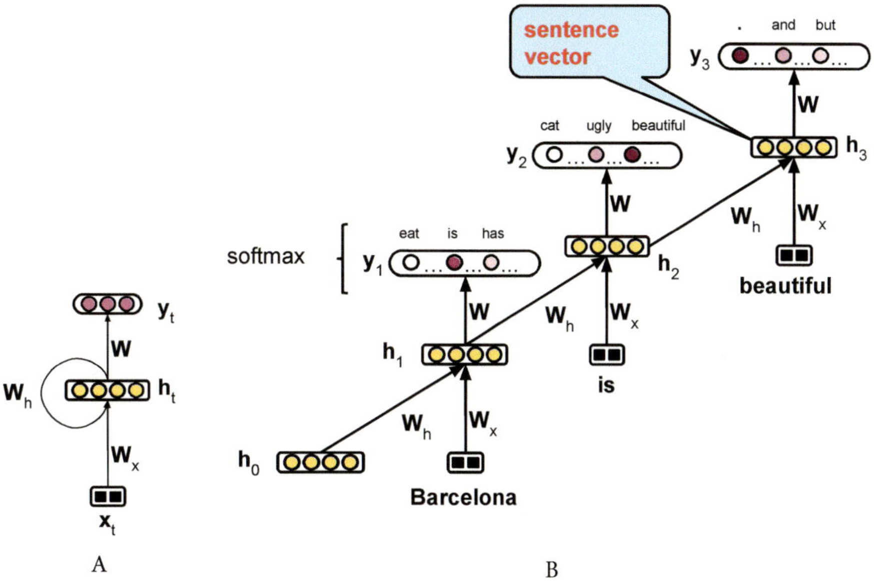

FIGURE 23.9 SRN (A) and the SRN when we unfold it (B). Notice that $\mathbf{h}_0$ can be simply $\vec{0}$.

2.2. Sequential Models of Sentence Semantics

2.2.1. Approach 4: Recurrent neural networks A fourth approach tries to compute vectors for sentences without (explicitly) using syntactic structures as glue to combine subcomponents. Successful versions of this approach are recurrent network and convolutional networks (which we will not discuss here, but see Collobert et al., 2011; Kalchbrenner, Grefenstette, & Blunsom, 2014; Kim, 2014).

A neural network is *recurrent* if it has at least one directed ring in its structure. In the simple recurrent neural network (SRN) model proposed by Elman (1990) (see figure 23.9A), an input $\mathbf{x}_t$ is fed to the network at each time t. The hidden layer $\mathbf{h}$, which has activation $\mathbf{h}_{t-1}$ right before $\mathbf{x}_t$ comes in, plays a role as a memory capturing the whole history $(\mathbf{x}_0, \ldots, \mathbf{x}_{t-1})$. When $\mathbf{x}t$ comes in, the hidden layer updates its activation by:

$$\mathbf{h}_t = g(\mathbf{W}_x\mathbf{x}_t + \mathbf{W}_h \mathrm{h}_{t-1} + \mathbf{b})$$

where $\mathrm{W}_x \in \mathbb{R}^{n\times dim(\mathrm{x}_t)}$, $\mathrm{W}_h \in \mathbb{R}^{n\times n}$, $\mathrm{b} \in \mathbb{R}^n$ are weight matrices and bias vector g is an activation function. This network model thus, in theory, can be used to estimate conditional probabilities with long histories.

Because the SRN can handle sequences with different lengths, we can use it to process sentences. For instance, figure 23.9B shows how we can use the SRN model for language modeling: the idea is to try to predict which word will come next. In this way, when we reach the end of a sentence, the hidden layer activation captures the whole sentence. In other words, we can use the activation of the hidden layer at that time point to represent the whole sentence. Figure 23.9B also shows that the SRN model is in fact a special case of the RxNN model: the input tree is always left-branching.

The network is efficiently trained using the backpropagation through time algorithm (Werbos, 1990). Although in theory recurrent neural networks are, at least, as expressiveness as Turing machines (Hyötyniemi, 1996), training them is difficult because of the vanishing gradient problem and the problem of how to capture long-range dependencies (Hochreiter, Bengio, Frasconi, & Schmidhuber, 2001). Two alternatives, which have become popular recently, are the *Long short-term memory architecture* (Hochreiter & Schmidhuber, 1997) and *gated recurrent neural networks* (Chung, Gulcehre, Cho, & Bengio, 2014).

3. Conclusion

How can we represent the meaning of words and sentences in a computational model? In this chapter we have discussed foundational work in distributional semantics, where word meanings are represented with (high-dimensional) numerical vectors, and recent extensions of this approach that generalize the vector representations to also apply to sentences. We have briefly contrasted these approaches to alternative modeling traditions. We have shown that vectorial models are surprisingly versatile and account for a wide variety of semantic phenomena often claimed to necessitate symbolic formalisms, while offering the additional benefits of being much more compatible with both neurobiological modeling and existing machine learning frameworks.

NOTES

1. See, for instance, http://esslli2016.unibz.it/?page_id=242.
2. Tensors describe linear relations among vectors, scalars, and other tensors. An n-dim vector is a first-order n-shape tensor, and an $n \times m$ matrix is a second-order $n \times m$-shape tensor. A third-order $(n \times m) \times k$–shape tensor **T** describes a linear map from a k-dim vector **v** to an $n \times m$ matrix **M**

$$\mathbf{M} = \mathbf{T} \times \mathbf{v}$$

where

$$M_{m,n} = \sum_k T_{m,n,k^{v}k}$$

Generally, a $(I_1 \times \ldots \times I_m) \times (J_1 \times \ldots \times J_n)$–shape tensor **T** describes a linear map from a $J_1 \times \ldots \times J_n$-shape tensor **V** to an $I_1 \times \ldots \times I_m$-shape tensor **M** by

$$M_{i_1,\ldots,i_m} = \sum_{j_1,\ldots,j_n} T_{i_1,\ldots,i_m,j_1,\ldots,j_n} V_{j_1,\ldots,j_n}$$

REFERENCES

Baroni, M. (2016). Grounding distributional semantics in the visual world. *Language and Linguistics Compass, 10*(1), 3–13.

Baroni, M., Bernardi, R., & Zamparelli, R. (2013). Frege in space: A program for compositional distributional semantics. *LiLT (Linguistic Issues in Language Technology), 9*, 241–346.

Baroni, M., & Lenci, A. (2010). Distributional memory: A general framework for corpus-based semantics. *Computational Linguistics, 36*(4), 673–721.

Bengio, Y., Ducharme, R., Vincent, P., & Janvin, C. (2003). A neural probabilistic language model. *Journal of Machine Learning Research, 3*, 1137–1155.

Blacoe, W., & Lapata, M. (2012). A comparison of vector-based representations for semantic composition. In J. Tsujii, J. Henderson, & M. Paşca (Eds.), *Proceedings of the 2012 Joint Conference on Empirical Methods in Natural Language Processing and Computational Natural Language Learning* (pp. 546–556). Stroudsburg, PA: Association for Computational Linguistics.

Chung, J., Gulcehre, C., Cho, K., & Bengio, Y. (2014). Empirical evaluation of gated recurrent neural networks on sequence modeling. *ArXiv Preprint*. Advanced online publication. arXiv:1412.3555.

Collobert, R., Weston, J., Bottou, L., Karlen, M, Kavukcuoglu, K., & Kuksa, P. (2011). Natural language processing (almost) from scratch. *Journal of Machine Learning Research, 12*, 2493–2537.

Cotterell, R., Schütze, H., & Eisner, J. (2016). Morphological smoothing and extrapolation of word embeddings. In K. Erk & N. A. Smith (Eds.), *Proceedings of the 54th Annual Meeting of the Association for Computational Linguistics (Volume 1: Long Papers)* (pp. 1651–1660). Stroudsburg, PA: Association for Computational Linguistics.

Cybenko, G. (1989). Approximation by superpositions of a sigmoidal function. *Mathematics of Control, Signals and Systems, 2*(4), 303–314.

Dinu, G., & Lapata, M. (2010). Measuring distributional similarity in context. In H. Li & L. Màrquez (Eds.), *Proceedings of the 2010 Conference on Empirical Methods in Natural Language Processing* (pp. 1162–1172). Stroudsburg, PA: Association for Computational Linguistics.

Elman, J. L. (1990). Finding structure in time. *Cognitive Science, 14*(2), 179–211.

Erk, K. (2012). Vector space models of word meaning and phrase meaning: A survey. *Language and Linguistics Compass, 6*(10), 635–653.

Firth, J. R. (1957). A synopsis of linguistic theory, 1930–1955. In *Studies in Linguistic Analysis (special volume of the Philological Society), Vol. 1952–59* (pp. 1–32). Oxford: The Philological Society.

Frermann, L., & Lapata, M. (2016). A Bayesian model of diachronic meaning change. *Transactions of the Association of Computational Linguistics, 4*, 31–45.

Goller, C., & Küchler, A. (1996). Learning task-dependent distributed representations by backpropagation through structure. In *IEEE International Conference on Neural Networks* (pp. 347–352). New York: IEEE Press.

Grefenstette, E., Dinu, G., Zhang, Y.-Z., Sadrzadeh, M., & Baroni, M. (2013). Multi-step regression learning for compositional distributional semantics. In K. Erk & A. Koller (Eds.), *Proceedings of the 10th International Conference on*

Computational Semantics (IWCS). Stroudsburg, PA: Association for Computational Linguistics.

Hamilton, W. L., Leskovec, J., & Jurafsky, D. (2016). Diachronic word embeddings reveal statistical laws of semantic change. In K. Erk & N. A. Smith (Eds.), *Proceedings of the 54th Annual Meeting of the Association for Computational Linguistics (Volume 1: Long Papers)* (pp. 1489–1501). Stroudsburg, PA: Association for Computational Linguistics.

Henderson, J., & Popa, D. N. (2016). A vector space for distributional semantics for entailment. *ArXiv Preprint*. Advanced online publication: arXiv:1607.03780.

Hochreiter, S., Bengio, Y., Frasconi, P., & Schmidhuber, J. (2001). Gradient flow in recurrent nets: The difficulty of learning long-term dependencies. In J. F. Kolen & S. C. Kolen (Eds.), *A field guide to dynamical recurrent neural networks* (pp. 237–244). New York: IEEE Press.

Hochreiter, S., & Schmidhuber, J. (1997). Long short-term memory. *Neural Computation, 9*(8), 1735–1780.

Hyötyniemi, H. (1996). Turing machines are recurrent neural networks. In J. Alander, T. Honkela, and M. Jakobsson (Eds.), *Proceedings of STeP'96, Finnish Artificial Intelligence Conference*. Vaasa, Finland: Finnish Artificial Intelligence Society.

Irsoy, O., & Cardie, C. (2013). Bidirectional recursive neural networks for token-level labeling with structure. *ArXiv Preprint*. Advanced online publication. arXiv:1312.0493.

Kalchbrenner, N., Grefenstette, E., & Blunsom, P. (2014). A convolutional neural network for modelling sentences. In K. Toutanova & H. Wu (Eds.), *Proceedings of the 52nd Annual Meeting of the Association for Computational Linguistics (Volume 1: Long Papers)* (pp. 655–665). Stroudsburg, PA: Association for Computational Linguistics.

Kim, Y. (2014). Convolutional neural networks for sentence classification. In A. Moschitti, B. Pang, & W. Daelemans (Eds.), *Proceedings of the 2014 Conference on Empirical Methods in Natural Language Processing (EMNLP)* (pp. 1746–1751). Stroudsburg, PA: Association for Computational Linguistics.

Landauer, T. K., & Dumais, S. T. (1997). A solution to Plato's problem: The latent semantic analysis theory of acquisition, induction, and representation of knowledge. *Psychological Review, 104*(2), 211.

Le, P., & Zuidema, W. (2014). Inside-outside semantics: A framework for neural models of semantic composition. In *Proceedings of the NIPS 2014 Workshop on Deep Learning and Representation Learning*. Montreal, Canada.

Le, P. & Zuidema, W. (2015a). Compositional distributional semantics with long short term memory. In C. Gardent, R. Bernardi, & I. Titov (Eds.), *Proceedings of the Joint Conference on Lexical and Computational Semantics (*SEM)*. Stroudsburg, PA: Association for Computational Linguistics.

Le, P., & Zuidema, W. (2015b). The forest convolutional network: Compositional distributional semantics with a neural chart and without binarization. In L. Màrquez, C. Callison-Burch, & J. Su (Eds.), *Proceedings of the 2015 Conference on Empirical Methods in Natural Language Processing* (pp. 1155–1164). Stroudsburg, PA: Association for Computational Linguistics.

Levy, O., & Goldberg, Y. (2014). Neural word embedding as implicit matrix factorization. *Advances in Neural Information Processing Systems, 27*, 2177–2185.

Liu, A. (2016). Word to Vec JS demo. Retrieved from http://turbomaze.github.io/word2vecjson/.

Lowe, W. (2001). Towards a theory of semantic space. In J. D. Moore & K. Stenning (Eds.), *Proceedings of the 23rd Annual Meeting of the Cognitive Science Society*. (pp. 576–581). Mahwah, NJ: Lawrence Erlbaum Associates.

Marelli, M., & Baroni, M. (2015). Affixation in semantic space: Modeling morpheme meanings with compositional distributional semantics. *Psychological Review, 122*(3), 485–515.

Mikolov, T., Chen, K., Corrado, G., & Dean, J. (2013). Efficient estimation of word representations in vector space. *ArXiv Preprint*. Advanced online publication. arXiv:1301.3781.

Mikolov, T., Karafiát, M., Burget, L., Černocký, J, & Khudanpur, S. (2010). Recurrent neural network based language model. In *Proceedings of the 11th Annual Conference of the International Speech Communication Association (INTERSPEECH 2010)* (pp. 1045–1048). Curran Associates, Inc.: Red Hook, NY.

Mikolov, T., Sutskever, I., Chen, K., Corrado, G. S., & Dean, J. (2013). Distributed representations of words and phrases and their compositionality. *Advances in Neural Information Processing Systems, 26*, 3111–3119.

Mikolov, T., Yih, W., & Zweig, G. (2013). Linguistic regularities in continuous space word representations. In L. Vanderwende, H. Daumé III, & K. Kirchhoff (Eds.), *Proceedings of the 2013 Conference of the North American Chapter of the Association for Computational Linguistics: Human Language Technologies* (pp. 746–751). Stroudsburg, PA: Association for Computational Linguistics.

Mitchell, J., & Lapata, M. (2008). Vector-based models of semantic composition. In *Proceedings of 46th Annual Meeting of the Association for Computational Linguistics: Human Language Technologies* (pp. 236–244). Stroudsburg, PA: Association for Computational Linguistics.

Mitchell, T. M., Shinkareva, S. V., Carlson, A., Chang, K.-M., Malave, V. L., Mason, R. A., & Just, M. A. (2008). Predicting human brain activity associated with the meanings of nouns. *Science, 320*(5880), 1191–1195.

Murphy, B., Talukdar, P., & Mitchell, T. (2012). Selecting corpus-semantic models for neurolinguistic decoding. In E. Agirre, J. Bos, M. Iab, S. Manandhar, Y. Marton, & D. Yuret (Eds.),**SEM 2012: The First Joint Conference on Lexical and Computational Semantics—Volume 1: Proceedings of the main conference and the shared task, and Volume 2: Proceedings of the Sixth International Workshop on Semantic Evaluation (SemEval 2012)* (pp. 114–123). Stroudsburg. PA: Association for Computational Linguistics.

Partee, B. (1995). Lexical semantics and compositionality. In L. R. Gleitman & M. Liberman (Eds.), *An invitation to cognitive science. Vol. 1: Language* (pp. 311–360). Cambridge, MA: MIT Press.

Pennington, J., Socher, R., & Manning, C. (2014). GloVe: Global vectors for word representation. In A. Moschitti, B. Pang, & W. Daelemans (Eds.), *Proceedings of the 2014 Conference on Empirical Methods in Natural Language Processing (EMNLP)* (pp. 1532–1543). Stroudsburg, PA: Association for Computational Linguistics.

Pereira, F., Gershman, S., Ritter, S., & Botvinick, M. (2016). A comparative evaluation of off-the-shelf distributed semantic representations for modelling behavioural data. *Cognitive Neuropsychology, 33*(3–4), 175–190.

Pollack, J. B. (1990). Recursive distributed representations. *Artificial Intelligence, 46*(1), 77–105.

Socher, R., Huval, B., Manning, C. D., & Ng, A. Y. (2012). Semantic compositionality through recursive matrix-vector spaces. In J. Tsujii, J. Henderson & M. Pasca (Eds.), *Proceedings of the 2012 Joint Conference on Empirical Methods in Natural Language Processing and Computational Natural Language Learning* (pp. 1201–1211). Stroudsburg, PA: Association for Computational Linguistics.

Socher, R., Manning, C. D., & Ng, A. Y. (2010). Learning continuous phrase representations and syntactic parsing with recursive neural networks. In *Proceedings of the NIPS 2014 Workshop on Deep Learning and Representation Learning*. Montreal, Canada.

Socher, R., Pennington, J., Huang, E. H., Ng, A. Y., & Manning, C. D. (2011). Semi-supervised recursive autoencoders for predicting sentiment distributions. In R. Barzilay & M. Johnson (Eds.), *Proceedings of the Conference on Empirical Methods in Natural Language Processing* (pp. 151–161). Stroudsburg, PA: Association for Computational Linguistics.

Socher, R., Perelygin, A., Wu, J. Y., Chuang, J., Manning, C. D., Ng, A. Y., & Potts, C. (2013). Recursive deep models for semantic compositionality over a sentiment treebank. In D. Yarowsky, T. Baldwin, A. Korhonen, K. Livescu, & S. Bethard (Eds.), *Proceedings of the 2013 Conference on Empirical Methods in Natural Language Processing (EMNLP)* (pp. 1631–1642). Stroudsburg, PA: Association for Computational Linguistics.

Tai, K. S., Socher, R., & Manning. C. D. (2015). Improved semantic representations from tree-structured long short-term memory networks. In C. Zong & M. Strube (Eds.), *Proceedings of the 53rd Annual Meeting of the Association for Computational Linguistics and the 7th International Joint Conference on Natural Language Processing (Volume 1: Long Papers)* (pp. 1556–1566). Stroudsburg, PA: Association for Computational Linguistics.

Werbos, P. (1990). Backpropagation through time: What it does and how to do it. *Proceedings of the IEEE, 78*(10), 1550–1560.

24 Language Processing in the Brain: Mapping Neural Activity to Language Meaning

LEILA WEHBE, ALONA FYSHE, AND TOM M. MITCHELL

The question of meaning has long interested philosophers, linguists, and scientists. In the 19th century, scientists began studying the brain's representation of language and meaning with case studies of patients with lesions. This led to the classical model of the neurobiology of language (Hagoort, 2013), in which Broca's area in the left inferior frontal gyrus is assumed to be the main seat of language production and the Wernicke area in the left temporal cortex is thought to be responsible for language comprehension. Analysis of brain lesions was limited to the reported cases of language deficit and required researchers to wait for postmortem analyses. The advent of noninvasive brain imaging techniques in the 20th century has since allowed language researchers to formulate myriad hypotheses about the representation of language and design experiments that probe the brain to test these hypotheses. Consequently, the language system has proven to be much more complex than initially thought.

One very common tool to study language representation in the brain is functional MRI (fMRI). fMRI measures brain activity indirectly: it is sensitive to the change in the level of oxygenated hemoglobin, which is a delayed consequence of the firing of neurons in an area. fMRI has good spatial resolution (2–3 mm) and allows us to obtain fine-grained maps of the areas involved in language tasks. However, fMRI acquires images at a slow rate (1–2 s per volume) and it measures a sluggish effect of neural activity. Language processing occurs at a faster rate (we read or speak multiple words per second, with individual words requiring a few hundreds of milliseconds to be perceived and understood). To study these fast processes, researchers also use electrophysiological techniques that record neural activity more directly. Electroencephalography (EEG) measures the change of voltage on the scalp due to neural activity. It has high time resolution (one sample per millisecond), but suffers from low spatial resolution, largely because of the distortion of the signal due to the skull. Magnetoencephalography (MEG) records the change in magnetic field outside the head due to neural activity. It has the same time resolution as EEG and suffers from less spatial distortion, therefore achieving better spatial resolution, while still remaining much lower than fMRI (~15 mm). Here we describe the use of fMRI and MEG to model the representation of language and meaning in the brain.

1. Modeling Neural Activity

There are a few main approaches that researchers typically use to study language processing (Friederici, 2011), most of which consist of establishing contrasts between two conditions. Researchers typically search for regions of the brain where neural activity varies across the conditions or time windows over which it varies.

- One type of contrast is between correct and erroneous language stimuli. This approach consists of breaking the normal processing of a sentence, for example by introducing a semantic incongruence or a grammatical mistake, and identifying resulting changes in activity. For example, the N400 event-related potential is a well-studied response to semantically surprising words (Kutas and Hillyard, 1980; Kutas & Federmeier, 2011), characterized by an increased negativity in the EEG/MEG signal around 400 ms after the onset of the incongruous word.
- A second type of contrast is achieved by varying the presence of information (Friederici; Fedorenko, Nieto-Castanon, & Kanwisher, 2012). For example, to study semantic representation, researchers compare neural activity responding to word versus nonword stimuli (i.e., strings that resemble words but have no meaning). To study syntax processing, researchers can compare sentences to word lists.
- A third type of contrast consists of presenting stimuli with different levels of difficulty and finding

the time or location of consequent increase in brain activity. Researchers can present to the participant sentences with difficult or ambiguous syntax or sentences with difficult or ambiguous meaning and compare them to easier sentences (Constable et al., 2004).

- Another type of contrast that is used to study language is to compare word lists that differ along the dimension of interest. For example, researchers have compared nouns to verbs (Li, Jin, & Hai Tan, 2004), abstract to concrete words (Wang, Conder, Blitzer, & Shinkareva, 2010), and animate to inanimate words (Grossman et al., 2002).

These different types of binary contrasts have been used extensively to study language in the brain. Usually, contrast-based analysis involves a two-sided statistical test to identify regions or time windows with statistically significant differences in activity. More recently, multivoxel pattern analysis (MVPA) (Norman, Polyn, Detre, & Haxby, 2006) has been used to detect differences in patterns of activity of brain regions, by finding the regions that can be used by a classifier to accurately classify between two conditions. While MVPA can detect regions with subtle differences of activity for multiple conditions, it cannot generalize its knowledge to classes that are not seen during training.

To the researcher interested in studying how the brain represents an infinite diversity of meanings, these binary contrast approaches have important limitations. It can be argued that methods that break the natural processing of language, and measure the brain's correcting mechanisms, do not reflect the processing of language by the brain in its natural state. It can also be argued that methods that measure the presence or absence of semantic processing (or the increased level of processing due to task difficulty) might detect *where* meaning is processed, but cannot find how different meanings are stored and represented (e.g., different regions might be involved in representing the roundness, manipulability, color and taste associated with the word *apple*). Designing specific contrasts to test each one of those dimensions of meaning (manipulable vs. not manipulable, edible vs. nonedible) is a possibility; however, the space of such binary divisions is very large and therefore such a task is arduous and extremely costly.

2. Intermediate Feature Space Modeling

In this chapter, we advocate a data analysis method for brain imaging experiments, which we call Intermediate Feature Space (IFS) modeling. Unlike earlier analysis methods, IFS modeling yields a model capable of making experimentally testable predictions for stimuli beyond those encountered during the model building phase. The essence of this approach is to first represent each experimental stimulus as a vector (point) in some predefined meaning space and then to use machine learning methods to train models that predict the neural activity from stimuli represented by points in the meaning space. Alternatively, the model can be trained to predict the stimulus point in meaning space from the neural activity. Instead of one dichotomy of interest (e.g., small vs. big objects), this vector encoding of meaning can capture many dimensions of meaning simultaneously. By testing the ability of these trained models to successfully cover test data held out from training, we can measure the statistical significance of the learned mappings and of corresponding hypotheses about where and when neural activity encodes different stimulus properties.

There are several advantages of the IFS method when it is compared to earlier approaches. First, compared to early univariate analysis methods that search for individual voxels whose activity is significantly different across conditions, IFS modeling is a multivariate approach that (like MVPA) can discover significant patterns in the joint activity across multiple components of neural activity. Second, IFS modeling goes beyond the abilities of MVPA because it naturally extrapolates beyond the stimuli on which it was trained to yield a model capable of making experimentally testable predictions for additional stimuli. For example, after training on neural activity collected while subjects read a specific set of word stimuli, the IFS approach yields a trained model that makes predictions of the neural activity that will be generated by arbitrary additional words not included in the experiment. In this sense, the IFS approach yields a model whose generality is much closer to the generality of the hypotheses sought by cognitive neuroscientists.

Given a set of stimuli and observed neural activity for those stimuli, the IFS approach involves the following steps:

1. Choose a vector representation for the stimulus set used in the experiment. That is, define the dimensions of the meaning space.
2. Choose a spatiotemporal region of neural activity to analyze. Represent this activity as a vector with each space-time point contributing one element to the vector.
3. Apply multivariate regression, or other machine learning algorithms, to learn a mapping between the stimulus vector (from step 1) and the neural activity vector (from step 2). The mapping can

be learned in both directions. The decoding direction refers to training a model to predict the stimulus vector (from step 1) from the neural activity vector (from step 2). The encoding direction refers to training a model to predict the neural activity vector (from step 2) from the stimulus vector (from step 1).

4. Test this trained model by applying it to held out test data, to evaluate both whether the model captures a statistically significant relationship between the neural activity and one or more components of the stimulus vector and the accuracy (effect size).

Figure 24.1 illustrates how IFS modeling operates in a meaning space. Consider the case where the experimental stimuli are individual nouns (e.g., *cat, desk*), as in Mitchell et al. (2008). In figure 24.1, words are points in a semantic space composed of two semantic features (dimensions): animacy and size. We see that similar words cluster together in this semantic space, and as we add additional dimensions (e.g., edibility), the word clusters can become more meaningful.

Given a predefined vector encoding of the stimuli (as in figure 24.1) and a brain image for each stimulus, IFS modeling can be applied using multivariate regression. Under this formulation, a multivariate decoding model is a set of weights that, when given a new brain image, can predict the value for one dimension of the meaning space. We train a model for every dimension of the meaning space, and the output of all models is combined to produce a final predicted point (white points in figure 24.1). IFS modeling can also be formulated to take a word vector as input and produce a predicted fMRI image as output, as in de Heer, Huth, Griffiths, Gallant, and Theunissen (2017); Huth, de Heer, Griffiths, Theunissen, and Gallant (2016); Mitchell et al. (2008); and Wehbe, Murphy, et al. (2014). This direction is also referred to as an encoding model (Naselaris, Kay, Nishimoto, & Gallant, 2011). Encoding models have also been used for studying visual perception in fMRI (Nishimoto et al., 2011).

To test that this model is sound, we train the model using only a subset of the available data (black points in figure 24.1). We then test the model by its performance on the remaining data (gray points in figure 24.1), which were held out during training. One common approach to testing (which we refer to as the *2 versus 2 test*) is to present the model with two held-out brain images, and their two held-out stimuli, but to omit the information about which of the two stimuli gave rise to which of the two brain images. If the trained model can assign the correct stimuli to the correct brain images, one has evidence of the correctness of the model on data beyond the training set. In greater detail, the

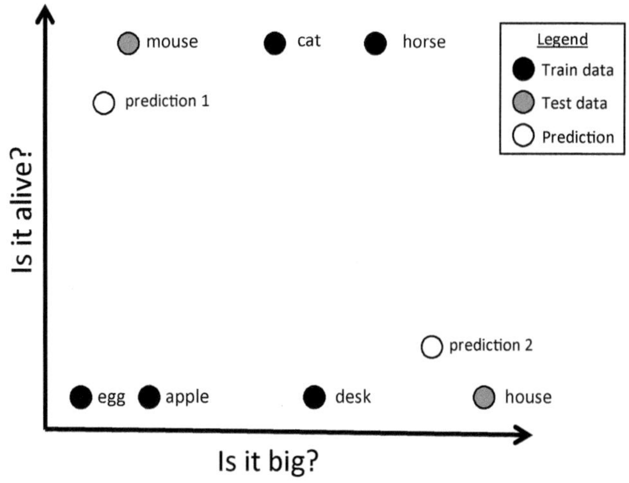

FIGURE 24.1 An example semantic space with only two dimensions: one corresponding to animacy and one to size. Words are represented as points (vectors) in the continuous space. In this example, each word's vector has two dimensions. For the 2 versus 2 test, we would train a model on brain images and word vectors for the black points and test it by providing brain images for the gray points. The trained model produces as output word vectors corresponding to the white points. We would assign prediction 1 to the word *mouse* and prediction 2 to the word *house*, as those are the closest observed test words to the predicted points.

trained model first predicts the word-meaning vector for each of the two brain images (e.g., prediction 1 and prediction 2 in figure 24.1). It then assigns the candidate words (*mouse* and *house* in this example) to predictions 1 and 2 based on which vectors are closest in the meaning vector space (in this case, prediction 1 is assigned the word *mouse* and prediction 2 is assigned *house*). By iteratively training on different subsets of the available data, a large set of such tests can be made and statistical significance can be evaluated.

Note that whereas univariate analysis and MVPA approaches typically require experimental designs using many repetitions of each stimulus or each condition, this IFS method places no such demands. Univariate methods require multiple repetitions of conditions in order to obtain statistically significant estimates of whether neural activity is modulated by the condition. Similarly, MVPA approaches require multiple repetitions of each class label (e.g., each word) to ensure a sufficiently large set of training examples for each class label. However, in IFS modeling, the parameters learned during training are not predicting individual words—they are instead predicting individual dimensions in the vector space that encodes the stimuli. Therefore, the need for repetitions in stimuli is replaced in IFS modeling by the naturally occurring repetition of dimensions of meaning across different stimuli—because every word stimulus has a value for "size" dimension, there are as many training examples of *size* as there are stimulus presentations.

Note that if the trained model is able to decode correctly the stimulus word from neural activity in a particular region of cortex, or a particular window of time, then we can hypothesize that this portion of neural activity is, in fact, modulated by the stimulus. Furthermore, we can hypothesize that the neural activity is also sensitive to the dimensions of meaning we have built into our definition of the meaning vector space. The observation that the model can successfully decode these dimensions of meaning from neural activity does not in and of itself tell us what neural processes are at work in the brain or what exact role these neural representations are playing in those processes. Note also that the implication flows only in one direction. We cannot claim that the inability of the model to decode the stimulus word from neural activity at a particular time or cortical location is proof that the property is not represented there. A failure to decode could be due to many factors, including noise and the limitations of brain imaging technology, not necessarily to a property of the brain. In addition, the relationship we detect with our modeling task could actually be the neural representation of some correlated property. Even with these caveats, our modeling approach is useful for exploring information processing in the brain.[1]

The following two sections present in greater detail our approach of training models that capture the relationship between language stimuli and the neural activity that encodes them. Subsequent sections then describe the application of IFS modeling to a study of semantic composition in adjective-noun phrases and a study of story reading.

3. Word Representations

In section 2, we described IFS modeling, a flexible approach that uses vector representations of word meaning to model the brain's response to words. These vector representations were not originally developed specifically to model brain activation, but rather for the more general task of representing word meaning (semantics).

The computational study of semantics has adopted the continuous meaning space and built various semantic models by automatically analyzing the word usage across many thousands of documents (a corpus). The assumption is that word meaning can be inferred by the context of a given word, averaged over many examples of the word's usage. For example, we might see the word *ball* occur frequently with verbs like *kick, throw, catch*; with adjectives like *bouncy*; or with nouns like *save* and *goal*. Large collections of text, often gathered from the Internet, are used to compile statistics about word usage, which can then be used to create a model of word meaning. In Vector-Space Models of semantics, also called Distributional Semantics, each word is assigned a vector, and the elements of the vector correspond to corpus statistics collected about the word. These statistics can include word-document co-occurrence (e.g., the word *ball* was seen 10 times in document 400), word-word co-occurrence within some window (e.g., the word *ball* and the word *goal* appear together within a five-word window 20 times in the corpus), or word-word-dependency triples (e.g., the word *ball* was the subject of the verb *kick* 19 times in the corpus). Each one of these statistics could become an element in the word vector representation of *ball*.

Especially when word-dependency-word triples (e.g., *ball-noun subj-kick*) are used, the number of elements in each word's vector can become very large, and often very sparse (e.g., the word *lettuce* may never occur in a sentence with the word *ball*, nor even in the same documents as the word *ball*). For this reason, compression of the word vectors is often performed with techniques such as singular value decomposition. This compression of word statistics is the basis of latent semantic analysis (Landauer & Dumais, 1997), one of the seminal works in distributional semantics. Each one of the dimensions captures some aspect of word meaning, but they can no longer be assigned simple labels, like those in figure 24.1. However, as we will see in section 5, there are ways to construct vectors from corpus statistics so that they retain interpretable meanings.

4. Zero-Shot Learning

A vector representation of word meaning is particularly powerful when it comes to studying language processing in the brain. Once we have word vectors, we can create IFS models to predict the individual dimensions of a word vector, rather than the discrete word. An IFS model is able to predict a new point in meaning space (e.g., prediction 1 in figure 24.1) and then identify the word that is closest to the predicted point (e.g., the word mouse in figure 24.1). This ability to predict new words (classes) is called zero-shot learning (ZSL) (Palatucci & Carlson, 2008).

Recall the distinction we made between MVPA and IFS modeling. MVPA detects patterns that distinguish between classes of things. For example, MVPA can be trained to distinguish the neural activity in response to viewing a house versus viewing a car. If we were to gather new brain images for people viewing motorhomes (recreational vehicles), our MVPA-trained model would be forced to predict either a house or a car, when in fact neither is correct. Upon receiving the neural activity in response to a motorhome, ZSL will allow the model to predict a new point on the landscape of word meaning that is somewhere between a car and a house. We can look this new point up in a dictionary of word vectors and (hopefully) find that the newly predicted point corresponds to a new concept that is a blend of a house and a car.

With the aid of word vectors, we can study language representations in the brain by moving through a *latent* vector representation of meaning, as shown in figure 24.2. There is a strong assumption here; we are assuming that brain activity observed when someone reads a word is somehow correlated with the meaning spaces we produce using word usage patterns. We are assuming there is a relationship between closeness in our latent meaning space and closeness in brain-activation space. This need not be the case, and until quite recently, there was only limited evidence that this might be true.

5. Zero-Shot Learning for Words and Phrases

The first test of the IFS hypothesis used 60 concrete nouns and fMRI (Mitchell et al., 2008). Word vectors were constructed using the co-occurrence probability for each concrete noun (e.g., dog) and one of 25 verbs (e.g., run, eat). An IFS model was used to learn a mapping from these 25 verb statistics to the activation of one voxel of fMRI data. Mitchell et al. used a ZSL test of their models in the 2 versus 2 fashion we described in section 2; they held out two words (and fMRI images) from the model training process. At test time, the word vectors for the two held-out words were used to create predicted fMRI images. The similarity between these predicted fMRI images and the two observed fMRI images was computed, and the assignment of true to observed image that maximized the similarity was the predicted word mapping.

Using the 2 versus 2 test, Mitchell et al. tested their algorithm on nine subjects and found that they were able to differentiate pairs of words with accuracy much higher than chance. This shows that the brain's activation (for at least this set of concrete nouns) can indeed be modeled with a continuous meaning space built with word usage statistics.

Once we have a weight vector that transforms the values of 25 verb dimensions to a predicted value for an fMRI voxel, we can view those weights superimposed on the fMRI image. The weight at each voxel represents the contribution from each verb dimension to the activation in a particular brain area. This allows us to view the brain's activation through the lens of a verb. For example, in figure 24.3, we see that the verb *push* is associated with high activity in the postcentral gyrus, an area of the brain known for premotor planning.[2]

6. Phrases in the Brain

What happens when we ask the brain to combine words to build more complex meaning? This process, called semantic composition, is one of the more fundamental cognitive tasks required for language comprehension. It is a skill that children acquire with amazing speed, and that adults perform with little effort. Semantic

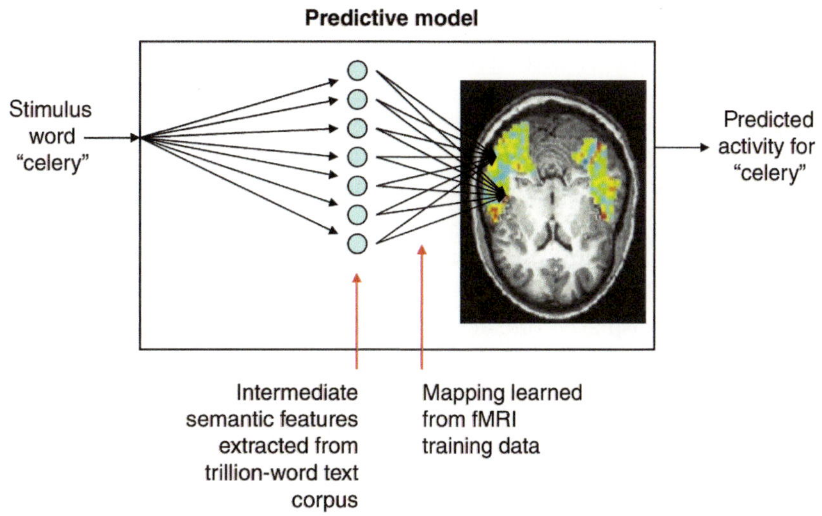

FIGURE 24.2 The ZSL approach. An intermediate (latent) representation for each word is created using word usage patterns in a large text corpus. From this intermediate representation, the model can predict the activation observed in the voxels of the fMRI image. From Mitchell et al. (2008).

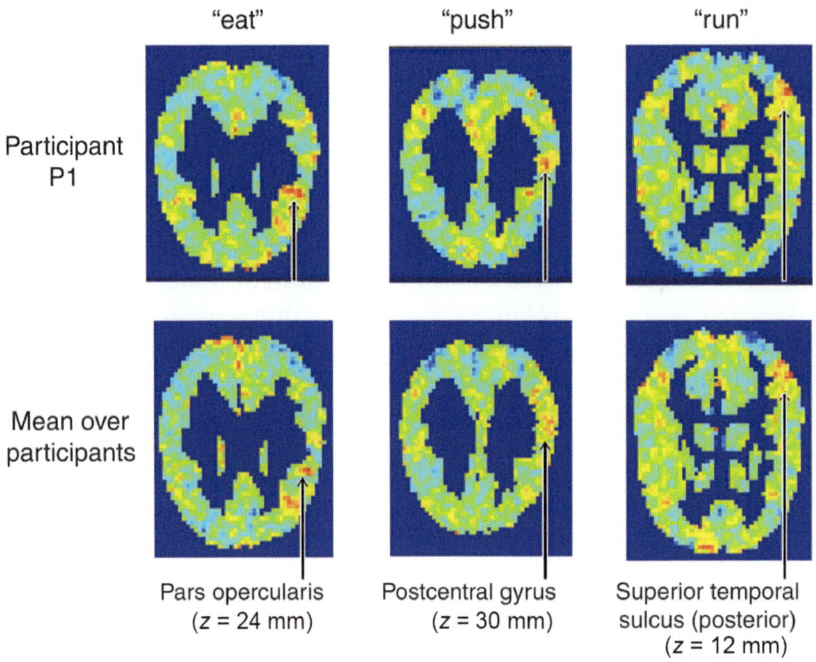

FIGURE 24.3 The predicted neural activity, according to the trained model, for several verb dimensions of the noun stimulus vectors. In this trained model, each of the verb features predicts high neural activation in an area of the brain previously shown to be related to the actions associated with that verb. For example, *eat* is associated with activity in right pars opercularis, part of the gustatory cortex; *push* is mapped to the postcentral gyrus known for premotor planning; and *run* is associated with posterior portion of the right superior temporal sulcus, an area associated with perceiving biological motion. From Mitchell et al. (2008).

composition in the brain has been studied previously in semantically anomalous sentences (Kuperberg, 2007; Kutas & Hillyard, 1980), as well as in simple phrases (Baron, 2012; Bemis & Pylkkänen, 2011). Still, relatively little is known about the neural processes governing the creation and representation of composed meaning.

To study the neural encoding of adjective, noun, and phrasal semantics, we visually presented adjective-noun phrases in isolation and recorded neural activity of nine human subjects using MEG (Fyshe, 2015). For this study, we chose six concrete nouns that were previously shown to be easily decodable (*carrot, tomato, bear, dog, shovel, hammer*) (Sudre et al., 2012). A set of eight adjectives were chosen to modulate the most decodable semantic qualities of the nouns (e.g., edibility, manipulability, size). In total, there are 30 adjective-noun phrases. Each word of the phrase is displayed one at a time for 500 ms, with 300-ms fixation between words of a phrase and 3 s total time between the onset of sequential phrases.

We will again use a vector representation of word meaning for the individual words of the phrase. Like the 25-dimensional vectors used for the 60 concrete nouns experiment, these vectors are built using word usage statistics, but here the dimensions correspond to frequently observed word-dependency-word triples (e.g., *dog* appearing as the subject of the verb *eat* would become *dog-nsubj-eat*) and are compressed using singular value decomposition. Because we are using MEG, we can chart the time course of decodability during the presentation of the adjective and noun, as seen in figure 24.4. Measures were taken to ensure that positive results were not due to the correlations between words of a phrase (e.g., to handle the fact that the word *rotten* is always followed by a food word). During the time the adjective is displayed (0–0.5 s) the model can decode the identity of the adjective using the 2 versus 2 test, and during the time the noun is visible (0.8–1.3 s) the model can decode the identity of the noun. However, the model can also decode the identity of the adjective while the noun is visible, and even as late as 2 s after the onset of the adjective stimuli (false discovery rate corrected for multiple comparisons). Thus, the model can detect that the brain is retaining the semantics of the adjective even while reading the noun, as well as during the probable window of composition, when the noun stimulus is no longer visible. Fyshe also attempted to detect the composed adjective-noun meaning, represented by human ratings of phrasal semantics, and there was some evidence of that composed representation around 2 s, but further investigation is needed.

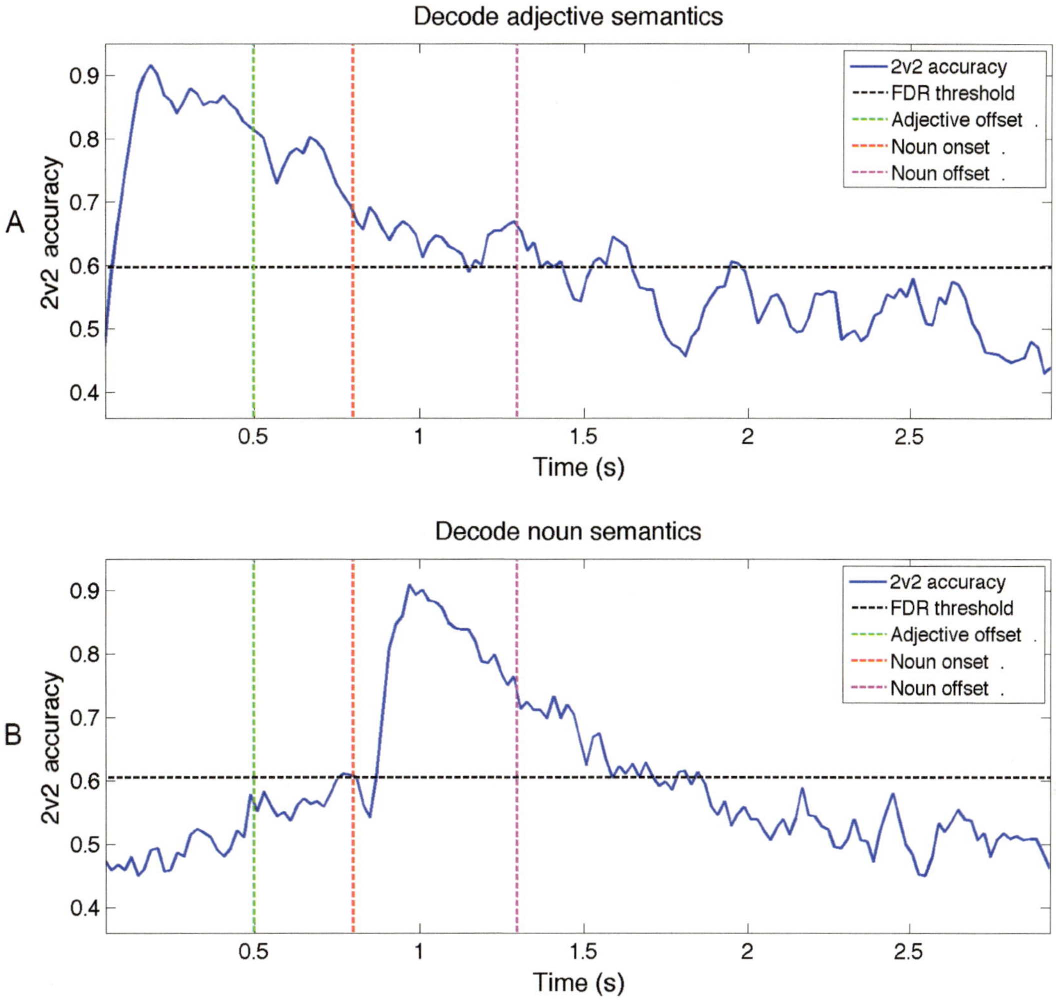

FIGURE 24.4 The 2 versus 2 (2v2) accuracy is shown as a function of time for the task of decoding the adjective or noun from MEG signal, based on its corpus-derived semantic vector. Time windows are 100-ms wide and overlap by 80 ms with adjacent windows. Time 0 is the onset of the adjective, green lines indicate the offset of the adjective, red lines the onset of the noun, and magenta lines the offset of the noun. (A) The 2 versus 2 accuracy is shown for decoding the semantics of the adjective, as a function of time. To avoid confounds, only 2 versus 2 pairs where the noun is shared are used to calculate accuracy. (B) The 2 versus 2 accuracy is shown for decoding the semantics of the noun, as a function of time. To avoid confounds, only 2 versus 2 pairs where the adjective is shared are used to calculate accuracy. FDR = false discovery rate.

7. *Natural Language Paradigms*

The IFS approach we are advocating for has important advantages. It offers us the opportunity to design experiments that not only are not restricted to a small number of categories, but that also use natural stimulus text such as sentences or even stories. This can be achieved due to multiple properties of this approach.

One of the main advantages of feature spaces is that it can be defined for multiple types of information representation and therefore can be used for studying the perception of real, complex text. Consider all the different processes involved in reading or listening to a story. First of all, there are processes involved in perceiving the words (involving different systems such as the visual or the auditory systems). There are other processes involved in representing the meanings of those words, combining them together, and constructing and representing the syntactic structure of the sentence to understand the conveyed message. Furthermore, a story has a narrative structure with scenes to imagine, events that one can predict, characters with point of views that one has to interpret, and so on.

Hypothetically, every one of the above-mentioned processes can be represented by a feature space. Every space would constitute an annotation of the events that occur at every time point, when considering only one type of processing. Instead of just focusing on the semantic properties of the words at a given time point, one could instead build a feature space that describes the part of speech of the words at that time point, the grammatical role of those words in their respective sentences, or their phonetical properties. One can then use these different feature spaces to predict the activity in the brain. In keeping with our previous assumptions, when we find that a specific feature space (e.g., the part of speech of a word) predicts the activity in a voxel with greater than chance performance, then we can say that the data support the hypothesis that this voxel is representing information related to that feature space (e.g., this voxel processes part of speech or related information).

Another advantage to the feature space method is that it frees the experimenter from having multiple repetitions for each stimulus. Brain-imaging methods are inherently noisy, and therefore when estimating brain responses to specific categories, one needs to repeat these categories. We have seen that the IFS approach allows us to jointly explore multiple dimensions along which the stimulus words differ. By presenting a large amount of text without repetitions, the researcher can maximize the ability of the model to detect the contribution of individual features to the brain signal, because these features will occur in different configurations with the other features. Instead of maximizing the occurrence of given stimuli, one instead maximizes the span of the stimulus space.

Additionally, the IFS approach allows the experimenter to have a data-driven approach that is not restricted by a previously defined hypothesis. This is not intended to replace hypothesis-driven experiments, but to complement them. The experimenter can gather a data set while subjects process complex text and then afterward use this data set to evaluate contrasting theories of brain representation (Murphy et al., 2011). For instance, to contrast two meaning composition hypotheses that place a different weight on the importance of previous words, one can construct the respective feature spaces and evaluate which predicts brain activity better.

Finally, this naturalistic experimental design allows the researcher to study the brain as it processes language in real life, without introducing artificial experimental tasks or breaking the normal process of language. We will take the example of a story reading experiment to illustrate these points: eight subjects read a chapter of *Harry Potter and the Sorcerer's Stone* in the fMRI scanner (Wehbe, Murphy, et al., 2014). The chapter was presented without modification, one word at a time, for 0.5 s each.

Multiple feature spaces were constructed to represent the different types of information processed by the brain (Wehbe, Murphy, et al., 2014). A non-negative and sparse semantic feature space was obtained by reducing the dimensionality of corpus concurrence counts (Murphy, Talukdar, & Mitchell, 2012). Another feature space was constructed to represent syntax: it includes the part of speech (e.g., noun, verb, adjective) and grammatical role (e.g., subject, object, noun modifier) of each word, as well as the numeric position of the word in the sentence (an approximate measure of sentence complexity at the point where the word is encountered). Narrative feature spaces were also built, indicating the appearance of the story characters, their emotions, their motions, and whether they were participating in directly quoted dialog.

A predictive model of brain activity was built using each of these feature spaces separately. On an unseen test data set, segments of brain activity were classified as corresponding to one of two possible story passages. This was done by first predicting the brain activity corresponding to the two passages using their feature representation and then choosing the passage with the smallest predicted activity to the real, unlabeled passage. This analysis is repeated for every feature space. The ability of a model derived with a specific feature space to predict the activity in a location in the brain suggests that the brain represents a related type of information in that location.

In order to increase classification accuracy so that it could be easier to distinguish chance performance from real performance (i.e., to improve power), the classification was done using 20 brain images at a time. More specifically, a segment of 40 s of brain activity was classified as corresponding to one of two 40-s text passages. This analysis was repeated for a $5 \times 5 \times 5$ voxel cube centered on each voxel location in the brain. For every feature space, the areas in which the average classification accuracy is higher than chance was found. In figure 24.5, these areas are colored according to the feature space that the data suggest is being represented by those areas. These areas span the areas referred to as language network (Fedorenko, Berh, & Kanwisher, 2011) that are usually responsive to the presence of meaningful language stimuli. This approach was able to suggest a decomposition for this area in terms of the different types of computations it performs. Additionally, this approach identifies narrative processing areas that seem to be the same as the regions that are used for social interaction: for example, characters seem to be represented in the right posterior superior temporal cortex, which is a center implicated in interpreting other people's intentions (Wehbe, Murphy, et al., 2014).

As we mentioned previously, MEG can be used to study the dynamics of language processing at the subword level. MEG recordings were collected when three

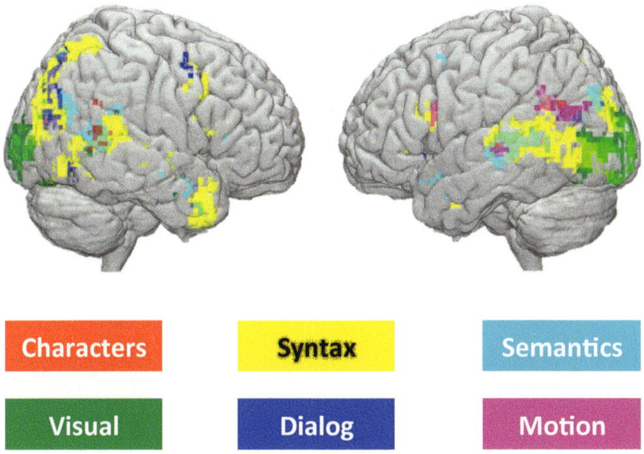

FIGURE 24.5 Brain map of the different reading subprocesses computed from combing data from multiple subjects. Each region is colored by the type of information it represents when subjects read a complex text. Details in Wehbe, Murphy, et al. (2014).

of the subjects underwent the same task: the same chapter was read, one word at a time, each for 0.5 s (Wehbe, Vaswani, Knight, & Mitchell, 2014). MEG creates the possibility of identifying the different stages of continuous meaning construction as subjects are reading each word. To study this, a recurrent neural network language model, by Mikolov, Kombrink, Deoras, Burget, and Cernocky (2011), was used as a source of feature spaces. The recurrent neural network language model predicts the probability of the incoming word as a function of the preceding word. To do so, it learns during training and embedding (vector representing the static properties) for each word in its dictionary. It simultaneously learns functions to combine the representations of successive words into a context vector representation. The context vector is recomputed at every word and represents information combined from the previous words. Consider now the similarity with the brain: the context vector can be used as a feature space for the representation of the previous words; the embedding vector can be used as a feature space for the properties of the current word; and the probability of each word given the previous words can be used as a measure of the integration process: more surprising words might incur more brain activity in regions responsible for integration.

The same type of analysis as with fMRI was applied to this data. The difference is that the modeling and classification was done not only by location, but also by time window (e.g., in the time window 0–100 ms after word onset). The results revealed that the context vector predicts brain activity before word onset (see figure 24.6). Once the word is on the screen, the context vector (computed before the current word) slowly becomes less useful for predicting brain activity, until its performance becomes indistinguishable from chance. This probably corresponds to the brain updating its representation of context. Furthermore, the results suggested a time line of how the brain updates its context representation. The results also depict the incremental perception of every new word starting early in the visual cortex, moving next to the temporal lobes and finally to the frontal regions. Lastly, the results suggested that the integration process occurs in the temporal lobes after the new word has been perceived.

8. Conclusion

In this chapter, we described IFS modeling, a method of exploring information flow in the brain. IFS requires the representation of stimuli in a vector space (e.g., of word meanings) and leverages the information encoded in the dimensions of that space to predict new and novel stimuli. We presented case studies that examined the neural representation of words, phrases, and whole stories. IFS modeling allows us to use a wide variety of stimuli and to create experimental paradigms that need not have multiple repetitions of stimuli in order to detect semantics in the brain.

The case studies we covered here use text corpora to build either all or some part of the IFS vectors. However, feature spaces can be built from any source that captures some information about the stimuli's properties. We saw this in section 7, where some dimensions of the vectors represented parts of speech. There are also examples where behavioral data have been used to define a feature space (Sudre et al., 2012) and even brain images themselves have been incorporated into the vector representations of words (Fyshe, Talukdar, Murphy, & Mitchell, 2014).

We focused exclusively on the application of IFS modeling to language experiments, but IFS modeling is flexible enough to be used in a variety of settings. For example, IFS modeling has been used to model the visual information in movies (Nishimoto et al., 2011) or their semantic content (Huth, Nishimoto, Vu, & Gallant, 2012). One could imagine using IFS modeling to study the neural processes involved in reading music or solving algebra problems, which would require different vector representations. For example, in music reading, we might want to include the identity and duration of the notes and the current beat of the bar. An algebraic formula could be encoded with digit identities and the functor involved (e.g., multiplication, addition). Even everyday activities such as baking cookies or driving to work could be encodable under an IFS modeling regime.

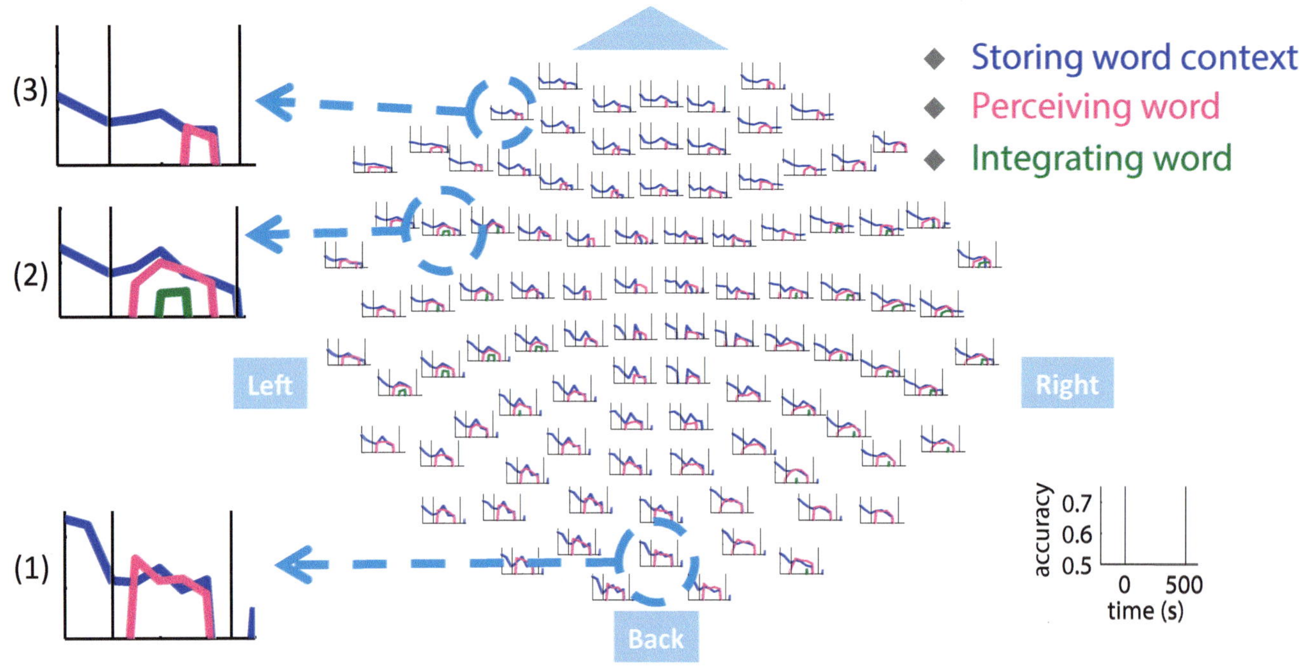

FIGURE 24.6 Adapted from Wehbe, Vaswani, et al. (2014). The time line of word integration across the MEG helmet is shown. For each of the 102 locations, the average accuracy at that location is plotted versus time. The axes are defined in the rightmost, empty plot. Accuracy should be seen as an indicator for the underlying process. Three plots have been magnified to show the increasing delay in processing the incoming word as information travels through the brain from the visual cortex (1) to the temporal (2) and frontal (3) cortices. Furthermore, the integration process is shown to occur in the left and right temporal cortices after the word perception begins. The word context is maintained before the word is shown and is gradually updated as the word is perceived.

IFS modeling is not without its limitations. It assumes that the neural activity related to a stimulus is time-locked to the onset of that stimuli. However, there are many examples of scenarios where that might not be the case. For example, in solving an algebraic formula, the mental processes for addition might not be active until the second number is viewed and then the onset of that activity might not be perfectly aligned to the onset of the second number. For example, if the numbers for different problems have different magnitudes (e.g., 5 + 5 vs. 5445 + 7835), it may take a variable amount of time for the first stage of addition to begin. IFS modeling also requires that we enumerate all of the features of the stimuli, which may be difficult or impossible. For example, in the story reading case study in section 7, we are certainly missing key components of the brain's story reading algorithm, as many of the higher-order aspects of the story (theme, story arc) are omitted from the feature space.

Still, IFS modeling is a powerful method for modeling brain activation. It allows us to represent stimuli in a very flexible way and to leverage that flexibility to predict for stimuli not encountered during the model building process. This allows for a more diverse set of stimuli and frees the experimenter from the burden of presenting stimuli many times. We hope that this chapter will inspire researchers to apply IFS modeling in new ways and ultimately lead to a better understanding of brain function.

NOTES

1. It is also important to specify that one should be careful when building a decoder that uses brain images as input: the weights that are given to different voxels or channels cannot be interpreted as a measure of the involvement of these voxels or channels in the representation of interest (Haufe et al., 2014). This is not the approach we are advocating for here.
2. We are able to perform this comparison because each model has been fit with a multivariate regression with the 25-dimensional feature vector as input. Had we been training the opposite model that predicts the feature value as a function of the voxels in the brain image, we would not be able to interpret the weights (Huth, Lee, et al., 2016).

REFERENCES

Baron, G. (2012). The neural basis of compositionality: Functional magnetic resonance imaging studies of conceptual combination (Unpublished doctoral dissertation). Princeton.

Bemis, D. K., & Pylkkänen, L. (2011). Simple composition: A magnetoencephalography investigation into the comprehension of minimal linguistic phrases. *Journal of Neuroscience, 31*(8), 2801–2814.

Constable, R. T., Pugh, K. R., Berroya, E., Mencl, W. E., Westerveld, M., Ni, W., & Shankweiler, D. (2004). Sentence complexity and input modality effects in sentence comprehension: An fMRI study. *Neuroimage, 22*(1), 11–21.

de Heer, W. A., Huth, A. G., Griffiths, T. L., Gallant, J. L., & Theunissen, E. (2017). The hierarchical cortical organization of human speech processing. *Journal of Neuroscience, 37*(27), 6539–6557.

Fedorenko, E., Behr, M. K., & Kanwisher, N. (2011). Functional specificity for high-level linguistic processing in the human brain. *Proceedings of the National Academy of Sciences, 108*(39), 16428–16433.

Fedorenko, E., Nieto-Castanon, A., & Kanwisher, N. (2012). Lexical and syntactic representations in the brain: An fMRI investigation with multi-voxel pattern analyses. *Neuropsychologia, 50*(4), 499–513.

Friederici, D. (2011). The brain basis of language processing: From structure to function. *Physiological Reviews, 91*(4), 1357–1392.

Fyshe, A. (2015). Corpora and cognition: The semantic composition of adjectives and nouns in the human brain (Unpublished doctoral dissertation), Carnegie Mellon University.

Fyshe, A., Talukdar, P. P., Murphy, B., & Mitchell, M. (2014). Interpretable semantic vectors from a joint model of brain- and text-based meaning. In K. Toutanova & H. Wu (Eds.), *Proceedings of the 52nd Annual Meeting of the Association for Computational Linguistics* (pp. 489–499). NIH Public Access. Bethesda, MD: National Institutes of Health.

Grossman, M., Koenig, P., DeVita, C., Glosser, G., Alsop, D., Detre, J., & Gee, J. (2002). The neural basis for category-specific knowledge: An fMRI study. *Neuroimage, 15*(4), 936–948.

Hagoort, P. (2013). MUC (memory, unification, control) and beyond. *Frontiers in Psychology, 4*, 416.

Haufe, S., Meinecke, F., Görgen, K., Dähne, S., Haynes, J.-D., Blankertz, B., & Bießmann, F. (2014). On the interpretation of weight vectors of linear models in multivariate neuroimaging. *NeuroImage, 87*, 96–110.

Huth, A. G., de Heer, W. A., Griffiths, T. L., Theunissen, F. E., & Gallant, L. (2016). Natural speech reveals the semantic maps that tile human cerebral cortex. *Nature, 532*(7600), 453–458.

Huth, A. G., Lee, T., Nishimoto, S., Bilenko, N. Y., Vu, A. T., & Gallant, J. L. (2016). Decoding the semantic content of natural movies from human brain activity. *Frontiers in Systems Neuroscience, 10*, 81.

Huth, A. G., Nishimoto, S., Vu, A. T., & Gallant, L. (2012). A continuous semantic space describes the representation of thousands of object and action categories across the human brain. *Neuron, 76*(6), 1210–1224.

Kuperberg, R. (2007). Neural mechanisms of language comprehension: Challenges to syntax. *Brain Research, 1146*, 23–49.

Kutas, M., & Federmeier, K. D. (2011). Thirty years and counting: Finding meaning in the N400 component of the event related brain potential (ERP). *Annual Review of Psychology, 62*, 621.

Kutas, M., & Hillyard, S. A. (1980). Reading senseless sentences: Brain potentials reflect semantic incongruity. *Science, 207*(4427), 203–205.

Landauer, T. K., & Dumais, T. (1997). A solution to Plato's problem: The latent semantic analysis theory of acquisition, induction, and representation of knowledge. *Psychological Review, 1*(2), 211–240.

Li, P., Jin, Z., & Hai Tan, L. (2004). Neural representations of nouns and verbs in Chinese: An fMRI study. *NeuroImage, 21*(4), 1533–1541.

Mikolov, T., Kombrink, S., Deoras, A., Burget, L., & Cernocky, J. (2011). RNNLM-recurrent neural network language modeling toolkit. In *Proceedings of the IEEE Workshop on Automatic Speech Recognition and Understanding* (pp. 196–201). New York: IEEE Press.

Mitchell, T. M., Shinkareva, S. V., Carlson, A., Chang, K.-M., Malave, V. L., Mason, R. A., & Just, M. A. (2008). Predicting human brain activity associated with the meanings of nouns. *Science, 320*(5880), 1191–1195.

Murphy, B., Poesio, M., Bovolo, F., Bruzzone, L., Dalponte, M., & Lakany, H. (2011). EEG decoding of semantic category reveals distributed representations for single concepts. *Brain and Language, 117*(1), 12–22.

Murphy, B., Talukdar, P. P., & Mitchell, M. (2012). Learning effective and interpretable semantic models using non-negative sparse embedding. In M. Kay & C. Boitet (Eds.), *Proceedings of the 24th International Conference on Computational Linguistics (COLING 2012)* (pp. 1933–1950). Mumbai, India: Indian Institute of Technology Bombay.

Naselaris, T., Kay, K. N., Nishimoto, S., & Gallant, L. (2011). Encoding and decoding in fMRI. *NeuroImage, 56*, 400–410.

Nishimoto, S., Vu, A. T., Naselaris, T., Benjamini, Y., Yu, B., & Gallant, J. L. (2011). Reconstructing visual experiences from brain activity evoked by natural movies. *Current Biology, 21*(19), 1641–1646.

Norman, K. A., Polyn, S. M., Detre, G. J., & Haxby, V. (2006). Beyond mind-reading: Multi-voxel pattern analysis of fMRI data. *Trends in Cognitive Sciences, 10*(9), 424–430.

Palatucci, M., & Carlson, A. (2008). On the chance accuracies of large collections of classifiers. In *Proceedings of the 25th International Conference on Machine Learning* (pp. 744–751). Boston.

Sudre, G., Pomerleau, D., Palatucci, M., Wehbe, L., Fyshe, A., Salmelin, R., & Mitchell, T. (2012). Tracking neural coding of perceptual and semantic features of concrete nouns. *NeuroImage, 62*(1), 463–451.

Wang, J., Conder, J. A., Blitzer, D. N., & Shinkareva, S. V. (2010). Neural representation of abstract and concrete concepts: A meta-analysis of neuroimaging studies. *Human Brain Mapping, 31*(10), 1459–1468.

Wehbe, L., Murphy, B., Talukdar, P., Fyshe, A., Ramdas, A., & Mitchell, T. (2014). Simultaneously uncovering the patterns of brain regions involved in different story reading subprocesses. *PLoS ONE, 9*(11), e112575.

Wehbe, L., Vaswani, A., Knight, K., & Mitchell, T. (2014). Aligning context-based statistical models of language with brain activity during reading. In A. Moschitti, B. Pang, & W. Daelemans (Eds.), *Proceedings of the 2014 Conference on Empirical Methods in Natural Language Processing* (pp. 233–243). Stroudsburg, PA: Association for Computational Linguistics.

25 The Robot Writer

EMIEL KRAHMER

1. What Are Robot Writers?

Robot writers are computer programs that can automatically produce coherent natural language text based on input data. The term is perhaps somewhat misleading, since, unlike, say, robot tutors (e.g., Kanda, Hirano, Eaton, & Ishiguro, 2004), robot writers are not *actual*, physically embodied robots. The robot terminology came into vogue with the emergence of applications that automatically generate news reports based on data, which are often referred to as *robot journalists* (e.g., Carlson, 2015; Clerwall, 2014). While the basic technology required for building robot journalists has been around for some time now, the quickly growing volumes of available data and the increase of digital publishing has paved the way for this new application, which allows for fast, large-scale, personalized, and on-demand information presentation.

Robot journalists typically rely on input in the form of numerical data, in well-defined domains such as sports, the weather, or finances. These applications have been studied extensively in the field of *Natural Language Generation (NLG)* (Reiter & Dale, 1997, 2000). NLG has been characterized as "the subfield of artificial intelligence and computational linguistics that is concerned with the construction of computer systems than can produce understandable texts in English or other human languages from some underlying non-linguistic representation of information" (Reiter & Dale, 1997, p. 1). The systems covered by this definition are also referred to as *data-to-text* systems (see, for example, Gatt et al., 2009; Goldberg, Driedger, & Kittredge, 1994). However, more recently, researchers have started exploring the possibilities of generating text based on different kinds of input, for example, building robot writers that can generate descriptions of photos or movies (*vision-to-text*; e.g., Bernardi et al., 2016; Mitchell et al., 2012) and those that take language as input (*text-to-text*; e.g., Mani, 2001; Nenkova & McKeown, 2011), in which case it would be more appropriate to talk of a robot *re*writers (with robot summarizers as a prime example). As a result of these recent developments, the NLG field has been changing considerably in recent years, in terms of methodology, application, and evaluation.

Importantly, for our current purposes, there is also another way to look at robot writers, namely as a *computational model of human language production*.[1] Human language production—going from "intention to articulation" (Levelt, 1989)—is a highly complex phenomenon, but our understanding of this process is slowly increasing (e.g., Goldrick, Ferreira, & Miozzo, 2014; Roelofs & Ferreira, chapter 4 of this volume). Interestingly, NLG researchers have long been inspired by psycholinguistic research on speech production (see, e.g., Reiter, 1994),[2] but in recent years, there has also been an increased interest in using robot writers as a research tool to better understand human language production, by developing cognitively plausible and computationally explicit models trying to model this process (see, e.g., van Deemter, 2016). After all, when we are capable of building a computationally explicit model of (some aspect of) human language behavior, this may greatly help our understanding of it (e.g., Lewandowsky & Farrell, 2010).

Trying to do this raises many interesting new questions: for example, while existing robot writers often tend to be deterministic (always producing the same output for the same input data), their human counterparts are generally *nondeterministic*: different speakers typically differ in how they would express a given piece of information in words, and even an individual speaker will presumably not always express the same information in exactly the same way. Modeling this kind of inter- and intraindividual variation may help us to better understand human language production and, at the same time, could lead to better (less predictable) robot writers. In a similar vein, human language production can be characterized by its *creativity*. Human speakers, for example, use language to create metaphors or puns; elements whose production is still poorly understood and sadly lacking from robot-written text.

The goal of the current chapter is to introduce robot writing for students and researchers interested in linguistics, both as an interesting application domain for theories of language production *and* as a research tool for studying human language production. We start in section 2 by briefly surveying earlier work on NLG, offering a how-to-guide of sorts for building robot writers. Next, in section 3, we zoom in on the robot writer as a research tool, discussing two case studies in the production of referring expressions: the choice of a referential form (when does a writer use a proper name, a pronoun, or a definite description to refer to something?) and conceptualization of descriptions (how does a writer decide on which properties to include in a description?). Finally, we ask what the future holds for robot writing, discussing recent forays in creative language generation, with a focus on affective, humorous, and poetic language use.

2. *How to Build a Robot Writer?*

If one wants to build a robot writer, there is a lot of earlier research one can build on, although there are not many ready-made solutions yet. For a robot writer in the data-to-text vein, typically six tasks are distinguished (Reiter & Dale, 2000) that roughly range from early planning tasks (deciding what to say) to later ones (deciding how to say it)—a distinction that is common both in various models of human speech production and in traditional approaches to NLG. Next, we briefly introduce these core tasks and discuss how they are integrated in architectures for robot writers. Two broad and related trends will emerge. First, as we shall see, the early tasks generally are highly domain- and application-dependent, whereas the later tasks are often more independent and generic. Second, the connection with human speech production is evident for all stages but more so for the later tasks. Interestingly, one could argue that robot writers put relatively more emphasis on early language production (what to say), and psycholinguistics more on late language production (how to say it), a theme to which we will return.

2.1. Which Tasks Does a Robot Writer Need to Address? The first commonly distinguished task for any robot writer is known as *content determination*. Typically, the input data contains more information or more detail than can effectively be expressed in text. Hence the robot writer needs to decide which information units ("messages") to include in the output and at which level of abstraction, which may depend on the communicative goal and the intended audience. In many instances, the task of deciding which information to include is closely tied to a specific application; content determination for a soccer report is obviously different from, say, that for weather information. As a result, very few general approaches to this problem exist. One of the few exceptions is Guhe (2007), who explicitly aims to model how human speakers deal with the task of content determination.

After the robot writer has determined which messages it wants to express, it needs to decide in which order these should be presented to the reader. This task is known as *text structuring* (and sometimes also as document/discourse, structuring/planning). Like content determination, text structuring is often closely linked to the application domain—in a soccer report, it makes sense to follow a chronological order, describing events such as attempts to score and actual goals as they occurred, while in an economics report, messages are often ordered with respect to their importance. Early approaches, such as McKeown (1985), often relied on handcrafted, domain-specific rules, but also sometimes build on more general models such as Rhetorical Structure Theory of how messages in a discourse relate to each other (e.g., Hovy, 1993). More recently, however, approaches that can automatically learn how to order "information units" have become increasingly popular (e.g., Barzilay & Lee, 2004; Lapata, 2006).

Once all messages are ordered in a text structure plan, the robot writer can decide which messages should end up in which sentences. Not every message needs to correspond with a single sentence (e.g., Hovy, 1988; Dalianis, 1999); instead, by combining different messages into a single sentence, the resulting text can become more fluid and readable (see what we did there?). The process of merging multiple messages into a single sentence is known as *aggregation*, and various approaches have been proposed that can automatically learn aggregation rules from data (e.g., Walker, Park, Rambow, & Rogati, 2001; Barzilay & Lapata, 2006). Aggregation is also studied by, for instance, educational psychologists interested in teaching pupils how to write better texts (see, e.g., Zimmerman & Kitsantas, 2002).

At this point, the robot writer "knows" what it wants to say in sufficient detail, after which it can start actually expressing it in natural language. Usually, this starts by deciding which words can be used to express the core message components, a process often referred to as *lexicalization*. Lexicalization can be difficult, because the input data is often very precise (e.g., 23:59:59 hours), while language is much more imprecise (should this timestamp be lexicalized as "late evening," "night," or "midnight"?). As it turns out, human speakers show considerable variation when faced with this kind of choice (Reiter et al., 2005), a theme to which we will return in section 3. Interestingly, work on lexicalization is closely related to psycholinguistic work on lexical access, a core ingredient of cognitive models of speech production (Levelt, 1989), which has also been extensively studied using computational cognitive modeling (e.g., Levelt, Roelofs, & Meyer, 1999).

A specific task, somewhat related to lexicalization, is how domain entities should be referred to. This is

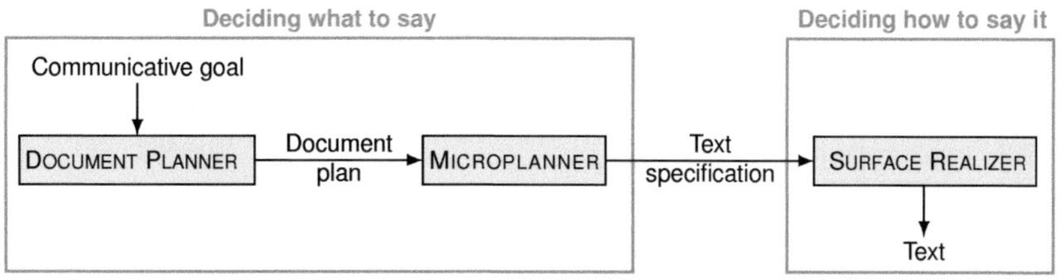

Figure 25.1 Classic pipeline architecture for a robot writer (after Mellish et al., 2006).

commonly referred to as *referring expression generation (REG)* (Dale & Reiter, 1995; Krahmer & van Deemter, 2012) and is among the most studied subtasks within NLG, perhaps because it can be studied in a mostly task-independent way. REG is often characterized as a two-step choice problem. When a robot writer wants to refer to an entity or object, it first needs to choose a referential *form* (which might, for example, be a proper name, a pronoun, or a definite description). Often, this will be followed by a second stage, in which the *contents* of the reference are decided on (e.g., which properties to include in a description?). The production of referring expressions also has a long history within psycholinguistics (e.g., Olson, 1970; Pechmann, 1989, among many others) and, in fact, these early psycholinguistic insights have been an important source of inspiration for developers of REG algorithms (Dale & Reiter, 1995).

Finally, when the robot writer has decided on all relevant words, phrases, and references, it needs to combine these into a coherent, grammatically correct sentence, a task that is known as *linguistic realization*. Traditional realizers were often grammar-based and did not scale up well. More recently, substantial progress has been made, especially with respect to probabilistic models, originated by, for example, Langkilde and Knight (1998) and Langkilde (2000). Their approach introduced the combination of a small, handcrafted grammar used to generate multiple sentence structures with a stochastic ranking module, which selects the optimal candidate. In more recent years, this approach has been extended, by looking at more advanced, automatically generated grammars and improved language modeling techniques. An example is OpenCCG (White, Rajkumar, & Martin, 2007), a wide-coverage English realizer based on representations in the manner of combinatory categorical grammar and derived from the Penn Treebank (Hockenmaier & Steedman, 2007), combined with a statistical language model. Again, there is also related (computational) psycholinguistic work, such as, for example, Chang, Dell, and Bock (2006), who offer a connectionist model of sentence production, modeling how speakers learn to produce sentences based on abstract meanings.

2.2. Architectures for Robot Writers In the previous section, we briefly surveyed six classic tasks that any data-to-text robot writer probably needs to address (Reiter & Dale, 2000), and indeed most existing NLG systems tackle all of them (Mellish et al., 2006). Different architectures have been proposed for dealing with the different tasks in an integrated manner.

Traditionally, NLG researchers often opted for a pipeline architecture, consisting of three modules: The first, usually referred to as the *Text or Document Planner* (sometimes also as *Macroplanner*), combines content selection and document planning and produces a text/document plan, which can be seen as a structured, abstract representation of messages. This plan is then fed into a second module, known as the *Sentence Planner* (or *Microplanner*), which typically performs aggregation, lexicalization, and referring expression generation. Finally, the resulting text specification plan needs to be turned into actual sentences, which is done by the *Linguistic* or *Surface Realizer*. Note that the first two modules are dedicated to deciding what to say, while the third module addresses the how-to-say-it question. Figure 25.1 (after Mellish et al., 2006) summarizes this architecture.

When Reiter (1994) described this three-stage pipeline architecture as the "consensus" architecture for NLG systems, he explicitly compared it to models of human speech production, most notably to the blueprint for the speaker, proposed by Levelt (1989), see figure 25.2. Even a cursory inspection of the two models reveals clear similarities: both assume a strict, modular (Fodor, 1983) separation between different stages, and in both there is an obvious distinction between modules relevant for deciding what to say and deciding how to say it (before actually saying it). There are also differences: Levelt's model, unlike the "consensus" NLG architecture, allows for a certain amount of self-monitoring via a feedback loop, and indeed such loops have sometimes been added to the NLG architecture by researchers as well (see, e.g., Mann & Moore, 1981, for an early example).

It is also interesting to note that for text-to-text robot writers, such as automatic summarizers, a comparable

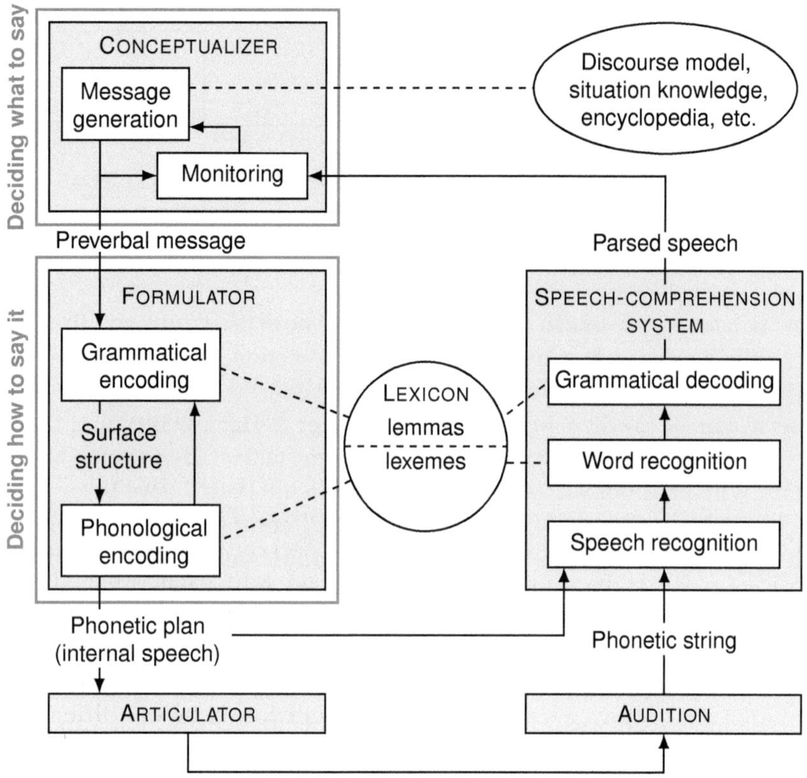

FIGURE 25.2 Blueprint for the speaker (after Levelt, 1989). Notice the distinction between deciding what to say and deciding how to say it.

tripartite architecture has been proposed (e.g., Mani, 2001; Nenkova & McKeown, 2011). Typically, *automatic summarizers* take as input a (set of related) document(s) and produce a shorter, coherent text containing the most important information in the source document(s). Summaries can either be *extractive*, meaning that the summary consists of sentences that are extracted verbatim from the input texts, or *abstractive*, in which case the summary may contain original sentences not found in the input. In practice, abstractive summarizers often rely on extractive methods to first find the most important sentences in the source document(s), which are then revised, for instance, by rewriting the referring expressions in extracted sentences to improve coherence (for example, in the case where the extracted sentence contains a pronoun, while the sentence containing the antecedent for this pronoun has not been selected for inclusion in the summary).

Traditionally, summarizers often consist of three separate modules, starting with *analysis* of the sources texts (deciding what to include in the summary), followed by *transformation* of the selected elements (deciding how to include it), and finally *synthesis* of the elements in the summary (realization of the summary) (Mani, 2001).

More recently, text generation researchers, just like some psycholinguists, have started to look into alternative, more integrated architectures, with less strict boundaries between modules. This can be seen, for instance, in planning-based architectures, such as those of Koller and Stone (2007) and Garoufi and Koller (2014), in which different tasks are often solved in tandem. But also the relatively recent shift to more data-driven methods in robot writing has been an important impetus for this change (e.g., Chen & Mooney, 2008; Konstas & Lapata, 2013; Mairesse & Young, 2014). Mairesse and Young, to give but one example, developed an end-to-end generation system, which automatically learns what the most likely sequences of concepts and linguistic realizations are according to factored language models (Bilmes & Kirchhoff, 2003), on the basis of a data set that aligns concepts (database records) and realizations (texts).

Vision-to-text robot writers also increasingly favor integrated architectures (see Bernardi et al., 2016, for a recent survey). In this particular application, robot writers typically take visual data as input and produce a textual description. Most often the input is a single image and the output is a single sentence, but there is also an increasing interest in generating descriptions from video (see, for example, the youtube2text system of Guadarrama et al., 2013). In general, *automatic image description* systems combine computer vision and NLG techniques—the former are required for analyzing the image and "understanding" it (including localization of the relevant, salient persons and objects in the image, retrieving their attributes, and detecting their relations and interactions), after which NLG techniques are applied to generate the description. This can involve all the tasks described herein (Bernardi et al., 2016), and existing systems rely on a range of techniques, varying from templates (e.g., Yang, Teo, Daume, & Aloimonos, 2011) to statistical language models (e.g., Kulkarni et al., 2013) and recurrent neural networks (e.g., Socher, Karpathy, Le, Manning, & Ng, 2014). A linguistically more advanced NLG approach is presented by Mitchell et al. (2012), who relied on syntactic trees akin to Tree Substitution Grammars during the generation process. Automatic image description is currently receiving a lot of scholarly attention, which at least partly is due to the availability of large data sets of images paired with human-produced descriptions, often obtained via crowdsourcing.

2.3. Is My Robot Writer Doing Well? Evaluation of robot writing is important and has become a central methodological concern within NLG in recent years (e.g., Reiter & Belz, 2009). Evaluation of NLG has proven to be complicated, in part because there is substantial variation, both within and between systems, and in terms of both inputs and outputs. As a result, there is no single evaluation methodology that is generally applicable (e.g., Mellish & Dale, 1998).

Perhaps for these reasons, one important group of evaluation metrics are subjective *human judgments*, in which participants are presented with system outputs and asked to rate certain aspects of the output, such as their *fluency*, *coherence*, and *comprehensibility*. Often, these are combined with some measure of *corpus similarity* (assuming that a suitable reference corpus is available). Arguably, corpus-based methods are a proxy for *humanlikeness*, since they indicate to what extent the outputs produced by a robot writer are similar to those produced by human authors. Frequently used measures include BLEU (Papineni, Roukos, Ward, & Zhu, 2002) and NIST (Doddington, 2002), which measure precision in terms of *n*-gram overlap and have their origins in the field of Machine Translation. Also metrics from automatic summarization are sometimes used, such as ROUGE (Lin, Hovy, & Rey, 2003) measuring *n*-gram overlap between automatically and human-produced reference summaries. Alternatively, so-called *extrinsic evaluation* metrics can be used, where the impact of an automatically generated text for a particular, real-life task is measured (for example, is the system helpful in changing undesired user behavior such as smoking?; Reiter, Robertson, & Osman, 2003).

Ideally, one would hope that different (kinds of) measures yield results that are consistent with each other (for example, that a higher human rating on the *fluency* dimension is accompanied by a higher BLEU score), so that different metrics converge for a particular system or so that comparative rankings of competing systems reveal a consistent picture. Unfortunately, it turns out that this often is not the case. For example, Gatt and Belz (2010), in a campaign comparing different REG algorithms, observed that human judgments, corpus-based metrics, and task-based metrics often did not correlate. Similarly, in a study on referential form selection (Generating Referring Expressions in Context [GREC]), Belz, Kow, Vieten, and Gatt (2010) found that different groups of measures did not necessarily correlate with each other. This phenomenon is not unique for data-to-text NLG tasks; in other areas of computational linguistics, such as Automatic Summarization and Machine Translation, a similar lack of correspondence between different measures has been observed (e.g., Callison-Burch, Osborne, & Koehn, 2006). The exact causes for this phenomenon are unclear (e.g., it remains to be seen whether other ways of measuring would lead to more convergence between metrics or whether there indeed is no clear underlying relation between different metrics). For now, this state of affairs raises the question: which metrics should be given preference during evaluation (what to do if subjective scores are low, but corpus-similarity is high, or vice versa)?

A further complication for NLG evaluation is that there usually are many different ways in which a particular input can be expressed into natural language and that it is still unclear how we can decide, in general, whether one output is better than the next. In practice, researchers often assume that an NLG system does well when it closely mimics the outputs found in some corpus. But, in truth, it is not clear whether an output that is different from a gold standard found in a corpus should be rated lower than an output that is found in the corpus (e.g., Reiter, Sripada, Hunter, Yu, & Davy, 2005). We will return to this issue.

3. The Robot Writer as a Research Tool

So far, in this chapter we have given a brief introduction to robot writers, discussing how one can build and evaluate the quality of a robot writer. As we have seen, the development of robot writers is firmly rooted in the field of NLG (and Artificial Intelligence, more generally), but in many cases comparable issues have also been studied by psycholinguists. In fact, we have seen various instances where NLG researchers took their inspiration from psycholinguistic theories or findings.

Accordingly, one way to look at robot writers is as technical applications of theories of human language production. In this section, however, we will argue that they may also serve as a test bed for new theories of human language production itself. After all, by developing robot writers and asking to what extent they are psychologically plausible, we may be able to further our understanding of human language production. In recent years, various researchers have developed computational models of language production, sometimes even based on algorithms originally developed for NLG. In this section, we discuss two case studies, both related to the production of referring expression.

3.1. Case Study 1: Modeling Human Referential Form Selection We have noted that robot writers generally approach the generation of referring expressions as a two-step procedure. First, a robot writer has to decide which form a particular referring expression should have; should it, for example, be a pronoun, a proper name, or a definite description? Then, in a second stage, the content of the reference is determined; if, say, a description is to be generated, which properties of the target should be mentioned? Both steps have been studied at length by psycholinguists, but comprehensive and explicit accounts have been lacking.

Psycholinguists have shown that the choice for a particular referential form can depend on factors such as *salience* of the antecedent and the *utterance* in which it occurs (e.g., Brennan, 1995; Gordon & Hendrick, 1998), and these factors have been integrated in theories of referential text coherence, such as *Centering Theory* (Grosz, Weinstein, & Joshi, 1995), However, such key notions as "salience" and "utterance" were never fully specified, and as a result, the theory offered an incomplete account of referential form selection (Poesio, Stevenson, Di Eugenio, & Hitzeman, 2004). Over the years, various NLG researchers have also studied referential form selection, developing computationally explicit models of this task (e.g., Callaway & Lester, 2002; Greenbacker & McCoy, 2009, among others), which often incorporate insights from psycholinguistically motivated accounts. Can these computational models offer us insights in human referential form selection?

To address this question, at least two issues need to be tackled. First, an important feature of many traditional NLG models is that they are *deterministic*, producing the same output for a given input context; for example, always predicting a pronoun when referring to a previously introduced entity at a particular point in a text. It is unlikely that this is what human speakers do, who instead can be expected to reveal substantial variation, both within and between speakers (van Deemter, Gatt, van Gompel, & Krahmer, 2012). The second issue, which arguably is somewhat related to the first one, is that we do not know exactly to what extent human speakers are nondeterministic, because data for this are generally missing. Notice that as a result of this, it is also difficult to say how well existing computational models of referential form selection actually perform. Typically, these are evaluated on text corpora, which only have one gold standard: the references actually produced by the writer. Now consider a situation in which a model predicts that a particular reference should take the form of a pronoun, while the corpus has a proper name at this specific location. We could treat this simply as an incorrect prediction, but the mere fact that a writer of a corpus text used a name in a particular context does not necessarily mean that using a pronoun there would have been incorrect or wrong. In fact, other writers might have used a pronoun as well, certainly when we take into account that human language production is nondeterministicc.

Ideally, we would like to know where and to what extent different writers agree in their choice of referential forms and where they do not. Based on this, one could develop a nondeterministic model of referential form selection, approximating human referential form selection.

To start addressing this, Castro Ferreira, Krahmer, and Wubben (2016a) collected a new corpus of texts, consisting of multiple gold standards for the referential forms. To achieve this, they selected 36 English texts from three different genres: news texts, product reviews, and encyclopedic texts, the latter randomly selected from the aforementioned GREC corpus (Belz et al., 2010). Next, all referring expressions to the main topic of the text (for example, the mountain, river, or country described in the encyclopedic texts) were removed and participants were asked to fill all the resulting gaps with references to the topic. This resulted in approximately 20 different referring expressions per gap.

To quantify the amount of variation for a particular gap, Castro Ferreira at al. computed the *normalized entropy*. This measure gives a value between 0 and 1, where 0 indicates complete agreement between the different authors (all use the same referential form in a gap) and 1 represents maximal variation. Next, these were averaged over all gaps, as depicted in figure 25.3, which shows the amount of individual variation, in terms of entropy, as a function of text genre. If there

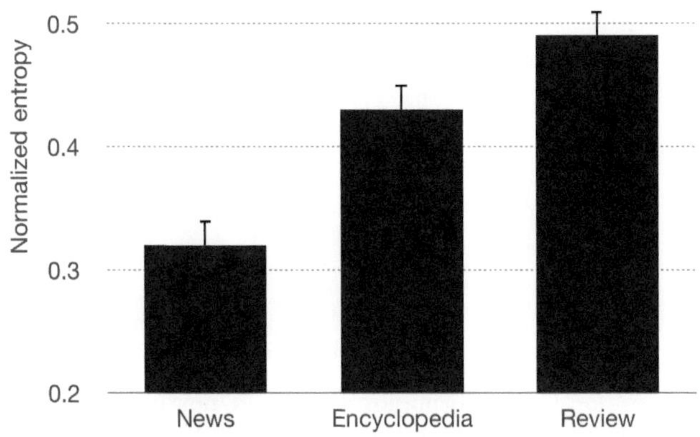

FIGURE 25.3 Average normalized entropy as a function of text genre (error bars are 95% confidence intervals). Figure adapted from Castro Ferreira et al. (2016a).

would be no individual variation in referential choices, the entropy should be 0 in all three genres. This is clearly not the case. Instead, Castro Ferreira et al. found substantial individual variation (entropy scores that are clearly above 0), which is lowest in the news texts and highest in the product reviews.

Consistent with earlier psycholinguistic findings (e.g., Brennan, 1995; Grosz et al., 1995; Gundel, Hedberg, & Zacharski, 1993), Castro Ferreira et al. observed that variation in referential form is higher when the reference is the syntactic object of the current sentence than when it is the subject, that variation is higher for repeated mentions than for first mentions in the text, and that variation tends to increase when the previous reference is further away. These trends are visualized in figure 25.4.

Inspired by these observations, Castro Ferreira, Krahmer, and Wubben (2016b) developed a probabilistic model, based on features such as syntactic position, referential status (old or new), and distance to previous reference, that predicts the probability of a certain referential form. In other words, for each reference, it predicts a *probability distribution* over all referential forms, rather than a single form. This paves the way for a nondeterministic robot writer, sampling referential forms during text generation by applying a so-called *roulette-wheel generation strategy* (Belz, 2008). In this way, referential forms in a text can be selected nondeterministically, but in accordance with the overall probability distribution.

So, how well does this strategy perform? To find out, an online experiment using CrowdFlower (a variant of Mechanical Turk) was conducted, in which 30 judges (all proficient in English, the language of the experiment) were asked to rate different texts in terms of their *coherence* and *comprehensibility* on a five-point scale, ranging from 1 (very bad) to 5 (very good). Castro Ferreira et al. (2016b) selected nine texts from the GREC 2.0 corpus and for each text automatically generated two variants of the original: one using the roulette-wheel strategy (Generated version) and one which randomly selected a referential form (Random version). All variants were distributed over three lists, so that each judge saw one variant of each text, and saw an equal number of Original, Generated, and Random texts. The results of this study are depicted in figure 25.5 and clearly show that the Generated texts are judged to be as coherent and comprehensible as the Original ones and are rated higher than the Random versions.

This shows that it is possible to mimic human, nondeterministic production of referential forms. Interestingly, this can be done without a loss of quality. Notice also that this paves the way for generating more varied texts (at least where referential forms are concerned), since regenerating the same text multiple times would lead to different sequences of referential forms in each case.

3.2. CASE STUDY 2: MODELING HUMAN CONCEPTUALIZATION OF DESCRIPTIONS Once a referential form has been chosen, some further choices may be necessary, which is most clearly an issue when the selected form is a definite description. The production of definite descriptions (i.e., a noun phrase describing some target object such as, for example, "the large cat") is often assumed to consist of at least two steps in itself: first, a speaker needs to decide which information to include—does she decide to conceptualize said cat as large or brown or some combination of these or other properties? And, second, once a selection of properties has been made, she has to decide how to express these properties in words (is the cat's brownish color referred to as being "brown" or "light brown" or "amber"). These two steps echo the distinction between deciding what to say and deciding how to say it, that was made in section 2. It is interesting to observe that psycholinguists studying language production generally had more to say about the latter than the former (Roelofs & Ferreira, chapter 4 of this volume), while the opposite seems the case for NLG researchers working on the generation of descriptions (e.g., Dale & Reiter, 1995; Krahmer & van Deemter, 2012).

Typically, NLG researchers studying the production of descriptions assume that some target object (like our large brown cat) needs to be distinguished form a set of one or more distractor objects (other domestic animals of possible different colors, sizes, owners, and so on). The question, then, is how to decide ("conceptualize") which properties of the target to include in a

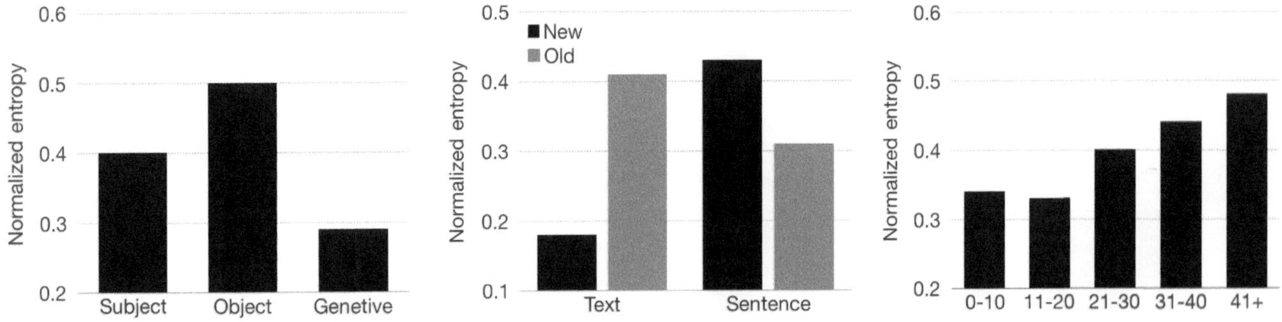

FIGURE 25.4 Average normalized entropy as a function of syntactic position (*left*), information status (old/new in text and sentence; *middle*) and recency of last mention (in words; *right*). Figure adapted from Castro Ferreira et al. (2016a).

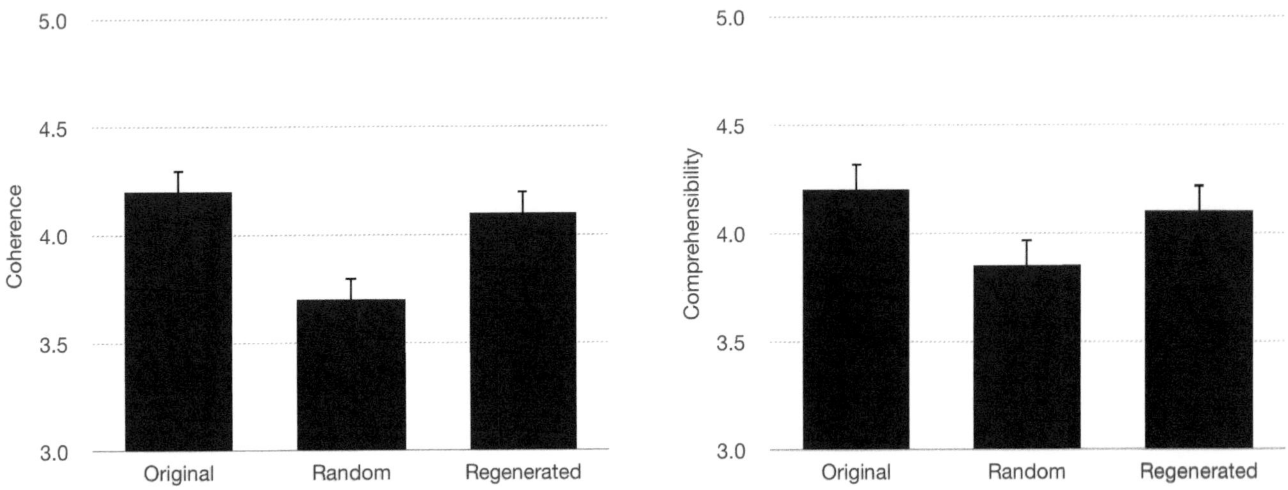

FIGURE 25.5 Average human rating of *coherence* (*left*) and *comprehensibility* (*right*) of Original, Random, and Generated texts on a five-point scale ranging from 1 (very bad) to 5 (very good) (error bars are 95% confidence intervals). Figure adapted from Castro Ferreira et al. (2016a).

description under construction. Various computational proposals have been put forward concerning how to make this choice. For example, Dale (1989) suggested that selection of properties is driven by their discriminatory power (or "informativeness" or "utility"), where the first property to be selected should be the one that rules out most distractors. This selection strategy has been dubbed the *greedy heuristic* approach and can be contrasted with a different strategy, which postulates that properties are selected incrementally, based on how preferred they are rather than their discriminatory power (Dale & Reiter, 1995). For example, based on psycholinguistic evidence (e.g., Pechmann, 1989), we might assume that the color of a target should be considered before its size, generally selecting a property if it rules out at least one distractor not previously ruled out. The question that concerns us here is whether these kinds of models can offer us insights in human conceptualization of descriptions?

As before, a *prima facie* hurdle for applying computational models such as those just mentioned to human conceptualization is that they are all *deterministic*—the greedy and incremental models make different predictions when the most discriminatory property is not the most preferred one, but crucially both always predict the same description in the same context, and this does not sit well with what we know of human reference production. For example, a well-studied phenomenon in the production of descriptions is the *overspecification* (terminology from Pechmann), which is used to refer to situations in which speakers select more properties than are necessary for identification (e.g., producing "the large brown cat" when all distractors are small cats, and hence "the large cat" would have sufficed for identification purposes). Studies of human reference production routinely find overspecification rates between 40% and 80% (see, e.g., Belke & Meyer, 2002; Engelhardt, Bailey, & Ferreira, 2006; Engelhardt, Demiral, & Ferreira, 2011; Koolen, Gatt, Goudbeek, & Krahmer, 2011). Moreover, it has been found that external factors can lead to an increase in overspecification rates, for example, when the costs of misunderstandings are high (Arts, Maes, Noordman, & Jansen, 2011) or when domains are more complex (Paraboni & van Deemter, 2014). Crucially, the aforementioned computational models predict, for each individual referential context, that a description is either always or never overspecified.

To find out to what extent models, such as the greedy and incremental ones, compare to human production strategies, van Gompel, van Deemter, Gatt, and Krahmer (2019) described a series of experiments in which participants were asked to refer to visually presented targets together with two distractors, both of which only differed from the target in their color, in their size, or in both their color and size. They then analyzed all descriptions that were produced by the different participants checking whether these mentioned the target's color, size, or both. A typical result, for English speakers, is depicted in figure 25.6.

A few observations can be made about this result. First, the Greedy Heuristic strategy cannot account for

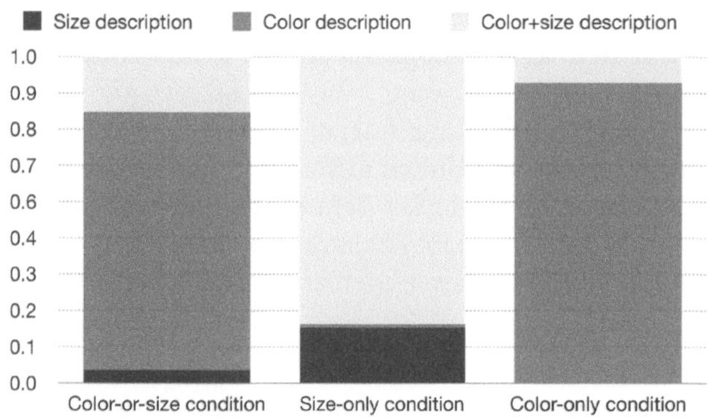

FIGURE 25.6 Proportions of color, size, and color+size descriptions produced by English speakers as a function of condition. Figure adapted from van Gompel et al. (2019).

the frequent use of color and size descriptions in the size-only condition. After all, since size has the highest discriminatory power in this condition, and size is sufficient to distinguish the target, the algorithm would always predict size-only descriptions in this case. Instead, speakers overwhelmingly produce overspecified color+size descriptions in the size-only condition. In fact, in all conditions, the majority of speakers produces a description that is in accordance with the predictions of the Incremental algorithm. However, it is also clear from the results that the predictions of the Incremental algorithm do not offer a good match with the actual results either, because the results clearly are not "all or nothing" as the Incremental strategy would predict (as visualized in the left panel of figure 25.7). In a nutshell, one could say that the Greedy Heuristic strategy does not fare well because it has no account for overspecification and the Incremental strategy does not fare well because it has no account for nondeterminism.

Still, this does not mean that computational models of conceptualization cannot be helpful for understanding human reference production. Gatt and colleagues discussed a different model called *PRO* (short for Probabilistic Referential Overspecification), which integrates core insights from the Greedy Heuristic and the Incremental models in a new probabilistic setting (Gatt, van Gompel, van Deemter, & Krahmer, 2013). In particular, the PRO model selects a first property if it is fully discriminating (in accordance with the Greedy Heuristic strategy). If there is no single property that rules out all distractors, it selects color with probability p (and size with probability $1-p$). Here, p can be thought of as the preference of color over size (in accordance with the Incremental strategy). Finally, in a second step, an additional property can be included, based on what the Gatt et al. call the "overspecify-eagerness" parameter r. Both p and r are estimated on held-out data, and

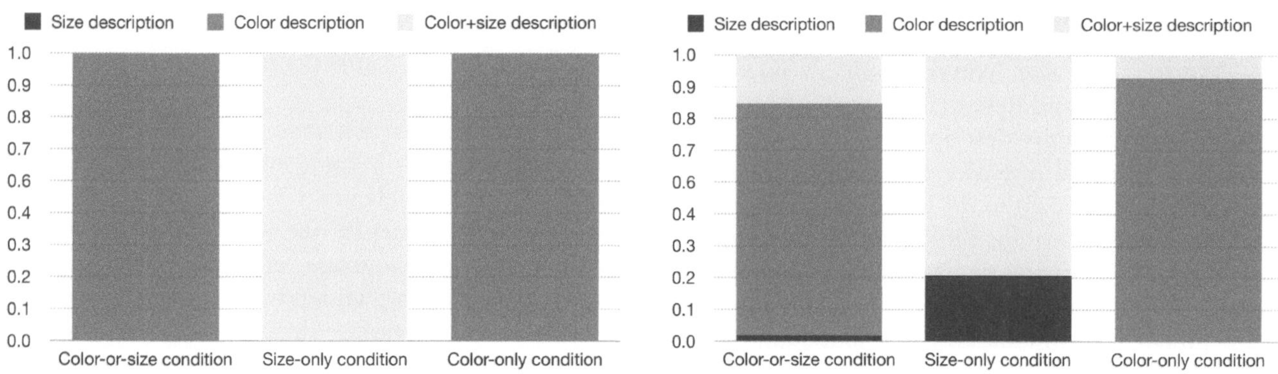

FIGURE 25.7 Predictions of the Incremental model (*left*) and the PRO model (*right*). Figure adapted from van Gompel et al. (2019).

both play a role for the selection of the second property, thereby accounting for the predominance of overspecification with preferred properties. Over a series of experiments, contrasting different pairs of properties in different languages (English and Dutch), Gatt et al. showed that PRO accounts for the human data rather well.

The PRO model, as outlined, is exceedingly simple, just accounting for whether speakers conceptualize an object's description in terms of, for example, its color or its size, depending on the color and size of the distractors. Still, it is one of the first computationally explicit models of the early stages of language production (deciding what to say, rather than how to say it). In this respect, it can serve as a simple model of human reference production, showing the potential of techniques from robot writing for understanding human language production.

4. Beyond Factual (and Boring) Robot Written Texts

Much of the earlier work on automatic text generation has been concerned with the production of factual texts, written with the intention to inform the reader. Even though robot writers are by now perfectly capable of producing coherent texts, typically based on close inspection of human-written example texts, reading a number of computer-generated texts in a row also reveals the limitations of this approach, in particular the repetitive nature of the generated texts. As a result, there is a growing interest in attempts to automatically create texts that are more *engaging*, for example, by adding affect or personality to the output. Similarly, researchers wonder whether it is possible to produce texts that are more *creative*, for example, by adding metaphors or even jokes to the text, or that are more *poetic*, by adding rhythm and rhyme. In this section, we present a sample of these recent developments to indicate where robot writing might be heading in the future.

4.1. Affective Robot Writing In an interesting study, Mahamood and Reiter (2011) proposed a system that generates affective descriptions for parents whose child is in neonatal intensive care. They argued that the system should adapt to the level of stress of the parents, which was determined based on the medical condition of their child and depending on this could add more or fewer qualifications such as "this is a positive development for your child." In an evaluation with parents that previously had a child in neonatal care, Mahamood and Reiter found that participating parents had a strong preference for affective texts over nonaffective, factual ones, both in low- and in high-stress scenarios. This clearly illustrates the potential benefits of generating affective texts, potentially tailored to individual audiences. It seems likely that this benefit is not restricted to the medical setting, but generalizes to different genres, such as sports reporting.

Some forays have been made into automatically generating tailored, emotional reports (e.g., de Rosis & Grasso, 2000; Mairesse & Walker, 2010; Van der Sluis & Mellish, 2010). However, none of these have developed fully functional systems that can generate emotional texts. In part, this may be due to a lack of suitable data sets, revealing how human writers express data in text as a function of their emotional state. To address this, Braun, Goudbeek, and Krahmer (2016) started collecting a corpus of affective sports reports (in Dutch, English, and German), based on the insight that the reports for the different teams in a match are often written in different affective states (e.g., sad or disappointed after losing, but happy after winning a game). Consider, for instance, the Burton Albion–Petersborough United match (2-1), and compare how this match was described on the Burton Albion website (1) and the Petersborough United website (2), respectively:

(1) If all League One games at the Pirelli Stadium this season are going to be like this it is going to be an entertaining if nerve jangling season.
(2) Peterborough United suffered a 2-1 defeat at Burton Albion in Sky Bet League One action and lost defender Gabi Zakuani to a straight red card during a nightmare spell at the Pirelli Stadium.

Notice that, for example, while the former seems to consider the match as "entertaining," the latter refers to the same event as "a nightmare spell." Braun et al. scraped the websites of all teams in the two highest regions of the Dutch, English, and German soccer competitions and in this way collected a parallel corpus of soccer reports. Combined with publicly available game statistics, this paves the way for learning how game statistics may trigger affective language production, which in turn may facilitate affective robot writing.

4.2. Creative Robot Writing Another way in which generated texts may become more engaging to read is by adding creative elements to them. In many cases, existing robot writers in one way or another rely on human-produced language, either because they are trained on previously written human texts (in the case of text-to-text) or because they rely on grammar rules or templates that were developed by a human researcher (in the case of data-to-text). Is it conceivable that robot writers produce a text that was not, in some way, anticipated by its developers? Addressing this question brings us into the realm of computational creativity, an area of research that is increasingly related to text generation. When discussing computational creativity in text, Veale (2015) drew an interesting parallel with the so-called *cut-up* technique, popularized by, most notably, the author William S. Burroughs. In this technique, an existing text is literally cut into pieces after which it is randomly put together again, thereby creating "new and unexpected meanings" as a result of the new combinations of words and sentences that arise in this way. Veale pointed out that the cut-up technique lends itself well for computational implementations, and indeed various early approaches to computational creativity can be seen as instantiations of this approach.

A good example is the *punning riddle* generator developed by Binsted and Ritchie (1994). It can automatically produce word riddles such the following:

(3) What's the difference between leaves and a car?
One you brush and rake, the other you rush and brake.

(4) What do you get when you cross a bag and a human?
A purse-on.

Essentially, the system developed by Binsted and Ritchie "cuts-up" phrases and words to create new combinations fitting a predefined punning riddle scheme. The starting point of the riddles are well-known templates, such as "What's the difference between X and Y?" or "What do you get when you cross X and Y?," after which an extensive search for interesting *X*'s and *Y*'s starts, using resources such as Wordnet (to find out, for instance, that *person* and *human* are related terms) and pronunciation dictionaries (to find out that *person* and *purse-on* sound similar).

Interestingly, Binsted, Pain, and Ritchie (1997) evaluated the performance of their "robot-joker" with the typical target audience of riddles: children. They presented 120 children between 8 and 11 years old with automatically generated riddles such as (3) and (4), interspersed with selected riddles from joke books as well as nonjoke controls ("What do you call a person who studies? A student."). Children were asked to indicate, for each sequence, whether they thought this was a joke and whether they had heard it before. In addition, they had to score its funniness on a five-point scale, ranging from 1 = not funny at all, to 5 = very funny. Of all riddles, the automatically generated (3) was rated as most funny ($M = 4.33$), with all children reporting this as a joke and one they had not heard before (to be fair, the next 13 on the list were human-written riddles).

While riddles have only limited applicability for text generation, similar techniques have also been proposed for the automatic generation of *metaphors*, which are an important example of creative language use in texts (Veale & Hao, 2007; Veale, 2013). Veale, for example, searched through a large text corpus for positive and negative associations for a user-provided input term, allowing for the creation (cut-up) of new metaphors. For example, through positive associations with "robot," one can generate novel metaphors such as "robots are like clocks" (emphasizing "their accuracy") or "robots are like scientists" (emphasizing their "cleverness"), while negative associations facilitate metaphors such as "robots are zombies" or "robots are soulless."

4.3. Poetic Robot Writing If robot writers can produce new metaphors, could they also produce full-fledged creative texts like poems? In an insightful paper, Greene, Bodrumlu, and Knight (2010) compared human and robot poets, pointing out that the former have a massive advantage over the latter (for instance, because they "experience joy and heartbreak"). However, they point out that the latter might also have a few small things going for them (such as, "a machine can easily come up with a five-syllable word that starts with *p* and rhymes with *early*"). In fact, the automatic generation of poems has a long tradition, and in recent years, various researchers have addressed this topic (Colton, Goodwin, & Veale, 2012; Gervas, 2001; Manurung, Ritchie, & Thompson, 2012; Netzer, Gabay, Goldberg, & Ehladad, 2009). Perhaps the relative popularity of this task may be explained by the observation that many poetic forms (like haikus and sonnets) have a very strictly defined structure in terms of allowed number of syllables per line as well as combinations of rhyme and meter, which provide useful constraints on the search process of the robot poet.

Greene et al. presented a probabilistic romantic poetry generation system, trained on a corpus of English love poems. It incorporates a discourse model, as well as probabilities for how a given user-provided stress pattern is likely to be realized (the meter model). It generates poems one line at a time, starting from a seed line, after which the model is ran again with additional rhyming constraints. This procedure yields couplets like the following:

(5) Of the bed to trust me
Around her twists the string
But I will not tell thee
Fire changes everything.

This output is fairly typical for robot-written poems: it sounds good, when read aloud, but there is no coherent meaning. This is not surprising: the model is all about "deciding how to say it" but has no idea "what to say."

Interestingly, Greene et al. argued that this makes their model very suitable for automatic poetry *translation*. To make this point more explicitly, Greene et al. described a case study of automatically translating Dante's *Divina Commedia* from Italian to English. They train a standard Phrase-Based Machine Translation system (Moses PBMT, Koehn et al., 2007) on an extensive parallel Italian-English corpus, including multiple English translations of Dante's epic poem (at first, this may sound a bit like cheating, but notice that human translators also typically inspect earlier translations for inspiration). They compare the performance of the standard PBMT system with a variant that additionally includes the aforementioned meter-model. An example illustrates the potential of this method:

(6) *Original:*
nel mezzo del cammin di nostra vita
mi ritrovai per una selva oscura
che la via diritta era smarrita.

(7) *Phrase-based translation (PBMT):*
midway in the journey of our life
i found myself within a forest dark
for the straight way was lost.

(8) *PBMT+ meter model:*
midway upon the journey of our life
i found myself within a forest dark
for the straightforward pathway had been lost.

While the standard PBMT model in its translation (7) only relies on the probabilities of translating one phrase into another, the addition of the meter model causes the system to reproduce a much better sounding earlier human translation (from Longfellow), determining, for example, that "upon" in the first line is a better translation than "in" due to the meter (8). Even though for other verses the system fares less well, sometimes even failing to find a suitable translation at all, the method offers an intriguing first look into the automatic translation of poetry.

5. Wrapping Up

In this chapter, we presented an introduction to robot writers for anyone interested in (psycho)linguistics. Robot writing can be characterized as the automatic generation of coherent text based on numeric, textual, or visual data and has been studied extensively in the area of NLG. The fact that computers can quickly produce large volumes of text, potentially tailored toward many individual readers, is one of the reasons why these techniques are increasingly used for applications in a wide range of domains including journalism, finance, and medicine.

We have presented a bird's eye perspective on robot writing, discussing a number of core tasks and common architectures. We have asked how the output of robot writers can be evaluated and seen that this is a knotty problem. We have also briefly looked into recent attempts to expand the repertoire of robot writers, toward more engaging and entertaining texts. Given that the goal of our chapter was to introduce readers to a new field (and given that the space for this was limited), our discussion of previous work in this area could not be exhaustive. If readers want to read more about machine-written texts (and we hope they do!), we refer them to Reiter and Dale (2000), which is still the most comprehensive account of the field. Gatt and Krahmer (2018) provided an extensive overview of NLG research of the years post-2000.

A central theme of this chapter was how previous research on robot writing has been inspired by psycholinguistic work. We have seen that "late" generation tasks ("deciding how to say it") such as lexicalization, referring expression generation, and linguistic realization all have counterparts in (computational) psycholinguistics and that insights from the latter have been important inspirations for the former (perhaps most evidently in the area of reference). We hope to have shown that, conversely, insights from robot writing, especially about early generation tasks ("deciding what to say") can also be helpful for psycholinguistic theories of human language production. We have discussed two case studies, both in the realm of reference, illustrating how explicit computational models can contribute to our understanding of human language.

The two cases addressed admittedly simple tasks: Should we use a pronoun or a proper name? Should we mention the color or the size of an object? While scaling up these choice problems to more referential forms and more object properties is straightforward, it is also clear that we are nowhere near a complete, computational model of the entire human speech production process. Typically, existing robot writers still take many shortcuts (Mellish, 2000), simply because the linguistic insights that would render these shortcuts superfluous are lacking. By building more advanced robot writers, doing away with one shortcut at a time, we stand to learn a lot more about human language.

Acknowledgments

Our general perspective on robot writing has been greatly influenced by many discussions with Albert Gatt. The two case studies in section 3 are based on joint work with Thiago Castro Ferreira, Sander Wubben, Albert Gatt, Roger van Gompel, and Kees van Deemter, and we thank them all for the pleasant collaboration. Financial support was received from The Netherlands Organization for Scientific Research (grants NWO 360-89-050, NWO 314-99-121, 2014-01-51PRO), which is gratefully acknowledged. Thanks to Thiago Castro Ferreira and Antal van den Bosch for their detailed comments on a previous version of this text. Thanks to Emiel van Miltenburg for his help in creating figures 25.1 and 25.2.

NOTES

1. While computational linguists tend to talk about *generating* language, psycholinguists generally prefer *producing* language. We use the two terms more or less interchangeably.
2. Interestingly, NLG researchers—developing systems that can automatically write language—have been inspired by models of human *speech* production and generally not by models of human *writing*; presumably this is because the latter models tend to focus more on planning and revision, which perhaps are perceived as lending themselves less well for producing a text in one go.

REFERENCES

Arts, A., Maes, A., Noordman, L., & Jansen, C. (2011). Overspecification in written instruction. *Linguistics, 49*(3), 555–574.

Barzilay, R., & Lapata, M. (2006). Aggregation via set partitioning for Natural Language Generation. In *Proceedings of the Main Conference on Human Language Technology Conference of the North American Chapter of the Association of Computational Linguistics* (pp. 359–366). Stroudsburg, PA: Association for Computational Linguistics.

Barzilay, R., & Lee, L. (2004). Catching the drift: Probabilistic content models, with applications to generation and summarization. In *Proceedings of the Main Conference on Human Language Technology Conference of the North American Chapter of the Association of Computational Linguistics* (pp. 113–120). Stroudsburg, PA: Association for Computational Linguistics.

Belke, E., & Meyer, A. (2002). Tracking the time course of multidimensional stimulus discrimination: Analysis of viewing patterns and processing times during same-different decisions. *European Journal of Cognitive Psychology, 14*, 237–266.

Belz, A. (2008). Automatic generation of weather forecast texts using comprehensive probabilistic generation-space models. *Natural Language Engineering, 14*(04), 431–455.

Belz, A., Kow, E., Vieten, J., & Gatt, A. (2010). Generating referring expressions in context: The GREC task evaluation challenges. In E. Krahmer & M. Theune (Eds.), *Empirical methods in Natural Language Generation* (pp. 294–327). Berlin: Springer.

Bernardi, R., Cakici, R., Elliott, D., Erdem, A., Erdem, E., Ikizler-Cinbis, N., ... Plank, B. (2016). Automatic description generation from images: A survey of models, datasets, and evaluation measures. *Journal of Artificial Intelligence Research, 55*, 409–442.

Bilmes, J. A., & Kirchhoff, K. (2003). Factored language models and generalized parallel backoff. In *Proceedings of the 2003 Conference of the North American Chapter of the Association for Computational Linguistics on Human Language Technology* (pp. 4–6). Stroudsburg, PA: Association for Computational Linguistics.

Binsted, K., Pain, H., & Ritchie, G. D. (1997). Children's evaluation of computer-generated punning riddles. *Pragmatics and Cognition, 5*(2), 305–354.

Binsted, K., & Ritchie, G. D. (1994). An implemented model of punning riddles. In *Proceedings of the Twelfth National Conference on Artificial Intelligence (AAAI-94)* (pp. 633–638). Palo Alto, CA: AAAI Press.

Braun, N., Goudbeek, M., & Krahmer, E. (2016). The Multilingual Affective Soccer Corpus (MASC): Compiling a biased parallel corpus on soccer reportage in English, German and Dutch. In *Proceedings of the 9th International Natural Language Generation (INLG) Conference* (pp. 74–78). Stroudsburg, PA: Association for Computational Linguistics.

Brennan, S. E. (1995). Centering attention in discourse. *Language and Cognitive Processes, 10*(2), 137–167.

Callaway, C. B., & Lester, J. C. (2002). Pronominalization in generated discourse and dialogue. In *Proceedings of the 40th Annual Meeting on Association for Computational Linguistics* (pp. 88–95). Stroudsburg, PA: Association for Computational Linguistics.

Callison-Burch, C., Osborne, M., & Koehn, P. (2006). Re-evaluating the role of BLEU in machine translation research. In *Proceedings of the 11th Conference of the European Chapter of the Association for Computational Linguistics* (pp. 249–256). Stroudsburg, PA: Association for Computational Linguistics.

Carlson, M. (2015). The robotic reporter: Automated journalism and the redefinition of labor, compositional forms, and journalistic authority. *Digital Journalism, 3*(3), 416–431.

Castro Ferreira, T., Krahmer, E., & Wubben, S. (2016a). Individual variation in the choice of referential form. In K. Knight, A. Nenkova, & O. Rambow (Eds.), *Proceedings of 2016 Conference of the North American Chapter of the Association for Computational Linguistics: Human Language Technologies* (pp. 423–427). Stroudsburg, PA: Association for Computational Linguistics.

Castro Ferreira, T., Krahmer, E., & Wubben, S. (2016b). Towards more variation in text generation: Developing and evaluating variation models for choice of referential form. In *Proceedings of the 54th Annual Meeting of the Association for Computational Linguistics* (Vol. 1: Long Papers, pp. 568–577). Stroudsburg, PA: Association for Computational Linguistics.

Chang, F., Dell, G. S., & Bock, J. K. (2006). Becoming syntactic. *Psychological Review*, 113, 234–272.

Chen, D. L., & Mooney, R. J. (2008). Learning to sportscast: A test of grounded language acquisition. In A. McCallum & S. Roweis (Eds.), *25th International Conference on Machine Learning (ICML 2008)* (pp. 128–135). Madison, WI: Omnipress.

Clerwall, C. (2014). Enter the robot journalist. *Journalism Practice, 8*(5), 519–531.

Colton, S., Goodwin, J., & Veale, T. (2012). Full-FACE poetry generation. In M. L. Maher, K. Hammond, A. Pease,

R. Pérez y Pérez, D. Ventura, & G. Wiggins (Eds.), *Proceedings of the Third International Conference on Computational Creativity* (pp. 95–102). Dublin, Ireland: University College Dublin.

Dale, R. (1989). Cooking up referring expressions. In *Proceedings of the 27th Annual Meeting of the Association for Computational Linguistics (ACL)* (pp. 68–75). Stroudsburg, PA: Association for Computational Linguistics.

Dale, R., & Reiter, E. (1995). Computational interpretations of the Gricean maxims in the generation of referring expressions. *Cognitive Science, 18,* 233–263.

Dalianis, H. (1999). Aggregation in Natural Language Generation. *Computational Intelligence, 15*(4), 384–414.

De Rosis, F., & Grasso, F. (2000). Affective Natural Language Generation. In A. Paiva (Ed.), *Affective interactions* (pp. 204–218). Berlin: Springer.

Doddington, G. (2002). Automatic evaluation of machine translation quality using n-gram co-occurrence statistics. In *Proceedings of the ARPA Workshop on Human Language Technology* (pp. 128–132). San Diego, CA.

Engelhardt, P. E., Bailey, K. G., & Ferreira, F. (2006). Do speakers and listeners observe the Gricean maxim of quantity? *Journal of Memory and Language, 54,* 554–573.

Engelhardt, P. E., Demiral, S. B., & Ferreira, F. (2011). Over-specified referring expressions impair comprehension: An ERP study. *Brain and Cognition, 77,* 304–314.

Fodor, J. A. (1983). *The modularity of mind: An essay on faculty psychology.* Cambridge; MA: MIT Press.

Garoufi, K., & Koller, A. (2014). Generation of effective referring expressions in situated context. *Language, Cognition, and Neuroscience, 29*(8), 986–1001.

Gatt, A., & Belz, A. (2010). Introducing shared task evaluation to NLG: The TUNA shared task evaluation challenges. In E. Krahmer, & M. Theune (Eds.), *Empirical methods in Natural Language Generation* (pp. 264–293). Berlin: Springer.

Gatt, A., & Krahmer, E. (2018). Automatic text generation: A survey of the state of the art in Natural Language Generation. *Journal of Artificial Intelligence Research, 61,* 65–170.

Gatt, A., Portet, F., Reiter, E., Hunter, J., Mahamood, S., Moncur, W., & Sripada, S. (2009). From data to text in the neonatal intensive care unit: Using NLG technology for decision support and information management. *AI Communications, 22*(3), 153–186.

Gatt, A., van Gompel, R. P. G., van Deemter, K., & Krahmer, E. (2013). Are we Bayesian referring expression generators? In *Proceedings of the 3rd CogSciWorkshop on Production of Referring Expressions.* Berlin.

Gervas, P. (2001). An expert system for the composition of formal Spanish poetry. *Knowledge-Based Systems, 14*(3–4), 181–188.

Goldberg, E., Driedger, N., & Kittredge, R. I. (1994). Using natural language processing to produce weather forecasts. *IEEE Expert, 2,* 45–53.

Goldrick, M., Ferreira, V., & Miozzo, M. (2014). *The Oxford handbook of language production.* Oxford: Oxford University Press.

Gordon, P. C., & Hendrick, R. (1998). The representation and processing of coreference in discourse. *Cognitive Science, 22,* 389–424.

Greenbacker, C. F., & McCoy, K. F. (2009). Feature selection for reference generation as informed by psycholinguistic research. In *Proceedings of the 3rd CogSci Workshop on Production of Referring Expressions.* Berlin.

Greene, E., Bodrumlu, T., & Knight, K. (2010). Automatic analysis of rhythmic poetry with applications to generation and translation. In *Proceedings of the 2010 Conference on Empirical Methods in Natural Language Processing (EMNLP)* (pp. 524–533). Stroudsburg, PA: Association for Computational Linguistics.

Grosz, B. J., Weinstein, S., & Joshi, A. K. (1995). Centering: A framework for modelling the local coherence of discourse. *Computational Linguistics, 21*(2), 203–225.

Guadarrama, S., Krishnamoorthy, N., Malkarnenkar, G., Venugopalan, S., Mooney, R., Darrell, T., & Saenko, K. (2013). Youtube2text: Recognizing and describing arbitrary activities using semantic hierarchies and zero-shot recognition. In *Proceedings of the 2013 IEEE International Conference on Computer Vision* (pp. 2712–2719). New York: IEEE.

Guhe, M. (2007). *Incremental conceptualization for language production.* Mahwah, NJ: Lawrence Erlbaum.

Gundel, J. K., Hedberg, N. & Zacharski, R. (1993). Cognitive status and the form of referring expressions in discourse. *Language, 69*(2), 274–307.

Hockenmaier, J., & Steedman, M. (2007). CCGbank: A corpus of CCG derivations and dependency structures extracted from the Penn Treebank. *Computational Linguistics, 33*(3), 355–396.

Hovy, E. H. (1988). *Generating natural language under pragmatic constraints.* Hillsdale, NJ: Lawrence Erlbaum Associates.

Hovy, E. H. (1993). Automated discourse generation using discourse structure relations. *Artificial Intelligence, 63*(1–2), 341–385.

Kanda, T., Hirano, T., Eaton, D., & Ishiguro, H. (2004). Interactive robots as social partners and peer tutors for children: A field trial. *Human-Computer Interaction, 19*(1), 61–84.

Koehn, P., Hoang, H., Birch, A., Callison-Burch, C., Federico, M., Bertoldi, N., ... Herbst, E. (2007). Moses: Open source toolkit for statistical machine translation. In A. Zaenen & A. van den Bosch (Eds.), *Proceedings of the 45th Annual Meeting of the Association for Computational Linguistics* (pp. 177–180). Stroudsburg, PA: Association for Computational Linguistics.

Koller, A., & Stone, M. (2007). Sentence generation as a planning problem. In A. Zaenen & A. van den Bosch (Eds.), *Proceedings of the 45th Annual Meeting of the Association for Computational Linguistics* (pp. 336–343). Stroudsburg, PA: Association for Computational Linguistics.

Konstas, I., & Lapata, M. (2013). A global model for concept-to-text generation. *Journal of Artificial Intelligence Research, 48,* 305–346.

Koolen, R., Gatt, A., Goudbeek, M., & Krahmer, E. (2011). Factors causing overspecification in definite descriptions. *Journal of Pragmatics, 43,* 3231–3250.

Krahmer, E., & van Deemter, K. (2012). Computational generation of referring expressions: A survey. *Computational Linguistics, 38,* 173–218.

Kulkarni, G., Premraj, V., Ordonez, V., Dhar, S., Li, S., Choi, Y., ... Berg, T. L. (2013). Baby talk: Understanding and generating simple image descriptions. *IEEE Transactions on Pattern Analysis and Machine Intelligence, 35*(12), 2891–2903.

Langkilde, I. (2000). Forest-based statistical sentence generation. In *Proceedings of the 1st North American Chapter of the Association for Computational Linguistics Conference* (pp. 170–177). Stroudsburg, PA: Association for Computational Linguistics.

Langkilde, I., & Knight, K. (1998). Generation that exploits corpus-based statistical knowledge. In *Proceedings of the 36th Annual Meeting of the Association for Computational Linguistics and 17th International Conference on Computational Linguistics* (pp. 704–710). Stroudsburg, PA: Association for Computational Linguistics.

Lapata, M. (2006). Automatic evaluation of information ordering: Kendall's tau. *Computational Linguistics, 32*(4), 471–484.

Levelt, W. J. M. (1989). *Speaking: From intention to articulation.* Cambridge, MA: MIT Press.

Levelt, W. J. M., Roelofs, A., & Meyer, A. S. (1999). A theory of lexical access in speech production. *Behavioural and Brain Sciences, 22*(1), 1–38.

Lewandowsky, S., & Farrell, S. (2010). *Computational modelling in cognition: Principles and practice.* Thousand Oaks, CA: Sage.

Lin, C.-Y., Hovy, E., & Rey, M. (2003). Automatic evaluation of summaries using n-gram co-occurrence statistics. In *Proceedings of the 2003 Conference of the North American Chapter of the Association for Computational Linguistics on Human Language Technology—Volume 1* (pp. 71–78). Stroudsburg, PA: Association for Computational Linguistics.

Mahamood, S., & Reiter, E. (2011). Generating affective natural language for parents of neonatal infants. In C. Gardent & K. Striegnitz (Eds.), *Proceedings of the 13th European Workshop on Natural Language Generation* (pp. 12–21). Stroudsburg, PA: Association for Computational Linguistics.

Mairesse, F., & Walker, M. A. (2010). Towards personality-based user adaptation: Psychologically informed stylistic language generation. *User Modelling and User-Adapted Interaction, 20*(3), 227–278.

Mairesse, F., & Young, S. (2014). Stochastic language generation in dialogue using factored language models. *Computational Linguistics, 4*(4), 763–799.

Mani, I. (2001). *Automatic summarization.* Amsterdam: John Benjamins.

Mann, W. C., & Moore, J. A. (1981). Computer generation of multiparagraph text. *American Journal of Computational Linguistics, 7*(1), 17–29.

Manurung, R., Ritchie, G., & Thompson, H. (2012). Using genetic algorithms to create meaningful poetic text. *Journal of Experimental and Theoretical Artificial Intelligence, 24*(1), 43–64.

McKeown, K. (1985). *Text generation.* Cambridge: Cambridge University Press.

Mellish, C. (2000). Understanding shortcuts in NLG systems. In T. Becker & S. Busemann (Eds.), *Proceedings of Impacts in Natural Language Generation: NLG between Technology and Applications. Workshop at Schloss Dagstuhl, Germany* (pp. 43–50). Kaiserslautern, Germany: Deutsches Forschungszentrum für Künstliche Intelligenz.

Mellish, C., & Dale, R. (1998). Evaluation in the context of natural language generation. *Computer Speech and Language, 12*(4), 349–373.

Mellish, C., Scott, D., Cahill, L., Paiva, D., Evans, R., & Reape, M. (2006). A reference architecture for natural language generation systems. *Natural Language Engineering, 12,* 1–34.

Mitchell, M., Han, X., Dodge, J., Mensch, A., Goyal, A., Berg, A., … Daumé III, H. (2012). Midge: Generating image descriptions from computer vision detections. In W. Daelemans (Ed.), *Proceedings of the 13th Conference of the European Chapter of the Association for Computational Linguistics* (pp. 747–756). Stroudsburg, PA: Association for Computational Linguistics.

Nenkova, A., & McKeown, K. (2011). Automatic summarization. *Foundations and Trends in Information Retrieval, 5*(2–3), 103–233.

Netzer, Y., Gabay, D., Goldberg, Y., & Elhadad, M. (2009). Gaiku: Generating haiku with word associations norms. In *Proceedings of the NAACL HLT Workshop on Computational Approaches to Linguistics Creativity* (pp. 32–39). Stroudsburg, PA: Association for Computational Linguistics.

Olson, R. (1970). Language and thought: Aspects of a cognitive theory of semantics. *Psychological Review, 77*(4), 257–273.

Papineni, K., Roukos, S., Ward, T., & Zhu, W. J. (2002). BLEU: A method for automatic evaluation of machine translation. In *Proceedings of the ACL-02 Conference on Empirical Methods in Natural Language Processing* (pp. 311–318). Stroudsburg, PA: Association for Computational Linguistics.

Paraboni, I., & van Deemter, K. (2014). Reference and the facilitation of search in spatial domains. *Language and Cognitive Processes, 29,* 1002–1007.

Pechmann, T. (1989). Incremental speech production and referential overspecification. *Linguistics, 27,* 98–110.

Poesio, M., Stevenson, R., Di Eugenio, B., & Hitzeman, J. (2004). Centering: A parametric theory and its instantiations. *Computational Linguistics, 30*(3), 309–363.

Reiter, E. (1994). Has a consensus NL generation architecture appeared and is it psycholinguistically plausible? In *Proceedings of the Seventh International Natural Language Generation Workshop (INLG 1994)* (pp. 163–170). Stroudsburg, PA: Association for Computational Linguistics.

Reiter, E., & Belz, A. (2009). An investigation into the validity of some metrics for automatically evaluating Natural Language Generation systems. *Computational Linguistics, 35*(4), 529–558.

Reiter, E., & Dale, R. (1997). Building natural language generation systems. *Natural Language Engineering, 3,* 57–87.

Reiter, E., & Dale, R. (2000). *Building Natural Language Generation systems.* Cambridge: Cambridge University Press.

Reiter, E., Robertson, R., & Osman, L. M. (2003). Lessons from a failure: Generating tailored smoking cessation letters. *Artificial Intelligence, 144*(1–2), 41–58.

Reiter, E., Sripada, S., Hunter, J., Yu, J., & Davy, I. (2005). Choosing words in computer-generated weather forecasts. *Artificial Intelligence, 167*(1–2), 137–169.

Socher, R., Karpathy, A., Le, Q. V., Manning, C. D., & Ng, A. Y. (2014). Grounded compositional semantics for finding and describing images with sentences. *Transactions of the Association for Computational Linguistics (TACL), 2,* 207–218.

van Deemter, K. (2016). *Computational models of referring: A study in cognitive science.* Cambridge, MA: MIT Press.

van Deemter, K., Gatt, A., van Gompel, R. P. G., & Krahmer, E. (2012). Towards a computational psycholinguistics of reference production. *Topics in Cognitive Science, 4,* 166–184.

Van der Sluis, I., & Mellish, C. (2010). Towards empirical evaluation of affective tactical NLG. In E. Krahmer & M. Theune (Eds.), *Empirical methods in natural language generation* (pp. 242–263). Berlin: Springer.

van Gompel, R. P. G., van Deemter, K., Gatt, A., Snoeren, R., & Krahmer, E. J. (2019). Conceptualization in reference production: Probabilistic modeling and experimental testing. *Psychological Review, 126*(3), 345–373.

Veale, T. (2013). Once more, with feeling! Using Creative affective metaphors to express information needs. In M. L. Maher, T. Veale, R. Saunders, & O. Bown (Eds.), *Proceedings of the Fourth International Conference on Computational Creativity* (pp. 16–23). Sydney, Australia: The University of Sydney.

Veale, T. (2015). Computational approaches to language and creativity: Creativity ex machina. In R. Jones (Ed.), *Routledge handbook of language and creativity* (pp. 353–366). New York: Routledge.

Veale, T., & Hao, Y. (2007). Comprehending and generating apt metaphors: A web-driven, case-based approach to figurative language. In R. C. Holte & A. Howe (Eds.), *Proceedings of the Twenty-Second AAAI Conference on Artificial Intelligence* (pp. 1471–1476). Menlo Park, CA: AAAI Press.

Walker, M. A., Park, F., Rambow, O., & Rogati, M. (2001). SPoT: A trainable Sentence Planner. In R. C. Moore, J. Bilmes, J. Chu-Carroll, & M. Sanderson (Eds.), *Proceedings of the Second Meeting of the North American Chapter of the Association for Computational Linguistics on Language technologies* (pp. 1–8). Stroudsburg, PA: Association for Computational Linguistics.

White, M., Rajkumar, R., & Martin, S. (2007). Towards broad coverage surface realization with CCG. In *Proceedings of the Workshop on Using Corpora for NLG: Language Generation and Machine Translation (UCNLG+MT)*. Copenhagen, Denmark.

Yang, Y., Teo, C. L., Daume, H., & Aloimonos, Y. (2011). Corpus-guided sentence generation of natural images. In R. Barzilay & M. Johnson (Eds.), *Proceedings of the Conference on Empirical Methods in Natural Language Processing* (pp. 444–454). Stroudsburg, PA: Association for Computational Linguistics.

Zimmerman, B. J., & Kitsantas, A. (2002). Acquiring writing revision and self-regulatory skill through observation and emulation. *Journal of Educational Psychology, 94*(4), 660–668.

26 Key Issues and Future Directions: Models of Human Language and Speech Processing

WILLEM ZUIDEMA AND HARTMUT FITZ

Computer programming and mathematics are not typically major components of educational programs in linguistics, and modeling papers only constitute a small fraction of the scientific literature that addresses the nature of languages and the cognitive apparatus to learn and use them. Nevertheless, it is striking how central mathematical and computational models have been and continue to be in many of the big debates in the language sciences, including those about the nature of the cognitive representations underlying language (such as discussed, for instance, in the "past tense debate"; see Pinker & Ullman, 2002, and references therein), innateness (e.g., Elman, Bates, & Johnson, 1998), learnability (e.g., Johnson, 2004), language change (e.g., Gray & Atkinson, 2003) and language evolution (e.g., Fitch, de Boer, Mathur, & Ghazanfar, 2016).

This central role of modeling in many different debates raises a number of questions, including: (a) why are there so many alternative schools of thought in modeling? (b) What are the advantages and attractions of modeling approaches that enable them to be taken so seriously? And (c) why has modeling, despite these advantages, not led to more consensus? In this closing chapter of the part *Modeling Language*, we will briefly consider these questions. We start describing the great variety of modeling approaches, as evidenced in the previous chapters and elsewhere. We will then highlight the advantages of having such a rich modeling toolbox at our disposal and discuss some of the inherent and some of the more easily avoidable causes of disagreement. We end with a number of suggestions for ways forward, including more systematic research on model comparison and a focus on developing models of the neurobiological basis of language.

1. *Why Are There So Many Different Modeling Paradigms?*

The primary medium of language is speech. Spoken utterances arrive at the ears of the receiver as patterns of vibrating air. The inner ear converts the stream of input into spike trains in the auditory nerve, and the receiver's brain somehow discovers (or imposes) segments in the input, recognizes segments as members of a category (phonemes, syllables, words), recognizes relations between segments in the input, assigns meaning to them, and decides whether the input is well-formed, incomplete, or ungrammatical.

In modeling the neural and cognitive processes involved in interpreting a spoken utterance (and similarly, in production, acquisition, and evolution of written or signed languages), modelers have to make a series of choices and simplify the unorderly, complex reality. Do we focus on the physical reality of the speech signal, on the neural reality of processing in the brain, or on the psychological reality of understanding a received message? The different research traditions reviewed in the previous chapters—including symbolic and neural network modeling paradigms that are often presented as incompatible—start from different answers to these questions.

For instance, the models of syntax and sentence processing reviewed in Demberg and Keller (chapter 22 of this volume), as well as many of the models of language generation reviewed in Krahmer (chapter 25) take abstract syntactic categories and hierarchical structure as a starting point, while greatly simplifying the nature of the signals. The various neural network models discussed in Frank, Monaghan, and Tsoukala (chapter 21) and Zuidema and Le (chapter 23), on the other hand, aim to account for how the empirical observations of linguistic behavior, with what at least to some extent looks like discrete categories and hierarchy, might emerge from the interaction between nodes with continuous activation values in a network. The work discussed in Wehbe, Fyshe, and Mitchell (chapter 24), then, addresses explicitly the relation of these kind of models with detectable activity in the human brain. Finally, the models of speech production and the vocal tract reviewed in de Boer (chapter 20), stay close to the physical signal.

These chapters already cover an enormous variety of approaches, but the variety of modeling paradigms is even larger than we have been able to represent in the chapters of part IV. One prominent framework not covered is that of Bayesian modeling.[1] Typically, Bayesian models are "rational" models, informing researchers about the *optimal strategy* under the assumed levels of uncertainty, without making direct claims about the actual cognitive processes that humans use. These models thus represent yet another idealization, orthogonal to the ones we already discussed (a good introduction of such Bayesian rational models, applied to the domain of language acquisition, can be found in Pearl & Goldwater, 2016).

We conclude that the great variety of modeling approaches that can be identified in language and speech research is to a large extent the unavoidable consequence of (a) the enormous complexity of language and speech processing, in combination with (b) the necessity of explanatory modeling to simplify. This richness could be an asset for research on language and speech, and rather than interpreting alternative models as rival accounts of language cognition, we would like to stress their complementary strengths. Differences among modeling paradigms are best viewed as different but defensible hard choices on simplifications necessary to reach a deeper understanding of how language and cognition work.

2. What Are the Advantages and Attractions of Modeling Approaches That Enable Them to Be Taken So Seriously?

Our second question concerns the advantages of models that have given them their prominent role in the cognitive and language sciences. An important part of the answer to this question is again the overwhelming complexity of speech and language. With so many interacting components, and so many crucial simplifications to keep track of, researchers need tools to clarify exactly what component in a theory plays what role, where, and when, and moreover they need tools to help derive (potentially counterintuitive) consequences of given assumptions.

The formalization and automation that computational models bring provide exactly these tools. There are endlessly many examples of models of speech and language that illustrate that formalization forces researchers to make theories much more precise, and as a result easier to criticize (which is a good thing). For instance, Demberg and Keller discuss garden path effects and related difficulties in language processing; the many models that have been developed to account for these difficulties have given increasingly precise characterizations of previously multi-interpretable notions such as "incompatibility," "reanalysis," and "semantic plausibility."

There is also a long tradition of using formal models to *automatically* generate predictions. In phonology, morphology, and syntax, the key paradigm of "generative linguistics" even reflects this feature in its name: it is named after a style of model building that is generative, that is, the automatic generation of new linguistic items that can be tested against native speaker intuition. A very different example can be found in Wehbe, Fyshe, and Mitchell, where neural networks are used to predict brain activation given a linguistic stimulus. Although the math for computing the activation of each individual node in these neural networks, and for updating the weights on connections between nodes, are well understood, obtaining the full prediction would be unfeasible without the automation that the implementation of computational models bring.

3. Why Has Modeling, Despite These Advantages, Not Led to More Consensus?

The answer to our third question, about continuing controversy, is more complex, and, ironically, more controversial—even the two authors of this chapter point to different causes for the ongoing controversies. Some of the disagreement about what the right modeling framework is, is unavoidable, due to the vastly different goals and interests of modelers. But some of the disagreements can be traced back to confusions among modelers themselves.

The first author of this chapter points to one particular cause of confusion that can be called "mission creep": The original goals of a modeling framework are gradually forgotten and simplifications that might have been defensible in earlier versions become problematic when models are repurposed to answer new questions. Important areas where we think mission creep plays a role, in many different subdomains of the language sciences, are the issues of how to deal with time (in particular whether and how to abstract out the time dimension), whether and how to convert the continuous physical signal into discrete objects, and how to represent structure in the input stream or sequence of units, including what could be called combinatorial structure, hierarchical structure, and/or slot-filler structure.

For instance, consider the practice of *abstracting out time* in neural network models, such as those discussed in the preceding chapters. These models are all so-called rate-coding networks: the "neurons" in these networks take continuous values, often between −1.0 and +1.0, representing the degree of activity of a single neuron or a group of neurons over a certain time window. The networks are typically organized in layers, and layers are updated one after the other, but all neurons in one layer in one go ("synchronous updating"; this implicitly imposes a global clock that regulates layer-by-layer updating). The exact timing of spikes, differences in oscillation phases, bursting, and other temporal relations between neurons are thus not represented in these models.

But time and timing are central to the study of language processing, which is reflected in many dependent measures used in psycholinguistics (e.g., reading and reaction times, fixation and regression in eye tracking, temporal characteristics of electroenceophalography and magnetoencephalography signals). To put time back into the models, a standard approach is to make the networks *recurrent* (Elman, 1990), but the relation between model time in recurrent networks and physical time in the systems they are meant to model is largely lost.

All this can still be a legitimate simplification, depending on the research questions. However, the simplification becomes problematic when the mission of modelers shifts away from providing an existence proof that neural networks can discover some structure in sentences (Elman, 1990) to questions that directly involve the way the human brain keeps track of time or to questions about supposed fundamental inabilities of neural networks. One example of the latter can be found in the literature on the "binding problem" in neural networks (see Kiela, 2011, for a discussion that requires more space than we have here); models that aim to solve the binding problem are forced to change the way time is represented, such that for instance binding-by-synchrony, or other uses of temporal information, become available.

The second author of this chapter emphasizes another source of confusion, which can be labeled "lack of constraints." Different computational modeling approaches—whether they are connectionist, symbolic, Bayesian, or other—have primarily been concerned with modeling the *output* of the language system, rather than the system itself. The aim of this approach is to redescribe speech and language processing data using some formalism or computational mechanism.

Clearly, however, any finite set of processing data can be captured by an infinity of formalisms, limited only be the imagination of the modeler. Thus, the choice between computational primitives, memory components, representational spaces, input encodings, learning algorithms, and such, is largely unconstrained. This toolbox approach to data recoding has yielded a vast array of distinct models that are based on different assumptions and abstractions. As a consequence, they are often difficult to compare, and rarely has it been possible to decide between architectures in a principled way. As we will discuss in section 4, apart from behavioral data, the neurobiology of the human language system provides a rich set of constraints that can inform the design of computational language models.

4. Ways Forward

The recommendations for the ways forward of the field "models of language" depend on the diagnosis. For the diagnosis "mission creep," the key issue is to continuously reevaluate the simplifications made, including implicit simplifications that modelers are hardly aware off. The remedy we recommend is a model comparison approach, where we develop very different models for the same phenomenon, using very different modeling paradigms and, through careful comparison, highlight the role of simplifications made.

For the diagnosis "lack of constraints," the key issue is to identify underappreciated sources of constraints that stem from characterizing the system itself. We argue that more attention should be paid to the neurobiological basis of language: models should be informed, first and foremost, by the properties of the neurobiological infrastructure that supports language and speech processing. We refer to this approach as *causal modeling* (Fitz et al., 2019).

4.1. Model Comparison Model comparison in the broad sense is standard practice in the modeling literature—every new version of a model is compared with an earlier version—but this broad sense is not what we have in mind here. Rather, we argue that modelers of language and speech should more often systematically compare models from very different modeling paradigms for the same phenomenon: spiking networks versus rate-coding networks, logic-based models of inference versus recursive neural networks (e.g., Bowman, Potts, & Manning, 2015), formal grammar-based models of grammaticality versus recurrent neural networks (e.g., Gulordava, Bojanowski, Grave, Linzen, & Baroni, 2018). Such cross-paradigm comparisons require taking both paradigms seriously, but they shed light on the implicit and explicit assumptions that both paradigms embody and thus yield an opportunity to reevaluate those assumptions.

An early example of such cross-paradigm comparisons yielding important insights is the discovery that context-free grammars, push-down automata, and augmented transition networks are really alternative ways at processing context-free languages and can be translated into each other. Another is the series of discoveries made about the relation between rival syntactic frameworks, including the finding that minimalism, combinatory categorical grammar, and lexicalized tree adjoining grammar have essentially the same binding theory (Steedman & Baldridge, 2011). Zuidema (2003) provided another example, implementing the differential equation model of Nowak, Komarova, and Niyogi (2001) in an agent-based simulation. Detailed comparison of the two models revealed that Nowak et al.'s result of the necessity of a finite and small size of the number of possible grammars (the "size of Universal Grammar") was crucially dependent on the problematic assumption of a uniform probability distribution over possible grammars that the learner is exposed to.

All these examples illustrate the benefits of seriously comparing models from different modeling traditions, accounting for the same phenomenon: Some disagreements disappear, as the supposed contradictions disappear on closer inspection. Other disagreements might remain, but are traced back to differences in more fundamental assumptions.

4.2. Causal Modeling Human language is a neurobiological system implemented as a sparsely connected, recurrently coupled network of highly dynamic neurons and synapses. Existing computational models of language processing have made little or no contact with the detailed biophysical properties of this system and have focused mainly on the reproduction of behavioral data.

Models of behavior start at the computational or algorithmic level of description (in the terminology of David Marr) and attempt to reverse engineer the language system from input-output relations. Recent experimental work, however, has shown that it is difficult to reverse engineer even simple computational systems (e.g., a microprocessor) whose functionality is completely known (Jonas & Kording, 2017). This methodological issue is exacerbated with increased system complexity and the noisiness of measurements made from the human language system.

A causal modeling approach, on the other hand, starts out at the implementational level of description and attempts to synthesize language function in the brain from first principles. These include, for instance, the principle of state-dependence (Buonomano & Maass, 2009) and the principle of information processing as computation over high-dimensional transients (Rabinovich, Huerta, & Laurent, 2008). Moreover, causal modeling attempts to capture the characteristics of the biophysical system itself, rather than to reproduce a particular aspect of behavior. The aim is to understand, through simulation and theoretical insight, how neural computation and memory in cortical circuits support language behavior. The extent to which a causal model approximates the neurobiological infrastructure of the real system determines how humanlike it will behave. Reproduction of behavior, however, is viewed as an independent outcome, not the primary goal of modeling.

Causal models also differ from models of behavior in that parameters have physical units of measurement that need to fall within physiological bounds. This places strong constraints on the model space and reduces degrees of freedom. In addition, many of the arbitrary design choices in models of behavior turn into empirical questions (see, e.g., Bartol et al., 2015). While models of behavior often attempt to fit data with as few parameters as possible, the challenge for causal models is to deal with the abundance of parameters provided by the neurobiological system (e.g., $\sim 10^{14}$ synaptic conductances).

Also the issue of time, which we discussed in section 3, can be dealt with from the causal modeling perspective. Here, network time corresponds to real physical time, since it arises from biophysical models of spike generation and the dynamics of synaptic transmission (Gerstner, Kistler, Naud, & Paninski, 2014). Due to this nomological relation, causal models allow us, in principle, to investigate how language and speech processing unfold over time at any desired grain size (e.g., on a millisecond scale).

Taking a causal modeling approach to language does not mean that we ought to replicate the neurobiological substrate at all levels of detail. Even causal models can (and should) be high-level abstractions of the underlying physiology in that they are composed of mathematically reduced but phenomenologically adequate parts. Many such parts have been described in computational neuroscience, for instance, the adaptive-exponential neuron (Brette & Gerstner, 2005), models of short-term synaptic facilitation and depression (Markram, Wang, & Tsodyks, 1998), mechanisms of homeostatic plasticity

(Vogels, Sprekeler, Zenke, Clopath, & Gerstner, 2011), long-term potentiation (Clopath, Büsing, Vasilaki, & Gerstner, 2010), and principles of synaptic consolidation (Clopath, Ziegler, Vasilaki, Büsing, & Gerstner, 2008). Using these experimentally validated components ensures that models will gradually begin to approximate the dynamic properties of the human language system.

As this chapter has attempted to highlight, it has been difficult to unify computational approaches to language and speech processing into a coherent framework. On the view outlined here, we should strive to replace data modeling by models of the neurobiological language system itself. This would mark a much needed paradigm shift in computational language modeling. The long-term goal of this approach is not to describe processing in terms of interacting neurons and synapses, but to distill an algorithmic abstraction from the neurobiological substrate that characterizes cognitive function.

There are many challenges involved in causal language modeling. For instance, complex networks of spiking neurons with plastic synapses are computationally costly to simulate, and their behavior is often difficult to interpret. These challenges can be overcome through the development of suitable neuromorphic architectures, novel approaches such as event-driven simulation, and large-scale team efforts to analyze and understand the computational role of component parts through model comparisons. Eventually, this approach might be able to bridge descriptions of brain function across all levels of explanation (Carandini, 2012) and yield a causal, mechanistic understanding of the human capacity for language.

NOTE

1. Models in the Bayesian framework define probability distributions over potentially quite rich structures (ranging from phonetic structure, as in de Boer, to syntactic structure, as in Demberg & Keller). The probability distributions are defined indirectly, by setting up a stochastic process (a "generative model") that generates the linguistic structures of interest. Using a rich toolbox of computational techniques, the probability distribution over possible data is used to compute the probability of observed data given a model and, using Bayes' law, to find the model or models that make the data maximally probable.

REFERENCES

Bartol, T. M., Jr., Bromer, C., Kinney, J., Chirillo, M. A., Bourne, J. N., Harris, K. M., & Sejnowski, T. J. (2015). Nanoconnectomic upper bound on the variability of synaptic plasticity. *eLife, 4,* e10778.

Bowman, S. R., Potts, C., & Manning, C. D. (2015). Recursive neural networks can learn logical semantics. In A. Allauzen, E. Grefenstette, K. M. Hermann, H. Larochelle, S. Wen-tau Yih (Eds.), *Proceedings of the 3rd Workshop on Continuous Vector Space Models and their Compositionality (CVSC)* (pp. 12–21). Stroudsburg, PA: Association for Computational Linguistics.

Brette, R., & Gerstner, W. (2005). Adaptive exponential integrate-and-fire model as an effective description of neuronal activity. *Journal of Neurophysiology, 94,* 3637–3642.

Buonomano, D. V., & Maass, W. (2009). State-dependent computations: Spatiotemporal processing in cortical networks. *Nature Reviews Neuroscience, 10,* 113–125.

Carandini, M. (2012). From circuits to behavior: A bridge too far? *Nature Neuroscience, 15,* 507–509.

Clopath, C., Büsing, L., Vasilaki, E., & Gerstner, W. (2010). Connectivity reflects coding: A model of voltage-based STDP with homeostasis. *Nature Neuroscience, 11,* 344–352.

Clopath, C., Ziegler, L., Vasilaki, E., Büsing, L., & Gerstner, W. (2008). Tag-trigger-consolidation: A model of early and late long-term-potentiation and depression. *PLoS Computational Biology, 4,* 1–14.

Elman, J. L. (1990). Finding structure in time. *Cognitive Science, 14*(2), 179–211.

Elman, J. L., Bates, E. A., & Johnson, M. H. (1998*). Rethinking innateness: A connectionist perspective on development* (Vol. 10). Cambridge, MA: MIT Press.

Fitch, W. T., de Boer, B., Mathur, N., & Ghazanfar, A. A. (2016). Monkey vocal tracts are speech-ready. *Science Advances, 2*(12), e1600723.

Fitz, H., van den Broek, D., Uhlmann, M., Duarte, R., Hagoort, P., & Petersson, K. M. (2019). Neuronal memory for language processing. *bioRxiv.* https://doi.org/10.1101/546325.

Gerstner, W., Kistler, W. M., Naud, R., & Paninski, L. (2014). *Neuronal dynamics: From single neurons to networks and models of cognition.* Cambridge: Cambridge University Press.

Gray, R. D., & Atkinson, Q. D. (2003). Language-tree divergence times support the Anatolian theory of Indo-European origin. *Nature, 426*(6965), 435.

Gulordava, K., Bojanowski, P., Grave, E., Linzen, T., & Baroni M. (2018). Colorless green recurrent networks dream hierarchically. In M. Walker, H. Ji & A. Stent (Eds.), *Proceedings NAACL-HLT 2018* (pp. 1195–1205). Stroudsburg, PA: Association for Computational Linguistics.

Johnson, K. (2004). Gold's theorem and cognitive science. *Philosophy of Science, 71*(4), 571–592.

Jonas, E., & Kording, K. P. (2017). Could a neuroscientist understand a microprocessor? *PLoS Computational Biology, 13,* e1005268.

Kiela, D. (2011). Variable binding in biologically plausible neural networks (Unpublished master's thesis). University of Amsterdam. Retrieved from https://eprints.illc.uva.nl/id/document/1994.

Markram, H., Wang, Y., & Tsodyks, M. (1998). Differential signaling via the same axon of neocortical pyramidal neurons. *Proceedings of the National Academy of Sciences, 95,* 5323–5328.

Nowak, M. A., Komarova, N. L., & Niyogi, P. (2001). Evolution of universal grammar. *Science, 291*(5501), 114–118.

Pearl, L., & Goldwater, S. (2016). Statistical learning, inductive bias, and bayesian inference in language acquisition. In J. Lidz, W. Snyder, & J. Pater (Eds.), *Oxford handbook of developmental linguistics* (pp. 664–695). Oxford: Oxford University Press.

Pinker, S., & Ullman, M. T. (2002). The past-tense debate: The past and future of and weaknesses of connectionist and rule-based models of language and cognition. *Trends in Cognitive Sciences, 6*(11), 456–463.

Rabinovich, M., Huerta, R., & Laurent, G. (2008). Transient dynamics for neural processing. *Science, 321*, 48–50.

Steedman, M., & Baldridge, J. (2011). Combinatory categorial grammar. In R. Borsley & K. Borjars (Eds.), *Non-transformational syntax: Formal and explicit models of grammar* (pp. 181–224). Hoboken, NJ: Wiley-Blackwell.

Vogels, T. P., Sprekeler, H., Zenke, F., Clopath, C., & Gerstner, W. (2011). Inhibitory plasticity balances excitation and inhibition in sensory pathways and memory networks. *Science, 334*, 1569–1573.

Zuidema, W. H. (2003). How the poverty of the stimulus solves the poverty of the stimulus. In S. Becker, S. Thrun, & K. Obermayer (Eds.), *Advances in neural information processing systems 15* (pp. 51–58). Cambridge, MA: MIT Press.

V THE FUNCTIONAL NEUROBIOLOGY OF LANGUAGE

SECTION EDITOR: DAVID POEPPEL

27 The Cortical Processing of Speech Sounds in the Temporal Lobe

MATTHIAS J. SJERPS AND EDWARD F. CHANG

Speech perception is a complex process that transforms the continuous stream of clicks, hisses, and vibrations that make up speech sounds into meaningful linguistic representations. This process unfolds at a remarkable speed, as naturally spoken speech typically contains around five syllables per second (Ding et al., 2017; Miller, Grosjean, & Lomanto, 1984). The cortical processing of spoken language involves a network of regions in the temporal, parietal, and frontal lobes in which the specific involvement of regions may vary depending on the task demands or goals of the listener (Hickok & Poeppel, 2004). It is widely recognized, however, that the posterior portions of the superior temporal gyrus (STG) and superior temporal sulcus (STS; see figure 27.1) play a pivotal role in early processing of speech sounds (e.g., Belin, Zatorre, Lafaille, Ahad, & Pike, 2000; Hickok & Poeppel, 2004; 2007; 2015; Rauschecker & Scott, 2009).

Indeed, local disruption of neural activity with focal electrical stimulation of the STG leads to sensory errors and/or phonemic errors (see, e.g., Boatman, 2004; Boatman, Hall, Goldstein, Lesser, & Gordon, 1997; Leonard, Cai, Babiak, Ren, & Chang, 2016; Quigg & Fountain, 1999; Roux et al., 2015). Furthermore, damage to the posterior part of the superior temporal lobe (STL, i.e., STS and STG combined) has been repeatedly associated with speech-perception deficits (Buchman, Garron, Trost-Cardamone, Wichter, & Schwartz, 1986; Buchsbaum, Baldo, et al., 2011; Rogalsky et al., 2015; Wilson et al., 2015). The STL is thus thought to play a critical role in the transformation of acoustic information into phonetic and prelexical representations.

One of the major questions that drives current research on early speech sound processing is the actual nature of speech representations in the STL (the STL is defined here as the *lateral parabelt auditory cortex*, including parts of Brodmann areas 41, 42, and 22; Hackett, 2011). Does this region mostly represent acoustic features (i.e., a responsiveness to energy at specific frequencies or perhaps to sounds for which the dominant frequencies change over time)? Or does this region mostly represent linguistic units such as phonemes and/or syllables? And how does the brain arrive at phonetic representations (or another form of prelexical representation) that allows for lexical access independently of how or by whom the speech sound was produced (i.e., abstract representations)? These questions are of particular importance for understanding the processing of spoken language as a whole because the representations in the STL constitute a critical link in processing, receiving direct input from primary input areas as well as interacting with associative auditory areas with higher-level representations (DeWitt & Rauschecker, 2012; Hickok & Poeppel, 2004; 2007; Lerner, Honey, Silbert, & Hasson, 2011; Rauschecker & Scott, 2009; Scott & Johnsrude, 2003; Steinschneider et al., 2011).

The current chapter provides a review of several concepts and recent findings that have informed our understanding of the role of the STL in early speech sound processing. Because this field of research is broad and highly active, we will focus our discussion by especially highlighting research that addresses the nature of speech sound representations in the STL. This approach, focusing on representations as distributed patterns of activation, has been especially informed by noninvasive imaging methods such as functional MRI (fMRI) and magnetoencephalography. In addition, invasive methods such as electrocorticography (ECoG) recordings, the main method used in our work, have also contributed meaningfully to research.

In section 1, we will briefly discuss speech sound processing in the primary auditory cortex (PAC), the main source of input for the STL with regard to acoustic information (chapter 35 by Formisano in this volume provides a more in-depth description of the language-relevant dominant properties of PAC organization). Subsequent sections will discuss the representation of speech sounds as acoustic phonetic features, the emergence of categorical/abstract representations, and how these representations are influenced by visual cues and other "contextual information" such as phoneme sequencing and lexical-semantic representations.

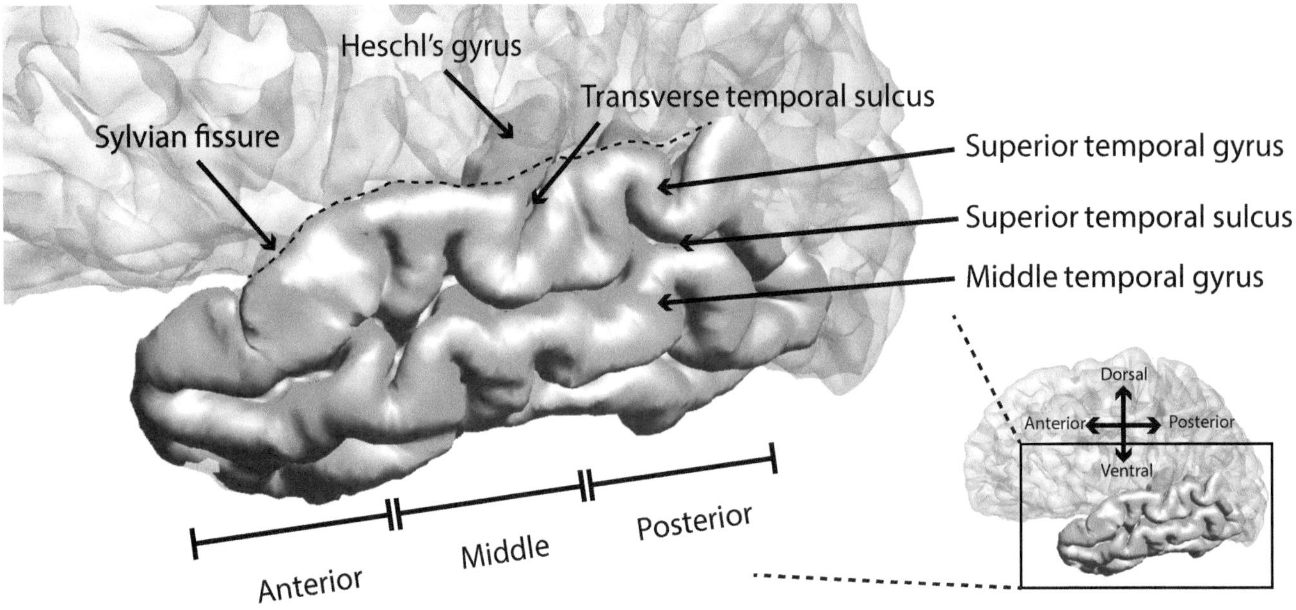

FIGURE 27.1 Anatomical landmarks of the temporal lobe on and around the regions involved in early speech sound processing. Regions outside the temporal lobe are displayed as transparent, allowing for the visualization of Heschl's gyrus, which is located inside the Sylvian fissure.

The research discussed here stresses the role of the STL as a highly versatile auditory association cortex that displays sensitivity to acoustic patterns at multiple levels of granularity (i.e., from acoustic features to phoneme sequences) but is also robustly influenced by concurrent visual information and lexical-semantic context. Moreover, abstraction, the property that allows for categorical and context-invariant mapping, seems to be an emergent but distributed property of processing in the STL.

1. *From Acoustics to Prelexical Abstraction*

1.1. REPRESENTATIONS IN PAC AND CLOSELY SURROUNDING REGIONS It is important to understand the functional pathway through which key speech auditory regions receive most of their input. The ascending auditory pathway projects to PAC through afferent input from the medial geniculate complex, which is part of the thalamus. Processing at these subcortical levels is subject to important transformations and is already influenced by linguistic and musical exposure (Bidelman, Gandour, & Krishnan, 2011; Krishnan, Gandour, & Bidelman, 2012; Weiss & Bidelman, 2015). Important for the current review, however, is that the representations also largely transmit the time-frequency properties of the sound waveform (Shamma & Lorenzi, 2013; Weiss & Bidelman; Young, 2008). This information is transmitted in a partly nonlinear fashion especially along the frequency axis (i.e., frequency resolution follows the so-called *mel scale*, which is a loglike scale, overrepresenting lower frequencies). PAC in humans is mostly confined to the bilateral transverse temporal gyrus (Heschl's gyrus; see figure 27.1). Its organization is traditionally characterized as having neuronal populations that display very fine frequency tuning, with at least two mirror-symmetric tonotopic frequency gradients (Bauman, Petkov, & Griffiths, 2013; Bitterman, Mukamel, Malach, Fried, & Nelken, 2008; Humphries, Liebenthal, & Binder, 2010; Moerel, De Martino, & Formisano, 2012; Saenz & Langers, 2014). As a result, sound representations in PAC allow for the transmission of acoustic cues that are critical for the perception of speech such as formants, formant transitions and amplitude modulations (e.g., Young, 2008). In addition to tonotopic representations, however, studies in animal models have also demonstrated more complex properties in PAC, such as tuning for temporal and spectral modulations rather than specific frequency representations per se (e.g., Schreiner, Froemke, & Atencio, 2011).

Secondary auditory areas such as the planum temporale (PT; located posterior to Heschl's gyrus) and the lateral STG largely depend on inputs from PAC (Hackett, 2011). This flow of information is facilitated by (bidirectional) functional connections between parts of PAC and its closely surrounding region, as well as direct projections from auditory thalamus. This has been demonstrated, for example, by activity in the laterally exposed STG that is observed at very short latencies after electrical stimulation in the PAC (Brugge,

Volkov, Garell, Reale, & Howard, 2003). Functionally, the regions immediately surrounding PAC, both within the Sylvian fissure and on the lateral part of the STG, display both tuning to narrow frequency ranges and sensitivity to increasingly complex spectrotemporal information. To exemplify, parts of the lateral STL display fairly low-level acoustic response properties. For example, Nourski et al. (2012) observed strong responses to simple pure tone stimuli in a restricted region surrounding the laterally exposed part of the transverse temporal sulcus (see figure 27.1), which runs parallel along the posterior side of Heschl's gyrus. The observation that this region inherits some amount of tonotopic organization is further supported by a body of research (Humphries et al. 2010; Moerel et al., 2012; Nourski et al.; Striem-Amit, Hertz, & Amedi, 2011; Talavage et al., 2004). In addition to these tonotopic characteristics, however, the regions surrounding PAC also display spectral preferences that become more complex, with more widespread tuning at octave intervals and harmonically related frequency intervals (Moerel et al., 2013; Ohl & Scheich, 1997). The characteristics of auditory representations in the PAC, on the one hand, and the lateral STG, on the other, are, thus, partly overlapping. However, the dominant representation in PAC is one of tonotopic distributions, whereas the more dominant forms of representation outside of PAC are of a more complex spectrotemporal nature (Hullett, Hamilton, Mesgarani, Schreiner, & Chang, 2016).

1.2. STL: From Spectrotemporal Receptive Fields to Speech Sound Representations The predominant preference for more complex spectrotemporal patterns shows that large portions of the medial and posterior STL transform relatively basic acoustic properties such as pure tones and sweeps into combined representations (Hickok & Poeppel, 2015; Peelle, Johnsrude, & Davis, 2010; Rauschecker & Scott, 2009). A popular approach in research on the involvement of the STL in speech sound processing has been to compare levels of activation to clear speech sounds with degraded speech sounds or nonspeech sounds (Belin et al., 2000; Binder et al., 2000; Davis & Johnsrude, 2003; Liebenthal, Binder, Spitzer, Possing, & Medler, 2005; Obleser, Eisner, & Kotz, 2008; Obleser, Zimmermann, Van Meter, & Rauschecker, 2007; Rosen, Wise, Chadha, Conway, & Scott, 2011; Scott, Blank, Rosen, & Wise, 2000; Takeichi, et al., 2010; Turkeltaub & Coslett, 2010; Zaehle, Geiser, Alter, Jancke, & Meyer, 2008). The general view that arises from this body of research is a hierarchy of responsiveness to increasingly speech-specific signal characteristics as activation spreads to more anterior and ventral regions (see Obleser & Eisner, 2009; Price, 2012, for general review, and Liebenthal, Desai, Humphries, Sabri, & Desai, 2014; Turkeltaub & Coslett; DeWitt & Rauschecker, 2012, for fMRI- and positron-emission tomography [PET]-based Activation Likelihood Estimation [ALE] meta-analyses). Turkeltaub and Coslett, for example, performed two ALE meta-analyses on studies that compared sublexical speech versus nonspeech signals. In a first analysis, they compared listening to speech with listening to relatively simple nonspeech signals (i.e., listening to isolated vowels or consonant-vowel sequences, compared to a variety of nonspeech signals such as pure tones, band-passed noise, music). Their analysis revealed large clusters in the bilateral STG extending into the STS that respond more strongly to speech, suggesting that these regions are involved in the processing of more complex acoustic properties of speech sounds. A subsequent analysis compared sublexical speech sounds with nonspeech stimuli that were closely matched to speech in terms of their spectrotemporal properties. This second analysis revealed a much smaller region of speech specificity, mostly located in the left STS but extending somewhat into the ventral bank of the left STG (see Desai, Liebenthal, Waldron, & Binder, 2008; DeWitt & Rauschecker; Jäncke, Wüstenberg, Scheich, & Heinze, 2002; Liebenthal et al., 2005, 2010, 2014; and Price, 2012, for very similar results), suggesting that only this more ventral portion of the left STL was involved in speech processing per se.

To further exemplify, a recent fMRI study demonstrated that cortical regions in the posteromedial STS reveal a preference for speechlike sounds over sounds that are matched by a number of spectrotemporal characteristics (Overath, McDermott, Zarate, & Poeppel, 2015). Overath et al. divided natural speech signals into short sound segments that were then randomly reshuffled but which adhered to local speechlike statistics. They found that the STS is increasingly sensitive to speechlike sounds when those sounds consisted of increasingly longer segments (ranging from 30 ms to ~1 s). The STS thus seemed to prefer speech sound cues, but increasingly so if they appear in a longer sequence. That is, STL prefers sequences that conform to the way that speech input is typically heard in more everyday listening situations. This was despite the fact that all stimuli (both long and short sequences) were equally meaningless to the listeners. Similarly, Canolty et al. (2007) compared ECoG responses in patients listening to both clear speech stimuli and to nonphonemic speechlike sounds (speech sounds for which specific formant details were removed, rendering complex but unintelligible sounds). Cortical responses in the STG were larger for speech than for the nonphonemic sounds. Interestingly, the differences in activation arose in a serial manner, with differences arising in posterior STG at ~120 ms, in the mid-STG at ~193 ms, and in the mid-STS at ~268 ms. This suggests a similar gradient across the STL (posterior to anterior/medial) of both the temporal progression of information and of increasing speech specificity.

Despite the demonstration of a gradient across the STL displaying selectivity for increasingly speechlike sounds, a fundamental question that remains is what properties of speech sounds are reflected by this neural activity. The dominant representations could be, among others, low-level spectrotemporal parameters, acoustic-phonetic features, or phonemes. In addressing this question, however, it is important to note that receptive fields and response properties of neurons are tightly matched to the statistics of natural input. This property has been demonstrated in both animal- and human-based research (David, Vinje, & Gallant, 2004; Hsu, Woolley, Fremouw, & Theunissen, 2004; Rieke, Bodnar, & Bialek, 1995; Talebi & Baker, 2012; Theunissen, Sen, & Doupe, 2000; Young, 2008). To obtain a detailed picture of how the human STL integrates auditory features into some form of higher-level representations, it is thus important to rely on natural or ecologically valid stimuli.

A powerful approach that has been developed in animal research to investigate auditory representations is *spectrotemporal receptive field (STRF) estimation.* STRFs are computed by first recording activity from a neural site in response to acoustic input. Then, through a procedure such as reverse correlation (e.g., Klein, Depireux, Simon, & Shamma, 2000; Theunissen et al., 2000; see, e.g., Hullett et al., 2016, who describe another estimation method called maximally informative dimension analysis), properties of the acoustic spectrogram are established that are found to either excite or inhibit neural activity (i.e., specific frequency bands at particular time lags). This method is distinguished from other measures by its broader descriptive power for encompassing both dynamics and spectral selectivity and for not requiring much prior knowledge such as frequency tuning or threshold. Moreover, this method has some advantages for analyzing responses to using natural stimuli.

Recently, STRF models have been used to describe the encoding of specific stimulus features in human STG for both speech and nonspeech input (e.g., Hullett et al., 2016; Mesgarani, Cheung, Johnson, & Chang, 2014; see figure 27.2A). Hullett et al., for example, observed that the human STG displays an anterior to posterior organization of different types of spectrotemporal tuning. Using ECoG, they demonstrated that sites toward the posterior STG were found to be increasingly tuned for speech sounds that have relatively constant energy across the frequency range (low spectral modulation; see figure 27.2B) but which are temporally changing at a fast rate. In contrast, sites toward the anterior STG were found to be increasingly tuned for speech sounds that show a high degree of spectral variation across the frequency range (high spectral modulation) and which are temporally changing at a slow rate (see Santoro et al., 2014 for corroborating findings from fMRI). This sensitivity to two types of modulations seems to be an important property of processing in the auditory processing stream that has been observed in animal models as well (e.g., Woolley, Fremouw, Hsu, & Theunissen, 2005; Nagel & Doupe, 2008). Of the regions in STG that Hullett et al., found to be responsive to basic auditory properties (these were mostly confined to our definition of posterior and medial STG in figure 27.1), on average about 23% of the variance in neural activity at specific sites could be explained by the patterns described in STRFs (Hullett et al., 2016; Pasley et al., 2012). This shows that a significant component of the information represented in the posterior and medial STG is closely related to acoustic features rather than higher-level (e.g., lexical or semantic) ones.

Despite this sensitivity to low-level acoustic properties of sound, it is clear that processing in the STG is strongly related to the behavioral relevance of the input. For example, when pure-tone selectivity is observed in the STG, it is generally confined to the low-frequency portions of the tonotopic map, which are the frequencies that are predominant in human voice sounds and speech, in particular (Moerel et al., 2012). Furthermore, auditory-based predictions of activity in the STG such as those already described (i.e., figure 27.2A) perform best for the ranges of spectrotemporal modulations most critical to speech intelligibility (Chi, Gao, Guyton, Ru, & Shamma, 1999; Elliott & Theunissen, 2009; Pasley et al., 2012). As another example, dynamic ripple stimuli (combinations of fluctuating sine tones) contain the same basic spectrotemporal modulations that are reflected in the speech, yet, because they occur in nonbehaviorally relevant auditory objects, they don't excite regions in the STG the way that speech does (Hullett et al., 2016). These findings suggest that the auditory stimulus preferences for regions on the lateral STL are closely related to the general acoustic properties of speech sounds.

In an investigation of how tuning to spectrotemporal properties in the STG is related to the processing of speech sounds, Mesgarani et al. (2014) presented listeners with a large number of naturally spoken sentences. Their participants were patients undergoing surgical monitoring for medically refractory epilepsy and were

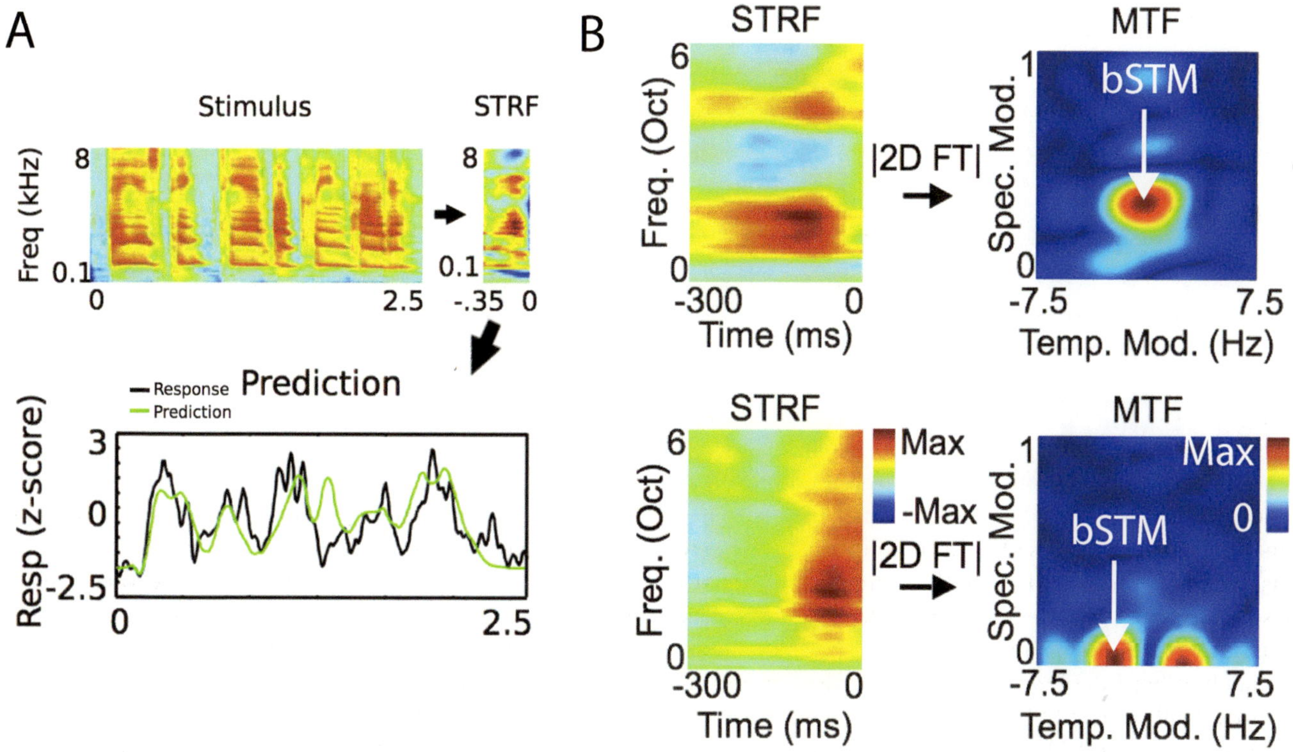

FIGURE 27.2 (A) STRF mapping. The spectrogram of a spoken sentence, an STRF, and the predicted and measured response for the sentence. Predicted responses are obtained by convolving the stimulus spectrogram with the STRF and are proportional to the similarity between the spectrotemporal content in the stimulus and the receptive field. (B) Computation of the modulation transfer function (MTF). The MTF is derived as the magnitude of the two-dimensional Fourier Transform (2D FT) of the STRF. It characterizes spectrotemporal modulation tuning for each site. Like the "best frequency" of a frequency tuning curve, the peak of the MTF defines the "best spectrotemporal modulation" (bSTM). For the site with the STRF shown on the top row, the MTF indicates that high spectral modulations and low temporal modulations drive activity at that site (i.e., prototypical of anterior sites). In contrast, the site on the bottom row has a bSTM at high temporal modulations and low spectral modulations, indicating that the site is driven by changes in temporal and not spectral energy (i.e., prototypical of posterior sites; figure reproduced from Hullet et al., 2016).

implanted with ECoG grids directly over the perisylvian cortex for clinical purposes (figure 27.3A). They listened to speech samples from the TIMIT corpus (figure 27.3B) covering a wide range of different sentences and speakers. Focal patterns of activity on the cortex of individual patients revealed selectivity for phonetic features, but not for individual phonemes. For example, one electrode (e1; figure 27.3C) displayed a reliable response to plosive phonemes /d/, /b/, /g/, /p/, /k/, and /t/. E2 displayed a reliable response to sibilant fricatives: /ʃ/, /z/, and /s/. E3 displayed a reliable response to low-back vowels (e.g., /a/ and /aʊ/). E4 displayed a reliable response to high-front vowels and glides (/i/ and /j/). And e5 was selective for nasals (/n/, /m/, and /ŋ/).

Mesgarani et al. then used unsupervised hierarchical clustering analyses on an electrode-specific measure of phoneme selectivity to find groups of electrodes with similar response characteristics. The clusters of electrodes that emerged from this procedure revealed very specific and speech-relevant STRFs. For example, a first cluster revealed an STRF (top row of figure 27.3D) displaying tuning for broadband excitation, a spectral property that is indicative of plosives (first panel in bottom row of figure 27.3D). A second cluster revealed an STRF that was tuned to a high-frequency component that is a defining feature of sibilant fricatives. Further clusters displayed STRFs indicative of other classes of speech sounds such as tuning for characteristic formants that define low-back, low-front, and high-front vowels, and a cluster revealed tuning for low acoustic frequencies, conforming to a general property of nasal speech sounds. These observations suggest that, at least on the lateral surface of the STG, speech sounds are most dominantly encoded as acoustic-phonetic features such as manner of articulation (e.g., /ta/ vs. /sa/) and voicing (e.g., /ba/ vs. /pa/). Moreover, features with acoustically very distinct cues such as manner of articulation were very strong determinants of selectivity, while acoustically weaker distinctions such as place of

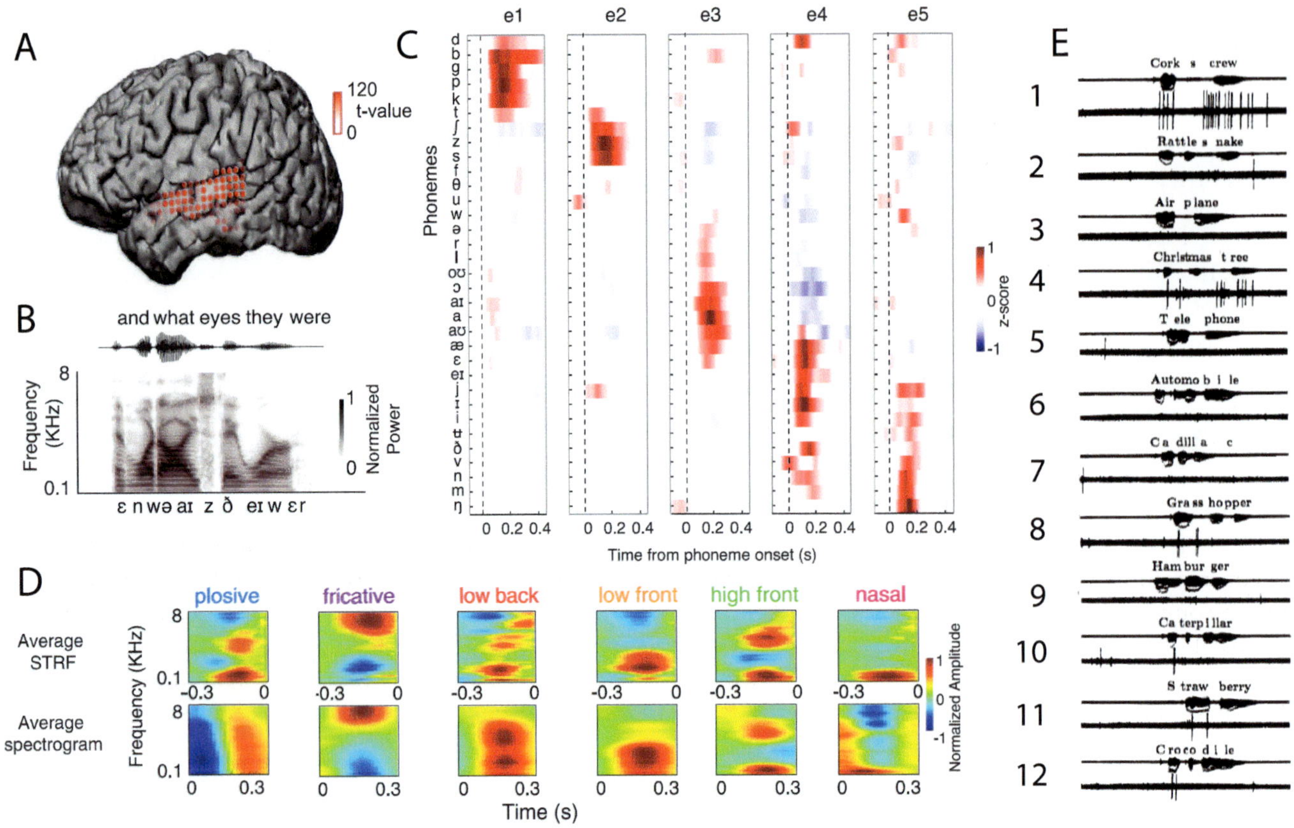

FIGURE 27.3 Human STG cortical selectivity to speech sounds. (A) MRI surface reconstruction of one participant's cerebrum. Auditory-responsive electrodes (red) are plotted with opacity signifying the *t*-test value when comparing responses to silence and speech. (B) Example sentence and its acoustic waveform, spectrogram, and phonetic transcription. (C) Average responses at five example electrodes to all English phonemes. (D, *top row*) Weighted average STRFs of main electrode clusters; (*bottom row*) average acoustic spectrograms for phonemes associated with each population cluster. (E) Oscillograms and corresponding responses of a single neuron in the right STG during passive listening to a list of 12 words. Panels A–D reproduced from Mesgarani et al. (2014); panel E reproduced from Creutzfeldt et al. (1989).

articulation (e.g., /pa/ vs. /ta/) were much less discriminable. It is important to note that single neuron recordings also failed to show selectivity to single phonemes (Chan et al., 2014; Creutzfeldt, Ojeman, & Lettich, 1989), suggesting that this phonetic feature organization is not simply a confound of the meso-scale ECoG sampling of thousands of neurons (see figure 27.3E, which displays activation after obstruent consonants such as /k/ and /t/, especially when these appear in obstruent clusters, e.g., *st* or *sk*). The data presented by Mesgarani et al. reveal that feature-level selectivity is a dominant property for STG processing of speech sounds (see also Arsenault & Buchsbaum, 2015; Steinschneider et al., 2011). This demonstrates that the organization of speech sounds in the STL is tightly linked to acoustic/phonetic cues and not to discrete phonemic or even articulatory ones.

The findings discussed in this section demonstrate that processing across the STG is dedicated to complex spectrotemporal events. The spectral and temporal modulation ranges that are covered by the STG closely align to the spectrotemporal properties that are important for the processing of natural speech sounds. This observation helps to better understand why processing across the STL becomes increasingly speech-specific: only those stimuli that involve longer sequences that adhere to speechlike statistics evoke activity in the more ventral and anterior regions of the STL. Moreover, it is clear that the dominant form of speech sound representations is one that reflects speech sound features rather than specific phonemes per se. That is, representations in the STL are closely related to the acoustic properties of the natural classes of speech sounds, such as fricatives, vowels, plosives.

1.3. ABSTRACT REPRESENTATIONS AND CONTEXTUAL INVARIANCE One of the main challenges of speech perception is that there is no one-to-one mapping between sounds and words or even between sounds and some form of prelexical representation such as features, phonemes, or syllables. This lack of a one-to-one mapping has many origins, such as differences in listening conditions, speakers with differently sized vocal tracts and speaker accents. To uniquely select lexicosemantic representations, however, there must exist a form of abstraction in the neural hierarchy that allows listeners to rely on some form of contextually invariant code. One way in which such abstraction has been demonstrated behaviorally is through categorical perception experiments where tokens on an acoustically linear continuum (e.g., a linear continuum spanning from the speech sound *ba* to *pa*, a distinction primarily cued by Voice Onset Time) show a nonlinear, sigmoidal, pattern of categorization by listeners (Harnad, 1987; Liberman, Harris, Kinney, & Lane, 1961). Abstraction in this case, then, requires some form of "warping" of neural space in the sense that ambiguous tokens (which can be considered as nonoptimal instances of prototypical speech sound representations) are assimilated to one's native category structure (Kuhl, 1993; Kuhl et al., 2008). In the following, we will describe how such forms of abstraction have indeed been observed to be an emergent property of processing across the STL.

Abstraction is a property that is useful for all behaviorally relevant sound classes, not only speech sounds. In fact, in some cases, meaningless nonspeech sounds may be an interesting alternative to investigate the acquisition of abstract representations because their representation has not yet been adjusted by lifelong exposure. Ley et al. (2012) presented participants with a set of complex (nonspeech) sound categories, both before and after training participants to distinguish them into two sound classes (based on pitch). An analysis of the distributed activation patterns during listening both before and after a training session demonstrated that sound category could only be decoded from the distributed blood oxygenation level-dependent (BOLD) response patterns after training. This suggests that some form of functional cortical reorganization had taken place as a result of learning that allowed for the enhanced processing of those aspects of the stimuli that are relevant for behavioral classification. Furthermore, the patterns of neural activity across the testing continuum after training revealed a nonlinear (i.e., sigmoidal) pattern of similarity between items on the pitch continuum thereby showing tight correspondence to participants' categorical behavior (Harnad, 1987). Interestingly, the regions that most strongly contributed to decoding involved a wide region spanning Heschl's gyrus, the PT, and portions of both posteromedial STG and STS. This suggests that this rather distributed region contributed to the emergence of more abstract representations of sound classes.

The distributed nature of the emergence of categorical representations has also been observed for the encoding of speech sounds, in particular for vowel and speaker representations. Formisano, De Martino, Bonte, and Goebel (2008) presented participants with speech sounds consisting of three vowels recorded from three different speakers. Statistical models were then trained to label the associated multivoxel BOLD responses according to vowel identity or according to speaker identity. The models could accurately classify novel trials (i.e., trials that were not in the training data) for both features. However, these features relied on separate, distributed regions in the STL. Importantly, the representations of these features were independent of specific acoustics and generalized across speakers and vowels. Correct classification of vowel identity was based on regions that included large portions of bilateral posteromedial STG and STS, along with a left-sided part of the PT. Speaker identity was mostly decoded based on a portion of the right STS. Similarly, Chang et al. (2010) presented ECoG patients with sounds on a place-of-articulation continuum spanning from /ba/ to /da/ to /ga/, with intermediate steps between the unambiguous speech sounds as well (a distinction that is cued by the F2 onset trajectory; cf. Liberman, Harris, Hoffman, & Griffith, 1957). It was observed that distinct spatiotemporal patterns of activity occurred across the posterior STG when patients listened to the different speech sounds. Importantly, however, they observed a nonlinearity in the neural encoding of the acoustically linearly spaced sound continuum. Tokens that were close to the unambiguous speech sounds (e.g., clear /da/) gave rise to patterns of activity that were very similar to those of the clear ones themselves. These signatures of abstraction arose within a latency range as short as 110–150 ms (see Tsunada, Lee, & Cohen, 2011, for closely related findings in macaque auditory belt regions; and Okada et al., 2010; and Altmann et al., 2014, for speech sound abstraction, especially in the STS).

Finally, converging evidence for the distributed effects of categorical representations have come from fMRI adaptation, or repetition suppression, paradigms (Grill-Spector, Henson & Martin, 2006; Grill-Spector & Malach, 2001). Adaptation can be used to reveal cortical regions that are sensitive to a particular characteristic that remains constant across a set of repeated items but differs in an oddball stimulus. For example, using speech sound stimuli on a continuum from /ga/ to /da/, Joanisse, Zevin, and McCandliss (2007) compared adaptation responses to pairs of stimuli that lay on either the same side of the phoneme-category boundary or stimuli that straddled the category boundary.

They found adaptation effects that were specific to phonemic (as opposed to acoustic) content in left mid-STG, suggesting abstraction in relatively early speech sound processing (but see Chevillet, Jiang, Rauschecker, & Riesenhuber, 2013, for conflicting results). Furthermore, in a similar approach, Leaver and Rauschecker (2010) showed adaptation effects for phonetic categories in left mid-STS (see also Humphries, Sabri, Lewis, & Liebenthal, 2014, for similar results).

As demonstrated, signatures of categorical representations seemed to be widely distributed (e.g., Chang et al., 2010; Formisano et al., 2008), including regions such as PT and posterior STG, regions typically thought to perform more basic acoustic integration. This observation demonstrates that variable levels of representation show a fair amount of overlap. In addition, one may speculate that abstraction can be a processing characteristic that is not restricted to phonemic and postphonemic levels of processing (Mitterer, Scharenborg, & McQueen, 2013), but may occur at the level of speech sound features as well. Abstraction at prelexical levels of representation is important because it allows listeners to understand the meaning of speech spoken by different speakers, despite their differences in pronunciation. Furthermore, when a listener comes across a speaker with an idiosyncratic pronunciation of, say, the phoneme /s/ (perhaps a speaker who lisps), the abstract nature of prelexical units allow the listener to apply their knowledge of this speaker's /s/ also to words that they have not heard this speaker produce before (e.g., McQueen, Cutler, & Norris, 2006; Sjerps & McQueen, 2010). An important addition to these observations, however, is the fact that neural abstraction is often not complete (e.g., Chang et al., 2010). That is, although these representations enhance between-category distinctiveness, "behaviorally irrelevant" auditory detail is not completely removed from the neural representations: within class items do not become neurally identical, just more similar. This property is important because it allows listeners to have access to fine phonetic detail as well when necessary, which can be extremely useful in reanalysis in ambiguous sentences and in the integration of speech cues over longer stretches of time (e.g., Andruski, Blumstein, & Burton, 1994; McMurray, Tanenhaus, Aslin, & Spivey, 2003; McMurray, Tanenhaus, & Aslin, 2009).

2. *Integrating Speech Sound Representations with Context*

Bottom-up processing of auditory information in speech perception is heavily influenced by various forms of context (e.g., Leonard & Chang, 2014). Such context includes factors such as concurrent visual information, phoneme sequence probabilities, and lexicosemantic properties. In the following, we will discuss recent findings that demonstrate how processing in the posteromedial STG and STS is affected by these factors during speech perception. In a final section, we will briefly describe the main regions in the human cortex that receive information from the STL as it forms part of the larger perisylvian language network.

2.1. Multimodal Integration Despite the focus on auditory processing of speech so far in this review, a typical setting for speech perception is a situation where the listener achieves comprehension through the incorporation of both auditory and visual signals. In fact, in people with extreme hearing loss, lip-reading alone can sometimes provide them with enough information to understand speech (Bernstein, Tucker, & Demorest, 2000; Suh, Lee, Kim, Chung, & Oh, 2009). Typically, however, Audiovisual (AV) perception will involve the integration of convergent signals, where listeners can utilize the relative importance of both signals by weighting their contribution depending on how informative they are (e.g., Ross, Saint-Amour, Leavitt, Javitt, & Foxe, 2007). In the following, we will describe how the visual and auditory flows of information interact in the STL and how their relative contributions can be adjusted to situation-specific demands.

Visual information has been shown to influence auditory processing throughout most of its cortical processing. For example, fMRI-based research has shown that both primary and secondary auditory cortices can be activated by visual (lip-read) information alone (Paulesu et al., 2003; Calvert et al., 1997). In addition, interactions between auditory and visual information occur across a number of these regions (Okada, Venezia, Matchin, Saberi, & Hickock, 2013; Skipper, van Wassenhove, Nusbaum, & Small, 2007; Miller & D'Esposito, 2005). In specific situations, auditory and visual information may conflict, as demonstrated in the well-known McGurk effect (e.g., auditory /ba/ presented with visual /ga/ often merge to a /da/ percept; McGurk & MacDonald, 1976). In an ALE meta-analysis of fMRI research on AV speech perception, Erickson, Heeg, Rauschecker, & Turkeltaub (2014) demonstrated the involvement of a large set of regions in resolving conflicting AV information. In addition, however, there was a (much smaller) set of regions involved when auditory and visual signals were in agreement. In the temporal lobe, Erickson et al. found that the posteromedial STL, especially the STS, was involved in both "validating" and conflicting AV situations, with a larger, more ventrally/posteriorly located region involved in situations of conflict. To further investigate the role of the STS in AV integration, Beauchamp, Nath, and Pasalar (2010) presented participants with McGurk stimuli and, on some trials, applied Transcranial Magnetic Stimulation (TMS) during presentation. Without TMS, most trials elicited the well-known McGurk fusion. In contrast, when TMS was applied to the STS, the proportion of reported fusions was significantly lower. In a further study, Nath and Beauchamp (2011) investigated functional connectivity between the AV integration area on the posterior STS (extending onto the lateral STG), with two regions: (i) a region on and around PAC and(ii) a primarily visual region. They then manipulated the reliability of the visual or the auditory signal (a clear auditory signal with a blurred visual signal, or a clear visual signal with an auditory signal in noise) and observed that when the auditory signal was more reliable, functional connectivity increased between the auditory region and the STS. When the visual signal was more reliable, functional connectivity increased between the visual region and the STS. This observation provides a crucial insight into the mechanism behind listeners' ability to flexibly change their dependence on one or the other channel to the most informative one (Ma, Zhou, Ross, Foxe, & Parra, 2009; Nath & Beauchamp, 2011).

To gain an insight into the time course of AV interactions, Rhone et al. (2016) presented ECoG patients with AV stimuli that consisted of the possible combinations of a speech/ nonspeech sound and a speech/ nonspeech lip movement. Rhone et al. recorded cortical signals from the STG, Heschl's gyrus (and premotor cortex, not discussed here). They observed that initial processing of sound in Heschl's gyrus was mostly unaffected by initial visual information (although they do report a small modulation of low-frequency oscillations), while STG activity was influenced by visual input. That is, in STG, stronger (additive) cortical responses were observed when both the visual and auditory signals consisted of speech than when they consisted of nonspeech. Their findings support a model of audiovisual processing in which visual information is integrated with auditory information in the STG, mostly beyond the PAC (but see Pekkola et al., 2005, for fMRI findings in favor of PAC integration). In a related approach, also using ECoG, Besle et al. (2008) reported AV interactions predominantly in secondary auditory cortex, almost immediately (~30 ms) after sound onset (see Reale et al., 2007, for further evidence of the involvement of STG in AV integration).

The findings reviewed here suggest the posterior STS, extending into the lateral STG, is a major site for the integration of auditory and visual speech signals. Recently, Peelle and Sommers (2015) proposed a multistage integration process where, at a first stage, visual information may aid early auditory processing (as early as primary auditory cortex) by predicting the timing of upcoming acoustic events, potentially by resetting ongoing oscillations (see, e.g., Schroeder, Lakatos, Kajikawa, Partan, & Puce, 2008; Stekelenburg & Vroomen, 2007; see Arnal, Poeppel, & Giraud, 2015; and Gross & Poeppel, chapter 29 of this volume, for recent discussions of the broader role of oscillations in speech perception). It should be noted, however, that the extent to which visual information may influence PAC processing beyond providing information about timing remains an active and hotly debated issue. In addition to influences in PAC, however, in regions involving mostly STS (and other higher-level regions outside the temporal lobe) visual information could help to constrain lexical processing, for example in noisy environments. That is, visual information could potentially contribute in distinguishing contrasts based on place of articulation (e.g., *bet* vs. *get*), which, as also described by Mesgarani et al. (2014), are not very clearly represented in the auditory signal and subsequent processing in the STG. Such integration, especially for auditorily weak signals, would thus involve the relative weighting of auditory and visual inputs (Peelle & Sommers; Ross, et al., 2007) through some of the mechanisms described here. Future research should focus on the representational form of the visually elicited activations on the posterior STL to further elucidate what the dominant code is for visual information that is integrated with auditory information on the STL.

2.2. STG Sensitivity to Phoneme Sequences In natural languages, words are not formed by random concatenations of phonemes or syllables. Instead, speech sounds are sequentially organized based on language-specific constraints (termed *phonotactics*). One such statistical regularity is the probability of a specific sound (e.g., *po*) being followed by another (e.g., *te*), called *transition probability* (or, p(B|A) in a sequence AB). Syllable transition probabilities tend to be higher within words than across word boundaries. Listeners can therefore use these probabilities, for example to segment continuous speech into words (e.g., McQueen, 1998), despite the fact that word boundaries are rarely marked by silence. Indeed, listeners, from a very young age, have been shown to be highly sensitive to such regularities (e.g., Pelucchi, Hay, & Saffran, 2009; Saffran, Aslin, & Newport, 1996; Tremblay, Baroni, & Hasson, 2013), despite typically not being consciously aware of them.

To understand at what stages of perceptual analysis transition probabilities affect ongoing cortical processing of speech sounds, McNealy, Mazziotta, and Dapretto (2006) examined the functional neuroanatomical correlates of speech processing while listeners became familiar with recurring patterns of syllables (these sequences, however, were all nonwords). During fMRI recording, the McNealy et al. presented listeners with syllables that appeared in either sequences with statistical regularities (i.e., some regularly recurring syllable triplets) or sequences with no statistical regularities (a third condition, involving stressed syllables, is not discussed here). They found that the bilateral (but left dominant) posterior STG was more active when listeners were presented with one of the statistically regular sequences than those which contained no regularities. Interestingly, this pattern displayed a buildup across the duration of the experiment as participants became more familiar with the sequences (see McNealy, Mazziotta, & Dapretto, 2010, for a replication in children; and Karuza et al., 2013, for a related study, again implicating bilateral STG). Sound processing in the STG is thus sensitive to local statistical relations, and it appears to be able to learn such relations even over the duration of a single experiment.

In addition to regularities at the syllabic level, similar constraints exist at the phoneme level (i.e., within syllables). For example, in English, hearing the sound /k/ followed by /uw/ (*koo*) is more common than hearing /k/ followed by /iy/ *kee* (we will refer to this relation as *forward probabilities* [*Pfwd*]; see figure 27.4A). Native speakers of English are behaviorally sensitive to these probabilities (e.g., Vitevitch & Luce, 1999), and recent work has begun to characterize the neural basis of these effects. Leonard, Bouchard, Tang, and Chang (2015) presented ECoG patients with a range of consonant-vowel-consonant (CVC) sounds of which the transitions between the CV and the VC parts had variable forward probabilities (and backward probabilities, not discussed here for brevity) based on patterns in spoken English. Firstly, in line with the literature discussed in section 1.3, Leonard et al. found that some electrodes displayed clear phoneme selectivity: Figure 27.4B displays the response of an example electrode, revealing that across the stimulus set this particular electrode had a clear preference for syllable-initial /n/ (blue lines in the left panel). This preference was related to the electrode's STRF which revealed sensitivity to low-frequency components that are characteristic of /n/ (figure 27.4C; STRF is estimated on independent data).

After controlling for the portion of the neural response explained by its acoustic sensitivity (the STRF), the electrode displays a strong effect of forward probability (figure 27.4D). That is, when the vowel was predictable based on the initial consonant, this electrode's response was significantly attenuated. Across electrodes, Leonard et al. observed both attenuating and facilitating effects of forward probability on processing of both the vowel and the final consonant. Time courses of the linear weights for STRF-based models and transition probability models (figure 27.4E) show that transition probabilities affect speech sound processing at a slight delay compared to auditory influences, as expected. These findings demonstrate that even when listening to individual CVC syllables, spoken English-based local transition probabilities have a strong effect on speech sound processing.

The two studies described here thus demonstrate that processing in the STG is highly sensitive to local statistical probabilities of sound patterns at a level beyond individual features of phonemes. These influences were observed both as emerging across the duration of an experiment (McNealy et al., 2006) and as a result of lifelong exposure on short CVC syllables (Leonard et al., 2015; see also Tremblay, Deschamps, Baroni, & Hasson, 2016). Furthermore, effects of local sequence probabilities at the word level have been argued to additionally influence cortical responses in the STG such that words that are statistically unlikely (given the just preceding speech) give rise to stronger activation (Willems, Frank, Nijhof, Hagoort, & Van den Bosch, 2015). Furthermore, a recent study has related listeners' abilities to learn syllable-wise statistical regularities to cortical thickness in a number of regions among which is the bilateral STG (Deschamps, Hasson, & Tremblay, 2016). These combined observations reveal that the human STG functions as an acoustic phonetic pattern recognizer that operates over a range of levels of granularity, from feature sequences to syllable sequences and possibly even word sequences.

2.3. Integration with Lexicosemantic Representations in STG and STS Going beyond sequences of features, a crucial step in language comprehension is the activation of lexical representations. Lexical representations allow for the linking of incoming information to stored semantic representations, which are themselves thought to be widely distributed across the cortex (e.g., Huth, de Heer, Griffiths, Theunissen & Gallant, 2016; Nastase et al., 2017; Ralph, Jefferies, Patterson, & Rogers, 2017, and references therein). Adult native speakers of English have access to somewhere between 25,000 and 75,000 lexical entries (e.g., Altmann, 1997; McMurray, 2007). These words are highly variable in terms of their frequency of use and both their semantic and phonological similarities among each other. To build up this massive lexical inventory,

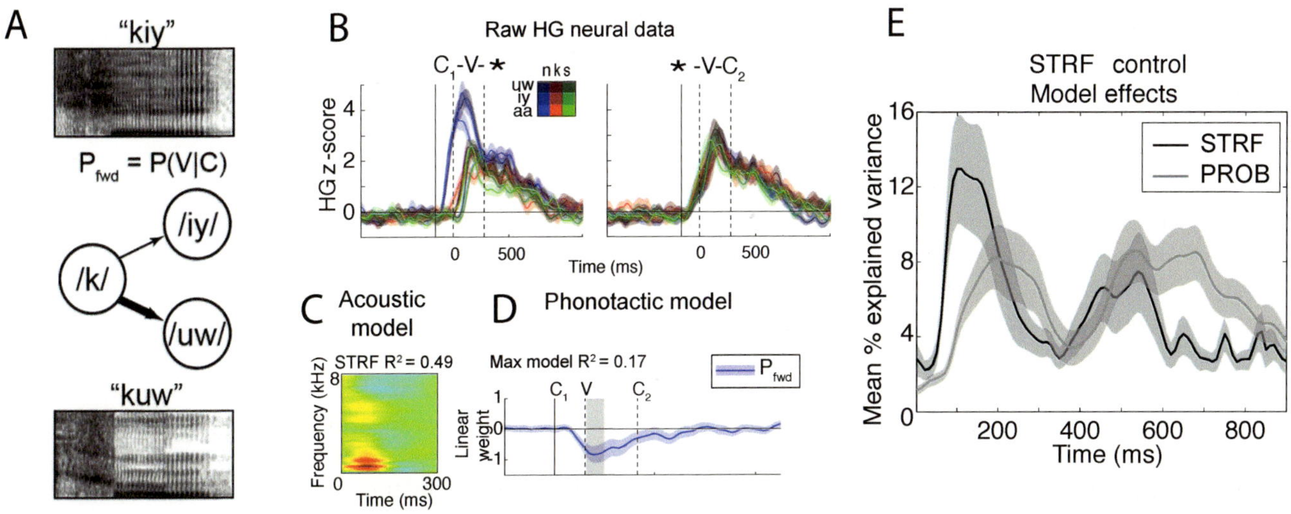

FIGURE 27.4 Phonotactic transition probabilities affect processing in the STG. (A) Two sequences, /k/-/iy/ (*top spectrogram*) and /k/-/uw/ (*bottom spectrogram*), have different transition probabilities between the consonant and the vowel (normalized by the marginal probabilities of the individual elements). The probability that /k/ is followed by /iy/ or /uw/ in English (Pfwd) is indicated by the thickness of the arrows in the *left diagram*. (B) Mean (± standard error of the mean [SEM]) High-Gamma (HG) cortical responses to all stimulus-CVC combinations in an example STG electrode. The electrode shows preferential responses to specific speech sounds in specific positions (see color grid; the electrode is most active when /n/ is the first phoneme). Dashed vertical lines indicate average V and C2 phoneme onsets. (C) The linear STRF of the example electrode in B demonstrates that selectivity is partly driven by phoneme acoustics. (D) Transition probabilities can modulate neural activity not explained by the linear STRF for the example electrode in B: time courses of linear weights show a significant effect during the vowel for Pfwd ($P<0.05$; gray shading). (E) Mean (±SEM) R^2 time courses across all significant electrodes show that auditory (STRF-model) effects generally precede phonotactic probability (PROB-model) effects. Figure reproduced from Leonard et al. (2015).

between birth and adulthood, people are thought to learn up to 10 new words a day on average. Even in adulthood, learning does not stop, as wordlike forms start to be processed as potential real words after only limited exposure (De Vaan, Schreuder, & Baayen, 2007; Lindsay, Sedin, & Gaskell, 2012), allowing the ongoing introduction of new words into a language's repertoire (e.g., *blog*, *selfie*, or *emoji*). In the following section, we will discuss some of the findings that suggest that speech representations in the STG are strongly affected by lexical- and semantic-level linguistic representations.

One influential approach to study lexical processing has been to compare processing of words and nonwords and especially where in the processing stream the two types of stimuli evoke different responses. Recently, Cibelli, Leonard, Johnson, and Chang (2015) presented ECoG patients with a list of auditory words and nonwords in an overt repetition task. Cibelli et al. observed that processing of both words and nonwords involved a temporal progression of peak latency high-gamma activity from more posterior-dorsal to more anterior-ventral temporal lobe sites, consistent with previous findings. In addition, they observed stronger responses to nonwords over words across the length of STG, an effect that was increasingly larger for more anterior sites. This finding aligns with a recent ALE meta-analysis of PET and fMRI research comparing activation between words and nonwords (Davis & Gaskell, 2009). In that meta-analysis, more activation was found for nonwords than words in a large region of the STG but, interestingly, the opposite was found for a number of downstream regions among which were the middle temporal gyrus (MTG) and a large area covering the supramarginal gyrus and adjacent regions of the most posterior portions of the STG. The increased activation for nonwords in the STG appears to reflect additional processing that is necessary while no lexical item has been selected. The repeated presentation of these nonwords, however, can lead to rapid changes in nonword processing. Davis, Di Betta, Macdonald, and Gaskell (2009) have shown, for example, that BOLD responses in the STG after hearing words and nonwords become increasingly similar once listeners have become familiar with the nonwords (and consolidated learning through sleep). The relative dominance of activation for words over nonwords for regions outside the STG probably reflects more semantic-level processing that fails to activate for nonwords.

A considerable behavioral literature has demonstrated that the ease with which a word is recognized is influenced by the number of words that are phonologically similar to it in a person's lexicon (Luce & Large,

2001; Luce & Pisoni, 1998; Vitevitch & Luce, 1999). That is, words that have many "phonological neighbors" are relatively hard to access at later stages of speech perception because of increased competition (although note that early on in processing having many frequent neighbors may be facilitatory). This pattern has been demonstrated, for example, with increased reaction times in lexical decision tasks or picture-naming latencies (Luce & Pisoni, 1998). To further understand the nature of lexical representations in the cortex, a number of researchers have manipulated these more subtle lexical properties. Although reports of regions that are affected by these manipulations appear to be somewhat variable, lexical statistics have been observed to affect processing in both the STG and STS. Cibelli et al. (2015) observed that for the processing of words, small and low-frequency cohorts (i.e., the number of words matching the phonetic input at each time point) led to increased activity in more anterior sites (also, see Zhuang, Randall, Stamatakis, Marslen-Wilson, & Tyler, 2011 for further effects of cohort on STG processing). Others have observed stronger activation in posterior STS for words that have a high phonological neighborhood density (Okada & Hickok, 2006). In a number of reports, however, researchers have failed to observe robust effects in STG/STS, but they have observed greater BOLD response for high-density neighborhood words than to low-density words in the left supramarginal gyrus (see also Righi, Blumstein, Mertus, & Worden, 2010, for the involvement of this region in phonological-lexical competition), and greater activation for high-frequency words in both anterior and posterior left MTG (Prabhakaran, Blumstein, Myers, Hutchison, & Britton, 2006). These findings suggest that among words, the ease of lexical access as governed by cohort size and neighborhood density impact processing in the STG/STS, but the variability in observed topography suggests that these effects may be relatively dependent on specific task requirements.

Further in the hierarchy from sound to meaning are lexicosemantic relations. The influence of lexicosemantic relations on speech processing has often been investigated with semantic priming paradigms (i.e., how does the prime *nurse* affect the processing of the subsequent target word *hospital*). Although effects of semantic relations are typically observed for MTG (see, e.g., Binder, Desai, Graves, & Conant, 2009, for review; Copland et al., 2003; Giesbrecht, Camblin, & Swaab, 2004; Guediche, Reilly, Santiago, Laurent, & Blumstein, 2016; Rissman, Eliassen, & Blumstein, 2003; Wible et al., 2006; and see Dronkers, Wilkins, Van Valin, Redfern, & Jaeger, 2004, for related research on lesions), a number of fMRI studies have observed similar modulations of STG activation (Matsumoto et al., 2005; Minicucci, Guediche, & Blumstein, 2013; Rissman, Eliassen, & Blumstein, 2003; Wible et al., 2006; but see Guediche et al.). For example, Wible et al. presented participants with (i) highly semantically related, (ii) mildly related, or (iii) unrelated prime-target pairs. They observed that activity in a large region in the posterior STG (along with a portion of the MTG) was strongly dependent on the semantic relation between the prime and target. That is, levels of activation in the STG are reduced when a particular target word has a semantic relation to a just-preceding prime stimulus (see also Rissman, Eliassen, & Blumstein, 2003, for corroborating evidence).

To investigate the temporal properties of the influence of lexicosemantic factors on STG processing, Travis et al. (2013) presented ECoG patients with a picture, followed by a spoken word or a noise sound. The spoken word was either congruent or incongruent with the picture. They observed early (within ~60 ms) differences between the processing of the speech sounds and the noise sounds, demonstrating the type of speech preference for the STG electrodes that was discussed in section 1.3. At a later time window (after ~217 ms), however, in the same region and in some cases the same electrodes, they observed differences between the congruent and incongruent prime-target relations. Their results suggest that early auditory processing in the posterior STG was unaffected by semantic-based expectations, whereas activity in the same region revealed lexicosemantic-dependent processing in a later time window (with, typically, larger responses to the incongruous words). Similar results were observed in a combined EEG-magnetoencephalography experiment (Sohoglu, Peelle, Carlyon, & Davis, 2012), suggesting that semantic-level information can robustly influence information processing in the STL as part of a top-down flow of information. These two studies suggested an initial window of information processing in the STG that was unaffected by semantic information, and a later window where it was. The extent and the time course over which higher-level representations such as semantics can affect early speech sound processing are of particular interest because of the important role that online lexical-phonetic interactions have played in the "autonomous" versus "feedback" debate in formal models of speech perception (see Fox & Blumstein, 2016; McClelland & Elman, 1986; Norris, McQueen, & Cutler, 2000, 2016, and references therein). A recent contribution to this debate was the demonstration that the phenomenon of *phoneme restoration* (the perceptual filling in of occluded phonemes) involves a neural reconstruction of the acoustic/phonetic events of the missing phoneme. This finding thus suggests that higher-level information may indeed feedback during online processing to activate acoustic/phonemic representations. A crucial next question, however, is what the role of such patterns of activation is (i.e., a role in online perception or in learning; Norris et al., 2016). Further research using the high-spatial and -temporal resolution as offered by ECoG is likely to provide important contributions to this ongoing debate.

The findings presented in this section demonstrate that both lexical- and semantic-level factors have a strong influence on processing in the posteromedial STL. These findings reveal that information typically thought to be represented in regions outside the STL can affect STL processing over short timescales.

2.4. The Role of the STL in the Larger Speech Perception Hierarchy The current chapter has discussed some of the core processing characteristics of the human STL in speech sound processing. The STL is just one part of a vastly interconnected language network in the broader perisylvian region, of which each of the components display processing characteristics that are too complex and detailed to describe here. However, some brief description of the regions that receive information from the STL and the functions that have been ascribed to them is in order.

As mentioned throughout this chapter, it is often suggested that speech sound processing becomes more speech-specific as activity spreads toward regions further away from PAC, especially on the posterior/dorsal to anterior/ventral axis. It appears that lexical and semantic levels of representations become activated in the anterior temporal lobe, the MTG, and the inferior temporal lobe (Binder et al., 1997; DeWitt & Rauschecker, 2012; Mesulam, Thompson, Weintraub, & Rogalski, 2015; Patterson & Johnsrude, 2008; Rissman et al., 2003; Rodd, Davis, & Johnsrude, 2005; Turken & Dronkers, 2011). In addition to this stream of processing in the temporal lobe, however, information is thought to relay from posteromedial STL to the supramarginal gyrus (Obleser & Eisner, 2009; Turkeltaub & Coslett, 2010), a region presumed to be involved in phonological working memory (see also Buchsbaum, Padmanabhan, & Berman, 2011, for the involvement of closely-situated portions of the posterior STS), but which may also play a role in the activation of lexical representations, especially in situations of phonological/lexical competition (Blumstein, 2009). There are also strong connections from the STL to premotor regions, especially as auditory information is important for self-monitoring in speech production (e.g., Chang, Niziolek, Knight, Nagarajan, & Houde, 2013). It has also been suggested that the motor region may play a role in the perception of speech produced by others, although this remains a hotly debated topic (see, e.g., Cheung, Hamilton, Johnson, & Chang, 2016; Galantucci, Fowler, & Turvey, 2006; Pulvermüller & Fadiga, 2010, and references therein). Finally, one of the main regions involved in speech sound perception and language tasks in general has been the inferior frontal gyrus (IFG; including Broca's area). The IFG is heavily involved in language processing more generally (e.g., Hagoort, Baggio, & Willems, 2009), but also in speech perception tasks, for example in resolving competition between task-relevant alternatives, especially when listeners make decisions about noisy or underspecified signals (e.g., Prabhakaran et al., 2006; Snyder, Feigenson, & Thompson-Schill, 2007; Swaab, Brown, & Hagoort, 1998; Utman, Blumstein, & Sullivan, 2001), and in tasks like phonological target detection (Chang et al., 2011). The higher-level role of IFG in phonetic speech perception tasks is supported by the observation that it is strongly dependent on attention (Alho et al., 2016).

The role of this network in speech perception has been conceptualized as following two parallel streams specialized for analyzing different aspects of the speech signal, which both originate from initial processing in parts of the STL. Although specific interpretations differ, it has been broadly suggested that information follows a ventral stream (involving the STL, MTG, anterior TL, [anterior] IFG) of which the dominant function involves lexicosemantic access and/or comprehension, and a dorsal stream (involving STL; supramarginal gyrus; sensory-motor cortex; [posterior] IFG) of which the primary function involves sensory–motor integration and phonological working memory (Hickok & Poeppel, 2007; Rauschecker & Scott, 2009; Scott & Johnsrude, 2003). Other chapters in part V of this volume provide a more detailed description of the several parts of this network.

3. Conclusion

The findings and concepts discussed in this chapter provide some important descriptions of the processing characteristics of the STL and its role in speech sound processing. Processing in the STL is partly characterized by sensitivity to relatively basic auditory properties, especially in regions close to PAC, but the dominant preference is one for more complex spectrotemporal stimulus properties, especially for those features that occur in speech (e.g., Hullett et al., 2016). For example, large portions of the STG were shown to be sensitive to specific spectrotemporal patterns that are associated with natural classes of speech sounds such as vowels, plosives, or fricatives. This observation shows that the representation of phonetic features is an important processing property of the STL (e.g., Arsenault & Buchsbaum, 2015; Mesgarani et al., 2014; Steinschneider et al., 2011). In addition, categorical representations were observed as an emergent property of processing in the STL. That is, acoustically continuous sequences of sounds become neurally "warped" depending on which parts of these continua are most behaviorally relevant (Chang et al., 2010; Ley et al., 2012). This property of the STL is fundamental to speech processing. It allows for the selection of representations (e.g., lexical and semantic) that are independent of idiosyncrasies in how particular speech sounds are uttered on different occasions or by different speakers (i.e., also known as the invariance problem).

In addition to processing related to immediate auditory stimulus properties, a number of observations demonstrate that processing in the STL is strongly influenced by contextual factors. Visual and auditory information were found to be rapidly integrated in the STL (especially for the posterior parts, e.g., Beauchamp et al., 2010; Erickson et al., 2014), in a way that allows for the selective weighting of visual and auditory input, depending on their reliability (e.g., Nath & Beauchamp, 2011). In addition, processing in the STL is affected by the local probabilities of perceived acoustic events given preceding and subsequent cues at multiple levels of granularity (i.e., sequence probabilities from acoustic features to phoneme sequences; Leonard et al., 2015; McNealy et al., 2006). Furthermore, evidence was presented for semantic influences affecting processing in the STL (Rissman et al., 2003; Wible et al., 2006), albeit potentially with a slight temporal delay relative to initial bottom-up processes (e.g., Sohoglu et al., 2012; Travis et al., 2013). These observations stress the role of the STL as a highly versatile auditory association cortex. Future research should be aimed at further investigating how representations in the STL are influenced by factors such as visual and semantic information. That is, what aspects of these featural representations are affected by visual and semantic representations? And what forms do these "contextual" representations have themselves that allow them to directly influence feature-based representations in the STL? We believe that such questions are likely to be informed by approaches that focus on speech sounds represented as distributed patterns of activity. That is, representation as distinctive patterns of activation, even in absence of changes in overall changes in signal level in broadly defined regions of interest. We think that this approach in combination with increased availability of invasive and noninvasive imaging techniques that allow for the measurement of cortical activity at high-spatial and -temporal resolution ushers in an exciting period for research on the neural processing of speech.

Acknowledgments

The first author received funding from the People Programme (Marie Curie Actions) of the European Union's Seventh Framework Programme FP7 2007–2013 under REA grant agreement nr. 623072. The authors are grateful to Matthew Leonard, Neal Fox, Nina Dronkers and David Poeppel for their comments on earlier versions of the manuscript.

REFERENCES

Alho, J., Green, B. M., May, P. J., Sams, M., Tiitinen, H., Rauschecker, J. P., & Jääskeläinen, I. P. (2016). Early-latency categorical speech sound representations in the left inferior frontal gyrus. *NeuroImage, 129*, 214–223.

Altmann, C. F., Uesaki, M., Ono, K., Matsuhashi, M., Mima, T., & Fukuyama, H. (2014). Categorical speech perception during active discrimination of consonants and vowels. *Neuropsychologia, 64*, 13–23.

Altmann, G. T. M. (1997). *The ascent of Babel: An exploration of language, mind, and understanding.* Oxford: Oxford University Press.

Andruski, J. E., Blumstein, S. E., & Burton, M. (1994). The effect of subphonetic differences on lexical access. *Cognition, 52*(3), 163–187.

Arnal, L. H., Poeppel, D., & Giraud, A. L. (2015). Temporal coding in the auditory cortex. *Handbook of Clinical Neurology, 129*, 85–98.

Arsenault, J. S., & Buchsbaum, B. R. (2015). Distributed neural representations of phonological features during speech perception. *Journal of Neuroscience, 35*(2), 634–642.

Baumann, S., Petkov, C. I., & Griffiths, T. D. (2013). A unified framework for the organization of the primate auditory cortex. *Frontiers in Systems Neuroscience, 7*, 11.

Beauchamp, M. S., Nath, A. R., & Pasalar, S. (2010). fMRI-guided transcranial magnetic stimulation reveals that the superior temporal sulcus is a cortical locus of the McGurk effect. *Journal of Neuroscience, 30*(7), 2414–2417.

Belin, P., Zatorre, R. J., Lafaille, P., Ahad, P., & Pike, B. (2000). Voice-selective areas in human auditory cortex. *Nature, 403*(6767), 309–312.

Bernstein, L. E., Tucker, P. E., & Demorest, M. E. (2000). Speech perception without hearing. *Perception and Psychophysics, 62*(2), 233–252.

Besle, J., Fischer, C., Bidet-Caulet, A., Lecaignard, F., Bertrand, O., & Giard, M. H. (2008). Visual activation and audiovisual interactions in the auditory cortex during speech perception: Intracranial recordings in humans. *Journal of Neuroscience, 28*(52), 14301–14310.

Bidelman, G. M., Gandour, J. T., & Krishnan, A. (2011). Cross-domain effects of music and language experience on the representation of pitch in the human auditory brainstem. *Journal of Cognitive Neuroscience, 23*(2), 425–434.

Binder, J. R., Desai, R. H., Graves, W. W., & Conant, L. L. (2009). Where is the semantic system? A critical review and meta-analysis of 120 functional neuroimaging studies. *Cerebral Cortex, 19*(12), 2767–2796.

Binder, J. R., Frost, J. A., Hammeke, T. A., Bellgowan, P. S., Springer, J. A., Kaufman, J. N., & Possing, E. T. (2000). Human temporal lobe activation by speech and nonspeech sounds. *Cerebral Cortex, 10*(5), 512–528.

Binder, J. R., Frost, J. A., Hammeke, T. A., Cox, R. W., Rao, S. M., & Prieto, T. (1997). Human brain language areas identified by functional magnetic resonance imaging. *Journal of Neuroscience, 17*(1), 353–362.

Bitterman, Y., Mukamel, R., Malach, R., Fried, I., & Nelken, I. (2008). Ultra-fine frequency tuning revealed in single neurons of human auditory cortex. *Nature, 451*(7175), 197–201.

Blumstein, S. E. (2009). Auditory word recognition: Evidence from aphasia and functional neuroimaging. *Language and Linguistics Compass, 3*(4), 824–838.

Boatman, D. (2004). Cortical bases of speech perception: Evidence from functional lesion studies. *Cognition, 92*(1), 47–65.

Boatman, D., Hall, C., Goldstein, M. H., Lesser, R., & Gordon, B. (1997). Neuroperceptual differences in consonant and vowel discrimination: As revealed by direct cortical electrical interference. *Cortex, 33*(1), 83–98.

Brugge, J. F., Volkov, I. O., Garell, P. C., Reale, R. A., & Howard, M. A. (2003). Functional connections between auditory cortex on Heschl's gyrus and on the lateral superior temporal gyrus in humans. *Journal of Neurophysiology, 90*(6), 3750–3763.

Buchman, A. S., Garron, D. C., Trost-Cardamone, J. E., Wichter, M. D., & Schwartz, M. (1986). Word deafness: One hundred years later. *Journal of Neurology, Neurosurgery and Psychiatry, 49*(5), 489–499.

Buchsbaum, B. R., Baldo, J., Okada, K., Berman, K. F., Dronkers, N., D'Esposito, M., & Hickok, G. (2011). Conduction aphasia, sensory-motor integration, and phonological short term memory an aggregate analysis of lesion and fMRI data. *Brain and Language, 119*(3), 119–128.

Buchsbaum, B. R., Padmanabhan, A., & Berman, K. F. (2011). The neural substrates of recognition memory for verbal information: Spanning the divide between short and long-term memory. *Journal of Cognitive Neuroscience, 23*, 978–991.

Calvert, G. A., Bullmore, E. T., Brammer, M. J., Campbell, R., Williams, S. C., McGuire, P. K., ... David, A. S. (1997). Activation of auditory cortex during silent lipreading. *Science, 276*(5312), 593–596.

Canolty, R. T., Soltani, M., Dalal, S. S., Edwards, E., Dronkers, N. F., Nagarajan, S. S., ... Knight, R. T. (2007). Spatiotemporal dynamics of word processing in the human brain. *Frontiers in Neuroscience, 1*, 14.

Chan, A. M., Dykstra, A. R., Jayaram, V., Leonard, M. K., Travis, K. E., Gygi, B., ... Cash, S. S. (2014). Speech-specific tuning of neurons in human superior temporal gyrus. *Cerebral Cortex, 24*(10), 2679–2693.

Chang, E. F., Edwards, E., Nagarajan, S. S., Fogelson, N., Dalal, S. S., Canolty, R. T., ... Knight, R. T. (2011). Cortical spatio-temporal dynamics underlying phonological target detection in humans. *Journal of Cognitive Neuroscience, 23*(6), 1437–1446.

Chang, E. F., Niziolek, C. A., Knight, R. T., Nagarajan, S. S., & Houde, J. F. (2013). Human cortical sensorimotor network underlying feedback control of vocal pitch. *Proceedings of the National Academy of Sciences, 110*(7), 2653–2658.

Chang, E. F., Rieger, J. W., Johnson, K., Berger, M. S., Barbaro, N. M., & Knight, R. T. (2010). Categorical speech representation in human superior temporal gyrus. *Nature Neuroscience, 13*(11), 1428–1432.

Cheung, C., Hamilton, L. S., Johnson, K., & Chang, E. F. (2016). The auditory representation of speech sounds in human motor cortex. *eLife, 5*, e12577.

Chevillet, M. A., Jiang, X., Rauschecker, J. P., & Riesenhuber, M. (2013). Automatic phoneme category selectivity in the dorsal auditory stream. *Journal of Neuroscience, 33*(12), 5208–5215.

Chi, T., Gao, Y., Guyton, M. C., Ru, P., & Shamma, S. (1999). Spectro-temporal modulation transfer functions and speech intelligibility. *Journal of the Acoustical Society of America, 106*(5), 2719–2732.

Cibelli, E. S., Leonard, M. K., Johnson, K., & Chang, E. F. (2015). The influence of lexical statistics on temporal lobe cortical dynamics during spoken word listening. *Brain and Language, 147*, 66–75.

Copland, D. A., De Zubicaray, G. I., McMahon, K., Wilson, S. J., Eastburn, M., & Chenery, H. J. (2003). Brain activity during automatic semantic priming revealed by event-related functional magnetic resonance imaging. *NeuroImage, 20*(1), 302–310.

Creutzfeldt, O., Ojemann, G., & Lettich, E. (1989). Neuronal activity in the human lateral temporal lobe. *Experimental Brain Research, 77*(3), 451–475.

David, S. V., Vinje, W. E., & Gallant, J. L. (2004). Natural stimulus statistics alter the receptive field structure of v1 neurons. *Journal of Neuroscience, 24*(31), 6991–7006.

Davis, M. H., Di Betta, A. M., Macdonald, M. J., & Gaskell, M. G. (2009). Learning and consolidation of novel spoken words. *Journal of Cognitive Neuroscience, 21*(4), 803–820.

Davis, M. H., & Gaskell, M. G. (2009). A complementary systems account of word learning: Neural and behavioural evidence. *Philosophical Transactions of the Royal Society of London B: Biological Sciences, 364*(1536), 3773–3800.

Davis, M. H., & Johnsrude, I. S. (2003). Hierarchical processing in spoken language comprehension. *Journal of Neuroscience, 23*(8), 3423–3431.

Desai, R., Liebenthal, E., Waldron, E., & Binder, J. R. (2008). Left posterior temporal regions are sensitive to auditory categorization. *Journal of Cognitive Neuroscience, 20*(7), 1174–1188.

Deschamps, I., Hasson, U., & Tremblay, P. (2016). The structural correlates of statistical information processing during speech perception. *PLOS ONE, 11*(2), e0149375.

De Vaan, L., Schreuder, R., & Baayen, R. H. (2007). Regular morphologically complex neologisms leave detectable traces in the mental lexicon. *Mental Lexicon, 2*(1), 1–24.

DeWitt, I., & Rauschecker, J. P. (2012). Phoneme and word recognition in the auditory ventral stream. *Proceedings of the National Academy of Sciences, 109*(8), E505–E514.

Ding, N., Patel, A. D., Chen, L., Butler, H., Luo, C., & Poeppel, D. (2017). Temporal modulations in speech and music. *Neuroscience and Biobehavioral Reviews 81*(Pt. B), 181–187.

Dronkers, N. F., Wilkins, D. P., Van Valin, R. D., Redfern, B. B., & Jaeger, J. J. (2004). Lesion analysis of the brain areas involved in language comprehension. *Cognition, 92*(1), 145–177.

Elliott, T. M., & Theunissen, F. E. (2009). The modulation transfer function for speech intelligibility. *PLOS Computational Biology, 5*(3), e1000302.

Erickson, L. C., Heeg, E., Rauschecker, J. P., & Turkeltaub, P. E. (2014). An ALE meta-analysis on the audiovisual integration of speech signals. *Human Brain Mapping, 35*(11), 5587–5605.

Formisano, E., De Martino, F., Bonte, M., & Goebel, R. (2008). "Who" is saying "what"? Brain-based decoding of human voice and speech. *Science, 322*(5903), 970–973.

Fox, N. P., & Blumstein, S. E. (2016). Top-down effects of syntactic sentential context on phonetic processing. *Journal of Experimental Psychology: Human Perception and Performance, 42*(5), 730.

Galantucci, B., Fowler, C. A., & Turvey, M. T. (2006). The motor theory of speech perception reviewed. *Psychonomic Bulletin and Review, 13*(3), 361–377.

Giesbrecht, B., Camblin, C. C., & Swaab, T. Y. (2004). Separable effects of semantic priming and imageability on word processing in human cortex. *Cerebral Cortex, 14*(5), 521–529.

Grill-Spector, K., Henson, R. N., & Martin, A. (2006). Repetition and the brain: Neural models of stimulus specific effects. *Trends in Cognitive Science, 10,* 14–25.

Grill-Spector, K., & Malach, R. (2001). fMR-adaptation: A tool for studying the functional properties of human cortical neurons. *Acta Psychologica, 107,* 293–321.

Guediche, S., Reilly, M., Santiago, C., Laurent, P., & Blumstein, S. E. (2016). An fMRI study investigating effects of conceptually related sentences on the perception of degraded speech. *Cortex, 79,* 57–74.

Hackett, T. A. (2011). Information flow in the auditory cortical network. *Hearing Research, 271*(1), 133–146.

Hagoort, P., Baggio, G., & Willems, R. M. (2009). Semantic unification. In M. S. Gazzaniga (Ed.), *The cognitive neurosciences* (4th ed., pp. 819–836). Cambridge, MA: MIT Press.

Harnad, S. (Ed.) (1987). *Categorical perception: The groundwork of cognition.* Cambridge: Cambridge University Press.

Hickok, G., & Poeppel, D. (2004). Dorsal and ventral streams: A framework for understanding aspects of the functional anatomy of language. *Cognition, 92*(1), 67–99.

Hickok, G., & Poeppel, D. (2007). The cortical organization of speech processing. *Nature Reviews Neuroscience, 8*(5), 393–402.

Hickok, G., & Poeppel, D. (2015). Neural basis of speech perception. *Human Auditory System: Fundamental Organization and Clinical Disorders, 129,* 149.

Hsu, A., Woolley, S. M., Fremouw, T. E., & Theunissen, F. E. (2004). Modulation power and phase spectrum of natural sounds enhance neural encoding performed by single auditory neurons. *Journal of Neuroscience, 24*(41), 9201–9211.

Hullett, P. W., Hamilton, L. S., Mesgarani, N., Schreiner, C. E., & Chang, E. F. (2016). Human superior temporal gyrus organization of spectrotemporal modulation tuning derived from speech stimuli. *Journal of Neuroscience, 36*(6), 2014–2026.

Humphries, C., Liebenthal, E., & Binder, J. R. (2010). Tonotopic organization of human auditory cortex. *NeuroImage. 50,* 1202–1211.

Humphries, C., Sabri, M., Lewis, K., & Liebenthal, E. (2014). Hierarchical organization of speech perception in human auditory cortex. *Frontiers in Neuroscience, 8,* 406.

Huth, A. G., de Heer, W. A., Griffiths, T. L., Theunissen, F. E., & Gallant, J. L. (2016). Natural speech reveals the semantic maps that tile human cerebral cortex. *Nature, 532*(7600), 453–458.

Jäncke, L., Wüstenberg, T., Scheich, H., & Heinze, H. J. (2002). Phonetic perception and the temporal cortex. *NeuroImage, 15*(4), 733–746.

Joanisse, M. F., Zevin, J. D., & McCandliss, B. D. (2007). Brain mechanisms implicated in the preattentive categorization of speech sounds revealed using fMRI and a short-interval habituation trial paradigm. *Cerebral Cortex, 17*(9), 2084–2093.

Karuza, E. A., Newport, E. L., Aslin, R. N., Starling, S. J., Tivarus, M. E., & Bavelier, D. (2013). The neural correlates of statistical learning in a word segmentation task: An fMRI study. *Brain and Language, 127*(1), 46–54.

Klein, D. J., Depireux, D. A., Simon, J. Z., & Shamma, S. A. (2000). Robust spectrotemporal reverse correlation for the auditory system: Optimizing stimulus design. *Journal of Computational Neuroscience, 9*(1), 85–111.

Krishnan, A., Gandour, J. T., & Bidelman, G. M. (2012). Experience-dependent plasticity in pitch encoding: From brainstem to auditory cortex. *NeuroReport, 23*(8), 498.

Kuhl, P. K. (1993). Early linguistic experience and phonetic perception: Implications for theories of developmental speech perception. *Journal of Phonetics, 21,* 125–139.

Kuhl, P. K., Conboy, B. T., Coffey-Corina, S., Padden, D., Rivera-Gaxiola, M., & Nelson, T. (2008). Phonetic learning as a pathway to language: New data and native language magnet theory expanded (NLM-e). *Philosophical Transactions of the Royal Society of London B: Biological Sciences, 363*(1493), 979–1000.

Leaver, A. M., & Rauschecker, J. P. (2010). Cortical representation of natural complex sounds: Effects of acoustic features and auditory object category. *Journal of Neuroscience, 30*(22), 7604–7612.

Leonard, M. K., Bouchard, K. E., Tang, C., & Chang, E. F. (2015). Dynamic encoding of speech sequence probability in human temporal cortex. *Journal of Neuroscience, 35*(18), 7203–7214.

Leonard, M. K., Cai, R., Babiak, M. C., Ren, A., & Chang, E. F. (2016). The peri-Sylvian cortical network underlying single word repetition revealed by electrocortical stimulation and direct neural recordings. *Brain and Language,* S0093-934X(15)30194-2. Advance online publication. doi:10.1016/j.bandl.2016.06.001

Leonard, M. K., & Chang, E. F. (2014). Dynamic speech representations in the human temporal lobe. *Trends in Cognitive Sciences, 18*(9), 472–479.

Lerner, Y., Honey, C. J., Silbert, L. J., & Hasson, U. (2011). Topographic mapping of a hierarchy of temporal receptive windows using a narrated story. *Journal of Neuroscience, 31*(8), 2906–2915.

Ley, A., Vroomen, J., Hausfeld, L., Valente, G., De Weerd, P., & Formisano, E. (2012). Learning of new sound categories shapes neural response patterns in human auditory cortex. *Journal of Neuroscience, 32*(38), 13273–13280.

Liberman, A. M., Harris, K. S., Hoffman, H. S., & Griffith, B. C. (1957). The discrimination of speech sounds within and across phoneme boundaries. *Journal of Experimental Psychology, 54*(5), 358.

Liberman, A. M., Harris, K. S., Kinney, J. A., & Lane, H. (1961). The discrimination of relative onset-time of the components of certain speech and non-speech patterns. *Journal of Experimental Psychology, 61*(5), 379.

Liebenthal, E., Binder, J. R., Spitzer, S. M., Possing, E. T., & Medler, D. A. (2005). Neural substrates of phonemic perception. *Cerebral Cortex, 15*, 1621–1631.

Liebenthal, E., Desai, R., Ellingson, M. M., Ramachandran, B., Desai, A., & Binder, J. R. (2010). Specialization along the left superior temporal sulcus for auditory categorization. *Cerebral Cortex, 20*(12), 2958–2970.

Liebenthal, E., Desai, R., Humphries, C., Sabri, M., & Desai, A. (2014). The functional organization of the left STS: A large scale meta-analysis of PET and fMRI studies of healthy adults. *Frontiers in Neuroscience, 8*, 289.

Lindsay, S., Sedin, L. M., & Gaskell, M. G. (2012). Acquiring novel words and their past tenses: Evidence from lexical effects on phonetic categorisation. *Journal of Memory and Language, 66*(1), 210–225.

Luce, P. A., & Large, N. R. (2001). Phonotactics, density, and entropy in spoken word recognition. *Language and Cognitive Processes, 16*(5–6), 565–581.

Luce, P. A., & Pisoni, D. B. (1998). Recognizing spoken words: The neighborhood activation model. *Ear and Hearing, 19*(1), 1.

Ma, W. J., Zhou, X., Ross, L. A., Foxe, J. J., & Parra, L. C. (2009). Lip-reading aids word recognition most in moderate noise: A Bayesian explanation using high-dimensional feature space. *PLOS ONE, 4*(3), e4638.

Matsumoto A., Iidaka T., Haneda K., Okada T., & Sadato N (2005). Linking semantic priming effect in functional MRI and event-related potentials. *Neuro-image, 24*(3), 624–634.

McClelland, J. L., & Elman, J. L. (1986). The TRACE model of speech perception. *Cognitive Psychology, 18*(1), 1–86.

McGurk, H., & MacDonald, J. (1976). Hearing lips and seeing voices. *Nature, 264*, 746–748.

McMurray, B. (2007). Defusing the childhood vocabulary explosion. *Science, 317*(5838), 631.

McMurray, B., Tanenhaus, M. K., & Aslin, R. N. (2009). Within-category VOT affects recovery from "lexical" garden-paths: Evidence against phoneme-level inhibition. *Journal of Memory and Language, 60*(1), 65–91.

McMurray, B., Tanenhaus, M. K., Aslin, R. N., & Spivey, M. J. (2003). Probabilistic constraint satisfaction at the lexical/phonetic interface: Evidence for gradient effects of within-category VOT on lexical access. *Journal of Psycholinguistic Research, 32*(1), 77–97.

McNealy, K., Mazziotta, J. C., & Dapretto, M. (2006). Cracking the language code: Neural mechanisms underlying speech parsing. *Journal of Neuroscience, 26*(29), 7629–7639.

McNealy, K., Mazziotta, J. C., & Dapretto, M. (2010). The neural basis of speech parsing in children and adults. *Developmental Science, 13*(2), 385–406.

McQueen, J. M. (1998). Segmentation of continuous speech using phonotactics. *Journal of Memory and Language, 39*(1), 21–46.

McQueen, J. M., Cutler, A., & Norris, D. (2006). Phonological abstraction in the mental lexicon. *Cognitive Science, 30*(6), 1113–1126.

Mesgarani, N., Cheung, C., Johnson, K., & Chang, E. F. (2014). Phonetic feature encoding in human superior temporal gyrus. *Science, 343*(6174), 1006–1010.

Mesulam, M. M., Thompson, C. K., Weintraub, S., & Rogalski, E. J. (2015). The Wernicke conundrum and the anatomy of language comprehension in primary progressive aphasia. *Brain, 138*(8), 2423–2437.

Miller, J. L., Grosjean, F., & Lomanto, C. (1984). Articulation rate and its variability in spontaneous speech: A reanalysis and some implications. *Phonetica, 41*(4), 215–225.

Miller, L. M., & D'Esposito, M. (2005). Perceptual fusion and stimulus coincidence in the cross-modal integration of speech. *Journal of Neuroscience, 25*(25), 5884–5893.

Minicucci, D., Guediche, S., & Blumstein, S. E. (2013). An fMRI examination of the effects of acoustic-phonetic and lexical competition on access to the lexical-semantic network. *Neuropsychologia, 51*(10), 1980–1988.

Mitterer, H., Scharenborg, O., & McQueen, J. M. (2013). Phonological abstraction without phonemes in speech perception. *Cognition, 129*(2), 356–361.

Moerel, M., De Martino, F., & Formisano, E. (2012). Processing of natural sounds in human auditory cortex: Tonotopy, spectral tuning, and relation to voice sensitivity. *Journal of Neuroscience, 32*(41), 14205–14216.

Moerel, M., De Martino, F., Santoro, R., Ugurbil, K., Goebel, R., Yacoub, E., & Formisano, E. (2013). Processing of natural sounds: Characterization of multipeak spectral tuning in human auditory cortex. *Journal of Neuroscience, 33*(29), 11888–11898.

Nagel, K. I., & Doupe, A. J. (2008). Organizing principles of spectro-temporal encoding in the avian primary auditory area field L. *Neuron, 58*(6), 938–955.

Nastase, S. A., Connolly, A. C., Oosterhof, N. N., Halchenko, Y. O., Guntupalli, J. S., Visconti di Oleggio Castello, M., ... Haxby, J. V. (2017). Attention selectively reshapes the geometry of distributed semantic representation. *Cerebral Cortex, 27*(8), 4277–4291.

Nath, A. R., & Beauchamp, M. S. (2011). Dynamic changes in superior temporal sulcus connectivity during perception of noisy audiovisual speech. *Journal of Neuroscience, 31*(5), 1704–1714.

Norris, D., McQueen, J. M., & Cutler, A. (2000). Merging information in speech recognition: Feedback is never necessary. *Behavioral and Brain Sciences, 23*(03), 299–325.

Norris, D., McQueen, J. M., & Cutler, A. (2016). Prediction, Bayesian inference and feedback in speech recognition. *Language, Cognition and Neuroscience, 31*(1), 4–18.

Nourski, K. V., Steinschneider, M., Oya, H., Kawasaki, H., Jones, R. D., & Howard, M. A. (2012). Spectral organization of the human lateral superior temporal gyrus revealed by intracranial recordings. *Cerebral Cortex, 24*(2), 340–352.

Obleser, J., & Eisner, F. (2009). Pre-lexical abstraction of speech in the auditory cortex. *Trends in Cognitive Sciences, 13*(1), 14–19.

Obleser, J., Eisner, F., & Kotz, S. A. (2008). Bilateral speech comprehension reflects differential sensitivity to spectral and temporal features. *Journal of Neuroscience, 28*(32), 8116–8123.

Obleser, J., Zimmermann, J., Van Meter, J., & Rauschecker, J. P. (2007). Multiple stages of auditory speech perception reflected in event-related FMRI. *Cerebral Cortex, 17*(10), 2251–2257.

Ohl, F. W., & Scheich, H. (1997). Orderly cortical representation of vowels based on formant interaction. *Proceedings of the National Academy of Sciences, 94*(17), 9440–9444.

Okada, K., & Hickok, G. (2006). Identification of lexical-phonological networks in the superior temporal sulcus using functional magnetic resonance imaging. *Neuro-Report, 17*(12), 1293–1296.

Okada, K., Rong, F., Venezia, J., Matchin, W., Hsieh, I. H., Saberi, K., ... Hickok, G. (2010). Hierarchical organization of human auditory cortex: Evidence from acoustic invariance in the response to intelligible speech. *Cerebral Cortex, 20*(10), 2486–2495.

Okada, K., Venezia, J. H., Matchin, W., Saberi, K., & Hickok, G. (2013). An fMRI study of audiovisual speech perception reveals multisensory interactions in auditory cortex. *PLOS ONE, 8*(6), e68959.

Overath, T., McDermott, J. H., Zarate, J. M., & Poeppel, D. (2015). The cortical analysis of speech-specific temporal structure revealed by responses to sound quilts. *Nature Neuroscience, 18*(6), 903–911.

Pasley, B. N., David, S.V., Mesgarani, N., Flinker, A., Shamma, S.A., Crone, N. E., ... Chang, E. F. (2012). Reconstructing speech from human auditory cortex. *PLOS Biology, 10*(1), e1001251.

Patterson, R. D., & Johnsrude, I. S. (2008) Functional imaging of the auditory processing applied to speech sounds. *Philosophical Transactions of the Royal Society B: Biological Sciences, 363*(1493), 1023–1035.

Paulesu, E., Perani, D., Blasi, V., Silani, G., Borghese, N. A., De Giovanni, U., ... Fazio, F. (2003). A functional-anatomical model for lipreading. *Journal of Neurophysiology, 90*(3), 2005–2013.

Prabhakaran, R., Blumstein, S. E., Myers, E. B., Hutchison, E., & Britton, B. (2006). An event-related fMRI investigation of phonological-lexical competition. *Neuropsychologia, 44*(12), 2209–2221.

Peelle, J. E., Johnsrude, I., & Davis, M. H. (2010). Hierarchical processing for speech in human auditory cortex and beyond. *Frontiers in Human Neuroscience, 4*, 51.

Peelle, J. E., & Sommers, M. S. (2015). Prediction and constraint in audiovisual speech perception. *Cortex, 68*, 169–181.

Pekkola, J., Ojanen, V., Autti, T., Jääskeläinen, I. P., Möttönen, R., Tarkiainen, A., & Sams, M. (2005). Primary auditory cortex activation by visual speech: An fMRI study at 3 T. *NeuroReport, 16*(2), 125–128.

Pelucchi, B., Hay, J. F., & Saffran, J. R. (2009). Statistical learning in a natural language by 8-month-old infants. *Child Development, 80*(3), 674–685.

Price, C. J. (2012). A review and synthesis of the first 20 years of PET and fMRI studies of heard speech, spoken language and reading. *NeuroImage, 62*(2), 816–847.

Pulvermüller, F., & Fadiga, L. (2010). Active perception: Sensorimotor circuits as a cortical basis for language. *Nature Reviews Neuroscience, 11*(5), 351–360.

Quigg, M., & Fountain, N. (1999). Conduction aphasia elicited by stimulation of the left posterior superior temporal gyrus. *Journal of Neurology, Neurosurgery, and Psychiatry, 66*(3), 393.

Ralph, M. L., Jefferies, E., Patterson, K., & Rogers, T. T. (2017). The neural and computational bases of semantic cognition. *Nature Reviews Neuroscience, 18*, 42–55.

Rauschecker, J. P., & Scott, S. K. (2009). Maps and streams in the auditory cortex: Nonhuman primates illuminate human speech processing. *Nature Neuroscience, 12*(6), 718–724.

Reale, R. A., Calvert, G. A., Thesen, T., Jenison, R. L., Kawasaki, H., Oya, H., ... Brugge, J. F. (2007). Auditory-visual processing represented in the human superior temporal gyrus. *Neuroscience, 145*(1), 162–184.

Rhone, A. E., Nourski, K. V., Oya, H., Kawasaki, H., Howard III, M. A., & McMurray, B. (2016). Can you hear me yet? An intracranial investigation of speech and non-speech audiovisual interactions in human cortex. *Language, Cognition and Neuroscience, 31*(2), 284–302.

Rieke, F., Bodnar, D. A., & Bialek, W. (1995). Naturalistic stimuli increase the rate and efficiency of information transmission by primary auditory afferents. *Proceedings of the Royal Society of London B: Biological Sciences, 262*(1365), 259–265.

Righi, G., Blumstein, S. E., Mertus, J., & Worden, M. S. (2010). Neural systems underlying lexical competition: An eye tracking and fMRI study. *Journal of Cognitive Neuroscience, 22*(2), 213–224.

Rissman, J., Eliassen, J. C., & Blumstein, S. E. (2003). An event related FMRI investigation of implicit semantic priming. *Journal of Cognitive Neuroscience, 15*(8), 1160–1175.

Rodd, J. M., Davis, M. H., & Johnsrude, I. S. (2005). The neural mechanisms of speech comprehension: FMRI studies of semantic ambiguity. *Cerebral Cortex, 15*, 1261–1269.

Rogalsky, C., Poppa, T., Chen, K. H., Anderson, S. W., Damasio, H., Love, T., & Hickok, G. (2015). Speech repetition as a window on the neurobiology of auditory–motor integration for speech: A voxel-based lesion symptom mapping study. *Neuropsychologia, 71*, 18–27.

Rosen, S., Wise, R. J., Chadha, S., Conway, E. J., & Scott, S. K. (2011). Hemispheric asymmetries in speech perception: Sense, nonsense and modulations. *PLOS ONE, 6*(9), e24672.

Ross, L. A., Saint-Amour, D., Leavitt, V. M., Javitt, D. C., & Foxe, J. J. (2007). Do you see what I am saying? Exploring visual enhancement of speech comprehension in noisy environments. *Cerebral Cortex, 17*(5), 1147–1153.

Roux, F. E., Minkin, K., Durand, J. B., Sacko, O., Réhault, E., Tanova, R., & Démonet, J. F. (2015). Electrostimulation mapping of comprehension of auditory and visual words. *Cortex, 71*, 398–408.

Saenz, M., & Langers, D. R. (2014). Tonotopic mapping of human auditory cortex. *Hearing Research, 307*, 42–52.

Saffran, J. R., Aslin, R. N., & Newport, E. L. (1996). Statistical learning by 8-month-old infants. *Science, 274*(5294), 1926–1928.

Santoro, R., Moerel, M., De Martino, F., Goebel, R., Ugurbil, K., Yacoub, E., & Formisano, E. (2014). Encoding of natural sounds at multiple spectral and temporal resolutions in the human auditory cortex. *PLOS Computational Biology, 10*(1), e1003412.

Schreiner, C. E., Froemke, R. C., & Atencio, C. A. (2011). Spectral processing in auditory cortex. In J. A. Winer & C. E. Schreiner (Eds.), *The auditory cortex* (pp. 275–308). New York: Springer.

Schroeder, C. E., Lakatos, P., Kajikawa, Y., Partan, S., & Puce, A. (2008). Neuronal oscillations and visual amplification of speech. *Trends in Cognitive Sciences, 12*(3), 106–113.

Scott, S. K., Blank, C. C., Rosen, S., & Wise, R. J. (2000). Identification of a pathway for intelligible speech in the left temporal lobe. *Brain, 123*(12), 2400–2406.

Scott, S. K., & Johnsrude, I. S. (2003). The neuroanatomical and functional organization of speech perception. *Trends in Neurosciences, 26*(2), 100–107.

Shamma, S., & Lorenzi, C. (2013). On the balance of envelope and temporal fine structure in the encoding of speech in the early auditory system. *Journal of the Acoustical Society of America, 133*(5), 2818–2833.

Sjerps, M. J., & McQueen, J. M. (2010). The bounds on flexibility in speech perception. *Journal of Experimental Psychology: Human Perception and Performance, 36*(1), 195.

Skipper, J. I., van Wassenhove, V., Nusbaum, H. C., & Small, S. L. (2007). Hearing lips and seeing voices: How cortical areas supporting speech production mediate audiovisual speech perception. *Cerebral Cortex, 17*(10), 2387–2399.

Snyder, H. R., Feigenson, K., & Thompson-Schill, S. L. (2007). Prefrontal cortical response to conflict during semantic and phonological tasks. *Journal of Cognitive Neuroscience, 19*(5), 761–775.

Sohoglu, E., Peelle, J. E., Carlyon, R. P., & Davis, M. H. (2012). Predictive top-down integration of prior knowledge during speech perception. *Journal of Neuroscience, 32*(25), 8443–8453.

Steinschneider, M., Nourski, K. V., Kawasaki, H., Oya, H., Brugge, J. F., & Howard, M. A. (2011). Intracranial study of speech-elicited activity on the human posterolateral superior temporal gyrus. *Cerebral Cortex, 21*(10), 2332–2347.

Stekelenburg, J. J., & Vroomen, J. (2007). Neural correlates of multisensory integration of ecologically valid audiovisual events. *Journal of Cognitive Neuroscience, 19*(12), 1964–1973.

Striem-Amit, E., Hertz, U., & Amedi, A. (2011). Extensive cochleotopic mapping of human auditory cortical fields obtained with phase-encoding fMRI. *PLOS ONE, 6*(3), e17832.

Suh, M. W., Lee, H. J., Kim, J. S., Chung, C. K., & Oh, S. H. (2009). Speech experience shapes the speechreading network and subsequent deafness facilitates it. *Brain, 132*(10), 2761–2771.

Swaab, T. Y., Brown, C., & Hagoort, P. (1998). Understanding ambiguous words in sentence contexts: Electrophysiological evidence for delayed contextual selection in Broca's aphasia. *Neuropsychologia, 36*(8), 737–761.

Takeichi, H., Koyama, S., Terao, A., Takeuchi, F., Toyosawa, Y., & Murohashi, H. (2010). Comprehension of degraded speech sounds with m-sequence modulation: An fMRI study. *NeuroImage, 49*(3), 2697–2706.

Talavage, T. M., Sereno, M. I., Melcher, J. R., Ledden, P. J., Rosen, B. R., & Dale, A. M. (2004). Tonotopic organization in human auditory cortex revealed by progressions of frequency sensitivity. *Journal of Neurophysiology, 91*, 1282–1296.

Talebi, V., & Baker, C. L. (2012). Natural versus synthetic stimuli for estimating receptive field models: A comparison of predictive robustness. *Journal of Neuroscience, 32*(5), 1560–1576.

Theunissen, F. E., Sen, K., & Doupe, A. J. (2000). Spectral-temporal receptive fields of nonlinear auditory neurons obtained using natural sounds. *Journal of Neuroscience, 20*(6), 2315–2331.

Travis, K. E., Leonard, M. K., Chan, A. M., Torres, C., Sizemore, M. L., Qu, Z., ... Halgren, E. (2013). Independence of early speech processing from word meaning. *Cerebral Cortex, 23*(10), 2370–2379.

Tremblay, P., Baroni, M., & Hasson, U. (2013). Processing of speech and non-speech sounds in the supratemporal plane: Auditory input preference does not predict sensitivity to statistical structure. *NeuroImage, 66*, 318–332.

Tremblay, P., Deschamps, I., Baroni, M., & Hasson, U. (2016). Neural sensitivity to syllable frequency and mutual information in speech perception and production. *NeuroImage, 136*, 106–121.

Tsunada, J., Lee, J. H., & Cohen, Y. E. (2011). Representation of speech categories in the primate auditory cortex. *Journal of Neurophysiology, 105*(6), 2634–2646.

Turkeltaub, P. E., & Coslett, H. (2010). Localization of sublexical speech perception components. *Brain and Language, 114*(1), 1–15.

Turken, A. U., & Dronkers, N. F. (2011). The neural architecture of the language comprehension network: Converging evidence from lesion and connectivity analyses. *Frontiers in Systems Neuroscience*, 5, 1.

Utman, J. A., Blumstein, S. E., & Sullivan, K. (2001). Mapping from sound to meaning: Reduced lexical activation in Broca's aphasics. *Brain and Language, 79*(3), 444–472.

Vitevitch, M. S., & Luce, P. A. (1999). Probabilistic phonotactics and neighborhood activation in spoken word recognition. *Journal of Memory and Language, 40*(3), 374–408.

Weiss, M. W., & Bidelman, G. M. (2015). Listening to the brainstem: Musicianship enhances intelligibility of subcortical representations for speech. *Journal of Neuroscience, 35*(4), 1687–1691.

Wible, C. G., Han, S. D., Spencer, M. H., Kubicki, M., Niznikiewicz, M. H., Jolesz, F. A., ... Nestor, P. (2006). Connectivity among semantic associates: An fMRI study of semantic priming. *Brain and Language, 97*(3), 294–305.

Willems, R. M., Frank, S. L., Nijhof, A. D., Hagoort, P., & Van den Bosch, A. (2015). Prediction during natural language comprehension. *Cerebral Cortex, 26*(6), 2506–2516.

Wilson, S. M., Lam, D., Babiak, M. C., Perry, D. W., Shih, T., Hess, C.P., ... Chang, E. F. (2015). Transient aphasias after left hemisphere resective surgery. *Journal of Neurosurgery, 123*(3), 581–593.

Woolley, S. M., Fremouw, T. E., Hsu, A., & Theunissen, F. E. (2005). Tuning for spectro-temporal modulations as a mechanism for auditory discrimination of natural sounds. *Nature Neuroscience, 8*(10), 1371–1379.

Young, E. D. (2008). Neural representation of spectral and temporal information in speech. *Philosophical Transactions of the Royal Society of London B: Biological Sciences, 363*(1493), 923–945.

Zaehle, T., Geiser, E., Alter, K., Jancke, L., & Meyer, M. (2008). Segmental processing in the human auditory dorsal stream. *Brain Research, 1220*, 179–190.

Zhuang, J., Randall, B., Stamatakis, E. A., Marslen-Wilson, W. D., & Tyler, L. K. (2011). The interaction of lexical semantics and cohort competition in spoken word recognition: An fMRI study. *Journal of Cognitive Neuroscience, 23*(12), 3778–3790.

28 Functional Anatomy of Speech Production: From Words to Motor Control

GREGORY HICKOK

Speech production in its broadest scope refers to the process of mapping conceptual information—a thought or message—onto a set of respiratory, laryngeal, and vocal tract gestures that yield a sequence of sounds—hums, chirps, buzzes, and whooshes—that we recognize as speech. This is a complex mapping involving a number of transformations between multiple hierarchical and interacting levels of representation. It's so complex, in fact, that its study requires several disciplines. Some of these representations and transformations are examined under the purview of linguistics and psycholinguistics and cover topics such as semantics, syntax, morphology, phonology, lexical organization and access, ambiguity resolution, working memory, sequencing, and more. Other parts of the process are studied under the purview of fields such as kinematics, biomechanical engineering, articulatory phonetics, and motor control and cover topics such as elasticity and damping, source-filter models, motor equivalence, state feedback control, sensorimotor learning and adaptation, and more.

Our goal here is not to provide a systematic overview of what has been learned to date regarding the concept to sound transformation process, but to introduce what we know about its neural basis.

1. *The Speech Production Network*

The neural basis of speech production has been studied from a variety of perspectives and with a variety of methods (Chang et al., 2017; Indefrey, 2011; Indefrey & Levelt, 2004; Mirman et al., 2015; Price, 2012; Wilson, Isenberg, & Hickok, 2009). There is general agreement from this body of work that speech production involves a large, distributed network of brain regions including portions of Broca's region, premotor cortex, supramarginal gyrus, primary sensorimotor cortex, auditory cortex, posterior superior temporal cortex, the temporal-parietal junction, middle temporal cortex, portions of the cerebellum, and the basal ganglia. There is less consensus on the functional organization of these regions, although significant progress is being made, as we will review.

1.1. Partitioning the Network: The View from Psycholinguistics The transformation from message to movement in speech production is not a one-step computation. Decades of research on slips of the tongue, chronometric picture-naming studies, and other tasks has revealed at least two major stages just in the mapping from concept to the phonological form of a word (Dell, 1986, 1995; Fromkin, 1971; Levelt, 1989, 1992; Levelt, Roelofs, & Meyer, 1999). The first major stage is *lexical selection*, the process of selecting an abstract word form, often called a *lemma*. The second major stage is *phonological selection* or *phonological encoding*, the process of selecting the phonological form of the word. Other linguistic processes or subprocesses include grammatical encoding, syllabification, and phonetic encoding. Beyond these linguistic stages and closer to the motor periphery lies the process of *articulation*, typically assumed to be the domain of motor control models.

There are a couple of obvious questions we might ask about the neural system for speech production: Can we identify neural correlates of the major stages in the mapping from message to movement? Can we identify systems involved in lexical selection, phonological selection, and articulation? Researchers have tried to do so using various methods. Indefrey and Levelt, for example, published a series of meta-analyses of functional neuroimaging studies on speech production in which they analyzed the task and control conditions to attempt to identify the various psycholinguistic stages of naming (Indefrey, 2011; Indefrey & Levelt, 2000; Indefrey & Levelt, 2004). They concluded that lexical selection implicates the middle temporal gyrus; phonological processes implicate the posterior superior temporal gyrus (STG)/posterior middle temporal gyrus for phonological selection and the inferior frontal gyrus (IFG) for syllabification; and articulation implicates lower sensorimotor cortex (pre- and postcentral gyri; see also Basilakos, Smith, Fillmore, Fridriksson, & Fedorenko, 2018; Okada, Matchin, & Hickok, 2018).

Lesion-based studies have also targeted the stages of naming. One study reported that damage to left anterior and middle temporal lobe regions is associated with semantic errors in naming, thus likely implicating prephonological processing stages (Walker et al., 2011), whereas another study reported that phonological errors in naming are associated with damage to premotor cortex, pre- and postcentral gyri, and posterior parietal/temporal-parietal junction regions (Schwartz, Faseyitan, Kim, & Coslett, 2012). Similar results were obtained (i) using a larger behavioral data set to assess phonological and lexical-semantic processes, which additionally implicated the posterior IFG in lexical-semantic processes (Mirman et al., 2015) (see also Schnur et al., 2009), and (ii) using level-associated parameters of a computational model of naming (phonological and semantic connection strength weight parameters) rather than error types (Dell, Schwartz, Nozari, Faseyitan, & Branch Coslett, 2013).

While the details of the implicated networks vary somewhat, a broad consensus is emerging for a hierarchical neural organization of lexical-semantic and phonological processes reflecting the psycholinguistic hierarchy. Specifically, lexical semantic access in naming implicates middle temporal regions as well as higher-order speech-related regions in the IFG, whereas phonological processes in naming implicate lower-level sensorimotor processes in the posterior STG, inferior parietal, premotor cortex, and sensorimotor cortex (figure 28.1).

The connectivity of these networks is not yet understood with debates centered on whether posterior-frontal connectivity courses through anterior temporal to frontal pathways (e.g., the uncinate fasciculus) or via posterior-dorsal pathways (e.g., the arcuate fasciculus or various branches of the superior longitudinal fasciculus) or both (Roelofs, 2014; Roelofs & Ferreira, chapter 4 of this volume; Ueno, Saito, Rogers, & Lambon Ralph, 2011). Most agree that sensorimotor connections underlying phonological processes involve posterior-dorsal pathways, although the details regarding which are involved are still under discussion.

1.2. Partitioning the Network: The View from Motor Control As noted at the beginning of this chapter, motor control scientists are also interested in the process and neural basis of speech production but their starting point for investigation is rather different. Instead of starting from a communicative message that needs to be transformed into a series of articulatory gestures and working their way "forward" toward the motor periphery, motor control scientists start with the need to control and coordinate dozens of muscles and work their way "backward" toward the things that do the controlling. In theory, these two approaches should eventually meet in the middle and hopefully arrive at similar conclusions (in fact, I suggest they have, at least in part, as we'll see later in this section and the next). Because of the different starting points, the two subfields of speech production research have latched on to different theoretical machinery to bootstrap their investigation. Psycholinguists naturally looked to

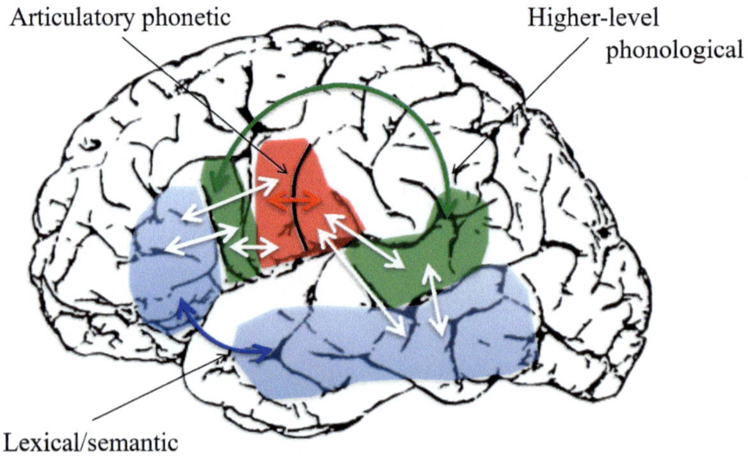

Figure 28.1 A broad schematic model of the hierarchical organization of lexical-semantic (blue), phonological (green), and articulatory processes (red) in speech production. White arrows indicate between-level connections. Location of arrows is not intended to reflect white matter pathways, only functional connections.

linguistic theory to guide the investigation into how concepts are translated into lexical and then phonological forms, whereas motor control–oriented speech scientists looked to motor control theory.

What are the big ideas in motor control theory? The most important one for our purposes here is feedback control, the idea that the goal of a motor act is to achieve a sensory outcome and to do so by minimizing error between the desired outcome and the actual sensory consequences of an action (Fairbanks, 1954). Consider the action of grasping a cup. The desired sensory outcome is for the hand to be wrapped around roughly the middle of the object, which is the best location for a stable grasp. As the reach is progressing, visual and somatosensory feedback can be used to determine whether there is any error in the movement—is the reach on the right trajectory to achieve the goal?—and this feedback information can be used to make corrections if an error is detected. But there's a problem, actual sensory feedback is delayed, noisy, and sometimes intermittent. The delay is a systemic issue: consider that it takes a nontrivial amount of time for neural signals to traverse from peripheral sensory organs to central processing networks, then error computations must be made, then correction signals must be sent back out the motor system, which also takes time. Meanwhile, the arm is no longer in the position at which the error was detected. In fact, by the time the brain detects the position of the limb in the first place, the limb is *already* not at that position anymore due to the time lag. Thus, the brain has the task of controlling the body without any direct information about the current state of what it is controlling. All it has is direct information about where the body was 100 ms or so in the past.

Modern theories of motor control argue that the brain solves this engineering problem by maintaining an *internal model* of the body. The idea is that through much experience with generating motor commands and observing their (delayed) sensory consequences, the brain can build a predictive model of where the effector should be given the current motor command and the most recent detected state of the effector. Predicting the consequences of a movement is sometimes called a *forward prediction* or *forward model* because it is predicting information about a state forward in time (Wolpert & Miall, 1996). It is assumed that this is achieved by sending a copy of the motor command, often termed an *efference copy* or *corollary discharge*, to the system that instantiates the internal model. Now, given that the brain can predict the consequences of a motor command before actual feedback occurs, it is capable of detecting and correcting errors near instantaneously based on the internal model. Actual overt feedback is used as a reference to make sure the predictions are correct and to keep the model tuned. These approaches are called *state feedback control (SFC)* models (Shadmehr & Krakauer, 2008; Todorov, 2004). Feedback control is still in play in SFC, but now much of the online control work is being done by internal rather than exclusively external feedback. Feedback control models have been proposed for speech (e.g., DIVA; Guenther, Hampson, & Johnson, 1998) as have SFC models (Hickok, Houde, & Rong, 2011; Houde & Nagarajan, 2011).

Notice that because feedback control and SFC models emphasize sensory targets as the goals of motor acts, these models in the speech domain are pitched at a fairly low representational level with "auditory" and "somatosensory" targets and motor commands aiming to hit those targets. Some models talk primarily about auditory-motor circuits in speech motor control (Hickok et al., 2011; Houde & Nagarajan, 2011), whereas others acknowledge the importance of the somatosensory system as an additional layer of sensory targets (Guenther, 2006; Hickok, 2012; Tremblay, Shiller, & Ostry, 2003). But all of these models are fundamentally sensorimotor in nature. Because linguistic representations are typically assumed to be abstract, modality independent objects, this reinforces the idea that motor control approaches to speech production are dealing with different processes compared to psycholinguistic models.

Some researchers have investigated the neural correlates of (state) feedback control in speech production. One interesting observation is that the electromagnetic response of auditory cortex to self-generated speech is attenuated compared to playback of that same speech (Aliu, Houde, & Nagarajan, 2009; Heinks-Maldonado, Nagarajan, & Houde, 2006; Houde, Nagarajan, Sekihara, & Merzenich, 2002; Tian & Poeppel, 2015). This is consistent with SFC models because forward predictions are typically modeled as suppression signals that tend to reduce our ability to detect our own actions when they match the predicted outcome—the neural basis for why we can't tickle ourselves (Blakemore, Wolpert, & Frith, 2000). In a recent study, researchers compared neural activation measured with functional MRI (fMRI) during imagined speech and silently articulated speech (Okada et al., 2018). Both conditions activated higher-level motor-speech areas in the IFG (among other regions), but only the silently articulated speech activated primary motor and somatosensory cortex, as expected. What was interesting was that silently articulated speech also activated auditory cortex quite robustly compared to imagined speech, even though neither condition involved any auditory speech stimulation. This result has a natural explanation in a forward prediction framework: the process of planning for overt speech articulation generates a forward prediction for acoustic consequences in auditory cortex. The increased activity in auditory cortex could be a result of inputs related to the prediction, increased processing due to the mismatch error between prediction and actual (lack of) input, or likely both.

Things start to get even more interesting when one looks more broadly at the proposed neural correlates of the various components of motor control–based models, which overlap those tied to phonological processes in psycholinguistic models, including IFG, ventral premotor cortex, posterior STG, temporal-parietal junction, and primary somatosensory-motor cortex (Golfinopoulos, Tourville, & Guenther, 2010; Guenther, 2006; Houde & Nagarajan, 2011). This raises the question, are these two approaches talking about distinct levels of processing or are they just using different terminology and architectures for describing the same processes?

2. *An Integrated Approach*

There are several additional reasons beyond the neural overlap to suspect that the two approaches are indeed studying the same processes, at least in part. One is that both motor control and linguistic representations and processes are hierarchical. This is a hallmark feature of language, but perhaps it is less appreciated in the motor domain from outside that field (Botvinick, Niv, & Barto, 2009; Crump & Logan, 2010; Lashley, 1951; Rosenbaum, Cohen, Jax, Weiss, & van der Wel, 2007). The use of hierarchical processes in motor control is a solution to a combinatorial explosion problem given the degrees of freedom an organism has in achieving a motor goal (Balleine, Dezfouli, Ito, & Doya, 2015). The idea is that higher levels of motor planning can operate over abstract motor goals that are spelled out by learned sequences of subgoals. This is structurally identical to the linguistic case where, say, an abstract word can be planned and then later spelled out in terms of its phonological subgoals. If both motor and linguistic processes rely on the same computational tricks, vis-à-vis the use of hierarchies, perhaps within the speech domain, then, the hierarchies in the two fields are one and the same.

There is reason to think these hierarchies indeed overlap. Consider the internal composition of the abstract phoneme according to the dominant theory (Chomsky & Halle, 1968): phonemes are decomposable into articulatory feature bundles (voicing, place of articulation, manner of articulation). These are precisely the effectors that motor control theories of speech production aim to control. Moreover, a major topic on motor control concerns how sequences of actions are coordinated efficiently (Balleine et al., 2015), which adds to the combinatorial problem. In the speech domain, sequences of vocal tract gestures correspond to sequences of phonemes, which correspond to the linguistic syllable, thus pushing motor control models deep into linguistic territory (e.g., see Bohland, Bullock, & Guenther, 2010; Bohland & Guenther, 2006).

Finally, both linguistic and motor control models have independently uncovered similar processes. We discussed the notion of feedback control, both external and internal varieties. Psycholinguistic research has also uncovered evidence for both types of feedback, called internal and external *monitoring*, and these loops are incorporated into psycholinguistic models (Levelt, 1983). Presumably the existence of feedback loops in motor control and psycholinguistics models of speech production is no coincidence.

For these reasons, work in my own lab has recently been aimed at integrating the two main approaches to speech production and its instantiation in the auditory dorsal stream. Before describing that work, we turn to a more detailed investigation into the functional anatomy of the dorsal stream system.

2.1. Anatomy and Physiology of the Auditory Dorsal Stream The auditory dorsal stream for speech comprises a network of regions, identified primarily on the basis of fMRI studies, that includes auditory-related cortices in the STG/superior temporal sulcus (STS), motor speech areas in the posterior lateral frontal lobe, and a region in the posterior Sylvian fissure at the parietal-temporal boundary (area *Spt*; figure 28.2) (Buchsbaum, Hickok, & Humphries, 2001; Hickok, Buchsbaum, Humphries, & Muftuler, 2003; Hickok, Okada, & Serences, 2009). Spt is hypothesized to serve as the interface between auditory and motor representations of speech (Hickok et al., 2011) in much the same way as regions of the posterior parietal cortex of macaques and humans function as a visuomotor interface (Andersen, 1997; Gallese, Fadiga, Fogassi, Luppino, & Murata, 1997; Grefkes & Fink, 2005; Milner & Goodale, 1995). Accordingly, Spt has been shown to exhibit auditory-motor response properties, activating both during the perception and (covert) production of speech; indeed, this is a defining property of the functional circuit. Spt has also been found to be relatively selective for vocal compared to manual auditory-motor actions. One fMRI study compared Spt activation during listening to simple novel melodies and then either silently humming them back or imagining playing them on a keyboard (participants were skilled pianists): Spt activated more strongly during the humming

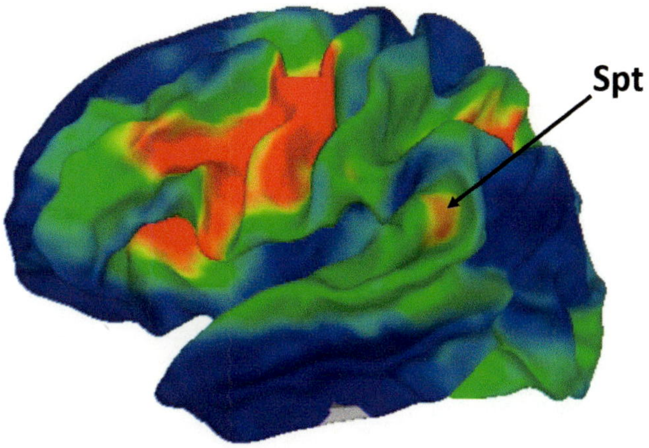

FIGURE 28.2 Location of area Spt based on an aggregate analysis of fMRI localization in 105 subjects (Buchsbaum et al., 2011).

condition despite identical auditory input (Pa & Hickok, 2008). Spt also appears to be functionally (Buchsbaum et al., 2001; Buchsbaum, Olsen, Koch, & Berman, 2005) and anatomically (Isenberg, Vaden, Saberi, Muftuler, & Hickok, 2012) connected to premotor regions involved in speech production, consistent with what is found for visuomotor integration areas in the parietal lobe (Grefkes & Fink, 2005). It also exhibits distinct activation patterns to auditory and motor phases of the task (Hickok et al., 2009), which argues against a purely auditory or purely motor explanation for its activation pattern.

Damage to this region is associated with conduction aphasia, a syndrome characterized by frequent sound-based errors, which are readily detected by the patient, in otherwise fluent speech production and difficulty with repeating speech verbatim, particularly nonwords; auditory comprehension is largely preserved (Goodglass, 1992). Functionally one can conceptualize conduction aphasia as an auditory-motor integration deficit (Buchsbaum et al., 2011; Hickok et al., 2011). Patients are fluent because frontal lobe motor speech networks are intact. They have good auditory comprehension and can detect their own errors because the speech perception system is intact. What's going wrong is the integration of the auditory and motor systems, leading to phonemic errors and difficulty repeating speech. (We take up the details of how this works in the next section.) If it is true that conduction aphasia is a deficit of auditory-motor integration and that Spt is a hub in the auditory-motor system, then the lesions in conduction aphasia should involve the Spt region, which is what has been reported (Buchsbaum et al., 2011). A more recent large-scale lesion analysis of word and nonword repetition more specifically found that repetition impairment is associated with the Spt region (Rogalsky et al., 2015).

As noted, Spt is located within the Sylvian fissure at its posterior most extent and, in individual subjects, typically involves the posterior-medial portions of the planum temporale and/or the parietal operculum, but Spt can extend laterally toward the crown of the superior temporal and/or supramarginal gyri. Cytoarchitectonically, the distribution range of Spt corresponds to area Tpt, a region *not* considered part of unimodal auditory cortex (Galaburda & Sanides, 1980; Smiley et al., 2007; Sweet, Dorph-Petersen, & Lewis, 2005), and therefore is consistent with the claim that Spt is not solely auditory but auditory-motor.

The anatomical connection between Spt and posterior motor areas is presumed to involve dorsal temporo-parietal-frontal white matter bundles including the arcuate fasciculus and the superior longitudinal fasciculus (Isenberg et al., 2012; Yagmurlu, Middlebrooks, Tanriover, & Rhoton, 2016). Some models, and one in particular, dubbed *Lichtheim 2*, have suggested a role for ventral temporofrontal pathways, such as the uncinate fasciculus, in speech production (Ueno et al., 2011). Although this proposal has been questioned empirically (Roelofs, 2014), it is not inconsistent with the view that the dorsal speech stream and its dorsal white matter connections are a critical part of the auditory-motor speech production network. Specifically, the dual stream model has never assumed that the dorsal stream is the only pathway for speech production. It is quite possible, therefore, that ventral stream temporofrontal pathways play a role in speech planning, perhaps serving to activate high-level action plans via lexical-semantic or combinatorial-semantic representations that are then fine-tuned by lower-level auditory-motor circuits. We turn now to the details of how this might work.

2.2. AN INTEGRATED (HIERARCHICAL) STATE FEEDBACK CONTROL MODEL OF SPEECH PRODUCTION The SFC model of speech production was a first attempt at developing a model of speech production that integrates the various traditions (Hickok et al., 2011) (figure 28.3). The model builds on previous work by Guenther et al. (1998), Tian and Poeppel (2010) and Houde and Nagarajan (2011). The essence of the model is the assumption that the feedback control architecture applies not only at low levels of motor control but at higher, "phonological levels," as well; see the discussion regarding what "phonological" might mean in this context. And it is at the phonological level (broadly speaking) that the model is primarily cast. In motor control terms, this phonological level is the "internal model," which is divided into two representational components, a motor-phonological

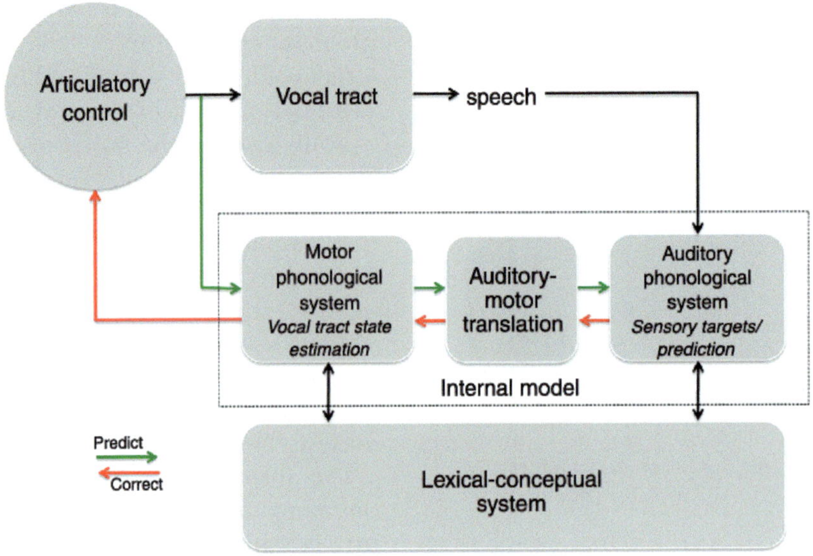

FIGURE 28.3 An integrated SFC model of speech production (Hickok et al., 2011).

component (which approximates a high-level internal model of the motor effector) and an auditory-phonological component (where the auditory consequences of actions are coded).

Consistent with psycholinguistic models, input to the phonological level of processing comes from a lexical level (the lexical and conceptual levels are collapsed into one box in the figure but assumed to be distinct), thus instantiating the familiar two-stage architecture. Different from many psycholinguistic models, but consistent with some neuropsychological models (Howard & Nickels, 2005; Jacquemot, Dupoux, & Bachoud-Levi, 2007; Martin, Lesch, & Bartha, 1999; Shelton & Caramazza, 1999), the phonological processing level is split into two components, an auditory-based and a motor-based system (input and output lexicons, in other terminology).

In the SFC model, the auditory- and motor-phonological systems are wired up via an interface network, which performs a kind of coordinate transform, linking phonological codes in auditory space to corresponding phonological codes in articulatory sequence space and vice versa. The anatomical correlate of this interface is hypothesized to be area Spt.

As noted, the phonological level receives input from a lexical-conceptual level. One unique aspect of the SFC model is the parallel input from the lexical-conceptual level to both auditory and motor phonological systems. This feature is not characteristic of motor control architectures (which typically start with the execution of a motor act) or two-stage psycholinguistic models, but it does have roots in classical 19th-century models of the neural organization of language (Lichtheim, 1885; Wernicke, [1874] 1977) and has similarities to more recent dual-route models of speech repetition (Hanley, Dell, Kay, & Baron, 2004). The parallel input assumption is key to explaining conduction aphasia as we will see.

What does it mean computationally to require input from the auditory target units? In typical visuomotor experiments, targets are defined by a sensory stimulus, such as an object in a spatial location, and the action is directed at that external target. Natural speech production (cf. shadowed or repeated speech) is different in the sense that the target is not defined in the immediate sensory environment but is an *internal* representation of previously heard auditory stimuli. But in both cases, it is still the sensory target that grounds the action. Motor control models often assume a kind of canceling operation as the basis for comparing motor predictions with overt sensory feedback, with the motor prediction implemented as an inhibitory signal. If the prediction (–) and the overt sensory input (+) cancel, then this indicates an accurate prediction. Prediction error, then, is sensory activation that is not canceled. Consistent with this general approach, the present model assumes excitatory inputs from the lexical-conceptual level to both the auditory- and motor-phonological units, excitatory inputs from auditory- to motor-phonological units, and inhibitory inputs from motor- to auditory-phonological units. The excitatory connections between auditory- and motor-phonological units can be conceptualized as an "error signal" in the sense described, although in this case input to sensory units are coming not from the external environment but from an internal top-down source. The prediction error signal computation is identical: if an auditory target is activated and there is no activity in the motor units, then the excitatory inputs from auditory to motor will correct this nonactivation "error" by activating the corresponding motor units. If an auditory target is activated and the corresponding motor units are also activated, then inhibitory motor-to-auditory inputs will cancel the auditory-to-motor excitation, thus squashing the error signal. If an auditory target is activated and the wrong (noncorresponding) motor units are activated, the motor-to-sensory feedback will inhibit nontarget auditory units and therefore "allow" the actual auditory target to continue sending excitatory inputs to the corresponding motor units, thus correcting the error. In this way, the same prediction/error detection calculation that is used for external feedback is used for internal error detection and correction. All of this is going on prior to speech articulation as part of the motor programming process. So, while the auditory targets are always activated, they only play a major role in the process when something goes wrong, either because the wrong motor units were activated or because no motor units were activated.

3. Conduction Aphasia

Conduction aphasia is one of the strongest sources of empirical validation for the proposed model. Explaining of the co-occurrence of the features of conduction aphasia—that is, generally fluent output, impaired phonemic planning, and preserved speech perception—has proven difficult. In models that assume one phonological level network for perception and production, a central phonological deficit could yield phonemic output problems but would also be expected to affect perception. Alternatively, assuming that separate phonological input and output systems exist, impairment to a phonological output system could explain the paraphasias but should also cause dysfluency. Furthermore, the lesions in conduction aphasia are in auditory-related temporal-parietal cortex (Baldo, Klostermann, & Dronkers, 2008; Buchsbaum et al., 2011; Damasio & Damasio, 1980), not in frontal cortex where one would expect to find motor-related systems. Damage to a phonological input system is more consistent with the lesion location, explains the preserved fluency because the motor phonological system is still intact, and could explain paraphasias if one assumes a role for the input system in speech production. However, again there is no explanation for why the system can easily recognize errors perceptually that it fails to prevent them in production. For this reason, some authors have attempted to explain conduction aphasia as a deficit to a buffer in phonological working memory, separate from the phonological input and output systems (Baldo et al., 2008), but then one loses the explanation of paraphasic errors (Buchsbaum et al., 2011).

Wernicke's original hypothesis that conduction aphasia is caused by a disconnection between sensory and motor speech systems is a viable solution (Hickok et al., 2000): fluency is preserved because the motor system is intact, perception is preserved because the sensory system is intact, and paraphasias occur because the sensory system can no longer play its role in speech production once the systems are disconnected (see also Buchsbaum et al., 2011; and Jacquemot et al., 2007, for similar arguments). What was lacking from Wernicke's account, though, was a principled explanation for why the sensory system plays a role in production. Internal state feedback control (as included in the SFC model) provides such a principled explanation: the sensory speech system is involved in production because the sensory system defines the targets of speech actions, and without access to information about the targets, actions will sometimes miss their mark. The only other modern adjustment that is needed to Wernicke's account is the anatomy. He proposed a white matter tract as the source of the disconnection, for which there is little evidence (Anderson et al., 1999; Dronkers & Baldo, 2009; Hickok, 2000). Modern findings instead implicate a cortical system that computes a sensorimotor coordinate transformation, which we have identified as area Spt (Buchsbaum et al., 2001; Hickok et al., 2003; Hickok et al., 2009) and which lies in the lesion distribution of conduction aphasia (Buchsbaum et al., 2011).

So far the discussion has been fairly vague about what kind of phonological information is coded in the auditory and motor networks and it has neglected a large literature showing the involvement of the somatosensory system in speech production. Here, I will outline an extension of the integrated SFC model, the *hierarchical* SFC model (figure 28.4) that partially remedies this situation. In the hierarchical SFC model, there are two hierarchically organized levels of SFC, which are similar to the levels proposed by Gracco and Lofqvist (Gracco, 1994; Gracco & Lofqvist, 1994). The higher level codes speech information roughly at the syllable level (that is, complete vocal tract opening and closing cycles) and involves a sensory-motor loop that includes sensory targets in auditory cortex, motor programs coded in the Brodmann area (BA) 44 portion of Broca's area and/or lower BA6, and area Spt, which computes a coordinate transform between the sensory and motor areas (Hickok, 2012; Hickok et al., 2003; Hickok et al., 2009). This is the loop described in the earlier SFC model (Hickok et al., 2011). The lower level

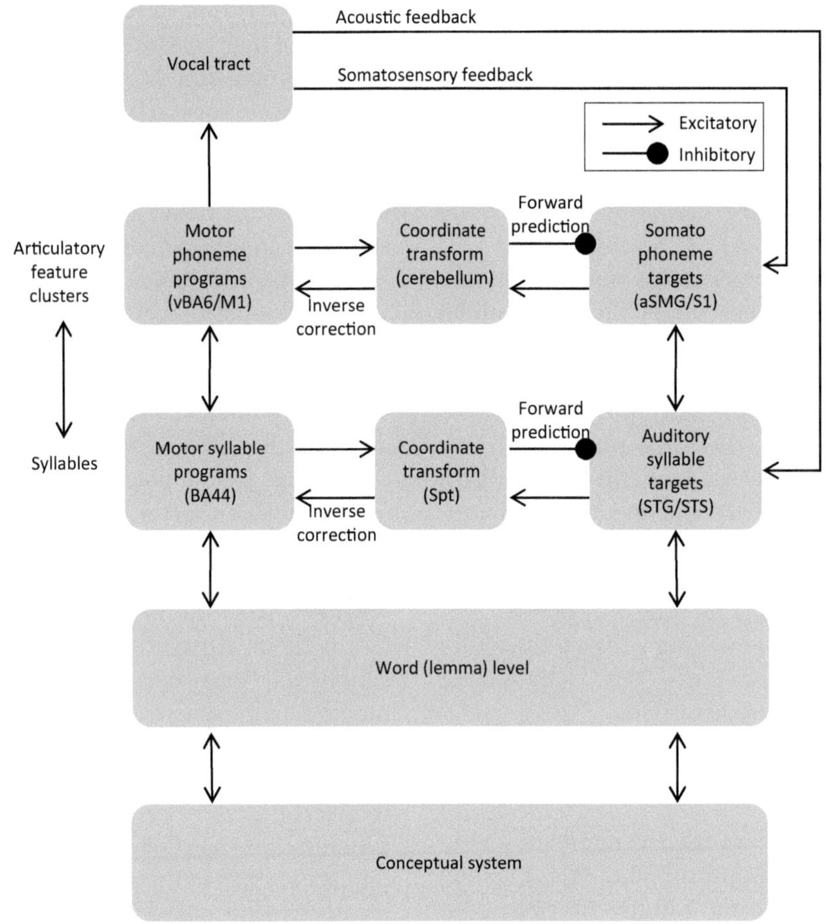

FIGURE 28.4 A hierarchical SFC model of speech production (Hickok, 2012).

of feedback control codes speech information at the level of articulatory feature clusters, that is, the collection of feature values that are associated with the targets of a vocal tract opening or closing gesture. Given that phonemes are defined in generative linguistics in terms of articulatory feature clusters (Chomsky & Halle, 1968), one can conceptualize this level of the speech motor control circuit as the most closely aligned to the theoretical construct of the phoneme (Gracco, 1994; Gracco & Lofqvist, 1994). The circuit is hypothesized to include sensory targets coded primarily in somatosensory cortex (as suggested by V. L. Gracco, personal communication, 2011), motor programs coded in lower primary motor cortex (M1), and a cerebellar circuit mediating the relation between the two.

As discussed, a large-scale lesion study provides some support for these claims. Schwartz et al. (2012) studied 106 unselected aphasic patients and documented 1,718 phonological errors on a naming task. Error rates in these patients were correlated with damaged regions in the auditory dorsal stream including prominent involvement of somatosensory cortex (postcentral gyrus) and a more posterior supramarginal gyrus region that included area Spt and surrounds. This pattern of findings is consistent with the hypothesis that auditory-motor as well as somatosensory-motor loops play a substantial role in the control of speech production at the phonological level(s).

4. A Computational Test of One Component of the SFC Architecture

From a psycholinguistic standpoint, the idea that the phonological system is split into two components is atypical (Dell, 1986; Dell, Schwartz, Martin, Saffran, & Gagnon, 1997; Levelt, 1999; Levelt, Roelofs, & Meyer, 1999). There is some precedent, however, in the neuropsychological literature where patterns of dissociation have led to the claim that there exist distinct phonological input and output lexicons (Jacquemot et al., 2007). Still, one would like to see additional psycholinguistic/computational evidence for the proposed architecture. One approach is to modify existing successful computational models of speech

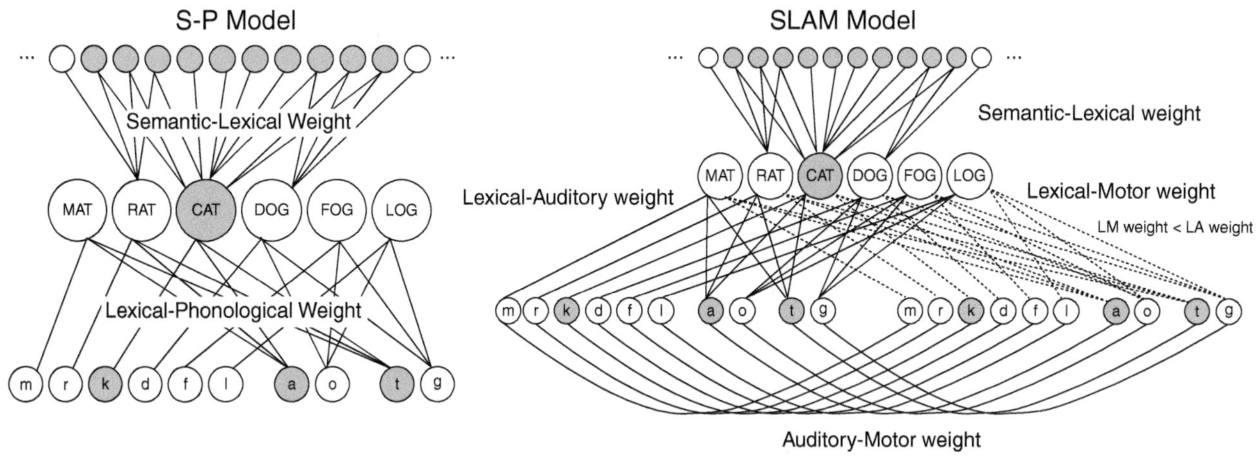

FIGURE 28.5 Graphic depiction of the S-P and SLAM models. LA = lexical-auditory; LM = lexical-motor.

production so that they approximate the architecture of the SFC model, that is, so that they contain separate auditory- and motor-phonological components, and then assess whether the modified model better explains the data.

The recently developed semantic lexical auditory motor (SLAM) computational model does just this (Walker & Hickok, 2016a). SLAM is a modification of the semantic-phonological (S-P) model (figure 28.5), which has proven quite successful in accounting for distributions of naming errors in aphasia (Dell, Martin, & Schwartz, 2007; Dell, Schwartz, Martin, Saffran, & Gagnon, 1997; Nozari, Kittredge, Dell, & Schwartz, 2010). The S-P model is a two-step model of word production where the first step involves lexical or lexeme selection and the second step involves phonological selection. The model has been implemented in a connectionist network composed of three layers, an input semantic feature layer, a middle lexical selection layer, and an output phonological layer. By adjusting only two connection strength parameters, the semantic-lexical weight and the lexical-phonological weight, the model can approximate the pattern of errors in individual aphasic patients with reasonably high accuracy.

SLAM made one modification to the S-P model: it split (actually duplicated) the phonological layer into an auditory- and motor-phonological layer, with the latter serving as the output layer. The question that was asked is whether SLAM provided a better fit to the aphasia naming data than S-P did. The answer was yes. Overall, SLAM did a better job of fitting the data than S-P and substantially outperformed S-P for patients with conduction aphasia, which fits nicely with the discussion in section 3 regarding the importance of auditory-motor interaction in speech production and its link to conduction aphasia (Walker & Hickok, 2016a). This finding provides an additional source of support for the architectural assumptions of the SFC framework.

As with any new theory, SLAM has its detractors. Goldrick (2016) argued that SLAM's improvement over S-P is merely a result of approximating a different model of production in which the phonological layer is broken down, not into auditory and motor components, but into lexical (higher-level) and postlexical (lower-level) layers (Goldrick & Rapp, 2007). This is an important point that highlights the fact that SLAM is not a complete model of speech production in that it does not model all of the hierarchical complexity of the system. For example, SLAM does not even attempt to model all the layers proposed in the hierarchical version of the SFC model (Hickok, 2012). That said, a computational evaluation of Goldrick's claims showed that simply adding an additional postlexical layer to the S-P model does not yield the same fit improvement found with SLAM (Walker & Hickok, 2016b); it is the additional layer arranged in a particular *architecture* that provides the boost.

5. *Conclusion*

The task of identifying the neural regions involved in speech production is largely complete to a first approximation, at least for word-level utterances. The field has now turned to understanding the connectivity and functional/computational roles of the network. A critical goal in this effort is to understand the relation between various hypotheses in both cognitive-computational and neural domains. Where do motor control and psycholinguistic models of speech make contact? What is the relation between sensorimotor speech systems and sensorimotor systems more broadly? How are linguistic/sensorimotor systems implemented in neurocomputational networks? We've surveyed some initial steps toward answering these questions and the results look promising but much more work needs to be done.

REFERENCES

Aliu, S. O., Houde, J. F., & Nagarajan, S. S. (2009). Motor-induced suppression of the auditory cortex. *Journal of Cognitive Neuroscience, 21*(4), 791–802.

Andersen, R. (1997). Multimodal integration for the representation of space in the posterior parietal cortex. *Philosophical Transactions of the Royal Society of London B Biological Sciences, 352*, 1421–1428.

Anderson, J. M., Gilmore, R., Roper, S., Crosson, B., Bauer, R. M., Nadeau, S., ... Hughes, J. D. (1999). Conduction aphasia and the arcuate fasciculus: A reexamination of the Wernicke-Geschwind model. *Brain and Language, 70*, 1–12.

Baldo, J. V., Klostermann, E. C., & Dronkers, N. F. (2008). It's either a cook or a baker: Patients with conduction aphasia get the gist but lose the trace. *Brain and Language, 105*(2), 134–140.

Balleine, B. W., Dezfouli, A., Ito, M., & Doya, K. (2015). Hierarchical control of goal-directed action in the cortical-basal ganglia network. *Current Opinion in Behavioral Sciences, 5*, 1–7.

Basilakos, A., Smith, K. G., Fillmore, P., Fridriksson, J., & Fedorenko, E. (2018). Functional characterization of the human speech articulation network. *Cerebral Cortex, 28*(5), 1816–1830.

Blakemore, S. J., Wolpert, D., & Frith, C. (2000). Why can't you tickle yourself? *NeuroReport, 11*(11), R11–R16.

Bohland, J. W., Bullock, D., & Guenther, F. H. (2010). Neural representations and mechanisms for the performance of simple speech sequences. *Journal of Cognitive Neuroscience, 22*(7), 1504–1529.

Bohland, J. W., & Guenther, F. H. (2006). An fMRI investigation of syllable sequence production. *NeuroImage, 32*(2), 821–841.

Botvinick, M. M., Niv, Y., & Barto, A. C. (2009). Hierarchically organized behavior and its neural foundations: A reinforcement learning perspective. *Cognition, 113*(3), 262–280.

Buchsbaum, B. R., Baldo, J., Okada, K., Berman, K. F., Dronkers, N., D'Esposito, M., & Hickok, G. (2011). Conduction aphasia, sensory-motor integration, and phonological short-term memory: An aggregate analysis of lesion and fMRI data. *Brain and Language, 119*(3), 119–128.

Buchsbaum, B. R., Hickok, G., & Humphries, C. (2001). Role of left posterior superior temporal gyrus in phonological processing for speech perception and production. *Cognitive Science, 25*, 663–678.

Buchsbaum, B. R., Olsen, R. K., Koch, P., & Berman, K. F. (2005). Human dorsal and ventral auditory streams subserve rehearsal-based and echoic processes during verbal working memory. *Neuron, 48*(4), 687–697.

Chang, E. F., Breshears, J. D., Raygor, K. P., Lau, D., Molinaro, A. M., & Berger, M. S. (2017). Stereotactic probability and variability of speech arrest and anomia sites during stimulation mapping of the language dominant hemisphere. *Journal of Neurosurgery, 126*(1), 114–121.

Chomsky, N., & Halle, M. (1968). *The sound pattern of English*. New York: Harper & Row.

Crump, M. J., & Logan, G. D. (2010). Hierarchical control and skilled typing: Evidence for word-level control over the execution of individual keystrokes. *Journal of Experimental Psychology: Learning, Memory, and Cognition, 36*(6), 1369.

Damasio, H., & Damasio, A. R. (1980). The anatomical basis of conduction aphasia. *Brain, 103*, 337–350.

Dell, G. S. (1986). A spreading-activation theory of retrieval in sentence production. *Psychological Review, 93*(3), 283–321.

Dell, G. S. (1995). Speaking and misspeaking. In L. R. Gleitman & M. Liberman (Eds.), *An invitation to cognitive science: Language* (2nd ed., Vol. 1, pp. 183–208). Cambridge, MA: MIT Press.

Dell, G. S., Martin, N., & Schwartz, M. F. (2007). A case-series test of the interactive two-step model of lexical access: Predicting word repetition from picture naming. *Journal of Memory and Language, 56*(4), 490–520.

Dell, G. S., Schwartz, M. F., Martin, N., Saffran, E. M., & Gagnon, D. A. (1997). Lexical access in aphasic and nonaphasic speakers. *Psychological Review, 104*, 801–838.

Dell, G. S., Schwartz, M. F., Nozari, N., Faseyitan, O., & Branch Coslett, H. (2013). Voxel-based lesion-parameter mapping: Identifying the neural correlates of a computational model of word production. *Cognition, 128*(3), 380–396.

Dronkers, N., & Baldo, J. (2009). Language: Aphasia. In L. R. Squire (Ed.), *Encyclopedia of neuroscience* (Vol. 5, pp. 343–348). Oxford: Academic Press.

Fairbanks, G. (1954). Systematic research in experimental phonetics: I. A theory of the speech mechanism as a servosystem. *Journal of Speech and Hearing Disorders, 19*(2), 133–139.

Fromkin, V. (1971). The non-anomalous nature of anomalous utterances. *Language, 47*, 27–52.

Galaburda, A., & Sanides, F. (1980). Cytoarchitectonic organization of the human auditory cortex. *Journal of Comparative Neurology, 190*, 597–610.

Gallese, V., Fadiga, L., Fogassi, L., Luppino, G., & Murata, A. (1997). A parietal-frontal circuit for hand and grasping movements in the monkey: Evidence from reversible inactivation experiments. In P. Thier & H.-O. Karnath (Eds.), *Parietal lobe contributions to orientation in 3D space* (pp. 255–270). Heidelberg: Springer-Verlag.

Goldrick, M. (2016). Integrating SLAM with existing evidence: Comment on Walker and Hickok (2015). *Psychonomic Bulletin and Review, 23*(2), 648–652.

Goldrick, M., & Rapp, B. (2007). Lexical and post-lexical phonological representations in spoken production. *Cognition, 102*(2), 219–260.

Golfinopoulos, E., Tourville, J. A., & Guenther, F. H. (2010). The integration of large-scale neural network modeling and functional brain imaging in speech motor control. *NeuroImage, 52*(3), 862–874.

Goodglass, H. (1992). Diagnosis of conduction aphasia. In S. E. Kohn (Ed.), *Conduction aphasia* (pp. 39–49). Hillsdale, NJ: Lawrence Erlbaum.

Gracco, V. L. (1994). Some organizational characteristics of speech movement control. *Journal of Speech, Language, and Hearing Research, 37*(1), 4–27.

Gracco, V. L., & Lofqvist, A. (1994). Speech motor coordination and control: Evidence from lip, jaw, and laryngeal movements. *Journal of Neuroscience, 14*(11), 6585–6597.

Grefkes, C., & Fink, G. R. (2005). The functional organization of the intraparietal sulcus in humans and monkeys. *Journal of Anatomy, 207*(1), 3–17.

Guenther, F. H. (2006). Cortical interactions underlying the production of speech sounds. *Journal of Communication Disorders, 39*(5), 350–365.

Guenther, F. H., Hampson, M., & Johnson, D. (1998). A theoretical investigation of reference frames for the planning of speech movements. *Psychological Review, 105*, 611–633.

Hanley, J. R., Dell, G. S., Kay, J., & Baron, R. (2004). Evidence for the involvement of a nonlexical route in the repetition of familiar words: A comparison of single and dual route models of auditory repetition. *Cognitive Neuropsychology, 21*(2), 147–158.

Heinks-Maldonado, T. H., Nagarajan, S. S., & Houde, J. F. (2006). Magnetoencephalographic evidence for a precise forward model in speech production. *NeuroReport, 17*(13), 1375–1379.

Hickok, G. (2000). Speech perception, conduction aphasia, and the functional neuroanatomy of language. In Y. Grodzinsky, L. Shapiro, & D. Swinney (Eds.), *Language and the brain* (pp. 87–104). San Diego: Academic Press.

Hickok, G. (2012). Computational neuroanatomy of speech production. *Nature Reviews Neuroscience, 13*(2), 135–145.

Hickok, G., Buchsbaum, B. R., Humphries, C., & Muftuler, T. (2003). Auditory-motor interaction revealed by fMRI: Speech, music, and working memory in area Spt. *Journal of Cognitive Neuroscience, 15*, 673–682.

Hickok, G., Erhard, P., Kassubek, J., Helms-Tillery, A. K., Naeve-Velguth, S., Strupp, J. P., … Ugurbil, K. (2000). A functional magnetic resonance imaging study of the role of left posterior superior temporal gyrus in speech production: Implications for the explanation of conduction aphasia. *Neuroscience Letters, 287*, 156–160.

Hickok, G., Houde, J. F., & Rong, F. (2011). Sensorimotor integration in speech processing: Computational basis and neural organization. *Neuron, 69*(3), 407–422.

Hickok, G., Okada, K., & Serences, J. T. (2009). Area Spt in the human planum temporale supports sensory-motor integration for speech processing. *Journal of Neurophysiology, 101*(5), 2725–2732.

Houde, J. F., & Nagarajan, S. S. (2011). Speech production as state feedback control. *Frontiers in Human Neuroscience, 5*, 82.

Houde, J. F., Nagarajan, S. S., Sekihara, K., & Merzenich, M. M. (2002). Modulation of the auditory cortex during speech: An MEG study. *Journal of Cognitive Neuroscience, 14*(8), 1125–1138.

Howard, D., & Nickels, L. (2005). Separating input and output phonology: Semantic, phonological, and orthographic effects in short-term memory impairment. *Cognitive Neuropsychology, 22*, 42–77.

Indefrey, P. (2011). The spatial and temporal signatures of word production components: A critical update. *Frontiers in Psychology, 2*, 255.

Indefrey, P., & Levelt, W. J. M. (2000). The neural correlates of language production. In M. S. Gazzaniga (Ed.), *The new cognitive neurosciences* (pp. 845–865). Cambridge, MA: MIT Press.

Indefrey, P., & Levelt, W. J. M. (2004). The spatial and temporal signatures of word production components. *Cognition, 92*(1–2), 101–144.

Isenberg, A. L., Vaden, K. I., Jr., Saberi, K., Muftuler, L. T., & Hickok, G. (2012). Functionally distinct regions for spatial processing and sensory motor integration in the planum temporale. *Human Brain Mapping, 33*(10), 2453–2463.

Jacquemot, C., Dupoux, E., & Bachoud-Levi, A. C. (2007). Breaking the mirror: Asymmetrical disconnection between the phonological input and output codes. *Cognitive Neuropsychology, 24*(1), 3–22.

Lashley, K. S. (1951). The problem of serial order in behavior. In L. A. Jeffries (Ed.), *Cerebral mechanisms in behavior* (pp. 112–131). New York: Wiley.

Levelt, W. J. M. (1983). Monitoring and self-repair in speech. *Cognition, 14*(1), 41–104.

Levelt, W. J. M. (1989). *Speaking: From intention to articulation.* Cambridge, MA: MIT Press.

Levelt, W. J. M. (1992). Accessing words in speech production: Stages, processes and representations. *Cognition, 42*(1–3), 1–22.

Levelt, W. J. M. (1999). Models of word production. *Trends in Cognitive Sciences, 3*, 223–232.

Levelt, W. J. M., Roelofs, A., & Meyer, A. S. (1999). A theory of lexical access in speech production. *Behavioral and Brain Sciences, 22*(1), 1–75.

Lichtheim, L. (1885). On aphasia. *Brain, 7*, 433–484.

Martin, R. C., Lesch, M. F., & Bartha, M. C. (1999). Independence of input and output phonology in word processing and short-term memory. *Journal of Memory and Language, 41*, 3–29.

Milner, A. D., & Goodale, M. A. (1995). *The visual brain in action.* Oxford: Oxford University Press.

Mirman, D., Chen, Q., Zhang, Y., Wang, Z., Faseyitan, O. K., Coslett, H. B., & Schwartz, M. F. (2015). Neural organization of spoken language revealed by lesion-symptom mapping. *Nature Communications, 6*, ncomms7762.

Nozari, N., Kittredge, A. K., Dell, G. S., & Schwartz, M. F. (2010). Naming and repetition in aphasia: Steps, routes, and frequency effects. *Journal of Memory and Language, 63*(4), 541–559.

Okada, K., Matchin, W., & Hickok, G. (2018). Neural evidence for predictive coding in auditory cortex during speech production. *Psychonomic Bulletin and Review, 25*, 423–430.

Pa, J., & Hickok, G. (2008). A parietal-temporal sensory-motor integration area for the human vocal tract: Evidence from an fMRI study of skilled musicians. *Neuropsychologia, 46*, 362–368.

Price, C. J. (2012). A review and synthesis of the first 20 years of PET and fMRI studies of heard speech, spoken language and reading. *NeuroImage, 62*(2), 816–847.

Roelofs, A. (2014). A dorsal-pathway account of aphasic language production: The WEAVER++/ARC model. *Cortex, 59*, 33–48.

Rogalsky, C., Poppa, T., Chen, K. H., Anderson, S. W., Damasio, H., Love, T., & Hickok, G. (2015). Speech repetition as a window on the neurobiology of auditory-motor integration for speech: A voxel-based lesion symptom mapping study. *Neuropsychologia*, 71, 18–27.

Rosenbaum, D. A., Cohen, R. G., Jax, S. A., Weiss, D. J., & Van Der Wel, R. (2007). The problem of serial order in behavior: Lashley's legacy. *Human Movement Science, 26*(4), 525–554.

Schnur, T. T., Schwartz, M. F., Kimberg, D. Y., Hirshorn, E., Coslett, H. B., & Thompson-Schill, S. L. (2009). Localizing interference during naming: Convergent neuroimaging and neuropsychological evidence for the function of Broca's area. *Proceedings of the National Academy of Sciences, 106*(1), 322–327.

Schwartz, M. F., Faseyitan, O., Kim, J., & Coslett, H. B. (2012). The dorsal stream contribution to phonological retrieval in object naming. *Brain, 135*(12), 3799–3814.

Shadmehr, R., & Krakauer, J. W. (2008). A computational neuroanatomy for motor control. *Experimental Brain Research, 185*(3), 359–381.

Shelton, J. R., & Caramazza, A. (1999). Deficits in lexical and semantic processing: Implications for models of normal language. *Psychonomic Bulletin and Review, 6*, 5–27.

Smiley, J. F., Hackett, T. A., Ulbert, I., Karmas, G., Lakatos, P., Javitt, D. C., & Schroeder, C. E. (2007). Multisensory convergence in auditory cortex: I. Cortical connections of the caudal superior temporal plane in macaque monkeys. *Journal of Comparative Neurology, 502*(6), 894–923.

Sweet, R. A., Dorph-Petersen, K. A., & Lewis, D. A. (2005). Mapping auditory core, lateral belt, and parabelt cortices in the human superior temporal gyrus. *Journal of Comparative Neurology, 491*(3), 270–289.

Tian, X., & Poeppel, D. (2010). Mental imagery of speech and movement implicates the dynamics of internal forward models. *Frontiers in Psychology, 1*, 166.

Tian, X., & Poeppel, D. (2015). Dynamics of self-monitoring and error detection in speech production: Evidence from mental imagery and MEG. *Journal of Cognitive Neuroscience, 27*(2), 352–364.

Todorov, E. (2004). Optimality principles in sensorimotor control. *Nature Neuroscience, 7*(9), 907–915.

Tremblay, S., Shiller, D. M., & Ostry, D. J. (2003). Somatosensory basis of speech production. *Nature, 423*(6942), 866–869.

Ueno, T., Saito, S., Rogers, T. T., & Lambon Ralph, M. A. (2011). Lichtheim 2: Synthesizing aphasia and the neural basis of language in a neurocomputational model of the dual dorsal-ventral language pathways. *Neuron, 72*(2), 385–396.

Walker, G. M., & Hickok, G. (2016a). Bridging computational approaches to speech production: The semantic-lexical-auditory-motor model (SLAM). *Psychonomic Bulletin and Review, 23*(2), 339–352.

Walker, G. M., & Hickok, G. (2016b). Evaluating quantitative and conceptual models of speech production: How does SLAM fare? *Psychonomic Bulletin and Review, 23*(2), 653–660.

Walker, G. M., Schwartz, M. F., Kimberg, D. Y., Faseyitan, O., Brecher, A., Dell, G. S., & Coslett, H. B. (2011). Support for anterior temporal involvement in semantic error production in aphasia: New evidence from VLSM. *Brain and Language, 117*(3), 110–122.

Wernicke, C. (1977). Der aphasische symptomencomplex: Eine psychologische studie auf anatomischer basis. In G. H. Eggert (Ed.), *Wernicke's works on aphasia: A sourcebook and review* (pp. 91–145). The Hague: Mouton (Reprint of original work of 1874, Breslau: Max Cohn & Weigert).

Wilson, S. M., Isenberg, A. L., & Hickok, G. (2009). Neural correlates of word production stages delineated by parametric modulation of psycholinguistic variables. *Human Brain Mapping, 30*(11), 3596–3608.

Wolpert, D. M., & Miall, R. C. (1996). Forward models for physiological motor control. *Neural Networks, 9*(8), 1265–1279.

Yagmurlu, K., Middlebrooks, E. H., Tanriover, N., & Rhoton, A. L., Jr. (2016). Fiber tracts of the dorsal language stream in the human brain. *Journal of Neurosurgery, 124*(5), 1396–1405.

29 Neural Oscillations and Their Role in Speech and Language Processing

JOACHIM GROSS AND DAVID POEPPEL

1. Introduction

In what way is a discussion of neural oscillations relevant in a book on speech and language processing? The first observation of neural oscillatory activity has been made almost 100 years ago. It lies, in some sense, at the very origin of electrophysiological recordings in humans, with Berger's 1929 discovery of rhythmic low-frequency brain activity with the electroencephalogram. Moreover, this form of rhythmic brain activity has been recorded and interpreted for decades in animal neurophysiology, very notably, for example, in the context of memory and hippocampal physiology (reviewed, e.g., in Buzsaki, 2002), as well as in vigorous discussions about the binding problem in visual perception (Singer & Gray, 1995). In light of the fact that neural oscillations are so ubiquitously observed in the central nervous system, it is reasonable to ask whether this form of brain activity has "causal force" and aids in shaping perception or cognition—or whether it constitutes functionally inert "exhaust fumes of cortical computation." Oscillatory activity arising in the cortex has been argued to play a role in virtually all aspects of perception and cognition, from regulating attention (cf. Cravo, Rohenkohl, Wyart, & Nobre, 2013; Lakatos et al., 2013; Landau & Fries, 2012; Thut, Miniussi, & Gross, 2012; Thut, Nietzel, Brandt, & Pascual-Leone, 2006) to encoding memory (Buzsaki, 2002; Hanslmayr, Staresina, & Bowman, 2016) to parsing running speech (cf. Bastiaansen & Hagoort, 2015; Ghitza, 2013; Giraud & Poeppel, 2012; Gross et al., 2013; Luo, Liu, & Poeppel, 2010; Zion-Golumbic, Ding, et al., 2013) to underpinning foundational aspects of temporal perception across domains (VanRullen, Zoefel, & Ilhan, 2014). The various hypotheses have been received skeptically, often legitimately so, and we raise some of the constructive discussions on the issues. However, the study of neural oscillations in human perception and cognition is a vibrant area of cognitive neuroscience research, and we will focus here specifically on how speech and language processing may exploit oscillatory activity as part of its processing architecture. The experiments provide a new window onto our understanding of different aspects of perceptual analysis, and the careful engagement with these phenomena stands a good chance at deepening our mechanistic understanding of the brain basis of human behavior.

Before focusing more exclusively on the phenomena at the center of this chapter, it is worth introducing relevant terminology and pointing to some broader background. For the purpose of this chapter, we define brain oscillations as rhythmic fluctuations in the local field potential (LFP). The focus of this chapter will be mostly on the spatial summation of LFPs that can be recorded noninvasively with magnetoencephalography (MEG) or electroencephalography (EEG, for the remainder of this chapter we will denote this with MEEG). In a strict sense, oscillations imply a sinusoidal shape of the recorded brain signals. To what degree brain activity can be regarded as sinusoidal and strictly oscillatory is a contentious issue that we discuss in more detail in section 3. Here, we adopt a more lenient definition of brain oscillations that includes nonsinusoidal components that reoccur at a certain rate. The literature suggests that neural (and specifically cortical) oscillations (or rhythms) are widespread in terms of their origin, characteristics, and function. They can be recorded from virtually all brain areas, and some studies suggest that the oscillatory profile differs between brain areas (Rosanova et al., 2009), and might even represent a spectral fingerprint that is characteristic for a given brain area (Keitel & Gross, 2016). These spatially specific brain oscillations are constrained and shaped by the anatomical substrate of each brain area, which includes local and global anatomical connectivity of diverse classes of neurons, neurotransmitter levels, and others.

Importantly, a common hallmark and requirement of oscillations (biological or physical [e.g., pendulum]) is the existence of two opposing forces. For neuronal oscillations these two forces are excitation and inhibition. Together—finely balanced and accurately timed with a specific delay—they lead to rhythmic changes in synchronized postsynaptic potentials (Cannon et al., 2014; Fries, 2009) that can be recorded noninvasively with MEEG. Depending on the exact physiological mechanisms and their respective time constants, these oscillations are expressed at frequencies ranging from below 1 Hz to above 150 Hz (Buzsaki & Draguhn, 2004; Buzsaki, Logothetis, & Singer, 2013; Wang, 2010).

The functional relevance of brain rhythms relies critically on the fact that they represent rhythmic excitability changes of neuronal populations. Neuronal firing rates are modulated by the phase of LFP oscillations (Montemurro, Rasch, Murayama, Logothetis, & Panzeri, 2008; Lakatos, Chen, O'Connell, Mills, & Schroeder, 2007; Panzeri, Macke, Gross, & Kayser, 2015) and this phase represents information over and above what is available from spike trains (Belitski et al., 2008; Kayser, Montemurro, Logothetis, & Panzeri, 2009). The cyclic excitability changes make brain rhythms a prime candidate for supporting information processing tasks that require accurate temporal coordination—a prerequisite for any form of human behavior. Buzsaki et al. (2013, p. 751) summarized this nicely by stating that brain oscillations "form a hierarchical system that offers a syntactical structure for the spike traffic within and across circuits at multiple time scales." This dynamic functional structure complements the more static anatomical structure and allows the flexible task-dependent routing and gating of information flow within anatomically constrained networks.

It is therefore not surprising that brain oscillations and their task-dependent modulations have been linked to a wide range of cognitive tasks such as working memory, spatial and temporal attention, perception, categorization, language, and so on (Buzsaki & Draguhn, 2004; Panzeri et al., 2015; Schroeder & Lakatos, 2009; Thut et al., 2012; Wang, 2010).

Before we delve into a more detailed discussion about how cyclic excitability changes in neuronal populations support speech processing, we clarify some of the methodological advances that made these discoveries possible.

2. *Methods to Study Oscillations in Continuous Speech*

Typical MEEG experiments rely on the repeated presentation of stimuli followed by averaging of brain signals time locked to these stimuli to compute event-related signals. However, this approach is not suitable to understand how the brain processes continuous speech. In recent years, several techniques have been developed to address this problem. All of these developments are applied—in one way or another—to continuous MEEG signals—typically in the absence of any temporally localized events. In the simplest case, power spectra can be computed over the time period of stimulation and power changes can be quantified between two experimental conditions (e.g., intelligible versus unintelligible speech). However, this approach does not account for the dynamic nature of speech. One strategy to address this has been used by Koskinen and Seppa (2014). They use canonical correlation to extract components from the MEEG signal that are maximally similar across participants listening to the same continuous speech stimulus. The analysis can be performed across different frequency bands of the MEEG signal and can therefore be used to identify and characterize components in the brain signal that are similar across participants. These similarities in brain activity across listeners will likely reflect the acoustic and linguistic structure of the processed continuous speech but can also be affected by other processes such as emotional processing or working memory demands.

Recently, other approaches have been developed that quantify more directly statistical dependencies between continuous auditory or visual speech and oscillatory brain activity. They were inspired by reports demonstrating that presentation of a given speech stimulus leads to a phase pattern that is specific to this stimulus (Luo & Poeppel, 2007). Repeated presentation of a stimulus will consistently evoke a similar phase pattern in the MEEG signal. These studies emphasize the importance of oscillatory phase (as compared to the amplitude) for representing stimulus identity. But they don't directly relate oscillatory components in the stimulus to oscillatory components in brain signals. This can be done by employing techniques that have previously been developed and used for studying corticomuscular interactions (Gross et al., 2001). These studies often use coherence, as a frequency domain equivalent to correlation, to quantify the bivariate dependency between continuous speech and brain signals (Peelle, Gross, & Davis, 2013). This analysis is typically based on the broadband envelope of the continuous speech signal (Chandrasekaran, Trubanova, Stillittano, Caplier, & Ghazanfar, 2009) that captures rhythmic changes in sound energy corresponding to syllables (3–7 Hz) or intonation and prosody (about 1 Hz; Giraud & Poeppel, 2012) (see also section 3.1).

Similarly, phase locking between the speech envelope and brain signals can be quantified in a given frequency band using linear or nonlinear measures such as coherence or mutual information, respectively (Cogan & Poeppel, 2011; Gross et al., 2013). In the time domain, simple cross-correlation analysis across different delays of both (band-pass filtered) time series also quantifies speech–brain interactions (Ahissar et al., 2001; Crosse & Lalor, 2014; Ding & Simon, 2012; Giraud & Poeppel, 2012). This analysis also identifies the time delay at which brain signals are most strongly correlated with the speech envelope.

When conducting research on the role of brain oscillations in continuous speech processing, scientists need to be aware of several pitfalls. The existence of a significant effect after band-pass filtering in a certain frequency band does not imply that brain oscillations at this frequency exist and that they are related to this effect. It could well be that the effect is broadband in nature. Similarly, a peak in a power spectrum is no proof that an oscillation exists at this frequency. Any (transient nonoscillatory) pattern in MEEG signals that is repeated at a certain frequency will lead to a peak in the power spectrum.

Still, the analysis of brain oscillations has proven very useful and has—in combination with relevant experimental paradigms—significantly improved our understanding of the different ways in which brain oscillations support speech and language processing. This is the main topic of the following section.

3. Oscillations in Speech and Language Processing

3.1. Rhythms and Speech Production Normal spoken language is a continuously varying signal with an enormous amount of structure over different time scales. The speech signal is aperiodic—we do not speak isochronously, timed like a metronome—but the signal nevertheless contains a compelling quasi-rhythmic structure. Indeed, temporal regularities in speech occur over multiple timescales, and these scales are associated with different types of perceptual information (Rosen 1992). Relatively rapidly modulated information, say in the range of 20 to 50 Hz, is associated with the spectrotemporal fine structure of a signal and is critical for establishing the order of rapid events, including phonemic information.

Amplitude modulation in the signal at rates between 2 and 9 Hz is associated with speech envelope fluctuations correlating with typical syllabic rates. Modulations at even slower rates, say 1 to 3 Hz or below, typically correlate with prosodic aspects of utterances, including intonation contours and phrasal attributes.

What is the origin of the temporal modulations in speech? The most visible and cross-linguistically evident signal modulation derives from the opening and closing of the mouth, at a given speaking rate. As the jaw, mouth, and lips open and close, the emanating signal increases and decreases in its amplitude. This amplitude modulation is at its peak during the vocalic portion of utterances (mouth most open); the signal is at its smallest during the generation of stop consonants (e.g., /p/, /t/, /k/), as the airflow is interrupted during the brief closure of lips or velum. In typical speech at typical rates, the consonant–vowel alternations occur in a relatively regular manner. The changes in signal energy that contribute the most to speech intelligibility are below 20 Hz and typically peak between 4 and 6 Hz.

The shorthand designation for the signal that represents these amplitude fluctuations is the *envelope.* The speech envelope is a way to obtain quantitative information about the rate of change of energy in the speech signal and reflects neural responses in auditory cortex. Specifically, the speech signal is decomposed into narrow frequency bands in the cochlea. The auditory cortex receives energy fluctuations in these narrow frequency bands. The frequency of energy fluctuations can be characterized by the *modulation spectrum* (Ding et al., 2017). For speech produced at a natural rate, the modulation spectrum across languages peaks between 4 and 6 Hz (Ding et al., 2017; Elliott & Theunissen, 2009). The remarkable crosslinguistic temporal regularity suggests that the peak and distribution of the modulation spectrum are likely to correlate with some important perceptual information (figure 29.1).

As it turns out, the energy modulations captured by the modulation spectrum correspond roughly to the syllabic rate of speech. A peak in the modulation spectrum between 4 and 6 Hz reflects periods of 160 to 250 ms. The mean syllable duration across languages lies between 150 and 300 ms (Pellegrino, Coupe, & Marsico, 2011), aligning well with the modulation peak. This alignment between the physics of the signal (modulation spectrum) and the linguistics related to this (syllabic units) invites the *linking hypothesis* that neural tracking of the envelope and its modulations provides crucial perceptual information that forms the basis for speech decoding and initiates the suite of operations underlying comprehension. Specifically, the syllabic-size alternations, as reflected by the energy envelope over time, are critical because syllabic rate (a) signals the speaking rate, (b) carries stress and tonal contrasts, and (c) can be viewed crosslinguistically as the carrier of the linguistic (question, statement, and so on) or affective (happy, sad, and so on) prosody of an utterance. As a consequence, a high sensitivity to envelope

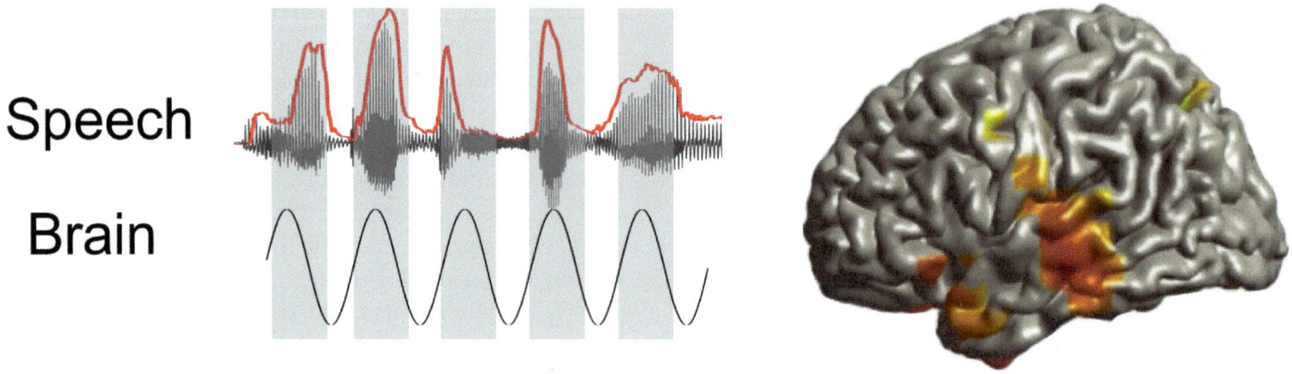

FIGURE 29.1 Schematic representation of speech entrainment. The red line in the 1-s long speech signal represents the broadband amplitude envelope. The five peaks correspond to syllables at a frequency of 5 Hz. The brain signal represents MEEG recordings from auditory cortex filtered around 5 Hz. The gray shaded areas illustrate the hallmark of speech entrainment—the temporal alignment of brain signals and fluctuations in the speech envelope.

structure and envelope dynamics appears critical for the successful perception of spoken language.

Before turning from speech production and its output (an amplitude modulated signal with a prominent peak in the 4- to 5-Hz range) to corresponding properties of the auditory system, it is important to note that the modulations at other timescales are also perceptually important and likely to carry essential aspects of the communication signal. Low frequency modulations, for example, in the delta range from 1 to 3 Hz, correlate with information in the signal that can reflect (suprasyllabic) lexical units, lexical stress, phrasal "chunks," or even sentential-level intonation contours. These lower frequency signal changes are less prominent (in the acoustic signal) than the 4- to 6-Hz envelope peak but can nevertheless be carriers of significant linguistic information (Meyer, Henry, Gaston, Schmuck, & Friederici, 2017). Similarly, higher frequency information, such as in the range of 20- to 50-Hz modulations, has been argued to correlate with featural and phonemic signal properties and may be necessary for speech decoding. That being said, we will now focus primarily on the largest obvious signal property of the speech envelope, its quasi-periodic modulation at 4 to 5 Hz, because it has been studied extensively in the last years. We turn next to properties of the auditory system that are relevant in developing the argument further.

3.2. RHYTHMS IN THE AUDITORY SYSTEM Given that the speech signal is modulated at a quasi-rhythmic rate, it is important to ask in which manner the auditory system deals with temporally modulated signals and whether there is a special sensitivity for analyzing amplitude modulated signals at that rate.

The response of the auditory system to temporally structured signals has been investigated for a long time, using a number of different techniques, from recording spike trains in the auditory nerve to recording MEG signals from auditory cortex. It is well established that the afferent auditory pathway from cochlea to cortex is extremely sensitive to temporal information (for review, see, e.g., Giraud & Poeppel, 2012). In studying the human auditory system, several response types have been used to investigate how the system responds to temporal regularities, including evoked responses mostly measured from cortex, the frequency following response recorded from brainstem structures, the auditory steady-state response (aSSR) recorded from auditory cortex, and others.

One neurophysiological response type that is germane to understanding how speech modulations can be investigated is the aSSR (Galambos, Makeig, & Talmachoff, 1981; Picton, John, Dimitrijevic, & Purcell, 2003). Suppose the system is presented with an extended acoustic signal, say >1 s duration. Whereas the frequency following response is a steady-state response that tracks the carrier frequency of an acoustic signal (assuming the carrier is <1,000 Hz), the aSSR is a steady-state response to the amplitude modulation of the signal. If one presents listeners with, say, an extended tone of frequency 80 Hz that is amplitude modulated at 4 Hz, the neural response analyzed in the frequency (Fourier) domain will reflect a structure very much like the stimulus, namely a peak at the carrier frequency and two side peaks at ± the modulation frequency. The aSSR is elicited by many different types of acoustic stimuli, including amplitude-modulated noise, sinusoidally amplitude-modulated tones, frequency-modulated tones, and even amplitude and frequency co-modulated signals that are more like speech (e.g., Luo, Wang, Poeppel, & Simon, 2006). The aSSR response can be recorded using MEEG, is effectively analyzed in the frequency domain, and provides evidence that the auditory cortex is very sensitive to amplitude modulation. This is a proof of principle, then, that signals such as speech—comprised of a carrier (which is approximately the fine structure of the signal) and its modulation (which is the envelope)—can be measured with high sensitivity using such techniques. Other experimental approaches have also shown a high sensitivity to low amplitude-modulation rates, for example, using functional MRI (Giraud et al., 2000; Overath, Zhang, Sanes, & Poeppel, 2012; Santoro et al., 2014), but electrophysiological approaches have the distinct advantage of providing data about the frequency, amplitude, and phase of the neural responses to such acoustic information.

In light of the sensitivity of auditory cortex to modulation, in particular temporally structured modulation (e.g., Teng, Tian, Rowland, & Poeppel, 2017), and in light of the fact that speech production yields quasi-rhythmic output, we now have—as mentioned at the outset—the ingredients to study naturalistic continuous speech (e.g., Ding & Simon, 2012; Luo & Poeppel, 2007; Park, Ince, Schyns, Thut, & Gross, 2015). Moreover, we now have a compelling linking hypothesis—amplified in the next section—among the physics of speech (the modulation spectrum, as shown in figure 29.1), the neurophysiological substrate to encode the physics via low-frequency (especially theta) neural activity, and the linguistic interpretation of the neural encoding, namely as acoustic segments or chunks that underpin syllabic analysis (e.g., Ghitza, 2011, 2013).

4. The Role of Oscillations in Communication at Different Spatial Scales

4.1. Entrainment in Auditory Cortex We have described how auditory cortex follows modulated signals faithfully, as measured by the aSSR. We now turn to the concept of *entrainment* (figure 29.2). Neural entrainment refers to the capacity of neurons or neural circuits to synchronize their response to the temporal structure (e.g., periodicity) of external stimuli, for example, auditory or visual inputs. How does this entrainment occur? There are two widely discussed possibilities. First, the entrainment of the neural responses could be reflecting a concatenation of stimulus-evoked responses, perhaps elicited by some feature in the input. Second, entrainment could be a consequence of intrinsic neural oscillations (hypothesized to be an architectural attribute of cortex) adjusting their frequency and phase to align with a stimulus. It is important to be cautious about these two alternative mechanisms. Both are viable mechanisms, and both are likely to occur under certain circumstances. There are analytic and experimental approaches to distinguish between stimulus-induced responses that reflect entrainment versus the entrainment of intrinsic oscillations (Thut, Schyns, & Gross, 2011). In what follows, we will focus on the oscillatory hypothesis.

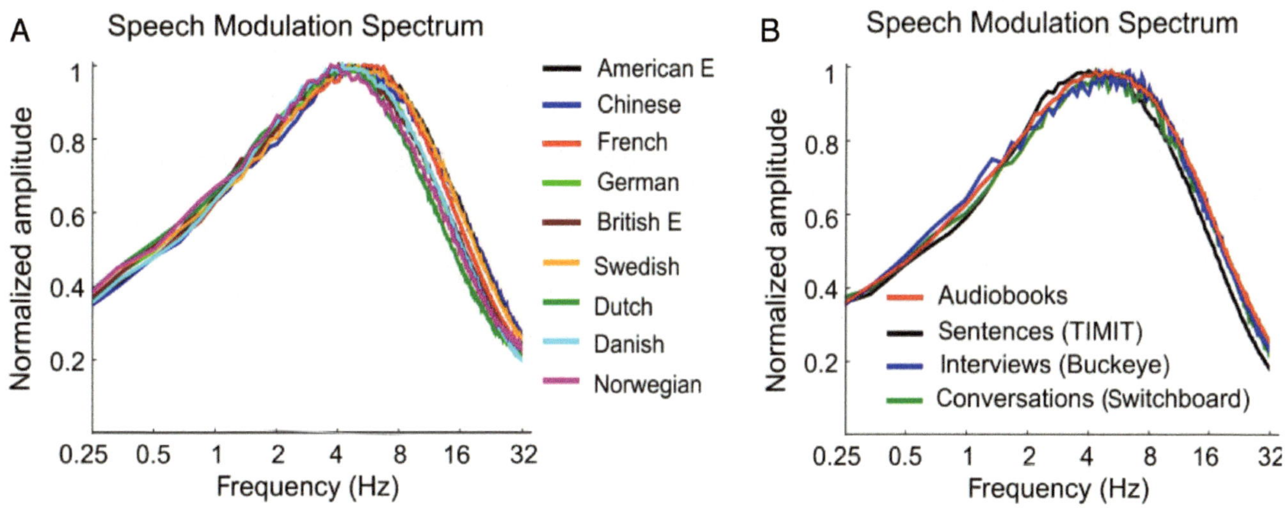

Figure 29.2 Speech modulation spectrum (modified from Ding et al., 2017). (A) The modulation spectrum across nine languages consistently shows a prominent peak at about 4 Hz. The spectrum for each language is normalized by the peak amplitude. (B) The modulation spectrum peaks at the same frequency of about 4 Hz for speech produced in different contexts. E = English.

The entrainment of intrinsic cortical oscillations to auditory stimuli has been studied for both nonspeech and speech sounds. For example, research in nonhuman primates by Schroeder, Lakatos, and colleagues (Lakatos, Karmos, Mehta, Ulbert, & Schroeder, 2008; Lakatos et al., 2013; Schroeder & Lakatos, 2009) demonstrates low-frequency phase entrainment to various forms of nonspeech acoustic stimulation, such as in trains of tones spaced rhythmically versus randomly (e.g., Ten Oever et al., 2017). A range of investigators have now also measured entrainment to continuous speech (e.g., Ahissar et al., 2001; Cogan & Poeppel, 2011; Ding & Simon, 2012; Doelling, Arnal, Ghitza, & Poeppel, 2014; Gross et al., 2013; Howard & Poeppel, 2012; Koskinen & Seppa, 2014; Luo et al., 2010; Luo & Poeppel, 2007; Peelle et al., 2013; Zion-Golumbic & Schroeder, 2012).

One especially noteworthy aspect in this context, as intimated previously, is that some oscillatory regimes link closely to the physical signal. The most compelling example is that the peak in the modulation spectrum of speech (and, concomitantly, mean syllable rate) is closely aligned with the frequency of theta oscillations (4–8 Hz). More generally, there is a fortuitous alignment of the mean period of some neural oscillations and the units of speech that require perceptual analysis. The temporal structure of speech reflects fluctuations at rates between 2 and 50 Hz (Drullman, Festen, & Plomp, 1994a, 1994b; Greenberg & Ainsworth, 2006; Greenberg, Carvey, Hitchcock, & Chang, 2003; Poeppel, 2003; Rosen, 1992). The temporal envelope itself contains variations at multiple distinguishable rates that convey different speech features. These include phonetic features or segments, which typically occur at a rate of 20 to 50 Hz (and therefore align with low gamma band neural activity); syllables that occur in a range of 3 to 9 Hz (aligning with theta oscillations); and phrases whose rate is usually between 0.5 and 3 Hz (aligning with the delta band). On the basis of these close alignments between the properties of neural oscillations and the temporal properties of speech, the hypothesis has been developed that oscillations play a central role in the segmentation and decoding of speech (Ghitza, 2011; Giraud & Poeppel, 2012; Poeppel, 2001, 2003). These models assume that there exist intrinsic oscillations in restricted regimes (that is to say, with independent physiological generators), that is, delta (1–3 Hz), theta (4–8 Hz), alpha (9–12 Hz), beta (15–25 Hz), low gamma (25–50 Hz), high gamma (70–200 Hz), and that some of these oscillatory regimes align closely with the to-be-analyzed signal (e.g., theta, delta, gamma), whereas other oscillations may play a key role in processing but are not necessarily linked to obvious physical properties (e.g., alpha and beta oscillations).

Sticking first to the oscillations more closely linked to the speech signal, for example, theta oscillations, how is entrainment of this neural activity achieved? The underlying physiology has been discussed and reviewed recently (Ding & Simon, 2014; Giraud & Poeppel, 2012; Peelle & Davis, 2012; Zion-Golumbic, Poeppel, & Schroeder, 2012). The entrainment is thought to be initiated by acoustic edges in the speech waveform (such as onsets or, alternatively, peaks in vowel energy) that lead to a phase resetting of ongoing oscillations in auditory cortex. As a result of the phase resetting, theta oscillations will be temporally aligned to the quasi-rhythmic syllable structure in speech. In MEEG data this is reflected in significant cross-correlation or coherence between the speech envelope and the theta component of brain activity in auditory cortex. Since theta oscillations in MEEG data represent cyclic excitability changes in underlying neuronal populations, this entrainment is thought to result in a rhythmic sampling or discretization of the continuous input stream into linguistically meaningful segments. Interestingly, auditory cortical brain oscillations may be organized in a hierarchy of nested rhythms, at frequencies matching relevant components in continuous speech (e.g., delta, theta, and gamma). These nested oscillations may therefore facilitate optimal parsing of speech components at different temporal scales (such as prosodic information, words, syllables, phonemes), while preserving their nested hierarchical structure (Gross et al., 2013).

These effects could exist to some degree even in a passive system that is merely driven by a quasi-rhythmic speech signal. However, there is evidence that the stimulus-driven brain activity in early auditory areas is effectively modulated by top-down signals from higher order areas. First, several studies demonstrate that higher intelligibility is reflected in stronger entrainment even when physical properties of the speech stimuli are similar (Peelle et al., 2013), which suggests that speech entrainment is not simply caused by the acoustic structure of the stimulus. Second, in a cocktail party scenario with multiple speakers, brain activity entrains strongest to the attended speaker (Ding & Simon, 2012; Zion-Golumbic & Schroeder, 2012). Finally, MEEG causality analysis has revealed directed causal effects from left inferior frontal and motor areas to auditory cortex, particularly at low frequencies that are relevant for speech processing (Park et al., 2015). These results indicate that the strength of speech brain entrainment in part reflects an active mechanism for attentional sampling and selection guided by prior information.

4.2. Entrainment in Audiovisual Brain Areas So far our discussion of the role of oscillations in speech processing has been restricted to auditory speech and its processing in the auditory cortex. However, during natural human communication, auditory signals are often complemented by highly informative visual signals, such as lip movements and gestures.

We have already noted that the local entrainment of activity in auditory brain areas by speech is not purely stimulus-driven, but is modulated by contextual factors such as attention and speech intelligibility. So how is the auditory processing of continuous speech modulated by congruent visual information? Several studies have demonstrated that congruent visual information significantly enhances auditory entrainment (Crosse, Butler, & Lalor, 2015; Zion-Golumbic, Cogan, Schroeder, & Poeppel, 2013).

But does rhythmic auditory entrainment also generalize to the visual cortex, in other words can we also observe entrainment in the visual cortex? It is well known that simple rhythmic visual stimulation leads to steady-state responses in the visual system (Herrmann, 2001). In addition, Spaak, de Lange, and Jensen (2014) demonstrated entrainment of activity in the visual cortex to rhythmic stimulation at alpha frequency (about 10 Hz). This precise phase locking between stimulus and brain activity (underlying entrainment) is even preserved during quasi-rhythmic visual stimulation where the frequency of stimulation changes over time (Keitel, Thut, & Gross, 2017). Interestingly, similar results have been demonstrated for more complex stimuli including speech. Luo et al. (2010) showed that the identity of different audiovisual movie stimuli can be robustly decoded from delta-theta phase in occipital MEG sensors. A recent MEG study directly investigated the existence of visual speech entrainment (Park, Kayser, Thut, & Gross, 2016). The dynamically changing time series of lip area of the speaker observed by the participants was used, in addition to the auditory speech envelope, to quantify the brain response to the input. Indeed, phase locking between lip area and brain activity in the visual cortex demonstrated visual speech entrainment. These results have recently been extended by demonstrating that visual areas encode visemes in addition to motion (O'Sullivan, Crosse, Di Liberto, & Lalor, 2016) and that sign language entrains activity in occipital and frontal brain areas (Brookshire, Lu, Nusbaum, Goldin-Meadow, & Casasanto, 2017). Together, these studies indicate that speech entrainment is a multisensory phenomenon (Giordano et al., 2017; Schroeder, Lakatos, Kajikawa, Partan, & Puce, 2008).

4.3. Neuronal Communication with Higher Order Areas Beyond supporting local processing in sensory cortical areas, brain oscillations likely play a more general role in the facilitation of communication between brain areas. Recent research has uncovered exciting insights about general principles that seem to underlie the organization of brain oscillations, particularly in relation to hierarchically organized processing streams in the brain. This has been most thoroughly studied in the visual system, where converging evidence from anatomical, invasive animal, and noninvasive human studies indicates that top-down and bottom-up signals are communicated via different frequency channels (Bastos et al., 2015; Michalareas et al., 2016; van Kerkoerle et al., 2014). Communication from higher order to lower order areas (top-down) seems to be preferentially communicated in the alpha/beta frequency bands (~8–12 Hz and ~15–25 Hz), whereas communication from lower to higher order areas (bottom-up) is most strongly reflected in the low gamma frequency band (~30–50 Hz, see figure 29.3). Interestingly, this functional asymmetry is supported by a structural asymmetry. Recent quantitative analysis of retrograde tracers in the macaque visual cortex revealed that feedforward projections typically start in supragranular layers and terminate in layer 4, whereas feedback projections predominantly start in infragranular layers and terminate in layers other than layer 4 (Markov et al., 2014). Together, these findings suggest that bottom-up and top-down communication across cortical hierarchies is anatomically and functionally segregated, where functional segregation is achieved through communication in different frequency channels. The predictive coding framework of brain communication further specifies the type of information that is transferred across these segregated routes. It postulates that top-down signals largely convey predictions to downstream areas whereas bottom-up signals represent prediction errors that are communicated upstream (Bastos et al., 2012). These principles have been most clearly argued for in the visual system. Similar effects exist in the auditory system.

Invasive recordings in human primary auditory cortex and auditory association cortex showed stronger Granger causality in the gamma band for the bottom-up direction and stronger Granger causality in the beta band for top-down direction (Fontolan, Morillon, Liegeois-Chauvel, & Giraud, 2014). Sedley et al. (2016) presented tone sequences to patients implanted with electrocorticography electrodes. They used computational models to extract estimates of three perceptually relevant measures—surprise, predictions, and precision. Linking these measures to the simultaneously recorded brain rhythms revealed results consistent with the predictive coding theory: surprise was reflected in gamma

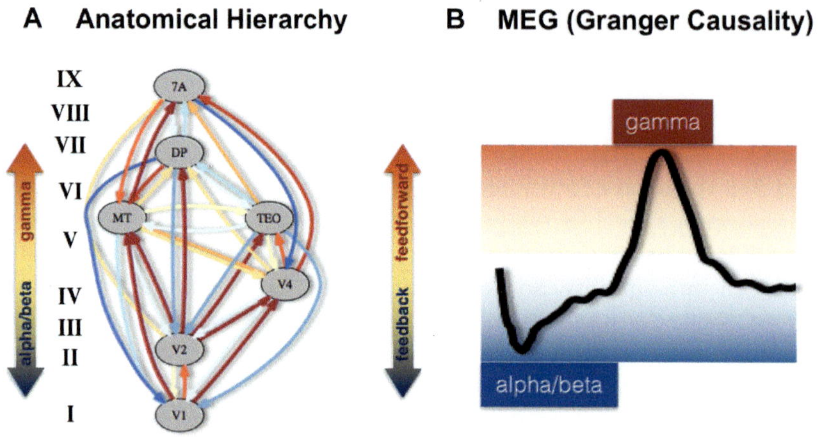

FIGURE 29.3 (A) Hierarchical model of seven selected areas along the visual pathway. Directed anatomical connectivity can be estimated from the fraction of supragranular labeled neurons. Color coding of arrows is based on supragranular labeled neuron values from Markov et al. (2014) where blue indicates feedback and red indicates feedforward direction. (B) Granger causality spectrum reconstructed from Michalareas et al. (2016). The spectrum represents the difference of Granger causality of anatomically defined feedforward and feedback connections averaged across both hemispheres. Positive values in this difference spectrum for gamma frequencies indicate dominance in feedforward direction. Negative values for alpha/beta frequencies indicate dominance in the feedback direction. Together, this suggests a segregation of feedforward and feedback signals in different frequency bands. 7A = angular gyrus; DP = dorsal aspect of prelunate gyrus; MT = motion-sensitive area, V5; TEO = posterior part of inferotemporal cortex.

oscillations, predictions in the beta band, and precision in alpha rhythms. We are just beginning to understand how these phenomena extend to language processing and how far these oscillation-based computational principles generalize across all levels of the auditory hierarchy.

Recent studies do provide converging evidence that oscillations play a role in the communication between two key areas of the dorsal auditory stream (Hickok & Poeppel, 2007)—namely early auditory areas and (pre) motor areas. A recent study aimed to identify the cortical generators of top-down signals to auditory areas during continuous auditory speech processing (Park et al., 2015). Interestingly, top-down signals originated from left inferior frontal gyrus and left (pre)motor area and modulated the entrainment of auditory brain activity to auditory speech and correlated with speech comprehension accuracy. Auditory-to-speech-motor coupling is strongest at low frequencies, corresponding to intonation and syllable rates, and seems to break down when syllable rate is lower or higher than the typical rate of about 4 to 5 Hz (Assaneo & Poeppel, 2018). Another recent MEG study nicely demonstrates that this coupling likely reflects temporal prediction of anticipated sensory input (Morillon & Baillet, 2017). However, the exact nature of computational processes during speech perception supported by (pre)motor areas is still the topic of a lively ongoing debate.

4.4. HUMAN COMMUNICATION Finally, recent studies suggest that brain oscillations not only facilitate communication between areas within a human brain but also have a role in supporting communication between humans. Interestingly, it seems that the same principles apply to interbrain communication as to intrabrain communication. Enhanced temporal synchronization of brain rhythms reflects (and possibly implements) improved communication. This has been described as brain-to-brain coupling (Hasson, Ghazanfar, Galantucci, Garrod, & Keysers, 2012), which can be seen as a special case of interactive alignment (Pickering & Garrod, 2013), and can be studied in the framework of two-person neuroscience (Hari & Kujala, 2009).

This relatively new research field is based on the idea that brain activity that is temporally synchronized between the speaker's and the listener's brains will improve the efficiency of information transfer. Temporally aligning brain oscillations would establish a communication channel for information exchange, not unlike two people communicating via a two-way radio device who agree on a common radio channel. But how could this alignment facilitate interpersonal communication?

One answer to this question is offered by the framework of predictive coding and its implications for human communication as outlined by Friston and Frith (2015). Friston and Frith model communication between interlocutors as two coupled dynamic systems that operate on active inference principles. This means that each system constantly generates predictions using an internal generative model. The model is optimized and adapted in order to minimize deviations between predictions and incoming sensory stimuli. Importantly, in the context of human communication, the same generative model is used to generate predictions about self-generated speech and also speech generated by the interlocutor. To be precise, as the speaker's motor system generates motor commands, this information is used by the speaker's generative model to predict the sensory consequences of these motor commands (i.e., the to-be-expected auditory input from your own speech). This feedback loop can be used to control, for example, speech rate and loudness. However, the generative model is also active in the listener's brain and constantly generates predictions based on the incoming sensory information from the speaker. Brain oscillations that are synchronized during communication provide important (temporal) constraints for these predictions. An interesting behavioral consequence of these predictions that are likely supported by synchronized brain oscillations is the typically short turn-taking gap between interlocutors with a median duration of about 200 ms (Levinson, 2016; M. Wilson & T. P. Wilson, 2005) (which happens to correspond to the period of the neural theta rhythm that is known to be robustly entrained during continuous speech). In addition, brain oscillations can facilitate human communication through other mechanisms. We have already discussed that brain oscillations represent cyclic changes in the excitability of neuronal populations. Obviously, there is a computational advantage to sample incoming sensory information at the same frequency that is used to produce it.

Some aspects of brain-to-brain coupling are not surprising, given the fact that brain activity is entrained by continuous speech. After all, in a dialogue, the speaker's brain generates the (rhythmic) motor commands that produce the continuous visual and auditory speech signals. The speech signal entrains brain activity in both the speaker's and the listener's auditory systems. This, in turn, will lead to a temporal alignment of activity in the speaker's and listener's auditory systems (mediated by the auditory speech signal). Indeed, this has been demonstrated experimentally (Kawasaki, Yamada, Ushiku, Miyauchi, & Yamaguchi, 2013; Perez, Carreiras, & Dunabeitia, 2017). But, interestingly, it seems that some aspects of brain-to-brain coupling are not directly mediated by the sensory speech signals. Perez et al. reported oscillatory brain coupling at frequencies in the alpha and beta ranges that appeared to be independent of the speech stimulus.

5. Problems and Prospects

The hypothesis that neural oscillations play a foundational role in perceptual analysis, in general, and in speech and language processing, in particular, remains vigorously debated—but it has also stimulated a variety of behavioral and neural experiments. There are some substantive concerns, but at the same time, there are thriving research programs examining the causal force that excitability cycles may have on perceptual processing and the extent to which one can measure neural oscillatory activity for nonobvious properties of the input (i.e., attributes that the observer/listener must infer, such as sentence structure). We discuss each issue in turn.

5.1. Problems and Challenges A criticism that is regularly raised in this literature is that the relation between rhythmic neural activity and perception and cognition is artifactual or epiphenomenal. It is, plausibly, pointed out that many of the phenomena can be explained by models that assume sequences of evoked responses or impulse responses, and that what looks like oscillatory activity derives from other, well-established mechanisms underlying transient neural activity. Indeed, evoked response analyses of ongoing speech reveal a stream of phase-locked, evoked responses (Sanders, Newport, & Neville, 2002). Moreover, responses to continuous speech can be modeled as signal-plus-noise additive response profiles (Howard & Poeppel, 2012).

There are, however, considerations that make these concerns not entirely compelling. First, although this might be viewed as a banal point, it is worth remembering that there is no reason to assume a single, monolithic mechanism. The range of phenomena can surely accommodate evoked, oscillatory, and hybrid mechanisms (e.g., Howard & Poeppel, 2012); there seems little motivation (other than impulsive Occam's razorism) to stipulate that one mechanism alone covers all phenomena, and it has indeed not been argued that oscillatory mechanisms must supersede or replace other mechanisms. Second, there are some effects that are problematic to explain without excitability cycles, namely prestimulus effects. It has been shown in a range of studies (e.g., Arnal, Doelling, & Poeppel, 2015; Busch, Dubois, & VanRullen, 2009; Fiebelkorn et al., 2013; Lakatos et al., 2013; Henry & Obleser, 2012) that the processing (e.g., detection, discrimination) of a stimulus depends precisely on when in the phase of a prestimulus oscillation it appears. Such dependence on prestimulus phase is naturally accounted for by assuming that oscillatory phase plays a critical role, whereas these effects are cumbersome to account for in models denying a role for excitability cycles. Likewise, there are data showing oscillatory phenomena after stimulus offset—again very awkward to account for based solely on impulse responses. Third, if processing depends entirely on evoked responses elicited by some form of an "edge" in a stimulus, one would expect every edge (within sensible physiological limits) to elicit an evoked response that is subsequently processed, that is, *all edges are equal.* In contrast, when presenting listeners with materials that range over modulation rates from slower (approximately theta range) to intermediate (approximately alpha range) to faster (approximately gamma range), only acoustic edges within the theta and gamma regimes generate entrainment responses; stimuli within the alpha range do not, even though they lie comfortably within the temporal range for entrainment (Luo & Poeppel, 2012; Teng et al., 2017). If we hypothesize that oscillatory regimes reflect distinct neural computations, the results are compatible with a view that theta and low gamma neural activity sample the input signal, whereas alpha activity could reflect functional inhibition and attentional gating. If we assume a strictly evoked model, these phenomena are a puzzle and anomaly: namely, why would an acoustic stimulus with edges at those rates not elicit responses? Fourth, there exists convergent evidence regarding a role for oscillations and entrainment from numerous domains: attention, working memory, vision, audition, speech, and language processing. In light of so much convergence, it seems at the very least impractical to re-explain so broad a set of phenomena when a rather straightforward oscillation-based analysis exists? There are further arguments to consider, and while the skeptical arguments are legitimate and deserve careful consideration—and may in some cases constitute the better model—it is safe to say that accounts of perception and cognition based on excitability cycles and the concomitant rhythmic computation are providing a fruitful basis for inquiry.

5.2. Experiments on the Causal Force of Entrainment by Oscillations A central question regarding oscillations, in general, and specifically neural entrainment to stimuli based on oscillatory activity (e.g., via phase resetting) concerns the causal force. It is nontrivial to demonstrate causality, in particular in human cognitive neuroscience experimentation, but various experiments/designs aim to establish, at the very least, a direct role for entrainment.

An early landmark study in this field was the MEG experiment by Ahissar et al. (2001). They presented listeners with sentences presented at different rates (compression ratios). The study showed (predictably) that compression by more than a factor of 2 compromises speech intelligibility. Critically, they were able to demonstrate that intelligibility of speech broke down precisely at rates at which phase locking between the cortex and stimulus envelope broke down because the cortex cannot entrain to the envelope at faster rates. This was one early demonstration pointing to the tight link between successful entrainment and successful speech perception.

This line of argumentation was extended in Luo and Poeppel (2007). There, several acoustic manipulations were performed that reduce intelligibility (e.g., presenting sentences using few channels, for instance, 4). The experiment revealed, again, that phase tracking correlated with intelligibility: the lower the tracking of the speech, reflected in theta band activity, the lower the intelligibility.

More recent experiments have sought to establish even more direct connections between oscillation-based entrainment and (aspects of) speech intelligibility. These studies are complicated by the fact that changes in intelligibility are almost always accompanied with changes in acoustic properties of the speech signal. This makes it difficult to unambiguously interpret changes of entrainment as being caused by differences in intelligibility (reviewed in Kösem & van Wassenhove, 2017). However, Ghitza (2012), using a psychophysical design, and Doelling et al. (2014), using a parallel MEG design, demonstrated that entrainment can be up- and down-regulated, in parallel to perceptual performance. In these experiments, the cues to entrainment were eliminated by filtering materials in distinct ways, for example, by band-rejecting (in each auditory channel) the modulation information between 2 and 9 Hz, which is precisely in the syllabic range. When such modulation information (and its associated acoustic cues) is artificially eliminated, there is no more phase tracking—and also essentially no intelligibility. The reinstatement of simple cues (e.g., clicks at peaks corresponding to vowel peaks), allowing for entrainment, up-regulated intelligibility, even though the quality of the speech acoustics was still drastically compromised.

Peelle et al. (2013) performed an influential MEG study that reinforces the conjecture that entrainment is required to mediate comprehension. They presented listeners with vocoded materials manipulated to change speech intelligibility in a controlled manner. They show that coherence between the acoustics of the signal and neural entrainment (cerebro-acoustic coherence) correlates closely with intelligibility and is unlikely to be explained by differences in acoustic properties of the speech stimulus. A related recent study expands on the coupling between speech acoustics and low-frequency oscillations. Park et al. (2015) demonstrated on the basis of MEG that the coupling is under influence of frontal cortical fields, and that intelligibility interacts with this coupling. They suggest that low-frequency brain oscillations play a role in implementing predictive top-down control during continuous speech perception.

New experiments seek to go beyond the largely correlational evidence summarized so far and aim to manipulate putatively causal mechanisms more directly. Herrmann and colleagues, in a series of reports, show that cortical modulation of intrinsic oscillations using transcranial direct-current stimulation and transcranial alternating-current stimulation can yield phase-dependent perceptual effects (e.g., Neuling, Rach, Wagner, Wolters, & Herrmann, 2012; for review, see Herrmann, Rach, Neuling, & Struber, 2013). Using such a stimulation approach, Riecke, Formisano, Sorger, Başkent, and Gaudrain (2017) were able to show that electrically stimulating auditory cortex modulates speech brain entrainment and speech intelligibility in the context of different experimental designs.

Cumulatively, the data briefly outlined here provide evidence for a compelling role for oscillation-based entrainment in speech. But what role? There are various experiments on entrainment that show strong neural effects without speech intelligibility, for example, on unintelligible reverse speech (Howard & Poeppel, 2010), nonspeech sequences (Ten Oever et al., 2017), or music (Doelling & Poeppel, 2015). The data invite the hypothesis that entrainment via low-frequency oscillations plays an intermediate role. Entrainment likely enables temporal segmentation and is necessary—but not sufficient—for speech comprehension. Entrainment (in the theta band) yields acoustic chunks of approximately syllable duration. These form the basis for subsequent decoding. Their causal force, therefore, lies in creating temporal primitives that constitute intermediate levels of representation that are necessary to drive perceptual and cognitive processing, in interaction with stimulus properties and top-down processes.

5.3. Extending Oscillations and Entrainment to Higher Order Structural Features The research programs and results we have summarized address primarily speech processing, a necessary step in spoken language comprehension. We have argued that oscillatory activity, and especially entrainment via oscillations, provides a critical substrate that underpins different operations in auditory perception. Language comprehension is composed of many operations beyond creating intelligible representations from speech input. Interestingly, the mechanistic and conceptual framework of oscillatory activity has been used to investigate higher order processing, that is, language processing proper.

Staying close to the concept of entrainment, recent experimentation has provided evidence that one can observe entrainment to inferred (i.e., nonobvious, not in the overt stimulus) structural properties of language. In order to understand—that is to say syntactically parse and semantically compose—a sentence, we take advantage of numerous cues: acoustic-prosodic cues to signal different landmarks such as phrase boundaries or ending, statistical cues to help predict contextually conditioned information, and purely grammatical cues. Ding, Melloni, Zhang, Tian, and Poeppel (2016) crafted sentential materials in which acoustic and statistical cues were eliminated or manipulated, and they showed that one can measure entrainment to purely grammar-dependent information. That is to say, the brain entrains (whether via evoked or rhythmic mechanisms) to information that must be internally constructed based on linguistic knowledge. Based on a novel computational model, Martin and Doumas (2017) were able to replicate and generalize this finding, demonstrating that oscillatory architectures of a certain form can capture the data recorded from humans and suggest a more general mechanism for the cortical computation of hierarchical structure, specifically for language.

Other important results on sentence-level processing in the present context capitalize on higher (beta band, gamma band) and lower (delta band) neural activity and correlate these neural signatures with other aspects of processing. For example, Bastiaansen, Magyari, and Hagoort (2010) measured neural activity to correct and incorrect sentences. They suggested that *syntactic unification*—a concept advanced by Hagoort to align syntactic, semantic, and phonological information—is reflected by neuronal synchronization in the low-beta frequency range. Indeed, Lewis and Bastiaansen (2015) linked this notion to predictive coding frameworks, proposing that beta band activity reflects the maintenance of the ongoing "model" as one processes the incoming linguistic signal, whereas gamma activity reflects the prediction error processing, as is commonly argued.

A related approach has been pursued by Meyer and colleagues (e.g., Meyer, 2018; Meyer et al., 2017), who have used EEG in different psycholinguistic paradigms and demonstrated that delta band activity can index linguistic biases in comprehension and provides a way to track sentence-level linguistic interpretation. Likewise, Molinaro and Lizarazu (2018) argued that delta band activity during sentence processing is reflecting speech-specific activity, unlike theta. In sum, a rich variety of new experiments are building on the oscillation framework to interrogate phenomena and mechanisms across the hierarchy of language processing.

6. Conclusion

The oscillation-based framework has generated a range of productive research programs. The approach has been applied to classic phenomena in experimental audition, such as the cocktail party effect (Zion-Golumbic, Cogan, et al., 2013; Zion-Golumbic, Ding, et al., 2013). It has been extended to processing regularities in other domains, such as music (Doelling & Poeppel, 2015) or trains of sounds (Ten Oever et al., 2017). In fact, the framework has also extended successfully across sensory modalities: recent experimental evidence shows that sign language processing is equally effective at eliciting these kinds of effects (Brookshire et al., 2017). Importantly, there is now a growing body of research using computational modeling techniques to investigate these topics. Modeling low-level oscillation effects to sound (Shamir, Ghitza, Epstein, & Kopell, 2009), effects of interaction across frequencies (e.g., Hyafil, Fontolan, Kabdebon, Gutkin, & Giraud, 2015), and higher order computational modeling of cognitive phenomena (e.g., Martin & Doumas, 2017) are all laying the groundwork for important new research directions that promise to elucidate the neural mechanisms.

REFERENCES

Ahissar, E., Nagarajan, S., Ahissar, M., Protopapas, A., Mahncke, H., & Merzenich, M. M. (2001). Speech comprehension is correlated with temporal response patterns recorded from auditory cortex. *Proceedings of the National Academy of Sciences, 98*(23), 13367–13372.

Arnal, L. H., Doelling, K. B., & Poeppel, D. (2015). Delta-beta coupled oscillations underlie temporal prediction accuracy. *Cerebral Cortex, 25*(9), 3077–3085.

Assaneo, F., & Poeppel, D. (2018). Listening to speech induces coupling between auditory and motor cortices in an unexpectedly rate-restricted manner. *Science Advances, 4*(2), eaao3842.

Bastiaansen, M., & Hagoort, P. (2015). Frequency-based segregation of syntactic and semantic unification during online sentence level language comprehension. *Journal of Cognitive Neuroscience, 27*(11), 2095–2107.

Bastiaansen, M., Magyari, L., & Hagoort, P. (2010). Syntactic unification operations are reflected in oscillatory dynamics during on-line sentence comprehension. *Journal of Cognitive Neuroscience, 22*(7), 1333–1347.

Bastos, A. M., Usrey, W. M., Adams, R. A., Mangun, G. R., Fries, P., & Friston, K. J. (2012). Canonical microcircuits for predictive coding. *Neuron, 76*(4), 695–711.

Bastos, A. M., Vezoli, J., Bosman, C. A., Schoffelen, J. M., Oostenveld, R., Dowdall, J. R., ... Fries, P. (2015). Visual areas exert feedforward and feedback influences through distinct frequency channels. *Neuron, 85*(2), 390–401.

Belitski, A., Gretton, A., Magri, C., Murayama, Y., Montemurro, M. A., Logothetis, N. K., & Panzeri, S. (2008). Low-frequency local field potentials and spikes in primary visual cortex convey independent visual information. *Journal of Neuroscience, 28*(22), 5696–5709.

Berger, H. (1929). Über das elektrenkephalogramm des menschen. *European Archives of Psychiatry and Clinical Neuroscience, 87*(1), 527–570.

Brookshire, G., Lu, J., Nusbaum, H. C., Goldin-Meadow, S., & Casasanto, D. (2017). Visual cortex entrains to sign language. *Proceedings of the National Academy of Sciences, 114*(24), 6352–6357.

Busch, N. A., Dubois, J., & VanRullen, R. (2009). The phase of ongoing EEG oscillations predicts visual perception. *Journal of Neuroscience, 29*(24), 7869–7876.

Buzsaki, G. (2002). Theta oscillations in the hippocampus. *Neuron, 33*(3), 325–340.

Buzsaki, G., & Draguhn, A. (2004). Neuronal oscillations in cortical networks. *Science, 304*(5679), 1926–1929.

Buzsaki, G., Logothetis, N., & Singer, W. (2013). Scaling brain size, keeping timing: evolutionary preservation of brain rhythms. *Neuron, 80*(3), 751–764.

Cannon, J., McCarthy, M. M., Lee, S., Lee, J., Borgers, C., Whittington, M. A., & Kopell, N. (2014). Neurosystems: brain rhythms and cognitive processing. *European Journal of Neuroscience, 39*(5), 705–719.

Chandrasekaran, C., Trubanova, A., Stillittano, S., Caplier, A., & Ghazanfar, A. A. (2009). The natural statistics of audiovisual speech. *PLOS Computational Biology, 5*(7), e1000436.

Cogan, G. B., & Poeppel, D. (2011). A mutual information analysis of neural coding of speech by low-frequency MEG phase information. *Journal of Neurophysiology, 106*(2), 554–563.

Cravo, A. M., Rohenkohl, G., Wyart, V., & Nobre, A. C. (2013). Temporal expectation enhances contrast sensitivity by phase entrainment of low-frequency oscillations in visual cortex. *Journal of Neuroscience, 33*(9), 4002–4010.

Crosse, M. J., Butler, J. S., & Lalor, E. C. (2015). Congruent visual speech enhances cortical entrainment to continuous auditory speech in noise-free conditions. *Journal of Neuroscience, 35*(42), 14195–14204.

Crosse, M. J., & Lalor, E. C. (2014). The cortical representation of the speech envelope is earlier for audiovisual speech than audio speech. *Journal of Neurophysiology, 111*(7), 1400–1408.

Ding, N., Melloni, L., Zhang, H., Tian, X., & Poeppel, D. (2016). Cortical tracking of hierarchical linguistic structures in connected speech. *Nature Neuroscience, 19*(1), 158–164.

Ding, N., Patel, A. D., Chen, L., Butler, H., Luo, C., & Poeppel, D. (2017). Temporal modulations in speech and music. *Neuroscience and Biobehavioral Reviews, 81,* 181–187. doi:10.1016/j.neubiorev.2017.02.011.

Ding, N., & Simon, J. Z. (2012). Neural coding of continuous speech in auditory cortex during monaural and dichotic listening. *Journal of Neurophysiology, 107*(1), 78–89.

Ding, N., & Simon, J. Z. (2014). Cortical entrainment to continuous speech: Functional roles and interpretations. *Frontiers in Human Neuroscience, 8,* 311.

Doelling, K. B., Arnal, L. H., Ghitza, O., & Poeppel, D. (2014). Acoustic landmarks drive delta-theta oscillations to enable speech comprehension by facilitating perceptual parsing. *NeuroImage, 85*(2), 761–768.

Doelling, K. B., & Poeppel, D. (2015). Cortical entrainment to music and its modulation by expertise. *Proceedings of the National Academy of Sciences, 112*(45), E6233–E6242.

Drullman, R., Festen, J. M., & Plomp, R. (1994a). Effect of reducing slow temporal modulations on speech reception. *Journal of the Acoustical Society of America, 95*(5), 2670–2680.

Drullman, R., Festen, J. M., & Plomp, R. (1994b). Effect of temporal envelope smearing on speech reception. *Journal of the Acoustical Society of America, 95*(2), 1053–1064.

Elliott, T. M., & Theunissen, F. E. (2009). The modulation transfer function for speech intelligibility. *PLOS Computational Biology, 5*(3), e1000302.

Fiebelkorn, I. C., Snyder, A. C., Mercier, M. R., Butler, J. S., Molholm, S., & Foxe, J. J. (2013). Cortical cross-frequency coupling predicts perceptual outcomes. *NeuroImage, 69*, 126–137.

Fontolan, L., Morillon, B., Liegeois-Chauvel, C., & Giraud, A. L. (2014). The contribution of frequency-specific activity to hierarchical information processing in the human auditory cortex. *Nature Communications*, 5, 4694.

Fries, P. (2009). Neuronal gamma-band synchronization as a fundamental process in cortical computation. *Annual Review of Neuroscience, 32*, 209–224.

Friston, K. J., & Frith, C. D. (2015). Active inference, communication and hermeneutics. *Cortex, 68*, 129–143.

Galambos, R., Makeig, S., & Talmachoff, P. J. (1981). A 40-Hz auditory potential recorded from the human scalp. *Proceedings of the National Academy of Sciences, 78*(4), 2643–2647.

Ghitza, O. (2011). Linking speech perception and neurophysiology: Speech decoding guided by cascaded oscillators locked to the input rhythm. *Frontiers in Psychology, 2*, 130.

Ghitza, O. (2012). On the role of theta-driven syllabic parsing in decoding speech: Intelligibility of speech with a manipulated modulation spectrum. *Frontiers in Psychology, 3*, 238.

Ghitza, O. (2013). The theta-syllable: A unit of speech information defined by cortical function. *Frontiers in Psychology, 4*, 138.

Giordano, B. L., Ince, R. A. A., Gross, J., Schyns, P. G., Panzeri, S., & Kayser, C. (2017). Contributions of local speech encoding and functional connectivity to audio-visual speech perception. *Elife, 6*. doi:10.7554/eLife.24763.

Giraud, A. L., Lorenzi, C., Ashburner, J., Wable, J., Johnsrude, I., Frackowiak, R., & Kleinschmidt, A. (2000). Representation of the temporal envelope of sounds in the human brain. *Journal of Neurophysiology, 84*(3), 1588–1598.

Giraud, A. L., & Poeppel, D. (2012). Cortical oscillations and speech processing: Emerging computational principles and operations. *Nature Neuroscience, 15*(4), 511–517.

Greenberg, S., & Ainsworth, W. A. (2006). *Listening to speech: An auditory perspective*. Mahwah, NJ: Erlbaum.

Greenberg, S., Carvey, H., Hitchcock, L., & Chang, S. (2003). Temporal properties of spontaneous speech—A syllable-centric perspective. *Journal of Phonetics, 31*(3), 465–485.

Gross, J., Hoogenboom, N., Thut, G., Schyns, P., Panzeri, S., Belin, P., & Garrod, S. (2013). Speech rhythms and multiplexed oscillatory sensory coding in the human brain. *PLOS Biology, 11*(12), e1001752.

Gross, J., Kujala, J., Hamalainen, M., Timmermann, L., Schnitzler, A., & Salmelin, R. (2001). Dynamic imaging of coherent sources: Studying neural interactions in the human brain. *Proceedings of the National Academy of Sciences, 98*(2), 694–699.

Hanslmayr, S., Staresina, B. P., & Bowman, H. (2016). Oscillations and episodic memory: Addressing the synchronization/desynchronization conundrum. *Trends in Neurosciences, 39*(1), 16–25.

Hari, R., & Kujala, M. V. (2009). Brain basis of human social interaction: From concepts to brain imaging. *Physiological Reviews, 89*(2), 453–479.

Hasson, U., Ghazanfar, A. A., Galantucci, B., Garrod, S., & Keysers, C. (2012). Brain-to-brain coupling: A mechanism for creating and sharing a social world. *Trends in Cognitive Sciences, 16*(2), 114–121.

Henry, M. J., & Obleser, J. (2012). Frequency modulation entrains slow neural oscillations and optimizes human listening behavior. *Proceedings of the National Academy of Sciences, 109*(49), 20095–20100.

Herrmann, C. S. (2001). Human EEG responses to 1–100 Hz flicker: Resonance phenomena in visual cortex and their potential correlation to cognitive phenomena. *Experimental Brain Research, 137*(3–4), 346–353.

Herrmann, C. S., Rach, S., Neuling, T., & Struber, D. (2013). Transcranial alternating current stimulation: A review of the underlying mechanisms and modulation of cognitive processes. *Frontiers in Human Neuroscience, 7*, 279.

Hickok, G., & Poeppel, D. (2007). The cortical organization of speech processing. *Nature Reviews Neuroscience, 8*(5), 393–402.

Howard, M. F., & Poeppel, D. (2010). Discrimination of speech stimuli based on neuronal response phase patterns depends on acoustics but not comprehension. *Journal of Neurophysiology, 104*(5), 2500–2511.

Howard, M. F., & Poeppel, D. (2012). The neuromagnetic response to spoken sentences: Co-modulation of theta band amplitude and phase. *NeuroImage, 60*(4), 2118–2127.

Hyafil, A., Fontolan, L., Kabdebon, C., Gutkin, B., & Giraud, A. L. (2015). Speech encoding by coupled cortical theta and gamma oscillations. *Elife, 4*, e06213.

Kawasaki, M., Yamada, Y., Ushiku, Y., Miyauchi, E., & Yamaguchi, Y. (2013). Inter-brain synchronization during coordination of speech rhythm in human-to-human social interaction. *Scientific Reports, 3*, 1692.

Kayser, C., Montemurro, M. A., Logothetis, N. K., & Panzeri, S. (2009). Spike-phase coding boosts and stabilizes information carried by spatial and temporal spike patterns. *Neuron, 61*(4), 597–608.

Keitel, A., & Gross, J. (2016). Individual human brain areas can be identified from their characteristic spectral activation fingerprints. *PLOS Biology, 14*(6), e1002498.

Keitel, C., Thut, G., & Gross, J. (2017). Visual cortex responses reflect temporal structure of continuous quasi-rhythmic sensory stimulation. *NeuroImage, 146*, 58–70.

Kösem, A., & Van Wassenhove, V. (2017). Distinct contributions of low-and high-frequency neural oscillations to speech comprehension. *Language, Cognition and Neuroscience, 32*(5), 536–544.

Koskinen, M., & Seppa, M. (2014). Uncovering cortical MEG responses to listened audiobook stories. *NeuroImage, 100*, 263–270.

Lakatos, P., Chen, C. M., O'Connell, M. N., Mills, A., & Schroeder, C. E. (2007). Neuronal oscillations and multisensory interaction in primary auditory cortex. *Neuron, 53*(2), 279–292.

Lakatos, P., Karmos, G., Mehta, A. D., Ulbert, I., & Schroeder, C. E. (2008). Entrainment of neuronal oscillations as a mechanism of attentional selection. *Science, 320*(5872), 110–113.

Lakatos, P., Musacchia, G., O'Connel, M. N., Falchier, A. Y., Javitt, D. C., & Schroeder, C. E. (2013). The spectrotemporal filter mechanism of auditory selective attention. *Neuron, 77*(4), 750–761.

Landau, A., & Fries, P. (2012). Attention samples stimuli rhythmically. *Current Biology, 22*(11), 1000–1004.

Levinson, S. C. (2016). Turn-taking in human communication—Origins and implications for language processing. *Trends in Cognitive Sciences, 20*(1), 6–14.

Lewis, A. G., & Bastiaansen, M. (2015). A predictive coding framework for rapid neural dynamics during sentence-level language comprehension. *Cortex, 68*, 155–168.

Luo, H., Liu, Z., & Poeppel, D. (2010). Auditory cortex tracks both auditory and visual stimulus dynamics using low-frequency neuronal phase modulation. *PLOS Biology, 8*(8), e1000445.

Luo, H., & Poeppel, D. (2007). Phase patterns of neuronal responses reliably discriminate speech in human auditory cortex. *Neuron, 54*(6), 1001–1010.

Luo, H., & Poeppel, D. (2012). Cortical oscillations in auditory perception and speech: Evidence for two temporal windows in human auditory cortex. *Frontiers in Psychology, 3*, 170.

Luo, H., Wang, Y., Poeppel, D., & Simon, J. Z. (2006). Concurrent encoding of frequency and amplitude modulation in human auditory cortex: MEG evidence. *Journal of Neurophysiology, 96*(5), 2712–2723.

Markov, N. T., Ercsey-Ravasz, M. M., Ribeiro Gomes, A. R., Lamy, C., Magrou, L., Vezoli, J., ... Kennedy, H. (2014). A weighted and directed interareal connectivity matrix for macaque cerebral cortex. *Cerebral Cortex, 24*(1), 17–36.

Martin, A. E., & Doumas, L. A. (2017). A mechanism for the cortical computation of hierarchical linguistic structure. *PLOS Biology, 15*(3), e2000663.

Meyer, L. (2018). The neural oscillations of speech processing and language comprehension: State of the art and emerging mechanisms. *European Journal of Neuroscience, 48*, 2609–2621

Meyer, L., Henry, M. J., Gaston, P., Schmuck, N., & Friederici, A. D. (2017). Linguistic bias modulates interpretation of speech via neural delta-band oscillations. *Cerebral Cortex, 27*(9), 4293–4302.

Michalareas, G., Vezoli, J., van Pelt, S., Schoffelen, J. M., Kennedy, H., & Fries, P. (2016). Alpha-beta and gamma rhythms subserve feedback and feedforward influences among human visual cortical areas. *Neuron, 89*(2), 384–397.

Molinaro, N., & Lizarazu, M. (2018). Delta(but not theta)-band cortical entrainment involves speech-specific processing. *European Journal of Neuroscience, 48*, 2642–2650.

Montemurro, M. A., Rasch, M. J., Murayama, Y., Logothetis, N. K., & Panzeri, S. (2008). Phase-of-firing coding of natural visual stimuli in primary visual cortex. *Current Biology, 18*(5), 375–380.

Morillon, B., & Baillet, S. (2017). Motor origin of temporal predictions in auditory attention. *Proceedings of the National Academy of Sciences, 114*(42), E8913–E8921.

Neuling, T., Rach, S., Wagner, S., Wolters, C. H., & Herrmann, C. S. (2012). Good vibrations: Oscillatory phase shapes perception. *NeuroImage, 63*(2), 771–778.

O'Sullivan, A. E., Crosse, M. J., Di Liberto, G. M., & Lalor, E. C. (2016). Visual cortical entrainment to motion and categorical speech features during silent lipreading. *Frontiers in Human Neuroscience, 10*, 679.

Overath, T., Zhang, Y., Sanes, D. H., & Poeppel, D. (2012). Sensitivity to temporal modulation rate and spectral bandwidth in the human auditory system: fMRI evidence. *Journal of Neurophysiology, 107*(8), 2042–2056.

Panzeri, S., Macke, J. H., Gross, J., & Kayser, C. (2015). Neural population coding: Combining insights from microscopic and mass signals. *Trends in Cognitive Sciences, 19*(3), 162–172.

Park, H., Ince, R. A., Schyns, P. G., Thut, G., & Gross, J. (2015). Frontal top-down signals increase coupling of auditory low-frequency oscillations to continuous speech in human listeners. *Current Biology, 25*(12), 1649–1653.

Park, H., Kayser, C., Thut, G., & Gross, J. (2016). Lip movements entrain the observers' low-frequency brain oscillations to facilitate speech intelligibility. *Elife, 5*, e14521.

Peelle, J. E., & Davis, M. H. (2012). Neural oscillations carry speech rhythm through to comprehension. *Frontiers in Psychology, 3*, 320.

Peelle, J. E., Gross, J., & Davis, M. H. (2013). Phase-locked responses to speech in human auditory cortex are enhanced during comprehension. *Cerebral Cortex, 23*(6), 1378–1387.

Pellegrino, F., Coupe, C., & Marsico, E. (2011). Across-language perspective on speech information rate. *Language, 87*(3), 539–558.

Perez, A., Carreiras, M., & Dunabeitia, J. A. (2017). Brain-to-brain entrainment: EEG interbrain synchronization while speaking and listening. *Scientific Reports, 7*(1), 4190.

Pickering, M. J., & Garrod, S. (2013). An integrated theory of language production and comprehension. *Behavioral and Brain Sciences, 36*(4), 329–347.

Picton, T. W., John, M. S., Dimitrijevic, A., & Purcell, D. (2003). Human auditory steady-state responses. *International Journal of Audiology, 42*(4), 177–219.

Poeppel, D. (2001). Pure word deafness and the bilateral processing of the speech code. *Cognitive Science, 25*(5), 679–693.

Poeppel, D. (2003). The analysis of speech in different temporal integration windows: Cerebral lateralization as "asymmetric sampling in time." *Speech Communication, 41*, 245–255.

Riecke, L., Formisano, E., Sorger, B., Başkent, D., & Gaudrain, E. (2018). Neural Entrainment to Speech Modulates Speech Intelligibility. *Current Biology, 28*(2), 161–169.

Rosanova, M., Casali, A., Bellina, V., Resta, F., Mariotti, M., & Massimini, M. (2009). Natural frequencies of human corticothalamic circuits. *Journal of Neuroscience, 29*(24), 7679–7685.

Rosen, S. (1992). Temporal information in speech: Acoustic, auditory and linguistic aspects. *Philosophical Transactions of the Royal Society London B: Biological Sciences, 336*(1278), 367–373.

Sanders, L. D., Newport, E. L., & Neville, H. J. (2002). Segmenting nonsense: An event-related potential index of perceived onsets in continuous speech. *Nature Neuroscience, 5*(7), 700–703.

Santoro, R., Moerel, M., De Martino, F., Goebel, R., Ugurbil, K., Yacoub, E., & Formisano, E. (2014). Encoding of natural sounds at multiple spectral and temporal resolutions in the human auditory cortex. *PLOS Computational Biology, 10*(1), e1003412.

Schroeder, C. E., & Lakatos, P. (2009). Low-frequency neuronal oscillations as instruments of sensory selection. *Trends in Neurosciences, 32*(1), 9–18.

Schroeder, C. E., Lakatos, P., Kajikawa, Y., Partan, S., & Puce, A. (2008). Neuronal oscillations and visual amplification of speech. *Trends in Cognitive Sciences, 12*(3), 106–113.

Sedley, W., Gander, P. E., Kumar, S., Kovach, C. K., Oya, H., Kawasaki, H., ... Griffiths, T. D. (2016). Neural signatures of perceptual inference. *Elife, 5*, e11476.

Shamir, M., Ghitza, O., Epstein, S., & Kopell, N. (2009). Representation of time-varying stimuli by a network exhibiting oscillations on a faster time scale. *PLOS Computational Biology, 5*(5), e1000370.

Singer, W., & Gray, C. M. (1995). Visual feature integration and the temporal correlation hypothesis. *Annual Review of Neuroscience, 18*, 555–586.

Spaak, E., de Lange, F. P., & Jensen, O. (2014). Local entrainment of alpha oscillations by visual stimuli causes cyclic modulation of perception. *Journal of Neuroscience, 34*(10), 3536–3544.

Teng, X., Tian, X., Rowland, J., & Poeppel, D. (2017). Concurrent temporal channels for auditory processing: Oscillatory neural entrainment reveals segregation of function at different scales. *PLOS Biology, 15*(11), e2000812.

Ten Oever, S., Schroeder, C. E., Poeppel, D., van Atteveldt, N., Mehta, A. D., Megevand, P., ... Zion-Golumbic, E. (2017). Low-frequency cortical oscillations entrain to subthreshold rhythmic auditory stimuli. *Journal of Neuroscience, 37*(19), 4903–4912.

Thut, G., Miniussi, C., & Gross, J. (2012). The functional importance of rhythmic activity in the brain. *Current Biology, 22*(16), R658–R663.

Thut, G., Nietzel, A., Brandt, S. A., & Pascual-Leone, A. (2006). α-Band electroencephalographic activity over occipital cortex indexes visuospatial attention bias and predicts visual target detection. *Journal of Neuroscience, 26*(37), 9494–9502.

Thut, G., Schyns, P. G., & Gross, J. (2011). Entrainment of perceptually relevant brain oscillations by non-invasive rhythmic stimulation of the human brain. *Frontiers in Psychology, 2*, 170.

van Kerkoerle, T., Self, M. W., Dagnino, B., Gariel-Mathis, M. A., Poort, J., van der Togt, C., & Roelfsema, P. R. (2014). Alpha and gamma oscillations characterize feedback and feedforward processing in monkey visual cortex. *Proceedings of the National Academy of Sciences, 111*(40), 14332–14341.

VanRullen, R., Zoefel, B., & Ilhan, B. (2014). On the cyclic nature of perception in vision versus audition. *Philosophical Transactions of the Royal Society of London B: Biological Sciences, 369*(1641), 20130214.

Wang, X. J. (2010). Neurophysiological and computational principles of cortical rhythms in cognition. *Physiological Reviews, 90*(3), 1195–1268.

Wilson, M., & Wilson, T. P. (2005). An oscillator model of the timing of turn-taking. *Psychonomic Bulletin and Review, 12*(6), 957–968.

Zion-Golumbic, E. M., Cogan, G. B., Schroeder, C. E., & Poeppel, D. (2013). Visual input enhances selective speech envelope tracking in auditory cortex at a "cocktail party." *Journal of Neuroscience, 33*(4), 1417–1426.

Zion-Golumbic, E. M., Ding, N., Bickel, S., Lakatos, P., Schevon, C. A., McKhann, G. M., ... Schroeder, C. E. (2013). Mechanisms underlying selective neuronal tracking of attended speech at a "cocktail party." *Neuron, 77*(5), 980–991.

Zion-Golumbic, E. M., Poeppel, D., & Schroeder, C. E. (2012). Temporal context in speech processing and attentional stream selection: A behavioral and neural perspective. *Brain and Language, 122*(3), 151–161.

Zion-Golumbic, E., & Schroeder, C. E. (2012). Attention modulates "speech-tracking" at a cocktail party. *Trends in Cognitive Sciences, 16*(7), 363–364.

30 Explaining Speech Comprehension: Integrating Electrophysiology, Evolution, and Cross-Linguistic Diversity

WILLIAM D. MARSLEN-WILSON

1. Introduction

Nearly 40 years ago we described the scientific challenge of spoken language comprehension as follows:

> A fundamental problem in understanding the perception of spoken language is to characterize the role of knowledge in the on-going perceptual processing of the sensory input. Recent studies of sentence perception show that the listener's immediate perceptual decisions about the incoming material represent a balance between two distinct sources of information. On the one hand there is the acoustic input itself, and on the other there is a set of internally generated structural constraints on the possible readings of this input (cf. Marslen-Wilson, 1975). These constraints are available to the listener as a result of his on-line syntactic and semantic interpretation of the sentence (Marslen-Wilson, 1973, 1975; Marslen-Wilson & Tyler, 1975; Tyler & Marslen-Wilson, 1977).[1]

Forty years on, the field has made some progress in meeting this challenge. Most saliently, it has moved from accounts solely in terms of cognitive theories based on behavioral data to neurocognitive theories based on combinations of behavioral, neuropsychological, and neuroimaging data. These data allow us to infer the broad outlines of a neurocognitive framework for real-time spoken language comprehension, instantiated in a generally accepted neuroanatomical framework.

The acoustic properties of the speech input are analyzed in bilateral Heschl's gyrus, propagating to bilateral superior temporal cortex where higher level phonological analyses are carried out along a posterior to anterior axis, cascading to early stages of lexical access in middle temporal cortex (MTC) (Dronkers, Wilkins, Van Valin, Redfern, & Jaeger, 2004; Hickok & Poeppel, 2007). Since word candidates are activated and decay over time as the speech unfolds, the continuous process of lexical activation may be underpinned bilaterally by recurrent activity between superior temporal cortex and MTC. Relatively early in this process (<200 ms from word onset), different types of lexically specified information will start to be activated, with resulting activations spreading from left MTC to posterior parietal regions and anteriorly to left inferior frontal cortex, as incremental processes lead to the development of higher level syntactic and semantic representations.

Incremental combinatorial processes resulting in morphological and structural syntactic representations primarily engage a left hemisphere (LH) frontotemporal network while sentential semantic incremental processing engages a bilateral temporal network (Wright, Stamatakis, & Tyler, 2012). The white matter tracts that transmit information via direct dorsal and ventral connections between frontal and posterior regions—currently thought to include extreme capsule fibers and the arcuate fasciculus—play a central role in all these processes (e.g., Friederici & Gierhan, 2013; Rolheiser, Stamatakis & Tyler, 2011), complemented by major interhemispheric pathways across the corpus callosum (Turken & Dronkers, 2011). Specific language functions may emerge synergistically from the integration across regions of this constant flow of different kinds of linguistic information (Rolheiser, Stamatakis, & Tyler, 2011).

Beyond this broad outline, however, surprisingly little is known about the specific spatiotemporal patterning and the specific neurocomputational properties of the wide range of processing operations that underpin the dynamic transition from sound to meaning. It is attractive and plausible to describe the process of spoken language comprehension as a multidimensional dynamic system instantiated in a complex neurobiological network of interacting brain networks, but this remains not much more than a descriptive promissory note. We are far from being able to spell out such an account at the necessary levels of quantitative and qualitative specificity. We still do not know what is going on.

My goal in this chapter is to make some proposals for what is required to construct a neurobiological account of language comprehension that will be more fully explanatory. Among the many ways in which most current accounts are lacking, I will focus on three main requirements. The first concerns the real-time flow of neurocomputational activity that underpins the comprehension process: How can we access this electrophysiological activity and how can we relate it to functional models of speech comprehension? The second seeks to situate language comprehension as an evolutionary, neurobiological process, viewing it as the overlay of potentially human-specific capacities on a more general primate substrate for conspecific communication in complex pragmatic contexts. The third concerns the need for systematic cross-linguistic investigation of the systems and processes supporting language comprehension across the immense diversity of extant human languages.

I do not claim that this is an exhaustive list of the methods and questions that will need to be assembled. Salient additional requirements include the "connectomics" of the neural systems in question, the functional connectivity within and between the resulting networks, and the role of neural oscillatory processes in driving these relations. Some of these topics are touched on in other chapters of this volume (e.g., Przeździk, Haak, Beckmann, & Bartsch, chapter 34; Gross & Poeppel, chapter 29).

2. *Capturing and Explaining Real-Time Neurocognition: What, When, and Where*

The most compelling property of speech comprehension is its immediacy and rapidity. We seem to understand speech literally as we hear it, with multilevel interpretations of the speech signal emerging into the consciousness of the listener within 250 ms of the arrival at their eardrum of the relevant speech input (Marslen-Wilson, 1973, 1975; Marslen-Wilson & Tyler, 1975, 1980). It is possible to capture the basic temporal dynamics of this remarkable sound-to-meaning transfer function, directly or indirectly, through speech shadowing paradigms and through a variety of reaction-time tasks. But like all behavioral measures, these tasks and paradigms reflect only the properties of the output of the system. It is not possible to infer from this output information what are the properties of the brain systems that generate these behaviors.

But it is precisely (and only) in terms of these brain systems that we will be able to develop explanatory, neurobiologically rooted, and evolutionarily intelligible accounts of complex human cognitive capacities such as speech comprehension. This account, furthermore, will have to be rooted in the real-time electrophysiological activity of the brain. It is this activity that is the causal substrate for the dynamic neural computations that mediate our perceptual experience and interpretation of spoken language. This poses the fundamental challenge of identifying these computations—to answer the question of *what* is being computed—and of locating these computations in neural space and time—to answer, in the incremental processing environment of speech comprehension, the *where* and the *when* questions.

Critically, we need to be able to answer all three of these questions (*what, where, when*) simultaneously. Any one (or two) of them on their own is not adequate to enable sufficiently direct inferences about the neurocognitive systems under investigation. These are inferences not only about the general spatiotemporal properties of these systems but also about the qualitative properties of the computations they support. Without this confluence of constraints, the data and their interpretation remain at too great a distance from the specific processes we seek to account for in a causally explicit account of higher cognitive functions—and in this respect not materially different from descriptive accounts based just on the observation of behavior.

For nearly a century, for example, it has been possible to measure the electrical activity of the brain through scalp electrodes—electroencephalography (EEG). This is in principle excellent for addressing *when* questions, with millisecond-level temporal resolution, but it is typically opaque with respect to *where* or *what* questions, especially when we move to higher order aspects of the EEG response. Early components of the EEG response driven by immediate sensory analyses, such as the N100, can be well localized (bilateral auditory cortices) and can be closely linked to cortical speech analysis processes. But the workhorses of psycholinguistically relevant EEG research—event-related potential components such as the N400, the P600, and so forth—while they have played an important role in the development of the field, are much more limited in so far as the *where* and *what* questions are involved—and indeed somewhat diffuse even where the *when* question is involved (the N400 should perhaps be renamed the N250–550). Thus, while suggestive constraints can be drawn, for example, from the relative ordering in a given stimulus environment of two such components, very little can be concluded about the specifics of the neural systems generating these components—especially given the uncertainty, even now, about the neural generators of these components. Similar caveats apply to studies using magnetoencephalography (MEG) where these are focused on the relative timing and presence or absence of distinctive event-related potential components.

The other great standby of modern cognitive neuroscience, functional MRI (fMRI), is potentially highly spatially resolved (down to single millimeter voxel size in high-field systems) but with notoriously poor temporal resolution. fMRI not only does not give a real-time readout of the fine-grained dynamic activity of the system, but it also blurs the temporal ordering of processes falling within each analysis time window. This is a critical disadvantage for the study of processes such as language comprehension, where the relative timing of processing events within and between different cortical sites is critical to determining the flow of information over time through different brain structures (e.g., Klimovich-Gray, Tyler, Randall, Kocagoncu, Devereux, & Marslen-Wilson, 2019; Kocagoncu, Clarke, Devereux, & Tyler, 2017; Pulvermüller, Shtyrov, & Ilmoniemi, 2003; Tyler, Cheung, Devereux, & Clarke, 2013; Whiting, Shtyrov, & Marslen-Wilson, 2014).

Both these types of imaging modalities, whatever their separate strengths in *when* and *where* domains, have been consistently weak in the *what* domain—in their ability to tell us specifically what neurocomputational content is being computed at a given time or place. For some effect that can be located in neural time or in neural space, there is still the question of what neurocomputational process (or fragment thereof) these results are indexing. Historically these questions have been answered by the use of univariate statistics focusing on differences in the average amounts of activity elicited by different experimental conditions. In this respect, the analysis strategy has remained conceptually yoked to the analysis strategies that have long dominated behavioral studies. The problem here is not simply that a univariate test typically discards information that is present in the original signal, but also that difference in amplitude is intrinsically opaque and interpretatively ambiguous, especially when computed on the basis of non-time-resolved (typically fMRI) data. In particular, these types of analyses are blind to ways in which the distribution of activity in each condition is potentially informative about differences in the computations being performed.

These limitations began to be mitigated with the introduction of multivariate pattern analysis (MVPA) techniques into neuroimaging (e.g., Haxby et al., 2001), where machine learning techniques were used to discriminate whether and where different patterns of activity were associated with different conditions. This standard classifier-based MVPA, though often more powerful and more sensitive than classic univariate measures, can nonetheless also be explanatorily opaque. The finding that there is a difference between distributions does not necessarily clarify the nature of this difference.

Much more powerful as an MVPA technique is *Representational Similarity Analysis* (RSA), introduced by Kriegeskorte and colleagues (Kriegeskorte, Goebel, & Bandettini, 2006; Nili et al., 2014; Su, Fonteneau, Marslen-Wilson, & Kriegeskorte, 2012). RSA is also based on comparison of multivariate patterns of activity, but it does so, critically, by comparing a representation of the correlational structure of the brain data under consideration with the correlational structure predicted by the relevant cognitive theory. The technique is not looking for *any* difference that is significant but for the specific difference that is predicted by a particular cognitive theory. Thus, in the example shown in figure 30.1, the theoretical model being tested, in the form of the correlational structure represented in the Model Representational Dissimilarity Matrix (RDM), hypothesizes that two conditions (SS and SI) should have similar neural activation patterns, while the other four conditions should have dissimilar activation patterns, both to each other and to conditions SS and SI. This predicted pattern is then tested against the pattern of correlations seen in the neuroimaging data generated by each condition, as illustrated by the Data RDM in figure 30.1. These data can be collected using either a searchlight process moving across the brain volume (as in the Kocagoncu et al., 2017, research illustrated in figure 30.3) or on a

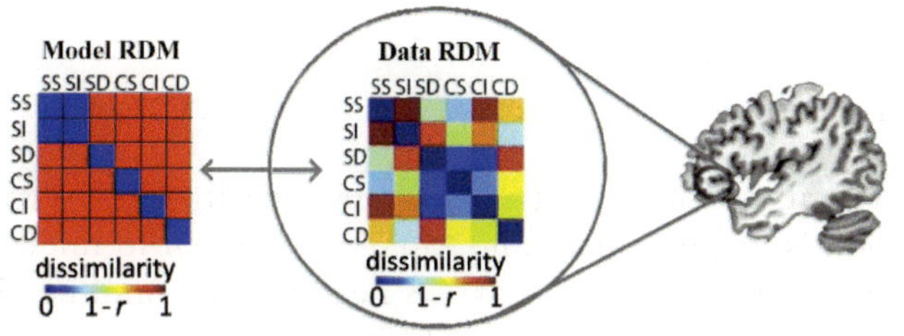

FIGURE 30.1 RSA: Sample analysis. (*Left*) A Model RDM, expressed as a 6×6 symmetrical matrix of predicted correlations between condition pairs, hypothesizes selective sensitivity to common processes in conditions [SS] and [SI]. Blue represents the predicted presence of a correlation between each pair while red represents its predicted absence. (*Center and right*) A brain Data RDM from a single searchlight location (in left inferior frontal cortex), again expressed as a 6×6 symmetrical matrix of correlation distances between conditions. Comparison of Model RDMs with Data RDMs (Spearman correlation) across the brain volume produces a brain-wide map of model fit for each model tested (for examples see figures 30.2 and 30.3). Condition labels are arbitrary and for illustration only.

region of interest (ROI) basis (as in the Bozic, Fonteneau, Su, & Marslen-Wilson, 2015, study illustrated in figure 30.2).

An early example of this approach in the language domain is an fMRI study (Bozic, Fonteneau, Su, & Marslen-Wilson, 2015) that looked at the distribution over bilateral frontal and temporal cortex of processing activity relevant to inflectional complexity on the one hand (as in forms like *plays* or *eats*) and phrasal constituent structure complexity on the other (as in *the play* or *I eat*). In a context where classical univariate analyses failed to find informative differences in levels of neural responses to inflectionally complex and phrasally complex forms, an RSA approach successfully separated out brain areas sensitive to the common and distinct elements of these different types of processing complexity. An inflectional "detector model" looking for brain regions where neural activations to inflected stimuli were highly correlated (see figure 30.2), produced selective LH model fit in left inferior frontal gyrus (LIFG) in Brodmann's area (BA)44 and left temporal cortex. The phrasal detector model, in contrast, provided bihemispheric temporal fit, but no selective engagement of LIFG.

Such analyses, however, do not provide an account of the real-time dynamics of the comprehension processes elicited by these phrasal and inflectional stimuli. fMRI data, coupled with RSA analyses, can go some way to providing answers to the *where* and *what* questions, but they are blind to the specific properties of the temporal sequence of different processes elicited by the presence, for example, of the inflectional [-s] morpheme. To achieve this, we need to use neuroimaging techniques that allow direct access to the real-time flow of electrophysiological activity of the brain (cf. Fonteneau, Bozic, & Marslen-Wilson, 2015) and to develop analysis techniques that allow the RSA method to be applied to time-resolved data streams.

Building on a pioneering combined MEG and RSA study by Tyler et al. (2013), recent research by Kocagoncu et al. (2017) achieved exactly this, working with MEG data recorded while listeners heard a series of isolated spoken words. This study used MRI-constrained

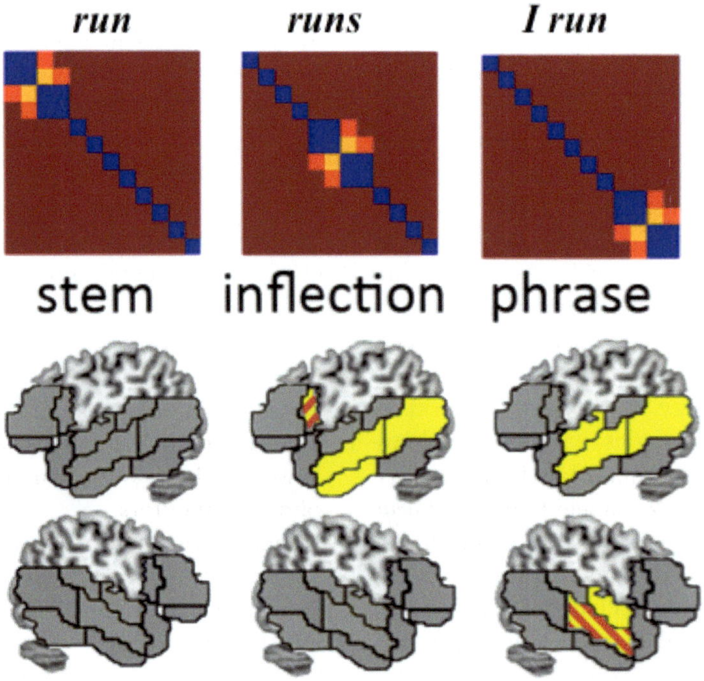

FIGURE 30.2 RSA models testing for inflectional and phrasal processing activity in a set of language-area ROIs. The upper row shows three Model RDMs, testing for differential sensitivity to processing activity related to each of three complexity types: simple spoken stems (*run*), inflected forms (*runs*), and phrases (*I run*). These are 12 × 12 matrices generated by the 12 conditions in the experiment (three complexity types—stem, inflection, and phrase—crossed by four verb/noun dominance levels). Dark red regions indicate no predicted correlation. Blue indicates correlated activation patterns due to a shared property (a specific type of processing complexity), with degree of similarity modulated by verb dominance (yellow, orange, red cells). *Middle* (LH) and *lower* (RH) rows show the significant correlations (in yellow) for each model ($P < .05$), with no model fit seen for the simple stems. Red bars indicate ROIs with significant differences in fit between the phrase and the inflection models. Figure modified from Bozic et al. (2015).

minimum norm source reconstruction methods (Gramfort et al., 2014) to record MEG signals at the scalp and project them back to the whole brain cortical surface (defined as the white matter/gray matter boundary). While necessarily noisy and imperfect, these noninvasive methods are nonetheless able to capture the dynamic electrophysiological events that underpin real-time brain function with millisecond-level temporal resolution and potentially subcentimeter spatial resolution. Developments in RSA methodology (spatiotemporal searchlight RSA [ssRSA]: Su et al., 2012; Su, Zulfiqar, Jamshed, Fonteneau, & Marslen-Wilson, 2014) make it possible to interrogate these MEG source space representations, using searchlight techniques, to determine where and when specific RSA models match the dynamic flow of neural activity across cortex.

Critically, these RSA models capture different aspects of the types of computational process that are hypothesized to underpin the basic operations of spoken word recognition, as laid out in the well-established cohort model of these processes (Marslen-Wilson & Welsh, 1978; Marslen-Wilson, 1987). The Kocagoncu et al. (2017) study focused on the core properties of the model—its claim for early activation of multiple word candidates as a word starts to be heard, with immediate access of both phonological and semantic information related to these word candidates, followed by the convergence of accumulating sensory (and contextual) information on a single unique candidate. This point of convergence is labeled the *Uniqueness Point* (UP), leading to the selective activation of the target word's semantics.

To test these claims of the cognitive model, Kocagoncu et al. constructed three distinct RSA models, tapping respectively into (i) early phonological competition at the level of lexical form between members of the cohort of word candidates (labeled the *LexComp model*), (ii) early semantic competition between this initial set of word candidates (labeled the *SemComp model*), and (iii) the semantic properties of the target word identified at the UP—labeled the *SemDis model.* The contents of these models and the location of the UP were determined using gating tasks to establish the membership of the early word cohorts (Grosjean, 1980) and using corpus-based distributional and feature-based models (Baroni & Lenci, 2010; McRae, Cree, Seidenberg, & McNorgan, 2005) to determine the semantic properties of the competitor words and of the target word. Using the ssRSA searchlight procedures, each of these models can be used to determine where and when the correlational structure of RSA models built on these three different principles can be significantly related to the correlational structure of the listeners' neural responses over time. This illustrates one of the important advantages of ssRSA in the MEG source space environment—it allows the same spatiotemporal stream of neural activity to be assessed from the perspective of multiple different theoretical claims about potentially relevant neurocomputational dimensions.

Kocagoncu et al. (2017) tested these models on the data obtained from a MEG study in which participants listened to familiar, concrete spoken words with a mean duration of 602 ms. The results of these analyses are summarized in figure 30.3 and show distinctive spatiotemporal patterns of model fit for each of the three models. We will discuss these results in sufficient detail to bring out the qualitative advances in explanatory specificity that become available when we can combine real-time neural *when* and *where* measures with the qualitative information about *what* is being computed that is provided by the RSA technique.

Overall, the results of this study were strongly consistent with the predictions of the cohort approach (Marslen-Wilson & Welsh, 1978) for a transition from competitive processes before the UP, involving early short-lived parallel activation of candidate lexical and semantic representations, to the activation of the unique semantics of the target word after the UP (see figure 30.3). Lexical competition effects, detected by the LexComp model, were observed as early as 400 ms before the UP and initially engaged the left superior temporal and superior marginal gyri, regions that are involved in acoustic and phonological analysis (Hickok & Poeppel, 2007; Rauschecker & Scott, 2009). These effects subsequently spread to left middle temporal gyrus which is associated with processing the semantic properties of words (e.g., Binder, Desai, Graves, & Conant, 2009; Dronkers et al., 2004), and were followed by model fit in BA45, which earlier studies (Zhuang, Tyler, Randall, Stamatakis, & Marslen-Wilson, 2014) showed to be involved in selecting among activated cohort competitors.

There were equally early effects, again detectable at around 400 ms before UP, of semantic competition between the multiple cohort word candidates, though with a quite different spatiotemporal distribution to the lexical form-based effects. Significant early fit of the SemComp model was seen in LIFG (BA47), spreading into the left anterior temporal lobe (LATL), followed by effects in the left angular gyrus and then the LATL again as the UP neared and the number of semantic competitors declined. The left angular gyrus is associated with a variety of semantic-based processes (Binder et al., 2009) such as semantic feature integration (Price, Peelle, Bonner, Grossman, & Hamilton, 2016), while the LATL is thought to be a core region

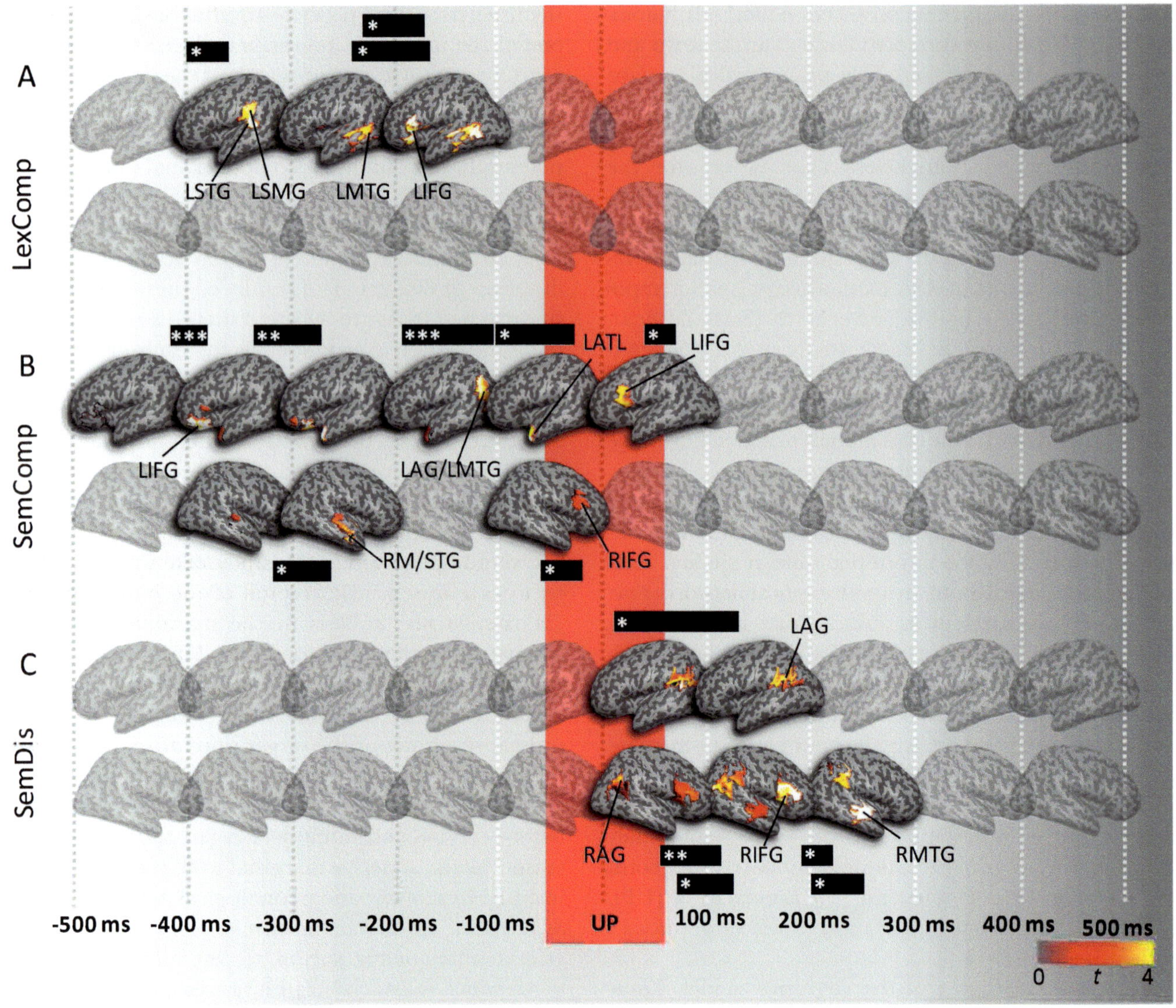

FIGURE 30.3 Kocagoncu et al. (2017) study: ssRSA results for the LexComp, SemComp, and SemDis RSA models (see section 2). The corrected *t*-maps show model fit for each model type, plotted at 100 ms intervals relative to the word's UP. The UP is marked by the vertical red bar. **P*<.05, ***P*<.01, ****P*<.001. Figure from Kocagoncu et al. (2017). LAG = left angular gyrus; LMTG = left middle temporal gyrus; LSMG = left superior marginal gyrus; LSTG = left superior temporal gyrus; RAG = right angular gyrus; RIFG = right inferior frontal gyrus; RMTG = right middle temporal gyrus; STG = superior temporal gyrus.

within an amodal semantic hub (Patterson, Nestor, & Rogers, 2007). The finding that semantic competition effects flow between these regions over time suggests their joint contribution to semantic analyses in a dynamically modulated network. Finally, around the UP, the SemComp model showed model fit in bilateral IFG (BA45), most likely reflecting the resolution of semantic competition and selection of the appropriate candidate (Grindrod, Bilenko, Myers, & Blumstein, 2008; Zhuang et al., 2014).

Although effects of cohort competition were seen at both the lexical form and semantic levels, semantic competition effects seemed more robust and sustained than lexical competitive effects. The SemComp model showed initial effects primarily in LIFG BA47, a region associated with cohort competition effects in previous studies of spoken word recognition (Zhuang et al., 2014) and later effects in more dorsal inferior frontal regions (BA44/BA45) associated with cohort selection processes (Zhang et al., 2014). These results suggest that the processes of cohort competition and selection, while modulated by incremental bottom-up constraints provided by the phonological input, mainly involve the semantic representations of cohort competitors (Devereux, Taylor, Randall, Geertzen, & Tyler, 2016; Gaskell & Marslen-Wilson, 1997). After the UP, there were no competitive effects of either cohort phonology or cohort semantics, supporting the claim that the UP marks the point at which a word can be reliably differentiated from its cohort competitors (Marslen-Wilson & Tyler, 1980; Marslen-Wilson & Welsh, 1978; Tyler, 1984). Consistent with this, it is also only at the UP that the SemDis model, capturing the semantic space defined by the properties of successfully recognized target words, begins to show significant model fit, in brain regions (bilateral angular gyrus, middle temporal gyrus) that have been associated previously with different dimensions of lexical semantic processing (e.g., Bonnici, Richter, Yazar, & Simons, 2016; Lau, Phillips, & Poeppel, 2008; Price et al., 2016, Seghier, 2013).

Overall, the results show that sound-to-meaning mapping during natural speech comprehension involves rapid dynamic computations aligned around the UP. By testing quantifiable cognitive models of key cognitive processes as they occur in real time in the brain, we can begin to map out the dynamic cortical networks that operate on the partially activated lexical and semantic representations of cohort candidates before the UP, the gradual resolution of competitive processes as speech information accrues over time, and the emergence of the target word's conceptual representation as computationally dominant at the UP. The salient characteristics of the cohort model—a cognitive theory initially built solely on behavioral data—are shown to correspond to the actual real-time patterning of neurocomputational activity across the brain as spoken words are heard and recognized.

More generally, these results suggest that behavioral methods—which reflect underlying neural organization only opaquely and ambiguously—are likely to be increasingly superseded by neuroimaging methods that can access the dynamic spatiotemporal patterning of neural activity, while at the same time allowing this activity to be neurocomputationally interpreted in terms of multiple higher level cognitive dimensions. In this connection, while I have focused here on noninvasive combined EEG and MEG (EMEG) methods for probing dynamic brain activity, important and revealing findings (e.g., Cheung, Hamilton, Johnson, & Chang, 2016) are emerging from studies using electrocorticographic methods, recording directly from the brains of epilepsy patients (for review, see, Leonard & Chang, 2014; Sjerps & Chang, chapter 27 of this volume).

3. Language Comprehension in Its Evolutionary Neurobiological Context

The preceding section argues for an approach to sentence comprehension that is embedded in the real-time electrophysiological activity of the brain and interpreted in terms of functional neurocognitive theories. Such an approach absolutely requires the study of language as a real-time system to be situated in its neurobiological and evolutionary context. The brain systems supporting all aspects of human neurocognitive performance—even those, such as language acquisition, comprehension, and production, that may have human-unique properties—can only be fully understood in the context of the nonhuman primate (NHP) systems from which they have emerged.

This reflects the foundational assumption across the neurosciences that there is a deep and explanatory continuity between the structural and functional architecture of the major cortical and subcortical systems seen in NHP species and those that underpin the corresponding systems in the modern human. In the domain of visual perception, for example, the concept of dorsal and ventral processing pathways is taken as fundamental to a neuroscientific account of the human perceptual system just as it is in the macaque and other primates (Mishkin, Ungerleider, & Macko, 1983). Similarly, pioneering work originally done in the macaque, describing the neurofunctional architecture of cortical auditory processing, with its division into a hierarchy of core, belt, and parabelt systems (Kaas & Hackett, 2000), forms the foundation for current understanding of human cortical auditory function. Such parallels can be found and straightforwardly explored in any domain—from attention and memory to emotion and motor control—where humans and their NHP relatives can be thought to be facing similar challenges and adopting similar solutions (e.g., Neubert, Mars, Thomas, Sallet, & Rushworth, 2014).

The critical question where human language and communication are concerned is how far the neurobiological systems supporting these capacities can also be traced back to systems shared with our primate ancestors. Historically, in the context of linguistically dominated approaches to language that emphasize the grammatical and combinatorial aspects of language function, these neurobiological linkages have not been systematically pursued. Indeed, in the biolinguistic framework (e.g., Berwick, Friederici, Chomsky, & Bolhuis, 2012), the causal evolutionary event driving the emergence of human language is claimed to be the sudden appearance in the human lineage, around 100,000 years ago, of the neurocomputational capacity for *recursion*. This key mutation is seen as bestowing unique combinatorial capacities on the human species and to be fundamentally and formally distinct from the multiple roles of language in conspecific communication.

Here we advocate a different approach to human language and communication, based on a *Dual Neurobiological Systems* (DNS) framework (Marslen-Wilson & Tyler, 2011; Marslen-Wilson, Bozic, & Tyler, 2014; Marslen-Wilson & Bozic, 2018). This places the neurobiological and evolutionary origins of language center stage and sees the communicative and combinatorial capacities of the modern human as reflecting a dynamic coalition of two intersecting but evolutionarily and functionally distinguishable sets of systems. Strong evolutionary continuity between humans and their primate relatives is provided by a distributed, bihemispheric set of capacities that support the dynamic interpretation of multimodal sensory inputs, most relevantly in the context of social communication between members of the same species. These systems are the object of intensive study in monkeys and apes, and we will argue that this basic architecture underpins related communicative functions in the human.

An influential example of this is the extension into the auditory perceptual domain, by Rauschecker and Tian (2000), of the dorsal/ventral distinction already established for visual perception. This distinction, developed in the macaque in terms of *what* and *where* auditory pathways, has come to dominate current thinking about human processing of complex auditory inputs, and of speech and language in particular (Friederici, 2012; Hickok & Poeppel, 2007; Poeppel, 2014; Rauschecker & Scott, 2009). Another important domain, critically relevant for conspecific social communication, relates to the intricate neural systems that support the integration of auditory and visual signals in the perception (and production) of NHP communicative signals (e.g., Ghazanfar, Chandrasekaran & Logothetis, 2008; Ghazanfar & Eliades, 2014). Still other research suggests that bilateral frontotemporal interactions in the macaque brain, elicited by the presence of conspecific calls and signals, serve to underpin the wider communicative function of these calls (e.g., Gil-da-Costa et al., 2006).

Our hypothesis here, in a DNS framework, is that these (and other) systems, already well established in the macaque and other primates to support their ecological niche as intelligent primates living in dynamically structured social groupings, largely form the neural substrate for parallel functions in the modern human. In the specific context of language comprehension, these bihemispheric systems support not only the ability to identify words in the speech input—integrating visual and auditory cues in face-to-face interaction—but they also provide the basis for our ability to make sense of these word meanings in the general context of the listener's knowledge of the world and of the specific context of speaking, allowing rapid and appropriate action in a complex social and physical environment. Primatologists (e.g., Seyfarth & Cheney, 2016) have highlighted the complexity of social communication in large NHP social communities, and the dynamic role of pragmatic inference in the interpretation of vocal calls in these complex contexts. Related issues are explored in chapter 14 by Levinson and chapter 15 by Rossano in the current volume.

At the same time, the bihemispheric, whole-brain systems inherited from our primate ancestors are by themselves not sufficient to support modern human language. Primate modes of communication with their conspecifics do not seem remotely comparable to human language in their expressive capacities (Tomasello, 2008). While there are many disparate factors feeding into this—symbolic representation, fine-grained speech motor control, cooperative communicative intent, and so forth—an important area of divergence is undoubtedly the domain of combinatorial grammatical function. Human language is much more than a set of signs that stand for things. It constitutes a powerful and flexible set of grammatical devices for organizing the flow of linguistic information and its interpretation, allowing us to represent and combine abstract linguistic elements (words and morphemes) that not only convey meaning but also the structural cues that indicate how these elements are linked together to drive dynamic processes of semantic and pragmatic interpretation.

A distinct LH frontotemporal system, seemingly human-specific, provides the principal neurobiological substrate for these core grammatical capacities, and forms the second component of the DNS approach. This LH system, furthermore, depends on neuroanatomical developments that are significantly more extensive in the human than in even our closest extant primate relatives. There are substantial increases in the size and complexity of the frontal and temporal areas critical for language function, coupled with the expansion of the dorsal white matter tracts connecting these areas (Rilling et al., 2008). While the functional consequences of these changes are still not fully understood, explaining how they provide the neural substrate on which certain grammatical functions depend, it is clear that their presence is essential. When the LH system is damaged, the parallel right hemisphere (RH) regions cannot fully take over these grammatical functions (e.g., Tyler, Wright, Randall, Marslen-Wilson, & Stamatakis, 2010; Tyler et al., 2011).

3.1. Functional Dissociability A key claim of the DNS approach is that the emerging LH system neither replaces nor displaces the bihemispheric system for social communication and action, so that we are dealing in the modern human with two evolutionarily distinct biological systems, however well integrated they may be in ordinary language use. Direct evidence for this comes from the functional separability of the two systems. Using a combination of behavioral and neuroimaging techniques, it is possible to demonstrate this dissociation both in unimpaired young adults (e.g., Bozic, Tyler, Ives, Randall, & Marslen-Wilson, 2010) and in patients with LH brain damage (Griffiths, Marslen-Wilson, Stamatakis, & Tyler, 2012; Marslen-Wilson & Tyler, 1997; Tyler et al., 2011; Wright et al., 2012). We focus here on the neuropsychological evidence for the global separability of a morphosyntactically focused left-lateralized system and a bilateral lexically and semantically oriented system.

The observation that intact left frontotemporal cortex is required for full command of syntax has a long history (e.g., Caplan, Hildebrand, & Makris, 1996; Dronkers et al., 2004; Friederici, Ruschemeyer, Hahne, & Fiebach, 2003; Snijders et al., 2009). The critical point here, however, is not just to confirm the importance of specific regions of the LH in supporting key aspects of language function but also to reveal what is still intact—specifically, those aspects of language function that may not be impaired even in the context of extensive LH damage.

In the case of syntax, direct support for the neuropsychological dissociability of distinct structural-functional networks supporting syntactic and semantic aspects of speech comprehension comes from studies that focus on the patterns of preservation and loss in patients with a variety of LH lesions and considerable variability in their language abilities (e.g., Tyler et al., 2011; Wright et al., 2012). These are correlational studies, where brain-wide patterns of neural damage, assessed on a voxel-by-voxel basis, are correlated both with performance on a range of tests of syntactic and semantic function in spoken sentence comprehension and with functional neuroimaging activity related to syntactic and semantic processing.

Using a classic measure of sentence comprehension (the reversible sentences test), patients were asked to match different spoken sentences to sets of pictures. Shown three pictures—a woman pushing a girl, a girl pushing a woman, and a woman teaching a girl—patients with impaired syntax will correctly match the sentence *The woman pushed the girl* to the first picture but will incorrectly match the passive sentence *The woman is being pushed by the girl* to the same picture. The second sentence requires the use of syntactic cues to extract the right meaning—just using the order of words is not sufficient. Patients do not match either sentence to the third picture (of the woman teaching the girl), reflecting their intact ability to identify the words being heard and to interpret their relationship on semantic and pragmatic grounds.

Performance on these and other tests of syntactic comprehension correlate with tissue integrity in LIFG (most strongly in BA45), in left posterior superior and middle temporal gyri and in the major white matter tracts connecting them, as well as with neural activity in these regions (and the functional connectivity between them) during the processing of syntactically complex materials. This combination of structural and functional effects strengthens the claim that complex syntax depends on a LH frontotemporal network with both dorsal and ventral components (Griffiths et al., 2012; Friederici, 2011; Papoutsi, Stamatakis, Griffiths, Marslen-Wilson & Tyler, 2011; Rolheiser et al., 2011) and where, notably, recruitment of RH activity is unable to support recovery after damage to the left (Tyler et al., 2010).

Critically, however, the amount of LH damage, and the extent to which it involved key LH frontotemporal circuits, did not affect the patients' ability to identify the words being spoken or to understand the messages being communicated—so long as syntactic cues were not required to do so. The Wright et al. (2012) study, using *Joint Independent Component Analysis* (JICA) to distinguish separable components in the structural and functional data attributable to syntax and to semantics, identified right temporal regions that supported semantic processing. Activity in these regions was not enhanced in patients with LH damage, but was at similar levels to control subjects, suggesting that it reflected continuing function in the residue of a normally bilateral network (Bozic et al., 2010). This is consistent with the observation that nonfluent LH patients often have preserved semantics (Crinion & Price, 2005; Hagoort, Wassenaar, & Brown, 2003; Tyler & Marslen-Wilson, 2008) and with evidence for RH involvement in the semantic and pragmatic interpretation of spoken language (Indefrey & Cutler, 2004). Overall, these findings demonstrate the functional dissociability of contrasting left lateralized and bilateral neural systems supporting core linguistic and general communicative processes.

These and other considerations lead to a view of human language and communication as being based on a *coalition* of underlying neural systems and networks. This means that the output of these systems, which we experience as the outcome of a single process, is in fact the integrated product of different sets of subsystems—analogous, indeed, to current thinking about the generation of visual perceptual experience.

This, in turn, raises the question of whether grammatically driven structure building in language comprehension is solely the province of the LH perisylvian system or whether ancestral bihemispheric systems can also contribute. The research by Tyler and Wright and their colleagues (e.g., Tyler et al., 2011; Wright et al., 2012) provides neuropsychological evidence that LH damage to systems supporting complex syntax does not eradicate the capacity to interpret sequences requiring only computation of linear adjacency relations. Further studies of unimpaired populations, combining fMRI with multivariate RSA analyses (Bozic et al., 2015; Yang, Marslen-Wilson, & Bozic, 2017) in a natural listening paradigm, shows that simple phrases (*I play, the smile*) engage bilateral temporal regions with no evidence of selective LIFG involvement, indicating that some aspects of sequential grammatical analysis may indeed be supported outside the LH perisylvian system.

In the evolutionary context, these basic sequence interpretation capacities have surfaced from a number of different perspectives (Dehaene, Meyniel, Wacongne, Wang, & Pallier, 2015; Wilson, Marslen-Wilson, & Petkov, 2017). Speculations that language in the modern human was preceded by the emergence of a protolanguage (e.g., Bickerton, 1981, 2009) presupposes capacities of this sort, while the proposals of Klein and Perdue (1997) for a "Basic Variety"—the spoken communication systems developed in untutored second language acquisition—similarly imply sentence-level capacities for sequential interpretation in the absence of hierarchical syntax or inflectional morphology. These properties of the Basic Variety may in fact represent the limits—but also the functional effectiveness—of a language built on the sequential processing capacities of the ancestral bihemispheric system.

A different kind of evidence for a common neurobiological substrate for basic sequence analysis capacities comes from studies using fMRI in macaques and in humans to examine the brain regions involved in processing the ordering relationships between auditory nonsense words in rule-based sequences (Wilson et al., 2015; Wilson et al., 2017). These studies suggest that key regions in the human bilateral ventral frontal and opercular cortex have functional counterparts in the monkey brain. These regions are also claimed to be associated with initial stages of human syntactic processing (cf. Friederici, Bahlmann, Heim, Schubotz, & Anwander, 2006). This raises the possibility that certain bilateral neural systems, which play a significant role in language function in modern humans, originally evolved to support domain-general abilities involved in sequence processing, and that these sit outside the core LH perisylvian system.[2]

4. Cross-Linguistic Variation in a Distributed Neurobiological Context

One reason that human language presents such a complex scientific challenge is that languages themselves exhibit such remarkable diversity. Even in an era of accelerating worldwide language loss, there are more than 6,500 languages still spoken, falling into some 250 different language families that vary along multiple linguistic dimensions. This greatly complicates the study of language as a neurobiological system. If we are studying, for example, the properties of visual perceptual systems, mapping from retinal input to underlying conceptual representation (e.g., Clarke & Tyler, 2015), we can assume that these systems will be essentially the same across the human population, both in terms of neural architecture and in terms of computational content—and with good evidence that these parallels cross species boundaries as well (e.g., Kriegeskorte et al., 2008).

In the case of human language, we are faced with potential variation, on a language-by-language basis, both in the overall neural architecture of the system and in the specific properties of the neural computations that the individual language requires. This becomes especially salient in the DNS framework, where the conception of language as a dynamic coalition of interacting brain systems directly raises the question of how basic linguistic functions are distributed across these brain systems and whether this distribution of functions varies between languages. To evaluate how far they are concentrated within a single core system (as proposed on the biolinguistic account), as opposed to reflecting the joint contribution of sets of distributed systems, it is essential to evaluate the neural expression of different linguistic functions across typologically different languages.

Here we summarize research that has explored this question for one of the most salient linguistic functions, the *generative capacity* of human language, whereby an infinite range of communicative meanings can be expressed using the combination and recombination of a finite set of linguistic elements. This is not just a property of syntactic combination across phrases and clauses, but it is also a property of combinatorial processes realized at lexical and morphological levels. This generative function is standardly defined in terms of recursion (Hauser, Chomsky, & Fitch, 2002), where two or more basic linguistic elements combine (in the operation known as *Merge*) to form a hierarchy of larger units that themselves enter into further combinatorial operations. At the lexical level this stands in contrast to the property of whole-form "storedness," which not only applies to the basic elements of linguistic analysis (words and morphemes), but also to stored combinations of these units as morphologically complex words or even as phrases (cf. Yang et al., 2017).

As discussed in section 3.1, generative functions and their implied combinatorial machinery are closely associated with the evolutionarily more recent left perisylvian system in humans. In a series of cross-linguistic studies, we have explored the generality of this observation, while simultaneously asking whether storedness is more characteristic of the linguistic operations of the bihemispheric subsystems. To do this, we have focused on the contrast between *inflectional* and *derivational* combinatorial word-formation processes, where inflectional morphology subserves primarily grammatical functions, while derivational morphology is associated with the creation and representation of new words (or *lexemes*) in the language. These distinctions relate directly to our questions about the respective roles of stored representation and decompositional processing in the neurobiological systems that handle morphemic and morphosyntactic combination in the world's languages. At the same time, they offer an experimentally tractable means of conducting linguistically comparable neuroimaging studies across a variety of languages.

4.1. English Inflection and Derivation For the neuroscientifically most intensively studied example of regular inflectional morphology (the English [-ed] past tense), there is compelling evidence for selective engagement of the left perisylvian system for decompositional access processes, where inflectionally complex forms (such as *jumped* or *agreed*) are not neurocognitively stored as full forms, and where lexical access involves the parsing, or decomposition, of the input stream into stems and affixes (Bozic et al., 2010; Marslen-Wilson, 2007; Marslen-Wilson & Tyler, 1997, 2007). Access to stems (e.g., *jump, agree*) is driven primarily by bilateral temporal processes, while access to inflectional affixes (and their grammatical implications) selectively engages left inferior frontal regions (primarily LIFG) as part of the broader left perisylvian network linking LIFG and posterior superior and middle temporal cortex (Bozic et al., 2010; Tyler, Stamatakis, Post, Randall, & Marslen-Wilson, 2005). These results, in turn, are consistent with extensive neuropsychological and neuroimaging research that demonstrates the critical role of this LH perisylvian network in supporting combinatorial syntactic functions more generally (e.g., Friederici, 2011; Hagoort, 2005, 2014; Hagoort & Indefrey, 2014; Tyler et al., 2011). In terms of the spatiotemporal dynamics of real-time speech processing, these contingencies are demonstrated most directly by Fonteneau, Bozic, and Marslen-Wilson (2015) in a study that combined analyses in EMEG source space with measures of functional connectivity between frontotemporal brain regions. These showed selective and early coactivation of the LIFG (BA44 and BA45) and left temporal regions as evidence became available that an inflectional suffix (as in forms like *jumped*) was present in the speech signal.

The selective activation of the LIFG, in conjunction with both left-lateralized and bilateral posterior temporal activation, is taken to be theoretically critical in these studies, as reflecting LH processing activities potentially specific to combinatorial morphosyntax (cf. Klimovich-Gray, Bozic, & Marslen-Wilson, 2016). This stands in contrast to bilateral IFG activation, which is thought to reflect more domain-general cognitive control processes, and triggered in the word-recognition context by increases in perceptual complexity that are not specifically linguistic in nature (cf. Bozic et al., 2010; Zhuang et al., 2014).

The neurocognitive properties of regular inflectional morphology (as seen in English) are consistent with the defining properties of inflectional morphology as linguistically defined (Bickel & Nichols, 2007; Marslen-Wilson, 2007). First, across languages, inflectional morphology does not create new words with new meanings that require new lexical entries. Inflectional variants like *cat* and *cats* or *walk* and *walked* are not listed as separate headwords in standard dictionaries, and their meaning is fully predictable from the meaning of the stem combined with the grammatical properties of the inflection. Second, inflections are responsive, in a regular and predictable way, to the properties of the syntactic environment in which they occur (e.g., Bickel & Nichols, 2007). The information carried by the inflection is not just about the stem itself, but also about the phrasal and sentential interpretation to which that stem relates. These properties of regularly inflected forms are consistent with a neurocognitive account where inflectional morphemes are stripped from their stems early in processing and where the grammatical information they carry engages combinatorial linguistic mechanisms supported by LH perisylvian brain regions working in close linkage with bilateral temporal areas (Marslen-Wilson & Tyler, 2007).

Derivationally complex forms, though also constructed by combining a stem with a grammatical morpheme, have quite different properties. Whereas inflectional morphemes result in different forms of the same word, responsive to its current syntactic environment, derivational processes produce new words—in linguistic terms, new lexemes with new lexical entries (Matthews, 1991)—whose meaning and syntactic function is much more context-independent, and where the meaning of these words is less predictable from the simple combination of the meaning of their constituent morphemes. This lack of compositionality—and the range of variation from opaque forms like *department* to transparent forms like *bravely*—has prompted long-standing psycholinguistic controversies as to whether derived words are represented and processed on a decompositional morphemic basis, or as whole forms with no underlying representation of their morphological structure (cf. Clahsen, Sonnenstuhl, & Blevins, 2003; Marslen-Wilson, Tyler, Waksler, & Older, 1994).

In this broader DNS context, Bozic, Tyler, Su, Wingfield, and Marslen-Wilson (2013) specifically addressed the question of whether derivationally complex words in English selectively engage the LH system in the same way as inflected forms—implying decompositional access processes—or whether they engage the distributed bihemispheric system, consistent with whole-form, noncompositional accounts. The study varied the semantic transparency and affix productivity of a set of English-derived and pseudo-derived words, forming a gradient from nondecomposable opaque forms like *breadth* and *archer*, to potentially decomposable transparent forms (varying in affix productivity) like *warmth* and *farmer*. The fMRI results provided no evidence for selective activation of the LH frontotemporal system for derived words, even for the highly decomposable *farmer* forms. Instead, derivationally complex forms, independent of whether they were synchronically transparent or opaque, activated a bilaterally distributed set of frontotemporal processes, with bilateral IFG involvement being driven by perceptual competition between the derived whole form and its onset-embedded stem or pseudo-stem (e.g., *breadth/bread*), modulated by the semantic transparency of the whole form-stem relationship (Bozic, Tyler, et al., 2013).[3] These results are far more consistent with the view that derived words are stored as whole forms, though they do not exclude the possibility that these stored representations are morphemically structured, especially if this structure (stem + affix) is synchronically semantically transparent (Clahsen et al., 2003; Marslen-Wilson, 2007).

The pattern of results for English suggests a strong distinction between inflectional morphology and derivational morphology, with the decomposition/storage distinction aligning with the different neural substrates proposed by the DNS approach. At the same time, this was a set of results that cried out for cross-linguistic investigation, given the idiosyncrasies of English inflectional morphology.

4.2. Slavic Inflection and Derivation Several further studies have examined the distinction between derivational and inflectional forms from a DNS perspective in a set of languages that contrast strongly both with each other and with English. Polish, a member of the Slavic language family, is typologically similar to English in the distinction it makes between inflectional and derivational processes, and in its use of *concatenative* word-formation processes (where stems and affixes are separate phonological forms that are strung together in a sequence to construct complex surface forms). It contrasts with English both in the prevalence of morphological complexity and in the richness and diversity of its inflectional system.

All content words in Polish are morphologically complex, realized within rich inflectional paradigms. Every noun is inflectionally marked for case, number, and gender, and similarly for adjectives, numerals, and pronouns. The neuter nominal form *badanie* 'inspection,' for example, from *badać* 'to inspect' breaks down into the derived stem *bad-ani*, made up of the verbal root [bad-] and the derivational suffix [-ani-], which in turn combines with the inflectional morpheme [-e] that marks the nominative case for a neuter singular noun. Polish verbal inflection is even more complex and heterogeneous, marking three tenses, with six person/number categories, together with affixes that express mood, voice, tense, and aspect.

Despite these major distributional and linguistic differences between Polish and English, there are close similarities in how the two languages interface with a dual-processing framework. Szlachta, Bozic, Jelowicka, and Marslen-Wilson (2012) found that inflectionally complex words selectively activated the LIFG (together with bilateral temporal regions) while Bozic, Szlachta, and Marslen-Wilson (2013), focusing on derivational complexity, found no evidence for selective left-lateralized activation, even for transparent and productive derived forms like *żabka* 'little frog' (from *żaba* 'frog') or *czytanie* 'reading' (from *czytać* 'to read'). Activation associated with the presence of derivational complexity was bilateral and primarily middle temporal. These conclusions are reinforced for Slavic languages more generally by a subsequent fMRI study in Russian (Klimovich-Gray et al., 2016), which found strong selective LIFG activation for simple and complex inflected words and phrases, but only bilateral temporal effects for derived words.[4]

Across both English and two Slavic languages, regular inflectional morphemes, carrying structurally relevant grammatical information, and requiring morphophonological parsing to separate these morphemes from their stems, selectively engage a LH circuit that includes the LIFG, as well as temporal regions bilaterally. Derivationally complex forms, in contrast, do not selectively engage the LIFG and are processed primarily by bilateral temporal structures, already identified as central to the mapping of stems onto meanings.

4.3. Mandarin Compounds and Aspectual Markers Two further studies address these issues in two typologically very different languages from the Indo-European languages examined so far. At one pole, we have Mandarin Chinese, which is notable for its virtual absence of combinatorial affixing morphology. Nominal marking for plural, verbal markings for person, and marking for number, all of which are commonly observed in Indo-European languages, are not found in Mandarin (Li & Thompson, 1981). Given that Mandarin also uses fewer linguistic combinatorial structures, this raises the possibility that the underlying neural organization of Mandarin may be less dependent on a left-lateralized frontotemporal network, specializing in morphosyntactic processing, and tilted more toward a bihemispheric system where semantic and pragmatic constraints play a larger role.

The only clear examples of affixing inflectional morphemes in Mandarin are arguably the small set of aspect markers, such as the perfective affix [-le] and the continuous affix [-zhe], which are seen as functionally and morphophonologically equivalent to grammatical affixes in Indo-European languages (Packard, 2000). Unlike compounding or word-forming morphemes, which change the meaning or the word category of the stem, these aspectual affixes, similar to tense marking in English, only add temporal information about the verb stem to which they attach. For example, when *ku* 'to cry' is followed by the perfective affix [-le], the meaning is 'have cried'; when it is followed by [-zhe], which marks continuous aspect, the meaning is 'to be crying'. The question of interest here is whether these affixes, despite their sparseness in the language, and the generally minimal combinatorial environment in which they occur, will nonetheless elicit selective left perisylvian activation as seen in our Indo-European examples. Earlier research in English (Fonteneau et al., 2015) showed that analyses in EMEG source space can track with high spatiotemporal precision the presence of selective left perisylvian activity when an inflectional suffix is present in the speech signal.

We tested for potential parallels in Mandarin in a study that combined source-localized EMEG data and multivariate RSA analyses (as in Kocagoncu et al., 2017) to examine the potential real-time involvement of the LH system in the perceptual processing of inflected as opposed to noninflected complex word forms (Huang, 2015; Huang et al., 2014). The predictions for the aspectually inflected items contrasted with those for bisyllabic compound words in Mandarin. Compounding is by far the dominant word-formation process in Mandarin, most frequently in the form of disyllabic compounds created by combining two existing morphemes. There is strong behavioral evidence that this word-formation process generates whole-form lexemic representations of the resulting compound (e.g., Zhou & Marslen-Wilson, 1994, 1995). In the current context, we would expect Mandarin compounds to primarily activate bilateral temporal regions, similarly to derivational forms in languages like English or Polish, which also seem to be stored and accessed as whole forms.

The critical contrasts fall at the onset of the second morpheme of the inflected forms (e.g., *ku+zhe* 'crying') and the compound forms (e.g., *hu+xi* 'to breathe'), where it begins to become clear what type of complex form the listener is hearing. Searchlight RSA analysis for the compound model reveals significant bilateral model fit in posterior temporoparietal regions from ~120 to 250 ms after second morpheme onset, but with no model fit in inferior temporal regions in either LH or RH. The inflection model, in contrast, shows selective LIFG fit in BA44 from 200 ms and is joined by BA45 and anterior temporal regions from around 240 ms.

This result, strikingly congruent with the patterns seen for English, emphasizes the pervasive engagement of LH frontotemporal circuits by inflectional morphemes. It is consistent with a primary role for the LH system in supporting combinatorial processes in language interpretation and the specialized function of inflectional morphemes in guiding this process. The bilateral temporal distribution of neural activity related to the access and interpretation of compound words, seen as nondecompositional and requiring storage of their meaning, is in turn consistent with the roles assigned to bihemispheric systems in supporting storedness—learned stable mappings between forms and meanings.

4.4. Arabic Roots, Word Patterns, and Inflectional Suffixes A quite different challenge to the DNS approach is provided by Arabic, a member of the Semitic language family. Arabic presents a morphological system that differs strongly from the languages we have investigated so far, in several respects. The first is its dependence on *nonconcatenative* word formation (Boudelaa & Marslen-Wilson, 2015). In all the languages discussed so far the predominant mechanism is *concatenative*, where complex forms are constructed by chaining together (concatenating) separate morphemes (stems and affixes) in linear sequence—as in the English word *unhappiness* (un+happy+ness), the Russian word *perewozshik* 'long distance lorry driver' (pere+woz+shik), or the Mandarin word *yanguang* 'judgment' (yan+guang). In Arabic, in contrast, surface word forms are constructed by interleaving an abstract consonantal *root* that conveys semantics with an abstract *word pattern*, primarily made up of vowels, which conveys phonological and morphosyntactic information (Holes, 1995; Versteegh, 1997). Thus the word *katam* 'to conceal' is built by interleaving the triconsonantal root [ktm], which has the semantic meaning of *concealing* or *hiding*, with the verbal word pattern [-a-a-], which has the morphosyntactic meaning of *active, perfective, verb*. These Arabic nonconcatenative processes interface, in turn, with typologically more standard concatenative mechanisms that attach inflectional affixes to these stems to generate more complex forms (e.g., *katam*+*~uu*= *katamuu* 'they hid'; *qalb* 'heart' + *hu* 'his' = *qalbuhu* 'his heart'). These forms are not only inflectionally complex, in terms of their sequential morphophonological structure, but also have a phrasal constituent structure as noun phrases and verb phrases.

Arabic also differs sharply from the languages we have considered so far in that its key word-formation morpheme—the word pattern—has both derivational and inflectional functions (Boudelaa & Marslen-Wilson, 2015; McCarthy, 1981). Consider the verbal word pattern [-a-a-] that generates verbs like *katam* 'hide', *katab* 'write', and *lakam* 'punch'. In each case, the word pattern is interleaved with a consonantal root ([ktm], [ktb], [lkm]) to generate a new lexeme with the specific grammatical category of *verb*, thereby fulfilling a core derivational word-formation function. At the same time, this morpheme conveys the morphosyntactic valence of *active mood* and *perfective aspect* for each form, standardly regarded as core inflectional dimensions (Bybee, 1985). If, in the languages considered so far, inflectional and derivational functions are aligned with different neurobiological dimensions of the substrate for language, where does this leave the nonconcatenative Arabic word pattern morpheme in relation to the patterns seen in other languages?

The final property of Arabic is the absence of a contrast in storedness between inflectionally and derivationally complex forms. There is strong behavioral evidence (Boudelaa & Marslen-Wilson, 2015) that all nonconcatenative forms in Arabic are obligatorily decomposed into their constituent roots and word patterns.[5] These decompositional effects—as reflected in numerous priming experiments—are driven solely by morphological variables. Pairs of words prime each other if they share a morpheme (a root or a word pattern), unaffected by the semantic variables that strongly condition morphological priming in languages like English (Marslen-Wilson et al., 1994). Our argument so far has been that combinatorial morphological processes driven by the grammatical properties of the morphemes involved—as in the case of inflectional affixes in their morphosyntactic contexts—selectively engage the LH perisylvian system. This raises the question of whether Arabic, where the nonconcatenative-derived forms exhibit comparable decompositional properties, will also elicit similarly left-lateralized processing.

These multiple divergences from the basic morphological characteristics of the languages investigated so far make Arabic a particular challenge for the DNS approach. To explore this challenge, we conducted an fMRI study (Carota, Boudelaa, Bozic, Su, & Marslen-Wilson, 2014) where participants heard Arabic words contrasting in degree and type of morphosyntactic complexity:

a. morphological simple words (chiefly function words) that are not analyzable into roots and word patterns, and are analogous to simple words in languages like English;
b. purely nonconcatenatively complex nouns and verbs, formed by adding either a nominal or a verbal word pattern to a consonantal root—for example, the noun *kitaab* 'book', formed by combining a nominal word pattern with the root [ktb], or the verb *kitaba* 'write', formed by combining the same root with a verbal word pattern; and
c. nonconcatenative complex nouns and verbs combined with a concatenative inflectional affix, as in forms like *katamuu* 'they hid', generated by combining the verbal stem *katam* 'hide' with a plural suffix, producing forms that are inflectionally complex and that have a phrasal constituent structure.

The results show that, despite the major divergences in its surface properties as a linguistic system, Arabic seems to fit the underlying dual neurobiological substrate remarkably well. The Arabic *morphologically simple* words showed similar activation to simple words in English, with univariate subtractions revealing bilateral temporal activity but no significant LIFG effects. Turning to the morphologically complex words, a combination of univariate and multivariate analyses allows us to distinguish two separate components in the neural responses to *nonconcatenative* (root+word-pattern) items. Selective LIFG engagement (BA44, BA45, BA47) is seen only for the verbal stimuli (e.g., *katam, kitaba*) and is arguably driven by word pattern *verb argument structure*, which is consistent with earlier imaging studies (Longe, Randall, Stamatakis, & Tyler, 2007; Tyler, Bright, Fletcher, & Stamatakis, 2004). RSA analyses focusing on the nonconcatenative nominal forms show a distinctive pattern of *bilateral IFG activation* (BA45), likely to be root-based and morphosemantic in nature, and consistent with previous research showing bilateral IFG involvement in implicit lexical and semantic control processes (Bozic, Tyler, et al., 2013; Raposo, Moss, Stamatakis, & Tyler, 2006; Zhuang et al., 2014).

This evidence for a two-way distinction between word-pattern–based left perisylvian effects and bilateral frontal root–related effects is supported by an earlier EEG experiment, using the Mismatch Negativity (MMN) paradigm, which also explored the distribution of neural responses to roots and word patterns in nonconcatenative Arabic words (Boudelaa, Pulvermüller, Hauk, Shtyrov, & Marslen-Wilson, 2009). When the deviant ("oddball") stimulus involved a change in the consonantal root, the resulting MMN response had a symmetric frontocentral distribution, whereas the word pattern MMN was significantly left-lateralized and centered over left perisylvian areas.

Finally, the concatenatively complex forms (as in *katamuu* 'they hid'), which have both inflectional and phrasal complexity, show an almost identical pattern of model fit to earlier research (Bozic et al., 2015), capturing the effects of inflectional and phrasal complexity in English (as described in section 2—see figure 30.2). The English study found that when a combined complexity RSA model was run, sensitive to the processing demands of both inflectional suffixes (as in *jumped*) and phrasal structure (as in *I jump*), this generated significant bilateral temporal model fit, selective LIFG effects (BA44, BA45), and no right IFG fit (see figure 2c in Bozic et al., 2015). The RSA results for the Arabic concatenatively complex forms (with the effects of nonconcatenative complexity partialled out) revealed a very similar pattern, with selective LIFG model fit in the same BA44 and BA45 ROIs, and substantial bilateral temporal model fit, somewhat more extensive on the left, as we also saw in English.

These results demonstrate, first, that there is a great deal of commonality across languages as diverse as Arabic and English in how similar underlying linguistic functions map onto similar neural structures across the major DNS domains. Morphologically simple words and forms with a simple phrasal structure activate bilateral temporal regions, in both English and Arabic, while morphemic elements with morphosyntactic functions selectively engage left inferior frontal regions in all the languages tested—in Arabic this would include both the concatenative inflectional morphemes in the *katamuu* forms and the verb features triggered by the verbal word patterns.

Second, however, Arabic has additional processing features associated with its typologically unusual nonconcatenative word-formation procedure, where the word pattern morpheme has both inflectional and derivational functions. These dual roles seem to be reflected in the two separable components we see in the neural responses to the nonconcatenative words. We see selective changes in LIFG activity that are specific to the verbal word patterns and a separate pattern of bilateral IFG model fit (BA45) related to the root, arguably reflecting the morphosemantics of its interpretation in the given word pattern context. Notably, this suggests that Arabic root and pattern morphology directly combines LH processes supporting morphosyntactic combination with bilateral domain-general semantic control processes. This finding—though requiring further research using time-resolved imaging methods—tells us that the mechanisms supporting basic morphological combinatorial processes in a language like Arabic can integrate across neurocognitive systems with divergent evolutionary and neurobiological origins, inconsistent with claims for any absolute distinction between them, even in the core domain of generative combination.

5. *Overview and Conclusions*

The journey taken in this chapter has three main staging points. It starts from the recognition that speech comprehension is remarkable for its immediacy and rapidity and reports on recent research that has begun to capture the real time flow of multilevel neurocomputational activity that underpins this process. These advances point the way to an explanatory account of sentence comprehension that is embedded in the dynamic electrophysiological activity of the brain, using multivariate analysis techniques to interpret this activity in terms of functional neurocognitive theories.

This is a scientific strategy that requires the study of language to be situated in its neurobiological and evolutionary context. The brain systems supporting human neurocognitive performance can only be understood in the context of the NHP systems from which they have emerged. This is reflected in the DNS approach, which distinguishes between a left perisylvian system that supports combinatorial morphosyntax, and a bihemispherically distributed set of conserved systems and processes that have deeper evolutionary roots. We cite a broad range of neuropsychological and neuroimaging evidence that argues both for the presence of these two systems in the modern human and for their separability.

At the third stage, we turn to the cross-linguistic domain, where the DNS conception of language, as an evolutionary coalition of interacting brain systems with contrasting neurocognitive properties, provides a unified framework for investigating how basic linguistic functions are distributed across these brain systems and whether this distribution varies between languages. Here I report a series of coordinated studies evaluating the neural expression of the same generative linguistic functions across six different languages.

Consistent with the DNS approach, we find a striking degree of alignment between the suggested dual neurobiological substrates for language function, the decomposition/storage dimension, and the inflection/derivation contrast. One axis of this alignment, involving inflectional morphology, decompositional representation and access, and the left perisylvian system, is very similar in its neural realization across all the languages tested. The second axis involving derivational morphology, full-form storage, and the bihemispheric subsystem, is equally consistent across languages, but allows for some intriguing variations in its realization—as seen in the bilateral frontal network engaged in the processing of Arabic derivation.

Overall, these outcomes suggest that there is a specific but broadly distributed repertoire of neurocomputational capacities underlying human language function. These capacities are not restricted to a classical left perisylvian system, even in its more expanded forms, and reflect instead a complex coalition of different neural systems of different evolutionary depths and origins. It is in terms of this complex evolutionary coalition that we should look for an explanation of how human language evolved and how it is neurally instantiated.

Acknowledgments

The extensive body of research covered in this chapter was supported, at different times, by an Advanced Investigator grant to W.D.M-W. from the European Research Council (AdG 230570 NEUROLEX), by MRC Cognition and Brain Sciences Unit (CBSU) funding to W.D.M-W (U.1055.04.002.00001.01), and by a European Research Council Advanced Investigator grant to Professor Lorraine Tyler under the European Community's Horizon 2020 Research and Innovation Programme (2014–2020 ERC Grant 669820 LANGDYN). I thank Lorraine Tyler for her many and substantial inputs to this chapter, for her warm support in the research we have done together, and for her unfailingly pioneering and critical spirit.

NOTES

1. From Marslen-Wilson and Welsh (1978), p. 29.
2. For further discussion of the DNS evolutionary framework and its implications, see Marslen-Wilson and Bozic (2018).
3. For a detailed examination of these issues in the rich context of Italian derivational morphology, see Carota, Bozic, and Marslen-Wilson (2016).
4. See also the results for derivational forms in German, where a MEG Mismatch Negativity study also points to whole form access for these words (Hanna & Pulvermüller, 2014).
5. Hebrew is markedly different from Arabic in these respects (cf. Boudelaa & Marslen-Wilson, 2015) and should be treated quite separately from Arabic in any discussions of Semitic languages.

REFERENCES

Baroni, M., & Lenci, A. (2010). Distributional memory: A general framework for corpus-based semantics. *Computational Linguistics, 36,* 673–721.

Berwick, R. C., Friederici, A. D., Chomsky, N., & Bolhuis, J. (2012). Evolution, brain, and the nature of language. *Trends in Cognitive Science, 17,* 89–98.

Bickel, B., & Nichols, J. (2007). Inflectional morphology. In T. Shopen (Ed.), *Language typology and syntactic description* (pp. 150–201). Cambridge: Cambridge University Press.

Bickerton, D. (1981). *Roots of language.* Ann Arbor, MI: Karoma.

Bickerton, D. (2009). *Adam's tongue: How humans made language, how language made humans.* New York: Hill and Wang.

Binder, J. R., Desai, R. H., Graves, W. W., & Conant, L. L. (2009). Where is the semantic system? A critical review and meta-analysis of 120 functional neuroimaging studies. *Cerebral Cortex, 19,* 2767–2796.

Bonnici, H. M., Richter, F. R., Yazar, Y., & Simons, J. S. (2016). Multimodal feature integration in the angular gyrus during episodic and semantic retrieval. *Journal of Neuroscience, 36,* 5462–5471.

Boudelaa, S., & Marslen-Wilson, W. D. (2015). Structure, form, and meaning in the mental lexicon: Evidence from Arabic. *Language, Cognition, and Neuroscience, 30*(8), 955–992.

Boudelaa, S., Pulvermüller, P., Hauk, O., Shtyrov, Y., & Marslen-Wilson, W. D. (2009). Arabic morphology in the neural language system: A mismatch negativity study. *Journal of Cognitive Neuroscience, 22,* 998–1010.

Bozic, M., Fonteneau, E., Su, L., & Marslen-Wilson W. D. (2015). Grammatical analysis as a distributed neurobiological function. *Human Brain Mapping, 36*(3), 1190–1202.

Bozic, M., Szlachta, Z., & Marslen-Wilson, W. D. (2013). Cross-linguistic parallels in processing derivational morphology: Evidence from Polish. *Brain and Language, 127,* 533–538.

Bozic, M., Tyler, L. K., Ives, D. T., Randall, B., & Marslen-Wilson, W. D. (2010). Bihemispheric foundations for human speech comprehension. *Proceedings of the National Academy of Sciences USA, 107*(40), 17439–17444.

Bozic, M., Tyler, L. K., Su, L., Wingfield, C., & Marslen-Wilson, W. D. (2013). Neurobiological systems for lexical representation and analysis in English. *Journal of Cognitive Neuroscience, 25*(10), 1678–1691.

Bybee, J. L. (1985). *Morphology: A study of the relation between meaning and form.* Amsterdam: John Benjamins.

Caplan, D., Hildebrandt, N., & Makris, N. (1996). Location of lesions in stroke patients with deficits in syntactic processing in sentence comprehension. *Brain, 119,* 933–949.

Carota, F., Boudelaa, S., Bozic, M., Su, L., & Marslen-Wilson, W. D. (2014). *Neurocognitive properties of concatenative and non-concatenative morphology in Arabic: Evidence from multivariate fMRI analyses.* Paper presented at the 21st Annual Meeting of the Cognitive Neuroscience Society, Boston, MA.

Carota, F., Bozic, M., & Marslen-Wilson, W. D. (2016). Decompositional representation of morphological complexity: Multivariate fMRI evidence from Italian. *Journal of Cognitive Science, 28*(12), 1878–1896.

Cheung, C., Hamilton, L. S., Johnson, K., & Chang, E. F. (2016). The auditory representation of speech sounds in human motor cortex. *eLIFE, 5*, e12577.

Clahsen, H., Sonnenstuhl, I., & Blevins, J. P. (2003). Derivational morphology in the German mental lexicon: A dual mechanism account. In Baayen, R. H. & Schreuder, R. (Eds.), *Morphological structure in language processing* (pp. 125–155). Berlin: Mouton de Gruyter.

Clarke, A., & Tyler, L. K. (2015). Understanding what we see: How we derive meaning from vision. *Trends in Cognitive Sciences, 19*, 677–687.

Crinion, J., & Price, C. J. (2005). Right anterior superior temporal activation predicts auditory sentence comprehension following aphasic stroke. *Brain, 128*, 2858–2871.

Dehaene, S., Meyniel, F., Wacongne, C., Wang, L., & Pallier, C. (2015). The neural representation of sequences: From transition probabilities to algebraic patterns and linguistic trees. *Neuron, 88*, 2–19.

Devereux, B. J., Taylor, K. I., Randall, B., Geertzen, J., & Tyler, L. K. (2016). Feature statistics modulate the activation of meaning during spoken word processing. *Cognitive Science, 40*, 325–350.

Dronkers, N. F., Wilkins, D. P., Van Valin, R. D. Jr., Redfern, B. B., & Jaeger, J. J. (2004). Lesion analysis of the brain areas involved in language comprehension. *Cognition, 92*, 145–177.

Fonteneau, E., Bozic, M., & Marslen-Wilson, W. D. (2015). Brain network connectivity during language comprehension: Interacting linguistic and perceptual subsystems. *Cerebral Cortex, 25*(10), 3962–3976.

Friederici, A. D. (2011). The brain basis of language processing: From structure to function. *Physiological Reviews, 91*, 1357–1392.

Friederici, A. D. (2012). The cortical language circuit: From auditory perception to sentence comprehension. *Trends in Cognitive Sciences, 16*, 262–268.

Friederici, A. D., Bahlmann, J., Heim, S., Schubotz, R. I., & Anwander, A. (2006). The brain differentiates human and non-human grammars: Functional localisation and structural connectivity. *Proceedings of the National Academy of Sciences, 103*, 2458–2463.

Friederici, A. D., & Gierhan, S. M. E. (2013). The language network. *Current Opinion in Neurobiology, 23*, 250–254.

Friederici, A. D., Ruschemeyer, S. A., Hahne, A., & Fiebach, C. J. (2003). The role of left inferior frontal and superior temporal cortex in sentence comprehension: Localizing syntactic and semantic processes. *Cerebral Cortex, 13*, 170–177.

Gaskell, G. M., & Marslen-Wilson, W. D. (1997). Integrating form and meaning: A distributed model of speech perception. *Language and Cognitive Processes, 12*, 613–656.

Ghazanfar, A. A., Chandrasekaran, C., & Logothetis, N. K. (2008). Interactions between the superior temporal sulcus and auditory cortex mediate dynamic face/voice integration in rhesus monkeys. *Journal of Neuroscience, 28*, 4457–4469.

Ghazanfar, A. A., & Eliades, S. J. (2014). The neurobiology of primate vocal communication. *Current Opinion in Neurobiology, 28*, 128–135.

Gil-da-Costa, R., Martin, A., Lopes, M. A., Munoz, M., Fritz, J. B., & Braun, A. R. (2006). Species-specific calls activate homologs of Broca's and Wernicke's areas in the macaque. *Nature Neuroscience, 9*, 1064–1070.

Gramfort, A., Luessi, M., Larson, E., Engemann, D. A., Strohmeier, D., Brodbeck, C., ... Hämäläinen M. S. (2014). MNE software for processing MEG and EEG data. *NeuroImage, 86*, 446–460.

Griffiths, J. D., Marslen-Wilson, W. D., Stamatakis, E., & Tyler, L. K. (2012). Functional organization of the neural language system: Dorsal and ventral pathways are critical for syntax. *Cerebral Cortex, 23*(1), 139–147.

Grindrod, C. M., Bilenko, N. Y., Myers, E. B., & Blumstein, S. E. (2008). The role of the left inferior frontal gyrus in implicit semantic competition and selection: An event-related fMRI study. *Brain Research, 1229*, 167–178.

Grosjean, F. (1980). Spoken word recognition processes and the gating paradigm. *Perception and Psychophysics, 28*, 267–283.

Hagoort, P. (2005). On Broca, brain, and binding: A new framework. *Trends in Cognitive Science, 9*, 416–423.

Hagoort, P. (2014). Nodes and networks in the neural architecture for language: Broca's region and beyond. *Current Opinion in Neurobiology, 28*, 136–141.

Hagoort, P., & Indefrey, P. (2014). The neurobiology of language beyond single words. *Annual Review of Neuroscience, 37*, 347–362.

Hagoort, P., Wassenaar, M., & Brown, C. (2003). Real-time semantic compensation in patients with agrammatic comprehension: Electrophysiological evidence for multiple-route plasticity. *Proceedings of the National Academy of Sciences, 100*, 4340–4345.

Hanna, J., & Pulvermüller, F. (2014). Neurophysiological evidence for whole form retrieval of complex derived words: A mismatch negativity study. *Frontiers in Human Neuroscience, 8*, 886.

Hauser, M. D., Chomsky, N., & Fitch, W. T. (2002). The faculty of language: What it is, who has it, and how did it evolve? *Science, 298*, 1569–1579.

Haxby, J. V., Gobbini, M. I., Furey, M. L., Ishai, A., Schouten, J. L., & Pietrini, P. (2001). Distributed and overlapping representations of faces and objects in ventral temporal cortex. *Science, 293*, 2425–2430.

Hickok, G., & Poeppel, D. (2007). The cortical organization of speech processing. *Nature Reviews Neuroscience, 8*, 393–402.

Holes, C. (1995). *Modern Arabic.* London: Longman.

Huang, Y-H. (2015). A cross-linguistic investigation of the neural mechanisms underlying linguistic combinatorial processing (unpublished doctoral dissertation). University of Cambridge, Department of Psychology.

Huang, Y-H., Fonteneau, E., Whiting, C., Su, L., Cai, Q., & Marslen-Wilson, W. D. (2014). *Spatiotemporal patterns of neural activity underlying morphological processing in a minimally inflected language: An EEG-MEG study in Mandarin.* Paper presented at 21st Annual Meeting of the Cognitive Neuroscience Society. Boston, MA.

Indefrey, P., & Cutler, A. (2004). Pre-lexical and lexical processing in listening. In M. S. Gazzaniga (Ed.), *The cognitive neurosciences* (3rd ed., pp. 759–774). Cambridge, MA: MIT Press.

Kaas, J., & Hackett, T.A. (2000). Subdivisions of auditory cortex and processing streams in primates. *Proceedings of the National Academy of Sciences, 97,* 11793–11799.

Klein, W., & Perdue, C. (1997). The basic variety (or: Couldn't natural languages be much simpler?). *Second Language Research, 13,* 301–347.

Klimovich-Gray, A., Bozic, M., & Marslen-Wilson, W. D. (2016). Domain-specific and domain-general processing in left perisylvian cortex: Evidence from Russian. *Journal of Cognitive Neuroscience, 29*(2), 382–397.

Klimovich-Gray, A., Tyler, L. K., Randall, B., Kocagoncu, E., Devereux, B. J., & Marslen-Wilson, W. D. (2019). Balancing prediction and sensory input in speech comprehension: The Spatiotemporal dynamics of word-recognition in context. *Journal of Neuroscience, 39*(3), 519–527.

Kocagoncu, E., Clarke, A., Devereux, B. J., & Tyler, L. K. (2017). Decoding the cortical dynamics of sound-meaning mapping. *Journal of Neuroscience, 37,* 1312–1319.

Kriegeskorte, N., Goebel, R., & Bandettini, P. (2006). Information-based functional brain mapping. *Proceedings of the National Academy of Sciences USA, 103,* 3863–3868.

Kriegeskorte, N., Mur, M., Ruff, D. A., Kiani, R., Bodurka, J., Esteky, H., ... Bandettini, P. A. (2008). Matching categorical object representations in inferior temporal cortex of man and monkey. *Neuron, 60*(6), 1126–1141.

Lau, E. F., Phillips, C., & Poeppel, D. (2008). A cortical network for semantics: (de)constructing the N400. *Nature Neuroscience Reviews, 9,* 920–933.

Leonard, M. K., & Chang, E. F. (2014). Dynamic speech representations in the human temporal lobe. *Trends in Cognitive Science, 18*(9), 472–479.

Li, C. N., & Thompson, S. A. (1981). *Mandarin Chinese: A functional reference grammar.* Berkeley: University of California Press.

Longe, O., Randall, B., Stamatakis, E. A., & Tyler, L. K. (2007). Grammatical categories in the brain: The role of morphological structure. *Cerebral Cortex, 17*(8), 1812–1820.

Marslen-Wilson, W. D. (1973). Linguistic structure and speech shadowing at very short latencies. *Nature, 244,* 522–523.

Marslen-Wilson, W. D. (1975). Sentence perception as an interactive parallel process. *Science, 189,* 226–228.

Marslen-Wilson, W. D. (1987). Functional parallelism in spoken word-recognition. *Cognition, 25,* 71–102.

Marslen-Wilson, W. D. (2007). Morphological processes in language comprehension. In G. Gaskell (Ed.), *Oxford handbook of psycholinguistics* (pp. 495–524). Oxford: Oxford University Press.

Marslen-Wilson, W. D., & Bozic, M. (2018). Dual neurobiological systems underlying language evolution: Inferring the ancestral state. *Current Opinion in Behavioral Sciences, 21,* 176–181.

Marslen-Wilson, W. D., Bozic, M., & Tyler, L. K. (2014). Morphological systems in their neurobiological contexts. In M. S. Gazzaniga & G. R. Mangun (Eds.), *The cognitive neurosciences* (5th ed., pp. 639–648). Cambridge, MA: MIT Press.

Marslen-Wilson, W. D., & Tyler, L. K. (1975). Processing structure of sentence perception. *Nature, 257,* 784–786.

Marslen-Wilson, W. D., & Tyler, L. K. (1980). The temporal structure of spoken language understanding. *Cognition, 8,* 1–71.

Marslen-Wilson, W. D., & Tyler, L. K. (1997). Dissociating types of mental computation. *Nature, 387,* 592.

Marslen-Wilson, W. D., & Tyler, L. K. (2007). Morphology, language and the brain: the decompositional substrate for language comprehension. *Philosophical Transactions of the Royal Society B: Biological Sciences, 362,* 823–836.

Marslen-Wilson, W. D., & Tyler, L. K. (2011). The communicative brain. *Research Horizons, 16,* 8–9.

Marslen-Wilson, W. D., Tyler, L. K., Waksler, R., & Older, L. (1994). Morphology and meaning in the English mental lexicon. *Psychological Review, 101,* 3–33.

Marslen-Wilson, W. D., & Welsh, A. (1978). Processing interactions and lexical access during word-recognition in continuous speech. *Cognitive Psychology, 10,* 29–63.

Matthews, P. H. (1991). *Morphology* (2nd ed.). Cambridge: Cambridge University Press.

McCarthy, J. J. (1981). A prosodic theory of non-concatenative morphology. *Linguistic Inquiry, 12,* 373–418.

McRae, K., Cree, G. S., Seidenberg, M. S., & McNorgan, C. (2005). Semantic feature production norms for a large set of living and nonliving things. *Behavior Research Methods, 37,* 547–559.

Mishkin, M., Ungerleider, L. G., & Macko, K. A. (1983). Object vision and spatial vision: Two cortical pathways. *Trends in Neurosciences, 6,* 414–417.

Neubert, F-X, Mars, R. B., Thomas. A. G., Sallet, J., & Rushworth, M. F. S. (2014). Comparison of human ventral frontal cortex areas for cognitive control and language with areas in monkey frontal cortex. *Neuron, 81,* 700–713.

Nili, H., Wingfield, C., Walther, A., Su, L., Marslen-Wilson, W. D., & Kriegeskorte, N. (2014). A toolbox for representational similarity analysis. *PLoS Computational Biology,* 10, e1003553.

Packard, J. L. (2000). *The morphology of Chinese: A linguistic and cognitive approach.* Cambridge: Cambridge University Press.

Papoutsi, M., Stamatakis, E. A., Griffiths, J. D., Marslen-Wilson, W. D., & Tyler, L. K. (2011). Is left fronto-temporal connectivity essential for syntax? Effective connectivity, tractography and performance in left-hemisphere damaged patients. *NeuroImage, 58,* 656–664.

Patterson, K., Nestor, P. J., & Rogers, T. T. (2007). Where do you know what you know? The representation of semantic knowledge in the human brain. *Nature Reviews Neuroscience, 8,* 976–987.

Poeppel, D. (2014). The neuroanatomic and neurophysiological infrastructure for speech and language. *Current Opinion in Neurobiology, 28,* 142–149.

Price, A. R., Peelle, J. E., Bonner, M. F., Grossman, M., & Hamilton, R. H. (2016). Causal evidence for a mechanism of semantic integration in the angular gyrus as revealed by high-definition transcranial direct current stimulation. *Journal of Neuroscience, 36,* 3829–3838.

Pulvermüller, F., Shtyrov, Y., & Ilmoniemi, I. J. (2003). Spatiotemporal dynamics of neural language processing: An MEG study using minimum-norm current estimates. *NeuroImage, 20,* 1020–1025.

Raposo, A., Moss, H. E., Stamatakis, E. A., & Tyler, L. K. (2006). Repetition suppression and semantic enhancement: An investigation of the neural correlates of priming. *Neuropsychologia, 44,* 2284–2295.

Rauschecker, J. P., & Scott, S. K. (2009). Maps and streams in the auditory cortex: nonhuman primates illuminate human speech processing. *Nature Neuroscience, 12,* 718–724.

Rauschecker, J. P., & Tian, B. (2000). Mechanisms and streams for processing of "what" and "where" in auditory cortex. *Proceedings of the National Academy of Sciences, 97,* 11800–11806.

Rilling, J. K., Glasser, M. F., Preuss, T. M., Ma, X., Zhao, T., Hu, X., & Behrens, T. E. J. (2008). The evolution of the arcuate fasciculus revealed with comparative DTI. *Nature Neuroscience, 11,* 426–428.

Rolheiser, T., Stamatakis, E. A., & Tyler, L. K. (2011). Dynamic processing in the human language system: Synergy between the arcuate fascicle and extreme capsule. *Journal of Neuroscience, 31*(47), 16949–16957.

Seghier, M. L. (2013). The angular gyrus: Multiple functions and multiple subdivisions. *Neuroscientist, 19,* 43–61.

Seyfarth, R. M., & Cheney, D. L. (2016). Precursors to language: Social cognition and pragmatic inference. *Psychonomic Bulletin and Review, 24*(1), 79–84.

Snijders, T. M., Vosse, T., Kempen, G., van Berkum, J. J., Petersson, K. M., & Hagoort, P. (2009). Retrieval and unification of syntactic structure in sentence comprehension: An fMRI study using word-category ambiguity. *Cerebral Cortex, 19,* 1493–1503.

Su, L., Fonteneau, E., Marslen-Wilson, W. D., & Kriegeskorte, N. (2012). Spatiotemporal searchlight representational similarity analysis in EMEG source space. *IEEE 2nd International Workshop on Pattern Recognition in Neuroimaging,* London, 97–100.

Su, L., Zulfiqar, I., Jamshed, F., Fonteneau, E., & Marslen-Wilson, W. D. (2014). Mapping tonotopic organization in human temporal cortex: Representational similarity analysis in EMEG source space. *Frontiers in Neuroscience, 8,* 368.

Szlachta, J., Bozic, M., Jelowicka, A., & Marslen-Wilson, W. D. (2012). Neurocognitive dimensions of lexical complexity in Polish. *Brain and Language, 121,* 219–225.

Tomasello, M. (2008). *Origins of human communication.* Cambridge, MA: MIT Press.

Turken, A. U., & Dronkers, N. F. (2011). The neural architecture of the language comprehension network: Converging evidence from lesion and connectivity analyses. *Frontiers in Systems Neuroscience, 5,* 1.

Tyler, L. K. (1984). The structure of the initial cohort: Evidence from gating. *Perception and Psychophysics, 36,* 417–427.

Tyler, L. K., Bright, P., Fletcher, P., & Stamatakis, E. A. (2004). Neural processing of nouns and verbs: The role of inflectional morphology. *Neuropsychologia, 42*(4), 512–523.

Tyler, L. K., Cheung, T. P., Devereux, B. J., & Clarke, A. (2013). Syntactic computations in the language network: Characterising dynamic network properties using representational similarity analysis. *Frontiers in Psychology, 4,* 271.

Tyler, L. K., & Marslen-Wilson, W. D. (1977). The on-line effects of semantic context on syntactic processing. *Journal of Verbal Learning and Verbal Behavior, 16,* 683–692.

Tyler, L. K., & Marslen-Wilson, W. (2008). Fronto-temporal brain systems supporting spoken language comprehension. *Philosophical Transactions of the Royal Society B: Biological Sciences, 363,* 1037–1054.

Tyler, L. K., Marslen-Wilson, W. D., Randall, B., Wright, P. Devereux, B. J., Zhuang, J., ... Stamatakis, E. A. (2011). Left inferior frontal cortex and syntax: Function, structure and behaviour in patients with left hemisphere damage. *Brain, 134,* 415–431.

Tyler, L. K., Stamatakis, E. A., Post, B., Randall, B., & Marslen-Wilson, W. D. (2005). Temporal and frontal systems in speech comprehension: An fMRI study of past tense processing. *Neuropsychologia, 43,* 1963–1974.

Tyler, L. K., Wright, P., Randall, B., Marslen-Wilson, W. D., & Stamatakis, E. A. (2010). Reorganization of syntactic processing following left-hemisphere brain damage: Does right-hemisphere activity preserve function? *Brain, 133,* 3396–3408.

Versteegh, K. (1997). *The Arabic language.* Edinburgh: Edinburgh University Press.

Whiting, C., Shtyrov, Y., & Marslen-Wilson, W. D. (2014). Real-time functional architecture of visual word recognition. *Journal of Cognitive Neuroscience, 27*(2), 246–265.

Wilson, B., Kikuchi, Y., Sun, L., Hunter, D., Dick, F., Smith, K., ... Petkov, C. I. (2015). Auditory sequence processing reveals evolutionarily conserved regions of frontal cortex in macaques and humans. *Nature Communications, 6,* 8901.1.

Wilson, B., Marslen-Wilson, W. D., & Petkov, C. I. (2017). Conserved sequence processing in primate frontal cortex. *Trends in Neurosciences, 40,* 72–82.

Wright, P., Stamatakis, E. A., & Tyler, L. K. (2012). Differentiating hemispheric contributions to syntax and semantics in patients with left-hemisphere lesions. *Journal of Neuroscience, 32,* 8149–8157.

Yang, Y-H., Marslen-Wilson, W. D., & Bozic, M. (2017). Syntactic complexity and frequency in the neurocognitive language system. *Journal of Cognitive Neuroscience, 29*(9), 1605–1620.

Zhou, X, & Marslen-Wilson, W. D. (1994). Words, morphemes, and syllables in the Chinese mental lexicon. *Language and Cognitive Processes, 9,* 393–423.

Zhou, X., & Marslen-Wilson, W. D. (1995). Morphological structure in the Chinese mental lexicon. *Language and Cognitive Processes, 10,* 545–601.

Zhuang, J., Tyler, L. K., Randall, B., Stamatakis, E. A., & Marslen-Wilson, W. D. (2014). Optimally efficient neural systems for processing spoken language. *Cerebral Cortex, 24*(4), 908–918.

31 A Cognitive Neuroscience Perspective on Language Comprehension in Context

JOS J. A. VAN BERKUM AND MANTE S. NIEUWLAND

1. Beyond the Code

Language is a great tool, an infinitely extendable discrete combinatorial coding system that has amplified the communicative capabilities of *Homo sapiens* far beyond that of other Great Apes or any other species on this planet. The beauty of this system has been documented by linguists in great detail, and, as evidenced by many chapters in this book, we are beginning to unravel how the human brain supports its acquisition and use. But what do we use it *for*? How does using the code mesh with, and amplify, whatever our particular animal species is up to *anyway* in life, as we navigate the physical and social world?

What the human brain shares with that of many other species is a capability to represent the external environment in great detail, such that we know what is around us, and what our options are—vast areas of the human neocortex are map-building areas (Damasio, 2010). What we also share with many other species is a set of affective control systems that make us care about the environment, such that we act on opportunities and threats in appropriate ways, and, if we don't do anything too stupid, learn from it along the way (Panksepp & Biven, 2012). Finally, like all species that can learn, we are forward-looking creatures, extrapolating from the here and now, and from what we have learned in the past, toward how the world might soon be. So, with virtually all of this in place well before language entered the scene, what did language add? And, given that all of those pre-existing capabilities and inclinations are still with us today and control everything we do, what are the implications for real-time language processing? In this chapter, we review some of our cognitive neuroscience research on how language comprehension unfolds in the context of our map-making, affective, and forward-looking nature.

The most obvious bonus of language is its referential power, the fact that it allows speakers to draw other people's attention to very precisely delineated aspects of the world, as such enriching their mental map of the here and now, as well as of possible, past, or future states of affairs, in infinitely detailed ways. Language helps us to convey that berries can be found in some particular spot down this and that track, but only in late summer, and that if we coordinate our attack well and come up from behind, the elephant on the left is probably our best target. And, as evidenced by examples like *I believe that this nation should commit itself to achieving the goal, before this decade is out, of landing a man on the moon and returning him safely to the earth*, language also allows us to draw other people's attention to our own mental representations. Because understanding reference is so central to the amplifying power of language, we start our review by examining how addressees (or overhearers) infer the speaker's referential intention, that is, relate apparently simple referential expressions such as *he*, *this nation*, or *itself* to elements of their mental representation of the real or imaginary environment.

Mapping what people refer to, and what they want to achieve by doing so, is only half of the story, though. What matters in the end is how all that relates to what addressees and overhearers *care about*. As argued extensively elsewhere (van Berkum, 2018, 2019a), language users are not dispassionate code-cracking and inference-generating information processors. We usually *care* about what people refer to and what their social intentions are. And, just like in other domains of mental life, it is these *affective* (emotional, evaluative) responses that typically move us (Damasio, 2010; Panksepp & Biven, 2012). That is, much of what makes language do real work, in other words, generate "perlocutionary effects" out there in the world, is how people feel about it, and what they think or do as a result. In the second part of this review, we discuss some of our cognitive neuroscience work on how the unfolding linguistic code makes contact with what people care about, and we introduce a model for thinking about the processes involved.

Understanding language comprehension in context requires an analysis of how the unfolding linguistic code rapidly makes contact with representations of the referential context, and of how the code can in the end affectively move comprehenders. However, language comprehension is not just about how we *react* to bits of code. Our vast capacity to learn about the world, and about how we communicate about that world, also allows us to *anticipate* upcoming communication (van Berkum, 2013). In the third part of our review, we briefly discuss some of our research on such anticipation.

2. *The Referential Context for Language*

The fact that language is about the world and refers to the world is what makes language so incredibly useful. The referential link between language and the world is perhaps the single most discussed issue throughout the philosophy of language (Martinich, 1985; Recanati, 1993). In contemporary cognitive science and psycholinguistics, the understanding of referential expressions, or *reference resolution*, also takes center stage by virtue of its importance in language acquisition and in the computation of speaker meaning (e.g., Gibson & Pearlmutter, 2011; Trueswell & Tanenhaus, 2005). The need to understand reference develops as early as language development itself, when children learn the meaning of words through understanding the referential intention of a speaker (Bloom, 2000; Tomasello, 2009). Already at a very young age (one to two years), children are sensitive to cues about referential intent such as the speaker's gaze, pointing, and prosody, and they use these cues to constrain the meaning of novel or ambiguous words (Bloom, 2000). And as early as age five, children can take into account the private perspective of a speaker when interpreting the speaker's referential expressions (Nadig & Sedivy, 2002), demonstrating a highly complex cognitive skill associated with Theory of Mind (Frith & Frith, 2005). Perhaps unsurprisingly, then, adults typically resolve reference without any conscious effort, whether reference is made to a nonlinguistic context (e.g., an object in the physical surroundings or real-world knowledge) or to a linguistic context (e.g., something that was mentioned before). This section zooms in on the latter, *anaphoric reference* to a linguistic antecedent.

2.1. Understanding Anaphoric Reference Anaphoric reference is one of the pillars of discourse cohesion (Sanford & Garrod, 1989), the stuff that makes sentences part of the same discourse. Referring to a previously mentioned concept is also the most basic way to stay on topic during conversation. How do people understand anaphora? In psychological theories of anaphora comprehension, comprehenders reactivate the intended antecedent from their memory representation of the discourse and subsequently integrate this antecedent information into the overall representation of the narrated event (e.g., Gernsbacher, 1989; Gerrig & McKoon, 1998; Myers & O'Brien, 1998; Sanford & Garrod, 1989). Antecedent reactivation is thus considered to be a memory-based process, in which the anaphor serves as a memory cue to reactivate or access the antecedent (e.g., Lewis & Vasishth, 2005; see also Martin, 2016). It follows that ease of anaphor resolution depends on the extent to which the antecedent is already activated before the anaphor appears and on how strongly the intended antecedent is uniquely cued by the semantic and syntactic content of the anaphor. Antecedents that are in the current focus of the discourse, and therefore relatively activated, are easier to access than those that are not. Focused antecedents can then be referred to with reduced-form anaphora like pronouns (e.g., *The construction worker smiled when he got his paycheck*). Pronouns may be very efficient referential devices, but their shallow content makes them particularly prone to ambiguity in interpretation (e.g., *The construction worker smiled at the foreman when he got his paycheck*, where *he* could refer to either worker). Referential ambiguity or misinterpretation may be a specific case where regular interpretation hampers, but this situation is not uncommon in everyday language interactions. Temporary ambiguity as in *That lady with the blue hat*, referring to one lady among other ladies (and thus disambiguated at *blue*), is particularly common. Temporary ambiguity simply arises because speakers do not always avoid ambiguity by mentioning the disambiguating information first (e.g., *that blue-hatted lady*). Having to briefly deal with ambiguity may just be that price to pay, at least occasionally, for having a referential communication system that works quickly as well as very flexibly.

Fortunately for language researchers, referential ambiguity offers a useful paradigm to investigate the time course and neural machinery associated with anaphor comprehension (for a review, see Nieuwland & van Berkum, 2008b; van Berkum, Koornneef, Otten & Nieuwland, 2007). This typically involves a comparison of behavioral or brain responses to referentially ambiguous expressions with those to nonambiguous expressions. One prominent example is the effect of referential ambiguity on people's looking patterns. This can be demonstrated with the *visual world paradigm* (see Trueswell & Tanenhaus, 2005), where participants look at objects in a display while receiving spoken instructions to touch or move those objects. Studies with this paradigm show that participants evenly distribute their fixations across potential referents for ambiguous expressions or quickly fixate on a single referent for nonambiguous expressions, all within 200 to 300 ms after the acoustic onset of a referential expression. A similar effect onset of referential ambiguity is observed in event-related brain potentials as measured by electroencephalography (EEG) recordings during language comprehension. Compared to nonambiguous anaphors, ambiguous anaphors elicit a sustained, negative voltage deflection with a frontocentral maximum that starts at about 200 to 300 ms after anaphor onset, called the *Nref effect* (van Berkum et al., 2007). Findings from a series of experiments suggest that this effect indexes a cognitively effortful attempt to deal with a referential problem and to establish a referentially coherent interpretation. The Nref effect has been observed for written and spoken noun phrase anaphora such as *girl* in a story context that contains two girls (van Berkum, Brown, & Hagoort, 1999; van Berkum, Brown, Hagoort, & Zwitserlood, 2003), for written and spoken pronouns such as *he* in a sentence context with two male characters (Nieuwland & van Berkum, 2006a; van Berkum, Zwitserlood, Bastiaansen, Brown, & Hagoort, 2004), as well as for elliptical expressions such as *another* (Martin, Nieuwland & Carreiras, 2012, 2014). The Nref effect appears when the context contains two equally plausible antecedents for an anaphor (Nieuwland, Otten & van Berkum, 2007; Nieuwland & van Berkum, 2006a) or when an anaphoric pronoun refers to an unknown referent (Nieuwland, 2014; see figure 31.1, left graph). The Nref effect is not elicited when only one of two antecedents is a plausible referent, in which case there is no genuine ambiguity, or when the anaphoric noun phrase itself is a semantically anomalous sentence continuation (Nieuwland & van Berkum, 2008a).

These observations demonstrate some key aspects of anaphor comprehension. The Nref effect onset, the time point at which event-related potentials (ERPs) elicited by ambiguous and nonambiguous words first start to diverge, shows that anaphor comprehension is rapidly impacted by the presence of suitable antecedents in the context, well before the onset of the subsequent word. Moreover, the spoken language results suggest that this effect of referential context starts off well before the complete anaphor has been heard (van Berkum, Brown, & Hagoort, 1999; Nieuwland, Otten, & van Berkum, 2007), after having heard perhaps only one or two phonemes. This shows anaphor comprehension to be a highly incremental process, meaning that people use relevant information as quickly and efficiently as possible to establish the intended referent. In terms of time course, this puts anaphor comprehension on a par with other influences on the computation of speaker meaning (van Berkum, 2009, 2012), such as whether an expression is plausible with respect to what was said before (van Berkum, van den Brink, Tesink, Kos, & Hagoort, 2008), who said it (van Berkum, Zwitserlood, et al., 2003), or knowledge about the world (Hagoort, Hald, Bastiaansen, & Petersson, 2004).

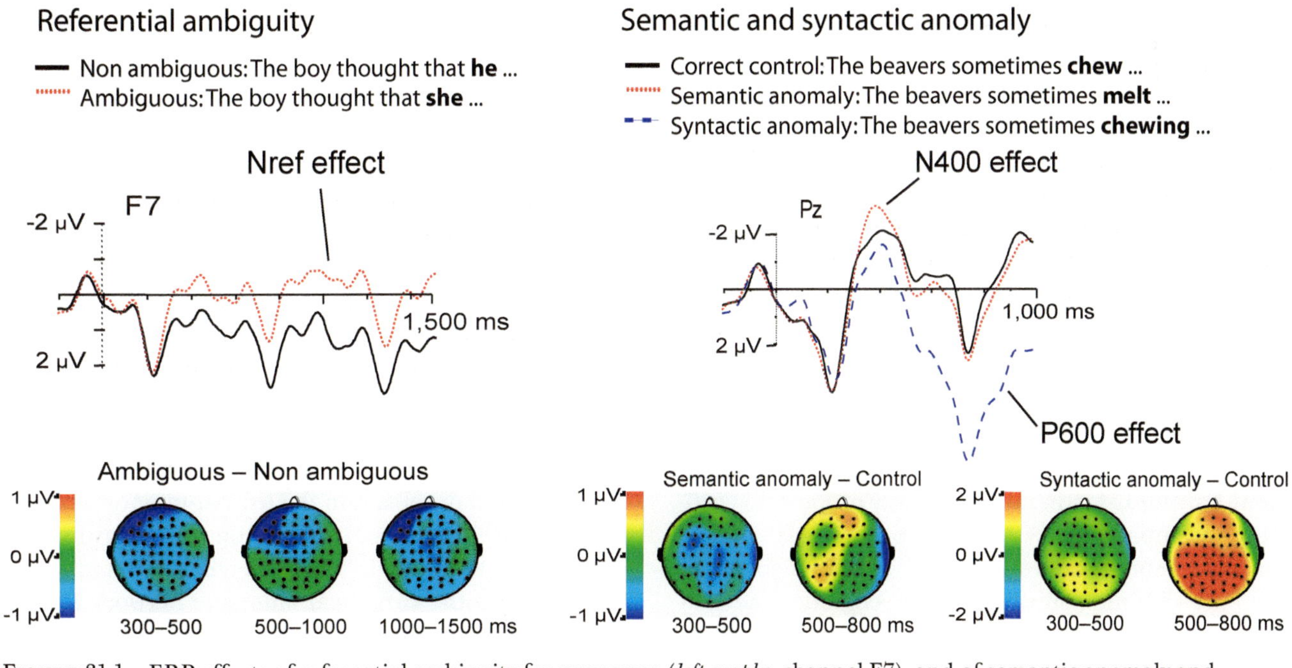

FIGURE 31.1 ERP effects of referential ambiguity for pronouns (*left graphs*, channel F7), and of semantic anomaly and syntactic anomaly (*right graphs*, channel Pz). This figure is adapted from Nieuwland (2014).

Irrespective of its time course, the Nref effect morphology and scalp distribution suggests that referential ambiguity entails processing costs that are qualitatively distinct from those imposed by semantically or syntactically problematic words (see figure 31.1, right graph), which elicit N400 and P600 effects, respectively. Referential ambiguity, unlike semantic ambiguity, does not impact the retrieval of lexical-semantic information associated with the anaphor, as arguably would be reflected in N400 ERP modulation (Nieuwland & van Berkum, 2008a). Building on memory-based accounts of language comprehension, referential ambiguity can be construed as a case of memory retrieval interference from overlapping content (e.g., Lewis & Vasishth, 2005). The observed ERP effects only point toward a qualitative distinction between Nref effect activity and N400/P600 activity, which is consistent with different sets of neural generators as has been revealed through functional MRI (fMRI) (Nieuwland, Petersson, & van Berkum, 2007).

Thus far, this section has focused strongly on referential ambiguity and on what its processing consequences may say about anaphor comprehension. But nonambiguous, *unproblematic* anaphor comprehension is arguably the default situation in natural language use. In the next section, we discuss a proposal for how this may work in the brain.

2.2. A Nascent Corticohippocampal Theory of Reference While the neurobiology of language has made major advances in recent years (Hickok & Small, 2015), it is still relatively unknown how the brain achieves the important feat of reference. Extant neurobiological accounts of sentence-level language comprehension primarily focus on syntactic and semantic processing in the traditional temporal-frontal language network (e.g., Friederici, 2012; Hagoort & Indefrey, 2014). Such accounts localize syntactic and semantic combinatorial operations in left prefrontal cortex (cf. Bemis & Pylkkänen, 2011), which use syntactic and lexical-semantic representations that are stored in temporal cortex and the angular gyrus (parietal cortex). To date, these accounts have had very little to say about reference.

As a first step toward a fully articulated neurobiological theory of language that captures reference, Nieuwland and Martin (2017) recently proposed a corticohippocampal theory of reference. This coarse-grained theory bridges psycholinguistic theories of reference with the neurobiology of recognition memory and of language. It builds on the broad consensus about antecedent reactivation based on the content features of the anaphor (e.g., Myers & O'Brien, 1998; Sanford & Garrod, 1989), such that a referential link can be established between multiple instantiations of the same concept despite linguistic form differences. The antecedent reactivation process equates to, or, minimally, involves recognition memory function, which allows people to distinguish new and old information. In recognition memory research, successful retrieval (i.e., recognition) is associated with increased activity in and connectivity between the medial temporal lobe and posterior parietal cortex (Aggleton & Brown, 2006; Eichenbaum, Yonelinas, & Ranganath, 2007; Wagner, Shannon, Kahn, & Buckner, 2005), which reflects the reinstatement of previously encoded items (e.g., Rissman & Wagner, 2012; Staresina, Henson, Kriegeskorte, & Alink, 2012). Building on these findings, the core claim of the corticohippocampal theory of reference is that anaphor comprehension draws on the interaction between the brain's recognition memory network (medial temporal lobe, including the hippocampus, and posterior parietal cortex) and the canonical frontal-temporal language network, with the former being chiefly responsible for antecedent reactivation and the latter primarily responsible for bringing the unfolding sentence representation to bear on situation model construction.

Available support for the corticohippocampal theory of reference comes from neuropsychological studies and neuroimaging. Hippocampal amnesia patients, who suffer from episodic memory dysfunction due to a hippocampal lesion, show impairments in pronoun comprehension (Kurczek, Brown-Schmidt, & Duff, 2013) and sometimes produce indefinite reference when definite reference is more appropriate (e.g., *a camel* where one would use *the camel* to refer to a camel that had been discussed before; Duff, Gupta, Hengst, Tranel, & Cohen, 2011). The hippocampus also shows increased blood oxygenation level–dependent fMRI activity for pronouns that match a given, "old" antecedent in the sentence context compared to pronouns that refer to an unknown, "new" antecedent (Nieuwland, Petersson, & van Berkum, 2007; see figure 31.2, left graph). This comparison is also associated with increased gamma-band neural oscillatory power, as measured in EEG activity (Nieuwland & Martin, 2017; van Berkum et al., 2004; see figure 31.2, right graph). Nieuwland and Martin showed that these increases in power take place in posterior parietal cortex at ~400 to 600 ms after pronoun onset and in the frontal-temporal cortex at ~500 to 1000 ms after pronoun onset, suggesting the involvement of the brain's parietal recognition memory network[1] and the canonical language network in successful reference resolution.

Several important predictions from the corticohippocampal theory of reference remain to be tested. For

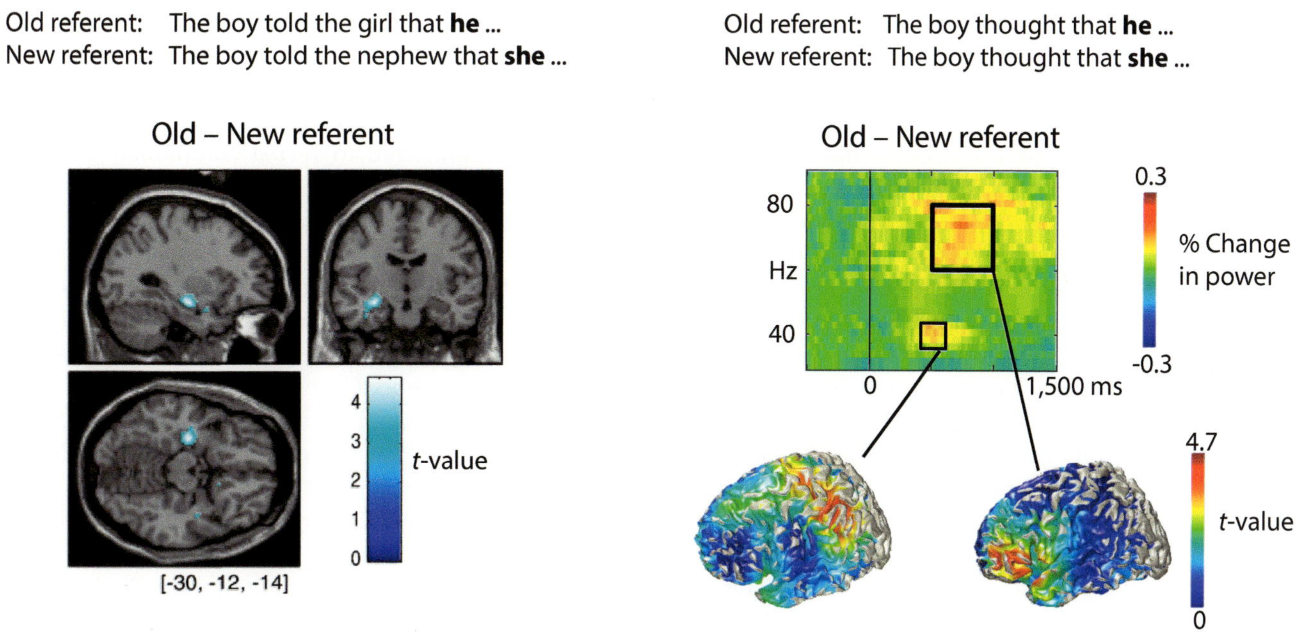

FIGURE 31.2 Relative increase in hippocampal blood oxygenation level–dependent fMRI activity for old referents compared to new referents (*left graph*, adapted from Nieuwland, Petersson, & van Berkum, 2007). Stronger gamma-band oscillatory EEG activity (measured as relative change in power) for old referents compared to new referents (*right graph*, adapted from Nieuwland & Martin, 2017). An increase at ~40 Hz at ~400 to 600 ms after pronoun onset was localized to left posterior parietal cortex, and an increase at ~60 to 80 Hz at ~500 to 1000 ms after pronoun onset was localized to left inferior frontal cortex.

example, successful reference resolution should lead to increased connectivity between the recognition memory network and language network, but activity in the recognition memory network may be of shorter duration or may even take place before activity in the language network, reflecting the processing phases of antecedent reactivation and sentence-level integration (e.g., Sanford & Garrod, 1989). In addition, the respective contributions of the hippocampus and posterior parietal cortex to antecedent reactivation need to be determined. Perhaps the initial binding operations between anaphor and antecedent take place in the hippocampus (e.g., Duff & Brown-Schmidt, 2012), but full reactivation, by bringing antecedent information back into the focus of attention, takes place in posterior parietal cortex (Wagner et al., 2005).

The corticohippocampal theory of reference could be a step in the right direction, but it only really covers anaphoric reference. Other aspects of reference are equally important in order to understand how the brain deals with language in context. Another common form of reference, nonanaphoric reference to things that are perceptually available, presumably does not engage recognition memory, but, instead may require multimodal integration of information from linguistic and perceptual modalities (e.g., Brodbeck, Gwilliams, & Pylkkänen, 2016; Willems, Özyürek, & Hagoort, 2008; see Martin, 2016, for a sensory integration-based process model). Moreover, neuroimaging research is only just starting to reveal the neurocognitive mechanisms for comprehension of reference to hypothetical worlds as opposed to real-life events (e.g., Menenti, Petersson, Scheeringa, & Hagoort, 2009; Nieuwland, 2012). For a full understanding of the various linguistic phenomena that are considered to be *referential* by nature, there are more than enough challenges ahead.

3. *The Affective Context for Language*

As we have seen, language greatly amplifies the mapmaking capabilities of our species, via arbitrarily precise reference to the world out there, and, as part of that world, our beliefs or imaginations about what is, was, or might be. As revealed by careful analyses in pragmatics (Tomasello, 2008; Clark, 1996), however, drawing attention to a particular referent or state of affairs is often just the beginning: communicators use language to inform, but also to move people around or share feelings. Saying things like *I forgot to buy cat food* to one's spouse upon returning home from the supermarket might be intended as a purely informative and helpful reference to some state of affairs in the world, but the speaker's intention in such cases is usually to share frustration and/or to elegantly get somebody else in the household to go back and buy a few cans. Language is used in the context of social intentions, as a means of achieving those intentions.

In a way, inferring the social intentions behind an utterance like *I forgot to buy cat food* is just another instance of mapping the world: depending on context, the listener not only learns about the speaker's shopping intention, its incomplete realization, and the implied current shortage of cat food, but, via additional inferences, also that the speaker would like the addressee to go to the store and buy some. However, whether all this additional information will have any effect on the listener beyond enriching his mental map of the world with somebody else's referential and social intentions depends on something else. Research on emotion and associated phenomena (evaluations, preferences, moods) has revealed that map making only takes us that far, and that in order to be moved, what is relevant is how the result *relates to our interests* (Damasio, 2010; Frijda, 2008; Lazarus, 1991; Panksepp & Biven, 2012; Scherer, 2005; Zajonc, 1980; see van Berkum, 2018, 2019a, 2019b, for language-relevant reviews). Emotions, evaluations, preferences and moods are *motive states*, urging or nudging us to act on entities or events out there in the world in certain ways, all because of how those entities or events *relate to our interests.* For example, because I care for my cat, the speaker's indirect reference to the cat food shortage may well by itself prompt me to go to the store, and if not, the fact that I care about my spouse's desires, or about being a cooperative partner, may well send me there anyway. I need not be conscious of such interests, nor of the emotional responses and associated motive states they induce. But without them, an utterance like *I forgot to buy cat food* would most likely leave me inert.

3.1. Understanding Affective Language People care about lots of things and they can talk about all of it. At the referential level, therefore, the possibilities for language that is affectively relevant are already infinite. However, because we are a highly social species, much of what we in fact care about is about how conspecifics relate to us and others, what they did or did not do, and whether they did the right thing (Greene, 2014). The interests at stake here are high. For a species heavily invested in cooperation, for example, being respected is critical: our public face (Goffman, 1959) determines whether people will trust and select us for collaboration, and, as in other primate species, respect also guarantees a comfortable position in social hierarchies, with access to resources.

These interests make for a context in which a canonically emotion-relevant piece of language, the verbal insult, can hit hard, to such an extent that we can see traces of this even in laboratory studies. In a recent EEG study, for example, Otten, Mann, van Berkum, and Jonas (2016) had people read insults such as "You are such a loser" or "Your opinions are too ridiculous for words." Compared to verbal compliment controls such as "You are so nice" or "Your opinion is often very interesting and illuminating," critical words in verbal insults elicited a Late Positive Potential (LPP) effect between 600 and 900 ms, a globally distributed positivity that is typically observed with emotional stimuli, and that is believed to index emotion-induced enhanced processing (Cacioppo, Crites, Berntson, & Coles, 1993; Sabatinelli, Keil, Frank, & Lang, 2013; see Hajcak, MacNamara, & Olvet, 2010, for review). Insulting words also elicited a negativity at ~300 to 400 ms, which, if interpreted as an N400 effect,[2] suggests that the insult was at some level unexpected and as such interfered with context-sensitive lexical retrieval (Kutas & Federmeier, 2011; van Berkum, 2009). Interestingly, when the insulting language was accompanied with cues to *public* humiliation (including recorded laughter and the outline of a crowd), the insult-induced LPP effect became considerably stronger and emerged much earlier, at some 300 ms, whereas the insult-induced negativity disappeared. We take this increase in the LPP to reflect that, because of their connection to public face and one's social hierarchy position, insults are usually much more threatening when delivered publically. In line with other observations (e.g., van Berkum, Holleman, Nieuwland, Otten, & Murre, 2009) of potential N400-LPP overlap, an increased LPP effect due to cues to public humiliation may perhaps have eliminated the N400-like negativity at ~300 to 400 ms.

In another recent EEG study (Struiksma, De Mulder, & van Berkum, 2019), we used a repetition design to explore whether people adapt to insults, and if so, in what way. Participants were exposed to massively repeated verbal insults such as "___ is a bitch" or "___ is ugly," presented in blocks that were mixed with blocks of linguistically matched compliments, like "___ is a sweetheart" or "___ is beautiful." Relative to compliments, insulting words elicited the same two effects as observed by Otten et al. (2016): an LPP effect, preceded by an enhanced N400-like negativity. In addition, however, insulting words elicited a very early modulation of the P2, indicative of enhanced attention. In contrast to the LPP and N400-like effects, the P2 effect was extremely robust to repetition, and also emerged when the participant's name was replaced by that of another person. Interestingly, post hoc analyses revealed that these effects hinged on those insults in the item set that had used a taboo word (e.g., *bitch*, but not, e.g., *ugly*). Because insults with taboo words were randomly intermixed with other insults, this also reveals that the ERP responses observed are not due to block-based anticipation. We suspect that what is happening here is that such swearwords very rapidly signal a taboo transgression, possibly as soon as they are retrieved from long-term memory, and that this is the source of the very rapid and "victim-insensitive" P2 effect.

This brings us to another domain where the interests guarded by emotion are high: morality. There is growing awareness that our species is (at least partly) moral by nature rather than "by culture," with morality and the associated emotions considered as biological adaptations that foster within-group collaboration and between-group competition (Greene, 2014; Haidt, 2012). If, as the relevant accounts hold, our perception of conspecifics (and ourselves) is intrinsically moral, language that touches on morality should have very rapid effects in processing (much like other forms of "moral perception"; Gantman & van Bavel, 2015). The early taboo-conditioned effects reported by Struiksma et al. (2019) are compatible with this: the use of an interpersonally violent taboo word seems to attract attention much like a slap in the face would do (Irvine, 2013), in a robust way that does not adapt with repetition, and is also very salient if *somebody else* is the victim.

Research on other types of morally loaded language also reveal a strong and rapid sensitivity to morally objectionable language. In one EEG study (van Berkum et al., 2009), participants with radically different (strict Christian or non-Christian) moral backgrounds read attitude survey statements that did or did not clash with their value system, such as "I think abortion is an acceptable/unacceptable course of action." Relative to value-consistent words, words that rendered the unfolding statement morally objectionable for the participant at hand (e.g., in the preceding example, "acceptable" for a strict Christian) elicited an early positivity at ~200 to 250 ms, a small centroparietally distributed N400 effect peaking at 400 ms, and an increased LPP between 500 and 650 ms, with the offending word defined by the specific moral background (see figure 31.3). Although they do not always occur together in the same study, each of these three morally induced EEG effects has been replicated, partly by other labs, and in one case also outside of the attitude survey context (Hundrieser & Stahl, 2016; Leuthold, Kunkel, Mackenzie, & Filik, 2015; Struiksma, Pantazi, de Mulder, & van Berkum, 2013).

Furthermore, we literally frown on language that describes morally objectionable behavior. In recent work ('t Hart, Struiksma, van Boxtel, & van Berkum, 2018; 2019), we have turned to facial electromyography to assess subtle—and not necessarily visible—activity of the frowning muscle (corrugator supercilli), as an index of language-induced moral indignation. In two studies, participants read short narratives where protagonists behaved in morally objectionable or prosocial ways (e.g., deliberately accelerating one's car through a

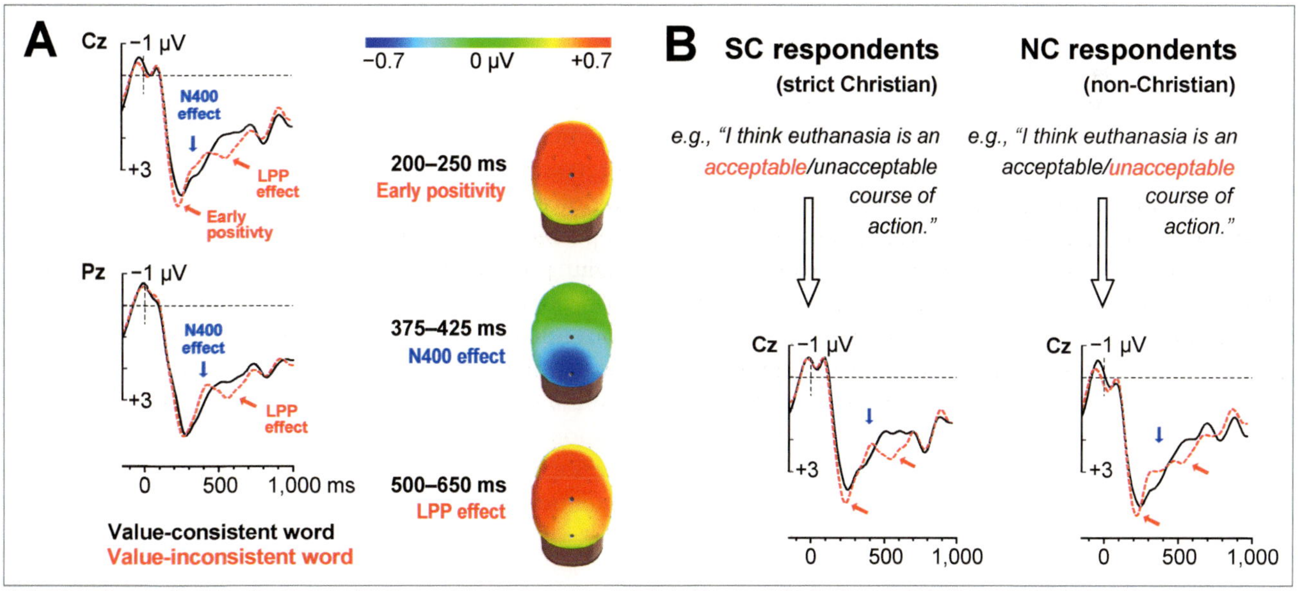

FIGURE 31.3 Triphasic response to morally objectionable language. (A) ERPs to value-consistent and value-inconsistent critical words, pooled across two groups of respondents, with scalp distributions for three differential effects. (B) ERPs and example item for two groups separately. Value-inconsistency is rapidly detected (early positivity), is conceptually unexpected (N400 effect), and induces additional processing (LPP effect). Scalp distributions are rendered on a three-dimensional head model (with Cz and Pz marked). See van Berkum et al. (2009) for details.

puddle to soak a pedestrian or deliberately slowing down instead). Relative to descriptions of prosocial behavior, descriptions of morally *objectionable* behavior induced a strong activation of the corrugator, which began to emerge at about 1 s into the offending passage. Furthermore, good or bad events that subsequently befell these protagonists also induced rapid differential corrugator responses, in ways that are again compatible with rapid moral (in this case fairness-based) evaluation.

In all, our findings indicate that as soon as unfolding language relates to strong interests of ours as affective creatures with things at stake, clear processing consequences emerge, in EEG as well as facial electromyography. As already mentioned, we believe that these effects at least partly reflect the response of our affective systems, whose job it is to watch over our interests and evaluate the environment accordingly. Those systems, and the interests they serve, provide the context for any bit of language coming in.

3.2. A Model for Affective Language Comprehension The findings discussed merely sample the neurocognitive exploration of how language moves people. For example, there is neurocognitive research on emotional prosody (e.g., Heilman, Bowers, Speedie, & Coslett, 1984; Pell, 1999), on the processing of face-saving indirectness and irony (e.g., Bašnákova, van Berkum, Weber, & Hagoort, 2015; Bašnákova, Weber, Petersson, van Berkum, & Hagoort, 2014; Spotorno, Koun, Prado, van der Henst, & Noveck, 2012), and on the impact of strongly valenced words and phrases (e.g., Citron, 2012; Fischler & Bradley, 2006; Foroni & Semin, 2009; Holt, Lynn, & Kuperberg, 2009; Wang & Bastiaansen, 2014). What has been missing, though, is a general perspective on how to conceive of these various effects—in line with general biases in the language sciences, extant models of language processing are fiercely nonaffective.

The *Affective Language Comprehension* (ALC) model (van Berkum, 2018, 2019a) depicted in figure 31.4 is a first attempt to remedy this, in a way that goes beyond the crude idea of "emotion words" and that respects the complexity of language. Building on sentence-oriented as well as more generic sentence-oriented and gesture-oriented analyses of human communication (e.g., Jackendoff, 2007; Tomasello, 2008), the model combines standard psycholinguistic and experimental pragmatic conceptions of what is needed to understand language with knowledge of how emotion can be triggered, consciously or unconsciously. The result is an explicit framework of processes and the associated representations, ranging from lexical retrieval and phrase-level linguistic processing to recovering the speaker's stance, referential intention, and social intention, plus whatever else the speaker did not intend but can be inferred anyway. Importantly, the ALC model predicts that an utterance can generate emotional responses in the addressee (or, for that matter, an overhearer) at *all levels of analysis*, from the ultimately relevant construal of the speaker's social intention all the way "down" to the level where a sign's meaning is retrieved from the mental lexicon. As such, the ALC model provides us with a detailed map of where and how the "perlocutionary effects" of an utterance arise, and a context within which we can also discuss how various phenomena involving rhetoric and persuasion (e.g., framing) might come about, as language is processed in real time. Along the way, the model provides a principled account of different forms of *word valence* (see van Berkum, 2018).

To give just one example related to the Struiksma et al. (2019) study, consider the potential emotional impact of hearing *you are a bitch/asshole!* uttered toward you in a way that is clearly aggressive. At the level of the social move, the speaker overtly expresses contempt for you, a clear trigger for emotion. However, beyond this obvious observation, the ALC model makes explicit various other potentially evocative ingredients that emerge in the language processing stream. One is the speaker's *referential* intention: although the situation described by the speaker (you being a bitch or asshole) need not be a correct characterization of the state of affairs in the world, *imagining* this state as an unavoidable part of language processing may nevertheless *still* induce an affective response (this is presumably part of why "benign," "playful" insults can still sting). Another additional ingredient involves experience-based emotional connotations of particular words or other signs. If a word like *bitch* has often been used to implement aggressive social moves or to describe very negative situations, or has often been witnessed to elicit strong negative responses in others, simple "Pavlovian" conditioning will lay down negative affective connotations as part of the memory trace for that particular word. As with other emotionally conditioned stimuli (e.g., a light that sufficiently reliably predicts a shock), the *mere use of the word* can then evoke a rapid automatic affective response, independent of the precise sentence-level message, the stance, the referential, and the social intention of the speaker.

4. Context-Dependent Anticipation

We have reviewed our cognitive neuroscience work on how incoming bits of linguistic code are mapped onto available referents and on how they are related to the comprehender's conscious or unconscious interests.

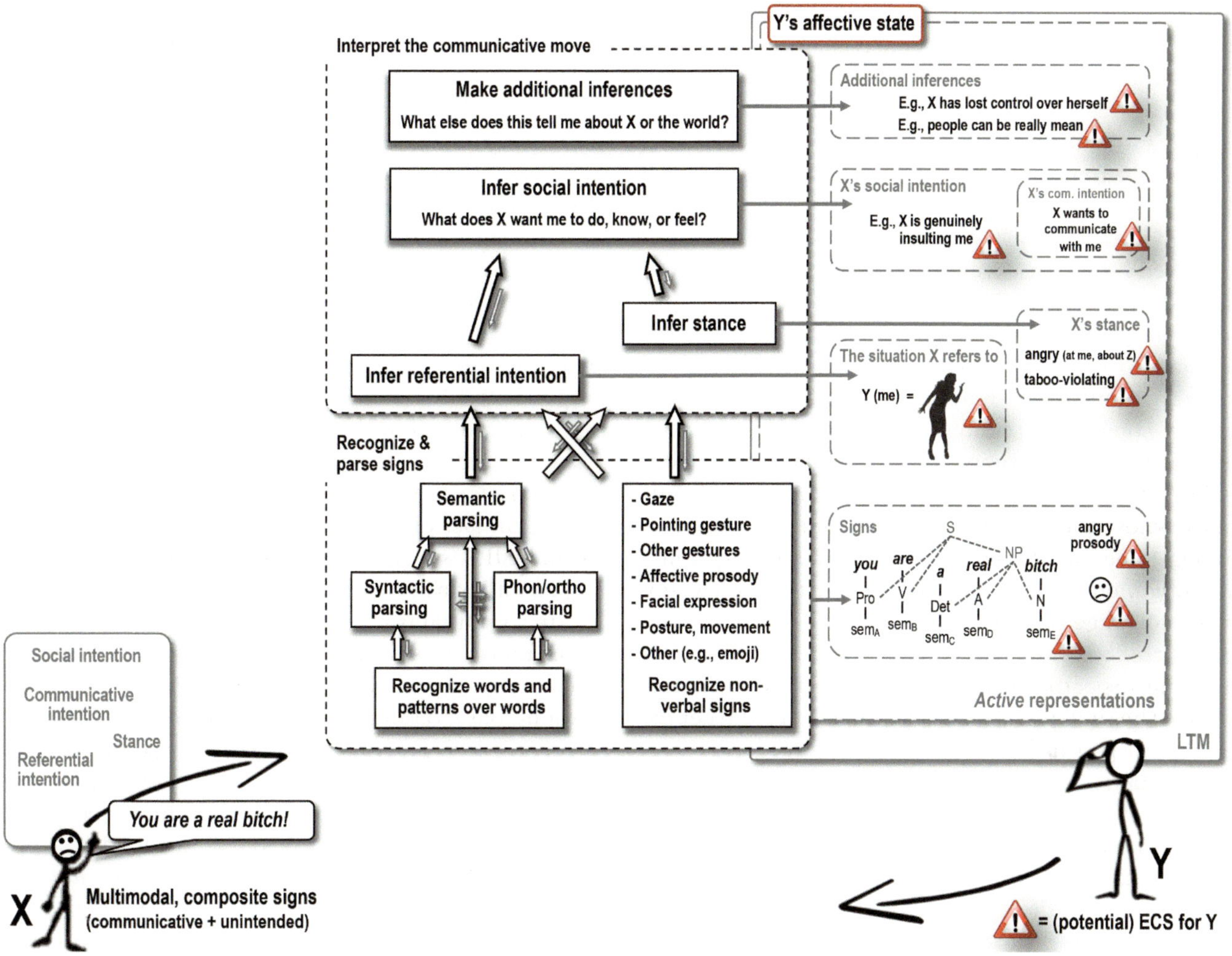

Figure 31.4 The ALC model. Mental processes and the associated retrieved or computed representations are expanded for addressee Y only. Y's computational processes draw upon (and add to) long-term memory (LTM) traces and involve currently active dynamic representations that reflect what is currently retrieved from LTM, composed from elements thereof and/or inferred from context, in response to the current communicative move. Y's active representations can be conscious or unconscious. ECS = emotionally competent stimulus; Phon/ortho parsing = phonological/orthographic parsing; X's com. intention = X's communicative intention. See van Berkum (2018, 2019a) for explanation. Adapted from van Berkum (2018, fig. 28.2) and van Berkum (2019a, fig. 29.1).

But language processing is not simply reactive. As argued elsewhere (van Berkum, 2013), people can and will have expectations about what speakers or authors will do at each of the levels relevant to language comprehension: the social intention, the referential intention, and the code used to realize them. In addition, as delineated in van Berkum (2009) in the context of the *Multiple-cause Intensified Retrieval* account of the N400, such expectations can be based on a wide variety of information sources, including, for example, associatively or semantically related prime words, scenario-based knowledge about the world activated by one or more words in the preceding text, a mental model of the situation being discussed, the immediately present physical situation, and such things as the discourse register and genre, the identity of the speaker, and the social goals believed to be pursued by him or her. Furthermore, everyday experience suggests that one's expectations can also be shaped by how we feel about what has been said before. With unfolding language, context is a very rich thing.

One early EEG study that illustrated the prediction of *specific upcoming bits of code* (van Berkum, Brown, Zwitserlood, Kooijman, & Hagoort, 2005) probed for traces of word anticipation *before* the anticipated word came along (see also DeLong, Urbach & Kutas, 2005; Wicha, Moreno & Kutas, 2004; but see Nieuwland et al., 2018, for a failure to replicate the effect reported by DeLong et al., 2005). Participants listened to (Dutch) mini-stories such as "The burglar had no trouble locating the secret family safe. Of course, it was situated behind a ___," which in a paper-and-pencil cloze test were predominantly completed with one particular critical noun (in this case, *painting*, the Dutch translation of which is a neuter-gender word). To test whether such discourse-based lexical prediction would also occur "on-line" as part of real-time language comprehension, the EEG participants at this point first heard a gender-inflected adjective, whose syntactic gender either agreed with the anticipated noun, as in (a) "$\text{big}_{\textbf{neu}}$ but rather unobtrusive $\text{painting}_{\textbf{neu}}$" or did at that point not agree with this expected noun, as in (b) "$\text{big}_{\textbf{com}}$ but rather unobtrusive $\text{bookcase}_{\textbf{com}}$." Relative to the gender-congruent prenominal adjective in (a), the gender-incongruent adjective in (b) elicited a small but reliable ERP effect right at the inflection. Because this prediction effect hinged on the syntactic gender of an expected but not yet presented noun ($\textit{painting}_{\textbf{neu}}$ in the example), and because the effect collapsed when the prior context sentence was removed, it suggested that discourse-level information can indeed lead people to anticipate specific upcoming words as the text unfolds. Follow-up experiments with the same gender-sensitive paradigm confirmed that such discourse-based lexical predictions cannot be reduced to the effects of simple or convergent lexical priming (Otten & van Berkum, 2008, 2009; Otten, Nieuwland, & van Berkum, 2007).[3]

Evidence for the anticipation of *upcoming reference* comes from an ERP study in which we examined the use of verb-based implicit causality information (van Berkum et al., 2007). When asked to complete a sentence fragment such as "David praised Linda because," readers and listeners will be inclined to continue the sentence with something about Linda, for example, "because she had done well." However, after "David apologized to Linda because," people tend to continue with something about *David* instead. In *person-1 verb-ed person-2 because* constructions, interpersonal verbs like *praise* and *apologize* thus supply information about whose behavior or state is the more likely immediate cause of the event at hand. Because this information is conveyed implicitly as part of the meaning of the interpersonal verb, it is usually referred to as *implicit causality*. In our experiment, we tested how rapidly readers were using this heuristic information as a context for anticipation, this by continuing the sentence with a bias-inconsistent pronoun ("David praised Linda because he"), and comparing the processing at this pronoun to its bias-consistent control ("Linda praised David because he"). Bias-inconsistent pronouns elicited a P600 effect, suggesting that as readers encountered the verb, their expectation for the sentence to continue with something about the person being praised was so strong that an expectation-inconsistent pronoun was momentarily taken as a processing—possibly morphosyntactic—error. The effect was replicated in a later study (van Berkum, de Goede, van Alphen, Mulder, & Kerstholt, 2013) and disappeared when readers were in a bad mood (consistent with other work on how mood affects cognitive processing; see e.g., Clore & Huntsinger, 2007).

In line with dominant interpretations of the N400 (Brouwer & Hoeks, 2013; Federmeier, 2007; Kutas & Federmeier, 2011; Lau, Phillips, & Poeppel, 2008; van Berkum, 2009), we think that the various discourse-dependent N400 effects obtained in many of our studies (e.g., Nieuwland, 2013, 2015; Nieuwland & van Berkum, 2005, 2006b; Otten & van Berkum, 2007; van Berkum et al., 2008; van Berkum, Hagoort, & Brown, 1999; van Berkum, Zwitserlood, et al., 2003) are best construed—at least partly—as instances of context-based anticipation as well. This includes our peanuts study (Nieuwland & van Berkum, 2006b), where, for example, a phrase like "The peanut was in love" elicited a large N400 effect on *in love* when presented in isolation, but *not* in the context of a cartoon-like story about a singing and dancing peanut—in the latter case, the precise mental model of the situation discussed prepared readers for *in love*, rather than for more typical features of peanuts. Context-based anticipation is presumably also why, for example, a word like *tattoo* in "I have a big tattoo on my back" elicits a larger N400 when the speaker has a stereotypical "upper-class" accent than when he or she has a stereotypical "lower-class" accent (van Berkum et al., 2008). Anticipations also arise during counterfactual language comprehension, which why a word that renders a counterfactual sentence false, like *USA* in "If NASA had not developed its Apollo Project, the first country to land on the moon would have been the USA," elicits a larger N400 compared to a word that renders the sentence true (*USSR*), despite the fact that *USA* is more strongly associated with all relevant concepts in the context (*NASA*, *Apollo Project*, *moon landing*) than *USSR* is (Nieuwland & Martin, 2012; see also Kulakova & Nieuwland, 2016).

In all, the evidence from our EEG studies on language processing confirms what should perhaps have been obvious anyway: we are memory-based forward-looking creatures, anticipating how the world might soon be—not just when we cross the street, but also when we process language.

5. Conclusion

In this chapter, we have reviewed our cognitive neuroscience research on how language is processed in context. We have explored how the unfolding linguistic code is related to the referential context, as well as how the code is related to the affective context, defined by the interests of the comprehender. In addition, we have examined some of the evidence for context-based anticipation during language processing.

We note that *context* has a wide range of interpretations that also depend on the grain size of the linguistic elements it is a context *for*, ranging from a single phoneme providing a context for the next, to one's interests and values providing a context for an entire discourse. We have also occasionally interpreted *context* differently, not as the synchronic or *enchronic* (Enfield, 2013) context for the processing or anticipation of a particular bit of code coming in, but as phylogenetically older systems that provide the basis for language, and allow language to *amplify* those older systems. These evolutionary and real-time contexts are related, though: since our map-making, affective, and anticipatory systems are amplified rather than replaced, they too provide the context for language processing as it unfolds in real time. Models such as the corticohippocampal theory of reference and the ALC model are models of how real-time language processing is interwoven with referential memory and affect.

The biggest addition brought by language may well be what people have focused on all along: it allows us to very precisely refer to states of affairs and refine our mental representations of what is, was, might, or could never be. However, that increased referential precision does its work in the same way that a simple pointing gesture can do its work (Tomasello, 2008; van Berkum, 2018, 2019a): by working out the exact referent, by inferring why the speaker wants to draw our attention to this in the first place, by relating both to what we care about as we navigate the physical and social world, and by responding affectively in appropriate ways. The increased referential precision doesn't just allow us to state that the elephant on the left is probably the easier target or that berries can be found in some particular spot near the river. The referential power of language also allows us to, for fun or science, create fictional worlds where peanuts can be in love. It allows us to define institutions, declare marriage, and denominate some individuals as presidents. And it allows us to discuss people's thoughts, feelings, and behavior, so as to realize some of our social intentions by evaluating—possibly even gossiping about—those of others. As such, the increased referential power that comes with language has not only massively amplified what we can achieve as the map-making, affective, and forward-looking animals that we are, but it is also deeply interwoven with our social and moral nature.

Acknowledgments

Supported by NWO Vici grant no. 277-89-001 to J.v.B.

NOTES

1. Neural oscillatory evidence for hippocampal involvement has yet to be found, but hippocampal oscillatory activity may not be detectable in EEG recordings (da Silva, 2013).
2. The negativity observed by Otten et al. (2016) had a slightly more anterior distribution than is typical for an N400 effect; this might reflect a different set of neural generators, but may also arise from overlap with a posteriorly distributed LPP effect.
3. These later studies also indicated that the exact ERP signature of prediction mismatch varied, for reasons we do not fully understand. A large-scale pre-registered replication study of van Berkum et al. (2005) is currently ongoing (supervised by M.S.N.) to explore the stability of that specific ERP effect.

REFERENCES

Aggleton, J. P., & Brown, M. W. (2006). Interleaving brain systems for episodic and recognition memory. *Trends in Cognitive Sciences, 10*(10), 455–463.

Bašnákova, J., van Berkum, J. J. A., Weber, K., & Hagoort, P. (2015). A job interview in the MRI scanner: How does indirectness affect addressees and overhearers? *Neuropsychologia, 76*, 79–91.

Bašnákova, J., Weber, K., Petersson, K. M., van Berkum, J. J. A., & Hagoort, P. (2014). Beyond the language given: The neural correlates of inferring speaker meaning. *Cerebral Cortex, 24*(10), 2572–2578.

Bemis, D. K., & Pylkkänen, L. (2011). Simple composition: A magnetoencephalography investigation into the comprehension of minimal linguistic phrases. *Journal of Neuroscience, 31*(8), 2801–2814.

Bloom, P. (2000). *How children learn the meanings of words*. Cambridge, MA: MIT Press.

Brodbeck, C., Gwilliams, L., & Pylkkänen, L. (2016). Language in context: MEG evidence for modality-general and-specific responses to reference resolution. *eNeuro, 3*(6), ENEURO-0145.

Brouwer, H., & Hoeks, J. C. (2013). A time and place for language comprehension: Mapping the N400 and the P600 to a minimal cortical network. *Frontiers in Human Neuroscience, 7*, 758.

Cacioppo, J. T., Crites, S. L., Berntson, G. G., & Coles, M. G. (1993). If attitudes affect how stimuli are processed, should they not affect the event-related brain potential? *Psychological Science, 4*(2), 108–112.

Citron, F. M. (2012). Neural correlates of written emotion word processing: A review of recent electrophysiological and hemodynamic neuroimaging studies. *Brain and Language, 122*(3), 211–226.

Clark, H. H. (1996). *Using language*. Cambridge: Cambridge University Press.

Clore, G. L., & Huntsinger, J. R. (2007). How emotions inform judgment and regulate thought. *Trends in Cognitive Sciences, 11*(9), 393–399.

Damasio, A. (2010). *Self comes to mind: Constructing the conscious mind.* New York: Pantheon.

da Silva, F. L. (2013). EEG and MEG: Relevance to neuroscience. *Neuron, 80*(5), 1112–1128.

DeLong, K. A., Urbach, T. P., & Kutas, M. (2005). Probabilistic word pre-activation during language comprehension inferred from electrical brain activity. *Nature Neuroscience, 8*(8), 1117–1121.

Duff, M. C., & Brown-Schmidt, S. (2012). The hippocampus and the flexible use and processing of language. *Frontiers in Human Neuroscience, 6,* 69.

Duff, M. C., Gupta, R., Hengst, J. A., Tranel, D., & Cohen, N. J. (2011). The Use of definite references signals declarative memory evidence from patients with hippocampal amnesia. *Psychological Science, 22*(5), 666–673.

Eichenbaum, H., Yonelinas, A. R., & Ranganath, C. (2007). The medial temporal lobe and recognition memory. *Annual Review of Neuroscience, 30,* 123–152.

Enfield, N. J. (2013). *Relationship thinking: Agency, enchrony, and human sociality.* New York: Oxford University Press.

Federmeier, K. D. (2007). Thinking ahead: The role and roots of prediction in language comprehension. *Psychophysiology, 44*(4), 491–505.

Fischler, I., & Bradley, M. (2006). Event-related potential studies of language and emotion: words, phrases, and task effects. *Progress in Brain Research, 156,* 185–203.

Foroni, F., & Semin, G. R. (2009). Language that puts you in touch with your bodily feelings the multimodal responsiveness of affective expressions. *Psychological Science, 20*(8), 974–980.

Friederici, A. D. (2012). The cortical language circuit: From auditory perception to sentence comprehension. *Trends in Cognitive Sciences, 16*(5), 262–268.

Frijda, N. H. (2008). *The psychologists' point of view.* In M. Lewis, J. M. Haviland-Jones, & L. Feldman Barrett (Eds.), *Handbook of emotions* (pp. 68–87). New York: Guilford.

Frith, C., & Frith, U. (2005). Theory of mind. *Current Biology, 15*(17), R644–R645.

Gantman, A. P., & van Bavel, J. J. (2015). Moral perception. *Trends in Cognitive Sciences, 19*(11), 631–633.

Gernsbacher, M. A. (1989). Mechanisms that improve referential access. *Cognition, 32*(2), 99–156.

Gerrig, R. J., & McKoon, G. (1998). The readiness is all: The functionality of memory-based text processing. *Discourse Processes, 26*(2–3), 67–86.

Gibson, E. A., & Pearlmutter, N. J. (2011). *The processing and acquisition of reference.* Cambridge, MA: MIT Press.

Goffman, E. (1959). *The presentation of self in everyday life.* New York: Doubleday.

Greene, J. (2014). *Moral tribes: Emotion, reason and the gap between us and them.* London: Atlantic Books.

Hagoort, P., Hald, L., Bastiaansen, M., & Petersson, K. M. (2004) Integration of word meaning and world knowledge in language comprehension. *Science, 304,* 438–441.

Hagoort, P., & Indefrey, P. (2014). The neurobiology of language beyond single words. *Annual Review of Neuroscience, 37,* 347–362.

Haidt, J. (2012). *The righteous mind: Why good people are divided by politics and religion.* London: Allen Lane.

Hajcak, G., MacNamara, A., & Olvet, D. M. (2010). Event-related potentials, emotion, and emotion regulation: An integrative review. *Developmental Neuropsychology, 35*(2), 129–155.

Heilman, K. M., Bowers, D., Speedie, L., & Coslett, H. B. (1984). Comprehension of affective and nonaffective prosody. *Neurology, 34*(7), 917–921.

Hickok, G., & Small, S. L. (2015). *The neurobiology of language.* Amsterdam: Elsevier.

Holt, D. J., Lynn, S. K., & Kuperberg, G. R. (2009). Neurophysiological correlates of comprehending emotional meaning in context. *Journal of Cognitive Neuroscience, 21*(11), 2245–2262.

Hundrieser, M., & Stahl, J. (2016). How attitude strength and information influence moral decision making: Evidence from event-related potentials. *Psychophysiology, 53,* 678–688.

Irvine, W. B. (2013). *A Slap in the face: Why insults hurt—And why they shouldn't.* Oxford: Oxford University Press.

Jackendoff, R. (2007). A parallel architecture perspective on language processing. *Brain Research, 1146,* 2–22.

Kulakova, E., & Nieuwland, M. S. (2016). Pragmatic skills predict online counterfactual comprehension: Evidence from the N400. *Cognitive, Affective, and Behavioral Neuroscience, 16,* 814–824.

Kurczek, J., Brown-Schmidt, S., & Duff, M. (2013). Hippocampal contributions to language: Evidence of referential processing deficits in amnesia. *Journal of Experimental Psychology: General, 142*(4), 1346–1354.

Kutas, M., & Federmeier, K. D. (2011). Thirty years and counting: Finding meaning in the N400 component of the event related brain potential (ERP). *Annual Review of Psychology, 62,* 621–647.

Lau, E. F., Phillips, C., & Poeppel, D. (2008). A cortical network for semantics: (De) constructing the N400. *Nature Reviews Neuroscience, 9*(12), 920–933.

Lazarus, R. S. (1991). *Emotion and adaptation.* Oxford: Oxford University Press.

Leuthold, H., Kunkel, A., Mackenzie, I. G., & Filik, R. (2015). Online processing of moral transgressions: ERP evidence for spontaneous evaluation. *Social Cognitive and Affective Neuroscience, 10*(8), 1021–1029.

Lewis, R. L., & Vasishth, S. (2005). An activation-based model of sentence processing as skilled memory retrieval. *Cognitive Science, 29*(3), 375–419.

Martin, A. E. (2016). Language processing as cue integration: Grounding the psychology of language in perception and neurophysiology. *Frontiers in Psychology, 7,* 120.

Martin, A. E., Nieuwland, M. S., & Carreiras, M. (2012). Event-related brain potentials index cue-based retrieval interference during sentence comprehension. *NeuroImage, 59,* 1859–1869.

Martin, A. E., Nieuwland, M. S., & Carreiras, M. (2014). Agreement attraction during comprehension of grammatical sentences: ERP evidence from ellipsis. *Brain and Language, 135,* 42–51.

Martinich, A. P. (1985). *The philosophy of language.* Oxford: Oxford University Press.

Menenti, L., Petersson, K. M., Scheeringa, R., & Hagoort, P. (2009). When elephants fly: Differential sensitivity of right and left inferior frontal gyri to discourse and world knowledge. *Journal of Cognitive Neuroscience, 21*(12), 2358–2368.

Myers, J. L., & O'Brien, E. J. (1998). Accessing the discourse representation during reading. *Discourse Processes, 26*(2–3), 131–157.

Nadig, A. S., & Sedivy, J. C. (2002). Evidence of perspective-taking constraints in children's on-line reference resolution. *Psychological Science, 13*(4), 329–336.

Nieuwland, M. S. (2012). Establishing propositional truth-value in counterfactual and real-world contexts during sentence comprehension: Differential sensitivity of the left and right inferior frontal gyri. *NeuroImage, 59*(4), 3433–3440.

Nieuwland, M. S. (2013). "If a lion could speak...": Online sensitivity to propositional truth-value of unrealistic counterfactual sentences. *Journal of Memory and Language, 68*(1), 54–67.

Nieuwland, M. S. (2014). "Who is he?" Event-related brain potentials and unbound pronouns. *Journal of Memory and Language, 76,* 1–28.

Nieuwland, M. S. (2015). The truth before and after: Brain potentials reveal automatic activation of event knowledge during sentence comprehension. *Journal of Cognitive Neuroscience, 27*(11), 2215–2228.

Nieuwland, M. S., & Martin, A. E. (2012). If the real world were irrelevant, so to speak: The role of propositional truth-value in counterfactual sentence comprehension. *Cognition, 122*(1), 102–109.

Nieuwland, M. S., & Martin, A. E. (2017). Neural oscillations and a nascent corticohippocampal theory of reference. *Journal of Cognitive Neuroscience, 29*(5), 896–910.

Nieuwland, M. S., Otten, M., & van Berkum, J.J.A. (2007). Who are you talking about? Tracking discourse-level referential processes with ERPs. *Journal of Cognitive Neuroscience, 19*(2), 1–9.

Nieuwland, M. S., Petersson, K. M., & van Berkum, J. J. A. (2007). On sense and reference: Examining the functional neuroanatomy of referential processing. *NeuroImage, 37*(3), 993–1004.

Nieuwland, M. S., Politzer-Ahles, S., Heyselaar, E., Segaert, K., Darley, E., Kazanina, N., ... Huettig, F. (2018). Limits on prediction in language comprehension: A multi-lab failure to replicate evidence for probabilistic pre-activation of phonology. *eLife,* 7: e33468. doi:10.7554/eLife.33468.

Nieuwland, M. S., & van Berkum, J. J. A. (2005). Testing the limits of the semantic illusion phenomenon: ERPs reveal temporary change deafness in discourse comprehension. *Cognitive Brain Research, 24*(3), 691–701.

Nieuwland, M. S., & van Berkum, J. J. A. (2006a). Individual differences and contextual bias in pronoun resolution: Evidence from ERPs. *Brain Research, 1118,* 155–167.

Nieuwland, M. S., & van Berkum, J. J. A. (2006b). When peanuts fall in love: N400 evidence for the power of discourse. *Journal of Cognitive Neuroscience, 18*(7), 1098–1111.

Nieuwland, M. S., & van Berkum, J. J. A. (2008a). The interplay between semantic and referential aspects of anaphoric noun phrase resolution: Evidence from ERPs. *Brain and Language, 106,* 119–131.

Nieuwland, M. S., & van Berkum, J. J. (2008b). The neurocognition of referential ambiguity in language comprehension. *Language and Linguistics Compass, 2*(4), 603–630.

Otten, M., Mann, L., van Berkum, J. J. A., & Jonas, K. J. (2016). No laughing matter: How the presence of laughing witnesses changes the perception of insults. *Social Neuroscience 12*(2), 182–193.

Otten, M., Nieuwland, M. S., & van Berkum, J. J. A. (2007). Great expectations: Specific lexical anticipation influences the processing of spoken language. *BMC Neuroscience, 8*(89).

Otten, M., & van Berkum, J. J. A. (2007). What makes a discourse constraining? Comparing the effects of discourse message and scenario fit on the discourse-dependent N400 effect. *Brain Research, 1153,* 166–177.

Otten, M., & van Berkum, J. J. A. (2008). Discourse-based word anticipation during language processing: Prediction of priming? *Discourse Processes, 45,* 464–496.

Otten, M., & van Berkum, J. J. A. (2009). Does working memory capacity affect the ability to predict upcoming words in discourse? *Brain Research, 1291,* 92–101.

Panksepp, J., & Biven, L. (2012). *The archaeology of mind: Neuroevolutionary origins of human emotions.* New York: Norton and Company.

Pell, M. D. (1999). Fundamental frequency encoding of linguistic and emotional prosody by right hemisphere-damaged speakers. *Brain and Language, 69*(2), 161–192.

Recanati, F. (1993). *Direct reference.* Oxford, UK: Blackwell Press.

Rissman, J., & Wagner, A. D. (2012). Distributed representations in memory: Insights from functional brain imaging. *Annual Review of Psychology, 63,* 101–128.

Sabatinelli, D., Keil, A., Frank, D. W., & Lang, P. J. (2013). Emotional perception: Correspondence of early and late event-related potentials with cortical and subcortical functional MRI. *Biological Psychology, 92*(3), 513–519.

Sanford, A. J., & Garrod, S. C. (1989). What, when, and how?: Questions of immediacy in anaphoric reference resolution. *Language and Cognitive Processes, 4*(3–4), SI235–SI262.

Scherer, K. R. (2005). What are emotions? And how can they be measured? *Social Science Information, 44*(4), 695–729.

Spotorno, N., Koun, E., Prado, J., van der Henst, J. B., & Noveck, I. A. (2012). Neural evidence that utterance-processing entails mentalizing: The case of irony. *NeuroImage, 63*(1), 25–39.

Staresina, B. P., Henson, R. N., Kriegeskorte, N., & Alink, A. (2012). Episodic reinstatement in the medial temporal lobe. *Journal of Neuroscience, 32*(50), 18150–18156.

Struiksma, M. E., De Mulder, H. N. M., & van Berkum, J. J. A. (2019). *The impact of verbal insults: The effects of person insulted, repetition, and taboo words.* Manuscript submitted for publication.

Struiksma, M. E., Pantazi, M., de Mulder, H. N. M., & van Berkum, J. J. A. (2013). The time-course of dealing with morally objectionable language: What happens when we read about bad stuff? *Proceedings of the 6th Annual Meeting of the Social & Affective Neuroscience Society,* San Francisco, CA.

't Hart, B., Struiksma, M. E., van Boxtel, A., & van Berkum, J. J. A. (2018). Emotion in stories: Facial EMG evidence for both mental simulation and moral evaluation. *Frontiers in Psychology, 9,* 613.

't Hart, B., Struiksma, M.E., van Boxtel, A. & van Berkum, J.J.A. (2019). Tracking affective language comprehension: Simulating and evaluation character affect in morally loaded narratives. *Frontiers in Psychology, 10,* 318.

Tomasello, M. (2008). *Origins of human communication.* Cambridge: MIT Press.

Tomasello, M. (2009). *Constructing a language: A usage-based theory of language acquisition.* Cambridge, MA: Harvard University Press.

Trueswell, J. C., & Tanenhaus, M. K. (2005). *Approaches to studying world-situated language use: Bridging the language-as-product and language-as-action traditions.* Cambridge, MA: MIT Press.

van Berkum, J. J. A. (2009). *The neuropragmatics of "simple" utterance comprehension: An ERP review.* In K. Yatsushiro & U. Sauerland (Eds.), *Semantics and pragmatics: From experiment to theory* (pp. 276–316). Basingstroke, UK: Palgrave Macmillan.

van Berkum, J. J. A. (2012). The electrophysiology of discourse and conversation. In M. J. Spivey, K. McRae, & M. F. Joanisse (Eds.), *The Cambridge handbook of psycholinguistics* (pp. 589–614). New York: Cambridge University Press.

van Berkum, J. J. A. (2013). Anticipating communication. *Theoretical Linguistics, 39*(1–2), 75–86.

van Berkum, J. J. A. (2018). Language comprehension, emotion, and sociality: Aren't we missing something? In S. A. Rueschemeyer & G. Gaskell (Eds.), *Oxford handbook of psycholinguistics* (2nd ed.), (pp. 644–669). Oxford: Oxford University Press.

van Berkum, J. J. A. (2019a). Language comprehension and emotion: Where are the interfaces, and who cares? In G. de Zubicaray & N. O. Schiller (Eds.), *Oxford handbook of neurolinguistics* (pp. 736–766). Oxford: Oxford University Press.

van Berkum, J. J. A. (2019b). What is emotion, and why is it relevant to language researchers? Manuscript submitted for publication, to appear in Schiewer, G.L., Altarriba, J., & Ng, B.C. (Eds.), *Language and Emotion: An international handbook*. Berlin: De Gruyter Mouton.

van Berkum, J. J. A., Brown, C. M., & Hagoort, P. (1999). Early referential context effects in sentence processing: Evidence from event-related brain potentials. *Journal of Memory and Language, 41*(2), 147–182.

van Berkum, J. J. A., Brown, C. M., Hagoort, P., & Zwitserlood, P. (2003). Event-related brain potentials reflect discourse-referential ambiguity in spoken language comprehension. *Psychophysiology, 40*(2), 235–248.

van Berkum, J. J. A., Brown, C. M., Zwitserlood, P., Kooijman, V., & Hagoort, P. (2005). Anticipating upcoming words in discourse: Evidence from ERPs and reading times. *Journal of Experimental Psychology: Learning, Memory, and Cognition, 31*, 443–467.

van Berkum, J. J. A., de Goede, D., van Alphen, P. M., Mulder, E. R., & Kerstholt, J. H. (2013). How robust is the language architecture? The case of mood. *Frontiers in Psychology, 4*, 505.

van Berkum, J. J. A., Hagoort, P., & Brown, C. M. (1999). Semantic integration in sentences and discourse: Evidence from the N400. *Journal of Cognitive Neuroscience, 11*(6), 657–671.

van Berkum, J. J. A., Holleman, B., Nieuwland, M., Otten, M., & Murre, J. (2009). Right or wrong? The brain's fast response to morally objectionable statements. *Psychological Science, 20*(9), 1092–1099.

van Berkum, J. J. A., Koornneef, A. W., Otten, M., & Nieuwland, M. S. (2007). Establishing reference in language comprehension: An electrophysiological perspective. *Brain Research, 1146*, 158–171.

van Berkum, J. J. A., van den Brink, D., Tesink, C. M. J. Y., Kos, M., & Hagoort, P. (2008). The neural integration of speaker and message. *Journal of Cognitive Neuroscience, 20*(4), 580–591.

van Berkum, J. J. A., Zwitserlood, P., Bastiaansen, M. C. M., Brown, C. M., & Hagoort, P. (2004). So who's "he" anyway? Differential ERP and ERSP effects of referential success, ambiguity and failure during spoken language comprehension. *Supplement to the Journal of Cognitive Neuroscience, 16*, 70.

van Berkum, J. J. A., Zwitserlood, P., Hagoort, P., & Brown, C. M. (2003). When and how do listeners relate a sentence to the wider discourse? Evidence from the N400 effect. *Cognitive Brain Research, 17*(3), 701–718.

Wagner, A. D., Shannon, B. J., Kahn, I., & Buckner, R. L. (2005). Parietal lobe contributions to episodic memory retrieval. *Trends in Cognitive Sciences, 9*(9), 445–453.

Wang, L., & Bastiaansen, M. (2014). Oscillatory brain dynamics associated with the automatic processing of emotion in words. *Brain and Language, 137*, 120–129.

Wicha, N. Y. Y., Moreno, E. M., & Kutas, M. (2004). Anticipating words and their gender: An event-related brain potential study of semantic integration, gender expectancy, and gender agreement in Spanish sentence reading. *Journal of Cognitive Neuroscience, 16*, 1272–1288.

Willems, R. M., Özyürek, A., & Hagoort, P. (2008). Seeing and hearing meaning: ERP and fMRI evidence of word versus picture integration into a sentence context. *Journal of Cognitive Neuroscience, 20*(7), 1235–1249.

Zajonc, R. B. (1980). Feeling and thinking: Preferences need no inferences. *American Psychologist, 35*(2), 151–175.

32 Key Issues and Future Directions: Where Have We Been and Where Are We Going?

DAVID POEPPEL

1. *The Harder We Look, the More Pieces We Find*

If we have learned anything about the neurobiological foundations of speech and language processing, it is that our textbooks must change! The field has lived comfortably and for a long time under the regime of left hemisphere imperialism. Whether approaching the area of research from the perspective of neurology, neuropsychology, cognitive neuroscience, or language research, we have proceeded in a calm and undisturbed manner with our familiar friends in the left inferior frontal gyrus (Broca's region) and the left posterior superior temporal lobe (Wernicke's region). While it is not incorrect to celebrate the relevance and historical significance of these brain regions and their roles for speech and language processing, it is now indisputable that we have vastly underestimated the number and distribution of brain regions that contribute in critical ways to what we achieve so effortlessly when we comprehend and produce language.

To start with, there is now essentially consensus that right hemisphere regions are instrumental in supporting particular subroutines in perception and comprehension (e.g., Friederici & Alter, 2004; Hickok & Poeppel, 2000, 2007; Marslen-Wilson, chapter 30 of this volume) and, under certain circumstances, sensorimotor aspects of speech processing as well (Cogan et al., 2014). But it is not just the rehabilitation—or rather the integration—of right hemisphere mechanisms that we need to address in current models. The next issue to remember is that it is now well established that the brain regions we have previously discussed coarsely and in a rather monolithic manner are subdivided into numerous fields that likely subserve different computational functions. Broca's region, for example, encompasses numerous cytoarchitecturally separable fields of tissue (Amunts et al., 2010), the functions of which remain under intensive investigation—although access to these different regions in the human brain is still challenging. Likewise, the superior temporal cortex, which we label informally as *auditory cortex*, is composed of a range of fields that are only now beginning to be characterized in their anatomic organization, neurophysiological response profile, and connectivity (e.g., Sjerps & Chang, chapter 27). Some of these cortical fields are more or less responsive to speech, some show multisensory sensitivity, some are tightly coupled to motor regions, and so on. Again, we dramatically (if understandably) underestimated the number, internal complexity, and connectivity of regions. Moreover, the tools we now have available have identified additional areas, both cortical and subcortical (e.g., Kotz & Schwartze, 2010) that are instrumental in language processing. A clear and compelling example of the expanded functional anatomy is provided by the case of speech production (Hickok, chapter 28), where the network that underlies this complex function is more or less agreed upon and involves distributed (at least) different regions in frontal, several distinct regions in parietal, and additional regions in temporal cortex, parts of the cerebellum, as well as the basal ganglia. In light of so extensive a network, new functional theories can and should capitalize on the rich neural infrastructure.

In short, the data of the last 20 years have expanded quite dramatically the amount of "brain real estate" that underpins speech perception and production and language processing. Not just are the operations bilaterally organized, but they extend over many areas that are both cortical and subcortical, the specific physiological and computational functions of which we do not yet understand in any satisfactory way. What, then, is the lesson of the last years? That with every new recording technique and every new analysis approach, we discover more components of what constitutes the total system. The harder we look, the more pieces we find. And we can perhaps expect this trend to continue. With novel instruments able to interrogate the human brain at higher and higher resolutions, we are likely to find circuit, cellular, and perhaps even subcellular structures implicated in language. The implementational level of description (in the words of David Marr) may indeed turn out to be a biological reductionist's dream. The studies of molecular genetics (see Fisher, chapter 43 of this volume) could yield unanticipated mechanisms about both storage and computation of the primitives of language.

2. The Toolkits, Old and New

How did we end up with new maps of language in the brain that look more like the Holy Roman Empire in 1618 (see, e.g., Huth, de Heer, Griffiths, Theunissen, & Gallant, 2016) than the simple map that populates our textbooks? Both older and established machinery as well as new recording technologies have led to ever more fine-grained measurements. But very importantly, significant changes in our data analysis approaches have been transformational in this area of inquiry, as well.

The experimental tools continue to include the usual suspects of systems and cognitive neuroscience. The last decade has seen exciting data generated by more traditional technologies such as electroencephalography (see, e.g., van Berkum & Nieuwland, chapter 31 of this volume) or functional MRI (fMRI; see, e.g., Hickok, chapter 28; Marslen-Wilson, chapter 30) as well as newer tools such as near infrared spectroscopy or cortical cooling during surgical intervention (Long et al., 2016). Stimulation paradigms using Transcranial Magnetic Stimulation and transcranial direct-current stimulation/transcranial alternating-current stimulation, as well, have generated both informative data and exciting controversy (Vöröslakos et al., 2018). The field has at its disposal recording techniques that range across scales: near infrared spectroscopy and fMRI are entirely noninvasive and yield, in the case of fMRI, extremely high-resolution spatial data, albeit at a lower temporal resolutions. Magnetoencephalography and electroencephalography are also entirely noninvasive and yield extremely high temporal resolution data. The recording of neural data in clinical contexts is more restricted, but it has been extremely influential in recent years; from local field potential data derived from electrocorticography recordings to multiple unit activity to single unit activity, patient studies have been a valuable resource enriching our current understanding, in part because such data combine very high spatial and very high temporal resolution. It is noteworthy that all these different recording techniques have been applied to aspects of language processing across the hierarchy of putative representations: experiments range from the relevance of features in speech perception (e.g., Sjerps & Chang, chapter 27) to lexical processing (Marslen-Wilson, chapter 30) to the construction of constituent-based syntactic representations (Gross & Poeppel, chapter 29). The methodological pluralism that is now possible—and often already practiced—makes it much easier to *pair an appropriate methodology with the specific question at hand.* We are definitely no longer restricted by techniques that impose narrow experimental designs; refreshingly, *we can now let the question be the epistemologically most significant factor* and let the technique be selected because it can contribute to answering a question—rather than dictating what question to ask to begin with.

With respect to research on the functional neurobiology of language, it is important to note that a growing body of work is using electrophysiological techniques. (The chapters in this section of the book exemplify this trend well.) In part, the increasing use of electrophysiological techniques reflects the necessity to study with high temporal resolution processes that are remarkably *fast.* The well-known speed of language processing provides not just a challenge to the neural system (how does the brain integrate so many different kinds of information in so short a time over such spatially distributed systems?), but that processing speed also makes certain experimental requirements patently transparent: methods that cannot capture neural activity at roughly a millisecond-level resolution are unlikely to be able to contribute sufficiently comprehensive and detailed mechanistic insights toward the models we seek to construct. Language processing is fast—so the methods used to study it need to be fast, and the data analysis techniques used to investigate all of this need to be sensitive to "fastness."

In that context, one major new development is the use of data analytic approaches that are significantly more sophisticated, both from the perspective of signal processing and from the perspective of statistics. For example, the data recorded by electrocorticography in presurgical epilepsy patients is now profitably analyzed using an approach adopted from neurophysiology in animal preps: spectrotemporal receptive fields. This (reverse correlation-based) way of looking at recordings during spoken language processing has provided compelling evidence for the kinds of representations situated in the superior temporal cortex of listeners, including features—a type of representation forcefully argued for by phonologists for decades on the basis of cross-linguistic data and the structure of phonological rules (Mesgarani, Cheung, Johnson, & Chang, 2014). Similarly, increasingly sensitive spectral analysis methods have revealed oscillatory structure in the data that is now being tested with respect to different processing models, both at the level of speech perception and at the level of sentential structure building (Gross & Poeppel, chapter 29). A data analysis strategy that is relatively new and has been used very effectively both on hemodynamic data and electrophysiological data is representational similarity analysis, a technique that allows one to query brain data vis-à-vis explicitly formulated predictions from cognitive models (Marslen-Wilson, chapter 30). Lurking behind many of these advances are the substantive new opportunities afforded by machine learning. Because the sheer size and complexity of the data sets generated in this research, machine learning approaches offer effective ways to look for regularities. But beyond that, machine learning techniques are being used relatively aggressively to contribute to our understanding of the structure of the domain. That is to say, machine learning–based approaches such as, say, *temporal generalization* (King & Dehaene, 2014) are posited to contribute to our understanding of processing models, for example, by providing online evidence for the degree to which the distinct levels of representation (say phonology and lexical levels) overlap during the recognition of spoken language, when representations are activated, reactivated, and so on. At the intersection of machine learning methods and computational linguistics lie many advances we can look forward to.

3. The Interdisciplinary Challenge at Its Core

There are many challenges that we face as we try to unravel the neurobiological foundations of speech and language. Let me highlight three questions that are discussed, in part, in the chapters of this section—but that, in any case, require significant additional study.

First, there is the issue of cerebral asymmetry and lateralization of function—a question that might be considered the "gift that keeps on giving," in light of how many careers have been constructed on the basis of issues concerning brain asymmetry. These questions, historically, lie at the very foundation of the field. It remains hard to explain, neurobiologically, why for some functions we observe such clear bilateral organization, whereas for others there appears to be compelling asymmetry. The data are now quite well characterized: while we previously thought of most aspects of language processing as lateralized to the dominant (typically left) hemisphere, it is now uncontroversial that, across levels of processing, there is a bilateral contribution. This is true at the level of speech perception (e.g., Hickok & Poeppel, 2000, 2007) and lexical level comprehension (e.g., Marslen-Wilson, chapter 30), at the level of sensorimotor mapping of speech (Cogan et al., 2014), and at the level of semantic organization, at least with respect to lexical semantics (e.g., Huth et al., 2016). On the other hand, the construction of grammatical representations that require more than simple adjacency appears to be lateralized (e.g., Friederici, 2011; Marslen-Wilson, chapter 30). *How* and *where* and *why* do such asymmetries arise? The combinatorics and hierarchical structure building that underlie syntactic processing do appear consistently to be asymmetrically mediated. There are speculations about connectivity and so on, but truth be told, we have no idea why the neural infrastructure of the left hemisphere and its component parts are suited for certain kinds of computations, whereas—in the healthy adult speaker/listener—the right hemisphere no longer appears to execute these functions. This requires explanation, including the fact that the right hemisphere can to a large degree assume such computational supports in the face of severe pathology, for example, in the context of hemispherectomies in children with epilepsy. All in all, the partly bilateral and partly lateralized brain representation of speech and language remains a deep puzzle, perhaps because we do not yet have a successful computational decomposition of the processes that map on to the specialized circuitry in the left versus right brain areas.

Second, how do we store words, or roots, or morphemes—or for that matter anything at all (cf. Gallistel & King, 2011)? While there appears to be some progress on questions of computation (say, for example, on questions of constituency or structure building; Ding, Melloni, Zhang, Tian, & Poeppel, 2016; Martin & Doumas, 2017; Nelson et al., 2017), there is very little mechanistic progress on the question of storage. This is surprising, given that the very basis of language processing is that we have a set of stored primitives—let us for the sake of argument call them *words*—that provide the terminal elements of the generative system that yields an infinite set of possible expressions. Estimates of vocabulary size in English range from 25,000 to 100,000. How is this internal dictionary organized, where does it live, are there multiple copies of it, and how do you even store even a single bit of information such that you can retrieve it for online processing in a tenth of a second? A satisfactory model of how lexical information is stored in a way to successfully interface with different sensory and motor modalities (auditory, visual for reading and sign, tactile for Braille) at an extremely rapid rate is not even remotely in sight.

Third, and most clearly at the intersection of the distinct relevant research areas, what are actually the objects of study? Adopting the perspective advocated, among others, by David Marr, what are the elementary computations, representations and algorithms, and implementational structures that form the basis of speech and language processing? At the level of description typically called *implementation*, we can turn to neurobiological research. The *parts list* provided by biological approaches is increasingly well described, although each part's role with respect to language is not well understood, beyond correlational speculations. For example, are dendritic spines, apical dendrites, single neurons, small ensembles, convolutional kernels/columns, interactions of areas, and so on the *functionally relevant* elements? If we take seriously the success of the neurosciences, we should endorse the set of primitive structures (say dendrite) and primitive operations (say coincidence detection) that have been identified. But likewise, the same status is due to the elements discovered by language research! There are, to be sure, theory-driven differences; nevertheless, there are *kinds* of representations that appear to be atomic and necessary. These include representations such as distinctive features, phonemes, syllables, morphemes, lexical items, constituents; operations such as concatenation, labeling; long distance relationships, for example, between anaphors and their antecedents, reference resolution operations, and so on. How are we to map the primitives of neurobiology onto the primitives of language research? Both "parts lists" have considerable evidence in support of their existence; that is to say, neither domain has epistemological priority. One goal therefore must be to identify plausible linking hypotheses that provide formal, ideally computationally specific, relations between the parts list of language on the parts list of the brain. Establishing such linking hypotheses is extremely challenging (see Poeppel & Embick, 2005, and Embick & Poeppel, 2015, for detailed discussion of the challenges; see Hagoort, 2005, for a proposed anatomic solution to the linking problem).

4. *Scaling Up or Scaling Down? Reconciling Artificial and Naturalistic Experimentation*

With rather few exceptions, experiments on the neurobiological foundations of speech and language processing are typically based on carefully designed materials deployed in highly controlled studies, serving to adjudicate between well-defined, detailed hypotheses. Whether testing speech perception by using single vowels or syllables, or using lexical decision in studies investigating aspects of word representation, or contrasting precisely crafted sentence pairs to identify structure-building operations, our conventional notions of how the brain mediates language processing build principally on "small" experiments that deconstruct language into constituent elements such as features, phonemes, syllables, morphemes, phrases, discourse representations, and other relevant and theoretically motivated basic parts. This has been a highly successful research program and, for the most part, forms the basis of our current understanding, whether one considers extant models sophisticated or relatively coarse. Indeed, many—if not most—of the studies discussed in the chapters of this section exemplify such an approach. The experiments use, say, categorical perception speech paradigms, lexical decision, anomaly detection of semantically unusual sentences, and so on. And it is unquestionable that this laboratory-based approach with "reduced" materials continues to be highly successful. In fact, this type of approach is simply adopting and adapting the research techniques commonly employed in the sciences: breaking the problem into smaller problems (e.g., heredity, genes), working with idealizations (forces, inclined frictionless plane), and testing parts of a broader scientific question in order to make incremental progress.

One of the aspects of these research programs that is typically unspoken is that we assume that our findings will "scale up." That is to say, we harbor a tacit belief that our findings generated in idealized reductionist, laboratory contexts will ultimately be compatible with the phenomena under investigation when considered in a more ecologically valid and naturalistic situation. This is an incredibly successful strategy in the sciences, but it makes assumptions about how results generalize, license extrapolation, and relate in a straightforward manner to the types of phenomena that are ultimately of more interest than laboratory-based simplifications and idealizations. The implicit presupposition that this is a productive research strategy that ultimately leads to explanatory (and not merely descriptive) mechanistic accounts reflects a particular epistemological stance.

In contrast to this more canonical approach embraced by the natural sciences, in the last years, increasing emphasis has been placed on using naturalistic materials. In the context of our area of investigation, such an approach has been used from low level aspects of speech perception (e.g., data discussed in Sjerps & Chang, chapter 27) to high-level aspects of discourse model construction (e.g., data discussed in van Berkum & Nieuwland, chapter 31). Rather than confront participants with experimentally crafted materials, it has become common to query the brain's responses to ecologically (more) valid materials. This kind of experimental approach comes with its own epistemological assumptions. Sticking with the theme, will the results from this type of experimentation "scale down"? There are two strategies that one can discern in these more recent study designs. One approach might be called the *neoclassical naturalistic experiment*. Although the experimental materials are, say, naturalistic spoken language presented to participants, the data are analyzed from the perspective that the "parts list" that linguistics, psycholinguistics, and computational linguistics have provided is the best available guide to query the recorded data. Although the materials are presented in as ecological a manner as possible, one is still searching for the neural reflex of, for example, decomposition into morphemes, competition between possible lexical candidates, the construction of constituent representation. The research strategy is "neo" because of the use of naturalistic spoken materials rather than deconstructed laboratory idealizations; the strategy is "conventional" because the analysis builds on the assumption that the primitives (or the ontological structure) of the domain under investigation has been provided by the relevant other fields, say linguistics.

A more radical approach might be called "data-driven language decomposition." In such a strategy, no overt assumptions are made about the infrastructure of speech and language; rather, the data are subjected to analysis pipelines (typically machine learning approaches) and the—again implicit—assumption is that certain units may "emerge" from analyzing the patterns in the data, by whatever algorithm finds regularity. The three approaches sketched out here—(i) the classical, laboratory-based decompositional approach,(ii) the neoclassical approach that builds on accrued knowledge but uses naturalistic materials, and(iii) the data-driven approach—are not just reflections of different researchers' distinct experimental aesthetics. Choosing these different approaches reflects epistemological presuppositions that cannot be ignored, because the results of such analyses and the ensuing *interpretations* may or may not yield explanatory models that meet with success or are doomed to failure. It behooves us as researchers to be aware of the advantages, pitfalls, and tacit epistemological assumptions.

REFERENCES

Amunts, K., Lenzen, M., Friederici, A. D., Schleicher, A., Morosan, P., Palomero-Gallagher, N., & Zilles, K. (2010). Broca's region: Novel organizational principles and multiple receptor mapping. *PLOS Biology, 8*(9), e1000489.

Cogan, G. B., Thesen, T., Carlson, C., Doyle, W., Devinsky, O., & Pesaran, B. (2014). Sensory-motor transformations for speech occur bilaterally. *Nature, 507*(7490), 94.

Ding, N., Melloni, L., Zhang, H., Tian, X., & Poeppel, D. (2016). Cortical tracking of hierarchical linguistic structures in connected speech. *Nature Neuroscience, 19*(1), 158.

Embick, D., & Poeppel, D. (2015). Towards a computational(ist) neurobiology of language: Correlational, integrated and explanatory neurolinguistics. *Language, Cognition, and Neuroscience, 30*(4), 357–366.

Friederici, A. D. (2011). The brain basis of language processing: From structure to function. *Physiological Reviews, 91*(4), 1357–1392.

Friederici, A. D., & Alter, K. (2004). Lateralization of auditory language functions: A dynamic dual pathway model. *Brain and Language, 89*, 267–276.

Gallistel, C. R., & King, A. P. (2011). *Memory and the computational brain: Why cognitive science will transform neuroscience* (Vol. 6). New York: John Wiley & Sons.

Hagoort, P. (2005). On Broca, brain, and binding: A new framework. *Trends in Cognitive Sciences, 9*(9), 416–423.

Hickok, G., & Poeppel, D. (2000). Towards a functional neuroanatomy of speech perception. *Trends in Cognitive Sciences, 4*(4), 131–138.

Hickok, G., & Poeppel, D. (2007). The cortical organization of speech perception. *Nature Reviews Neuroscience, 8*(5), 393.

Huth, A. G., de Heer, W. A., Griffiths, T. L., Theunissen, F. E., & Gallant, J. L. (2016). Natural speech reveals the semantic maps that tile human cerebral cortex. *Nature, 532*(7600), 453–458.

King, J. R., & Dehaene, S. (2014). Characterizing the dynamics of mental representations: The temporal generalization method. *Trends in Cognitive Sciences, 18*(4), 203–210.

Kotz, S. A., & Schwartze, M. (2010). Cortical speech processing unplugged: A timely subcortico-cortical framework. *Trends in Cognitive Sciences, 14*(9), 392–399.

Long, M. A., Katlowitz, K. A., Svirsky, M. A., Clary, R. C., Byun, T. M., Majaj, N., … Greenlee, J. D. (2016). Functional segregation of cortical regions underlying speech timing and articulation. *Neuron, 89*(6), 1187–1193.

Martin, A. E., & Doumas, L. A. (2017). A mechanism for the cortical computation of hierarchical linguistic structure. *PLOS Biology, 15*(3), e2000663.

Mesgarani, N., Cheung, C., Johnson, K., & Chang, E. F. (2014). Phonetic feature encoding in human superior temporal gyrus. *Science, 343*, 1006–1010.

Nelson, M. J., El Karoui, I., Giber, K., Yang, X., Cohen, L., Koopman, H., … Dehaene, S. (2017). Neurophysiological dynamics of phrase-structure building during sentence processing. *Proceedings of the National Academy of Sciences, 114*, E3669–E3678.

Poeppel, D., & Embick, D. (2005). Defining the relation between linguistics and neuroscience. In A. Cutler (Ed.), *Twenty-first century psycholinguistics: Four cornerstones* (pp. 103–118). London: Routledge.

Vöröslakos, M., Takeuchi, Y., Brinyiczki, K., Zombori, T., Oliva, A., Fernández-Ruiz, A., … Berényi, A. (2018). Direct effects of transcranial electric stimulation on brain circuits in rats and humans. *Nature Communications, 9*(1), 483.

VI NEUROANATOMY OF LANGUAGE

SECTION EDITORS: PETER HAGOORT AND CHRISTIAN F. BECKMANN

33 Receptor and Cytoarchitecture of Language-Related Cortical Areas

NICOLA PALOMERO-GALLAGHER AND KARL ZILLES

1. Introduction

The neuroanatomical network underlying language and speech processing has been studied for over 100 years. In his pioneering work, Broca (1865) laid the foundation of the localization of speech functions. He linked the fact that two of his patients had lost the ability to speak to the existence of a lesion caused by stroke in the posterior part of their left inferior frontal gyrus (for a comprehensive description of the macroanatomy of these brains, see Dronkers, Plaisant, Iba-Zizen, & Cabanis, 2007). Specifically, in his report to the Anatomical Society, Broca (1861) stated that "the integrity of the third frontal convolution (and perhaps of the second) seems indispensable to the exercise of the faculty of articulate language." Subsequent clinical observations enabled Broca to more specifically pinpoint speech production abilities to the inferior frontal gyrus of the left cerebral hemisphere (Broca, 1865). Thus, this lesioned part of the brain is now known as Broca's region, which was the first brain region to which a circumscribed function, overt speech, had been related: the deficit in language production has often been termed Broca's aphasia.

A further crucial milestone in understanding the neural basis of language was achieved by Wernicke's clinical and pathoanatomical studies. Wernicke (1874) proposed the first theory of language. He conceptualized an anterior, "motor" speech center (i.e., Broca's region), which is connected via the arcuate fascicle to a posterior "sensory" brain region (i.e., Wernicke's region). He considered the latter region to be involved in the comprehension of written and spoken language. The anatomical counterpart of Wernicke's region was originally thought to be restricted to the superior temporal gyrus of the left hemisphere (Wernicke, 1874). To date Wernicke's region often includes the auditory cortex and areas (or parts thereof) located on the superior and middle temporal gyri, as well as on the angular and supramarginal gyri (Hickok & Poeppel, 2007; Binder, 2015; DeWitt & Rauschecker, 2013).

The identification and characterization of the microstructural (cytoarchitecture and myeloarchitecture) and molecular organization of language-relevant areas cannot be analyzed in vivo using available MRI studies because of their low spatial resolution compared to microstructural features. However, cyto- and receptor architecture are crucial prerequisites for further understanding the neurobiological basis of language. The importance of the microstructural ground truth is caused by the fact that functional and clinical studies point to the existence of numerous distinct areas exceeding by far the concept of a homogenous Broca or Wernicke region.

Most commonly applied cytoarchitectonic criteria include the number and delineability/visibility of cortical layers, the packing density and size of the cell bodies in cortical layers, the occurrence of specialized cell types, the thickness of the cortical ribbon, and the width of each layer relative to other layers. Particularly, the correlation between Brodmann's (1909) map and the functional data is not studied in sufficient detail in the Wernicke region.

Classical cytoarchitectonic maps are the result of observations of stained histological sections by pure visual microscopic inspection and qualitative descriptions of regional differences in cell morphology and laminar distribution patterns. Since this approach requires experience and pattern recognition ability, which are highly variable between observers, the resulting maps are different between different observers (tables 33.1 and 33.2). To solve this problem caused by an observer-dependent method, a quantitative approach to cytoarchitectonic mapping was introduced, that is, a statistically testable method for the definition of borders between cortical areas. This approach is based on measurements at microscopic resolution of the regional and laminar distribution of cell bodies in histological serial sections of postmortem brains (Schleicher, Amunts, Geyer, Morosan, & Zilles, 1999; Schleicher et al., 2000; Schleicher et al., 2005; Zilles, 1978). The identification of borders with this observer-independent

TABLE 33.1

Comparison of several parcellation schemes of Broca's region with that of Brodmann's (1909) map

Author(s)	BA44	BA45
von Economo and Koskinas (1925)	FCBm	FDΓ
Stengel (1930)	FCBm	FDΓ
Riegele (1931)	56, part of 57	Part of 57, 58, 59
Ngowyang (1934)	56, part of 57	Part of 57, 58, 59
Kononowa (1935)	44	45
Kreht (1935)	56, part of 57	Part of 57, 58, 59
Sarkissov, Filimonoff, and Preobrashenskaya (1949)	44	45, 45a
Sanides (1962, 1964)	56, 57	58, 59
Petrides and Pandya (1994)	44	45A, 45B
Amunts et al. (1999)	44	45
Amunts et al. (2010)	44d, 44v	45a, 45p

Note: Brodmann's map is currently the most widely used anatomical reference for the interpretation of functional imaging data, frequently via the atlas of Talairach and Tournoux (1988). Brodmann identified one cytoarchitectonic area on the opercular part of the inferior frontal gyrus (i.e., area BA44) and a second area on the triangular part of the inferior frontal gyrus (i.e., area BA45). Correspondences are based on descriptions by von Economo and Koskinas (1925), and Amunts and Zilles (2012). The map of Sanides (1962, 1964) is the result of cyto- and myeloarchitectonic observations, by Amunts et al. (2010), on a receptor architectonic analysis. The remaining studies examined the cytoarchitecture of Broca's region. FCBm = cytoarchitcctonic arca FCBm; FDΓ = cytoarchitcctonic arca FDΓ.

TABLE 33.2

Comparison of several nomenclature systems, based on cytoarchitectonic or receptor architectonic observations, with that of Brodmann's (1909) areas 41, 42, 52, 22, 21, 39, and 40

Author(s)	Staining	BA41	BA42	BA52	BA22	BA21	BA39	BA40
von Economo and Koskinas (1925)	C	TC, part of TD	TB	Part of TD	TA1, TA2	TE1	PG	PF
von Economo and Horn (1930)	C	TC, part of TD	TBa, TBp	TC, part of TD				
Sarkissov et al. (1955)	C	41c, 41b, 41bc	42	52	22	21	39	40, 40op
Morosan et al. (2001); Morosan, Rademacher, et al. (2005); Morosan, Schleicher, et al. (2005)	C, R	Te1	Te2.1, Te2.2	TI1	Te3, Te4 (parts of BA22)			
Eickhoff, Schleicher, et al. (2006)	C							OP1 (probably also OP2; parts of BA40)
Caspers et al. (2006); S. Caspers et al. (2013)	C, R						PGa, PGp	PF, PFcm, PFm, PFop, PFt (parts of BA40)
Present study	C, R			Te2.3, TI	Te5, Te6 (parts of BA22)			

C = cytoarchitecture; n.d. = not determined; R = receptor architecture; TB = cytoarchitectonic area TB; TBa = cytoarchitectonic area TBa; TBp = cytoarchitectonic area TBp; TC = cytoarchitectonic area TC; TD = cytoarchitectonic area TD.

approach in 10 or more individual brains and their registration to a common reference space or single reference brain leads to the so-called continuous cytoarchitectonic probabilistic maps. They provide information about variability between individuals in position of cytoarchitectonically defined borders with respect to surrounding sulci and enable quantification of intersubject differences in the extent and location of cortical borders (Zilles & Amunts, 2010). These continuous cytoarchitectonic probabilistic maps, and the maximum probability maps, which are computed from them, can also be used to test the degree of overlap between activation clusters obtained by functional imaging studies and thus enable the highest currently available localization precision (Eickhoff et al., 2005; Eickhoff, Heim, Zilles, & Amunts, 2006; Eickhoff et al., 2007). Furthermore, such observations in the Broca region and the application of receptor architectonic criteria led to the detection of subdivisions of Brodmann's areas (BA)44 and BA45 as well as to the identification of previously undescribed cytoarchitectonic areas (Amunts et al., 2010).

Despite such observer-independent mapping procedures, cytoarchitectonic maps only reveal part of the microstructural ground truth, since they are based on a single staining technique, and thus reveal only one aspect of the tissue properties, for example, distribution of cell bodies. Therefore, the application of in vitro receptor autoradiography provided a powerful tool to analyze an important aspect of the molecular organization of the cerebral cortex, since this method is quantitative, can be applied to serial sections through entire brains and can visualize numerous different receptor types, that is, multimodality in one and the same brain. Since the density of receptors is regionally heterogeneous, variations in regional and laminar receptor densities not only corroborate previously identified cytoarchitectonic borders, but they also reveal hitherto unknown subdivisions within cytoarchitectonically defined areas or new areas (e.g., Amunts et al., 2010; Geyer et al., 1996; Scheperjans, Grefkes, Palomero-Gallagher, Schleicher, & Zilles, 2005). Furthermore, the regional-specific balance between the densities of multiple receptor types reflects hierarchical and systemic aspects of cortical functional organization (Zilles, Palomero-Gallagher, et al., 2002; Zilles & Amunts, 2009; Zilles, Bacha-Trams, et al., 2015).

Aims of the present chapter are to provide a comprehensive multimodal description of the cyto- and receptor architecture in Broca's and Wernicke's regions and to describe how the differential distribution of transmitter receptors in language-relevant areas not only highlight their borders, but also reveal the existence of a common molecular basis for interaction in the functionally defined sentence comprehension network.

2. *Broca's Region*

The most widely accepted parcellation scheme, and the first description of cytoarchitectonically distinct entities within Broca's region, is that proposed by Brodmann (1909), who identified his areas as BA44 and BA45 (figure 33.1). The borders of BA44 and BA45 were confirmed by Amunts et al. (1999) in their quantitative cytoarchitectonic study. The most conspicuous cytoarchitectonic feature that distinguishes BA44 and BA45 from the surrounding cortex is the occurrence of large pyramidal cells in deep layer III (figure 33.1; Amunts et al., 1999). Furthermore, both areas present a diffuse transition from layer II to layer III, as well as a relatively low cell packing density in layer VI compared to surrounding areas. BA44 and BA45 can be distinguished mainly from each other due to differences in layer IV, which is dysgranular (i.e., an interrupted layer IV [Petrides, 2013] invaded by pyramidal cells from layers III and V and layer IV) in the former area (figure 33.1). Although BA45 contains a clearly visible layer IV, it is still less distinct than that in the rostrally adjoining lateral prefrontal cortex (Amunts et al., 1999).

Although BA44 and BA45 are often considered synonymous with opercular and triangular parts of the inferior frontal gyrus, these macroscopical landmarks cannot be used as reliable markers for the cytoarchitectonic borders due to a high intersubject variability in both the sulcal pattern of Broca's region (Keller et al., 2007; Ono, Kubik, & Abermathey, 1990) and the relationship of areal borders to surrounding sulci (Amunts et al., 2004; Amunts et al., 1999; Amunts, Schleicher, Ditterich, & Zilles 2003; Kononowa, 1935, 1938; Riegele, 1931). For instance, although the ascending branch of the lateral fissure does at times serve as an anatomical landmark to pinpoint the border between BA44 and BA45, it can also be found within either of these two areas (Amunts et al., 2004). Likewise, the inferior rostral sulcus and the horizontal branch of the lateral fissure do not constitute reliable markers for the dorsal and ventral borders of BA45, respectively (Kononowa, 1935, 1938). This intersubject variability in the shape, size, and position of BA44 and BA45 relative to the sulcal and gyral anatomy of Broca's region has been quantified and visualized by means of cytoarchitectonic probabilistic maps (Amunts et al., 2004). Figure 33.2A shows maximum probability maps of BA44 and BA45.

Functional imaging studies have not only shown that different parts of BA44 and BA45 are involved in different aspects of language processing, but also that the

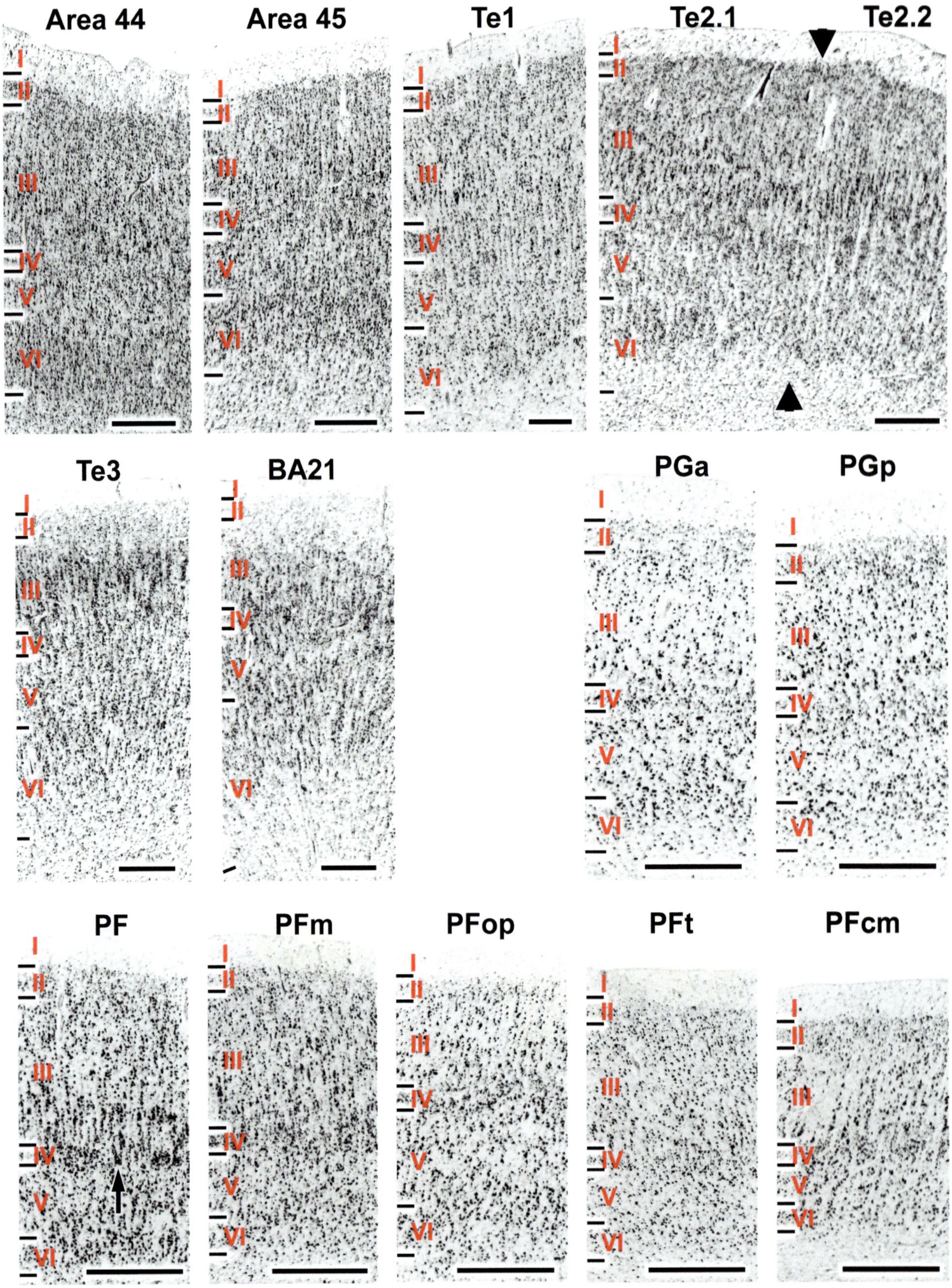

FIGURE 33.1 Cytoarchitecture of areas 44 and 45 of the Broca region as well as of primary auditory area Te1 (corresponds to BA41), secondary auditory areas Te2.1 and 2.2 (parts of BA42), temporal areas Te3 (part of BA22) and BA21, and parietal areas PF, PFcm, PFm, PFop, PFt, PGa, and PGp of the Wernicke region. Arrowheads highlight the position of the border between Te2.1 and Te2.2. Arrow in PF highlights a string of layer III pyramids that extend through layer IV into layer V. Micrograph of BA21 was taken from the dome of the middle temporal gyrus. Roman numerals indicate cortical layers. Scale bars = 0.5 mm.

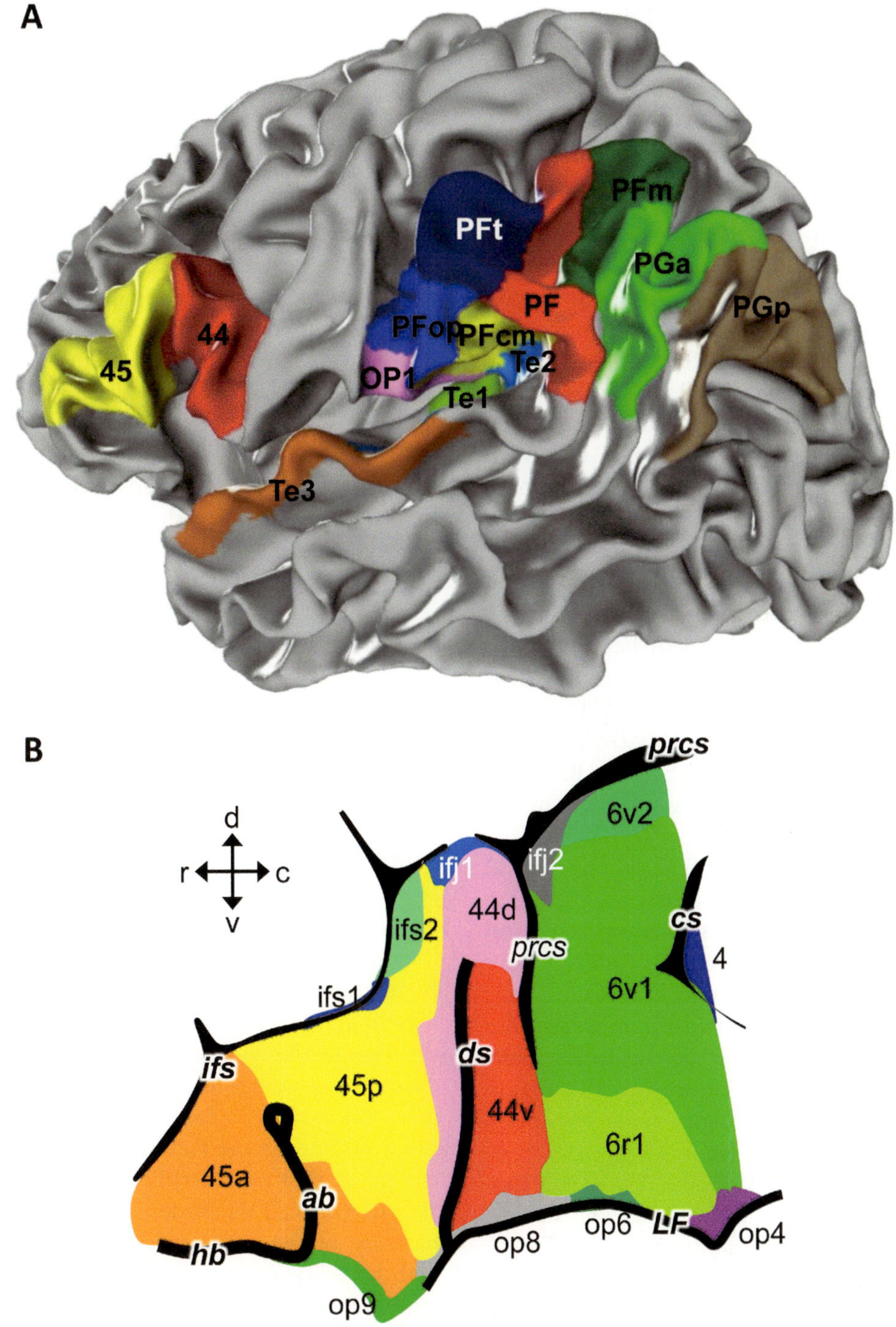

FIGURE 33.2 (A) Maximum probabilistic maps of cytoarchitectonic areas in Broca's and Wernicke's region. Data for BA44 and BA45 from Brodmann's (1909); Te1, Te2, and Te3 from Morosan et al. (2001); Morosan, Rademacher, et al., (2005); and Morosan, Schleicher, et al. (2005); opercular area OP1 from Eickhoff, Schleicher, et al. (2006) and Eickhoff, Amunts, et al. (2006); and parietal areas PFop, PFt, PFcm, PF, PFm, PGa, and PGp from Caspers et al. (2006). (B) Receptor architectonic map of Broca's areas 44 and 45 and of the surrounding cortex. 4 = rostral part of the primary motor cortex BA4; 6r1 = transitional area found between areas BA6 and BA44; 6v1, 6v2 = ventral subdivisions of BA6; 44d, 44v = dorsal and ventral subdivisions of BA44; 45a, 45p = anterior and posterior subdivisions of BA45; ab = ascending branch of the lateral fissure; c = caudal; cs = central sulcus; d = dorsal; ds = diagonal sulcus; hb = horizontal branch of the lateral fissure; ifj1, ifj2 = areas 1 and 2 at the junction between the inferior frontal and precentral sulci; ifs = inferior frontal sulcus; ifs1, ifs2 = areas 1 and 2 on the ventral bank of the inferior frontal sulcus; LF = lateral fissure; op4, op6, op8, op9 = opercular areas 4, 6, 8, and 9; prcs = precentral sulcus; r = rostral; v = ventral. Modified from Amunts et al. (2010).

concept of Broca's region as encompassing only opercular and triangular portions of the inferior frontal gyrus is too restrictive (Ardila, Bernal, & Rosselli, 2016; Hagoort, 2014). Under this assumption, a recent multimodal *postmortem* study revised the parcellation of Broca's region and of the adjoining areas on the frontal operculum, as well as of the ventral premotor cortex based on differences in transmitter receptor distribution patterns (Amunts et al., 2010). The resulting receptor map is shown in figure 33.2B. One of the main findings of this study is the existence of receptor architectonically distinct subdivisions within BA44 and BA45 as well as the detection of additional new areas in Broca's region. Multiple receptors (e.g., kainate, gamma-aminobutyric acid ionotropic receptor family B [$GABA_B$], muscarinic acetylcholine receptor M2 [M_2], alpha-1 adrenergic receptor [α_1], 5-hydroxytryptamine $[HT]_{1A}$, and adenosine A1 receptor [A_1]) segregate the cytoarchitectonically identified area BA44 into dorsal (44d) and ventral (44v) parts (figure 33.3; Amunts et al., 2010). Likewise, area BA45 can be subdivided into receptor architectonically distinct anterior (45a) and posterior (45p) areas. Area 45p presents lower kainate, *N*-methyl-D-aspartate receptor (NMDA), metabotropic glutamate receptor (mGluR)2/3, $GABA_B$, M_3, and 5-HT_2 densities than 45a does (figure 33.3; Amunts et al., 2010).

Several small areas were described on the ventral bank of the inferior frontal sulcus (areas 1 and 2 of the inferior frontal sulcus, ifs1 and ifs2, respectively), and at its junction with the precentral sulcus (inferior frontal junction areas 1 and 2, ifj 1 and ifj2, respectively; Amunts et al., 2010). Areas ifj1 and ifj2 represent the receptor architectonic correlates of the functionally identified inferior frontal junction region. This region is consistently activated by paradigms requiring cognitive control (Amunts & von Cramon, 2006; Derrfuss, Brass, & von Cramon, 2004; Derrfuss, Brass, von Cramon, Lohmann, & Amunts, 2009). Area 45p can be distinguished from ifs1 by its higher kainate and M_3, but lower NMDA, and 5-HT_2 receptor densities (figure 33.3; Amunts et al., 2010). At more dorsal and ventral levels, 45a is delimited by BA46 and BA47, respectively. These latter areas are thought to be part of the extended Broca region (Ardila et al., 2016). Area 45a contains lower alpha-amino-3-hydroxy-5-methyl-4-isoxazolepropionic acid receptor (AMPA) and $GABA_A$, but higher NMDA and M_2 receptor densities than either BA46 or BA47 do (Zilles, Bacha-Trams, et al., 2015). Ventral to areas BA44 and BA45, two additional dysgranular areas were identified within the deep frontal operculum: the more caudal op8 and more rostral op9 (Amunts et al., 2010).

Amunts et al. (2010) also identified and delineated a novel area, namely 6r1, located in the ventral precentral sulcus between 44v and the most ventral portion of BA6. Cytoarchitectonically, 6r1 is characterized by a rather indistinct laminar differentiation, comparable to that of BA6, and by the presence of a layer IV, though considerably thinner and even less continuous than that of dysgranular BA44. Therefore, 6r1 is a transitional area between the "agranular" premotor cortex and areas of the Broca region (Amunts et al., 2010). The receptor architectonic border between 6r1 and BA6 is revealed by differences in M_2 and α_1 receptor densities, and that between 6r1 and BA44 is highlighted by kainate, $GABA_A$, and α_1 receptors (Amunts et al., 2010). Finally, two further receptor architectonically distinct areas were identified on the precentral gyrus at a position occupied by the most the ventrolateral portion of BA6 and were thus named 6v1 and 6v2 (figure 33.2B; Amunts et al., 2010).

3. Wernicke's Region

Wernicke's region is currently thought to encompass the primary and secondary auditory cortices, the superior temporal gyrus, as well as parts of the middle temporal, supramarginal, and angular gyri (Hickok & Poeppel, 2007; Binder, 2015; DeWitt & Rauschecker, 2013).

3.1. Primary Auditory Cortex BA41 constitutes the cytoarchitectonic correlate of the primary auditory cortex (Brodmann, 1907) and is located mainly on the Heschl gyrus (accompanied by the first transverse and the Heschl sulcus), though in some brains it extends caudally onto the planum temporale, whereas in other cases the border is found on the wall of the Heschl gyrus and not in the fundus of the first transverse sulcus or in that of Heschl's sulcus (Brodmann, 1909; Morosan et al., 2001; Rademacher et al., 2001). For comparisons of the various delineations in the literature and the present map, see table 33.2. BA41 corresponds to temporal area 1 (Te1) of Morosan et al. (2001).

The primary auditory cortex is characterized by its koniocortical appearance (figure 33.1). That is, it presents a broad and densely packed layer IV with notably small granule cells and an invasion of small granular cells in the supragranular layers, which makes it difficult to distinguish the border between layers II and IV (Morosan et al., 2001; Sarkissov, Filimonoff, Kononowa, & Preobrachenskaja, 1955; von Economo & Koskinas, 1925). Furthermore, layer III neurons are arranged in distinct radial columns known as *rain shower* or *organ pipe* formations of von Economo and Koskinas (1925).

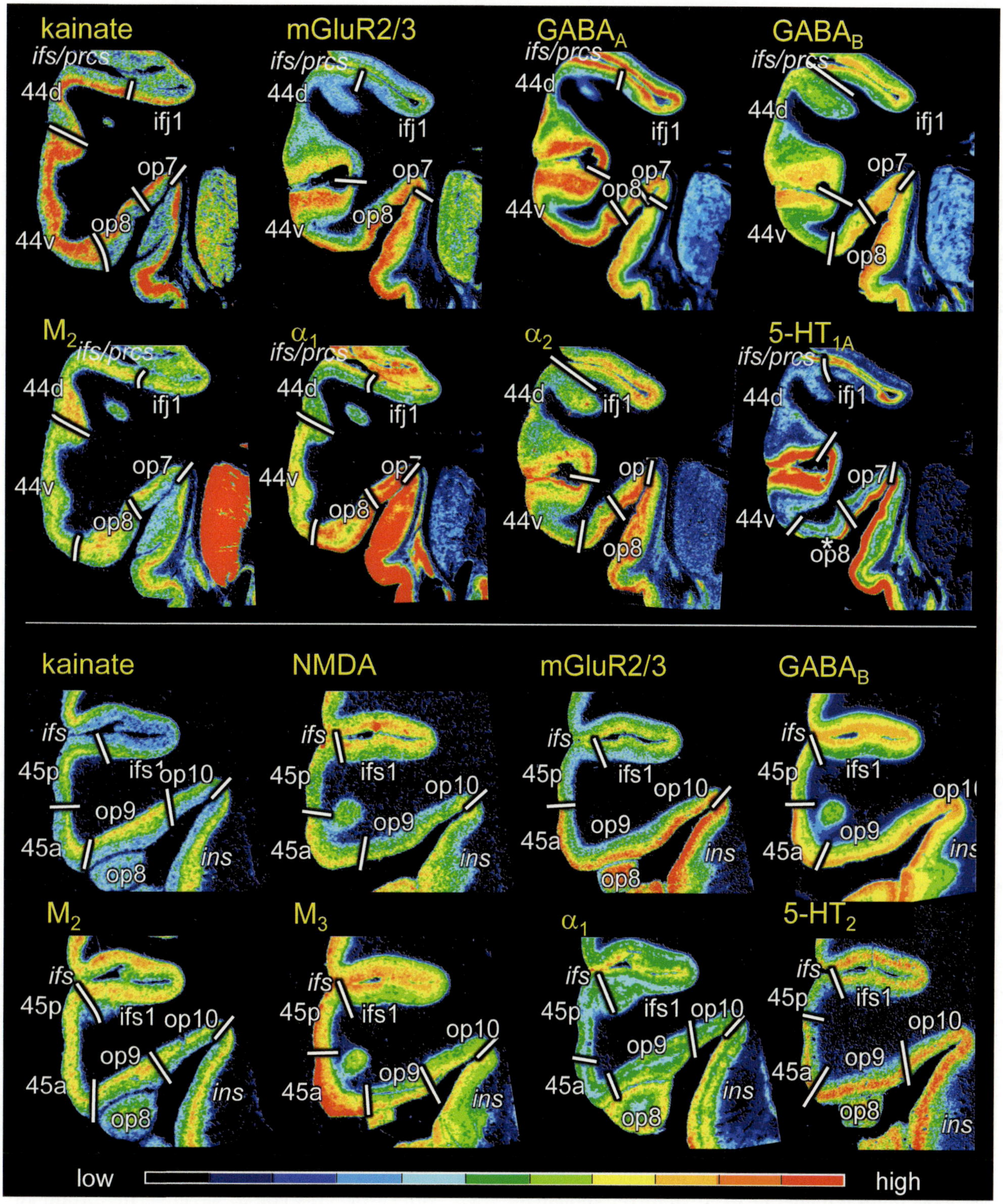

FIGURE 33.3 Neighboring coronal sections through the opercular (*top*) and triangular (*bottom*) parts of the inferior frontal gyrus showing the distribution of receptors for glutamate (ionotropic kainate, NMDA, metabotropic mGluR2/3), GABA ($GABA_A$, $GABA_B$), acetylcholine (muscarinic M_2, M_3), noradrenalin (α_1, β_2), and serotonin ($5\text{-}HT_{1A}$, $5\text{-}HT_2$) in cytoarchitectonically identified areas 44 and 45 and the surrounding cortex. The M_2 do not reveal the border between 45a and 45p. 44d, 44v = receptor architectonically identified dorsal and ventral subdivisions of 44, respectively; 45a, 45p = receptor architectonically identified anterior and posterior subdivisions of 45, respectively; ifj1 = area 1 at the junction between the inferior frontal and precentral sulci; ifs/prcs = inferior frontal sulcus/precentral sulcus; ifs1 = area 1 of the inferior frontal sulcus; ins = insular cortex; op7, op8, op9, op10 = opercular areas 7, 8, 9, and 10, respectively. Asterisks mark artifacts (artificial folds or lost tissue). Color bar codes for receptor densities. Autoradiographs of kainate, $GABA_A$, M_2, and α_1 receptors in area 44 and of α_1 receptors in area 45 were taken from Amunts et al. (2010).

Like all primary sensory areas, the primary auditory cortex is characterized by a distinct receptor architecture that clearly demarcates it from the adjacent secondary auditory and temporoinsular cortices (Zilles & Palomero-Gallagher, 2017). In particular, the muscarinic cholinergic M_2 receptors are expressed at considerably higher levels in the primary auditory cortex than in the surrounding cortex (figure 33.4; Morosan, Rademacher, Palomero-Gallagher, & Zilles, 2005; Zilles, Schleicher, Palomero-Gallagher, & Amunts, 2002; Zilles & Palomero-Gallagher, 2017). The NMDA, kainate, $GABA_A$, α_2, and 5-HT_2 receptors, as well as the nicotinic $\alpha_4\beta_2$ receptor, which presents a conspicuous local maximum in layer IV, also highlight Te1 by their high densities (figure 33.4).

3.2. Secondary Auditory Cortex BA42, which is thought to be the cytoarchitectonic correlate of the secondary auditory cortex (Brodmann, 1907), lies on the planum temporale, adjacent to BA41, and encroaches laterally onto the free surface of the superior temporal gyrus, where it surrounds the most lateral portion of BA41 (Brodmann, 1909; Morosan et al., 2001; Morosan, Rademacher, et al., 2005). For a correspondence of the various nomenclatures in the literature, see table 33.2. The cortex of the planum temporale reaches a thickness of 3 to 3.2 mm and is therefore surprisingly broad, particularly considering the fact that it constitutes the dorsal wall of the superior temporal gyrus, and that walls of gyri are generally quite thin (von Economo & Koskinas, 1925).

Since Brodmann did not provide a detailed description of his area 42, we here refer to von Economo and Koskinas (1925) and to Morosan, Rademacher, et al. (2005) and adhere to the nomenclature of the latter authors. The cytoarchitecture of temporal area 2 (Te2) of Morosan, Rademacher, et al. (2005) displays intermediate features between those of the primary and higher auditory cortices. As in the primary auditory cortex Te1, pyramids in layer III of Te2 are arranged in organ pipe formations. The most conspicuous difference between Te1 and Te2 is the appearance of medium-sized pyramidal neurons in layer IIIc of the latter area (figure 33.1; Morosan, Rademacher, et al., 2005; von Economo & Koskinas, 1925). The packing density of layer V pyramids is higher in Te2 than in Te1, though it is lower than in the laterally adjacent higher auditory cortex on the superior temporal gyrus (Galaburda & Sanides, 1980; Morosan, Rademacher, et al., 2005; von Economo & Koskinas, 1925). Differences in neuronal packing density in layer V, as well as in the width of layer III, enable the subdivision of Te2 into two cytoarchitectonically distinct areas (figure 33.1): Area Te2.1 abuts the primary auditory cortex and is demarcated laterally and anteriorly by Te2.2 (Morosan, Rademacher, et al., 2005; von Economo & Horn, 1930). Area Te2.1 presents a thinner layer III and more densely packed layer V than Te2.2 (Morosan, Rademacher, et al., 2005; von Economo & Horn, 1930). The border between Te2.1 and Te2.2 is also revealed by the higher NMDA, $GABA_A$, M_2, M_3, α_2, and 5-HT_2, but lower AMPA, kainate, mGluR2/3, $GABA_B$, nicotinic $\alpha_4\beta_2$, α_1, 5-HT_{1A}, and dopamine 1 (D_1) receptor densities in the former than in the latter area (figure 33.4; Morosan, Rademacher, et al., 2005). Medial to the primary auditory cortex, area BA52 constitutes a transitional area to the insular region (Brodmann, 1907) and is characterized by a wider layer III, thinner layer IV, and more densely packed layers V and VI than those of the primary auditory cortex (Morosan, Rademacher, et al., 2005).

The present receptor architectonic observations reveal the existence of subdivisions within this transition region BA52 (figure 33.4). We identified a novel area Te2.3, which directly abuts the primary auditory cortex, and would thus constitute a medial part of the secondary auditory cortex. Te2.3 is separated from the insula by the temporoinsular area TI. Te2.3 contains higher AMPA, mGluR2/3, M_1, M_3, α_1, 5-HT_{1A}, and D_1, but lower NMDA, kainate, $GABA_A$, M_2, nicotinic $\alpha_4\beta_2$, α_2, and 5-HT_2 receptor densities than Te1. The $GABA_B$ receptor does not highlight the border between these two areas (figure 33.4). Te2.3 contains higher NMDA, $GABA_A$, M_2, nicotinic $\alpha_4\beta_2$, and α_2, but lower AMPA, kainate, $GABA_B$, M_1, M_3, α_1, 5-HT_{1A}, and D_1 receptor densities than medially adjacent TI (figure 33.4). The mGluR2/3 and 5-HT_2 receptors do not highlight the border between Te2.3 and TI. Finally, TI presents higher AMPA, $GABA_A$, M_2, and M_3, but lower NMDA, kainate (particularly in the supragranular layers), mGluR2/3, M_1, and D_1 receptor densities than the adjacent insular cortex (figure 33.4). This border, however, was not revealed by the $GABA_B$, nicotinic $\alpha_4\beta_2$, α_1, α_2, 5-HT_{1A}, or 5-HT_2 receptors (figure 33.4).

The various areas of the Te2 family (i.e., Te2.1, Te2.2, and Te2.3) seem to correspond to the areas of the auditory belt region in primates (Kaas & Hackett, 2000; Kaas, Hackett, & Tramo, 1999).

3.3. Superior Temporal Gyrus In Brodmann's (1909) map, the superior temporal gyrus is mostly occupied by BA22, the superior temporal area, which forms an arc around BA41, BA42, and BA52. Rostrally and caudally, BA22 reaches dorsally into the planum polare and temporale, respectively. Ventrally, the border is located within the superior temporal sulcus at most rostral and caudal levels of BA22, but along its

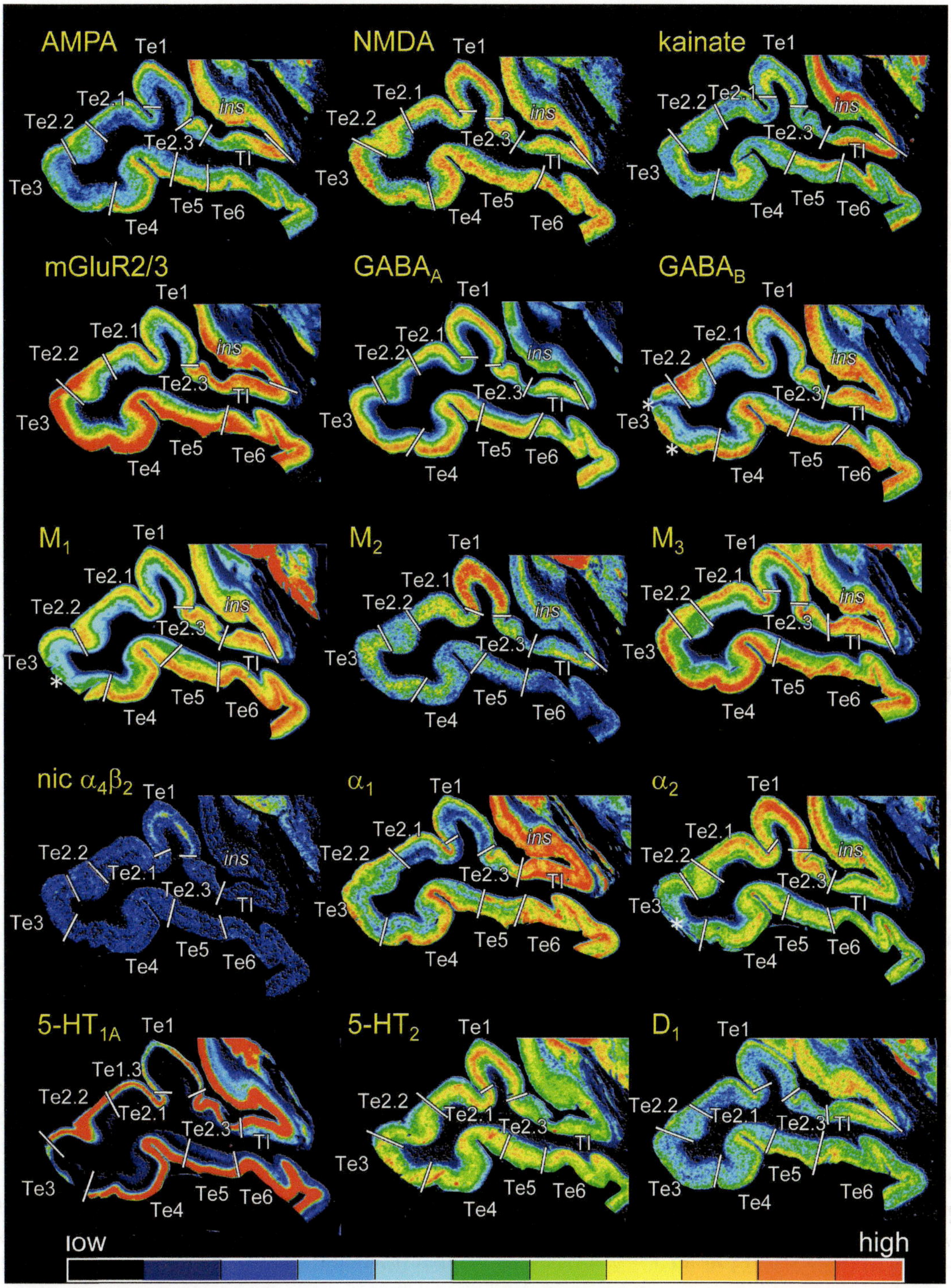

FIGURE 33.4 Receptor architectonic parcellation of the superior temporal gyrus. ins = insular cortex, Te1 = primary auditory cortex (corresponds to BA41); Te2.1, Te2.2 (parts of BA42) and Te2.3 (part of BA52) secondary auditory areas; and Te3 to Te6 = receptor architectonic areas within BA22. Color bar codes for receptor densities.

central portion the border is found on the free surface of the middle temporal gyrus (Brodmann, 1909).

The cortex of the superior temporal area is of average thickness, reaching values around 3 mm (von Economo & Koskinas, 1925). It can be generally characterized by somewhat narrow layers II and IV, a relatively broad layer III with medium-sized neurons, the presence of the organ pipe formation, and a poor demarcation between layer VI and the white matter (figure 33.1; Morosan, Schleicher, Amunts, & Zilles, 2005; von Economo & Koskinas, 1925). However, subsequent studies (including the present observations) have shown that BA22 is not homogeneous, but can be subdivided into several areas (table 33.2). For example, von Economo and Koskinas (1925) described two areas within BA22: area TA_1 occupies the caudal portion of BA22, TA_2 its rostral one. Furthermore, Morosan and colleagues (Morosan, Rademacher, et al., 2005; Morosan, Schleicher, et al., 2005) identified a subdivision along the dorsoventral axis of TA_1 using multimodal and statistically testable approaches. Thus, their temporal area 3 (Te3) is found mainly on the free surface of the superior temporal gyrus, though it encroaches onto the dorsal wall of the superior temporal sulcus (figure 33.2A), where it is followed by temporal area 4 (Te4). Te3 is probably part of the parabelt region (Kaas & Hackett, 2000; Kaas et al., 1999).

A detailed analysis of the receptor architecture of the superior temporal gyrus revealed that area Te4 is restricted to the dorsal wall of the superior temporal sulcus, where it is followed by temporal area 5 (Te5), which is in turn delimited medially by temporal area 6 (Te6; figure 33.4). Area Te4 can be distinguished from laterally adjacent Te3 by its higher AMPA, NMDA, kainate, $GABA_A$, $GABA_B$, M_1, α_1, α_2, $5\text{-}HT_{1A}$, and D_1, but lower M_2, nicotinic $\alpha_4\beta_2$, and $5\text{-}HT_2$ receptor densities. The mGluR2/3 and M_3 receptors do not highlight the border between these two areas. Medially, Te4 presents higher $GABA_A$, M_2, M_3, nicotinic $\alpha_4\beta_2$, and α_2, but lower (particularly in infragranular layers) AMPA, kainate, $GABA_B$, M_1, α_1, $5\text{-}HT_{1A}$, $5\text{-}HT_2$, and D_1 receptor densities than Te5 (figure 33.4). These two areas do not differ significantly in their NMDA or mGluR2/3 receptor densities. Finally, the border between Te5 and Te6 is revealed by the higher mGluR2/3, $GABA_A$, $GABA_B$, M_2, and α_2, but lower AMPA, NMDA, kainate, M_1, nicotinic $\alpha_4\beta_2$, α_1, $5\text{-}HT_{1A}$, $5\text{-}HT_2$, and D_1 receptor densities in the former area (figure 33.4). This border was not shown by the M_3 receptors (figure 33.4).

3.4. Middle Memporal Gyrus Brodmann identified his area 21 on the middle temporal gyrus and specified that its borders cannot be consistently associated with the superior or inferior temporal sulci and that the changes in cytoarchitecture that enable its demarcation from rostrally and caudally adjacent areas are gradual in nature (Brodmann, 1909). Based on its macroanatomical location, BA21 is the equivalent of von Economo and Koskinas's (1925) area TE1 (Zilles, Palomero-Gallagher, Bludau, Mohlberg, & Amunts, 2015).

The cortex of middle temporal gyrus, and in particular of its rostral portion, is one of the thickest of the entire brain, reaching values between 3 and 3.8 mm depending on whether measurements were taken from the walls or the dome of the middle temporal gyrus, respectively (von Economo, 1927; von Economo & Koskinas, 1925). This brain region is characterized by an arrangement of cell bodies into vertical columns in layers III to VI, and a diffuse layer VI/white matter border (area BA21 in figure 33.1; von Economo & Horn, 1930). Layer III is relatively thin and cell-sparse and presents a superficial-to-deep increase in the size of its pyramids, whereas layer IV is relatively broad (von Economo & Koskinas, 1925). Layers V and VI contain numerous large cells and are also particularly prominent due to both their thickness relative to total cortical width and their higher cell packing density compared to that of supragranular layers.

Future quantitative studies of the cyto- and receptor architecture of the middle temporal gyrus are needed to (a) determine whether it is indeed a homogeneous brain region, (b) identify the position of its borders (and those of possible subdivisions) by means of multivariate statistical analysis, and (c) create a cytologically informed probabilistic three-dimensional map in stereotaxic space that will enable analysis of intersubject variability and can be applied as template for localization of functional imaging data.

3.5. Angular Gyrus Classical cytoarchitectonic parcellation schemes identified a single area on the angular gyrus, that is, BA39 (table 33.2). BA39 is largely coincident with area PG of von Economo and Koskinas (1925). However, a recent quantitative approach identified two cytoarchitectonically distinct areas (areas PGa and PGp) within BA39 (Caspers et al., 2006), and the resulting probabilistic maps not only serve as templates for localization of functional imaging data, but quantify intersubject variability in the size and location of PGa and PGp (figure 33.2A; Caspers et al., 2008).

Cortex on the angular gyrus is relatively broad, with a cortical thickness ranging from 2.9 to 3.3 mm (von Economo & Koskinas, 1925). Areas PGa and PGp are both characterized by a well-developed layer IV which divides the cortex into a supragranular and an infragranular portion of comparable width and a very low cell-packing density in layer V (Caspers et al., 2006). Pyramidal neurons in layer III of PGa form vertical columns and show a clear superficial-to-deep increase in size. In contrast, layer III of PGp contains medium-sized and homogeneously distributed pyramids, with only a very slight increase in size toward the border with layer IV. Layer VI is more cell dense in PGa than in PGp. Finally, PGp shows a sharp layer VI/white matter border, whereas that of PGa is indistinct.

Most conspicuous receptor architectonic differences between PGa and PGp were found for kainate, α_1, 5-HT_{1A}, 5-HT_2, and D_1 (not shown here) receptors. Densities of 5-HT_{1A} and α_1 receptors are lower (particularly in layers V and VI), and those of kainate, 5-HT_2, and D_1 receptors are higher in PGa than in PGp (S. Caspers et al., 2013). Furthermore, PGa presents higher AMPA, $GABA_A$, nicotinic $\alpha_4\beta_2$, α_2, and D_1, but lower kainate, 5-HT_{1A}, and 5-HT_2 receptor densities as well as benzodiazepine binding site densities than rostrally adjacent PFt (S. Caspers et al., 2013).

3.6. Supramarginal Gyrus Existing parcellation schemes differ considerably in the number of areas found on the supramarginal gyrus (table 33.2). Brodmann (1909) identified a single area (BA40) that is now known to encompass opercular area OP1 (and probably also area OP2), as well as areas PFop, PFt, PFcm, PF, and PFm, all of which were identified by means of quantitative cytoarchitectonic analyses and brought into stereotaxic space to enable comparison with results from functional imaging studies (figure 33.2A; Eickhoff, Amunts, Mohlberg, & Zilles, 2006; Eickhoff, Schleicher, Zilles, & Amunts, 2006; Caspers et al., 2006; Caspers et al., 2008).

The cortex of the supramarginal gyrus is relatively thick, with values ranging from 3 to 3.5 mm (von Economo & Koskinas, 1925). General cytoarchitectonic characteristics of areas OP1, PFop, PFt, PFcm, PF, and PFm are the extremely small layer II granule cells, a superficial-to deep increase in the size of layer III pyramids, and a relatively cell-sparse layer V, which appears as a pale stripe between layers IV and VI (figure 33.1; Caspers et al., 2006; von Economo & Koskinas, 1925). Cytoarchitectonic features specific to each area are as follows:

- OP1: distinct columnar arrangement of cell bodies in layers III to VI and a relatively diffuse layer VI/white matter border (Eickhoff, Schleicher, et al., 2006).
- PFop: layer VI is broad and can be easily demarcated from the white matter (Caspers et al., 2006). PFop also presents the most undifferentiated lamination of all areas identified on the supramarginal gyrus.
- PFt: layer IV is relatively thin. Layer V can be subdivided into superficial and deep parts, with a lower cell density in the latter one (Caspers et al., 2006).
- PFcm: columnar arrangement of cell bodies in layers III to VI and a very sharp layer VI/white matter border. PFcm is also extremely thin and overall rather cell-sparse compared to surrounding areas (Caspers et al., 2006).
- PF: conspicuous layer III due to its considerable width relative to total cortical thickness and arrangement of its pyramids into relatively wide vertical radial columns (see arrow in figure 33.1; Caspers et al., 2006).
- PFm: layer III is relatively wide and has a particularly pronounced superficial-to-deep increase in the size of its neurons. Layer IV can be clearly demarcated from layers III and V (Caspers et al., 2006).

The subdivision of BA40 is also supported by differences in the regional and/or laminar distribution patterns of receptors for classical neurotransmitters (S. Caspers et al., 2013). Area PFop contains higher AMPA, kainate, $GABA_B$, M_1, M_2, and α_1 receptor densities as well as benzodiazepine binding site concentrations, but lower NMDA, α_2, 5-HT_2, and D_1 receptor densities than OP1. Furthermore, infragranular layers of PFop present higher 5-HT_{1A} receptor densities than those of OP1. The border between PFt and caudally adjacent PF is revealed by the clearly higher AMPA and M_2 receptor densities and benzodiazepine binding site concentrations, but lower NMDA, α_2, and 5-HT_2 receptor densities in the former area (S. Caspers et al., 2013). The border between PF and PFcm is most clearly highlighted by differences in the densities of AMPA, kainate, $GABA_A$, and D_1 receptors, which are all higher in the former than in the latter area (S. Caspers et al., 2013).

4. Receptors and the Sentence Comprehension Network

Differences in receptor densities not only corroborate previously identified borders of cytoarchitectonic areas, but reveal hitherto unknown subdivisions within these areas as exemplarily described herein for the subdivisions within BA44 and BA45 (Amunts et al., 2010). Furthermore, receptors can also group two or more cytoarchitectonically identified areas into a family of neurochemically related areas. Therefore, the combined analysis of multiple receptor types in a specific brain region provides insights into the functional organization of the cortex (Zilles, Palomero-Gallagher, et al., 2002; Zilles & Amunts, 2009).

The concept of a "receptor fingerprint" was introduced to define the balance of multiple receptor types from different neurotransmitter systems in a specific brain region (Zilles, Palomero-Gallagher, et al., 2002; e.g., fingerprints of areas 44v, 44d, 45a and 45p in figure 33.5A). Receptor fingerprints, which can differ in shape and/or size, represent the molecular default organization of the regionally specific local information processing in each cortical area. Similarities in fingerprints are indicative of comparable multireceptor balances and enable identification of neurochemically related cortical areas, which may also participate in the same functional system (Zilles, Palomero-Gallagher, et al., 2002; Zilles & Amunts, 2009). The degree of (dis)similarity between fingerprints can be objectively examined by means of multimodal statistical tests such as multidimensional scaling or hierarchic cluster analyses, which have been applied to demonstrate the existence of hierarchical progression within areas of the visual, parietal, and cingulate cortices (Caspers et al., 2015; S. Caspers et al., 2013; Palomero-Gallagher et al., 2009; Scheperjans, Palomero-Gallagher, Grefkes, Schleicher, & Zilles, 2005).

A recent study examined whether areas involved in language processing share a common multireceptor expression, that is, they can be characterized by similar receptor fingerprints and whether these fingerprints differ from those of areas subserving nonlanguage functions (Zilles, Bacha-Trams, et al., 2015). The densities of 15 different receptors were quantified in eight areas belonging to the large-scale frontotemporal language network activated during sentence comprehension and in 18 areas involved in motor control, sensory perception, higher visual functions, or diverse cognitive or emotion-related functions. The anatomical sites from which receptor densities were extracted were identified based on the localization of functional activations in three of the areas examined, that is, ifs1/ifj, posterior superior temporal gyrus (pSTG)/superior temporal sulcus (STS), and 44d. All other areas were localized based on their cyto- and/or receptor architectonic characteristics. Areas ifs1/ifj, pSTG/STS, and 44d are involved in the processing of syntactically complex, embedded sentences and have been identified by subtracting activation for syntactically simple sentences from that for syntactically complex sentences (Friederici, 2009; Makuuchi, Bahlmann, Anwander, & Friederici, 2009). Additionally, area 44d, which can also be delineated based on its receptor architecture (Amunts et al., 2010), is activated during the processing of noncanonical object-first sentences (Friederici, Fiebach, Schlesewsky, Bornkessel, & von Cramon, 2006; Grewe et al., 2005). Further language-related areas examined were: areas 44v, 45a, and 45p, which are receptor architectonically identified subdivisions of Broca's region (Amunts et al., 2010); Brodmann's (1909) cytoarchitectonic area 47, located rostral to Broca's region and involved in language comprehension (Dronkers, Wilkins, van Valin, Redfern, & Jaeger, 2004; Hagoort, Hald, Bastiaansen, & Petersson, 2004; Turken & Dronkers, 2011); area Te2, which can be cyto- and receptor architectonically segregated from adjacent areas on the temporal lobe (Morosan, Rademacher, et al., 2005) and has been found to be activated by speech stimuli and language processing (Kubanek, Brunner, Gunduz, Poeppel, & Schalk, 2013). The nonlanguage-related areas included were the hand (4d) and mouth (4v) representation regions within the primary motor area 4 (Geyer et al., 1996), primary visual area V1 (Amunts, Malikovic, Mohlberg, Schormann, & Zilles, 2000), primary auditory cortex Te1 (Morosan, Schleicher, et al., 2005), primary somatosensory area 3b (Geyer, Schleicher, & Zilles, 1997), extrastriate higher visual areas FG1 and FG2 on the fusiform gyrus (J. Caspers et al., 2013; Caspers et al., 2015), cingulate area 32 (Palomero-Gallagher et al., 2009), prefrontal areas 9 and 46 (Brodmann, 1909), area 7 of the superior parietal lobule (Scheperjans, Palomero-Gallagher, et al., 2005; Scheperjans et al., 2008), as well as areas PF, PFcm, PFm, PFop, PFt, PGa, and PGp of the inferior parietal lobule (Caspers et al., 2006). Hierarchical cluster and multidimensional scaling analyses were carried out separately for receptor fingerprints extracted from these areas in the left and right hemispheres.

The hierarchical cluster analysis of the receptor fingerprints extracted from the left hemisphere (figure 33.5B; Zilles, Bacha-Trams, et al., 2015) revealed a tight clustering of all language-relevant areas, despite their cytoarchitectonic diversity and their topological widespread distribution throughout the brain, and segregated from the cortical areas subserving other functions. The mouth representation region within the primary motor cortex more closely resembled the language-related areas in its neurochemical composition than it did the hand representation region. Although activations within lateral prefrontal areas 46 and 9 have been reported during language processing (Chow, Kaup, Raabe, & Greenlee, 2008; Romero, Walsh, & Papagno, 2006; Turken & Dronkers, 2011), they cluster with parietal area 7 and cingulate area 32. The latter two areas are components of cortical networks active during working memory tasks (Garn, Allen, & Larsen, 2009; Hernandez, Martinez, & Kohnert, 2000; Kan & Thompson-Schill, 2004; Whitney et al., 2009). Since areas 46 and 9 have also been involved in memory functions (D'Esposito, Postle, &

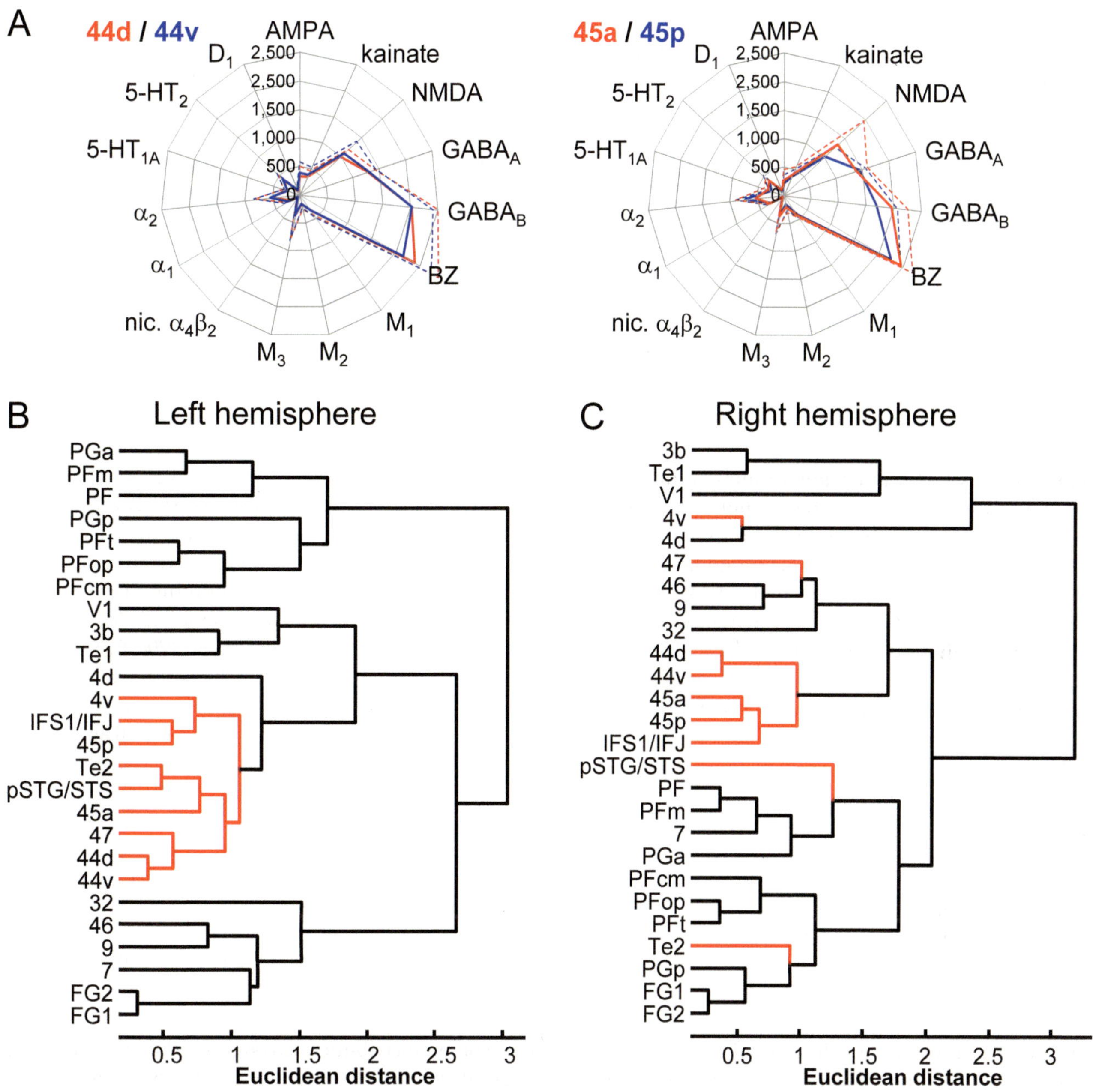

FIGURE 33.5 (A) Receptor fingerprints of areas 44d, 44v, 45a, and 45p. Axis scaling is identical in all four plots and indicates absolute receptor densities in fmol/mg protein. Continuous lines represent the mean absolute receptor densities and dashed lines the standard error of the mean. Data from Zilles, Bach-Trams, et al. (2015). (B) Hierarchical cluster tree resulting from analysis of receptor fingerprints measured in the left hemisphere. (C) Hierarchical cluster tree resulting from analysis of receptor fingerprints measured in the right hemisphere. 3b = primary somatosensory cortex, part of BA3; 4d, 4v = hand and mouth representation regions of the primary motor cortex, BA4; 7 = superior parietal BA7; 9 = dorsal prefrontal BA9; 32 = cingulate BA32; 44d, 44v = dorsal and ventral subdivisions of BA44; 45a, 45p = anterior and posterior subdivisions of BA45; 46 = ventrolateral prefrontal BA46; 47 = ventrolateral prefrontal BA47; FG1, FG2 = cyto- and receptor architectonically defined extrastriate visual areas on the fusiform gyrus; ifs1/ifj = cyto- and receptor architectonically defined areas in the inferior frontal sulcus and at the junction between the inferior frontal and precentral sulci; PF, PFm, PFcm, PFop, and PFt = cyto- and receptor architectonically defined areas located within BA40; PGa, PGp = cyto- and receptor architectonically defined areas located within BA39; Te1 = primary auditory cortex, BA41; Te2 = auditory belt area, BA42; V1 = primary visual cortex, BA17. BZ = $GABA_A$ associated benzodiazepine binding site. Figure adapted from Zilles, Bacha-Trams, et al. (2015).

Rypma, 2000; Levy & Goldman-Rakic, 2000; Marshuetz, Smith, Jonides, DeGutis, & Chenevert, 2000), this clustering pattern highlights their preferential involvement in the memory-related processes associated with language. Analysis of the receptor fingerprints extracted from the right hemisphere (figure 33.5C; Zilles, Bacha-Trams, et al., 2015) resulted in a strikingly different clustering pattern. Here only the receptor architectonically identified components of Broca's region, that is, areas 45a, 45p, 44d, 44v, and ifs1/ifj clustered together. The other three language-related areas, namely temporal areas pSTG, STS, and Te2, as well as prefrontal area 47 were each located in a different cluster. Interestingly, the three primary sensory areas (V1, Te1, and 3b) built an independent cluster regardless of the hemisphere from which receptor densities were extracted (figure 33.5B and C), thus highlighting their unique receptor organization (Zilles & Palomero-Gallagher, 2017).

Therefore, Zilles, Bacha-Trams, et al. (2015) not only provided the first evidence for the existence of a common molecular basis of interaction in the functionally defined sentence comprehension network, but they also further emphasized that the multireceptor balance of each cortical area enables and subserves distinct local information processing abilities.

Acknowledgments

This project has partly been funded by the European Union's Horizon 2020 Research and Innovation Programme under Grant Agreement No. 7202070 (HBP SGA1).

REFERENCES

Amunts, K., Lenzen, M., Friederici, A. D., Schleicher, A., Morosan, P., Palomero-Gallagher, N., & Zilles, K. (2010). Broca's region: Novel organizational principles and multiple receptor mapping. *PLOS Biology, 8*, e1000489. doi: 10.1371/journal.pbio.1000489.

Amunts, K., Malikovic, A., Mohlberg, H., Schormann, T., & Zilles, K. (2000). Brodmann's areas 17 and 18 brought into stereotaxic space-where and how variable? *NeuroImage, 11*, 66–84.

Amunts, K., Schleicher, A., Bürgel, U., Mohlberg, H., Uylings, H. B. M., & Zilles, K. (1999). Broca's region revisited: Cytoarchitecture and intersubject variability. *Journal of Comparative Neurology, 412*, 319–341.

Amunts, K., Schleicher, A., Ditterich, A., & Zilles, K. (2003). Broca's region: Cytoarchitectonic asymmetry and developmental changes. *Journal of Comparative Neurology, 465*, 72–89.

Amunts, K., & von Cramon, D. Y. (2006). The anatomical segregation of the frontal cortex: What does it mean for function? *Cortex, 42*, 525–528.

Amunts, K., Weiss, P. H., Mohlberg, H., Pieperhoff, P., Eickhoff, S. B., Gurd, J. M., ... Zilles, K. (2004). Analysis of neural mechanisms underlying verbal fluency in cytoarchitectonically defined stereotaxic space—The roles of Brodmann areas 44 and 45. *NeuroImage, 22*, 42–56.

Amunts, K., & Zilles, K. (2012). Architecture and organizational principles of Broca's region. *Trends in Cognitive Sciences, 16*, 418–426.

Ardila, A., Bernal, B., & Rosselli, M. (2016). How localized are language brain areas? A review of Brodmann areas involvement in oral language. *Archives of Clinical Neuropsychology, 31*, 112–122.

Binder, J. R. (2015). The Wernicke area: Modern evidence and a reinterpretation. *Neurology, 85*, 2170–2175.

Broca, P. (1861). Remarques sur le siège de la faculté du langage articulé, suivies d'une observation d'aphémie (perte de la parole). *Bulletin de la Société d'anthropologie de Paris, 6*, 330–357.

Broca, P. (1865). Sur le siege de la faculte du langage articule. *Bulletin de la Société d'anthropologie de Paris, 6*, 337–393.

Brodmann, K. (1907). Beiträge zur histologischen Lokalisation der Grosshirnrinde. VI. Mitteilung: Die Cortexgliederung der Großhirnrinde. *Journal für Psychologie und Neurologie, 10*, 232–246.

Brodmann, K. (1909). *Vergleichende Lokalisationslehre der Großhirnrinde in ihren Prinzipien dargestellt auf Grund des Zellbaues*. Leipzig: Johann Ambrosius Barth.

Caspers, J., Palomero-Gallagher, N., Caspers, S., Schleicher, A., Amunts, K., & Zilles, K. (2015). Receptor architecture of visual areas in the face and word-form recognition region of the posterior fusiform gyrus. *Brain Structure and Function, 220*, 205–219.

Caspers, J., Zilles, K., Eickhoff, S. B., Schleicher, A., Mohlberg, H., & Amunts, K. (2013). Cytoarchitectonical analysis and probabilistic mapping of two extrastriate areas of the human posterior fusiform gyrus. *Brain Structure and Function, 218*, 511–526.

Caspers, S., Eickhoff, S. B., Geyer, S., Scheperjans, F., Mohlberg, H., Zilles, K., & Amunts, K. (2008). The human inferior parietal lobule in stereotaxic space. *Brain Structure and Function, 212*, 481–495.

Caspers, S., Geyer, S., Schleicher, A., Mohlberg, H., Amunts, K., & Zilles, K. (2006). The human inferior parietal cortex: Cytoarchitectonic parcellation and interindividual variability. *NeuroImage, 33*, 430–448.

Caspers, S., Schleicher, A., Bacha-Trams, M., Palomero-Gallagher, N., Amunts, K., & Zilles, K. (2013). Organization of the human inferior parietal lobule based on receptor architectonics. *Cerebral Cortex, 23*, 615–628.

Chow, H. M., Kaup, B., Raabe, M., & Greenlee, M. W. (2008). Evidence of fronto-temporal interactions for strategic inference processes during language comprehension. *NeuroImage, 40*, 940–954.

D'Esposito, M., Postle, B. R., & Rypma, B. (2000). Prefrontal cortical contributions to working memory: Evidence from event-related fMRI studies. *Experimental Brain Research, 133*, 3–11.

Derrfuss, J., Brass, M., & von Cramon, D. Y. (2004). Cognitive control in the posterior frontolateral cortex: Evidence from common activations in task coordination, interference control, and working memory. *NeuroImage, 23*, 604–612.

Derrfuss, J., Brass, M., von Cramon, D. Y., Lohmann, G., & Amunts, K. (2009). Neural activations at the junction of the inferior frontal sulcus and the inferior precentral sulcus: Interindividual variability, reliability, and association with sulcal morphology. *Human Brain Mapping, 30,* 299–311.

DeWitt, I., & Rauschecker, J. P. (2013). Wernicke's area revisited: Parallel streams and word processing. *Brain and Language, 127,* 181–191.

Dronkers, N. F., Plaisant, O., Iba-Zizen, M. T., & Cabanis, E. A. (2007). Paul Broca's historic cases: High resolution MR imaging of the brains of Leborgne and Lelong. *Brain, 130,* 1432–1441.

Dronkers, N. F., Wilkins, D. P., Van Valin, R. D. J., Redfern, B., & Jaeger, J. (2004). Lesion analysis of the brain areas involved in language comprehension. *Cognition, 92,* 145–177.

Eickhoff, S. B., Amunts, K., Mohlberg, H., & Zilles, K. (2006). The human parietal operculum. II. Stereotaxic maps and correlation with functional imaging results. *Cerebral Cortex, 16,* 268–279.

Eickhoff, S. B., Heim, S., Zilles, K., & Amunts, K. (2006). Testing anatomically specified hypotheses in functional imaging using cytoarchitectonic maps. *NeuroImage, 32,* 570–582.

Eickhoff, S. B., Paus, T., Caspers, S., Grosbras, M. H., Evans, A. C., Zilles, K., & Amunts, K. (2007). Assignment of functional activations to probabilistic cytoarchitectonic areas revisited. *NeuroImage, 36,* 511–521.

Eickhoff, S. B., Schleicher, A., Zilles, K., & Amunts, K. (2006). The human parietal operculum. I. Cytoarchitectonic mapping of subdivisions. *Cerebral Cortex, 16,* 254–267.

Eickhoff, S. B., Stephan, K. E., Mohlberg, H., Grefkes, C., Fink, G. R., Amunts, K., & Zilles, K. (2005). A new SPM toolbox for combining probabilistic cytoarchitectonic maps and functional imaging data. *NeuroImage, 25,* 1325–1335.

Friederici, A. D. (2009). Pathways to language: Fiber tracts in the human brain. *Trends in Cognitive Sciences, 13,* 175–181.

Friederici, A. D., Fiebach, C. J., Schlesewsky, M., Bornkessel, I. D., & von Cramon, D. Y. (2006). Processing linguistic complexity and grammaticality in the left frontal cortex. *Cerebral Cortex, 16,* 1709–1717.

Galaburda, A. M., & Sanides, F. (1980). Cytoarchitectonic organization of the human auditory cortex. *Journal of Comparative Neurology, 190,* 597–610.

Garn, C. L., Allen, M. D., & Larsen, J. D. (2009). An fMRI study of sex differences in brain activation during object naming. *Cortex, 45,* 610–618.

Geyer, S., Ledberg, A., Schleicher, A., Kinomura, S., Schormann, T., Bürgel, U., … Roland, P. E. (1996). Two different areas within the primary motor cortex of man. *Nature, 382,* 805–807.

Geyer, S., Schleicher, A., & Zilles, K. (1997). The somatosensory cortex of human: Cytoarchitecture and regional distributions of receptor-binding sites. *NeuroImage, 6,* 27–45.

Grewe, T., Bornkessel, I., Zysset, S., Wiese, R., von Cramon, D. Y., & Schlesewsky, M. (2005). The emergence of the unmarked: a new perspective on the language-specific function of Broca's area. *Human Brain Mapping, 26,* 178–190.

Hagoort, P. (2014). Nodes and networks in the neural architecture for language: Broca's region and beyond. *Current Opinion in Neurobiology, 28,* 136–141.

Hagoort, P., Hald, L., Bastiaansen, M., & Petersson, K. M. (2004). Integration of word meaning and world knowledge in language comprehension. *Science, 304*(5669), 438–441.

Hernandez, A. E., Martinez, A., & Kohnert, K. (2000). In search of the language switch: An fMRI study of picture naming in Spanish-English bilinguals. *Brain and Language, 73,* 421–431.

Hickok, G., & Poeppel, D. (2007). The cortical organization of speech processing. *Nature Reviews Neuroscience, 8,* 393–402.

Kaas, J. H., & Hackett, T. A. (2000). Subdivisions of auditory cortex and processing streams in primates. *PNAS, 97,* 11793–11799.

Kaas, J. H., Hackett, T. A., & Tramo, M. J. (1999). Auditory processing in primate cerebral cortex. *Current Opinion in Neurobiology, 9,* 164–170.

Kan, I. P., & Thompson-Schill, S. L. (2004). Effect of name agreement on prefrontal activity during overt and covert picture naming. *Cognitive, Affective, and Behavioral Neuroscience, 4,* 43–57.

Keller, S., Highley, J. R., Garcia-Finana, M., Sluming, V., Rezaie, R., & Roberts, N. (2007). Sulcal variability, stereological measurement and asymmetry of Broca's area on MR images. *Journal of Anatomy, 211,* 534–555.

Kononowa, E. P. (1935). Structural variability of the cortex cerebri: Inferior frontal gyrus in adults (Russian). In S. A. Sarkissov & I. N. Filimonoff (Eds.), *Annals of the Brain Research Institute* (pp. 49–118). Moscow-Leningrad: State Press for Biological and Medical Literature.

Kononowa, E. P. (1938). Variability in the structure of the human cerebral cortex: Frontal lobe of adult man (Russian). In S. A. Sarkissov & I. N. Filimonoff (Eds.), *Annals of the Brain Research Institute* (pp. 213–274). Moscow-Leningrad: State Press for Biological and Medical Literature.

Kreht, H. (1935). Cytoarchitektonik und motorisches Sprachzentrum. *Zeitschrift für mikroskopisch-anatomische Forschung, 39,* 331–354.

Kubanek, J., Brunner, P., Gunduz, A., Poeppel, D., & Schalk, G. (2013). The tracking of speech envelope in the human cortex. *PLOS One, 8,* e53398.

Levy, R., & Goldman-Rakic, P. S. (2000). Segregation of working memory functions within the dorsolateral prefrontal cortex. *Experimental Brain Research, 133,* 23–32.

Makuuchi, M., Bahlmann, J., Anwander, A., & Friederici, A. D. (2009). Segregating the core computational faculty of human language from working memory. *PNAS, 106,* 8362–8367.

Marshuetz, C., Smith, E., Jonides, J., DeGutis, J., & Chenevert, T. L. (2000). Order information in working memory: fMRI evidence for parietal and prefrontal mechanisms. *Journal of Cognitive Neuroscience, 12*(Supplement 2), 130–144.

Morosan, P., Rademacher, J., Palomero-Gallagher, N., & Zilles, K. (2005). Anatomical organization of the human auditory cortex: Cytoarchitecture and transmitter receptors. In P. Heil, E. König, & E. Budinger (Eds.), *Auditory cortex—Towards a synthesis of human and animal research* (pp. 27–50). Mahwah, NJ: Lawrence Erlbaum.

Morosan, P., Rademacher, J., Schleicher, A., Amunts, K., Schormann, T., & Zilles, K. (2001). Human primary auditory cortex: Cytoarchitectonic subdivisions and mapping into a spatial reference system. *NeuroImage, 13,* 684–701.

Morosan, P., Schleicher, A., Amunts, K., & Zilles, K. (2005). Multimodal architectonic mapping of human superior temporal gyrus. *Anatomy and Embryology, 210,* 401–406.

Ngowyang, G. (1934). *Die Zytoarchitektonik des menschlichen Stirnhirns: Academia Sinica.* Taipei: The National Research Institute of Psychology.

Ono, M., Kubik, S., & Abernathey, C. D. (1990). *Atlas of the cerebral sulci.* Stuttgart: Thieme Verlag.

Palomero-Gallagher, N., Vogt, B. A., Schleicher, A., Mayberg, H. S., Schleicher, A., & Zilles, K. (2009). Receptor architecture of human cingulate cortex: evaluation of the four-region neurobiological model. *Human Brain Mapping, 30,* 2336–2355.

Petrides, M. (2013). *Neuroanatomy of language regions of the human brain.* Amsterdam: Elsevier.

Petrides, M., & Pandya, D. N. (1994). Comparative architectonic analysis of the human and the macaque frontal cortex. In F. Boller & J. Grafman (Eds.), *Handbook of neuropsychology* (pp. 17–58). Amsterdam: Elsevier.

Rademacher, J., Morosan, P., Schormann, T., Schleicher, A., Werner, C., Freund, H. J., & Zilles, K. (2001). Probabilistic mapping and volume measurement of human primary auditory cortex. *NeuroImage, 13,* 669–683.

Riegele, L. (1931). Die Cytoarchitektonik der Felder der Brocaschen Regionen. *Journal of Psychology and Neurology, 42,* 496–514.

Romero, L., Walsh, V., & Papagno, C. (2006). The neural correlates of phonological short-term memory: A repetitive transcranial magnetic stimulation study. *Journal of Cognitive Neuroscience, 18,* 1147–1155.

Sanides, F. (1962). *Die Architektonik des menschlichen Stirnhirns.* Berlin: Springer.

Sanides, F. (1964). The cyto-myeloarchitecture of the human frontal lobe and its relation to phylogenetic differentiation of the cerebral cortex. *Journal für Hirnforsch, 7,* 269–282.

Sarkissov, S. A., Filimonoff, I. N., Kononowa, E. P., & Preobrachenskaja, I. S. (1955). *Atlas of the cytoarchitectonics of the human cerebral cortex.* Moscow: Medgiz.

Sarkissov, S. A., Filimonoff, I. N., & Preobrashenskaya, N. S. (1949). *Cytoarchitecture of the human cortex cerebri (Russian).* Moscow: Medgiz.

Scheperjans, F., Grefkes, C., Palomero-Gallagher, N., Schleicher, A., & Zilles, K. (2005). Subdivisions of human parietal area 5 revealed by quantitative receptor autoradiography: A parietal region between motor, somatosensory and cingulate cortical areas. *NeuroImage, 25,* 975–992.

Scheperjans, F., Hermann, K., Eickhoff, S. B., Amunts, K., Schleicher, A., & Zilles, K. (2008). Observer-independent cytoarchitectonic mapping of the human superior parietal cortex. *Cerebral Cortex, 18,* 846–867.

Scheperjans, F., Palomero-Gallagher, N., Grefkes, C., Schleicher, A., & Zilles, K. (2005). Transmitter receptors reveal segregation of cortical areas in the human superior parietal cortex: Relations to visual and somatosensory regions. *NeuroImage, 28,* 362–379.

Schleicher, A., Amunts, K., Geyer, S., Kowalski, T., Schormann, T., Palomero-Gallagher, N., & Zilles, K. (2000). A stereological approach to human cortical architecture: Identification and delineation of cortical areas. *Journal of Chemical Neuroanatomy, 20,* 31–47.

Schleicher, A., Amunts, K., Geyer, S., Morosan, P., & Zilles, K. (1999). Observer-independent method for microstructural parcellation of cerebral cortex: A quantitative approach to cytoarchitectonics. *NeuroImage, 9,* 165–177.

Schleicher, A., Palomero-Gallagher, N., Morosan, P., Eickhoff, S. B., Kowalski, T., de Vos, K., … Zilles, K. (2005). Quantitative architectural analysis: A new approach to cortical mapping. *Anatomy and Embryology, 210,* 373–386.

Stengel, E. (1930). Morphologische und cytoarchitektonische Studien über den Bau der unteren Frontalwindung bei Normalen und Taubstummen. Ihre individuellen und Seitenunterschiede. *Zeitschrift für die gesamte Neurologie und Psychiatrie, 130,* 631–677.

Talairach, J., & Tournoux, P. (1988). *Co-planar stereotaxic atlas of the human brain.* Stuttgart, Germany: Thieme.

Turken, A. U., & Dronkers, N. F. (2011). The neural architecture of the language comprehension network: Converging evidence from lesion and connectivity analyses. *Frontiers in Systems Neuroscience, 5,* 1.

von Economo, C. (1927). *Zellaufbau der Grosshirnrinde des Menschen.* Berlin: Verlag Julius Springer.

von Economo, C., & Horn, L. (1930). Über Windungsrelief, Masen und Rindenarchitektonik der Supratemporalfläche, ihre individuellen und Seitenunterschiede. *Zeitschrift für die gesamte Neurologie und Psychiatrie, 130,* 678–755.

von Economo, C., & Koskinas, G. N. (1925). *Die Cytoarchitektonik der Hirnrinde des erwachsenen Menschen.* Berlin: Springer.

Wernicke, C. (1874). *Der aphasische Symptomencomplex Eine psychologische Studie auf anatomischer Basis.* Breslau, Germany: M. Cohn & Weigert.

Whitney, C., Weis, S., Krings, T., Huber, W., Grossman, M., & Kircher, T. (2009). Task-dependent modulations of prefrontal and hippocampal activity during intrinsic word production. *Journal of Cognitive Neuroscience, 21,* 697–712.

Zilles, K. (1978). A quantitative approach to cytoarchitectonics. I. The areal pattern of the cortex of *Tupaia belangeri. Anatomy and Embryology, 153,* 195–212.

Zilles, K., & Amunts, K. (2009). Receptor mapping: Architecture of the human cerebral cortex. *Current Opinion in Neurology, 22,* 331–339.

Zilles, K., & Amunts, K. (2010). Centenary of Brodmann's map—Conception and fate. *Nature Reviews Neuroscience, 11,* 139–145.

Zilles, K., Bacha-Trams, M., Palomero-Gallagher, N., Amunts, K., & Friederici, A. D. (2015). Common molecular basis of the sentence comprehension network revealed by neurotransmitter receptor fingerprints. *Cortex, 63,* 79–89.

Zilles, K., & Palomero-Gallagher, N. (2017). Comparative analysis of receptor types that identify primary cortical sensory areas. In J. H. Kaas (Ed.), *Evolution of nervous systems* (pp. 225–245). Oxford: Elsevier.

Zilles, K., Palomero-Gallagher, N., Bludau, S., Mohlberg, H., & Amunts, K. (2015). Cytoarchitecture and maps of the human cerebral cortex. In A. W. Toga (Ed.), *Brain mapping: An encyclopedic reference* (pp. 115–135). New York: Elsevier.

Zilles, K., Palomero-Gallagher, N., Grefkes, C., Scheperjans, F., Boy, C., Amunts, K., & Schleicher, A. (2002). Architectonics of the human cerebral cortex and transmitter receptor fingerprints: Reconciling functional neuroanatomy and neurochemistry. *European Neuropsychopharmacology, 12,* 587–599.

Zilles, K., Schleicher, A., Palomero-Gallagher, N., & Amunts, K. (2002). Quantitative analysis of cyto- and receptorarchitecture of the human brain. In A. W. Toga & J. C. Mazziotta (Eds.), *Brain mapping: The methods* (pp. 573–602). Amsterdam: Elsevier.

34 The Human Language Connectome

IZABELA PRZEŹDZIK, KOEN V. HAAK, CHRISTIAN F. BECKMANN, AND ANDREAS BARTSCH

1. From Localization to a Distributed Systems View

Speech and language, writing and reading included, are probably the most distinctive cognitive abilities of humans. By means of a complex and far from completely understood apparatus within our brains, these cognitive skills endow us with a unique ability to express ourselves and to communicate with each other by spoken and written words.

Historically, investigations into the human language system focused on representations of localized effects and behavioral associations. Initial discoveries of such representations of speech and language functions in the human brain started with (a) the recognition of hemispheric dominance/lateralization for speech and language (M.G. Dax, 1865a; M.G. Dax, 1865b) and (b) the localization of Broca's speech area in the opercular part of the left inferior frontal gyrus (IFG=F1, for first frontal convolution) based on the study of brain lesions, which lead to nonfluent motor aphasia (*aphemia*; Broca, 1861, 1865; Levelt, 2013).

In following years, further cortical "centers" for other speech and language functions, including reading and writing, were identified, such as Wernicke's sensory area in the posterior superior temporal gyrus/superior temporal sulcus (STG=T1/STS=t1; Meynert, 1866, Wernicke, 1874), the inferior parietal language area (IPL=P2; later termed *Geschwind's territory*; Geschwind, 1970) of the angular gyrus (ANG) and supramarginal gyrus (SMG), the basotemporal language area (Mills & Martin, 1912) or visual word form area (VWFA; Cohen et al., 2000) in the lateral occipitotemporal sulcus (= t3) and posterior fusiform gyrus (= T4/O4), and Exner's graphical motor area in the posterior middle frontal gyrus (= F2; Exner, 1881). In fact, Exner (1881) conducted the first systematic brain lesion-symptom mapping study and concluded that the cortical field for speech and language follows a relative, not absolute representation with respect to the underlying anatomy of the cortical surface.

The realization that damage to certain brain areas does not inevitably lead to aphasic symptoms and that the resulting clinical symptoms vary considerably across patients prompted strong criticisms on a plain localization of human speech and language functions in the brain. In a famous debate at Berlin's Society for Anthropology, Ethnology, and Ancient History (*Berliner Sprachstreit*) between Hitzig (1874) and Steinthal (1874), who argued for precise descriptions of linguistic deficits in aphasic patients, Westphal (1874) finally refuted to assume simple "centers" for speech and language. Similarly, Freud (1891) refused "the assumptions that brain organs for speech (*Sprachapparat*) would consist of segregated centers" and conceptualized "these centers as edges of a field of speech and language."[1] This can be viewed as one of the earliest neuroscientific precursors to today's concepts of a functional human language connectome in terms of a distributed system.

Von Monakow (1905) reasoned that associations of aphasic deficits with particular brain lesions, which are not confined by strict borders anyway, would not indicate that speech and language functions are indeed seated within these brain areas. Instead, disruptions of functional connectivity are expected to lead to remote effects that better explain the broad spectrum of aphasic semiology after cerebral lesions and, with their waning during recovery from an insult, the brain's plasticity. Thus, his concept of intercortical *diaschisis* can be interpreted as another forerunner to modern views on distributed *connectomic* approaches to understanding how the faculty of language is represented in the human brain.

In this chapter, we review current characterizations of the language system in terms of the structural and functional associations between areas of the human language connectome at rest and under a task condition. Speech and language cannot be localized by pure anatomic criteria alone, and advances in neuroimaging now provide unprecedented access to in vivo measurements of brain function, structure, and connectivity. Diffusion-weighted MRI (dMRI) and functional MRI (fMRI) are, despite their limitations, particularly attractive tools for investigating the human language connectome for their ability to characterize the brain

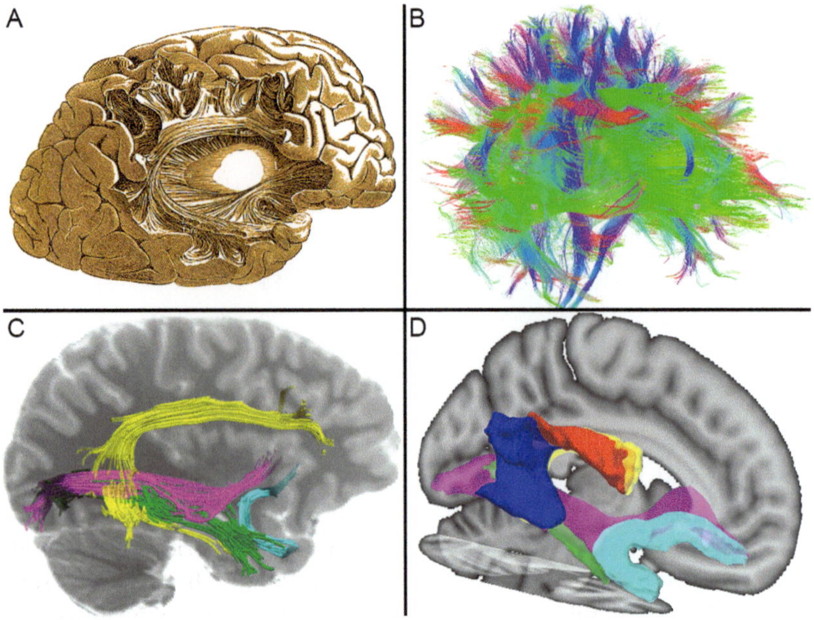

FIGURE 34.1 Association fiber pathways underlying intrahemispheric structural connectivity between cortical areas involved in speech and language processing. (A) Based on postmortem dissections according to Meynert (1884). (B) In vivo whole-brain tractography based on dMRI. (C) Visualization of probabilistic dMRI tractography streamlines reconstructing the arcuate (yellow; see [A], inferior longitudinal (green), inferior occipitofrontal (pink), and the uncinate fasciculi (cyan; see [A]) in a single subject (data of probabilistic streamlines courtesy of N. Voets). (D) Group parcellation across 491 subjects from the Human Connectome Project (van Essen et al., 2013) of these tracts in Montreal Neurological Institute 152 standard space further separating the arcuate fasciculus into a long (yellow), anterior (red), and posterior (blue) segment. The latter is the vertical branch of the AF, corresponds to the SLF-tp, and overlaps with the anterior VOF of Wernicke.

structure and function in vivo. Here, we describe how these tools can be used to characterize the structural and functional connectome and how connectivity methods can be used to investigate specific models of language processing in the human brain.

2. *The Structural Language Connectome*

White matter has, for a long time, been primarily considered as occupying (*Belegungs-*) or filling matter (*Ausfüllungsmasse*) in the brain (Arnold, 1838; Burdach, 1822). However, with the advent of postmortem fiber dissections, fiber connections between cortical areas and intrahemispheric association fiber systems, in particular, were considered more important (Meynert, 1884; cf. figure 34.1A).

Major association fiber pathways linking cortical areas involved in speech and language processing are shown in figure 34.1. These represent the structural connectome of the human language system (Friederici & Gierhan, 2013) and can be subdivided into (a) inferior and (b) superior perisylvian fiber systems. The main inferior fiber pathways are the inferior longitudinal fasciculus (ILF), the inferior occipitofrontal (or fronto-occipital) fasciculus (IOFF) or inferior fronto-occipital fasciculus, and the uncinate fasciculus (UF). The main superior fiber pathway is the arcuate fasciculus (AF; or fourth portion of the superior longitudinal fasciculus, SLF-IV). The AF can be subdivided into a long (brachium anterius of the SLF; Nieuwenhuys, Voogd, & van Huijzen, 2008), anterior and posterior (vertical/SLF-temporoparietal [tp]) segment (figure 34.1D).

Note that figures 34.1A and 34.1B reveal a fanning continuum of the ILF, IOFF, and UF (Ebeling & von Cramon, 1992) while figures 34.1C and 34.1D virtually separate these tracts by using a priori seed, target, and waypoint masks for in vivo tractography. UF and IOFF, for example, are compacted and intertwined at the anterior base of the external capsule, with the UF extending more laterocaudally and the IFO more mediocranially. Note further, that fiber pathways in the central nervous system are not covered by perineural sheaths, which confine and separate peripheral nerves. Often, separation of contiguous and possibly crossing or interwoven fiber pathways by diffusion tractography is limited and will not necessarily represent the underlying course of axons.

The distinction of the anterior segment of the AF and the third portion of the SLF (SLF-III) as well of the posterior segment of the AF and the tp branch of the SLF (SLF-tp) is not clear and, at least in part, arbitrary. The IOFF seems to have more frontal terminations in the triangular and orbital part of the IFG (Yagmurlu,

Middlebrooks, Tanriover, & Rhoton, 2016) while the AF seems to have more in the opercular part (i.e., Broca's area proper; Petrides, Tomaiuolo, Yeterian, & Pandya, 2012; Schmahmann & Pandya, 2006). This very likely reflects known functional specializations within the IFG.

Within the dual-stream model of cortical speech and language processing (Hickok & Poeppel, 2007; see figure 34.4 below), the anterior segment of the AF can be conceptualized as the primary structural connection of a strongly lateralized, dorsal phonological-articulatory stream while the ILF, IOFF, and UF would represent structural connections of less hemispheric dominance within a ventral lexical-semantic stream, which is assumed to integrate cross-modal content and conceptual meaning. Indeed, connectional asymmetries of the dorsal stream may largely underlie functional language lateralization (Powell et al., 2006) while the ventral stream is more bilaterally organized even though the semantic system also exhibits lateralization (Binder, Desai, Graves, & Conant, 2009; Wright, Stamatakis, & Tyler, 2012). Syntax and morphology seem to be processed partly by both and synergistically (Friederici, 2012; Griffiths, Marslen-Wilson, Stamatakis, & Tyler, 2013; Rolheiser, Stamatakis, & Tyler, 2011). The posterior (and parts of the long) segment of the AF may be viewed to interconnect these two streams of information processing.

Others have suggested that it is the long AF segment, terminating/originating mainly around the middle temporal gyrus (MTG=T2), that transmits lexical-semantic content in the left hemisphere as opposed to prosodic content in the right (Glasser & Rilling, 2008). The left MTG is considered important for language comprehension and has terminations of AF, ILF, and IOFF fibers (Turken & Dronkers, 2011). According to the classical Wernicke-Lichtheim-Geschwind model, disruption of the dominant AF should lead to conduction (i.e., repetition) aphasia. Lichtheim's (1885) original model built a connectionist language model based on clinical presentations of specific deficits found in patients. Implicitly, thereby, it derived a model of normal language functioning under the assumption that aphasiological deficits would reveal the "normal flow of information" by its interruptions. This is, for conduction aphasia according to the later, topographically refined Wernicke-Lichtheim-Geschwind model concept, a disconnection within the AF.

The anterior and the long segment of the AF run in the suprainsular white matter above the semicircular sulcus insulae, the anterior segment more laterally, and the long segment more medially (figure 34.1). The frontal aslant and the anterior portion of Wernicke's perpendicular fasciculus, or vertical occipital fasciculus (VOF), are not shown in the figure. The frontal aslant tract is a structural connection among the supplementary motor area (SMA)/pre-SMA, the precentral area, and IFG (Broca's area, in particular; Vergani et al., 2014). Its disconnection may be associated with (usually transient) mutism in SMA syndromes (Rostomily, Berger, Ojemann, & Lettich, 1991) due to disrupted speech initiation and orofacial movement control. The anterior portion of the VOF interconnects the VWFA with the IPL (ANG/SMG). Its disconnection may be associated with pure "preangular" alexia without agraphia and hemianopsia (Bartsch, Jbabdi, & Geletneky, 2013; Greenblatt, 1990).

Fiber pathways connecting the IPL with the middle and posterior temporal cortex are considered relevant for language given that their surgical or ischemic damage predicts sustained aphasic deficits (Caverzasi et al., 2016; Yourganov, Fridriksson, Rorden, Gleichgerrcht, & Bonilha, 2016). The posterior segment of the AF has originally been described as its vertical branch and corresponds to SLF-tp (Catani & Jones, 2005; Chang, Raygor, & Berger, 2015). The trajectory of these fibers is lateral to the long segment of the AF and overlap with the anterior VOF. Their existence has been confirmed by dissections (Lawes et al., 2008; Fernandez-Miranda et al., 2008; Martino et al., 2013), and Direct Subcortical Electrical Stimulation of these fiber pathways may cause phonological and repetition errors (Chang et al., 2015).

While the initial interpretation of brain lesions leading to different types of aphasias focused on the cortex, subcortical structural connections of the language connectome clearly received less attention. Wernicke (1874, 1881) theoretically established conduction aphasia (*Leitungsaphasie*) as a subcortical disconnection syndrome of structural connections between his presumed sensory and Broca's cortical motor language center. Semiologically, he expected paraphasias and correction attempts (*conduite d'approche*) to dominate conduction aphasia, in other words, not a primarily deficient repetition. Structurally, he identified inferior (sub)insular fibers with the fiber pathway connecting Broca's motor speech production center and his sensory language perception center. Yet, the presumed inferior (= ventral) pathway of the structural language connectome got rapidly out of focus in favor of a different superior (= dorsal) fiber tract: the AF (Weiller, Bormann, Saur, Musso, & Rijntjes, 2011). Both Freud (1891) and Heilbronner (1908) identified conduction aphasia with the core symptom of impaired repetition. Later, the term *repetition aphasia* was introduced; though it was simultaneously questioned that repetition aphasia is the result of a true disconnection syndrome (Kleist, 1934). While Freud denied the possibility of a pure repetition deficit and a separate entity of conduction aphasia entirely (because the tract connecting Broca's with Wernicke's area should, in his view, be utilized not just during repetition but also for spontaneous speech), Heilbronner (1908) ascertained the existence. Given that the AF was considered the main association tract between the temporal and frontal lobe, conduction aphasia was suggested to result from lesions to this dorsal candidate pathway—a view that then persisted into the modern era (Geschwind, 1965). Speech and language were, in these models, primarily seen as neurological functions and analyzed by the corresponding deficits.

Interestingly, tract tracer studies in primates have challenged the existence of direct connections between IFG and STG via the AF. Instead, a complex system of multiple fiber bundles was identified (Petrides et al., 2012; Schmahmann & Pandya, 2006). Von Monakow (1905) had already suggested that the ILF and AF may consist of fewer long-ranging than short between-relay connections—a notion that has recently been revived to explain conduction aphasias by considering bidirectional (e.g., precentral) cortical relays or downstream or convergence nodes within the AF (Bernal & Ardila, 2009; Ellmore, Beauchamp, O'Neill, Dreyer, & Tandon, 2009). Furthermore, precentral AF terminations seem to be present at birth while more rostral IFG terminations of the AF mature later during development (Brauer, Anwander, Perani, & Friederici, 2013).

Direct Subcortical Electrical Stimulation along the arcuate in the dominant hemisphere has been shown to routinely disrupt speech production and to induce phonemic paraphasias, occasionally with repetitions (Duffau, 2015; Duffau, Peggy Gatignol, Mandonnet, Capelle, & Taillandier, 2008), while ischemic strokes affecting AF fibers have been shown to disrupt phonological processing and repetition, in particular (Kümmerer et al., 2013; Rolheiser et al., 2011), and to reduce speech fluency (Marchina et al., 2011). According to these data, the AF of the dominant hemisphere indeed constitutes a "dorsal phonological route" of speech monitoring, planning, and execution in the human brain.

Historically, Freud (1891) did not only question the dissociation of intact spontaneous speech from primary repetition impairments by lesioning the AF. In addition to proposing more of a functional network approach (as opposed to segregated cortical speech centers), he also re-emphasized the importance of a ventral pathway without specifying a clinical syndrome that would result from its damage. Thus, a dual-stream hypothesis of cortical speech and language functions relying on a ventral and dorsal loop of the structural connectome can be traced back to the seminal studies and theoretical conceptualizations of aphasiology.

It is generally presumed that structural connectivity between cortical areas underlies, to an important extent, their functional connectivity (cf., for example, Vincent et al., 2007, and Wang et al., 2013). This should not be mistaken to indicate that the functional language connectome can be uniquely determined by the structural language connectome. However, one structural network may form many functional network configurations depending on what task the network nodes (i.e., cortical areas) are engaged in. When it comes to understanding the mapping between the structural and functional language connectomes, there are also the challenges that (a) there is no fully satisfactory, empirically founded model of subcortical speech and language processing along the association fiber systems outlined, and that (b) virtually all models of the structural language connectome have been related to those of cortical speech and language processing for which no comprehensive, universally accepted theory exists either.

Therefore, this chapter will, in the following, focus on empirical characterizations of functional connectivities. In the future, further insights can be expected in combining functional and structural connectivity information. Evidence suggests that structural connectivity predicts, to some extent, functional activations (Homol, Jbabdi, Beckmann, & Bartsch, 2012; Jbabdi & Behrens, 2012; Saygin et al., 2012, Tavor et al., 2016). If the pattern of functional activations (F) within two language areas can be predicted by the structural connectivity (S) between them, the association tract connecting these two areas is likely to mediate a direct interaction. However, all parts of such existing structural connections do not necessarily have to be involved in each and every particular language task or language processing, in general, and functional peaks of activation should, for that reason, be predictable by peaks of structural connectivity but not vice versa.

3. The Functional Language Connectome

Investigations into the functional neuroanatomy of language have been dominated by activation (localization) studies. This trend parallels the advances in noninvasive neuroimaging technology and the emergence of new techniques such as fMRI. These tools are an attractive addition to lesion studies, and they enable studying brain function noninvasively by measuring changes in local blood oxygenation (the blood oxygenation level-dependent or BOLD signal), which reflects local brain activity.

Through these new technologies, activation (localization) studies can now be designed to sensitively focus on changes in the BOLD signal in response to externally controlled tasks. By virtue of the technique's high spatial resolution, we can localize brain regions associated with particular cognitive functions, and the resulting activation maps display task-dependent functional coactivations in response to the task. That is, the images show the BOLD dynamics that are evoked by the task condition(s). This sensitivity, however, comes at the cost of the scope of these characterizations: the brain activity patterns and resulting localizations obtained during the experiment are inevitably dependent on the particular task and the associated cognitive function of interest as well as on supporting cognitive functions, such as general modes of attention or working memory or sensory aspects involved in stimulus processing. It remains a formidable challenge, and indeed may be impossible, to design task experiments in which numerous inherently linked processes can be fully disentangled.

Many different tasks have been designed to actively probe different language functions, for example, figure 34.2 shows BOLD activations in response to four different language tasks. In the auditory detection of words task, that is, processing at the word level only with lexical recognition (figure 34.2A), we can see (a) left hemispheric SMG activations (that refer to area *Spt* in the dual-stream model, see figure 34.4), (b) left-lateralized opercular and triangular IFG as well as peri-Rolandic activations, and (c) less lateralized STG/STS and MTG/ANG activations. In turn, the auditory description-cued naming task (figure 34.2C), which requires processing at the sentence level and lexical retrieval, also recruits aforementioned (a) and (b) with relatively more left triangular and bilateral opercular involvement and stronger left-lateralized upper MTG/lower ANG activations. Passive listening to nonfinal embedded clause sentences (figure 34.2B), which is phonologically not very challenging, evoked the weakest activations with hardly any SMG or IFG contributions. Furthermore, the covert reading of nonfinal embedded clause sentences (figure 34.2D) leads to the most extensive and strongest left angular (and VWFA) activations, and the orbital contributions to the left and right IFG activations are most profound. This is consistent with the known IPL and VWFA contributions of reading, as well as proposed sound/(letter)/word/sentence/meaning processing gradients in the IFG (see the *Memory, Unification, and Control* (MUC) model, figure 34.4; Kleist, 1934.)

Overall, this shows that different experimental paradigms evoke different language-related activation patterns. Thus, while task-fMRI has a clear advantage when controlled experiments are of interest, it is inherently limited in its ability to measure the full range of functions that a brain region or a network might subserve. In other words, by using task-fMRI, we are able to probe specific aspects of language processing; however, we can never get a full system's level characterization in a single experiment. As a complementary approach, resting-state fMRI can be used as a useful tool for studying the more generic aspects of brain function in the absence of specific cognitive manipulation and sensory stimulation, providing a description of coactivation (*functional connectivity*) within broad systems and may be a suitable approach for testing overarching cognitive models.

Additionally, as it will be discussed later, these task-fMRI examples show that several tasks together can accommodate different aspects of the overarching models of language processing. For example, here some aspects of both the dual-stream model (Spt/SMG lateralization, bilateral/less lateralized representation of STG/MTG activations) as well as the MUC model (figures 34.2A and 34.2D; SMG-Triangular/ANG-Orbital gradients) can be observed. Therefore, this shows that the task-fMRI results can bias the interpretation depending on the data and how we read it.

3.1. From Activity to Connectivity The image produced by task-fMRI experiments can be thought of as a functional connectivity "map," because all activated areas coactivated in response to the task. This means that these areas exhibited correlated BOLD signal time courses. Another way to characterize the functional language connectome is by considering the coactivations among brain areas when a subject is not engaged in any particular task. This approach to characterize functional connectivity is known as resting-state fMRI (rfMRI). It has been shown that the ensuing coactivation maps (resting-state networks) contain, to a large extent, the same information in terms of spatial configuration of coactivations as can be derived through meta-analysis from thousands of separate task-activation maps (Smith et al., 2009). This means that the functional connectivity characterizations obtained with rfMRI provide, in some sense, a comprehensive picture of brain function across the entire functional repertoire. Therefore, functional connectivities extracted from rfMRI data appear particularly suited to investigate and test the overarching neurobiological models of complex cognitive processes involved in language processing.

Over the past decade, several approaches have been developed to characterize resting-state functional connectivity. For example, seed-based connectivity analysis (Biswal, Zerrin Yetkin, Haughton, & Hyde, 1995) measures the resting-state correlations of a predefined single seed region with every voxel in the brain, while Independent Component Analysis (Beckmann, 2012; Beckmann, DeLuca, Devlin, & Smith, 2005; Smith et al., 2013) takes all the available fMRI data and

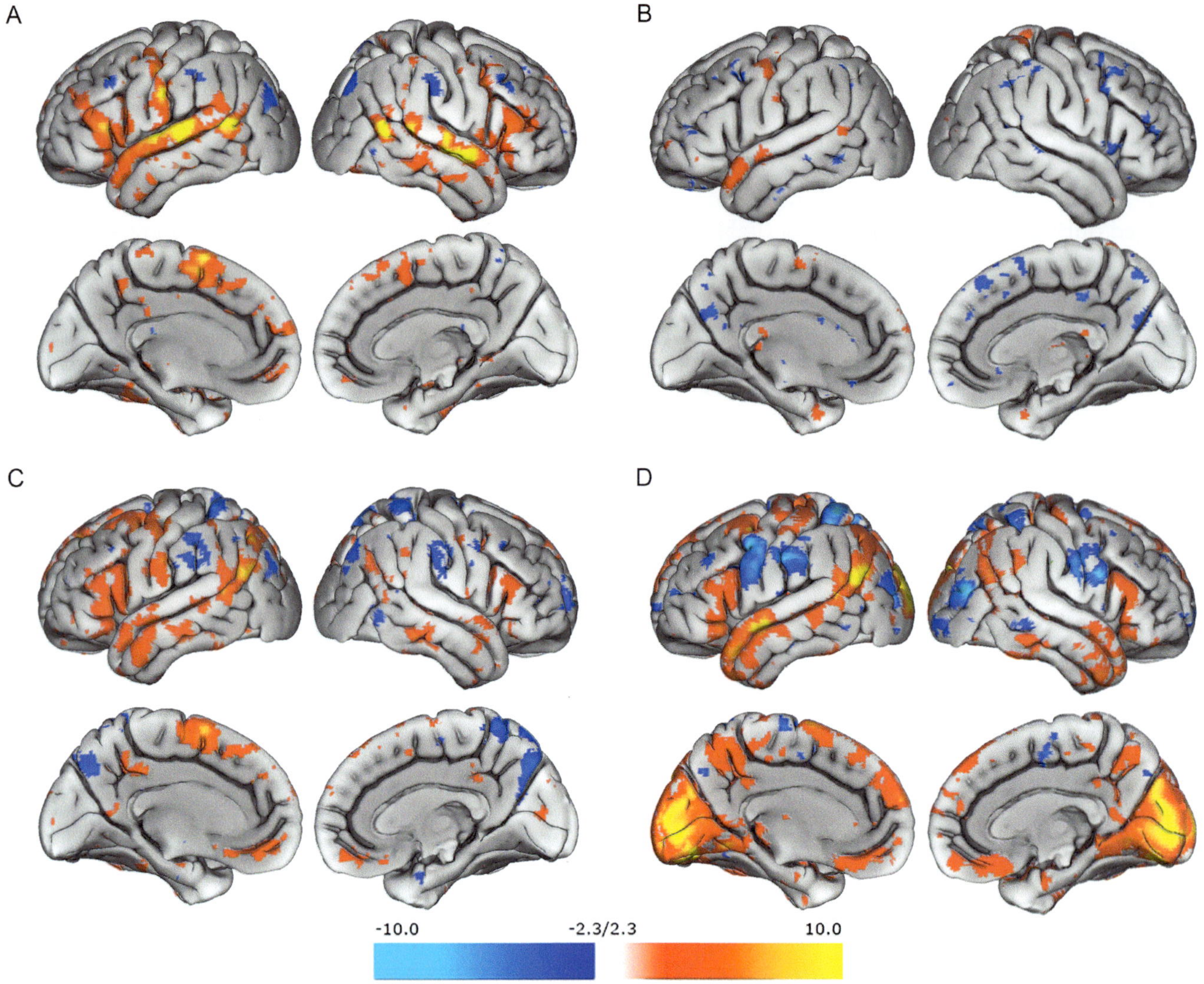

FIGURE 34.2 Task-dependent fMRI activations. Depending on the experimental paradigm, we observe strikingly different language-related activation patterns: (A) auditory detection of words among pseudowords versus listening to sound beeps of different tones (frequencies matched to human speech), (B) listening to nonfinal embedded clause sentences played forward versus backward, (C) auditory description cued naming played versus listening to non- and pseudowords, and (D) reading the same nonfinal embedded clause sentences as in (B) versus consonant strings presented in a sentence-like structure. While single aspects of certain language models can be tested using appropriate task-fMRI experiments, the approach is not very suitable for testing overarching models of language.

decomposes it into temporally coherent, spatially independent components. Figure 34.3 shows the results of the seed-based resting-state functional connectivity analysis where the seed was defined as Broca's complex (i.e., Brodmann's areas [BA] 44 and 45). The results of this analysis show functional connectivity (statistically significant correlations between the fMRI BOLD signal time series) between the seed regions and middle frontal gyrus, SMG, SMG/STG, superior/inferior parietal lobule, planum temporale, posterior temporal lobe (BA39/BA37/BA21/BA22), posterior MTG/posterior inferior temporal gyrus, and the temporal pole. These resting-state functional connectivity patterns are largely consistent with the patterns of structural connectivity we've described. Independent Component Analysis can also be used to reveal language networks based on the rfMRI data. For instance, Tie et al. (2014) reported two resting-state language networks (independent components) at the group level: a frontal language network including regions in the left frontal cortex that are weakly connected with their homologues in right cerebral hemisphere, as well as left temporal areas; and a temporal language network, consisting of regions in the left temporal/parietal cortex and to a lesser extent their homologues in the right hemisphere. Together, these two networks correspond well to the

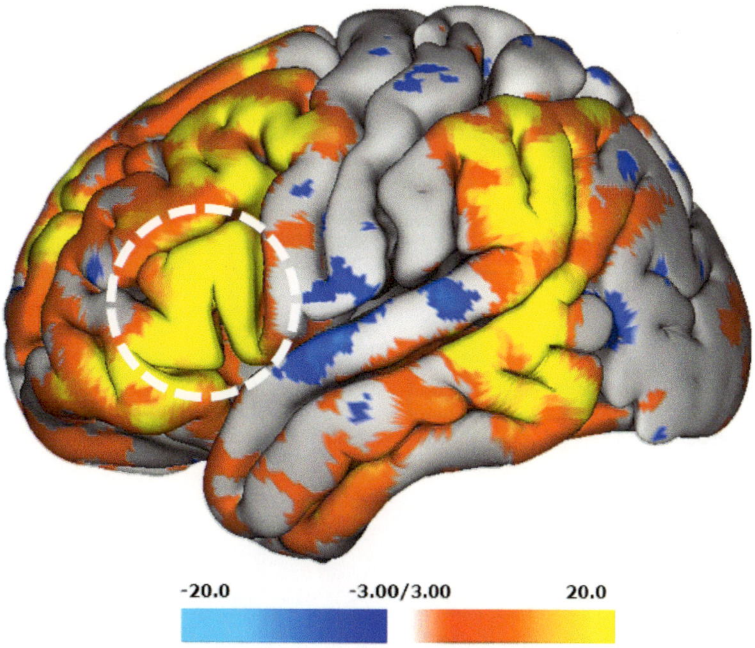

FIGURE 34.3 Seed-based resting-state functional connectivity of Broca's area. The image shows of the Fisher-transformed correlation (z) between the rfMRI time courses in Broca's area (defined as BA44 and BA45, marked in white) and the rest of the left hemisphere of the brain. These functional connectivity patterns largely correspond to the cortical terminations of the white matter fibers associated with language processing.

results obtained with the seed-based analysis. Thus, while these functional connectivity methods are conceptually very different, both analyses principally quantify the strength of functional association between different brain areas, which can in turn be used to characterize the functional language connectome in a task-free and therefore relatively unbiased manner.

4. Using Connectomics and Connectopics to Test Cognitive Models of Language Processing

Models of language processing typically involve network diagrams of processing nodes and the connections between them. One way to falsify these models is by separately testing, with task-based experiments, its single aspects (one particular node or connection). However, this approach can only falsify those language models that are incorrect in terms of the function of one particular node or connection. It cannot falsify those that are incorrect in terms of the overall organization defined by all nodes and connections. This is, however, quite important, as the function of the entire language network is more than just a simple sum of its parts. This, in turn, means that any cognitive model of language or some other cognitive function, must also be tested in its entirety against the available neurobiological data. Furthermore, good models will adhere to both the structural and functional language connectomes. In what follows, we will therefore provide a few examples of possible approaches to this objective.

4.1. CONNECTOMIC EVIDENCE FOR THE DUAL-STREAM HYPOTHESIS In order to test models of language against connectomes, it can be helpful to create a simplified network representation that is still faithful to the measured connectomics data, but can be more easily compared to the theoretical model. One way to create such a representation is a mathematical technique called Multidimensional scaling (MDS). The input data of the MDS is usually a measure of global similarity (here the functional connectivity strength) between the objects (here brain regions). The main output of the MDS is a spatial representation (a graph or network) depicting the geometric configuration of the similarity between the objects. The higher the similarity between two objects as indicated by the initial similarity measure, the closer together the objects will be grouped in the spatial map. Objects that are more dissimilar will, in turn, be placed further away from each other (Kruskal & Wish, 1978). To illustrate this approach, we applied MDS to the rfMRI data obtained from 470 Human Connectome Project subjects (van Essen et al., 2013, S-500 release).[2] Forty-two language-related brain regions (figure 34.5) were defined using a digital brain atlas (Destrieux, Fischl, Dale, & Halgren, 2010). We extracted the mean fMRI time courses within each of these regions

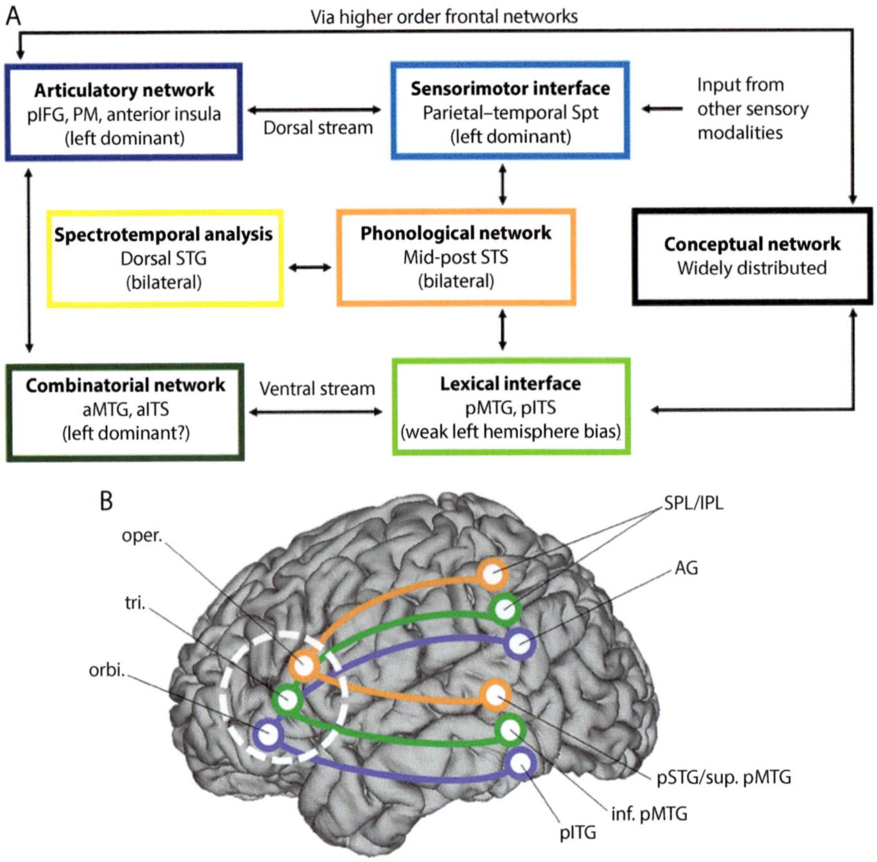

FIGURE 34.4 Neurobiological models of language processing: (A) The dual-stream model of the functional anatomy of language proposed by Hickok and Poeppel (2007) emphasizes the role of two broad streams: a ventral and a dorsal pathway. Before these diverge, the earliest stage of speech processing involving spectrotemporal computations and phonological processing is thought to happen in the dorsal STG and the middle to posterior parts of STS (the most central units in the schematic). After this point, the system separates into the two streams. The ventral stream is bilateral (although possible differences in computational operations between the two hemispheres are indicated) and it involves superior (sup) and middle parts of temporal lobe: the posterior (p) MTG and posterior inferior temporal sulcus (pITS) (*lexical interface*) as well as more anterior (a) parts: the aMTG and aITS (*combinatorial network*). The dorsal stream, on the other hand, is strongly left-lateralized and includes the area Spt (*sensorimotor interface* at the Sylvian fissure/parietotemporal boundary) and several more anterior regions including pIFG, dorsal premotor area, and anterior insula (all involved in the *articulatory network*). Overall, the ventral stream is assumed to be mapping sensory or phonological representations onto lexical conceptual representations, whereas the dorsal stream is thought to be mapping sensory or phonological representations onto articulatory motor representations. The schematic diagram was recreated after Hickok and Poeppel (2007), and colors are adjusted to match the color-coding in figures 34.5A and 34.5C. (B) In the MUC model, Hagoort (2005) proposed that the distribution of the activations of multiple brain regions, while different types of information are being unified, create a spatial gradient in the left inferior (inf) frontal cortex (marked in white). According to this model, language processing can be divided into three functional components: memory, unification, and control. The memory component refers to the linguistic knowledge, which is encoded and consolidated in neocortical memory structures when we acquire language. The unification component reflects our ability to derive these building blocks from memory and then to combine them in novel ways. This unification happens at syntactic, semantic, and phonological levels. Finally, the control component of the model relates to the usage of language in social interactions. It is important for both understanding the context in a conversation and using appropriate registers for particular social situations. Regions related to the memory component are, depending on the type of knowledge, located in different parts of temporal cortex as well as in parietal cortex (angular gyrus [AG]). They subserve the knowledge representations. In turn, frontal regions (Broca's area) and adjacent cortex are required for the unification processes to happen. Here, in the inferior frontal cortex (so-called Unification space), a spatial activation gradient can be observed. As we've mentioned, different types of unification can be distinguished depending on the type of information that is unified. For example, BA47 and BA45 are recruited during semantic unification; BA45 and BA44 are involved in syntactic unification; and BA44 and ventral parts of BA6 are important for phonological unification. The dorsolateral prefrontal, anterior cingulate, and parts of the parietal cortex involved in attention are associated with the control element of the model. The above-mentioned distribution of labor is not absolute, and overall the language processing is subserved by dynamic networks of brain regions. The contribution of a particular region depends on the interaction with other brain areas that are part of the network. The topology of connectivity patterns is therefore highlighted here as an important aspect of the model. The schematic diagram was recreated after Hagoort (2014), and adjusted to highlight the topographic connectivity pattern in left inferior frontal cortex. ITG = inferior temporal gyrus; oper = opercular; orbit = orbital; PM = premotor cortex; SPL = superior parietal language area; tri = triangular.

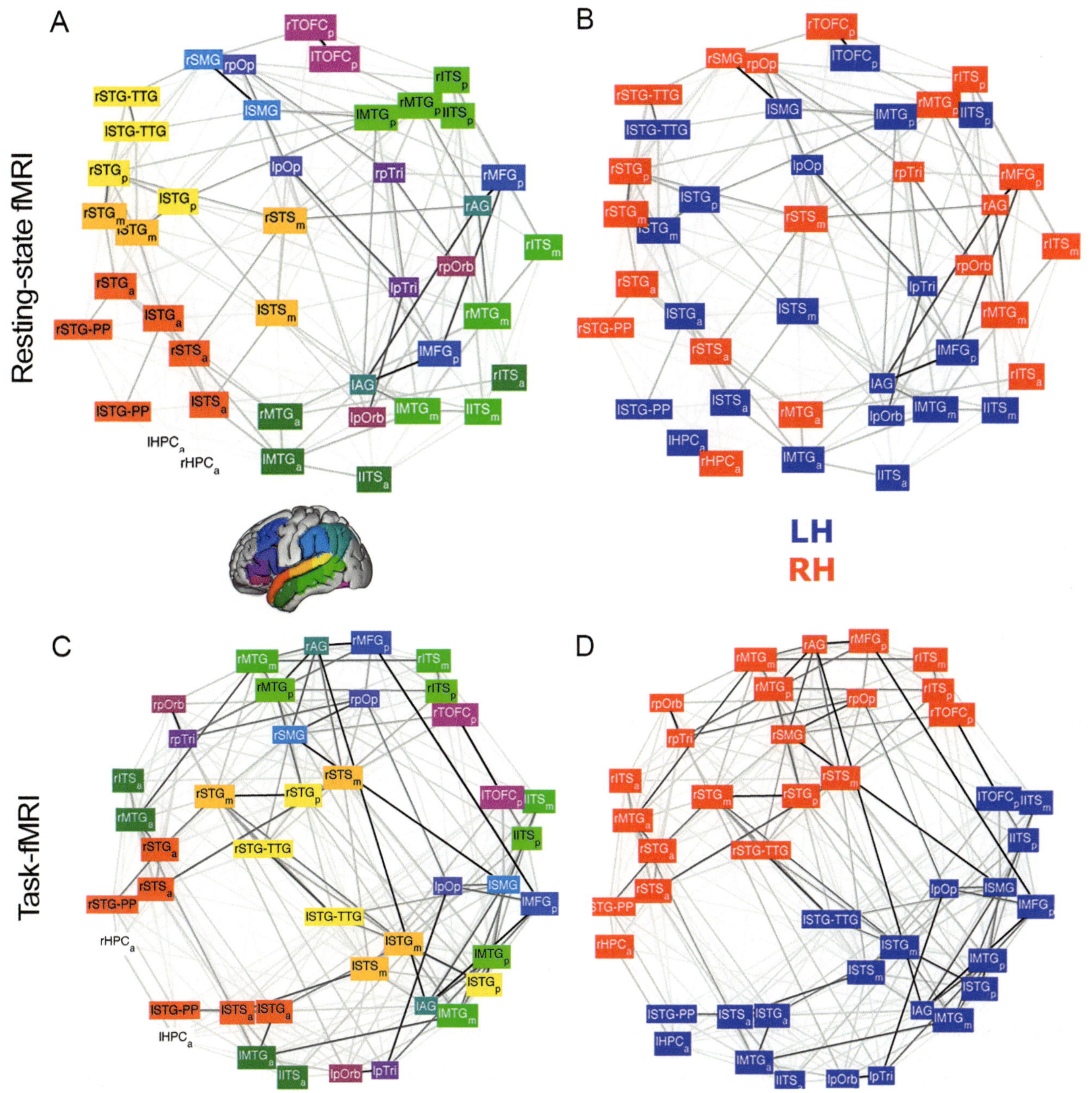

FIGURE 34.5 MDS of the language system: (A) The MDS results of the resting-state functional connectivity analysis are largely consistent with the compartmentalization of different subnetworks as described in the Hickok and Poeppel (2007) model. (B) The resting-state functional connectivity MDS results color-coded according to the left hemisphere (LH) and right hemisphere (RH). (C) The MDS results of the fMRI language task do not capture the predicted groupings of regions as consistently as the MDS results of the resting-state functional connectivity analysis. (D) MDS of the fMRI language task (Binder et al., 2011) indicates that language lateralization is task dependent. (*Middle*) Schematic of the main ROIs used in the MDS analysis. a = anterior; HPC = hippocampus; l = left; m = middle; MFG = middle frontal gyrus; Op = operculum; Orb = orbital; p = posterior; PP = planum polare; r = right; TOFC = temporo-occipital fusiform cortex; TTG = transverse temporal gyrus.

and used these to compute the partial correlation between each possible time-series pair (i.e., the correlation between each pair of regions controlled for the activity in the 40 remaining regions; this is a standard approach for obtaining estimates of the direct functional connectivity strength between 2 brain regions). In this way, we obtained a matrix of partial correlations, which was then submitted to MDS.

The results are shown in figure 34.5, which we can now compare to, for example, the influential dual-stream model of language processing by Hickok and Poeppel (2007) (see figure 34.4 for details). It becomes immediately clear that the two streams of information processing according to the model are not reflected by MDS of the data even though a similar MDS approach captured separate processing streams in the visual system in both human and nonhuman primates (Haak & Beckmann, 2017; Young, 1992). In particular, lateralization of the dorsal stream component of the Spt/SMG area does not become apparent at rest and lateralization of the opercular and triangular part of the IFG is no stronger than for the orbital part. Thus, it appears that this aspect of the model does not adhere to the data. Other aspects of the model, however do appear to adhere to the data. Language regions that form specific networks within the dual-stream model are similarly grouped together by the MDS. For instance, according to the dual-stream model, the spectrotemporal analysis network (marked yellow in figure 34.4A) includes the dorsal STG bilaterally. When we compare it with the MDS graph (figure 34.5A), we can clearly see that left and right STG–transverse temporal gyrus and posterior STG (all marked yellow) are very close to each other.

Other language regions, such as more middle parts of the temporal cortex including bilateral STS (the posterior STS was not part of the MDS) form a separate group, in both the MDS and the model (the phonological network marked orange). It is important to note here that the regions of interest (ROIs) in the MDS analysis are not an exact match with the ROIs proposed in the theoretical model by Hickok and Poeppel (2007). The regions largely correspond to the dual-stream schematic although some additional regions were included and an additional distinction between the sulci and gyri was made in the MDS analysis. As a result, the ROIs marked as orange include separately the middle portions of STG and the middle portions of STS (figure 34.5A). We can see that these middle STGs are much closer to other portions of the STGs (the yellow spectrotemporal analysis group), whereas the orange middle STSs create a clearly separate group (the phonological network), which is in line with the theoretical model.

A similar pattern can be observed when we consider regions associated with the model's ventral stream. In the MDS (marked bright green), bilateral posterior MTG and inferior temporal sulcus regions are grouped very closely together. These are the exact regions that, according to the dual-stream model, make up the posterior part of the ventral stream, the lexical interface network (marked bright green). The same is true for more anterior (including middle) parts of the ventral stream, in the graph depicted as dark green/green. They create a separate group and correspond to the combinatorial network (marked dark green). Overall, although the two main streams are not very clearly separated by MDS, the grouping of brain regions within the language network according to the dual-stream model appears to be largely consistent with the resting-state connectomics data.

Additionally, according to the dual-stream model, there appear to be differences in the representation of the two streams; the ventral stream is thought to be bilateral, whereas the dorsal stream is thought to be strongly left-lateralized (Hickok & Poeppel, 2007). These features partly appear in the MDS, although not as consistently as one could have expected based on the theoretical model. For instance, the MDS results on the basis of resting-state data (figures 34.5A and 34.5B) show that regions in the ventral stream (particularly green/dark green combinatorial network)—although clustering consistently into left versus right ROIs—exhibit distances between homologue areas of comparable magnitude to those observed between left and right ROIs from the articulatory network/sensorimotor interface (dorsal stream dark/light blue areas in figure 34.5A). As such, the MDS embedding shows lateralization effects in resting data that are similar between the dorsal and ventral stream and no particularly strong lateralization exclusively in the dorsal stream areas. Indeed, consistent with our labeling in Figure 34.4A, ventral stream areas show weak lateralization according to the data-driven analysis. Furthermore, and consistent with the model, areas forming part of the phonological network and the spectrotemporal analysis regions (orange/yellow) show smaller distances between homologue ROIs, suggesting that these regions are indeed more bilateral than either dorsal or ventral stream regions.

To better understand the apparent inconsistency between the model and the data in terms of a high degree of lateralization, it is worth considering the fact that the dual-stream model represents a synthesis of myriad task-based experimental results. As such, it is possible that the lateralization of the human language system is not an inherent characteristic of the human language system but rather a feature imposed by or manifested within specific language tasks. Indeed, when we do not use rfMRI data but the fMRI data recorded while the same subjects were engaged in a language comprehension task (Binder et al., 2011), we obtain highly lateralized MDS results (figure 34.5D). However, while the task-based MDS results recapitulate much of the lateralization predicted under the dual-stream model, it does not capture the predicted groupings of regions as consistently as before (figure 34.5C). Thus, it appears that the overall organization of the dual-stream model into groups of brain regions, but not the predicted large-scale lateralization, which is itself task-dependent, reflects an inherent property of the human language network.

Overall, this provides an example of how—through applying connectomic analysis on empirical data—we can test models of the language processing system, in the absence of task experiments. Although the main two streams were not as clearly visible as portrayed by Hickok and Poeppel (2007), by using the MDS technique, we were able to recapture the compartmentalization of different subnetworks as well as reveal representations of lateralization largely consistent with the model.

4.2. Connectopic Evidence for the MUC Model Other models of information processing in the brain transcend mere descriptions of functional nodes and associated connections between these, to additionally make predictions about finer-grained topographic organization of functional "nodes" and about the preservation of topological organization across processing hierarchies. These models move away from the idea that the language connectome is comprised of strictly segregated processing centers and the connectivity between them. Instead, theories such as the MUC model (Hagoort, 2005, 2013; see figure 34.4B) form a different class of neurobiological models of language processing that proposes that the different language areas are connected with each other in an orderly, highly topographic manner.

Here, we apply a novel data analysis, connectopic mapping (Haak, Marquand, & Beckmann, 2018; Marquand, Haak, & Beckmann, 2017), to investigate the connectivity changes and gradual transitions of the connectivity in one analysis. The connectopic mapping analysis produces maps of a predefined region of interest in which similar values (colors) indicate similar connectivity profiles (i.e., subregions with similar colors are connected with similar parts of the rest of the brain, while subregions with very different colors are connected with very different parts of the brain). As before, we used the rfMRI data of 470 subjects Human Connectome Project subjects, and our ROI corresponded to Broca's area in the left hemisphere of the brain (BA44 and BA45).

With regard to Broca's area, the MUC model proposes that connectivity is topographically organized along a smooth gradient that follows the rostral-caudal axis of the left IFG. The results of the connectopic mapping analysis within Broca's area (figure 34.6B) indicate that there is indeed such a gradual change of connectivity within the left IFG. We can further examine the predictions of the MUC model by projecting the connectopic map shown in figure 34.6A onto the rest of the brain. This can be done simply by color-coding all cortical voxels according to the color of the voxels in Broca's area that they correlate the most with. The results of this analysis are shown in figure 34.6C, indicating that the gradual connectivity pattern in Broca's area is preserved in most other language-related regions (these maps were thresholded such that only significant connectivity is shown). This preservation of the gradient across various language-related areas indicates that these areas are topographically connected, very much as predicted by

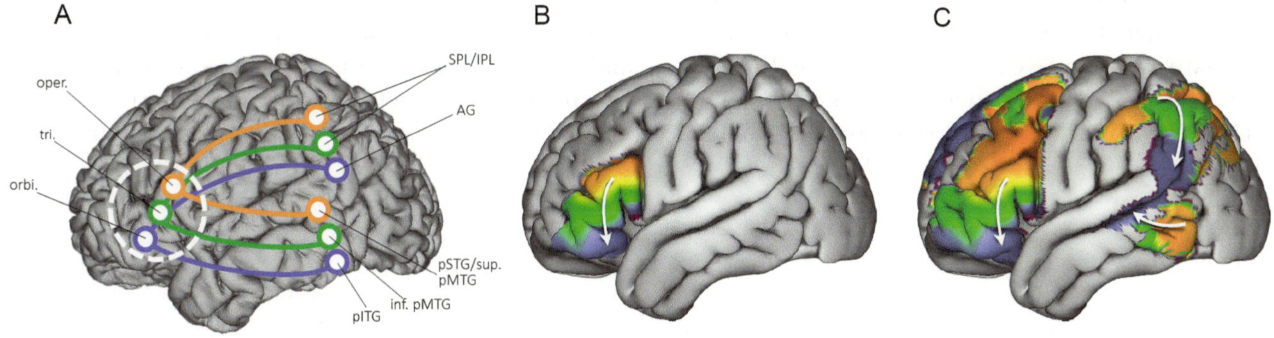

Figure 34.6 Connectopic mapping of the main language regions: (A) Spatial gradient described in the MUC model (Hagoort, 2005) across BA44/BA45 and preserved topographic organization in posterior temporal and parietal regions. (B) A gradual transition of the connectivity as we move from BA44 to BA45 in the IFG captured by connectopic mapping analysis. (C) Preservation of the connectivity change (the connectivity gradient) in other language-related regions. The arrows indicate the trajectory of the gradual change of connectivity within the main language areas.

the MUC model (but note the difference between the predicted and empirical gradient directions in posterior MTG). Overall, by using the connectopic mapping approach, we were able to provide empirical evidence of topographic patterns of the connectivity in the left IFG, an idea that corresponds to the unification gradient described in the MUC model.

5. Conclusion

In this chapter we reviewed current characterizations of the structural and functional associations between areas of the language connectome and provided some examples of how, by using connectomic and connectopic analysis, we can test some of the well-known neurobiological models of speech and language on empirical data.

We showed that over the years, different concepts and methods including representations of localized effects and behavioral associations allowed us to study language function. The move away from strict cortical localization toward network connectivity measures and representations enables us to investigate the complexity of the language system and understand its mechanisms to an even greater extent. By means of recent connectivity methods, such as diffusion-weighted imaging (dMRI) we are now able to characterize main association fibers and whole pathways linking cortical areas involved in speech and language processing. Although we still lack entirely satisfactory, empirically founded models of subcortical gray matter structures along the association fiber systems, these new approaches provide us with new information and add to the investigations of this highly distributed system. The dMRI results show that by using this structural imaging technique we can identify particular tracts, such as the anterior segment of the AF, that can be conceptualized as the primary structural connection of the dorsal stream, whereas the ILF, IOFF, and UF represent structural connections of the ventral stream described by Hickok and Poeppel (2007).

Similarly, the possibility of characterizing the functional connectome as a whole allows for a dynamic functional description of the language system. By using connectomic and connectopic analyses approaches applied to rfMRI, we are able to show the strength of functional associations between the regions involved in language processing. This allows for more independent, as compared to task-fMRI, characterization of the whole system. The results of the novel data analysis approaches that were used here to test the functional connectome against the overarching neurobiological models showed an overlap between the empirical data and some aspects of the theoretical models. For instance, the MDS analysis recaptured the compartmentalization of different subnetworks as well as revealed a lateralized representation across these subnetworks as described by the Hickok and Poeppel (2007) model, whereas with the connectopic mapping approach we were able to identify a gradient in left IFG, much as previously proposed in the MUC model by Hagoort (2005).

Disentangling the different neurobiological substrates of language is not a trivial task. Despite all the knowledge that we have gained over the past decades by virtue of different approaches and methodological advancements, there is still a lot to be done. Given that structural connections underlie the functional connectivity, integrating both aspects of processing with language systems and validating theoretical models against the empirical data provides a promising approach to more fully understand language-specific and other neurobiological systems. New data analysis approaches, combining information across different imaging modalities (e.g., magnetoencephalography, electroencephalography), are required to deliver meticulous empirical validations of existing models of the different aspects of language processing and will aid in better understanding this very distinct function of the human brain.

NOTES

1. Translated by Andreas Bartsch.
2. Human Connectome Project, https://www.humanconnectome.org.

REFERENCES

Arnold, F. (1838). *Bemerkungen über den Bau des Hirns und Rückenmarks, nebst Beiträgen zur Physiologie des zehnten und eilften Hirnnerven, mehrern kritischen Mittheilungen, so wie verschiedenen pathologischen und anatomischen Beobachtungen.* Zurich: Höhr.

Bartsch, A. J., Jbabdi, S., & Geletneky, K. (2013). The temporo-parietal fiber intersection area (TPFIA) and Wernicke's perpendicular fasciculus (WpF). *Neurosurgery, 73*(2), E381–E382.

Beckmann, C. F. (2012). Modelling with independent components. *NeuroImage, 62*(2), 891–901.

Beckmann, C. F., DeLuca, M., Devlin, J. T., & Smith, S. M. (2005). Investigations into resting-state connectivity using independent component analysis. *Philosophical Transactions of the Royal Society B: Biological Sciences, 360*(1457), 1001–1013.

Bernal, B., & Ardila, A. (2009). The role of the arcuate fasciculus in conduction aphasia. *Brain, 132*(Pt 9), 2309–2316.

Binder, J. R., Desai, R. H., Graves, W. W., & Conant, L. L. (2009). Where is the semantic system? A critical review and meta-analysis of 120 functional neuroimaging studies. *Cerebral Cortex, 19*(12), 2767–2796.

Binder, J. R., Gross, W. L., Allendorfer, J. B., Bonilha, L., Chapin, J., Edwards, J. C., ... Weaver, K. E. (2011). Mapping anterior temporal lobe language areas with FMRI: A multicenter normative study. *NeuroImage, 54*(2), 1465–1475.

Biswal, B., Zerrin Yetkin, F., Haughton, V. M., & Hyde, J. S. (1995). Functional connectivity in the motor cortex of resting human brain using echo-planar MRI. *Magnetic Resonance in Medicine, 34*(4), 537–541.

Brauer, J., Anwander, A., Perani, D., & Friederici, A. D. (2013). Dorsal and ventral pathways in language development. *Brain and Language, 127*(2), 289–295.

Broca, P. (1861). Remarques sur le siège de la faculté du langage articulé, suivies d'une observation d'aphémie (perte de la parole). *Bulletins et memoires de la Société Anatomique de Paris, 36*, 330–357.

Broca, P. (1865). Sur le siege de la faculte du langage articule. *Bulletin de la Société d'anthropologie de Paris, 6*, 337–393.

Burdach, K. F. (1822). *Vom Baue und Leben des Gehirns. 2. Band.* Leipzig: Dyk'sche Buchhandlung.

Catani, M., & Jones, D. K. (2005). Perisylvian language networks of the human brain. *Annals of Neurology, 57*, 8–16.

Caverzasi, E., Hervey-Jumper, S. L., Jordan, K. M., Lobach, I. V., Li, J., Panara, V., ... Henry, R. G. (2016). Identifying preoperative language tracts and predicting postoperative functional recovery using HARDI q-ball fiber tractography in patients with gliomas. *Journal of Neurosurgery, 125*(1), 33–45.

Chang, E. F., Raygor, K. P., & Berger, M. S. (2015). Contemporary model of language organization: An overview for neurosurgeons. *Journal of Neurosurgery, 122*(2), 250–261.

Cohen, L., Dehaene, S., Naccache, L., Lehericy, S., Dehaene-Lambertz, G., Henaff, M. A., & Michel, F. (2000). The visual word form area: Spatial and temporal characterization of an initial stage of reading in normal subjects and posterior split-brain patients. *Brain, 123 (Pt 2)*, 291–307.

Dax, M. G. (1865a). Lesions de la moitie gauche de l'encephale coincident avec l'ouble des signes de la pensee. *Gazette Hebdomadaire de Médecine et de Chirurgie, 2*(2), 259–260.

Dax, M. G. (1865b). Sur le même sujet. *Gazette Hebdomadaire de Médecine et de Chirurgie, 2*(2), 260–262.

Destrieux, C., Fischl, B., Dale, A., & Halgren, E. (2010). Automatic parcellation of human cortical gyri and sulci using standard anatomical nomenclature. *NeuroImage, 53*(1), 1–15.

Duffau, H. (2015). Stimulation mapping of white matter tracts to study brain functional connectivity. *Nature Reviews Neurology, 11*(5), 255–265.

Duffau, H., Peggy Gatignol, S. T., Mandonnet, E., Capelle, L., & Taillandier, L. (2008). Intraoperative subcortical stimulation mapping of language pathways in a consecutive series of 115 patients with grade II glioma in the left dominant hemisphere. *Journal of Neurosurgery, 109*(3), 461–471.

Ebeling, U., & von Cramon, D. (1992). Topography of the uncinate fascicle and adjacent temporal fiber tracts. *Acta Neurochirurgica (Wien), 115*(3–4), 143–148.

Ellmore, T. M., Beauchamp, M. S., O'Neill, T. J., Dreyer, S., & Tandon, N. (2009). Relationships between essential cortical language sites and subcortical pathways. *Journal of Neurosurgery, 111*(4), 755–766.

Exner, S. (1881). *Untersuchungen über die Localisation der Functionen in der Großhirnrinde des Menschen.* Vienna: Braumüller.

Fernandez-Miranda, J. C., Rhoton, A. L., Jr., Alvarez-Linera, J., Kakizawa, Y., Choi, C., & de Oliveira, E. P. (2008). Three-dimensional microsurgical and tractographic anatomy of the white matter of the human brain. *Neurosurgery, 62* (6 Suppl 3), 989–1028.

Freud, S. (1891). *Zur Auffassung der Aphasien: Eine kritische Studie.* Leipzig: Deuticke.

Friederici, A. D. (2012). The cortical language circuit: From auditory perception to sentence comprehension. *Trends in Cognitive Sciences, 16*(5), 262–268.

Friederici, A. D., & Gierhan, S. M. E. (2013). The language network. *Current Opinion in Neurobiology, 23*(2), 250–254.

Geschwind, N. (1965). Disconnexion syndromes in animals and man. *Brain, 88*(3), 585–644.

Geschwind, N. (1970). The organization of language and the brain. *Science, 170*(3961), 940–944.

Glasser, M. F., & Rilling, J. K. (2008). DTI tractography of the human brain's language pathways. *Cerebral Cortex, 18*(11), 2471–2482.

Greenblatt, S. H. (1990). Left occipital lobectomy and the preangular anatomy of reading. *Brain and Language, 38*(4), 576–595.

Griffiths, J. D., Marslen-Wilson, W. D., Stamatakis, E. A., & Tyler, L. K. (2013). Functional organization of the neural language system: Dorsal and ventral pathways are critical for syntax. *Cerebral Cortex, 23*(1), 139–147.

Haak, K. V., & Beckmann, C. F. (2017). Objective analysis of the topological organization of the human cortical visual connectome suggests three visual pathways. *Cortex, 98*, 73–83.

Haak, K. V., Marquand, A. F., & Beckmann, C. F. (2018). Connectopic mapping with resting-state fMRI. *NeuroImage, 170*, 83–94.

Hagoort, P. (2005). Broca's complex as the unification space for language. In A. Cutler (Ed.), *Twenty-first century psycholinguistics: Four cornerstones* (pp. 157–173). Mahwah, NJ: Erlbaum.

Hagoort, P. (2013). MUC (Memory, Unification, Control) and beyond. *Frontiers in Psychology, 4*, 416.

Hagoort, P. (2014). Nodes and networks in the neural architecture for language: Broca's region and beyond. *Current Opinion in Neurobiology, 28*, 136–141.

Heilbronner, K. (1908). Zur Symptomatologie der Aphasie mit besonderer Berücksichtigung zwischen Sprachverständnis, Nachsprechen und Wortfindung. *Archiv für Psychiatrie und Nervenkrankheiten, 43*, 234–298.

Hickok, G., & Poeppel, D. (2007). The cortical organization of speech processing. *Nature Reviews Neuroscience, 8*(5), 393–402.

Hitzig, E. (1874). *Verhandlungen der Berliner Gesellschaft für Anthropologie, Ethnologie und Urgeschichte.* Berlin: Berliner Gesellschaft für Anthropologie, Ethnologie, and Urgeschichte. (Specifically, discussions on pp. 47–50, 131–134, 138, 139, 140.)

Homola, G. A., Jbabdi, S., Beckmann, C. F., & Bartsch, A. J. (2012). A brain network processing the age of faces. *PLOS ONE, 7*, e49451.

Jbabdi, S., & Behrens, T. E. (2012). Specialization: The connections have it. *Nature Neuroscience, 15*(2), 171–172.

Kleist, K. (1934). Kriegsverletzungen des Gehirns in ihrer Bedeutung für die Hirnlokalisation und Hirnpathologie. In K. Bonhoeffer & O. von Schjerning (Eds.), *Handbuch der Ärztlichen Erfahrungen im Weltkriege 1914/1918. Band IV. Geistes- und Nervenkrankheiten.* Leipzig: Johann Ambrosius Barth.

Kruskal, J. B., & Wish, M. (1978). *Multidimensional scaling.* Newbury Park, CA: Sage.

Kümmerer, D., Hartwigsen, G., Kellmeyer, P., Glauche, V., Mader, I., Kloppel, S., ... Saur, D. (2013). Damage to ventral and dorsal language pathways in acute aphasia. *Brain, 136*(Pt 2), 619–629.

Lawes, I. N., Barrick, T. R., Murugam, V., Spierings, N., Evans, D. R., Song, M., & Clark, C. A. (2008). Atlas-based segmentation of white matter tracts of the human brain using

diffusion tensor tractography and comparison with classical dissection. *NeuroImage, 39*(1), 62–79.
Levelt, W. J. M. (2013). *A history of psycholinguistics: The pre-Chomskyan era.* Oxford: Oxford University Press.
Lichtheim, L. (1885). On aphasia. *Brain, 7,* 433–484.
Marchina, S., Zhu, L. L., Norton, A., Zipse, L., Wan, C. Y., & Schlaug, G. (2011). Impairment of speech production predicted by lesion load of the left arcuate fasciculus. *Stroke, 42*(8), 2251–2256.
Marquand, A. F., Haak, K. V., & Beckmann, C. F. (2017). Functional corticostriatal connection topographies predict goal-directed behaviour in humans. *Nature Human Behaviour, 1*(8), 0146.
Martino, J., da Silva-Freitas, R., Caballero, H., Marco de Lucas, E., Garcia-Porrero, J. A., & Vazquez-Barquero, A. (2013). Fiber dissection and diffusion tensor imaging tractography study of the temporoparietal fiber intersection area. *Neurosurgery, 72*(1 Suppl Operative), 87–97.
Meynert, Th. (1866). Ein Fall von Sprachstörung anatomisch begründet. *Medizinische Jahrbücher der Zeitschrift der K.K. Gesellschaft der Ärzte in Wien, 12,* 152–189.
Meynert, Th. (1884). *Psychiatrie. Klinik der Erkrankungen des Vorderhirns begründet auf dessen Leistungen und Ernährung.* Vienna: Braumüller.
Mills, C. K., & Martin, E. (1912). Aphasia and agraphia in some practical surgical relations. *JAMA, 59,* 1513–1518.
Nieuwenhuys, R., Voogd, J., & van Huijzen, C. (2008). *The human central nervous system: A synopsis and atlas.* New York: Springer.
Petrides, M., Tomaiuolo, F., Yeterian, E. H., & Pandya, D. N. (2012). The prefrontal cortex: Comparative architectonic organization in the human and the macaque monkey brains. *Cortex, 48*(1), 46–57.
Powell, H. W. R., Parker, G. J. M., Alexander, D. C., Symms, M. R., Boulby, P. A., Wheeler-Kingshott, C. A. M., ... Duncan, J. S. (2006). Hemispheric asymmetries in language-related pathways: A combined functional MRI and tractography study. *NeuroImage, 32*(1), 388–399.
Rolheiser, T., Stamatakis, E. A., & Tyler, L. K. (2011). Dynamic processing in the human language system: Synergy between the arcuate fascicle and extreme capsule. *Journal of Neuroscience, 31*(47), 16949–16957.
Rostomily, R. C., Berger, M. S., Ojemann, G. A., & Lettich, E. (1991). Postoperative deficits and functional recovery following removal of tumors involving the dominant hemisphere supplementary motor area. *Journal of Neurosurgery, 75*(1), 62–68.
Saygin, Z. M., Osher, D. E., Koldewyn, K., Reynolds, G., Gabrieli, J. D., & Saxe, R. R. (2012). Anatomical connectivity patterns predict face selectivity in the fusiform gyrus. *Nature Neuroscience, 15*(2), 321–327.
Schmahmann, J. D., & Pandya, D. N. (2006). *Fiber pathways of the brain.* Oxford: Oxford University Press.
Smith, S. M., Fox, P. T., Miller, K. L., Glahn, D. C., Fox, P. M., Mackay, C. E., ... Beckmann, C. F. (2009). Correspondence of the brain's functional architecture during activation and rest. *Proceedings of the National Academy of Sciences, 106*(31), 13040–13045.
Smith, S. M., Vidaurre, D., Beckmann, C. F., Glasser, M. F., Jenkinson, M., Miller, K. L., ... van Essen, D. C. (2013). Functional connectomics from resting-state fMRI. *Trends in Cognitive Sciences, 17*(12), 666–682.
Steinthal, H. (1874). *Verhandlungen der Berliner Gesellschaft für Anthropologie, Ethnologie und Urgeschichte.* Berlin: Berliner Gesellschaft für Anthropologie, Ethnologie, and Urgeschichte. (Specifically discussions on pp. 130, 131, 138.)
Tavor, I., Parker Jones, O., Mars, R. B., Smith, S. M., Behrens, T. E., & Jbabdi, S. (2016). Task-free MRI predicts individual differences in brain activity during task performance. *Science, 352*(6282), 216–220.
Tie, Y., Rigolo, L., Norton, I. H., Huang, R. Y., Wu, W., Orringer, D., ... Golby, A. J. (2014). Defining language networks from resting-state fMRI for surgical planning: A feasibility study. *Human Brain Mapping, 35*(3), 1018–1030.
Turken, A. U., & Dronkers, N. F. (2011). The neural architecture of the language comprehension network: Converging evidence from lesion and connectivity analyses. *Frontiers in Systems Neuroscience, 5,* 1.
van Essen, D. C., Smith, S. M., Barch, D. M., Behrens, T. E. J., Yacoub, E., Ugurbil, K., & WU-Minn HCP Consortium. (2013). The WU-Minn Human Connectome Project: An overview. *NeuroImage, 80,* 62–79.
Vergani, F., Lacerda, L., Martino, J., Attems, J., Morris, C., Mitchell, P., ... Dell'Acqua, F. (2014). White matter connections of the supplementary motor area in humans. *Journal of Neurology, Neurosurgery, and Psychiatry, 85*(12), 1377–1385.
Vincent, J. L., Patel, G. H., Fox, M. D., Snyder, A. Z., Baker, J. T., van Essen, D. C., ... Raichle, M. E. (2007). Intrinsic functional architecture in the anaesthetized monkey brain. *Nature, 447*(7140), 83–86.
Voets, N. (2017). [Probabilistic streamline data]. Unpublished data.
Von Monakow, C. (1905). *Gehirnpathologie, 2. Auflage.* Vienna: Hölder.
Wang, Z., Chen, L. M., Négyessy, L., Friedman, R. M., Mishra, A., Gore, J. C., & Roe, A. W. (2013). The relationship of anatomical and functional connectivity to resting-state connectivity in primate somatosensory cortex. *Neuron, 78*(6), 1116–1126.
Weiller, C., Bormann, T., Saur, D., Musso, M., & Rijntjes, M. (2011). How the ventral pathway got lost: And what its recovery might mean. *Brain and Language, 118*(1–2), 29–39.
Wernicke, C. (1874). *Der aphasische Symptomencomplex; Eine psychologische Studie auf anatomischer Basis.* Breslau, Germany: Cohn and Weigert.
Wernicke, C. (1881). *Lehrbuch der Gehirnkrankheiten für Ärzte und Studierende. Band 1.* Kassel, Germany: Fischer.
Westphal, C. (1874). *Verhandlungen der Berliner Gesellschaft für Anthropologie, Ethnologie und Urgeschichte.* Berlin: Berliner Gesellschaft für Anthropologie, Ethnologie, and Urgeschichte. (Specifically discussions on pp. 94-102, 134, 135, 138, 139.)
Wright, P., Stamatakis, E. A., & Tyler, L. K. (2012). Differentiating hemispheric contributions to syntax and semantics in patients with left-hemisphere lesions. *Journal of Neuroscience, 32*(24), 8149–8157.
Yagmurlu, K., Middlebrooks, E. H., Tanriover, N., & Rhoton, A. L. Jr. (2016). Fiber tracts of the dorsal language stream in the human brain. *Journal of Neurosurgery, 124*(5), 1396–1405.
Young, M. P. (1992). Objective analysis of the topological organization of the primate cortical visual system. *Nature, 358*(6382), 152–155.
Yourganov, G., Fridriksson, J., Rorden, C., Gleichgerrcht, E., & Bonilha, L. (2016). Multivariate connectome-based symptom mapping in post-stroke patients: Networks supporting language and speech. *Journal of Neuroscience, 36*(25), 6668–6679.

35 The "Speech Ready" Auditory Cortex

ELIA FORMISANO

1. Introduction

Decoding speech is arguably the most complex and relevant function the human auditory cortex contributes to. It is thus conceivable that in the human brain, the circuitry and computational mechanisms devoted to auditory perception and cognition are optimized to ensure efficient and robust analysis of speech signals. So far, a large number of cognitive neuroimaging studies on speech perception have focused on discovering what is "special" about the processing of speech in the human brain compared to the processing of all other sounds. This is comprehensible as deriving linguistic information from speech certainly involves additional and unique processing steps compared to those required for the analysis of other complex sounds. On the other side, the analysis of speech, especially at the early processing stages, also relies on general purpose auditory mechanisms that are equivalently relevant for the analysis of any other sound (Diehl, Lotto, & Halt, 2004; Moore, 2008; Young, 2008). In this chapter, I will examine several of these basic auditory processing mechanisms, with an emphasis on the underlying neurophysiological bases and coding principles in the (human) auditory cortex. I will consider the hypothesis that the human auditory cortex is "speech ready," that is, that—in the human brain—even basic and general purpose auditory mechanisms are fine-tuned to analyze optimally the acoustic components of speech that are most relevant for behavior.

In the long-standing tradition of psychoacoustics, the processing mechanisms underlying auditory perception have been thoroughly characterized and modeled using psychophysical methods (Moore, 2003). More recently, their possible neural implementation could be investigated with noninvasive functional imaging in humans. These investigations are progressing with an ever improving spatial (functional MRI [fMRI], Formisano, Moerel, & Bonte, 2015) and temporal (electroencephalography/magnetoencephalography, Poeppel & Hickok, 2015) resolution and an ever increasing sophistication in terms of signal analysis and interpretation. Gaining a mechanistic insight into the neural coding mechanisms underlying the analysis of sounds, however, requires examining the activity of single neurons or small neuronal populations at a fine-grained spatiotemporal scale. Despite the advances in noninvasive methods, these types of investigations can only be achieved using intracortical electrophysiological recordings in animal models (Eggermont, 2001) and, more rarely, in human patients (see chapter 27 by Sjerps & Chang in this volume).

After a brief description of the anatomy of the human auditory system, in section 2, I will discuss major findings on the spectral and temporal analysis of sounds from psychophysics, functional neuroimaging, and neurophysiology. In the past years, each of these research fields has produced valuable insights into the mechanisms of sound analysis from the auditory periphery (cochlea) to the central auditory pathway and the cortex. However, the information obtained in the separate fields is of very different nature and relates to diverse spatiotemporal scales. Whereas a detailed discussion of the results from the different disciplines is beyond the scope of this chapter, my goal will be to highlight established or potential links between perceptual phenomena (as studied in psychoacoustics) and neurophysiological mechanisms (as investigated with invasive and noninvasive methods). Although necessarily tentative in several aspects, this integrative effort appears obligatory for gaining a coherent understanding of the sound processing mechanisms in the human brain.

A typical experimental paradigm in both psychophysical and neurophysiological research on audition consists of measuring behavior (e.g., detection thresholds) and/or neural activity (e.g., single neuron spiking, fMRI responses) in response to simple synthetic sounds (e.g., tones, noises) that vary in a single acoustic feature (e.g., frequency, bandwidth). The advantage of this approach is that it provides a means to assess the covariation of behavioral/neuronal responses and the stimulus manipulation, while having strict experimental control over all other features, which are typically kept constant across conditions. In real life, however, sounds continuously vary along several acoustic dimensions and scales. The auditory processing system thus requires mechanisms to analyze these processes simultaneously. For example, in speech, formants occur at a coarser spectral scale, that is, they are more widely spaced, than the harmonics of the speaker's fundamental frequency. Likewise, temporal modulations at different time scales co-occur; for example, modulations associated with syllabic events are characterized by a longer temporal scale than modulations associated with phonemic transitions (Rosen, 1992). In section 3, I will discuss results from psychophysics and neurophysiology that examined the mechanisms of combined spectrotemporal processing using more complex and naturalistic stimuli, such as speech, human voices, and other environmental sounds. This experimental approach is complementary to the traditional approach employing simplified stimuli. Its distinct advantage is that the auditory system is probed with ecologically valid and behaviorally relevant stimuli, whereas the disadvantage is that it remains difficult to control for the intrinsic complexity of natural stimuli and to model their effects. Convergent results from experiments employing simple, synthetic versus complex, realistic sounds strongly mutually reinforce the conclusions on the mechanisms being studied. Divergent results, on the other side, provide a good indication of the limits of applicability of simplified designs to explain the complexity of real-life perception.

Finally, in the section 4, I will describe a recent series of studies combining high spatial resolution fMRI, naturalistic stimuli, and computational modeling to examine the encoding of real-life sounds (including speech) in the auditory cortex of humans and macaque monkeys. And in section 5, I will argue that the results of these studies provide empirical support to the hypothesis that the human auditory cortex is "speech ready."

2. Anatomy of the Human Auditory Pathway and Cortex

Sounds enter our ears in the form of air pressure waves; at the cochlea, these waves produce spatiotemporal patterns of displacement of the basilar membrane, which are then transduced into spatiotemporal patterns of neural responses, transmitted to the cortex via the auditory nerves and the ascending auditory pathway. Before reaching the auditory cortex, the sound-related neuronal response patterns pass through a number of relays and nuclei, including the cochlear nucleus, the superior olivary complex in the brain stem, the inferior colliculus in the midbrain, and the middle geniculate body in the thalamus (figure 35.1A). The auditory cortex occupies the upper surface of the superior temporal gyrus (STG), which runs parallel to the lateral sulcus (Sylvian fissure) (figure 35.1B). The superior surface of the STG is typically divided—from front to back—into three partitions: the planum polare (PP), the transverse temporal gyrus or gyrus of Heschl (HG), and the planum temporale (PT). HG runs obliquely across the surface of the STG and is separated from the PP by the first transverse sulcus and from the PT by the Heschl's sulcus. Note that the HG is not present in macaques, while it may be present—often in a rudimentary form—in chimpanzees (Hackett, Preuss, & Kaas, 2001). In humans, the gross morphology of the Heschl's region is highly variable, both among individuals and between hemispheres (Abdul-Kareem & Sluming, 2008; Campain & Minckler, 1976). This variability includes complete and incomplete HG duplications, and even (incomplete) triplications may occur. In the case of an incomplete duplication or triplication, an intermediate sulcus divides the HG (Leonard, Puranik, Kuldau, & Lombardino, 1998). The HG is located more anteromedially in the right, as compared to the left hemisphere and there is a consistent asymmetry (left greater than right) in the white matter volume of the auditory cortex (Penhune, Zatorre, MacDonald, & Evans, 1996). The PT is a triangular region located posterior to the HG (figure 35.1B). Medially, the PT is delimited by the posterior extremity of the Sylvian fissure, and laterally it is bordered by the rim of the supratemporal plane (Kim et al., 2000). Similar to the HG, the PT is asymmetrical, with a larger surface in the left hemisphere (Geschwind & Levitsky, 1968). The PP is located anterior to HG and is delimited by the insula and the frontal operculum (Kim et al., 2000). Besides in HG, PT, and PP, strong responses to sounds are consistently observed along the STG, extending into the superior temporal sulcus (STS), which separates the superior from the middle temporal gyrus (figure 35.1B) and is deeper in the right than in the left hemisphere (Bonte et al., 2013). Auditory responsive regions further extend posteriorly toward the temporal-parietal junction and the superior operculum, anteriorly toward the temporal pole, and inferiorly toward the middle temporal gyrus.

Beyond this macroanatomical description, several parcellation schemes have been proposed for the human auditory cortex on the basis of cyto- or myeloarchitectonic properties (see chapter 33 by Palomero-Gallagher & Zilles in this volume). In the commonly used Brodmann cytoarchitectonic classification of brain areas (or Brodmann's area [BA]; Brodmann, 1909), the auditory cortex is subdivided into a primary area 41

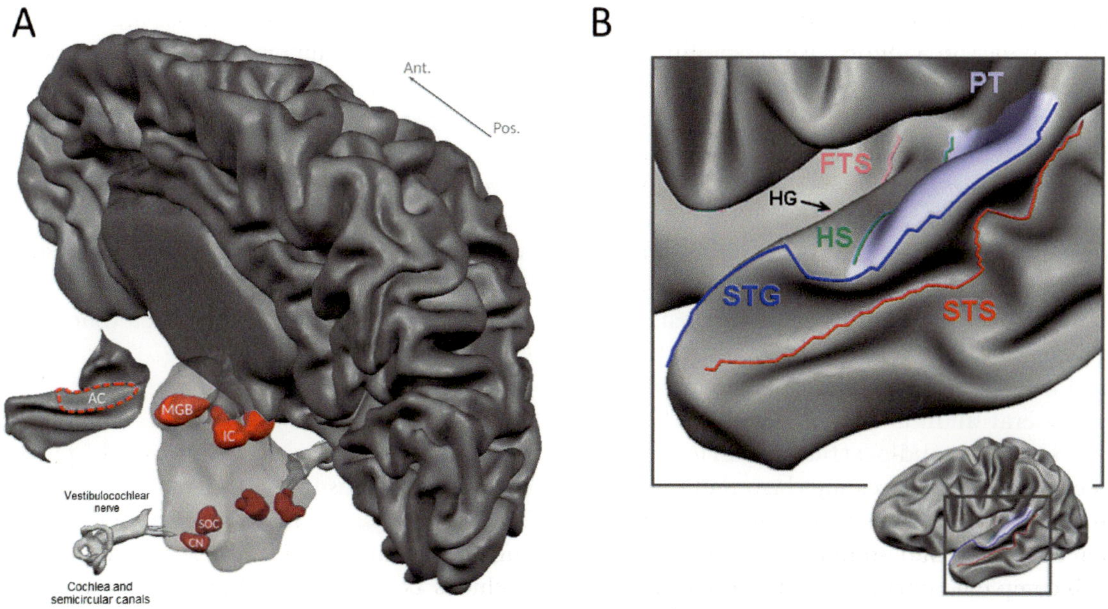

FIGURE 35.1 The human auditory pathway and cortex. (A) Reconstruction of the auditory pathway and auditory cortex (AC) as obtained from the high-resolution structural and functional MR imaging in an individual participant. (B) Zoom-in of the auditory cortex with highlighted relevant macro-anatomical landmarks and regions. Ant = anterior; CN = cochlear nucleus; FTS = first transverse sulcus; HS = Heschl's sulcus; IC = inferior colliculus; MGB = medial geniculate body; Pos = posterior; SOC = superior olivary complex; STS = superior temporal sulcus. Panel A courtesy of Omar Faruk Gulban and Federico De Martino; panel B is from Bonte et al., 2013.

(BA41) and secondary association areas 42 (BA42) and 22 (BA22). BA22 includes Wernicke's area, a brain area that has been associated with the processing of spoken language. In subsequent anatomical studies, however, it became clear that the superior temporal cortex entails a larger number of subdivisions than those originally identified by Brodmann (1909). Yet, results varied across studies with respect to number of areas, reference to anatomical landmarks and cyto- and myeloarchitectonic properties of the areas (e.g., Galaburda & Sanides, 1980; Hackett et al., 2001; Morosan et al., 2001; Rademacher et al., 2001; Rivier & Clarke, 1997). Moerel, De Martino, and Formisano (2014) described a tentative synthesis of results from several *postmortem* anatomical studies, with the goal to derive a coherent framework for reporting auditory cortical activations. Concerning Heschl's region, there is consensus that the primary auditory cortex (PAC) occupies the most posterior-medial aspects of the HG. In case multiple gyri are present, PAC occupies the most anterior of the HGs, occasionally extending posterior-medially into Heschl's sulcus and the second gyrus (Galaburda & Sanides, 1980; Hackett et al., 2001; Morosan et al., 2001; Rademacher et al., 2001; Rivier & Clarke, 1997). Importantly, recent advancements in MRI acquisition techniques have made it possible to obtain high resolution anatomical images whose contrast is related to the intracortical myelin content. Consistent with the notion that the primary auditory area is more densely myelinated than surrounding areas (Nieuwenhuys, 2013), several studies have shown that the primary auditory area can be delineated in vivo and in individual subjects through the analysis of the MR images with myelin-related contrast (Dick et al., 2012; De Martino et al., 2014).

3. Spectral and Temporal Processing of Sounds

The auditory periphery transforms the input sound from its one-dimensional temporal representation (sound wave) to a two-dimensional time-frequency representation. This spectrogram-like representation describes the time-varying amplitude of each acoustic frequency and highlights the rich spectrotemporal dynamics typical of natural sounds. For simplicity, I will first examine separately the mechanisms for analyzing spectral and temporal sound components and then consider more complex computational mechanisms that enable the representation and processing of a sound spectrogram at multiple temporal and spectral scales.

3.1. Spectral Processing

3.1.1. Frequency selectivity and the cochlear filters Human listeners are able of distinguishing a target frequency within a complex sound containing multiple frequencies, at a threshold referred to as *frequency selectivity* or *resolution.* This property of the auditory system is relevant for many common auditory tasks, including listening to music, speech, and many other relevant real-life sounds. Although the intelligibility of speech is largely preserved even after spectral degradation of the sounds (Shannon, Zeng, Kamath, Wygonski, & Ekelid, 1995), a good frequency resolution is especially required for complex speech recognition tasks and in noisy conditions (Shannon, Fu, & Galvin, 2004).

The frequency selectivity of the auditory system has been characterized in numerous studies using psychoacoustics. A central finding is that auditory processing at the peripheral auditory system (cochlea) can be modeled as a bank of band-pass filters, with logarithmically spaced center frequencies and overlapping passbands (Moore, 2003). The auditory filter bank model is closely anchored to the properties of the cochlea. Sounds at a given frequency produce a larger displacement in a specific location of the basilar membrane. The so-called Equivalent Rectangular Bandwidth (ERB) scale (Glasberg & Moore, 1990) is often used to describe the logarithmic spacing of the cochlear filters' center frequencies on a frequency scale related to the bandwidth of the filters. It has been suggested that one unit on the ERB scale corresponds approximately to a constant spacing of 0.89 mm on the basilar membrane (which has a total length of about 30 mm).

Psychophysical tuning curves describing the shape of the cochlear filters have been obtained using different variants of masking paradigms. The filters are asymmetrical around the center frequency, with a steeper slope for higher than for lower frequencies. Estimates of the "critical" bandwidth obtained with simultaneous masking (i.e., presenting masking noise simultaneously to a target tone) indicate that the bandwidth of auditory filters increases with the center frequency (from ~30 Hz to 900 Hz for center frequencies from 50 Hz to 8 kHz, respectively). A function introduced by Glasberg and Moore (1990) describes the relation between center frequency and bandwidth and provides an approximation of the ERB of the auditory filter for a given center frequency. As the slope of the linear equation is less than 1, the sharpness of tuning (i.e., the filter selectivity: bandwidth/center frequency) of the auditory filters is higher for high center frequencies than for low center frequencies (~0.11 octaves at 8 kHz, ~0.13 octaves at 1 kHz, and ~0.35 octaves at 100 Hz; see figure 35.4A). Note, however, that the experimental procedures and the sound level at which the psychophysical tuning curves are calculated can significantly affect the estimates of shape and bandwidth of the auditory filters. For instance, Oxenham and Shera (2003) showed that ERBs calculated using forward masking instead of simultaneous masking are substantially narrower, which may be due to the fact that nonlinear effects may be more prominent for simultaneous masking. Furthermore, this same study showed that the sharpness of tuning doubles over the range of center frequencies tested (1–8 kHz). Importantly, the estimates obtained with forward masking are in agreement with those obtained with an objective method based on the measurement of stimulus-frequency otoacoustic emissions produced by the cochlea (Shera, Guinan, & Oxenham, 2002).

3.1.2. Neuronal frequency tuning and frequency discrimination The decomposition of sounds by frequency and the relation between frequency and space described by the cochlear filters is preserved in the ascending auditory pathway and in the auditory cortex, although both the complexity of the relation and the frequency selectivity changes across levels of the processing hierarchy. Neurophysiological measures of frequency selectivity in the auditory nerves and in auditory neurons can be obtained by plotting the threshold sound level of the stimulus that is required to evoke a discharge rate for each of a range of stimulus frequencies. The resulting tuning curves show that the auditory nerves as well as auditory neurons in the subcortical relays and in the cortex exhibit a "characteristic" or preferred frequency. At the characteristic frequency, the sound level that is required to evoke a discharge is lower than for all other frequencies. In many animal species, measurements of frequency tuning of single neurons at different stages of the auditory nerve, the ascending auditory pathway, and in the cortex, found bandwidth values comparable to or larger than those obtained for the cochlear filter (e.g., see Bartlett, Sadagopan, & Wang, 2011, table 1). In contrast with this notion, however, Bitterman, Mukamel, Malach, Fried, and Nelken (2008) reported that frequency tuning in single neurons recorded from human auditory cortex is substantially narrower than that typically described in any other mammalian species (besides bats). In fact, Bitterman et al. (2008) reported auditory cortical neurons with a selectivity of 0.08 octaves, which reflects better frequency selectivity than that attributed to the human auditory periphery. The discrepancy between the estimates in human auditory cortex with those reported in previous literature based on recordings in animals may be due to the fact that the latter were mostly obtained in anesthetized preparations. In support of this hypothesis, recordings in awake marmoset monkeys found that neurons in "the primary auditory cortex exhibit frequency tuning that is finer than the typical frequency tuning of the auditory nerve and substantially finer than previously reported cortical data obtained from anesthetized animals" (Bartlett et al., 2011, p. 849).

Even the reported ultra-fine tuning of single neurons in the auditory cortex, however, cannot fully explain how the auditory system can give rise to the behavioral performance levels observed for fine-grained frequency discrimination. In fact, for human listeners, the smallest detectable frequency changes (or just noticeable differences) for two tones presented in rapid succession are in the order of 0.1% to 0.3% of the carrier, such that—for a reference tone at 1,000 Hz—changes of 1 to 3 Hz can be resolved. Frequency just noticeable differences are thus substantially smaller than the bandwidth of the peripheral sensors and of cortical neurons, a phenomenon that is referred to as *frequency hyperacuity*. Furthermore, the frequency discrimination threshold is frequency-dependent (and task-dependent) and is smaller for lower frequencies than for higher frequencies, even when using ERB units (Moore & Sek, 1995). This frequency dependence is not predicted by the auditory filter models, as one would expect for a constant threshold across frequencies represented on an ERB scale. Both these psychophysical observations indicate that frequency discrimination does not rely on a single frequency channel, but it involves the analysis of the activation patterns across multiple channels. A possible explanation for this frequency dependence may be that temporal mechanisms (e.g., phase locking to the stimulus, see section 3.2.2) contribute relevant information for frequency discrimination for low but not high frequencies (Sek & Moore, 1995).

3.1.3. Tonotopic maps in the auditory pathway and cortex A consistent finding of electrophysiology in numerous animal species is that neurons with similar frequency preference cluster together and form tonotopic (or cochleotopic) maps (Merzenich & Brugge, 1973; Merzenich, Knight, & Roth, 1973). Tonotopy is a prominent organizational principle of the auditory system, which defines the topography and connectivity of the subcortical and cortical pathways in humans and other mammalians. In macaque monkeys, the PAC has been subdivided in three areas, which include area A1, the rostral area R, and rostrotemporal area RT, that forms a so-called *primary core*. In these areas, neurons respond best to tones at specific frequencies and exhibit narrow frequency-response curves. Analogously to retinotopic maps in early areas of the visual cortex, frequency-selective neurons in the core subdivisions of the auditory cortex form tonotopic maps with a mirror symmetry (see inset in figure 35.2). In A1, the most caudal of these fields, the tonotopic gradient from high to low frequencies is caudorostrally oriented. In R, located rostrally to A1, the tonotopic gradient is reversed such that A1 and R share a low-frequency border that appears to coincide with the histologically defined border (Kaas & Hackett, 2000; Merzenich & Brugge, 1973; Morel, Garraghty, & Kaas, 1993). The primary core is surrounded by nonprimary areas forming the so-called *belt*. Areas of the lateral belt also display tonotopic organization, even though their neurons are less responsive to tones and prefer increasingly complex stimuli (Kosaki, Hashikawa, He, & Jones 1997; Kusmierek & Rauschecker, 2009; Rauschecker, Tian, & Hauser, 1995). A similar functional organization has been found in the human auditory cortex. Formisano et al. (2003) combined high spatial resolution fMRI, "sparse" imaging, and an analysis of the functional images that accounted for the cortical folding to reveal a mirror-symmetric frequency map occupying HG. A quantitative analysis of the spatial frequency allowed defining the reversal in the best-frequency gradient in fine detail, delineating two subdivisions that each contained a complete frequency gradient. The spatial arrangement, the selectivity, and the mirror symmetry of the responses suggested that these subdivisions reflected the human homologues of monkey areas A1 and R (Kaas & Hackett, 2000). Since this report (Formisano et al., 2003), many other studies obtained fMRI-based tonotopic maps of the human PAC (see, e.g., Da Costa et al., 2011; Striem-Amit, Hertz, & Amedi, 2011; Talavage et al., 2004). Tonotopic maps shown in these studies are generally consistent with the presence of at least two mirror-symmetric tonotopic maps sharing a low-frequency border and delineating the primary areas (figure 35.2, left column). However, the exact orientation of the tonotopic gradients with respect to the HG anatomy remains debated (see Baumann, Petkov, & Griffiths, 2013; Leaver & Rauschecker, 2016; Moerel et al., 2014; Saenz & Langers, 2014, for a detailed discussion). Additional information on functional (e.g., frequency selectivity, Moerel, De Martino, & Formisano, 2012), anatomical (e.g., myelin content, Dick et al., 2012; De Martino et al., 2014; figure 35.2, left column), and physiological properties (e.g., blood flow, Gardumi, Ivanov, Havlicek, Formisano, & Uludag, 2017) can be used to disambiguate the exact location of the primary subdivisions in individual subjects.

3.1.4. Tonotopic maps and their relation to speech processing A relevant finding of tonotopic mapping experiments conducted using frequency-modulated sweeps (Striem-Amit et al., 2011) or natural sounds in combination with fMRI encoding (Moerel et al., 2012) is that tonotopic responses extend well beyond the primary areas, into the nonprimary belt and parabelt

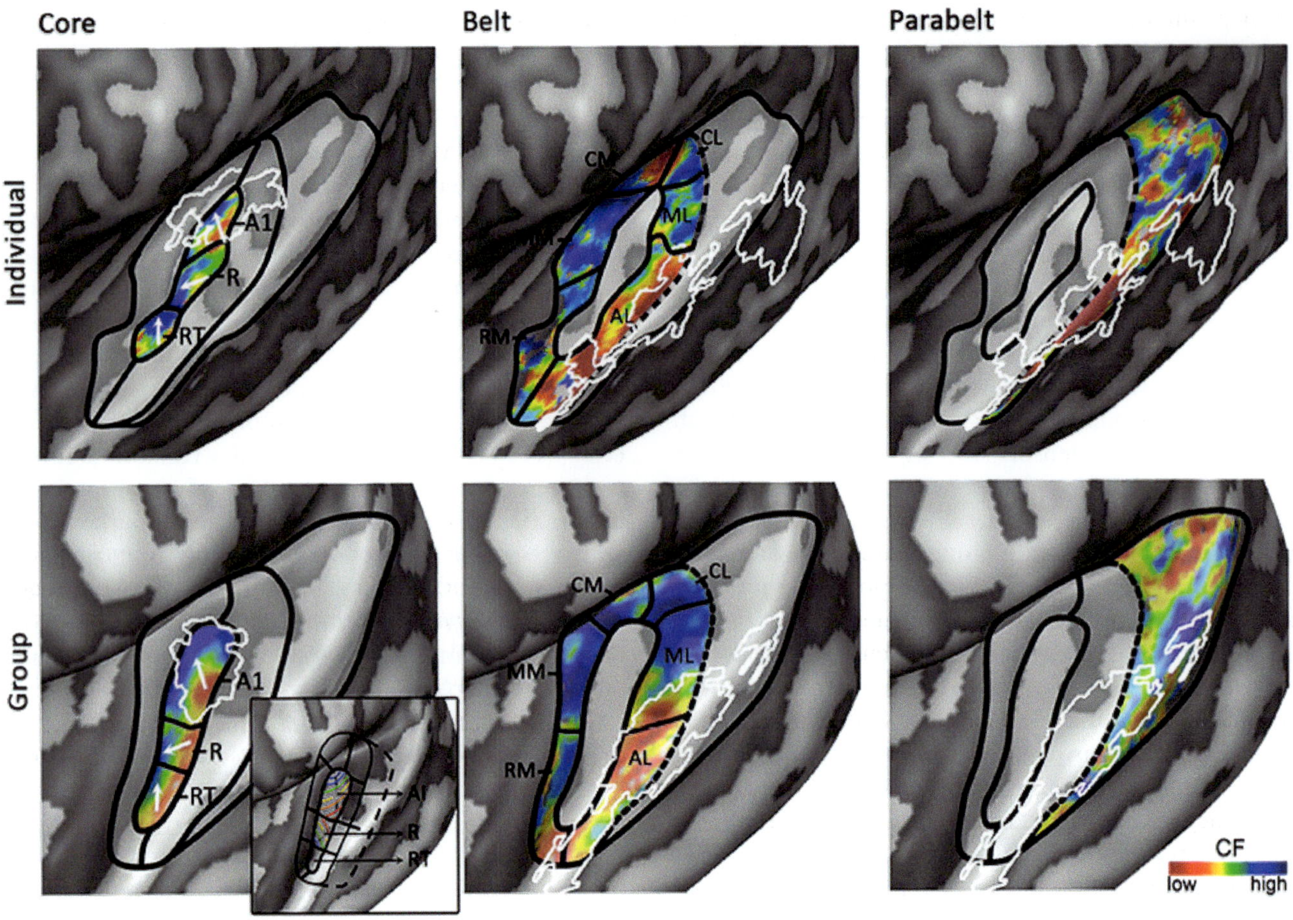

FIGURE 35.2 Subdivisions of the human auditory cortex. Parcellation of the auditory cortex into core (*left*), medial and lateral belt (*middle*), and parabelt (*right*) based on individual (*top*) and group (*bottom*) data. Solid black lines indicate boundaries between auditory fields based on maps of tonotopy, tuning width, or myelin. Dotted black lines indicate boundaries assumed based on literature, but for which no objective measure is available. White outlines indicate maps of myelin (*left column*) or speech/voice selectivity (*middle and right columns*). The white arrows in the first column show the main direction of the frequency gradient within the three core fields. The inset in the panel "Group-Core" shows the primate model of the auditory cortex as interpreted and adapted from Kaas and Hackett (2000). The model is rotated so that the core aligns with HG. Isofrequency gradients are color-coded to match the color scale of the human tonotopy maps. Images are from Moerel et al. (2014). A1 = primary auditory area; AL = anterolateral area; CF = characteristic frequency; CL = caudolateral area; CM = caudomedial area; ML = middle lateral area; MM = middle medial area; RM = rostromedial.

areas. Interestingly, overlaying statistical maps of preference for speech (and voice) sounds on tonotopic maps, reveals an intrinsic relation between these two maps (figure 35.2, center and right columns). That is, speech-/voice-sensitive regions display a bias toward low frequencies. This low-frequency bias is preserved even when the responses to simple tones are considered (Moerel et al., 2012). Results from behavioral experiments are also consistent with this close link between general acoustic mechanisms and speech and voice perception. For instance, Laing, Liu, Lotto, and Holt (2012) reported significant contextual influences of the long-term average spectrum of nonspeech stimuli on subsequent processing and perception of speech/voice (Laing et al., 2012).

Together, these findings suggest a general organizational principle for higher level auditory areas, which predicts that the topographic arrangement of "specialized" areas is determined by their (residual) sensitivity to low-level acoustic properties. This is similar to that observed in the visual cortex where face selective areas occupy the foveal part of a single eccentricity map in the occipitotemporal cortex (Hasson, Harel, Levy, & Malach, 2003; Rajimehr, Devaney, Bilenko, Young, & Tootell, 2011). Speech and voice sounds contain most energy in relatively low-frequency ranges (<1 kHz; Crandall & MacKenzie, 1922). Assuming that regions in the STG/superior temporal sulcus are the site of relevant computations and transformations from low-level to high-level sound representation of speech (Overath,

McDermott, Zarate, & Poeppel, 2015) and vocal sounds (Belin, Zatorre, Lafaille, Ahad, & Pike, 2000), the observed low-frequency bias may reflect the input stage of these transformations. Understanding the entire computational chain underlying these transformations remains an intriguing experimental challenge for current research.

3.2. Temporal Processing

3.2.1. Temporal resolution and the modulation filters The capability of the auditory system to detect small, transient energy changes over time (i.e., its temporal resolution) is pivotal to the extraction and analysis of information from any sound in the environment. A good sensitivity to temporal changes in sound signals is essential for speech perception (Kluender, Coady, & Kiefte, 2003). In speech, as well as in many other complex sounds, temporal changes are present at multiple dimensions. When considering the mechanisms of temporal processing by the auditory system, it is thus useful to distinguish between the processing of the fine temporal structure of the sound, which is related to the temporal variations in the carrier, and the processing of the sound envelope, which determines the overall amplitude shape of the sound. For modulated sounds, *temporal resolution* typically refers to the detection of changes in the sound envelope. In psychophysics, estimates of the temporal resolution of the auditory system are obtained by measuring the threshold of detecting a small gap in an otherwise continuous broadband noise (gap detection threshold). These estimates suggest that humans are unable to detect gaps smaller than 2 to 3 ms. A more complete characterization of the properties of temporal processing by the auditory system is obtained by calculating the *temporal modulation transfer function* (tMTF). This is done with a modulation detection (or discrimination) task where the modulation depth (i.e., the peak-to-valley amplitude distance) is manipulated in order to measure the listener's modulation detection threshold, that is, the minimum modulation depth required to detect signal amplitude changes. Detection thresholds are measured over a range of modulation frequencies to yield the tMTF. For broadband noise, detection threshold is smallest and relatively constant for slow temporal modulations (up to 10–15 Hz) but with a minimum around 4 Hz; detection threshold then increases monotonically with rate (Moore, 2003, 2008). Above 100 Hz, transient modulation changes become basically undetectable (Bacon & Viemeister, 1985; O'Connor et al., 2011; Viemeister, 1979).

Similar to the analysis of frequency by the cochlear filter bank, it has been suggested that the analysis of amplitude changes of the sound envelope by the auditory system can be modeled by a bank of modulation-tuned band-pass filters (*modulation filter bank*). In the range 0 to 10 Hz, the filters have a constant bandwidth of 5 Hz. Above 10 Hz, a logarithmic scaling is assumed (Dau, Kollmeier, & Kohlrausch, 1997). This model explains well the tMTF estimated experimentally using broadband noise. For narrow band noise and sinusoidal carriers, results whether the carrier frequency has an influence on the tMTF beyond the influence of spectral cues are controversial (Dau et al., 1997; Kohlrausch, Fassel, & Dau, 2000; Moore & Glasberg, 2001; Moore & Sek, 1995). Simpson, Reiss, and McAlpine (2013) estimated the sensitivity for rates of 0.5, 5, 10, and 33 Hz, ranging across a large number of frequency carriers (100 Hz–10 kHz). Results suggested that at low carrier frequencies, modulation filters are most sensitive to slow modulation rates in the 1 to 5 Hz range. Vice versa, toward higher carrier frequencies, the filters appear to be more sensitive to faster modulation rates (10 Hz). Simpson et al. (2013) speculated that this frequency-dependent modulation tuning is related to the neural processing of characteristic acoustic properties of speech, namely to the processing of prosodic, syllabic, and phonemic information (Giraud & Poeppel, 2012). However, this relation remains to be tested empirically.

3.2.2. Neuronal coding of temporal modulation The temporal modulation of a sound is encoded along the auditory pathway by means of both temporal and place/firing rate coding mechanisms. The temporal code is implemented through the phase locking of the neuronal responses to the envelope of the sounds (envelope phase locking). Instead, the firing rate code is implemented through changes in average neuronal firing rate as a function of the envelope modulation rate. A thorough overview of the electrophysiological findings on the coding of temporal modulation is provided by Joris, Schreiner, and Rees (2004). The major conclusions can be summarized in the following points. At peripheral stages, the coding is predominantly driven by phase locking (temporal code); in the auditory nerve fibers, phase locking is observed up to 4 to 5 kHz (in the cat). The proportion of neurons exhibiting phase-locked responses and the upper frequency limit of phase-locking rate both decrease along the auditory pathway. Results further indicate a partial shift from temporal coding to rate coding moving up from the auditory nerve to the cortex. In primates, the upper limit of phase locking that has been observed in auditory cortex is 250 Hz for auditory intracortical potentials measured in response to pure tones using depth electrodes (100 Hz for multiunit activity) (Steinschneider, Arezzo, & Vaughan Jr., 1980). In humans, phase locking up to 200 to 250 Hz has been observed with intracortical recordings and click trains (Nourski et al., 2013). However, in the majority of cases, the phase locking of neuronal populations in the cortex occurs for much lower modulation rates, whereas higher modulation rates are encoded by spike rate changes (Lu, Liang, & Wang, 2001).

Additionally, studies that examined large neuronal population responses (fMRI) to temporal rates with noise bursts (Harms & Melcher, 2002), amplitude modulated broadband noise (Giraud et al., 2000), and narrowband noise (Overath, Zhang, Sanes, & Poeppel, 2012) have reported that in the cortex there is a shift of selectivity toward low modulation rates. As it has been argued that an ordered spatial distribution of modulation rates might exist in addition to tonotopy (Schreiner & Winer, 2007), several fMRI studies also examined this possibility. However, while all studies reported clusters of neuronal populations responding selectively to different modulation rates, their topographical arrangement has not been consistent (Barton & Venezia, 2012; Giraud et al., 2000; Herdener et al., 2013; Langers, Backes, & van Dijk, 2003). A recent fMRI study in the macaque monkey (Baumann et al., 2015) described a topographical arrangement of selective responses for temporal modulations progressing from high rates in the posterior core to low rates in the anterior core and lateral belt cortex. This is in agreement with other results obtained in humans using naturalistic stimuli (see Santoro et al., 2014, and section 5).

4. *Spectrotemporal Analysis of Speech and Real-Life Sounds*

So far, I have considered the spectral and temporal analysis of sound in the context of simple synthetic stimuli. However, the processing mechanisms considered in section 3 are highly relevant for the analysis of more complex sounds that we commonly encounter in daily environments. As illustrated in figure 35.3, spectral and temporal modulations are characteristically present in many real-life sounds. In addition, the spectrogram of vocalization calls typically exhibit simultaneous spectral and temporal changes that may be tightly coupled. Furthermore, in natural sounds, modulations are not specific to a given temporal or spectral scale, but rather they occur over multiple scales (Singh & Theunissen, 2003).

4.1. Sensitivity to Spectrotemporal Modulations In section 3, I have described the results of psychoacoustic studies that estimated the sensitivity of human listeners to temporal modulations (i.e., the tMTF) based on detection thresholds. A similar approach has also been adopted to measure the sensitivity to spectral modulations (Eddins & Bero, 2007; Green, 1986) and to combined spectrotemporal modulations. Chi, Gao, Guyton, Ru, and Shamma (1999) measured spectrotemporal psychophysical MTFs using dynamic ripples. Dynamic ripples are broadband sounds that are simultaneously modulated both in time and frequency (see figure 35.3B). With dynamic ripples, two parameters can be changed to control for temporal and spectral modulations, respectively: velocity (in hertz) controls the temporal variation, and ripple density (in cycles/octave) controls the spectral variations. Results showed that normal-hearing listeners have the highest sensitivity between 1 and 8 Hz range for temporal modulations and up to 1 cycle/octave for spectral modulations. For both temporal and spectral modulations, the sensitivity decreases at higher ranges of the parameter (depending on the stimulus type, the cutoff is between 10 and 64 Hz for temporal modulations and 2 and 4 cycles/octave for spectral modulations) (Chi et al., 1999).

Several psychoacoustic studies examined how performance on auditory perceptual tasks correlates with a systematic degradation of the stimulus and provided evidence that the analysis of spectral, temporal, and spectrotemporal modulations is relevant for the perception of complex sounds. A common approach of these studies is to measure task performance in relation to stimuli that have been filtered to remove spectral and/or temporal modulations at specific frequencies. This procedure results in a task-specific MTF that indicates which modulation frequencies are relevant for the task of interest. When applied to speech intelligibility, this method has revealed that speech comprehension is significantly impaired when low-frequency spectral or temporal modulations are removed (Drullman, Festen, & Plomp, 1994; Elliott & Theunissen, 2009; Liu & Eddins, 2008; Shannon et al., 1995). Specifically, the spectrotemporal MTF for speech intelligibility has been shown to be bandpass for temporal modulations and low-pass for spectral modulations, with temporal modulations between 1 and 7 Hz and spectral modulations <1 cycle/kHz being the most relevant (Elliott & Theunissen, 2009).

4.2. Neuronal Tuning to Spectrotemporal Modulations Dynamic ripples have also been used in electrophysiology to measure the sensitivity of neurons to spectrotemporal modulations. By examining the responses to many ripples with different velocity and density values, the spectrotemporal response field (STRF) of the neuron (or equivalently its MTF) or population of neurons (in fMRI and electrocorticography

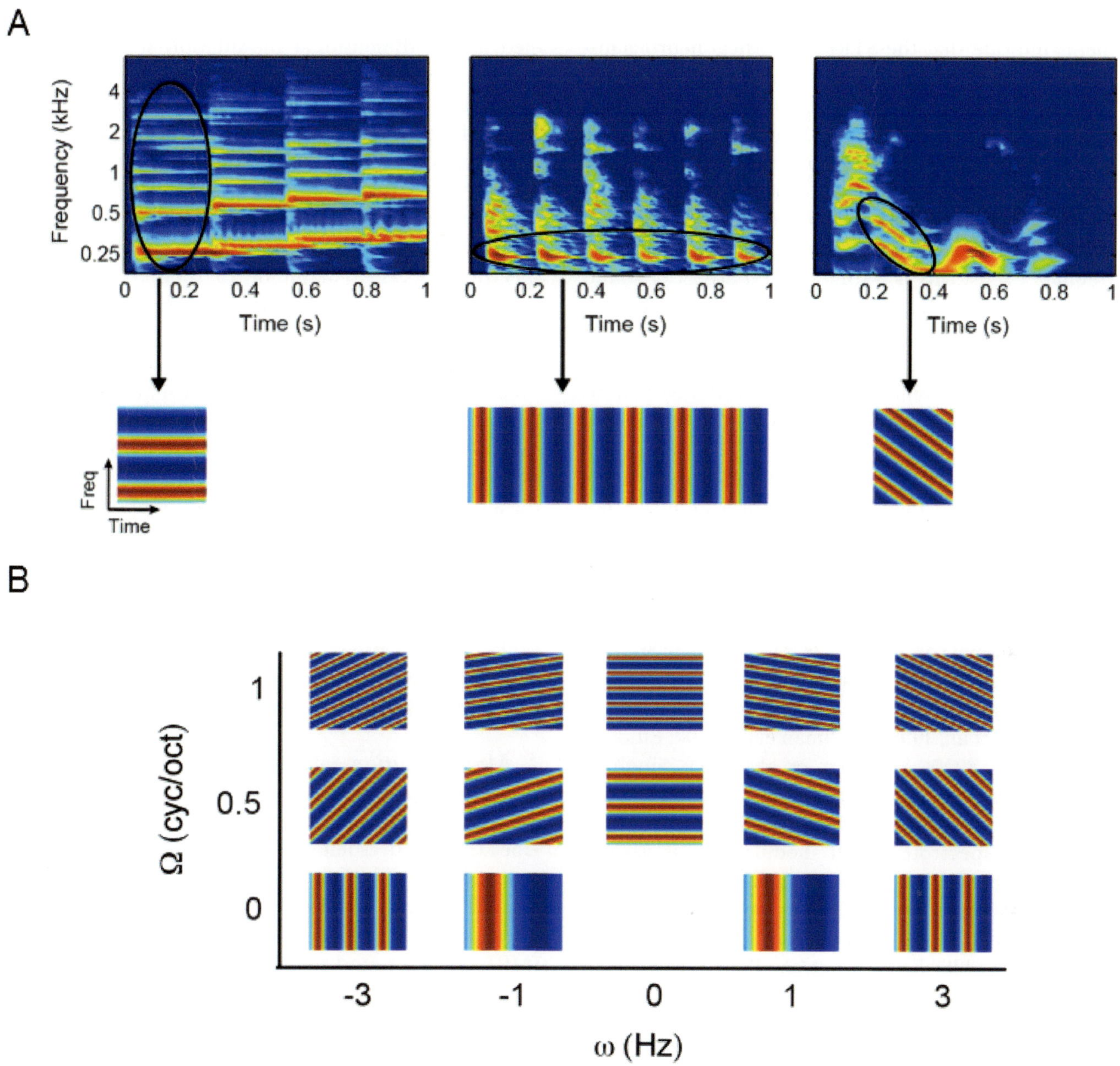

FIGURE 35.3 Examples of spectrograms and of spectral and temporal modulations. (A) Spectrogram of three exemplary natural sounds. Spectrograms are computed using a computational model mimicking early auditory processing (Chi et al., 2005). Black ellipses highlight examples of spectral (*left*), temporal (*middle*), and spectrotemporal (*right*) modulations. Schematics of these modulation patterns are shown under the spectrograms. Spectral modulations are amplitude oscillations along the frequency axis (e.g., speech formants), while temporal modulations are oscillations along the temporal axis (e.g., syllabic structure in speech). Spectrotemporal modulations are combined oscillations over both time and frequency (Freq) (e.g., formant transitions). (B) Spectral and temporal modulations are defined by the frequency of the oscillation, which determines the spacing between consecutive amplitude peaks. Schematic spectrograms illustrate how the modulation pattern changes with spectral (Ω) and temporal (ω) modulation frequencies. Purely temporal and spectral modulations are defined for $\Omega = 0$ and $\omega = 0$, respectively. By convention, positive and negative temporal modulation frequencies distinguish between downward and upward spectrotemporal modulations. Artificial sounds whose spectrogram resembles the spectral and spectrotemporal modulations depicted here are referred to as static and dynamic ripples, respectively. Images from Santoro (2015).

[ECoG]) can be calculated. Results from such experiments indicate that the STRFs of cortical neurons display a wide range of properties from temporally fast to slow and from spectrally sharp to broad (e.g., Depireux, Simon, Klein, Shamma, & Ding, 2001; Kowalski, Depireux, & Shamma, 1996; Shamma, Versnel, & Kowalski, 1995). Importantly, some neurons display oriented STRFs (i.e., with preference only for upward or downward modulations), indicating that the combined time-frequency sensitivity is a fundamental aspect of sound processing (Chi, Ru, & Shamma, 2005; deCharms, 1998). Consistently, a study examining the fMRI responses to dynamic ripples in humans reported selective tuning to combined spectrotemporal modulations in the primary and secondary auditory cortexes (Schönwiesner & Zatorre, 2009).

Neuronal tuning to spectrotemporal modulations has also been investigated using natural stimuli. Ethological considerations have led to the hypothesis that brain processing of sounds is optimized for spectrotemporal modulations that are characteristically present in ecologically relevant sounds (Theunissen & Elie, 2014), such as in animal vocalizations (e.g., zebra finch: Singh & Theunissen, 2003; macaque monkeys: Fukushima, Doyle, Mullarkey, Mishkin, & Averbeck, 2015), and speech (Chi et al., 1999; Drullman et al., 1994; Elliott & Theunissen, 2009; Shannon et al., 1995; Singh & Theunissen, 2003). Electrophysiological investigations in several species have reported single neurons tuned to specific spectrotemporal modulations at various stages of the auditory pathway (e.g., inferior colliculus: Rodríguez, Read, & Escabí, 2010; auditory thalamus: Miller, Escabi, Read, & Schreiner, 2001) and in PAC (deCharms, 1998; Kowalski et al., 1996). It has been suggested that this tuning represents a mechanism for optimally discriminating ecologically relevant stimuli (Woolley, Fremouw, Hsu, & Theunissen, 2005).

4.3. Computational Modeling of Spectrotemporal Modulation Analysis A computational model described in Chi et al. (2005) accounts for these psychophysical and electrophysiology observations and generalizes the concept of modulation filter banks to combinations of spectral and temporal modulations. Their auditory model consists of two main components: an early stage that accounts for the transformations that acoustic signals undergo in the early auditory system, from the cochlea to the midbrain (figure 35.4A) and a cortical stage that simulates the processing at the level of the (primary) auditory cortex (figure 35.4B). The auditory spectrogram generated by the early stage is further analyzed by the cortical stage, which explicitly encodes the modulation content of the spectrogram. Cortical neuronal units are modeled as two-dimensional modulation selective filters that are tuned to a specific combination of spectral and temporal modulations, and operate over a limited range of frequencies along the tonotopic axis. These filters have constant bandwidth and are directional, that is, they respond either to upward or downward frequency sweeps. Computationally, the cortical filter bank performs a complex wavelet decomposition of the auditory spectrogram. The magnitude of such decomposition yields a phase-invariant measure of modulation content. Ultimately, the model's output is a multiresolution representation of the spectrogram envelope as a function of time, frequency, spectral and temporal modulations, as well as directionality (upward or downward frequency sweeps). Just as the cochlea decomposes the acoustic waveform into its frequency components, in this model, the auditory cortex is postulated to decompose the spectrogram into multiple views via arrays of filters that are tuned to different combinations of spectral and temporal modulations. Channels tuned to low modulation frequencies extract a smooth version of the spectrogram, while channels tuned to high modulation frequencies capture its fine-grained spectral and temporal amplitude variations (figure 35.4B).

5. *Multiresolution Analysis in Human Auditory Cortex*

Does this multiresolution model accurately reflect the acoustic analysis of (real-life) complex sounds in the human auditory cortex? Recently, we examined this question in two studies that combined high-resolution fMRI, natural sound stimuli, and computational modeling (Santoro et al., 2014; Santoro et al., 2017). We also examined the homology and differences between human and macaque auditory cortexes by comparing the results obtained in humans with those obtained in macaque monkeys using fMRI and identical stimuli and methods (Erb et al., 2018). Sections 5.1 to 5.3 discuss the results of these studies in some detail.

5.1. Encoding of Spectrotemporal Modulations in Human Auditory Cortex In Santoro et al. (2014), we used single-voxel encoding (Kay, Naselaris, Prenger, & Gallant, 2008) to compare different computational models of auditory cortical processing. In particular, we considered several variants of the modulation filter model described in section 4.3 and derived a sensitivity profile (or voxel response function [vRF]) per voxel and per model variant from the observed fMRI responses. Similar to response functions obtained in single neurons, the vRF describes the response properties of a

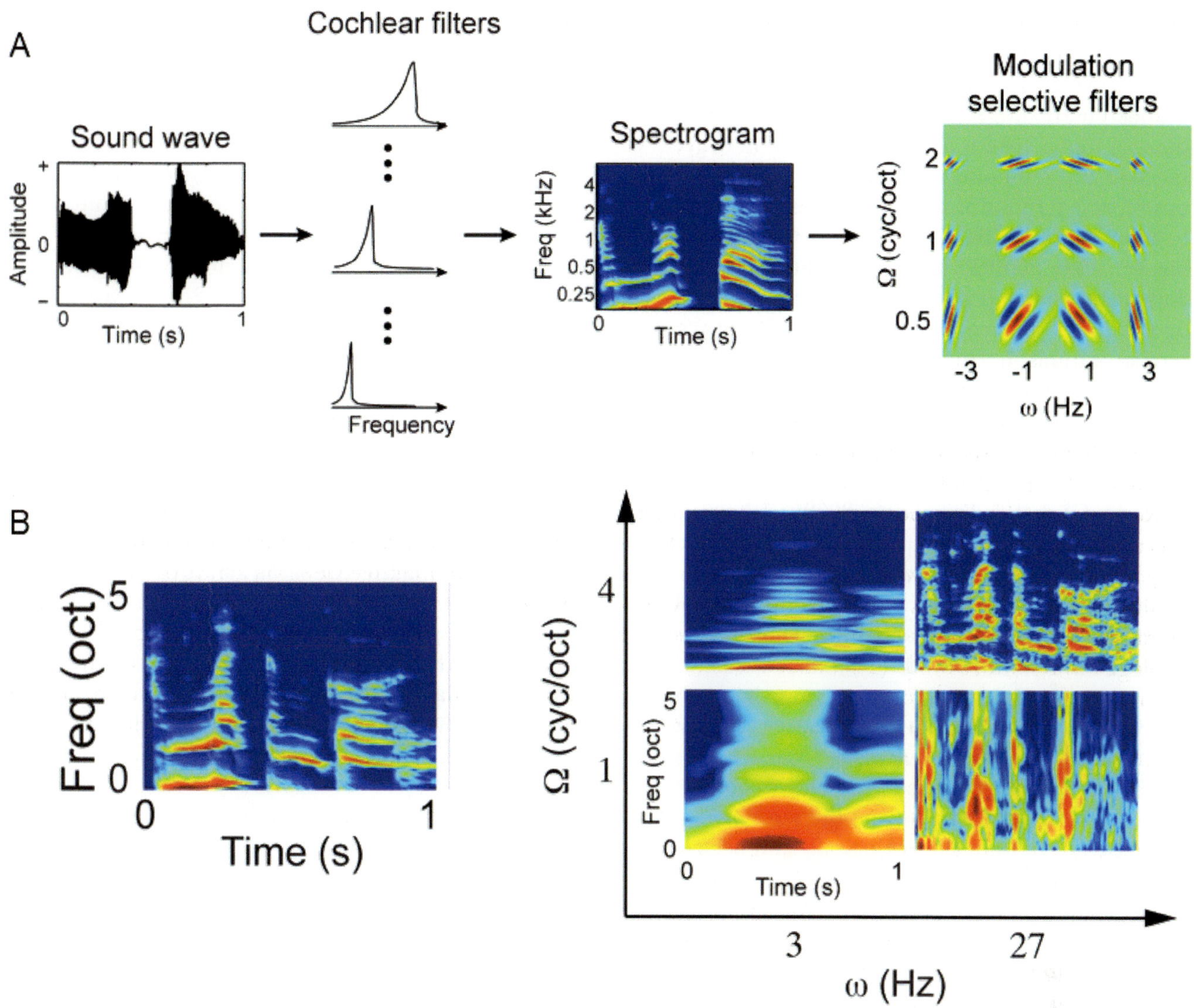

FIGURE 35.4 Computational model of auditory processing. (A) The sound wave enters the peripheral stage, where it is analyzed by a bank of band-pass cochlear-like filters (a schematic of the filters' transfer function is depicted in the figure). Further transformations mimicking hair cells and midbrain mechanisms (not shown here) generate the auditory spectrogram. Finally, the cortical stage computes the modulation content of the spectrogram using an array of modulation selective filters tuned to different values of spectral and temporal modulations frequencies. Impulse responses are shown for a selection of cortical filters. (B) Spectrogram and multiresolution representation of a speech sound. Different "views" are obtained as output of modulation channels tuned to specific spectral modulation (Ω) and temporal modulation (ω) frequencies. Each channel represents the spectrogram with a different combination of spectral and temporal detail. Panel A is from Santoro (2015) and panel B is from Santoro et al. (2014).

neuronal population (voxel) and can be mono- or multidimensional depending on the sound representation model being considered. For example, in case of a model variant that considers *frequency* as the only relevant feature driving the fMRI responses, the vRF has a single dimension (axis) and corresponds to the voxel frequency tuning curve. On the contrary, in the case of a model that considers the voxel responses being driven by *frequency* as well as *spectral and temporal modulations*, the vRF has three dimensions. In Santoro et al. (2014), we compared different vRF models based on their capability to predict the fMRI responses to stimuli that were not used for their estimation. The best performing vRF was that derived from the sound representation model as described in Chi et al. (2005). In other words, modeling a vRF as a (linear) function of frequency-specific filters tuned to specific combinations of spectral and temporal modulations resulted in highly accurate predictions of the fMRI responses to natural sounds. These predictions were significantly better than those obtained considering representation models based on frequency alone or on separate processing of frequency, spectral, and temporal modulations. In particular, these results provided empirical evidence for the notion that the cortical encoding of natural sounds entails the formation of multiple representations of sound spectrograms with different degrees of spectral and temporal resolution. Multiple simultaneous representations of the same incoming sounds may be relevant for enabling efficient processing and flexible behavior, as different goal-oriented and context-dependent sound processing (e.g., sound localization or identification) may benefit from different types of representations. Furthermore, it is consistent with distributed accounts of sound processing in auditory cortex (Bonte, Hausfeld, Scharke, Valente, & Formisano, 2014; Formisano, De Martino, Bonte, & Goebel, 2008; Ley, Vroomen, & Formisano, 2014; Staeren, Renvall, De Martino, Goebel, & Formisano, 2009) as the representations of sounds at multiple resolutions may provide the computational basis analyzing the acoustic elements that are required to form auditory objects (Formisano et al., 2008) and solve complex auditory scenes (Elhilali & Shamma, 2008).

5.2. Cortical Topographies and Spectrotemporal Trade-off The methodological approach of single voxel encoding provides the possibility to obtain maps of multiple sound features from the same set of fMRI responses and within the ecologically and behaviorally relevant context of natural sounds processing. Therefore, we could analyze the fMRI responses to natural sounds to examine not only the spatial organization of frequency (tonotopy), but also that of temporal and spectral modulation processing. Figure 35.5 describes the resulting maps of voxels' characteristic spectral modulation, temporal modulation, and characteristic frequency. These maps were obtained by averaging the estimated vRF across irrelevant dimensions (i.e., spectral and temporal modulations for characteristic frequency), assigning the point of maximum of the marginal sum as the voxel's preferred feature value, and color-coding along the relevant dimension (figure 35.5). Maps of characteristic frequency confirmed the presence of multiple tonotopic gradients in primary auditory regions (HG) and surrounding superior temporal cortex (Moerel et al., 2012). The spatial distribution of voxels' characteristic spectral and temporal modulations suggested distinct regional sensitivities to modulation frequencies. The most evident characteristic was that—in both the hemispheres—regions located posterior-medially to HG preferably encoded coarse spectral and fast temporal components in the sounds (i.e., fast temporal rates) while regions located along HG or on the lateral superior temporal plane preferably encoded fine-grained spectral and slow temporal components in the sounds (see figure 35.5, middle and right columns). Supporting this spatial dissociation between preference for spectral and temporal modulations at map level, we found a significant negative correlation between voxels' characteristic spectral and temporal modulations (Santoro et al., 2014).

These results thus suggest a "spectrotemporal trade-off" between neuronal processing in posterior-lateral and anterior-ventral auditory regions and that the multiresolution analysis of natural sounds is actually realized by partially distinct auditory neuronal networks. An analogous spectrotemporal resolution trade-off has been described previously for neurons in the inferior colliculus of the cat (Rodríguez et al., 2010) and in the cortex of the squirrel monkey (Bendor & Wang,

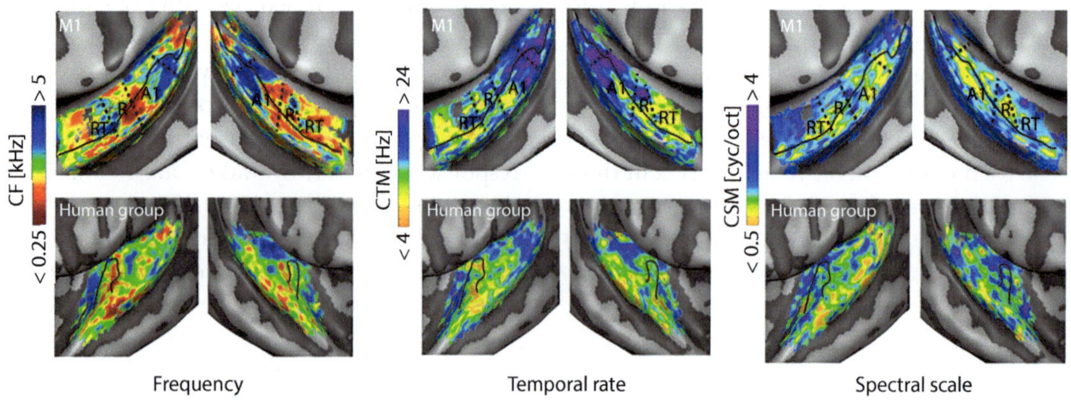

Figure 35.5 Topographic maps of characteristic frequency, temporal rate (CTM), and spectral scale (CSM) in the macaque (*upper row*) and human (*lower row*) auditory cortexes. Maps were obtained from the analysis of fMRI responses to natural sounds (fMRI encoding). Maps are for a single macaque (monocrystalline iron oxide nanoparticle-enhanced fMRI) and a group of five human subjects (7 T fMRI). Images are courtesy of Erb et al. (2018).

2008). It is thus possible that these effects found in the cortex are inherited from previous processing stages along the auditory pathway. In the context of neuroimaging investigations in humans, the proposal of distinct systems for the temporal and spectral analysis of sounds in the auditory cortex is long-standing. Zatorre and Belin (2001) suggested the existence of a trade-off between auditory processing in the two hemispheres, with a left hemispheric specialization for the processing of fine-grained temporal details of sounds and a right hemispheric specialization for the analysis of spectral details (and pitch), respectively. Hemispheric differences in the processing of sounds are also assumed in the *Asymmetric Sampling in Time* scheme proposed in Poeppel (2003). In this latter scheme, however, only the temporal aspects are considered, with the left and right auditory cortices differing in the size of the temporal windows over which they preferentially sample acoustic information (small for the left hemisphere, large for the right hemisphere). Many of the asymmetrical findings in neuroimaging research have been interpreted in terms of these proposals, which, however, have also been criticized (McGettigan & Scott, 2012).

In our studies with natural sounds, we specifically tested the hypothesis of hemispheric functional asymmetries and did not find significant differences between the two hemispheres, but rather a medial to lateral (around HG) and posterior to anterior (along STG) dissociation of vRF characteristics (Santoro et al., 2014). We found the very same dissociation in the auditory cortex of macaque monkeys in a recent study employing fMRI and the same stimuli and analysis methods as those described for the study in humans (see figure 35.5 and Erb et al., 2018). Interestingly, a recent study investigating the representation of speech using ECoG in humans reached the same conclusions and suggested a posterior to anterior gradient of temporal versus spectral sensitivity along the STG (Hullett, Hamilton, Mesgarani, Schreiner, & Chang, 2016). Further research should sample a much larger number of participants to clarify the relation between our recent findings with those pointing to the left-right hemispheric differences. In addition, it should be taken into account that anatomical asymmetries may be central to explaining this apparent discrepancy. For example, in regions posterior to HG, the PT is larger on the left than on the right, whereas the region anterior to HG is more pronounced on the right than on the left. These anatomical asymmetries may interact in complex ways with the specific methodological approach used to perform the group-level statistical analyses and influence the outcome.

5.3. Is the Human Auditory Cortex Speech Ready?

In the introduction, I have put forward the hypothesis that,—as a result of evolution and/or lifelong experience—in the human auditory cortex even general purpose auditory mechanisms may be shaped by the spectrotemporal characteristics of speech. Do empirical findings support this hypothesis? Pasley et al. (2012) used ECoG to investigate which acoustic information in speech sounds is encoded in population neural activity from non-PAC activity. They addressed this question using the approach of model-based stimulus reconstruction (or decoding). Unlike the more common classification-based decoding, which only allows discriminating between a small set of stimulus categories, brain-based stimulus reconstruction embeds a representational model of the stimuli in terms of elementary features, thereby enabling the reconstruction of single features (or whole stimuli) from brain-response patterns. Results indicated that slow and intermediate temporal fluctuations, such as those corresponding to syllable rate, were accurately reconstructed using a linear model based on the auditory spectrogram. However, reconstruction of fast temporal fluctuations, such as syllable onsets and offsets, required a sound representation based on temporal modulation energy. Importantly, reconstruction accuracy was highest within the range of spectrotemporal fluctuations that have been found to be critical for speech intelligibility. This suggests that the representation of acoustic information relevant to the decoding of speech is particularly enhanced in the human auditory cortex.

As Pasley et al. (2012) only examined the brain responses to speech, no conclusion could be derived on whether their results reflect a general or a speech-specific mechanism. In a recent study (Santoro et al., 2017), we could address this question by combining high field MRI (7 T) with model-based decoding of stimuli from many categories, including speech and vocal sounds but also animal cries and other environmental sounds. Compared to single-voxel encoding, model-based fMRI decoding exploits the information in spatial (multivoxel) response patterns (Miyawaki et al., 2008). First, we found that accurate and specific reconstructions of modulations (and spectrograms) of real-life sounds (and speech) could be obtained from auditory fMRI response patterns in the human auditory cortex. Reconstructions based on the sound representation model including frequency-specific spectrotemporal modulations were significantly better than those based on a frequency-only model. This result brings further support to the notion of multiresolution analysis in the human auditory cortex. Second, consistent with the hypothesis of topographic arrangement of spectrotemporal modulations,

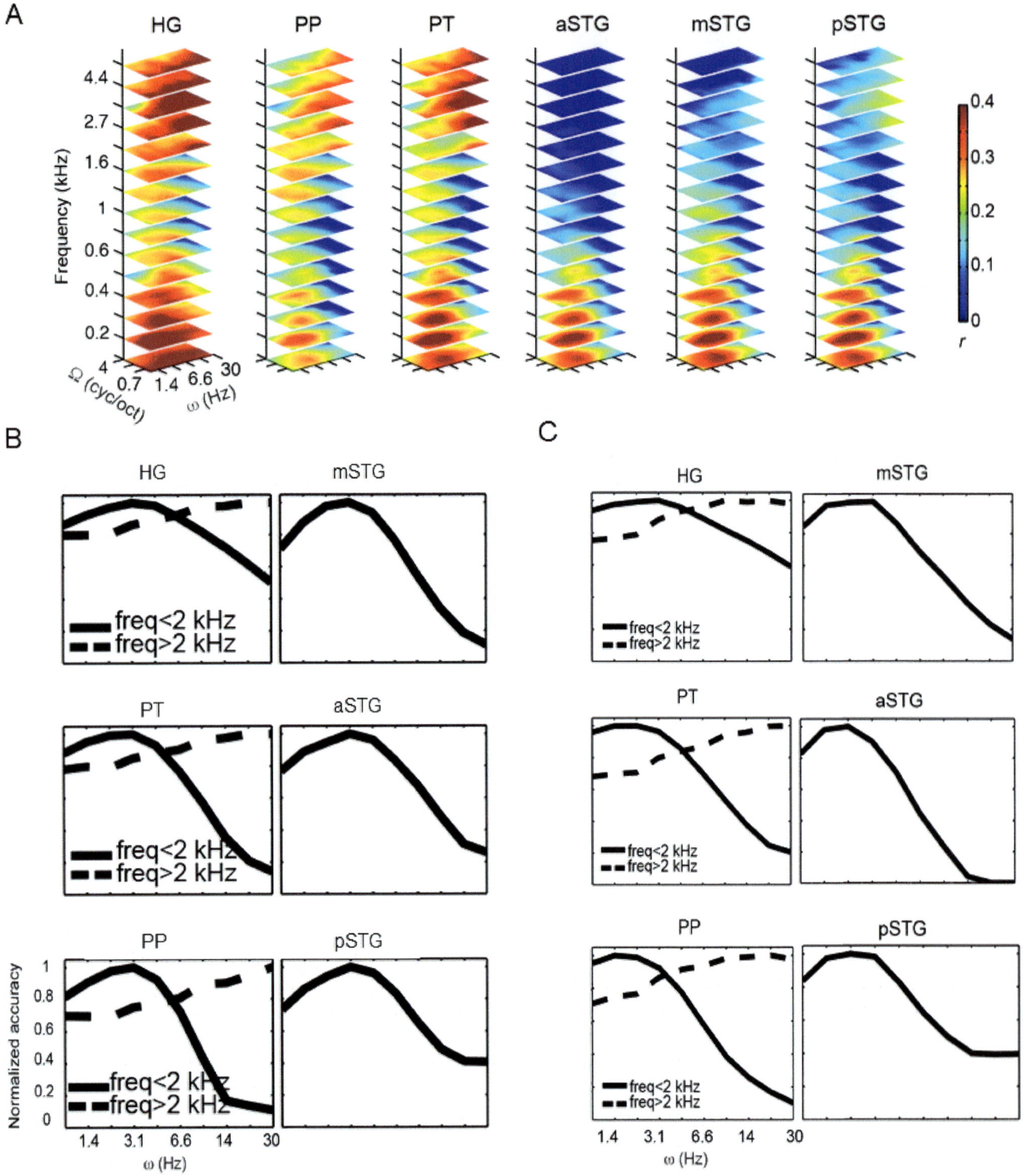

FIGURE 35.6 Reconstruction of spectrotemporal modulation of real-life sounds from fMRI response patterns. (A) MTFs for different auditory regions of interest obtained by color-coding the accuracy of the brain-based reconstruction of acoustic features (group average: $n = 10$; Pearson's r between reconstructed and original features). Each slice represents the MTF at a given combination of frequency, spectral (Ω) and temporal (ω) modulation (only 15 frequencies are shown). Features with a nonsignificant r are depicted in dark blue. The average MTF across hemispheres is shown. MTFs have been interpolated for display purposes. (B, C) Region of interest–based marginal profiles for temporal modulation obtained with a stimulus set including (B) or excluding (C) speech. Each profile was obtained by averaging the MTFs along the other two dimensions (frequency, spectral scale) and normalized by its maximum value. Based on the visual inspection of the full MTF (A) for the HG, PP, and PT, distinct marginal tuning functions were computed for frequencies below and above 2 kHz. Figure from Santoro et al. (2017). a = anterior; m = middle; p = posterior.

the region-based analyses indicated that separate portions of the auditory cortex encode distinctive sets of spectrotemporal modulations. Most relevantly, results revealed that in early auditory regions (HG, PP, PT), and progressively more in surrounding regions (STG), temporal modulations in a range relevant for speech analysis (~2–4 Hz) were reconstructed more faithfully than other temporal modulations. In early auditory regions, this effect was frequency-dependent and only present for lower frequencies (<~2 kHz), whereas for higher frequencies, reconstruction accuracy was higher for faster temporal modulations (figure 35.6). Several additional analyses and results are relevant for interpreting these findings. First, a control analysis showed that neither in primary regions nor in STG regions were the profiles of reconstruction accuracy for temporal modulations affected by the removal of speech/voice sounds from the entire stimulus set (figure 35.6B). This is important as it indicates that the estimated profiles reflect an actual property of the regions and that the specific shape of the profile does not simply reflect an artifact of, for example, enhanced response to speech stimuli. Second, a stimulus-discrimination analysis suggested that the observed enhancement of acoustic components favors the fine-grained discrimination of speech (and voices) rather than for natural sounds per se. Finally, the preferential tuning, which was evident in humans even in primary areas along the HG, could not be observed in the macaque auditory cortex, where in fact reconstruction accuracy was most accurate at fast rates (~30 Hz; Erb et al., 2018). Importantly, these results are in line with psychoacoustic findings showing that the macaques' sensitivity to amplitude modulations is highest at fast modulation rates of approximately 30 to 60 Hz, whereas human listeners perform best in the discrimination of slow temporal modulations in the 2 to 4 Hz range (O'Connor et al., 2011).

In sum, speech is a complex sound requiring the simultaneous analysis of spectral and temporal components at multiple scales. The set of analyses and results reviewed are consistent with the hypothesis that, in the human brain, neuronal populations display a unique sensitivity to those acoustic components of speech that are most relevant for behavior and that enable optimal discrimination between speech tokens. This is true not only in higher level auditory cortex but also in early auditory areas, where neuronal populations are similarly fine-tuned to the characteristic acoustic properties of speech. The absence of such tuning to speech in the auditory cortex of macaque monkeys suggests that this is a uniquely human characteristic that is acquired through lifelong learning or may have emerged as a critical step in the evolution of speech and language.

Acknowledgments

This work is supported by the Maastricht University, the Dutch Province of Limburg, and the Netherlands Organization for Scientific Research (NWO grant 453-12-002). The author wishes to thank Milene Bonte, Federico De Martino, Julia Erb, Michelle Moerel, Roberta Santoro, Isma Zulfiqar, and the other members of the Auditory Perception and Cognition unit at the Maastricht Brain Imaging Center for their contribution to the original work on which this chapter is based.

REFERENCES

Abdul-Kareem, I. A, & Sluming, V. (2008). Heschl gyrus and its included primary auditory cortex: Structural MRI studies in healthy and diseased subjects. *Journal of Magnetic Resonance Imaging, 28*, 287–299.

Bacon, S. P., & Viemeister, N. F. (1985). Temporal modulation transfer functions in normal-hearing and hearing-impaired listeners. *Audiology, 24*, 117–134.

Bartlett, E. L., Sadagopan, S., & Wang, X. (2011). Fine frequency tuning in monkey auditory cortex and thalamus. *Journal of Neurophysiology, 106*, 849–859.

Barton, B., & Venezia, J. (2012). Orthogonal acoustic dimensions define auditory field maps in human cortex. *Proceedings of the National Academy of Sciences, 109*, 20738–20743.

Baumann, S., Joly, O., Rees, A., Petkov, C. I., Sun, L., Thiele, A., & Griffiths, T. D. (2015). The topography of frequency and time representation in primate auditory cortices. *Elife, 4*, e03256.

Baumann, S., Petkov, C. I., & Griffiths, T. D. (2013). A unified framework for the organization of the primate auditory cortex. *Frontiers in Systems Neuroscience, 7*, 11.

Belin, P., Zatorre, R. J., Lafaille, P., Ahad, P., & Pike, B. (2000). Voice-selective areas in human auditory cortex. *Nature, 403*, 309–312.

Bendor, D., & Wang, X. (2008). Neural response properties of primary, rostral, and rostrotemporal core fields in the auditory cortex of marmoset monkeys. *Journal of Neurophysiology, 100*, 888–906.

Bitterman, Y., Mukamel, R., Malach, R., Fried, I., & Nelken, I. (2008). Ultra-fine frequency tuning revealed in single neurons of human auditory cortex. *Nature, 451*, 197–201.

Bonte, M., Frost, M. A., Rutten, S., Ley, A., Formisano, E., & Goebel, R. (2013). Development from childhood to adulthood increases morphological and functional inter-individual variability in the right superior temporal cortex. *NeuroImage, 83*, 739–750.

Bonte, M., Hausfeld, L., Scharke, W., Valente, G., & Formisano, E. (2014). Task-dependent decoding of speaker and vowel identity from auditory cortical response patterns. *Journal of Neuroscience, 34*, 4548–4557.

Brodmann, K. (1909). *Vergleichende Localisationslehre der Grosshirnrinde in ihren Prinzipien dargestellt auf Grund des Zellenhaues.* Leipzig: Verlag von Johann Ambrosius Barth.

Campain, R., & Minckler, J. (1976). A note on the gross configurations of the human auditory cortex. *Brain and Language, 3*(2), 318–323.

Chi, T., Gao, Y., Guyton, M. C., Ru, P., & Shamma, S. (1999). Spectro-temporal modulation transfer functions and speech intelligibility. *Journal of the Acoustical Society of America, 106,* 2719–2732.

Chi, T., Ru, P., & Shamma, S. (2005). Multiresolution spectrotemporal analysis of complex sounds. *Journal of the Acoustical Society of America, 118,* 887–906.

Crandall, I. B., & MacKenzie, D. (1922). Analysis of the energy distribution in speech. *Bell Labs Technical Journal, 1*(1), 116–128.

Da Costa, S., van der Zwaag, W., Marques, J. P., Frackowiak, R. S. J., Clarke, S., & Saenz, M. (2011). Human primary auditory cortex follows the shape of Heschl's gyrus. *Journal of Neuroscience, 31,* 14067–14075.

Dau, T., Kollmeier, B., & Kohlrausch, A. (1997). Modeling auditory processing of amplitude modulation. I. Detection and masking with narrow-band carriers. *Journal of the Acoustical Society of America, 102,* 2892–2905.

deCharms, R. C. (1998). Optimizing sound features for cortical neurons. *Science, 280*(5368), 1439–1444.

De Martino, F., Moerel, M., Xu, J., van de Moortele, P.-F., Ugurbil, K., Goebel, R., ... Formisano, E. (2014). High-resolution Mapping of myeloarchitecture in vivo: Localization of auditory areas in the human brain. *Cerebral Cortex, 25*(10), 3394–3405.

Depireux, D. A., Simon, J. Z., Klein, D. J., Shamma, S. A., & Ding, N. (2001). Spectro-temporal response field characterization with dynamic ripples in ferret primary auditory cortex. *Journal of Neurophysiology, 85,* 1220–1234.

Dick, F., Tierney, A. T., Lutti, A., Josephs, O., Sereno, M. I., & Weiskopf, N. (2012). In vivo functional and myeloarchitectonic mapping of human primary auditory areas. *Journal of Neuroscience, 32,* 16095–16105.

Diehl, R. L., Lotto, A. J., & Holt, L. L. (2004). Speech perception. *Annual Review of Psychology, 55,* 149–179.

Drullman, R., Festen, J., & Plomp, R. (1994). Effect of temporal envelope smearing on speech reception. *Journal of the Acoustical Society of America, 95,* 1053–1064.

Eddins, D. A., & Bero, E. M. (2007). Spectral modulation detection as a function of modulation frequency, carrier bandwidth, and carrier frequency region. *Journal of the Acoustical Society of America, 121,* 363–372.

Eggermont, J. J. (2001). Between sound and perception: Reviewing the search for a neural code. *Hearing Research, 157,* 1–42.

Elhilali, M., & Shamma, S. (2008). A cocktail party with a cortical twist: How cortical mechanisms contribute to sound segregation. *Journal of the Acoustical Society of America, 124,* 3751–3771.

Elliott, T. M., & Theunissen, F. E. (2009). The modulation transfer function for speech intelligibility. *PLOS Computational Biology, 5*(3), e1000302.

Erb, J., Armendariz, M., De Martino, F., Goebel, R., Vanduffel, W., & Formisano, E. (2018). *Homology and specificity of natural sound-encoding in human and monkey auditory cortex.*

Formisano, E., De Martino, F., Bonte, M., & Goebel, R. (2008). "Who" is saying "what"? Brain-based decoding of human voice and speech. *Science, 322*(5903), 970–973.

Formisano, E., Kim, D. S., Di Salle, F., van de Moortele, P. F., Ugurbil, K., & Goebel, R. (2003). Mirror-symmetric tonotopic maps in human primary auditory cortex. *Neuron, 40,* 859–869.

Formisano, E., Moerel, M., & Bonte, M. (2015). Functional MRI of the auditory cortex. In K. Uludag, K. Uğurbil, & L. Berliner (Eds.), *fMRI: From Nuclear Spins to Brain Function* (pp. 473–478). Boston: Springer.

Fukushima, M., Doyle, A. M., Mullarkey, P., Mishkin, M., & Averbeck, B. B. (2015). Distributed acoustic cues for caller identity in macaque vocalization. *Royal Society Open Science, 2*(12), 150432.

Galaburda, A., & Sanides, F. (1980). Cytoarchitectonic organization of the human auditory cortex. *Journal of Comparative Neurology, 190,* 597–610.

Gardumi, A., Ivanov, D., Havlicek, M., Formisano, E., & Uludag, K. (2017). Tonotopic maps in human auditory cortex using arterial spin labeling. *Human Brain Mapping, 38,* 1140–1154.

Geschwind, N., & Levitsky, W. (1968). Human brain: Left-right asymmetries in temporal speech region. *Science, 161*(3837), 186–187.

Giraud, A. L., Lorenzi, C., Ashburner, J., Wable, J., Johnsrude, I., Frackowiak, R., & Kleinschmidt, A. (2000). Representation of the temporal envelope of sounds in the human brain. *Journal of Neurophysiology, 84,* 1588–1598.

Giraud, A.-L., & Poeppel, D. (2012). Cortical oscillations and speech processing: Emerging computational principles and operations. *Nature Neuroscience, 15,* 511–517.

Glasberg, B. R., & Moore, B. C. J. (1990). Derivation of auditory filter shapes from notched-noise data. *Hearing Research, 47,* 103–138.

Green, D. (1986). "Frequency" and the detection of spectral shape change. In B. C. J. Moore & R. Patterson (Eds.), *Auditory Frequency Selectivity* (pp. 351–359). Boston: Springer.

Hackett, T. A., Preuss, T. M., & Kaas, J. H. (2001). Architectonic identification of the core region in auditory cortex of macaques, chimpanzees, and humans. *Journal of Comparative Neurology, 441,* 197–222.

Harms, M. P., & Melcher, J. R. (2002). Sound repetition rate in the human auditory pathway: Representations in the waveshape and amplitude of fMRI activation. *Journal of Neurophysiology, 88,* 1433–1450.

Hasson, U., Harel, M., Levy, I., & Malach, R. (2003). Large-scale mirror-symmetry organization of human occipito-temporal object areas. *Neuron, 37,* 1027–1041.

Herdener, M., Esposito, F., Scheffler, K., Schneider, P., Logothetis, N. K., Uludag, K., & Kayser, C. (2013). Spatial representations of temporal and spectral sound cues in human auditory cortex. *Cortex, 49,* 2822–2833.

Hullett, P. W., Hamilton, L. S., Mesgarani, N., Schreiner, C. E., & Chang, E. F. (2016). Human superior temporal gyrus organization of spectrotemporal modulation tuning derived from speech stimuli. *Journal of Neuroscience, 36,* 2014–2026.

Joris, P. X., Schreiner, C. E., & Rees, A. (2004). Neural processing of amplitude-modulated sounds. *Physiological Review, 84,* 541–577.

Kaas, J. H., & Hackett, T. A. (2000). Subdivisions of auditory cortex and processing streams in primates. *Proceedings of the National Academy of Sciences, 97,* 11793–11799.

Kay, K. N., Naselaris, T., Prenger, R. J., & Gallant, J. L. (2008). Identifying natural images from human brain activity. *Nature, 452,* 352–355.

Kim, J. J., Crespo-Facorro, B., Andreasen, N. C., O'Leary, D. S., Zhang, B., Harris, G., & Magnotta, V. A. (2000). An

MRI-based parcellation method for the temporal lobe. *NeuroImage, 11,* 271–288.

Kluender, K. R., Coady, J. A., & Kiefte, M. (2003). Sensitivity to change in perception of speech. *Speech Communication, 41,* 59–69.

Kohlrausch, A., Fassel, R., & Dau, T. (2000). The influence of carrier level and frequency on modulation and beat-detection thresholds for sinusoidal carriers. *Journal of the Acoustical Society of America, 108,* 723.

Kosaki, H., Hashikawa, T., He, J., & Jones, E. G. (1997). Tonotopic organization of auditory cortical fields delineated by parvalbumin immunoreactivity in macaque monkeys. *Journal of Comparative Neurology, 386,* 304–316.

Kowalski, N., Depireux, D. A., & Shamma, S. A. (1996). Analysis of dynamic spectra in ferret primary auditory cortex. I. Characteristics of single-unit responses to moving ripple spectra. *Journal of Neurophysiology, 76,* 3503–3523.

Kusmierek, P., & Rauschecker, J. P. (2009). Functional specialization of medial auditory belt cortex in the alert rhesus monkey. *Journal of Neurophysiology, 102,* 1606–1622.

Laing, E. J., Liu, R., Lotto, A. J., & Holt, L. L. (2012). Tuned with a tune: Talker normalization via general auditory processes. *Frontiers in Psychology, 3,* 203.

Langers, D. R. M., Backes, W. H., & van Dijk, P. (2003). Spectrotemporal features of the auditory cortex: The activation in response to dynamic ripples. *NeuroImage, 20,* 265–275.

Leaver, A. M., & Rauschecker, J. P. (2016). Functional topography of human auditory cortex. *Journal of Neuroscience, 36,* 1416–1428.

Leonard, C. M., Puranik, C., Kuldau, J. M., & Lombardino, L. J. (1998). Normal variation in the frequency and location of human auditory cortex landmarks. Heschl's gyrus: Where is it? *Cerebral Cortex, 8*(5), 397–406.

Ley, A., Vroomen, J., & Formisano, E. (2014). How learning to abstract shapes neural sound representations. *Frontiers in Neuroscience, 8,* 132.

Liu, C., & Eddins, D. A. (2008). Effects of spectral modulation filtering on vowel identification. *Journal of the Acoustical Society of America, 124,* 1704–1715.

Lu, T., Liang, L., & Wang, X. (2001). Temporal and rate representations of time-varying signals in the auditory cortex of awake primates. *Nature Neuroscience, 4,* 1131–1138.

McGettigan, C., & Scott, S. K. (2012). Cortical asymmetries in speech perception: What's wrong, what's right and what's left? *Trends in Cognitive Sciences, 16,* 269–276.

Merzenich, M. M., & Brugge, J. F. (1973). Representation of the cochlear partition of the superior temporal plane of the macaque monkey. *Brain Research, 50,* 275–296.

Merzenich, M. M., Knight, P. L., & Roth, G. L. (1973). Cochleotopic organization of primary auditory cortex in the cat. *Brain Research, 63,* 343–346.

Miller, L. M., Escabi, M. A., Read, H. L., & Schreiner, C. E. (2001). Functional convergence of response properties in the auditory thalamocortical system. *Neuron, 32,* 151–160.

Miyawaki, Y., Uchida, H., Yamashita, O., Sato, M. A., Morito, Y., Tanabe, H. C., ... Kamitani, Y. (2008). Visual image reconstruction from human brain activity using a combination of multiscale local image decoders. *Neuron, 60,* 915–929.

Moerel, M., De Martino, F., & Formisano, E. (2012). Processing of natural sounds in human auditory cortex: Tonotopy, spectral tuning, and relation to voice sensitivity. *Journal of Neuroscience, 32,* 14205–14216.

Moerel, M., De Martino, F., & Formisano, E. (2014). An anatomical and functional topography of human auditory cortical areas. *Frontiers in Neuroscience, 8,* 225.

Moore, B. C. J. (2003). *An Introduction to the Psychology of Hearing.* London: Academic Press.

Moore, B. C. J. (2008). Basic auditory processes involved in the analysis of speech sounds. *Philosophical Transactions of the Royal Society B: Biological Sciences, 363,* 947–963.

Moore, B. C. J., & Glasberg, B. R. (2001). Temporal modulation transfer functions obtained using sinusoidal carriers with normally hearing and hearing-impaired listeners. *Journal of the Acoustical Society of America, 110,* 1067–1073.

Moore, B. C. J., & Sek, A. (1995). Effects of carrier frequency, modulation rate, and modulation waveform on the detection of modulation and the discrimination of modulation type (amplitude modulation versus frequency modulation). *Journal of the Acoustical Society of America, 97,* 2468–2478.

Morel, A., Garraghty, P. E., & Kaas, J. H. (1993). Tonotopic organization, architectonic fields, and connections of auditory cortex in macaque monkeys. *Journal of Comparative Neurology, 335,* 437–459.

Morosan, P., Rademacher, J., Schleicher, A., Amunts, K., Schormann, T., & Zilles, K. (2001). Human primary auditory cortex: Cytoarchitectonic subdivisions and mapping into a spatial reference system. *NeuroImage, 13,* 684–701.

Nieuwenhuys, R. (2013). The myeloarchitectonic studies on the human cerebral cortex of the Vogt-Vogt school, and their significance for the interpretation of functional neuroimaging data. *Brain Structure and Function, 218,* 303–352.

Nourski, K. V., Brugge, J. F., Reale, R. A., Kovach, C. K., Oya, H., Kawasaki, H., ... Howard, M. A. (2013). Coding of repetitive transients by auditory cortex on posterolateral superior temporal gyrus in humans: An intracranial electrophysiology study. *Journal of Neurophysiology, 109,* 1283–1295.

O'Connor, K. N., Johnson, J. S., Niwa, M., Noriega, N. C., Marshall, E. A., & Sutter, M. L. (2011). Amplitude modulation detection as a function of modulation frequency and stimulus duration: Comparisons between macaques and humans. *Hearing Research, 277,* 37–43.

Overath, T., McDermott, J. H., Zarate, J. M., & Poeppel, D. (2015). The cortical analysis of speech-specific temporal structure revealed by responses to sound quilts. *Nature Neuroscience, 18,* 903–911.

Overath, T., Zhang, Y., Sanes, D. H., & Poeppel, D. (2012). Sensitivity to temporal modulation rate and spectral bandwidth in the human auditory system: fMRI evidence. *Journal of Neurophysiology, 107,* 2042–2056.

Oxenham, A. J., & Shera, C. A. (2003). Estimates of human cochlear tuning at low levels using forward and simultaneous masking. *Journal of the Association for Research in Otolaryngology, 4,* 541–554.

Pasley, B. N., David, S. V., Mesgarani, N., Flinker, A., Shamma, S. A., Crone, N. E., ... Chang, E. F. (2012). Reconstructing speech from human auditory cortex. *PLOS Biology, 10,* e1001251.

Penhune, V. B., Zatorre, R. J., MacDonald, J. D., & Evans, A. C. (1996). Interhemispheric anatomical differences in human primary auditory cortex: Probabilistic mapping and volume measurement from magnetic resonance scans. *Cerebral Cortex, 6,* 661–672.

Poeppel, D. (2003). The analysis of speech in different temporal integration windows: Cerebral lateralization as "asymmetric sampling in time." *Speech Communication, 41,* 245–255.

Poeppel, D., & Hickok, G. (2015). Electromagnetic recording of the auditory system. *Handbook of Clinical Neurology, 129,* 245–255.

Rademacher, J., Morosan, P., Schormann, T., Schleicher, A., Werner, C., Freund, H. J., & Zilles, K. (2001). Probabilistic mapping and volume measurement of human primary auditory cortex. *NeuroImage, 13,* 669–683.

Rajimehr, R., Devaney, K. J., Bilenko, N. Y., Young, J. C., & Tootell, R. B. (2011). The "parahippocampal place area" responds preferentially to high spatial frequencies in humans and monkeys. *PLOS Biology, 9,* e1000608.

Rauschecker, J. P., Tian, B., & Hauser, M. (1995). Processing of complex sounds in the macaque nonprimary auditory cortex. *Science, 268*(5207), 111–114.

Rivier, F., & Clarke, S. (1997). Cytochrome oxidase, acetylcholinesterase, and NADPH-diaphorase staining in human supratemporal and insular cortex: Evidence for multiple auditory areas. *NeuroImage, 6,* 288–304.

Rodríguez, F. A., Read, H. L., & Escabí, M. A. (2010). Spectral and temporal modulation tradeoff in the inferior colliculus. *Journal of Neurophysiology, 103,* 887–903.

Rosen, S. (1992). Temporal information in speech: Acoustic, auditory and linguistic aspects. *Philosophical Transactions of the Royal Society B: Biological Sciences, 336,* 367–373.

Saenz, M., & Langers, D. R. (2014). Tonotopic mapping of human auditory cortex. *Hearing Research, 307,* 42–52.

Santoro, R. (2015). The computational architecture of the human auditory cortex (Doctoral dissertation). Maastricht University. Retrieved from https://cris.maastrichtuniversity.nl/portal/en/publications/the-computational-architecture-of-the-human-auditory-cortex(f51d20e2-bbde-4b56-b04e-fb264abdc2c6).html (from an institutional database).

Santoro, R., Moerel, M., De Martino, F., Goebel, R., Ugurbil, K., Yacoub, E., & Formisano, E. (2014). Encoding of natural sounds at multiple spectral and temporal resolutions in the human auditory cortex. *PLOS Computational Biology, 10,* e1003412.

Santoro, R., Moerel, M., De Martino, F., Valente, G., Ugurbil, K., Yacoub, E., & Formisano, E. (2017). Reconstructing the spectrotemporal modulations of real-life sounds from fMRI response patterns. *Proceedings of the National Academy of Sciences, 114*(18), 4799–4804.

Schönwiesner, M., & Zatorre, R. J. (2009). Spectro-temporal modulation transfer function of single voxels in the human auditory cortex measured with high-resolution fMRI. *Proceedings of the National Academy of Sciences, 106,* 14611–14616.

Schreiner, C. E., & Winer, J. A. (2007). Auditory cortex mapmaking: Principles, projections, and plasticity. *Neuron, 56,* 356–365.

Sek, A., & Moore, B. C. J. (1995). Frequency discrimination as a function of frequency, measured in several ways. *Journal of the Acoustical Society of America, 97,* 2479–2486.

Shamma, S., Versnel, H., & Kowalski, N. (1995). Ripple analysis in ferret primary auditory cortex. I. Response characteristics of single units to sinusoidally rippled spectra. Technical research report. *Auditory Neuroscience, 1,* 233–254.

Shannon, R. V., Fu, Q.-J., & Galvin J., III. (2004). The number of spectral channels required for speech recognition depends on the difficulty of the listening situation. *Acta Oto-Laryngologica, 124,* 50–54.

Shannon, R. V., Zeng, F. G., Kamath, V., Wygonski, J., & Ekelid, M. (1995). Speech recognition with primarily temporal cues. *Science, 270*(5234), 303–304.

Shera, C. A., Guinan, J. J., & Oxenham, A. J. (2002). Revised estimates of human cochlear tuning from otoacoustic and behavioral measurements. *Proceedings of the National Academy of Sciences, 99,* 3318–3323.

Simpson, A. J. R., Reiss, J. D., & McAlpine, D. (2013). Tuning of human modulation filters is carrier-frequency dependent. *PLOS ONE, 8,* 1–9.

Singh, N. C., & Theunissen, F. E. (2003). Modulation spectra of natural sounds and ethological theories of auditory processing. *Journal of the Acoustical Society of America, 114,* 3394–3411.

Staeren, N., Renvall, H., De Martino, F., Goebel, R., & Formisano, E. (2009). Sound categories are represented as distributed patterns in the human auditory cortex. *Current Biology, 19,* 498–502.

Steinschneider, M., Arezzo, J., & Vaughan Jr., H. G. (1980). Phase-locked cortical responses to a human speech sound and low-frequency tones in the monkey. *Brain Research, 198,* 75–84.

Striem-Amit, E., Hertz, U., & Amedi, A. (2011). Extensive cochleotopic mapping of human auditory cortical fields obtained with phase-encoding FMRI. *PLOS ONE, 6,* e17832.

Talavage, T. M., Sereno, M. I., Melcher, J. R., Ledden, P. J., Rosen, B. R., & Dale, A. M. (2004). Tonotopic organization in human auditory cortex revealed by progressions of frequency sensitivity. *Journal of Neurophysiology, 91,* 1282–1296.

Theunissen, F. E., & Elie, J. E. (2014). Neural processing of natural sounds. *Nature Reviews Neuroscience, 15,* 355–366.

Viemeister, N. F. (1979). Temporal modulation transfer functions based upon modulation thresholds. *Journal of the Acoustical Society of America, 66,* 1364–1380.

Woolley, S. M. N., Fremouw, T. E., Hsu, A., & Theunissen, F. E. (2005). Tuning for spectro-temporal modulations as a mechanism for auditory discrimination of natural sounds. *Nature Neuroscience, 8,* 1371–1379.

Young, E. D. (2008). Neural representation of spectral and temporal information in speech. *Philosophical Transactions of the Royal Society B: Biological Sciences, 363,* 923–945.

Zatorre, R. J., & Belin, P. (2001). Spectral and temporal processing in human auditory cortex. *Cerebral Cortex, 11,* 946–953.

36 The Speech-Ready Organization of the Motor Cortex

SERENA BIANCHI AND KRISTINA SIMONYAN

Spoken language is a complex form of human communication that is unparalleled in the animal kingdom. For a speaker to convey a message in a linearized sequence of speech sounds, it is necessary that concepts are retrieved, phonological segments are combined with abstract syllabic templates, and articulatory gestures are produced to sound out the intended words. The neurobiological bases of these processes have intrigued philosophers, psychologists, anthropologists, neuroscientists, and linguists for centuries, but it has been only in the last decade that our understanding of the speech- and language-ready brain has seen remarkable advances. While historically the majority of studies focused on the function and structure of the key language-related brain regions, such as Broca's and Wernicke's areas, more recent methodological advances in the conduct of neuroanatomical, neurophysiological, and brain imaging studies have allowed for expansion of our knowledge on complex neural networks controlling specific sensorimotor aspects of speech production. In the recent years, speech production has also rekindled the interest of the evolutionary-minded community seeking to understand comparative neurobiological origins of unique human behaviors.

It is worthwhile to mention that the terms *language* and *speech* are often used interchangeably, although they underlie different processes. The former refers to more cognitive aspects of the human communication system and is independent of the modality whereby it is conveyed (e.g., sign language); the latter mostly constitutes the sensorimotor component of language and relies specifically on auditory-vocal modality of transmission. In this chapter, we will focus on neurobiology behind the last step of the cascaded set of speaking operations that is the sensorimotor control of speech production, which depends on organization and integrity of the primary motor cortex as the final cortical output structure. We will first discuss the structural and functional organization of the laryngeal motor cortex (LMC) and then examine evolutionary specializations that may have contributed to the emergence of unique speech motor control in the hominin lineage.

1. The Speech-Ready Motor Cortex: Somatotopic Organization

Production of speech is a complex motor behavior, which requires coordinated activity of over 100 respiratory, laryngeal, and orofacial muscles. To orchestrate such a level of complexity, the nervous system must engage in a degree of motor control that is practically unparalleled among other motor systems in the human body (Kent, 2004).

First explorations into the anatomical organization of the primary motor cortex relevant to speech articulators date back to the seminal work of Leyton and Sherrington (1917) in the anthropoid apes. Using direct electrical stimulation and ablation of the precentral gyrus, this study mapped discrete representations of body parts, with the lower limbs and trunk located most dorsally; the hand and fingers in the middle; followed by the face, mouth, and larynx ventrally (figure 36.1). Movements of the vocal folds, which are necessary for voice and speech production, were evoked more dorsally and anteriorly within the ventral motor cortex, whereas extralaryngeal muscles, which move the larynx up and down, were localized more ventrally along the central sulcus. The larynx representation within the motor cortex showed high variability within and across species and was not identified in each experimental subject. Specifically, in the chimpanzee, this area was mapped within the ventral portion of precentral sulcus but intermingled with the representations of other orofacial articulators (figure 36.1a) (Leyton & Sherrington, 1917). Electrical stimulation of this region evoked an unspecified sound (Hines, 1940), although this instance occurred only in one individual and could not be further replicated. In the gorilla, the vocal folds were mapped to a small field most anterior and ventral to precentral sulcus (figure 36.1b) (Leyton & Sherrington, 1917), while later studies in Old and New World monkeys (i.e., rhesus and squirrel monkeys) have moved the laryngeal representation even further anterior and ventral within the motor cortex (figure 36.1c) (Coude et al., 2011; Hast, Fischer, Wetzel, & Thompson,

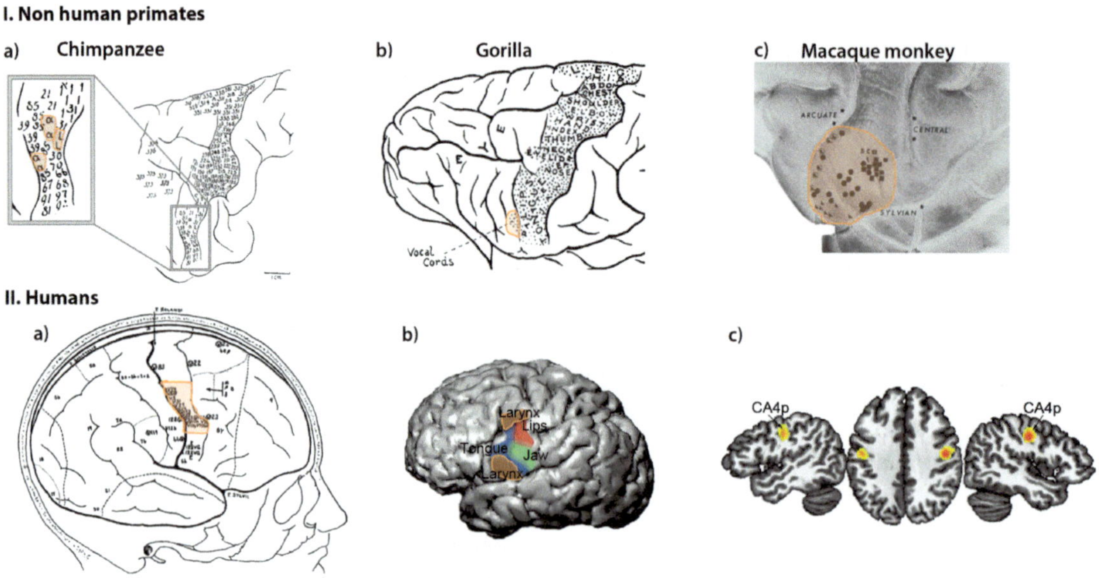

FIGURE 36.1 The location of the LMC in nonhuman primates (I) and humans (II). The LMC is circled in orange. In panel I, (a) α indicates a site of vocal cord adduction and L indicates external movement of the larynx, presumably due to contraction of the extralaryngeal muscles, in chimpanzees, (b) a site of vocal cord adduction in gorillas, and (c) LMC in macaque monkeys (*Macaca mulatta*), where triangles indicate sites eliciting movements of the cricothyroid muscles, circles indicate tyroarythenoid muscles, encircled triangles indicate combined responses of the cricothyroid and tyroarythenoid muscles, and squares indicate the activity of extrinsic laryngeal muscles. In panel II, (a) the "vocalization" area, from where voicing was elicited in six neurosurgical subjects; (b) the dorsal and ventral locations of the larynx representation identified using electrocorticography mapping; (c) the location of the LMC in bilateral area 4p of the primary motor cortex based on a meta-analysis of functional neuroimaging studies from 2000 to 2013. The images were adapted as follows: panels Ia and Ib from Leyton and Sherrington (1917), Ic from Hast et al. (1974a), IIa from Penfield and Boldrey (1937), IIb from Conant et al. (2014), and IIc from Simonyan (2014).

1974a; Jurgens, 1974; Simonyan & Jurgens, 2002, 2005b).

The first report of laryngeal representation in humans should be credited to Foerster (1936), who described bilateral movements of vocal folds while electrically stimulating the motor cortex in neurosurgical patients. Later influential work by Penfield and colleagues (especially Penfield & Boldrey, 1937) mapped discrete representations of body parts along the dorsal-ventral axis of the precentral gyrus, a so-called motor homunculus, which is much similar to somatotopical representations in apes and monkeys. However, the human motor homunculus lacked the larynx representation. Instead, a *vocalization* representation was assigned to a larger region, which elicited prolonged vocal sounds, such as *ahhh*, spanning from an area dorsal to the lip representation to a ventral area closer to the Sylvian fissure, which also elicited grunts while electrically stimulated. The vocalization representation, thus, was not a segregated region but largely overlapped with orofacial articulators (figure 36.1-IIa). While speech production does require integration among laryngeal, orofacial, and respiratory behaviors, the location of the larynx itself (not vocalization as a behavior) remained unknown for more than half a century following Penfield's work.

Due to advances in modern electrophysiology and brain imaging approaches, the interest in laryngeal motocortical representation was renewed in early 2000s. Using single-pulse cortical transcranial magnetic stimulation (TMS), Rödel et al. (2004) were first to localize and segregate not-overlapping representations of laryngeal muscles in the primary motor cortex. Two distinct sites with fast response of 7.3 to 11.3 ms were identified as somatotopic representations of the cricothyroid and thyroarytenoid muscles, which participate in vocal fold tension and closure (Rödel et al., 2004). Combining TMS with MRI, Espadaler et al. (2012) further confirmed the location of this site by evoking movements of the crycothyroid muscle lateral to the lip representation. This location of laryngeal activity was further validated in an electrocorticography (ECoG) study in neurosurgical patients that leveraged the high temporal and spatial resolutions of this methodology to measure cortical activity of different speech articulators (Bouchard, Mesgarani, Johnson, & Chang, 2013). Consistent with previous observations, Bouchard et al. (2013) confirmed a dorsal and segregated location of the laryngeal representation in the precentral gyrus (figure 36.1-IIb). While Penfield's earlier studies were hampered by the relatively large currents delivered for cortical mapping, finer-grained ECoG studies have also suggested that the representation of speech effectors is subject to interindividual variations and a greater spatial overlap (Breshears, Molinaro, & Chang, 2015; Takai, Brown, & Liotti, 2010) as initially observed in earlier studies in great apes (Leyton & Sherrington, 1917). In addition, Bouchard et al. (2013) reported a second site of laryngeal evoked activity, which was located more ventrally, above the Sylvian fissure. While electrical stimulation of the dorsal laryngeal representation elicited prolonged vowel sounds, stimulation of the ventral region led to throat movements and the sensation of being unable to breathe, speak, and swallow (Breshears et al., 2015; Herve, Zago, Petit, Mazoyer, & Tzourio-Mazoyer, 2013). The latter region was also similar to the site that Penfield and Boldrey (1937) once described as eliciting grunts. Furthermore, a more recent magnetoencephalography study revealed the same ventral area to contain the sensory representation of the larynx (Miyaji et al., 2014). It, thus, appears that "true motor" representation of laryngeal muscles is located in the dorsal primary motor cortex, whereas its sensory representation spans the most ventral region, which, in fact, cytoarchitectonically represents the somatosensory (not motor) cortex, extending rostrally from the postcentral gyrus underneath the central sulcus.

The findings from these electrophysiological studies found their confirmation in numerous functional brain imaging studies that mapped the larynx representation within the motor cortex based on its activity during speech and nonspeech tasks. Loucks et al. (2007) described a dorsal region in the ventral primary motor cortex that was active during both phonatory tasks (syllable production designed to elicit laryngeal but not orofacial movements) and exhalation (which physiologically is critical for voice and speech production). The location of this dorsal region was similar to the one found with TMS and ECoG as we've already discussed. This location was further confirmed in the studies that separated brain activation during movements of the larynx in the presence or absence of phonation and other orofacial (e.g., lips, tongue) tasks (Brown, Ngan, & Liotti, 2008) as well as during other voluntary laryngeal behaviors, such as coughing, sniffing, and voluntary breathing (Simonyan, Saad, Loucks, Poletto, & Ludlow, 2007). Summarizing the last few decades of neuroimaging effort in two independent meta-analyses, the larynx representation was at last mapped to the dorsal portion of the ventral primary motor cortex in area 4p (Brown et al., 2009; Simonyan, 2014) (see figure 36.1-IIc).

2. *The Speech-Ready Motor Cortex: Connectivity*

As we mentioned earlier, speech production relies on coordinated activity and integration among laryngeal, respiratory, and orofacial articulators, which is thought to rely on a vast network of primary motor cortex, with input and output cortical and subcortical structures, forming a so-called *speech connectome* (Fuertinger, Horwitz, & Simonyan, 2015).

In order to understand how the speech connectome is formed, it is necessary to understand the neural processes of voice production for speech. Human vocalizations are considered to be either innate (i.e., not depending on audiomotor learning) or voluntary (i.e., learned during development through exposure to human speech). Examples of innate vocalizations include cries, groans, and laughter that are acoustically simple. Voluntary learned vocalizations include speech and song that are acoustically complex and combine discrete elements (i.e., phonemes, notes). In the animal kingdom, most auditory-vocal communications rely on the production of innate, genetically preprogrammed calls, while learned vocalizations that have some resemblance to human speech are found only in rare instances, such as in songbirds, parrots, hummingbirds, bats, dolphins, sea lions, and elephants (Simonyan, Horwitz, & Jarvis, 2012).

Production of innate and learned vocalizations appears to be regulated by distinct neural networks (Jurgens, 2002, 2009; Simonyan & Horwitz, 2011). Innate vocalizations are produced at the level of brain stem structures, relying on phonatory motor and sensory nuclei and reticular formation. This is why anencephalic infants, who lack the entire forebrain, are still able to react with groans to painful stimuli (Monnier & Willi, 1953). Parallel to child's development, innate vocalizations come under more voluntary control, with children being able to voluntarily initiate, suppress, or modulate their cries and laughter. This is achieved due to proper functioning and maturity of higher order structures, such as periaqueductal gray and cingulate cortex. Lesions in the cingulate cortex lead to a loss of voluntary control of emotional intonations during speaking, while destruction of the periaqueductal gray results in mutism (Barrett, Pike, & Paus, 2004; von Cramon & Jurgens, 1983). The reason why speech is preserved in the presence of cingulate damage is because it is directly controlled by the orofacial and laryngeal primary motor cortex. The developmental maturity of the latter structure allows learning and controlling new vocalizations for speech and song, independent but in association with a separate cingulate-periaqueductal-brain stem pathway that is responsible for production of innate vocalizations. As a result of this separation, bilateral lesions of the LMC abolish the ability to speak and sing while preserving the ability to produce innate vocalizations, such as cries and laughter.

Anatomical tract tracing studies in nonhuman primates and probabilistic diffusion tractography studies in humans revealed that, to fulfill its function, the LMC forms a widespread structural network of cortical and subcortical connections. With its bilateral projections, the LMC is connected with dorsal and ventral areas of the primary sensorimotor cortex, premotor and supplementary motor area (SMA), inferior frontal gyrus, ventrolateral and dorsolateral prefrontal cortex, insula, cingulate cortex, angular gyrus, supramarginal gyrus, superior and middle temporal gyri, claustrum, thalamus, and brain stem nuclei, including the parabrachial nucleus, mesencephalic nucleus, and locus coeruleus (Jurgens, 1976; Simonyan & Jurgens, 2002, 2003, 2005a). Unilateral efferents from the LMC project to the inferior parietal cortex, posterior parietal operculum, intraparietal sulcus, caudate nucleus, putamen, globus pallidus, reticular formation, solitary tract nucleus, facial nucleus, spinal trigeminal nucleus; unilateral afferents are received from the orbital cortex, ventral tegmental area, substantia nigra, and the raphe nucleus. These connections exist in both humans and nonhuman primates, with the exception of projections from the LMC to nucleus ambiguus, the brain stem site of laryngeal motoneurons that appears to exist in humans only (Iwatsubo, Kuzuhara, Kanemitsu, Shimada, & Toyokura, 1990; Kuypers, 1958a) (see discussion in section 3 on human evolutionary specializations).

While the exact role of each of these connections within the speech connectome still needs to be elucidated, several clues exist about the potential contribution of specific brain regions within this extensive anatomical network. Connections with premotor regions may regulate aspects of movement preparation and sequencing, which may be important to prearrange the correct sequence of movements necessary to articulate and combine individual phonemes before their execution is performed at the level of the primary motor cortex (Riecker et al., 2005). The SMA is a key region for the production of learned vocalizations such as speech, and, similar to the primary motor cortex, electrical stimulation of this area elicits vocalizations in humans, but not in nonhuman primates (Jurgens, 2002; Kirzinger & Jurgens, 1982). In humans, electrical stimulation of this region evokes perseveration of speech (Brickner, 1940), as well as production of a vowel sound, syllable/word repetition, and movements of the mouth and jaw (Penfield & Welch, 1951). By contrast, SMA lesions do not affect the acoustic structure of innate calls in nonhuman primates, although damage to its most anterior portion, which may share some functional similarities in affective/volitional vocal control with the anterior cingulate cortex, reduces the overall number of isolation calls produced (Kirzinger & Jurgens, 1982).

Outside the motor areas, LMC connections with the adjacent primary somatosensory and association regions of the parietal lobe are important in providing sensory feedback to guide motor actions (Guenther, 2006) as well as higher order level sensorimotor integration (Kumar, Croxson, & Simonyan, 2016). In particular, connections with areas regulating orofacial and respiratory activity may allow for the degree of interregional communication necessary to orchestrate the coupling of laryngeal and orofacial components that uniquely characterize speech production. Projections with posterior parietal and temporal areas may play a role in specific aspects of auditory-motor mapping, which are fundamental to speaking. Accordingly, frontal, parietal, and temporal regions and their underlying pathways are proposed to form the so-called *dorsal stream* of language control, which provides the interface for articulatory functions (Hickok & Poeppel, 2004). In particular, it has been suggested that an area at the conjunction of the temporal and parietal lobes (Sylvian parietal-temporal, Spt) may be a key site of auditory-motor integration through its connections with ventral primary motor cortex, surrounding the LMC (Hickok, 2016). Strong connectivity between frontal motor and posterior temporal areas has been also demonstrated based on greater coherence between these regions during voice production relative to playback (Kingyon et al., 2015). The connectivity between the motor cortex and the superior temporal gyrus may be especially important in processes of auditory monitoring to detect and correct errors during speaking (Behroozmand et al., 2015; Chang, Niziolek, Knight, Nagarajan, & Houde, 2013; Greenlee et al., 2013).

Furthermore, the LMC is functionally connected with inferior frontal gyrus (IFG) (Greenlee et al., 2004), a region that has been traditionally implicated in a variety of speech and language processes (Price, 2000). With regard to its role in the speech motor network, an ECoG study that monitored brain activity during word repetition has shown that IFG activity preceded the activity of premotor and motor regions, which, by contrast, went "online" before and during speech was uttered (Flinker et al., 2015). These results indicate a prearticulatory function of IFG, which may serve as a relay region through which auditory "images" are propagated to motor cortices and transformed into articulated sounds. Using a brain surface cooling technique in awake neurosurgical patients, Long et al. (2016) were able to dissociate the function of the ventral primary motor cortex and IFG, further pointing to importance of the former in maintaining the speech quality and the latter in speech rate.

The LMC also exhibits strong and reciprocal connections with the cingulate cortex, thus forming a point of convergence for both evolutionary conserved (innate vocalizations) and novel (speech/song) pathways (Simonyan & Jurgens, 2005a). This potentially contributes to integration between motivational and affective aspects of voice production for speech at higher levels of voluntary control (Paus, 2001; Simonyan & Horwitz, 2011).

In addition to corticocortical projections, the LMC is directly connected with a range of subcortical structures. Among these, the putamen is the largest target of the LMC outputs, which project according to somatotopical organization, with the laryngeal efferents overlapping the orofacial efferents in the ventral postcommissural portion of the putamen and forming a segregated laryngeal region in the rostral associative putamen (Simonyan, Herscovitch, & Horwitz, 2013; Simonyan & Jurgens, 2003). Lesions of the striatum result in various speech motor deficits, including dysarthria and dysphonia (Liegeois et al., 2003; Vargha-Khadem, Watkins, Alcock, Fletcher, & Passingham, 1995; Watkins, Gadian, & Vargha-Khadem, 1999), but leave spontaneous vocalizations intact in nonhuman primates (Jurgens, Kirzinger, & von Cramon, 1982). This suggests that putamen may have an exclusive role in the production of learned but not innate vocalizations. Its role was also proposed in the evolution of the neuroanatomical bases of vocal learning in modern humans (Enard, 2011; Lai, Gerrelli, Monaco, Fisher, & Copp, 2003), with recent studies showing that the associative putamen is involved in learning new articulatory patterns for non-native words (Simmonds, Leech, Iverson, & Wise, 2014). The role of the putamen in speech production may be mediated by neurotransmitter mechanisms such as nigrostriatal dopamine release and its coupling with neural activity in the associative putamen (Simonyan et al., 2013), which further drives the left-hemispheric lateralization of speech and language networks (Fuertinger, Zinn, & Simonyan, 2015).

While tract tracing, lesioning and neuroimaging studies provide a comprehensive view on the regions that are connected with the LMC, functional dynamics of speech motor control remains less well understood. Recent series of studies employed seed-based and graph theoretical analyses to identify the patterns of motocortical connectivity formed during real-life speaking (Fuertinger, Horwitz, & Simonyan, 2015; Simonyan & Fuertinger, 2015) (see figure 36.2). By seeding functional networks from seven key regions involved in speech sensorimotor control and examining the interactions between these neural networks, the LMC was found to be the most homogeneous region of the speech network with a complete overlap with all other speech-controlling networks (Simonyan & Fuertinger, 2015). On the other hand, a distinct involvement of the inferior parietal cortex and cerebellum in the speech but not resting-state network appears to be one of the hallmarks of neural organization controlling speech production. These results were extended by a follow-up study that examined the complexities of the large-scale speech connectome using graph theoretical analysis (Fuertinger, Horwitz, & Simonyan, 2015). Compared to other tasks (i.e., finger tapping, auditory discrimination, and syllable production), speech was characterized by an extensive and complex functional network of highly connected hubs (i.e., pivotal regions of high information transfer) in the primary sensorimotor and parietal cortex. Transition from resting to speaking required recruitment of the parietal and cingulate cortex, thalamus and cerebellum, while transition from syllable production (the closest motor task to word production) to real-life speaking required additional involvement of prefrontal cortex, insula, and putamen by sensorimotor hubs. Finally, unique topological architecture of the large-scale functional speech network depended on reorganization and specialization of the network hubs specifically centered in and around the primary motor cortex (area 4) where the LMC is represented (Fuertinger, Horwitz, & Simonyan, 2015). Taken together, these studies indicate that the LMC is a pivotal region of the speech network that establishes extensive structural and functional connections to integrate sensorimotor, auditory, and cognitive information flow for speech production.

3. The Speech-Ready Motor Cortex: Human Evolutionary Specializations

The last few decades have witnessed renewed interest in the question of the evolution of speech and language, and, in particular, the origins of their neurobiological bases. An increasing number of comparative studies have been conducted, many of which focused on examining interspecies differences in the structure of traditional language regions, such as Broca's and Wernicke's areas (Cantalupo & Hopkins, 2001; Rilling et al., 2008; Schenker et al., 2010; Sherwood, Holloway, Erwin, &

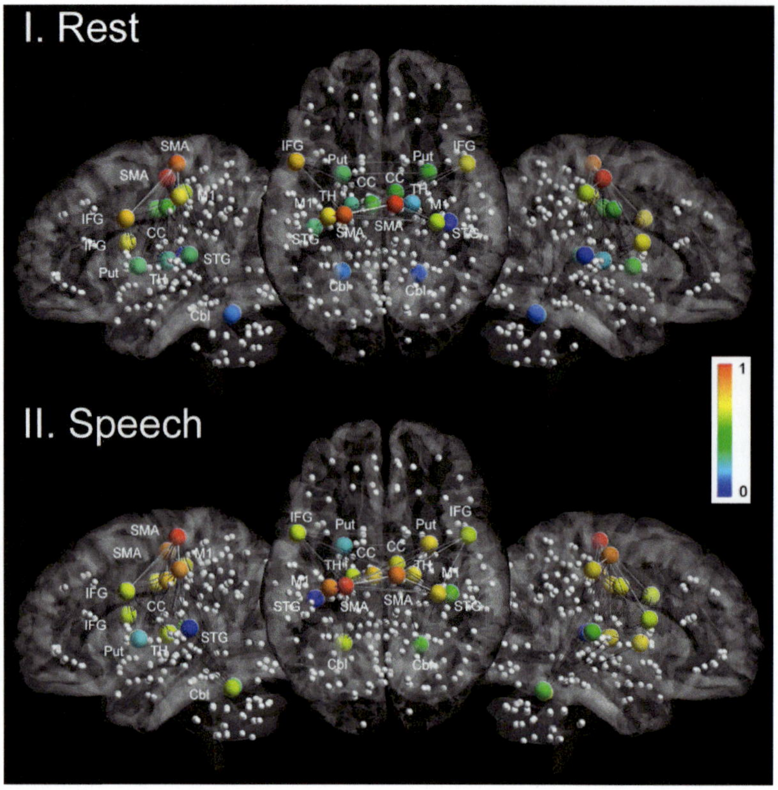

FIGURE 36.2 Illustration of the interaction between sensorimotor network components during resting (I) and speech production (II). Sphere color indicates nodal strength, and line color indicates the strength of nodal connectivity. Color bar indicates nodal strength. The images are adapted from Simonyan and Fuertinger (2015).

Hof, 2004; Spocter et al., 2010), as well as white matter tracts supporting speech and language in humans (Kumar et al., 2016; Rilling et al., 2008). These studies have shown that the representation of orofacial musculature in the motor cortex is similarly organized between primate species (Graziano, Taylor, & Moore, 2002; Huang, Sirisko, Hiraba, Murray, & Sessle, 1988), with no major cellular differences among macaque, baboon, orangutan, gorilla, and human (Sherwood, Holloway, Erwin, Schleicher, et al., 2004). As a general trend, however, layer III of this region showed greater thickness and reduced cell density in great apes relative to Old World monkeys, thus suggesting increased space for corticocortical connections. From a functional perspective, enhanced connectivity of this region may underlie greater motor control of orofacial muscles, which may be important for more complex facial expressions and communicative gestures in this species (Sherwood, Holloway, Erwin, Schleicher, et al., 2004).

Some of the most remarkable differences between humans and nonhuman primates have been found in relation to LMC connectivity. While in humans this area has been identified in area 4p, dorsally and caudally buried along the central sulcus (Simonyan, 2014), movements of the laryngeal muscles in Old and New World monkeys were elicited from a more anterior-ventral area in the precentral gyrus, as discussed earlier (Coude et al., 2011; Hast et al., 1974a; Jurgens, 1974; Simonyan & Jurgens, 2002, 2005b). Cytoarchitectonically, crycothyroid and thyroarytenoid laryngeal muscles were mapped to a segregated region within the premotor cortex (area 6), in an area bordering the arcuate sulcus, in the macaque monkey (Hast et al., 1974a). A more recent study showed that the laryngeal representation may also be intermingled with orofacial representation within the primary motor cortex (Cerkevich & Strick, 2015), but without forming an isolated laryngeal region, which is in contrast to mapping of an isolated area in area 6 in nonhuman primates (Hast, Fischer, Wetzel, & Thompson, 1974b) or in area 4p in humans (Rödel et al., 2004). Similarly, the larynx representation appears to be intermingled with orofacial representation in both primary motor and premotor cortex in the chimpanzee, whereas in the gorilla it appears to be mostly positioned in the premotor cortex. Such intermixed representation of laryngeal and orofacial muscles explains why previous electrical stimulation studies have failed to identify a segregated region within the primary motor cortex in nonhuman primates (Coude et al., 2011; Hage & Nieder, 2013; Jurgens, 1976; Simonyan & Jurgens, 2002, 2005b). This evidence further points to the fact that the LMC has undergone not only a dorsal-caudal shift in humans from its ventral-anterior location in nonhuman primates, but also somatotopic maturation that led to the formation of a specialized and functionally segregated region within the primary motor cortex for fine speech motor control. Nevertheless, because stimulation of both premotor cortex in the nonhuman primates/great apes and the primary motor cortex in humans elicits isolated adduction of the vocal folds necessary for voice and speech production, these regions are considered to be functional homologies (Kumar et al., 2016).

On another hand, as a consequence of somatotopic reorganization in the course of evolution, the LMC connectivity, too, appears to have undergone remarkable changes (Simonyan, 2014). In primates, including humans, the ventral primary motor cortex sends monosynaptic connections to the motor nuclei of the pons and medulla, controlling orofacial movements (Kuypers, 1958a, 1958b). These direct connections provide fine motor control over the orofacial muscles for the production of facial expressions and communicative orofacial gestures (e.g., lip smacks, raspberries) that do not involve laryngeal movements and are commonly used among anthropoids. Only in humans do connections between the LMC and laryngeal motoneurons of the nucleus ambiguus appear to exist, albeit based on a limited number of studies (Iwatsubo et al., 1990; Kuypers, 1958a). By contrast, in nonhuman primates, descending connections between the LMC and ambiguus motoneurons are indirect and synapse first in the surrounding reticular formation as well as brain stem sensory phonatory nuclei (Jurgens, 1976; Kuypers, 1958b; Simonyan & Jurgens, 2003). In light of this evidence, it has been hypothesized that direct corticobulbar connections originating from the phylogenetically new, caudally shifted, and functionally matured LMC in humans underlie the faster and finer control of laryngeal movements that are necessary for speech production (Simonyan, 2014; Simonyan & Jurgens, 2003). The retention of indirect connections through the reticular formation as well as the presence of a phylogenetically old LMC in the premotor cortex may instead provide an additional support and subserve less complex and/or non-speech-related laryngeal movements. It is uncontroversial that one major specialization of the motor cortex during mammalian and, especially, primate evolution has entailed the progressive increase in direct motoneuronal projections for finer motor control of complex and learned behaviors (Fitch, 2010; Kuypers, 1958a; Rathelot & Strick, 2009; Simonyan, 2014; Striedter, 2005). Yet, the mechanisms whereby these changes have occurred remain little understood.

On the other hand, direct motor connections with laryngeal motoneurons in the brain stem are potentially not the only pathway that emerged to support speech production. As previously discussed, speech relies on timely orchestration of manifold muscles to coordinate laryngeal, respiratory, and orofacial movements. A recent diffusion tractography study confirmed the presence of a largely comparable LMC network between nonhuman primates and humans despite the differences in the cytoarchitectonic location of LMC (Kumar et al., 2016). It also identified that human LMC establishes nearly seven-fold enhanced connectivity with the somatosensory and inferior parietal cortex in humans than in the macaque monkey (Kumar et al., 2016), providing yet another structural evidence for the critical importance of sensorimotor integration for speech production.

Other major descending LMC pathways that run in parallel to the direct corticobulbar tract include connections with (a) the reticular formation, which is densely interconnected and passes information to phonatory motor and sensory nuclei of the brain stem; (b) basal ganglia, especially the putamen and substantia nigra pars reticulata, and (c) cingulate cortex, its anterior and middle portions (Jurgens, 1976; Simonyan & Jurgens, 2002, 2003, 2005a, 2005b). Both the basal ganglia and cingulate pathways collide in the reticular formation, thus indirectly, and additionally, connecting the LMC to nucleus ambiguus. Such enhanced organization of the LMC network ensures the presence of an extremely robust and nearly damage-proof hardwired framework for the fine-grained control of speech. It may partly be for this very reason why unilateral damage to the human LMC has little, if any, effect on the ability to speak.

Taken together, our current state of knowledge suggests that the LMC has undergone remarkable reorganization in structure and connectivity during hominin evolution, which potentially involved the establishment of direct connections with laryngeal motorneurons in the brain stem, as well as enhanced LMC-parietal connectivity, supporting primary and secondary levels of sensorimotor integration. The chance that these evolutionary developments specifically supported the emergence of greater motor control to regulate fast and coordinated laryngeal and orofacial movements for speech is probable. More research is needed to further elucidate the mechanisms and adaptive reasons why humans, yet not other primate species, underwent this remarkable evolution.

4. Future Directions and Challenges

Although the recent decades have witnessed huge strides toward better understanding of the neurobiological mechanisms and origins of speech, many questions still remain open. In order to understand the evolution of our remarkable capacity for speech production, future research should elucidate the neural bases of flexibly learned vocal behavior that has been recently described in great apes, humans' closest living relatives. This would entail, in particular, the characterization of the patterns of connectivity of the motor cortex and test the hypothesis of direct corticobulbar pathway in these species with contemporary, noninvasive methodologies. Yet, such an endeavor is likely to be hampered by the difficulty of current imaging methods to identify specific brain stem structures, such as nucleus ambiguus, and to differentiate its projections from those of the surrounding reticular formation. Furthermore, if a full understanding of the speech-ready motor cortex is to be achieved, it would be necessary to shed light on the mechanisms whereby the representation of the LMC may have shifted from the premotor cortex and matured in the primary motor cortex. It would also be important to understand the LMC's adaptive function(s), which collectively would answer the questions of how such evolutionary "tinkering" led to a brain that seemingly and effortlessly engages in one of the most complex motor behaviors that is speech production.

REFERENCES

Barrett, J., Pike, G. B., & Paus, T. (2004). The role of the anterior cingulate cortex in pitch variation during sad affect. *European Journal of Neuroscience, 19*(2), 458–464.

Behroozmand, R., Shebek, R., Hansen, D. R., Oya, H., Robin, D. A., Howard, M. A., 3rd, & Greenlee, J. D. (2015). Sensory-motor networks involved in speech production and motor control: An fMRI study. *NeuroImage, 109,* 418–428.

Bouchard, K. E., Mesgarani, N., Johnson, K., & Chang, E. F. (2013). Functional organization of human sensorimotor cortex for speech articulation. *Nature, 495*(7441), 327–332.

Breshears, J. D., Molinaro, A. M., & Chang, E. F. (2015). A probabilistic map of the human ventral sensorimotor cortex using electrical stimulation. *Journal of Neurosurgery, 123*(2), 340–349.

Brickner, R. (1940). A human cortical area producing repetitive phenomena when stimulated. *Journal of Neurophysiology, 3,* 128–130.

Brown, S., Laird, A. R., Pfordresher, P. Q., Thelen, S. M., Turkeltaub, P., & Liotti, M. (2009). The somatotopy of speech: Phonation and articulation in the human motor cortex. *Brain and Cognition, 70*(1), 31–41.

Brown, S., Ngan, E., & Liotti, M. (2008). A larynx area in the human motor cortex. *Cerebral Cortex, 18*(4), 837–845.

Cantalupo, C., & Hopkins, W. D. (2001). Asymmetric Broca's area in great apes. *Nature, 414*(6863), 505.

Cerkevich, C. M., & Strick, P. L. (2015). *A representation of the larynx in macaque monkey M1.* Paper presented at the Annual Meeting of the Society for Neuroscience, Chicago, IL.

Chang, E. F., Niziolek, C. A., Knight, R. T., Nagarajan, S. S., & Houde, J. F. (2013). Human cortical sensorimotor network underlying feedback control of vocal pitch. *Proceedings of the National Academy of Sciences, 110*(7), 2653–2658.

Conant, D., Bouchard, K. E., & Chang, E. F. (2014). Speech map in the human ventral sensory-motor cortex. *Current Opinion in Neurobiology, 24,* 63–67.

Coude, G., Ferrari, P. F., Roda, F., Maranesi, M., Borelli, E., Veroni, V., ... Fogassi, L. (2011). Neurons controlling voluntary vocalization in the macaque ventral premotor cortex. *PLOS ONE, 6*(11), e26822.

Enard, W. (2011). FOXP2 and the role of cortico-basal ganglia circuits in speech and language evolution. *Current Opinion in Neurobiology, 21*(3), 415–424.

Espadaler, J., Rogic, M., Deletis, V., Leon, A., Quijada, C., & Conesa, G. (2012). Representation of cricothyroid muscles at the primary motor cortex (M1) in healthy subjects, mapped by navigated transcranial magnetic stimulation (nTMS). *Clinical Neurophysiology, 123*(11), 2205–2211.

Fitch, W. T. (2010). *The evolution of language.* Cambridge: Cambridge University Press.

Flinker, A., Korzeniewska, A., Shestyuk, A. Y., Franaszczuk, P. J., Dronkers, N. F., Knight, R. T., & Crone, N. E. (2015). Redefining the role of Broca's area in speech. *Proceedings of the National Academy of Sciences, 112*(9), 2871–2875.

Foerster, O. (1936). Motorische Felder und Bahnen. In O. Bumke & O. Foerster (Ed.), *Handbuch der Neurologie* (Vol. 6, pp. 1–357). Berlin: Springer.

Fuertinger, S., Horwitz, B., & Simonyan, K. (2015). The functional connectome of speech control. *PLOS Biology, 13*(7), e1002209.

Fuertinger, S., Zinn, J. C., & Simonyan, K. (2015). *Dopamine modulates lateralization of motocortical activity during speech: An fMRI and modeling study.* Paper presented at the Annual Meeting of the Organization for Human Brain Mapping, Honolulu, HI.

Graziano, M. S., Taylor, C. S., & Moore, T. (2002). Complex movements evoked by microstimulation of precentral cortex. *Neuron, 34*(5), 841–851.

Greenlee, J. D., Behroozmand, R., Larson, C. R., Jackson, A. W., Chen, F., Hansen, D. R., ... Howard III, M. A. (2013). Sensory-motor interactions for vocal pitch monitoring in non-primary human auditory cortex. *PLOS ONE, 8*(4), e60783.

Greenlee, J. D., Oya, H., Kawasaki, H., Volkov, I. O., Kaufman, O. P., Kovach, C., ... Brugge, J. F. (2004). A functional connection between inferior frontal gyrus and orofacial motor cortex in human. *Journal of Neurophysiology, 92*(2), 1153–1164.

Guenther, F. H. (2006). Cortical interactions underlying the production of speech sounds. *Journal of Communication Disorders, 39*(5), 350–365.

Hage, S. R., & Nieder, A. (2013). Single neurons in monkey prefrontal cortex encode volitional initiation of vocalizations. *Nature Communications, 4,* 2409.

Hast, M. H., Fischer, J. M., Wetzel, A. B., & Thompson, V. E. (1974a). Cortical motor representation of the laryngeal muscles in *Macaca mulatta. Brain Research, 73*(2), 229–240.

Hast, M. H., Fischer, J. M., Wetzel, A. B., & Thompson, V. E. (1974b). Cortical motor representation of laryngeal muscles in *Macaca mulatta*. *Federation Proceedings, 33*(3), 399–399.

Herve, P. Y., Zago, L., Petit, L., Mazoyer, B., & Tzourio-Mazoyer, N. (2013). Revisiting human hemispheric specialization with neuroimaging. *Trends in Cognitive Sciences, 17*(2), 69–80.

Hickok, G. (2016). A cortical circuit for voluntary laryngeal control: Implications for the evolution language. *Psychonomic Bulletin and Review, 24*(1), 56–63.

Hickok, G., & Poeppel, D. (2004). Dorsal and ventral streams: A framework for understanding aspects of the functional anatomy of language. *Cognition, 92*(1–2), 67–99.

Hines, M. (1940). Movements elicited from precentral gyrus of adult chimpanzees by stimulation with sine wave currents. *Journal of Neurophysiology, 3*, 442–466.

Huang, C. S., Sirisko, M. A., Hiraba, H., Murray, G. M., & Sessle, B. J. (1988). Organization of the primate face motor cortex as revealed by intracortical microstimulation and electrophysiological identification of afferent inputs and corticobulbar projections. *Journal of Neurophysiology, 59*(3), 796–818.

Iwatsubo, T., Kuzuhara, S., Kanemitsu, A., Shimada, H., & Toyokura, Y. (1990). Corticofugal projections to the motor nuclei of the brainstem and spinal cord in humans. *Neurology, 40*(2), 309–312.

Jurgens, U. (1974). On the elicitability of vocalization from the cortical larynx area. *Brain Research, 81*(3), 564–566.

Jurgens, U. (1976). Projections from the cortical larynx area in the squirrel monkey. *Experimental Brain Research, 25*(4), 401–411.

Jurgens, U. (2002). Neural pathways underlying vocal control. *Neuroscience and Biobehavioral Reviews, 26*(2), 235–258.

Jurgens, U. (2009). The neural control of vocalization in mammals: A review. *Journal of Voice, 23*(1), 1–10.

Jurgens, U., Kirzinger, A., & von Cramon, D. (1982). The effects of deep-reaching lesions in the cortical face area on phonation: A combined case report and experimental monkey study. *Cortex, 18*(1), 125–139.

Kent, R. D. (2004). The uniqueness of speech among motor systems. *Clinical Linguistics and Phonetics, 18*(6–8), 495–505.

Kingyon, J., Behroozmand, R., Kelley, R., Oya, H., Kawasaki, H., Narayanan, N. S., & Greenlee, J. D. (2015). High-gamma band fronto-temporal coherence as a measure of functional connectivity in speech motor control. *Neuroscience, 305*, 15–25.

Kirzinger, A., & Jurgens, U. (1982). Cortical lesion effects and vocalization in the squirrel monkey. *Brain Research, 233*(2), 299–315.

Kumar, V., Croxson, P. L., & Simonyan, K. (2016). Structural organization of the laryngeal motor cortical network and its implication for evolution of speech production. *Journal of Neuroscience, 36*(15), 4170–4181.

Kuypers, H. G. (1958a). Corticobular connexions to the pons and lower brain-stem in man: An anatomical study. *Brain, 81*(3), 364–388.

Kuypers, H. G. (1958b). Some projections from the pericentral cortex to the pons and lower brain stem in monkey and chimpanzee. *Journal of Comparative Neurology, 110*(2), 221–255.

Lai, C. S., Gerrelli, D., Monaco, A. P., Fisher, S. E., & Copp, A. J. (2003). FOXP2 expression during brain development coincides with adult sites of pathology in a severe speech and language disorder. *Brain, 126*(Pt 11), 2455–2462.

Leyton, A. S. F., & Sherrington, C. S. (1917). Observations on the excitable cortex of the chimpanzee, orang-utan and gorilla. *Quarterly Journal of Experimental Physiology, 11*, 135–222.

Liegeois, F., Baldeweg, T., Connelly, A., Gadian, D. G., Mishkin, M., & Vargha-Khadem, F. (2003). Language fMRI abnormalities associated with FOXP2 gene mutation. *Nature Neuroscience, 6*(11), 1230–1237.

Long, M. A., Katlowitz, K. A., Svirsky, M. A., Clary, R. C., Byun, T. M., Majaj, N., ... Greenlee, J. D. (2016). Functional segregation of cortical regions underlying speech timing and articulation. *Neuron, 89*(6), 1187–1193.

Loucks, T. M., Poletto, C. J., Simonyan, K., Reynolds, C. L., & Ludlow, C. L. (2007). Human brain activation during phonation and exhalation: Common volitional control for two upper airway functions. *NeuroImage, 36*(1), 131–143.

Miyaji, H., Hironaga, N., Umezaki, T., Hagiwara, K., Shigeto, H., Sawatsubashi, M., ... Komune, S. (2014). Neuromagnetic detection of the laryngeal area: Sensory-evoked fields to air-puff stimulation. *NeuroImage, 88*, 162–169.

Monnier, M., & Willi, H. (1953). The integrative activity of the nervous system of a meso-rhombencephalic anencephalus. II. Anatomical part. *Monatsschrift für Psychiatrie und Neurologie, 126*(4–5), 259–273.

Paus, T. (2001). Primate anterior cingulate cortex: Where motor control, drive and cognition interface. *Nature Reviews Neuroscience, 2*(6), 417–424.

Penfield, W., & Boldrey, E. (1937). Somatic motor and sensory representation in the cerebral cortex of man as studied by electrical stimulation. *Brain, 60*, 389–443.

Penfield, W., & Welch, K. (1951). The supplementary motor area of the cerebral cortex; a clinical and experimental study. *AMA Archives of Neurology and Psychiatry, 66*(3), 289–317.

Price, C. J. (2000). The anatomy of language: Contributions from functional neuroimaging. *Journal of Anatomy, 197* (Pt 3), 335–359.

Rathelot, J. A., & Strick, P. L. (2009). Subdivisions of primary motor cortex based on cortico-motoneuronal cells. *Proceedings of the National Academy of Sciences, 106*(3), 918–923.

Riecker, A., Mathiak, K., Wildgruber, D., Erb, M., Hertrich, I., Grodd, W., & Ackermann, H. (2005). fMRI reveals two distinct cerebral networks subserving speech motor control. *Neurology, 64*(4), 700–706.

Rilling, J. K., Glasser, M. F., Preuss, T. M., Ma, X., Zhao, T., Hu, X., & Behrens, T. E. (2008). The evolution of the arcuate fasciculus revealed with comparative DTI. *Nature Neuroscience, 11*(4), 426–428.

Rödel, R. M., Olthoff, A., Tergau, F., Simonyan, K., Kraemer, D., Markus, H., & Kruse, E. (2004). Human cortical motor representation of the larynx as assessed by transcranial magnetic stimulation (TMS). *Laryngoscope, 114*(5), 918–922.

Schenker, N. M., Hopkins, W. D., Spocter, M. A., Garrison, A. R., Stimpson, C. D., Erwin, J. M., ... Sherwood, C. C. (2010). Broca's area homologue in chimpanzees (Pan troglodytes): Probabilistic mapping, asymmetry, and comparison to humans. *Cerebral Cortex, 20*(3), 730–742.

Sherwood, C. C., Holloway, R. L., Erwin, J. M., & Hof, P. R. (2004). Cortical orofacial motor representation in Old World monkeys, great apes, and humans. II. Stereologic analysis of chemoarchitecture. *Brain, Behavior and Evolution, 63*(2), 82–106.

Sherwood, C. C., Holloway, R. L., Erwin, J. M., Schleicher, A., Zilles, K., & Hof, P. R. (2004). Cortical orofacial motor representation in Old World monkeys, great apes, and humans. I. Quantitative analysis of cytoarchitecture. *Brain, Behavior and Evolution, 63*(2), 61–81.

Simmonds, A. J., Leech, R., Iverson, P., & Wise, R. J. (2014). The response of the anterior striatum during adult human vocal learning. *Journal of Neurophysiology, 112*(4), 792–801.

Simonyan, K. (2014). The laryngeal motor cortex: Its organization and connectivity. *Current Opinions in Neurobiology, 28,* 15–21.

Simonyan, K., & Fuertinger, S. (2015). Speech networks at rest and in action: Interactions between functional brain networks controlling speech production. *Journal of Neurophysiology, 113*(7), 2967–2978.

Simonyan, K., Herscovitch, P., & Horwitz, B. (2013). Speech-induced striatal dopamine release is left lateralized and coupled to functional striatal circuits in healthy humans: A combined PET, fMRI and DTI study. *NeuroImage, 70,* 21–32.

Simonyan, K., & Horwitz, B. (2011). Laryngeal motor cortex and control of speech in humans. *Neuroscientist, 17*(2), 197–208.

Simonyan, K., Horwitz, B., & Jarvis, E. D. (2012). Dopamine regulation of human speech and bird song: A critical review. *Brain and Language, 122*(3), 142–150.

Simonyan, K., & Jurgens, U. (2002). Cortico-cortical projections of the motorcortical larynx area in the rhesus monkey. *Brain Research, 949*(1–2), 23–31.

Simonyan, K., & Jurgens, U. (2003). Efferent subcortical projections of the laryngeal motorcortex in the rhesus monkey. *Brain Research, 974*(1–2), 43–59.

Simonyan, K., & Jurgens, U. (2005a). Afferent cortical connections of the motor cortical larynx area in the rhesus monkey. *Neuroscience, 130*(1), 133–149.

Simonyan, K., & Jurgens, U. (2005b). Afferent subcortical connections into the motor cortical larynx area in the rhesus monkey. *Neuroscience, 130*(1), 119–131.

Simonyan, K., Saad, Z. S., Loucks, T. M., Poletto, C. J., & Ludlow, C. L. (2007). Functional neuroanatomy of human voluntary cough and sniff production. *NeuroImage, 37*(2), 401–409.

Spocter, M. A., Hopkins, W. D., Garrison, A. R., Bauernfeind, A. L., Stimpson, C. D., Hof, P. R., & Sherwood, C. C. (2010). Wernicke's area homologue in chimpanzees (Pan troglodytes) and its relation to the appearance of modern human language. *Proceedings of the Royal Society of London B: Biological Sciences, 277*(1691), 2165–2174.

Striedter, G. (2005). *Principles of brain evolution.* Sunderland, MA: Sinauer Associates.

Takai, O., Brown, S., & Liotti, M. (2010). Representation of the speech effectors in the human motor cortex: Somatotopy or overlap? *Brain and Language, 113*(1), 39–44.

Vargha-Khadem, F., Watkins, K., Alcock, K., Fletcher, P., & Passingham, R. (1995). Praxic and nonverbal cognitive deficits in a large family with a genetically transmitted speech and language disorder. *Proceedings of the National Academy of Sciences, 92*(3), 930–933.

von Cramon, D., & Jurgens, U. (1983). The anterior cingulate cortex and the phonatory control in monkey and man. *Neuroscience and Biobehavioral Reviews, 7*(3), 423–425.

Watkins, K. E., Gadian, D. G., & Vargha-Khadem, F. (1999). Functional and structural brain abnormalities associated with a genetic disorder of speech and language. *American Journal of Human Genetics, 65*(5), 1215–1221.

37 Insights into the Neurobiology of Language from Individuals Born Blind or Deaf

MARINA BEDNY AND MAIRÉAD MACSWEENEY

In this chapter we discuss insights into theories of the neurobiology of language from work with individuals who are born blind or deaf. Why should language acquired under atypical circumstances be of interest to those striving toward a general theory of language? An answer to this question was articulated by Lila Gleitman and her colleagues twenty years ago in the context of behavioral research on language acquisition (e.g., Landau & Gleitman, 1985; Gleitman & Newport, 1995). Children acquiring language with different access to the environment offer unique insights into the contribution of nature and nurture to the language faculty. Landau and Gleitman (1985) showed that children born blind acquire the meanings of words such as *red* and *look* at around the same time as sighted children do. Deaf children growing up without access to either a signed or spoken language develop language-like communication systems called homesign (Goldin-Meadow, chapter 16 of this volume). Language acquisition turns out to be surprisingly robust to dramatic changes in the environment.

An analogous argument applies to the neurobiology of language. The anatomical substrates of language are remarkably consistent across languages and cultures. Left-lateralized perisylvian cortices are a core neurobiological mechanism across users of English, German, Russian, and Mandarin Chinese (Chee et al., 1999; Rüschemeyer, Fiebach, Kempe, & Friederici, 2005). What predisposes language toward these cortical systems? Has evolution uniquely programmed the left perisylvian cortices for language processing? Are there any circumstances under which regions outside the classic network take on language functions? Alternatively, since many aspects of the acquisition experience are shared across cultures, do these commonalities give rise to a similar neural phenotype? For example, most people learn language through audition. We might therefore suppose that lateral temporal areas become involved in language because of their proximity to auditory cortices. Research with those born deaf allows us to test this hypothesis. Conversely, studies with individuals who are born blind have the potential to uncover how and whether absence of vision changes the neurobiological instantiation of concrete concepts. As it turns out, studies of blindness also reveal a surprising capacity of language to colonize unused cortical territory in the visual cortex. Research with individuals who grow up with distinctive experiences thus provides unique insights into the contributions of nature and nurture to the neurobiology of language.

1. The Cognitive and Neural Basis of Language in Blind Individuals

1.1. Representation of Word Meanings in Blindness: Insights into the Relationship of Meaning and Sensory Experience When philosophers and psychologists consider the effects of blindness on language, they often focused on the meanings of concrete words, supposing such meanings to be very different in people who are blind from birth (Berkeley, 1709; Hume, 1748; Locke, 1690). Vision is a key source of information about objects, actions, events, properties, and agents. Sighted people use information from the eyes to distinguish lions from hippos and rolling from spinning. Prominent philosophical, cognitive, and neural theories of concepts posit that understanding concrete words involves reactivating recollections of modality-specific experiences with referents. For example, the meaning of the word *run* includes visual memories of seeing others run, motor memories of running, and the sound of running footsteps. At the neurobiological level, these different modality-specific aspects of meaning are said to be stored in distinct neural systems (e.g., visual image of running stored in visual motion areas, e.g., middle temporal complex (MT+), and the motor program for running in motor cortex) (Barsalou, Simmons, Barbey, & Wilson, 2003; Kable, Kan, Wilson,

Thompson-Schill, & Chatterjee, 2005; Martin, 2007; Martin, Haxby, Lalonde, Wiggs, & Ungerleider, 1995; Patterson, Nestor, & Rogers, 2007; Pulvermüller, 2001; Warrington & Shallice, 1984). Different versions of such embodied theories have been proposed (see Meteyard, Cuadrado, Bahrami, & Vigliocco, 2012, for a review). Many current versions of such theories posit additional binding mechanisms or domain-general convergence zones that retrieve and weigh the modality-specific representations (Patterson, Nestor, & Rogers, 2007; Damasio, 1989). Some further claim that words and concepts depend not on primary sensory cortices but rather on secondary sensory areas or even sensory-derived representations in regions that are immediately adjacent to sensory cortices. Nevertheless, what makes all such theories "embodied" is the shared claim that modality-specific aspects of experience are causally relevant to the cognitive and neural instantiation of the meanings of words.

Such theories predict that changes in sensory experience should translate to changes in language comprehension. Some evidence consistent with this idea comes from studies of expertise. For example, hockey players activate different neural systems when listening to descriptions of hockey actions than nonhockey players do (Beilock, Lyons, Mattarella-Micke, Nusbaum, & Small, 2008), as do left- as opposed to right-handed individuals (Willems, Hagoort, & Casasanto, 2010). One interpretation of these results is that action experiences change the representation of action-verb meanings. Theories that predict changes to hockey players' concept of actions due to sensory-motor experience also predict that in blindness representations of concrete language should be substantially different. We would expect changes to the meanings for visual words such as *blue* and *sparkle*, as well as names of concrete objects (e.g., *lion*) and concrete events (e.g., *rolling*).

Contrary to this straightforward prediction, studies of linguistic behavior in blind children and adults reveal that blind individuals have rich "visual" concepts. As noted, Landau and Gleitman (1985) showed that the acquisition of color adjectives and verbs of perception (look and see) is preserved in blind children. For example, a blind five-year-old understood that a car could be yellow but an idea could not. Blind adults also know something about which colors are more similar to each other (e.g., orange to red more than to blue) (Connolly, Gleitman, & Thompson-Schill, 2007; Marmor, 1978; Shepard & Cooper, 1992). We recently found that blind adults' knowledge of visual perception and light emission verbs is indistinguishable from that of the sighted. Like the sighted, blind individuals distinguish between prolonged and brief acts of looking (e.g., *stare* vs. *glance*) and represent light emission verbs along dimensions of intensity (*blaze* vs. *shimmer*) and temporal stability (*blaze* vs. *flash*). Together these studies demonstrate that direct first-person sensory access is not necessary for the acquisition of concrete word meanings. Either first-person sensory experience is not relevant or at the very least equivalent information is redundantly present in nonsensory sources (e.g., language).

These behavioral data nevertheless leave open the possibility that representations of concrete meanings are qualitatively different in format across blind and sighted individuals, albeit behaviorally equivalent. For example, the meaning of *red* might be modality-specific and visual in the sighted, while abstract, linguistic, or represented in some other modality for people who are blind. Studies of the neural basis of word meanings can provide insight into this problem.

A key signature of semantic retrieval is that different categories of concepts depend on partially nonoverlapping neural mechanisms (Hart, Berndt, & Caramazza, 1985; Warrington & Shallice, 1984). The cognitive and neural interpretation of these category-specific differences have long been debated. According to the embodied theories outlined above, such neural differences reflect the degree to which different types of sensory experiences (e.g., visual, auditory, or motor) contribute to different categories of words. For example, retrieving the meanings of motion verbs depends disproportionately on regions within the left posterior temporal lobe that are partially overlapping with or near visual motion perception areas (MT+ and RSTS, right superior temporal sulucus). Such activation is said to reflect retrieval of visual motion features associated with verbs such as *roll* (Kable et al., 2005; Kemmerer, Castillo, Talavage, Patterson, & Wiley, 2008; Martin et al., 1995). Contrary to this interpretation, congenitally blind individuals activate the same posterior temporal regions when making semantic judgments about motion verbs (Bedny, Caramazza, Pascual-Leone, & Saxe, 2012; Noppeney, 2003). Crucially, it is not merely that blind individuals activate the same cortical area, rather the response profile of this region is identical across blind and sighted speakers across multiple different categories of words (Bedny, Caramazza, Pascual-Leone, & Saxe, 2012). As it turns out both in the sighted and blind participants, this region responds more to all verbs than to all nouns, including verbs like *believe* that have no apparent motion associations (Bedny, Caramazza, Grossman, Pascual-Leone, & Saxe, 2008).

Analogous evidence comes from studies of object nouns. Different categories of objects activate distinct parts of the ventral temporal cortices, and it has been suggested that these neural differences reflect the retrieval of modality-specific visual representations of object color and shape (Martin, 2007; Martin et al., 1995). Yet ventral temporal category-specific responses to objects names are also preserved in blindness. Large objects (e.g., *castle*) activate a region of medial ventral temporal cortex that overlaps with the parahippocampal place area (Bi, Wang, & Caramazza, 2016; He et al., 2013). Names of tools activate a medial region in the ventral stream and a region of inferior parietal cortex (Mahon, Anzellotti, Schwarzbach, Zampini, & Caramazza, 2009; Mahon, Schwarzbach, & Caramazza, 2010) and spatial propositions activate the supramarginal gyrus (Struiksma, Noordzij, Neggers, Bosker, & Postma, 2011). Further studies are needed to characterize the details of the information represented within these brain regions in blind individuals. Nevertheless, the few studies that have attempted to do so find similar content across blind and sighted speakers. Peelen, He, Han, Caramazza, and Bi (2014) used multivoxel pattern analysis with concrete object nouns to show similar spatial patterns of activation among blind and sighted individuals.

In sum, all of the behavioral and neural evidence from blindness suggests that dramatic changes to sensory experience leave word meanings unaltered. This does not imply the phenomenological experiences, memories, and knowledge of sighted and blind people do not differ. On the contrary, there is ample evidence that cortical systems that typically store and process sensory visual representations do change quite dramatically in blindness. What is remarkable is that these changes do not carry forward into conceptual systems. Evidence from blindness thus challenges the idea that sensory motor representations play a major role in semantics. Word meaning representations appear to develop independent of modality-specific experiences.

It remains possible that future studies will reveal some differences not only in blind individual's sensory perception but also in semantic knowledge. Indeed such differences seem quite possible when it comes to knowledge of appearance information (e.g., the color differences between *deer* and *kangaroo*). Such differences are of interest and could reveal the unique contribution of vision to information gathering. The available evidence suggests, however, that such differences, if they do exist, are a small change to the overall semantic knowledge about concrete categories. They do not lead to any fundamental change in the neural or cognitive architecture of word meanings or concepts. As such, they are no different from the typical variation in expertise across domains in the population. We therefore make two claims about semantics. First, sensory experience per se plays a rather peripheral role. Abstract aspects of experience are more likely to yield semantic differences, that is, the hockey-related concepts of a hockey player are different not merely because the player has different muscular and sensory motor memory, but because they have much richer knowledge about hockey. Second, the meanings of words are not faithful reflections of our own first-person histories but rather shared cultural constructs. It is this property of words that enables us to communicate with each other, whether we are sighted, blind, musicians, or athletes. This property of words also makes their gross neural signatures robust to idiosyncratic variation in individual experience.

1.2. Frontotemporal Language System in Blindness During language acquisition, blind children have reduced access to extralinguistic information such as object referents and visual social cues (e.g., eye-gaze). There is some evidence that this causes subtle and temporary delays in the earliest milestones of language acquisition: blind children are slightly late to produce first words and first multiword utterances (Bigelow, 1987; Landau & Gleitman, 1985; Urwin, 1987). There is also some recent evidence of transient delays in the functional maturation of cortical systems that support language processing. We found that in blind children, the degree of inferior frontal selectivity for language is reduced relative to that for sighted children. Interestingly, no such delay is observed in lateral temporal cortices (Bedny, Richardson, & Saxe, 2015). This finding is consistent with reports that development of prefrontal language regions is sensitive to early deafness (MacSweeney, Waters, Brammer, Woll, & Goswami, 2008) and variability in socioeconomic status (Raizada, Richards, Meltzoff, & Kuhl, 2008).

Importantly, both the neural and behavioral delays described resolve. Older blind children and adults have similar linguistic competence as sighted individuals do (Lane, Kanjlia, Omaki, & Bedny, 2015). We recently found that blind adults actually perform better on comprehension of grammatically complex sentences (Loiotile, Omaki, & Bedny 2019) and multiple studies show superior verbal working memory performance in blindness (Amedi, Raz, Pianka, Malach, & Zohary, 2003; Raz, Striem, Pundak, Orlov, & Zohary, 2007). Paralleling preserved linguistic competence, the functional signatures of the frontotemporal language system are similar in blind and sighted adults. Prefrontal language selectivity increases in blind children between ages 4 and 17, and frontotemporal neural specialization for language is similar in blind and sighted adults (Bedny et al., 2015). In blind individuals, regions of the inferior frontal and lateral temporal cortices are active during sentence processing, word comprehension, reading, and production (Bedny, Pascual-Leone, Dodell-Feder, Fedorenko, & Saxe, 2011; Burton, 2002; Röder, Stock, Bien, Neville, & Rösler, 2002; Watkins et al., 2012). Activity in these frontotemporal areas is sensitive to phonological, lexical, and syntactic information, in the same way as that of sighted adults (e.g., sensitive to the grammatical complexity of sentences) (Bedny et al., 2011; Lane, et al., 2015; Röder et al., 2002). Specialization within the perisylvian language network is also preserved. Like sighted individuals, blind individuals show larger responses to verbs than nouns in the posterior lateral temporal and inferior frontal cortices (Bedny, Caramazza, Pascual-Leone, & Saxe, 2012).

One exception to this general pattern of preservation within the frontotemporal language network is lateralization. During auditory sentence processing, congenitally blind individuals are more likely to have right-lateralized or bilateral frontotemporal responses (Lane et al., 2015, Röder et al., 2002; Röder, Rösler, & Neville, 2000). Importantly, reduced left-lateralization in blindness does not appear to be associated with reduced language performance. An intriguing possibility is that reduced laterality is a neural vestige of early delays in language acquisition (see Bishop, 2013). It has been suggested that the maturational phase of cortex at the time of language acquisition affects lateralization (Bates et al., 1997; Bishop, 2013; Corballis & Morgan, 2010; Locke, 1997). If so, then delayed acquisition early in life could reduce the likelihood of left-lateralization in adulthood.

1.3. Visual Cortex Recruited for Language Processing in Blindness: Insights into Function Structure Mapping The biggest change to the neurobiology of language in blindness is not within the perisylvian language network but outside it. In blindness, parts of the so-called visual cortices appear to be incorporated into the language network. Visual cortices are active when blind individuals listen to spoken sentences, generate verbs from heard nouns, retrieve lists of words from long-term memory, and read Braille (Amedi et al., 2003; Bedny et al., 2011; Burton, 2002; Burton et al., 2002; Röder et al., 2002; Watkins et al., 2012). Language responses are observed in lateral and ventral occipital cortices, as well as V1 (Bedny et al., 2011). Visual cortex responses to spoken language and Braille are part of a broader phenomenon known as cross-modal plasticity, where regions of the brain that normally respond primarily to visual input come to respond to auditory and tactile stimuli (Pascual-Leone, Amedi, Fregni, & Merabet, 2005). However, visual cortex responses to language go beyond sensory responses to sound or touch. Visual cortices respond more to auditory and tactile language tasks than to nonlinguistic auditory and tactile control tasks that are matched on low-level sensory and working memory demands. Responses to language are anatomically separable from responses to nonlinguistic auditory tasks within visual cortices (Kanjlia, Lane, Feigenson, & Bedny, 2016). Most importantly, visual cortices are sensitive to the grammatical and lexical properties of linguistic stimuli. Visual cortex activity scales with amount of sentence-level structure and lexical information. As in frontotemporal language areas, highest responses are observed for sentence stimuli, followed by Jabberwocky and lists of words, and lowest to lists of nonwords (Bedny et al., 2011). Among sentences, larger responses are observed for those that are grammatically complex (Lane et al., 2015; Röder et al., 2002). Visual cortices are also sensitive to subtle semantic manipulations. For example, the N400 component has a more posterior distribution in blind as compared to sighted individuals (Röder et al., 2000). The functional profile of language-responsive visual areas is thus similar in several important respects to that of frontotemporal language regions (see figure 37.1).

There is also some evidence that in blindness visual cortex activity is functionally relevant to some linguistic tasks. Transiently disrupting visual cortex activity with transcranial magnetic stimulation causes semantic verb generation errors in blind but not sighted participants (Amedi, Floel, Knecht, Zohary, & Cohen, 2004). Transcranial magnetic stimulation to the occipital pole impairs reading Braille words and letters and abolishes repetition priming (Cohen et al., 1997; Kupers et al., 2007). Across blind individuals, a greater amount of visual cortex activity correlates with superior linguistic performance (Amedi et al., 2003; Lane et al., 2015).

In blindness, language-responsive visual cortices also become functionally connected with frontotemporal language networks. In resting state data, correlations between frontoparietal networks are higher in blind than in sighted individuals (Bedny et al., 2011; Deen, Saxe, & Bedny, 2015; Watkins et al., 2012). Those areas of visual cortex that are active during language tasks are also more correlated with prefrontal language areas. By contrast, areas of visual cortex that are involved in nonlinguistic processing correlate with different regions of prefrontal and parietal cortices (Kanjlia et al., 2016). Finally, in blind individuals, the laterality of occipital language responses follows the laterality of frontotemporal cortices. Blind individuals that have right-lateralized responses to spoken language in frontotemporal cortices also show larger responses on the

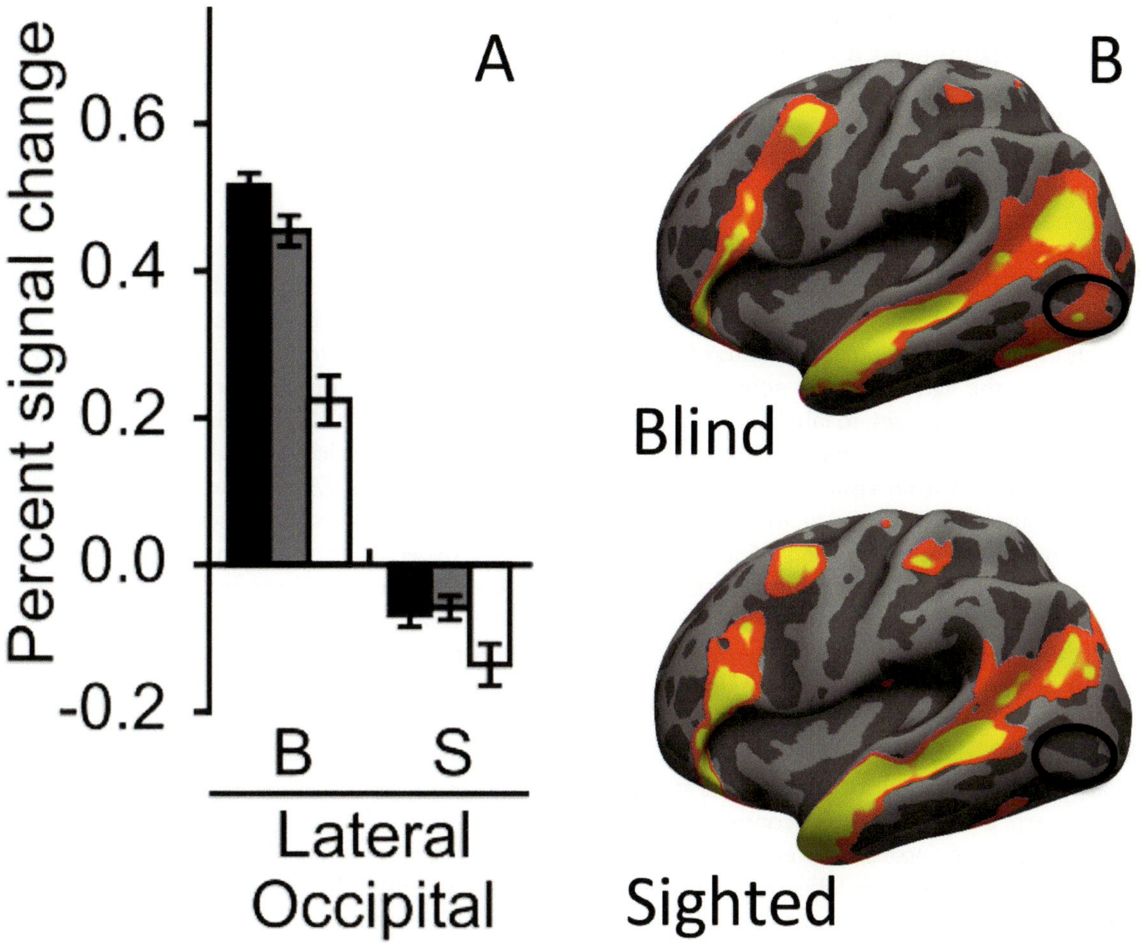

FIGURE 37.1 (A) Visual cortex responses to spoken sentences with syntactic movement (black) and without syntactic movement (light gray) as well as matched lists of nonwords (white) in sighted (S) and blind (B) groups. (B) Whole-brain responses to sentences (average of +Move and −Move) greater than nonword lists in blind and sighted.

right in occipital cortices (Lane et al., 2015). Together, the available evidence suggests that parts of visual cortices become functionally incorporated into the language network in blindness.

How and when does language colonize the visual cortices of blind individuals? We hypothesize that in the absence of bottom-up visual input, signals from higher cognitive networks in frontoparietal and temporal cortices dominate activity in the visual system during development (Bedny et al., 2011; Deen et al., 2015; Watkins et al., 2012). In sighted individuals, these systems typically modulate activity in the visual cortices based on goals, attention, and conceptual content. In blindness, they incorporate the visual system into higher cognitive networks. There is evidence that language-related plasticity in the occipital cortex occurs early in life and follows a sensitive period. In blind children, occipital responses to spoken language are observed by four years of age and do not require Braille literacy (Bedny et al., 2015). Selective responses to language in the visual system do not appear to be present in adult-onset blind individuals, even after many years of blindness (Bedny, Pascual-Leone, Dravida, & Saxe, 2012; Burton, 2002).

The striking neural reorganization of the language system in blindness has implications for theories of language neurobiology. Frontotemporal networks are involved in language processing across variation in experience, including blindness and deafness, and therefore may well be predisposed for language processing. However, evidence from blindness also suggests that development of the language system is not well described as the gradual maturation of a preordained set of frontotemporal areas. Despite its usual consistency, the neural phenotype of language emerges as a result of a dynamic process. When other brain regions are available at the right time (i.e., not already doing something else), they can be incorporated into the language system. Studies of blindness further suggest that regions that did not evolve for language can nevertheless participate in language processing. The neural phenotype of language is influenced by the state of other neurocognitive systems during development (e.g. the visual system).

1.4. Neural Basis of Braille-Reading in Blindness: Insights into the Neurobiology of Reading The blind community uses a tactile reading system called Braille. Like visual reading systems, Braille translates spoken language into permanent written record. As such it enables blind readers to do all of the things that print makes possible for the sighted: to take notes, to scan text and reread important passages, to label and organize, to write down and pore over mathematical equations. Yet Braille is also different from visual print in interesting respects. Most obviously Braille is tactile, while print is visual. Each Braille cell is composed of a two-row-by-three-column grid of raised dots that are perceived by passing the palm of the finger over the dots. Each English Braille letter corresponds to a letter of the print English alphabet. For example, the Braille character for the letter K is <⠅> and for the letter T is <⠞>. Proficient Braille readers of English use a contracted form of Braille in which single characters and character combinations stand for larger units, such as frequent letter combinations and words. For example, the single character K <⠅> also stands for the whole word "knowledge." There are also single character Braille contractions for commonly used bound morphemes such as "ing" and "ed." The most commonly used form of English Braille text is a combination of contractions and letter-by-letter spelling.

There are some differences in the way Braille and print are recognized. A key challenge in the recognition of visual letters and words is the invariance problem. Sighted readers must recognize different instances of A a *a* as the same, while remaining sensitive to the subtle shape difference that distinguishes *e* from *c*. By contrast, Braille has no font variation. Braille dots are of a standardized size and spacing and even capital letters are indicated by a preceding dot and not by changing the character itself. Proficient Braille readers are believed to recognize letters through shear patterns on the finger (Millar, 2008). Like print readers, proficient readers of Braille represent orthographic sublexical structure beyond single letters (Fischer-Baum & Englebretson, 2016). However, while expert sighted print readers read words in parallel and do not show word-length effects on reading speed (Adelman, Marquis, & Sabatos-DeVito, 2010), even for highly skilled Braille readers, reading speed per word scales with letter length because the finger must pass in sequence over each character (Bertelson, Mousty, & Radeau, 1992; Daneman, 1988; Legge, Madison, & Mansfield, 1999). Thus, reading by touch has some consequences for the cognitive properties of orthography. Does it also influence the neurobiology of reading?

Sighted print readers recruit a consistent network of areas during reading. A key node this network is the so-called *visual word form area* (VWFA), located in the ventral occipitotemporal cortex. The VWFA is situated among object-selective regions such as the fusiform face area and the parahippocampal place area (Epstein & Kanwisher, 1998; Kanwisher, McDermott, & Chun, 1997). According to one prominent hypothesis, the VWFA supports the recognition of letter and word forms (Dehaene & Cohen, 2007, 2011; Tsapkini & Rapp, 2010; Vinckier et al., 2007; see Price & Devlin, 2011, for an alternative account). The VWFA is of particular theoretical importance within cognitive neuroscience because it is the only category-specific brain region that is known to acquire its function as a result of learning (Dehaene et al., 2010).

Different nonmutually exclusive hypotheses have been proposed regarding the causal mechanisms that lead this particular part of the ventral occipital temporal cortex to specialize for reading (Dehaene & Cohen, 2007, 2011; Price & Devlin, 2011). One possibility is that the VWFA develops at the nexus of visual and linguistic inputs, because reading involves extracting linguistic content from visual symbols (Dehaene & Dehaene-Lambertz, 2016; Saygin et al., 2016; Kravitz, Saleem, Baker, Ungerleider, & Mishkin, 2013). Alternatively, it has been suggested that the VWFA has an intrinsic preference for representing the types of forms found in written alphabets (e.g., line junctions) (Szwed, Cohen, Qiao, & Dehaene, 2009). At face value, all of these hypotheses predict that the VWFA should not become reading-selective in blindness. Braille has no line junction and does not require connecting vision and language.

Surprisingly, there is evidence that congenitally blind readers of Braille activate the VWFA during reading (Burton et al., 2002; Reich, Szwed, Cohen, & Amedi, 2011; Sadato et al., 1996). The occipitotemporal cortex responds more when blind individuals read Braille words than when they pass their fingers over strings of raised dots without linguistic content and the response to Braille words is larger than to spoken words (Reich et al., 2011). A VWFA-like response is also observed when blind subjects listen to letters through a sensory substitution device (Striem-Amit, Cohen, Dehaene, & Amedi, 2012). Remarkably, the peak response to Braille words within the occipitotemporal cortex is located in the same location as the peak response to printed words in sighted readers (Reich et al., 2011). Based on this finding, it has been suggested that like sighted print readers, blind Braille readers develop a VWFA reading-selective region in the ventral stream. In an effort to account for the

Braille findings, it was recently suggested that the VWFA is suited for processing shape, independent of modality (Hannagan, Amedi, Cohen, Dehaene-Lambertz, & Dehaene, 2015). This, along with connectivity to language networks, is said to account for both print and Braille reading-specialization in this part of the brain (Hannagan et al., 2015). It remains unclear, however, how well such an account squares with either the nature of Braille reading, which may depend on shear patterns more so than shape recognition, or the function of the occipitotemporal cortex, which is involved in view-invariant object recognition, rather than shape extraction per se (DiCarlo, Zoccolan, & Rust, 2012).

We recently tested an alternative prediction: Since the VWFA has strong anatomical connectivity to the frontotemporal language network prior to literacy, we predicted that in the absence of visual input, this region would become incorporated into the frontotemporal language network. Consistent with this possibility, we found that in individual blind Braille readers, the same occipitotemporal area that responds to Braille also responds to the grammatical complexity of spoken sentences (Kim, Kanjlia, Merabet, Bedny, 2017). By contrast, in sighted print readers, reading-responsive VWFA was insensitive to grammatical complexity. This observation is consistent with evidence that large parts of the visual system become incorporated into language networks in blindness. Indeed, the VWFA may serve as a gateway region for the language to enter the visual system. These data suggest that in the sighted, the VWFA develops where it does, in part because print is read through the visual modality. Thus, the neurobiology of tactile reading is at least in part different from that of reading visual print. Whether Braille depends on other, reading-specific neural mechanisms, or whether VWFA-like category specificity is unique to the visual system, remains to be determined.

The case of the VWFA in blindness is a powerful illustration that activating the same region during language tasks in two different populations does not necessary imply that the region is performing the same computations. Determining whether a region is functionally similar requires measuring its response across a range of cognitive manipulations. At the end of the chapter, we return to this issue and the flexibility of the mind to brain mapping more generally.

2. *The Cognitive and Neural Basis of Language in Individuals Who Are Deaf*

Children born severely or profoundly deaf offer unique insights into the neurobiology of language that cannot be gained from working with hearing people alone. First, we can ask whether the neurobiology of language changes when language is not auditory. Only 5% to 10% of deaf children are born to deaf parents and are therefore likely to be exposed to a fully accessible native language—a sign language. Sign languages are natural human languages that are perceived in the visual modality alone. Examining the neural systems that support sign language processing is interesting in its own right. However, sign languages are also a powerful tool with which to test constraints and plasticity of the language system. Determining commonalities between how signed and spoken languages are processed allows identification of language processes and language networks that are recruited *independent of modality.* First, in section 2.1, we will therefore review the current knowledge about the neural systems supporting language in deaf native signers and the insights that have been gained about plasticity of the language system from this group. Second, in section 2.2, we consider the impact of absence of auditory input on plasticity of sensory cortices and specifically whether auditory regions play a role in linguistic or nonlinguistic processing in those born deaf.

Finally, in section 2.3, we discuss the impact of impoverished early access to language. As many as 90% to 95% of deaf children are born to hearing parents. These children do not have early access to sign language. Furthermore, as is the case for all deaf children, they have impoverished access to the auditory component of *spoken language.* Unlike the subtle consequences of impoverished access to visual aspects of speech for blind children, the impact of deafness on spoken language processing can be severe. There have been tremendous improvements in recent years in digital hearing aids and in cochlear implants. However, there are large individual differences in the extent to which deaf children's spoken language benefits from these forms of amplification. There is a broad continuum from those who learn the spoken language around them well to those who do not. Hearing parents may or may not decide to learn a sign language. Furthermore, the age at which a deaf child of hearing parents encounters a signed language in an educational or social environment can vary greatly. Thus deaf children of hearing parents have extremely heterogeneous early language experiences. Here, we consider sign language processing in these individuals who have learned a sign language late, building on impoverished early access to a spoken language. These studies provide insights into the consequences of delayed access to language for the development of the language system in the brain. Research with deaf individuals with different language histories provides unique insights into the environmental constraints that influence the development of the language system.

2.1. Using Sign Languages as a Tool to Identify Modality-Independent Language Networks: Studies with Native Signers One of the goals within the field of the neurobiology of language is to identify the neural systems supporting higher order language processing *regardless of modality*. Studies with deaf signers provide a unique opportunity to answer this question because it is only in this population that we can rule out activation of auditorily derived representations (see MacSweeney, Capek, Campbell, & Woll, 2008, for review). Sign languages arise wherever Deaf[1] communities come together, and they differ across countries (e.g., American Sign Language [ASL], British Sign Language [BSL]). Children born to signing parents grow up to be native signers. Studies have clearly shown that deaf children learning a sign language from deaf parents show the same developmental milestones in their language acquisition as hearing children learning a spoken language do (Morgan & Woll, 2002). By identifying regions of overlapping activation during tasks performed in signed and spoken languages, we can then propose the testable hypothesis that the linguistic computations and processes being carried out in these regions are truly modality-independent.

As in spoken language users, lesion studies in signers overwhelmingly indicate that left hemisphere damage leads to severely impaired language processing (aphasia) while right hemisphere damage does not (e.g., Klima & Bellugi, 1979; Marshall, Atkinson, Woll, & Thacker, 2005). Neuroimaging studies also indicate a critical role for the left hemisphere in sign language processing. Despite differences in the articulators used, both covert and overt sign production rely predominantly on the left hemisphere (Gutierrez-Sigut et al., 2015; Gutierrez-Sigut, Payne, & MacSweeney, 2016) and specifically on the left inferior frontal gyrus, as does speech production. Critically, whether the right or left hand is used to produce signs has little impact on the recruitment of this left-lateralized network (Corina, San Jose-Robertson, Guillemin, High, & Braun, 2003). Furthermore, although sign languages are well suited to represent the visual features of referents (i.e., to be iconic) this appears to have little direct impact on the neural systems supporting sign language production (Emmorey et al., 2004). Finally, hearing people who have grown up as bilinguals, having learned a signed language from their deaf parents and a spoken language from the broader hearing community, engage the same left frontoparietal network during both sign and speech production (Emmorey, McCullough, Mehta, & Grabowski, 2014).

With regard to language perception, we used conjunction analysis across different groups, to identify the regions involved in processing audiovisual speech in hearing people and BSL in deaf native signers (MacSweeney, Woll, Campbell, McGuire et al., 2002). We identified a primarily left frontotemporal network involving the superior temporal gyrus and sulcus as well as the left inferior frontal gyrus, extending into the prefrontal gyrus, that was involved in processing both sign and speech (see also Sakai, Tatsuno, Suzuki, Kimura, & Ichida, 2005). Numerous studies of sign language perception alone have also identified a primarily left-lateralized frontotemporal network involved in sign language perception when contrasted with non-sense movement (MacSweeney et al., 2004), gesture (Newman, Supalla, Fernandez, Newport, & Bavelier, 2015), or transitive actions (Corina et al., 2007). Together these studies suggest that the classic left-lateralized perisylvian network is resilient to change in the sensory modality of language.

In summary, in native users, signed and spoken languages engage very similar left-perisylvian networks for perception and production of sign and speech. This similarity appears to extend to metalinguistic judgments regarding the phonological structure of sign and speech, which have been shown to engage a left frontoparietal network (MacSweeney, Waters, et al., 2008).

The fact that signed languages, like spoken languages, are predominantly processed in the left hemisphere is an important finding that should be taken into account in theories of hemispheric lateralization of language processing. It has been argued by some that the left hemisphere shows a predisposition to process certain temporal aspects of auditory information that are critical to speech processing (see McGettigan & Scott, 2012; Scott & McGettigan, 2013, for reviews and comment). The inference is then made, explicitly or implicitly, that this is the cause of left hemisphere lateralization for language processing. That sign languages are also predominantly processed in the left hemisphere poses a problem for any purely auditory-based account. Although the finding with sign languages does not answer *why* language is predominantly left-lateralized in most people, it should prompt the field to generate hypotheses that are modality-independent and can account for left hemisphere lateralization of both sign and speech.

2.2. Beyond Commonalities: Modality-/Language-Dependent Differences between Signed and Spoken Language Processing Networks Although the overlap between the networks supporting sign and speech processing is extensive, there are some differences. Not surprisingly, direct contrasts have highlighted differences reflecting early sensory processing. Audiovisual speech elicits greater activation than sign language in auditory processing regions in superior temporal cortices. In contrast, sign languages elicit greater activation than audiovisual speech does, in biological motion processing regions of the posterior middle temporal gyri, bilaterally (Emmorey, McCullough, & Weisberg, 2014; MacSweeney, Woll, Campbell, McGuire, et al., 2002; Söderfeldt et al., 1997). Above and beyond sensory demands of visual motion processing, the posterior middle temporal gyri also appear to be recruited when visual movement is specifically linguistic, such as in the perception of classifiers representing movement of a referent (MacSweeney, Woll, Campbell, Calvert, et al., 2002; McCullough, Saygin, Korpics, & Emmorey, 2012). This finding is in line with the proposal from the blindness literature that regions primarily involved in visual processing can also be engaged in linguistic processing.

There is also growing evidence suggesting that the left inferior and superior parietal lobules appear to play a greater role in sign language processing than spoken language processing (see MacSweeney, Capek, et al., 2008, for review). Greater activation has been reported in the inferior and superior parietal lobules during sign language compared to spoken language production (Emmorey, Mehta, & Grabowski, 2007) and during short-term memory tasks for sign compared to speech (e.g., Bavelier, 2008). It has been suggested that the left superior parietal lobule may be involved in motor rehearsal during memory tasks (Buchsbaum et al., 2005) and/or in proprioceptive monitoring during sign production (Emmorey, McCullough, & Weisberg, 2016). Online monitoring of language production is, by necessity, different for sign and speech. The feedback received by a hearing person during their own speech production is almost equivalent to that heard by their interlocutor. In contrast, a signer's perception of their own hands while signing is very different to that perceived by their communication partner. There are likely to be important implications of this difference between languages in the online feedback loop for the neural systems supporting language production. The current state of knowledge suggests that the superior parietal lobe plays an important role in this monitoring for sign production (see Emmorey et al., 2016).

In contrast to the superior parietal lobule, the left supramarginal gyrus in the inferior parietal lobule appears to play a particularly important role in phonological processing of sign language. Corina et al. (1999) found that direct stimulation to the left supramarginal gyrus elicited errors in selecting appropriate handshapes and movements during sign production. Furthermore, phonological similarity judgments for signs elicit greater activation in the superior portion of the left supramarginal gyrus, extending into the superior parietal lobule, than phonological similarity judgments for speech (MacSweeney, Waters, et al., 2008). The inferior parietal lobule is reliably recruited during the imitation and production of hand configurations or movements and also when such stimuli are passively viewed or imagined (Rizzolatti & Craighero, 2004). The need to attend to the location and configuration of the hands in space in any task involving sign language processing could explain the enhanced involvement of this region. The focus on these features during a sign language phonological task is likely to increase these processing demands further.

A critical question regarding the influence of sensory experience on language processing in deafness that relates to the blindness literature is whether deaf individuals additionally recruit auditory cortices during sign language processing, as blind individuals recruit visual cortices for spoken and written language. Unlike in the case of blindness, the evidence is mixed regarding whether sign language, or any other visual stimuli, activates the primary auditory cortices (see Cardin et al., 2016; Scott, Karns, Dow, Stevens, & Neville, 2014). However, there are now numerous reports of increased activation in secondary auditory and auditory association cortices in the superior temporal cortex (STC) in deaf individuals that is greater than that in hearing individuals during sign language perception, even when sign language experience is matched across deaf and hearing groups (Capek et al., 2010; MacSweeney et al., 2004). In addition to the influence of deafness, however, there is a further influence of sign language experience. For example, the left STC has been shown to be more active in deaf signers than in *deaf nonsigners* during sign language perception (Cardin et al., 2013). It has also been shown that responses in left STC in deaf signers are larger for sign language stimuli than for other nonlinguistic visual stimuli (e.g., MacSweeney et al., 2004).

Unlike in the case of blindness, it is not clear whether responses to sign language in auditory cortices of deaf signers are language-specific. Studies in deaf cats suggest that portions of the STC, that are typically auditory cortex in hearing cats, are recruited for visuospatial tasks. For example, Lomber, Meredith, & Kral (2010) demonstrated that regions involved in auditory localization in hearing cats are recruited during visual localization in deaf cats. It is therefore possible that larger auditory responses to sign than nonsign stimuli in signers are partially due to the visual complexity of the signed input and subsequent increased visual processing demands. There is also evidence that portions of the posterior STC, that typically participate in higher order auditory processing, are recruited by deaf individuals in visual tasks that do not involve language, such as a visual rhythm discrimination task (Bola et al., 2017) and a visuospatial working memory (Ding et al., 2015). Ding et al. (2015) found that auditory cortex activity is active not only when complex visual stimuli are displayed, but also during the maintenance phase of a visual working memory task, during which only a static crosshair was visible on the screen. Furthermore, the effects were load-dependent, with larger responses when more information was maintained in working memory during the delay. Ding et al. (2015) also reported correlations between amplitude of response in STG and task performance in deaf but not hearing participants. They argue therefore that auditory association cortices play an important role in visuospatial working memory in those born deaf (see MacSweeney & Cardin, 2015, for commentary). Furthermore, in a recent study, we find that these STC regions are particularly involved in sign language tasks in which the visual working memory and visual imagery demands are high—a picture-based BSL phonological judgment task (Twomey, Waters, Price, Evans, & MacSweeney, 2017). We hypothesize that plasticity within the left STC is most likely an interaction between deafness and sign language knowledge. As discussed in section 2.3, this interaction is likely to be further influenced by age of sign language acquisition and sign language proficiency (MacSweeney, Waters, et al., 2008; Mayberry, Chen, Witcher, & Klein, 2011).

Future studies are needed to determine whether the linguistic and nonlinguistic cognitive processes recruit distinct portions of auditory association cortices in deaf individuals, as they appear to do in the visual cortices of blind adults. Or instead whether activity during sign language tasks is related to nonlinguistic cognitive functions. We would expect that some general principles of plasticity in sensory loss apply across deafness and blindness. For example, as reviewed, in both deaf and blind humans, there is evidence that sensory cortices are recruited for higher cognitive functions (e.g., language and working memory in deafness). However, the specific patterns of functional reorganization in deafness and blindness will depend in part on the connectivity patterns of auditory and visual cortices, respectively. Furthermore, other aspects of experience that differ among deaf as opposed to blind individuals may lead to differences in cortical repurposing. Language-related plasticity in sensory cortices could be less likely in cases of reduced language experience early in life (e.g., deaf individuals who lack access to spoken and signed language). Uncovering the similarities and differences between patterns of functional reorganization in deafness and blindness is an important future avenue for understanding plasticity mechanisms.

2.3. Consequences of Impoverished Access to Early Language: Evidence from Deaf Late Learners of a Sign Language The vast majority of deaf children are not exposed to a sign language from birth. They are born to hearing parents, who may or may not decide to learn a sign language. Often these children are exposed to a sign language at school or on leaving school, at an age past the point that would normally be considered the critical period for language development (see Mayberry & Lock, 2003; Mayberry, Lock, & Kazmi, 2002). These children have extremely heterogeneous language experiences and can provide unique insights into the influence of timing on the language system.

The impact of late sign language acquisition on sign language processing has been investigated extensively at the behavioral level. Studies have reported poorer syntactic (Morford, 2003; Johnson & Newport, 1989) and phonological (Mayberry & Fischer, 1989) performance by this group (see Mayberry, 2007, for review). The impact of late sign language acquisition on the neural systems supporting language has yet to be fully explored. Newman, Bavelier, Corina, Jezzard, and Neville (2002) reported greater right angular gyrus activation in hearing native than hearing non-native signers while watching ASL. Newman et al. (2002) argued that this reflected the spatial-linguistic processing requirements of a sign language. However, hearing late learners of sign have already successfully acquired a first (spoken) language; deaf late learners of a sign language have not. When a deaf person learns a sign language later in life, it is typically built on impoverished early access to a spoken language. That is, it cannot always be considered a second language, as is clearly the case for hearing late learners of a signed language. As a consequence, it is unclear to what extent findings from hearing late learners of sign can be extrapolated to late sign language learners who are deaf.

To date only a handful of studies have examined the impact of late sign language acquisition on the neural systems supporting sign language processing in those born deaf. MacSweeney, Waters, et al. (2008) tested deaf native and non-native signers. The non-native signers had learned BSL between the ages of 4 and 21 years and importantly the deaf native and non-native signers did not differ significantly on measures of English proficiency (reading, speech reading [lipreading], and English vocabulary). Participants were asked to make phonological judgments about signs (same location?) and speech (rhyme?) in response to picture pairs. We found increased activation in the left inferior frontal gyrus in the non-native compared to in the native signers (MacSweeney, Waters, et al., 2008). Critically, this was the case not only for BSL, which was learned late, but also for English, of which both groups had similar experience and had also shown equal levels of performance on English online (rhyme task) and offline tasks. The deaf late learners of sign language, needed to call on left frontal cortices to a greater extent for both sign and speech tasks, even when their performance on the tasks was equivalent to that of the native signers. Such recruitment might reflect more effortful processing. One interpretation of these data, therefore, is that having a robust first language (here a signed language) provides a solid basis on which to learn a second language (here English). These data support behavioral data underlining the critical importance of early language experience, in any modality, for later language development (see Mayberry, 2007, for review).

Mayberry et al. (2011) also investigated the influence of age of sign language acquisition by testing participants whose age of onset of ASL acquisition ranged from birth to 14 years old. Participants were tested on phonemic and grammatical judgments in response to ASL sentences. In contrast to the findings of MacSweeney, Waters, et al. (2008), Mayberry et al. (2011) found *decreased* recruitment of left frontal regions in late compared to early signers. Late signers, in contrast, showed enhanced recruitment of occipital cortices. There were a number of stimulus (ASL video/static pictures) and task differences between the MacSweeney, Waters, et al. (2008) and Mayberry et al. (2011) studies that may have contributed to the different pattern of results. One key difference appears to be language skills of those tested. In the MacSweeney, Waters, et al. (2008) study, the deaf late signers had a relatively high level of English, as measured by reading and speech reading tests. In contrast, the deaf late learners of ASL tested by Mayberry et al. (2011, p. 18) rated their spoken language communication skills as not able to "understand spoken language sufficiently for functional communication." The discrepancy in language skills of the participants tested could explain the different patterns observed across studies. Future studies should examine how proficiency and task demands interact with age-of-language acquisition in predicting inferior frontal recruitment during sign language processing.

Whatever the cause for the difference in results between these two studies, it is clear that the left inferior frontal cortices are sensitive (in one direction or another) to the age of sign language acquisition and/or to the consequences of impoverished first language input. These findings align well with data from blind children reviewed in section 1, suggesting that the left frontal cortices are sensitive to early language experience (Bedny et al., 2011). Future studies are needed to dissociate those effects that are related to age of sign language exposure and those related to sign language proficiency.

Mayberry and colleagues (Ramirez et al., 2014; Ferjan Ramirez et al., 2016) have also had the opportunity to examine ASL processing in two deaf adolescents who moved to the United States from Central America and who are described as having no first language (spoken language) before encountering ASL at the age of 14 years (Ramirez et al., 2014). Critically, only in cases of extreme deprivation could such cases be argued to be found in the hearing population. These case studies therefore offer unique insights into the consequences of severe early language deprivation.

Using magnetoencephalography, Ferjan Ramirez et al. (2016) showed that even after three years of exposure to ASL, the teenager's responses to single signs were highly atypical, engaging right dorsal frontoparietal regions, rather than the typical left-lateralized frontotemporal network. When followed up just over a year later, these cases still showed atypical neural processing for less familiar signs. However, interestingly, for more familiar signs, they started to show activation in the typical left perisylvian network. Ferjan Ramirez et al. (2016) argued that even though timing of language experience inevitably affects the organization of neural language processing, language representation in the human brain can continue to evolve with experience, even into adolescence. Continuing to study the language development of these individuals and testing them on more complex language input will provide unique insights into the consequences of extremely impoverished early language experience on the neural bases of language processing.

3. Conclusion

Studies of blindness and deafness highlight that language is an abstract system of phonological, semantic, and syntactic representations that can be realized as speech sounds, combinations of hand shapes and movements, printed words or tactile Braille. Even in the absence of optimal sensory access, children acquire language by hook or by crook, as long as linguistic information is available to them in some format. Blind children acquire language without seeing colors, without pointing, eye-gaze, or the sight of moving lips. Children who are born deaf can acquire a visuospatial signed language, provided they have access to the language early in life. Like the behavioral phenotype of language, the neuroanatomical layout and functional profile of the perisylvian language system is robust to dramatic changes in sensory experience. This suggests that whatever predisposes perisylvian systems to take on language processing, these predispositions are related to the computational demands of language as a system of thought and communication and not to a particular sensory instantiation.

By contrast to sensory change per se, reduced or delayed access to language in childhood impacts language competence and changes language neurobiology. In blindness, delays to language acquisition are subtle and do not affect ultimate language competence, but nevertheless, may modify the time course of functional maturation in prefrontal language areas during childhood and reduce the left-lateralization of frontotemporal language systems. In deaf children of hearing parents, language access can be severely reduced. Impoverished exposure to signed and spoken language early in life can have significant consequences for later language proficiency and affects the neural systems supporting language. Reports of extreme case studies with deaf individuals who had no access to sign language, and limited spoken language exposure, during childhood suggest that severe delays in first language acquisition can lead to large-scale and long-term functional changes in frontotemporal language networks (e.g., Ferjan Ramirez et al., 2016). By contrast, early access to a sign language leads to a preserved frontotemporal system and provides robust foundations for later spoken language processing. An interesting question is whether, in the absence of access to language early in life, frontotemporal systems merely diminish in their language capabilities or whether in the absence of language experience the perisylvian cortices are colonized by nonlinguistic functions, akin to the colonization of sensory systems of deaf and blind individuals by nonsensory functions. Such colonization, if it occurs, could affect the capacity of perisylvian cortices to support language.

Conversely, an important insight from studies of blindness is that there is more than one neural solution to the computational problem of language. Blind adults perform the same or better on language tasks as compared to matched groups of sighted individuals. Nevertheless, blind individuals show reduced left-lateralization of frontotemporal language areas and recruit "visual cortices" during language processing. It is therefore an oversimplification to consider the infant brain as a miniature version of the adult language system (Bates et al., 1997; Karmiloff & Karmiloff-Smith, 2009). Rather the language-ready brain contains predispositions, which are then amplified and transformed during acquisition. As a result there is potential for nonverbal experience and the maturation of other brain systems to influence the neural basis of language.

Studies of blindness and deafness furthermore demonstrate that cortical location and cognitive computation are not inextricably linked at birth. When experience is dramatically changed, brain regions can assume drastically different functional profiles (e.g., visual cortex in blindness). This could also be true on a smaller scale across speakers of different languages. For example, computing sentence structure depends to different degrees on word order as opposed to morphology depending on the language. Perisylvian cortices participate in sentence processing across speakers of different languages but the precise information they represent is different. What are the limits on such cortical flexibility? Studies with adult-onset blind individuals suggest that once the cognitive role of a cortical area is established during development, there are limits on how much it can change. For example, even after years of blindness, the visual cortices are limited in their ability to take on linguistic functions in those who become blind later in life (e.g., Bedny, Pascual-Leone, Dravida, et al., 2012). On the other hand, smaller scale cognitive reorganization is possible in adult learning (e.g., Merzenich et al., 1984). Such flexibility may also exist within an individual across contexts. For example, it has been suggested that the same cortical region could contribute to syntactic or semantic processing depending on the other areas and networks engaged during a given language task (Hagoort, 2014). We might say then that a cortical region has a cognitive range that depends on its connectivity and microcircuitry. At birth this range is wide and discontinuous, encompassing cognitively unrelated functions. With experience and age the range narrows and becomes more cognitively homogenous, but nevertheless it does not boil down to a single cognitive operation.

Many further opportunities remain open for insights into the neurobiology of language from blindness and deafness and recent methodological advances continue to open new avenues of inquiry. When individuals with different developmental histories activate the same brain regions (e.g., the VWFA or perisylvian cortices) during similar tasks, are the same representations being processed in the same way for both groups (Amedi, Hofstetter, Maidenbaum, & Heimler, 2017)? If not, what are the differences? Conversely, when different cortical areas are activated by groups with different developmental histories, do different cortical areas nevertheless perform the same computations across groups? An important goal for future research is to use multivariate approaches to examine in greater detail the linguistic representations and computations of cortical systems involved in language processing across populations. This would provide insight into the degree to which the information represented and processed within language-responsive cortical systems is similar across blind, sighted, hearing, and deaf individuals.

There is also a need for research into the neurobiology of language in blind and deaf *children*. Only a single imaging study has looked at the neurobiology of language development in blind children, and there are no published studies with deaf children using techniques with high spatial resolution such as functional MRI. Examining the developing neural system supporting language, rather than only the end state, will provide unique insights into how language and sensory experience influences development of language networks in the human brain. These, and many other open questions, remain to be answered. Studying language in individuals with diverse developmental experiences continues to be an important source of insight into human language.

Acknowledgments

We would like to acknowledge the blind and deaf communities for their support. Without their generous willingness to give their time to benefit science, this research would not be possible. We would particularly like to acknowledge the US National Federation of the Blind. We also thank the funding agencies that have supported the work reported here: Johns Hopkins Science of Learning Institute (to MB); Johns Hopkins Catalyst Award (to MB); National Eye Institute 1R01EY027352-01A1 (to MB); Wellcome Trust Fellowship 100229/Z/12/Z (to MMacS) and the UK Economic and Social Research Council—Deafness Cognition and Language Research Centre Grant RES-620-28-0002 (to MMacS).

NOTE

1. In line with convention, "Deaf" is used when referring to those who use a signed language as their primary language.

REFERENCES

Adelman, J. S., Marquis, S. J., & Sabatos-DeVito, M. G. (2010). Letters in words are read simultaneously, not in left-to-right sequence. *Psychological Science, 21*(12), 1799–1801.

Amedi, A., Floel, A., Knecht, S., Zohary, E., & Cohen, L. G. (2004). Transcranial magnetic stimulation of the occipital pole interferes with verbal processing in blind subjects. *Nature Neuroscience, 7*(11), 1266–1270.

Amedi, A., Hofstetter, S., Maidenbaum, S., & Heimler, B. (2017). Task selectivity as a comprehensive principle for brain organization. *Trends in Cognitive Sciences, 21*(5), 307–310.

Amedi, A., Raz, N., Pianka, P., Malach, R., & Zohary, E. (2003). Early visual cortex activation correlates with superior verbal memory performance in the blind. *Nature Neuroscience, 6*(7), 758–766.

Barsalou, L. W., Simmons, W. K., Barbey, A. K., & Wilson, C. D. (2003). Grounding conceptual knowledge in modality-specific systems. *Trends in Cognitive Sciences, 7*(2), 84–91.

Bates, E., Thal, D., Trauner, D., Fenson, J., Aram, D., Eisele, J., & Nass, R. (1997). From first words to grammar in children with focal brain injury. *Developmental Neuropsychology, 13*(3), 275–343.

Bavelier, D. (2008). Encoding, rehearsal, and recall in signers and speakers: Shared network but differential engagement. *Cerebral Cortex, 18*(10), 2263–2274.

Bedny, M., Caramazza, A., Grossman, E., Pascual-Leone, A., & Saxe, R. (2008). Concepts are more than percepts: The case of action verbs. *Journal of Neuroscience, 28*(44), 11347–11353.

Bedny, M., Caramazza, A., Pascual-Leone, A., & Saxe, R. (2012). Typical neural representations of action verbs develop without vision. *Cerebral Cortex, 22*(2), 286–293.

Bedny, M., Pascual-Leone, A., Dodell-Feder, D., Fedorenko, E., & Saxe, R. (2011). Language processing in the occipital cortex of congenitally blind adults. *Proceedings of the National Academy of Sciences, 108*(11), 4429–4434.

Bedny, M., Pascual-Leone, A., Dravida, S., & Saxe, R. (2012). A sensitive period for language in the visual cortex: Distinct patterns of plasticity in congenitally versus late blind adults. *Brain and Language, 122*(3), 162–170.

Bedny, M., Richardson, H., & Saxe, R. (2015). "Visual" cortex responds to spoken language in blind children. *Journal of Neuroscience, 35*(33), 11674–11681.

Beilock, S. L., Lyons, I. M., Mattarella-Micke, A., Nusbaum, H. C., & Small, S. L. (2008). Sports experience changes the neural processing of action language. *Proceedings of the National Academy of Sciences, 105*(36), 13269–13273.

Berkeley, G. (1709). *Essay towards a new theory of vision.* Dublin: Printed for Jeremy Pepyat.

Bertelson, P., Mousty, P., & Radeau, M. (1992). The time course of Braille word recognition. *Journal of Experimental Psychology: Learning, Memory, and Cognition, 18*(2), 284–297.

Bi, Y., Wang, X., & Caramazza, A. (2016). Object domain and modality in the ventral visual pathway. *Trends in Cognitive Sciences, 20*(4), 282–290.

Bigelow, A. (1987). Early words of blind children. *Journal of Child Language, 14*, 47–56.

Bishop, D. V. M. (2013). Cerebral asymmetry and language development: Cause, correlate, or consequence? *Science, 340*(6138), 1230531.

Bola, Ł., Zimmermann, M., Mostowski, P., Jednoróg, K., Marchewka, A., Rutkowski, P., & Szwed, M. (2017). Task-specific reorganization of the auditory cortex in deaf humans. *Proceedings of the National Academy of Sciences, 114*, E600–E609.

Buchsbaum, B., Pickell, B., Love, T., Hatrak, M., Bellugi, U., & Hickok, G. (2005). Neural substrates for verbal working memory in deaf signers: fMRI study and lesion case report. *Brain and Language, 95*(2), 265–272.

Burton, H. (2002). Adaptive changes in early and late blind: A fMRI study of verb generation to heard nouns. *Journal of Neurophysiology, 88*(6), 3359–3371.

Burton, H., Snyder, A. Z., Conturo, T. E., Akbudak, E., Ollinger, J. M., & Raichle, M. E. (2002). Adaptive changes in early and late blind: A fMRI study of Braille reading. *Journal of Neurophysiology, 87*(1), 589–607.

Capek, C. M., Woll, B., MacSweeney, M., Waters, D., McGuire, P. K., David, A. S., ... Campbell, R. (2010). Superior temporal activation as a function of linguistic knowledge: Insights from deaf native signers who speechread. *Brain and Language, 112*(2), 129–134.

Cardin, V., Orfanidou, E., Ronnberg, J., Capek, C. M., Rudner, M., & Woll, B. (2013). Dissociating cognitive and sensory neural plasticity in human superior temporal cortex. *Nature Communications, 4,* 1473.

Cardin, V., Smittenaar, R. C., Orfanidou, E., Ronnberg, J., Capek, C. M., Rudner, M., & Woll, B. (2016). Differential activity in Heschl's gyrus between deaf and hearing individuals is due to auditory deprivation rather than language modality. *NeuroImage, 124*(Pt A), 96–106.

Chee, M. W. L., Caplan, D., Soon, C. S., Sriram, N., Tan, E. W. L., Thiel, T., & Weekes, B. (1999). Processing of visually presented sentences in Mandarin and English studied with fMRI. *Neuron, 23*(1), 127–137.

Cohen, L. G., Celnik, P., Pascual-Leone, A., Corwell, B., Faiz, L., Dambrosia, J., ... Hallett, M. (1997). Functional relevance of cross-modal plasticity in blind humans. *Nature, 389*(6647), 180–183.

Connolly, A. C., Gleitman, L. R., & Thompson-Schill, S. L. (2007). Effect of congenital blindness on the semantic representation of some everyday concepts. *Proceedings of the National Academy of Sciences, 104*(20), 8241–8246.

Corballis, M. C., & Morgan, M. J. (2010). On the biological basis of human laterality: I. Evidence for a maturational left-right gradient. *Behavioral and Brain Sciences, 1*(02), 261–269.

Corina, D., Chiu, Y. S., Knapp, H., Greenwald, R., San Jose-Robertson, L., & Braun, A. (2007). Neural correlates of human action observation in hearing and deaf subjects. *Brain Research, 1152*(1), 111–129.

Corina, D. P., McBurney, S. L., Dodrill, C., Hinshaw, K., Brinkley, J., & Ojemann, G. (1999). Functional roles of Broca's area and SMG: Evidence from cortical stimulation mapping in a deaf signer. *NeuroImage, 10*(5), 570–581.

Corina, D. P., San Jose-Robertson, L., Guillemin, A., High, J., & Braun, A. R. (2003). Language lateralization in a bimanual language. *Journal of Cognitive Neuroscience, 15*(5), 718–730.

Damasio, A.R. (1989). Time-locked multiregional retroactivation: A systems-level proposal for the neural substrates of recall and recognition. *Cognition, 33*(1–2), 25–62.

Daneman, M. (1988). How reading Braille is both like and unlike reading print. *Memory and Cognition, 16*(6), 497–504.

Deen, B., Saxe, R., & Bedny, M. (2015). Occipital cortex of blind individuals is functionally coupled with executive control areas of frontal cortex. *Journal of Cognitive Neuroscience, 27*(8), 1–15.

Dehaene, S., & Cohen, L. (2007). Cultural recycling of cortical maps. *Neuron, 56*(2), 384–398.

Dehaene, S., & Cohen, L. (2011). The unique role of the visual word form area in reading. *Trends in Cognitive Sciences, 15*(6), 254–262.

Dehaene, S., & Dehaene-Lambertz, G. (2016). Is the brain prewired for letters? *Nature Neuroscience, 19*(9), 1192–1193.

Dehaene, S., Pegado, F., Braga, L. W., Ventura, P., Nunes Filho, G., Jobert, A., ... Cohen, L. (2010). How learning to read changes the cortical networks for vision and language. *Science, 330*(6009), 1359–1364.

DiCarlo, J. J., Zoccolan, D., & Rust, N. C. (2012). How does the brain solve visual object recognition? *Neuron, 73*(3), 415–434.

Ding, H., Qin, W., Liang, M., Ming, D., Wan, B., Li, Q., & Yu, C. (2015). Cross-modal activation of auditory regions during visuo-spatial working memory in early deafness. *Brain, 138*(9), 2750–2765.

Emmorey, K., Grabowski, T., McCullough, S., Damasio, H., Ponto, L., Hichwa, R., & Bellugi, U. (2004). Motor-iconicity of sign language does not alter the neural systems underlying tool and action naming. *Brain and Language, 89*(1), 27–37.

Emmorey, K., McCullough, S., Mehta, S., & Grabowski, T. J. (2014). How sensory-motor systems impact the neural organization for language: Direct contrasts between spoken and signed language. *Frontiers in Psychology, 5,* 484.

Emmorey, K., McCullough, S., & Weisberg, J. (2014). Neural correlates of fingerspelling, text, and sign processing in deaf American Sign Language–English bilinguals. *Language, Cognition and Neuroscience, 30*(6), 749–767.

Emmorey, K., McCullough, S., & Weisberg, J. (2016). The neural underpinnings of reading skill in deaf adults. *Brain and Language, 160,* 11–20.

Emmorey, K., Mehta, S., & Grabowski, T. J. (2007). The neural correlates of sign versus word production. *NeuroImage, 36*(1), 202–208.

Epstein, R., & Kanwisher, N. (1998). A cortical representation of the local visual environment. *Nature, 392*(6676), 598–601.

Ferjan Ramirez, N., Leonard, M. K., Davenport, T. S., Torres, C., Halgren, E., & Mayberry, R. I. (2016). Neural language processing in adolescent first-language learners: Longitudinal case studies in American Sign Language. *Cerebral Cortex, 26*(3), 1015–1026.

Fischer-Baum, S., & Englebretson, R. (2016). Orthographic units in the absence of visual processing: Evidence from sublexical structure in Braille. *Cognition, 153,* 161–174.

Gleitman, L. R., & Newport, E. L. (1995). The invention of language by children: Environmental and biological influences on the acquisition of language. In L. R. Gleitman & M. Liberman (Eds.), *An invitation to cognitive science* (2nd ed., Vol. 1, pp. 1–24). Cambridge, MA: MIT Press.

Gutierrez-Sigut, E., Daws, R., Payne, H., Blott, J., Marshall, C., & MacSweeney, M. (2015). Language lateralization of hearing native signers: A functional transcranial Doppler sonography (fTCD) study of speech and sign production. *Brain and Language, 151,* 23–34.

Gutierrez-Sigut, E., Payne, H., & MacSweeney, M. (2016). Examining the contribution of motor movement and language dominance to increased left lateralization during sign generation in native signers. *Brain and Language, 159,* 109–117.

Hagoort, P. (2014). Nodes and networks in the neural architecture for language: Broca's region and beyond. *Current Opinion in Neurobiology, 28,* 136–141.

Hannagan, T., Amedi, A., Cohen, L., Dehaene-Lambertz, G., & Dehaene, S. (2015). Origins of the specialization for letters and numbers in ventral occipitotemporal cortex. *Trends in Cognitive Sciences, 19*(7), 374–382.

Hart, J., Berndt, R. S., & Caramazza, A. (1985). Category-specific naming deficit following cerebral infarction. *Nature, 316*(6027), 439–440.

He, C., Peelen, M. V., Han, Z., Lin, N., Caramazza, A., & Bi, Y. (2013). Selectivity for large non-manipulable objects in scene-selective visual cortex does not require visual experience. *NeuroImage, 79*(C), 1–9.

Hume, D. (1748). *An enquiry concerning human understanding.* London: Printed for A. Millar.

Johnson, J. S., & Newport, E. L. (1989). Critical period effects in second language learning: The influence of maturational state on the acquisition of English as a second language. *Cognitive Psychology, 21*(1), 60–99.

Kable, J. W., Kan, I. P., Wilson, A., Thompson-Schill, S. L., & Chatterjee, A. (2005). Conceptual representations of action in the lateral temporal cortex. *Journal of Cognitive Neuroscience, 17*(12), 1855–1870.

Kanjlia, S., Lane, C., Feigenson, L., & Bedny, M. (2016). Absence of visual experience modifies the neural basis of numerical thinking. *Proceedings of the National Academy of Sciences, 113*(40), 11172–11177.

Kanwisher, N., McDermott, J., & Chun, M. M. (1997). The fusiform face area: A module in human extrastriate cortex specialized for face perception. *Journal of Neuroscience, 17*(11), 4302–4311.

Karmiloff, K., & Karmiloff-Smith, A. (2009). *Pathways to language: From fetus to adolescent.* Cambridge, MA: Harvard University Press.

Kemmerer, D., Castillo, J. G., Talavage, T., Patterson, S., & Wiley, C. (2008). Neuroanatomical distribution of five semantic components of verbs: Evidence from fMRI. *Brain and Language, 107*(1), 16–43.

Kim, J. S., Kanjlia, S., Merabet, L. B., & Bedny, M. (2017). Development of the visual word form area requires visual experience: Evidence from blind Braille readers. *Journal of Neuroscience, 37*(47), 11495–11504.

Klima, E. S., & Bellugi, U. (1979). *The signs of language.* Cambridge, MA: Harvard University Press.

Kravitz, D. J., Saleem, K. S., Baker, C. I., Ungerleider, L. G., & Mishkin, M. (2013). The ventral visual pathway: An expanded neural framework for the processing of object quality. *Trends in Cognitive Sciences, 17*(1), 26–49.

Kupers, R., Pappens, M., de Noordhout, A. M., Schoenen, J., Ptito, M., & Fumal, A. (2007). rTMS of the occipital cortex abolishes Braille reading and repetition priming in blind subjects. *Neurology, 68*(9), 691–693.

Landau, B., & Gleitman, L. R. (1985). *Language and experience: Evidence from the blind child.* Cambridge, MA: Harvard University Press.

Lane, C., Kanjlia, S., Omaki, A., & Bedny, M. (2015). "Visual" cortex of congenitally blind adults responds to syntactic movement. *Journal of Neuroscience, 35*(37), 12859–12868.

Legge, G. E., Madison, C. M., & Mansfield, J. S. (1999). Measuring Braille reading speed with the MNREAD test. *Visual Impairment Research, 1*(3), 131–145.

Locke, J. (1690). *An essay concerning human understanding.* London: Thomas Bassett.

Locke, J. L. (1997). A theory of neurolinguistic development. *Brain and Language, 58*(2), 265–326.

Loiotile, R. E., & Bedny, M. (2018). "Visual" cortices of congenitally blind adults respond to executive demands. *bioRxiv 390450.*

Loiotile, R., Omaki, A., & Bedny, M. (2019). Enhanced sentence processing abilities among congenitally blind adults. PsyArXiv. https://doi.org/10.31234/osf.io/3vn4a.

Lomber, S. G., Meredith, M. A., & Kral, A. (2010). Cross-modal plasticity in specific auditory cortices underlies visual compensations in the deaf. *Nature Neuroscience, 13*(11), 1421.

MacSweeney, M., Campbell, R., Woll, B., Giampietro, V., David, A. S., McGuire, P. K., ... Brammer, M. J. (2004). Dissociating linguistic and nonlinguistic gestural communication in the brain. *NeuroImage, 22*(4), 1605–1618.

MacSweeney, M., Capek, C. M., Campbell, R., & Woll, B. (2008). The signing brain: The neurobiology of sign language. *Trends in Cognitive Sciences, 12*(11), 432–440.

MacSweeney, M., & Cardin, V. (2015). What is the function of auditory cortex without auditory input? *Brain, 138*(Pt 9), 2468–2470.

MacSweeney, M., Waters, D., Brammer, M. J., Woll, B., & Goswami, U. (2008). Phonological processing in deaf signers and the impact of age of first language acquisition. *NeuroImage, 40*(3), 1369–1379.

MacSweeney, M., Woll, B., Campbell, R., Calvert, G., McGuire, P., David, A., ... Brammer, M. (2002). Neural correlates of British Sign Language comprehension: Spatial processing demands of topographic language. *Journal of Cognitive Neuroscience,* 14, (7), 1064–1075.

MacSweeney, M., Woll, B., Campbell, R., McGuire, P. K., David, A. S., Williams, S. C. R., ... Brammer, M. J. (2002). Neural systems underlying British Sign Language and audio-visual English processing in native users. *Brain, 125*(7), 1583–1593.

Mahon, B. Z., Anzellotti, S., Schwarzbach, J., Zampini, M., & Caramazza, A. (2009). Category-specific organization in the human brain does not require visual experience. *Neuron, 63*(3), 397–405.

Mahon, B. Z., Schwarzbach, J., & Caramazza, A. (2010). The representation of tools in left parietal cortex is independent of visual experience. *Psychological Science, 21*(6), 764–771.

Marmor, G. S. (1978). Age at onset of blindness and the development of the semantics of color names. *Journal of Experimental Child Psychology, 25*(2), 267–278.

Marshall, J., Atkinson, J., Woll, B., & Thacker, A. (2005). Aphasia in a bilingual user of British Sign Language and English: Effects of cross-linguistic cues. *Cognitive Neuropsychology, 22*(6), 719–736.

Martin, A. (2007). The representation of object concepts in the brain. *Annual Review of Psychology, 58*(1), 25–45.

Martin, A., Haxby, J. V., Lalonde, F. M., Wiggs, C. L., & Ungerleider, L. G. (1995). Discrete cortical regions associated with knowledge of color and knowledge of action. *Science, 270*(5233), 102–105.

Mayberry, R. I. (2007). When timing is everything: Age of first-language acquisition effects on second-language learning. *Applied Psycholinguistics, 28*(3), 537–549.

Mayberry, R. I., Chen, J. K., Witcher, P., & Klein, D. (2011). Age of acquisition effects on the functional organization of language in the adult brain. *Brain and Language, 119*(1), 16–29.

Mayberry, R. I., & Fischer, S. D. (1989). Looking through phonological shape to lexical meaning: The bottleneck of non-native sign language processing. *Memory and Cognition, 17*(6), 740–754.

Mayberry, R. I., & Lock, E. (2003). Age constraints on first versus second language acquisition: Evidence for linguistic plasticity and epigenesis. *Brain and Language, 87*(3), 369–384.

Mayberry, R. I., Lock, E., & Kazmi, H. (2002). Development: Linguistic ability and early language exposure. *Nature, 417*(6884), 38.

McCullough, S., Saygin, A. P., Korpics, F., & Emmorey, K. (2012). Motion-sensitive cortex and motion semantics in American Sign Language. *NeuroImage, 63*, 111–118.

McGettigan, C., & Scott, S. K. (2012). Cortical asymmetries in speech perception: What's wrong, what's right and what's left? *Trends in Cognitive Science, 16*(5), 269–276.

Merzenich, M. M., Nelson, R. J., Stryker, M. P., Cynader, M. S., Schoppmann, A., & Zook, J. M. (1984). Somatosensory cortical map changes following digit amputation in adult monkeys. *Journal of Comparative Neurology, 224*(4), 591–605.

Meteyard, L., Cuadrado, S. R., Bahrami, B., & Vigliocco, G. (2012). Coming of age: A review of embodiment and the neuroscience of semantics. *Cortex, 48*(7), 788–804.

Millar, S. (2008). *Space and sense.* New York: Psychology Press.

Morford, J. (2003). Grammatical development in adolescent first-language learners. *Linguistics 41.* 10.1515/ling.2003.022.

Morgan, G., & Woll, B. (2002). *Directions in sign language acquisition.* Amsterdam: John Benjamins.

Newman, A. J., Bavelier, D., Corina, D., Jezzard, P., & Neville, H. J. (2002). A critical period for right hemisphere recruitment in American Sign Language processing. *Nature Neuroscience, 5*(1), 76–80.

Newman, A. J., Supalla, T., Fernandez, N., Newport, E. L., & Bavelier, D. (2015). Neural systems supporting linguistic structure, linguistic experience, and symbolic communication in sign language and gesture. *Proceedings of the National Academy of Sciences, 112*(37), 11684–11689.

Noppeney, U. (2003). Effects of visual deprivation on the organization of the semantic system. *Brain, 126*(7), 1620–1627.

Pascual-Leone, A., Amedi, A., Fregni, F., & Merabet, L. B. (2005). The plastic human brain cortex. *Annual Review of Neuroscience, 28*(1), 377–401.

Patterson, K., Nestor, P. J., & Rogers, T. T. (2007). Where do you know what you know? The representation of semantic knowledge in the human brain. *Nature Reviews Neuroscience, 8*(12), 976–987.

Peelen, M. V., He, C., Han, Z., Caramazza, A., & Bi, Y. (2014). Non-visual and visual object shape representations in occipitotemporal cortex: Evidence from congenitally blind and sighted adults. *Journal of Neuroscience, 34*(1), 163–170.

Price, C. J., & Devlin, J. T. (2011). The interactive account of ventral occipitotemporal contributions to reading. *Trends in Cognitive Sciences, 15*(6), 246–253.

Pulvermüller, F. (2001). Brain reflections of words and their meaning. *Trends in Cognitive Sciences, 5*(12), 517–524.

Raizada, R. D. S., Richards, T. L., Meltzoff, A., & Kuhl, P. K. (2008). Socioeconomic status predicts hemispheric specialisation of the left inferior frontal gyrus in young children. *NeuroImage, 40*(3), 1392–1401.

Ramirez, N. F., Leonard, M. K., Torres, C., Hatrak, M., Halgren, E., & Mayberry, R. I. (2014). Neural language processing in adolescent first-language learners. *Cerebral Cortex, 24*(10), 2772–2783.

Raz, N., Striem, E., Pundak, G., Orlov, T., & Zohary, E. (2007). Superior serial memory in the blind: A case of cognitive compensatory adjustment. *Current Biology, 17*(13), 1129–1133.

Reich, L., Szwed, M., Cohen, L., & Amedi, A. (2011). A ventral visual stream reading center independent of visual experience. *Current Biology, 21*(5), 363–368.

Rizzolatti, G., & Craighero, L. (2004). The mirror-neuron system. *Annual Review of Neuroscience, 27*, 169–192.

Röder, B., Rösler, F., & Neville, H. J. (2000). Event-related potentials during auditory language processing in congenitally blind and sighted people. *Neuropsychologia, 38*(11), 1482–1502.

Röder, B., Stock, O., Bien, S., Neville, H., & Rösler, F. (2002). Speech processing activates visual cortex in congenitally blind humans. *European Journal of Neuroscience, 16*(5), 930–936.

Rüschemeyer, S. A., Fiebach, C. J., Kempe, V., & Friederici, A. D. (2005). Processing lexical semantic and syntactic information in first and second language: fMRI evidence from German and Russian. *Human Brain Mapping, 25*(2), 266–286.

Sadato, N., Pascual-Leone, A., Grafman, J., Ibañez, V., Deiber, M. P., Dold, G., & Hallett, M. (1996). Activation of the primary visual cortex by Braille reading in blind subjects. *Nature, 380*(6574), 526–528.

Sakai, K. L., Tatsuno, Y., Suzuki, K., Kimura, H., & Ichida, Y. (2005). Sign and speech: Amodal commonality in left hemisphere dominance for comprehension of sentences. *Brain, 128*(6), 1407–1417.

Saygin, Z. M., Osher, D. E., Norton, E. S., Youssoufian, D. A., Beach, S. D., Feather, J., ... Kanwisher, N. (2016). Connectivity precedes function in the development of the visual word form area. *Nature Neuroscience, 19*(9), 1250–1255.

Scott, G. D., Karns, C. M., Dow, M. W., Stevens, C., & Neville, H. J. (2014). Enhanced peripheral visual processing in congenitally deaf humans is supported by multiple brain regions, including primary auditory cortex. *Frontiers in Human Neuroscience, 8*, 177.

Scott, S. K., & McGettigan, C. (2013). Do temporal processes underlie left hemisphere dominance in speech perception? *Brain and Language, 127*(1), 36–45.

Shepard, R. N., & Cooper, L. A. (1992). Representation of colors in the blind, color-blind, and normally sighted. *Psychological Science, 3*(2), 97–104.

Söderfeldt, B., Ingvar, M., Rönnberg, J., Eriksson, L., Serrander, M., & Stone-Elander, S. (1997). Signed and spoken language perception studied by positron emission tomography. *Neurology, 49*(1), 82–87.

Striem-Amit, E., Cohen, L., Dehaene, S., & Amedi, A. (2012). Reading with sounds: Sensory substitution selectively activates the visual word form area in the blind. *Neuron, 76*(3), 640–652.

Struiksma, M. E., Noordzij, M. L., Neggers, S. F. W., Bosker, W. M., & Postma, A. (2011). Spatial language processing in the blind: Evidence for a supramodal representation and cortical reorganization. *PLOS ONE, 6*(9), e24253-12.

Szwed, M., Cohen, L., Qiao, E., & Dehaene, S. (2009). The role of invariant line junctions in object and visual word recognition. *Vision Research, 49*(7), 718–725.

Tsapkini, K., & Rapp, B. (2010). The orthography-specific functions of the left fusiform gyrus: Evidence of modality and category specificity. *Cortex, 46*(2), 185–205.

Twomey, T., Waters, D., Price, C.J., Evans, S., & MacSweeney, M. (2017). How auditory experience differentially influences the function of left and right superior temporal cortices. *Journal of Neuroscience, 37*(39), 9564–9673.

Urwin, C. (1987). Dialogue and cognitive functioning in the early language development of three blind children. In A. E. Mills (Ed.), *Language acquisition in the blind child: Normal and deficient* (pp. 142–161). San Diego, CA: College-Hi Press.

Vinckier, F., Dehaene, S., Jobert, A., Dubus, J. P., Sigman, M., & Cohen, L. (2007). Hierarchical coding of letter strings in the ventral stream: Dissecting the inner organization of the visual word-form system. *Neuron, 55*(1), 143–156.

Warrington, E. K., & Shallice, T. (1984). Category specific semantic impairments. *Brain, 107*(Pt 3), 829–854.

Watkins, K. E., Cowey, A., Alexander, I., Filippini, N., Kennedy, J. M., Smith, S. M., ... Bridge, H. (2012). Language networks in anophthalmia: Maintained hierarchy of processing in "visual" cortex. *Brain, 135*(5), 1566–1577.

Willems, R. M., Hagoort, P., & Casasanto, D. (2010). Body-specific representations of action verbs: Neural evidence from right- and left-handers. *Psychological Science, 21*(1), 67–74.

38 Key Issues and Future Directions: The Neural Architecture for Language

PETER HAGOORT AND CHRISTIAN F. BECKMANN

An adequate understanding of the human capacity for language requires a merging of constraints from functional accounts (the cognitive architecture) and from specifications of the neural infrastructure that supports it (the neural architecture). This part focuses mainly on the latter. Which aspects of brain organization have made our species language-ready? Which building blocks of brain organization are especially critical for the cognitive computations that are involved in speaking, listening, reading, and related language skills? Clearly, language is a multicomponent system, minimally involving levels of phonology, syntax, and semantics (Hagoort, 2005, 2013; Jackendoff, 2002). Each of these components might need to recruit the contribution of a different set of neuronal circuits. The degree to which brain areas and brain networks are specialized for aspects of language is an issue of debate. Another unresolved issue is which level of brain organization is most critical for supporting cognitive computations.

Palomero-Gallagher and Zilles (chapter 33) argue that insights into cytoarchitecture and receptor architecture are crucial prerequisites for our understanding of the neurobiological basis of language. They discuss these architectural features in quite some detail for the language-relevant areas. In addition, these authors argue that especially the receptor fingerprints represent the molecular default organization of the regionally specific local information processing in each cortical area. Neurochemically, related cortical areas may participate in the same functional system. Interestingly, the left hemisphere receptor fingerprints revealed a tight clustering of all language-relevant areas.

Another central aspect of brain organization is its organization of distinct building blocks into networks, forming structural and functional systems extending across the brain. This is the focus of the contribution of Przeździk, Haak, Beckmann, and Bartsch (chapter 34). They discuss current descriptions of structural and functional organization between distinct areas that form the "language connectome" and further argue that neurobiological models of language ought to be validated on empirical data, considering the mesoscopic variations in brain connectivity patterns as well as more macroscopic connectivity and the topographic organization thereof. The observed level of coactivity across different language-relevant parts of the brain provide an important set of benchmark boundary conditions for investigating the complexity of interactions within the language system.

Listening and speaking are the evolutionary primordial speech and language skills. Although tightly linked, they link differently to the underlying neural infrastructure. For listening, it is conceivable that in the human brain the circuitry devoted to auditory perception is optimized to ensure efficient and robust analysis of the complex speech signals (Formisano, chapter 35). The online analysis of an incoming speech signal requires the simultaneous analysis of spectral and temporal components at multiple scales. As Formisano reviews, not only in higher-level auditory cortex but also in early auditory areas neuronal populations are fine-tuned to the characteristic properties of speech. The absence of these tuning characteristics in the auditory cortex of the macaque monkey suggests that human auditory areas express some unique features that are likely acquired through lifelong learning and have emerged as a critical step in the evolution of speech and language.

Likewise, for speaking, the motor cortex seems to have developed a speech-ready organization (Bianchi & Simonyan, chapter 36). Speaking is one of the most complicated motor acts we perform, requiring the coordinated activity of more than 100 respiratory, laryngeal, and orofacial muscles. This necessitates a degree of motor control unmatched by any other motor system in the body. The laryngeal motor cortex (LMC) is a key component of the motor system for supporting speech. Compared to the macaque brain, the human LMC establishes a nearly sevenfold increased connectivity with the somatosensory and inferior parietal cortex, putatively to enable sensorimotor integration needed for speech production. An additional reorganization of LMC in the course of hominin evolution is the establishment of direct connections with laryngeal motor neurons in the brain stem. In short, both auditory and motor cortices have undergone important adaptations to realize a neural architecture that is speech- and language-ready.

Despite these sizeable brain adaptations to requirements of speech and language, there remains an important riddle for our understanding of the mapping relation between the cognitive and the neural architecture for language. This is discussed chapter 37 by Bedny and MacSweeney about the changes in the neural architecture in individuals that are born blind or deaf. Their contribution suggests that there is more than one neuronal solution to the computational challenges of language. For instance, the visual cortex of the congenitally blind is sensitive to the grammatical and lexical properties of linguistic stimuli, which prima facie have quite different computational requirements than the visual input in the seeing brain. This brings us back to the issue of what are the computational degrees of freedom over variations in the underlying neural architecture.

1. The Neuroanatomical Stance

Among experts in the field of neuroanatomy and neurophysiology there are, in broad strokes, two opposing views (cf. Hagoort, 2018). This we will illustrate at the example of the role of cytoarchitectonics. A prime example of the contribution of neuroanatomy is the famous map by Korbinian Brodmann (1869–1918). This map consisted originally of 52 different areas, usually referred to by expressions such as BA44 for Broadmann area 44. The labeling of the Brodmann areas was determined by the order in which Brodmann went through the brain, analyzing one area after the other. Brodmann's classification is based on the cytoarchitectonics of the brain, which refers to the structure, form, and position of the cells in the six layers of the cortex, the quantification of which was done by Brodmann based on a small number of postmortem brains. These were sectioned into slices of 5- to 10-micron thickness that underwent Nissl staining and were then inspected under the microscope. In this way, the distribution of different cell types across cortical layers and brain areas could be determined. Even today, Brodmann's map, published in 1905, is seen as a hallmark achievement in the history of neuroscience. Brodmann's work reveals that the composition of the six cortical layers, both in terms of cell types and cell numbers, varies substantially across the brain. The primary visual cortex, for instance, has about twice as many neurons per cortical column as other brain areas (Amaral, 2000). In recent times, the work of Brodmann and others from the classical era of neuroanatomy has been substantially modernized by Zilles, Amunts, and colleagues (see Palomero-Gallagher & Zilles, chapter 33 of this volume).

Importantly, the classical view among neuroanatomists is that these architectural differences in brain structure are indicative of functional differences, and, conversely, that functional differences demand differences in architecture (Bartels & Zeki, 2005; Brodmann, 1905; Vogt & Vogt, 1919; Von Economo & Koskinas, 1925). Following the classical view, through different ways of characterizing brain structure (i.e., cyto-, myelo-, and receptorarchitectonics; Amunts & Zilles, 2015; Zilles & Palomero-Gallagher, 2001), brain areas can be identified, for which differences in structural characteristics imply functional differences. From this view, it follows that one should look for the structural features that determine why a particular brain area can support, for instance, morphological decomposition or syntactic encoding.

In contrast to the classical view in neuroanatomy, more recent accounts have argued that, from a computational perspective, different brain areas are very similar. For instance, Douglas and Martin (2004) argued that "the same basic laminar and tangential organization of the excitatory neurons of the neocortex, the spiny neurons, is evident wherever it has been sought. The inhibitory neurons similarly show a characteristic morphology and patterns of connections throughout the cortex ... all things considered, many crucial aspects of morphology, laminar distribution, and synaptic targets are very well conserved between areas and between species" (p. 439; see also Douglas & Martin, 2018). Bastos et al. (2012) therefore spoke of a canonical microcircuit that has the same computational organization across all of the neocortex, despite the cytoarchitectonic differences that can be observed between, for instance, the sensory and motor cortexes. Functional differences between brain areas arise in this perspective mainly due to variability of the input signals in forming functional specializations. The functional contribution of a particular piece of cortex might thus primarily not be determined by heterogeneity of brain tissue, but rather by the way in which its functional characteristics are shaped by the afferent connections.

2. Key Issues and Future Directions

The two opposing views discussed in section 1 have consequences for how one envisions to identify the relevant neural building block for cognitive computations. At the same time, functional and structural neuroimaging studies can provide some relevant insights into this matter. Importantly, a number of remarkable manifestations of neural plasticity have been reported in recent years. For instance, Amedi, Raz, Pianka, Malach, and Zohary (2003) reported on congenitally blind subjects with increased activation in primary visual cortex (V1) during a verbal memory task. Moreover, the stronger the activation in V1, the better the memory performance. If the structural properties of V1 had been exclusively decisive for its functional capacities, then it would be hard to see how the same neurons that in subjects with intact vision support basic visual processing could be recruited in the blind for verbal memory, assuming that major cytoarchitectonic features of visual cortex are not shaped by the specifics of the sensory experience. Bedny, Pascual-Leone, Dodell-Feder, Federenko, and Saxe (2011; see Bedny & MacSweeney, chapter 37) reported language processing in the occipital cortex of congenitally blind individuals. This even includes high-level language function such as the computation of sentence-level combinatorial structure. These findings demonstrate that the cytoarchitectonic constraints for specifications of cognitive function are rather loose. Presumably, the patterns of connectivity between areas and ensuing differences in inputs are a more relevant functional parameter than the differences between the cytoarchitectonic details of specific cortical areas. For another part of brain, Bola et al. (2017) reported that in deaf humans the high-level auditory cortex gets involved in vision. Based on these and similar findings, Bedny (2017, p. 637) concluded that "human cortices are cognitively pluripotent, that is, capable of assuming a wide range of cognitive functions. Specialization is driven by input during development, which is constrained by connectivity and experience."

All this does not imply that one should deny the great importance of cytoarchitectonic structures for human cognition. Clearly, without these basic building blocks of the brain, cognition would not be possible. Without different types of neurons, glia, and axons, the cognitive machinery would not work. Nevertheless, the issue here is that these building blocks enter into processes of functional specialization and segregation. It might be that the exact nature of these functional specializations are more easily inferred from an analysis of input/output and connectivity than by looking at the detailed cytoarchitectonic characteristics.

The conundrum that remains for this view is what to do with the concept of a brain area. How should one define borders independently of cytoarchitecture? One solution is to specify brain areas themselves in functional instead of structural terms. An example is parcellations based on resting-state functional MRI (e.g., Zhang et al., 2015; see Eickhoff, Thirion, Varoquaux, & Bzdok, 2015, for an overview). Alternative options are a parcellation of cerebral cortex on the basis of a multimodal combination of structural and functional measures (Glasser et al., 2016). Connectivity-based parcellations are predicated on the assumption of a correspondence between a cortical brain area and its connectivity fingerprint, as derived by diffusion-weighted MRI and resting-state functional MRI (Eickhoff et al., 2015). In either case, however, current approaches for characterizing elementary building blocks of computation will require further advances in imaging methodology to capture the rich functional variations across different parts of the brain. These provide further constraints in terms of the necessity to capture both functional variation within brain areas as well as functional multiplicity. The former refers to the fact that evidence from sensory cortices suggest a multiscale hierarchical functional organization while the latter acknowledges the possibility that the same patch of cortex can exhibit multiple modes of organization simultaneously. For instance, V1 can be probed for orientation selectivity as well as for foveal versus high eccentricity, giving rise to multiple modes of area description. Indeed, recent work suggests that various areas of interest might be characterized by multiple modes of organization, with overlapping topographically organized modes of connectivity (Haak, Marquand, & Beckmann, 2018), further challenging the concept of a brain area with piecewise homogeneous and unitary structure.

What are the consequences for accounts of language learning and language processing? We don't think that at the moment there is much evidence for the claim that all hinges on whether a particular activation is found in, say, BA44 rather than BA45. Patterns of activity have been shown to vary depending on the precise experimental context (see Przeździk et al., chapter 34, figure 34.2). An approach based on reversed inferences from structural anatomy or functional activation to cognitive function seems in the light of our current knowledge, not well-constrained enough. It is functional anatomy that counts, and that might provide stronger constraints than structural anatomy for specifications of the different forms of human cognition. This implies that for an evolutionary perspective on the neural architecture for language, we should look at patterns of associations and dissociations in measures of brain activity, and do comparative studies of, for example, tractography in different species (e.g., Rilling et al., 2008; Rilling & van den Heuvel, 2018). The differences in the arrangements of network elements in the nervous system among related species might be key to understanding the evolutionary origins of language and other cognitive functions (Petersen & Sporns, 2015). The network topologies might provide more direct insights into the neural instantiations of cognitive functions than of the classical anatomical measures (Xiang, Fonteijn, Norris, & Hagoort, 2010). In the words of Park and Friston (2013, p. 7), "Function may deviate from structure to exhibit dynamic and contextualized behavior. Such divergence of function from structure is perhaps the most intriguing property of the brain."

3. Conclusion

A cognitive neuroscience approach to language takes information and constraints from different levels of analysis into consideration, in the service of a full account of the neurobiology of language. The assumption hereby is that these different levels become mutually informative and can be connected in a transparent way. At the same time, not all constraints have the same impact. Here it is argued that the constraints provided by the classical anatomical measures (cyto- and myeloarchitectonics) are–in our current understanding–not very tight constraints for detailed specifications of cognitive functions, including language learning and language processing. Maybe, measures of the computational features of brain tissue might provide stronger constraints. For understanding cognitive specialization, future research should pay close attention to measures of functional and structural connectivity.

REFERENCES

Amaral, D. G. (2000). The anatomical organization of the central nervous system. In E. R. Kandel, J. H. Schwartz, & T. M. Jessell (Eds.), *Principles of neural science* (4th ed., pp. 317–336). New York: McGraw-Hill.

Amedi, A., Raz, N., Pianka, P., Malach, R., & Zohary, E. (2003). Early "visual" cortex activation correlates with superior verbal memory performance in the blind. *Nature Neuroscience, 6,* 758–766.

Amunts, K., & Zilles, K. (2015). Architectonic mapping of the human brain beyond Brodmann. *Neuron, 88*(6), 1086–1107.

Bartels, A., & Zeki, S. (2005). The chronoarchitecture of the cerebral cortex. *Philosophical Transactions of the Royal Society of London (Series B-Biological Sciences), 360,* 733–750.

Bastos, A. M. W., Usrey, M., Adams, R. A., Mangun, G. R., Fries, P., & Friston, K. J. (2012). Canonical microcircuits for predictive coding. *Neuron, 76,* 695–711.

Bedny, M. (2017). Evidence from blindness for a cognitively pluripotent cortex. *Trends in Cognitive Science, 21,* 637–648.

Bedny, M., Pascual-Leone, A., Dodell-Feder, D., Federenko, E., & Saxe, R. (2011). Language processing in the occipital cortex of congenitally blind adults. *Proceedings of the National Academy of Sciences, 15,* 4429–4434.

Bola, L., Zimmermann, M., Mostowski, P., Jednorog, K., Marchewska, A., Rutkowski, P., & Szwed, M. (2017). Task-specific reorganization of the auditory cortex in deaf humans. *Proceedings of the National Academy of Sciences, 114,* E600–E609.

Brodmann, K. (1909). *Vergleichende Lokalisationslehre der Grosshirnrinde in ihren Prinzipien dargestellt auf Grund des Zellenbaues.* Leipzig: Johann Ambrosius Barth.

Douglas, R. J., & Martin, K. A. (2004). Neuronal circuits of the neocortex. *Annual Review of Neuroscience, 27,* 419–451.

Douglas, R. J., & Martin, K. A. (2018). Canonical cortical circuits. In G. Shepard & S. Grillnes (Eds.), *Handbook of brain microcircuits* (pp. 23–33). Oxford: Oxford University Press.

Eickhoff, S. B., Thirion, B., Varoquaux, G., & Bzdok, D. (2015). Connectivity-based parcellation: Critique and implications. *Human Brain Mapping, 36,* 4771–4792.

Glasser, M. F., Coalson, T. S., Robinson, E. C., Hacker, C. D., Harwell, J., Yacoub, E., ... van Essen, D. C. (2016). A multimodal parcellation of human cerebral cortex. *Nature, 536,* 171–178.

Haak, K. V., Marquand, A. F., & Beckmann, C. F. (2018). Connectopic mapping with resting-state fMRI. *NeuroImage, 170,* 83–94.

Hagoort, P. (2005). On Broca, brain, and binding: A new framework. *Trends in Cognitive Sciences, 9*(9), 416–423.

Hagoort, P. (2013). MUC (memory, unification, control) and beyond. *Frontiers in Psychology, 4,* 416.

Hagoort, P. (2018). Prerequisites for an evolutionary stance on the neurobiology of language. *Current Opinion in Behavioral Sciences,* 21, 191–194.

Jackendoff, R. (2002). *Foundations of language: brain, meaning, grammar, evolution.* Oxford: Oxford University Press.

Park, H. J., & Friston, K. (2013). Structural and functional brain networks: From connections to cognition. *Science, 342*(6158), 1238411.

Petersen, S. E., & Sporns, O. (2015). Brain networks and cognitive architectures. *Neuron, 88,* 207–219.

Rilling, J. K., Glasser, M. F., Preuss, T. M., Ma, X., Zhao, T., Hu, X., & Behrens, T. E. J. (2008). The evolution of the arcuate fasciculus revealed with comparative DTI. *Nature Neuroscience, 11*(4), 426–428.

Rilling, J. K., & van den Heuvel, M. P. (2018). Comparative primate connectomics. *Brain, Behavior and Evolution, 91,* 170–179.

Vogt, C., & Vogt, O. (1919). Allgemeinere Ergebnisse unserer Hirnforschung. Vierte Mitteilung: Die physiologische Bedeutung der architektonischen Rindenfelderung auf Grund neuer Rindenreizungen. *Journal of Psychology and Neurology, 25,* 399–462.

Von Economo, C., & Koskinas, G. N. (1925). *Die Cytoarchitektonik der Hirnrinde des erwachsenen Menschen.* Berlin: Springer.

Xiang, H., Fonteijn, H. M., Norris, D. G., & Hagoort, P. (2010). Topographical functional connectivity pattern in the Perisylvian language networks. *Cerebral Cortex, 20,* 549–560.

Zhang, Y., Caspers, S., Fan, L., Fan, Y., Song, M., Liu, C., ... Jiang, T. (2015). Robust brain parcellation using sparse representation on resting-state fMRI. *Brain Structure and Function, 220,* 3565–3579.

Zilles, K., & Palomero-Gallagher, N. (2001). Cyto-, myelo-, and receptor architectonics of the human parietal cortex. *NeuroImage, 14,* S8–S20.

VII THE GENETICS OF LANGUAGE

SECTION EDITOR: SIMON E. FISHER

39 Mapping Genes Involved in Reading and Language Skills

MICHELLE LUCIANO AND TIMOTHY C. BATES

Reading and language skills are arguably the most valuable cognitive skills for any individual to acquire. They are required for navigating our social world and for broader learning. Problems with language development can limit the capacity for skilled reading, although difficulties with reading can still occur in the absence of language problems. There is large interindividual variation in people's reading and language abilities, with a major cause of this variation being a person's genes. This chapter will describe the gene mapping progress being made in the areas of reading and language research. It will draw on findings from research focused on reading/language ability (based on quantitative measures) and disability (based on dichotomous categorization), which suggest many of the same genes operate on all levels of ability. Early gene mapping efforts that used linkage methodology will be introduced, and association analysis (the most common approach used today) will be presented in the context of fine mapping and genome-wide studies. The chapter concludes with current directions that gene technologies and methodologies are taking, explaining how these may add to our understanding of the genetic etiology of reading and language disorders.

1. Measuring Reading and Language Skills

Like other cognitive abilities, reading and language proficiency can be measured by standardized tests, performance on which is distributed along a continuum (i.e., they are quantitative traits). Performance in reading is typically assessed by accuracy and/or fluency of reading aloud and by comprehension of text. Reading aloud requires individuals to recognize printed words: unfamiliar words are sounded out using grapheme-phoneme rules and words that do not follow letter-sound correspondence rules (which are common in English) can be retrieved from one's lexicon, or memory store for words. These respective processes can be distinguished by asking a person to read aloud nonwords that are not part of one's lexicon but can be pronounced using grapheme-phoneme rules (e.g., *yot*) versus irregular words that do not conform to grapheme-phoneme rules (e.g., *yacht*). Regular words (i.e., real words that follow grapheme-phoneme rules) can be pronounced using either phonological or lexical processing routes and are less informative for specific processes, although high-frequency words are more likely to be lexically accessed. Children with reading difficulties tend to have particular problems with phonological processing; letter knowledge, phoneme awareness, and rapid naming are all early predictors of reading skill/disability (Hulme & Snowling, 2016). Reading comprehension is a broader measure that taps word meaning, grammar, and their integration with existing knowledge. To diagnose a reading impairment (commonly known as developmental dyslexia), scores on tests of reading, writing, and spelling are typically evaluated against some predefined criterion (e.g., 1 standard deviation below the age-relevant population mean) indicative of reading difficulties.

Children with a specific language impairment (SLI)[1] are typically identified by their poor receptive and/or expressive language, that is, they have difficulties understanding others' language and/or expressing themselves through language despite otherwise normal development. Like reading disorder diagnoses, an arbitrary cutoff is used to categorize scores on language tasks as low and indicative of an SLI. Psycholinguistic markers of language skill such as linguistic tense marking knowledge, sentence, and nonword repetition are sensitive to current/resolved SLI (Conti, Ramsden, Botting, & Faragher, 2001) and also show variation in the normal population. Nonword repetition, one of the most commonly used tests in genetic studies of language skills, assesses the accuracy of an examinee's immediate repetition of a nonword after it is read aloud to them (nonwords in the list increase in syllable length). This is thought to reflect processes related to phonological working memory, phonological analysis, and articulatory output and is related to vocabulary acquisition, reading ability, and language comprehension (Gathercole, Willis, Baddeley, & Emslie, 1994).

Genetic studies of reading/language ability focus on quantitative measurement whereas studies of disability tend to focus on case-control comparisons. Historically, specific reading and language disorders were diagnosed only when these skills were discrepant with global cognitive ability, or IQ, but there is a strong shift away from this requirement (Siegel, 2003), which may have implications for genetic research. Reading/language disorder in the presence of low IQ may have additional genetic burden from general cognitive processes that influence reading/language skill. Indeed, there may be cases where general cognitive processes alone are sufficient to cause poor reading. Including such cases in a genetic study would reduce the power to identify genes that specifically affect reading/language processes. Ultimately, definitions of disorder are arbitrary and do not acknowledge potential heterogeneity in the underlying processes that contribute to poor reading/language, although it is likely that a deficit in more than one process is needed to break down typical reading/language function (Bishop, Laws, Adams, & Norbury, 2006). Genetic studies that focus on individual processes will be maximally powered to detect the genes influencing specific processes and therefore risk of reading/language impairment when combined.

2. *Reading and Language Skills and Their Relationship to Disability*

Interest in mapping genes for reading and language skills has largely been driven by the need to understand the biological causes of reading and language disability. Many early gene mapping attempts in the field of reading and language disabilities expected that these disorders would be primarily explained by genes of large effect and that these genes would influence disability but not individual variation in the normal range (e.g., Pennington et al., 1991). Based on findings of a bimodal distribution of reading scores in children, Rutter and Yule (1975) developed an influential concept of "specific reading retardation," arguing that specific reading disability reflected a qualitatively different phenotype to a "reading-backward" group in terms of sex ratio, neurodevelopmental profile, and impact on education. However, these authors suggested that this did *not* reflect a genetic discontinuity, but rather an artifact of the specific reading retardation group not differing (by diagnostic definition) from normal IQ. The Rutter and Yule (1975) finding then might be viewed more as evidence for speech and language deficits existing *independent* of variation in general IQ. There has been support for a model of reading and dyslexia occurring along a continuum (Rodgers, 1983; Shaywitz, Escobar, Shaywitz, Fletcher, & Makuch, 1992; Silva, Mcgee, & Williams, 1985). Similarly, the idea that SLI should be viewed as the low tail of a normal distribution in language skills is also influential (Dale & Cole, 1991; Leonard, 1991). It should be noted that this continuum view is compatible with the existence of some cases of low reading and language scores that have an etiology linked to rare genetic or neurological pathological events. Rather the prediction is that any level of dysfunction (or function) is possible, with these severe cases reflecting the rare combination of events leading to very low performance. Like other quantitative traits, a highly polygenic genetic architecture is likely to underlie the heritable variation in reading and language skills and, by extension, to explain most cases meeting diagnostic criteria for dyslexia and SLI.

3. *Genetic Variation in Reading and Language Abilities*

Dyslexia and SLI can be understood in terms of the *Common Disorders are Quantitative Traits Hypothesis*, which has been applied to learning difficulties more generally (Haworth & Plomin, 2010). The assumption is that the polygenic liability for common disorders is distributed normally in the population, so that genes influencing low levels of a trait (i.e., classified as disorder) operate over the entire range of the distribution. Twin and family studies are common methodologies to estimate the proportion of genetic influence on a trait, that is, its heritability. The population variance can be decomposed into genetic and environmental parts by modeling the observed phenotypic similarity between family members in relation to their expected genetic similarity. For example, identical twins share virtually all their genetic variants, whereas nonidentical twins share roughly half their genetic variants. Thus, when identical twins are observed to be more phenotypically similar than nonidentical twins are (i.e., stronger correlation for continuous traits or higher concordance—presence of the same trait for diagnoses), this provides evidence in favor of genetic influence. The heritability estimates for reading traits based on unselected general population samples (e.g., Bates, Castles, et al., 2007; Harlaar, Spinath, Dale, & Plomin, 2005) and samples ascertained based on affected probands (e.g., Defries, Fulker, & Labuda, 1987; Olson, Wise, Conners, Rack, & Fulker, 1989) have converged on similar estimates, typically in the 0.30 to 0.70 range (i.e., genetic factors explain 30% to 70% of phenotypic variation in these samples), depending on the trait. For language traits, fewer studies have focused on unselected samples. In a sample of 301 four-year-old twins with normal language development, assessed on a battery of 18 verbal and nonverbal tests, a language factor emerged in principal components analyses, and scores on this factor showed a heritability of 0.39 (Colledge et al., 2002). Estimates of heritability of over 0.5 have been reported for language impairment where the diagnosis was based on speech deficits, with research showing that heritability is lower for broader language deficits (Hayiou-Thomas, 2008). Nonword repetition scores, a psycholinguistic marker of SLI, have been argued to show greater genetic influence in children with language impairment than in those without (Bishop et al., 1999).

Evidence supporting continuity in the genes influencing disordered versus normal-range ability and variation among the normal range comes from DeFries-Fulker extremes analysis. This method is designed to correctly decompose variance in samples that consist of probands (ascertained on the basis of the presence of a disorder) and their co-twins. It estimates the degree to which trait differences between an extreme group (disordered probands) and the population (normal range of reading) is due to genetic variation (i.e., the group heritability). If we interpret the group heritability as a bivariate analysis between extreme scores and continuous scores, a nonzero estimate of heritability indicates a genetic link between disability and normal variation (Plomin & Kovas, 2005). Results from larger studies show group heritability estimates of 0.37 for language disability and ability (a composite of eight language tests) and 0.52 and 0.60 for reading disability and ability (using respective Peabody Individual Achievement Test composite and the Test of Word Reading Efficiency), indicating substantial genetic commonality (Plomin & Kovas, 2005).

4. Linkage Methodology

Human somatic cells are diploid, that is, they contain two sets of chromosomes consisting of DNA, regions of which constitute genes. One variant of a gene (an allele) is inherited from a mother and the other from a father, with the variant passed down to an offspring being random; this segregation occurs at gametogenesis. Individual genes do not always segregate independently, but they are often inherited as a group, with relatively long stretches of DNA in grandparents, parents, and offspring being identical. Linkage analysis uses this violation of the principle of segregation in combination with polymorphic DNA markers—regions of genetic sequence with population variance—to identify genomic regions containing these markers that are shared between family members (that is, inherited from the same common ancestor) and segregate (co-occur) with the trait being studied. Because of the relatively long segments of DNA involved, the resolution or ability of linkage to identify the specific causal gene variant is low; it instead identifies broad regions on a chromosome where a disease gene may lie. In addition, the approach is only sensitive to gene variants that have a large effect, explaining greater than 10% of phenotypic variance (Williams & Blangero, 1999).

Parametric linkage analysis requires specification of a model of inheritance (e.g., dominant, recessive, X-linked), disease prevalence, marker allele frequency, mutation rate, and penetrance—the proportion of mutation carriers who exhibit the trait (Bush & Haines, 2010). This type of approach is most effective for rare disorders/extreme traits that occur in multiple individuals within large multigenerational pedigrees. Using likelihood inference, it is possible to estimate recombination fractions, which represent the probability that two alleles at different positions on the same chromosome have been inherited from different parental chromosomes. Any nonrandom segregation of the disease allele and the marker allele indicates linkage between the disease and marker locus. The strength of linkage can be evaluated statistically by dividing the likelihood that the marker and disease alleles are linked (recombination fraction less than 0.5) by the likelihood that they are unlinked (recombination fraction of 0.5). The logarithm of this odds ratio (LOD score) is used, so that evidence from different families can be simply added. LOD scores greater than 3 are generally accepted as indicative of significant linkage in parametric studies, corresponding to one false positive result in every 20 genome-wide linkage scans, while a score of 2.2 can be considered as suggestive of linkage, corresponding to one false positive per genome scan (Lander & Kruglyak, 1995).

In the nonparametric linkage approach, minimal assumptions are made with respect to the model of disease inheritance or the marker allele frequencies (Ferreira, 2004). Here, geneticists test whether the extent of sharing between related individuals at a genetic marker is correlated with the similarity of these individuals on the measured phenotypic trait. In the case of pairs of siblings, at any locus they can share zero, one, or two allelic variants identical by descent (i.e., showing the same inheritance pattern). Under an additive model, linkage to a locus would be supported if sibling pairs sharing two alleles at that locus were more similar on the trait than those sharing one allele, who in turn, were more similar than those not sharing any alleles identical by descent. The same logic can be applied to dichotomous traits; in which case, the test assesses whether there is a relationship between allelic sharing and concordance for the trait.

5. *Association Methodology*

Whereas linkage analysis works well for single-gene disorders and for complex traits involving rare genetic variants with large effect sizes, for the majority of traits influenced by many common variants of small effect (Plomin & Kovas, 2005), gene finding requires association analysis. Historically, association studies have progressed from early studies of candidate genes, focusing on a specific genetic variant hypothesized to have a functional role in influencing the trait, to hypothesis-free genome-wide association scans (GWASs), where millions of genetic markers across the entire genome are systematically tested for association with the trait. Association analysis has also been used to refine a locus identified via linkage analysis, via dense genotyping of polymorphic markers in the linkage region (fine mapping) followed by association analysis with these markers.

Within a case-control context, association analysis compares the frequency of alleles in samples of unrelated cases (with the disorder) and controls (who are free from the disorder). Logistic regression can be used to accommodate more complex models, for example, including risk modifiers (see Morris & Cardon, 2007). For a quantitative trait, a regression approach is similarly used. Association analysis typically uses biallelic SNPs as genetic markers, which gives three possible diploid genotypes (combinations of maternal and paternal alleles) at any specific locus. For example, at a SNP with the alleles C and G, a person could have a genotype of CC (i.e., homozygous for the C allele), CG (heterozygous), or GG (homozygous for the G allele). Under an additive model, a linear relationship is tested where the heterozygotes of the quantitatively defined phenotype are expected to fall midway between the phenotypic scores of the two types of homozygotes. Dominance represents the deviation from additivity, where the heterozygote phenotype is similar to that of one of the homozygotes, but this effect is not often modeled in atheoretical GWAS studies. Genome-wide genotyping can be carried out using DNA microarray chips in which assays for millions of SNPs are arrayed on a single chip. Whereas these SNP arrays contain three orders of magnitude fewer polymorphisms than the three billion base pairs of a sequenced genome, many polymorphisms that are not genotyped directly can be imputed using a population-relevant human DNA reference sequence. Imputation takes advantage of the phenomenon of linkage disequilibrium (i.e., closely neighboring variants on a section of the genome tend to be inherited together), making it possible to deduce the likely sequence at loci that have not been genotyped. This yields information on up to 38 million SNPs, with around 8 million of these reliably imputed and having minor allele frequencies greater than 1%.

An important consideration in population-based association studies is the potential presence of population stratification, which can induce false positive findings. If a sample comprises groups of individuals from distinct populations (i.e., differing genetic background) who also differ on the trait being studied, then any SNPs showing different allele frequencies between populations will show significant association with the trait. There are various methods to control for this artifact; one of the most common is to adjust the analysis for population-structure principal components that are derived from genome-wide genetic data. Family-based association studies are free from population stratification confounding because cases and controls are drawn from the same family. Variations of family-based methods—transmission/disequilibrium tests, haplotype-based relative risk, logistic regression models, extension to quantitative traits—are discussed in detail in Dudbridge (2007). Such studies depend on multiple family members agreeing to participate, so they are slower and require greater investment in recruitment to achieve the large sample sizes needed for GWASs. However, they can be highly informative: whether associated markers are causal or in linkage disequilibrium with the causal variant can be evaluated by modeling combined linkage and association. Parent-of-origin effects can also be investigated because it is possible to resolve which allele is maternal (inherited from the mother) and which is paternal (inherited from the father). Whether the maternal or paternal gene influences a trait is often determined by genomic imprinting, an epigenetic process tied to DNA and histone methylation that affects gene expression in a parent-of-origin specific way. For example, if the paternal copy of a gene is "imprinted," this means that it is silenced, while the maternal copy of the gene will be normally expressed.

Linkage and association methods typically map genetic variation at a single nucleotide site or short sequence length, but structural genetic changes can also contribute to trait variation. Structural variants are much larger and involve variation in chromosomal structure that may arise, for example, during gamete formation, wherein stretches of DNA are rearranged within/between chromosomes. For example, studies of rare translocations have made important contributions to reading and language research; these are rearrangements where one stretch of chromosome breaks off and aligns to another chromosome. For a balanced translocation, no genetic material is either gained or lost in the cell, but it still might disturb gene function if the location of the breakpoint occurs within, or close to, a gene. Deletion and duplication of DNA that affects a substantial number of nucleotide pairs is known as a copy number variation; this type of variant has only recently been considered in relation to reading and language but is commonly associated with neurodevelopmental disorders.

6. Genetic Linkage of Reading and Language Abilities and Identification of Candidate Genes

In the 1990s, the only affordable gene mapping method able to interrogate hundreds of genetic variants across the genome was genetic linkage. At the time, there was still uncertainty regarding the size of the expected gene effects. The new millennium saw rapid progress in genotyping technology, and a shift to genome-wide association methods that could detect much smaller gene effects, consistent with multifactorial inheritance. But the most researched candidate genes for reading disability and language impairment have arisen from linkage analysis and subsequent fine mapping (incorporating association analyses), so these findings will be reviewed, with a brief summary presented in table 39.1.

6.1. Reading-Related Loci

6.1.1. DYX1 The first finding in this field showed dominant transmission of specific reading disability in native English–speaking families linked to chromosome 15 (Smith, Kimberling, & Pennington, 1983). Whereas this locus was not supported in future studies, there was convergence on a nearby region,15q15.1 to 15q21 (the DYX1 locus), from five independent studies of reading disability (Cardon et al., 1994; Chapman et al., 2004; Grigorenko et al., 1997; Nothen et al., 1999; Schumacher et al., 2008). These studies were based on native English or German speakers. Whole genome linkage scans in Finnish, Norwegian, and Dutch samples (de Kovel et al., 2004; Fagerheim et al., 1999; Kaminen et al., 2003; Nopola-Hemmi et al., 2001) and a targeted scan of chromosome 15 in a Danish sample (Bisgaard, Eiberg, Moller, Niebuhr, & Mohr, 1987) did not find linkage to DYX1, and three studies in native English samples also failed to replicate this region (Grigorenko et al., 2001; Kaplan et al., 2002; Marlow et al., 2003). Studies of two Finnish families with reading problems and balanced translocations (different position in each family) helped to refine the locus region (Nopola-Hemmi et al., 2000), and a candidate gene (*DYX1C1*[2], also now known as *DNAAF4*) therein was later discovered (Taipale et al., 2003).

Support for *DYX1C1* involvement in reading and language is equivocal. Meta-analyses of 9 and 10 reading disability case-control/family-based studies did not find support for association with two functional SNPs (-3G/A and 1249G/T polymorphisms, respectively) within *DYX1C1* (Tran et al., 2013). In 790 unselected Australian families comprising up to five adolescent/young adult siblings, association of these variants was not supported, but another exonic coding SNP (rs1781926) showed association to nonword reading (Bates et al., 2010). In a larger unselected sample of nine-year-olds from the United Kingdom, rs1781926 was not associated with a word reading measure or reading and spelling component (Luciano et al., 2013). A more recent finding shows an association with a SNP and haplotype (a set of neighboring SNPs on a single chromosome inherited from one parent and that tag unmeasured variants) in the *DYX1C1* regulatory promoter region (responsible for starting gene transcription) in an Indian dyslexic sample (Venkatesh, Siddaiah, Padakannaya, & Ramachandra, 2014). But across eight European countries (NeuroDys cohort) including 958 individuals with dyslexia versus 1,150 controls, meta-analysis found no association with four previously associated SNPs in *DYX1C1* (Becker et al., 2014). Nonreplication was argued to be a likely product of low power, with initial findings susceptible to the winner's curse (inflated effect size) that requires even larger samples to replicate the smaller true effect. The region containing *DYX1C1* has been linked to word reading scores in families with SLI probands and suggestively linked to categorical and quantitative measures of language skill in this same cohort (Rice, Smith, & Gayan, 2009). A variant within *DYX1C1* has been associated with SLI case-control status, but in the opposite direction to the original report (Newbury et al., 2011).

Another candidate gene, *CYP19A1*, was identified in the DYX1 locus because it was disrupted by the other translocation observed in the Finnish families (Anthoni et al., 2012). In four samples (*N* ranging from 22 to 268) from Finland, Germany, and the United States, where parental dyslexia status was also known, association between various SNPs/haplotypes and dyslexia was supported in three of the samples. Quantitative measures of reading, oral motor skills, vocabulary, and phonological processing were also associated with single SNPs situated within the coding region of the gene. In the Finnish and German samples, exonic variants within this gene were not associated with dyslexia (Matsson et al., 2015), but further studies are needed to confirm whether *CYP19A1* has a role.

6.1.2. DYX2 Motivated by a purported relationship between autoimmune disease and dyslexia (Pennington, Smith, Kimberling, Green, & Haith, 1987), Cardon et al. (1994) tested linkage of reading disability to the human leukocyte antigen region, a highly polymorphic gene complex encoding cell-surface proteins that regulate the human immune system. They reported significant linkage to 6p21.3 in two samples (*N* of 114 and 50) of sibling pairs (the latter comprising dizygotic twins). Linkage to this DYX2 locus was replicated in both targeted (Gayan et al., 1999; Grigorenko et al., 2003) and genome-wide (Fisher et al., 2002) linkage scans in native English–speaking cohorts. However, genome-wide linkage scans in Finnish, Norwegian, Dutch, and German samples did not replicate this finding (e.g., de Kovel et al., 2004; Fagerheim et al., 1999; Kaminen et al., 2003; Nothen et al., 1999). There is some evidence that this region is also involved in language impairment (Rice et al., 2009), but note that there have been null findings, too (Bartlett et al., 2002; SLI Consortium, 2002). Fine mapping of the region using high-density SNP maps was undertaken to refine the location of the quantitative trait locus. Deffenbacher et al. (2004) reported family-based association with SNPs in five genes—*VMP*, *DCDC2*, *KIAA0319*, *TTRAP*, and *THEM2*—displaying strong linkage disequilibrium. In subsequent studies, two of these genes, *DCDC2* and *KIAA0319*, showed heightened support (Cope et al., 2005; Francks et al., 2004; Schumacher et al., 2006).

Scerri and Schulte-Koerne (2010) summarized the association findings of SNPs and haplotypes in these DYX2 locus genes for six independent studies of dyslexia (native English– and German-speaking) and two studies of unselected samples (native English–speaking). All studies confirmed at least one association with a variant that they tested in either *DCDC2* or *KIAA0319*, although there was inconsistency between studies in which variant that was and occasionally the direction of the effect (e.g., Luciano et al., 2007; Scerri et al., 2011). Harold et al. (2006) found no evidence for association with *DCDC2* in their dyslexic sample, although an interaction between SNPs in *DCDC2* and *KIAA0319* was found and later supported in a German sample (Ludwig et al., 2008). Epistatic (gene by gene) interaction effects occur when the effect of a particular genetic variant is dependent on the status of one or more other genetic variants; that is, they do not combine in a purely additive fashion to influence the phenotype (risk of disorder or quantitative trait). Epistatic interactions can be modeled within a regression framework, but the power to detect such effects, especially where more than two genes are considered, is lower than for single gene effects.

In 191 US families, Brkanac et al. (2007) reported no association between dyslexia case-control children and the two SNPs that they tested in *KIAA0319*. A meta-analysis of the family-based and transmission disequilibrium test results (347 families from two studies) for the nonsynonymous (protein-changing) variant, 931C>T, in *KIAA0319*, reported an odds ratio of 0.87, with the T variant associated with greater risk of dyslexia (Zou et al., 2012). A similar meta-analysis including 941 cases and 1,183 controls from four studies and 1,119 families from five studies focused on five SNPs in *DCDC2* and found association for rs807701, with the C allele related to increased risk of dyslexia (Zhong et al., 2013). However, in the NeuroDys cohort (eight European samples comprising 958 dyslexia cases versus 1,150 controls), no association was found with *DCDC2* or *KIAA0319* (Becker et al., 2014). In a large Chinese study of dyslexic children (502 cases, 522 controls), several SNPs in *DCDC2* and *KIAA0319* (and a haplotype within it) showed nominal association, but these did not withstand a multiple testing correction (Sun et al., 2014). In a smaller Indian cohort of 210 cases and 256 controls aged 6 to 18 years, a SNP in *KIAA0319*, but none in *DCDC2*, was associated with dyslexia (Venkatesh et al., 2014). The generally inconsistent association findings can be due to many reasons, including variation in genetic effects between different populations, variation in genetic effects influencing different languages, heterogeneity in phenotypes (and definitions of cases) used to test association, varying levels of statistical power between studies (with low-powered studies plagued by false positive results), and variation in environmental modifiers of a genetic effect between studies.

It has also been investigated whether variation in the DYX2 locus influences language ability. In families with an SLI proband (86 probands, 134 siblings, 102 parents), evidence of linkage to the DYX2 locus was found and SNPs in *KIAA0319* (but not in *DCDC2*) were associated with language scores (Rice et al., 2009). Variation in *KIAA0319*, but not in *DCDC2*, was again supported for an expressive language quantitative trait and in SLI case-control analysis (Newbury et al., 2011). But null association between *KIAA0319* and language skills have been reported, too (Mascheretti et al., 2014). Eicher et al. (2014) found association between a marker in *DCDC2* and severe language impairment in the Avon Longitudinal Study of Parents And Children (ALSPAC), and this was replicated in an independent language impairment cohort from Iowa. In another study of ALSPAC, when reading disability and comorbid reading disability and language impaired cases were combined, a stronger association signal for SNPs in *DCDC2* and *KIAA0319* was observed compared to that for reading disability alone (Scerri et al., 2011). A short tandem repeat polymorphism within a *DCDC2* deletion site, one of the original *DCDC2* variants associated with reading disability (Meng et al., 2005), has also been reported to be associated with language impairment in ALSPAC (Powers et al., 2013). Interestingly, *regulatory element associated with dyslexia 1* (or *READ1*, as it is now known because of evidence that it may modulate activity of the *DCDC2* promoter) has been reported to interact with a haplotype in *KIAA0319*, yielding either a protective or harmful effect on reading and language traits (Powers et al., 2016).

6.1.3. DYX3 In one of the first genome-wide linkage studies of reading disability, linkage to region 2p15–16 (DYX3) was found in a large Norwegian pedigree (Fagerheim et al., 1999). This linkage was not replicated in Finnish or Dutch extended families (de Kovel et al., 2004; Nopola-Hemmi et al., 2001), but support was found in US, UK, and Canadian samples (Fisher et al., 2002; Francks et al., 2002; Petryshen, Kaplan, Hughes, Tzenova, & Field, 2002). In an unselected Australian sample of adolescents and young adults, the region showed suggestive linkage to nonword spelling and a nearby region, 2q22.3, to measures of irregular reading and regular spelling (Bates, Luciano, et al., 2007). It has proved difficult to identify reliably associated candidate genes in this region (Francks et al., 2002; Peyrard-Janvid et al., 2004). One study found within a 157 kb candidate region two co-regulated genes, *MRPL19* and *C2ORF3*, which were highlighted as important because their expression was associated with other dyslexia candidate genes and found in relevant brain regions (Anthoni et al., 2007). However, these genes may be more relevant for general cognitive ability, given that variants have been associated with IQ in multiple independent samples of children (ranging to adolescents and young adults) and to white matter integrity in a smaller sample of 76 children and young adults (Scerri et al., 2012).

6.1.4. DYX4 to DYX9 Among the other six loci (DYX4 to DYX9) with strongest evidence of linkage to reading disability, DYX5 (mapping to 3p12–q13), DYX6 (18p11.2), and DYX8 (1p36–p34) have produced candidate genes. A translocation involving DYX5 was observed in an individual with developmental dyslexia, with its breakpoint located in *roundabout, axon guidance receptor, homolog 1* (*ROBO1*), and a risk haplotype affecting expression of this gene co-segregated with dyslexia in 19 of 21 affected members of a large multigenerational family (Hannula-Jouppi et al., 2005). SNPs in *ROBO1* were subsequently associated with nonword repetition, but not reading and spelling measures, in an unselected sample of 1,111 individuals, most of whom were dizygotic twins (Bates et al., 2011), and a SNP correlated with low expression in brain tissue was associated with reading disability in 421 Canadian families (Tran et al., 2014). However, *ROBO1* association was not found with reading or language traits in 493 Italian families with a developmental dyslexia proband (Mascheretti et al., 2014). Linkage to the DYX5 region and preliminary associations of *ROBO1* SNPs have also been found for speech-sound disorder–related traits (Bates, Luciano, et al., 2007; Bates et al., 2011; Stein et al., 2004).

Linkage at DYX6 was originally found in UK and US developmental dyslexia samples (Fisher et al., 2002). Scerri et al. (2010) later genotyped almost 3,000 SNPs in the region in 187 UK families and found replicated association for markers in melanocortin 5 receptor (*MC5R*), dymeclin (*DYM*), and neural precursor cell expressed, developmentally down-regulated 4-like (*NEDD4L*). Variants in *NEDD4L* and *myosin VB* (*MYO5B*; initially identified, but not replicated, by Scerri et al., 2010) were associated with dyslexia in a German study of 388 cases and 364 controls (Mueller et al., 2014). Like DYX5 and DYX6, less linkage support has been found for DYX8 (cf. DYX1 to DYX3), but the *KIAA0319-like* (*KIAA0319L*) gene, originally investigated because of its similarity to *KIAA0319*, has been weakly associated with reading disability in family-based analyses and to a quantitative measure of word reading efficiency in this same sample (Couto et al., 2008). Two previously untested SNP variants within this gene were associated with developmental dyslexia under dominant and recessive models in a Chinese sample of 409 cases and 410 controls (Shao et al., 2015). Further replication studies of this gene (and especially the associated SNPs) are needed before it can be considered a reliable candidate.

In a relatively recent linkage study, which typed 133,165 SNP markers in 718 individuals from 101 families ascertained for dyslexia, five regions were implicated at a suggestive level of significance (Field et al., 2013). Three were replications of prior findings suggesting linkage at 4q13.1, 16p12.1, and 17q22 regions, and the two novel loci fell in regions linked to autism. Some replicated linkage regions (e.g., on 18p11) could only be confirmed in separate analyses of the larger pedigrees, underscoring the importance of major gene effects in individual families. Association analyses of the SNPs within the linkage intervals of interest were not significant.

6.2. Language-Related Loci The first locus linked to language impairment was discovered in a multigenerational pedigree (family KE) where a severe speech and language disorder (including orofacial dyspraxia) was segregating in an autosomal dominant way (Fisher, Vargha-Khadem, Watkins, Monaco, & Pembrey, 1998). In 27 members of this family (15 affected), strong evidence of linkage was found on chromosomal band

7q31, initially given the locus name *SPCH1*. A mutation that disrupted the *forkhead/winged helix transcription factor* (*FOXP2*) gene from this region was later implicated in this impairment for the KE family, as well as a translocation disrupting the gene in an unrelated individual with a similar disorder (Lai, Fisher, Hurst, Vargha-Khadem, & Monaco, 2001). Transcription factors are proteins that initiate and regulate the transcription of genes and thus affect expression of genes in specific cells throughout development. Rare mutations of *FOXP2* have been found in multiple independent cases with severe speech and language disorders, typically characterized by childhood apraxia of speech (Morgan, Fisher, Scheffer, & Hildebrand, 2017). This gene shows no evidence for association with variation in normal language abilities as indexed by a composite score of five language subtests (Mueller et al., 2016) or for more common language impairment, although investigations have been limited in testing only a small number of SNPs across this large gene, with one study testing only for the presence/absence of the single rare mutation that had first been found in the KE family (Meaburn, Dale, Craig, & Plomin, 2002; Newbury et al., 2002; O'Brien, Zhang, Nishimura, Tomblin, & Murray, 2003). Linkage to this region has not been supported in other studies that sampled less severe forms of language impairment (Bartlett et al., 2002; SLI Consortium, 2002; O'Brien et al., 2003). Another gene, *contactin-associated protein-like 2* (*CNTNAP2*), located in a more distal region of chromosome 7, at 7q35–q36, was investigated due to a FOXP2-bound fragment being found in one of its introns, that is, in a functional regulatory region with influence on gene expression (Vernes et al., 2008). SNPs in *CNTNAP2* were genotyped in 184 families collected by the SLI Consortium (a group of UK researchers studying the genetic etiology of SLI), and rs17236239 was the most significant SNP in a cluster associated with nonword repetition; a haplotype spanning this cluster also showed association. Variation in these same SNPs with the same direction of allelic effect has also been associated with age of first word in autism spectrum disorder (Alarcon et al., 2008) and with communicative behavior in toddlers from the general population (Whitehouse et al., 2011). A different SNP in *CNTNAP2*, rs6951437, has been associated with language impairment in the ALSPAC population cohort (Eicher et al., 2013).

The SLI Consortium conducted the first systematic linkage study of less severe (but persistent) language impairment using receptive and expressive language and nonword repetition measures collected in 219 sibling pairs from 98 families drawn from clinical and epidemiological studies (SLI Consortium, 2002). Two regions, 16q24 (SLI1) and 19q13 (SLI2), were significant, with the best evidence coming from nonword repetition linked to 16q24. By adding an additional 86 families to the analysis, the linkage on chromosome 16 increased (SLI Consortium, 2004), and in multivariate analyses (which can increase power) this locus was further linked to reading and spelling measures, and the chromosome 19 linkage peak plus a novel peak on chromosome 10 showed marginal significance (Monaco, 2007). In an independent sample of 93 SLI probands and 300 of their first-degree relatives, replication of the chromosome 16 and 19 linkage regions was sought and confirmed (Falcaro et al., 2008). However, in a genome-wide linkage study of 214 sibling pairs where 40% of the probands showed poor language ability, these regions were not replicated; novel peaks at 10q23.33 and 13q33.3 were significantly linked to nonword repetition (Evans, Mueller, Gamazon, Cox, & Tomblin, 2015). In follow-up work by the SLI Consortium, two candidate genes (*c-maf-inducing protein* [*CMIP*] and *calcium-transporting ATPase type 2C* [*ATP2C2*]) within the SLI1 locus were identified through fine mapping. SNPs in these genes were associated with nonword repetition (phonological short-term memory) and reading-related measures in language-impaired individuals (an expanded sample of the SLI cohort in which the SLI1 locus was first identified) but not to phonological memory in the ALSPAC population-based sample (Newbury et al., 2009; Newbury et al., 2011).

Taking a different analysis approach, Bartlett et al. (2002) investigated genome-wide linkage to categorical groupings of clinical SLI, language impairment based on a quantitative cutoff, and reading discrepancy with IQ, in five Canadian families (total of 86 individuals) with a strong history of language impairment. They did not replicate previous SLI loci, but found that a locus on 13q21 (SLI3) was linked to their reading discrepancy trait under a recessive model; evidence for this locus increased when a further 21 families from the United States were included (Bartlett et al., 2004). It is not clear whether this locus is more important for reading ability than language, but note that early genetic studies of autism also identified linkage to this locus with increased evidence for probands where language delay is present (Barrett et al., 1999).

In a Chilean population isolate sample with a high SLI prevalence (Robinson Crusoe Island), linkage analysis of four pedigrees found significant linkage to chromosome 7 (SLI4), in a broad region spanning *FOXP2* and *CNTNAP2* candidate genes (Villanueva, Jara, & Palomino, 2010; Villanueva et al., 2011). However, segregation analysis implicated alternative genes,

NOBOX and *TPK1*, as likely candidates. The SLI3 locus was replicated, but SLI1 and SLI2 were not. In a three-generational German family segregating language impairment, a very large region on chromosome 12 (containing around 600 genes) was linked to impairment, but no causative variants could be found in any of the six candidate genes sequenced at this locus (Addis et al., 2010). A deletion in 2q36.3 (SLI5), which removes exon 3 of the *transmembrane 4 L six family member 20* (*TM4SF20*) gene, was shown to be linked with language delay and accompanying white matter hyperintensities in Southeast Asian carriers (Wiszniewski et al., 2013).

As well as reviewing prior published literature, we report here findings from a new linkage study of SLI-related measures in a population sample comprising 396 families (286 sibling pairs, 98 trios, 11 quads, and 1 five-sibling family; 17.6% monozygotic twins, the remainder dizygotic twins or nontwin siblings). Adolescents and young adults (aged 14 to 26 years, mean 20.15 ± 2.95) from the Brisbane Adolescent Twin Study were measured on phonological memory (two nonword repetition tasks) and other measures of short-term (digit span) and working (letter-number sequencing) memory. A description of the tests and sample characteristics can be found in Bates et al. (2011). A genome-wide linkage scan incorporating 1,133 microsatellite and 15,872 autosomal SNP markers (see Middelberg et al., 2008, for genotyping information and quality control) was performed using the MERLIN program (Abecasis, Cherney, Cookson, & Cardon, 2000). The most significant linkage was found for the Wechsler Adult Intelligence Scale (WAIS) III forward digit span to 3q26.31 (peak marker was rs10486390) with a LOD score of 3.28. The score is just short of the 3.3 threshold for genome-wide significance that is commonly applied for this kind of analysis; no correction for multiple testing of different phenotypes was applied, given the moderate correlation among traits (see figure 39.1). All other traits showed some evidence of linkage in this same region, albeit with lower peaks, which suggests that general short-term memory processes are influenced by this locus. Suggestive linkage ($LOD > 2.2$) was also found for WAIS III letter-number sequencing (LOD of 2.50) to 7p21.1 (peak markers rs10486390 and rs10486391), with a lower peak in this region observed for backward digit span suggesting linkage to working memory processes. The Gathercole and Baddeley nonword repetition task showed suggestive linkage to 10q25.3 (LOD of 2.91; peak marker rs10509979) and 18q11.2 (LOD of 2.17; peak markers rs632301 and rs8094228). With regard to replication of previous SLI loci, we found no support for SLI1–SLI3, but observed replication of the SLI4 locus to the Donoghan and Campbell nonword repetition task (LOD of 1.48 with peak marker rs4726715 on 7q34) and composite nonword repetition score (LOD of 1.46 with peak marker rs10487338 on 7q31.1). Note, however, that, previously, language-related SNPs in *CNTNAP2* did not show significant association in our sample (Luciano et al., 2013). Support for linkage close to SLI5 (2q36) was also found for composite nonword repetition (LOD of 2.1 with peak marker rs10497958 on 2q34). Because linkage studies have been superseded by GWASs, it will be important to check whether the results of future, large GWAS studies of language ability align with the linkage regions we report here.

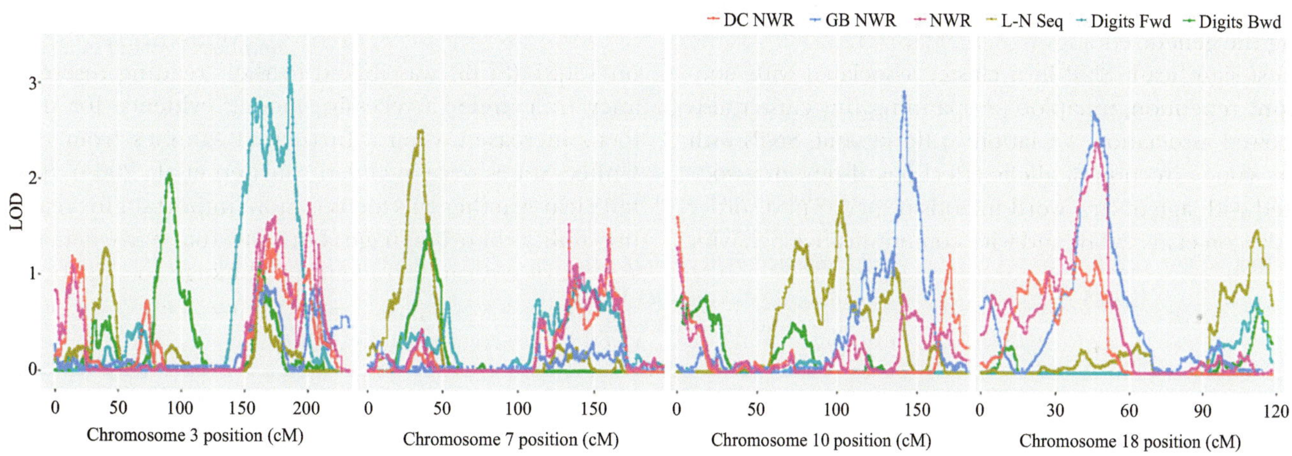

FIGURE 39.1 Linkage plots for four chromosomes (3, 7, 10, and 18) where suggestive linkage ($LOD > 2.2$) was observed for at least one of the SLI-related measures: Donoghan and Campbell's nonword repetition task (DC NWR); Gathercole and Baddeley's nonword repetition task (GB NWR); composite nonword repetition score (NWR); WAIS III Letter-Number Sequencing (L-N Seq); WAIS III Digit span forward (Digits Fwd); WAIS III Digit span backward (Digits Bwd).

Table 39.1

Summary of the linkage regions to reading disability (RD) and SLI and the candidate genes within them

Locus name	Chromosomal region	Original linkage report	Candidate gene	Associated phenotype
Dyslexia				
DYX1	15q21	RD (Smith et al., 1983)	*DYX1C1*	RD (Taipale et al., 2003)
DYX2	6p23–21.3	RD (Cardon et al., 1994)	*DCDC2, KIAA0319*	RD (Kaplan et al., 2002; Deffenbacher et al., 2004)
		Early grammatical impairment in SLI ascertained cohort (Rice et al., 2009)		
DYX3	2p15–16	RD (Fagerheim et al., 1999)	*MRPL19, C2ORF3*	RD (Anthoni et al., 2007)
	2p11	RD (Kaminen et al., 2003)		
DYX4	6q11.2–q12	RD (Petryshen et al., 2001)		
DYX5	3p12–q13	RD (Nopola-Hemmi et al., 2001)	*ROBO1*	RD (Hannula-Jouppi et al., 2005)
DYX6	18p11.2	RD (Fisher et al., 2002)		
DYX7	11p15.5	RD (Fisher et al., 2002)		
DYX8	1p36	RD (Rabin et al., 1993)	*KIAA0319L*	RD (Couto et al., 2008)
		Omnibus language measure in SLI cohort (Rice et al., 2009)		
DYX9	Xq27.2–q28	RD (de Kovel et al., 2004)		
SLI				
SLI1	16q23.1–q24	SLI (SLI Consortium, 2004)	*CMIP, ATP2C2*	Phonological memory in an SLI cohort (Newbury et al., 2009)
SLI2	19q13.13–q13.41	SLI (SLI Consortium, 2004)		
SLI3	13q14.3–q31.1	SLI (SLI Consortium, 2004)		
		Reading-IQ discrepancy in an SLI cohort (Bartlett et al., 2002)		
SLI4	7q31–q36	SLI (Monaco, 2007)	*CNTNAP2*	Phonological memory in an SLI cohort including nonaffected family members (Vernes et al., 2008)
		RD (Kaminen et al., 2003)		
SLI5	2q36	Language delay (Wiszniewski et al., 2013); exon 3 deletion co-segregated with language delay in South-east Asian families	*TM4SF20*	
Other	2p22	SLI (Bartlett et al., 2002)		
	17q23	Reading-IQ discrepancy in an SLI cohort (Bartlett et al., 2002)		

Note that indicative (or original) study reference is given, for a review of positive and null findings for reading loci, see Scerri and Schulte-Koerne (2010), and for a review of language loci, see Reader, Covill, Nudel, and Newbury (2014).

7. Genome-Wide Association Studies of Reading and Language Abilities

Linkage mapping has been successful in identifying genes responsible for a large number of rare single gene disorders, but this approach lacks the power and precision needed to detect more subtle effects underlying complex polygenic traits. Linkage-based strategies have found very limited evidence of genes of large effect size influencing reading and language traits, with the major loci harboring private mutations only relevant to individual families, and equivocal replication support for candidate genes within linkage intervals. The recognized ability of association mapping to identify genes of small effect, together with the advent of low-cost SNP chips with genome-wide coverage led to a rapid shift in psychiatry and complex genetics toward hypothesis-free association scans. Precision GWASs come with a heavy price tag in the form of sample sizes orders of magnitude larger than those conventionally envisaged. To maximize power, it is standard to meta-analyze GWAS results from cohorts where the same or similar phenotypes have been collected. This becomes challenging for studies of reading and language traits where different measures can be used to index the same trait. The number of cohorts contributing to meta-analysis will be limited to those with comparable indices and any heterogeneity in measurement will weaken association signals if variants have heightened/specific effects on some cognitive processes. The meta-analysis of GWASs using case-control designs suffers similar decreases in power when different diagnostic criteria have been used between studies and thus create heterogeneity within case and control groupings.

GWASs of reading and language abilities have used quantitative traits (from the general population) and case-control designs (Eicher et al., 2013; Field et al., 2013; Harlaar, Meaburn, et al., 2014; Luciano et al., 2013; Nudel et al., 2014; St. Pourcain et al., 2014); the results of which are summarized in table 39.2. The largest GWAS on quantitative reading traits was performed using an Australian (Brisbane Adolescent Twin Study; aged 12 to 25 years) and a British sample (ALSPAC; age 9), giving a total of 6,255 participants for meta-analysis (Luciano et al., 2013). Association tests using imputed data (around 2.5 million SNPs) for both a whole-word reading task and a factor-score based on regular-word, irregular-word, and nonword reading and spelling tests, showed suggestive association of a SNP in *DAZAP1* with both these measures. Additional suggestive associations were found that were unique to one or other of the measures. Two SNPs (not located in genes) associated with whole-word reading were subsequently tested for replication in a Dutch population sample (N=1,020) measured on various reading measures, but uncorrected for general intelligence (Carrion-Castillo et al., 2016). These were not significant in either children or adults. In addition to testing single SNP associations, Luciano et al. (2013) tested whether associations of SNPs located in known genes and their regulatory regions were over-represented in particular genes. Around 18,000 genes were examined in this way, and four genes located on chromosome 1—*CDC2L1*, *CDC2L2*, *LOC728661*, and *RCAN3*—showed significant effects. These genes do not, however, have any obvious biological relevance to reading or language, in terms of currently known functions.

A second substantial quantitative trait design, this time including children with reading and language impairments and their nonaffected siblings was undertaken by Gialluisi et al. (2014). These authors tested association for around 5.5 million directly genotyped or imputed SNPs with a factor derived from a range of reading and language measures in three samples with a total of 1,862 (mean age 10–11 years) and combining data from the United Kingdom and the United States. Their analysis showed suggestive associations for SNPs in *CCDC136/FLNC* and *RBFOX2* genes. Both *FLNC* and *CCDC136* had previously shown nominal association with reading and language in the gene-based tests of the Luciano et al. (2013) GWAS. The SNP rs3734972 (located in *FLNC*) was associated in both the Luciano et al. (2013) and Gialluisi et al. (2014) studies, although with opposite allelic direction of effects in the different cohorts. The most significant SNP from the Gialluisi et al. (2014) study (in *RBFOX2*) was subsequently tested in the replication study of Carrion-Castillo et al. (2016), but no support was found.

In what would be considered a very underpowered analysis by current GWAS standards, Eicher et al. (2013) used performance on quantitative measures of reading and language in the ALSPAC cohort to classify 174 participants as having impairment in both reading and language. In addition to this case-control analysis using observed genotypes (~500,000 SNPs), they performed comparisons of those with "pure" reading (n=353) and language deficits (n=163), that is, excluding those with comorbid status. For comorbid reading and language deficits, four SNPs in *ZNF385D* showed suggestive association, the largest with an odds ratio of 1.811 (greater risk for those with the minor allele). Two of these SNPs were also associated with picture vocabulary test scores (but not oral reading performance) in an independent sample (N=440). This second sample had also been characterized using brain imaging, and

Table 39.2

Summary of the top genome-wide association signals from studies of reading and language traits

Chromosome	SNP	Closest known gene	Trait	SNPs tested, *N*	Total *N*
1p13.1	rs4839516		Single word reading (Luciano et al., 2013)	2,373,249 imputed	6,135 individuals (children and adolescents)
2q31	rs12474600	*CCDC141*	Language factor (Harlaar, Meaburn, et al., 2014)	1,724,317 imputed	2,329 children
3p12.3	**rs7642482**	***ROBO2***	**Expressive vocabulary in 15 month olds (St. Pourcain et al., 2014)**	**2,449,665 imputed/ genotyped**	**8,889 individuals (toddlers)**
4p15.3	rs11945798		Single word reading (Luciano et al., 2013)	2,373,249 imputed	6,135 individuals (children and adolescents)
4q26	rs482700		SLI (Eicher et al., 2013)	500,527 genotyped	163 SLI (based on test performance) and 4,117 controls (children)
5p13	rs1353835 (maternal effect)		SLI (Nudel et al., 2014)	614,937 genotyped	278 families (297 SLI children)
5q35.1	rs9313548	*FGF18*	RD (Field et al., 2013)	125,047 autosomal SNPs	718 families (400 with RD)
6p24	rs479526		Reading and spelling factors (Luciano et al., 2013)	2,373,249 imputed	6,127 individuals (children and adolescents)
7p21	rs6954796		Phonological memory (Luciano et al., 2013)	2,373,249 imputed	6,583 individuals (children and adolescents)
7p21	rs13307587		Reading and spelling factors; single word reading (Luciano et al., 2013)	2,373,249 imputed	6,127 individuals (children and adolescents)
7q32	rs59197085	*CCDC136*	Reading and language factor (Gialluisi et al., 2014)	5,518,496 imputed	1,862 SLI/RD children and their nonaffected siblings
8q24.1	rs2590673		RD (Eicher et al., 2013)	500 527 genotyped	353 SLI (based on test performance) and 4,117 controls (children)
10p15	rs17135159		Reading and spelling factors (Luciano et al., 2013)	2,373,249 imputed	6,127 individuals (children and adolescents)
10q23.3	rs1326167		Language factor (Harlaar, Meaburn, et al., 2014)	1,724,317 imputed	2,329 children

Chromosome	SNP	Closest known gene	Trait	SNPs tested, *N*	Total *N*
10q25.3	rs180950		RD (Eicher et al., 2013)	500,527 genotyped	353 SLI (based on test performance) and 4,117 controls (children)
10q21.2	rs3213056	*CDK1**	Reading and spelling factor (Luciano et al., 2013)	2,373,249 imputed	6,127 individuals (children and adolescents)
11p15.2	rs10734234	*INSC*	Expressive vocabulary in 15 month olds (St. Pourcain et al., 2014)	2,449,665 imputed/ genotyped	8,889 individuals (toddlers)
12q15	rs11176749	*CAND1*	Expressive vocabulary in 15 month olds (St. Pourcain et al., 2014)	2,449,665 imputed/ genotyped	8,889 individuals (toddlers)
13q21.3	rs1928007		Single word reading (Luciano et al., 2013)	2,373,249 imputed	6,135 individuals (children and adolescents)
14q12	**rs11158632 (paternal effect)**		**SLI (Nudel et al., 2014)**	**614,937 genotyped**	**278 families (297 SLI children)**
14q23	rs11158345	*PRKCH**	Single word reading (Luciano et al., 2013)	2,373,249 imputed	6,135 individuals (children and adolescents)
16q22	rs764255	*C16orf47**	Single word reading (Luciano et al., 2013)	2,373,249 imputed	6,135 individuals (children and adolescents)
16q23	rs7187223		Single word reading (Luciano et al., 2013)	2,373,249 imputed	6,135 individuals (children and adolescents)
19p13.3	rs1654584	*DAPK3**	Expressive vocabulary in 15 month olds (St. Pourcain et al., 2014)	2,449,665 imputed/ genotyped	8,889 individuals (toddlers)
19p13.3	rs4807927	*DAZAP1*	Reading and spelling factors; single word reading (Luciano et al., 2013)	2,373,249 imputed	6,127 individuals (children and adolescents)
21q11.2	rs2192161	*ABCC13*	Phonological memory (Luciano et al., 2013)	2,373,249 imputed	6,583 individuals (children and adolescents)
22q12.3	rs5995177	*RBFOX2*	Reading and language factors (Gialluisi et al., 2014)	5,518,496 imputed	1,862 SLI/RD children and their nonaffected siblings

Bold entries achieved genome-wide significance.

Note that a comprehensive summary of the effect allele, effect size, and *P* value is given in Carrion-Castillo et al. (2016), along with replication results for many of these SNPs.

*Located within the gene.

several associations between the two SNPs and diffusion tensor imaging tract volumes were identified. Analyses in this sample did not confirm replication of any other SNPs, showing suggestive association in their discovery sample, and no gene-based tests were significant. In the subsequent replication study by Carrion-Castillo et al. (2016), rs12636438 (in *ZNF385D*) was associated with their reading measures in children, but in the direction opposite to that predicted. The analyses of isolated reading and language impairments by Eicher et al. (2013) showed no overlap in the suggestively significant SNP associations, and of those tested for replication in that study, none were supported.

Field et al. (2013), in their linkage study of dyslexia using SNP markers, also performed GWA using a family-based association method (the transmission disequilibrium test). By GWAS standards, the number of variants tested (125,047) was small and so too was the sample (N=718), however, one association (rs9313548) was just short of genome-wide significance and was located near a relevant candidate gene (*FGF18*) involved in asymmetry and neocortex development.

Other neurodevelopmental disorders are potentially affected by epigenetic mechanisms, that is, heritable or environmentally induced chemical modifications that influence gene expression. Therefore, the role of parent-of-origin gene effects (which can indicate genetic imprinting, one type of epigenetic phenomenon) was investigated in a GWAS of language impairment. Nudel et al. (2014) examined SLI in 278 families of language-impaired children. Whereas they found no significant associations for the child genotypes without considering parent-of-origin, they did find paternal and maternal parent-of-origin effects on respective chromosomes 14q12 (in *NOP9*, which has been associated with schizophrenia) and 5p13 (near genes *PTGER4* and *DAB2*, which are implicated in autism and attention deficit). The paternal effect reached genome-wide significance. In an independent analysis (mothers from the ALSPAC cohort) a maternal effect was again found but in the opposite direction (i.e., the risk allele appeared to be protective). Although Nudel et al. (2014) had found a paternal effect for rs4280164 (in *NOP9*) in their parent-of-origin GWAS, a maternal effect at this SNP was subsequently reported for a variety of quantitative language measures in an attempted replication using 106 families (including those with language difficulties; Pettigrew et al., 2016). For two reading measures, a paternal effect of this SNP was shown, but the allelic effect was in the opposite direction to the original finding. Thus, the Pettigrew et al. (2016) replication study raised inconsistencies both with respect to direction of effect and parent-of-origin (paternal or maternal). Clearly, larger studies are needed to explore these inconsistent findings. Indeed, a reanalysis of the original Nudel et al. (2014) data using a newly developed method for more reliable haplotype estimation found reduced support for parent-of-origin associations (Howey et al., 2015), raising the possibility that the findings reflected type 1 error in an underpowered sample.

Harlaar, Trzaskowski, Dale, and Plomin (2014) reported on receptive language ability in the UK Twins Early Development Study (TEDS) sample (aged 12 years; N=2,329), assessed over the internet for vocabulary, semantics, syntax, and pragmatics. Analysis of a composite score against DNA data imputed to one million SNPs yielded no significant associations, and a test of the strongest SNP associations found in the remainder of the cohort did not replicate. Heritability of a trait due to common SNPs can be estimated using genetic array data. In this method, information across SNPs is combined to produce a genetic relationship matrix between unrelated individuals. The genetic similarity between individuals is related to their phenotypic similarity to estimate the heritability for the trait that is due to SNPs across the entire genome (Visscher et al., 2014). Using this method, a SNP heritability of 0.29 for receptive language and estimates of 0.20 and 0.29 for reading fluency at 7 and 12 years have been reported (Plomin et al., 2013). Given that these estimates are lower than those from twin studies, this suggests that much larger GWAS samples—on the order of tens of thousands rather than the thousands studied to date would both be needed and should yield genome-wide significant SNP associations.

The largest GWAS of a language-related trait was carried out in toddlers measured on expressive vocabulary at around 18 months (N=8,889) and 2 years (N=10,819) (St. Pourcain et al., 2014). Sampled from the European Early Genetics and Lifecourse Epidemiology Consortium, the study included cohorts whose reading and language measures have been analyzed in childhood (e.g., TEDS, ALSPAC) and a cohort from the Netherlands and one from Australia. The GWAS yielded one genome-wide significant result (for rs7642482, near *ROBO2*) for expressive vocabulary at 18 months, but this was not among the top associated SNPs at 24 months. In the TEDS sample (N~ 5,500), rs7642482 was nominally associated with reading speed, but not with phonological memory or reading comprehension at age 10. Nor was it associated with reading and language measures in an independent study of 483 children and 505 adults from the Netherlands (Carrion-Castillo et al., 2016). St. Pourcain et al. (2014) estimated SNP heritability for expressive vocabulary at each time point within each cohort; meta-analysis yielded a heritability of ~0.13

(standard error of 0.05) at both ages. This represented over half of the genetic variance estimated from a twin analysis of the TEDS sample.

The biggest barrier to gene discovery via GWAS is power, and this is fundamentally resolvable only by much larger sample numbers: Progress in discovery for reading and language studies, then, requires that phenotyping very large samples must be the first priority going forward. An example of the kind of breakthrough that can be made with such large aggregations, even of imperfect phenotypes, is educational attainment, a measure that is easily collected via questionnaire, unlike reading and language traits that require psychometric testing (Lee et al., 2018). The Social Science Genetics Consortium members have aggregated 1.1 million individuals for whom years of education are available. Analyses of these cohorts have yielded 1,271 genome-wide significant hits, with SNPs disproportionately located in genomic regions involved in regulating gene expression in the fetal brain and showing strong enrichment for biological pathways involved in neural development. The results are highly replicable (for instance in open data from the UK Biobank study) and can be used to derive polygenic scores in independent samples with genome-wide data. To calculate polygenic scores for a trait, each allele is weighted by its effect (e.g., beta or odds ratio) on the trait (estimated from a large GWAS) then summed to form a single score, typically across all independent SNPs in the genome (Wray et al., 2014). A polygenic educational attainment score has been reported to explain 5% of variance in reading achievement at age 14 (Selzam et al., 2017), suggesting that part of the signal in the GWAS of educational attainment reflects genes influencing reading ability. Likewise, polygenic reading ability scores (based on a much smaller GWAS sample of children and adolescents) have been shown to predict variance in adult socioeconomic and cognitive outcomes (Luciano et al., 2017).

8. *Other Gene Mapping Techniques*

The effect of chromosomal abnormalities on reading and language disabilities has been considered in the context of linkage and cytogenetics (studies of chromosomal number and structure using microscopy techniques). For example, translocations in particular chromosomal regions were shown to co-segregate in families with dyslexia and severe speech delay (Froster, Schultekorne, Hebebrand, & Remschmidt, 1993; Nopola-Hemmi et al., 2000), and a deletion on 21q22.3 was found to co-segregate with dyslexia (Poelmans et al., 2009). SNP array and DNA sequencing technology now allows structural genetic variants, such as copy number variants (CNVs; representing deletions and duplications of stretches of DNA), to be investigated on a genome-wide level. In a study of large CNVs (>1 Mbp) no difference in burden was found between dyslexics ($n=376$) and controls ($n=337$), even when restricted to rarer CNVs (Girirajan et al., 2011). However, two potential candidate genes (disrupted by a duplication) were identified, *AUTS2* and *WBSCR17*, and a deletion involved *IMMP2L*. For SLI, a case-control study found higher CNV burden in cases: their CNVs were more frequent, longer in overall length, of larger average size, and spanning an increased number of genes (Simpson et al., 2015). This increased burden was not due to large or de novo CNVs. (A *de novo* variant is one that newly arises in a child without being present in either parent, e.g., due to mutation.) In 749 participants from 343 unrelated families (with a proband with either reading disability or attention deficit/hyperactivity disorder), a genome-wide CNV analysis confirmed the lack of association between CNVs (number and total length), and a general factor derived from language and reading scores (Gialluisi et al., 2016). Furthermore, this study investigated CNV disruption of biological pathways considered important in reading disability and of candidate genes for reading and language abilities; no support for heightened involvement of these CNVs was found.

Association studies of rare genetic variants in reading and language abilities are lagging behind those of common variants due to the increased cost of sequencing and lower statistical power. To date, although most rare variant studies have been targeted to chromosomal/genetic regions of interest, there are a couple of exceptions. One example is a next-generation sequencing study of SLI cases from Robinson Crusoe Island, the same Chilean population isolate that had shown linkage to SLI3 and SLI4 loci in an earlier study (discussed in section 6.2). In this more recent investigation, Villanueva et al. (2015) sequenced the protein coding parts of the genome (the exome) of five SLI individuals and searched for novel nonsynonymous variants (i.e., variants that change protein sequence) as well as splice site variants (i.e., variants positioned at intron-exon boundaries, which can affect processing of RNA and thus protein production). When variants were observed in at least three of the five sequenced SLI individuals, they were genotyped in a larger validation cohort from the population. A nonsynonymous variant in the *NFXL1* gene was significantly associated with language impairment although this did not fall within the previously identified linkage regions. The coding region of this gene was sequenced in an independent British sample of children with SLI ($N=117$), and heterozygous variants predicted to be functionally important were found in four individuals, more than expected by chance. *NFXL1* is predicted to encode a transcription factor; it is expressed in different tissues, but in the brain it is most abundant in the cerebellum (Nudel, 2016).

By focusing on an extended Swedish family where developmental dyslexia was modeled as an autosomal dominant disorder with high to complete penetrance, Einarsdottir et al. (2015) compared the exome sequences of family members with and without developmental dyslexia. The authors discovered a mutation in a novel candidate gene, *CEP63*, encoding the centrosomal protein of 63 kDa and localized in the centrosome, an organelle they purport to be relevant to developmental dyslexia given the links between other candidate genes (exogenous *DYX1C1*, *DCDC2*) and centrosome and cilia functions. *CEP63* is expressed in the brain and plays a role in cell division; in this study a common *CEP63* variant was associated with brain white matter volume, in a region (left temporoparietal) previously reported to be influenced by variants of *DYX1C1* and *KIAA0319* (Darki, Peyrard-Janvid, Matsson, Kere, & Klingberg, 2012). Whole genome sequencing studies have not yet been described for dyslexia or SLI and neither have rare variant studies of reading and language abilities in the general population.

9. Candidate Gene Function

Of the most convincing candidate genes associated with reading abilities, the function has been best described for those in the DYX2 locus. The function of *KIAA0319* has been explored by characterizing expression patterns related to the variants (including haplotypes) associated with dyslexia susceptibility; the risk variants were associated with lower expression of the gene (Dennis et al., 2009; Paracchini et al., 2006). Knocking down expression of Kiaa0319 in embryonic rat neocortex disrupted neuronal migration, with potentially lasting effects on particular brain structures (Szalkowski et al., 2012). However, in mouse knockouts there was no evidence of disrupted neuronal migration, although protein expression was localized to brain (Martinez-Garay et al., 2017). Structural and cellular function similarities between *DCDC2* and *DCX*, a gene with known function, suggested neuronal migration to be affected by variation in *DCDC2*; downregulation of DCDC2 via in utero RNA interference in rats, confirmed changes in neuronal migration in topologically homologous brain regions that are activated in fluent and dyslexic readers (Meng et al., 2005). However, it is possible that migration effects observed in acute knockdown experiments are caused by unintended off-target or inflammatory responses (Baek et al., 2014). In *Dcdc2* mutant mice, increased glutamatergic synaptic transmission has been demonstrated (Che, Truong, Fitch, & LoTurco, 2015). This alteration was partly mediated by activated presynaptic *N*-methyl-D-aspartate receptors, which can cause variability in neuronal signaling. Deficits in behavioral phenotypes, such as visuospatial working memory and rapid auditory processing, have also been shown in *Dcdc2* mutant mice (Gabel et al., 2011; Truong et al., 2014). Relatedly, higher levels of glutamate have been associated with poorer reading and language skills in emergent readers from the general population (Pugh et al., 2014).

ROBO1 encodes a receptor that is a member of the neural cell adhesion molecule family of receptors. It is highly expressed in the developing and adult brain, and in *Robo1* knockout mice, a role in cell migration, axon growth, and guidance has been shown (Andrews et al., 2006). Complementary expression patterns have been observed in their ligands, the Slit molecules, with Slit-Robo signaling appearing to be important for interneuron migration (Andrews et al., 2008). A single nucleotide variant in the *ROBO1* region associated with a dyslexia susceptibility haplotype is bound by the LIM homeobox 2 protein, with support that it regulates *ROBO1* in humans (Massinen et al., 2016). In 10 dyslexic individuals carrying a rare ROBO1 haplotype that is weakly expressed, deficits in binaural suppression were observed, suggesting a role in auditory pathway crossing (Lamminmäki, Massinen, Nopola-Hemmi, Kere, & Hari, 2012).

The likely relevance of disordered neuronal migration in the etiology of reading disability has been highlighted by Poelmans, Buitelaar, Pauls, and Franke (2011) who used bioinformatics techniques to draw similarities in biological pathways for 14 dyslexia candidate genes (*DYX1C1*, *DCDC2*, *KIAA0319*, *C2ORF3*, *MRPL19*, *ROBO1*, *KIAA0319L*, *FMR1*, *PCNT*, *DIP2A*, *S100B*, *PRMT2*, *DOCK4*, and *GTF2I*) based on mining of the existing literature. Enrichment in the following gene ontology terms was found: generation of neurons, neurogenesis, axonogenesis, neuron projection morphogenesis, neuron projection development, and neuron development. *DCDC2*, *ROBO1*, and *S100B* were linked to neurite outgrowth, but when considering proposed protein function of the genes, seven other candidates could be linked to processes involved in neurite outgrowth and/or neuronal migration. There is some evidence that the analysis of interactions between genes in the neuronal migration and neurite outgrowth network might also influence reading disability (Harold et al., 2006; Shao et al., 2015), but this needs to be followed up by testing an increased number of relevant genes that have experimental evidence to support their function in the pathways.

There are fewer candidate genes for language ability, although *FOXP2* has been extensively characterized, with animal and cellular models contributing to its known functions in brain (French et al., 2012; Groszer et al., 2008; Shu et al., 2005; Wohlgemuth, Adam, & Scharff, 2014). FOXP2 is expressed in fetal and adult brain and several other organs; as an evolutionary conserved transcription factor it is involved in regulating gene expression. Lower levels of FoxP2 protein have been associated with compromised vocalization in songbirds and motor-skill learning deficits in mice (Fisher & Scharff, 2009). Area X, a basal ganglia structure with striatal characteristics is important for song learning in zebra finch, with increased FoxP2 expression shown in zebra finches actively learning to sing, and when FoxP2 in this region was reduced by RNA interference, errors in song imitation were observed (Haesler et al., 2007). Reduced FoxP2 levels in zebra finch disrupt dopaminergic modulation of vocal variability (Murugan, Harward, Scharff, & Mooney, 2013) and manipulation of FoxP2 expression also alters the effects of song practice on vocal variability (Heston & White, 2015), a behavior contributing to accurate song imitation. In mice that have two human-specific amino acid substitutions from *FOXP2* introduced into their endogenous *Foxp2* gene, learning and striatal neuroplasticity is affected, potentially as a result of differential tuning of the corticostriatal systems influencing declarative and procedural learning (Schreiweis et al., 2014). See Vernes (chapter 41 in this volume), for a comprehensive description of progress in studying neuromolecular mechanisms contributing to language, including current understanding of *FOXP2* function to the wider role of larger networks. One such network is the *FOXP2-CNTNAP2* regulatory pathway (Vernes et al., 2008). *CNTNAP2* codes for a neuronal cell adhesion protein and it lies in the same pathway as *FOXP2*, which regulates its expression. Other candidate genes for language ability, such as *CMIP*, *ATP2C2*, and *CYP19A1*, are functionally diverse, although the expression of *CYP19A1* in the human brain is similar to *DYX1C1* and *ROBO1* (Anthoni et al., 2012). While preliminary, the diversity of gene function and potential overlap with function for reading disability suggests that the genetic mechanisms underlying language ability are not specific. At present, no clear biological pathway has been highlighted as particularly important, although Foxp2 has been implicated in neurite outgrowth, neuronal migration, and synaptic plasticity (Devanna, Middelbeek, & Vernes, 2014; Vernes et al., 2011).

10. Moving Forward

A reasonable summary of the position to date from candidate and genome-wide studies is that the individual genetic effects influencing reading and language abilities (and conversely, disability) are very small, but that common genetic variants will together explain at least half of their genetic variance. This position is better established for reading than for language, and reading candidate genes appear to be linked to a molecular network involved in neuronal migration and neurite outgrowth (Poelmans et al., 2011), although better controlled RNA interference experiments (as well as data from complementary experimental approaches) are needed to confirm this. Novel candidate genes arising from GWASs of normal populations have not yet been validated, and given the relatively small sample sizes used thus far, many of these findings are likely to be false positives, which will not be reproducible at the individual SNP level. However, large GWASs of normal variation in other cognitive traits have been successful in demonstrating the importance of common SNP variation (Davies et al., 2016) and suggest that large-scale efforts directed to reading and language skills will be hugely informative. The larger sample sizes afforded by sampling quantitative measures in the general population may be preferable to a case-control design of reading/language disability and ability. Both designs present challenges in data collection: quantitative measures are labor intensive to collect (although the development of online reading and language tests would substantially increase sample sizes), and recruitment of individuals with clinical diagnoses requires ties with health and school authorities. Other quantitative traits, such as height—where self-report is fairly reliable—or clinical conditions (e.g., schizophrenia) that have a long history of psychiatric genetic research are clearly advantaged in terms of amassing the huge sample sizes needed for GWASs.

Going forward, it may also be useful to prioritize genes identified in existing large GWASs of reading/language endophenotypes, such as functional and structural brain MRI measures. In targeted studies of candidate genes, temporoparietal white matter structure has been suggested to be affected by *DYX1C1*, *DCDC2*, and *KIAA0319* (Darki et al., 2012), and a SNP in *THEM2* (near *KIAA0319*) has been associated with asymmetry in functional activation of the superior temporal sulcus (Pinel et al., 2012). Note, however, that neuroimaging studies that focus on reading/language have so far suffered from small sample size and few, if any, independent replications exist.

Beyond the main effect of genes, gene-by-gene interactions will be important to evaluate; targeted gene-by-gene investigation of existing candidate genes show some evidence for epistatic effects among *DYX1C1*, *KIAA0319/TTRAP*, and *GRIN2B* genes with developmental dyslexia traits (Mascheretti, Bureau, Trezzi, Giorda, & Marino, 2015) and between *DCDC2* and *KIAA0319* in developmental dyslexia case-control studies and developmental dyslexia-related quantitative traits in ALSPAC (Harold et al., 2006; Ludwig et al., 2008; Powers et al., 2013). Similarly, investigation of gene-by-environment interaction might also indicate moderation of a gene's effect on reading and language abilities by specific environmental variables. One study has implicated maternal smoking during pregnancy, socioeconomic class, and birth weight as potential environmental moderators of a variant in *DYX1C1* on reading and memory composite scores (Mascheretti et al., 2013). These results await replication and care should be exercised in interpreting such moderator effects as environmental because these might be confounded with genetic effects. Large samples will enable reliable SNP-level discovery. For example, a GWAS of educational attainment identified 74 significant SNPs, with 72 of these having a consistent direction of effect and 52 showing nominal significance in an independent cohort (Okbay et al., 2016). Genetic mapping approaches building on this work may be enhanced by gene-level analysis and biological pathway analysis, which can target cellular functions of genes (Hill, de Leeuw, et al., 2014) or groupings by biological role of genes (Hill, Davies, et al., 2014). Aggregation of information into SNP sets (Luciano, Montgomery, Martin, Wright, & Bates, 2011) and using genome-wide prediction functions to understand pleiotropy (where the same gene affects multiple different traits at the same time; Luciano et al., 2017) will also be of great value for maximizing information across samples, for translating results between samples, and disentangling the causes of comorbidity (Wray et al., 2014).

NOTES

1. The phrase specific language impairment (or SLI) has been in use for many decades. A recent Delphi survey of international experts now recommend use of the phrase developmental language disorder (or DLD) instead. However, since the studies cited here were conducted when SLI was the accepted term, we retain this term in the current chapter.
2. Following standard nomenclature, genes are denoted in italics, proteins in regular font. Uppercase letters denote the human version of the gene (i.e., *FOXP2*), lowercase the mouse version of the gene (i.e., *Foxp2*).

REFERENCES

Abecasis, G. R., Cherney, S. S., Cookson, W. O., & Cardon, L. R. (2002). Merlin-rapid analysis of dense genetic maps using sparse gene flow trees. *Nature Genetics, 30*, 97–101.

Addis, L., Friederici, A. D., Kotz, S. A., Sabisch, B., Barry, J., Richter, N., ... Monaco, A. P. (2010). A locus for an auditory processing deficit and language impairment in an extended pedigree maps to 12p13.31-q14.3. *Genes Brain and Behavior, 9*(6), 545–561.

Alarcon, M., Abrahams, B. S., Stone, J. L., Duvall, J. A., Perederiy, J. V., Bomar, J. M., ... Geschwind, D. H. (2008). Linkage, association, and gene-expression analyses identify *CNTNAP2* as an autism-susceptibility gene. *American Journal of Human Genetics, 82*(1), 150–159.

Andrews, W., Barber, M., Hernadez-Miranda, L. R., Xian, J., Rakic, S., Sundaresan, V., ... Parnavelas, J. G. (2008). The role of Slit-Robo signaling in the generation, migration and morphological differentiation of cortical interneurons. *Developmental Biology, 313*(2), 648–658.

Andrews, W., Liapi, A., Plachez, C., Camurri, L., Zhang, J. Y., Mori, S., ... Richards, L. J. (2006). Robol regulates the development of major axon tracts and interneuron migration in the forebrain. *Development, 133*(11), 2243–2252.

Anthoni, H., Sucheston, L. E., Lewis, B. A., Tapia-Paez, I., Fan, X. T., Zucchelli, M., ... Kere, J. (2012). The aromatase gene *CYP19A1*: Several genetic and functional lines of evidence supporting a role in reading, speech and language. *Behavior Genetics, 42*(4), 509–527.

Anthoni, H., Zucchelli, M., Matsson, H., Muller-Myhsok, B., Fransson, I., Schumacher, J., ... Peyrard-Janvid, M. (2007). A locus on 2p12 containing the co-regulated *MRPL19* and *C2ORF3* genes is associated to dyslexia. *Human Molecular Genetics, 16*(6), 667–677.

Baek, S. T., Kerjan, G., Bielas, S. L., Lee, J. E., Fenstermaker, A. G., Novarino, G., & Gleeson, J. G. (2014). Off-target effect of doublecortin-family shRNA on neuronal migration associated with endogenous microRNA dysregulation. *Neuron, 82*(6), 1255–1262.

Barrett, S., Beck, J. C., Bernier, R., Bisson, E., Braun, T. A., Casavant, T. L .... Gilman, S. (1999). An autosomal genomic screen for autism: Collaborative linkage study of autism. *American Journal of Medical Genetics, 88*(6), 609–615.

Bartlett, C. W., Flax, J. F., Logue, M. W., Smith, B. J., Vieland, V. J., Tallal, P., & Brzustowicz, L. M. (2004). Examination of potential overlap in autism and language loci on chromosomes 2, 7, and 13 in two independent samples ascertained for specific language impairment. *Human Heredity, 57*(1), 10–20.

Bartlett, C. W., Flax, J. F., Logue, M. W., Vieland, V. J., Bassett, A. S., Tallal, P., & Brzustowicz, L. M. (2002). A major susceptibility locus for specific language impairment is located on 13q21. *American Journal of Human Genetics, 71*(1), 45–55.

Bates, T. C., Castles, A., Luciano, M., Wright, M. J., Coltheart, M., & Martin, N. G. (2007). Genetic and environmental bases of reading and spelling: A unified genetic dual route model. *Reading and Writing, 20*(1–2), 147–171.

Bates, T. C., Lind, P. A., Luciano, M., Montgomery, G. W., Martin, N. G., & Wright, M. J. (2010). Dyslexia and DYX1C1: Deficits in reading and spelling associated with a missense mutation. *Molecular Psychiatry, 15*(12), 1190–1196.

Bates, T. C., Luciano, M., Castles, A., Coltheart, M., Wright, M. J., & Martin, N. G. (2007). Replication of reported linkages for dyslexia and spelling and suggestive evidence for novel regions on chromosomes 4 and 17. *European Journal of Human Genetics, 15*(2), 194–203.

Bates, T. C., Luciano, M., Medland, S. E., Montgomery, G. W., Wright, M. J., & Martin, N. G. (2011). Genetic variance in a component of the language acquisition device: ROBO1 polymorphisms associated with phonological buffer deficits. *Behavior Genetics, 41*(1), 50–57.

Becker, J., Czamara, D., Scerri, T. S., Ramus, F., Csepe, V., Talcott, J. B., ... Schumacher, J. (2014). Genetic analysis of dyslexia candidate genes in the European cross-linguistic NeuroDys cohort. *European Journal of Human Genetics, 22*(5), 675–680.

Bisgaard, M. L., Eiberg, H., Moller, N., Niebuhr, E., & Mohr, J. (1987). Dyslexia and chromosome-15 heteromorphism: Negative lod score in a Danish material. *Clinical Genetics, 32*(2), 118–119.

Bishop, D. V. M., Bishop, S. J., Bright, P., James, C., Delaney, T., & Tallal, P. (1999). Different origin of auditory and phonological processing problems in children with language impairment: Evidence from a twin study. *Journal of Speech Language and Hearing Research, 42*(1), 155–168.

Bishop, D. V. M., Laws, G., Adams, C., & Norbury, C. F. (2006). High heritability of speech and language impairments in 6-year-old twins demonstrated using parent and teacher report. *Behavior Genetics, 36*(2), 173–184.

Brkanac, Z., Chapman, N. H., Matsushita, M. M., Chun, L., Nielsen, K., Cochrane, E., ... Raskind, W. H. (2007). Evaluation of candidate genes for DYX1 and DYX2 in families with dyslexia. *American Journal of Medical Genetics Part B-Neuropsychiatric Genetics, 144B*(4), 556–560.

Bush, W. S., & Haines, J. (2010). Overview of linkage analysis in complex traits. *Current Protocols in Human Genetics, 64*(1), 1–9.

Cardon, L. R., Smith, S. D., Fulker, D. W., Kimberling, W. J., Pennington, B. F., & Defries, J. C. (1994). Quantitative trait locus for reading-disability on chromosome-6. *Science, 266*(5183), 276–279.

Carrion-Castillo, A., van Bergen, E., Vino, A., van Zuijen, T., de Jong, P. F., Francks, C., & Fisher, S. E. (2016). Evaluation of results from genome-wide studies of language and reading in a novel independent dataset. *Genes Brain and Behavior, 15*(6), 531–541.

Chapman, N. H., Igo, R. P., Thomson, J. B., Matsushita, M., Brkanac, Z., Holzman, T., ... Raskind, W. H. (2004). Linkage analyses of four regions previously implicated in dyslexia: Confirmation of a locus on chromosome 15q. *American Journal of Medical Genetics Part B-Neuropsychiatric Genetics, 131B*(1), 67–75.

Che, A., Truong, D. T., Fitch, R. H., & LoTurco, J. J. (2015). Mutation of the dyslexia-associated gene *Dcdc2* enhances glutamatergic synaptic transmission between layer 4 neurons in mouse neocortex. *Cerebral Cortex, 26*(9), 3705–3718.

Colledge, E., Bishop, D. V. M., Koeppen-Schomerus, G., Price, T. S., Happe, F. G. E., Eley, T. C., ... Plomin, R. (2002). The structure of language abilities at 4 years: A twin study. *Developmental Psychology, 38*(5), 749–757.

Conti-Ramsden, G., Botting, N., & Faragher, B. (2001). Psycholinguistic markers for specific language impairment (SLI). *Journal of Child Psychology and Psychiatry, 42*(6), 741–748.

Cope, N., Harold, D., Hill, G., Moskvina, V., Stevenson, J., Holmans, P., ... Williams, J. (2005). Strong evidence that KIAA0319 on chromosome 6p is a susceptibility gene for developmental dyslexia. *American Journal of Human Genetics, 76*(4), 581–591.

Couto, J. M., Gomez, L., Wigg, K., Cate-Carter, T., Archibald, J., Anderson, B., ... Barr, C. L. (2008). The KIAA0319-like (*KIAA0319L*) gene on chromosome 1p34 as a candidate for reading disabilities. *Journal of Neurogenetics, 22*(4), 295–313.

Dale, P. S., & Cole, K. N. (1991). What's normal? Specific language impairment in an individual differences perspective. *Language, Speech, and Hearing Services in Schools, 22*(2), 80–83.

Darki, F., Peyrard-Janvid, M., Matsson, H., Kere, J., & Klingberg, T. (2012). Three dyslexia susceptibility genes, *DYX1C1*, *DCDC2*, and *KIAA0319*, affect temporo-parietal white matter structure. *Biological Psychiatry, 72*(8), 671–676.

Davies, G., Marioni, R. E., Liewald, D. C., Hill, W. D., Hagenaars, S. P., Harris, S. E., ... Deary, I. J. (2016). Genome-wide association study of cognitive functions and educational attainment in UK Biobank (N = 112 151). *Molecular Psychiatry, 21*(6), 758–767.

Deffenbacher, K. E., Kenyon, J. B., Hoover, D. M., Olson, R. K., Pennington, B. F., DeFries, J. C., & Smith, S. D. (2004). Refinement of the 6p21.3 quantitative trait locus influencing dyslexia: linkage and association analyses. *Human Genetics, 115*(2), 128–138.

Defries, J. C., Fulker, D. W., & Labuda, M. C. (1987). Evidence for a genetic etiology in reading-disability of twins. *Nature, 329*(6139), 537–539.

de Kovel, C. G. F., Hol, F. A., Heister, J. G. A. M., Willemen, J. J. H. T., Sandkuijl, L. A., Franke, B., & Padberg, G. W. (2004). Genomewide scan identifies susceptibility locus for dyslexia on Xq27 in an extended Dutch family. *Journal of Medical Genetics, 41*(9), 652–657.

Dennis, M. Y., Paracchini, S., Scerri, T. S., Prokunina-Olsson, L., Knight, J. C., Wade-Martins, R., ... Monaco, A. P. (2009). A common variant associated with dyslexia reduces expression of the *KIAA0319* gene. *PLOS Genetics, 5*(3), e1000436.

Devanna, P., Middelbeek, J., & Vernes, S. C. (2014). FOXP2 drives neuronal differentiation by interacting with retinoic acid signaling pathways. *Frontiers in Cellular Neuroscience, 8*, 305.

Dudbridge, F. (2007). Family-based association. In D. J. Balding, M. Bishop, & C. Cannings (Eds.), *Handbook of statistical genetics* (pp. 1264–1285). Chichester, UK: John Wiley & Sons.

Eicher, J. D., Powers, N. R., Miller, L. L., Akshoomoff, N., Amaral, D. G., Bloss, C. S., ... Gruen, J. R. (2013). Genome-wide association study of shared components of reading disability and language impairment. *Genes, Brain and Behavior, 12*(8), 792–801.

Eicher, J. D., Powers, N. R., Miller, L. L., Mueller, K. L., Mascheretti, S., Marino, C., ... Gruen, J. R. (2014). Characterization of the DYX2 locus on chromosome 6p22 with reading disability, language impairment, and IQ. *Human Genetics, 133*(7), 869–881.

Einarsdottir, E., Svensson, I., Darki, F., Peyrard-Janvid, M., Lindvall, J. M., Ameur, A., ... Matsson, H. (2015). Mutation in CEP63 co-segregating with developmental dyslexia in a Swedish family. *Human Genetics, 134*(11–12), 1239–1248.

Evans, P. D., Mueller, K. L., Gamazon, E. R., Cox, N. J., & Tomblin, J. B. (2015). A genome-wide sib-pair scan for quantitative language traits reveals linkage to chromosomes 10 and 13. *Genes, Brain and Behavior, 14*(5), 387–397.

Fagerheim, T., Raeymaekers, P., Tonnessen, F. E., Pedersen, M., Tranebjaerg, L., & Lubs, H. A. (1999). A new gene (DYX3) for dyslexia is located on chromosome 2. *Journal of Medical Genetics, 36*(9), 664–669.

Falcaro, M., Pickles, A., Newbury, D. F., Addis, L., Banfield, E., Fisher, S. E., ... Conti-Ramsden, G. (2008). Genetic and phenotypic effects of phonological short-term memory and grammatical morphology in specific language impairment. *Genes, Brain and Behavior, 7*(4), 393–402.

Ferreira, M. A. R. (2004). Linkage analysis: Principles and methods for the analysis of human quantitative traits. *Twin Research, 7*(5), 513–530.

Field, L. L., Shumansky, K., Ryan, J., Truong, D., Swiergala, E., & Kaplan, B. J. (2013). Dense-map genome scan for dyslexia supports loci at 4q13, 16p12, 17q22; suggests novel locus at 7q36. *Genes, Brain and Behavior, 12*(1), 56–69.

Fisher, S. E., Francks, C., Marlow, A. J., MacPhie, I. L., Newbury, D. F., Cardon, L. R., ... Monaco, A. P. (2002). Independent genome-wide scans identify a chromosome 18 quantitative-trait locus influencing dyslexia. *Nature Genetics, 30*(1), 86–91.

Fisher, S. E., & Scharff, C. (2009). FOXP2 as a molecular window into speech and language. *Trends in Genetics, 25*(4), 166–177.

Fisher, S. E., Vargha-Khadem, F., Watkins, K. E., Monaco, A. P., & Pembrey, M. E. (1998). Localisation of a gene implicated in a severe speech and language disorder. *Nature Genetics, 18*(2), 168–170.

Francks, C., Fisher, S. E., Olson, R. K., Pennington, B. F., Smith, S. D., DeFries, J. C., & Monaco, A. P. (2002). Fine mapping of the chromosome 2p12-16 dyslexia susceptibility locus: quantitative association analysis and positional candidate genes SEMA4F and OTX1. *Psychiatric Genetics, 12*(1), 35–41.

Francks, C., Paracchini, S., Smith, S. D., Richardson, A. J., Scerri, T. S., Cardon, L. R., ... Monaco, A. P. (2004). A 77-kilobase region of chromosome 6p22.2 is associated with dyslexia in families from the United Kingdom and from the United States. *American Journal of Human Genetics, 75*(6), 1046–1058.

French, C. A., Jin, X., Campbell, T. G., Gerfen, E., Groszer, M., Fisher, S. E., & Costa, R. M. (2012). An aetiological Foxp2 mutation causes aberrant striatal activity and alters plasticity during skill learning. *Molecular Psychiatry, 17*(11), 1077–1085.

Froster, U., Schultekorne, G., Hebebrand, J., & Remschmidt, H. (1993). Cosegregation of balanced translocation (1,2) with retarded speech development and dyslexia. *Lancet, 342*(8864), 178–179.

Gabel, L. A., Marin, I., LoTurco, J. J., Che, A., Murphy, C., Manglani, M., & Kass, S. (2011). Mutation of the dyslexia-associated gene *Dcdc2* impairs LTM and visuo-spatial performance in mice. *Genes, Brain and Behavior, 10*(8), 868–875.

Gathercole, S. E., Willis, C. S., Baddeley, A. D., & Emslie, H. (1994). The children's test of nonword repetition: A test of phonological working memory. *Memory, 2*(2), 103–127.

Gayan, J., Smith, S. D., Cherny, S. S., Cardon, L. R., Fulker, D. W., Brower, A. M., ... DeFries, J. C. (1999). Quantitative-trait locus for specific language and reading deficits on chromosome 6p. *American Journal of Human Genetics, 64*(1), 157–164.

Gialluisi, A., Newbury, D. F., Wilcutt, E. G., Olson, R. K., DeFries, J. C., Brandler, W. M., ... Fisher, S. E. (2014). Genome-wide screening for DNA variants associated with reading and language traits. *Genes, Brain and Behavior, 13*(7), 686–701.

Gialluisi, A., Visconti, A., Willcutt, E. G., Smith, S. D., Pennington, B. F., Falchi, M., ... Fisher, S. E. (2016). Investigating the effects of copy number variants on reading and language performance. *Journal of Neurodevelopmental Disorders, 8*(1), 17.

Girirajan, S., Brkanac, Z., Coe, B. P., Baker, C., Vives, L., Vu, T. H., ... Eichler, E. E. (2011). Relative burden of large CNVs on a range of neurodevelopmental phenotypes. *PLOS Genetics, 7*(11), e1002334.

Grigorenko, E. L., Wood, F. B., Golovyan, L., Meyer, M., Romano, C., & Pauls, D. (2003). Continuing the search for dyslexia genes on 6p. *American Journal of Medical Genetics Part B-Neuropsychiatric Genetics, 118B*(1), 89–98.

Grigorenko, E. L., Wood, F. B., Meyer, M. S., Hart, L. A., Speed, W. C., Shuster, A., & Pauls, D. L. (1997). Susceptibility loci for distinct components of developmental dyslexia on chromosomes 6 and 15. *American Journal of Human Genetics, 60*(1), 27–39.

Grigorenko, E. L., Wood, F. B., Meyer, M. S., Pauls, J. E. D., Hart, L. A., & Pauls, D. L. (2001). Linkage studies suggest a possible locus for developmental dyslexia on chromosome 1p. *American Journal of Medical Genetics, 105*(1), 120–129.

Groszer, M., Keays, D. A., Deacon, R. M. J., de Bono, J. P., Prasad-Mulcare, S., Gaub, S., ... Fisher, S. E. (2008). Impaired synaptic plasticity and motor learning in mice with a point mutation implicated in human speech deficits. *Current Biology, 18*(5), 354–362.

Haesler, S., Rochefort, C., Georgi, B., Licznerski, P., Osten, P., & Scharff, C. (2007). Incomplete and inaccurate vocal imitation after knockdown of FoxP2 in songbird basal ganglia nucleus area X. *PLOS Biology, 5*(12), 2885–2897.

Hannula-Jouppi, K., Kaminen-Ahola, N., Taipale, M., Eklund, R., Nopola-Hemmi, J., Kaariainen, H., & Kere, J. (2005). The axon guidance receptor gene ROBO1 is a candidate gene for developmental dyslexia. *PLOS Genetics, 1*(4), 467–474.

Harlaar, N., Meaburn, E. L., Hayiou-Thomas, M. E., Davis, O. S., Docherty, S., Hanscombe, K. B., ... & Plomin, R. (2014). Genome-wide association study of receptive language ability of 12-year-olds. *Journal of Speech, Language, and Hearing Research, 57*(1), 96–105.

Harlaar, N., Spinath, F. M., Dale, P. S., & Plomin, R. (2005). Genetic influences on early word recognition abilities and disabilities: A study of 7-year-old twins. *Journal of Child Psychology and Psychiatry, 46*(4), 373–384.

Harlaar, N., Trzaskowski, M., Dale, P. S., & Plomin, R. (2014). Word reading fluency: Role of genome-wide single-nucleotide polymorphisms in developmental stability and correlations with print exposure. *Child Development, 85*(3), 1190–1205.

Harold, D., Paracchini, S., Scerri, T., Dennis, M., Cope, N., Hill, G., ... Monaco, A. P. (2006). Further evidence that the KIAA0319 gene confers susceptibility to developmental dyslexia. *Molecular Psychiatry, 11*(12), 1085–1091.

Haworth, C. M. A., & Plomin, R. (2010). Quantitative genetics in the era of molecular genetics: Learning abilities and disabilities as an example. *Journal of the American Academy of Child and Adolescent Psychiatry, 49*(8), 783–793.

Hayiou-Thomas, M. E. (2008). Genetic and environmental influences on early speech, language and literacy development. *Journal of Communication Disorders, 41*(5), 397–408.

Heston, J. B., & White, S. A. (2015). Behavior-linked FoxP2 regulation enables zebra finch vocal learning. *Journal of Neuroscience, 35*(7), 2885.

Hill, W. D., Davies, G., van de Lagemaat, L. N., Christoforou, A., Marioni, R. E., Fernandes, C. P. D., ... Deary, I. J. (2014). Human cognitive ability is influenced by genetic variation in components of postsynaptic signalling complexes assembled by NMDA receptors and MAGUK proteins. *Translational Psychiatry, 4*, e341.

Hill, W. D., de Leeuw, C., Davies, G., Liewald, D. C., Payton, A., Craig, L. C., ... Deary, I. J. (2014). Functional gene group analysis indicates no role for heterotrimeric G proteins in cognitive ability. *PLOS ONE, 9*(3), e91690.

Howey, R., Mamasoula, C., Topf, A., Nudel, R., Goodship, J. A., Keavney, B. D., & Cordell, H. J. (2015). Increased power for detection of parent-of-origin effects via the use of haplotype estimation. *American Journal of Human Genetics, 97*(3), 419–434.

Hulme, C., & Snowling, M. J. (2016). Reading disorders and dyslexia. *Current Opinion in Pediatry, 28*(6), 731–735.

Kaminen, N., Hannula-Jouppi, K., Kestila, M., Lahermo, P., Muller, K., Kaaranen, M., ... Kere, J. (2003). A genome scan for developmental dyslexia confirms linkage to chromosome 2p11 and suggests a new locus on 7q32. *Journal of Medical Genetics, 40*(5), 340–345.

Kaplan, D. E., Gayan, J., Ahn, J., Won, T. W., Pauls, D., Olson, R. K., ... Gruen, J. R. (2002). Evidence for linkage and association with reading disability, on 6p21.3-22. *American Journal of Human Genetics, 70*(5), 1287–1298.

Lai, C. S. L., Fisher, S. E., Hurst, J. A., Vargha-Khadem, F., & Monaco, A. P. (2001). A forkhead-domain gene is mutated in a severe speech and language disorder. *Nature, 413*(6855), 519–523.

Lamminmäki, S., Massinen, S., Nopola-Hemmi, J., Kere, J., & Hari, R. (2012). Human *ROBO1* regulates interaural interaction in auditory pathways. *Journal of Neuroscience, 32*(3), 966–971.

Lander, E., & Kruglyak, L. (1995). Genetic dissection of complex traits: Guidelines for interpreting and reporting linkage results. *Nature Genetics, 11*(3), 241–247.

Lee, J. J., Wedow, R., Okbay, A., Kong, E., Maghzian, O., Zacher, M., ... Cesarini, D. (2018). Gene discovery and polygenic prediction from a genome-wide association study of educational attainment in 1.1 million individuals. *Nature Genetics, 50*(8), 1112–1121.

Leonard, L. B. (1991). Specific language impairment as a clinical category. *Language, Speech, and Hearing Services in Schools, 22*(2), 66–68.

Luciano, M., Evans, D. M., Hansell, N. K., Medland, S. E., Montgomery, G. W., Martin, N. G., ... Bates, T. C. (2013). A genome-wide association study for reading and language abilities in two population cohorts. *Genes, Brain and Behavior, 12*(6), 645–652.

Luciano, M., Hagenaars, S. P., Cox, S. R., Hill, W. D., Davies, G., Harris, S. E., ... Bates, T. C. (2017). Single nucleotide polymorphisms associated with reading ability show connection to socio-economic outcomes. *Behavior Genetics, 47*(5), 469–479.

Luciano, M., Lind, P. A., Duffy, D. L., Castles, A., Wright, M. J., Montgomery, G. W., ... Bates, T. C. (2007). A haplotype spanning KIAA0319 and TTRAP is associated with normal variation in reading and spelling ability. *Biological Psychiatry, 62*(7), 811–817.

Luciano, M., Montgomery, G. W., Martin, N. G., Wright, M. J., & Bates, T. C. (2011). SNP sets and reading ability: Testing confirmation of a 10-SNP set in a population sample. *Twin Research and Human Genetics, 14*(3), 228–232.

Ludwig, K. U., Roeske, D., Schumacher, J., Schulte-Korne, G., Konig, I. R., Warnke, A., ... Hoffmann, P. (2008). Investigation of interaction between DCDC2 and KIAA0319 in a large German dyslexia sample. *Journal of Neural Transmission, 115*(11), 1587–1589.

Marlow, A. J., Fisher, S. E., Francks, C., MacPhie, I. L., Cherny, S. S., Richardson, A. J., ... Cardon, L. R. (2003). Use of multivariate linkage analysis for dissection of a complex cognitive trait. *American Journal of Human Genetics, 72*(3), 561–570.

Martinez-Garay, I., Guidi, L. G., Holloway, Z. G., Bailey, M. A. G., Lyngholm, D., Schneider, T., ... Velayos-Baeza, A. (2017). Normal radial migration and lamination are maintained in dyslexia-susceptibility candidate gene homolog Kiaa0319 knockout mice. *Brain Structure and Function, 222*(3), 1367–1384.

Mascheretti, S., Bureau, A., Battaglia, M., Simone, D., Quadrelli, E., Croteau, J., ... Marino, C. (2013). An assessment of gene-by-environment interactions in developmental dyslexia-related phenotypes. *Genes, Brain and Behavior, 12*(1), 47–55.

Mascheretti, S., Bureau, A., Trezzi, V., Giorda, R., & Marino, C. (2015). An assessment of gene-by-gene interactions as a tool to unfold missing heritability in dyslexia. *Human Genetics, 134*(7), 749–760.

Mascheretti, S., Riva, V., Giorda, R., Beri, S., Lanzoni, L. F. E., Cellino, M. R., & Marino, C. (2014). *KIAA0319* and *ROBO1*: Evidence on association with reading and pleiotropic effects on language and mathematics abilities in developmental dyslexia. *Journal of Human Genetics, 59*(4), 189–197.

Massinen, S., Wang, J., Laivuori, K., Bieder, A., Tapia Paez, I., Jiao, H., & Kere, J. (2016). Genomic sequencing of a dyslexia susceptibility haplotype encompassing ROBO1. *Journal of Neurodevelopmental Disorders, 8*(1), 4.

Matsson, H., Huss, M., Persson, H., Einarsdottir, E., Tiraboschi, E., Nopola-Hemmi, J., ... Kere, J. (2015). Polymorphisms in *DCDC2* and *S100B* associate with developmental dyslexia. *Journal of Human Genetics, 60*(7), 399–401.

Meaburn, E., Dale, P. S., Craig, I. W., & Plomin, R. (2002). Language-impaired children: No sign of the FOXP2 mutation. *NeuroReport, 13*(8), 1075–1077.

Meng, H. Y., Smith, S. D., Hager, K., Held, M., Liu, J., Olson, R. K., ... Gruen, J. R. (2005). *DCDC2* is associated with reading disability and modulates neuronal development in the brain. *Proceedings of the National Academy of Sciences, 102*(47), 17053--17058.

Middelberg, R. P., Gordon, S. D., Zhu, G., Mcrae, A., Montgomery, G. W., Martin, N. G., & Whitfield, J. B. (2008). Linkage and association analyses of longitudinally measured lipid phenotypes in adolescence. *Twin Research and Human Genetics, 11*(6), 603–620.

Monaco, A. P. (2007). Multivariate linkage analysis of specific language impairment (SLI). *Annals of Human Genetics, 71*, 660–673.

Morgan, A., Fisher, S. E., Scheffer, I., & Hildebrand, M. (2017). FOXP2-related speech and language disorders. In M. P. Adam, H. H. Ardinger, R. A. Pagon et al. (Eds.), *GeneReviews®* [Internet]. Seattle: University of Washington, Seattle; 1993–2019. Retrieved from https://www.ncbi.nlm.nih.gov/books/NBK368474/.

Morris, A. P., & Cardon, L. R. (2007). Whole genome association. In D. J. Balding, M. Bishop, & C. Cannings (Eds.),

Handbook of statistical genetics (pp. 1238–1263). Chichester, UK: John Wiley & Sons.

Mueller, B., Ahnert, P., Burkhardt, J., Brauer, J., Czepezauer, I., Quente, E., ... Kirsten, H. (2014). Genetic risk variants for dyslexia on chromosome 18 in a German cohort. *Genes, Brain and Behavior, 13*(3), 350–356.

Mueller, K. L., Murray, J. C., Michaelson, J. J., Christiansen, M. H., Reilly, S., & Tomblin, J. B. (2016). Common genetic variants in FOXP2 are not associated with individual differences in language development. *PLOS ONE, 11*(4), e0152576.

Murugan, M., Harward, S., Scharff, C., & Mooney, R. (2013). Diminished FoxP2 levels affect dopaminergic modulation of corticostriatal signaling important to song variability. *Neuron, 80*(6), 1464–1476.

Newbury, D. F., Bonora, E., Lamb, J. A., Fisher, S. E., Lai, C. S., Baird, G., ... & Bolton, P. F. (2002). FOXP2 is not a major susceptibility gene for autism or specific language impairment. *American Journal of Human Genetics, 70*(5), 1318–1327.

Newbury, D. F., Paracchini, S., Scerri, T. S., Winchester, L., Addis, L., Richardson, A. J., ... Monaco, A. P. (2011). Investigation of dyslexia and SLI risk variants in reading- and language-impaired subjects. *Behavior Genetics, 41*(1), 90–104.

Newbury, D. F., Winchester, L., Addis, L., Paracchini, S., Buckingham, L. L., Clark, A., ... Monaco, A. P. (2009). *CMIP* and *ATP2C2* modulate phonological short-term memory in language impairment. *American Journal of Human Genetics, 85*(2), 264–272.

Nopola-Hemmi, J., Myllyluoma, B., Haltia, T., Taipale, M., Ollikainen, V., Ahonen, T., ... Widen, E. (2001). A dominant gene for developmental dyslexia on chromosome 3. *Journal of Medical Genetics, 38*(10), 658–664.

Nopola-Hemmi, J., Taipale, M., Haltia, T., Lehesjoki, A. E., Voutilainen, A., & Kere, J. (2000). Two translocations of chromosome 15q associated with dyslexia. *Journal of Medical Genetics, 37*(10), 771–775.

Nothen, M. M., Schulte-Korne, G., Grimm, T., Cichon, S., Vogt, I. R., Muller-Myhsok, B., ... Remschmidt, H. (1999). Genetic linkage analysis with dyslexia: Evidence for linkage of spelling disability to chromosome 15. *European Child and Adolescent Psychiatry, 8*, 56–59.

Nudel, R. (2016). An investigation of NFXL1, a gene implicated in a study of specific language impairment. *Journal of Neurodevelopmental Disorders, 8*, 13.

Nudel, R., Simpson, N. H., Baird, G., O'Hare, A., Conti-Ramsden, G., Bolton, P. F., ... Newbury, D. F. (2014). Genome-wide association analyses of child genotype effects and parent-of-origin effects in specific language impairment. *Genes, Brain and Behavior, 13*(4), 418–429.

O'Brien, E. K., Zhang, X., Nishimura, C., Tomblin, J. B., & Murray, J. C. (2003). Association of specific language impairment (SLI) to the region of 7q31. *American Journal of Human Genetics, 72*(6), 1536–1543.

Olson, R., Wise, B., Conners, F., Rack, J., & Fulker, D. (1989). Specific deficits in component reading and language skills: Genetic and environmental influences. *Journal of Learning Disabilities, 22*(6), 339–348.

Paracchini, S., Thomas, A., Castro, S., Lai, C., Paramasivam, M., Wang, Y., ... Monaco, A. P. (2006). The chromosome 6p22 haplotype associated with dyslexia reduces the expression of *KIAA0319*, a novel gene involved in neuronal migration. *Human Molecular Genetics, 15*(10), 1659–1666.

Pennington, B. F., Gilger, J. W., Pauls, D., Smith, S. A., Smith, S. D., & Defries, J. C. (1991). Evidence for major gene transmission of developmental dyslexia. *JAMA, 266*(11), 1527–1534.

Pennington, B. F., Smith, S. D., Kimberling, W. J., Green, P. A., & Haith, M. M. (1987). Left-handedness and immune disorders in familial dyslexics. *Archives of Neurology, 44*(6), 634–639.

Petryshen, T. L., Kaplan, B. J., Fu Liu, M., de French, N. S., Tobias, R., Hughes, M. L., & Leigh Field, L. (2001). Evidence for a susceptibility locus on chromosome 6q influencing phonological coding dyslexia. *American Journal of Medical Genetics, 105*(6), 507–517.

Petryshen, T. L., Kaplan, B. J., Hughes, M. L., Tzenova, J., & Field, L. L. (2002). Supportive evidence for the DYX3 dyslexia susceptibility gene in Canadian families. *Journal of Medical Genetics, 39*(2), 125–126.

Pettigrew, K. A., Frinton, E., Nudel, R., Chan, M. T., Thompson, P., Hayiou-Thomas, M. E., ... Paracchini, S. (2016). Further evidence for a parent-of-origin effect at the NOP9 locus on language-related phenotypes. *Journal of Neurodevelopmental Disorders, 8*, 24.

Peyrard-Janvid, M., Anthoni, H., Onkamo, P., Lahermo, P., Zucchelli, M., Kaminen, N., ... Kere, J. (2004). Fine mapping of the 2p11 dyslexia locus and exclusion of TACR1 as a candidate gene. *Human Genetics, 114*(5), 510–516.

Pinel, P., Fauchereau, F., Moreno, A., Barbot, A., Lathrop, M., Zelenika, D., ... Dehaene, S. (2012). Genetic variants of FOXP2 and KIAA0319/TTRAP/THEM2 locus are associated with altered brain activation in distinct language-related regions. *Journal of Neuroscience, 32*(3), 817–825.

Plomin, R., Haworth, C. M., Meaburn, E. L., Price, T. S., Wellcome Trust Case Control Consortium 2, & Davis, O. S. (2013). Common DNA markers can account for more than half of the genetic influence on cognitive abilities. *Psychological Science, 24*(4), 562–568.

Plomin, R., & Kovas, Y. (2005). Generalist genes and learning disabilities. *Psychological Bulletin, 131*(4), 592–617.

Poelmans, G., Buitelaar, J. K., Pauls, D. L., & Franke, B. (2011). A theoretical molecular network for dyslexia: Integrating available genetic findings. *Molecular Psychiatry, 16*(4), 365–382.

Poelmans, G., Engelen, J. J. M., Van Lent-Albrechts, J., Smeets, H. J., Schoenmakers, E., Franke, B., ... Schrander-Stumpel, C. (2009). Identification of novel dyslexia candidate genes through the analysis of a chromosomal deletion. *American Journal of Medical Genetics Part B-Neuropsychiatric Genetics, 150B*(1), 140–147.

Powers, N. R., Eicher, J. D., Butter, F., Kong, Y., Miller, L. L., Ring, S. M., ... Gruen, J. R. (2013). Alleles of a polymorphic ETV6 binding site in *DCDC2* confer risk of reading and language impairment. *American Journal of Human Genetics, 93*(1), 19–28.

Powers, N. R., Eicher, J. D., Miller, L. L., Kong, Y., Smith, S. D., Pennington, B. F., ... Gruen, J. R. (2016). The regulatory element *READ1* epistatically influences reading and language, with both deleterious and protective alleles. *Journal of Medical Genetics, 53*(3), 163–171.

Pugh, K. R., Frost, S. J., Rothman, D. L., Hoeft, F., Del Tufo, S. N., Mason, G. F., ... Fulbright, R. K. (2014). Glutamate and choline levels predict individual differences in reading ability in emergent readers. *Journal of Neuroscience, 34*(11), 4082–4089.

Rabin, M., Wen, X. L., Hepburn, M., Lubs, H. A., Feldman, E., & Duara, R. (1993). Suggestive linkage of developmental dyslexia to chromosome 1p34-p36. *Lancet, 342*(8864), 178.

Reader, R. H., Covill, L. E., Nudel, R., & Newbury, D. F. (2014). Genome-wide studies of specific language impairment. *Current Behavioral Neuroscience Reports*, 1(4), 242–250.

Rice, M. L., Smith, S. D., & Gayan, J. (2009). Convergent genetic linkage and associations to language, speech and reading measures in families of probands with specific language impairment. *Journal of Neurodevelopmental Disorders, 1*(4), 264–282.

Rodgers, B. (1983). The identification and prevalence of specific reading retardation. *British Journal of Educational Psychology, 53*(3), 369–373.

Rutter, M., & Yule, W. (1975). Concept of specific reading retardation. *Journal of Child Psychology and Psychiatry and Allied Disciplines, 16*(3), 181–197.

Scerri, T. S., Darki, F., Newbury, D. F., Whitehouse, A. J., Peyrard-Janvid, M., Matsson, H., ... Paracchini, S. (2012). The dyslexia candidate locus on 2p12 is associated with general cognitive ability and white matter structure. *PLOS ONE, 7*(11), e50321.

Scerri, T. S., Morris, A. P., Buckingham, L. L., Newbury, D. F., Miller, L. L., Monaco, A. P., ... Paracchini, S. (2011). *DCDC2, KIAA0319* and *CMIP* are associated with reading-related traits. *Biological Psychiatry, 70*(3), 237–245.

Scerri, T. S., Paracchini, S., Morris, A., MacPhie, I. L., Talcott, J., Stein, J., ... Monaco, A. P. (2010). Identification of candidate genes for dyslexia susceptibility on chromosome 18. *PLOS ONE, 5*(10), e13712.

Scerri, T. S., & Schulte-Koene, G. (2010). Genetics of developmental dyslexia. *European Child and Adolescent Psychiatry, 19*(3), 179–197.

Schreiweis, C., Bornschein, U., Burguiere, E., Kerimoglu, C., Schreiter, S., Dannemann, M., ... Graybiel, A. M. (2014). Humanized Foxp2 accelerates learning by enhancing transitions from declarative to procedural performance. *Proceedings of the National Academy of Sciences, 111*(39), 14253–14258.

Schumacher, J., Anthoni, H., Dahdouh, F., König, I. R., Hillmer, A. M., Kluck, N., ... Remschmidt, H. (2006). Strong genetic evidence of *DCDC2* as a susceptibility gene for dyslexia. *American Journal of Human Genetics, 78*(1), 52–62.

Schumacher, J., Konig, I. R., Schroder, T., Duell, M., Plume, E., Propping, P., ... Nothen, M. M. (2008). Further evidence for a susceptibility locus contributing to reading disability on chromosome 15q15-q21. *Psychiatric Genetics, 18*(3), 137–142.

Selzam, S., Dale, P. S., Wagner, R. K., DeFries, J. C., Cederlof, M., O'Reilly, P. F., ... Plomin, R. (2017). Genome-wide polygenic scores predict reading performance throughout the school years. *Scientific Studies of Reading, 21*(4), 334–349.

Shao, S., Kong, R., Zou, L., Zhong, R., Lou, J., Zhou, J., ... Song, R. (2015). The roles of genes in the neuronal migration and neurite outgrowth network in developmental dyslexia: Single- and multiple-risk genetic variants. *Molecular Neurobiology, 53*(6), 3967–3975.

Shaywitz, S. E., Escobar, M. D., Shaywitz, B. A., Fletcher, J. M., & Makuch, R. (1992). Evidence that dyslexia may represent the lower tail of a normal-distribution of reading-ability. *New England Journal of Medicine, 326*(3), 145–150.

Shu, W., Cho, J. Y., Jiang, Y., Zhang, M., Weisz, D., Elder, G. A., ... Buxbaum, J. D. (2005). Altered ultrasonic vocalization in mice with a disruption in the *Foxp2* gene. *Proceedings of the National Academy of Sciences, 102*(27), 9643–9648.

Siegel, L. S. (2003). IQ-discrepancy definitions and the diagnosis of LD: Introduction to the special issue. *Journal of Learning Disabilities, 36*(1), 2.

Silva, P. A., Mcgee, R., & Williams, S. (1985). Some characteristics of 9-year-old boys with general reading backwardness or specific reading retardation. *Journal of Child Psychology and Psychiatry and Allied Disciplines, 26*(3), 407–421.

Simpson, N. H., Ceroni, F., Reader, R. H., Covill, L. E., Knight, J. C., Nudel, R., ... Slonims, V. (2015). Genome-wide analysis identifies a role for common copy number variants in specific language impairment. *European Journal of Human Genetics, 23*(10), 1370–1377.

SLI Consortium. (2002). A genomewide scan identifies two novel loci involved in specific language impairment. *American Journal of Human Genetics, 70*(2), 384–398.

SLI Consortium. (2004). Highly significant linkage to the SLI1 locus in an expanded sample of individuals affected by specific language impairment. *American Journal of Human Genetics, 74*(6), 1225–1238.

Smith, S. D., Kimberling, W. J., & Pennington, B. F. (1983). Specific reading-disability: Identification of an inherited form through linkage analysis. *Science, 219*(4590), 1345–1347.

Stein, C. M., Schick, J. H., Taylor, H. G., Shriberg, L. D., Millard, C., Kundtz-Kluge, A., ... Iyengar, S. K. (2004). Pleiotropic effects of a chromosome 3 locus on speech-sound disorder and reading. *American Journal of Human Genetics, 74*(2), 283–297

St. Pourcain, B., Cents, R. A., Whitehouse, A. J., Haworth, C. M., Davis, O. S., O'Reilly, P. F., ... Davey Smith, G. (2014). Common variation near ROBO2 is associated with expressive vocabulary in infancy. *Nature Communications, 5*, 4831.

Sun, Y. M., Gao, Y., Zhou, Y. X., Chen, H., Wang, G. Q., Xu, J. Q., ... Tan, L. H. (2014). Association study of developmental dyslexia candidate genes *DCDC2* and *KIAA0319* in Chinese population. *American Journal of Medical Genetics Part B-Neuropsychiatric Genetics, 165*(8), 627–634.

Szalkowski, C. E., Fiondella, C. G., Galaburda, A. M., Rosen, G. D., LoTurco, J. J., & Fitch, R. H. (2012). Neocortical disruption and behavioral impairments in rats following in utero RNAi of candidate dyslexia risk gene Kiaa0319. *International Journal of Developmental Neuroscience, 30*(4), 293–302.

Taipale, M., Kaminen, N., Nopola-Hemmi, J., Haltia, T., Myllyluoma, B., Lyytinen, H., ... Kere, J. (2003). A candidate gene for developmental dyslexia encodes a nuclear tetratricopeptide repeat domain protein dynamically regulated in brain. *Proceedings of the National Academy of Sciences, 100*(20), 11553–11558.

Tran, C., Gagnon, F., Wigg, K. G., Feng, Y., Gomez, L., Cate-Carter, T. D., ... Barr, C. L. (2013). A family-based association analysis and meta-analysis of the reading disabilities candidate gene DYX1C1. *American Journal of Medical Genetics Part B-Neuropsychiatric Genetics, 162B*(2), 146–156.

Tran, C., Wigg, K. G., Zhang, K., Cate-Carter, T. D., Kerr, E., Field, L. L., ... Barr, C. L. (2014). Association of the *ROBO1* gene with reading disabilities in a family- based analysis. *Genes, Brain and Behavior, 13*(4), 430–438.

Truong, D. T., Che, A., Rendall, A. R., Szalkowski, C. E., LoTurco, J. J., Galaburda, A. M., & Holly Fitch, R. (2014). Mutation of *Dcdc2* in mice leads to impairments in auditory processing and memory ability. *Genes, Brain and Behavior, 13*(8), 802–811.

Venkatesh, S. K., Siddaiah, A., Padakannaya, P., & Ramachandra, N. B. (2014). Association of SNPs of DYX1C1 with developmental dyslexia in an Indian population. *Psychiatric Genetics, 24*(1), 10–20.

Vernes, S. C., Newbury, D. F., Abrahams, B. S., Winchester, L., Nicod, J., Groszer, M., ... Fisher, S. E. (2008). A functional genetic link between distinct developmental language disorders. *New England Journal of Medicine, 359*(22), 2337–2345.

Vernes, S. C., Oliver, P. L., Spiteri, E., Lockstone, H. E., Puliyadi, R., Taylor, J. M., ... Fisher, S. E. (2011). Foxp2 regulates gene networks implicated in neurite outgrowth in the developing brain. *PLOS Genetics, 7*(7), e1002145.

Villanueva, P., Jara, L., & Palomino, H. (2010). Association of D16S515 microsatellite with specific language impairment on Robinson Crusoe Island, an isolated Chilean population: A possible key to understanding language development. *Human Biology, 82*(4), 395–408.

Villanueva, P., Newbury, D. F., Jara, L., De Barbieri, Z., Mirza, G., Palomino, H. M., ... Palomino, H. (2011). Genome-wide analysis of genetic susceptibility to language impairment in an isolated Chilean population. *European Journal of Human Genetics, 19*(6), 687–695.

Villanueva, P., Nudel, R., Hoischen, A., Fernandez, M. A., Simpson, N. H., Gilissen, C., ... Newbury, D. F. (2015). Exome sequencing in an admixed isolated population indicates NFXL1 variants confer a risk for specific language impairment. *PLOS Genetics, 11*(3), e1004925.

Visscher, P. M., Hemani, G., Vinkhuyzen, A. A., Chen, G. B., Lee, S. H., Wray, N. R., ... Yang, J. (2014). Statistical power to detect genetic (co)variance of complex traits using SNP data in unrelated samples. *PLOS Genetics, 10*(4), e1004269.

Whitehouse, A. J., Bishop, D. V., Ang, Q. W., Pennell, C. E., & Fisher, S. E. (2011). CNTNAP2 variants affect early language development in the general population. *Genes, Brain and Behavior, 10*(4), 451–456.

Williams, J. T., & Blangero, J. (1999). Power of variance component linkage analysis to detect quantitative trait loci. *Annals of Human Genetics, 63*(6), 545–563.

Wiszniewski, W., Hunter, J. V., Hanchard, N. A., Willer, J. R., Shaw, C., Tian, Q., ... Lalani, S. R. (2013). *TM4SF20* ancestral deletion and susceptibility to a pediatric disorder of early language delay and cerebral white matter hyperintensities. *American Journal of Human Genetics, 93*(2), 197–210.

Wohlgemuth, S., Adam, I., & Scharff, C. (2014). FoxP2 in songbirds. *Current Opinion in Neurobiology, 28*, 86–93.

Wray, N. R., Lee, S. H., Mehta, D., Vinkhuyzen, A. A., Dudbridge, F., & Middeldorp, C. M. (2014). Research review: Polygenic methods and their application to psychiatric traits. *Journal of Child Psychology and Psychiatry, 55*(10), 1068–1087.

Zhong, R., Yang, B. F., Tang, H., Zou, L., Song, R. R., Zhu, L. Q., & Miao, X. P. (2013). Meta-analysis of the association between DCDC2 polymorphisms and risk of dyslexia. *Molecular Neurobiology, 47*(1), 435–442.

Zou, L., Chen, W., Shao, S. S., Sun, Z., Zhong, R., Shi, J. X., ... Song, R. R. (2012). Genetic variant in KIAA0319, but not in DYX1C1, is associated with risk of dyslexia: An integrated meta-analysis. *American Journal of Medical Genetics Part B-Neuropsychiatric Genetics, 159B*(8), 970–976.

40 Deciphering Genetics of Speech-Motor Disorders

ANGELA MORGAN

Acquisition of the complex act of speech begins in infancy, driven by genetic forces and shaped by the environment (Levelt, 1989). This skill takes all of childhood to develop and is refined into adolescence; there is arguably never a "steady state" for speaking (Levelt, 1989). The skills of speech and language are inextricably linked, but dissociations can occur, such as cases where speech is impaired in the presence of preserved language ability, which provides justification for considering speech sound disorders separately. The most severe and striking speech diagnoses are arguably speech-motor disorders, the focus of this chapter; including childhood apraxia of speech (CAS) and dysarthria.

Speech-motor disorders appear highly heritable, with increasing evidence of specific gene pathways leading to these conditions (Graham, Deriziotis, & Fisher, 2015). In fact, the very first gene finding in this field, which catalyzed study of the genetics of communication disorder, was identified in association with a speech-motor disorder. Specifically, mutations in *FOXP2*[1] were found to explain a rare monogenic form of CAS (Lai, Fisher, Hurst, Vargha-Khadem, & Monaco, 2001). This discovery provided the first insights into critical molecular pathways for speech disorder. Since this initial finding, technological advances of rapid DNA genotyping and sequencing have led to a proliferation of studies identifying further gene pathways to speech-motor disorders (Graham et al., 2015, for review).

This chapter focuses on the genetic bases of speech-motor disorders and in particular on genotype-phenotype association studies in the field. Our premise is that precise speech phenotyping is critical to lead to targeted genetic analyses to reveal further insights into the genetic bases of speech disorder. To this end, the first part of the chapter is dedicated to an overview of childhood speech sound disorders, with a focus on speech-motor disorders. The conditions of dysarthria and CAS are defined and the necessary steps required to determine a diagnosis of these disorders are elucidated.

The second part of this chapter is concerned with gene pathways implicated in speech-motor disorders. An exhaustive list of possible gene pathways to speech-motor disorders associated with long-standing neurogenetic syndromes is beyond the scope of this chapter (e.g., Kabuki syndrome due to *KMT2D, KDM6A* mutations, Morgan et al., 2015; Dravet syndrome due to *SCN1A* mutations, Turner et al., 2017). Rather, this chapter focuses on more recently identified novel gene pathways implicated in speech-motor disorders. Three core areas will be discussed in turn: single gene mutations, multigene deletion syndromes, and chromosomal rearrangements. The chapter emphasizes the important role of precise speech phenotyping for deciphering genetic contributions to childhood speech-motor disorder.

1. Childhood Speech Disorders

When speech development is atypical, it may be explained by an array of diagnoses. Articulation and phonological disorders represent the most prevalent, typically less severe and more tractable conditions. These occur in around 4% to 5% of children in preschool or at school-entry (Shriberg, Tomblin, & McSweeny, 1999; Eadie et al., 2015). Articulation disorder is an inability to produce a perceptually acceptable pronunciation of one or more speech sounds, in words or in isolation, at every attempt (Morgan & Guenther, 2017). These errors manifest as a phonetic-level distortion of sound. By contrast, a phonological disorder concerns a failure in language-specific knowledge about speech sounds that contrast meaning and how they may be sequenced, legally, to form words (Dodd, 2017). Natural history studies of community cohorts are lacking, yet existing evidence suggests that the majority of children eventually outgrow or resolve at least their surface articulation and phonological errors, achieving highly intelligible speech, between five and seven years of age (Dodd, Holm, Hua, & Crosbie, 2003).

Far more complex than commonplace articulation and phonological disorders, speech-*motor* disorders, are rare and much less tractable. Speech-motor disorders may occur in the presence of otherwise average or borderline nonverbal IQ with minimal or no other developmental concerns, or as part of a broader phenotype or syndrome of neurodevelopmental concerns. As noted, subtypes of speech-motor disorders include CAS and dysarthria. In either case, these speech deficits are typically chronic, leading to communication breakdown and a cycle of reduced confidence in speaking to others, particularly unfamiliar listeners, and in turn, social withdrawal. Longer term, children with speech-motor disorders are at increased risk for bullying, depression, and anxiety. The intractability of these conditions is due to their genetic or neurological etiological basis (Liégeois & Morgan, 2012). Abnormalities can be present at birth or emerge in later childhood after acquired neurologic injury or progressive disease. Stuttering can also be categorized as a speech-motor disorder, but does not commonly arise with CAS or dysarthria, the foci of the present chapter. For further detail on the genetics of stuttering, the reader is referred to Kang and Drayna (2012), Domingues et al. (2014), and Raza et al. (2015). This chapter presents gene pathways and the specific observable speech characteristics (phenotypes) associated with speech-motor disorders of CAS and/or dysarthria, conditions that frequently co-occur and hence lend themselves to be considered together here and will discussed in turn.

1.1. Speech-Motor Disorder CAS CAS is a rare disorder that begins in childhood but leaves a watermark of lifelong impacts. The most commonly cited prevalence estimate for CAS is approximately 1% (12 of 944 in Newcastle Upon-Tyne, UK [Morley, 1972]; 10 of 1,000 in Minnesota, US [Yoss, 1975]). More contemporaneous epidemiological data drawn from a population cohort is not available. The most recent prevalence estimate in the field is drawn from a US-based university clinic (Shriberg, 1994), that is, a clinical sample, which is inherently biased and unlikely to be representative of the general population. Nevertheless, this estimate is still in line with historical prevalence data, again suggesting that 1 to 2 children per thousand are likely to present with CAS (Shriberg, 1994).

CAS is a "deficit in speech-motor programming and/or planning that affects a child's ability to perform the spatiotemporal parameters of movement sequences, resulting in errors in speech sound production and prosody" (American Speech-Language-Hearing Association [ASHA], 2007, p. 1). Hence, the core deficit lies in the higher-order motor system as regards motor planning: transforming abstract phonological representations into speech-motor goals or commands, and motor programming: further specifying the articulatory and/or acoustic tactics or parameters within each articulator for speech production (Morgan & Murray, 2017). In "pure" CAS, the speech deficit is unrelated to orofacial structural impairments (such as cleft lip or palate) or physical strength, tone, or range of movement affecting speech production (associated with dysarthria). However, CAS may co-occur with orofacial structural deficits, dysarthria, or other more common subtypes of speech disorder such as articulation and phonological impairment. It is more challenging to reliably diagnose CAS when it co-occurs with these additional diagnoses. CAS is also known as developmental verbal dyspraxia in the United Kingdom (Royal College of Speech Language Therapists [RCSLT], 2011) and has historically been known as verbal dyspraxia, speech dyspraxia, and developmental apraxia of speech. This multifarious nosology mirrors the complexity of symptomatology associated with this diagnosis.

1.1.1. Symptomatology and diagnosis of CAS For over 50 years, the field of Speech Pathology debated whether a parsimonious set of speech deficits could be operationalized to signify a CAS diagnosis. Over the past decade there has been relative agreement and uptake of three core diagnostic features for CAS, as provided by an American Speech-Language-Hearing Association consensus panel (ASHA, 2007):

1. Inconsistent speech errors (e.g., producing the same syllable or word differently across repetitions of the same word such as *copta, upa, opta* for helicopter).
2. Lengthened and disrupted co-articulatory transitions, e.g., oral groping behaviors during speech; difficulty sequencing phonemes and syllables; difficulty maintaining syllable integrity; hypernasality (thought to be due to incoordination of velum for denoting oral/nasal contrasts, e.g., *a-m-u/a-m-u/a-m-u/a-m-u/* in Verbal Motor Production Assessment for Children test); slowed and disrupted diadochokinetic sequences (e.g., repetition of *pa-ta-ka*).
3. Inappropriate prosody (e.g., lexical stress errors, equal stress across words and sentences) (Morgan & Murray, 2017).

An important caveat is that, while these features are common across many cases with CAS, they are recognized not to be simply the "necessary" or "sufficient" features required to reach a diagnosis (ASHA, 2007, p. 2). This is because the features of CAS change across the life span and heterogeneity is seen across the population, making it challenging to define and operationalize a restricted list of diagnostic features.

Further common speech characteristics of children with CAS include limited babbling during development, delayed onset of first words, loss of words once mastered, and delayed and limited development of speech sound repertoires of consonants and vowels. Children with CAS use simplified syllable shapes, such as a single vowel [ɛ] for *leg*, deleting a syllable to use a single consonant-vowel syllable [ba] for *baby*, and reducing consonant clusters as in [ta] for *star*. Such errors reflect difficulties in motor planning and programming the sounds in their inventory together to form words (Morgan & Murray, 2017). Broader motor system difficulties may be associated with CAS including oral-motor dyspraxia (RCSLT, 2011) and fine or gross motor impairments, sometimes expounded under a diagnosis of Developmental Coordination Disorder (Newmeyer et al., 2009). Sensory integration impairments are also reported in some affected children. Nonverbal "performance" IQ is generally more preserved than verbal IQ. In the first few years of life, children with CAS typically have receptive language functioning that is more preserved than their expressive language functioning. In school-age children with CAS, language abilities are typically in line with a child's cognitive profile. Once literacy develops, difficulties with real word and non-word reading, spelling, and phonological awareness skills are reported (e.g., Gillon & Moriarty, 2007; Lewis, Freebairn, Hansen, Iyengar, & Taylor, 2004; Marion, Sussman, & Marquardt, 1993). In addition to the overt motor production deficits, difficulties with phonological encoding and short-term phonological memory are now also acknowledged as part of the condition (Shriberg, Lohmeier, Strand, & Jakielski, 2012).

1.1.2. Age of onset and prognosis of CAS Very early developmental features of CAS, such as hypotonia, restricted babbling, feeding difficulties, a restricted phonemic repertoire, and restricted language development are synonymous across a number of neurodevelopmental disorders where linguistic involvement is a core feature in the early years. That is, this developmental history is not specific for CAS (Morgan, Fisher, Scheffer, & Hildebrand, 2017). Specific signs of CAS are not typically probed until development of sufficient speech and language has occurred to allow systematic testing of core symptoms. Consequently, CAS is typically diagnosed at around three years of age, although earlier diagnosis is encouraged when the family history is positive.

Most children with "idiopathic"[2] CAS have speech that is unintelligible in the first years of life, even to familiar listeners such as friends and family. With effective treatment, children can become largely intelligible by the time they commence school. Toward the end of the first decade of life, most children with pure "idiopathic" CAS have resolved, apart from a few residual errors (e.g., articulation of /s, r/, consonant clusters production; Preston, Hull, & Edwards, 2013).

In terms of neurological involvement, routine clinical MRI scans are not currently sensitive enough to identify neural anomalies or correlates for most individual children with idiopathic CAS, even when the phenotype or clinical presentation is severe. A handful of reports exist on overt clinical MRI findings in children with known causes of CAS such as galactosemia due to disruption to *GALT1* (Hughes et al., 2009 and see Robertson, Singh, Guerrero, Hundley, & Elsas, 2000, for early discussion of *GALT1* genotype-CAS phenotype association), creatine transporter deficits (Battini et al., 2007), and rolandic epilepsy (Roll et al., 2006; Scheffer et al., 1995; see Liégeois & Morgan, 2012, for review of neural bases of speech-motor disorders). To date however, the most comprehensive quantitative MRI work in this area has been conducted in CAS associated with *FOXP2* mutation (Belton, Salmond, Watkins, Vargha-Khadem, & Gadian, 2003; Liégeois et al., 2003; Liégeois, Morgan, Connelly, & Vargha-Khadem, 2011; Liégeois et al., 2016; Vargha-Khadem et al., 1998; Watkins et al., 2002). Considering all of these study findings together, it appears that CAS is most commonly associated with brain anomalies of the inferior frontal cortices, rolandic cortices, basal ganglia, and cerebellum (Liégeois & Morgan, 2012; Liégeois et al., 2016).

1.2. Speech-Motor Disorder Dysarthria The term *dysarthria* is restricted to speech disorders with a known neurological origin, that is, damage to the central or peripheral nervous system. Most commonly in childhood, dysarthria is due to damage occurring almost anywhere along the corticobulbar and corticospinal tracts, the core white matter pathways responsible for controlling movement or production of speech (Morgan & Liégeois, 2012). Although peripheral nerve damage, lesion, or paralysis (e.g., Bell's facial palsy) may also result in dysarthria. Both central and peripheral neural damage result in aberrant tone that in turn manifest as weakness, incoordination, spasticity, involuntary movements, or reduced movements of the speech mechanism (i.e., lips, tongue, jaw, cheeks, soft palate, larynx, respiratory system) to result in dysarthria.

1.2.1. Symptomatology and diagnosis of dysarthria An operationalized definition for diagnosis of dysarthria is a disorder of the central or peripheral nervous system that impairs neuromuscular control and muscle tone (e.g., flaccidity, spasticity, fluctuating tone, or incoordinated and involuntary movements) and results in speech deficits across one or more features of phonation, articulation, resonance, prosody, or respiration impacting on naturalness (listener comfort) and/or intelligibility of speech.

The type of underlying tone or movement disturbance (e.g., spasticity, flaccidity, ataxia) is critical for determining the subtype of dysarthria (e.g., spastic dysarthria, flaccid dysarthria, ataxic dysarthria) when making a diagnosis using the Mayo Clinic Dysarthria Rating Scale (Duffy, 2013). This scale was developed for application to adult populations (Darley, Aronson & Brown, 1969), although in the absence of a pediatric-specific assessment (Morgan & Liégeois, 2010), this tool has been frequently applied to pediatric populations in both acquired (Liégeois et al., 2010; Morgan et al., 2011; Morgan, Masterton, Pigdon, Connelly, & Liégeois, 2013) and genetic conditions (Morgan et al., 2015; Turner et al., 2013; Turner, Mayes, et al., 2015; Turner et al., 2017). Speech features affected may include one, or a combination, of the following:

- voice quality or phonation, for example, strained, tremulous, or breathy voice
- poor breath support for speech, resulting in short phrases or bursts of speech
- resonance, most commonly realized as hypernasality
- distortion or "slurring" of speech sounds, leading to imprecise articulation
- speech rate, usually slower than typical, but sometimes fast
- speech rhythm, for example, too little or excessive variation in pitch or loudness

Significantly, children with neurogenetic syndromes and conditions such as cerebral palsy, where bilateral damage is apparent from birth, are at increased risk for a comorbid diagnosis of both dysarthria and CAS. Dysarthria may also co-occur with other acquired or developmental disorders (e.g., language impairment).

1.2.2. Age of onset and prognosis of dysarthria Symptoms of dysarthria are apparent from the onset of neurological involvement, whether at birth or as a later acquired disorder such as brain tumor, traumatic brain injury, or stroke. Dysarthria due to perinatal neural lesions (occurring around the time of birth) will be identifiable as soon as the child begins to speak. Children acquiring dysarthria due to a brain lesion later in life, after having developed typical speech and language, typically have a better prognosis than those with perinatal injury. Children with acquired brain injury tend to recover or compensate for their speech function relatively well, unless the brain injury is particularly severe and diffuse (Liégeois et al., 2010; Morgan et al., 2013; Morgan et al., 2011). This prognosis is understandable in the context of neuroplasticity of the developing brain (see Anderson, Spencer-Smith, & Wood, 2011, for review). As for CAS, persistent dysarthria is also most likely to occur when there is bilateral damage to the speech-motor pathways (Liégeois et al., 2010; Morgan et al., 2013). Given that most genetic conditions are also associated with bilateral neural involvement, children with neurogenetic forms of dysarthria typically have poor speech prognosis.

2. Genetic Bases of Speech-Motor Disorders

Speech and language disorders are highly heritable, evidenced by increased concordance rates for these conditions in monozygotic compared to dizygotic twins (Bishop, North, & Donlan, 1995; DeThorne et al., 2006; Lewis & Thompson, 1992; Locke & Mather, 1989; Matheny & Bruggemann, 1973). There is also evidence that speech abilities may be more heritable than other language-related traits (Stromswold, 2006). The more speech-akin domain of phonology, for example, is reportedly more heritable than linguistic domains of vocabulary or syntax (Stromswold, 2006). Furthermore, based on current evidence, it appears there will be many more monogenic pathways leading to speech-motor disorders in comparison to specific language disorders. For example, there has been no confirmed single gene mutation associated with specific language impairment (SLI).[3] SLI is a language disorder in the absence of intellectual disability or frank neurological lesion. Current findings suggest SLI has more complex genetic architecture than CAS, with a greater polygenic contribution and possible environmental contributions (Simpson et al., 2015; Villanueva et al., 2015). A recent exploratory genome-wide copy number variant (CNV) study on individuals with SLI compared to population controls found that SLI individuals had an increased CNV burden in terms of the average number of events, the total length of CNVs, the average CNV size, and the number of genes spanned (Simpson et al., 2015). Hence, the authors suggested the majority of risk conferred by CNVs in SLI is via common, inherited events within a "common disorder-common variant" model (Simpson et al., 2015). Importantly, the authors did not rule out de novo CNVs leading to pathology in individual cases, but they suggested that the risk conferred by CNVs in the majority of SLI cases likely depends on the combination of inherited events (both CNVs and SNPs[4]), genetic background, and environmental factors (Simpson et al., 2015). By contrast, there have been a number of single genes associated with both CAS and dysarthria over the past decade, as will be described further in section 2.1.

The first significant genetic finding for speech-motor disorders was the discovery of mutations of *FOXP2,* explaining a rare monogenic form of CAS (Lai et al., 2001). This provided our first insights into critical molecular pathways for speech disorder. The discovery could be considered "ahead of its time" because the work occurred before the first human genome sequence was completed. Fifteen years on, we now have next-generation technologies enabling rapid and relatively cost-efficient whole-genome sequencing. This recent technological advance is leading to a proliferation of studies identifying further gene pathways to speech-motor disorders and CAS in particular.

An exhaustive list of possible gene pathways to speech-motor disorders is beyond the scope of this chapter. For example, a number of long-standing neurogenetic syndromes are associated with speech-motor and language difficulties, including Floating Harbor syndrome due to mutations of *SRCAP* (White et al., 2010), Cri du chat syndrome due to 5p15.2-p15.3 deletion (Elmakky et al., 2014), galactosemia associated with mutations in *GALT1* (Potter, Nievergelt, & Shriberg, 2013), and Kabuki syndrome associated with mutations in *KMT2D* and *KDM6A* (Morgan et al., 2015) to name but a few. Rather than detail these well-recognized syndromes, this section will focus on more novel gene pathways identified in recent times to show an association with speech-motor disorders. These pathways and associated genotype-phenotype relationships are summarized in table 40.1 and discussed in the following section.

2.1. Single Gene Mutations

2.1.1. FOXP2 *FOXP2* was the first gene implicated in a monogenic form of specific speech and language disorder (Lai et al., 2001). As outlined earlier, the phrase *specific speech and language disorder* refers to a communication disorder phenotype occurring in the absence of frank neurological lesion or intellectual impairment (Reilly et al., 2014). Heterozygous *FOXP2* mutations that cause disorder have been reported to occur in both de novo and inherited forms (Lai et al., 2001; Reuter et al., 2017; Turner et al., 2013). The penetrance for this severe speech and language disorder is high, close to 100% based on reported cases (see Morgan et al., 2017, for review). The story of *FOXP2* began with identification of the KE family; a three generation family, in which approximately half of the members (n = 15) are affected with CAS (Vargha-Khadem, Gadian, Copp, & Mishkin, 2005), see figure 40.1. The family were first described as an example of a large family with an autosomal dominantly inherited speech disorder (Hurst et al., 1990). Genetic linkage analysis of the family then mapped the likely damaged gene to a critical genomic interval on 7q31 (Fisher, Vargha-Khadem, Watkins, Monaco, & Pembrey, 1998). Later, identification of CS, an unrelated child with a similar

Table 40.1

A summary of speech-motor phenotypes associated with specific gene pathways

Gene	Dysmorphology	Epilepsy	NVIQ impairment	Autism	ADHD	Motor deficits	Language deficits	Oral-motor disorder/ praxis	CAS	Dysarthria
FOXP2-only	–	–	–/Borderline	–	–	–	+	+	+	+
FOXP2-plus	–/ Nonspecific	–	+Mild-mod.	?	–	+	+	+	+	+
FOXP1	+	+	+Mild-mod.	+	–	+	+	+	+	+
CNTNAP2	+Mild	+	+ Mod-sev.	+	+	–	+	?	+?	–
BCL11A	+	–	+ Mild	–	–	+	+	+	+	+
ELKS/ERC1	–/nonspecific	–	+	+	+	–	+	+	+	–
2q:1q trans.	–	–	–	–	–	–	+*	+	+	–
4q:16q trans.	+	–	+	–	–	–	+	+	+	–
16p11.2 del.	+Mild	+	+	+	–	–	+	+	+	+Mild
GRIN2A	+Mild	+	+	–	–	–	+	+	+	+

Phenotypic heterogeneity is common, hence this table is intended to provide a representative summary of features often seen in association with the genotype in question.

*Expressive only.

ADHD = attention deficit/hyperactivity disorder; del = deletion; mod = moderate; NVIQ = nonverbal IQ; sev = severe; trans = translocation; + = feature present; – = feature absent; ? = unclear from literature to date.

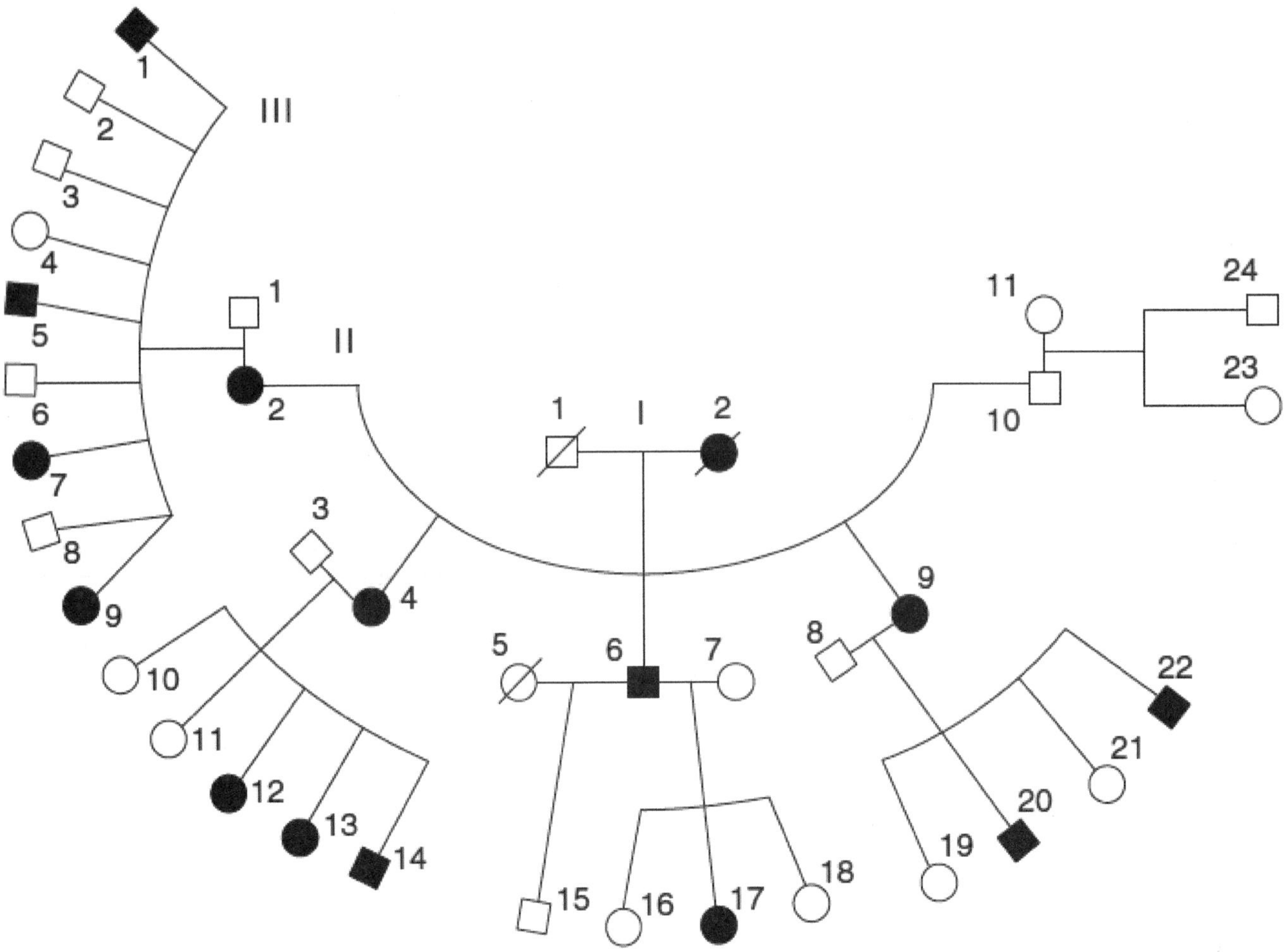

FIGURE 40.1 Pedigree of the KE family. I, II, and III mark the generations. Shaded symbols designate affected status; circles, females; squares, males (from Vargha-Khadem et al., 1998, ©1998 National Academy of Sciences USA). Participants in the present study are females II-2, II-9, III-7 and males III-5, III-20, III-22.

phenotype and a translocation involving this region led to identification of a mutation in *FOXP2*, explaining the phenotype in the KE family (Lai et al., 2001).

Since the discovery of *FOXP2* and detailed phenotyping in the KE family, a number of further cases with *FOXP2* disruption have been reported allowing further delineation of the associated phenotype and generation of a *GeneReview* chapter to guide clinical practice in this area (Morgan et al., 2017). Individuals with intragenic mutations disturbing *FOXP2* may be categorized as having "*FOXP2*-only"–related speech and language disorders, whereas cases with *FOXP2* disruption via translocation, deletion, or other complex variants such as pericentromeric inversion or uniparental disomy may be referred to as "*FOXP2*-plus"–related speech and language disorders (Morgan et al., 2017).

All *FOXP2*-related speech and language disorders—regardless of the underlying genetic alteration—have a core phenotype of CAS (Morgan et al., 2017). Similar to pure CAS cases described earlier in this chapter, additional findings across children with *FOXP2*-only– or *FOXP2*-plus–related speech and language conditions can include oral-motor dyspraxia, dysarthria, moderate-to-severe receptive and expressive language disorder, and reading and spelling impairments.

Of those individuals with disruption to *FOXP2*, cases with the underlying genetic alteration of *FOXP2*-only, have typically been found to have a relatively preserved nonverbal (performance) IQ compared to verbal IQ, appropriate social abilities, and both fine and gross motor skills that are typically normal. However a recent paper encompassing a further 14 individuals with *FOXP2* disruptions, from eight unrelated families, has extended the clinical spectrum in these cases, suggesting that mild cognitive impairment, mild motor impairments, autistic features (but not a formal diagnosis), and even mild dysmorphology (e.g., narrow palpebral fissures, mild finger pads, horizontal eyebrows, large ears) may be seen in some cases with *FOXP2*-only mutations (Reuter et al., 2017).

By contrast, individuals with *FOXP2*-plus genetic aberrations are more likely to have global developmental and behavioral issues with oral-motor deficits, global developmental delay, and autism spectrum disorder (ASD) as frequent concomitants (for review, see Morgan et al., 2017). These additional phenotypes may relate to disruptions of other genes neighboring *FOXP2* in that region of chromosome 7. Better understanding of the phenotype and awareness of *FOXP2*-related disorders will result in further *FOXP2*-affected cases reported in the literature in coming years. Continued refinement of the phenotype and more targeted therapeutic techniques will naturally follow. There is a groundswell of animal model and pharmacological research into therapeutics that may lead to exciting novel therapies targeting the underlying neurobiological pathways involved in *FOXP2*-related speech and language–related disorders (Chabout et al., 2016; Groszer et al., 2008; Haesler et al., 2007; Murugan, Harward, Scharff, & Mooney, 2013; Vernes et al., 2011).

2.1.2. FOXP1 *FOXP2* encodes a protein that belongs to a divergent subgroup of forkhead-box transcription factors (FOXP1–4), featuring a characteristic DNA-binding domain (Vernes, MacDermot, Monaco, & Fisher, 2009). These proteins act as regulators of gene expression. Among the FOXP subgroup, *FOXP1*, located at 3p13, was the next candidate gene hypothesized to influence speech and language–related pathways, by potentially working together with *FOXP2* (Vernes et al., 2009). This hypothesis was based on a body of evidence from functional gene studies, as well as mouse and songbird models (see Vernes et al., 2009). Studies of humans, mice and songbirds indicate several key sites of shared neural expression of FoxP1 and FoxP2 (Teramitsu, Kudo, London, Geschwind, & White, 2004), while cell biology investigations have demonstrated that the two proteins can directly interact to regulate target genes (e.g., Wang, Lin, Li, & Tucker, 2003). Studies of the significance of *FOXP1* to speech-motor disorders have been approached from both a phenotype-first and a genotype-first perspective as will be explained. A phenotype-first approach in the field of genotype-phenotype relationship studies is where cases are ascertained on the basis of their phenotype (e.g., CAS) and examined for possible genetic contributions to the condition. By contrast, a genotype-first approach is where cases are ascertained on the basis of genotype (e.g., *FOXP1, FOXP2, CNTNAP2* mutations) and then examined to study the phenotype.

Using a phenotype-led approach, Vernes et al. (2009) tested the hypothesis of *FOXP1* as a possible candidate for CAS by conducting mutation screening of this gene in a panel of 49 probands screened previously for *FOXP2* coding changes (MacDermot et al., 2005). Importantly, inclusion criteria for this CAS cohort included the absence of intellectual impairment. The authors identified a nonsynonymous coding change in one proband, yielding a proline-to-alanine change (P215A). A similar change was also found in a random control sample, however, and analyses of noncoding SNPs revealed no correlation with affection status (Vernes et al., 2009). Hence, the authors concluded that *FOXP1* mutations are unlikely to represent a major cause of CAS, at least in the absence of intellectual impairment (Vernes et al., 2009).

Many years later, next-generation sequencing enabled identification of multiple individuals with *FOXP1* mutations (Carr et al., 2010; Hamdan et al., 2010; Horn et al., 2010; Le Fevre et al., 2013; Lozano, Vino, Lozano, Fisher, & Deriziotis, 2015; Palumbo et al., 2013; Sollis et al., 2016). Such findings have allowed a genotype-led perspective on the contributions of this gene to disorder, yielding definition of a recognizable *FOXP1* syndrome. The syndrome involves mild-to-moderate intellectual disability (ID); features of ASD; behavioral problems such as obsessions, compulsions, aggression, and hyperactivity; and macrocephaly (see Sollis et al., 2016, for review). Moderate-to-severe speech and language impairments have also been reported, with expressive language often more affected than receptive language, and speech deficits possibly related to orofacial motor dysfunction (Sollis et al., 2016). The phenotype has been recently extended in three cases to include sensory integration disorder; hypotonia; hypertelorism; and abnormal facial features including a broad forehead, down-slanting palpebral fissures, a short nose with broad tip, and a frontal hair upsweep (see figure 40.2; Sollis et al., 2016). Speech and language impairments were also noted in all three new cases in Sollis et al. (2016), yet the phenotype was described as including articulation and pronunciation deficits rather than a diagnosis of CAS per se. Interestingly, "apraxia of the tongue" was expressly noted for one individual, suggesting an oral-motor praxis specific to that articulator. From a functional perspective, experiments in cellular models by Sollis et al. (2016) suggested that disruption of the transcriptional repression activity of *FOXP1* underlies the neurodevelopmental phenotype in *FOXP1*-related disorder (Sollis et al., 2016). The authors also noted that the three novel *FOXP1* variants reported in their paper retained the ability to interact with *FOXP2*, with potential relevance to the speech and/or language

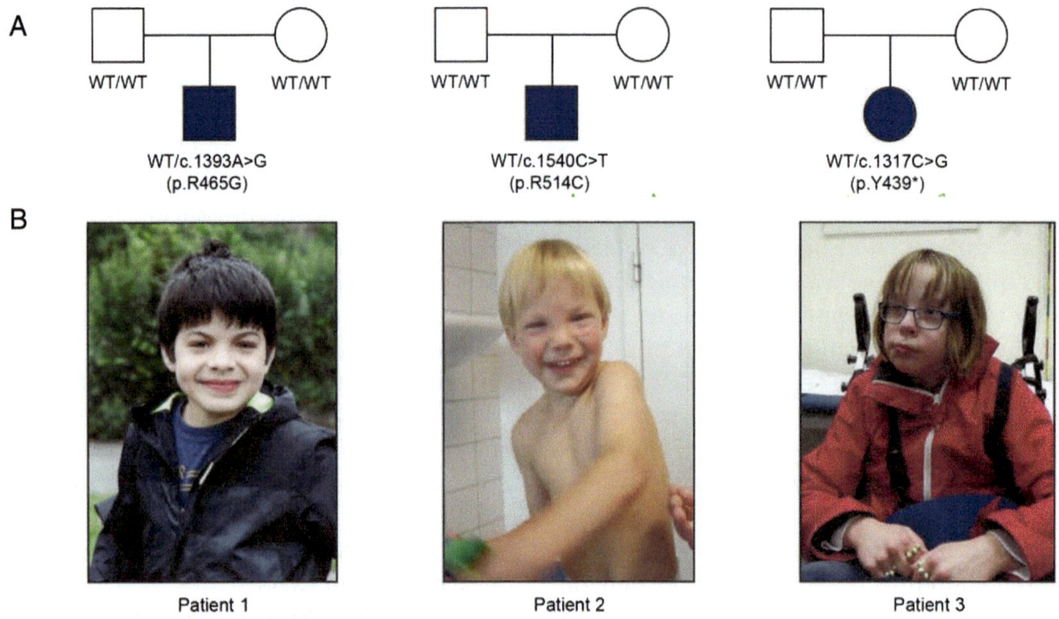

FIGURE 40.2 Identification of de novo *FOXP1* variants in three patients with neurodevelopmental disorder including reported speech impairment. (A) Pedigrees of the three patients. (B) Photographs of patients 1, 2, and 3 at 11, 3, and 12 years of age, respectively. Adapted from Sollis et al. (2016).

deficits observed in these cases. More systematic speech and language phenotyping is required in children with *FOXP1* mutations to enable greater understanding of the exact nature of communication difficulties in this group, including the potential significance of speech-motor disorders.

2.1.3. CNTNAP2 Following on from *FOXP2* and *FOXP1*, the next gene identified as a potential candidate for speech and language disorders was *CNTNAP2*. This connection was first discovered in a study hypothesizing that neural pathways downstream of *FOXP2* might influence more common phenotypes such as SLI (Vernes et al., 2008). Genomic screening was performed for regions bound by *FOXP2*, leading to identification of a *FOXP2*-bound fragment located within intron 1 of *CNTNAP2*. Further analyses in cellular models demonstrated that *FOXP2* can indeed regulate the expression of *CNTNAP2*, confirming the latter as a downstream target. *CNTNAP2* encodes CASPR2, a member of the neurexin superfamily of transmembrane proteins expressed in the human cortex (Rodenas-Cuadrado et al., 2016). Vernes et al. (2008) went on to examine associations between SNPs in *CNTNAP2* and a quantitative language measure of nonword repetition in a well-characterized set of 184 families affected with SLI. Intriguingly, *CNTNAP2* had also been linked to language delays in autism (Alarcón et al., 2008; Vernes et al., 2008). Hence, Vernes et al. (2008) suggested that the *FOXP2-CNTNAP2* pathway provides a mechanistic link between syndromes involving disrupted language. A further study supporting this apparent link found common variants in the exon region of *CNTNAP2* influence early language acquisition, as assessed at age two, in the general population (Whitehouse, Bishop, Ang, Pennell, & Fisher, 2011). Whitehouse et al. (2011) suggested that *CNTNAP2* risk variants increase susceptibility to SLI or autism when they occur together with other risk factors.

Based on genotype-driven phenotyping studies of individuals with mutations, the *CNTNAP2* story has since become a prime example of a single candidate gene being implicated across multiple neurodevelopmental disorders. In particular, mutations of *CNTNAP2* have been reported to impact ASD, schizophrenia, ID, cortical dysplasias, epilepsy, dyslexia, and language impairment (reviewed by Rodenas-Cuadrado, Ho, & Vernes, 2014). Most recently, Smogavec et al. (2016) identified homozygous mutations, compound heterozygous CNVs, or CNVs and mutations in *CNTNAP2* in eight individuals from six unrelated families, all of which were predicted to be deleterious for protein function. Unlike the de novo mutations of *FOXP1*, all *CNTNAP2* aberrations were inherited from healthy, heterozygous parents (Smogavec et al., 2016). Early onset epilepsy was a core feature, occurring in all affected individuals, with onset in the first 3.5 years of life. Further commonly seen features were severe ID (6 of 8), regression of speech development (5 of 8), and behavioral anomalies (7 of 8). The characterization of speech and language was limited, however, making it challenging to understand the exact nature of "regression" of speech development. Based on the increasing evidence from genotype-led studies in this area, however, it would appear that speech and language symptoms associated with *CNTNAP2* loss-of-function mutations are part of a broader syndrome rather than part of a specific causative pathway to speech-motor disorders.

Only one study has directly evaluated the association between *CNTNAP2* and the speech-motor disorder of CAS (Centanni et al., 2015). The authors examined genetic profiles of seven children with CAS and eight children with SLI who were reported to be "free" of comorbid speech, language, cognitive, or literacy impairments. They identified deletions in alternative intron 18 of *CNTNAP2* at 7q35 in two children with CAS but not in children with SLI. Based on this finding, the authors suggested CNVs in *CNTNAP2* occur with a higher rate than chance in children with CAS than in the general population (Centanni et al., 2015). Given the careful selection of participants without comorbid impairments, they further suggested that *CNTNAP2* variants may be associated with deficits in speech production in the absence of comorbid reading, language, and cognitive impairments. Importantly, however, each child who carried a *CNTNAP2* CNV in this study also carried additional variants elsewhere in the genome that met reportable criteria (Centanni et al., 2015). Furthermore, a pathogenic role for the *CNTNAP2* CNV identified in the CAS children was not proven via gene function studies or any other supporting forms of evidence. Hence, further studies examining the link between *CNTNAP2* and speech-motor function are warranted.

2.1.4. GRIN2A The identification of a role for *GRIN2A* has a long history, beginning 60 years ago with the discovery of epilepsy-associated speech and language disorders (Landau & Kleffner, 1957). Further clinical enquiry into children with epilepsy since this time has revealed a broad spectrum of epilepsy aphasias—which represents an association among epilepsy, speech and language disorders, and the EEG signature of centrotemporal spikes (Turner, Morgan, Perez, & Scheffer, 2015). At the mild end of the spectrum is benign childhood epilepsy with centrotemporal spikes, typically mild language impairment, and speech and oral-motor dyspraxia (Turner, Morgan, et al., 2015). The severe end of the spectrum by contrast, is signified by the epileptic encephalopathies of Landau-Kleffner Syndrome and epileptic encephalopathy with continuous spike-wave during sleep (ECSWS), in synchrony with severe speech and language deficits (Turner, Morgan, et al., 2015). While this spectrum focuses on speech and language deficits in the presence of epilepsy, just as for many neurodevelopmental syndromes, there is phenotypic variability. That is, other cognitive and motor difficulties can occur with ECSWS in the absence of speech and language problems. Finally, acquired epileptic opercular syndrome is a further condition associated with this spectrum, evoking oral-motor function (oral-motor dyspraxia, dysarthria, drooling) with CSWS.

Subsequent to the advent of next- generation sequencing, gene discovery in Landau-Kleffner syndrome and other epilepsy-aphasia spectrum (EAS) syndromes has intensified (Turner, Morgan, et al., 2015). Genes encoding proteins with diverse functions (ion channels, RNA binding, cell adhesion), as well as genomic regions associated with other neurodevelopmental disorders, have been implicated in EAS syndromes (see Turner, Morgan, et al., 2015, for review). This section focuses on the discovery of *GRIN2A* mutations as the first monogenic cause of EAS disorders.

In November 2013, three parallel studies (Carvill et al., 2013; Lemke et al., 2013; Lesca et al., 2013) reported mutations or very small deletions in *GRIN2A* in patients with focal epilepsy and speech and language dysfunction. While language functioning in individuals with these epilepsy aphasias was historically well-characterized, the speech phenotype has only recently been reported (Turner, Morgan, et al., 2015). Turner, Mayes, et al. (2015) examined 11 individuals with *GRIN2A* mutations from three families, A-C and D (see figure 40.3). Variable phenotypes may be associated with the *GRIN2A* genotype, as seen across these families (Turner, Mayes, et al., 2015).

Family A included five members with autosomal dominant rolandic epilepsy with speech dyspraxia previously reported in Scheffer et al. (1995). family C was related to family A and consisted of a father-son pair with ECSWS with the same *GRIN2A* mutation. Family D comprised two brothers with Intermediate Epilepsy Aphasia Disorder, a sister with ECSWS, and a mother with no history of seizures, previously reported in Tsai et al., (2013). In their examination of the speech phenotype, Turner, Mayes, et al. (2015) reported a profile of oral-motor dysfunction, dysarthria, and features of apraxia of speech across all three families regardless of the associated form of epilepsy. The study also highlighted that this speech phenotype may occur in the absence of a seizure disorder, suggesting an important role for *GRIN2A* in speech-motor function (Turner, Mayes, et al., 2015).

A growing body of work has considered the biological functions of *GRIN2A*, which encodes the NR2A subunit of the glutamate *N*-methyl-D-aspartate (NMDA) receptor. NMDA receptors are ligand-gated ion channels involved in brain development, synaptic plasticity, and memory. The receptors are tetrameric and

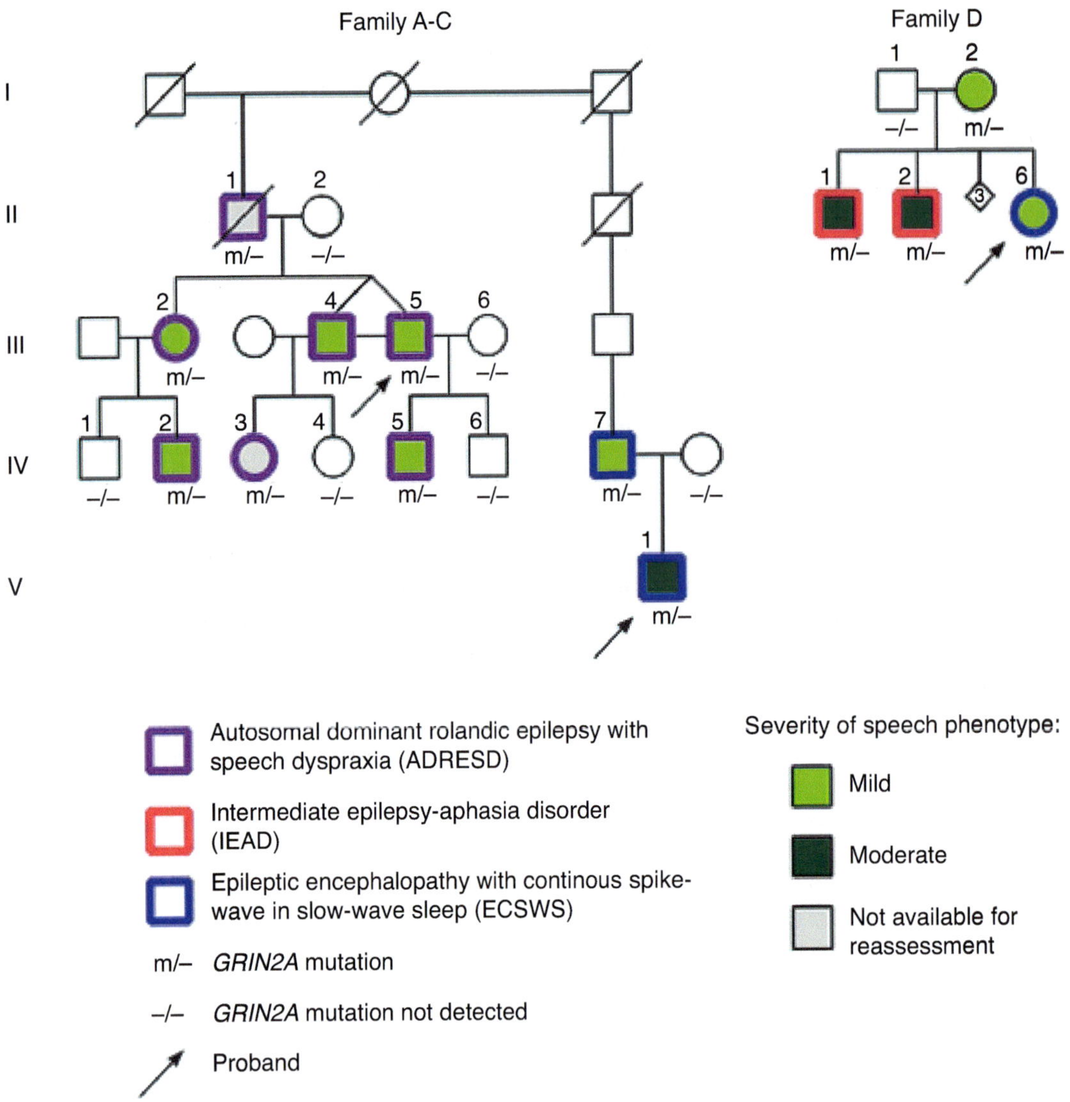

FIGURE 40.3 Pedigree of families A-C and D with mutations in *GRIN2A* from Turner, Mayes et al. (2015). Family A previously reported in Scheffer et al. (1995); family D is family 1 in Tsai et al. (2013).

comprise two NR1 subunits and two NR2 subunits (from NR2A to NR2D) (Kornau, Schenker, Kennedy, & Seeburg, 1995). The NR2A subunit is critical to NMDA receptor functioning, controlling cell surface expression and localization (Kornau et al., 1995), modifying channel properties (Monyer et al., 1992), and providing glutamate-binding sites (Laube, Hirai, Sturgess, Betz, & Kuhse, 1997). *GRIN2A* mutations disrupt the ligand-binding domain of the NR2A subunit and alter NMDA receptor gating (Lesca et al., 2013; Carvill et al., 2013). Mice expressing truncated NR2A show deficits in synaptic plasticity and reorganization and have impaired motor coordination (Sprengel et al., 1998).

In the human brain, NR2A is expressed in many cortical and subcortical structures including those relevant to speech and language (Akbarian et al., 1996; Conti, Barbaresi, Melone, & Ducati, 1999; Clinton & Meador-Woodruff, 2004; Hynd, Scott, & Dodd, 2004; Kosinski et al., 1998; Scherzer et al., 1998). Aberrant NMDA receptor functioning in the basal ganglia may contribute to impaired motor speech planning/programming and execution, with these structures implicated in both childhood dysarthria and CAS (Liégeois & Morgan, 2012). While we have some key insights into the biological functions of *GRIN2A*, further work is underway to better understand these mechanisms. For example, we have little understanding of NR2A expression in Broca's area, a key expressive language region implicated in a range of speech and language disorders including CAS (Liégeois & Morgan, 2012). Greater understanding of the biological function of *GRIN2A* will lead to more targeted therapies to improve speech, language, and seizure outcomes.

2.1.5. BCL11A Ten cases of 2p15p16.1 microdeletions had been reported worldwide up until 2014, with the sizes of deletions ranging from 0.3 to 5.7 Mb and spanning three genes. At that time, Peter, Matsushita, Oda, and Raskind (2014) described a de novo microdeletion of 2p15p16.1 that disrupted just a single gene, *BCL11A*, in an 11-year-old male. *BCL11A* encodes a transcription factor that is a putative member of the BAF swi/snf chromatin-remodeling complex (Dias et al., 2016). Prior to the work of Peter et al. (2014), descriptions of the 2p15p16.1 deletion syndrome had involved a broad ID profile with concomitant poor expressive language, growth retardation, craniofacial and skeletal dysmorphic traits, internal organ defects, abnormal muscle tone, and gross motor delays. The careful phenotyping of Peter et al. (2014) revealed the presence of these features in their case, occurring alongside CAS and dysarthria with hypotonia and general oral and gross motor dyspraxia. The presence of speech-motor disorder in this case with deletion limited to *BCL11A* led the authors to propose this gene as a potential new candidate for CAS.

More recently, Dias et al. (2016) characterized children with de novo heterozygous missense, nonsense, and frameshift mutations specific to *BCL11A*. They identified nine individuals with intellectual disability with de novo mutations in *BCL11A*: three missense and six loss of function. All individuals had delayed developmental milestones globally, including speech and language delay. Cognitive impairment varied from mild to severe but most had moderate ID. Two individuals had a diagnosis of ASD and four others had a spectrum of behavioral anomalies including sensory issues and repetitive behaviors (Dias et al., 2016). Commonly shared features of dysmorphology included joint laxity (87%), strabismus (100%), external ear abnormalities (62%), microcephaly (55%), and thin upper lip and flat midface (see figure 40.4A). The ear anomalies were reported to be more severe in individuals with nonsense and frameshift mutations (Dias et al., 2016). Fetal hemoglobin was significantly elevated in all affected individuals explicitly tested for this feature, including those with missense mutations in *BCL11A* (figure 40.4B).

The study provided multiple lines of functional evidence for phenotypic effects involving human cellular analyses coupled to mouse behavioral, neuroanatomical, and molecular phenotyping (Dias et al., 2016). Bcl11a haploinsufficiency in mice led to impaired cognition, abnormal social behavior, and microcephaly in accordance with the human phenotype (Dias et al., 2016). Anomalous transcriptional profiles were revealed in the cortex and hippocampus of these same mouse models. Hence, this work implicates *BCL11A* haploinsufficiency in neurodevelopmental disorders, including the presence of speech and language disorder, seen across all cases. Phenotyping delineating subtypes of speech disorder was not conducted, hence whether individuals presented expressly with CAS remains unknown. Nevertheless, *BCL11A* may be a promising new candidate gene for investigation in relation to speech and language development, and with further work, may be found to be more specifically associated with CAS.

2.2. Multigene Deletion Syndromes

2.2.1. The 16p11.2 deletion syndrome The 16p11.2 microdeletion represents one of the most well-studied CNVs in neurodevelopmental disorders. The proband often has a de novo deletion; however, the deletion can also be transmitted from a parent to a child. These deletions are relatively common and are seen in ~1 per 2,000 individuals (Jacquemont et al., 2011; Stefansson et al., 2014). Common features of the deletion syndrome include moderate ID, ASD, poor cognitive and language ability (as great as 2 standard deviations below familial noncarriers), epilepsy, macrocephaly (apparent by age two years), brain abnormalities such as chiari malformations or cerebellar ectopia, and even obesity (Miller et al., 2015). Speech deficits have also been proposed as an associated feature (Newbury et al., 2013; Raca et al., 2013). Specifically, authors conducting CNV screens of individuals with CAS identified three unrelated individuals with 16p11.2 deletions (Newbury et al., 2013; Raca et al., 2013). This finding, revealed by a phenotype-first approach, then begged the question as to whether CAS was prevalent across a broader population of children with 16p11.2 deletions. Hence, Fedorenko et al. (2016) took a genotype-first approach, conducting the first detailed, systematic, and replicable characterization and differential diagnosis of speech abilities in a group of 11 unrelated children with 16p11.2 deletion syndrome.

Two recruited individuals were largely nonverbal, preventing detailed speech analysis, whereas the remaining nine met the standard accepted diagnostic criteria for CAS (Fedorenko et al., 2016). Findings linked 16p11.2 deletions to a highly penetrant form of CAS and underlined the need for further precise characterization of speech and language profiles in larger groups of affected individuals to enhance our understanding of how genetic pathways contribute to human communication disorders. As noted by the authors, it will also be important to determine which of the specific genes implicated in the 16p11.2 deleted regions are most closely associated with the profile of CAS.

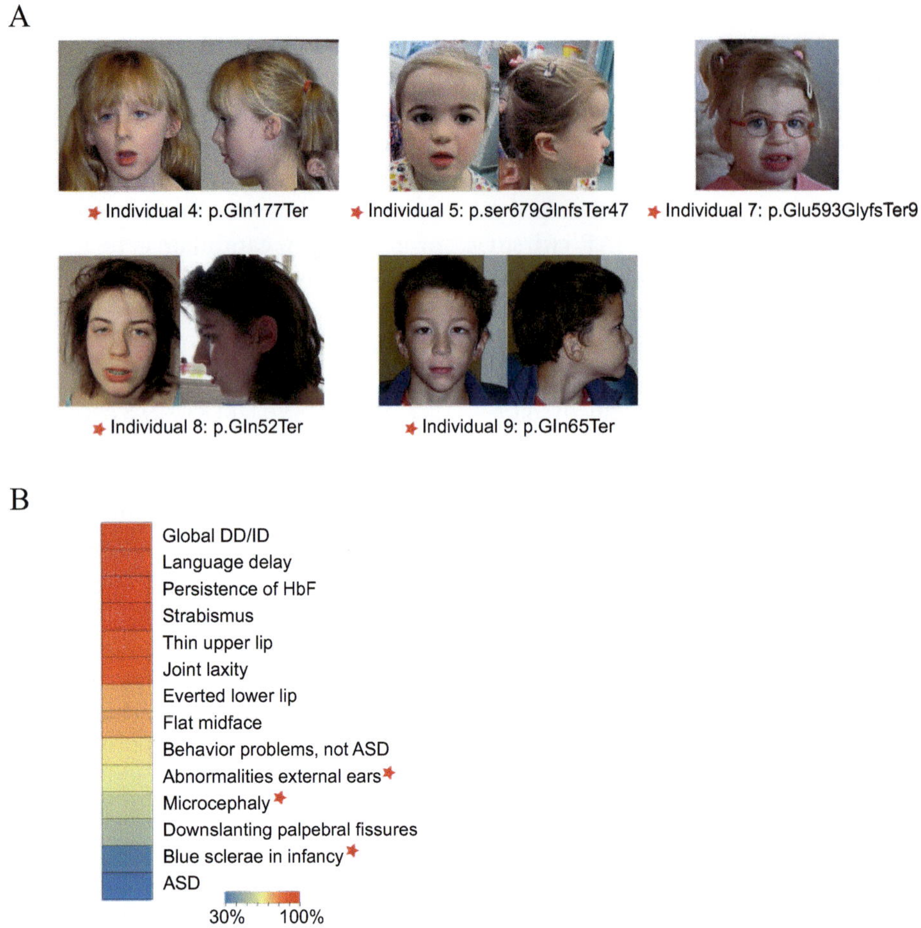

FIGURE 40.4 *BCL11A* mutations from Dias et al. (2016). (A) Dysmorphology associated with *BCL11A* mutations, including strabismus, down-slanting palpebral fissures, synophrys, flat midface, thin upper lip, and full lower lip. (B). Heatmap represents frequency of the feature seen, from 100% (red) to 30% (blue); frequency calculated based on information available for each feature. Predominant (microcephaly, abnormal external ears) or exclusive features (blue sclerae in infancy) in individuals with loss of function mutations are indicated by a red star. DD = developmental delay; HbF = fetal hemoglobin. Adapted from Dias et al. (2016).

Following on from this story, a more recent study reported on the speech phenotype of 136 individuals with 16p11.2 deletions (Steinman et al., 2016). "Speech articulation abnormalities" were noted in 79% of the cohort. Furthermore, articulation abnormalities differed significantly between proband and familial carriers (86% vs. 48% with articulation problems; Steinman et al., 2016). There was no mention of systematic assessment for CAS or any other subtypes of speech sound disorders in this report. It appears that the speech assessment was observation of speech during an informal neurological exam. A lack of detail on the stimuli or tools used to assess speech makes it challenging to interpret findings in relation to the report by Fedorenko et al. (2016) of increased penetrance for CAS in 16p11.2 deletion syndrome. Precise phenotyping, with the use of correct terminology, diagnostic tools, and criteria accessible and replicable to others is critical to elucidate genotype-phenotype relations in this relatively common deletion syndrome.

2.2.2. The 12p13.33 deletion and ELKS/ERC1 Cost-effective genetic screening methods have naturally led to the identification of many more cases with rare microdeletion syndromes. By collating and studying these cases together for genotype-phenotype associations, we may be able to eventually identify single genes of interest most closely aligned with specific aspects of behavior. The recent identification of *ELKS/ERC1* as a candidate for CAS provides a good example of this approach. Thevenon et al. (2013) identified nine participants across six families with rare deletions at 12p13.33. They noted that *ELKS/ERC1* was the only gene consistently deleted across the different affected people and hence hypothesized this gene was key to expression of the phenotype.

Until the report of Thevenon et al. (2013), only four cases of 12p13.33 microdeletion syndrome had been reported in the literature that were associated with phenotypic features that included ID, various psychiatric manifestations, and delayed first words in some (Abdelmoity, Hall, Bittel, & Yu, 2011; Baker et al., 2002; MacDonald et al., 2010; Rooryck et al., 2009). Thevenon et al. (2013) expanded this clinical phenotype through examination of their nine cases. Phenotyping was prospective and detailed in some cases, and retrospective in others, making it challenging to determine penetrance of specific features across the group. "Speech delay" was reported across all nine participants. However, on the basis of data presented (stating that first words did not occur until 36 to 42 months), it appears that the authors were in fact referring here to "language delay," instead of delayed phonetic or phonemic speech acquisition. Walking was also delayed in some cases. Absent to nonspecific dysmorphic features were reported with prominent ear lobes being a frequent feature. Five of the nine participants were assessed by a speech therapist and reported to have CAS. Language and reading difficulties were reported in some but not systematically assessed across all cases, making it challenging to interpret the significance of this gene for literacy and language problems. The participants also had a broader profile of neurodevelopmental disorder beyond speech and language, including intellectual impairment (5 of 9), psychiatric manifestations (5 of 9), behavioral difficulties (7 of 9), attention deficit/hyperactivity disorder (6 of 9), ASD (2 of 9), and reported mild nonspecific dysmorphic features (Thevenon et al., 2013).

Overall, it appears that deletions of 12p13.33 may be responsible for variable phenotypes including CAS, which is associated with a broader neurobehavioral syndrome. At a biological level, little is understood regarding gene function for *ELKS/ERC1*. Although *ELKS* protein is concentrated in the presynaptic active zone of neuron cells, it is not brain-specific and is also expressed at neuromuscular junctions (Held, Liu, & Kaeser, 2016; Juranek et al., 2006; Liu et al., 2014).

2.3. Chromosomal Rearrangements The majority of genetic disruptions associated with speech-motor disorders thus far have involved de novo or inherited autosomal dominant mutations or CNV syndromes. An additional form of genetic diagnosis associated with neurodevelopmental disorders, including speech, language, or literacy impairments (Froster, Schulte-Körne, Hebebrand, & Remschmidt, 1993; Rabin et al., 1993; Ozantürk et al., 2016; Moralli et al., 2015), is a chromosomal translocation or rearrangement. These genetic aberrations involve a section of a chromosome, or even an entire chromosome, breaking off and joining another chromosome. *Balanced translocation* refers to cases where the genetic material is neither added nor deleted. To date there have been only a handful of non-*FOXP2*-related chromosomal translocations reported alongside well-phenotyped speech-motor disorder and specifically CAS. Weistuch and Schiff-Myers (1996) reported on a case with a balanced location with a breakpoint at the long arm of chromosome 2 (2q) that was attached to the long arm of chromosome 1 (1q) (karyotype: 46, XY, t[2q:1q]). As far as could be detected at the time of examination in 1987, there was no loss of genetic material nor was there additional material. The phenotype of the child described by Weistuch and Schiff-Myers (1996) included verbal and oral-motor dyspraxia in the context of preserved nonverbal cognition, play, and receptive language skills. No dysmorphic features were noted, nor any other significant medical features. Hence, on the face of the evidence provided, this case of chromosomal rearrangement was associated with a relatively specific form of speech praxis.

Over a decade later, Shriberg, Jakielski, and El-Shanti (2008) reported in three siblings associated with CAS, an unbalanced 4q;16q translocation; cognitive and language deficits; and dysmorphology characterized by frontal bossing, epicanthal folds, posterior ear rotation, hypoplastic alae nasi, and crowded, unusual, or decreased number of teeth as well as hand and foot malformations (Shriberg et al., 2008; see figure 40.5). Mapping of the chromosomal breakpoints revealed that, in addition to potential contributions from duplicated genes on chromosome 16, haploinsufficiency of 1 or more of 11 genes deleted in the telomeric region of the long arm of chromosome 4 was the likely cause of the phenotype including CAS.

Shriberg et al. (2008) were early proponents of using genotype-phenotype studies of CAS associated with complex neurodevelopmental disorders to elucidate the pathophysiology of speech-motor impairment. There have been no further translocation reports in the literature with a core or obvious phenotype of CAS or dysarthria, nor any further delineation of which of the 11 genes noted here on the long arm of chromosome 4 may form a pathway to CAS. A number of new cases with chromosomal translocations associated with speech-motor disorders are likely to be reported in the future, given the current landscape of advancing genetic technologies, appreciation of accurate speech phenotyping to lead to genetic diagnoses, and intensified research efforts focused on explaining the genetic contribution for speech-motor disorders.

FIGURE 40.5 Siblings with an unbalanced 4q;16q translocation and CAS, arranged from A oldest to C youngest. Figure shows common dysmorphic features, including frontal bossing, epicanthal folds, posterior ear rotation, hypoblastic alae nasi, and crowded, unusual, or decreased number of teeth. Adapted from Shriberg et al. (2008).

3. Conclusion

While evidence is increasing for potential candidate genes associated with speech-motor disorders, the picture is also becoming more complex. A surprising level of comorbidity is seen across different communication disorders, indicating that distinct conditions may sometimes represent alternative manifestations of the same genetic risk variants (Graham et al., 2015). Hence, just as one phenotype may lead to many genotypes, we now recognize that one genotype may also lead to many phenotypes (Graham et al., 2015), underscoring the complexity of these gene-behavior relationships. Mutations of *FOXP2*-only often lead to a profile of CAS and language disorder in the absence of ID, but this has been the only candidate gene to date to show a selectivity of this kind. That is, other candidate genes or genotypes so far associated with speech-motor disorders generally lead to broader ID syndromes (e.g., *BCL11A, FOXP1, CNTNAP2*) and/or significant medical comorbidities such as epilepsy (e.g., *GRIN2A*), where CAS and/or dysarthria may occur as part of the broader spectrum of the condition. Next-generation sequencing methods and improved bioinformatic analysis techniques will continue to generate a proliferation of rare variants and new candidate genes for severe and intractable childhood speech-motor disorders. It is clear that deep phenotyping, carefully differentiating among oral-motor, speech, language, generalized motor, and nonverbal IQ at a minimum, will be necessary to lead to a more informed understanding of genotype-phenotype associations and to delineate the truly significant gene pathways to speech-motor disorders in this field. The potential clinical implications for this body of work are clearly significant in terms of personalized medicine, with the prospect of optimized diagnosis, genetic and prognostic counseling, and targeted therapeutics.

NOTES

1. Following standard nomenclature, genes are denoted in italics, proteins in regular font. Uppercase letters denote the human version of the gene (i.e., *FOXP2*), lowercase the mouse version of the gene (i.e., *Foxp2*).
2. Term "idiopathic" denoted in quotes as CAS is likely caused by a yet to be determined monogenic, polygenic, or complex multifactorial etiology rather than being a truly unknown cause.
3. The phrase *specific language impairment* has been in use for many decades (Reilly et al., 2014). A recent Delphi survey of international experts in this field (Bishop, Snowling, Thompson, Greenhalgh, & CATALISE Consortium, 2016, 2017), however, now suggests use of the phrase *developmental language disorder* in place of SLI. The term SLI is retained in this chapter as studies cited here were conducted when SLI was the accepted term.
4. A single nucleotide polymorphism, or SNP, is a single base-pair difference in the DNA sequence in individuals. SNPs do not necessarily result in a pathological mutation, but are commonly examined as a covarying marker within a complex disease phenotype.

REFERENCES

Abdelmoity, A. T., Hall, J. J., Bittel, D. C., & Yu, S. (2011). 1.39 Mb inherited interstitial deletion in 12p13.33 associated with developmental delay. *European Journal of Medical Genetics, 54,* 198–203.

Akbarian, S., Sucher, N. J., Bradley, D., Tafazzoli, A., Trinh, D., Hetrick, W. P., ... Jones, E. G. (1996). Selective alterations in gene expression for NMDA receptor subunits in prefrontal cortex of schizophrenics. *Journal of Neuroscience, 16,* 19–30.

Alarcón, M., Abrahams, B. S., Stone, J. L., Duvall, J. A., Perederiy, J. V., Bomar, J. M., ... Geschwind, D. H. (2008). Linkage, association, and gene-expression analyses identify *CNTNAP2* as an autism-susceptibility gene. *American Journal of Human Genetics, 82,* 150–159.

American Speech-Language-Hearing Association (ASHA). (2007). *Ad Hoc Committee on Childhood Apraxia of Speech.* Rockville, MD. Retrieved from http://www.asha.org/policy/PS2007-00277/.

Anderson, V., Spencer-Smith, M., & Wood, A. (2011). Do children really recover better? Neurobehavioural plasticity after early brain insult. *Brain, 134,* 2197–2221.

Baker, E., Hinton, L., Callen, D. F., Haan, E. A., Dobbie, A., & Sutherland, G. R. (2002). A familial cryptic subtelomeric deletion 12p with variable phenotypic effect. *Clinical Genetics, 61,* 198–201.

Battini, R., Chilosi, A., Mei, D., Casarano, M., Alessandrì, G. M., Leuzzi, V., ... Cioni, G. (2007). Mental retardation, verbal dyspraxia in a new patient with de novo creatine transporter (SLC6A8) mutation. *American Journal of Medical Genetics A*, 143A, 1771–1774.

Belton, E., Salmond, C. H., Watkins, K. E., Vargha-Khadem, F., & Gadian, D. G. (2003). Bilateral brain abnormalities associated with dominantly inherited verbal and orofacial dyspraxia. *Human Brain Mapping, 18*, 194–200.

Bishop, D. V. M., North, T., & Donlan C. (1995). Genetic basis of specific language impairment: Evidence from a twin study. *Developmental Medicine and Child Neurology, 37*, 56–71.

Bishop, D. V. M., Snowling, M. J., Thompson, P. A., Greenhalgh, T., & CATALISE Consortium. (2016). CATALISE: A multinational and multidisciplinary Delphi consensus study: Identifying language impairments in children. *PLoS One*, 11, e0158753.

Bishop, D. V. M., Snowling, M. J., Thompson, P. A., Greenhalgh, T., & CATALISE Consortium. (2017). CATALISE: a multinational and multidisciplinary Delphi consensus study: Phase 2: Terminology. *Journal of Child Psychology and Psychiatry, 58*(10), 1068–1080.

Carr, C. W., Moreno-De-Luca, D., Parker, C., Zimmerman, H. H., Ledbetter, N., Martin, C. L., ... Abdul-Rahman, O.A. (2010). Chiari I malformation, delayed gross motor skills, severe speech delay, and epileptiform discharges in a child with *FOXP1* haploinsufficiency. *European Journal of Human Genetics, 18*, 1216–1220.

Carvill, G. L., Regan, B. M., Yendle, S.C., O'Roak, B. J., Lozovaya, N., Bruneau, N., ... Mefford, H. C. (2013). GRIN2A mutations cause epilepsy-aphasia spectrum disorders. *Nature Genetics, 45*, 1073–1076.

Centanni, T. M., Sanmann, J. N., Green, J. R., Iuzzini-Seigel, J., Bartlett, C., Sanger, W. G., & Hogan, T. P. (2015). The role of candidate-gene *CNTNAP2* in childhood apraxia of speech and specific language impairment. *American Journal of Medical Genetics B Neuropsychiatric Genetics, 168*, 536–543.

Chabout, J., Sarkar, A., Patel, S. R., Radden, T., Dunson, D. B., Fisher, S. E., & Jarvis, E. D. (2016). A Foxp2 mutation implicated in human speech deficits alters sequencing of ultrasonic vocalizations in adult male mice. *Frontiers in Behavioural Neuroscience, 10*, 197.

Clinton, S. M., & Meador-Woodruff, J. H. (2004). Abnormalities of the NMDA receptor and associated intracellular molecules in the thalamus in schizophrenia and bipolar disorder. *Neuropsychopharmacology, 29*, 1353–1362.

Conti, F., Barbaresi, P., Melone, M., & Ducati, A. (1999). Neuronal and glial localization of NR1 and NR2A/B subunits of the NMDA receptor in the human cerebral cortex. *Cerebral Cortex*, 9, 110–120.

Darley, F. L., Aronson, A. E., & Brown, J. R. (1969). Differential diagnostic patterns of dysarthria. *Journal of Speech and Hearing Research, 12*, 246–269.

DeThorne, L. S., Hart, S. A., Petrill, S. A., Deater-Deckard, K., Thompson, L. A., Schatschneider, C., & Davison, M. D. (2006). Children's history of speech-language difficulties: Genetic influences and associations with reading-related measures. *Journal of Speech Language and Hearing Research, 49*, 1280–1293.

Dias, C., Estruch, S. B., Graham, S. A., McRae, J., Sawiak, S. J., Hurst, J. A., ... Logan, D. W. (2016). *BCL11A* haploinsufficiency causes an intellectual disability syndrome and dysregulates transcription. *American Journal of Human Genetics, 99*, 253–274.

Dodd, B. (2017). Clinical management of phonological impairment in children. In B. Dodd & A. Morgan (Eds.), *Intervention case studies of child speech impairment.* London: J&R Press.

Dodd, B., Holm, A., Hua, Z., & Crosbie, S. (2003). Phonological development: A normative study of British English-speaking children. *Clinical Linguistics and Phonetics, 17*, 617–643.

Domingues, C. E., Olivera, C. M., Oliveira, B. V., Juste, F. S., Andrade, C. R., Giacheti, C.M., ... Drayna, D. (2014). A genetic linkage study in Brazil identifies a new locus for persistent developmental stuttering on chromosome 10. *Genetics and Molecular Research, 13*, 2094–2101.

Duffy, J. R. (2013). *Motor speech disorders.* St. Louis, MO: Mosby.

Eadie, P., Morgan, A., Ukoumunne, O. C., Ttofari-Eecen, K., Wake, M., & Reilly, S. (2015). Speech sound disorder at 4 years: Prevalence, comorbidities, and predictors in a community cohort of children. *Developmental Medicine and Child Neurology, 57*, 578–584.

Elmakky, A., Carli, D., Lugli, L., Torelli, P., Guidi, B., Falcinelli, C., ... Percesepe, A. (2014). A three-generation family with terminal microdeletion involving 5p15.33-32 due to a whole-arm 5;15 chromosomal translocation with a steady phenotype of atypical cri du chat syndrome. *European Journal of Medical Genetics, 57*, 145–150.

Fedorenko, E., Morgan, A., Murray, E., Cardinaux, A., Mei, C., Tager-Flusberg, H., ... Kanwisher, N. (2016). A highly penetrant form of childhood apraxia of speech due to deletion of 16p11. 2. *European Journal of Human Genetics, 24*, 302–306.

Fisher, S. E., Vargha-Khadem, F., Watkins, K. E., Monaco, A. P., & Pembrey, M. E. (1998). Localisation of a gene implicated in a severe speech and language disorder. *Nature Genetics, 18*, 168–170.

Froster, U., Schulte-Körne, G., Hebebrand, J., & Remschmidt, H. (1993). Cosegregation of balanced translocation (1;2) with retarded speech development and dyslexia. *Lancet, 342*(8864), 178–179.

Gillon, G. T., & Moriarty, B. C. (2007). Childhood apraxia of speech: Children at risk for persistent reading and spelling disorder. *Seminars in Speech and Language, 28*, 48–57.

Graham, S. A., Deriziotis, P., & Fisher, S. E. (2015). Insights into the genetic foundations of human communication. *Neuropsychological Review, 25*, 3–26.

Groszer, M., Keays, D. A., Deacon, R. M., de Bono, J. P., Prasad-Mulcare, S., Gaub, S., ... Fisher S. E. (2008). Impaired synaptic plasticity and motor learning in mice with a point mutation implicated in human speech deficits. *Current Biology, 18*, 354–362.

Haesler, S., Rochefort, C., Georgi, B., Licznerski, P., Osten, P., & Scharff, C. (2007). Incomplete and inaccurate vocal imitation after knockdown of FoxP2 in songbird basal ganglia nucleus Area X. *PLOS Biology*, 5, e321.

Hamdan, F. F., Daoud, H., Rochefort, D., Piton, A., Gauthier, J., Langlois, M., ... Michaud, J. L. (2010). De novo mutations in *FOXP1* in cases with intellectual disability, autism, and language impairment. *American Journal of Human Genetics, 87*, 671–678.

Held, R. G., Liu, C., & Kaeser, P. S. (2016). *ELKS* controls the pool of readily releasable vesicles at excitatory synapses through its N-terminal coiled-coil domains. *eLife*, 5, e14862.

Horn, D., Kapeller, J., Rivera-Brugués, N., Moog, U., Lorenz-Depiereux. B., Eck. S., ... Strom, T. M. (2010). Identification of *FOXP1* deletions in three unrelated patients with mental retardation and significant speech and language deficits. *Human Mutation, 31,* E1851–E1860.

Hughes, J., Ryan, S., Lambert, D., Geoghegan, O., Clark, A., Rogers, Y., ... Treacy, E. P. (2009). Outcomes of siblings with classical galactosemia. *Journal of Pediatrics, 154,* 721–726.

Hurst, J.A., Baraitser, M., Auger, E., Graham, F., & Norell, S. (1990). An extended family with a dominantly inherited speech disorder. *Dev Med Child Neurol, 32,* 352–355.

Hynd, M. R., Scott, H. L., & Dodd, P. R. (2004). Differential expression of N-methyl-D-aspartate receptor NR2 isoforms in Alzheimer's disease. *Journal of Neurochemistry, 90,* 913–919.

Jacquemont, S., Reymond, A., Zufferey, F., Harewood, L., Walters, R. G., Kutalik, Z., ... Froguel, P. (2011). Mirror extreme BMI phenotypes associated with gene dosage at the chromosome 16p11.2 locus. *Nature, 478,* 97–102.

Juranek, J., Mukherjee, K., Rickmann, M., Martens, H., Calka, J., Südhof, T. C., & Jahn, R. (2006). Differential expression of active zone proteins in neuromuscular junctions suggests functional diversification. *European Journal of Neuroscience, 24,* 3043–3052.

Kang, C., & Drayna, D. (2012). A role for inherited metabolic deficits in persistent developmental stuttering. *Molecular Genetics and Metabolism, 107,* 276–280.

Kornau, H. C., Schenker, L. T., Kennedy, M. B., & Seeburg, P. H. (1995). Domain interaction between NMDA receptor subunits and the postsynaptic density protein PSD-95. *Science, 269,* 1737–1740.

Kosinski, C. M., Standaert, D. G., Counihan, T. J., Scherzer C. R., Kerner J. A., Daggett L. P., ... Landwehrmeyer, G. B. (1998). Expression of N-methyl-D-aspartate receptor subunit mRNAs in the human brain: Striatum and globus pallidus. *Journal of Computational Neurology, 390,* 63–74.

Lai, C. S., Fisher, S. E., Hurst, J. A., Vargha-Khadem, F., & Monaco, A. P. (2001). A forkhead-domain gene is mutated in a severe speech and language disorder. *Nature, 413,* 519–523.

Landau, W. M., & Kleffner, F. R. (1957). Syndrome of acquired aphasia with convulsive disorder in children. *Neurology, 7,* 523–530.

Laube, B., Hirai, H., Sturgess, M., Betz, H., & Kuhse, J. (1997). Molecular determinants of agonist discrimination by NMDA receptor subunits: Analysis of the glutamate binding site on the NR2B subunit. *Neuron, 18,* 493–503.

Le Fevre, A. K., Taylor, S., Malek, N. H., Horn, D., Carr, C. W., Abdul-Rahman, O. A., ... Hunter, M. F. (2013). *FOXP1* mutations cause intellectual disability and a recognizable phenotype. *American Journal of Medical Genetics Part A, 161A,* 3166–3175.

Lemke, J. R., Lal, D., Reinthaler, E. M., Steiner, I., Nothnagel, M., Alber, M., ... von Spiczak S. (2013). Mutations in GRIN2A cause idiopathic focal epilepsy with rolandic spikes. *Nature Genetics, 45,* 1067–1072.

Lesca, G., Rudolf, G., Bruneau, N., Lozovaya, N., Labalme, A., Boutry-Kryza, N., ... Szepetowski P. (2013). GRIN2A mutations in acquired epileptic aphasia and related childhood focal epilepsies and encephalopathies with speech and language dysfunction. *Nature Genetics, 45,* 1061–1066.

Levelt, W. J. M. (1989). *Speaking: From intention to articulation.* Cambridge, MA: MIT Press.

Lewis, B. A., Freebairn, L. A., Hansen, A. J., Iyengar, S. K., & Taylor, H. G. (2004). School-age follow-up of children with childhood apraxia of speech. *Language, Speech and Hearing Services in Schools, 35,* 122–140.

Lewis, B. A., & Thompson, L. A. (1992). A study of developmental speech and language disorders in twins. *Journal of Speech Hearing Research,* 35, 1086–1094.

Liégeois, F., Baldeweg, T., Connelly, A., Gadian, D. G., Mishkin, M., & Vargha-Khadem, F. (2003). Language fMRI abnormalities associated with FOXP2 gene mutation. *Nature Neuroscience, 6,* 1230–1237.

Liégeois, F. J., Hildebrand, M. S., Bonthrone, A., Turner, S. J., Scheffer, I. E., Bahlo, M., ... & Morgan, A. T. (2016). Early neuroimaging markers of *FOXP2* intragenic deletion. *Scientific Reports, 6,* 35192.

Liégeois, F. J., & Morgan, A. T. (2012). Neural bases of childhood speech disorders: Lateralization and plasticity for speech functions during development. *Neuroscience and Biobehavioural Reviews, 36,* 439–458.

Liégeois, F., Morgan, A. T., Connelly, A., & Vargha-Khadem, F. (2011). Endophenotypes of *FOXP2*: Dysfunction within the human articulatory network. *European Journal of Paediatric Neurology, 15,* 283–288.

Liégeois, F., Morgan, A. T., Stewart, L. H., Cross, H. J., Vogel, A. P., & Vargha-Khadem, F. (2010). Speech and oral motor profile after childhood hemispherectomy. *Brain and Language, 114,* 126–134.

Liu, C., Bickford, L. S., Held, R. G., Nyitrai, H., Südhof, T. C., & Kaeser, P. S. (2014). The active zone protein family *ELKS* supports Ca2+ influx at nerve terminals of inhibitory hippocampal neurons. *Journal of Neuroscience, 34,* 12289–12303.

Locke, J. L., & Mather, P. L. (1989). Genetic factors in the ontogeny of spoken language: Evidence from monozygotic and dizygotic twins. *Journal of Child Language, 16,* 553–559.

Lozano, R., Vino, A., Lozano, C., Fisher, S. E., & Deriziotis, P. (2015). A de novo *FOXP1* variant in a patient with autism, intellectual disability and severe speech and language impairment. *European Journal of Human Genetics, 23,* 1702–1707.

MacDermot, K. D., Bonora, E., Sykes, N., Coupe, A. M., Lai, C. S., Vernes, S. C., ... Fisher, S. E. (2005). Identification of *FOXP2* truncation as a novel cause of developmental speech and language deficits. *American Journal Human Genetics, 76,* 1074–1080.

MacDonald, A. H., Rodríguez, L., Aceña, I., Martínez-Fernández, M. L., Sánchez-Izquierdo, D., Zuazo, E., & Martínez-Frías, M. L. (2010). Subtelomeric deletion of 12p: Description of a third case and review. *American Journal of Medical Genetics Part A, 152,* 1561–1566.

Marion, M. J., Sussman, H. M., & Marquardt, T. P. (1993). The perception and production of rhyme in normal and developmentally apraxic children. *Journal of Communication Disorders, 26,* 129–160.

Matheny, A. P., & Bruggemann, C. E. (1973). Children's speech: Hereditary components and sex differences. *Folia Phoniatrica, 25,* 442–449.

Miller, D. T., Chung, W., Nasir, R., Shen, Y., Steinman, K. J., Wu, B. L., & Hanson, E. (2015). 16p11.2 recurrent microdeletion. In M. P. Adam, H. H. Ardinger, R. A. Pagon, et al.

(Eds.), *GeneReviews®* [Internet]. Seattle: University of Washington, Seattle; 1993–2019. Retrieved from https://www.ncbi.nlm.nih.gov/books/NBK11167/.

Monyer, H., Sprengel, R., Schoepfer, R., Herb, A., Higuchi, M., Lomeli, H., ... Seeburg, P. H. (1992). Heteromeric NMDA receptors: Molecular and functional distinction of subtypes. *Science, 256*, 1217–1221.

Moralli, D., Nudel, R., Chan, M. T., Green, C. M., Volpi, E. V., Benítez-Burraco, A., ... García-Bellido, P. (2015). Language impairment in a case of a complex chromosomal rearrangement with a breakpoint downstream of *FOXP2*. *Molecular Cytogenetics*, 8, 36.

Morgan, A. T., Fisher, S. E., Scheffer, I., & Hildebrand, M. (2017). FOXP2-related speech and language disorders. In M. P. Adam, H. H. Ardinger, R. A. Pagon et al. (Eds.), *GeneReviews®* [Internet]. Seattle: University of Washington, Seattle; 1993–2019. Retrieved from https://www.ncbi.nlm.nih.gov/books/NBK368474/.

Morgan, A. T., & Guenther, T. (2017). Clinical management of articulation impairment in children. In B. Dodd & A. T. Morgan (Eds.), *Intervention case studies of child speech impairment* (pp. 9–29). London: J&R Press.

Morgan, A. T., & Liégeois, F. (2010). Re-thinking diagnostic classification of the dysarthrias: A developmental perspective. *Folia Phoniatrica et Logopaedica, 62*, 120–126.

Morgan, A. T., Liégeois, F., Liederkerke, C., Vogel, A. P., Hayward, R., Harkness, W., ... Vargha-Khadem F. (2011). Role of cerebellum in fine speech control in childhood: Persistent dysarthria after surgical treatment for posterior fossa tumour. *Brain and Language, 117*, 69–76.

Morgan, A. T., Masterton, R., Pigdon, L., Connelly, A., & Liégeois, F. J. (2013). Functional magnetic resonance imaging of chronic dysarthric speech after childhood brain injury: Reliance on a left-hemisphere compensatory network. *Brain, 136*, 646–657.

Morgan, A. T., Mei, C., Da Costa, A., Fifer, J., Lederer, D., Benoit, V., ... White, S. M. (2015). Speech and language in a genotyped cohort of individuals with Kabuki syndrome. *American Journal of Medical Genetics Part A, 167*(7), 1483–1492.

Morgan, A., T. & Murray, E. (2017). Clinical management of motor speech disorders. In B. Dodd & A. T. Morgan (Eds.), *Intervention case studies of child speech impairment* (pp. 75–94). London: J&R Press.

Morley, M. E. (1972). *The development and disorders of speech in childhood* (3rd ed.). Baltimore, MD: Williams and Wilkins.

Murugan, M., Harward, S., Scharff, C., & Mooney, R. (2013). Diminished FoxP2 levels affect dopaminergic modulation of corticostriatal signaling important to song variability. *Neuron, 80*, 1464–1476.

Newbury, D. F., Mari, F., Sadighi Akha, E., MacDermot, K. D., Canitano, R., Monaco, A. P., ... Knight, S. J. (2013). Dual copy number variants involving 16p11 and 6q22 in a case of childhood apraxia of speech and pervasive developmental disorder. *European Journal of Human Genetics, 21*, 361–365.

Newmeyer, A. J., Aylward, C., Akers, R., Ishikawa, K., Grether, S., deGrauw, T., ... White, J. (2009). Results of the sensory profile in children with suspected childhood apraxia of speech. *Physical and Occupational Therapy Pediatrics, 29*, 203–218.

Ozantürk, A., Davis, E. E., Sabo, A., Weiss, M. M., Muzny, D., Dugan-Perez, S., ... Serdaroglu, E. (2016). A t (5; 16) translocation is the likely driver of a syndrome with ambiguous genitalia, facial dysmorphism, intellectual disability, and speech delay. *Molecular Case Studies, 2*(2), a000703.

Palumbo, O., D'Agruma, L., Minenna, A. F., Palumbo, P., Stallone, R., Palladino, T., ... Carella, M. (2013). 3p14.1 de novo microdeletion involving the *FOXP1* gene in an adult patient with autism, severe speech delay and deficit of motor coordination. *Gene, 516*, 107–113.

Peter, B., Matsushita, M., Oda, K., & Raskind, W. (2014). De novo microdeletion of *BCL11A* is associated with severe speech sound disorder. *American Journal of Medical Genetics Part A, 164A*, 2091–2096.

Potter, N. L., Nievergelt, Y., & Shriberg, L. D. (2013). Motor and speech disorders in classic galactosemia. *Journal of Inherited Metabolic Disorders Reports, 11*, 31–41.

Preston, J. L., Hull, M., & Edwards, M. L. (2013). Preschool speech error patterns predict articulation and phonological awareness outcomes in children with histories of speech sound disorders. *American Journal of Speech Language Pathology, 22*, 173–184.

Rabin, M., Wen, X. L., Hepburn, M., Lubs, H. A., Feldman, E., & Duara, R. (1993). Suggestive linkage of developmental dyslexia to chromosome 1p34-p36. *Lancet, 342*(8864), 178.

Raca, G., Baas, B. S., Kirmani, S., Laffin, J. J., Jackson, C. A., Strand, E. A., ... Shriberg, L. D. (2013). Childhood apraxia of speech (CAS) in two patients with 16p11.2 microdeletion syndrome. *European Journal of Human Genetics, 21*, 455–459.

Raza, M. H., Mattera, R., Morell, R., Sainz, E., Rahn, R., Gutierrez, J., ... Drayna, D. (2015). Association between rare variants in *AP4E1*, a component of intracellular trafficking, and persistent stuttering. *American Journal of Human Genetics, 97*, 715–725.

Reilly, S., Tomblin, B., Law, J., McKean, C., Mensah, F. K., Morgan, A., ... Wake, M. (2014). Specific language impairment: A convenient label for whom? *International Journal of Language and Communication Disorders, 49*, 416–451.

Reuter, M. S., Riess, A., Moog, U., Briggs, T. A., Chandler, K. E., Rauch, A., ... Zweier, C. (2017). FOXP2 variants in 14 individuals with developmental speech and language disorders broaden the mutational and clinical spectrum. *Journal of Medical Genetics, 54*, 64–72.

Robertson, A., Singh, R. H., Guerrero, N. V., Hundley, M., & Elsas, L. J. (2000). Outcomes analysis of verbal dyspraxia in classic galactosemia. *Genetics in Medicine, 2*, 142–148.

Rodenas-Cuadrado, P., Ho, J., & Vernes, S. C. (2014). Shining a light on *CNTNAP2*: Complex functions to complex disorders. *European Journal of Human Genetics, 22*, 171–178.

Rodenas-Cuadrado, P., Pietrafusa, N., Francavilla, T., La Neve, A., Striano, P., & Vernes, S. C. (2016). Characterisation of *CASPR2* deficiency disorder: A syndrome involving autism, epilepsy and language impairment. *BMC Medical Genetics, 17*, 8.

Roll, P., Rudolf, G., Pereira, S., Royer, B., Scheffer, I. E., Massacrier, A., & Szepetowski, P. (2006). *SRPX2* mutations in disorders of language cortex and cognition. *Human Molecular Genetics, 15*, 1195–1207.

Rooryck, C., Stef, M., Burgelin, I., Simon, D., Souakri, N., Thambo, J. B., ... Arveiler, B. (2009). 2.3 Mb terminal deletion in 12p13.33 associated with oculoauriculovertebral spectrum and evaluation of WNT5B as a candidate gene. *European Journal of Medical Genetics, 52*, 446–449.

Royal College of Speech Language Therapists (RCSLT). (2011). Policy statement—Developmental verbal dyspraxia. Retrieved from www.rcslt.org.

Scheffer, I. E., Jones, L., Pozzebon, M., Howell, R. A., Saling, M. M., & Berkovic, S. F. (1995). Autosomal dominant rolandic epilepsy and speech dyspraxia: A new syndrome with anticipation. *Annals of Neurology, 38*, 633.

Scherzer, C. R., Landwehrmeyer, G. B., Kerner, J. A., Counihan T. J., Kosinski C. M., Standaert D. G.... Young A. B. (1998). Expression of N-methyl-D-aspartate receptor subunit mRNAs in the human brain: Hippocampus and cortex. *Journal of Computational Neurology, 390*, 75–90.

Shriberg, L. D. (1994). Five subtypes of developmental phonological disorders. *Clinics in Communication Disorders, 4*, 38–53.

Shriberg, L. D., Jakielski, K. J., & El-Shanti, H. (2008). Breakpoint localization using array-CGH in three siblings with an unbalanced 4q;16q translocation and childhood apraxia of speech (CAS). *American Journal of Medical Genetics Part A, 146A*, 2227–2233.

Shriberg, L. D., Lohmeier, H. L., Strand, E. A., & Jakielski, K. J. (2012). Encoding, memory, and transcoding deficits in childhood apraxia of speech. *Clinical Linguistics and Phonetics, 26*, 445–482.

Shriberg, L. D., Tomblin, J. B., & McSweeny, J. L. (1999). Prevalence of speech delay in 6-year-old children and comorbidity with language impairment. *Journal of Speech Language Hearing Research, 42*, 1461–1481.

Simpson, N. H., Ceroni, F., Reader, R. H., Covill, L. E., Knight, J. C., Nudel, R., ... Slonims, V. (2015). Genome-wide analysis identifies a role for common copy number variants in specific language impairment. *European Journal of Human Genetics, 23*(10), 1370.

Smogavec, M., Cleall, A., Hoyer, J., Lederer, D., Nassogne, M. C., Palmer, E. E., ... Zweier, C. (2016). Eight further individuals with intellectual disability and epilepsy carrying bi-allelic *CNTNAP2* aberrations allow delineation of the mutational and phenotypic spectrum, *Journal of Medical Genetics, 53*, 820–827.

Sollis, E., Graham, S. A., Vino, A., Froehlich, H., Vreeburg, M., Dimitropoulou, D., ... Fisher, S. E. (2016). Identification and functional characterization of de novo FOXP1 variants provides novel insights into the etiology of neurodevelopmental disorder. *Human Molecular Genetics, 25*, 546–557.

Sprengel, R., Suchanek, B., Amico, C., Brusa, R., Burnashev, N., Rozov, A., ... Kim, J. J. (1998). Importance of the intracellular domain of NR2 subunits for NMDA receptor function in vivo. *Cell, 92*(2), 279–289.

Stefansson, H., Meyer-Lindenberg, A., Steinberg, S., Magnusdottir, B., Morgen, K., Arnarsdottir, S., ... Stefansson, K. (2014). CNVs conferring risk of autism or schizophrenia affect cognition in controls. *Nature, 505*, 361–366.

Steinman, K. J., Spence, S. J., Ramocki, M. B., Proud, M. B., Kessler, S. K., Marco, E. J., ... Simons VIP Consortium. (2016). 16p11.2 deletion and duplication: Characterizing neurologic phenotypes in a large clinically ascertained cohort. *American Journal of Medical Genetics Part A, 170*, 2943–2955.

Stromswold, K. (2006). Why aren't identical twins linguistically identical? Genetic, prenatal and postnatal factors. *Cognition, 101*, 333–384.

Teramitsu, I., Kudo, L. C., London, S. E., Geschwind, D. H., & White, S. A. (2004). Parallel FoxP1 and FoxP2 expression in songbird and human brain predicts functional interaction. *Journal of Neuroscience, 24*, 3152–3163.

Thevenon, J., Callier, P., Andrieux, J., Delobel, B., David, A., Sukno, S., ... Faivre, L. (2013). 12p13.33 microdeletion including ELKS/ERC1, a new locus associated with childhood apraxia of speech. *European Journal of Human Genetics, 21*, 82–88.

Tsai, M. H., Pardoe, H. R., Perchyonok, Y., Fitt, G. J., Scheffer, I. E., Jackson, G. D., & Berkovic, S. F. (2013). Etiology of hippocampal sclerosis: Evidence for a predisposing familial morphologic anomaly. *Neurology, 81*, 144–149.

Turner, S. J., Brown, A., Arpone, M., Anderson, V., Morgan, A. T., & Scheffer, I. E. (2017). Dysarthria and broader motor speech deficits in Dravet syndrome. *Neurology, 88*, 743–749.

Turner, S. J., Hildebrand, M. S., Block, S., Damiano, J., Fahey, M., Reilly, S., ... Morgan, A. T. (2013). Small intragenic deletion in FOXP2 associated with childhood apraxia of speech and dysarthria. *American Journal of Medical Genetics Part A, 161A*(9), 2321–2326.

Turner, S. J., Mayes, A. K., Verhoeven, A., Mandelstam, S. A., Morgan, A. T., & Scheffer, I. E. (2015). *GRIN2A*: An aptly named gene for speech dysfunction. *Neurology, 84*, 586–593.

Turner, S. J., Morgan, A. T., Perez, E. R., & Scheffer, I. E. (2015). New genes for focal epilepsies with speech and language disorders. *Current Neurology and Neuroscience Reports, 15*, 35.

Vargha-Khadem, F., Gadian, D. G., Copp, A., & Mishkin, M. (2005). *FOXP2* and the neuroanatomy of speech and language. *Nature Reviews Neuroscience, 6*, 131–138.

Vargha-Khadem, F., Watkins, K. E., Price, C. J., Ashburner, J., Alcock, K. J., Connelly, A., ... Passingham, R. E. (1998). Neural basis of an inherited speech and language disorder. *Proceedings National Academy of Science, 95*, 12695–12700.

Vernes, S. C., MacDermot, K. D., Monaco, A. P., & Fisher, S. E. (2009). Assessing the impact of FOXP1 mutations on developmental verbal dyspraxia. *European Journal of Human Genetics, 17*, 1354–1358.

Vernes, S. C., Newbury, D. F., Abrahams, B. S., Winchester, L., Nicod, J., Groszer, M., ... Fisher S. E. (2008). A functional genetic link between distinct developmental language disorders. *New England Journal of Medicine, 359*, 2337–2345.

Vernes, S. C., Oliver, P. L., Spiteri, E., Lockstone, H. E., Puliyadi, R., Taylor, J. M., ... Fisher, S. E. (2011). Foxp2 regulates gene networks implicated in neurite outgrowth in the developing brain. *PLOS Genetics, 7*, e1002145.

Villanueva, P., Nudel, R., Hoischen, A., Fernández, M. A., Simpson, N. H., Gilissen, C., ... Newbury D. F. (2015). Exome sequencing in an admixed isolated population indicates NFXL1 variants confer a risk for specific language impairment. *PLOS Genetics, 11*, e1004925.

Wang, B., Lin, D., Li, C., & Tucker, P. (2003). Multiple domains define the expression and regulatory properties of Foxp1 forkhead transcriptional repressors. *Journal of Biological Chemistry*, 278, 24259–24268.

Watkins, K. E., Vargha-Khadem, F., Ashburner, J., Passingham, R. E., Connelly, A., Friston, K. J., ... Gadian, D. G. (2002). MRI analysis of an inherited speech and language disorder: Structural brain abnormalities. *Brain, 125*, 465–478.

Weistuch, L., & Schiff-Myers, N. B. (1996). Chromosomal translocation in a child with SLI and apraxia. *Journal of Speech Hearing Research, 39,* 668–671.

White, S. M., Morgan, A., Da Costa, A., Lacombe, D., Knight, S. J., Houlston, R., ... Hurst, J. A. (2010). The phenotype of Floating-Harbor syndrome in 10 patients. *American Journal of Medical Genetics Part A, 152A,* 821–829.

Whitehouse, A. J., Bishop, D. V., Ang, Q. W., Pennell, C. E., & Fisher, S. E. (2011). *CNTNAP2* variants affect early language development in the general population. *Genes Brain Behaviour, 10,* 451–456.

Yairi, E., & Ambrose, N. (2013). Epidemiology of stuttering: 21st century advances. *Journal of Fluency Disorders, 38,* 66–87.

Yoss, K. A. (1975). *Developmental apraxia of speech in children: Familial patterns and behavioral characteristics.* Paper presented at the ASHA North Central Regional Conference, Minneapolis, MN.

41 Neuromolecular Approaches to the Study of Language

SONJA C. VERNES

That language has a strong genetic basis has been clearly established in the previous chapters of this book and elsewhere. The difficult task of finding genes that underlie language skills and disorders has also been comprehensively covered in the other chapters of this part. Thus the goal of the present chapter is to discuss the ways in which we can move beyond gene identification into a deeper understanding of how these genes function, how these functions relate to normal development, and why variations or mutations in these genes lead to altered language abilities or, at the extreme, language disorders.

Communication via language is a complex multimodal task that involves a so-called *system of systems* (Levinson & Holler, 2014). To perform this task we must attend to the signal by which language is conveyed in our environment, that is, text, speech, and/or gestures, via auditory or visual perception. The information we receive must be processed via neural circuitry, and then we must plan and enact a motor program intended to produce sounds, gestures, signs, and/or writing to convey meaningful information (Fitch, Hauser, & Chomsky, 2005; Hauser, Chomsky, & Fitch, 2002; Levinson & Holler, 2014). This is a gross oversimplification of how we communicate and each of these steps can be broken down into many more intricate and interacting components that contribute to how we communicate via language. What this description is intended to do, however, is give a little insight into the complexity of the problem that we are dealing with when we talk about the genetics of language. When we say "language genetics" are we talking about genes that help us hear, speak, or read; genes that help us process linguistic information; genes that give us the ability to use grammar or learn new languages; or genes that when malfunctioning can cause language disorders? The answer is yes. We are looking for genes that do all this and more. But what we are clearly not looking for is one gene that does all these things. Rather, we are looking for many genes, some of which are likely to have very subtle influences, that together give us our unique human capacity for language.

But once we have found these genes we do not hang up our lab coats and go home. Finding the candidate genes that are involved in language phenotypes is the first step in a long and fascinating road toward understanding how these strings of letters in our DNA are able to do such an amazing task as building a "language-ready brain" (Fisher & Marcus, 2006). Once we identify a candidate language-related gene, we need to answer a range of questions to understand the function of the gene and its relationship to the capacity for language. We can ask questions such as: What sort of RNA or protein does the gene encode? When does it function—that is, is it acting developmentally to set up the biological components needed for language or is it performing some function during the very act of using language, or both? Where is it needed—that is, is it present in specific brain regions or neural circuitry subserving language? Via which molecular mechanisms does this gene produce these effects? And why does disruption of this gene cause language disorder? To answer these questions, we need a battery of approaches and model systems that we can use to explore gene function. In the following sections I will present background on the types of genetic variation that can give us clues to the genetics of language, the model systems that are widely used in a neuromolecular approach to language, as well as case studies exploring the functions of genes linked to language-related phenotypes—focusing on neuromolecular mechanisms. Finally, I will discuss approaches for finding new language-related genes, including recent advances taking a genome-wide view of language genetics.

1. *Types of Genomic Variation Contributing to Language Phenotypes*

The presence of genetic differences allows us to determine which genes underlie a trait in the human population by linking the presence of DNA changes to the trait under study. Although we share 99.9% of our DNA, this still leaves a lot of room for variation to occur in the approximately three billion DNA "letters" of the human genome. Thus, each person carries approximately three million variants in their genome. Given the complexity of language, it is perhaps not surprising that many different types of genetic variation have been identified that contribute to language phenotypes. Before going into specific examples of language-related genes, I will outline some of the types of variation present in the genome that helps us find such genes. Broadly, I have classified genetic variation into classes based on size of the variation, its frequency in the population, and position in the genome.

1.1. Size of Genetic Changes The size of a genetic variant reflects how much of the genome it affects. Variation may take the form of large genomic rearrangements (copy number variants, translocations, or inversions) or small insertions or deletions (indels), or variation may change the identity of single letters in the DNA sequence. Importantly, the size of the change alone cannot be used as a proxy for the severity of the resulting phenotype, since a single base substitution in one gene could be more severe than the insertion of hundreds of bases in another part of the genome depending on whether any gene(s) are affected and what the affected genes do.

1.2. Frequency of Variation in the Population In terms of frequency, genetic variation can be broadly classed as either common—positions in the genome that are highly variable in the population, or rare—found in small numbers of people or even single individuals.

Common variation often has more subtle effects on phenotypes, as severely detrimental changes to the genome are usually not maintained in the population. The best-studied common variants are known as SNPs, which are single nucleotides in the genome that vary in large numbers of individuals, and thus can be used to track phenotypic variation in the population. The human genome has millions of these SNPs, and it is possible to look for associations between phenotypes and millions of SNPs simultaneously to determine whether one or more are linked to a trait—known as a genome-wide association study. By assaying common variation, it has been possible to identify genomic regions related to normal variation in language ability in the general population, as well as identify genetic risk factors related to language impairment or language-related disorders (Becker et al., 2016; Gialluisi et al., 2014; Newbury, Fisher, & Monaco, 2010; Newbury & Monaco, 2002, 2010; Reader, Covill, Nudel, & Newbury, 2014).

Rare changes may be inherited or de novo—that is, newly arising in the individual. These can be more severe and are often associated with disorders when they disrupt an essential gene. Rare genetic mutations can have severe consequences for language and many types of rare variation have been identified in individuals with language disorders, some examples of which will be discussed in subsequent sections (Deriziotis & Fisher, 2013; Fisher & Scharff, 2009; Kang & Drayna, 2011; Rodenas-Cuadrado, Ho, & Vernes, 2014; Vernes & Fisher, 2009, 2011).

1.3. Position of Variation in the Genome Genetic variation can be further classified into coding and noncoding variations. *Coding* DNA is the part of the genome containing genes, which encode proteins. Proteins compose much of the functional machinery of cells and therefore disruption of protein coding regions can have severe effects on phenotypes. Mutations in coding DNA might result in a change to the sequence identity of a protein, truncation of a protein (making the sequence shorter than it should be), or a complete loss of the protein.

Although crucial to phenotypes, only about 1% to 2% of the human genome codes for proteins. Much of the rest of the genome is so-called *noncoding DNA*, which has a range of functions, often having a regulatory role that controls when, where, and how much protein is produced from a gene. Although often more subtle than protein-coding variation, changes in regulatory regions of DNA can change the timing, location, or amount of gene expression and in this way have been implicated in disorders (Devanna et al., 2018; Turner et al., 2016; Vaishnavi, Manikandan, Tiwary, & Munirajan, 2013).

2. Models for Functional Testing

When attempting to understand functional links between genes and language phenotypes, there are multiple different ways to address this problem. We can ask about the general properties of the gene in its normal state. This will tell us how, when, and where the gene normally acts. We can also determine what happens when the gene is no longer present and compare this to its normal function. The effects of removing the gene from a system (via a genetic "knockout") or severely reducing its expression (via a genetic "knockdown") can tell us a lot about what it normally does. However, not all human variants or mutations completely destroy the function of a gene, so it is often of interest to also investigate the effects when a patient-identified mutation is introduced into the gene/protein.

In this section, I will briefly outline some key model systems used for neuromolecular investigations of

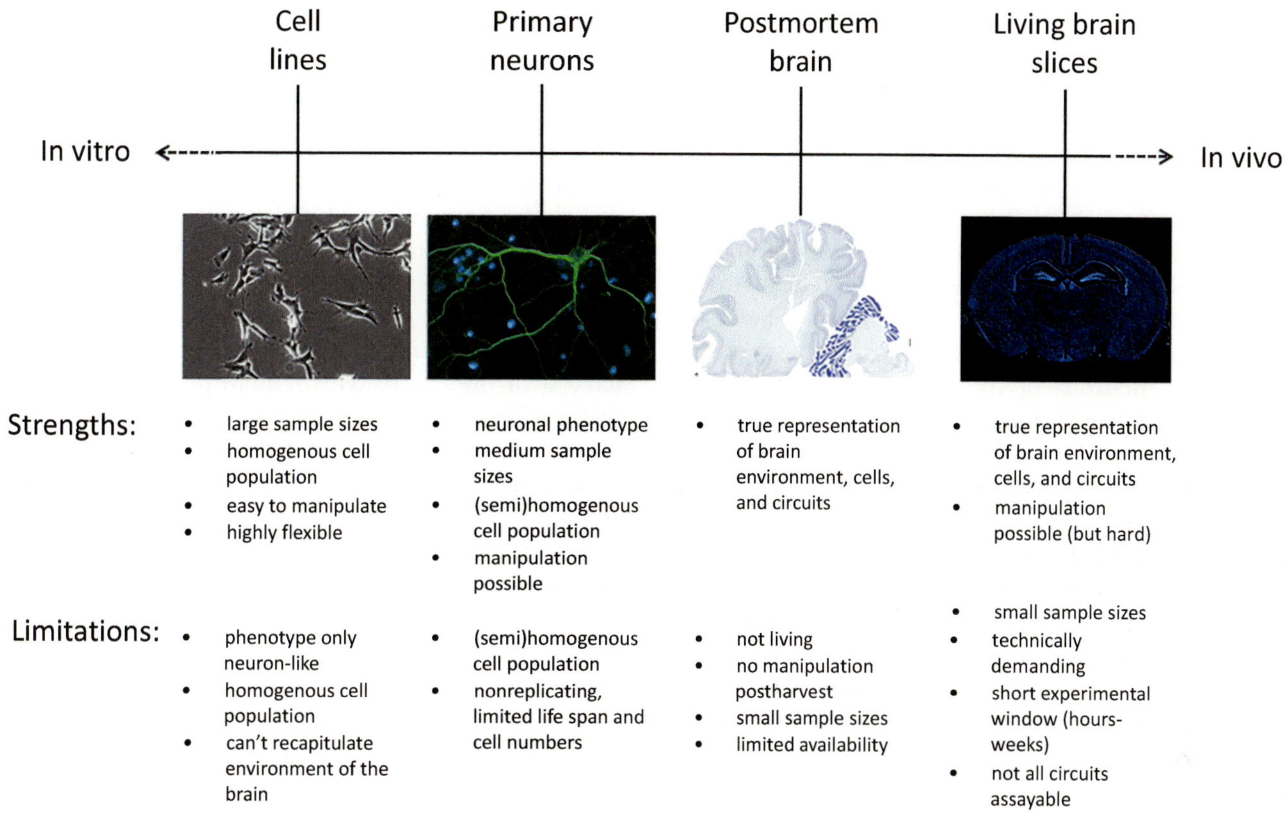

FIGURE 41.1 Summary of a selection of models used in neuromolecular approaches to studying language. Models are placed along a hypothetical scale from in vitro to more in vivo–like systems and the main strengths and limitations of each model are summarized. Postmortem human brain image adapted from Ding et al. (2016).

language-related genes. These range from in vitro systems, to models that more closely resemble the in vivo brain, and each has its strength and limitations (see figure 41.1 for summary). In sections 3–5, I will give specific examples of how these methods have been applied to understand three language-related genes.

2.1. CELL LINES One of the most simple and flexible models routinely used to explore gene function is an immortalized cell line. These cells are called *immortalized* as they can continue to divide in culture almost indefinitely and retain their properties as they divide,[1] meaning that very large quantities of highly homologous cells can be generated—a considerable advantage for experimentation at the molecular level. These cells can be frozen for long-term storage and thawed without harm. They also have the great advantage that it is relatively straightforward to manipulate their gene expression. DNA constructs that express the normal or mutant copy of a gene, or that express RNA molecules to switch off endogenous genes in the cell (known as short hairpin or small interfering RNA molecules) can be easily introduced into cell lines. Cell lines may come from almost any species (there are human, mouse, rat, bat, and fruit fly cell lines, to name just a few) and many different tissue sources (e.g., brain, skin, lung, kidney). Although these cells do not exactly match the phenotype of their source tissue, they show properties akin to cells of the original tissue including similar epigenetic marks, gene expression profiles, and morphology (Agholme, Lindstrom, Kagedal, Marcusson, & Hallbeck, 2010; Odom et al., 2007; Xie, Hu, & Li, 2010). Neurogenetics regularly makes use of human cell lines derived from neural tissue—the so-called *neuron-like cell lines* (Agholme et al., 2010; Cheung et al., 2009). An advantage of these neuron-like cells is that although they start off only moderately resembling neurons, with the addition of growth factors they can be differentiated into a state that is much more characteristic of a mature neuron and may even form active synapses and start signaling as a connected neural network (Agholme et al., 2010; Encinas et al., 2000; Jamsa, Hasslund, Cowburn, Backstrom, & Vasange, 2004; Joshi, Guleria, Pan, DiPette, & Singh, 2006). Because of these properties (easy to manipulate, able to get large quantities of homogenous cells, similarity to source tissue), these model systems facilitate a wide range of molecular experimentation aimed at understanding the core functions of language-related genes.

Exciting advances in cellular techniques are also now making it possible to derive human neurons from non-neuronal sources such as from human embryonic stem cell lines or patient tissue biopsies (e.g., induced pluripotent stem cell lines) (Chailangkarn, Acab, & Muotri, 2012). These emerging methods direct the differentiation of cells into neurons via reprogramming of gene expression and addition of growth factors. The resulting neurons closely resemble specific in vivo phenotypes (e.g., neurons of the cortex or even complex assemblies of neuron types such as organoids) (Lancaster et al., 2013; Wapinski et al., 2013) and allow study of gene-phenotype relationships within a human neuronal system. A great benefit of this approach is the opportunity to study human neurons directly (especially neurons generated from patient samples that could carry causative genetic mutations). However, generating these models requires substantial investment as they are currently technically very demanding and time-consuming, meaning that their use is not yet routine.

2.2. Primary Neurons Another cell model that can be grown in a dish in the lab, but that is closer to the in vivo brain than to immortalized cell lines are primary neurons. Primary neurons are made by dissecting out tissue from regions of interest in the developing brain (often from mouse or rat) and treating the tissue so the cells are gently dissociated into a single cell suspension. The neurons are then maintained alive in a rich growth environment in a petri dish. Primary neurons prepared in this way start off as single cells, but after only one day the cells start extending their neurites (axons and dendrites) and after approximately two weeks they have grown long, highly branched neurites to form connected, active, neuronal networks (Falk, Zhang, Erbe, & Sherman, 2006; Geissler & Faissner, 2012; Kim & Lee, 2012). This means that the effect of gene manipulation on neurite growth, synapse formation, and network activity can be studied in real time in these living neurons in a petri dish. Primary neurons are not immortal and do not divide, thus they have a limited life span once in a petri dish, however they can survive in culture for weeks or months with the proper care. Like cell lines, primary neurons can be genetically manipulated, but they are far less receptive to the introduction of DNA constructs, thus it is often necessary to use techniques such as packaging the DNA into viruses and infecting the cells to deliver DNA to the neurons (Holehonnur, Lella, Ho, Luong, & Ploski, 2015; Manfredsson, 2016).

2.3. Postmortem Brain Tissue Directly studying postmortem brain tissue moves closer toward an environment that reflects the in vivo brain. The greatest strength of this model is that it is possible to take slices through different regions of a human or animal brain and look directly at where and how strongly a gene/protein is expressed and how brain structure might be affected by differences. However, a major limitation with this system is that because the tissue is not living it is not possible to perform any further manipulations. Thus all genetic manipulations must be done prior to harvest of the tissue. With human postmortem tissue this is ethically out of the question; however, it is possible to compare brains of people with naturally occurring variation and look for common features (although still difficult given that tissue is scarce and thus sample size is necessarily small). In animal models, particularly in mouse models, this is more straightforward, as transgenic animal lines are now routinely created in which a gene is *knocked-out*—such that it can no longer be expressed; *knocked-down*—such that its expression is reduced; or *knocked-in*—where a change, for example, a patient-identified mutation, is added to the genome. Comparison of a knockout brain versus a matched control (normal) brain is a powerful way to determine the neuronal functions of a gene.

2.4. Brain Slices Finally, it is possible to harvest brain slices from animal models and keep them alive in a dish for hours (acute slices) or weeks (organotypic slices). This can be done with transgenic animal models to look at effects of genetic manipulations, but it is also possible to infect brain slices from normal animals with viruses carrying DNA constructs to alter gene expression or apply chemicals to manipulate specific molecular pathways. In this way it is possible to directly measure the effects of genetic mutations or chemical treatments on the activity of synapses and neural circuits in real time. A major benefit of using slices is that they closely reflect many aspects of the living brain. Prior to harvest, the neurons and circuits have formed in the complex three-dimensional environment of the brain and thus slices match the in vivo brain more closely than cell models do (e.g., primary neurons). However, there are some limitations that should be considered. Working with brain slices is technically demanding and time-consuming. Often, neurons are recorded one at a time, meaning that only a handful of neurons can be measured per brain slice. Slices can also only be worked with for a limited time window. This means that, for example, genetic or chemical treatments that take days or weeks to have an effect cannot be observed in acute slices and long-term experiments (longer than weeks) are sometimes not feasible. One way around this is to perform manipulations in the animal (by creating transgenic animals or administering chemicals to the living animal) and then harvest the brain to measure long-term effects. Another limitation is that these are, by nature, slices, thus it is not possible to measure all circuitry in the brain. Brain slices tend to be between 100 and 500 µm thick.[2] Only connections that are preserved after the slices are cut can be assayed, meaning that any connectivity in the brain extending more than a few hundred microns perpendicular to the cut will be lost.

To illustrate how the models outlined in this section can be utilized and what they have revealed about the neurogenetics of language, in the subsequent sections I will discuss two well-characterized examples and one new example of a putative language-related gene. For each gene (where the information is available), I will describe what the gene encodes; where it is expressed; what molecular, cellular, and neuronal functions it performs; and how this has informed our understanding of why mutations cause language-related disorders.

3. Function of Language-Related Genes: FOXP2

The first gene implicated in human language was *FOXP2*,[3] and it is for this gene that we have the most comprehensive investigations of how genes contribute to a language-ready brain. *FOXP2* was first identified as a monogenic cause of a rare and severe form of speech and language disorder in a large pedigree, known as the KE family (Lai, Fisher, Hurst, Vargha-Khadem, & Monaco, 2001). All affected members of this family carried a single letter DNA change in the protein coding region of one copy of this gene and that led to a single amino acid change in the protein sequence (Lai et al., 2001). Because humans have two copies of each gene in their genomes (one inherited from each parent), affected individuals in this family had one normal copy of *FOXP2* and one copy that carried the protein altering mutation—which is sufficient to cause the disorder. Since its initial identification, mutations affecting *FOXP2* (ranging from point mutations to large-scale multigene deletions) have been identified in a number of independent cases, each with a single copy of *FOXP2* disrupted and displaying similar speech and language disorders (Morgan, Fisher, Scheffer, & Hildebrand, 2017). These cases provide convincing evidence that mutation of a single copy of *FOXP2* is sufficient to cause speech/language disorder and that both copies of the *FOXP2* gene are required for normal language to develop in humans. Although it should be noted that *FOXP2* mutations are rare and can only account for the observed speech/language disorder in a small minority of cases.

3.1. When and Where Is *FOXP2* Expressed? Understanding the location and timing of *FOXP2* expression can give clues to its role in the brain.[4] Expression of *FOXP2* has been observed throughout the developing human brain including in the developing cortex, subcortical regions (e.g., thalamus, hypothalamus, caudate nucleus), midbrain, and cerebellum (Lai, Gerrelli, Monaco, Fisher, & Copp, 2003). Little is known about *FOXP2* expression in the postnatal human brain given the difficulty in obtaining samples; however, animal models have given us insight into what adult expression pattern might look like. Human *FOXP2* and mouse *Foxp2* are highly conserved at a sequence level and the expression pattern of *Foxp2/FOXP2* was highly similar in mouse and human embryonic brains (Ferland, Cherry, Preware, Morrisey, & Walsh, 2003; Lai et al., 2003). Thus it was predicted that in the adult brain, expression patterns would also be highly conserved. In adult mice, *Foxp2* expression was maintained in the same areas as in the developing mouse and human brains. In many regions, however, expression became more restricted to subsets of neurons in these regions in adult brains. As a result, adult mice displayed *Foxp2* expression in deep layer cortical neurons, medium spiny neurons of the striatum, a subset of nuclei of the thalamus, inferior olives in the brain stem, and Purkinje cells of the cerebellum (Ferland et al., 2003; Lai et al., 2003). This evidence from human and mouse studies suggests that *FOXP2/Foxp2* is expressed across multiple structures of the developing brain, and in the adult brain is found in areas involved in higher order cognition, sensory integration, and motor control/learning. Thus a 50% reduction in functional *FOXP2*, as found in the affected KE family members, might have subtle but specific effects on development and function of these circuits.

3.2. What Does *FOXP2* Encode? Given its links to language, it was of intense interest to discover the function of the protein encoded by the *FOXP2* gene. Analysis of the protein sequence encoded by *FOXP2* indicated that the protein contained a domain known as a forkhead-box (*FOX* for short), identifying it as part of the FOX family of genes (Katoh & Katoh, 2004; Lai et al., 2001).[5] FOX genes are known to act as transcription factors, suggesting that *FOXP2* was likely to have the same function, that is, to regulate the expression of other genes in the genome.

3.3. In Which Molecular Mechanisms Is *FOXP2* Involved? Around the same time the mutations in human *FOXP2* in the KE family were identified, researchers independently showed that the mouse Foxp2 protein was indeed acting as a transcription factor to regulate gene expression (Shu, Yang, Zhang, Lu, & Morrisey, 2001). The ability of human FOXP2 to directly regulate gene expression was also shown in human cell lines (Vernes et al., 2006), opening up the possibility to explore the genes it regulates and thus understand the downstream molecular mechanisms directed by FOXP2. Studies using human neuron-like cells, human fetal brain and developing mouse brain identified the genes that were regulated by FOXP2; its so-called *target genes* (Spiteri et al., 2007; Vernes et al., 2007; Vernes et al., 2011). These studies identified overlapping sets of genes that were involved in highly conserved molecular pathways. Across both human and mouse model systems there was a high proportion of target genes that played a role in neuronal development including such processes as neuronal migration, neurite outgrowth, and synapse function (Spiteri et al., 2007; Vernes et al., 2007; Vernes & Fisher, 2011). This gave first insight into the molecular pathways that were acting downstream of FOXP2 and furthermore suggested cellular processes and phenotypes that may be directly affected by this gene.

3.4. In Which Cellular/Neuronal Mechanisms Is *FOXP2* Involved? The molecular data described have suggested a role for FOXP2 in cellular processes including migration, neurite outgrowth, and synaptic activity—hypotheses that could be directly tested using model systems.

Precise control of neuronal migration is important to ensure that neurons reach the appropriate place in the brain at the right time during development. Neurons migrating too far, not far enough, or arriving at the right place but at the wrong time can all have important consequences for the formation of brain circuits (Marin, Valiente, Ge, & Tsai, 2010). Effects of Foxp2 on neuronal migration were first observed in the developing mouse brain, where overexpression of *Foxp2* led to reduced migration of neurons during cortical development (Clovis, Enard, Marinaro, Huttner, & De Pietri Tonelli, 2012). This study investigated the regulation of Foxp2 via microRNA molecules and its consequences for neuronal migration. However, it was also possible from this data to observe the effect of excess Foxp2 on cortical migration during development. In this study, *Foxp2* or a control DNA construct, along with a visible label, were injected into cells of the early embryonic cortex. Embryos were then left to grow and the effects were observed in the brains of postnatal animals. Mice that received excess *Foxp2*, compared to control animals, displayed more labeled cells in deep layers of the cortex, suggesting that the normal migration of neurons through the cortical layers had been disrupted.

Comparable effects were observed for human FOXP2 in a human neuron-like cell line model (Devanna, Middelbeek, & Vernes, 2014). An advantage of studying such questions in these cell lines is that is it possible to compare the behavior of pure populations of FOXP2-positive and FOXP2-negative cells in real time. Using this system to observe the migration of living cells using time-lapse photography, it was demonstrated that cells expressing FOXP2 migrated more slowly than cells lacking FOXP2 (see figure 41.2A; Devanna et al., 2014). Thus the evidence from both brain and cell models suggest that human and mouse FOXP2/Foxp2 acts to influence neuronal migration and that increasing *FOXP2/Foxp2* expression results in reduced cell migration.

Neurogenetic models were also used to investigate the role of FOXP2 in neurite outgrowth—the process by which axons and dendrites grow and branch to allow neurons to connect to each other. The developing mouse striatum is a site of high *Foxp2* expression and thus primary neurons from this region were generated to study how Foxp2 can influence the growth and development of neurites. Primary striatal neurons from mice with mutations in *Foxp2* that mirrored the mutation found in the KE family developed shorter neurites with fewer branches than did primary neurons from normal (so-called *wild-type*) animals (figure 41.2B). This showed that a normal function of Foxp2 is to promote the growth and branching of axons and dendrites in the striatum, pointing to a role for Foxp2 in establishing striatal circuitry (such as cortico-striatal-thalamic loops) in the brain (Vernes et al., 2011). Cell line models showed a comparable role for human FOXP2 in neurite outgrowth. Human neuron-like cells expressing FOXP2 grew longer and more highly branched neurites than FOXP2-negative cells did (Devanna et al., 2014), suggesting that both mouse Foxp2 and human FOXP2 normally act to promote neurite outgrowth and branching.

Once neurons have migrated to the right position in the brain and extended their neurites toward other neurons, they must make and maintain connections with other neurons via the formation of synapses. It is this signaling across synapses and the resulting strengthening or weakening of synapses (known as synaptic plasticity) that underlies the function of neural circuits. By using brain slices from wild-type or *Foxp2* mutant animals, it was possible to record directly from neurons and observe synaptic effects. Mutation of one

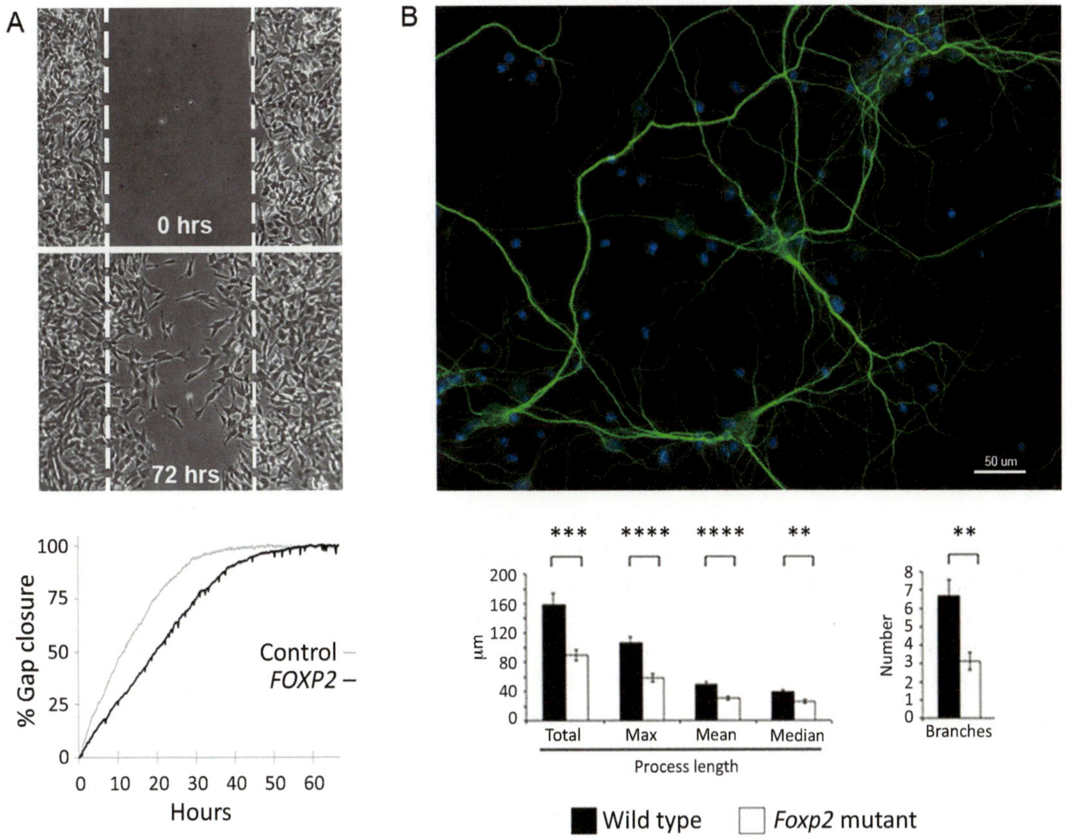

FIGURE 41.2 *FOXP2* affects neuronal migration and neurite outgrowth. (A) Observing the migration of human neuron-like cells over 72 hours showed that cells expressing *FOXP2* migrated more slowly than control cells without *FOXP2*. (B) Primary striatal neurons taken from normal wild-type mice had longer and more branched neurites than those from mice that are homozygous for mutations in *Foxp2* (*Foxp2* mutant) (**** = p=0.001, *** = p=0.001, ** = p=0.01). Figures adapted from Devanna et al. (2014) (A) and Vernes et al. (2011) (B).

copy of *Foxp2* altered synaptic plasticity in the mouse striatum, showing that *Foxp2* is involved in regulating the activity of striatal synapses and firing through striatal circuitry (Groszer et al., 2008). Recordings made by implanting electrodes in awake-behaving mice also showed aberrant striatal activity in mice with heterozygous mutations in *Foxp2* (French et al., 2012). Taken together, studies across human and mouse model systems demonstrate that *FOXP2*/*Foxp2* influences multiple aspects of brain development that can influence the formation of, and signaling through, specific neural circuitry.

3.5. Why Does Mutation of *FOXP2* Cause Disorder? Understanding the molecular and cellular functions of FOXP2, it is possible to start to build a picture of why mutations of this gene result in language-related disorders. At the most fundamental level it was shown that introducing patient mutations (such as the mutation found in the KE family) into *FOXP2* severely affected the transcription factor function of this protein in cell lines (Vernes et al., 2006). Normally, FOXP2 is restricted to the nucleus of cells, where it binds to DNA. However, introducing the KE family mutation into FOXP2 led to increased levels of the protein outside the nucleus in human cell lines (Vernes et al., 2006). Moreover, the mutant protein was no longer able to bind to target DNA or regulate gene expression (Vernes et al., 2006). As a result, in individuals carrying these mutations, the target genes normally regulated by FOXP2 are likely to be misregulated. Since affected individuals retain one normal copy of *FOXP2*, these effects may be subtle (i.e., an altered amount of target gene being expressed rather than a complete loss of gene regulation); however, this could still lead to important changes in how the brain develops. Indeed, both structural and functional abnormalities were observed in affected members of the KE family, in regions where *FOXP2* is expressed (Belton, Salmond, Watkins, Vargha-Khadem, & Gadian, 2003; Liégeois et al., 2003; Pinel et al., 2012). For example in affected KE family members (who all have a mutation in one copy of *FOXP2*), the caudate nucleus, putamen, and select cortical regions (including the inferior frontal gyrus) showed both structural (via voxel-based morphometry tests) and functional differences (via functional MRI) from those regions of the brains of unaffected family members (Belton et al., 2003; Liégeois et al., 2003; Pinel et al., 2012). Thus by understanding the role of FOXP2 in model systems to regulate target genes and cellular processes, together with the phenotypes observed in affected individuals, we can start to understand the mechanisms that may underlie the disorders of speech and language caused by *FOXP2* mutations.

4. Function of Language-Related Genes: CNTNAP2

Understanding FOXP2 function also led to the identification of new candidate language-related genes. Given that the function of FOXP2 is to regulate the expression of other genes, the phenotypic effects it has on cells, circuits, or behaviors must be mediated by these target genes. For this reason, it was hypothesized that the target genes acting downstream of FOXP2 would represent new candidate genes for language or language disorders. This was proven to be the case by the identification of the *CNTNAP2* gene as a target of FOXP2 (Vernes et al., 2008) and the association of common variation in this gene with disorders involving language (specific language impairment [SLI], autism spectrum disorder [ASD], dyslexia) and language-related phenotypes (early communicative behavior) (Alarcon et al., 2008; Anney et al., 2012; Arking et al., 2008; Ji et al., 2012; Newbury et al., 2011; Peter et al., 2011; Poot, 2014; Vernes et al., 2008; Whitehouse, Bishop, Ang, Pennell, & Fisher, 2011). Rare mutations of *CNTNAP2* have also been found in individuals with disorders affecting language (including language regression, ASD, and a small number of individuals with speech apraxia) and also more widespread deficits such as intellectual disability and epilepsy (Poot, 2015; Rodenas-Cuadrado et al., 2014; Rodenas-Cuadrado et al., 2016).

4.1. When and Where Is *CNTNAP2* Expressed? *CNTNAP2* expression has been well studied in mouse and human tissue and is highly expressed in a number of brain regions during development and increasingly in postnatal stages (Abrahams et al., 2007; Alarcon et al., 2008). In the human brain, expression is highest in the striatum, dorsal thalamus, and cortical areas. In the developing human cortex, expression is highest in layers II–V and enrichment in the inferior frontal gyrus and perisylvian regions has been noted (Rodenas-Cuadrado et al., 2014). The strong expression of *CNTNAP2* in brain regions involved in higher order cognitive processes, speech, and language supports a role for this gene in human language development.

4.2. What Does *CNTNAP2* Encode? The *CNTNAP2* gene encodes a protein (known as CASPR2) that is a member of the neurexin superfamily of proteins (Nakabayashi & Scherer, 2001; Rodenas-Cuadrado et al., 2014). Neurexins are well known to facilitate cell-cell interactions in the nervous system and play a role in synapse development and function, giving first hints to the likely function of CASPR2 (Dean & Dresbach, 2006; Siddiqui & Craig, 2011; Sudhof, 2008).

4.3. In Which Molecular Mechanisms Is *CNTNAP2* Involved? CASPR2 is a transmembrane protein (meaning that it sits at the cell surface) and most of the CASPR2 protein is projected outside the cell into the extracellular space (Poliak et al., 1999). Initially CASPR2 was observed at the surface of axons. The small intracellular component of CASPR2 helps to cluster ion channels together and the large extracellular component facilitates interactions with myelin (which surrounds and insulates nerve fibers), and these functions contribute to rapid conduction of nerve impulses (Horresh et al., 2008; Inda, 2006; Poliak et al., 1999; Poliak et al., 2003; Traka, 2003; Vabnick et al., 1999).

More recently, it has been shown that CASPR2 is also located at the synaptic membrane (Bakkaloglu et al., 2008; Varea et al., 2015). Like other members of the protein family to which it belongs (neurexins) it is thought to act as scaffold protein bridging the presynapse and postsynapse to facilitate synaptic stability (Bakkaloglu et al., 2008; Sudhof, 2008; Varea et al., 2015).

4.4. In Which Cellular/Neuronal Mechanisms Is *CNTNAP2* Involved? CASPR2 has been implicated in neuronal migration, network formation, and synaptic activity. Evidence for a role in neuronal migration came from patients with *CNTNAP2* mutations and a severe syndrome including epileptic seizures, autism, and language regression (Strauss et al., 2006). Following surgery aimed at controlling their epilepsy, brain biopsies were analyzed and the tissue was found to contain evidence of abnormal migration including poorly defined white and gray matter junctions, abnormal organization of neurons, and ectopic (misplaced) glia and neurons (Strauss et al., 2006). Migration defects were also observed in a mouse model of *Cntnap2*. Brain slices taken from *Cntnap2* knockout mice showed ectopic neurons in the corpus callosum and aberrant layering of neurons in the cortex, suggesting deficiencies in migration during development (Penagarikano et al., 2011). Brain tissue from these mice also showed reduced numbers of inhibitory interneurons in the cortex, striatum, and hippocampus, which could reflect either altered neuronal migration or neurogenesis (Penagarikano et al., 2011).

In primary cortical neurons from the developing mouse brain, Caspr2 affected neuronal network formation and connectivity. Reduction in Caspr2 expression (via knockdown) resulted in reduced growth and branching of dendrites (Anderson et al., 2012). Neurons cultured from transgenic *Cntnap2* knockout mice displayed reduced density of spines—the sites where synapses may form (Varea et al., 2015). In primary neurons, Caspr2 knockdown/knockout also affected synaptic properties and transmission of signals across the synapse by altering levels of receptors at the synapse (Varea et al., 2015) and altering the amplitude of synaptic responses (Anderson et al., 2012).

4.5. Why Does Mutation of *CNTNAP2* Cause Disorder? The links between *CNTNAP2* mutations and disorder are not as clear-cut as the ones for *FOXP2*. As noted, individuals with rare *CNTNAP2* mutations display a complex profile that is not restricted to language but includes ASD, speech/language impairments, epilepsy, and intellectual disability. These phenotypes are consistently present in individuals with homozygous *CNTNAP2* mutations; however, heterozygous changes seem to have a mixed penetrance, from severe to unaffected (Murdoch et al., 2015; Rodenas-Cuadrado et al., 2014; Rodenas-Cuadrado et al., 2016). It is not clear why some heterozygous individuals have more severe phenotypes than others do. It could be due to the genetic background of the individual, making some people more vulnerable to *CNTNAP2* mutation than others are (due to other genetic variants they carry). Conversely, it may be that some mutations of *CNTNAP2* produce more severe disruptions of CASPR2 function. Functional effects of patient mutations have been directly investigated for a small number of variants in human cell lines. Some nonsynonymous point mutations caused retention of the protein in the endoplasmic reticulum of the cell (where proteins are produced, modified, and folded into their mature shape) and in some cases degradation of the mutant protein. Introducing patient mutations into cells resulted in reduced amounts of functional protein to be present in human cell lines (Falivelli et al., 2012). A homozygous frameshift mutation causing a disorder involving cortical dysplasia, focal epilepsy, intellectual disability, attention deficit/hyperactivity disorder, and ASD in an Amish population was predicted to cause a truncated CASPR2 protein (Strauss et al., 2006). This mutation was found to truncate CASPR2 prior to its transmembrane domain, meaning that instead of being lodged at the cell surface/synapses, this protein is secreted into the extracellular space and lost (Falivelli et al., 2012).

Common variation in *CNTNAP2* has been associated with language disorder phenotypes and these variants have been linked to differences in the structure and function of brain circuits related to language processing. Structural changes included effects on gray matter volume (Tan, Doke, Ashburner, Wood, & Frackowiak, 2010; Udden, Snijders, Fisher, & Hagoort, 2016); structural and functional connectivity (Scott-van Zeeland et al., 2010; Dennis et al., 2011); and brain activation during sentence, artificial grammar, and syntax violation processing (Folia, Forkstam, Ingvar, Hagoort, & Petersson, 2011; Kos et al., 2012; Whalley et al., 2011). Why or how these common variants could be influencing brain structure or function is unknown, and further neuromolecular investigations are needed to understand these links.

Taken together these data support multiple roles for *CNTNAP2* in migration, neuronal connectivity, and synapse activity that may underlie the phenotypic differences observed in individuals carrying rare or common genetic variants in this gene.

5. Function of Language-Related Genes: ARHGEF39

Common variation in *ARHGEF39* was only recently associated with endophenotypes of language impairment (Devanna et al., 2018) and thus relatively little is known about the function of this gene. However, it warrants further discussion here as it represents an unusual example of a noncoding variant with clear functional consequences that may be relevant for language disorder.

As we have seen in sections 3 and 4, rare coding variants that disrupt protein sequence can have severe consequences and thus lead to disorder. However, noncoding variation may also contribute to disorder if it is located in a region of the noncoding genome that is responsible for regulating gene expression. Such noncoding regulatory variants are particularly good candidates for complex language disorders because these disorders are likely to be caused by the additive effects of subtle changes to multiple genes, rather than disruption of a single coding gene.

A recent study aimed to discover such noncoding regulatory variants in children with language impairment, focusing specifically on one type of regulatory region, known as a 3′-untranslated region (or *3′UTR*) of genes (Devanna et al., 2018). 3′UTRs control how much protein is expressed from a gene via multiple mechanisms including by interacting with regulatory molecules known as *microRNAs* (Grimson et al., 2007). From 43 affected children, a variant was identified in the 3′UTR of the *ARHGEF39* gene, which was predicted to disrupt this regulatory process. This variant was the alternative allele of a common SNP (rs72727021), which made it possible to determine whether it was associated with quantitative measures of language impairment in a larger cohort of 983 individuals from 285 families affected by language impairment (Devanna et al., 2018). The alternative allele of the rs72727021 SNP had a frequency of 12.3% in the language-impaired children in this cohort, as compared to a frequency of around 10.8% in unselected European populations and was significantly associated with reduced performance on tests of nonword repetition[6] in the language-impaired children (Devanna et al., 2018).

5.1. What Does *ARHGEF39* Encode? *ARHGEF39* is a member of the *ARHGEF* family of Rho guanine nucleotide exchange factors—enzymes that catalyze activation of guanosine triphosphatases and act as adapter proteins in a number of cellular reactions. Little is known regarding the specific function of ARHGEF39, however other members of this family of proteins regulate a range of processes such as gene expression, cell migration, cell growth and dendritic outgrowth (Goicoechea, Awadia, & Garcia-Mata, 2014; Newey, Velamoor, Govek, & van Aelst, 2005; van Aelst & Cline, 2004; van Aelst & D'Souza-Schorey, 1997).

5.2. Why Does Variation in *ARHGEF39* Associate with Disorder? The SLI-associated rs72727021 variant is located in the 3′UTR of the *ARHGEF39* gene and was predicted to disrupt the interaction between 3′UTR and a microRNA molecule (Devanna et al., 2018). MicroRNAs bind to 3′UTR regions of messenger RNAs, resulting in assembly of a protein complex that causes reduced translation of the messenger RNA and, as a consequence, lower protein levels (Chen & Rajewsky, 2007). Because this interaction is mediated via complementary base pairing, the sequences bound by microRNAs have been well defined. Thus, clear predictions of binding sites and effects of variants can be made *in silico* (Lewis, Burge, & Bartel, 2005). The rs72727021 SNP was located in a binding site for microRNA-215, and its presence was predicted to alter microRNA binding and protein expression (Devanna et al., 2018), however direct functional tests in cell lines were needed to confirm these effects.

By introducing the 3′UTR of *ARHGEF39* together with microRNA-215 into cell lines, it was possible to see that the reference sequence (the A allele that is common in the general population) was downregulated by microRNA-215. By contrast, the alternative 3′UTR sequence (carrying the SLI-associated alternative C allele) was not downregulated (see figure 41.3A) (Devanna et al., 2018). Comparable disruption of gene regulation and expression were also observed in the human brain (see figure 41.3B). In postmortem human tissue samples, individuals with the A allele had lower expression of *ARHGEF39* than individuals with the C allele (in heterozygous or homozygous state) across multiple regions of the brain, including the cortex (Devanna et al., 2018). Taken together, these data suggest that the presence of this single change in a noncoding DNA region disrupts regulation of *ARHGEF39* expression. Why elevated *ARHGEF39* expression may be related to language impairment is currently unknown, but functional studies into this gene in the future may shed light on this putative new language-related gene.

Interestingly, *ARHGEF39* is not the only member of the *ARHGEF* gene family that has been implicated in language impairment or neurodevelopmental phenotypes. A common variant thought to affect *ARHGEF19* expression is associated with language impairment (Nudel et al., 2014), *ARHGEF6* has been linked to intellectual disability (Kutsche et al., 2000), and *ARHGEF9* mutations have been found in individuals with developmental delay, intellectual disability, and seizures (de Ligt et al., 2012; Harvey et al., 2004; Kalscheuer et al., 2009; Lemke et al., 2012; Lesca et al., 2011; Marco et al., 2008; Shimojima et al., 2011), as well as ASD and speech delays (Bhat et al., 2016). These data may point to a larger role for the *ARHGEF* family in neurodevelopment and language-related circuitry and supports the call for greater studies to be performed on the neuromolecular functions of these genes.

6. Finding New Language-Related Genes via Neuromolecular Approaches

Although there are infrequent examples where mutations in a single gene like *FOXP2* can cause language impairments, it is clear that there is no single "gene for language," rather there are interconnected networks of genes that contribute to complex phenotypes underlying the human capacity for language. It is therefore essential that, rather than only considering individual genes, gene networks be considered as a whole to obtain a complete understanding of the molecular-genetic underpinnings of language. There are many ways to identify gene networks related to language. One way is to use known language-related genes, such as *FOXP2*,

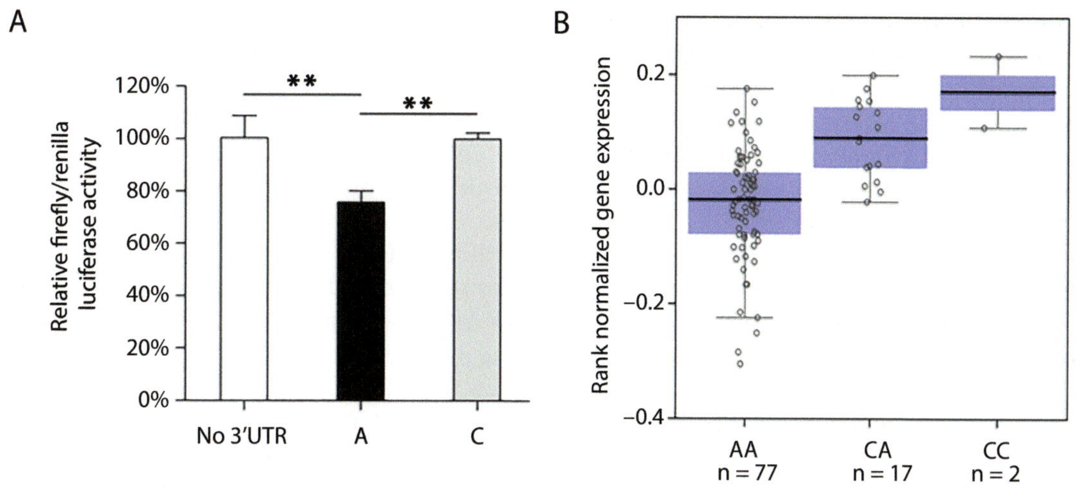

FIGURE 41.3 The presence of the rs72727021 SNP in the 3′UTR of the *ARHGEF39* gene results in altered regulation in human cell lines (** = p<0.01) (A) and postmortem human cortex tissue (B). The alternative allele was significantly associated with nonword repetition in language-impaired children and led to increased expression in both model systems. Figures adapted from Devanna et al. (2018).

as an entry point to identify the wider molecular networks within which they are embedded. Another way is to perform hypothesis-free discovery of networks by exploring the molecular properties of relevant circuits, tissues, and/or behaviors.

6.1. MOLECULAR WINDOWS INTO LANGUAGE Language-related genes can be used as a molecular window into language pathways by understanding their wider molecular networks. These networks include transcription factors that regulate expression, interacting proteins, and downstream molecular cascades (e.g., resulting from enzymatic reactions, signaling cascades, or gene regulatory mechanisms).

Exploration of the molecular networks related to *FOXP2* has shown the power of this approach (Fisher & Scharff, 2009). Proteins that interact with FOXP2 have proven strong candidates for language-related disorders. In performing its function as a transcription factor, FOXP2 interacts with other regulatory factors in order to bind DNA and regulate gene expression (Li, Weidenfeld, & Morrisey, 2004; Wang, Lin, Li, & Tucker, 2003). The closely related protein, FOXP1 (also a FOX transcription factor), is expressed in overlapping brain regions with FOXP2, including parts of the cortex (motor cortex layer IV) and striatum (caudate nucleus and putamen) where these proteins are able to dimerize to regulate gene expression (Ferland et al., 2003; Hisaoka, Nakamura, Senba, & Morikawa, 2010; Li et al., 2004). Mutations in *FOXP1* have been found in individuals displaying motor impairments, intellectual disability, autism, and speech/language impairments (Bacon & Rappold, 2012; Bacon et al., 2015; Carr et al., 2010; Chien et al., 2013; Hamdan et al., 2010; Horn, 2011; O'Roak et al., 2011; Sollis et al., 2016; Vernes, MacDermot, Monaco, & Fisher, 2009).

Identifying the downstream targets of FOXP2 has facilitated the identification of new candidate genes and identified molecular links with existing candidate genes for language-related disorders. *CNTNAP2* was the first example of a FOXP2 target gene used as a candidate for language disorder (Vernes et al., 2008). *CNTNAP2*'s status as a FOXP2 target led to its association with SLI and represented the first molecular link between language-related disorders (see section 4).

DISC1 was a well-studied candidate gene for schizophrenia when it was identified as a direct target of FOXP2 (Spiteri et al., 2007), suggesting a possible molecular link between speech/language disorder and schizophrenia (Farrell et al., 2015). Schizophrenia is a neuropsychiatric disorder with a range of clinical symptoms including auditory hallucinations, delusions, and social and communicative impairments (Kuperberg, 2010). Impaired verbal communication is a key feature of schizophrenia and may manifest in many ways, such as alogia (poverty of speech), disorganized speech pattern, misassociation of words and phrases, use of neologisms or nonwords, and production of meaningless sentences (Kuperberg, 2010). *DISC1* disruptions have also been linked to ASDs (Crepel et al., 2010; Kilpinen et al., 2008; Williams et al., 2009; Zheng et al., 2011). FOXP2 was found to bind the promoter of the *DISC1* gene in embryonic human basal ganglia tissue (Spiteri et al., 2007). In functional cell experiments, FOXP2 was shown to bind to the promoter to downregulate the expression of *DISC1*, and this repression was disrupted when patient mutations of *FOXP2* were introduced (Walker et al., 2012). Using induced pluripotent stem cell models, loss of *DISC1* expression was shown to affect synapse formation and synaptic activity and have knock-on consequences for large networks of neuronal genes (Wen et al., 2014).

Many more studies are needed to clarify how *FOXP2* genetic networks relate to language disorders and to elucidate the molecular links between diverse disorders involving language phenotypes, but these examples give good reasons to expect this approach to yield valuable insights in the coming years.

6.2. Hypothesis-Free Network Identification It is possible to identify language-related molecular networks without prior hypotheses about the genes involved. One way to do this is to survey the genes expressed in specific brain regions to understand the molecular pathways acting during development or function of these regions. This can be done by surveying known language-related regions of postmortem human brain (see parts V and VI of this volume), or circuitry in animal brains underlying language-relevant behaviors (see part VIII of this volume). Molecular networks can be surveyed by transcriptome analysis, that is, RNA-Sequencing, which will identify every gene being expressed in the tissue at a given time point, together with the levels of expression for each gene (from very high to very low). By surveying multiple individuals/animals and different time points or behaviors, relationships between genes can be extracted from these data following the principal that genes that are tightly coupled in their expression patterns across samples will be likely to act together (Langfelder & Horvath, 2008). Linking this to the known functions of these genes facilitates the building of large-scale, molecular networks of functional relevance to the tissue/behavior under study.

This approach has been used in vocal learning songbirds to identify gene networks regulated during singing (Hilliard, Miller, Fraley, Horvath, & White, 2012; Hilliard, Miller, Horvath, & White, 2012). By surveying key regions of the song circuitry in singing and silent brains using transcriptomics and *in silico* network building methods, large-scale molecular networks were built that were driven by, and involved in, singing behavior. This showed that specific parts of the song circuitry displayed distinct molecular profiles and identified new genes that may be involved in such behavior across species (Hilliard, Miller, Fraley, et al., 2012; Hilliard, Miller, Horvath, & White, 2012). More recently, bats have emerged as a promising mammalian model to study vocal learning. A comparable network building approach was used to identify molecular networks in vocal-related regions of the brain (see figure 41.4) (Rodenas-Cuadrado, Chen, Wiegrebe, Firzlaff, & Vernes, 2015). This hypothesis-free approach highlighted a molecular pathway underlying glutamatergic synaptic function in the periaqueductal gray of the bat brain—a region that, across mammals, is crucial for the control of vocalizations. Activation or blockade of this glutamatergic synaptic pathway in mammals can induce or prevent vocalizations in mammals, respectively (Rodenas-Cuadrado et al., 2015). As such, this unbiased network building approach made it possible to identify a gene network that may be mechanistically important for vocal-motor control in mammals and demonstrated the potential of such approaches to identify gene networks involved in communicative behavior.

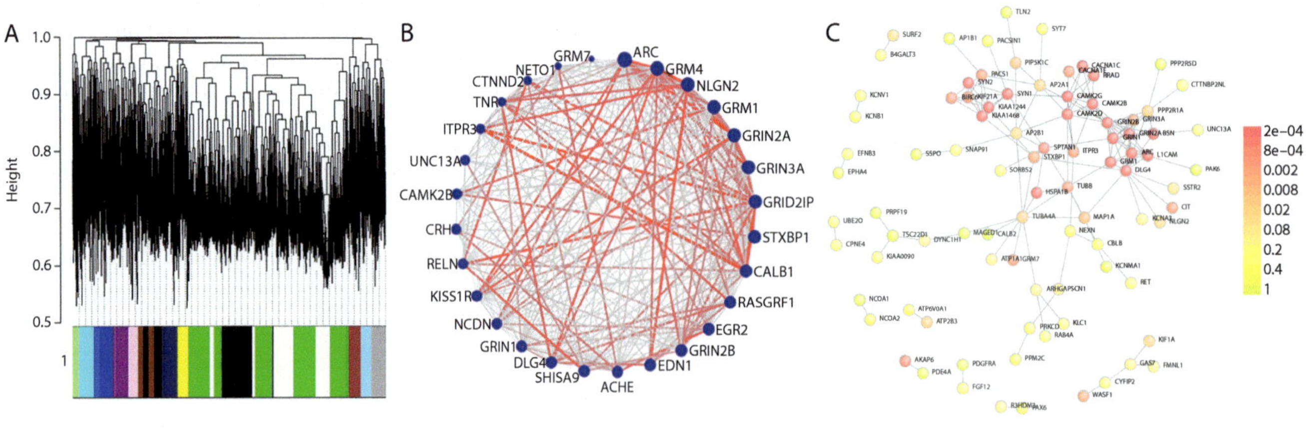

Figure 41.4 Molecular networks can be identified by surveying gene expression in specific brain regions and using network clustering methods (A) to identify highly interconnected gene networks (B) with known functional relevance (C). Example figures adapted from Rodenas-Cuadrado et al. (2015).

Further such studies in animal models linking gene networks to speech/language relevant behaviors or from postmortem human tissue (e.g., Konopka et al., 2012) are likely to reveal a genome level view of molecular networks and identify new candidate genes underlying speech, language, and related disorders (see also Johnson & White, chapter 8 of this volume).

7. Conclusion

The use of neuromolecular approaches as outlined in this chapter can help to elucidate how genes contribute to human speech and language and language-related disorders. By coupling such approaches with crucial research to determine brain networks involved in language, clinical studies of disorders, and animal behavior studies it will be possible to form an integrated view from gene to brain and behavior of the underlying biology of human language.

Acknowledgments

This work was funded by a Marie Curie Career Integration Grant (PCIG12-GA-2012-333978) and a Max Planck Research Group Grant both awarded to S.C.V. Thank you also to the members of my research group who provided data for many of the image panels, especially Nienke Hoeksema and Paolo Devanna.

NOTES

1. Although it should be noted that over time and after many divisions, the properties of the cells change and often look less like their source tissue if kept dividing for too long (e.g., years)
2. One micron is equivalent to one millionth of a meter. For perspective, the width of human hair is estimated to be between 20 and 200 µm.
3. Following standard nomenclature, genes are denoted in italics, proteins in regular font. Uppercase letters denote the human version of the gene (i.e., *FOXP2*), lowercase the mouse version of the gene (i.e., *Foxp2*).
4. *FOXP2* is also expressed in non-neuronal tissue such as the lungs and heart, see (Schroeder & Myers, 2008).
5. It is from its membership in the FOX family that *FOXP2* gets its name.
6. Children with language impairment commonly perform poorly on the nonword repetition task, making it a common metric in cohorts such as this.

REFERENCES

Abrahams, B. S., Tentler, D., Perederiy, J. V., Oldham, M. C., Coppola, G., & Geschwind, D. H. (2007). Genome-wide analyses of human perisylvian cerebral cortical patterning. *Proceedings of the National Academy of Sciences, 104*(45), 17849–17854.

Agholme, L., Lindstrom, T., Kagedal, K., Marcusson, J., & Hallbeck, M. (2010). An in vitro model for neuroscience: Differentiation of SH-SY5Y cells into cells with morphological and biochemical characteristics of mature neurons. *Journal of Alzheimer's Disease, 20*(4), 1069–1082.

Alarcon, M., Abrahams, B. S., Stone, J. L., Duvall, J. A., Perederiy, J. V., Bomar, J. M., ... Geschwind, D. H. (2008). Linkage, association, and gene-expression analyses identify *CNTNAP2* as an autism-susceptibility gene. *American Journal of Human Genetics, 82*(1), 150–159.

Anderson, G. R., Galfin, T., Xu, W., Aoto, J., Malenka, R. C., & Sudhof, T. C. (2012). Candidate autism gene screen identifies critical role for cell-adhesion molecule CASPR2 in dendritic arborization and spine development. *Proceedings of the National Academy of Sciences, 109*(44), 18120–18125.

Anney, R., Klei, L., Pinto, D., Almeida, J., Bacchelli, E., Baird, G., ... Devlin, B. (2012). Individual common variants exert weak effects on the risk for autism spectrum disorders. *Human Molecular Genetics, 21*(21), 4781–4792.

Arking, D. E., Cutler, D. J., Brune, C. W., Teslovich, T. M., West, K., Ikeda, M., ... Chakravarti, A. (2008). A common genetic variant in the neurexin superfamily member *CNTNAP2* increases familial risk of autism. *American Journal of Human Genetics, 82*(1), 160–164.

Bacon, C., & Rappold, G. A. (2012). The distinct and overlapping phenotypic spectra of *FOXP1* and *FOXP2* in cognitive disorders. *Human Genetics, 131*(11), 1687–1698.

Bacon, C., Schneider, M., Le Magueresse, C., Froehlich, H., Sticht, C., Gluch, C., ... Rappold, G. A. (2015). Brain-specific Foxp1 deletion impairs neuronal development and causes autistic-like behaviour. *Molecular Psychiatry, 20*(5), 632–639.

Bakkaloglu, B., O'Roak, B. J., Louvi, A., Gupta, A. R., Abelson, J. F., Morgan, T. M., ... State, M. W. (2008). Molecular cytogenetic analysis and resequencing of contactin associated protein-like 2 in autism spectrum disorders. *American Journal of Human Genetics, 82*(1), 165–173.

Becker, M., Guadalupe, T., Franke, B., Hibar, D. P., Renteria, M. E., Stein, J. L., ... Fisher, S. E. (2016). Early developmental gene enhancers affect subcortical volumes in the adult human brain. *Human Brain Mapping, 37*(5), 1788–1800.

Belton, E., Salmond, C. H., Watkins, K. E., Vargha-Khadem, F., & Gadian, D. G. (2003). Bilateral brain abnormalities associated with dominantly inherited verbal and orofacial dyspraxia. *Human Brain Mapping, 18*(3), 194–200.

Bhat, G., LaGrave, D., Millson, A., Herriges, J., Lamb, A. N., & Matalon, R. (2016). Xq11.1-11.2 deletion involving ARHGEF9 in a girl with autism spectrum disorder. *European Journal of Medical Genetics, 59*(9), 470–473.

Carr, C. W., Moreno-De-Luca, D., Parker, C., Zimmerman, H. H., Ledbetter, N., Martin, C. L., ... Abdul-Rahman, O. A. (2010). Chiari I malformation, delayed gross motor skills, severe speech delay, and epileptiform discharges in a child with *FOXP1* haploinsufficiency. *European Journal of Human Genetics, 18*(11), 1216–1220.

Chailangkarn, T., Acab, A., & Muotri, A. R. (2012). Modeling neurodevelopmental disorders using human neurons. *Current Opinion in Neurobiology, 22*(5), 785–790.

Chen, K., & Rajewsky, N. (2007). The evolution of gene regulation by transcription factors and microRNAs. *Nature Reviews Genetics, 8*(2), 93–103.

Cheung, Y. T., Lau, W. K., Yu, M. S., Lai, C. S., Yeung, S. C., So, K. F., & Chang, R. C. (2009). Effects of all-trans-retinoic acid on human SH-SY5Y neuroblastoma as in vitro model in neurotoxicity research. *Neurotoxicology, 30*(1), 127–135.

Chien, W. H., Gau, S. S., Chen, C. H., Tsai, W. C., Wu, Y. Y., Chen, P. H., & Shang, C. Y. (2013). Increased gene expression of *FOXP1* in patients with autism spectrum disorders. *Molecular Autism, 4*(1), 23.

Clovis, Y. M., Enard, W., Marinaro, F., Huttner, W. B., & De Pietri Tonelli, D. (2012). Convergent repression of *Foxp2* 3′UTR by miR-9 and miR-132 in embryonic mouse neocortex: Implications for radial migration of neurons. *Development, 139*(18), 3332–3342.

Crepel, A., Breckpot, J., Fryns, J. P., De la Marche, W., Steyaert, J., Devriendt, K., & Peeters, H. (2010). *DISC1* duplication in two brothers with autism and mild mental retardation. *Clinical Genetics, 77*(4), 389–394.

Dean, C., & Dresbach, T. (2006). Neuroligins and neurexins: Linking cell adhesion, synapse formation and cognitive function. *Trends in Neurosciences, 29*(1), 21–29.

de Ligt, J., Willemsen, M. H., van Bon, B. W., Kleefstra, T., Yntema, H. G., Kroes, T., ... Vissers, L. E. (2012). Diagnostic exome sequencing in persons with severe intellectual disability. *New England Journal of Medicine, 367*(20), 1921–1929.

Dennis, E. L., Jahanshad, N., Rudie, J. D., Brown, J. A., Johnson, K., McMahon, K. L., ... Thompson, P. M. (2011). Altered structural brain connectivity in healthy carriers of the autism risk gene, CNTNAP2. *Brain Connectivity, 1*(6), 447–459.

Deriziotis, P., & Fisher, S. E. (2013). Neurogenomics of speech and language disorders: the road ahead. *Genome Biology, 14*(4), 204.

Devanna, P., Chen, X. S., Ho, J., Gajewski, D., Smith, S. D., Gialluisi, A., ... Vernes, S. C. (2018). Next-gen sequencing identifies non-coding variation disrupting miRNA-binding sites in neurological disorders. *Molecular Psychiatry, 23*(5), 1375–1384.

Devanna, P., Middelbeek, J., & Vernes, S. C. (2014). *FOXP2* drives neuronal differentiation by interacting with retinoic acid signaling pathways. *Frontiers in Cellular Neuroscience, 8*, 305.

Ding, S., Royall, J. J., Sunkin, S. M., Ng, L., Facer, B. A. C., Lesnar, P., ... Lein, E. S. (2016). Comprehensive cellular-resolution atlas of the adult human brain. *Journal of Comparative Neurology, 524*, 3127–3481.

Encinas, M., Iglesias, M., Liu, Y., Wang, H., Muhaisen, A., Cena, V., ... Comella, J. X. (2000). Sequential treatment of SH-SY5Y cells with retinoic acid and brain-derived neurotrophic factor gives rise to fully differentiated, neurotrophic factor-dependent, human neuron-like cells. *Journal of Neurochemistry, 75*(3), 991–1003.

Falivelli, G., De Jaco, A., Favaloro, F. L., Kim, H., Wilson, J., Dubi, N., ... Comoletti, D. (2012). Inherited genetic variants in autism-related *CNTNAP2* show perturbed trafficking and ATF6 activation. *Human Molecular Genetics, 21*(21), 4761–4773.

Falk, T., Zhang, S., Erbe, E. L., & Sherman, S. J. (2006). Neurochemical and electrophysiological characteristics of rat striatal neurons in primary culture. *Journal of Comparative Neurology, 494*(2), 275–289.

Farrell, M. S., Werge, T., Sklar, P., Owen, M. J., Ophoff, R. A., O'Donovan, M. C., ... Sullivan, P. F. (2015). Evaluating historical candidate genes for schizophrenia. *Molecular Psychiatry, 20*(5), 555–562.

Ferland, R. J., Cherry, T. J., Preware, P. O., Morrisey, E. E., & Walsh, C. A. (2003). Characterization of *Foxp2* and *Foxp1* mRNA and protein in the developing and mature brain. *Journal of Comparative Neurology, 460*(2), 266–279.

Fisher, S. E., & Marcus, G. F. (2006). The eloquent ape: Genes, brains and the evolution of language. *Nature Reviews Genetics, 7*(1), 9–20.

Fisher, S. E., & Scharff, C. (2009). *FOXP2* as a molecular window into speech and language. *Trends in Genetics, 25*(4), 166–177.

Fitch, W. T., Hauser, M. D., & Chomsky, N. (2005). The evolution of the language faculty: Clarifications and implications. *Cognition, 97*(2), 179–210; discussion 211–125.

Folia, V., Forkstam, C., Ingvar, M., Hagoort, P., & Petersson, K. M. (2011). Implicit artificial syntax processing: Genes, preference, and bounded recursion. *Biolinguistics, 5*, 105–132.

French, C. A., Jin, X., Campbell, T. G., Gerfen, E., Groszer, M., Fisher, S. E., & Costa, R. M. (2012). An aetiological *Foxp2* mutation causes aberrant striatal activity and alters plasticity during skill learning. *Molecular Psychiatry, 17*(11), 1077–1085.

Geissler, M., & Faissner, A. (2012). A new indirect co-culture set up of mouse hippocampal neurons and cortical astrocytes on microelectrode arrays. *Journal of Neuroscience Methods, 204*(2), 262–272.

Gialluisi, A., Newbury, D. F., Wilcutt, E. G., Olson, R. K., DeFries, J. C., Brandler, W. M., ... Fisher, S. E. (2014). Genome-wide screening for DNA variants associated with reading and language traits. *Genes, Brain, and Behavior, 13*(7), 686–701.

Goicoechea, S. M., Awadia, S., & Garcia-Mata, R. (2014). I'm coming to GEF you: Regulation of RhoGEFs during cell migration. *Cell Adhesion and Migration, 8*(6), 535–549.

Grimson, A., Farh, K. K., Johnston, W. K., Garrett-Engele, P., Lim, L. P., & Bartel, D. P. (2007). MicroRNA targeting specificity in mammals: Determinants beyond seed pairing. *Molecular Cell, 27*(1), 91–105.

Groszer, M., Keays, D. A., Deacon, R. M., de Bono, J. P., Prasad-Mulcare, S., Gaub, S., ... Fisher, S. E. (2008). Impaired synaptic plasticity and motor learning in mice with a point mutation implicated in human speech deficits. *Current Biology, 18*(5), 354–362.

Hamdan, F. F., Daoud, H., Rochefort, D., Piton, A., Gauthier, J., Langlois, M., ... Michaud, J. L. (2010). De novo mutations in *FOXP1* in cases with intellectual disability, autism, and language impairment. *American Journal of Human Genetics, 87*(5), 671–678.

Harvey, K., Duguid, I. C., Alldred, M. J., Beatty, S. E., Ward, H., Keep, N. H., ... Harvey, R. J. (2004). The GDP-GTP exchange factor collybistin: An essential determinant of neuronal gephyrin clustering. *Journal of Neuroscience, 24*(25), 5816–5826.

Hauser, M. D., Chomsky, N., & Fitch, W. T. (2002). The faculty of language: What is it, who has it, and how did it evolve? *Science, 298*(5598), 1569–1579.

Hilliard, A. T., Miller, J. E., Fraley, E. R., Horvath, S., & White, S. A. (2012). Molecular microcircuitry underlies functional specification in a basal ganglia circuit dedicated to vocal learning. *Neuron, 73*(3), 537–552.

Hilliard, A. T., Miller, J. E., Horvath, S., & White, S. A. (2012). Distinct neurogenomic states in basal ganglia subregions relate differently to singing behavior in songbirds. *PLOS Computational Biology, 8*(11), e1002773.

Hisaoka, T., Nakamura, Y., Senba, E., & Morikawa, Y. (2010). The forkhead transcription factors, *Foxp1* and *Foxp2*, identify different subpopulations of projection neurons in the mouse cerebral cortex. *Neuroscience, 166*(2), 551–563.

Holehonnur, R., Lella, S. K., Ho, A., Luong, J. A., & Ploski, J. E. (2015). The production of viral vectors designed to express large and difficult to express transgenes within neurons. *Molecular Brain, 8*, 12.

Horn, D. (2011). Mild to moderate intellectual disability and significant speech and language deficits in patients with *FOXP1* deletions and mutations. *Molecular Syndromology, 2*(3–5), 213–216.

Horresh, I., Poliak, S., Grant, S., Bredt, D., Rasband, M. N., & Peles, E. (2008). Multiple molecular interactions determine the clustering of Caspr2 and Kv1 channels in myelinated axons. *Journal of Neuroscience, 28*(52), 14213–14222.

Inda, M. C. (2006). Voltage-gated ion channels in the axon initial segment of human cortical pyramidal cells and their relationship with chandelier cells. *Proceedings of the National Academy of Sciences, 103*(8), 2920–2925.

Jamsa, A., Hasslund, K., Cowburn, R. F., Backstrom, A., & Vasange, M. (2004). The retinoic acid and brain-derived neurotrophic factor differentiated SH-SY5Y cell line as a model for Alzheimer's disease-like tau phosphorylation. *Biochemical and Biophysical Research Communications, 319*(3), 993–1000.

Ji, W., Li, T., Pan, Y., Tao, H., Ju, K., Wen, Z., … Shi, Y. (2012). *CNTNAP2* is significantly associated with schizophrenia and major depression in the Han Chinese population. *Psychiatry Research, 207*(3), 225–228.

Joshi, S., Guleria, R., Pan, J., DiPette, D., & Singh, U. S. (2006). Retinoic acid receptors and tissue-transglutaminase mediate short-term effect of retinoic acid on migration and invasion of neuroblastoma SH-SY5Y cells. *Oncogene, 25*(2), 240–247.

Kalscheuer, V. M., Musante, L., Fang, C., Hoffmann, K., Fuchs, C., Carta, E., … Harvey, R. J. (2009). A balanced chromosomal translocation disrupting ARHGEF9 is associated with epilepsy, anxiety, aggression, and mental retardation. *Human Mutation, 30*(1), 61–68.

Kang, C., & Drayna, D. (2011). Genetics of speech and language disorders. *Annual Review of Genomics and Human Genetics, 12*, 145–164.

Katoh, M., & Katoh, M. (2004). Human FOX gene family. *International Journal of Oncology, 25*(5), 1495–1500.

Kilpinen, H., Ylisaukko-Oja, T., Hennah, W., Palo, O. M., Varilo, T., Vanhala, R., … Peltonen, L. (2008). Association of *DISC1* with autism and Asperger syndrome. *Molecular Psychiatry, 13*(2), 187–196.

Kim, Y. G., & Lee, Y. I. (2012). Differential expressions of synaptogenic markers between primary cultured cortical and hippocampal neurons. *Experimental Neurobiology, 21*(2), 61–67.

Konopka, G., Friedrich, T., Davis-Turak, J., Winden, K., Oldham, M. C., Gao, F., … Geschwind, D. H. (2012). Human-specific transcriptional networks in the brain. *Neuron, 75*(4), 601–617.

Kos, M., van den Brink, D., Snijders, T. M., Rijpkema, M., Franke, B., Fernandez, G., & Hagoort, P. (2012). *CNTNAP2* and language processing in healthy individuals as measured with ERPs. *PLOS ONE, 7*(10), e46995.

Kuperberg, G. R. (2010). Language in schizophrenia: Part 1: An introduction. *Language and Linguistics Compass, 4*(8), 576–589.

Kutsche, K., Yntema, H., Brandt, A., Jantke, I., Nothwang, H. G., Orth, U., … Gal, A. (2000). Mutations in ARHGEF6, encoding a guanine nucleotide exchange factor for Rho GTPases, in patients with X-linked mental retardation. *Nature Genetics, 26*(2), 247–250.

Lai, C. S., Fisher, S. E., Hurst, J. A., Vargha-Khadem, F., & Monaco, A. P. (2001). A forkhead-domain gene is mutated in a severe speech and language disorder. *Nature, 413*(6855), 519–523.

Lai, C. S., Gerrelli, D., Monaco, A. P., Fisher, S. E., & Copp, A. J. (2003). *FOXP2* expression during brain development coincides with adult sites of pathology in a severe speech and language disorder. *Brain, 126*(11), 2455–2462.

Lancaster, M. A., Renner, M., Martin, C. A., Wenzel, D., Bicknell, L. S., Hurles, M. E., … Knoblich, J. A. (2013). Cerebral organoids model human brain development and microcephaly. *Nature, 501*(7467), 373–379.

Langfelder, P., & Horvath, S. (2008). WGCNA: An R package for weighted correlation network analysis. *BMC Bioinformatics, 9*, 559.

Lemke, J. R., Riesch, E., Scheurenbrand, T., Schubach, M., Wilhelm, C., Steiner, I., … Biskup, S. (2012). Targeted next generation sequencing as a diagnostic tool in epileptic disorders. *Epilepsia, 53*(8), 1387–1398.

Lesca, G., Till, M., Labalme, A., Vallee, D., Hugonenq, C., Philip, N., … Sanlaville, D. (2011). De novo Xq11.11 microdeletion including ARHGEF9 in a boy with mental retardation, epilepsy, macrosomia, and dysmorphic features. *American Journal of Medical Genetics. Part A, 155A*(7), 1706–1711.

Levinson, S. C., & Holler, J. (2014). The origin of human multi-modal communication. *Philosophical Transactions of the Royal Society of London. Series B, Biological Sciences, 369*(1651), 20130302.

Lewis, B. P., Burge, C. B., & Bartel, D. P. (2005). Conserved seed pairing, often flanked by adenosines, indicates that thousands of human genes are microRNA targets. *Cell, 120*(1), 15–20.

Li, S., Weidenfeld, J., & Morrisey, E. E. (2004). Transcriptional and DNA binding activity of the Foxp1/2/4 family is modulated by heterotypic and homotypic protein interactions. *Molecular and Cellular Biology, 24*(2), 809–822.

Liégeois, F., Baldeweg, T., Connelly, A., Gadian, D. G., Mishkin, M., & Vargha-Khadem, F. (2003). Language fMRI abnormalities associated with *FOXP2* gene mutation. *Nature Neuroscience, 6*(11), 1230–1237.

Manfredsson, F. P. (2016). Introduction to viral vectors and other delivery methods for gene therapy of the nervous system. *Methods in Molecular Biology, 1382*, 3–18.

Marco, E. J., Abidi, F. E., Bristow, J., Dean, W. B., Cotter, P., Jeremy, R. J., … Sherr, E. H. (2008). ARHGEF9 disruption in a female patient is associated with X linked mental retardation and sensory hyperarousal. *Journal of Medical Genetics, 45*(2), 100–105.

Marin, O., Valiente, M., Ge, X., & Tsai, L. H. (2010). Guiding neuronal cell migrations. *Cold Spring Harbor Perspectives in Biology, 2*(2), a001834.

Morgan, A., Fisher, S. E., Scheffer, I., & Hildebrand, M. (2017). FOXP2-related speech and language disorders. In M. P. Adam, H. H. Ardinger, R. A. Pagon et al. (Eds.), *GeneReviews®* [Internet]. Seattle: University of Washington, Seattle; 1993–2019. Retrieved from https://www.ncbi.nlm.nih.gov/books/NBK368474/.

Murdoch, J. D., Gupta, A. R., Sanders, S. J., Walker, M. F., Keaney, J., Fernandez, T. V., ... State, M. W. (2015). No evidence for association of autism with rare heterozygous point mutations in contactin-associated protein-like 2 (*CNTNAP2*), or in other contactin-associated proteins or contactins. *PLOS Genetics, 11*(1), e1004852.

Nakabayashi, K., & Scherer, S. W. (2001). The human contactin-associated protein-like 2 gene (*CNTNAP2*) spans over 2 Mb of DNA at chromosome 7q35. *Genomics, 73*(1), 108–112.

Newbury, D. F., Fisher, S. E., & Monaco, A. P. (2010). Recent advances in the genetics of language impairment. *Genome Medicine, 2*(1), 6.

Newbury, D. F., & Monaco, A. P. (2002). Molecular genetics of speech and language disorders. *Current Opinion in Pediatrics, 14*(6), 696–701.

Newbury, D. F., & Monaco, A. P. (2010). Genetic advances in the study of speech and language disorders. *Neuron, 68*(2), 309–320.

Newbury, D. F., Paracchini, S., Scerri, T. S., Winchester, L., Addis, L., Richardson, A. J., ... Monaco, A. P. (2011). Investigation of dyslexia and SLI risk variants in reading- and language-impaired subjects. *Behavior Genetics, 41*(1), 90–104.

Newey, S. E., Velamoor, V., Govek, E. E., & van Aelst, L. (2005). Rho GTPases, dendritic structure, and mental retardation. *Journal of Neurobiology, 64*(1), 58–74.

Nudel, R., Simpson, N. H., Baird, G., O'Hare, A., Conti-Ramsden, G., Bolton, P. F., ... Newbury, D. F. (2014). Genome-wide association analyses of child genotype effects and parent-of-origin effects in specific language impairment. *Genes, Brain, and Behavior, 13*(4), 418–429.

Odom, D. T., Dowell, R. D., Jacobsen, E. S., Gordon, W., Danford, T. W., MacIsaac, K. D., ... Fraenkel, E. (2007). Tissue-specific transcriptional regulation has diverged significantly between human and mouse. *Nature Genetics, 39*(6), 730–732.

O'Roak, B. J., Deriziotis, P., Lee, C., Vives, L., Schwartz, J. J., Girirajan, S., ... Eichler, E. E. (2011). Exome sequencing in sporadic autism spectrum disorders identifies severe de novo mutations. *Nature Genetics, 43*(6), 585–589.

Penagarikano, O., Abrahams, B. S., Herman, E. I., Winden, K. D., Gdalyahu, A., Dong, H., ... Geschwind, D. H. (2011). Absence of *CNTNAP2* leads to epilepsy, neuronal migration abnormalities, and core autism-related deficits. *Cell, 147*(1), 235–246.

Peter, B., Raskind, W. H., Matsushita, M., Lisowski, M., Vu, T., Berninger, V. W., ... Brkanac, Z. (2011). Replication of *CNTNAP2* association with nonword repetition and support for *FOXP2* association with timed reading and motor activities in a dyslexia family sample. *Journal of Neurodevelopmental Disorders, 3*(1), 39–49.

Pinel, P., Fauchereau, F., Moreno, A., Barbot, A., Lathrop, M., Zelenika, D., ... Dehaene, S. (2012). Genetic variants of *FOXP2* and KIAA0319/TTRAP/THEM2 locus are associated with altered brain activation in distinct language-related regions. *Journal of Neuroscience, 32*(3), 817–825.

Poliak, S., Gollan, L., Martinez, R., Custer, A., Einheber, S., Salzer, J. L., ... Peles, E. (1999). *Caspr2*, a new member of the neurexin superfamily, is localized at the juxtaparanodes of myelinated axons and associates with K+ channels. *Neuron, 24*(4), 1037–1047.

Poliak, S., Salomon, D., Elhanany, H., Sabanay, H., Kiernan, B., Pevny, L., ... Peles, E. (2003). Juxtaparanodal clustering of Shaker-like K+ channels in myelinated axons depends on Caspr2 and TAG-1. *Journal of Cell Biology, 162*(6), 1149–1160.

Poot, M. (2014). A candidate gene association study further corroborates involvement of contactin genes in autism. *Molecular Syndromology, 5*(5), 229–235.

Poot, M. (2015). Connecting the *CNTNAP2* networks with neurodevelopmental disorders. *Molecular Syndromology, 6*(1), 7–22.

Reader, R. H., Covill, L. E., Nudel, R., & Newbury, D. F. (2014). Genome-wide studies of specific language impairment. *Current Behavioral Neuroscience Reports, 1*(4), 242–250.

Rodenas-Cuadrado, P., Chen, X. S., Wiegrebe, L., Firzlaff, U., & Vernes, S. C. (2015). A novel approach identifies the first transcriptome networks in bats: A new genetic model for vocal communication. *BMC Genomics, 16*(1), 836.

Rodenas-Cuadrado, P., Ho, J., & Vernes, S. C. (2014). Shining a light on *CNTNAP2*: Complex functions to complex disorders. *European Journal of Human Genetics, 22*(2), 171–178.

Rodenas-Cuadrado, P., Pietrafusa, N., Francavilla, T., La Neve, A., Striano, P., & Vernes, S. C. (2016). Characterisation of *CASPR2* deficiency disorder—A syndrome involving autism, epilepsy and language impairment. *BMC Medical Genetics, 17*(1), 8.

Schroeder, D. I., & Myers, R. M. (2008). Multiple transcription start sites for *FOXP2* with varying cellular specificities. *Gene, 413*(1–2), 42–48.

Scott-Van Zeeland, A. A., Abrahams, B. S., Alvarez-Retuerto, A. I., Shimojima, K., Sugawara, M., Shichiji, M., Mukaida, S., Takayama, R., Imai, K., & Yamamoto, T. (2011). Loss-of-function mutation of collybistin is responsible for X-linked mental retardation associated with epilepsy. *Journal of Human Genetics, 56*(8), 561–565.

Shu, W., Yang, H., Zhang, L., Lu, M. M., & Morrisey, E. E. (2001). Characterization of a new subfamily of winged-helix/forkhead (Fox) genes that are expressed in the lung and act as transcriptional repressors. *The Journal of Biological Chemistry, 276*(29), 27488–27497.

Siddiqui, T. J., & Craig, A. M. (2011). Synaptic organizing complexes. *Current Opinion in Neurobiology, 21*(1), 132–143.

Sollis, E., Graham, S. A., Vino, A., Froehlich, H., Vreeburg, M., Dimitropoulou, D., ... Fisher, S. E. (2016). Identification and functional characterization of de novo *FOXP1* variants provides novel insights into the etiology of neurodevelopmental disorder. *Human Molecular Genetics, 25*(3), 546–557.

Sonnenblick, L. I., Rudie, J. D., Ghahremani, D., ... Bookheimer, S. Y. (2010). Altered functional connectivity in frontal lobe circuits is associated with variation in the autism risk gene CNTNAP2. *Science Translational Medicine, 2*(56), 56ra80.

Spiteri, E., Konopka, G., Coppola, G., Bomar, J., Oldham, M., Ou, J., ... Geschwind, D. H. (2007). Identification of the transcriptional targets of *FOXP2*, a gene linked to speech and language, in developing human brain. *American Journal of Human Genetics, 81*(6), 1144–1157.

Strauss, K. A., Puffenberger, E. G., Huentelman, M. J., Gottlieb, S., Dobrin, S. E., Parod, J. M., ... Morton, D. H. (2006). Recessive symptomatic focal epilepsy and mutant contactin-associated protein-like 2. *New England Journal of Medicine, 354*(13), 1370–1377.

Sudhof, T. C. (2008). Neuroligins and neurexins link synaptic function to cognitive disease. *Nature, 455*(7215), 903–911.

Tan, G. C., Doke, T. F., Ashburner, J., Wood, N. W., & Frackowiak, R. S. (2010). Normal variation in fronto-occipital circuitry and cerebellar structure with an autism-associated polymorphism of *CNTNAP2*. *NeuroImage, 53*(3), 1030–1042.

Traka, M. (2003). Association of TAG-1 with Caspr2 is essential for the molecular organization of juxtaparanodal regions of myelinated fibers. *Journal of Cell Biology, 162*(6), 1161–1172.

Turner, T. N., Hormozdiari, F., Duyzend, M. H., McClymont, S. A., Hook, P. W., Iossifov, I., ... Eichler, E. E. (2016). Genome sequencing of autism-affected families reveals disruption of putative noncoding regulatory DNA. *American Journal of Human Genetics, 98*(1), 58–74.

Udden, J., Snijders, T. M., Fisher, S. E., & Hagoort, P. (2016). A common variant of the *CNTNAP2* gene is associated with structural variation in the left superior occipital gyrus. *Brain and Language, 172*, 16–21.

Vabnick, I., Trimmer, J. S., Schwarz, T. L., Levinson, S. R., Risal, D., & Shrager, P. (1999). Dynamic potassium channel distributions during axonal development prevent aberrant firing patterns. *Journal of Neuroscience, 19*(2), 747–758.

Vaishnavi, V., Manikandan, M., Tiwary, B. K., & Munirajan, A. K. (2013). Insights on the functional impact of microRNAs present in autism-associated copy number variants. *PLOS ONE, 8*(2), e56781.

van Aelst, L., & Cline, H. T. (2004). Rho GTPases and activity-dependent dendrite development. *Current Opinion in Neurobiology, 14*(3), 297–304.

van Aelst, L., & D'Souza-Schorey, C. (1997). Rho GTPases and signaling networks. *Genes and Development, 11*(18), 2295–2322.

Varea, O., Martin-de-Saavedra, M. D., Kopeikina, K. J., Schurmann, B., Fleming, H. J., Fawcett-Patel, J. M., ... Penzes, P. (2015). Synaptic abnormalities and cytoplasmic glutamate receptor aggregates in contactin associated protein-like 2/Caspr2 knockout neurons. *Proceedings of the National Academy of Sciences, 112*(19), 6176–6181.

Vernes, S. C., & Fisher, S. E. (2009). Unravelling neurogenetic networks implicated in developmental language disorders. *Biochemical Society Transactions, 37*, 1263–1269.

Vernes, S. C., & Fisher, S. E. (2011). Functional genomic dissection of speech and language disorders. *Genomics, Proteomics, and the Nervous System, 2*, 253–278.

Vernes, S. C., MacDermot, K. D., Monaco, A. P., & Fisher, S. E. (2009). Assessing the impact of *FOXP1* mutations on developmental verbal dyspraxia. *European Journal of Human Genetics, 17*(10), 1354–1358.

Vernes, S. C., Newbury, D. F., Abrahams, B. S., Winchester, L., Nicod, J., Groszer, M., ... Fisher, S. E. (2008). A functional genetic link between distinct developmental language disorders. *New England Journal of Medicine, 359*(22), 2337–2345.

Vernes, S. C., Nicod, J., Elahi, F. M., Coventry, J. A., Kenny, N., Coupe, A. M., ... Fisher, S. E. (2006). Functional genetic analysis of mutations implicated in a human speech and language disorder. *Human Molecular Genetics, 15*(21), 3154–3167.

Vernes, S. C., Oliver, P. L., Spiteri, E., Lockstone, H. E., Puliyadi, R., Taylor, J. M., ... Fisher, S. E. (2011). *Foxp2* regulates gene networks implicated in neurite outgrowth in the developing brain. *PLOS Genetics, 7*(7), e1002145.

Vernes, S. C., Spiteri, E., Nicod, J., Groszer, M., Taylor, J. M., Davies, K. E., ... Fisher, S. E. (2007). High-throughput analysis of promoter occupancy reveals direct neural targets of *FOXP2*, a gene mutated in speech and language disorders. *American Journal of Human Genetics, 81*(6), 1232–1250.

Walker, R. M., Hill, A. E., Newman, A. C., Hamilton, G., Torrance, H. S., Anderson, S. M., ... Evans, K. L. (2012). The *DISC1* promoter: Characterization and regulation by *FOXP2*. *Human Molecular Genetics, 21*(13), 2862–2872.

Wang, B., Lin, D., Li, C., & Tucker, P. (2003). Multiple domains define the expression and regulatory properties of Foxp1 forkhead transcriptional repressors. *Journal of Biological Chemistry, 278*(27), 24259–24268.

Wapinski, O. L., Vierbuchen, T., Qu, K., Lee, Q. Y., Chanda, S., Fuentes, D. R., ... Wernig, M. (2013). Hierarchical mechanisms for direct reprogramming of fibroblasts to neurons. *Cell, 155*(3), 621–635.

Wen, Z., Nguyen, H. N., Guo, Z., Lalli, M. A., Wang, X., Su, Y., ... Ming, G. L. (2014). Synaptic dysregulation in a human iPS cell model of mental disorders. *Nature, 515*(7527), 414–418.

Whalley, H. C., O'Connell, G., Sussmann, J. E., Peel, A., Stanfield, A. C., Hayiou-Thomas, M. E., ... Hall, J. (2011). Genetic variation in CNTNAP2 alters brain function during linguistic processing in healthy individuals. *American Journal of Medical Genetics Part B: Neuropsychiatric Genetics, 156*(8), 941–948.

Whitehouse, A. J., Bishop, D. V., Ang, Q. W., Pennell, C. E., & Fisher, S. E. (2011). *CNTNAP2* variants affect early language development in the general population. *Genes, Brain, and Behavior, 10*(4), 451–456.

Williams, J. M., Beck, T. F., Pearson, D. M., Proud, M. B., Cheung, S. W., & Scott, D. A. (2009). A 1q42 deletion involving *DISC1*, DISC2, and TSNAX in an autism spectrum disorder. *American Journal of Medical Genetics Part A, 149A*(8), 1758–1762.

Xie, H. R., Hu, L. S., & Li, G. Y. (2010). SH-SY5Y human neuroblastoma cell line: In vitro cell model of dopaminergic neurons in Parkinson's disease. *Chinese Medical Journal, 123*(8), 1086–1092.

Zheng, F., Wang, L., Jia, M., Yue, W., Ruan, Y., Lu, T., ... Zhang, D. (2011). Evidence for association between Disrupted-in-Schizophrenia 1 (*DISC1*) gene polymorphisms and autism in Chinese Han population: A family-based association study. *Behavioral and Brain Functions, 7*(1), 14.

42 The Genetic Bases of Brain Lateralization

CLYDE FRANCKS

1. Lateralizations of Brain and Behavior

Lateralization of the human central nervous system begins early in development (figure 42.1). In a study of 72 fetuses at 10 weeks' gestational age, using in utero ultrasound scanning, Hepper, McCartney, and Shannon (1998) reported that the majority (85%) moved their right arms more than their left arms. This is consistent with the adult predominance of right-handedness and points to an embryonic precursor of this landmark behavioral lateralization (Hepper, 2013; Hepper et al., 1998). At 11 weeks of gestation, the choroid plexuses, which are highly vascularized structures that control the composition of cerebrospinal fluid in the brain's lateral ventricles, show an average leftward asymmetry of size (Abu-Rustum, Ziade, & Abu-Rustum, 2013), again based on in utero ultrasound scanning. This asymmetry may affect broader lateralized development of the brain via the secretion of diffusible signaling molecules into the ventricles (Corballis, 2013; Lehtinen et al., 2013). Slightly later in development, in utero ultrasound scanning of 274 human fetuses aged from 15 weeks of gestation showed a population-level preference for sucking of the right thumb (Hepper, Shahidullah, & White, 1991). In fact, the individual preference for fetal thumb sucking was also shown to be predictive of handedness aged 12 years, in follow-up analysis of 75 of the same subjects (Hepper, 2013; Hepper, Wells, & Lynch, 2005). Furthermore, anatomical brain lateralizations, including of perisylvian regions of the cerebral cortex that are important for language, have been reported from the second trimester and onward through fetal and infant development, with methodologies that have included ultrasound, *postmortem* analysis, and MRI (Chi, Dooling, & Gilles, 1977; Dubois et al., 2008; Fagard, 2013; Hering-Hanit, Achiron, Lipitz, & Achiron, 2001; Holland et al., 2014; Kasprian et al., 2011; G. Li et al., 2013; Y. Liu et al., 2010; Wada, Clarke, & Hamm, 1975; Witelson & Pallie, 1973).

These lateralizations of central nervous system (CNS) and limb activity in utero point strongly to a lateralized genetic-developmental program. In addition, varying degrees of lateralized brain activity have been recorded in adults during the performance of diverse cognitive tasks, including those related to language (figure 42.2), visuospatial cognition, and hand motor control (Gotts

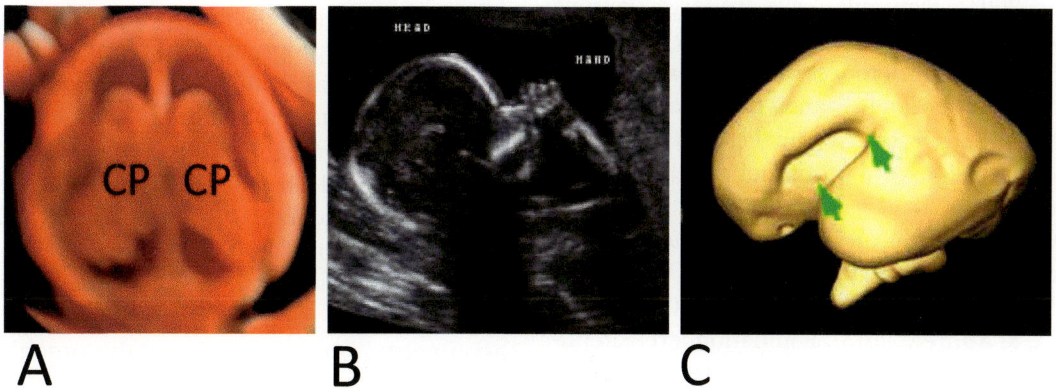

Figure 42.1 Asymmetrical development of the human brain and behavior in utero. (A) The left choroid plexus (CP) is 8% larger than the right one, on average, in fetuses at gestational age of 11 weeks. (B) At 15 weeks, most fetuses perform a right-handed "thumb sucking" behavior. (C) By mid-gestation, the anatomy around the sylvian fissure (highlighted by green arrows) has become left-right asymmetrical. See text for references. (A) adapted from Corballis (2013), (B) retrieved from https://www.pregmed.org/pregnancy-week-by-week/15-weeks-pregnant, (C) adapted from https://embryology.med.unsw.edu.au/embryology/index.php/Human_Sylvian_Fissure_Movie.

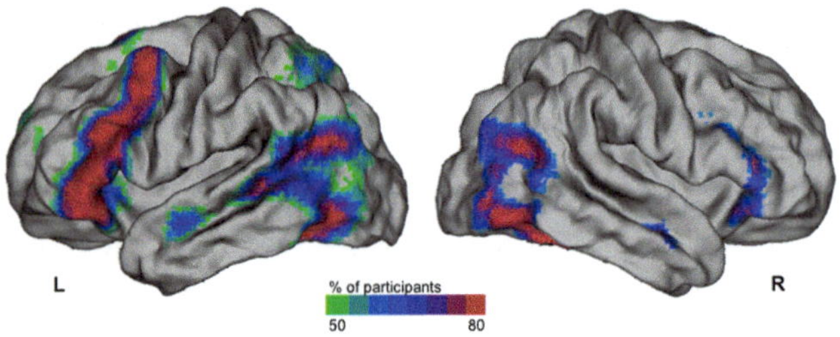

FIGURE 42.2 Laterality of brain activation during language production, based on 144 right-handed adults. Activation is more extensive in the left hemisphere. The task contrast used in this fMRI experiment was covert sentence-level versus word-list production. Figure reproduced, under the Creative Commons Attribution (CC BY) license, from Mazoyer et al. (2014).

et al., 2013; Hervé, Zago, Petit, Mazoyer, & Tzourio-Mazoyer, 2013; Rentería, 2012; Tomasi & Volkow, 2012). Lateralized functions imply left-right differences in the activity of proteins that modify the information-processing properties of neural circuits. Furthermore, syllabic speech rhythms have been reported to correspond to left-hemispheric neural oscillatory frequencies, in a manner that may preferentially support auditory and language processing in the left hemisphere (Morillon et al., 2010). Such neurophysiological lateralization is presumably supported at the molecular-genetic level by genes whose abundances determine the signaling properties of neuronal circuitry, including classes of genes involved in synaptogenesis, neurotransmission, and synaptic cell adhesion (Francks, 2011; Margeta & Shen, 2010). It has also been observed, by postmortem analysis of the auditory and other regions of the adult temporal lobe, that left superficial layers of the cortex contain a greater number of large pyramidal cells than right layers do (Hutsler, 2003). Pyramidal cells are large neurons involved in synaptic integration and plasticity (Spruston, 2008). Such left-right differences of microanatomy are a further indication that lateralization of gene activity is to be expected, since different neuronal classes are known to have their own signature profiles of gene expression (Zeisel et al., 2015).

In fact CNS lateralization is not unique to humans, being a feature of many vertebrate clades (Ocklenburg & Güntürkün, 2012; Rogers & Andrew, 2002; Rogers, Vallortigara, & Andrew, 2013). Population-level lateralization manifests, for example, in direction-biased turning behavior of schooling fish and reactions to visual stimuli in chicks (Ocklenburg & Güntürkün, 2012). Some crucial developmental events underlying lateralization have been elucidated in bird and fish species (Concha, Bianco, & Wilson, 2012; Concha, Signore, & Colombo, 2009; Ocklenburg & Güntürkün, 2012), and in zebrafish, in particular, characterization of CNS lateralized development at the molecular level is at a relatively advanced state (Concha et al., 2012). Asymmetrical development of the zebrafish forebrain involves the epithalamus, a structure of the dorsal posterior diencephalon, which migrates away from its embryonic origin at the midline, toward the left side (Concha et al., 2012). The epithalamus then innervates asymmetrically and influences broader CNS development in a lateralized manner (Concha et al., 2012). The genetic-developmental program that controls this process is linked to the same molecular factors that set up left-right lateralization of the viscera (e.g., heart, lungs), which include the Nodal signaling molecule (Concha et al., 2012). In *Xenopus* too, lateralization of tadpole swimming behavior is linked to visceral organ lateralization (Blackiston & Levin, 2013).

However, it is not clear how closely the mechanisms of lateralized brain development in fish, birds, and amphibians are related to those in humans. Humans with the rare genetic condition *situs inversus*, involving a mirror reversal of visceral asymmetries on the left-right axis, have shown normal population rates of right-handedness and left-lateralized language dominance, in the largest studies of these kinds to have been performed (McManus, Martin, Stubbings, Chung, & Mitchison, 2004; Tanaka, Kanzaki, Yoshibayashi, Kamiya, & Sugishita, 1999). Therefore an early developmental dissociation is suggested between visceral and brain asymmetries in humans, at least as regards handedness and language lateralization. This contrasts with the process of epithalamus-driven lateralized development of the zebrafish forebrain and means that the earliest developmental origins of human brain asymmetry remain mysterious. Regardless, the studies of zebrafish have clearly illustrated the principle that lateralized genetic-developmental programs can create brain structural and functional asymmetries, and therefore the search for such mechanisms in humans is strongly motivated. Note that here we are concerned with mechanisms of population-level lateralization, whereas there are many examples of lateralities that occur within species equally frequently in their leftward and rightward forms (Rogers et al., 2013; Vallortigara & Rogers, 2005). Individual-level lateralities presumably occur when asymmetry is advantageous for the organism but the direction is not important (Rogers et al., 2013; Vallortigara & Rogers, 2005).

Some mammalian species may prove to be useful models for understanding aspects of the genetics and development of human brain lateralization. Mice have not been widely reported to have population-level asymmetries of brain structure or function, but lateralization has recently been reported using imaging in vivo (Apostolova et al., 2012; Spring, Lerch, Wetzel, Evans, & Henkelman, 2010), as well as molecular-level asymmetries in the hippocampus, which affect learning and memory (Goto et al., 2010; Kawakami et al., 2003). Weak population-level paw preference has also been observed in inbred mice during reaching tasks, although these lateralizations required large samples to detect them, and they varied in leftward versus rightward direction depending on the task (Waters & Denenberg, 1994). Rats have been reported to show a stronger population-level bias (73% right paw preference) (Güven, Elalmis, Binokay, & Tan, 2003) than mice, as well as hemispheric differences in spatial cognition (LaMendola & Bever, 1997) and proteomic lateralization in the hippocampus (Samara et al., 2011). Great apes have also shown evidence for population-level handedness, albeit at levels much weaker than for humans, and also some structural brain lateralizations similar to those found in regions important for language in humans (Cantalupo et al., 2009; Hopkins, 2013; Lyn et al., 2011; Meguerditchian, Vauclair, & Hopkins, 2013). However, research with nonhuman primates, particularly apes, is restricted ethically and legally, while human brain lateralizations linked closely to language may not manifest sufficiently, or at all, in rodents. Therefore, genetic studies that have a direct focus on human tissues and traits are critical for making progress in this field.

2. *Asymmetrical Gene Expression in Human Brain Development*

The most well-studied lateralizations of the human brain involve the cerebral cortex, and several attempts have been made to identify genes that are asymmetrically active in this tissue, either during development or in adulthood (Hawrylycz et al., 2012; Johnson et al., 2009; Lambert et al., 2011; Pletikos et al., 2014; Sun et al., 2005). For these studies, postmortem tissue samples have been carefully dissected shortly after donor death, and the levels of messenger RNA (mRNA) of thousands of genes was measured simultaneously, using an approach called *transcriptomic profiling*. The level of mRNA of any specific gene within a tissue is an imperfect indication of the amount of protein encoded by that gene. Sun et al. (2005) studied the cerebral cortices of human fetuses at 12–19 weeks, using a transcriptomic technique called *serial analysis of gene expression*, and found higher right-than-left mRNA levels of the transcription factor *LMO4*[1] at the earlier developmental stages, which was not detected at 19 weeks. Transcription factors are proteins that regulate the mRNA expression of other genes and can influence many cellular and developmental processes. *LMO4* was more recently shown to affect neurogenesis and axonal projection (Q. Li et al., 2013). Induced unilateral knockdown of the *Lmo4* mRNA level in one embryonic mouse brain hemisphere, *in utero*, caused suppression of neurogenesis in that hemisphere, which resulted in asymmetries of neuronal production, functional area formation, and axonal projection (Q. Li et al., 2013). Later, 12-week-old mice that had been treated in this way as embryos showed behavioral lateralizations, including of turning during swimming, and paw preference (Q. Li et al., 2013). Nontreated mice showed no population-level lateralization of these behaviors, as well as little individual-level lateralization (Q. Li et al., 2013). However, asymmetrical developmental outcomes would seem likely to arise from unilateral manipulation of a cortically expressed transcription factor that affects processes such as neurogenesis and functional area formation, even if that transcription factor is not naturally important for lateralized development. Furthermore, lateralization of *LMO4* mRNA in the human fetus was statistically tentative in the original study and has yet to be replicated in the literature. This therefore remains a key finding that is in need of confirmation.

Using a more up-to-date technology, that is, microarray transcriptomics, Lambert et al. (2011) did not identify significant asymmetries of gene expression in frontal or temporal cortical tissue from human fetuses aged 17 and 19 gestational weeks. Pletikos et al. (2014) also used microarrays to study postmortem neocortical regions, this time all across the human life span from embryo to old age, but again they did not find significant evidence for differential left-right gene expression, either at the level of individual genes or in terms of the changes observed in gene expression over time. Another microarray-based postmortem study, by Johnson et al. (2009), also did not identify significant lateralization of cortical mRNA expression in tissue taken from mid-fetal human brains aged between 18 and 23 weeks of gestation. A recent expression-profiling study from adult brain tissue also did not identify asymmetrically expressed genes (Hawrylycz et al., 2012).

However, each of these transcriptomic studies was based on only tiny numbers of postmortem samples from any given stage of development. Such studies are severely limited by the availability of tissue samples, as well as the expense of transcriptomic profiling techniques. For the transcriptomic-screening stage of their study, Sun et al. (2005) used tissue from only two fetuses at 12 weeks, two at 14 weeks, and one at 19 weeks; Lambert et al. (2011) analyzed one fetus at 17 weeks and one fetus at 19 weeks; Johnson et al. (2009) analyzed four mid-fetal brains; and Hawrylycz et al. (2012) analyzed two adult brains for which data from both hemispheres

TABLE 42.1
Studies of gene expression in human embryonic and fetal cerebral cortex

Study	Age range (gestational weeks)	Sample size	Technology	Asymmetrical gene expression
Sun et al. (2005)	12–19*	5	SAGE	LMO4 and others
Lambert et al. (2011)	17–19	2	Microarray	None found
Johnson et al. (2009)	18–23	4	Microarray	None found
Pletikos et al. (2014)	14–40	20†	Microarray	None found

*It was not stated in the study whether the age range referred to postconception or gestational weeks (gestational weeks are measured from the last reported menstruation and therefore roughly two weeks prior to conception).

†Left-right differential expression analysis was performed within age-restricted subsets of samples.

SAGE = serial analysis of gene expression.

were available. The study by Pletikos et al. (2014) was the most substantial in terms of sample size, being based on 57 brains spanning the life span, but again the number of brains at any given stage of development averaged less than one for the prenatal material and was similarly low for the infant, childhood, and teenage samples. In addition, for testing lateralized gene expression, Pletikos et al. (2014) used a maximum sample size of four brains that were grouped by consecutive ages, including for the adult samples (see table 42.1).

It follows that none of these transcriptomic studies was well powered, in statistical terms, to detect subtle contrasts of left-right gene expression. In addition, these studies involved individually testing thousands of genes and performing false-discovery-rate correction, which necessarily meant that they were not powered to detect asymmetries of less than 1.5-fold expression level for a given gene, in the small sample sizes that were used. Yet, expression differences of 1.5-fold magnitude may be functionally relevant, especially when considered over multiple genes interacting together in networks to influence neuronal and circuit properties.

3. Lateralization of Cerebral Cortical Gene Expression in Adult Brains

In light of the limited power of previous studies of asymmetrical gene expression, Karlebach and Francks (2015) recently reanalyzed some of the adult cerebral cortical gene expression data of Pletikos et al. (2014) and Hawrylycz et al. (2012), using various techniques to increase the statistical power to detect left-right differences, which had not been previously applied. First, data from the posterior superior temporal cortex (corresponding to Brodmann's area [BA] 22), as well as from the primary auditory cortex (BA41) (see figure 42.3), were specifically targeted, as lateralizations of these regions have been reported in terms of function, neurophysiology, gross anatomy, and histological microanatomy (as discussed in section 1). Data from Pletikos et al. (2014) for all 13 adults within the age range 18–55 years were entered into a single analysis without subdividing by age, followed by meta-analysis with the data of Hawrylycz et al. (2012). Bayesian smoothing of gene expression variance estimates was used to aid statistical testing in these small data sets. Lateralization was tested at the level of individual genes, but also at the level of functional gene sets defined according to Gene Ontology classifications, by which gene products are grouped hierarchically according to molecular functions, biological processes, and cellular components (Ashburner et al., 2000). Through applying these data-analytic approaches, robust evidence for left-right differences were found for BA22, which manifested most strongly at the level of sets of genes that are involved in synaptic transmission,

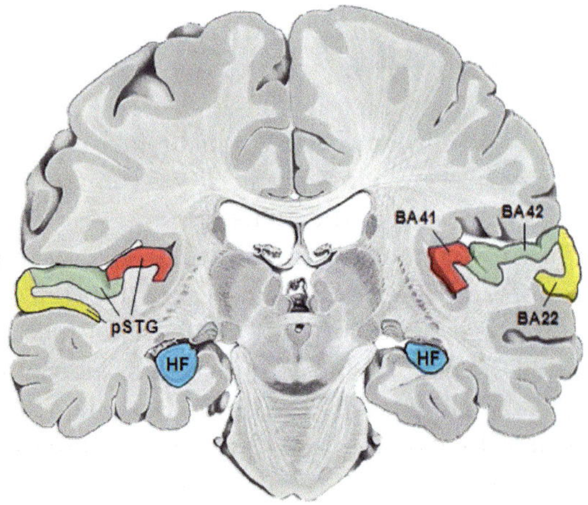

FIGURE 42.3 Regions of the posterior superior temporal gyrus (pSTG) of the adult human cerebral cortex, including BA22, which showed lateralization of gene expression levels in the study by Karlebach and Francks (2015). Lateralization was detected for genes involved in membrane receptor activity and synaptic functions (among others). A coronal section of the brain is shown. HF = hippocampal formation. Figure reproduced, under the Creative Commons Attribution 2.5 Generic license, from Talbot et al. (2011).

signal transduction, glutamate receptor activity, and transmission of nerve impulses, all showing relative up-regulation in the left-hemispheric region (Karlebach & Francks, 2015). The genes involved have neuronal functions that are likely to affect signaling, learning, and information-processing properties of circuitry. Interestingly, even though these data were from adult brains, there was also lateralization of gene sets that are defined for their developmental roles, which suggests that transcription factors and other developmentally important proteins have roles in maintaining lateralized function in the adult brain. Overall, the findings of Karlebach and Francks (2015) indicate that the combinatorial effects of subtle, quantitative left-right differences, over many genes, are likely to fine-tune neurophysiological outcomes differently in the two hemispheres and underpin lateralized cortical functions.

4. Heritability of Brain and Behavioral Asymmetries

A very different approach to identify genes involved in human brain lateralization is to correlate genetic polymorphisms in the population with interindividual differences in structural or functional brain asymmetries or lateralized behaviors. The summed effects of all such polymorphisms in the genome indicate the heritability, that is, the proportion of population variation in a trait that is caused by genetic differences. Heritability can be measured through studies of twins, in which monozygotic pairs of twins show how similar individuals are when they are genetically identical, by contrast with dizygotic pairs of twins who share, on average, half of their chromosomes identical-by-descent.

So far there have only been a small number of twin studies of asymmetries of human brain structure, function, or behavior, and they have generally shown evidence for zero-to-modest heritability (Badzakova-Trajkov, Häberling, & Corballis, 2010; Bishop, 2013; Häberling, Badzakova-Trajkov, & Corballis, 2013; Jahanshad et al., 2010; Medland et al., 2009; Steinmetz, Herzog, Schlaug, Huang, & Jäncke, 1995). The heritability of left-handedness, for example, was estimated at close to 24%, in a large meta-analysis study that involved data from more than 25,000 families with twins (Medland et al., 2009). In other words, when one twin in a pair was left-handed, the other twin was significantly (but only slightly) more likely also to be left-handed when the pair was monozygotic than when the pair was dizygotic. The large sample size of this study meant that the weak heritability was accurately measured, with a tight confidence interval. It is therefore clear that genomic variation has a weak effect on the probability of becoming left-handed, while it is also clear that environmental effects and/or random effects during development are largely responsible (additionally there was no evidence that shared rearing environments for given pairs of twins have an effect).

By contrast with handedness, heritability studies of brain structural and functional asymmetries have only been performed in low hundreds of twin pairs, rather than thousands, and therefore the heritability estimates have been much less accurately measured. In addition, a large number of different imaging-based measures of asymmetry have sometimes been derived within individual studies, leading to the issue of multiple testing, which is difficult to control adequately in smaller samples. Eyler et al. (2013) found significant heritabilities of regional cortical areas and thicknesses in a study of 130 monozygotic twin pairs, 97 dizygotic pairs, and 61 unpaired twins, using automated segmentation of MRI images. However, their data indicated that left-right homologous regions of the two hemispheres share most or all of the genetic contributions to their variances, and there was little evidence for genetic effects that were different between the hemispheres and that would contribute to asymmetry (Eyler et al., 2013). In a study of 374 human twins, Jahanshad et al. (2010) used diffusion tensor imaging (DTI) to study nerve fiber bundles. They found the heritability of asymmetry indexes to range from 0% to 47%, depending on the particular fiber tract and DTI-based metric of white matter integrity. Frontal and temporal regions showed the most significant population-level asymmetries (Jahanshad et al., 2010), and genetic factors accounted for 33% of the variance in asymmetry in the inferior fronto-occipital fasciculus, 37% of the variance in the anterior thalamic radiation, and 20% of the variance in the forceps major and the uncinate fasciculus (Jahanshad et al., 2010).

As regards measures of brain function, hemispheric language lateralization was shown to have a heritability of 31% using functional transcranial Doppler in families from the Netherlands (Somers et al., 2015). This study used families rather than monozygotic versus dizygotic twins, and it therefore remains possible in principle that part of the apparent heritability was driven by shared environmental influences on relatives from the same family, although this seems unlikely since shared environmental effects did not prove significant in the studies of handedness or brain asymmetries that we have discussed.

In general, large nonshared environmental or random effects on brain asymmetries are indicated by these studies, in addition to low-to-moderate heritabilities, as is the case for handedness. Larger imaging studies of twins will be required to more accurately assess the degree to which brain structural and functional asymmetries are heritable.

5. *Viscera and the Principle of Randomization on the Left-Right Axis*

Visceral lateralization (of the heart, lungs, and such) illustrates some important principles of asymmetrical development that are likely to be informative for studies of brain asymmetry. When early developmental mechanisms underlying visceral asymmetry are disrupted by certain genetic mutations, the direction of asymmetrical development can become randomized: half of mutation carriers develop visceral asymmetry in the normal orientation, and half develop the mirrored form situs inversus (Sharma, Berbari, & Yoder, 2008) (see figure 42.4). In humans this condition has a population frequency of roughly 1 in 10,000, and it can be induced by mutagenesis or gene knockouts in various vertebrate species. The typical pattern of lateralized visceral development likely has its origins in asymmetrical motions of protein cilia located on the ventral surface of very early mammalian embryos (Shinohara et al., 2012). Cilia rotate predominantly in only one of two theoretically possible orientations, due to their protein components being constructed by chiral amino acid molecules (Shinohara et al., 2012). Beating of the cilia apparently causes a unidirectional flow of fluid within a pitted embryonic structure called the node, resulting in mechanical and/or chemical differences between the left and right sides (Yoshiba et al., 2012) that are thought to trigger differential gene expression. Lateralization may even be initiated earlier than this, by molecular chirality of subcellular components such as cytoskeletal elements (Burdine & Caspary, 2013; Levin & Palmer, 2007). Primordial left-right differences are then amplified by differential gene activity into distinct developmental fates for the left and right sides of the embryonic viscera (Levin, 2005; Shinohara et al., 2012). In situs inversus with primary ciliary dyskinesia, mutations in genes encoding protein components of the nodal cilia, or other genes functionally related to these, result in a loss of unidirectional fluid flow or its detection, and thus a lack of consistency in the direction of asymmetrical development. Left-right differentiation of the viscera still proceeds in the embryo, but it is triggered with an equal probability in either orientation, probably by random and slight asymmetrical fluctuations of key developmental gene activities (Concha et al., 2009; Levin, 2005).

Randomization that results from the genetic loss of consistent, direction-giving mechanisms early in development has also been considered in relation to left-handedness and other aspects of human brain lateralization (Annett, 1985; Klar, 1999; McManus & Bryden, 1992; McManus, Davison, & Armour, 2013). However, there is no direct evidence for this, since core genes and mechanisms that give rise to human brain lateralization are unknown. As mentioned, people with situs inversus and primary ciliary dyskinesia have normal population proportions of left-handedness and left-lateralized language dominance, which suggests a dissociation of visceral lateralization from at least some aspects of cerebral lateralization, in terms of their embryonic development. Regardless, the concept of randomization in the direction of brain asymmetry, which arises from disruption of normally lateralized genetic-developmental programs, is consistent with the weak heritability estimates for brain asymmetries and lateralized behaviors that have been measured in studies of twins and families. In these studies, any random contribution to trait variability is confounded with the nonshared environmental component of variance that is estimated.

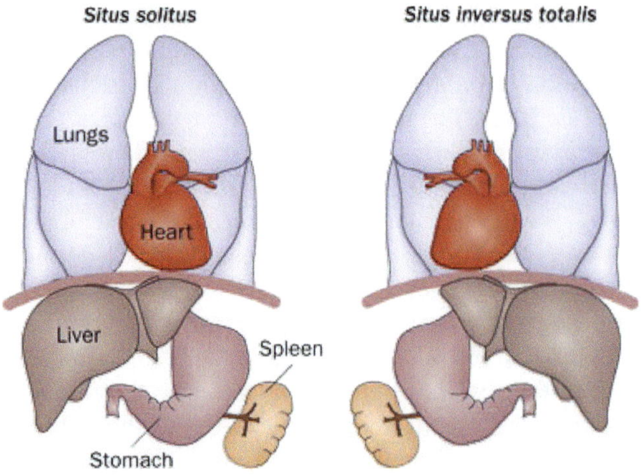

FIGURE 42.4 Mirror reversal of the viscera (situs inversus) can result from genetic mutations that disrupt the functions of cilia in the very early embryo. It is unclear whether the same fundamental mechanisms of left-right patterning are involved in all brain lateralizations, since people with situs inversus due to ciliary mutations appear to have similar proportions of right-handedness and left-hemisphere auditory language dominance as the general population (see text). Picture reproduced from Patel and Honoré (2010).

6. *Cognitive Performance and Plasticity*

Variability in lateralized brain structures, functions, or behaviors has been shown to weakly associate with cognitive or behavioral performance in some studies, including with verbal ability and scholastic achievement

(Badzakova-Trajkov, Häberling, & Corballis, 2011; Björk, Brus, Osika, & Montgomery, 2012; Boles, Barth, & Merrill, 2008; Catani et al., 2007; Gotts et al., 2013; Groen, Whitehouse, Badcock, & Bishop, 2013; Kikuchi et al., 2011; Leask & Crow, 2001; Mellet et al., 2013; Prichard, Propper, & Christman, 2013). For example, Björk et al. (2012) observed an association between mixed-handedness and slightly reduced performance on school tests, including tests of verbal ability and mathematics, in a British birth cohort of 10,612 children. This effect was limited to those children who scored within the lower third on a measure of right-hand motor performance, suggesting interactions between variances in motor skill, lateralization, and cognition. Catani et al. (2007) found that individuals with more symmetric patterns of white matter connections in the perisylvian language network were better at remembering words using semantic association, although this was a DTI-based study of only 50 participants. If such findings are correct, it appears that some specific aspects of cognition can benefit from relatively more bilateral organization, while general academic performance may benefit from relatively stronger lateralization. In general, however, reorganizations of lateralized brain structure and function, such as left-handedness and reversed language lateralization, can obviously occur developmentally without major consequences for cognitive or behavioral performance (Mazoyer et al., 2014; Mellet et al., 2013; Willems, der Haegen, Fisher, & Francks, 2014).

It has also become clear in recent years that the variances in different aspects of brain-asymmetrical structure and function are often largely uncorrelated with each other (Badzakova-Trajkov, Häberling, Roberts, & Corballis, 2010; Bishop, 2013; Guadalupe, Willems, et al., 2014; Guadalupe, Zwiers, et al., 2014; Knecht et al., 2002; Liu, Stufflebeam, Sepulcre, Hedden, & Buckner, 2009; Mazoyer et al., 2014; Rentería, 2012). For example, handedness and lateralized language dominance are only weakly related, which has been found using both functional MRI (fMRI) and functional transcranial Doppler sonography (Knecht et al., 2000; Mazoyer et al., 2014). Liu et al. (2009) analyzed intersubject variance in lateralized brain activity using resting-state fMRI in 300 participants and found four separately lateralized factors: systems involved in vision, internal thought, attention, and language. These observations, together with the generally weak associations of altered lateralization with cognitive performance, imply a high degree of developmental plasticity of lateralization on a brain-regional and process-specific basis. Either hemisphere is apparently able to become dominant for any given function, especially if the requirement to do so is initiated early enough in development. This plasticity is consistent with the data on lateralized gene expression that were already discussed, which indicated only subtle, quantitative variations on what are bilaterally homologous themes at the molecular level and which are likely to be developmentally readjustable.

At the same time, many studies have found cognitive and psychiatric disorders to be modestly associated with alterations of brain asymmetry and/or lateralized behavior (Eyler, Pierce, & Courchesne, 2012; Floris et al., 2013; Herbert et al., 2002; Herbert et al., 2005; Kawasaki et al., 2008; Lindell & Hudry, 2013; McCarley et al., 2002; Mock et al., 2012; Preslar, Kushner, Marino, & Pearce, 2014; Seidman et al., 2002; Shenton, Dickey, Frumin, & McCarley, 2001; Somers, Sommer, Boks, & Kahn, 2009; Sommer, Ramsey, Kahn, Aleman, & Bouma, 2001; Tsuang, Chen, Kuo, & Hsiao, 2013), including schizophrenia, autism, dyslexia, and language impairment, although not in all populations affected with these disorders (Deep-Soboslay et al., 2010; Preslar et al., 2013). A comprehensive meta-analysis study published in 2001 found that schizophrenia was associated with mixed- and left-handedness and also with reductions of structural lateralization of the planum temporale and sylvian fissure (Sommer et al., 2001). Both findings have subsequently received further support (Kawasaki et al., 2008; Somers et al., 2009; Tsuang et al., 2012). People with autism spectrum disorders have been reported to show changes of cortical structure, handedness, and functional lateralization for language (Lindell & Hudry, 2013). Genetic variations and environmental influences that contribute to these disorders may therefore affect brain lateralized development and function, although cause-effect relationships between altered lateralization and disorders are not currently understood.

7. *Genetic Association Studies*

Genetic association studies have identified individual polymorphisms within certain genes and genetic networks that may have modifying effects on brain or behavioral asymmetries (Arning et al., 2013; Brandler et al., 2013; Francks et al., 2002; Francks et al., 2003; Francks et al., 2007; Medland et al., 2005; Ocklenburg et al., 2013a, 2013b; Ocklenburg et al., 2011; Ocklenburg, Beste, & Güntürkün, 2013; Scerri et al., 2011). Measures used in these studies have included indices of lateralized hand motor skill (Francks et al., 2007), binary-trait hand preference (Medland et al., 2005), and lateralization of auditory language dominance as assessed by dichotic listening (Ocklenburg et al., 2011). The implicated genes have functions including steroid hormone biology (*AR*) (Medland et al., 2005), synaptic adhesion (*LRRTM1*) (Francks, 2011), glutamatergic neurotransmission (*GRIN2B*) (Ocklenburg et al., 2011), transcriptional regulation (*FOXP2*) (Ocklenburg et al., 2013b), dopamine release (*CCKAR*) (Ocklenburg et al., 2013a), and left-right lateralization of the viscera (*PCSK6*) (Brandler et al., 2013; Scerri et al., 2011). However, these studies were based on data sets of hundreds of individuals and were therefore not well powered to establish effects of individual, common polymorphisms on etiologically heterogeneous and complex traits (Altshuler, Daly, & Lander, 2008; Sham & Purcell, 2014). Each of the findings therefore remains tentative. Studies using thousands of participants will be required (Sham & Purcell, 2014) to reliably pinpoint individual, common genetic effects on brain asymmetry measures, especially given the generally low heritabilities of these traits (see section 4). Recent success has been achieved for various multifactorial human traits through the use of genetic association studies in tens of thousands of participants, including for human height and body mass index, and complex diseases such as diabetes and schizophrenia (J. Z. Liu et al., 2010; Welter et al., 2014). However, the only genetic studies of brain asymmetries performed at something approaching this scale were two recent genome-wide association study meta-analyses, which were both based on just over 3,000 subjects (Guadalupe, Zwiers, et al., 2014; Guadalupe et al., 2015) (see figure 42.5). One of these studies found that structural lateralization within and around the planum temporale is sexually dimorphic and associated with genes involved in steroid hormone biology (Guadalupe et al., 2015). The other study analyzed volume asymmetry of the caudate nucleus but did not find genetic polymorphisms associated with individual differences in this measure (Guadalupe, Zwiers, et al., 2014).

It is not clear how a reported association between hand motor skill asymmetry and polymorphisms within visceral asymmetry genes (Brandler et al., 2013) is consistent with the dissociation of situs inversus from handedness that was discussed in section 5. Some elements of visceral asymmetrical development may in fact be shared with those that set up human brain asymmetries, and indeed, situs inversus in mice has been reported to affect molecular lateralization in the hippocampus (Kawakami, Dobi, Shigemoto, & Ito, 2008). However, the human genetic association data still require replication in independent data sets.

When associations are discovered between common genetic polymorphisms and brain-asymmetrical traits or lateralized behaviors, the genes that are implicated can, in principle, be either slight modifiers of asymmetrical outcomes or else essential for patterning developmental lateralization. Genetic association studies using common polymorphisms, in which effects are expected to be small, should also be complemented by investigating families that show unusually high rates of atypical lateralization of structure, function, or behavior, for example, an unusually high rate of left-handedness (Willems et al., 2014). Such families may carry genetic

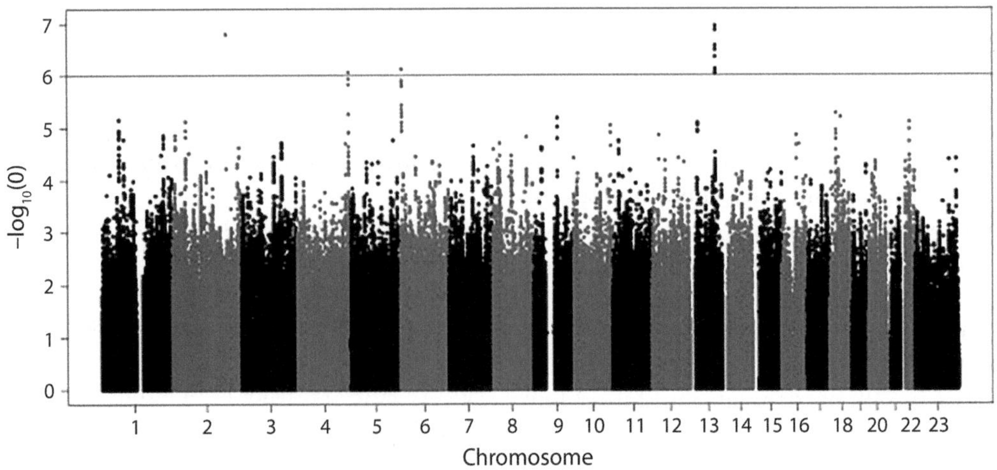

FIGURE 42.5 Genome-wide scan for polymorphisms associated with an index of gray matter asymmetry, measured within and around the planum temporale (cerebral cortical region). The *x*-axis represents the chromosomes laid end-to-end, from short to long arms, in ascending numerical order from left to right. The *y*-axis shows the significance of association for each of over 2.5 million individual polymorphisms located at unique points in the genome. Shading represents the different chromosomes. The horizontal line represents the threshold used for defining suggestive association ($P = 1 \times 10^{-6}$). No result reached genome-wide significance ($P = 5 \times 10^{-8}$), but genes involved in steroid hormone biology were found to have an enrichment of low association *P* values, within the overall distribution of signals. Figure reproduced from Guadalupe et al. (2015).

mutations that are rare in the population but that have large effects on lateralized brain development when they occur (McManus et al., 2013; Willems et al., 2014). Genes identified through studying such families would be likely to have key roles in setting up brain lateralization, rather than having downstream, modifying effects on lateralized traits. Epigenetic effects on brain lateralization, owing to variation in the structure and function of chromosomes that is not attributable to DNA polymorphisms, are also a possibility (Brucato, DeLisi, Fisher, & Francks, 2014). Epigenetic variation involves chromosomal properties such as DNA methylation (Gordon et al., 2012) or chemical modifications of proteins that associate with DNA in the cell nucleus (Gräff & Mansuy, 2008), and this variation can be caused heritably, environmentally, or randomly.

8. Reconciling Theme with Variations: Are Strong Genetic Effects on Lateralization Compatible with Weak Heritabilities?

As we've discussed, lateralization at the population level is characteristic of many aspects of human brain structure, function, microanatomy, neurophysiology, and behavior, right across the life span. For example, handedness and language dominance are lateralized at the population level at a rate of 90% or more. Lateralized genetic-developmental programs are required to create differently adjusted properties of neural circuitry in the two hemispheres. Such molecular programs have been described for the brains of fish and for the viscera of humans and other species. Furthermore, hemispheric differences in adult gene expression that involve multiple individual genes, such as those identified by Karlebach and Francks (2015), are likely to underlie functional lateralization for language and other aspects of cognition. It is therefore clear that genes have a major role in setting up and maintaining the brain lateralizations that are found in the majority of people.

A strong genetic effect on a lateralized trait's asymmetric mean in the population can be reconciled with a weak genetic effect on its variance (i.e., low heritability) if lateralized developmental programs become randomized in their directions in response to environmental or genetic disruptions. The notion of randomization in response to disruption is supported by the known effects of certain genetic mutations that cause situs inversus of the viscera with 50% probability. The picture can be further complicated by genetic mutations that affect processes downstream in development from the primary direction-setting mechanisms. Again the viscera can illustrate the various genetic mutations affecting visceral asymmetry that cause complex and partial disruptions of laterality known as *heterotaxias* (Peeters & Devriendt, 2006). In these conditions, specific organs or clusters of organs become located abnormally on the left-right axis and are sometimes malformed (Peeters & Devriendt, 2006). While the mirror reversal in situs inversus can have no direct medical consequences, misplacement or malformation of organs in heterotaxias often has health implications (Peeters & Devriendt, 2006). In terms of brain lateralization, although the key developmental mechanisms have not been identified, genetic variation is likely to affect the degree of lateralization in addition to its direction on the left-right axis (Arning et al., 2013). As for the viscera, functional consequences in the brain (for cognitive performance) are most likely to occur when lateralization is incomplete, or when certain functions become dissociated with respect to each other, rather than when lateralization develops completely in the reverse direction to the typical form. Observations regarding academic performance and mixed hand dominance support this notion (see section 6). Again by analogy with heterotaxias, distinct lateralized traits such as language dominance and hand preference, which develop consistently with respect to each other in the typical brain, might become dissociated as a result of disruptions of a lateralized genetic-developmental program. Such traits would then appear to be largely uncorrelated in the population, even though they may stem from overlapping mechanisms in typically developing people. Furthermore, it seems likely that some early developmental disruptions, whether they are environmentally mediated, genetic, or random in nature, might be localized to specific brain regions or networks. This could then give rise to an array of different outcomes, depending on how early or late they occur during development.

9. Future Research and Implications

For the reasons discussed herein, further research on the genetic bases of brain lateralization will require complementary approaches that are focused on both the *majority average form* and also on lateralized *trait variances*. A powerful method to study the mean form is transcriptomic analysis of left and right CNS regions in postmortem human tissue from a necessarily limited number of donors, whereas trait variance can best be studied in data sets of thousands of participants, using, for example, genome-wide association scanning and brain imaging.

Improved transcriptomic and proteomic studies will be required to measure lateralized gene activity more accurately than has been achieved to date. Given the findings of Karlebach and Francks (2015) on lateralization of gene expression in the adult cerebral cortex, the study of embryonic, fetal, and developing cerebral cortices should also benefit greatly from further studies that make use of larger numbers of postmortem samples than used previously, as well as methods such as functional gene set analysis and meta-analysis. Furthermore, new studies will ideally be based on the most accurate and up-to-date method of transcriptome analysis, which is currently RNA sequencing (Külahoglu & Bräutigam, 2014). This method has not yet been used to investigate lateralization of gene expression in the human brain.

Beyond the cerebral cortex, other regions of the brain have been less well investigated for genetic lateralization. Only the study by Johnson et al. (2009), which was based on four postmortem fetal brains, included cerebellar, subcortical, and hippocampal tissue in addition to cerebral cortical tissue. As noted, the epithalamus (subcortical) is a crucial site of CNS asymmetrical development in zebrafish, and the human dorsal thalamus is therefore worth investigating in this regard. The reported lateralization of human embryonic arm movements at 10 weeks of gestation, which occurs before most or all neural connections between the arms and forebrain are in place (Clowry, 2007), also suggests that more caudal regions of the CNS are functionally lateralized at early developmental stages. Future studies of asymmetrical gene activity in the human brain should therefore investigate brain regions beyond the cerebral cortex.

Through genetic studies of lateralization it may become possible to understand some of the properties of cerebral cortical regions that are especially suited to language perception (Morillon et al., 2010) by contrasting left-sided regions at the genetic level against their "natural control" homologue regions on the right. This approach can identify molecular profiles that are likely to specify the fine-tuning of neuronal circuitry for particular types of information processing. Genetic networks arising from this work may then be manipulated in animal models: for example, genetically modified rodents with more *human left-type* cortices may be compared to those with more *human right-type* cortices. Multilevel neurobiological analyses in model organisms may then deliver a new understanding of how one human hemisphere is preferentially adapted for carrying out particular functions. The data of Karlebach and Francks (2015) suggest that this will be challenging, however, because lateralization of gene expression in the adult superior temporal cortex involved subtle left-right differences over many genes. To model these effects in cell and animal models would therefore require the simultaneous and precise manipulation of many genes, and this may not be feasible with current technology. Lateralization of gene expression will also need to be studied at the level of individual neuronal and glial subtypes, rather than at the level of mRNA derived from whole tissue excisions.

As we've already discussed, psychiatric and cognitive disorders including schizophrenia and dyslexia have been associated with alterations of brain structural and/or functional asymmetries. Genes involved in brain lateralization may therefore influence these diseases, in terms of both individual susceptibility and disorder progression. Further progress toward understanding the genetic basis of human brain lateralization is needed in order to assess the relevance to these disorders.

Lateralization is also of considerable interest with respect to human evolution. Many of our higher cognitive functions show cerebral hemispheric dominance, while our language faculty is likely to have co-opted aspects of our motor circuitry (French & Fisher, 2014), perhaps even involving a hand-gestural component in its origins (Corballis, 2003). Genes that are found to be involved in human brain lateralization may therefore be analyzed informatively with respect to their comparative genomics: whether they show evidence for having undergone positive selection in human or primate evolution.

Acknowledgments

The author's research group is supported by the Max Planck Society of Germany, the Netherlands Organisation for Scientific Research, and Joint Transnational Call funding linked to the Human Brain Project.

NOTE

1. Following standard nomenclature, genes are denoted in italics, proteins in regular font. Uppercase letters denote the human version of the gene (i.e., *FOXP2*), lowercase the mouse version of the gene (i.e., *Foxp2*).

REFERENCES

Abu-Rustum, R. S., Ziade, M. F., & Abu-Rustum, S. E. (2013). Reference values for the right and left fetal choroid plexus at 11 to 13 weeks: An early sign of "developmental" laterality? *Journal of Ultrasound in Medicine*, *32*(9), 1623–1629.

Altshuler, D., Daly, M. J., & Lander, E. S. (2008). Genetic Mapping in Human Disease. *Science*, *322*(5903), 881–888.

Annett, M. (1985). *Left, right, hand and brain: The right shift theory*. London: Laurence Erlbaum Associates.

Apostolova, I., Wunder, A., Dirnagl, U., Michel, R., Stemmer, N., Lukas, M., ... Buchert, R. (2012). Brain perfusion SPECT in the mouse: Normal pattern according to gender and age. *NeuroImage*, *63*(4), 1807–1817.

Arning, L., Ocklenburg, S., Schulz, S., Ness, V., Gerding, W. M., Hengstler, J. G., ... Beste, C. (2013). *PCSK6* VNTR polymorphism is associated with degree of handedness but not direction of handedness. *PLOS ONE, 8*(6), e67251.

Ashburner, M., Ball, C. A., Blake, J. A., Botstein, D., Butler, H., Cherry, J. M., ... Sherlock, G. (2000). Gene Ontology: Tool for the unification of biology: The Gene Ontology Consortium. *Nature Genetics, 25,* 25–29.

Badzakova-Trajkov, G., Häberling, I. S., & Corballis, M. C. (2010). Cerebral asymmetries in monozygotic twins: An fMRI study. *Neuropsychologia, 48*(10), 3086–3093.

Badzakova-Trajkov, G., Häberling, I. S., & Corballis, M. C. (2011). Magical ideation, creativity, handedness, and cerebral asymmetries: A combined behavioral and fMRI study. *Neuropsychologia, 49*(10), 2896–2903.

Badzakova-Trajkov, G., Häberling, I. S., Roberts, R. P., & Corballis, M. C. (2010). Cerebral asymmetries: Complementary and independent processes. *PLOS ONE, 5*(3), e9682.

Bishop, D. V. M. (2013). Cerebral asymmetry and language development: Cause, correlate, or consequence? *Science, 340*(6138), 1230531.

Björk, T., Brus, O., Osika, W., & Montgomery, S. (2012). Laterality, hand control and scholastic performance: A British birth cohort study. *BMJ Open, 2*(2), e000314.

Blackiston, D. J., & Levin, M. (2013). Inversion of left–right asymmetry alters performance of Xenopus tadpoles in nonlateralized cognitive tasks. *Animal behavior, 86*(2), 459-466.

Boles, D. B., Barth, J. M., & Merrill, E. C. (2008). Asymmetry and performance: Toward a neurodevelopmental theory. *Brain and Cognition, 66*(2), 124–139.

Brandler, W. M., Morris, A. P., Evans, D. M., Scerri, T. S., Kemp, J. P., Timpson, N. J., ... Paracchini, S. (2013). Common variants in left/right asymmetry genes and pathways are associated with relative hand skill. *PLOS Genetics, 9*(9), e1003751.

Brucato, N., DeLisi, L. E., Fisher, S. E., & Francks, C. (2014). Hypomethylation of the paternally inherited LRRTM1 promoter linked to schizophrenia. *American Journal of Medical Genetics Part B: Neuropsychiatric Genetics, 165*(7), 555–563.

Burdine, R. D., & Caspary, T. (2013). Left-right asymmetry: Lessons from Cancún. *Development, 140*(22), 4465–4470.

Cantalupo, C., Oliver, J., Smith, J., Nir, T., Taglialatela, J. P., & Hopkins, W. D. (2009). The chimpanzee brain shows human-like perisylvian asymmetries in white matter. *European Journal of Neuroscience, 30*(3), 431–438.

Catani, M., Allin, M. P. G., Husain, M., Pugliese, L., Mesulam, M. M., Murray, R. M., & Jones, D. K. (2007). Symmetries in human brain language pathways correlate with verbal recall. *Proceedings of the National Academy of Sciences, 104*(43), 17163–17168.

Chi, J. G., Dooling, E. C., & Gilles, F. H. (1977). Left-right asymmetries of the temporal speech areas of the human fetus. *Archives of Neurology, 34*(6), 346–348.

Clowry, G. J. (2007). The dependence of spinal cord development on corticospinal input and its significance in understanding and treating spastic cerebral palsy. *Neuroscience and Biobehavioral Reviews, 31*(8), 1114–1124.

Concha, M. L., Bianco, I. H., & Wilson, S. W. (2012). Encoding asymmetry within neural circuits. *Nature Reviews Neuroscience, 13*(12), 832–843.

Concha, M. L., Signore, I. A., & Colombo, A. (2009). Mechanisms of directional asymmetry in the zebrafish epithalamus. *Seminars in Cell and Developmental Biology, 20*(4), 498–509.

Corballis, M. C. (2003). From mouth to hand: Gesture, speech, and the evolution of right-handedness. *Behavioral and Brain Sciences, 26*(2), 199–208; discussion 208–160.

Corballis, M. C. (2013). Early signs of brain asymmetry. *Trends in Cognitive Sciences, 17*(11), 554–555.

Deep-Soboslay, A., Hyde, T. M., Callicott, J. P., Lener, M. S., Verchinski, B. A., Apud, J. A., ... Elvevag, B. (2010). Handedness, heritability, neurocognition and brain asymmetry in schizophrenia. *Brain, 133*(10), 3113–3122.

Dubois, J., Benders, M., Cachia, A., Lazeyras, F., Ha-Vinh Leuchter, R., Sizonenko, S. V., ... Hüppi, P. S. (2008). Mapping the early cortical folding process in the preterm newborn brain. *Cerebral Cortex, 18*(6), 1444–1454.

Eyler, L. T., Pierce, K., & Courchesne, E. (2012). A failure of left temporal cortex to specialize for language is an early emerging and fundamental property of autism. *Brain, 135*(3), 949–960.

Eyler, L. T., Vuoksimaa, E., Panizzon, M. S., Fennema-Notestine, C., Neale, M. C., Chen, C.-H., ... Kremen, W. S. (2013). Conceptual and data-based investigation of genetic influences and brain asymmetry: A twin study of multiple structural phenotypes. *Journal of Cognitive Neuroscience, 26*(5), 1100–1117.

Fagard, J. (2013). The nature and nurture of human infant hand preference. *Annals of the New York Academy of Sciences, 1288*(1), 114–123.

Floris, D., Chura, L., Holt, R., Suckling, J., Bullmore, E., Baron-Cohen, S., & Spencer, M. (2013). Psychological correlates of handedness and corpus callosum asymmetry in autism: The left hemisphere dysfunction theory revisited. *Journal of Autism and Developmental Disorders, 43*(8), 1758–1772.

Francks, C. (2011). Leucine-rich repeat genes and the finetuning of synapses. *Biological Psychiatry, 69*(9), 820–821.

Francks, C., DeLisi, L. E., Fisher, S. E., Laval, S. H., Rue, J. E., Stein, J. F., & Monaco, A. P. (2003). Confirmatory evidence for linkage of relative hand skill to 2p12-q11. *American Journal of Human Genetics, 72*(2), 499–502.

Francks, C., Fisher, S. E., MacPhie, I. L., Richardson, A. J., Marlow, A. J., Stein, J. F., & Monaco, A. P. (2002). A genomewide linkage screen for relative hand skill in sibling pairs. *American Journal of Human Genetics, 70*(3), 800–805.

Francks, C., Maegawa, S., Lauren, J., Abrahams, B., Velayos-Baeza, A., Medland, S., ... Caffrey, T. (2007). LRRTM1 on chromosome 2p12 is a maternally suppressed gene that is associated paternally with handedness and schizophrenia. *Molecular Psychiatry, 12,* 1129–1139.

French, C. A., & Fisher, S. E. (2014). What can mice tell us about Foxp2 function? *Current Opinion in Neurobiology, 28,* 72–79.

Gene Ontology Consortium, The. (2017). Expansion of the Gene Ontology knowledgebase and resources. *Nucleic Acids Research, 45*(D1), D331–D338.

Gordon, L., Joo, J. E., Powell, J. E., Ollikainen, M., Novakovic, B., Li, X., ... Saffery, R. (2012). Neonatal DNA methylation profile in human twins is specified by a complex interplay between intrauterine environmental and genetic factors, subject to tissue-specific influence. *Genome Research, 22*(8), 1395–1406.

Goto, K., Kurashima, R., Gokan, H., Inoue, N., Ito, I., & Watanabe, S. (2010). Left–right asymmetry defect in the hippocampal circuitry impairs spatial learning and working memory in IV mice. *PLOS ONE, 5*(11), e15468.

Gotts, S. J., Jo, H. J., Wallace, G. L., Saad, Z. S., Cox, R. W., & Martin, A. (2013). Two distinct forms of functional lateralization in the human brain. *Proceedings of the National Academy of Sciences, 110*(36), E3435–E3444.

Gräff, J., & Mansuy, I. M. (2008). Epigenetic codes in cognition and behavior. *Behavioral Brain Research, 192*(1), 70–87.

Groen, M. A., Whitehouse, A. J. O., Badcock, N. A., & Bishop, D. V. M. (2013). Associations between handedness and cerebral lateralisation for language: A comparison of three measures in children. *PLOS ONE, 8*(5), e64876.

Guadalupe, T., Willems, R. M., Zwiers, M. P., Arias Vasquez, A., Hoogman, M., Hagoort, P., ... Francks, C. (2014). Differences in cerebral cortical anatomy of left- and right-handers. *Frontiers in Psychology, 5*, 261.

Guadalupe, T., Zwiers, M. P., Teumer, A., Wittfeld, K., Vasquez, A. A., Hoogman, M., ... Francks, C. (2014). Measurement and genetics of human subcortical and hippocampal asymmetries in large datasets. *Human Brain Mapping, 35*(7), 3277–3289.

Guadalupe, T., Zwiers, M. P., Wittfeld, K., Teumer, A., Vasquez, A. A., Hoogman, M., ... Francks, C. (2015). Asymmetry within and around the human planum temporale is sexually dimorphic and influenced by genes involved in steroid hormone receptor activity. *Cortex, 62*, 41–55.

Güven, M., Elalmis, D. D., Binokay, S., & Tan, Ü. (2003). Population-level right-paw preference in rats assessed by a new computerized food-reaching test. *International Journal of Neuroscience, 113*(12), 1675–1689.

Häberling, I. S., Badzakova-Trajkov, G., & Corballis, M. C. (2013). Asymmetries of the arcuate fasciculus in monozygotic twins: Genetic and nongenetic influences. *PLOS ONE, 8*(1), e52315.

Hawrylycz, M. J., Lein, E. S., Guillozet-Bongaarts, A. L., Shen, E. H., Ng, L., Miller, J. A., ... Jones, A. R. (2012). An anatomically comprehensive atlas of the adult human brain transcriptome. *Nature, 489*(7416), 391–399.

Hepper, P. G. (2013). The developmental origins of laterality: Fetal handedness. *Developmental Psychobiology, 55*(6), 588–595.

Hepper, P. G., McCartney, G. R., & Shannon, E. A. (1998). Lateralised behavior in first trimester human foetuses. *Neuropsychologia, 36*(6), 531–534.

Hepper, P. G., Shahidullah, S., & White, R. (1991). Handedness in the human fetus. *Neuropsychologia, 29*(11), 1107–1111.

Hepper, P. G., Wells, D. L., & Lynch, C. (2005). Prenatal thumb sucking is related to postnatal handedness. *Neuropsychologia, 43*(3), 313–315.

Herbert, M. R., Harris, G., Adrien, K., Ziegler, D., Makris, N., Kennedy, D., ... Hodgson, J. (2002). Abnormal asymmetry in language association cortex in autism. *Annals of Neurology, 52*, 588–596.

Herbert, M. R., Ziegler, D. A., Deutsch, C. K., O'Brien, L. M., Kennedy, D. N., Filipek, P. A., ... Caviness, V. S., Jr. (2005). Brain asymmetries in autism and developmental language disorder: A nested whole-brain analysis. *Brain, 128*(1), 213–226.

Hering-Hanit, R., Achiron, R., Lipitz, S., & Achiron, A. (2001). Asymmetry of fetal cerebral hemispheres: in utero ultrasound study. *Archives of Disease in Childhood-Fetal and Neonatal Edition, 85*(3), 194–196.

Hervé, P.-Y., Zago, L., Petit, L., Mazoyer, B., & Tzourio-Mazoyer, N. (2013). Revisiting human hemispheric specialization with neuroimaging. *Trends in Cognitive Sciences, 17*(2), 69–80.

Holland, D., Chang, L., Ernst, T. M., Curran, M., Buchthal, S. D., Alicata, D., ... Kuperman, J. M. (2014). Structural growth trajectories and rates of change in the first 3 months of infant brain development. *JAMA Neurology, 71*(10), 1266–1274.

Hopkins, W. D. (2013). Neuroanatomical asymmetries and handedness in chimpanzees (Pan troglodytes): A case for continuity in the evolution of hemispheric specialization. *Annals of the New York Academy of Sciences, 1288*(1), 17–35.

Hutsler, J. J. (2003). The specialized structure of human language cortex: Pyramidal cell size asymmetries within auditory and language-associated regions of the temporal lobes. *Brain and Language, 86*(2), 226–242.

Jahanshad, N., Lee, A. D., Barysheva, M., McMahon, K. L., de Zubicaray, G. I., Martin, N. G., ... Thompson, P. M. (2010). Genetic influences on brain asymmetry: A DTI study of 374 twins and siblings. *NeuroImage, 52*(2), 455–469.

Johnson, M. B., Kawasawa, Y. I., Mason, C. E., Krsnik, Z., Coppola, G., Bogdanovic, D., ... Sestan, N. (2009). Functional and evolutionary insights into human brain development through global transcriptome analysis. *Neuron, 62*(4), 494–509.

Karlebach, G., & Francks, C. (2015). Lateralization of gene expression in human language cortex. *Cortex, 67*, 30–36.

Kasprian, G., Langs, G., Brugger, P. C., Bittner, M., Weber, M., Arantes, M., & Prayer, D. (2011). The prenatal origin of hemispheric asymmetry: An in utero neuroimaging study. *Cerebral Cortex, 21*(5), 1076–1083.

Kawakami, R., Dobi, A., Shigemoto, R., & Ito, I. (2008). Right isomerism of the brain in inversus viscerum mutant mice. *PLOS ONE, 3*(4), e1945.

Kawakami, R., Shinohara, Y., Kato, Y., Sugiyama, H., Shigemoto, R., & Ito, I. (2003). Asymmetrical allocation of NMDA receptor epsilon2 subunits in hippocampal circuitry. *Science, 300*(5621), 990–994.

Kawasaki, Y., Suzuki, M., Takahashi, T., Nohara, S., McGuire, P. K., Seto, H., & Kurachi, M. (2008). Anomalous cerebral asymmetry in patients with schizophrenia demonstrated by voxel-based morphometry. *Biological Psychiatry, 63*(8), 793–800.

Kikuchi, M., Shitamichi, K., Yoshimura, Y., Ueno, S., Remijn, G. B., Hirosawa, T., ... Minabe, Y. (2011). Lateralized theta wave connectivity and language performance in 2- to 5-year-old children. *Journal of Neuroscience, 31*(42), 14984–14988.

Klar, A. J. S. (1999). Genetic models for handedness, brain lateralization, schizophrenia, and manic-depression. *Schizophrenia Research, 39*(3), 207–218.

Knecht, S., Drager, B., Deppe, M., Bobe, L., Lohmann, H., Floel, A., ... Henningsen, H. (2000). Handedness and hemispheric language dominance in healthy humans. *Brain, 123*(12), 2512–2518.

Knecht, S., Floel, A., Drager, B., Breitenstein, C., Sommer, J., Henningsen, H., ... Pascual-Leone, A. (2002). Degree of language lateralization determines susceptibility to unilateral brain lesions. *Nature Neuroscience, 5*(7), 695–699.

Külahoglu, C., & Bräutigam, A. (2014). Quantitative transcriptome analysis using RNA-seq. In D. Staiger (Ed.), *Plant circadian networks* (Vol. 1158, pp. 71–91). New York: Springer.

Lambert, N., Lambot, M.-A., Bilheu, A., Albert, V., Englert, Y., Libert, F., ... Vanderhaeghen, P. (2011). Genes expressed in specific areas of the human fetal cerebral cortex display distinct patterns of evolution. *PLOS ONE, 6*(3), e17753.

LaMendola, N. P., & Bever, T. G. (1997). Peripheral and cerebral asymmetries in the rat. *Science, 278*(5337), 483–486.

Leask, S. J., & Crow, T. J. (2001). Word acquisition reflects lateralization of hand skill. *Trends in Cognitive Sciences, 5*(12), 513–516.

Lehtinen, M. K., Bjornsson, C. S., Dymecki, S. M., Gilbertson, R. J., Holtzman, D. M., & Monuki, E. S. (2013). The choroid plexus and cerebrospinal fluid: Emerging roles in development, disease, and therapy. *Journal of Neuroscience, 33*(45), 17553–17559.

Levin, M. (2005). Left-right asymmetry in embryonic development: A comprehensive review. *Mechanisms of Development, 122*(1), 3–25.

Levin, M., & Palmer, A. R. (2007). Left–right patterning from the inside out: Widespread evidence for intracellular control. *BioEssays, 29*(3), 271–287.

Li, G., Nie, J., Wang, L., Shi, F., Lyall, A. E., Lin, W., ... Shen, D. (2013). Mapping longitudinal hemispheric structural asymmetries of the human cerebral cortex from birth to 2 years of age. *Cerebral Cortex, 24*(5), 1289–1300.

Li, Q., Bian, S., Liu, B., Hong, J., Toth, M., & Sun, T. (2013). Establishing brain functional laterality in adult mice through unilateral gene manipulation in the embryonic cortex. *Cell Research, 23*(9), 1147–1149.

Lindell, A. K., & Hudry, K. (2013). Atypicalities in cortical structure, handedness, and functional lateralization for language in autism spectrum disorders. *Neuropsychology Review, 23*(3), 257–270.

Liu, H., Stufflebeam, S. M., Sepulcre, J., Hedden, T., & Buckner, R. L. (2009). Evidence from intrinsic activity that asymmetry of the human brain is controlled by multiple factors. *Proceedings of the National Academy of Sciences, 106*(48), 20499–20503.

Liu, J. Z., Tozzi, F., Waterworth, D. M., Pillai, S. G., Muglia, P., Middleton, L., ... Marchini, J. (2010). Meta-analysis and imputation refines the association of 15q25 with smoking quantity. *Nature Genetics, 42*(5), 436–440.

Liu, Y., Balériaux, D., Kavec, M., Metens, T., Absil, J., Denolin, V., ... Aeby, A. (2010). Structural asymmetries in motor and language networks in a population of healthy preterm neonates at term equivalent age: A diffusion tensor imaging and probabilistic tractography study. *NeuroImage, 51*(2), 783–788.

Lyn, H., Pierre, P., Bennett, A. J., Fears, S., Woods, R., & Hopkins, W. D. (2011). Planum temporale grey matter asymmetries in chimpanzees (Pan troglodytes), vervet (Chlorocebus aethiops sabaeus), rhesus (Macaca mulatta) and bonnet (Macaca radiata) monkeys. *Neuropsychologia, 49*(7), 2004–2012.

Margeta, M. A., & Shen, K. (2010). Molecular mechanisms of synaptic specificity. *Molecular and Cellular Neuroscience, 43*(3), 261–267.

Mazoyer, B., Zago, L., Jobard, G., Crivello, F., Joliot, M., Perchey, G., ... Tzourio-Mazoyer, N. (2014). Gaussian mixture modeling of hemispheric lateralization for language in a large sample of healthy individuals balanced for handedness. *PLOS ONE, 9*(6), e101165.

McCarley, R. W., Salisbury, D. F., Hirayasu, Y., Yurgelun-Todd, D. A., Tohen, M., Zarate, C., ... Shenton, M. E. (2002). Association between smaller left posterior superior temporal gyrus volume on magnetic resonance imaging and smaller left temporal P300 amplitude in first-episode schizophrenia. *Archives of General Psychiatry, 59*(4), 321–331.

McManus, I. C., & Bryden, M. P. (1992). *The genetics of handedness, cerebral dominance and lateralization.* Amsterdam: Elsevier.

McManus, I. C., Davison, A., & Armour, J. A. L. (2013). Multilocus genetic models of handedness closely resemble single-locus models in explaining family data and are compatible with genome-wide association studies. *Annals of the New York Academy of Sciences, 1288*(1), 48–58.

McManus, I. C., Martin, N., Stubbings, G. F., Chung, E. M., & Mitchison, H. M. (2004). Handedness and situs inversus in primary ciliary dyskinesia. *Proceedings of the Royal Society of London B: Biological Sciences, 271*(1557), 2579–2582.

Medland, S., Duffy, D., Spurdle, A., Wright, M., Geffen, G., Montgomery, G., & Martin, N. (2005). Opposite effects of androgen receptor CAG repeat length on increased risk of left-handedness in males and females. *Behavior Genetics, 35*(6), 735–744.

Medland, S. E., Duffy, D. L., Wright, M. J., Geffen, G. M., Hay, D. A., Levy, F., ... Boomsma, D. I. (2009). Genetic influences on handedness: Data from 25,732 Australian and Dutch twin families. *Neuropsychologia, 47*(2), 330–337.

Meguerditchian, A., Vauclair, J., & Hopkins, W. D. (2013). On the origins of human handedness and language: A comparative review of hand preferences for bimanual coordinated actions and gestural communication in nonhuman primates. *Developmental Psychobiology, 55*(6), 637–650.

Mellet, E., Jobard, G., Zago, L., Crivello, F., Petit, L., Joliot, M., ... Tzourio-Mazoyer, N. (2013). Relationships between hand laterality and verbal and spatial skills in 436 healthy adults balanced for handedness. *Laterality: Asymmetries of Body, Brain and Cognition, 19*(4), 383–404.

Mock, J. R., Zadina, J. N., Corey, D. M., Cohen, J. D., Lemen, L. C., & Foundas, A. L. (2012). Atypical brain torque in boys with developmental stuttering. *Developmental Neuropsychology, 37*(5), 434–452.

Morillon, B., Lehongre, K., Frackowiak, R. S. J., Ducorps, A., Kleinschmidt, A., Poeppel, D., & Giraud, A.-L. (2010). Neurophysiological origin of human brain asymmetry for speech and language. *Proceedings of the National Academy of Sciences, 107*(43), 18688–18693.

Ocklenburg, S., Arning, L., Gerding, W. M., Epplen, J. T., Güntürkün, O., & Beste, C. (2013a). Cholecystokinin A receptor (CCKAR) gene variation is associated with language lateralization. *PLOS ONE, 8*(1), e53643.

Ocklenburg, S., Arning, L., Gerding, W. M., Epplen, J. T., Güntürkün, O., & Beste, C. (2013b). FOXP2 variation modulates functional hemispheric asymmetries for speech perception. *Brain and Language, 126*(3), 279–284.

Ocklenburg, S., Arning, L., Hahn, C., Gerding, W. M., Epplen, J. T., Güntürkün, O., & Beste, C. (2011). Variation in the NMDA receptor 2B subunit gene GRIN2B is associated with differential language lateralization. *Behavioral Brain Research, 225*(1), 284–289.

Ocklenburg, S., Beste, C., & Güntürkün, O. (2013). Handedness: A neurogenetic shift of perspective. *Neuroscience and Biobehavioral Reviews, 37*(10), 2788–2793.

Ocklenburg, S., & Güntürkün, O. (2012). Hemispheric asymmetries: The comparative view. *Frontiers in Psychology, 3*, 5.

Patel, A., & Honoré, E. (2010). Polycystins and renovascular mechanosensory transduction. *Nature Reviews Nephrology, 6*(9), 530–538.

Peeters, H., & Devriendt, K. (2006). Human laterality disorders. *European Journal of Medical Genetics, 49*(5), 349–362.
Pletikos, M., Sousa, A. M., Sedmak, G., Meyer, K. A., Zhu, Y., Cheng, F., ... Šestan, N. (2014). Temporal specification and bilaterality of human neocortical topographic gene expression. *Neuron, 81*(2), 321–332.
Preslar, J., Kushner, H. I., Marino, L., & Pearce, B. (2014). Autism, lateralisation, and handedness: A review of the literature and meta-analysis. *Laterality: Asymmetries of Body, Brain and Cognition, 19*(1), 64–95.
Prichard, E., Propper, R. E., & Christman, S. D. (2013). Degree of handedness, but not direction, is a systematic predictor of cognitive performance. *Frontiers in Psychology, 4*, 9.
Rentería, M. E. (2012). Cerebral asymmetry: A quantitative, multifactorial, and plastic brain phenotype. *Twin Research and Human Genetics, 15*(3), 401–413.
Rogers, L. J., & Andrew, R. (2002). *Comparative vertebrate lateralization*. Cambridge: Cambridge University Press.
Rogers, L. J., Vallortigara, G., & Andrew, R. (2013). *Divided brains: The biology and behavior of brain asymmetries*. Cambridge: Cambridge University Press.
Samara, A., Vougas, K., Papadopoulou, A., Anastasiadou, E., Baloyanni, N., Paronis, E., ... Tsangaris, G. T. (2011). Proteomics reveal rat hippocampal lateral asymmetry. *Hippocampus, 21*(1), 108–119.
Scerri, T. S., Brandler, W. M., Paracchini, S., Morris, A. P., Ring, S. M., Richardson, A. J., ... Monaco, A. P. (2011). PCSK6 is associated with handedness in individuals with dyslexia. *Human Molecular Genetics, 20*(3), 608–614.
Seidman, L. J., Faraone, S. V., Goldstein, J. M., Kremen, W. S., Horton, N. J., Makris, N., ... Tsuang, M. T. (2002). Left hippocampal volume as a vulnerability indicator for schizophrenia: A magnetic resonance imaging morphometric study of nonpsychotic first-degree relatives. *Archives of General Psychiatry, 59*(9), 839–849.
Sham, P. C., & Purcell, S. M. (2014). Statistical power and significance testing in large-scale genetic studies. *Nature Reviews Genetics, 15*(5), 335–346.
Sharma, N., Berbari, N. F., & Yoder, B. K. (2008). Ciliary dysfunction in developmental abnormalities and diseases. In K. Y. Bradley (Ed.), *Current topics in developmental biology* (Vol. 85, pp. 371–427). Waltham, MA: Academic Press.
Shenton, M. E., Dickey, C. C., Frumin, M., & McCarley, R. W. (2001). A review of MRI findings in schizophrenia. *Schizophrenia Research, 49*(1), 1–52.
Shinohara, K., Kawasumi, A., Takamatsu, A., Yoshiba, S., Botilde, Y., Motoyama, N., ... Hamada, H. (2012). Two rotating cilia in the node cavity are sufficient to break left-right symmetry in the mouse embryo. *Nature Communications, 3*, 622.
Somers, M., Ophoff, R. A., Aukes, M. F., Cantor, R. M., Boks, M. P., Dauwan, M., ... Sommer, I. E. (2015). Linkage analysis in a Dutch population isolate shows no major gene for left-handedness or atypical language lateralization. *Journal of Neuroscience, 35*(23), 8730–8736.
Somers, M., Sommer, I. E., Boks, M. P., & Kahn, R. S. (2009). Hand-preference and population schizotypy: A meta-analysis. *Schizophrenia Research, 108*(1), 25–32.
Sommer, I., Ramsey, N., Kahn, R., Aleman, A., & Bouma, A. (2001). Handedness, language lateralisation and anatomical asymmetry in schizophrenia: meta-analysis. *British Journal of Psychiatry, 178*(4), 344–351.
Spring, S., Lerch, J. P., Wetzel, M. K., Evans, A. C., & Henkelman, R. M. (2010). Cerebral asymmetries in 12-week-old C57Bl/6J mice measured by magnetic resonance imaging. *NeuroImage, 50*(2), 409–415.
Spruston, N. (2008). Pyramidal neurons: Dendritic structure and synaptic integration. *Nature Reviews Neuroscience, 9*(3), 206–221.
Steinmetz, H., Herzog, A., Schlaug, G., Huang, Y., & Jäncke, L. (1995). Brain (a)symmetry in monozygotic twins. *Cerebral Cortex, 5*(4), 296–300.
Sun, T., Patoine, C., Abu-Khalil, A., Visvader, J., Sum, E., Cherry, T. J., ... Walsh, C. A. (2005). Early asymmetry of gene transcription in embryonic human left and right cerebral cortex. *Science, 308*(5729), 1794–1798.
Talbot, K., Louneva, N., Cohen, J. W., Kazi, H., Blake, D. J., & Arnold, S. E. (2011). Synaptic dysbindin-1 reductions in schizophrenia occur in an isoform-specific manner indicating their subsynaptic location. *PLOS ONE, 6*(3), e16886.
Tanaka, S., Kanzaki, R., Yoshibayashi, M., Kamiya, T., & Sugishita, M. (1999). Dichotic listening in patients with situs inversus: Brain asymmetry and situs asymmetry. *Neuropsychologia, 37*(7), 869–874.
Tomasi, D., & Volkow, N. D. (2012). Resting functional connectivity of language networks: Characterization and reproducibility. *Molecular Psychiatry, 17*(8), 841–854.
Tsuang, H. C., Chen, W. J., Kuo, S. Y., & Hsiao, P. C. (2013). The cross-cultural nature of the relationship between schizotypy and mixed-handedness. *Laterality: Asymmetries of Body, Brain and Cognition, 18*(4), 476–490.
Vallortigara, G., & Rogers, L. J. (2005). Survival with an asymmetrical brain: Advantages and disadvantages of cerebral lateralization. *Behavioral and Brain Sciences, 28*(4), 575–589; discussion 589–633.
Wada, J. A., Clarke, R., & Hamm, A. (1975). Cerebral hemispheric asymmetry in humans: Cortical speech zones in 100 adult and 100 infant brains. *Archives of Neurology, 32*(4), 239–246.
Waters, N. S., & Denenberg, V. H. (1994). Analysis of two measures of paw preference in a large population of inbred mice. *Behavioral Brain Research, 63*(2), 195–204.
Welter, D., MacArthur, J., Morales, J., Burdett, T., Hall, P., Junkins, H., ... Parkinson, H. (2014). The NHGRI GWAS Catalog, a curated resource of SNP-trait associations. *Nucleic Acids Research, 42*(D1), D1001–D1006.
Willems, R. M., der Haegen, L. V., Fisher, S. E., & Francks, C. (2014). On the other hand: including left-handers in cognitive neuroscience and neurogenetics. *Nature Reviews Neuroscience, 15*(3), 193–201.
Witelson, S. F., & Pallie, W. (1973). Left hemisphere specialization for language in the newborn: Neuroanatomical evidence of asymmetry. *Brain, 96*(3), 641–646.
Yoshiba, S., Shiratori, H., Kuo, I. Y., Kawasumi, A., Shinohara, K., Nonaka, S., ... Hamada, H. (2012). Cilia at the node of mouse embryos sense fluid flow for left-right determination via Pkd2. *Science, 338*(6104), 226–231.
Zeisel, A., Muñoz-Manchado, A. B., Codeluppi, S., Lönnerberg, P., La Manno, G., Juréus, A., ... Linnarsson, S. (2015). Cell types in the mouse cortex and hippocampus revealed by single-cell RNA-seq. *Science, 347*(6226), 1138–1142.

43 Key Issues and Future Directions: Genes and Language

SIMON E. FISHER

Determining the molecular mechanisms that underlie human capacities for speech, language, and reading is a vibrant and exciting initiative. The prior chapters of this part comprehensively reviewed the state-of-the-art on mapping genetic factors that contribute to language- and reading-related skills (Luciano & Bates, chapter 39), discovery of mutations involved in speech disorders (Morgan, chapter 40), and deciphering the impacts of key genes and mutations on development and function of the nervous system, in different model systems (Vernes, chapter 41). Insights in this area come not only from language itself, but also by studying genetic influences on the associated neural architecture, most notably the lateralization of much of the relevant functional circuitry (Francks, chapter 42). Here, in this final chapter of the part, I lay out the promise and challenges of the coming years for tracing genetic pathways involved in language, in light of rapid improvements in the available research tools. In particular, I focus on the dramatically reduced costs and enhanced feasibility of large-scale genotyping and high-throughput DNA sequencing and what this will mean for the future landscape of studies in this area. I emphasize that the substantial advances in molecular technologies will need to be matched by increased sophistication in our approach to defining and characterizing language-related phenotypes. Thus, interdisciplinary connections among molecular biologists, psychologists, linguists, and neuroscientists are set to become ever more essential for progress in this fast-moving research endeavor. I will begin by considering rare mutations and the impact of next-generation DNA sequencing, before moving on to discuss common genetic variants and the prospect of systematic studies of interindividual variation in language proficiency.

1. Rare Gene Variants and How to Find Them

Rare gene variants that disrupt aspects of speech and language development offer powerful windows into neurobiological pathways (see Vernes, chapter 41). The fundamental scientific value afforded by the discovery of such mutations is clearly illustrated by the example of *FOXP2*[1] (Fisher & Scharff, 2009). The identification of *FOXP2* preceded the completion of the first full draft of the human genome sequence. It relied on analysis of an unusual family (the KE family) in which 15 relatives displayed apparent monogenic inheritance of developmental speech and language problems across three successive generations. Even with such a large family available to study, the process of uncovering the causative mutation was a lengthy one. It began with a screen for genomic regions linked to the disorder (Fisher, Vargha-Khadem, Watkins, Monaco, & Pembrey, 1998), followed by an intensive gene-by-gene search through one particular section of chromosome 7, containing many unknown genes that had never been sequenced before (Lai et al., 2000). Clues to limit the search came from an unrelated child with a similar disorder, along with a chromosomal rearrangement involving that same region. Eventually, researchers pinpointed the KE family mutation (Lai, Fisher, Hurst, Vargha-Khadem, & Monaco, 2001), a change of a single base in a crucial part of *FOXP2* (at that time, a novel human gene that had not yet been described). Subsequent studies could then look specifically at *FOXP2* in other smaller families. Although these targeted investigations successfully confirmed an etiological role, by finding other mutations of this gene in independent cases/families with a similar profile of speech and language deficits, they also illustrated that such mutations are extremely rare (MacDermot et al., 2005). Indeed, more than a decade and a half after discovering *FOXP2*, most cases of speech and language disorders remain unexplained at the genetic level (Deriziotis & Fisher, 2017; Graham & Fisher, 2015).

In the meantime, the broader landscape of genetic research has been dramatically transformed by the development of novel technologies for reading off a person's genomic sequence (Metzker, 2010). Traditional methods, such as Sanger sequencing, are highly laborious, time-consuming, and expensive. These standard approaches were used to assemble the first draft sequence of the human genome, work that took the best part of a decade, requiring the efforts of tens of thousands of researchers around the world, at a cost of billions of US dollars. So-called next-generation techniques (Goodwin, McPherson, & McCombie, 2016), massively parallel sequencing of a great many DNA fragments at the same time, allow for an almost entire human genome to be sequenced from a simple saliva sample in a matter of hours, at a cost of less than US$1,000 (based on estimates in 2018, at time of writing). Another upcoming generation of technologies involves direct sequencing of individual intact molecules of DNA from start to finish, without shearing up into fragments (Loose, 2017; Zaaijer et al., 2017). So, as it becomes ever cheaper and quicker to readout the genomes of people with speech and language disorders, how will this impact on our understanding of the genetic underpinnings?

In principle, the routine availability of whole genome sequencing should make it considerably easier to study pedigrees, such as the KE family, in which there are multiple relatives affected with speech-, language-, and/or reading-related disorders across several generations. Thus, one might expect a substantial acceleration in the discovery of novel causative mutations. In practice, even with these new molecular technologies, progress in uncovering susceptibility genes has been relatively slow, and it is worth considering some of the explanations for this. Crucially, the majority of cases of language-related disorders involve multifactorial etiology (Graham & Fisher, 2015), resulting from the combined effects of several, perhaps many, genetic risk factors of small effect (as will be discussed in section 2). Researchers rely on a certain amount of serendipity to successfully track down those families that have truly monogenic forms, in which the disorder can be fully explained by a single genetic mutation with a large effect size, also known as high penetrance. A pedigree may show what appears to be a simple inheritance pattern at the phenotypic level, but this surface level view will not necessarily be reflected in the underlying biology. Hence, it is not uncommon to come up empty-handed after searching for a high-penetrance genetic mutation in a family of interest, despite using the fastest and most powerful sequencing techniques.

Moreover, whole genome sequencing has revealed that each human individual carries a surprisingly large number of rare sequence variants that could potentially disrupt gene function. It can be challenging to distinguish causative mutations from more benign sequence variants in the genomic background (Deriziotis & Fisher, 2017). When a rare disruptive variant cosegregates with a disorder in multiple affected relatives, this finding is consistent with an etiological role, but often families with language-related disorders are not large enough to be sure that the observed pattern represents a statistically significant finding (i.e., that the cosegregation is beyond that expected by chance). The KE family is exceptional in this regard (Fisher et al., 1998). Sometimes a study will identify several different candidate mutations in a family, without concrete evidence to highlight a single true culprit. To have high confidence that a gene is implicated in a disorder, it is necessary to collate observations of this same gene being disrupted by independent mutations in different families and/or cases. These factors make it challenging to identify completely novel etiological genes that have not already been confirmed as potential contributors to language-related disorders.

To distinguish true causative mutations from benign variation in an affected family, cosegregation with the disorder is important, but not sufficient. We also need to draw inferences about the likely functional relevance of any variant of interest (i.e., determining whether it will disturb gene function). Although it is sometimes possible to make confident *in silico* predictions about the disruptive potential of a genetic variant, based on DNA sequence data alone, the ability to do this varies considerably for different parts of the genome. We are on firmest ground when it comes to the genomic regions that directly code for proteins. For example, a DNA sequence variant in a protein-coding gene may lead to substitution of one amino acid for another at a key point of the encoded protein, and if that substituted amino acid has different chemical properties it can distort the shape of the protein, preventing it from carrying out its normal cellular functions. (This is the kind of mutation that was found in *FOXP2* in the affected members of the KE family; Lai et al., 2001.) Other well-studied types of mutation lead to truncation of the encoded protein, so that it is missing crucial parts, or may even stop the protein from being expressed at all (MacDermot et al., 2005). Importantly, the protein-coding regions (collectively known as the *exome*) make up only 1% to 2% of the genetic makeup of a person—the vast majority of the genome comprises noncoding DNA. Variants in noncoding regions can also play etiological roles in a disorder, by disturbing the way that expression of a gene is regulated (Devanna et al., 2018), but our understanding of such effects is lagging behind that for coding variants. Thus, even as it becomes trivial to obtain whole genome sequences from people affected with language-related disorders, and to accurately identify all the rare DNA variants that they carry, we still face difficulties in interpreting the functional significance of the variants, and this will impede progress in the field. Chapter 41 by Vernes shows how experiments in model systems, such as neurons grown in the laboratory, will become increasingly important for making sense of the avalanche of sequence data we face.

Finding families with multiple affected relatives is not the only way to pinpoint rare mutations involved in neurodevelopmental disorders. Perhaps counterintuitively, we may sometimes learn about genetic underpinnings by investigating so-called *simplex* families in which there is just a single affected proband with unaffected parents and siblings and no obvious family history (Fischbach & Lord, 2010). In this kind of strategy, we compare the genome (or exome) sequence of the proband to those of the parents and search for any new mutations that have arisen de novo in his or her genome (i.e., changes that are present in the proband, but absent from both parents) (Veltman & Brunner, 2012). The natural mutation rate in humans is low, so that a typical child carries fewer than a hundred de novo changes, with only one or two of these being located in the protein-coding parts of the genome. In the rare cases where a de novo mutation disrupts an important gene in a substantive way, this may be sufficient to lead to a disorder in the child who carries it. Using next-generation sequencing, it has been shown that severe neurodevelopmental disorders such as autism and intellectual disability (ID) are enriched for these types of causative de novo mutations, explaining a significant proportion of cases (Gilissen et al., 2014; Iossifov et al., 2014). By studying large enough numbers of simplex families, it becomes possible to find genes that are recurrently mutated in independent probands. Interestingly, a number of genes thereby implicated in autism and/or ID phenotypes, such as *FOXP1* and *TBR1*, are connected to language-related genetic pathways through protein-protein interactions with *FOXP2* (Deriziotis et al., 2014; Sollis et al., 2016).

2. *Next-Generation Sequencing in Language-Related Disorders: The Story So Far*

It is early days for applying next-generation DNA sequencing to speech and language disorders (Deriziotis & Fisher, 2017). Only a handful of studies in this field have so far taken advantage of these techniques. Most have faced limitations in reliably determining etiological relevance of identified variants, in line with the issues discussed in section 1. For example, exome sequencing of people from Robinson Crusoe Island, an isolated population with a high incidence of specific language impairment (SLI, also known as developmental language disorder, DLD), revealed a rare variant of the *NFXL1* gene that was associated with the disorder, but within a complex model, since only 39% of affected cases carried the putative risk variant, and it was found in 10% of people with typical language ability (Villanueva et al., 2015). Another study of a geographical isolate with elevated levels of SLI/DLD, from a remote cluster of villages in the north of Russia, combined data from genome-wide association screening (see section 3) and exome sequencing in a subset of individuals, and again was unable to find a high-penetrant mutation accounting for the phenotype, although a pathway-based analysis identified enrichment for variants affecting *MEF2*-regulated genes (Kornilov et al., 2016). The emerging view that most cases of SLI/DLD involve complex genetic architecture is further supported by an exome sequencing study of rare variants in 43 unrelated probands, which found possible pathogenic variants in known speech-/language-related genes *ERC1*, *GRIN2A*, and *SRPX2*, as well as in novel candidates including *OXR1*, *SCN9A*, and *KMT2D*, and potential "multiple-hit" cases (i.e., rare variants in two different genes in the same proband) involving genes such as *AUTS2*, *STARD9*, *SEMA6D*, and *SYNPR* (X. S. Chen et al., 2017). However, it was difficult to draw firm conclusions about the etiological significance of the variants, without clear-cut evidence of segregation with disorder, recurrent independent mutations, or data from functional experiments.

Typically cases of SLI/DLD are not as severe as autism or ID and may not be enriched for high-penetrance mutations. Interestingly, the monogenic disorder caused by *FOXP2* mutation has a distinct phenotype from common forms of SLI/DLD, in that it involves childhood apraxia of speech (CAS) accompanied by impairments affecting most aspects of language development (Watkins, Dronkers, & Vargha-Khadem, 2002). Indeed, next-generation sequencing studies suggest that by targeting a disorder such as CAS it is possible to increase the yield of causative variants. A recent investigation used whole genome sequencing to analyze 19 unrelated probands with a robust diagnosis of CAS (Eising et al., 2018). For nine of the probands, whole genome sequences were also obtained from both unaffected parents, allowing systematic screening for de novo coding variants, the first time that this had been done for a primary speech disorder. Three probands carried de novo disruptions with high likelihood of being pathogenic, affecting the genes *CHD3*, *SETD1A*, and *WDR5*. Analyses of the genome sequences from the 10 isolated probands (for whom parental

DNA was not available) identified novel loss-of-function variants affecting *KAT6A*, *SETBP1*, *ZFHX4*, *TNRC6B*, and *MKL2*. The sample size in this study was too small for observation of recurrent mutations (same gene independently affected in different cases) within the cohort. However, these new potential CAS candidates are regulatory genes with known links to neurodevelopment, and analyses of RNA-sequencing data revealed that they are clustered in a single module of genes showing concordant spatiotemporal expression in the developing human brain (Eising et al., 2018). Another notable feature of these findings is that, despite the lack of recurrent mutation within the 19 CAS probands, most of the new candidate genes have been implicated in neurodevelopmental phenotypes in prior human studies.

Overall, the first data on rare variants from next-generation sequencing studies of speech-/language-related disorders are consistent with the emerging view of convergence in the neurogenetic pathways underlying multiple distinct brain-related disorders (e.g., ID, autism, epilepsy, schizophrenia) (Deriziotis & Fisher, 2017). Putative risk genes often do not seem to respect the clinical boundaries that we impose on definitions of disorder, which should not be surprising given the complex relationships among genes, brains, and behavior (Fisher, 2006). Thus, a major goal for the future is to gain a better understanding of the mappings between rare high-penetrance mutations and the potential cognitive and behavioral (as well as physiological and anatomical) outcomes observed in the children that carry them. Although the individual mutation events are by definition rare, through national and international networks of geneticists and clinicians, it is becoming easier to pull together data from multiple unrelated cases for a particular candidate gene of interest, enabling a genotype-driven approach to phenotype definition. For example, much of our initial knowledge on the consequences of *FOXP2* mutation came from studies of the etiological substitution found in members of one family (the KE family), demonstrating disproportionate effects on speech and language skills (Watkins, Dronkers, & Vargha-Khadem, 2002). Recent years have seen acceleration in the identification of new point mutations disrupting *FOXP2* (e.g., Reuter et al., 2017). Systematic in-depth studies of these independent cases, to characterize the range of profiles of speech, language, and other aspects of behavior and cognition, promise valuable new insights into the relationship between *FOXP2* disruptions and brain development. More broadly, this kind of genotype-driven strategy, applied to language-related candidate genes, is likely to enhance our studies of rare disruptive variants in the coming decade, and it will reshape concepts concerning the boundaries between neurodevelopmental disorders involving speech and language deficits.

3. Learning from Common Genetic Architecture

Rare mutations are just one source of evidence for connecting genotypes to phenotypes. Studies of common genetic variation provide a complementary strategy for dissecting biological pathways that are important for language skills. Moving beyond a view that language proficiency is a fixed unvarying trait across all human beings, the idea here is to capture variability in aspects of language proficiency and try to relate that to the well-documented interindividual variation in people's genomes. Again, technological advances are making the molecular side of this rather trivial to carry out—for less than US$100 per sample, DNA microarrays can simultaneously read out the alleles of hundreds of thousands of positions across the genome. Using these variants, known as SNPs, together with our knowledge of population genetics, it is possible to recover the majority of common genetic variation that a person carries, without the need to directly sequence the whole genome. (But note that rare variation is more difficult to reliably reconstruct with these techniques.) Hence, as summarized in the present volume (Luciano & Bates, chapter 39) and elsewhere (Deriziotis & Fisher, 2017; Graham & Fisher, 2015), the past few years have seen the first genome-wide association scans (GWASs) of speech-, language-, and reading-related traits, either studying disorders (Eicher et al., 2013; Field et al., 2013; Gialluisi et al., 2014; Kornilov et al., 2016; Nudel et al., 2014; Roeske et al., 2011) or normal variation (Harlaar et al., 2014; Luciano et al., 2013; St. Pourcain et al., 2014). Some interesting suggestive findings have emerged from these studies. However, beyond an association of variants in the *ROBO2* gene with a measure of early vocabulary, which didn't persist for language measures at later stages of development (St Pourcain et al., 2014), robust significant evidence of association has eluded researchers, and there are not yet any positive independent replications of GWAS hits. Moreover, it turns out that few, if any, of the claims for common variant association from earlier studies (prior to advancements in GWAS technologies) are borne out by these newer large-scale screens (e.g., Becker et al., 2014). Why have GWAS approaches failed to deliver major insights so far for the language sciences, and what are the prospects for turning this situation around in the future?

A key constraint here is genetic complexity. Most language-related traits are highly heritable, meaning that the combined effects of all genetic variations together account for a significant proportion of phenotypic variability. However, the underlying genetic architecture is nonetheless complex, involving multiple different loci (Graham & Fisher, 2015). For language phenotypes, just as with other complex traits, the individual effect size of any single common genetic variant is usually very small and can be difficult to detect. Moreover, a standard GWAS involves testing of many hundreds of thousands of largely independent SNPs spread across the genome. The value of this approach is that the search is unbiased and does not depend on prior assumptions about the nature of the genes involved (Visscher et al., 2017), which is important for language traits where we are only just beginning to glimpse the underlying biology. But the downside of a systematic search is that it imposes a huge multiple-testing burden and renders a traditional threshold of $P<.05$ useless, since it would yield a hugely inflated number of false positive associations. This issue is usually dealt with through a rigorously adjusted *P* value threshold for declaring genome-wide significance, of 5×10^{-8}, ideally along with an independent replication in another sample. Taken together, the small effect sizes of individual SNPs and the substantive multiple-testing involved, mean that cohorts of thousands of people (preferably tens of thousands of people) are required to give adequate basic power for a GWAS (Visscher et al., 2017). A classic and often-cited example of the impact of sample size is schizophrenia, where early GWAS efforts struggled to identify significant associations. Through combining available cohorts into a shared sample of almost 37,000 cases, the field moved from only around 30 potential associated loci to more than 128 robust independent genome-wide significant signals (Schizophrenia Working Group of the Psychiatric Genomics, 2014). As sample sizes continue to increase, more and more replicable loci are being uncovered. This is just one of many potentially compelling illustrations from the field over the past few years (Visscher et al., 2017). Note that it can be hard to predict the sample size at which one can be confident of capturing associations, since that depends on multiple unknown factors, including the precise nature of the genomic architecture. But it is clear that the sample sizes used thus far for language-related phenotypes are on the edge of what is desirable, to say the least. To give an extreme example, one study performed GWAS case-control analyses with only 163 language-impaired cases (Eicher et al., 2013), which is a highly underpowered study design for seeking out common variant associations in a genome-wide context.

Clearly, a single laboratory is unlikely to by itself have the capacity to recruit and study (phenotype and genotype) cohorts of tens of thousands of participants. The success of GWAS approaches for biomedical traits has largely stemmed from researchers embracing highly collaborative models of team science, in which data from multiple independent smaller cohorts are brought together for meta-analysis (or even mega-analysis, depending on issues over data alignment, sharing policies, and ethical approvals). To return to the much-cited schizophrenia GWAS example (Schizophrenia Working Group of the Psychiatric Genomics, 2014), this was only possible through the efforts of a large international network of clinical and research groups, known as the Psychiatric Genomics Consortium, which also supports genetic analyses of several other psychiatric phenotypes (Sullivan, 2010). To coordinate large-scale consortium investigations of this kind, which may depend on work of hundreds of clinicians and scientists at different sites around the world, is by no means a trivial task. It involves harmonized approaches to phenotypic definition, integrating data from different genotyping platforms, standardizing analytical procedures, and so on, to ensure optimal signal-to-noise ratios in any combined genetic analysis. Still, if the language sciences are to take full advantage of the promise offered by GWAS approaches, then this field will also have to move toward the team science model, otherwise we will continue to come up against limitations of insufficient power in our studies. Indeed, this move is already beginning. For example, the new GenLang Consortium (www.genlang.org) has been established to try to integrate existing cohorts (both disorder-based and epidemiological) that have available speech, language, and/or reading phenotype data coupled to genome-wide genotype information, and the first meta-GWAS studies are underway.

4. Toward Large-Scale Genomic Screens of Speech, Language, and Reading Skills

With the formation of consortium efforts focused on language and genetics, and advances in methods for analyzing and integrating high-throughput molecular data, there are good grounds for optimism. Nonetheless, it is worth recognizing that phenotypes related to speech, language, and reading pose some special challenges. Large-scale biomedical studies often tap into routine clinical diagnostic pipelines in order to recruit sufficient numbers of affected participants for genetic analyses of adequate power. For instance, a wealth of information about the genetic architecture of ID and autism (and many other more medically orientated phenotypes) has come through study of clinically ascertained cohorts. In contrast, language-related neurodevelopmental difficulties, such as dyslexia, SLI, and

CAS, are less likely to be considered within a clinical framework, especially when they occur against a background of preserved general cognitive function. This makes it harder to assemble large cohorts through opportunistic routes involving medical screening systems that are already in place (but see the following discussion for some possibilities). There is also a degree of resistance in certain quarters to "medicalization" of problems in this area. Even if considering speech and language therapists as a potential source, it is not yet common practice for therapists to posit a potential role for genetic factors, despite growing evidence of their relevance (see Morgan, chapter 40). It is sobering to note that while developmental disorders that disturb speech, language, and/or reading are estimated to be some of the most common problems facing children in the modern world (Newbury & Monaco, 2010), the available cohorts collected for genetic studies over the past few decades remain disappointingly small (Deriziotis & Fisher, 2017). Thus, even with an international consortium like GenLang bringing together the existing data sets for meta-/mega-analysis, the combined sample sizes still fall short of that which would be optimal for high-powered GWASs.

Another complication when characterizing language ability for large-scale genetic studies is the complex multifaceted nature of the relevant capacities. There is no single agreed test that can give a simple readout of a person's skills in this area, since to be linguistically competent one must draw on a wide range of distinct processes, acquiring and manipulating items at multiple levels from phonemes to words to sentences and discourse, using different modalities (sign, speech, writing), expressing, and comprehending. Test batteries do exist and have been used in prior genetic studies for a variety of purposes. Some batteries have been designed for diagnosing children with problems in acquiring aspects of language or for indexing impaired linguistic competence in adults with neurological lesions and or age-related decline. Other batteries were developed to identify proficiency of language use in adult second language learners, primarily as tools for teaching and assessment purposes. Few, if any, available batteries are effective for capturing interindividual differences across the full suite and spectrum of language-related abilities in the general population. Where particular tests are available and have already been used to collect phenotypes from different cohorts with matched genotype data, aligning the phenotypes is often not straightforward. For example, tests of "single-word reading" (typically used to diagnose dyslexia; Carrion-Castillo, Franke, & Fisher, 2013) exist in multiple different forms and can be coded in different ways for accuracy and speed, among other things. Moreover, consortium efforts usually depend on bringing together cohorts from different countries across the world—if your phenotype of interest is a biometric one such as blood pressure, height, or head circumference, then your test instruments can be easily aligned across international boundaries, regardless of the host language. However, if the phenotype of interest is itself a language-related skill, and there are different native languages spoken by the distinct cohorts being tested, then harmonization of test data becomes much more challenging, for reasons that are self-evident. This difficulty applies not only to tests of spoken language and grammar processing where the surface properties of the languages involved might be highly distinct, but also to assessments of reading and spelling, given obvious differences in orthographic depth of alternative writing systems (Ziegler & Goswami, 2006).

Thus, for the language sciences to fully take advantage of the potential for consortium-based efforts to perform productive large-scale genetic analyses across existing cohorts, there is much work to do in the area of phenotype alignment over the coming years. Success in studying genetic architecture that underpins language will depend on collaborations across disciplines, in particular interactions among geneticists, psychologists, and linguists. More crucially, perhaps, it could be valuable to develop novel test batteries that are designed to reliably capture the true extent of interindividual differences in language-related skills at multiple levels, in both expressive and receptive domains. The degree of variation in such skills in randomly selected people from the general population, the relationships between supposedly different skills, how much they draw on general cognitive processes, and so on, are all important unaddressed questions in their own right, worthy of attention by the field. Indeed, this is the focus of a major ongoing project by the Dutch Language in Interaction Consortium. If new batteries assessing language-related skills are designed in such a way that they can be reliably administered via remote testing, through web portals or apps, that could revolutionize genetic studies of such traits. There are more and more large-scale population cohorts available with genome-wide genotype data already collected (e.g., Ge, Chen, Neale, Sabuncu, & Smoller, 2017), but in which no language-relevant phenotypes are yet available. Web-/app-based collection of the missing phenotypes offers a highly cost-effective way to gather standardized data for association analyses with the pre-existing genetic information.

Another approach that is being used for clinical traits makes use of electronic health records (EHRs) in the medical databases of hospital biobanks (Wei &

Denny, 2015). These are biobanks in which hundreds of thousands of DNA samples have been collected from all patients, and genome-wide genotypes have been obtained. By searching through the matching EHRs from patients for particular medical codes that designate a particular disorder, it is possible to virtually assemble a sample of cases from the biobank and run a GWAS against suitable controls also taken from the same biobank. A number of investigations are currently underway using *International Statistical Classification of Diseases and Related Health Problems, 10th Revision* codes for disorders related to speech, language, and reading. These types of disorders are likely to be under-represented in hospital-based biobanks, and the EHR approach may lack the rigor of actively recruited samples, but it will be interesting to see how this strategy fares.

GWAS efforts will not only fuel the identification of significantly associated genetic variants that can give insights into the biological pathways that underlie language. The expanding tool set of molecular genetic epidemiology also offers innovative ways to answer important questions about how our various language-related skills change during development, how they relate to each other and to other traits, as well as their connections to evolutionary data (see section 6). In such studies, we can directly use molecular data from cohorts to empirically assess genetic correlations, that is, how much of the shared variance between two different traits is accounted for by shared effects at the genetic level. To give just one example, one can ask broadly whether the same genetic factors are contributing to a language-related skill when monitored at different stages of children's development (St. Pourcain, Eaves, et al., 2018; St. Pourcain, Robinson, et al., 2018). These kinds of empirical observations can then constrain future theories about how nature and nurture interact during language acquisition, which again indicates the increased need for interdisciplinary exchanges as this field matures.

5. Investigating Language Neurobiology with Neuroimaging Genetics

As discussed extensively in section 4, the small effect size of each individual genetic variant is a limiting factor for studies of any complex multifactorial trait. Researchers have attempted to overcome this problem in a variety of ways, highlighted in sections 3 and 4. One complementary avenue that initially held particular promise centered on the idea of moving from more peripheral traits of behavioral/cognitive performance toward the structural and functional neural correlates (Bigos & Weinberger, 2010). The original argument was that brain-based measures, for example collected using methods such as MRI, would be closer to the underlying biological mechanisms, such that the effect sizes of associated genetic variants would become larger and easier to detect in small samples. A new interdisciplinary field—now known as neuroimaging genetics—crystallized around this hope at the turn of the century. Early prominent studies targeted a handful of well-studied candidate genes in samples with only tens of participants and reported significant associations that, if true, would account for a substantial proportion of variation in the brain-based measures (e.g., Egan et al., 2003; Hariri et al., 2002). However, as neuroimaging genetics has become more systematic and rigorous and has moved away from candidate gene studies toward unbiased GWASs, early optimism about effect sizes has proved unfounded, at least for variations in neuroanatomy (de Zubicaray et al., 2008; Hibar et al., 2015). Sadly, genes really do not care about how much cost and effort it has taken researchers to collect a phenotype. It turns out that even if that phenotype has been measured using expensive sophisticated tools of MRI, it is likely to face just the same issues as other multifactorial traits.

To help illustrate this point, consider a fairly standard brain trait that can be reliably captured in living participants through MRI scanning, such as the volume of a subcortical structure. In 2015, GWAS work by the ENIGMA (Enhancing NeuroImaging Genetics through Meta-Analysis) Consortium analyzed the volumes of seven structures, including the hippocampus, nucleus accumbens, caudate, putamen, pallidum, amygdala, and thalamus, in almost 31,000 people from 50 different cohorts around the world and identified a number of significant associations (Hibar et al., 2015). The strongest association observed in that study, for a SNP with a significant replicable influence on the volume of the putamen, accounted for only 0.52% in the variance of the measured trait. Even with tens of thousands of people, most of the variance in subcortical volumes is still to be accounted for, but larger follow-on consortium studies (as yet unpublished) are uncovering additional significant replicable loci and moving toward a more complete account of the relevant genetic architecture. Perhaps for a consortium like ENIGMA, involving a great many different sites (Bearden & Thompson, 2017), it could be that the diversity of scanning machines and parameters, and of types of cohorts across the world, leads to increased heterogeneity in the genetic analyses, reducing power. However, there are now various neuroimaging genetics cohorts, such as that collected by UK Biobank, in which the same scanners and test paradigms have been used for relatively homogenous populations and the sample sizes are rapidly approaching similar levels to that for ENIGMA (Alfaro-Almagro et al., 2018). The first neuroimaging genetics analyses of resources such as UK Biobank are drawing the same conclusions as prior consortium studies; just as with behavioral and cognitive traits, interindividual variations in brain structure have a highly multifactorial genetic basis, involving a great many common variants spread across the genome, most with a tiny effect size.

Thus, to be successful, neuroimaging genetics studies need to confront the exact same issues as more basic genetic association studies, with designs that have adequate power for reliably detecting modest effect sizes, and that factor in sufficient replication and validation steps to guard against false positives. Thus far, most language-related neuroimaging genetics studies have failed to fulfil these requirements, and they may have already distorted the literature of this emerging field. A brief example, to highlight the issues, concerns *FOXP2*, the most well-studied of language-related candidate genes (Fisher & Scharff, 2009). Neuroimaging studies of people with rare high-penetrance mutations that cause severe speech and language disorders have revealed differences in brain structure and function; these alterations are subtle yet significant and they are distributed across various regions, with inferior frontal gyrus, striatum, and cerebellum being of particular interest (Watkins, Vargha-Khadem, et al., 2002). Subsequent to identification of rare etiological mutations that obviously disrupt *FOXP2*, research groups around the world sought evidence that common variants (of unknown molecular effect) might be associated with variability in aspects of behavior, brain structure or function in other disorders, or in the general population (Mueller et al., 2016; Padovani et al., 2010; Peter et al., 2011; Pinel et al., 2012; Premi et al., 2012; Wilcke et al., 2012). Despite the fact that common variants are expected to have (at best) much smaller effect sizes than the high-penetrance mutations causing monogenic disorders, the group sizes in neuroimaging genetics studies of common *FOXP2* SNPs were low—as few as 14 healthy participants in one report (Wilcke et al., 2012). A later study of 1,300 people from the general population targeted each of the *FOXP2* SNPs that had been claimed to show associations from the earlier smaller investigations and found no evidence of any effects on neuroanatomy in that larger independent sample (Hoogman et al., 2014). Moreover, no significant associations of *FOXP2* SNPs with aspects of brain structure have so far emerged in larger GWAS studies of tens of thousands of people, even when those studies included measurement of striatal volumes (e.g., Hibar et al., 2015), a key structure of interest from studies of rare mutations of this gene. This does not discount the possibility that effects of common *FOXP2* variants will eventually be found, perhaps depending on the granularity of the structural phenotypes that we are studying. But the main point here is that the prior positive claims of *FOXP2* SNP effects based on only tens of individuals are unlikely to be valid.

If high-powered cohorts with thousands of people fail to find effects, why are there some reports in the literature that show supposedly significant associations in much smaller samples? First, although it is not often appreciated, a study design that is severely underpowered is also more prone to spurious false positive findings; for discussion of these issues, see Button et al. (2013) and follow-up commentaries. Second, neuroimaging genetic approaches involve bringing together complex data sets of high multidimensionality in which there are many choice points regarding how the data are processed and tested. At the genetic level there are choices about which markers to choose from a gene of interest, which modes of inheritance to consider (additive, recessive/dominant), how to deal with rare alleles or markers with multiple different alleles, what direction of effect to expect, whether to test genders separately, and so on. At the neuroimaging level, there are choices about whether to look bilaterally or at individual hemispheres, which regions of the brain to test, how to best define those regions of interest (e.g., data-driven or atlas-based, which atlas to choose), which processed features to assess (e.g., gray matter or white matter density, volume, thickness, surface area, shape, some measure of connectivity). The excessive choice points of even the simplest neuroimaging genetics study makes this area of work especially susceptible to inflation bias due to implicit (or even explicit) selective reporting, without suitable levels of multiple-testing adjustment. Third, and in a related point, given the broad possibilities regarding the (typically unknown) relationship between a genetic variant of interest and its impact on brain measures, exploratory testing and hypothesis-driven testing are often conflated. Some studies engage in post hoc hypothesis testing using the same data set that was originally used for exploratory analyses, rather than carrying out a formal replication in an independent cohort. These, and other issues facing neurogenetic association studies, are part of the broader discussions currently underway in psychology and neuroscience around the so-called *replicability crisis* (Grabitz et al., 2018). Future neuroimaging genetics studies of language should pay close attention.

Notwithstanding the need for increased rigor moving forward, the prospects for this area of work remain exciting. There is little doubt that neuroimaging genetics will be a useful strategy for understanding effects of molecular factors on brain circuits involved in language, especially with the growing data sets that are becoming available, such as those from large-scale consortia like ENIGMA (Bearden & Thompson, 2017) and population resources such as UK Biobank (Alfaro-Almagro et al., 2018). It is already feasible to run high-powered GWAS studies for cortical thickness and surface area of language-related areas, such as the inferior frontal gyrus and superior temporal gyrus, and it will soon be possible to thoroughly investigate more sophisticated characterizations of language systems, such as structural and functional connectivity of the key circuits, aspects of gyral/sulcal morphology, and degree of lateralization of key phenotypes (Guadalupe et al., 2017). One area where it will remain difficult to obtain suitably high-powered data sets for genetic association studies is in task-based functional neuroimaging, tapping into particular aspects of language processing. Often, due to scanning time limitations, the task-based paradigms used in establishing large-scale public biobanks are not designed to tackle research questions about language (or to go deeply into any particular cognitive trait). It is an open question as to how much of the relevant genetic architecture involved in language processing could be deciphered through (for example) resting-state functional MRI data or some other more generic source. Nonetheless, several different groups around the world have been using language tasks for functional MRI and/or magnetoencephalography in hundreds of participants and have obtained DNA samples to match (Mahowald & Fedorenko, 2016; Schoffelen et al., 2017); it should be possible to harmonize data across these groups for meta-analyses. If these combined sample sizes are not large enough to support GWASs themselves, they could still attain sufficient power for targeted follow-up investigations of top SNPs (or a polygenic risk score—a weighted cumulative score based on SNP associations at multiple loci) from other GWAS efforts. Finally, as discussed by Francks (chapter 42), there is also considerable potential to integrate association findings with data from analyses of gene expression in different brain regions, another area that is rapidly expanding with advances in molecular technology.

6. *Concluding Remarks*

Due to limitations of space, this chapter has primarily focused on methods for capturing rare and common variation in the genomes of modern humans and relating that to brain-based and behavioral/cognitive measures. Remarkably, the impact of advanced molecular technologies is also extending its reach into the distant past, giving us novel means to answer fundamental questions about the evolution of our unusual communicative capacities. With next-generation sequencing, it has become possible to read the genome sequences not only of our extant primate cousins but also of extinct archaic hominins, such as Neanderthals and Denisovans (Meyer et al., 2012; Prufer et al., 2014; Slon et al., 2017). For more information on this, I refer the reader to recent reviews discussing the promise of molecular anthropology (Paabo, 2014) and how comparative genomics of different hominins can help shed light on language origins (Fisher, 2017). The researchers of tomorrow are incredibly fortunate to have access to catalogues of genetic differences between *Homo sapiens* and other extant and extinct apes, along with information about parts of the genome that have undergone selection, or introgressed from one hominin to another, at distinct stages of evolution (Fisher, 2017; Paabo, 2014). Perhaps the most exciting work of the future will be that which integrates these historical findings with the new wave of genome-wide screening data from studies of language-related phenotypes in modern human populations (Deriziotis & Fisher, 2017). Do genes that are implicated in speech and language disorders, or interindividual variability in language-related skills, overlap with those that changed during the emergence of our species, and how do such genes influence the structure and function of key brain circuits, in humans and other species? Moreover, with the rise of gene-editing tools, such as CRISPR/Cas9 (J. S. Chen et al., 2017) and novel systems for modeling human brain development in the lab, such as cerebral organoids (Lancaster et al., 2013), we are in an unprecedented position for tracing out pathways at multiple levels, unpicking effects of language-related gene variants on neurons and circuits, to help explain this elusive human trait.

NOTE

1. Following standard nomenclature, genes are denoted in italics, proteins in regular font. Uppercase letters denote the human version of the gene (i.e., *FOXP2*), lowercase the mouse version of the gene (i.e., *Foxp2*).

REFERENCES

Alfaro-Almagro, F., Jenkinson, M., Bangerter, N. K., Andersson, J. L. R., Griffanti, L., Douaud, G., ... Smith, S. M. (2018). Image processing and quality control for the first 10,000 brain imaging datasets from UK Biobank. *NeuroImage, 166*, 400–424.

Bearden, C. E., & Thompson, P. M. (2017). Emerging global initiatives in neurogenetics: The Enhancing Neuroimaging Genetics through Meta-analysis (ENIGMA) Consortium. *Neuron, 94*(2), 232–236.

Becker, J., Czamara, D., Scerri, T. S., Ramus, F., Csepe, V., Talcott, J. B., ... Schumacher, J. (2014). Genetic analysis of dyslexia candidate genes in the European cross-linguistic NeuroDys cohort. *European Journal of Human Genetics, 22*(5), 675–680.

Bigos, K. L., & Weinberger, D. R. (2010). Imaging genetics—Days of future past. *NeuroImage, 53*(3), 804–809.

Button, K. S., Ioannidis, J. P., Mokrysz, C., Nosek, B. A., Flint, J., Robinson, E. S., & Munafo, M. R. (2013). Power failure: Why small sample size undermines the reliability of neuroscience. *Nature Reviews Neuroscience, 14*(5), 365–376.

Carrion-Castillo, A., Franke, B., & Fisher, S. E. (2013). Molecular genetics of dyslexia: an overview. *Dyslexia, 19*(4), 214–240.

Chen, J. S., Dagdas, Y. S., Kleinstiver, B. P., Welch, M. M., Sousa, A. A., Harrington, L. B., ... Doudna, J. A. (2017). Enhanced proofreading governs CRISPR-Cas9 targeting accuracy. *Nature, 550*(7676), 407–410.

Chen, X. S., Reader, R. H., Hoischen, A., Veltman, J. A., Simpson, N. H., Francks, C., ... Fisher, S. E. (2017). Next-generation DNA sequencing identifies novel gene variants and pathways involved in specific language impairment. *Scientific Reports, 7*, 46105.

Deriziotis, P., & Fisher, S. E. (2017). Speech and language: Translating the genome. *Trends in Genetics, 33*(9), 642–656.

Deriziotis, P., O'Roak, B. J., Graham, S. A., Estruch, S. B., Dimitropoulou, D., Bernier, R. A., ... Fisher, S. E. (2014). De novo TBR1 mutations in sporadic autism disrupt protein functions. *Nature Communications, 5*, 4954.

Devanna, P., Chen, X. S., Ho, J., Gajewski, D., Smith, S. D., Gialluisi, A., ... Vernes, S. C. (2018). Next-gen sequencing identifies non-coding variation disrupting miRNA-binding sites in neurological disorders. *Molecular Psychiatry, 23*(5), 1375–1384.

de Zubicaray, G. I., Chiang, M. C., McMahon, K. L., Shattuck, D. W., Toga, A. W., Martin, N. G., ... Thompson, P. M. (2008). Meeting the challenges of neuroimaging genetics. *Brain Imaging Behavior, 2*(4), 258–263.

Egan, M. F., Kojima, M., Callicott, J. H., Goldberg, T. E., Kolachana, B. S., Bertolino, A., ... Weinberger, D. R. (2003). The BDNF val66met polymorphism affects activity-dependent secretion of BDNF and human memory and hippocampal function. *Cell, 112*(2), 257–269.

Eicher, J. D., Powers, N. R., Miller, L. L., Akshoomoff, N., Amaral, D. G., Bloss, C. S., ... Pediatric Imaging, Neurocognition and Genetics Study. (2013). Genome-wide association study of shared components of reading disability and language impairment. *Genes, Brain and Behavior, 12*(8), 792–801.

Eising, E., Carrion Castillo, A., Vino, A., Strand, E. A., Jakielski, K. J., Scerri, T. S., ... Fisher, S. E. (2018). A set of regulatory genes co-expressed in embryonic human brain is implicated in disrupted speech development. *Molecular Psychiatry*. Advance online publication. doi: 10.1038/s41380-018-0020-x.

Field, L. L., Shumansky, K., Ryan, J., Truong, D., Swiergala, E., & Kaplan, B. J. (2013). Dense-map genome scan for dyslexia supports loci at 4q13, 16p12, 17q22; suggests novel locus at 7q36. *Genes, Brain and Behavior, 12*(1), 56–69.

Fischbach, G. D., & Lord, C. (2010). The Simons Simplex Collection: A resource for identification of autism genetic risk factors. *Neuron, 68*(2), 192–195.

Fisher, S. E. (2006). Tangled webs: Tracing the connections between genes and cognition. *Cognition, 101*(2), 270–297.

Fisher, S. E. (2017). Evolution of language: Lessons from the genome. *Psychonomic Bulletin and Review, 24*(1), 34–40.

Fisher, S. E., & Scharff, C. (2009). FOXP2 as a molecular window into speech and language. *Trends in Genetics, 25*(4), 166–177.

Fisher, S. E., Vargha-Khadem, F., Watkins, K. E., Monaco, A. P., & Pembrey, M. E. (1998). Localisation of a gene implicated in a severe speech and language disorder. *Nature Genetics, 18*(2), 168–170.

Ge, T., Chen, C. Y., Neale, B. M., Sabuncu, M. R., & Smoller, J. W. (2017). Phenome-wide heritability analysis of the UK Biobank. *PLOS Genetics, 13*(4), e1006711.

Gialluisi, A., Newbury, D. F., Wilcutt, E. G., Olson, R. K., DeFries, J. C., Brandler, W. M., ... Fisher, S. E. (2014). Genome-wide screening for DNA variants associated with reading and language traits. *Genes, Brain and Behavior, 13*(7), 686–701.

Gilissen, C., Hehir-Kwa, J. Y., Thung, D. T., van de Vorst, M., van Bon, B. W., Willemsen, M. H., ... Veltman, J. A. (2014). Genome sequencing identifies major causes of severe intellectual disability. *Nature, 511*(7509), 344–347.

Goodwin, S., McPherson, J. D., & McCombie, W. R. (2016). Coming of age: Ten years of next-generation sequencing technologies. *Nature Reviews Genetics, 17*(6), 333–351.

Grabitz, C. R., Button, K. S., Munafo, M. R., Newbury, D. F., Pernet, C. R., Thompson, P. A., & Bishop, D. V. M. (2018). Logical and methodological issues affecting genetic studies of humans reported in top neuroscience journals. *Journal of Cognitive Neuroscience, 30*(1), 25–41.

Graham, S. A., & Fisher, S. E. (2015). Understanding language from a genomic perspective. *Annual Review of Genetics, 49*, 131–160.

Guadalupe, T., Mathias, S. R., vanErp, T. G. M., Whelan, C. D., Zwiers, M. P., Abe, Y., ... Francks, C. (2017). Human subcortical brain asymmetries in 15,847 people worldwide reveal effects of age and sex. *Brain Imaging Behavior, 11*(5), 1497–1514.

Hariri, A. R., Mattay, V. S., Tessitore, A., Kolachana, B., Fera, F., Goldman, D., ... Weinberger, D. R. (2002). Serotonin transporter genetic variation and the response of the human amygdala. *Science, 297*(5580), 400–403.

Harlaar, N., Meaburn, E. L., Hayiou-Thomas, M. E., Wellcome Trust Case Control, C., Davis, O. S., Docherty, S., ... Plomin, R. (2014). Genome-wide association study of receptive language ability of 12-year-olds. *Journal of Speech, Language, and Hearing Research, 57*(1), 96–105.

Hibar, D. P., Stein, J. L., Renteria, M. E., Arias-Vasquez, A., Desrivieres, S., Jahanshad, N., ... Medland, S. E. (2015). Common genetic variants influence human subcortical brain structures. *Nature, 520*(7546), 224–229.

Hoogman, M., Guadalupe, T., Zwiers, M. P., Klarenbeek, P., Francks, C., & Fisher, S. E. (2014). Assessing the effects of common variation in the FOXP2 gene on human brain structure. *Frontiers in Human Neuroscience, 8*, 473.

Iossifov, I., O'Roak, B. J., Sanders, S. J., Ronemus, M., Krumm, N., Levy, D., ... Wigler, M. (2014). The contribution of de novo coding mutations to autism spectrum disorder. *Nature, 515*(7526), 216–221.

Kornilov, S. A., Rakhlin, N., Koposov, R., Lee, M., Yrigollen, C., Caglayan, A. O., ... Grigorenko, E. L. (2016). Genome-wide association and exome sequencing study of language disorder in an isolated population. *Pediatrics, 137*(4), e20152469.

Lai, C. S., Fisher, S. E., Hurst, J. A., Levy, E. R., Hodgson, S., Fox, M., ... Monaco, A. P. (2000). The SPCH1 region on human 7q31: Genomic characterization of the critical interval and localization of translocations associated with speech and language disorder. *American Journal of Human Genetics, 67*(2), 357–368.

Lai, C. S., Fisher, S. E., Hurst, J. A., Vargha-Khadem, F., & Monaco, A. P. (2001). A forkhead-domain gene is mutated in a severe speech and language disorder. *Nature, 413*(6855), 519–523.

Lancaster, M. A., Renner, M., Martin, C. A., Wenzel, D., Bicknell, L. S., Hurles, M. E., ... Knoblich, J. A. (2013). Cerebral organoids model human brain development and microcephaly. *Nature, 501*(7467), 373-379.

Loose, M. W. (2017). The potential impact of nanopore sequencing on human genetics. *Human Molecular Genetics, 26*(R2), R202–R207.

Luciano, M., Evans, D. M., Hansell, N. K., Medland, S. E., Montgomery, G. W., Martin, N. G., ... Bates, T. C. (2013). A genome-wide association study for reading and language abilities in two population cohorts. *Genes, Brain and Behavior, 12*(6), 645–652.

MacDermot, K. D., Bonora, E., Sykes, N., Coupe, A. M., Lai, C. S., Vernes, S. C., ... Fisher, S. E. (2005). Identification of FOXP2 truncation as a novel cause of developmental speech and language deficits. *American Journal of Human Genetics, 76*(6), 1074–1080.

Mahowald, K., & Fedorenko, E. (2016). Reliable individual-level neural markers of high-level language processing: A necessary precursor for relating neural variability to behavioral and genetic variability. *NeuroImage, 139,* 74–93.

Metzker, M. L. (2010). Sequencing technologies—The next generation. *Nature Reviews Genetics, 11*(1), 31–46.

Meyer, M., Kircher, M., Gansauge, M. T., Li, H., Racimo, F., Mallick, S., ... Paabo, S. (2012). A high-coverage genome sequence from an archaic Denisovan individual. *Science, 338*(6104), 222–226.

Mueller, K. L., Murray, J. C., Michaelson, J. J., Christiansen, M. H., Reilly, S., & Tomblin, J. B. (2016). Common genetic variants in FOXP2 are not associated with individual differences in language development. *PLOS One, 11*(4), e0152576.

Newbury, D. F., & Monaco, A. P. (2010). Genetic advances in the study of speech and language disorders. *Neuron, 68*(2), 309–320.

Nudel, R., Simpson, N. H., Baird, G., O'Hare, A., Conti-Ramsden, G., Bolton, P. F., ... Newbury, D. F. (2014). Genome-wide association analyses of child genotype effects and parent-of-origin effects in specific language impairment. *Genes, Brain and Behavior, 13*(4), 418–429.

Paabo, S. (2014). The human condition—A molecular approach. *Cell, 157*(1), 216–226.

Padovani, A., Cosseddu, M., Premi, E., Archetti, S., Papetti, A., Agosti, C., ... Borroni, B. (2010). The speech and language FOXP2 gene modulates the phenotype of frontotemporal lobar degeneration. *Journal of Alzheimer's Disease, 22*(3), 923–931.

Peter, B., Raskind, W. H., Matsushita, M., Lisowski, M., Vu, T., Berninger, V. W., ... Brkanac, Z. (2011). Replication of CNTNAP2 association with nonword repetition and support for FOXP2 association with timed reading and motor activities in a dyslexia family sample. *Journal of Neurodevelopmental Disorders, 3*(1), 39–49.

Pinel, P., Fauchereau, F., Moreno, A., Barbot, A., Lathrop, M., Zelenika, D., ... Dehaene, S. (2012). Genetic variants of FOXP2 and KIAA0319/TTRAP/THEM2 locus are associated with altered brain activation in distinct language-related regions. *Journal of Neuroscience, 32*(3), 817–825.

Premi, E., Pilotto, A., Alberici, A., Papetti, A., Archetti, S., Seripa, D., ... Borroni, B. (2012). FOXP2, APOE, and PRNP: New modulators in primary progressive aphasia. *Journal of Alzheimer's Disease, 28*(4), 941–950.

Prufer, K., Racimo, F., Patterson, N., Jay, F., Sankararaman, S., Sawyer, S., ... Paabo, S. (2014). The complete genome sequence of a Neanderthal from the Altai Mountains. *Nature, 505*(7481), 43–49.

Reuter, M. S., Riess, A., Moog, U., Briggs, T. A., Chandler, K. E., Rauch, A., ... Zweier, C. (2017). FOXP2 variants in 14 individuals with developmental speech and language disorders broaden the mutational and clinical spectrum. *Journal of Medical Genetics, 54*(1), 64–72.

Roeske, D., Ludwig, K. U., Neuhoff, N., Becker, J., Bartling, J., Bruder, J., ... Schulte-Korne, G. (2011). First genome-wide association scan on neurophysiological endophenotypes points to trans-regulation effects on SLC2A3 in dyslexic children. *Molecular Psychiatry, 16*(1), 97–107.

Schizophrenia Working Group of the Psychiatric Genomics Consortium. (2014). Biological insights from 108 schizophrenia-associated genetic loci. *Nature, 511*(7510), 421–427.

Schoffelen, J. M., Hulten, A., Lam, N., Marquand, A. F., Udden, J., & Hagoort, P. (2017). Frequency-specific directed interactions in the human brain network for language. *Proceedings of the National Academy of Sciences, 114*(30), 8083–8088.

Slon, V., Hopfe, C., Weiss, C. L., Mafessoni, F., de la Rasilla, M., Lalueza-Fox, C., ... Meyer, M. (2017). Neandertal and Denisovan DNA from Pleistocene sediments. *Science, 356*(6338), 605–608.

Sollis, E., Graham, S. A., Vino, A., Froehlich, H., Vreeburg, M., Dimitropoulou, D., ... Fisher, S. E. (2016). Identification and functional characterization of de novo FOXP1 variants provides novel insights into the etiology of neurodevelopmental disorder. *Human Molecular Genetics, 25*(3), 546–557.

St. Pourcain, B., Cents, R. A., Whitehouse, A. J., Haworth, C. M., Davis, O. S., O'Reilly, P. F., ... Davey Smith, G. (2014). Common variation near ROBO2 is associated with expressive vocabulary in infancy. *Nature Communications, 5,* 4831.

St. Pourcain, B., Eaves, L. J., Ring, S. M., Fisher, S. E., Medland, S., Evans, D. M., & Davey Smith, G. (2018). Developmental changes within the genetic architecture of social communication behavior: A multivariate study of genetic variance in unrelated individuals. *Biological Psychiatry, 83*(7), 598–606.

St. Pourcain, B., Robinson, E. B., Anttila, V., Sullivan, B. B., Maller, J., Golding, J., ... Davey Smith, G. (2018). ASD and schizophrenia show distinct developmental profiles in common genetic overlap with population-based social communication difficulties. *Molecular Psychiatry, 23*(2), 263–270.

Sullivan, P. F. (2010). The psychiatric GWAS consortium: Big science comes to psychiatry. *Neuron, 68*(2), 182–186.

Veltman, J. A., & Brunner, H. G. (2012). De novo mutations in human genetic disease. *Nature Reviews Genetics, 13*(8), 565–575.

Villanueva, P., Nudel, R., Hoischen, A., Fernandez, M. A., Simpson, N. H., Gilissen, C., … Newbury, D. F. (2015). Exome sequencing in an admixed isolated population indicates NFXL1 variants confer a risk for specific language impairment. *PLOS Genetics, 11*(3), e1004925.

Visscher, P. M., Wray, N. R., Zhang, Q., Sklar, P., McCarthy, M. I., Brown, M. A., & Yang, J. (2017). 10 years of GWAS discovery: Biology, function, and translation. *American Journal of Human Genetics, 101*(1), 5–22.

Watkins, K. E., Dronkers, N. F., & Vargha-Khadem, F. (2002). Behavioural analysis of an inherited speech and language disorder: comparison with acquired aphasia. *Brain, 125*(3), 452–464.

Watkins, K. E., Vargha-Khadem, F., Ashburner, J., Passingham, R. E., Connelly, A., Friston, K. J., … Gadian, D. G. (2002). MRI analysis of an inherited speech and language disorder: structural brain abnormalities. *Brain, 125*(3), 465–478.

Wei, W. Q., & Denny, J. C. (2015). Extracting research-quality phenotypes from electronic health records to support precision medicine. *Genome Medicine, 7*(1), 41.

Wilcke, A., Ligges, C., Burkhardt, J., Alexander, M., Wolf, C., Quente, E., … Kirsten, H. (2012). Imaging genetics of FOXP2 in dyslexia. *European Journal of Human Genetics, 20*(2), 224–229.

Zaaijer, S., Gordon, A., Speyer, D., Piccone, R., Groen, S. C., & Erlich, Y. (2017). Rapid re-identification of human samples using portable DNA sequencing. *eLife, 6*, e27798.

Ziegler, J. C., & Goswami, U. (2006). Becoming literate in different languages: Similar problems, different solutions. *Developmental Science, 9*(5), 429–436.

VIII ANIMAL MODELS OF LANGUAGE

SECTION EDITORS: CAREL TEN CATE AND CONSTANCE SCHARFF

44 The Comparative Approach to the Biology and Evolution of Language

W. TECUMSEH FITCH

1. Introduction

The comparative method involves comparing and analyzing traits of interest as they appear in multiple different species and is one of the cornerstones of evolutionary biology and thus of biology more generally. Although some basic comparisons between human language and animal communication are probably as old as language itself, and Aristotle had already recorded some speculations on their relationship (Aristotle, 350 B.C./1970), the modern comparative approach began in earnest with Darwin, who had already identified key issues that continue to occupy researchers today (Darwin, 1871, 1872). The study of animal communication within an evolutionary framework took off in the 1950s, when the sound spectrograph gave a huge impetus to the study of acoustic communication and the ethological perspective of Karl von Frisch, Konrad Lorenz, and Niko Tinbergen became widely known and practiced. By 1960, the linguist Charles Hockett used comparisons between human language and the communication systems of diverse animal species, including learned birdsong and the honeybee dance "language," to inform his proposed "design principles" of language (Hockett, 1960). Thus began the modern era of comparative analysis of the biology and evolution of language, which has been a research area of steadily growing importance in our modern understanding of language.

What is a language? My simple definition, in essentially Aristotelian terms, is "a system that allows a signaler to express any cognitive state (thoughts and feelings) it is capable of, and allows a similarly-equipped receiver to process such expressions and thus understand those cognitive states." The key here is the word "any," since most communication systems allow their bearers to communicate only a narrow spectrum of their possible cognitive states (e.g., concerning food, mating, or alarm, but not color, space, or time). Even the honeybee dance system, which is remarkable in its capacity to inform bees about the location of distant objects that are not visible to the communicating bees, does not allow the signaler to communicate the color or shape of the flowers. Since we know that honeybees are well able to process and remember color and shape information (Chittka & Skorupski, 2001; Giurfa, Backhaus, & Menzel, 1995; Menzel, 2012), this means that this dance system is not a language by the preceding definition. Indeed, according to current knowledge, humans are the *only* living species with a true language as just defined. Accepting this uniqueness, how do we proceed to study human language comparatively?

The first and most crucial step is to recognize that despite its uniqueness in toto, language is a complex cognitive capacity comprising multiple separable mechanisms, and that many of these mechanisms are shared with other species. For example, learning to produce a novel acoustic signal requires auditory and vocal capabilities, the motor control capabilities to link these together, and a system of learning and memory to retain and access the learned signal. Some of these capabilities (auditory, learning, and memory mechanisms) are widely shared among vertebrates while others (linking audition to vocal motor control to yield signal learning) are unusual but nonetheless are found in other species (see section 5.1). Recognizing that many of the various mechanisms that underlie language (the so-called *faculty of language in a broad sense*) are shared with other species is the first key step in comparative analysis (Hauser, Chomsky, & Fitch, 2002) and is often termed the *multicomponent approach* to language study (Fitch, 2017).

The second key to a productive comparative approach is to recognize the essential role of phylogeny in deriving valid comparative insights. That is, we need not only to recognize similarities and differences in capabilities across a wide range of species, but also to interpret those observations in the context of the phylogenetic relationship among the species in question. Although in Darwin's time there was still considerable debate and confusion about how organisms are related, he provided the essential modern insight that our taxonomies should be constructed *phylogenetically*, based on common descent, essentially like family trees. Today, at least with respect to the vertebrate species most often discussed in connection with language, these issues have been resolved and we have a clear understanding of the framework of phylogenetic relatedness. Molecular-based trees provide the key armature or scaffolding for interpreting comparative cognitive or behavioral data. A tree reconstructed from DNA sequence data, independent of any cognitive traits, can thus provide the crucial context for studying the evolution of behavioral and cognitive features (Felsenstein, 1985, 2004).

In this chapter I will introduce the essential components of the modern comparative method, using examples relevant to language to illustrate the fundamental points (which apply to any trait, whether morphological, cognitive, or physiological). Then I will put these tools to work in a concise analysis of the biology and evolution of language that summarizes our current understanding of the different mechanisms underlying language, and their comparative status.

2. *Essentials of Comparative Analysis*

2.1 Homology—Traits Inherited from a Common Ancestor One of the most central concepts in all of evolutionary biology is homology. It was famously distinguished from analogy by the great comparative anatomist Richard Owen in deceptively simple terms: a homolog is "the same organ in different animals under every variation of form and function" while an analog is "a part or organ in one animal which has the same function as another part or organ in a different animal" (Owen, 1843). Thus, some organs with different functions are homologs (e.g., the bat's wing and human hand), but the bat's and fly's wing are analogs. Owen's definition leaves open what criteria we use to determine "sameness," but this gap was rapidly filled by Darwinian ideas about evolution, which provided the standard modern interpretation, based on phylogeny: homologous features are those features present in extant multiple species due to inheritance from the common ancestor of those species (Lankester, 1870b). While widely accepted, this evolutionary interpretation of homology is not without its difficulties (Donoghue, 1992), both theoretical and empirical, and there is still considerable debate about what precise criteria should be used to identify homology (and particularly the role of developmental or genetic processes in determining homology; discussed in detail by Hall, 1994).

Skirting these issues for a moment, let us first consider why homology, once identified, is so important in biology. Traditionally, homology provides the crucial tool for taxonomists attempting to understand phylogenetic relationships among organisms. For more than a century after Darwin, taxonomy was probably the main reason biologists worried about homology. But today, given the power and accuracy of modern systematics based on DNA sequences, this importance has waned. Instead, we mainly use homologies to try to *interpret* evolutionary patterns (based on a phylogenetic tree derived from DNA), and we use homologies identified in some set of existing related species to infer traits that were present in common ancestors of those species. In some cases, such as feathers in ancestral birds or hair in the ancestral mammals, these inferences can be verified by fossil discoveries (Ji et al., 2002; Qiang, Currie, Norell, & Shu-An, 1998), but the more interesting cases are those where no fossil evidence is available. In such cases, which include neural, cognitive, and behavioral mechanisms, homologies combined with clear phylogenies provide a kind of "time machine" allowing us to reconstruct the brain, behavior, or cognition of extinct ancestors: traits that do not fossilize. This is the central importance of homology for interpreting cognitive evolution.

It is clear that homology can apply at multiple levels of biological organization, from genes and molecules to organs or body parts to cognitive mechanisms to the behaviors they produce. It also applies at multiple taxonomic levels, from a few species to whole orders or phyla. This makes it difficult, if not impossible, to come up with a single definition of homology that provides clear criteria in all cases at all levels (Hall, 1994). Nonetheless, in a given clade (e.g., vertebrates) and level (e.g., neural and cognitive mechanisms) it is often quite clear when characters are homologous. Examples at the neural level include cell types such as neurons and glia, neural circuits like those of the mammalian neocortex or cerebellum, or the dopamine reward system across vertebrates. At the behavioral level some clear examples of behavioral homology would be REM sleep or the innate suckling reflex in placental mammals, the courtship calls of males among frogs and toads, or purring in house cats and mountain lions (Gingras, Mohandesan, Boko, & Fitch, 2013; Peters, 2002; Siegel, 1995). Perhaps more controversially, I will place many other cognitive mechanisms in the homology category (see section 4), including basic spatial awareness and navigation, color vision, social decision making, and many basic learning mechanisms from conditioning to sequence learning. These allow us to rebuild the cognitive abilities of our common ancestor with chimpanzees (which was *not* a chimpanzee), and thus to deduce what additional capabilities needed to evolve in our ancestors to support modern human language. But first, let us discuss analogy.

2.2. Analogy or Convergent Evolution Consider the wings of a bird, a bat, and a housefly: two flattened appendages attached to the thorax that when vigorously flapped lead to powered flight. We call them "wings" because of this shared function (and include airplane "wings" in the same category, by imperfect analogy). In all of these cases, the common ancestor of the species in question did not possess wings: the reptilian common ancestor of birds and mammals was a terrestrial quadruped, and the Ur-bilaterian ancestor of flies and vertebrates was a worm-like creature (De Robertis & Sasai, 1996). Thus despite their similarities, insect and vertebrate wings provide clear examples of analogies: traits with similar functions that evolved via convergent evolution.

Analogy plays an important complementary role to homology in comparative biology. In particular, because analogous traits by definition evolved independently, they provide independent data points for testing functional and evolutionary hypotheses (Harvey & Pagel, 1991). This is not the case for homologous traits, which, no matter how widespread, by definition evolved once, in the common ancestor of the species bearing them. Thus, as far as the evolution of powered flight is concerned, all birds represent a single evolutionary event, while birds, bats, and flies provide three independent points. Thus if we are interested in the evolution of bird flight, we are best served by taking a *broad* evolutionary approach, including bats and flies in our analysis (not to mention gliding animals like flying squirrels or extinct flying reptiles like pterosaurs). Similarly, if we are interested in the evolution of human bipedalism, we should include organisms that independently evolved bipedal locomotion in our analyses (e.g., birds, kangaroos, and various reptiles; Gatesy & Biewiener, 1991).

Convergent evolution thus takes center stage for testing hypotheses in modern comparative biology (Harvey & Pagel, 1991; Lockwood & Fleagle, 1999). We first seek cases where a given characteristic or ability (flight, bipedalism, color vision, vocal learning, male parental care, or a descended larynx) has convergently evolved multiple times, and we use those as the data points in a hypothesis-testing procedure often called the "method of independent contrasts" (Felsenstein, 1985; Pagel, 1992) that is implemented in common statistical packages (e.g., PGLS in R). Crucially, like homology, these analyses require an independent and accurate phylogeny for the clade in question (today typically stemming from DNA analysis). To include homologous traits as independent data points in this procedure would constitute a statistical error that inflates our sample size.

The crucial requirement here is that a trait *has* evolved multiple times, which may seem to invalidate the approach for "unique" traits such as language. But if we break such traits down into their components, we can often find that some of these components are shared (whether as analogies or homologies), as I will discuss. It is also worth noting that traits are often claimed to be uniquely human based on inadequate data: the descended human larynx was claimed to be unique to our species for decades (Lieberman, 1984; Negus, 1949), before anyone actually looked carefully at other species like deer or lions (Fitch & Reby, 2001) or even chimpanzees (Nishimura, Mikami, Suzuki, & Matsuzawa, 2006). Thus it behooves anyone interested in the evolution of language to stay informed about new developments in animal behavior or cognition (the primary point of this part of this book).

A technical term related to analogy is "homoplasy" (Lankester, 1870a; Sanderson & Hufford, 1996). Although sometimes treated as a synonym of analogy, homoplasy is essentially a "garbage can" term referring to *any* similar traits that are *not* due to homology, and in addition to convergence, a homoplasy can include parallelism, reversals, and other more obscure processes. Because of this polysemy I generally avoid using it.

2.3. The Comparative Method Applied to Behavior: The Ethological View Although Darwin clearly thought of behavior in comparative terms, scientific application of the comparative method to animal behavior was uncommon until the rise of ethology in the early 20th century. Ethology involves the mechanistic study of animals' natural behavior from an evolutionary and ecological viewpoint, and in its early days was mainly practiced in continental Europe, by such luminaries as the Austrians Karl von Frisch (who discovered the honeybee dance "language") and Konrad Lorenz (well-known for his work on imprinting) and the Dutch biologist Niko Tinbergen (a master of experimental methods in a field context). Ethology was initially seen as a competitor to the North American tradition of animal behavior (that focused more on laboratory studies of learning in a narrow set of "model" species such as rats and pigeons), and some early academic battles pitted these two disciplines against one another as mutually incompatible. Partly this was a matter of focus: the ethologists were mostly field biologists interested in species-typical behaviors with a strong innate component, while the behaviorists had a strong psychology background and were focused on learning and discovering general "laws" of learning that applied to all species. But after some angry debate

(Lehrman, 1953; Lorenz, 1965), Tinbergen pointed out that there are many interesting and valid questions in behavioral biology, and he clearly recognized that these apparent differences concerned research focus rather than any fundamental disagreement (Tinbergen, 1963).

This resolution of early debates led to the foundation of modern behavioral biology, which recognizes at least four issues of fundamental interest and validity (sometimes called *Tinbergen's Four Whys* or *Tinbergen's four questions*). These are *ontogeny* (development in the individual—learning would be included here), *phylogeny* (evolutionary history of a species), *mechanism* (the physiological or cognitive mechanisms that generate and control behavior), and adaptive *function* (the selective forces leading to particular behaviors being favored over evolutionary time) (Tinbergen, 1963). The comparative method has consistently played an important role in analyzing all of these questions.

Because homology was first defined in the context of anatomy (mostly the skeleton), there was initially considerable skepticism when the early ethologists began to use the homology concept in a behavioral context. Lorenz made two convincing and lasting points in this regard. First, particular behaviors must have some inherited genetic basis if it is to be meaningful to discuss their evolution. This is one reason that ethologists initially focused mostly on instinct and "fixed action patterns" rather than behaviors involving learning: instinctual behaviors have the clearest genetic basis, and when present at birth, they beg for evolutionary interpretation. Second, because of this, we can discuss phylogenetic patterns of behavior with the same rationale and justification as patterns of anatomy or physiology (Lorenz, 1965). Lorenz illustrated these points empirically with his masterful analysis of duck courtship behavior. This revealed a complex phylogenetic pattern, where certain components are shared widely among ducks and others are derived features shown only by one or a few species (Lorenz, 1953).

The delineation of homologous behaviors allows us to rebuild the behavior of extinct ancestors. But clearly, it requires that behavioral differences have some clear genetic basis (otherwise they cannot have evolved). Crucially, despite perennial confusion, these assertions are not in conflict with the idea that behaviors show individual learning (any more than the fact that muscles or bones grow larger when exercised conflicts with their having a genetic basis). An "instinct to learn" can be as innate as having five fingers or wings. Examples include the imprinting of young birds on their caregivers, or the instinct to learn their species-typical song shown by most songbirds (Marler, 1991; ten Cate, 1994). Thus there is no conflict between learning and instinct, and indeed any organism capable of learning does so using biologically given learning mechanisms that evolved in its ancestors.

Despite the clear acceptance of these points by behavioral biologists, a strict dichotomy between learning and instinct continues to bedevil psychology and linguistics, where the idea that complex cognitive mechanisms have *both* innate and learned components remains distressingly controversial (and rears its head periodically in the "nature vs. nurture" debate; cf. Bates & Elman, 1996). For example, a conceptual analysis of "innateness" in the cognitive sciences by Richard Samuels (2004, p. 134) states that the biological interpretation above "appears to have the highly implausible consequence that traits can be both learned and innate—a result that is strongly at variance with the notion of innateness as used by cognitive scientists." If so, such cognitive scientists need to revise their notion of innateness to avoid further fruitless debates (cf. Fitch, 2011; Griffiths, 2002; Marler, 1999). In the rest of this chapter, I will assume the standard biological interpretation of "homologous behavior," which assumes some genetic basis for behaviors without in any way denying a learned component for many linguistically relevant behaviors, or plasticity in their underlying neural mechanisms.

2.4. Distinguishing Homology from Convergence: It's Complicated Based on the preceding discussion, it should be clear that both homologous and analogous traits play important, but distinct, roles in evolutionary thinking and the comparative method. Often, particularly when considering humans and our close mammalian relatives, these two classes are relatively easy to distinguish. Our basic tool-using capacities are homologous with those of chimpanzees, but not those of New Caledonian crows. Similarly, our capacity for complex vocal learning is analogous, not homologous, to that of parrots.

Unfortunately, these distinctions become more difficult as the phylogenetic scope is broadened and the time depth to the last common ancestor (LCA) of relevance is accordingly deepened. One reason is that a complex organ in one species might be homologous at one level, and analogous at another, to a similar organ in a different species. Thus, although the wings of birds and bats are functionally analogous *as wings* (because the common ancestor of birds and mammals clearly lacked wings), they are homologous *as tetrapod forelimbs* (they exhibit the same skeletal structure, inherited from that common ancestor). There is a clear homology, at this level, between most parts of the skeleton in vertebrates. Thus many of the bones (e.g., humerus, ulna) and muscles in bird and bat wings are homologous, and they are also homologous to the human forelimb or dolphin's flipper for the same reason.

This issue is particularly important when considering components of human language. For example, the human capacity to rapidly acquire a large lexicon may be quantitatively distinct from anything in other animals, but nonetheless rely on both widely shared learning mechanisms (e.g., associative learning) and others more specifically relevant to word learning (e.g., fast mapping and mutual exclusivity; Kaminski, Call, & Fischer, 2004; Pepperberg & Wilcox, 2000; Schusterman & Krieger, 1984). While a basic capacity for associative learning is probably homologous across all of these species, mutual exclusivity may not be (cf. Fitch, 2010). It is thus quite important that our subdivision of any complex trait into components makes biological and evolutionary sense and "carves nature at the joints." This is potentially problematic because much comparative work relevant to language starts out by utilizing existing terms denoting subcomponents created by linguists or psychologists with little knowledge of or attention to which of these subcomponents is evolutionarily ancient or recent. Although this starting point is sensible (we have to start somewhere), as new comparative data comes in, we should welcome the opportunity to modify the multicomponent framework to fit biological and evolutionary reality more closely. Any particular multicomponent breakdown (including the one advanced in this chapter) should be treated as provisional and open for updating.

2.5. Deep Homology One of the most interesting recent "tweaks" on the classical comparative concepts discussed here occurred when molecular and developmental biology reached an adequate level of sophistication to sequence specific genes and determine the roles they play in development. Until the mid-1990s, it was widely assumed, based on differences in development, that similar traits in widely separated species (e.g., the wings of flies and birds) involved completely different genes (Mayr, 1963). Thus it was a big surprise when genetic developmental analysis repeatedly showed that the same genes are involved in regulating the development of features such as limbs, hearts, or eyes, even in groups as different as insects and vertebrates (Gehring & Ikeo, 1999; Shubin, Tabin, & Carroll, 1997; Tabin, Carroll, & Panganiban, 1999). This is true despite the uncontested fact that the common ancestor of insects and vertebrates lacked image-forming eyes or jointed limbs. This preservation of ancient genetic determinants of development is strikingly illustrated by the fact that an eyeless mutant fruit fly can be "rescued" by expression of the homologous *Pax6* gene taken from a mouse. This conservative pattern, termed "deep homology," now appears to be surprisingly common in evolution (Held, 2017; Shubin, Tabin, & Carroll, 2009; Wilkins, 2002), suggesting that the developmental genetic toolkit is far more conservative than previously suspected based on the diverse phenotypes seen in modern organisms.

An interesting example of deep homology relevant to human language concerns the capacity for vocal learning (see chapter 45 by Fischer & Hage and chapter 46 by Scharff, Knörnschild, & Jarvis in this volume, for more discussion). Although it appears to be unique to humans among primates, complex vocal learning (call production learning *sensu* Janik & Slater, 2000) is found in a variety of other birds and mammals including songbirds, parrots, hummingbirds, cetaceans, elephants, and at least some seals and bats (Fitch & Jarvis, 2013; Ralls, Fiorelli, & Gish, 1985; Stoeger et al., 2012). In most, if not all, of these clades, this capacity appears to have evolved independently, and thus to constitute an analogy, although there is some debate about parrots and songbirds (which are closely related). Nonetheless, after it was discovered that the *FOXP2* gene plays a role in complex oromotor control and vocal sequencing in humans, analysis of the role of this gene in vocal learning in birds showed that *Foxp2* plays a role in zebra finch song learning and production (Haesler et al., 2007; White, Fisher, Geschwind, Scharff, & Holy, 2006). This thus appears to be a clear case of deep homology in an analogous trait directly related to spoken language in humans (Fitch, 2009; Scharff & Petri, 2011).

The implications of such deep homology for the study of the genetic basis of apparently unique linguistic traits in humans are potentially quite profound. It means that if we can find analogous traits in other species, however distantly related, we might find that the genetic basis of these traits is shared and homologous despite many millions of years of evolutionary distance. This possibility brightens the prospect of detailed biological and genetic study of such traits and underscores the fact that examination of distant relatives (such as aquatic mammals, birds, or even invertebrates) can be as profitable as examining our closer relatives such as primates.

2.6. Preadaptation and Exaptation A final consideration in successfully deploying the comparative method concerns change in function. As Darwin tirelessly stressed, many complex traits also have a complex *functional* history, meaning that their function has changed one or more times (Darwin, 1859, 1871). A classic example from Darwin concerns the lungs of terrestrial vertebrates and the swim bladders found in many fish, which are clearly homologous structures with totally different functions (swim bladders allow buoyancy compensation in fish). Another clear example are the mammalian middle ear ossicles, known from both developmental biology and fossils to have a complex functional history (cf. Fitch, 2010). Initially, these ossicles (like the jaw and hyoid apparatus) were part of the gill bars of the jawless fish ancestral to all jawed vertebrates. They became part of the jaw joint in ancestral tetrapods and still play that role in reptiles and birds. But in the ancestral mammal, a new jaw joint developed and the bones of the older joint became superfluous and gradually reduced in size. Eventually, these former jaw bones became incorporated into the mammalian middle ear, where they now play a key role in conducting and controlling sound from the eardrum into the cochlea (Hopson, 1966; Luo, Crompton, & Sun, 2001). When an existing trait acquires a novel function, it is termed an "exaptation," and the precursor trait can be termed a "preadaptation" for this new function.

3. Comparative Analysis of Human Language

Let us turn to the topic of specific comparative interest: the biology and evolution of human language. As mentioned at the outset, the key steps are to first determine what the subcomponents of human language are and then examine a wide range of animal species experimentally to see which of these components are shared. Deriving a biologically and neurally informed subdivision of "language" is far from a trivial task, and the set of subcomponents proposed at present remains quite tentative. Most of the subcomponents identified in the current literature are derived from existing terms in cognitive psychology (e.g., working, episodic, declarative, and procedural memory) or from linguistics (phonetics, phonology, syntax, and semantics). Some of these terms seem terminologically inappropriate for animals (e.g., "declarative" memory is defined by our ability to say what we know—but how can this be investigated with nonhuman animals or prelinguistic infants?). But in some cases, for example, the "declarative/procedural" distinction just mentioned, these subcomponents can be linked to particular neural subsystems (hippocampal vs. basal ganglia in this case), and these can in turn be investigated in nonhuman species (Schreiweis et al., 2014; Terrace, Son, & Brannon, 2003; Ullman, 2001). Such reframing of cognitive/psychological distinctions in more biological and mechanistic terms is clearly the direction in which the field should be moving, but this goal remains far from complete at present.

4. The Broad Cognitive Foundations

A key piece of evidence for the idea that evolution involves descent with modification comes from comparative anatomy and physiology, where most components of the skeleton, muscular system, circulatory and respiratory systems, as well as the nervous system, are widely shared among vertebrates. For example, most of the bones of the human skeleton are clearly homologous to those of a newt and those of our skull with a rat's. As Darwin (1871) emphasized, most cognitive traits also fall into this class of broadly shared characteristics, so many that they are too numerous to discuss in detail (see Fitch, 2010, chapter 4, for more detail). What I have termed the *basic cognitive toolkit* that is shared among vertebrates quite broadly (and often beyond) includes such capacities as all of our perceptual modalities, categorization, all of the types of memory mentioned in section 3, a sense of time and space (often used in sophisticated navigation abilities), an ability to anticipate and plan for the future, and some basic inference abilities (e.g., transitive inference, where training that A>B and B>C yields, with no training, the inference that A>C). Learning can occur at all of these levels, from simple perceptual matching to categorization to high-level if/then rules, and social learning from conspecifics enables a basic form of animal "culture" in many species (Laland & Galef, 2009; Shettleworth, 2009; Vauclair, 1996; Whiten et al., 1999).

In terms of communication, such features as functional referentiality (calls that can be used by receivers to derive inferences about states of affairs in the world, e.g., the presence of food or predators) are widespread among both primates and birds (Dittus, 1984; Evans, Evans, & Marler, 1993; Macedonia, 1991; Manser, Seyfarth, & Cheney, 2002; Seyfarth, Cheney, & Marler, 1980). The vast majority of the apparatus for speech perception, from formant perception to categorical perception and prosody perception, is widely shared (cf. Kriengwatana & Beckers, chapter 48 of this volume). It also appears that basic syntactic ordering and rule learning, at the phonological level, are also broadly shared (at least with birds and mammals—see chapter 47 by ten Cate & Petkov in this volume). Further widely shared neural/cognitive traits include the vast majority of neural components, from the molecular level of neurotransmitters and receptors, to the neuronal level of pyramidal and Purkinje cells, to the circuit level of thalamocortical or hippocampal-medial temporal circuitry (Striedter, 2004)—along with their genetic bases. Finally, many traits initially suggested (or assumed) to be uniquely human, from tool use and tool making to social intelligence and theory of mind have been found in other species, even if only in a small subset of our animal relatives (cf. Fitch, Huber, & Bugnyar, 2010).

The conclusions that can be drawn from this broadly shared foundation of our cognition are many. Crucially, they indicate that most of the "cognitive toolkit" used by modern humans in our day-to-day life was already in place long before language evolved. This clearly falsifies the old notion that without language an organism cannot think (Descartes, 1982; Müller, 1861). Virtually every successful study in animal cognition refutes this notion (at least not without some serious rear-guard redefinition of "thought"). Equally, this research provides an empirical grounding for hypotheses about where and when abilities of humans, even those highly developed in our species, may have started. Finally, recognizing these broadly shared foundations of cognition simplifies our search for differences, since many intuitively plausible potential differences have already been explored and found to be shared. Thus, a study of animal cognition is an important start to a discussion of potential human linguistic differences.

Particularly important in understanding language evolution is a clear cognitive characterization of our LCA with chimpanzees. Crucially, this LCA was not a chimpanzee but an extinct species whose cognitive capabilities must be reconstructed using comparisons with extant apes, particularly our nearest living relatives, the chimpanzees (*Pan troglodytes*) and bonobos (*Pan paniscus*). Such comparisons need to be grounded in careful study of ape behavior, both in lab and field, of a sort that has made great advances in the last few decades (Byrne, 2000; McGrew, 2004; Tomasello & Call, 1997; Schmelz, Call, & Tomasello, 2013), but where much remains unknown (particularly about bonobos in the field). Even for chimpanzees, where field studies have been ongoing across Africa since the pioneering work of Jane Goodall in the 1960s (Goodall, 1968, 1986), there is much that remains unknown, and new data from new field sites continue to surprise (Sanz, Call, & Morgan, 2009). The cognitive abilities of apes in captivity seem to be pushed further every year, and old ideas (e.g., that apes have no "Theory of Mind") have been conclusively overturned by careful lab studies (Call & Tomasello, 2008; Schmelz, Call, & Tomasello, 2013).

Given this contemporary database, I have identified three areas of difference between modern humans and chimpanzees, and thus by inference from our common ancestor with chimpanzees, that I have called the "three Ss" of signal, syntax, and semantics/pragmatics (Fitch, 2010). These are novel human abilities that must have appeared in roughly the last six million years, since the evolutionary parting of ways between hominins and chimpanzees/bonobos. I will briefly review these novel traits, focusing on data appearing in the last decade, here; for more detailed explication and examination please see Fitch (2010).

5. What Evolved? Three Derived Components of Human Language

5.1 Vocal Production Learning

The most obvious and well-documented ability that humans have but other apes lack is our capacity for vocal production learning—the ability to learn new vocalizations from environmental input (cf. Fischer & Hage, chapter 45 of this volume; Fischer, Wheeler, & Higham, 2015; Fitch & Zuberbühler, 2013; Janik & Slater, 2000). This needs to be distinguished from the ability to modify innate calls from the species-typical vocal repertoire, which has been documented in numerous primate species (e.g., Crockford, Herbinger, Vigilant, & Boesch, 2004; Lemasson, Ouattara, Petit, & Zuberbühler, 2011; Watson et al., 2015) and is probably a common occurrence in vertebrates when social conditions change (e.g., maturation, finding a mate). The clearest illustration of the inflexibility of primate vocal production comes from experiments where infant apes are raised with humans. Despite learning to imitate many manual actions, apes never learn to produce even a few distinct words of speech in such cross-fostering experiments (Furness, 1916; Hayes, 1951; Kellogg, 1968; Yerkes & Yerkes, 1929). Similarly, macaques cross-fostered to another macaque species continue to produce their species-typical calls, not those of their adopted mothers (Owren, Dieter, Seyfarth, & Cheney, 1993). Although humans, like other primates, can learn to control our innate vocal repertoire (e.g., inhibiting crying or laughing), we are unique among primates in our ability to learn to produce novel vocalizations like speech or song that use all vocal tract components (and especially requiring voluntary control over the larynx).

There has been a long debate about the precise biological basis of this human vocal learning capacity, dating back at least to Darwin (1871). One hypothesis is that changes in human vocal anatomy (particularly the descent of the human larynx and the concomitant reconfiguration of the human vocal tract) gave us vocal control abilities not present in other primates. Although this hypothesis is often quoted as a textbook truth (e.g., Crystal, 2003; Harley, 2014; Hoff, 2008) recent data cast severe doubt on any strong form of this hypothesis. The ultimate source for this belief are anatomical studies (Kelemen, 1948; Negus, 1949) and in particular early work by Lieberman and colleagues (Lieberman, Crelin, & Klatt, 1972; Lieberman, Klatt, & Wilson, 1969), who modeled the vocal tract of a rhesus macaque and found that it could not produce all of the vowels of human speech, particularly the three "point vowels" /i/, /a/, and /u/. Note that even in these original papers the claim was *not* that vocal anatomy prevented monkeys from speaking, but rather from producing a few particular vowel sounds. Lieberman's most recent work makes this point clearly: "a nonhuman SVT [supralaryngeal vocal tract] can produce all vowels excepting quantal vowels" (Lieberman, 2012, p. 613). But even this weak version of the hypothesis is challenged by recent data showing that nonhuman primates (like other mammals) lower the larynx during calling (Fitch, 2000), thus attaining a flexible vocal tract that could, with proper neural control, make a large number of discriminable vowels (Fitch, Mathur, de Boer, & Ghazanfar, 2016), fully adequate for spoken linguistic communication (even if different in detail than those of modern humans).

The main alternative hypothesis is that changes in the neural control circuitry for the vocal tract, and in particular the gaining of cortical control over the larynx, were required for volitional vocal control (e.g., speech or singing). This idea builds on a considerable database (cf. Holstege & Subramanian, 2016; Simonyan, 2014) indicating that humans have direct (monosynaptic) connections between motor cortex and the motor neurons controlling the larynx (located in the nucleus ambiguus of the medulla in the brain stem) that are lacking in other primates (Jürgens, 2002). Because lesions in human face motor cortex prevent voluntary lip retraction but do not prevent normal smiling, it is widely believed that cortical control plays a key role in volitional movement (Holstege & Subramanian, 2016; Jürgens, Kirzinger, & von Cramon, 1982; Lemon, 2008). This is consistent with a huge amount of comparative data showing that direct connections from cortex onto motor neurons are important for limb and finger control (Kuypers, 1981). Also consistent with this hypothesis, because chimpanzees and other primates *do* have direct connections to the motor neurons controlling lip and tongue movements (Jürgens & Alipour, 2002; Jürgens et al., 1982), several studies show that great apes have excellent control over their mouth and lips and can produce voluntary sounds using these articulators (Marshall, Wrangham, & Arcadi, 1999; Reynolds Losin, Russell, Freeman, Meguerditchian, & Hopkins, 2008; Wich et al., 2009). Finally, in bird species with vocal production learning (e.g., songbirds and parrots), the analogous monosynaptic connections between motor "cortex" and vocal motor neurons exist (Wild, 1993) and these connections are lacking in non-vocal-learning birds (Striedter, 2004). All of these comparative data converge in indicating that it is neural control rather than peripheral anatomy that needed to change in our recent evolution to give us the rich voluntary control over our vocalizations needed for speech (cf. Fitch, 2010, 2018).

What selective forces could have driven the evolution of complex vocal control? Here, the comparative data can again offer insights, this time in the form of convergent evolution. Vocal learning has independently evolved in multiple clades, including among birds—songbirds, hummingbirds, parrots (Jarvis, 2007)—and among mammals—cetaceans, elephants, some pinnipeds, and some bats (Fitch & Jarvis, 2013; Reichmuth & Casey, 2014; Stoeger et al., 2012). In most of these clades, vocal learning is operative in some form of communication termed "song," often limited to males during the breeding season (seals, some bats, baleen whales, hummingbirds, and many songbirds). This lends support to Darwin's hypothesis that the origins of human vocal control might be found in a posited "musical protolanguage" that involved something resembling birdsong more than language (Darwin, 1871). However, in some clades such as parrots, toothed whales, and humans, vocal learning abilities are equally developed in both sexes and are widely used outside of a mating context; in such cases, some selection for increased communicative diversity and control may have been the driving force. These two hypotheses are not necessarily contradictory: early sexual selection may have initially driven the evolution of vocal control, only to be later co-opted or "exapted" for more general communicative purposes (e.g., among mated pairs or between parents and offspring). In any case, the acquisition of complex vocal production learning is one evolutionary transition that must have occurred sometime in the last six million years.

5.2. Complex Hierarchical Syntax and "Dendrophilia" A second, and more central, change from our LCA with chimpanzees was the acquisition of complex hierarchical syntax: our capacity not only to learn new words but to put them together into complex hierarchical structures (Bickerton, 1990; Hauser et al., 2002). While vocal control is relevant to only one mode of linguistic output (speech), syntax is equally relevant to all language modalities including sign and writing (Armstrong, Stokoe, & Wilcox, 1984; Stokoe, 1960). Indeed, similar brain regions are involved in syntactic processing in both sign and speech (Emmorey, 2002; Horwitz et al., 2003), consistent with the hypothesis that syntax is supported by a unitary, modality-independent neural subsystem. Thus, while full modern language can exist without speech, it cannot exist without syntax.

Recent comparative investigations of the syntactic abilities of nonhuman animals, although not without controversy, paint a relatively consistent picture. A wide variety of birds and mammals have basic syntactic capabilities, in the sense of being able to learn and generalize rules about signal sequences. For example, rats and various primates and birds in the lab can learn simple sequential rules about element order (Herbranson & Shimp, 2008; Ravignani, Sonnweber, Stobbe, & Fitch, 2013; Sonnweber, Ravignani, & Fitch, 2015; Spierings & ten Cate, 2016; Toro & Trobalón, 2005). Although this constitutes a basic form of syntax, all of the patterns learned in these cases can be formally characterized as belonging to the simplest level of the formal language hierarchy, comprising the so-called *finite state grammars* or *regular languages* (Jäger & Rogers, 2012). Similarly, the abilities of animals in more naturalistic settings are also at this level (Robinson, 1984; Whitney, 1985; Zuberbühler, 2002), as are the abilities of apes trained with signed languages (Savage-Rumbaugh et al., 1993; Yang, 2013). But hierarchical syntactic structures in human language require computational resources above this simplest level (because they are above the regular grammars, they can be termed *supra-regular* (Fitch, 2014); other common terms include *context-free* or *phrase-structure* grammars).

Because most empirical claims of animals perceiving syntax beyond the regular level (Abe & Watanabe, 2011; Gentner, Fenn, Margoliash, & Nusbaum, 2006; Rey, Perruchet, & Fagot, 2012) have been refuted in the sense that simpler strategies better explain the data (Beckers, Bolhuis, Okanoya, & Berwick, 2012; Poletiek, Fitz, & Bocanegra, 2016; van Heijningen, de Visser, Zuidema, & ten Cate, 2009), this regular/supra-regular contrast appears at present to be a well-defined characterization of the distinction between human and animal syntactic capacities (see also chapter 47 by ten Cate and Petkov in this volume). However, recent results indicate that, with intensive training, macaques can acquire a supra-regular mirror grammar in the visual-motor domain, so this distinction is not absolute (Jiang et al., 2018). I have elsewhere termed the human proclivity to impose treelike hierarchical structure on perceptual input "dendrophilia" (Fitch, 2014).

Turning to the neural mechanisms underlying hierarchical syntax, there is strong converging evidence that regions in the inferior frontal gyrus (IFG), along with their direct connections to other cortical areas, play a central role. This general region is often termed "Broca's area," although Broca's region strictly speaking is restricted to the left hemisphere, but considerable data suggest that IFG on both sides of the brain plays an important role in processing both speech and sign syntax (Emmorey, 2002; Horwitz et al., 2003; MacSweeney, Capek, Campbell, & Woll, 2008) as well as harmonic syntax in music (Koelsch, Maess, & Friederici, 2000; Maess, Koelsch, Gunter, & Friederici, 2001; Tillmann, Janata, & Bharucha, 2003). These areas, which can be called "Broca's regions" in a broad sense, are also present in other primates and involved in many linguistic tasks. In humans, they appear particularly sensitive to syntax (Fitch & Friederici, 2012; Friederici, 2012): for example even meaningless "jabberwocky" speech of increasing hierarchical complexity specifically activates the IFG (Pallier, Devauchelle, & Dehaene, 2011). Although there is some debate about precisely which subregions of the IFG are central in syntax processing (with both Brodmann's areas 44 and 45 as likely candidates), current brain imaging methods often do not account for possible interindividual differences in the localization of such defined anatomical regions in living humans (Fedorenko et al., 2011).

In any case, histological postmortem comparisons of human and chimpanzee brains indicate that both BA 44 and 45 are greatly expanded in the human brain: a five- to sevenfold expansion is observed that depends on the precise area and the hemisphere: left area 44 is 6.6 times larger in humans than in chimpanzees and left area 45 is 6.0 times larger (Schenker et al., 2010). Furthermore, these frontal regions show greatly expanded connectivity with the rest of the brain in humans, particularly the parietal and temporal lobes (Rilling et al., 2008). Consistent with the hypothesized importance of Broca's regions, comparative neuroimaging studies in macaques and humans show that although both species are sensitive to violations of sequential and numerical rules, in humans the IFG responds preferentially to their joint combination (Wang, Uhrig, Jarraya, & Dehaene, 2015). Thus, while many brain regions are obviously involved in processing language, the expansion of Broca's areas and its corresponding cortical connections seem likely to have played a key neural role in the evolution of hierarchical syntax.

5.3. Evolving the Social Mind: Pragmatics and Ostensive Signaling The acquisition of syntax or speech alone would not necessarily yield modern human language: linguistic communication also requires semantic and pragmatic complexity. While it is clear that single words referring to specific objects or actions can be rapidly learned by many species (Kaminski et al., 2004; Pepperberg & Wilcox, 2000; Pilley & Reid, 2011; Savage-Rumbaugh et al., 1993), it is semantic combinations of words that give language its open capacity to represent meaning. Syntax obviously plays an important role in this (Berwick, Friederici, Chomsky, & Bolhuis, 2013; Chomsky, 2010). But other capacities relevant to semantics and/or pragmatics have also been hypothesized to be lacking in chimpanzees. As mentioned, for decades "Theory of Mind"—the capacity to mentally represent the thoughts of other individuals—was thought to be one of these abilities (Povinelli, Nelson, & Boysen, 1990; Premack & Woodruff, 1978). This was based on the failure of chimpanzees to differentiate between knowledgeable and ignorant humans in various food-begging paradigms. The overturning of this consensus involved changing the task so that it involved competition for food between two chimpanzees (Hare, Call, Agnetta, & Tomasello, 2000; Hare, Call, & Tomasello, 2001)—arguably a much more "natural" situation for chimpanzees. Although this result has led to considerable debate, and ever more clever attempts to find simple "behavior reading" explanations of positive results, new paradigms have been able to control for and refute most of these killjoy explanations (Bräuer, Call, & Tomasello, 2007; Call & Tomasello, 2008), and current consensus is thus that chimpanzees (and ravens: Bugnyar, Reber, & Buckner, 2016) can indeed mentally represent what others know.

Two further pragmatic capacities have been hypothesized to be lacking in our nearest relatives and thus to have evolved during the last six million years. They are the drive to share information cooperatively and an ability for ostensive signaling. Apes trained to communicate with signs or keyboards could, in principle, use them to freely communicate about a huge variety of events and objects, but typically use these abilities only instrumentally, to make requests or to answer questions for a food reward. In contrast children acquiring language rapidly begin to use it to share information, and this "urge to share" continues into adulthood (at least among friends and family). Since there is no explicit English word for this urge, I have adopted the German word *Mitteilungsbedürfnis* for this strong and species-typical urge of our species (Fitch, 2010), hypothesizing that the evolution of a strong *Mitteilungsbedürfnis* was another requirement for modern human linguistic communication during our recent evolutionary history. Related to this, but not necessarily expressed verbally, is the capacity (and drive) for joint attention and shared intentionality that characterizes all cooperative and group activities in our species, again either lacking or underdeveloped in other primates (Tomasello, 2007; Tomasello, Carpenter, Call, Behne, & Moll, 2005). Finally, our drive (and ability) to read others' minds and inform them underlies human teaching—another aspect of our use of language that appears to go far beyond the basic forms of teaching seen in nonhuman animals (Csibra & Gergely, 2006, 2011; Thornton & McAuliffe, 2006).

Finally, recent theoretical work has highlighted the importance of ostensive signaling—signaling of signalhood—in human communication (and this includes gesture and pantomime as much as language; Scott-Phillips, 2014; Stegmann, 2013). While the degree to which this characteristic is absent in nonhuman animals remains contested (cf. Fitch, 2015; Moore, 2016), the capacity to indicate *that* one is attempting, with some act, to communicate is arguably fundamental to human communication in all of its forms, and there can be little doubt that this is another trait that has been greatly developed during recent human evolution (cf., Scott-Phillips, 2014).

In summary then, there are several plausible candidates for human specializations related to the communicative use of language, all of them possibly related and potentially important in the evolution of semantics and pragmatics. Although there are to my knowledge no clear hypotheses about the precise neural changes that might underlie these specializations, this stems mostly from our poor understanding of apes' "social brain" rather than a lack of knowledge of the social brain in humans, which is a very active research topic (Lamm, Decety, & Singer, 2011; Schilbach et al., 2010; Singer et al., 2004). Furthermore, it seems likely that if any "special" aspect of human linguistic cognition has strong roots in, and continuity with, primate social intelligence, it is this complex of socially central abilities (Barton & Dunbar, 1997; Dunbar, 2009; Shultz & Dunbar, 2007; Silk, 2007). Indeed, primatologists Dorothy Cheney and Robert Seyfarth have convincingly argued that primate social cognition provides the most coherent foundation for the subsequent evolution of many derived features of human language comprehension (Cheney & Seyfarth, 1998; Seyfarth & Cheney, 2014).

6. Conclusion

In this overview, I first gave a general perspective on how we can use the comparative approach to understand behavioral evolution in an empirically grounded evolutionary context and then briefly illustrated how such a general perspective can be put to work analyzing both shared and derived features of human language. It seems clear that, once we take the crucial first step of subdividing language into multiple components, we find that most of these components are shared, sometimes quite widely, with other vertebrates. We have seen examples of both homology and convergence, and seen how both of these can be useful in formulating and testing biological hypotheses. The virtues of Tinbergen's multifaceted four questions concerning ontogeny, phylogeny, mechanisms, and adaptive functions have been illustrated. While a brief chapter like this one cannot explore all of the data relevant to language evolution, I hope that it will inspire the reader to pursue other more detailed analyses applying the comparative method to problems in the biology and evolution of language, including problems discussed in the remaining chapters in this volume.

REFERENCES

Abe, K., & Watanabe, D. (2011). Songbirds possess the spontaneous ability to discriminate syntactic rules. *Nature Neuroscience, 14*(8), 1067–1074.

Aristotle. (1970). *The history of animals* (A. L. Peck, Trans.). London: Heinemann. (Translation and reprint of Aristotle's [The history of animals] written 350 B.C.)

Armstrong, D. F., Stokoe, W. C., & Wilcox, S. E. (1984). Signs of the origin of syntax. *Current Anthropology, 35*(4), 349–368.

Barton, R. A., & Dunbar, R. I. M. (1997). Evolution of the social brain. In A. Whiten & R. W. Byrne (Eds.), *Machiavellian intelligence II* (pp. 240–263). Cambridge: Cambridge University Press.

Bates, E., & Elman, J. (1996). Learning rediscovered. *Science, 274*(5294), 1849–1850.

Beckers, G. J. L., Bolhuis, J. J., Okanoya, K., & Berwick, R. C. (2012). Birdsong neurolinguistics: Songbird context-free grammar claim is premature. *NeuroReport, 23*(3), 139–145.

Berwick, R. C., Friederici, A. D., Chomsky, N., & Bolhuis, J. J. (2013). Evolution, brain, and the nature of language. *Trends in Cognitive Sciences, 17*(2), 89–98.

Bickerton, D. (1990). *Language and species.* Chicago: Chicago University Press.

Bräuer, J., Call, J., & Tomasello, M. (2007). Chimpanzees really know what others can see in a competitive situation. *Animal Cognition, 10,* 439–448.

Bugnyar, T., Reber, S. A., & Buckner, C. (2016). Ravens attribute visual access to unseen competitors. *Nature Communications, 7,* 10506.

Byrne, R. W. (2000). Evolution of primate cognition. *Cognitive Science, 24*(3), 543–570.

Call, J., & Tomasello, M. (2008). Does the chimpanzee have a theory of mind? 30 years later. *Trends in Cognitive Sciences, 12*(5), 187–192.

Cheney, D. L., & Seyfarth, R. M. (1998). Why monkeys don't have language. In G. Petersen (Ed.), *The Tanner lectures on human values* (Vol. 19, pp. 173–210). Salt Lake City: University of Utah Press.

Chittka, L., & Skorupski, P. (2001). Information processing in miniature brains. *Proceedings of the Royal Society B, Biological Sciences, 278*(1707), 885–888.

Chomsky, N. (2010). Some simple evo devo theses: How true might they be for language? In R. Larson, V. Deprez, & H. Yamakido (Eds.), *The evolution of human language: Biolinguistic perspectives* (pp. 45–62). Cambridge: Cambridge University Press.

Crockford, C., Herbinger, I., Vigilant, L., & Boesch, C. (2004). Wild chimpanzees produce group-specific calls: A case for vocal learning? *Ethology, 110,* 221–243.

Crystal, D. (2003). *The Cambridge encyclopedia of language.* Cambridge: Cambridge University Press

Csibra, G., & Gergely, G. (2006). Social learning and social cognition: The case for pedagogy. In Y. Munakata & M. H. Johnson (Eds.), *Attention and Performance XXI: Processes of change in brain and cognitive development* (pp. 249–274). Oxford: Oxford University Press.

Csibra, G., & Gergely, G. (2011). Natural pedagogy as evolutionary adaptation. *Philosophical Transactions of the Royal Society B, 366,* 1149–1157.

Darwin, C. (1859). *On the origin of species.* London: John Murray.

Darwin, C. (1871). *The descent of man and selection in relation to sex.* London: John Murray.

Darwin, C. (1872). *The expression of the emotions in man and animals.* London: John Murray.

De Robertis, E. M., & Sasai, Y. (1996). A common plan for dorsoventral patterning in Bilateria. *Nature, 380,* 37–40.

Descartes, R. (1982). Discours de la Méthode V. In E. Haldane (Ed.), *The philosophical works of Descartes.* Cambridge: Cambridge University Press.

Dittus, W. P. G. (1984). Toque macaque food calls: Semantic communication concerning food distribution in the environment. *Animal Behaviour, 32,* 470–477.

Donoghue, M. J. (1992). Homology. In E. F. Keller & E. A. Lloyd (Eds.), *Keywords in evolutionary biology* (pp. 170–179). Cambridge, MA: Harvard University Press.

Dunbar, R. I. M. (2009). The social brain hypothesis and its implications for social evolution. *Annals of Human Biology, 36*(5), 562–572.

Emmorey, K. (2002). *Language, cognition and the brain: Insights from sign language research.* London: Lawrence Erlbaum.

Evans, C. S., Evans, L., & Marler, P. (1993). On the meaning of alarm calls: Functional reference in an avian vocal system. *Animal Behaviour, 46,* 23–38.

Felsenstein, J. (1985). Phylogenies and the comparative method. *American Naturalist, 125,* 1–15.

Felsenstein, J. (2004). *Inferring phylogenies.* Sunderland, MA: Sinauer Associates.

Fischer, J., Wheeler, B. C., & Higham, J. P. (2015). Is there any evidence for vocal learning in chimpanzee food calls? *Current Biology, 25*(21), 1028–1029.

Fitch, W. T. (2000). The phonetic potential of nonhuman vocal tracts: Comparative cineradiographic observations of vocalizing animals. *Phonetica, 57,* 205–218.

Fitch, W. T. (2009). The biology and evolution of language: "Deep homology" and the evolution of innovation. In M. S. Gazzaniga (Ed.), *The cognitive neurosciences* (Vol. 4, pp. 873–883). Cambridge, MA: MIT Press.

Fitch, W. T. (2010). *The evolution of language.* Cambridge: Cambridge University Press.

Fitch, W. T. (2011). Innateness and human language: A biological perspective. In M. Tallerman & K. R. Gibson (Eds.), *The Oxford handbook of language evolution* (pp. 143–156). Oxford: Oxford University Press.

Fitch, W. T. (2014). Toward a computational framework for cognitive biology: Unifying approaches from cognitive neuroscience and comparative cognition. *Physics of Life Reviews, 11*(3), 329–364.

Fitch, W. T. (2015). Book review: Evolving pragmatics. *Current Biology, 25*(23), R1110–R1112.

Fitch, W. T. (2017). Empirical approaches to the study of language evolution. *Psychonomic Bulletin and Review, 24*(1), 3–33.

Fitch, W. T. (2018). The biology and evolution of speech: A comparative analysis. *Annual Review of Linguistics, 4*, 255–279.

Fitch, W. T., & Friederici, A. D. (2012). Artificial grammar learning meets formal language theory: An overview. *Philosophical Transactions of the Royal Society B, 367*, 1933–1955.

Fitch, W. T., Huber, L., & Bugnyar, T. (2010). Social cognition and the evolution of language: Constructing cognitive phylogenies. *Neuron, 65*, 795–814.

Fitch, W. T., & Jarvis, E. D. (2013). Birdsong and other animal models for human speech, song, and vocal learning. In M. A. Arbib (Ed.), *Language, music, and the brain: A mysterious relationship* (Vol. 10, pp. 499–539). Cambridge, MA: MIT Press.

Fitch, W. T., Mathur, N., de Boer, B., & Ghazanfar, A. A. (2016). Monkey vocal tracts are speech-ready. *Science Advances, 2*(12), e1600723.

Fitch, W. T., & Reby, D. (2001). The descended larynx is not uniquely human. *Proceedings of the Royal Society of London, B, 268*(1477), 1669–1675.

Fitch, W. T., & Zuberbühler, K. (2013). Primate precursors to human language: Beyond discontinuity. In E. Zimmerman, S. Schmidt, & E. Altenmüller (Eds.), *The evolution of emotional communication* (pp. 26–48). Oxford: Oxford University Press.

Friederici, A. D. (2012). The cortical language circuit: From auditory perception to sentence comprehension. *Trends in Cognitive Sciences, 16*(5), 262–268.

Furness, W. H. (1916). Observations on the mentality of chimpanzees and orang-utans. *Proceedings of the American Philosophical Society, 55*(3), 281–290.

Gatesy, S. M., & Biewiener, A. A. (1991). Bipedal locomotion: Effects of speed, size and limb posture in birds and humans. *Journal of Zoology, 224*(1), 127–147.

Gehring, W. J., & Ikeo, K. (1999). Pax6: Mastering eye morphogenesis and eye evolution. *Trends in Genetics, 15*, 371–377.

Gentner, T. Q., Fenn, K. M., Margoliash, D., & Nusbaum, H. C. (2006). Recursive syntactic pattern learning by songbirds. *Nature, 440*, 1204–1207.

Gingras, B., Mohandesan, E., Boko, D., & Fitch, W. T. (2013). Phylogenetic signal in the acoustic parameters of the advertisement calls of four clades of anurans. *BMC Evolutionary Biology, 13*, e134.

Giurfa, M., Backhaus, W., & Menzel, R. (1995). Color and angular orientation in the discrimination of bilateral symmetric patterns in the honeybee. *Naturwissenschaften, 82*(4), 198–201.

Goodall, J. (1968). The behaviour of free-living chimpanzees in the Gombe Stream Reserve, Tanzania. *Animal Behaviour Monographs, 1*, 161–311.

Goodall, J. (1986). *The chimpanzees of Gombe: Patterns of behavior.* Cambridge, MA: Harvard University Press.

Griffiths, P. E. (2002). What is innateness? *Monist, 85*(1), 70–85.

Haesler, S., Rochefort, C., Geogi, B., Licznerski, P., Osten, P., & Scharff, C. (2007). Incomplete and inaccurate vocal imitation after knockdown of FoxP2 in songbird basal ganglia nucleus Area X. *PLOS Biology, 5*(12), e321.

Hall, B. K. (Ed.) (1994). *Homology: The hierarchical basis of comparative biology.* San Diego: Academic Press.

Hare, B., Call, J., Agnetta, B., & Tomasello, M. (2000). Chimpanzees know what conspecifics do and do not see. *Animal Behaviour, 59*(4), 771–785.

Hare, B., Call, J., & Tomasello, M. (2001). Do chimpanzees know what conspecifics know? *Animal Behavior, 61*(1), 139–151.

Harley, T. (2014). *The psychology of language: From data to theory.* Sussex, UK: Psychology Press.

Harvey, P. H., & Pagel, M. D. (1991). *The comparative method in evolutionary biology.* Oxford: Oxford University Press.

Hauser, M., Chomsky, N., & Fitch, W. T. (2002). The language faculty: What is it, who has it, and how did it evolve? *Science, 298*, 1569–1579.

Hayes, C. (1951). *The ape in our house.* New York: Harper.

Held, L. I. (2017). *Deep homology?* Cambridge: Cambridge University Press.

Herbranson, W. T., & Shimp, C. P. (2008). Artificial grammar learning in pigeons. *Learning and Behavior, 36*(2), 116–137.

Hockett, C. F. (1960). Logical considerations in the study of animal communication. In W. E. Lanyon & W. N. Tavolga (Eds.), *Animal sounds and communication* (pp. 392–430). Washington, DC: American Institute of Biological Sciences.

Hoff, E. (2008). *Language development.* Belmont, CA: Wadsworth.

Holstege, G., & Subramanian, H. H. (2016). Two different motor systems are needed to generate human speech. *Journal of Comparative Neurology, 524*, 1558–1577.

Hopson, J. A. (1966). The origin of the mammalian middle ear. *American Zoologist, 6*, 437–450.

Horwitz, B., Amunts, K., Bhattacharyya, R., Patkin, D., Jeffries, K., Zilles, K., & Braun, A. R. (2003). Activation of Broca's area during the production of spoken and signed language: A combined cytoarchitectonic mapping and PET analysis. *Neuropsychologia, 41*, 1868–1876.

Jäger, G., & Rogers, J. (2012). Formal language theory: Refining the Chomsky hierarchy. *Philosophical Transactions of the Royal Society B, 267*(1598), 1956–1970.

Janik, V. M., & Slater, P. J. B. (2000). The different roles of social learning in vocal communication. *Animal Behaviour, 60*, 1–11.

Jarvis, E. D. (2007). Neural systems for vocal learning in birds and humans: A synopsis. *Journal of Ornithology, 148*(1), 35–44.

Ji, Q., Luo, Z.-X., Yuan, C.-X., Wible, J. R., Zhang, J.-P., & Georgi, J. A. (2002). The earliest eutherian mammal. *Nature, 416*, 816–822.

Jiang, X., T. Long, W. Cao, et al. 2018. Production of supraregular spatial sequences by macaque monkeys. *Current Biology, 28*, 1851–1859.

Jürgens, U. (2002). Neural pathways underlying vocal control. *Neuroscience and Biobehavioral Reviews, 26*(2), 235–258.

Jürgens, U., & Alipour, M. (2002). A comparative study on the cortico-hypoglossal connections in primates, using biotin dextranamine. *Neuroscience Letters, 328*(3), 245–248.

Jürgens, U., Kirzinger, A., & von Cramon, D. Y. (1982). The effects of deep-reaching lesions in the cortical face area on phonation: A combined case report and experimental monkey study. *Cortex, 18*(1), 125–139.

Kaminski, J., Call, J., & Fischer, J. (2004). Word learning in a domestic dog: Evidence for "fast mapping." *Science, 304*, 1682–1683.

Kelemen, G. (1948). The anatomical basis of phonation in the chimpanzee. *Journal of Morphology, 82*(2), 229–256.

Kellogg, W. N. (1968). Chimpanzees in experimental homes. *Psychological Record, 18*, 489–498.

Koelsch, S., Maess, B., & Friederici, A. D. (2000). Musical syntax is processed in the area of Broca: An MEG study. *NeuroImage, 11*, 56.

Kuypers, H. G. J. M. (1981). Anatomy of the descending pathways. In J. Brookhart & V. B. Mountcastle (Eds.), *Handbook of physiology: Section I: The nervous system* (pp. 597–666). Bethesda, MD: American Physiological Society.

Laland, K. N., & Galef, B. (Eds.). (2009). *The question of animal culture*. Cambridge, MA: Harvard University Press.

Lamm, C., Decety, J., & Singer, T. (2011). Meta-analytic evidence for common and distinct neural networks associated with directly experienced pain and empathy for pain. *NeuroImage, 54*(3), 2492–2502.

Lankester, E. (1870a). On the use of the term "homology." *Annals and Magazine of Natural History, 6*, 342.

Lankester, E. (1870b). On the use of the term homology in modern zoology, and the distinction between homogenetic and homoplastic agreements. *Annals and Magazine of Natural History, 6*, 34–43.

Lehrman, D. S. (1953). A critique of Konrad Lorenz's theory of instinctive behavior. *Quarterly Review of Biology, 28*(4), 337–363.

Lemasson, A., Ouattara, K., Petit, E. J., & Zuberbühler, K. (2011). Social learning of vocal structure in a nonhuman primate? *BMC Evolutionary Biology, 11*(1), 1.

Lemon, R. N. (2008). Descending pathways in motor control. *Annual Review of Neuroscience, 31*, 195–218.

Lieberman, P. H. (1984). *The biology and evolution of language*. Cambridge, MA: Harvard University Press.

Lieberman, P. H. (2012). Vocal tract anatomy and the neural bases of talking. *Journal of Phonetics, 40*(4), 608–622.

Lieberman, P.H., Crelin, E. S., & Klatt, D. H. (1972). Phonetic ability and related anatomy of the newborn and adult human, Neanderthal man, and the chimpanzee. *American Anthropologist, 74*(3), 287–307.

Lieberman, P. H., Klatt, D. H., & Wilson, W. H. (1969). Vocal tract limitations on the vowel repertoires of rhesus monkey and other nonhuman primates. *Science, 164*, 1185–1187.

Lockwood, C. A., & Fleagle, J. G. (1999). The recognition and evaluation of homoplasy in primate and human evolution. *Yearbook of Physical Anthropology, 42*, 189–232.

Lorenz, K. (1953). *Comparative studies on the behaviour of the anatinae*. London: Goodale Press.

Lorenz, K. (1965). *Evolution and modification of behavior*. Chicago: University of Chicago Press.

Luo, Z.-X., Crompton, A. W., & Sun, A.-L. (2001). A new mammaliaform from the Early Jurassic and evolution of mammalian characteristics. *Science, 292*, 1535–1540.

Macedonia, J. M. (1991). What is communicated in the antipredator calls of lemurs: Evidence from playback experiments with ring-tailed and ruffed lemurs. *Ethology, 86*, 177–190.

MacSweeney, M., Capek, C. M., Campbell, R., & Woll, B. (2008). The signing brain: The neurobiology of sign language. *Trends in Cognitive Sciences, 12*(11), 432–440.

Maess, B., Koelsch, S., Gunter, T. C., & Friederici, A. D. (2001). Musical syntax is processed in Broca's area: An MEG study. *Nature Neuroscience, 4*, 540–545.

Manser, M. B., Seyfarth, R. M., & Cheney, D. L. (2002). Suricate alarm calls signal predator class and urgency. *Trends in Cognitive Sciences, 6*(2), 55–57.

Marler, P. (1991). The instinct to learn. In S. Carey & R. Gelman (Eds.), *The epigenesis of mind: Essays on biology and cognition* (pp. 37–66). Hillsdale, NJ: Lawrence Erlbaum Associates.

Marler, P. (1999). On innateness: Are sparrow songs "learned" or "innate"? In M. D. Hauser & M. Konishi (Eds.), *The design of animal communication* (pp. 293–318). Cambridge, MA: MIT Press.

Marshall, A. J., Wrangham, R. W., & Arcadi, A. C. (1999). Does learning affect the structure of vocalizations in chimpanzees? *Animal Behaviour, 58*, 825–830.

Mayr, E. 1963. *Animal Species and Evolution*. Cambridge, Mass., Harvard University Press.

McGrew, W. C. (2004). *The cultured chimpanzee*. Cambridge: Cambridge University Press.

Menzel, R. (2012). The honeybee as a model for understanding the basis of cognition. *Nature Reviews Neuroscience, 13*, 758–768.

Moore, R. (2016). Meaning and ostension in great ape gestural communication. *Animal Cognition, 19*(1), 223–231.

Müller, F. M. (1861). The theoretical stage, and the origin of language. In *Lectures on the Science of Language* (pp. 287–328). London: Longman, Green, Longman, and Roberts.

Negus, V. E. (1949). *The comparative anatomy and physiology of the larynx*. New York: Hafner Publishing Company.

Nishimura, T., Mikami, A., Suzuki, J., & Matsuzawa, T. (2006). Descent of the hyoid in chimpanzees: Evolution of face flattening and speech. *Journal of Human Evolution, 51*, 244–254.

Owen, R. (1843). *Lectures on the comparative anatomy and physiology of the invertebrate animals*. London: Longman, Brown, Green and Longmans.

Owren, M. J., Dieter, J. A., Seyfarth, R. M., & Cheney, D. L. (1993). Vocalizations of rhesus (*Macaca mulatta*) and Japanese (*M. fuscata*) macaques cross-fostered between species show evidence of only limited modification. *Developmental Psychobiology, 26*, 389–406.

Pagel, M. D. (1992). A method for the analysis of comparative data. *Journal of Theoretical Biology, 156*, 434–442.

Pallier, C., Devauchelle, A.-D., & Dehaene, S. (2011). Cortical representation of the constituent structure of sentences. *Proceedings of the National Academy of Sciences, 108*(6), 2522–2527.

Pepperberg, I. M., & Wilcox, S. E. (2000). Evidence for a form of mutual exclusivity during label acquisition by grey parrots (*Psittacus erithacus*)? *Journal of Comparative Psychology, 114*(3), 219–231.

Peters, G. (2002). Purring and similar vocalizations in mammals. *Mammal Review, 32*, 245–271.

Pilley, J. W., & Reid, A. K. (2011). Border collie comprehends object names as verbal referents. *Behavioural Processes, 86*(2), 184–195.

Poletiek, F. H., Fitz, H., & Bocanegra, B. R. (2016). What baboons can (not) tell us about natural language grammars. *Cognition, 151*, 108–112.

Povinelli, D. J., Nelson, K. E., & Boysen, S. T. (1990). Inferences about guessing and knowing by chimpanzees (*Pan troglodytes*). *Journal of Comparative Psychology, 104*, 203–210.

Premack, D., & Woodruff, G. (1978). Does the chimpanzee have a theory of mind? *Behavioral and Brain Sciences, 4*, 515–526.

Qiang, J., Currie, P. J., Norell, M. A., & Shu-An, J. (1998). Two feathered dinosaurs from northeastern China. *Nature, 393*, 753–761.

Ralls, K., Fiorelli, P., & Gish, S. (1985). Vocalizations and vocal mimicry in captive harbor seals, *Phoca vitulina*. *Canadian Journal of Zoology, 63*, 1050–1056.

Ravignani, A., Sonnweber, R.-S., Stobbe, N., & Fitch, W. T. (2013). Action at a distance: Dependency sensitivity in a New World primate. *Biology Letters, 9*(6), 20130852.

Reichmuth, C. J., & Casey, C. (2014). Vocal learning in seals, sea lions, and walruses. *Current Opinion in Neurobiology, 28*, 66–71.

Rey, A., Perruchet, P., & Fagot, J. (2012). Centre-embedded structures are a by-product of associative learning and working memory constraints: Evidence from baboons (*Papio papio*). *Cognition, 123*(1), 180–184.

Reynolds Losin, E. A., Russell, J. L., Freeman, H., Meguerditchian, A., & Hopkins, W. D. (2008). Left hemisphere specialization for oro-facial movements of learned vocal signals by captive chimpanzees. *PLOS ONE, 3*(6), e2529.

Rilling, J. K., Glasser, M. F., Preuss, T. M., Ma, X., Zhao, T., Hu, X., & Behrens, T. E. J. (2008). The evolution of the arcuate fasciculus revealed with comparative DTI. *Nature Neuroscience, 11*(4), 426–428.

Robinson, J. G. (1984). Syntactic structures in the vocalizations of wedge-capped capuchin monkeys, *Cebus nigrivittatus*. *Behaviour, 90*, 46–79.

Samuels, R. (2004). Innateness in cognitive science. *Trends in Cognitive Sciences, 8*(3), 136–141.

Sanderson, M. J., & Hufford, L. (Eds.). (1996). *Homoplasy: The recurrence of similarity in evolution*. San Diego: Academic Press.

Sanz, C., Call, J., & Morgan, D. B. (2009). Design complexity in termite-fishing tools of chimpanzees. *Biology Letters, 5*(3), 293–296.

Savage-Rumbaugh, E. S., Murphy, J., Sevcik, R. A., Brakke, K. E., Williams, S. L., & Rumbaugh, D. M. (1993). Language comprehension in ape and child. *Monographs of the Society for Research in Child Development, 58*, 1–221.

Scharff, C., & Petri, J. (2011). Evo-devo, deep homology and FoxP2: Implications for the evolution of speech and language. *Philosophical Transactions of the Royal Society B, 366*(1574), 2124–2140.

Schenker, N. M., Hopkins, W. D., Spocter, M. A., Garrison, A. R., Stimpson, C. D., Erwin, J. M., ... Sherwood, C. C. (2010). Broca's area homologue in chimpanzees (*Pan troglodytes*): Probabilistic mapping, asymmetry and comparison to humans. *Cerebral Cortex, 20*(3), 730–742.

Schilbach, L., Wilms, M., Eickhoff, S. B., Romanzetti, S., Tepest, R., Bente, G., ... Vogeley, K. (2010). Minds made for sharing: Initiating joint attention recruits reward-related neurocircuitry. *Journal of Cognitive Neuroscience, 22*(12), 2702–2715.

Schmelz, M., Call, J., & Tomasello, M. (2013). Chimpanzees predict that a competitor's preference will match their own. *Biology Letters, 9*(1), 20120829.

Schreiweis, C., Bornschein, U., Burguière, E., Kerimoglu, C., Schreiter, S., Dannemann, M., ... Graybiel, A. M. (2014). Humanized Foxp2 accelerates learning by enhancing transitions from declarative to procedural performance. *Proceedings of the National Academy of Sciences, 111*(39), 14253–14258.

Schusterman, R. J., & Krieger, K. (1984). California sea lions are capable of semantic comprehension. *Psychological Record, 34*, 3–25.

Scott-Phillips, T. C. (2014). *Speaking our minds*. New York: Palgrave Macmillan.

Seyfarth, R. M., & Cheney, D. L. (2014). The evolution of language from social cognition. *Current Opinion in Neurobiology, 28*, 5–9.

Seyfarth, R. M., Cheney, D. L., & Marler, P. (1980). Vervet monkey alarm calls: Semantic communication in a free-ranging primate. *Animal Behaviour, 28*, 1070–1094.

Shettleworth, S. J. (2009). *Cognition, evolution, and behavior.* Oxford: Oxford University Press.

Shubin, N., Tabin, C., & Carroll, S. (1997). Fossils, genes and the evolution of animal limbs. *Nature, 388*, 639–648.

Shubin, N., Tabin, C., & Carroll, S. (2009). Deep homology and the origins of evolutionary novelty. *Nature, 457*, 818–823.

Shultz, S., & Dunbar, R. I. M. (2007). The evolution of the social brain: Anthropoid primates contrast with other vertebrates. *Proceedings of the Royal Society B, 274*(1624), 2429–2436.

Siegel, J. M. (1995). Phylogeny and the function of REM sleep. *Behavioural Brain Research, 69*(1–2), 29–34.

Silk, J. B. (2007). Social components of fitness in primate groups. *Science, 317*(5843), 1347–1351

Simonyan, K. (2014). The laryngeal motor cortex: Its organization and connectivity. *Current Opinion in Neurobiology, 28*, 15–21.

Singer, T., Seymour, B., O'Doherty, J., Kaube, H., Dolan, R. J., & Frith, C. D. (2004). Empathy for pain involves the affective but not sensory components of pain. *Science, 303*(5661), 1157–1162.

Sonnweber, R. S., Ravignani, A., & Fitch, W. T. (2015). Non-adjacent visual dependency learning in chimpanzees. *Animal Cognition, 18*(3), 733–745.

Spierings, M. J., & ten Cate, C. (2016). Budgerigars and zebra finches differ in how they generalize in an artificial grammar learning experiment. *Proceedings of the National Academy of Sciences, 113*, E3977–E3984.

Stegmann, U. E. (Ed.) (2013). *Animal communication theory: Information and influence.* Cambridge: Cambridge University Press.

Stoeger, A. S., Mietchen, D., Oh, S., de Silva, S., Herbst, C. T., Kwon, S., & Fitch, W. T. (2012). An Asian elephant imitates human speech. *Current Biology, 22*, 2144–2148.

Stokoe, W. C. (1960). *Sign language structure: An outline of the communicative systems of the American deaf.* Silver Spring, MD: Linstock Press.

Striedter, G. F. (2004). *Principles of brain evolution.* Sunderland, MA: Sinauer.

Tabin, C. J., Carroll, S. B., & Panganiban, G. (1999). Out on a limb: Parallels in vertebrate and invertebrate limb patterning and the origin of the appendages. *American Zoologist, 39*, 650–663.

ten Cate, C. (1994). Perceptual mechanisms in imprinting and song learning in birds. In J. A. Hogan & J. J. Bolhuis (Eds.), *Causal mechanisms of behavioural development* (pp. 116–146). Cambridge: Cambridge University Press.

Terrace, H. S., Son, L. K., & Brannon, E. M. (2003). Serial expertise of rhesus macaques. *Psychological Science, 14*, 66–73.

Thornton, A., & McAuliffe, K. (2006). Teaching in wild meerkats. *Science, 313*, 227–229.

Tillmann, B., Janata, P., & Bharucha, J. J. (2003). Activation of the inferior frontal cortex in musical priming. *Cognitive Brain Research, 16*(2), 145–161.

Tinbergen, N. (1963). On aims and methods of ethology. *Zeitschrift für Tierpsychologie, 20*, 410–433.

Tomasello, M. (2007). If they're so good at grammar, then why don't they talk? Hints from apes' and humans' use of gestures. *Language Learning and Development, 3*(2), 133–156.

Tomasello, M., & Call, J. (1997). *Primate cognition.* Oxford: Oxford University Press.

Tomasello, M., Carpenter, M., Call, J., Behne, T., & Moll, H. (2005). Understanding and sharing intentions: The origins of cultural cognition. *Behavioral and Brain Sciences, 28*, 675–735.

Toro, J. M., & Trobalón, J. B. (2005). Statistical computations over a speech stream in a rodent. *Perception and Psychophysics, 67*(5), 867–875.

Ullman, M. T. (2001). A neurocognitive perspective on language: The declarative/procedural model. *Nature Reviews Neuroscience, 2*, 717–726.

van Heijningen, C. A. A., de Visser, J., Zuidema, W., & ten Cate, C. (2009). Simple rules can explain discrimination of putative recursive syntactic structures by a songbird species. *Proceedings of the National Academy of Sciences, 106*, 20538–20543.

Vauclair, J. (1996). *Animal cognition: An introduction to modern comparative psychology.* London: Harvard University Press.

Wang, L., Uhrig, L., Jarraya, B., & Dehaene, S. (2015). Representation of numerical and sequential patterns in macaque and human brains *Current Biology, 25*(15), 1966–1974.

Watson, S. K., Townsend, S. W., Schel, A. M., Wilke, C., Wallace, E. K., Cheng, L., ... Slocombe, K. E. (2015). Vocal learning in the functionally referential food grunts of chimpanzees. *Current Biology, 25*, 495–499.

White, S. A., Fisher, S. E., Geschwind, D. H., Scharff, C., & Holy, T. E. (2006). Singing mice, songbirds, and more: Models for FOXP2 function and dysfunction in human speech and language. *Journal of Neuroscience, 26*(41), 10376–10379.

Whiten, A., Goodall, J., McGrew, W. C., Nishida, T., Reynolds, V., Sugiyama, Y., ... Boesch, C. (1999). Cultures in chimpanzees. *Nature, 399*, 682–685.

Whitney, C. L. (1985). Serial order in wood thrush song. *Animal Behaviour, 33*, 1250–1265.

Wich, S. A., Swartz, K. B., Hardus, M. E., Lameira, A. R., Stromberg, E., & Shumaker, R. W. (2009). A case of spontaneous acquisition of a human sound by an orangutan. *Primates, 50*(1), 56–64.

Wild, J. M. (1993). Descending projections of the songbird nucleus robustus archistriatalis. *Journal of Comparative Neurology, 338*, 225–241.

Wilkins, A. S. (2002). *The evolution of developmental pathways.* Sunderland, MA: Sinauer.

Yang, C. (2013). Ontogeny and phylogeny of language. *Proceedings of the National Academy of Sciences, 110*(16), 6323–6327.

Yerkes, R. M., & Yerkes, A. W. (1929). *The great apes.* New Haven, CT: Yale University Press.

Zuberbühler, K. (2002). A syntactic rule in forest monkey communication. *Animal Behaviour, 63*, 293–299.

45 Primate Vocalization as a Model for Human Speech: Scopes and Limits

JULIA FISCHER AND STEFFEN R. HAGE

1. Nonhuman Primate Vocalizations and the Origin of Speech and Language

The question of language evolution is as intriguing as it is elusive, since we neither know precisely when the first humans began to talk to each other, nor what specifically happened that enabled them to do so. This heady combination of intense interest and few constraints by available data has given rise to much speculation and a flurry of different scenarios about how and when the language faculty evolved. One productive research avenue in the last few decades has been the comparative approach (Fitch, 2010). By studying the communication of other animals, the aim is to shed light on the mechanisms that support linguistic abilities, as well as the evolutionary trajectory of language evolution.

The language faculty can be deconstructed into a variety of core capacities, including symbolic representation, intentionality to communicate, and combinatorial capacities that enable the generation and comprehension of syntax-like structures (Hauser, Chomsky, & Fitch, 2002; Hockett, 1960). Importantly, language can take different modes of expression (Hagoort & Poeppel, 2013), including sign language, speech, and other symbolic means of communication such as written language or Morse code. In this chapter, we will focus on the vocal communication of nonhuman primates (hereafter "primates") as the potential substrate from which speech evolved. We are aware, however, that gestural communication, or a combination of gestures, facial expressions, and vocalizations have also been proposed as possible routes in the emergence of the language faculty (Arbib, Liebal, & Pika, 2008; Corballis, 2002).

Almost 50 years of research into the vocal communication of primates has provided ample evidence that the human language faculty vastly outperforms primate vocal communication systems in scope and flexibility (Balter, 2010; Ghazanfar, 2008; Fischer, 2017b; Hammerschmidt & Fischer, 2008; Hage & Nieder, 2016). From an evolutionary point of view, the emerging language faculty most likely co-opted prelinguistic components instead of evolving de novo language-specific components (Fitch, 2010). These prelinguistic modules can be conceived as preadaptations or precursors to language (Fischer & Price, 2017; Hurford, 2003). Such possible preadaptations might be call usage learning and perception learning in general, which are distinct from vocal production learning (Fitch, 2010; Janik & Slater, 1997, 2000). Briefly, three types of vocal learning have been defined (Arriaga & Jarvis, 2013; Janik & Slater, 1997; Fischer & Hammerschmidt, 2011; Petkov & Jarvis, 2012; Gultekin & Hage, 2017): *Auditory comprehension learning* is characterized by the ability to associate a distinct auditory cue with an adequate behavioral response or objects in the environment and is broadly distributed among vertebrates, including primates. *Vocal usage learning* is defined by the ability to volitionally control when and where, but not how, to produce a specific vocalization in a specific cognitive, social, or environmental context. Finally, *vocal production learning* is the ability to generate new vocal patterns as a response to experience-based auditory feedback.

In this chapter, we will focus on vocal production learning, as it is considered to be one of the key prerequisites for the acquisition of speech. We will review the evidence from studies in the wild, as well as results from investigations of the neural foundations of primate vocal behavior. The ultimate aim is to clarify whether the observed degree of flexibility likely provided the substrate for the enormous flexibility and control of human vocal production. Furthermore, we will examine the extent to which the neural structures supporting vocal production learning are conserved or have reorganized in humans to allow for the expression of linguistic thought. As pointed out above, it is important to distinguish between learning to produce sounds with a specific structure and learning the appropriate usage of calls (Janik & Slater, 2000; Seyfarth & Cheney, 1997; see also section 2). Before we turn to the question of vocal flexibility, we provide some context by briefly delving into the debate what primate vocalizations mean.

2. *What Do Primate Vocalizations Mean?*

A core debate in the study of primate vocalizations is what the calls express: do primates refer to objects or events in the environment, or do they rather express their affective evaluation of specific situations (Fischer & Price, 2017). A classic notion advanced by Darwin (1872) is that primate (and other mammal) vocalizations should be considered as expressions of affective or motivational states (see also Ackermann, Hage, & Ziegler, 2014; Hammerschmidt & Fischer, 2008). Peter Marler, a British ethologist who pioneered the study of birdsong learning and primate vocal communication, challenged the view that primate vocalizations were merely affective. In a persuasive essay, he suggested that primate vocalizations may have a symbolic component (Marler, 1977).

Marler had been inspired by Tom Struhsaker's observation that vervet monkeys, now classified as *Chlorocebus pygerythrus*, produce different types of alarm calls in response to different types of predators (Struhsaker, 1967). Importantly, the monkeys also produce different types of adaptive escape responses when confronted with different predator classes: when a leopard is lurking, they jump up into a tree, while they run out of a tree and into thick bushes when an eagle circles in the sky. Snakes, in turn, prompt the monkeys to stand on their hind legs, scanning the ground and mobbing the snake with conspecifics (Struhsaker, 1967). To test whether conspecific listeners would be able to interpret the different types of calls correctly in the absence of any contextual information, Marler teamed up with Robert Seyfarth and Dorothy Cheney, to produce what turned out to be one of the most influential studies in the study of primate vocal behavior (Seyfarth, Cheney, & Marler, 1980a, 1980b). In this study, they played back the different alarm calls to the monkeys in the absence of the predators, and they observed that the monkeys were likely to choose the appropriate escape strategy in response to the respective alarm call category (Seyfarth et al., 1980a). From the listeners' perspective, these calls appeared to function as though the callers referred to specific events in the environment, and subsequently, they were termed as *functionally referential* (Macedonia & Evans, 1993). Yet, further analyses revealed some important differences between the referential communication of humans and the functionally referential communication of other animals. One important requirement for (symbolic) referential communication is that the signaler has a certain degree of volitional control over their vocalizations and is thus able to choose (and change) the "labels" for different objects or events in the environment. Another requirement for referential communication is that listeners are able to attribute the correct meaning to vocalizations, by establishing the link between a word or sound and what it stands for.

One indicator that the vervet monkey alarm call system differs from human symbolic communication was the observation that vervets may produce similar sounding calls not only in response to, for example, snakes, but also during intergroup encounters (Cheney, 1984; Struhsaker, 1967). A recent detailed acoustic analysis set out to examine the context-specificity of the alarm calls as well as calls given during aggression in more detail (Price et al., 2015).

First, this analysis corroborated that alarm calls given to leopards, snakes, and eagles are acoustically different (figure 45.1A). Second, the study confirmed that female vervet monkeys produce similar call types in response to snakes and during intergroup encounters and also similar call types in response to eagles and during within-group aggression (figure 45.1B). Furthermore, male vervet monkeys (recorded in South Africa by Price) uttered barks not only in response to leopards (and other dangerous predators), but also during intergroup encounters (Price et al., 2015). A second indicator that vervets do not fulfill the key criterion for symbolic communication (namely volitional choice of the "label" or "signifier") is that the vocal repertoires of different populations and species of vervet monkeys do not differ substantially (see section 4).

3. *Primate Vocalizations as Expressions of Internal States*

As noted, vervet monkeys may produce similar sounding calls not only in response to predators, but also during aggression (Cheney, 1984; Price et al., 2015; Struhsaker, 1967). These observations beg the question whether similar calls are tied to similar or rather to different internal states of the animals. Possibly, different contexts may elicit similar internal states, which in turn evoke the same calls (figure 45.2A). Alternatively, different contexts evoke different internal states that activate the same pattern generators (figure 45.2B).

Because it is difficult to distinguish between these two accounts (similar calls reflect similar internal states vs. specific internal states can give rise to a variety of calls) by analyzing the animals' overt behavior, independent assessments of the "internal state" are needed. Jürgens (1979) investigated the link between specific call types and internal state in squirrel monkeys, *Saimiri sciureus*, using electric stimulation of the brain. The animals were trained to control this stimulation of specific brain areas by switching between different

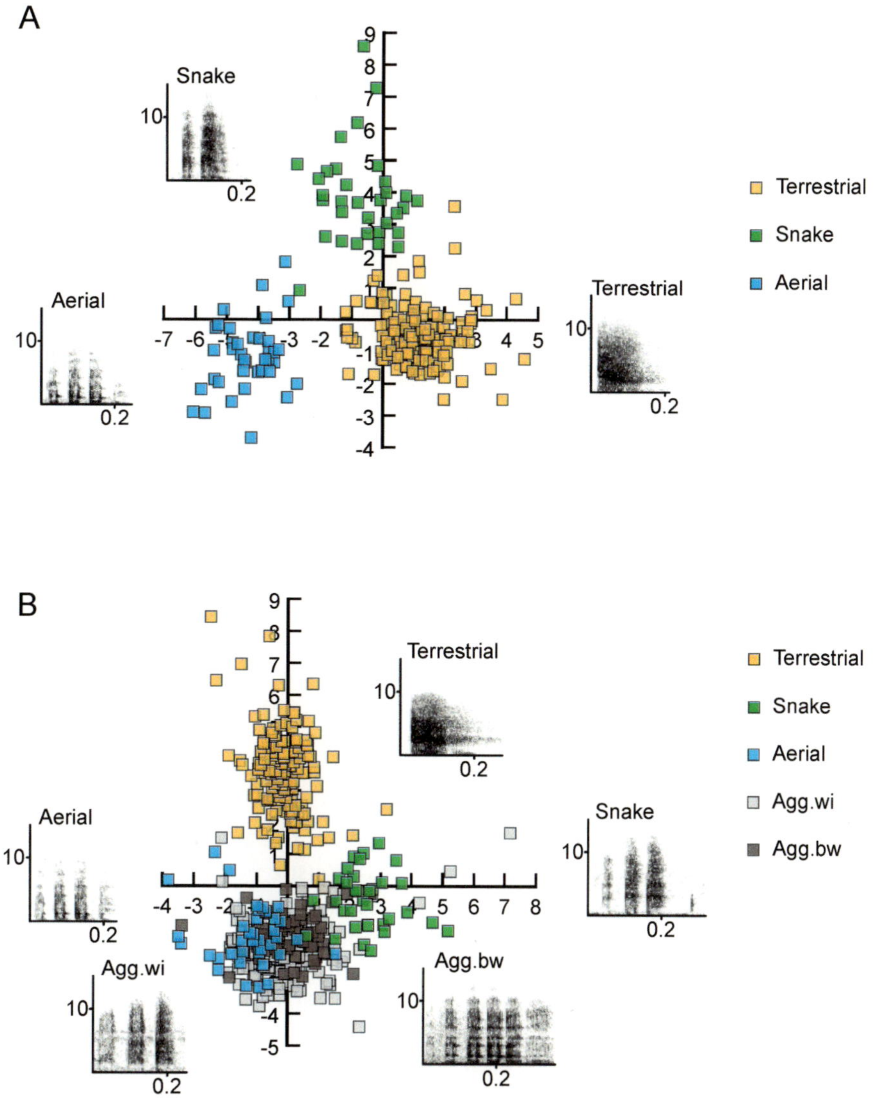

FIGURE 45.1 Structure and discriminability of female vervet vocalizations given in alarm and aggressive contexts. (A) Scatter plot of the discriminant scores with corresponding spectrograms of female alarm calls given in response to leopards, eagles, and snakes. (B) Scatter plot of the discriminant scores with corresponding spectrograms of female alarm calls given in response to leopards, eagles, and snakes, as well as during within- (Agg.wi) and between-group (Agg.bw) aggression. All spectrograms were made using the following settings in Avisoft: 256 fast Fourier transform frame size of 100% (Hamming window), frequency resolution 172 Hz; 50% window overlap, temporal resolution 2.9 ms. Adapted from Price et al. (2015), with permission.

compartments in a cage (Jürgens, 1979). Thus, they could either seek or avoid stimulation. In this way, it was possible to calculate an aversion score. Situations with high degree of aversion were associated with shriek cackles, shrieks, and alarm peeps, while in hedonistic situations, the monkeys produced twitters, groans, and chucks (Jürgens, 1979). While these results support the view that calls are tied to different affective states, the link between internal state and vocal output was not always consistent. In some instances, the stimulation of the same brain area could evoke different call types. Similarly, numerous studies reported that primates may produce very different call types in the same situation (e.g., Hammerschmidt & Fischer, 1998). Overall, however, these findings support the view of a close link between affective states and vocal output, similar to different degrees of emotional utterances like groans and shrieks of pain in humans, while they clearly contrast with the human ability to say *it hurts*, or *es tut weh*.

To date, we still do not fully understand the link between mental representation, associated affective states, and the resulting vocal behavior. The most parsimonious explanation appears to be that internal states are driving the activation of different pattern

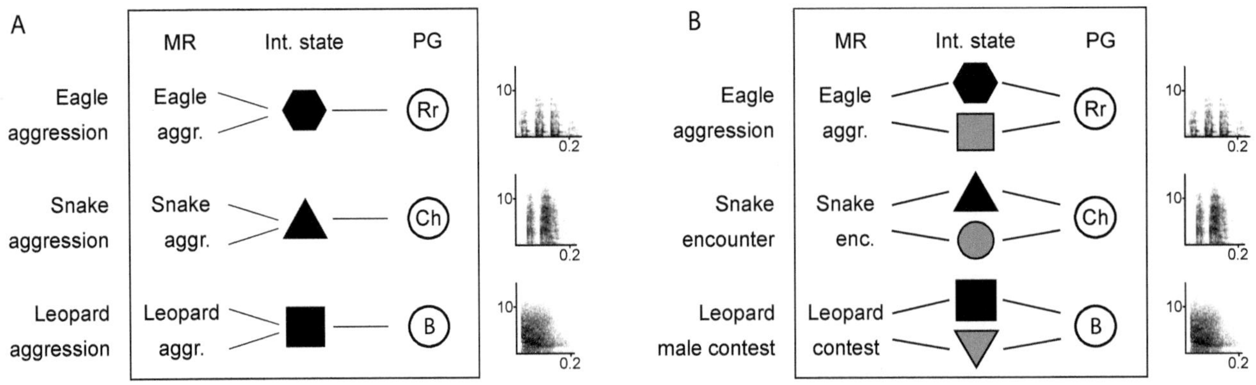

FIGURE 45.2 Different conceptions among the link of external events, putative mental representations (MRs) and affective evaluations, and specific vocalizations. (A) The occurrence of the same calls in different situations could be due to the fact that different situations evoke similar internal (Int) states, here represented by different shapes, which in turn activate specific pattern generators (PGs). (B) Alternatively, different situations may evoke different mental representations, as well as different internal states, which eventually converge onto the same pattern generators. Aggr = aggressions; B = bark; Ch = chutter; Contest = male contest; Enc = encounter; Rr = Rraup. From Fischer and Price (2017).

generators, which in turn give rise to specific vocalizations. In addition, there is ample evidence that variation in arousal affects the acoustic structure of calls. With increasing arousal, call rate and duration increase, and calls tend to become noisier, higher-pitched, and more variable (see Fischer & Price, 2017, for a review). At the same time, the animals have a certain degree of volitional control over the activation (use) of specific vocal patterns, that is, they may call on command (at least during specific stages in their lives) or remain silent. What is not possible, apparently, is to activate call patterns "freely," such as using alarm calls to obtain a food reward, for instance. Indeed, calls used in conditioning experiments all had an appetitive component, corroborating the view that the calls are broadly tied to specific affective states.

In summary, some call types appear to broadly correspond to specific internal states, but there is only a probabilistic mapping between events in the environment and specific calls, and this may also be true for the link between specific internal states and specific calls. This probabilistic link between event and signal, or state and signal poses specific challenges for listeners, who may need to draw on contextual information to disambiguate the information content of specific calls (Fischer, 2013; Fischer & Price, 2017).

The vervet monkey alarm call system, and indeed most primate vocal communication, thus rests on a combination of calls that are sufficiently specific on the one hand and listeners who are able to pick up on this variation to choose the appropriate responses, on the other hand. From the production side, this system bears little resemblance to human linguistic communication, however (Wheeler & Fischer, 2012). There is now ample evidence that a large number of primate species produce alarm calls that vary to a larger or lesser degree with predator type or context (Cäsar & Zuberbühler, 2012; Kirchhof & Hammerschmidt, 2006; Zuberbühler, 2000c, 2001). The commonalities between nonhuman primate and human communication appear to be rather at the side of the listeners, who are sensitive to this acoustic variation than at the levels of the signaler. Indeed, primates have been shown to categorize continuous variation in calls into distinct types, resembling categorical perception in human speech (Fischer, 1998, 2006). Furthermore, primates integrate information from different sources, such as caller reliability and acoustic variation (Cheney & Seyfarth, 1988), or contextual information and acoustic information when choosing different responses (Price & Fischer, 2014; Zuberbühler, 2000a). Thus, "smart" listeners who pick up on variation not only of their own species' calls but also of other species (Zuberbühler, 2000b) make the communication system as a whole relatively powerful. Yet, this also applies to other terrestrial mammal species, such as dogs. A striking example is Rico, the border collie who knew the names of more than 200 toys, and who was able to map new words onto novel toys (Kaminski, Call, & Fischer, 2004)—yet his vocal production was largely confined to barks and occasional growls.

4. *Between-Species Variation in Vocal Production*

Comparisons between species provide insights into the selective pressures and constraints that play a role in the evolution of vocal communication (Wilkins, Seddon, & Safran, 2013). Notably, the "barks" of male members of the genus *Chlorocebus* differ only very minimally between

East African and South African vervets (Price, Ndiaye, & Fischer, 2014), two subspecies of *Chlorocebus pygerythrus*, whose last common ancestor lived about 1.5 million years ago (Perelman et al., 2011; but see Warren et al., 2015 for markedly shorter divergence times). Even more strikingly, males of the West African congener *C. sabaeus*, whose last common ancestor with *Chlorocebus pygerythrus* lived around 2.1 million years ago, also produce barks with a highly similar call structure (figure 45.3). Further comparative analyses of the call structure of closely related species revealed that the acoustic distance correlates highly with phylogenetic distance. Within specific *call types*, such as the loud calls of leaf monkeys (Meyer et al., 2012), the alarm calls of macaques (Hauser, 1996), or the grunts of baboons, only minor differences in acoustic structure were observed (Maciej, Ndao, Hammerschmidt, & Fischer, 2013). A number of studies found high concordance between phylogenetic relationships and acoustic characteristics, indicating that genetic drift accounts for substantial parts of the variation (Geissmann, 1984; Meyer et al., 2012; Thinh, Hallam, Roos, & Hammerschmidt, 2011). The acoustic characteristics of the habitat, in contrast, do not appear to shape the structure of signals much, although the usage might be affected (Schneider, Hodges, Fischer, & Hammerschmidt, 2008). At the same time, there may be notable differences at the level of the *repertoire*: male Barbary macaques, *Macaca sylvanus*, do not produce loud calls, while male crested macaques, *Macaca nigra*, do (Neumann, Assahad, Hammerschmidt, Perwitasari-Farajallah, & Engelhardt, 2010). Furthermore, the vocal repertoire of geladas, *Theropithecus gelada*, differs markedly from that of baboons (Gustison, le Roux, & Bergman, 2012). Both environmental constraints and sexual selection appear to play a role here (Wilkins, Seddon, & Safran, 2013). In contrast, the vocal repertoires of different baboon species are largely similar (Maciej et al., 2013), despite the fact that their social systems vary substantially (Fischer et al., 2017).

Although systematic quantitative comparisons of the vocalizations of different chimpanzee, *Pan troglodytes*, subspecies have not been conducted yet, the verbal

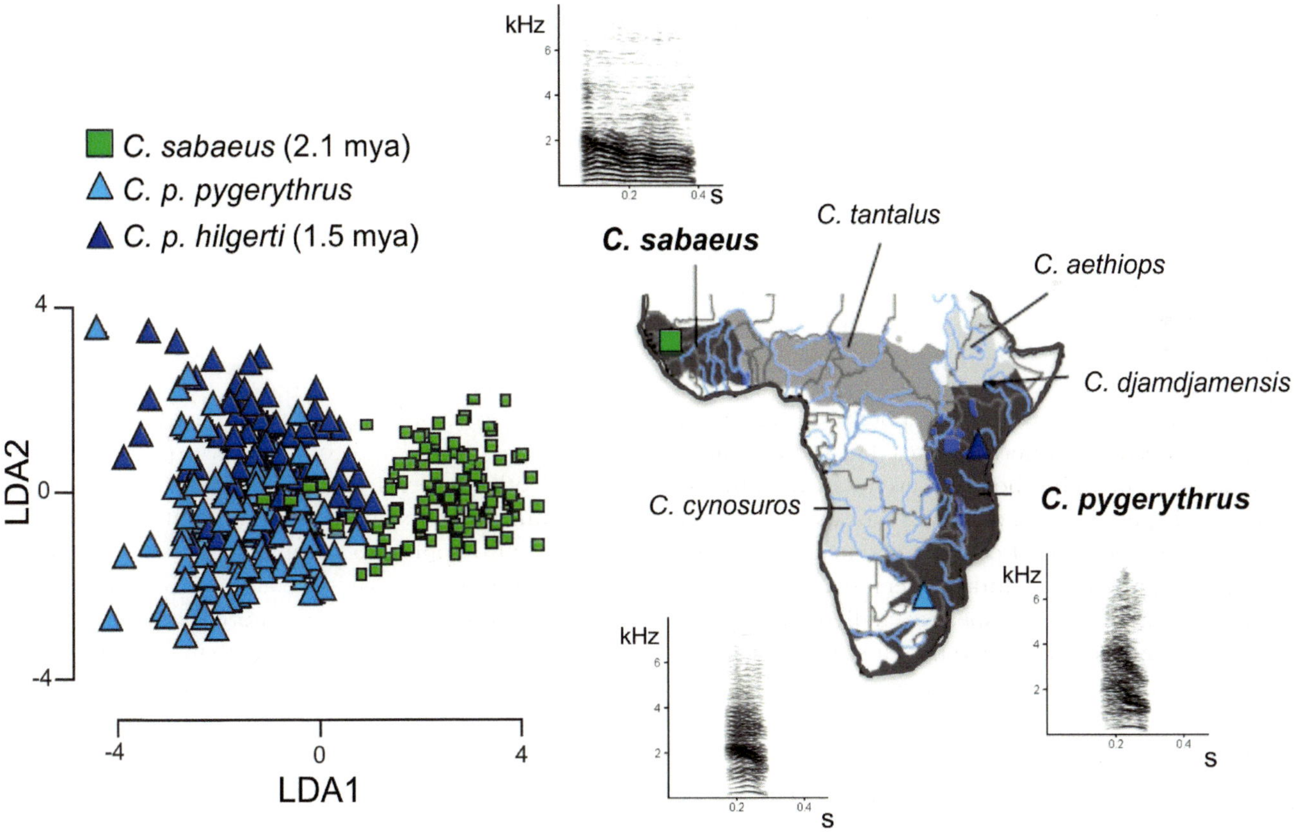

FIGURE 45.3 Scatterplot and spectrograms illustrating population differences in the acoustic structure of *C. sabaeus* (green monkey) from West Africa, *C. p. hilgerti*, and *C. p. pygerythrus* (vervet monkey) barks from East and South Africa, respectively. The scatterplot presents the distribution of the first and second linear discriminant analysis scores. Spectrograms illustrate a typical call exemplar for each call group, with typical calls defined as those that were most likely to be assigned by linear discriminant analysis to the correct caller/population. mya = millions of years ago. Modified from Price et al. (2014).

reports and spectrographic depictions of their vocalizations seem to be relatively similar. So-called *pant-hoots* as well as *screams, alert hoos,* and *rough grunts* have been reported for different subspecies (Crockford, Wittig, & Zuberbühler, 2014; Marler, 1969; Mitani & Macedonia, 1996; Slocombe & Zuberbühler, 2005a, 2005b). At the same time, the vocalizations of bonobos, *Pan paniscus* (Clay, Archbold, & Zuberbühler, 2015; Schamberg, Cheney, Clay, Hohmann, & Seyfarth, 2016), indicate some marked differences between the two members of this genus. To pin down the selective pressures shaping vocal repertoires more precisely, broad comparative analyses would be needed that encompass the entire repertoire of calls of a given species. Presently, such data are not available and given the huge sampling effort, this question mostly likely remains elusive.

Overall, the analysis of the vocal communication of primates thus suggests that the calls are species-specific and genetically fixed. Yet, within these constraints, some flexibility appears to be possible. We will first provide an overview over studies that investigated variation within species at the group and individual levels, before we turn to developmental studies that more explicitly addressed the question of auditory feedback.

5. Within-Species Variation in Vocal Production

Marshall, Wrangham, and Arcadi (1999) investigated whether the pant-hoots of two groups of captive chimpanzees differed with regard to their acoustic structure, and they also compared the calls of the captive chimpanzees to vocalizations recorded from three males of the Kanyawara community in Uganda. First, pant-hoots recorded in captivity and in the wild were largely similar with regard to the spectral structure of the elements of this multisyllable call, suggesting a strong innate component of the call type. Second, there was significant variation in temporal aspects, specifically the duration of the climax element and the number of build-up elements in the calls of the two captive groups (comprising 11 and 3 males, respectively). The authors deemed genetic differences between the two groups unlikely, although the provenance of the captive chimpanzees was mostly unclear. Marshall et al. (1999) also reported that one of the males in one enclosure had introduced a novel element into his pant-hoot that subsequently was also used by five other males from the same colony. This novel element ("Bronx cheer") is produced by blowing air through pursed lips, suggesting that chimpanzees have some control over their respiratory activity and their lips. Yet, it has to be noted that the novel element was not voiced, that is, it did not require volitional control over the voice box. Thus, this report did not provide convincing evidence for much voluntary control over the acoustic structure of voiced patterns in chimpanzees.

A more recent study by Watson et al. (2015) examined the structure of chimpanzee food grunts after individuals from two different groups ($n=7$ and 6, respectively) had been brought together to form a new group. The animals' calls were recorded immediately after introduction, and then one and three years later. At the same time, the social integration of the animals was studied using social network analysis. The results of the statistical analyses suggested that with increasing integration, the calls of the animals from the different groups became more similar to each other (Watson et al., 2015). Yet, the call structure of the animals from the two groups largely had overlapped from the start of the study on. Thus, the data could also be taken to support the opposite claim: the fact that subjects from two groups produce the same general call type in response to food strongly indicates that both the call types and the general usage are largely genetically fixed (Fischer, Wheeler, & Higham, 2015).

Studies on orangutans (*Pongo* spp.) suggest a larger degree of vocal flexibility compared to that of other apes. Orangutans were found to produce novel voiced and unvoiced sounds that are not part of the natural repertoire of vocalizations (Wich et al., 2009; Wich et al., 2012). A number of subjects living in captivity were reported to whistle (an unvoiced mode of sound production), and one subject, Tilda, a former entertainment orangutan, produced "clicks" and "faux-speech" (Lameira et al., 2013). These behaviors mainly occurred in the feeding context when Tilda aimed to attract the attention of the keeper. The faux-speech consisted of broad-band vowel-like elements with a fundamental frequency between 50 and 95 Hz, with clearly discernible formants.

Pygmy marmosets, *Cebuella pygmaea,* were found to modify their trill vocalizations when placed with a new partner such that the calls of both partners became more similar to each over time (Snowdon & Elowson, 1999). Ultimately, such social influences may give rise to group-specific differences ("dialects"). For instance, Green (1975) reported that three different populations of Japanese macaques, *Macaca fuscata,* produced acoustically different "coo" calls. The alarm calls recorded in two populations of Barbary macaques differed between sites, and playback experiments revealed that this variation was salient to the listeners (Fischer, Hammerschmidt, & Todt, 1998).

In addition, some monkey species seem to be capable to slightly modulate some call parameters. A few studies were able to show that a few parameters of call production, for example, call frequency, amplitude, and duration, are under volitional control in operant conditioning experiments in primates (Larson, Sutton, & Lindeman, 1973; Sutton, Larson, Taylor, & Lindeman, 1973; Trachy, Sutton, & Lindeman, 1981; Hihara, Yamada, Iriki, & Okanoya, 2003; Hage, Gavrilov, & Nieder, 2013). There is also evidence from the wild that Olive baboons, *Papio anubis*, increase the length of their grunts when they move in habitats with poor visibility compared to open habitats (Ey, Rahn, Hammerschmidt, & Fischer, 2009). Interestingly, the fundamental frequency of the grunts was not affected, tentatively suggesting that the change in call pattern was not simply due to increased arousal.

Taken together, these studies suggest that within the species-specific calls, some more or less systematic variation is possible, but this appears to be rather in the magnitude of an "accent" rather than an entirely different language. Thus, it has also been likened to the vocal accommodation in human speech (Hammerschmidt & Fischer, 2008). Remarkably, the mechanisms that underpin this "vocal accommodation" in primates are poorly understood (Fischer, 2017a).

6. Ontogenetic Development of Primate Vocalizations

So-called *Kaspar-Hauser* experiments, deafening, as well as lesion studies were conducted to examine the role of auditory feedback on vocal development. In the Kaspar-Hauser experiments, squirrel monkeys were raised without auditory feedback from conspecifics, by either hand-raising them or by muting the mother. These monkeys still produced the complete species-specific vocal repertoire and used it in most motivational and social contexts appropriately (Winter, Handley, Ploog, & Schott, 1973). Deaf-born or deafened monkeys showed similar call patterns as their normal-hearing conspecifics with only minor changes in call amplitude and pitch (Talmage-Riggs, Winter, Ploog, & Mayer, 1972; Hammerschmidt, Jürgens, & Freudenstein, 2001; see figure 45.4). Several studies in Old World and New World monkeys found no differences in vocal pattern production after lesioning homologue brain structures that are crucial for speech production in humans (Sutton, Larson, & Lindeman, 1974; Sutton, Trachy, & Lindeman, 1985; Aitken, 1981; Jürgens, Kirzinger, & von Cramon, 1982; Kirzinger & Jürgens, 1982). Changes in acoustic structure that can be observed during ontogeny can largely be attributed to growth, and sometimes also practice/increased motor control (Ey, Hammerschmidt, Seyfarth, & Fischer, 2007; Hammerschmidt, Newman, Champoux, & Suomi, 2000). In contrast to human speech, therefore, the development of the normal species-specific acoustic structure does not require auditory experience in primates (Ackermann et al., 2014; Deacon, 2010; Hage & Nieder, 2016).

Recent studies, however, reveal a certain role of auditory feedback in vocal development in New World monkeys. Marmosets and tamarins typically live in family groups. The mother normally gives birth to twins, and both the father and older siblings carry the young. When separated, both young and adult subjects emit vocalizations that play an important role in re-establishing contact. Contact calls by one party typically elicit counter-calling from the other party, which constitutes an excellent opportunity to study the effects of parental feedback on the vocal development of the young. During ontogeny, marmoset vocalizations undergo distinct changes in several call parameters, such as decreases in call frequency and entropy and increases in call duration and intercall intervals (Takahashi et al., 2015; Zhang & Ghazanfar, 2016; Pistorio, Vintch, & Wang, 2006). Contingent parental auditory feedback accelerates vocal development into mature vocalizations (Takahashi et al., 2015; Takahashi, Liao, & Ghazanfar, 2017). Another recent study investigated whether such parental auditory feedback is an obligate requirement for proper vocal development or whether it simply accelerates vocal development, without a detrimental effect if absent, by comparing the vocal behavior of two sets of offspring. One set of infants was normally raised, while the other was separated from the parents after the third postnatal month (Gultekin & Hage, 2017). All the monkeys eventually produced mature vocalizations. In contrast to normally raised monkeys, however, marmosets with limited parental feedback also produced infant-specific vocal behavior, which is characterized by specific call types, distributions, and transitions (figure 45.5). These results indicate that marmoset monkeys need contingent parental auditory feedback to learn which call types to use or not in specific behavioral contexts and, therefore, that such auditory feedback can prune repertoires of innate call types.

7. Preadaptations in Vocal Production

Even if the neuronal network that is involved in human speech production is more complex than the vocal motor system of primates (Hage & Nieder, 2016), evolutionary theories postulate preadaptations in the primate lineage (Darwin, 1859), no matter how exiguous they might be. A hot topic in this context is the question of the evolution of speech rhythm, that is the pattern with which voicing and articulation (and breathing) are entrained to generate the rapid temporal changes

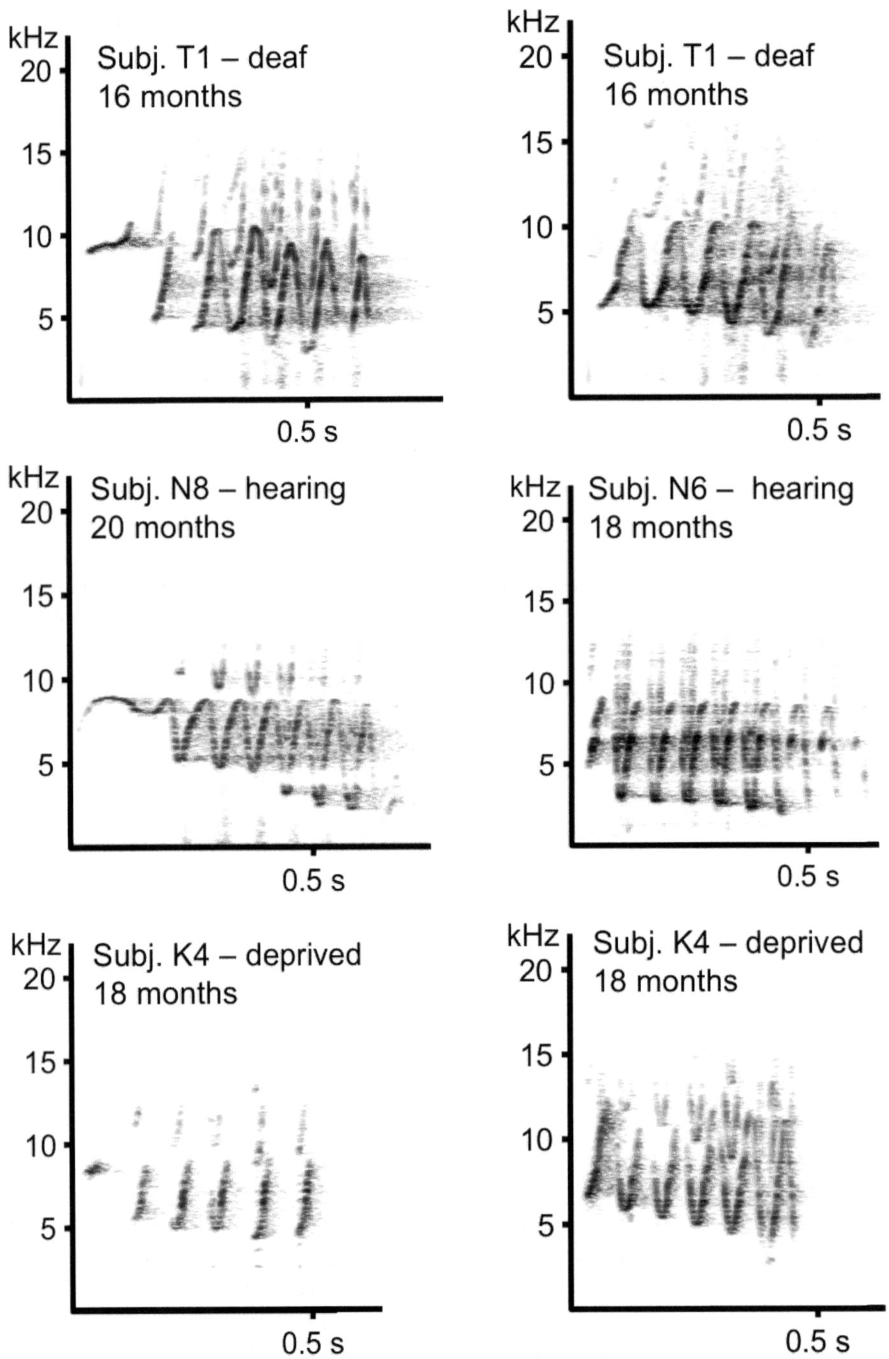

FIGURE 45.4 Spectrograms of trill vocalizations from four adult squirrel monkey, *Saimiri sciureus*, individuals. Recordings from a deaf subject (Subj) (*top*), two hearing subjects (*middle*), and an acoustically deprived subject (*lower*). The age in months and identification codes are given. Spectrograms kindly made available by Kurt Hammerschmidt; original recordings from Winter et al. (1973), and reanalyzed by Hammerschmidt et al. (2001).

that characterize speech. Across all languages that have been investigated to date, a speech rhythm of 3 to 8 Hz was observed. This rhythm is mostly reflected in the rate of syllable production (reviewed in Ghazanfar & Takahashi, 2014). MacNeilage and Davis (1999, 2001) posited that human speech evolved from the rhythmic facial movements in ancestral primates.

Intriguingly, rhythmic facial movements in the form of lip-smacking can be found in numerous primate taxa, including different macaque species (Ferrari et al., 2006; Fischer & Hammerschmidt, 2002; Preuschoft, 1992; Rowell & Hinde, 1962), different baboon species, *Papio* spp. (Smuts & Watanabe, 1990), and chimpanzees (Parr, Cohen, & de Waal, 2005).

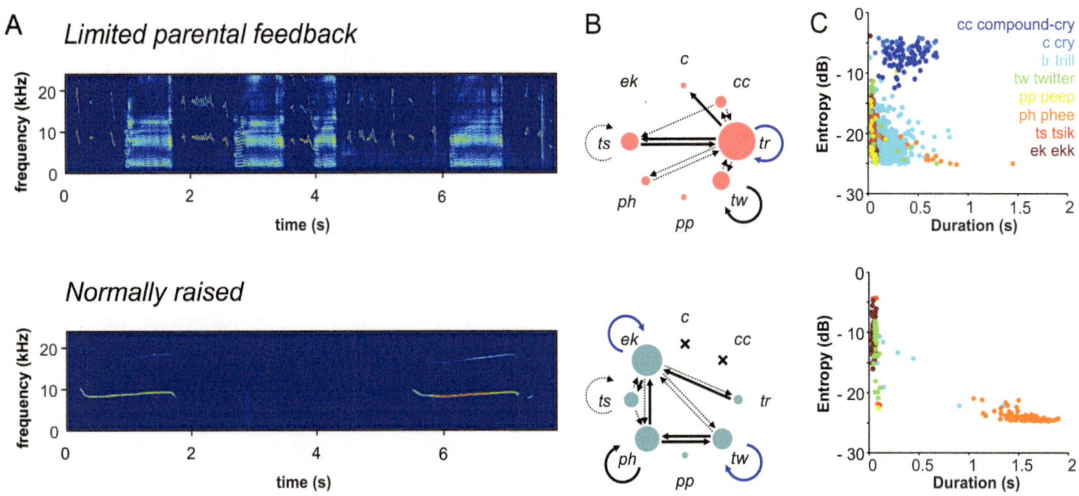

FIGURE 45.5 Vocal development of subadult marmoset monkeys depends on parental feedback. (A) Spectrograms of vocal sequences indicate differences in vocal behavior between a normally raised and a limited parental feedback monkey. (B) Transition diagrams visualizing vocal sequences of a normally raised and a limited parental feedback monkey. Each node in the diagram corresponds to a type of call, and the arrows correspond to the transitions between call types. Node size is proportional to the fraction of the call types, and edge size is proportional to the transition probability between calls. Blue arrows are transitions above 10% occurrences, black solid arrows between 5% and 10%, and dashed arrows between 1% and 5%. (C) Differences in call entropy and duration between a normally raised and a limited parental feedback monkey. c = cry; cc = compound-cry; ek = ekk; ph = phee; pp = peep; tr = trill; ts = tsik; tw = twitter. Modified from Gultekin and Hage (2017).

Lip-smacking involves chewing movements of the jaw and/or rhythmic parting of the lips. Ghazanfar, Takahashi, Mathur, and Fitch (2012) used cineradiography to assess the dynamics of rhesus monkey, *Macaca mulatta*, lip-smacking. They found that the lips, tongue, and parts of the voice box move with a 5 Hz rhythm when the animals lip-smack. This rate was substantially faster than chewing movements. Furthermore, during lip-smacking, a functional coordination of the involved structures was observed (Ghazanfar et al., 2012). Similar conclusions were drawn from the analysis of other primate species. For instance, geladas produce so-called wobbles, which are produced while lip-smacking (Bergman, 2013). The gelada vocal repertoire appears highly derived and remarkably distinct from their sister group, the members of the genus *Papio* who all reveal largely similar call types (Maciej et al., 2013). With 6–9 Hz, the rhythm of wobble production also falls into the range of human speech production. The faux-speech of Tilda, the orangutan, revealed peaks in lip movement and voicing between 4.5 and 6 Hz and substantial coherence between the two features (Lameira et al., 2015). These findings are compatible with the idea that such rhythmic movements provide the evolutionary motor substrate for speech production; it has been argued that these rhythms are driven by brain oscillations that also play a role in cognition, by segregating information (Giraud & Poeppel, 2012). Despite these intriguing links between motor patterns and cognitive processes, it is still unclear how volitional control over the vocal apparatus emerged and how the representational and sociocognitive aspects of language evolved (Arbib, 2017).

8. Volitional Control over Call Usage

Call usage learning—that is, the capability to volitionally control when and where, but not how, to produce a specific vocalization in a specific cognitive, social, or environmental context—is another important step in vocal learning (Arriaga & Jarvis, 2013). Conceptually, one might distinguish two types of call usage learning. First, the ability to withhold or initiate a specific vocalization, although it is still tied to the respective (motivational) context (Seyfarth et al., 1980a; Seyfarth & Cheney 1986; Price et al., 2015), and second, the more elaborate ability to decouple calls of their innate vocal repertoire from the accompanying motivational state, such that the subjects are able to use the calls in a novel context (Fitch, 2010; Ackermann et al., 2014; Hage & Nieder, 2016). In both types of vocal usage learning, vocal output does not have to depend on learning new call patterns. Vocal usage learning is deemed a critical, albeit insufficient, preadaptation for the development of vocal production learning, a flexible communicative system in the primate lineage (Balter, 2010; Ackermann et al., 2014).

Several behavioral studies report that monkeys are able to volitionally initiate vocal output and instrumentalize their calls in a goal-directed way (Fischer & Price, 2017). Chacma baboons, for instance, use their "grunt" calls in a strategic way to facilitate affiliative interactions with specific individuals (Silk, Seyfarth, & Cheney, 2016). Furthermore, primates appear to be able to take into account the attentional state of the recipient when signaling, as they are more likely to use acoustic signals when the recipient cannot see them, indicating a certain degree of control over the signaling modality (e.g., Liebal, Pika, & Tomasello, 2004). Primates may also produce or withhold alarm calls depending on the social context (Seyfarth et al., 1980a; Rendall, Seyfarth, Cheney, & Owren, 1999), a phenomenon associated with the umbrella term *audience effects* (Zuberbühler, 2008). Furthermore, they modulate vocal timing in response to external ambient noise (Egnor, Wickelgren, & Hauser, 2007; Roy, Miller, Gottsch, & Wang, 2011) or conspecific calls (Hage, 2013; Miller, Beck, Meade, & Wang, 2009). There is also evidence from operant conditioning tasks that primates are able to produce a vocalization or remain silent (Sutton et al., 1973; Aitken & Wilson, 1979; Hihara, Yamada, Iriki, & Okanoya, 2003; Koda, Oyakawa, Kato, & Masataka, 2007; Coude et al., 2011; Hage et al., 2013). These studies clearly indicate that several monkey species have volitional control on the timing of their vocal output in both natural situations as well as operant conditioning tasks.

Given that primates are equipped with a limited set of species-specific calls, but a certain degree of volitional control over call usage, their strategy might be to employ call combinations to broaden the scope of the messages that might be conveyed (Arnold & Zuberbühler, 2006b). Putty-nosed monkeys, *Cercopithecus nictitans martini*, possess two call types, "hacks" and "pyows," which are emitted in different contexts. In response to the appearance of an eagle, the monkeys predominantly (though not exclusively) produce a series of hacks or hacks followed by pyows. When they encounter a leopard, the animals typically give a series of pyows, which sometimes fade into hacks (Arnold & Zuberbühler, 2006a). Interestingly, before initiating group movement, the animals produce mixed series of pyows and hacks (Arnold & Zuberbühler, 2008). Yet there are no clear rules specifying how the succession of pyows and hacks may be generated. Instead, a probabilistic system is arguably at work here, where the presence of leopards is more likely to elicit pyows. Playbacks of such mixed sequences elicited group travel, while pure pyow series evoked cryptic behavior in female monkeys (Arnold & Zuberbühler, 2008). Thus, the different call combinations were meaningful to the listeners. In the meantime, a number of studies in other species corroborated the view that not only call structure but also the occurrence of calls may vary with context (e.g., Candiotti, Zuberbühler, & Lemasson, 2011). Yet, none of these studies found clear rules for the generation of calls (see also Scott-Phillips, Gurney, Ivens, Diggle, & Popat, 2014), rendering it unlikely that these combinations are a result of a computational process that constitutes a preadaptation for human syntax (see chapter 47 in this volume by ten Cate & Petkov). The exact processes underlying the production of different types of sequences remains unclear, but in light of the probabilistic variation with context, the most parsimonious explanation appears to be that they are mediated by variation in internal state.

So how much volitional control do monkeys indeed have over their vocal output? In a recent study, Hage et al. (2013) demonstrated that rhesus monkeys are capable to selectively emit different call types in response to distinct visual cues. At first, they showed that monkeys can be trained in a visual detection task to vocalize whenever a colored visual cue appeared on the screen, showing consistent vocal performance (figure 45.6A). In addition, this study revealed that rhesus monkeys can be trained to switch between two distinct vocalizations ("coos" and "grunts") on command on a trial-by-trial base in a visual discrimination task (figure 45.6B). These results indicate that rhesus monkeys are able to decouple their innate calls in a goal-directed way from the corresponding state of arousal and, therefore, to instrumentalize their vocal output to perform a specific task successfully.

The observed capability for volitional control on vocal output seems to change during development. In a longitudinal study, Hage, Gavrilov, and Nieder (2016) collected data over several years to investigate potential developmental changes in cognitive vocal behavior of macaques from the juvenile period to adulthood. They observed a systematic decline of cognitive vocal behavior, which could not be explained by a general loss of vocal output, environmental changes, lack of motivation, or a general loss of cognitive function. Since this loss of cognitive vocal behavior occurred at the transition between the juvenile and the adult period, the authors suggested that connections between the monkeys' neural central executive and the vocal motor network as, for example, the connections between prefrontal and cingulate cortexes might have gotten lost due to developmental reorganization in the monkey brain (Hage et al., 2016). These results might indicate that one important aspect of flexible communication might already have been present in the primate lineage and could be exploited during the emergence of functional flexibility of prelinguistic vocalizations of human infants (Oller et al., 2013).

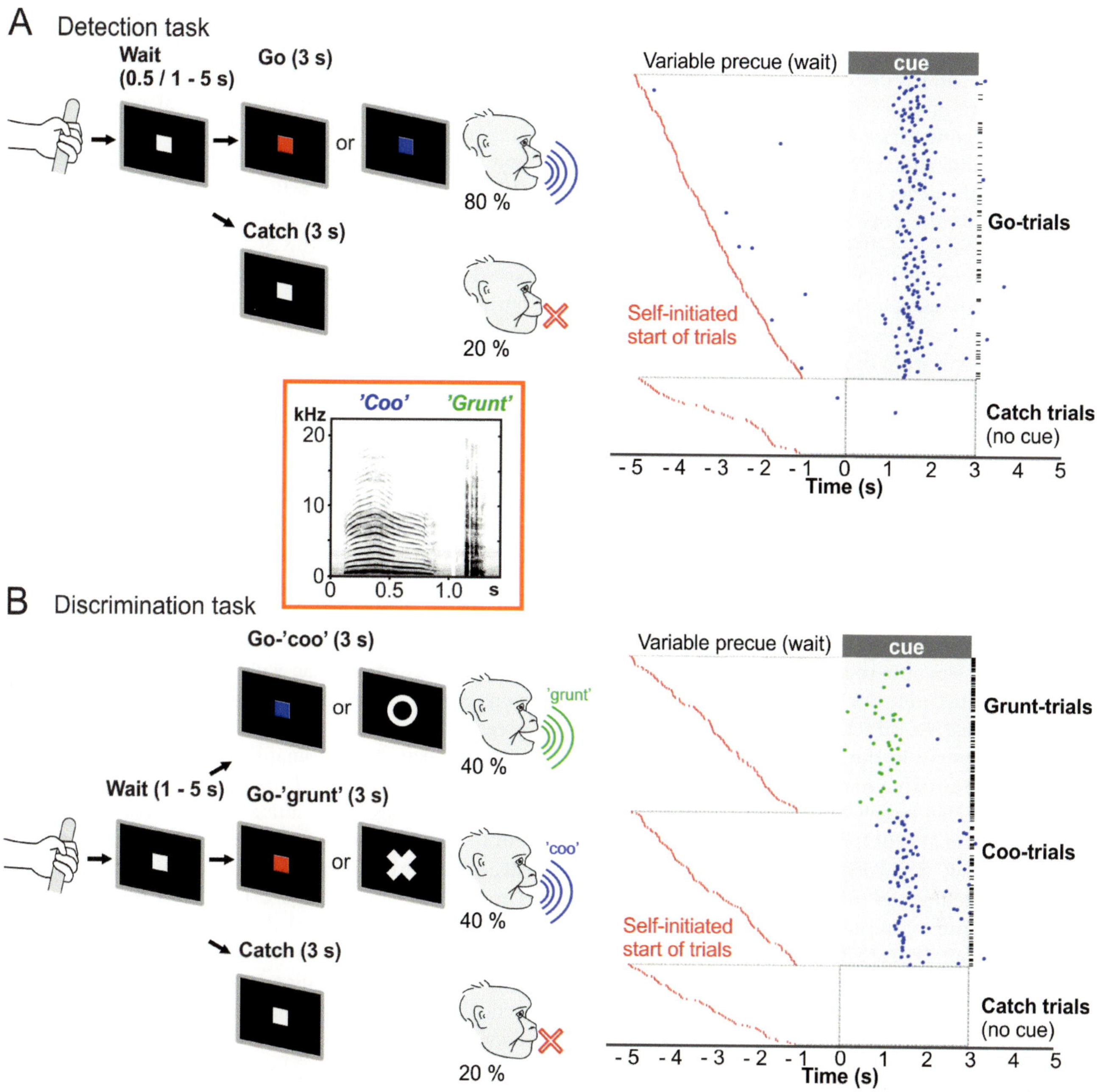

FIGURE 45.6 Two types of vocal usage learning in rhesus monkeys. (A) Experimental design (*left*) and vocal performance (*right*) in a detection task in which monkeys were trained in a go/no go protocol to vocalize whenever a visual cue (red or blue square) appeared (go trials) or to withhold vocalization whenever the precue (white square) remained (catch trials). (B) Experimental design (*left*) and vocal performance (*right*) in a discrimination task in which monkeys were trained in a successive training period to utter distinct vocalizations in response to specific visual cues. For both task types (A and B), examples of a single session indicate successful vocal performances. Trials were pseudorandomized but for better presentation, responses to go and catch trials (in A) as well as to grunt, coo, and catch trials (in B) are sorted according to the length of the precue signal (white square during the wait period) in the figure. Each line represents a single trial, blue circles indicate vocal onsets (in A), blue circles indicate coo call onsets and green circles grunt call onsets (in B). Go trials ignored by the monkey (misses) are marked with a horizontal black bar at trial end. Inset: Spectrograms of representative coo and grunt vocalizations. Spectral intensity is represented by different shades of gray. Modified from Hage et al. (2013).

In none of the cases where operant conditioning was employed to assess vocal flexibility did the animals produce a vocal pattern de novo. Instead, they shifted the median distribution of distinct call parameters within the natural boundaries of their innate repertoire. In summary, these studies indicate that monkeys possess preadaptations such as call usage learning that are crucial for the evolution of a learned vocal communication system but lack the ability to learn or imitate new vocal signals (Deacon, 2010; Fitch, 2010; Ackermann et al., 2014). Human speech, however, requires much more than just being able to volitionally control the onset of a distinct vocalization or shifting the duration of calls. Such limited vocal learning capabilities are far from sufficient for the production of human speech signals (Fitch, 2010). From all what is known so far, the ability to generate new vocal patterns and thus vocal pattern learning does not seem to exist in primates.

9. Neural Control Mechanisms Underlying Volitional Control of Vocal Behavior

All primates possess a vocal motor network that produces species-specific vocalizations with a largely fixed structure. This network consists of two structurally and functionally distinct parts. First, a vocal pattern-generating system in the brain stem with an upstream limbic vocal-initiating network driving the pattern generator controlled largely by the animal's affective state (Ackermann et al., 2014; Jürgens, 2002; Hage & Nieder, 2016). Second, a cortical network including several structures such as the monkey areas 44 and 45 (putative homologues of human Broca's area) in the ventrolateral prefrontal cortex (vlPFC), as well as the premotor and/or primary motor cortices, which are involved in the volitional control of vocal initiation and call modulation (Coude et al., 2011; Fukushima, Saunders, Fujii, Averbeck, & Mishkin, 2014; Gemba, Kyuhou, Matsuzaki, & Amino, 1999; Hage & Nieder, 2013; Simonyan & Horwitz, 2011). This dual network model suggests the incipient linking of the prefrontal central executive of the brain with the vocalization system, as a key neurobiological event and preadaptation for the evolution of speech and language in the primate lineage (Hage & Nieder, 2016). This hypothesis is supported by several comparative anatomical studies demonstrating that the basic architectonic plans of vlPFC are similar in both monkey and man, despite considerable development of the human vlPFC in the human brain.

The above-mentioned conditioning experiments where monkeys are trained to vocalize on cue opened the door to investigate these neuronal preadaptations of cognitive control of vocal behavior in the monkey brain. The first neurophysiological studies that have investigated conditioned vocal behavior indicate that the anterior cingulate cortex (ACC) might play a crucial role in the cognitive control of monkey vocal production. In humans, ACC is implicated to be involved in the initiation of nonverbal vocal utterances such as laughter (Wild, Rodden, Grodd, & Ruch, 2003). At the same time, it seems also to be important for human speech, since bilateral infarction of the ACC results in akinetic mutism (Barris & Schuman, 1953; Nielsen & Jacobs, 1951), but with the potential for restoration of speech characterized by monotonous intonation, suggesting that the ACC is involved in the emotional intonation of human speech (Jürgens, Kirzinger, & von Cramon, 1982; Paus, 2001). In primates, bilateral destruction of ACC abolished conditioned vocal behavior, but left spontaneous vocal output intact (Smith, 1945; Jürgens & Ploog, 1970; West & Larson, 1995). Recently, several studies suggested a putative monkey homologue of Broca's area in the vlPFC of rhesus monkeys that might play a major role in the cognitive control of vocal behavior (figure 45.7A). Electrical microstimulation in monkey area 44 elicited orofacial responses (Petrides, Cadoret, & Mackey, 2005), and auditory stimulation with species-specific vocalizations evoked distinct patterns of brain activity in monkey areas 44 and 45 in awake behaving rhesus monkeys (Gil-da-Costa et al., 2006; Romanski & Goldman-Rakic, 2002; Romanski, Averbeck, & Diltz, 2005) that were comparable to responses in the human Broca's area (Joly et al., 2012). In humans, Broca's area is one of the crucial structures involved in speech production (Broca, 1861; Dronkers & Baldo, 2010). Using direct cortical recordings, recent studies revealed that Broca's area in humans is predominantly activated before the utterance of a speech signal, but is silent during the corresponding articulation, indicating that Broca's area is involved in coordinating speech initiation rather than in producing speech output directly (Flinker et al., 2015).

Vocalization-related activity that specifically predicts the preparation of instructed vocalizations in response to a visual cue was recorded in monkey areas 44 and 45 (putative homologues of human Broca's area in the left hemisphere; Hage & Nieder, 2013) and the ventral premotor cortex (Coude et al., 2011; Hage & Nieder, 2013; figure 45.7B). Furthermore, neuronal responses during conditioned vocalizations were higher than during spontaneous vocalizations, suggesting a specific involvement of the putative monkey homologue of Broca's area during volitional monkey vocalizations (figure 45.7C). In addition, it has been shown that neurons within this area show audio-vocal response properties. These neurons showed auditory response properties combined with a vocalization-related activity preceding

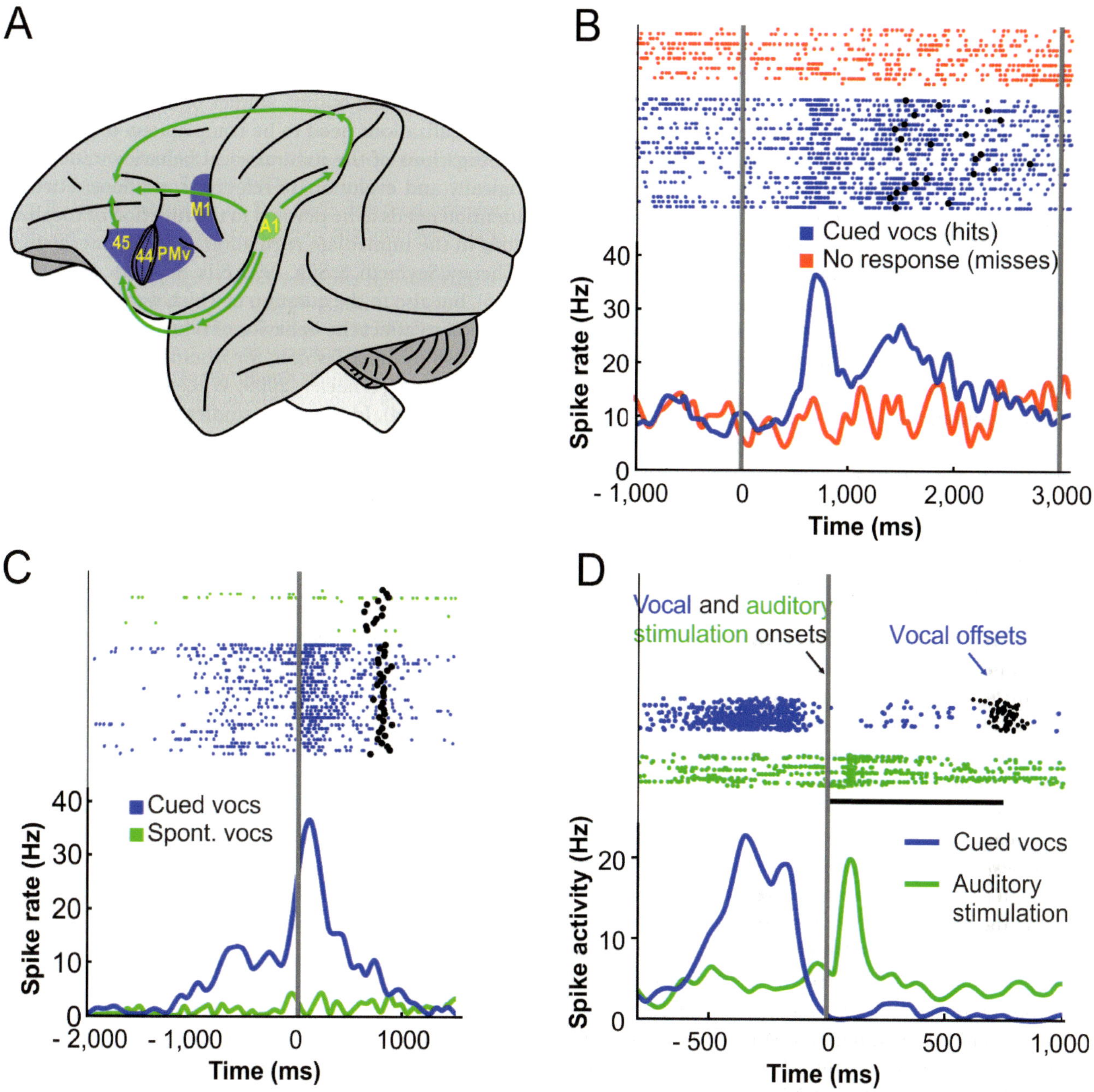

FIGURE 45.7 Vocalization-correlated and audio-vocal activity in single neurons of the monkey homologue of human Broca's area. (A) Lateral view of a macaque brain indicating the monkey homologue of human Broca's area (Brodmann's areas 44 and 45), ventral premotor (PMv), and motor cortex (M1) as well as the auditory cortex (A1) with a scheme of direct and indirect connections between A1 and monkey area 45. (B) Example neuron recorded in the monkey homologue of human Broca's area showing a significant increase in neuronal activity during trials with cued vocalizations (hits) in comparison to no response trials (misses). (C) Example neuron recorded in area 45 (within the putative monkey homologue of human Broca's area) showing significantly higher activity prior to vocal onset of volitional coo calls (cued vocs) compared to spontaneous coo calls (spont vocs). (D) Example neuron recorded in the monkey homologue of human Broca's area showing a phasic response during auditory stimulation and increased neuronal activity prior to volitional calls (cued vocs). (B–D) *Upper* panels show raster plots, black dots indicate vocal onset during hit trials in B and D and mark call offsets in C; *lower* panels represent the corresponding spike density histograms averaged and smoothed for illustration. The vertical gray lines indicate the onset of the go signal in D, the onset and offset of the go signal in B, and the vocal onset in C. Modified from Hage and Nieder (2016) (A) and Hage and Nieder (2013, 2015) (B–D).

Gemba, H., Kyuhou, S., Matsuzaki, R., & Amino, Y. (1999). Cortical field potentials associated with audio-initiated vocalization in monkeys. *Neuroscience Letters, 272*, 49–52

Ghazanfar, A. A. (2008). Language evolution: Neural differences that make the difference. *Nature Neuroscience, 11*, 382–384

Ghazanfar, A. A., & Takahashi, D. Y. (2014). Facial expressions and the evolution of the speech rhythm. *Journal of Cognitive Neuroscience, 26*(3), 1196–1207.

Ghazanfar, A. A., Takahashi, D. Y., Mathur, N., & Fitch, W. T. (2012). Cineradiography of monkey lip-smacking reveals putative precursors of speech dynamics. *Current Biology, 22*(13), 1176–1182.

Gil-da-Costa, R., Martin, A., Lopes, M. A., Munoz, M., Fritz, J. B., & Braun, A. R. (2006). Species-specific calls activate homologs of Broca's and Wernicke's areas in the macaque. *Nature Neuroscience, 9*, 1064–1070.

Giraud, A.-L., & Poeppel, D. (2012). Cortical oscillations and speech processing: Emerging computational principles and operations. *Nature Neuroscience, 15*(4), 511–517.

Green, S. (1975). Dialects in Japanese monkeys: Vocal learning and cultural transmission of locale-specific vocal behavior? *Zeitschrift Für Tierpsychologie, 38*, 304–314.

Gultekin, Y. B., & Hage, S. R. (2017). Limiting parental feedback disrupts vocal development in marmoset monkeys. *Nature Communications, 7*, 14046.

Gustison, M. L., le Roux, A., & Bergman, T. J. (2012). Derived vocalizations of geladas (*Theropithecus gelada*) and the evolution of vocal complexity in primates. *Philosophical Transactions of the Royal Society B: Biological Sciences, 367*(1597), 1847–1859.

Hage, S. R. (2013). Audio-vocal interactions in vocal communication of squirrel monkeys and their neurobiological implications. *Journal of Comparative Physiology A, 199*, 663–668.

Hage, S. R., Gavrilov, N., & Nieder, A. (2013). Cognitive control of distinct vocalizations in rhesus monkeys. *Journal of Cognitive Neuroscience, 25*, 1692–1701.

Hage, S. R, Gavrilov, N., & Nieder, A. (2016). Developmental changes of cognitive vocal control in monkeys. *Journal of Experimental Biology, 219*, 1744–1749.

Hage, S. R., & Nieder, A. (2013). Single neurons in monkey prefrontal cortex encode volitional initiation of vocalizations. *Nature Communications, 4*, 2409.

Hage, S. R., & Nieder, A. (2015). Audio-vocal interaction in single neurons of the monkey ventrolateral prefrontal cortex. *Journal of Neuroscience, 35*(18), 7030–7040.

Hage, S. R., & Nieder, A. (2016). Dual neural network model for the evolution of speech and language. *Trends in Neurosciences, 39*, 813–829.

Hagoort, P., & Poeppel, D. (2013). The infrastructure of the language-ready brain. In M. A. Arbib (Ed.), *Language, Music, and the Brain* (Vol. 10, pp. 233–255). Cambridge, MA: MIT Press.

Hammerschmidt, K., & Fischer, J. (1998). The vocal repertoire of Barbary macaques: A quantitative analysis of a graded signal system. *Ethology, 104*, 203–216.

Hammerschmidt, K., & Fischer, J. (2008). Constraints in primate vocal production. In D. K. Oller & U. Griebel (Eds.), *Evolution of communicative flexibility: Complexity, creativity, and adaptability in human and animal communication* (pp. 93–119). Cambridge, MA: MIT Press.

Hammerschmidt, K., Jürgens, U., & Freudenstein, T. (2001). Vocal development in squirrel monkeys. *Behaviour, 138*, 1179–1204.

Hammerschmidt, K., Newman, J. D., Champoux, M., & Suomi, S. J. (2000). Changes in rhesus macaque "coo" vocalizations during early development. *Ethology, 106*(10), 873–886.

Hauser, M. D. (1996). *The evolution of communication*. Cambridge, MA: MIT Press.

Hauser, M. D., Chomsky, N., & Fitch, W. T. (2002). The faculty of language: What is it, who has it, and how did it evolve? *Science, 298*(5598), 1569–1579.

Hihara, S., Yamada, H., Iriki, A., & Okanoya, K. (2003). Spontaneous vocal differentiation of coo-calls for tools and food in Japanese monkeys. *Neuroscience Research, 45*, 383–389.

Hockett, C. F. (1960). The origin of speech. *Scientific American, 203*, 88–111.

Holstege, G., & Subramanian, H. H. (2016). Two different motor systems are needed to generate human speech. *Journal of Comparative Neurology, 524*, 1558–1577.

Hurford, J. R. (2003). The language mosaic and its evolution. In M. H. Christiansen & S. Kirby (Eds.), *Language evolution* (pp. 38–57). Oxford: Oxford University Press.

Janik, V. M., & Slater, P. J. B. (1997). Vocal learning in mammals. *Advances in the Study of Behavior, 26*, 59–99.

Janik, V. M., & Slater, P. J. B. (2000). The different roles of social learning in vocal communication. *Animal Behaviour, 60*, 1–11.

Joly, O., Pallier, C., Ramus, F., Pressnitzer, D., Vanduffel, W., & Orban, G. A. (2012). Processing of vocalizations in human and monkeys: A comparative fMRI study. *NeuroImage, 62*, 1376–1389.

Jürgens, U. (1979). Vocalization as an emotional indicator: A neuroethological study in the squirrel monkey. *Behaviour, 69*(1/2), 88–117.

Jürgens, U. (2002). Neural pathways underlying vocal control. *Neuroscience and Biobehavioral Reviews, 26*(2), 235–258.

Jürgens, U., Kirzinger, A., & von Cramon, D. (1982). The effects of deep-reaching lesions in the cortical face area on phonation: A combined case report and experimental monkey study. *Cortex, 18*, 125–139.

Jürgens, U., & Ploog, D. (1970). Cerebral representation of vocalization in the squirrel monkey. *Experimental Brain Research, 10*, 532–554.

Kaminski, J., Call, J., & Fischer, J. (2004). Word learning in a domestic dog: Evidence for "fast mapping." *Science, 304*(5677), 1682–1683.

Kirchhof, J., & Hammerschmidt, K. (2006). Functionally referential alarm calls in tamarins (*Saguinus fuscicollis* and *Saguinus mystax*) : Evidence from playback experiments. *Ethology, 112*(4), 346–354.

Kirzinger, A., & Jürgens, U. (1982). Cortical lesion effects and vocalization in the squirrel monkey. *Brain Research, 233*, 299–315.

Koda, H., Oyakawa, C., Kato, A., & Masataka, N. (2007). Experimental evidence for the volitional control of vocal production in an immature gibbon. *Behaviour, 144*, 681–692.

Lameira, A. R., Hardus, M. E., Bartlett, A. M., Shumaker, R. W., Wich, S. A., & Menken, S. B. J. (2015). Speech-like rhythm in a voiced and voiceless orangutan call. *PLOS ONE, 10*(1), e116136.

Lameira, A. R., Hardus, M. E., Kowalsky, B., de Vries, H., Spruijt, B. M., Sterck, E. H. M., ... Wich, S. A. (2013). Orangutan (*Pongo* spp.) whistling and implications for the emergence of an open-ended call repertoire: A replication and extension. *Journal of the Acoustical Society of America, 134*(3), 2326–2335.

Larson, C. R., Sutton, D., & Lindeman, R. C. (1973). Cerebellar regulation of phonation in rhesus monkey (*Macaca mulatta*). *Experimental Brain Research, 33*, 1–18.

Liebal, K., Pika, S., & Tomasello, M. (2004). Social communication in siamangs (*Symphalangus syndactylus*): Use of gestures and facial expressions. *Primates, 45*(1), 41–57.

Macedonia, J. M., & Evans, C. S. (1993). Variation among mammalian alarm call systems and the problem of meaning in animal signals. *Ethology, 93*, 177–197.

Maciej, P., Ndao, I., Hammerschmidt, K., & Fischer, J. (2013). Vocal communication in a complex multi-level society: Constrained acoustic structure and flexible call usage in Guinea baboons. *Frontiers in Zoology, 10*(1), 58.

MacNeilage, P. F., & Davis, B. L. (1999). Evolution of the form of spoken words. *Evolution of Communication, 3*(1), 3–20.

MacNeilage, P. F., & Davis, B. L. (2001). Motor mechanisms in speech ontogeny: Phylogenetic neurobiological and linguistic implications. *Current Opinion in Neurobiology, 11*, 696–700.

Marler, P. (1969). Vocalizations of wild chimpanzees: An introduction. In *Proceedings 2nd International Congress Primatology, Atlanta 1968* (Vol. 1, pp. 94–100). Basel: Karger.

Marler, P. (1977). Primate vocalization: Affective or symbolic? In G. Bourne (Ed.), *Progress in ape research* (pp. 85–96). New York: Academic Press.

Marshall, A. J., Wrangham, R. W., & Arcadi, A. C. (1999). Does learning affect the structure of vocalizations in chimpanzees? *Animal Behaviour, 58*(4), 825–830.

Meyer, D., Hodges, K., Rinaldi, D., Wijaya, A., Roos, C., & Hammerschmidt, K. (2012). Acoustic structure of male loud-calls support molecular phylogeny of Sumatran and Javanese leaf monkeys (genus *Presbytis*). *BMC Evolutionary Biology, 12*(1), 16.

Miller, C. T., Beck, K., Meade, B., & Wang, X. (2009). Antiphonal call timing in marmosets is behaviorally significant: Interactive playback experiments. *Journal of Comparative Physiology A, 195*, 783–789.

Miller, C. T., Thomas, A. W., Nummela, S. U., & de la Mothe, L. (2015). Responses of primate frontal cortex neurons during natural vocal communication. *Journal of Neurophysiology, 114*, 1158–1171.

Mitani, J. C., & Macedonia, J. M. (1996). Selection for acoustic individuality within the vocal repertoire of wild chimpanzees. *International Journal of Primatology, 17*(4), 569–583.

Neumann, C., Assahad, G., Hammerschmidt, K., Perwitasari-Farajallah, D., & Engelhardt, A. (2010). Loud calls in male crested macaques, *Macaca nigra*: A signal of dominance in a tolerant species. *Animal Behaviour, 79*(1), 187–193.

Nielsen, J. M., & Jacobs, L. L. (1951). Bilateral lesions of the anterior cingulate gyri; report of case. *Bulletin of the Los Angeles Neurological Society, 16*(2), 231.

Oller, D. K., Buder, E. H., Ramsdell, H. L., Warlaumont, A. S., Chorna, L., & Bakeman, R. (2013). Functional flexibility of infant vocalization and the emergence of language. *Proceedings of the National Academy of Sciences, 110*, 6318–6323.

Owren, M. J., Amoss, R. T., & Rendall, D. (2011). Two organizing principles of vocal production: Implications for nonhuman and human primates. *American Journal of Primatology, 73*, 530–544.

Parr, L., Cohen, M., & de Waal, F. (2005). Influence of social context on the use of blended and graded facial displays in chimpanzees. *International Journal of Primatology, 26*(1), 73–103.

Paus, T. (2001) Primate anterior cingulate cortex: Where motor control, drive and cognition interface. *Nature Reviews Neuroscience, 2*, 417–424.

Perelman, P., Johnson, W. E., Roos, C., Seuánez, H. N., Horvath, J. E., Moreira, M. A. M., ... Pecon-Slattery, J. (2011). A molecular phylogeny of living primates. *Plos Genetics, 7*(3), e1001342.

Petkov, C. I., & Jarvis, E. D. (2012). Birds, primates, and spoken language origins: Behavioral phenotypes and neurobiological substrates. *Frontiers in Evolutionary Neuroscience, 4*, 12.

Petrides, M., Cadoret, G., & Mackey, S. (2005). Orofacial somatomotor responses in the macaque monkey homologue of Broca's area. *Nature, 435*, 1235–1238.

Pistorio, A.L., Vintch, B., & Wang, X. (2006). Acoustic analysis of vocal development in a New World primate, the common marmoset (*Callithrix jacchus*). *Journal of the Acoustical Society of America, 120*, 1655–1670.

Preuschoft, S. (1992). "Laughter" and "smile" in Barbary macaques (*Macaca sylvanus*). *Zeitschrift für Tierpsychologie, 91*, 220–236.

Price, T., & Fischer, J. (2014). Meaning attribution in the West African green monkey: Influence of call type and context. *Animal Cognition, 17*(2), 277–286.

Price, T., Ndiaye, O., & Fischer, J. (2014). Limited geographic variation in the acoustic structure of and responses to adult male alarm barks of African green monkeys. *Behavioural Ecology and Sociobiology, 68*(5), 815–825.

Price, T., Wadewitz, P., Cheney, D. L., Seyfarth, R. M., Hammerschmidt, K., & Fischer, J. (2015). Vervets revisited: A quantitative analysis of alarm call structure and context specificity. *Scientific Reports, 5*(13220), 1–11.

Rendall, D., Seyfarth, R. M., Cheney, D. L., & Owren, M. J. (1999). The meaning and function of grunt variants in baboons. *Animal Behaviour, 57*(3), 583–592.

Romanski, L. M., Averbeck, B. B., & Diltz, M. (2005). Neural representation of vocalizations in the primate ventrolateral prefrontal cortex. *Journal of Neurophysiology, 93*, 734–747.

Romanski, L. M., & Goldman-Rakic, P. S. (2002). An auditory domain in primate prefrontal cortex. *Nature Neuroscience, 5*, 15–16.

Rowell, T. E., & Hinde, R. A. (1962). Vocal communication by the rhesus monkey (*Macaca mulatta*). *Proceedings of the Zoological Society of London, 138*(2), 279–294.

Roy, S., Miller, C. T., Gottsch, D., & Wang, X. (2011). Vocal control by the common marmoset in the presence of interfering noise. *Journal of Experimental Biology, 214*, 3619–3629.

Roy, S., Zhao, L., & Wang, X. (2016). Distinct neural activities in premotor cortex during natural vocal behaviors in a New World primate, the common marmoset (*Callithrix jacchus*). *Journal of Neuroscience, 36*, 12168–12179.

Schamberg, I., Cheney, D. L., Clay, Z., Hohmann, G., & Seyfarth, R. M. (2016). Call combinations, vocal exchanges and interparty movement in wild bonobos. *Animal Behaviour, 122*, 109–116.

Schneider, C., Hodges, J. K., Fischer, J., & Hammerschmidt, K. (2008). Acoustic niches of Siberut primates. *International Journal of Primatology, 29,* 601–613.

Scott-Phillips, T. C., Gurney, J., Ivens, A., Diggle, S. P., & Popat, R. (2014). Combinatorial communication in bacteria: Implications for the origins of linguistic generativity. *PLOS ONE, 9*(4), e95929.

Seyfarth, R. M., & Cheney, D. L. (1986). Vocal development in vervet monkeys. *Animal Behaviour, 34,* 1640–1658.

Seyfarth, R. M., & Cheney, D. L. (1997). Some general features of vocal development in nonhuman primates. In C. T. Snowdon & M. Hausberger (Eds.), *Social influences on vocal development* (pp. 249–273). Cambridge: Cambridge University Press.

Seyfarth, R. M., Cheney, D. L., & Marler, P. (1980a). Monkey responses to three different alarm calls: Evidence of predator classification and semantic communication. *Science, 210,* 801–803.

Seyfarth, R. M., Cheney, D. L., & Marler, P. (1980b). Vervet monkey alarm calls: Semantic communication in a free-ranging primate. *Animal Behaviour, 28*(4), 1070–1094.

Silk, J. B., Seyfarth, R. M., & Cheney, D. L. (2016). Strategic use of affiliative vocalizations by wild female baboons. *PLOS ONE, 11*(10), e0163978.

Simonyan, K., & Horwitz, B. (2011). Laryngeal motor cortex and control of speech in humans. *Neuroscientist, 17*(2), 197–208.

Slocombe, K. E., & Zuberbühler, K. (2005a). Agonistic screams in wild chimpanzees (*Pan troglodytes schweinfurthii*) vary as a function of social role. *Journal of Comparative Psychology, 119*(1), 67–77.

Slocombe, K. E., & Zuberbühler, K. (2005b). Functionally referential communication in a chimpanzee. *Current Biology, 15*(19), 1779–1784.

Smith, W. K. (1945). The functional significance of the rostral cingulate cortex as revealed by its responses to electrical excitation. *Journal of Neurophysiology, 8,* 241–255.

Smuts, B., & Watanabe, J. M. (1990). Social relationships and ritualized greetings in adult male baboons (*Papio cynocephalus anubis*). *International Journal of Primatology, 11*(2), 147–172.

Snowdon, C. T., & Elowson, A. M. (1999). Pygmy marmosets modify call structure when paired. *Ethology, 105*(10), 893–908.

Struhsaker, T. T. (1967). Auditory communication among vervet monkeys (*Cercopitheceus aethiops*). In S. Altmann (Ed.), *Social communication among primates* (pp. 281–324). Chicago, IL: University of Chicago Press.

Sutton, D., Larson, C., & Lindeman, R. C. (1974). Neocortical and limbic lesion effects on primate phonation. *Brain Research, 71,* 61–75.

Sutton, D., Larson, C., Taylor, E. M., & Lindeman, R. C. (1973). Vocalization in rhesus monkeys: Conditionability. *Brain Research, 52,* 225–231.

Sutton, D., Trachy, R. E., & Lindeman, R. C. (1985). Discriminative phonation in macaques: Effects of anterior mesial cortex damage. *Experimental Brain Research, 59,* 410–413.

Takahashi, D. Y., Fenley, A. R., Teramoto, Y., Narayanan, D. Z., Borjon, J. I., Holmes, P., & Ghazanfar, A. A. (2015). The developmental dynamics of marmoset monkey vocal production. *Science, 349,* 734–738.

Takahashi, D. Y., Liao, D. A., & Ghazanfar, A. A. (2017). Vocal learning via social reinforcement by infant marmoset monkeys. *Current Biology, 27,* 1844–1852.e6.

Talmage-Riggs, G., Winter, P., Ploog, D., & Mayer, W. (1972). Effect of deafening on the vocal behavior of the squirrel monkey (*Saimiri sciureus*). *Folia Primatologica, 17,* 404–420.

Thinh, V. N., Hallam, C., Roos, C., & Hammerschmidt, K. (2011). Concordance between vocal and genetic diversity in crested gibbons. *BMC Evolutionary Biology, 11*(1), 36.

Trachy, R. E., Sutton, D., & Lindeman, R. C. (1981). Primate phonation: Anterior cingulate lesion effects on response rate and acoustical structure. *American Journal of Primatology, 1,* 43–55.

Warren, W. C., Jasinska, A. J., García-pérez, R., Svardal, H., Tomlinson, C., Rocchi, M., … Hughes, J. (2015). The genome of the vervet (*Chlorocebus aethiops sabaeus*). *Genome Research, 4,* 1921–1933.

Watson, S. K., Townsend, S. W., Schel, A. M., Wilke, C., Wallace, E. K., Cheng, L., … Slocombe, K. E. (2015). Vocal learning in the functionally referential food grunts of chimpanzees. *Current Biology, 25*(4), 495–499.

West, R. A., & Larson, C. R. (1995). Neurons of the anterior mesial cortex related to faciovocal activity in the awake monkey. *Journal of Neurophysiology, 74,* 1856–1869.

Wheeler, B. C. (2009). Monkeys crying wolf? Tufted capuchin monkeys use anti-predator calls to usurp resources from conspecifics. *Proceedings of the Royal Society B: Biological Sciences, 276,* 3013–3018.

Wheeler, B. C., & Fischer, J. (2012). Functionally referential signals: A promising paradigm whose time has passed. *Evolutionary Anthropology, 21,* 195–205.

Wheeler, B. C., Tiddi, B., & Heistermann, M. (2014). Competition-induced stress does not explain deceptive alarm calling in tufted capuchin monkeys. *Animal Behaviour, 93,* 49–58.

Wich, S. A., Krützen, M., Lameira, A. R., Nater, A., Arora, N., Bastian, M. L., … van Schaik, C. P. (2012). Call cultures in orang-utans? *PLOS ONE, 7*(5), e36180.

Wich, S. A., Swartz, K. B., Hardus, M. E., Lameira, A. R., Stromberg, E., & Shumaker, R. W. (2009). A case of spontaneous acquisition of a human sound by an orangutan. *Primates, 50*(1), 56–64.

Wild, B., Rodden, F. A., Grodd, W., & Ruch, W. (2003). Neural correlates of laughter and humour. *Brain, 126*(10), 2121–2138.

Wilkins, M. R., Seddon, N., & Safran, R. J. (2013). Evolutionary divergence in acoustic signals: Causes and consequences. *Trends in Ecology and Evolution, 28*(3), 156–166.

Winter, P., Handley, P., Ploog, D., & Schott, D. (1973). Ontogeny of squirrel monkey calls under normal conditions and under acoustic isolation. *Behaviour, 47,* 230–239.

Zhang, Y. S., & Ghazanfar, A. A. (2016). Perinatally influenced autonomic system fluctuations drive infant vocal sequences. *Current Biology, 26,* 1249–1260.

Zuberbühler, K. (2000a). Causal knowledge of predators' behaviour in wild Diana monkeys. *Animal Behaviour, 59,* 917–927.

Zuberbühler, K. (2000b). Interspecies semantic communication in two forest primates. *Proceedings of the Royal Society of London Series B—Biological Sciences, 267*(1444), 713–718.

Zuberbühler, K. (2000c). Referential labelling in Diana monkeys. *Animal Behaviour, 59,* 917–927.

Zuberbühler, K. (2001). Predator-specific alarm calls in Campbell's monkeys, *Cercopithecus campbelli. Behavioral Ecology and Sociobiology, 50*(5), 414–422.

Zuberbühler, K. (2008). Audience effects. *Current Biology, 18*(5), R189–R190.

46 Vocal Learning and Spoken Language: Insights from Animal Models with an Emphasis on Genetic Contributions

CONSTANCE SCHARFF, MIRJAM KNÖRNSCHILD, AND ERICH D. JARVIS

1. From Language and Speech to Biolinguistics

Language is a specialized human trait, tightly linked to our intellectual capacity (Hauser, Chomsky, & Fitch, 2002). Scores of philosophers, linguists, and biologists have commented on the astonishing complexity of language and proposed different scenarios of how it might have arisen (Christiansen & Kirby, 2003; Di Sciullo & Boeckx, 2011; Nowak, 2000). We learn language as infants and use it throughout our lives to express our thoughts. One feature that makes this possible is that spoken language combines a limited repertoire of sounds into an essentially unlimited combination of words and sentences. This fact already fascinated medieval Arab and Jewish religious scholars (Eco, 1995) and eventually, Wilhelm von Humboldt (1836, p. 106) put it succinctly as "the infinite use of finite means."

From a biolinguistic perspective (Boeckx & Grohmann, 2013; Hauser et al., 2002), one can ask whether language evolution required unique components, including human-unique genes, or whether it was the result of a unique combination or of a continuum of traits that exist to varying degrees in other animals (Fitch, 2010; Larson, Déprez, & Yamakido, 2010; Petkov & Jarvis, 2012; Scharff & Petri, 2011). Which of these traits are relevant to language evolution is a matter of debate (Arbib, Liebal, & Pika, 2008; Bolhuis, Tattersall, Chomsky, & Berwick, 2014; Tomasello, 2005). Speech is one feature suitable for comparative studies because many animals also communicate vocally and in a subset of those the production of those vocalizations are learned. How does the neural control of jaw, orofacial, and laryngeal muscles necessary for speech relate to sound production mechanisms in vocally communicating animals? These and other key components of spoken language, such as vocal learning, syntax, and semantics, which concern the acquisition, structure, and meaning of a signal, respectively, are increasingly the focus of comparative biolinguistic studies (Christiansen & Kirby, 2003; Fitch 2010; Larson et al., 2010; Petkov & Jarvis, 2012).

1.1. Definitions of Social Learning in Vocal Communication To understand the relevance of animal models to study genes relevant for language learning, we briefly review some relevant terms (Janik & Slater, 1997, 2000; Jarvis, 2004; Petkov & Jarvis, 2012). *Vocal learning,* or more precisely, *vocal production learning* (VPL), is the ability to learn to produce new vocalizations based on auditory feedback (vocal imitation) or reshape existing vocalizations based on social influences (vocal modification). *Vocal usage learning* (VUL) is the ability of a sender to learn which sound signal (whether it be an innate or learned sound) to produce in a specific context. *Auditory comprehension learning* is the ability of a receiver to learn which message a sound signal carries in a particular context (table 46.1).

Of these types of learning, VPL is the rarest. It is documented in only three orders of birds (songbirds, parrots, and hummingbirds) and in five orders of mammals (humans, cetaceans, bats, elephants, and pinnipeds) (Janik & Slater, 1997, 2000; Petkov & Jarvis, 2012). Songbirds and parrots are particularly accomplished in a special form of VPL, *vocal mimicry,* where even heterospecific sounds are copied, including human words and sentences (Garamszegi, Eens, Pavlova, Avilés, & Møller, 2007; Kelley, Coe, Madden, & Healy, 2008; Pepperberg, 2010). In these instances, mimicry of other species often occurs by socialization with them and absence of their own species. Except for parrots, imitation of human speech (Pepperberg, 1981) has been reported in an Asian elephant (Stoeger et al.,

TABLE 46.1

Definitions of terms used in the chapter with comparison to terms used in Janik and Slater (2000) and Petkov and Jarvis (2012)

Present chapter	Janik and Slater (2000)	Petkov and Jarvis (2012)
Auditory comprehension learning: "the ability of a receiver to learn which message a sound signal carries in a particular context."	*Comprehension learning*: "a receiver comes to extract a novel meaning from a signal as a result of experience with the usage of signals by other individuals" (table 1).	*Auditory learning*: "an animal learns to perceive something novel or behaviorally react to sounds differently as a result of experience" (p. 4).
Vocal usage learning (VUL): "the ability of a sender to learn which sound signal to produce in a specific context."	*Usage learning:* "an existing signal is produced in a new context as a result of experience with the usage of signals by other individuals" (table 1).	*Usage vocal learning*: "an animal learns to use acoustically innate or already learned vocalizations in a new context" (p. 4).
Vocal production learning (VPL): "the ability to learn to produce new vocalizations based on auditory feedback (*vocal imitation*) or reshape existing vocalizations based on social influences (*vocal modification*)."	*Production learning*: "signals are modified in form as a result of experience with those of other individuals. This can lead to signals that are either similar or dissimilar to the model" (table 1).	*Production vocal learning*: the ability of an animal to produce novel vocalizations.
Limited VPL: the ability to learn to produce new vocalizations is limited. The adult vocal repertoire of species with limited VPL may not need to be imitatively learned from an external source, but can be (see obligatory/ facultative VPL).		Limited vocal learning
Extensive VPL: the ability to learn to produce new vocalizations is extensive and adult vocalizations of species with extensive VPL are usually imitatively learned from an external source. Note that different vocalizations (song, calls) can differ in the degree of VPL.		High vocal learning
Oobligatory VPL: VPL is obligatory in species that cannot acquire their species-typical sound repertoire in the absence of an external source.		

2012), bottlenose dolphins (Lilly, 1965), a male harbor seal (Ralls, Fiorelli, & Gish, 1985), and a male beluga whale (Ridgway, Carder, Jeffries, & Todd, 2012). Heterospecific imitation was described in a young Risso's dolphin cross-fostered with bottlenose dolphins (Favaro et al., 2016), a juvenile free-ranging orca separated from its natal group that imitated the barks of sea lions (Foote et al., 2006), and a single African elephant housed with Asian elephants (Poole, Tyack, & Stoeger-Horwath, 2005). Cross-species VPL has also been noted in children deprived of human contact but fortunately, examples are few (Bettelheim, 1959; Ogburn, 1959).

In contrast, auditory comprehension learning is widespread among animals (Seyfarth & Cheney, 2010; Petkov & Jarvis, 2012). For example, dogs can learn to respond to the sound *sit* by sitting down (Jarvis, 2004; Kaminski, Call, & Fisher, 2004), but they cannot learn to produce the sound *sit*. VUL is also not uncommon. Some animals learn which vocalizations to use when faced with particular types of predators (Blumstein, 1999; Seyfarth & Cheney, 2003; Scarantino & Clay, 2015). For example, vervet monkeys utter different types of alarm calls for different predators, and conspecifics respond with the appropriate avoidance behavior. Juveniles need not learn how to produce these alarm calls but they must learn when to use them (Seyfarth, Cheney, & Marler, 1980a, 1980b). Rhesus macaques can be trained to produce specific calls in response to specific visual stimuli (Hage, Gavrilov, & Nieder, 2013; Hage & Nieder, 2013) and walruses in response to hand gestures (Schusterman & Reichmuth, 2008). The fact that through VUL such species can be taught to utter particular vocalizations to request particular items (Pepperberg, 1981; Richards, 1986) suggests that at least some animals make the connection between arbitrary sounds and objects (and in the case of Alex the parrot, among sounds and the shape, color, or material of objects; Pepperberg, 2010). Given that mapping of arbitrary sound to meaning is a prerequisite for the semanticity of language, claims about the unsuitability of animal communication to inform domains of language besides speech should be reconsidered (Berwick, Okanoya, Beckers, & Bolhuis, 2011; Mozzi et al., 2016).

Traditionally, VPL has been discussed in a binary way; some species are capable of VPL (e.g., humans and songbirds) and others are not (e.g., monkeys and mice). However, work showing modifications of vocalizations in species previously considered incapable of VPL has led to the continuum hypothesis of VPL (Arriaga & Jarvis, 2013; Petkov & Jarvis, 2012). In this framework, VPL is not a dichotomous trait but a continuum (figure 46.1), ranging from subtle modifications of vocalizations in mice (Arriaga, Zhou, & Jarvis, 2012), goats (Briefer & McElligott, 2012), and chimpanzees (Watson et al., 2015); to social influences on repertoire maturation in bats (Prat, Taub, & Yovel, 2015) and marmosets (Gultekin & Hage, 2017; Takahashi et al., 2015; Takahashi, Liao, & Ghazanfar, 2017); to full imitative acquisition of communication sounds in many songbirds and speech in humans (Doupe & Kuhl, 1999). Species capable of VPL do not always use it, it can be facultative (Hammerschmidt et al., 2012). For instance, laboratory mice can acquire normal song without external input (Hammerschmidt et al., 2012; Mahrt, Perkel, Tong, Rubel, & Portfors, 2013), but when available, external input can lead to modified song (Arriaga et al., 2012).

In the context of the continuum theory, it is noteworthy that among the thousands of species of songbirds there is a range of how much of the song is learned from an external auditory source (often an adult "tutor") and how much of the song is internally represented ("innate"). Humans and many songbirds develop abnormal vocal communication when deprived of external input (Fromkin, Krashen, Curtiss, Rigler, & Rigler, 1974; Price, 1979; Williams, Kilander, & Sotanski, 1993), making VPL obligatory for developing species-typical communication sounds. However, some songbirds can develop normal sounding song without exposure to song models. As long as they can hear themselves they use this to guide their own vocalizations toward an internal model. Examples are European sedge warblers (Leitner, Nicholson, Leisler, DeVoogd, & Catchpole, 2002), gray catbirds (Kroodsma, Houlihan, Fallon, & Wells, 1997), and chipping sparrows (Liu & Nottebohm, 2007). The different strengths of internal predispositions toward species-typical vocalizations are often only uncovered in the absence of the normally occurring external, auditory, or other input (Hammerschmidt et al., 2012; Mets & Brainard, 2018). For instance, Fehér, Wang, Saar, Mitra, and Tchernichovski (2009) found that zebra finches, which typically copy their tutor's song with high fidelity, do not faithfully copy experimentally induced "bad" zebra finch songs. Instead, they seem to be internally biased toward species-typical song features and improve the "bad" tutor model toward more species-typical song features.

In summary, the ability to modify communicative vocalizations in response to auditory feedback occurs on a continuum; on one end are species that communicate with sounds that mostly have been imitatively learned from external models (hereafter *extensive VPL*). At the other end are species that can develop their sound repertoire without reference to an external model, but those sounds can still be modified in limited ways (hereafter *limited VPL*). Importantly, both extensive and limited VPL require auditory feedback, whereas changes to vocalizations through arousal do not (Arriaga & Jarvis, 2013; Janik & Slater, 1997; Konishi, 1985; Petkov & Jarvis, 2012).

1.2. Occurrence of Vocal Production Learning in Birds and Mammals Extensive VPL is widespread in 3 of the 42 avian orders (Kroodsma, 1982; figure 46.2A, table 46.2) and 5 of the 28 mammalian orders mentioned earlier (figure 46.2B, table 46.3). Among the three avian orders, oscine songbirds are the most numerous, with more than 4,000 species, followed by ~300 species each of parrots and hummingbirds, of the over 10,500 species of birds. VPL in songbirds has been studied in more detail in songbirds (Catchpole & Slater, 1995; Kroodsma & Miller, 1996) than in parrots (Farabaugh, Linzenbold, & Dooling, 1994; Pepperberg, 2010) or hummingbirds (Baptista & Schuchmann, 1990; Ferreira, Smulders, Sameshima, Mello, & Jarvis, 2006; Gahr, 2000; Gaunt, Baptista, Sánchez, & Hernandez, 1994). Songbirds' close relatives, the suboscines, are generally regarded as not capable of VPL (Kroodsma, 1989; Kroodsma & Konishi, 1991), but bellbirds are (Kroodsma et al., 2013; Saranathan, Hamilton, Powell, Kroodsma, & Prum, 2007). Experimentally demonstrated absence of extensive VPL has also been reported for several nonpasserine birds such as domestic chickens (Konishi, 1963), ring doves (Nottebohm & Nottebohm, 1971), barn owls, and gulls (Gahr, 2000); these species can develop normal vocalizations without auditory feedback (Gahr, 2000; Jarvis; 2004; Petkov & Jarvis, 2012), but it is possible that more detailed analysis would reveal limited VPL. Although most avian orders have not been analyzed for the absence or presence of extensive VPL in detail so far, there are no

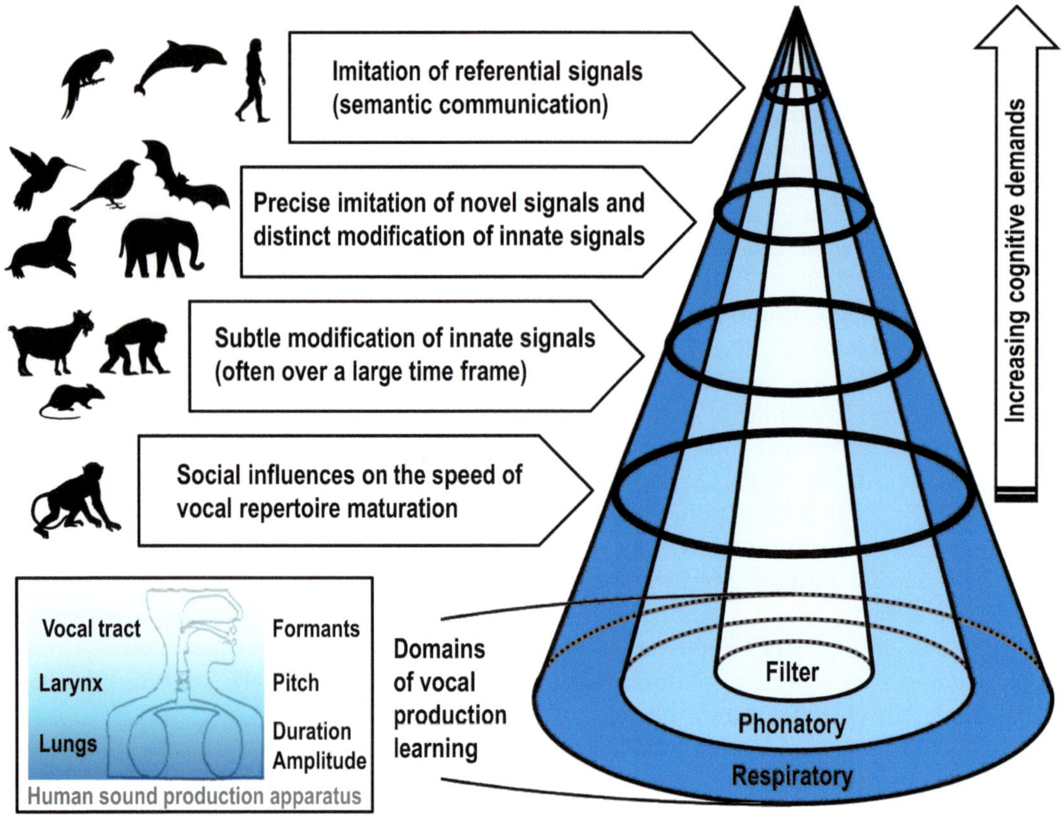

FIGURE 46.1 Different forms of VPL are arranged on a continuum of ability. Basic forms of VPL, such as social influences on repertoire maturation or subtle modifications of innate vocalizations over a long time period, constitute one end of the continuum, while complex forms of VPL, such as the imitation of referential signals, constitute the other end. Three domains of vocal production learning (respiratory, phonatory, and filter learning), their association with the sound producing apparatus, and the resulting signal characteristics are depicted as well. Sketches provide graphic references to avian and mammalian vocal production learners (on the order level).

known obvious vocal mimics such as songbirds and parrots.

Among mammals, VPL of varying degrees has been described in multiple species of cetaceans (Janik, 2014), pinnipeds (e.g., harbor seal: Ralls, Fiorelli, & Gish, 1985; walrus: Schusterman & Reichmuth, 2008), and bats (Knörnschild, 2014; Prat, Azoulay, Dor, & Yovel, 2017; Prat et al., 2015). Limited VPL exists in three mammalian orders: even-toed ungulates (Briefer & McElligott, 2012); nonhuman primates (chimpanzees: Watson et al., 2015; common marmosets: Gultekin & Hage, 2017; Takahashi et al., 2015; Takahashi et al., 2017; pygmy marmosets: Snowdon & Elowson, 1999, but also see Fischer, Wheeler, & Higham, 2015; Mahrt et al., 2013), and murids (Arriaga et al., 2012, but see Hammerschmidt et al., 2012). Concerning the capacity for VPL in mice, it is conceivable that VPL plays a bigger role in wild mice than in inbred laboratory mice for lack of sexual selection pressure in the latter (Chalfin et al., 2014; von Merten, Hoier, Pfeifle, & Tautz, 2014).

1.3. DOMAINS OF VOCAL PRODUCTION THAT CAN BE SHAPED BY LEARNING Vocalizations are defined by temporal and spectral parameters, among them duration, pitch, timbre, and the order in which sound elements are produced. Those parameters are shaped by different components of the body's sound production apparatus (Fitch, 2006; Kriengwatana & Beckers, chapter 48 of this volume) and their neural control by the brain (Arriaga & Jarvis, 2013; Jürgens, 2009; Mackevicius & Fee, 2017; Petkov & Jarvis 2012). Because the development and function of this system involve the concerted activity of many genes, we will point out the parts of the sound production system associated with different domains of VPL (Janik & Slater, 1997, 2000; figure 46.1).

The respiratory domain of vocal production is associated with the lungs and shapes the duration, interval, and amplitude of vocalizations. The phonatory domain of vocal production is associated with the larynx in mammals and the syrinx in birds, the sound source that shapes the pitch and some other acoustic features

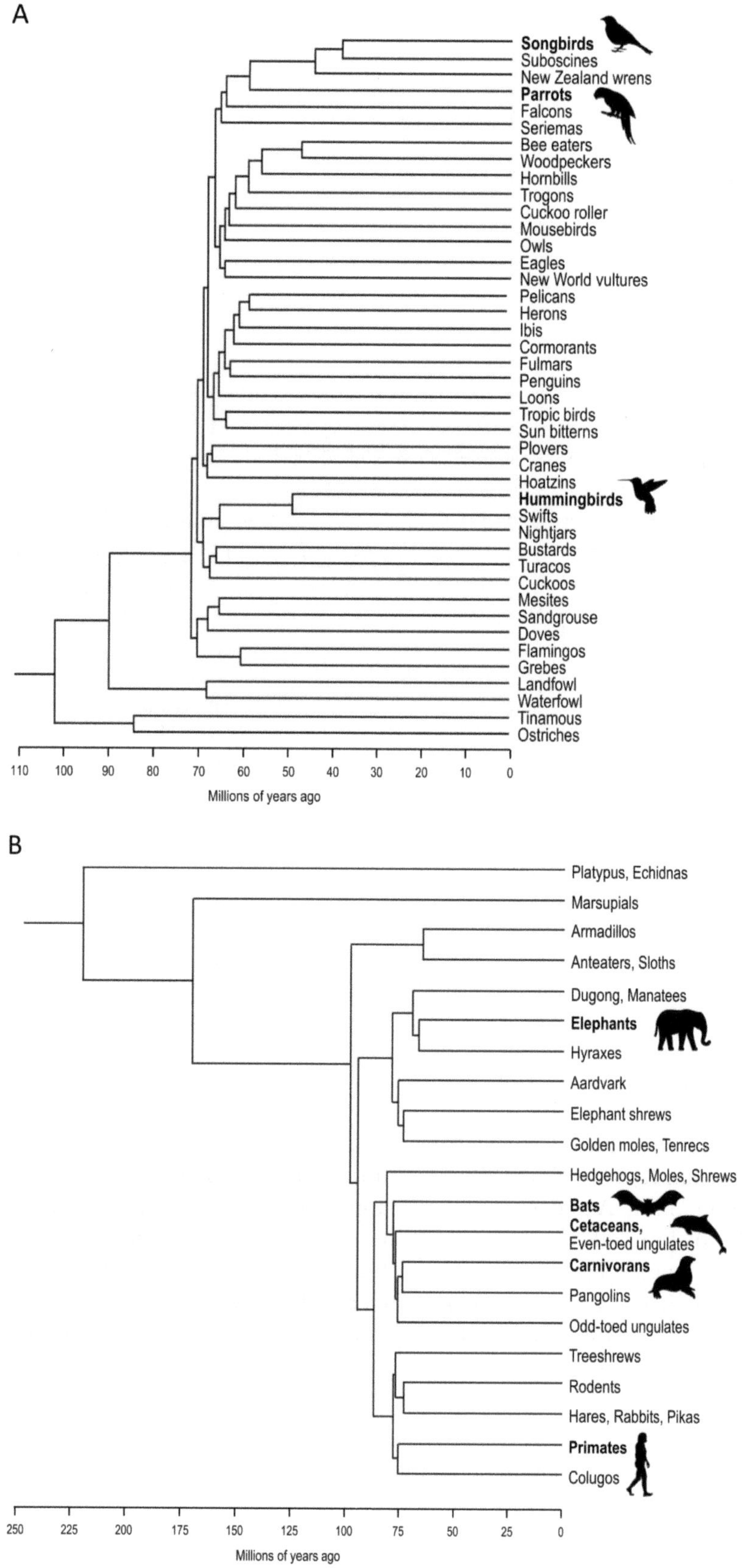

FIGURE 46.2 Phylogenetic tree of extant avian and mammalian orders highlighting the occurrence of VPL. (A) The avian tree has three orders capable of extensive VPL (designated by bold), namely songbirds, parrots, and hummingbirds. The evidence for extensive VPL in suboscines is currently limited to bellbirds. (B) The mammalian tree has five orders, which have at least one species capable of extensive VPL, namely humans among primates, bats, pinnipeds, cetaceans, and elephants. Four additional taxa (murids, apes, marmosets, goats) have at least one species with limited VPL capacities. Trees modified from Jarvis et al. (2014) and Foley et al. (2011).

Table 46.2

Selected cases of avian VPL

English name	Latin name	Family	Order	VPL type	Vocalization type	Description	References
Zebra finch	*Taeniopygia guttata*	Estrildidae	Passeriformes (oscines)	Imitation	Male song	Slightly imprecise copying of adult tutor song (normally father's or, rarer, sibling's song) results in an individual song type per male; song learning possible in adulthood when no suitable tutor was available before; species-specific song type can be established de novo over several generations	Immelmann (1969); Böhner (1983, 1990); Slater et al. (1988); Tchernichovski et al. (2001); Derégnaucourt and Gahr (2013); Eales (1985); Morrison and Nottebohm (1993); Jones et al. (1996); Fehér et al. (2009)
Canary	*Serinus canaria*	Fringillidae	Passeriformes (oscines)	Imitation	Male song	Male song is copied from adult tutors; open-ended learning; improvisation in the absence of tutors is also possible	Poulsen (1959); Waser and Marler (1977); Nottebohm & Nottebohm (1978); Metfessel (1935)
White-crowned sparrow	*Zonotrichia leucophrys*	Emberizidae	Passeriformes (oscines)	Imitation	Male song	Improvisation and precise imitation of adult tutor songs, regional dialects; song overproduction during development is followed by selective attrition	Marler and Tamura (1962); Marler (1970); Baptista (1977); Nelson et al. (2004); Nelson (2000)
Swamp sparrow	*Melospiza georgiana*	Emberizidae	Passeriformes (oscines)	Imitation	Male song	Male song is copied from adult conspecifics, but not heterospecifics; song overproduction during development is followed by selective attrition	Marler and Peters (1977, 1988); Marler and Peters (1982)
Song sparrow	*Melospiza melodia*	Emberizidae	Passeriformes (oscines)	Imitation	Male song	Male song is copied from adult conspecific tutors **(and in captivity also from heterospecific tutors); young males copy song types from their neighbors**	Marler and Peters (1987); Beecher et al. (1994, **1996); Nordby et al. (2001)**

Chipping sparrow	*Spizella passerina*	Emberizidae	Passeriformes (oscines)	Imitation	Male song	Songs are learned by imitating a single adult tutor (neighbor after dispersal)	Liu and Kroodsma (1999, 2006)
Savannah sparrow	*Passerculus sandwichensis*	Emberizidae	Passeriformes (oscines)	Imitation	Male song	Specific song elements are copied from different adult tutors and combined into one song type per male; regional dialects; long-term cultural changes affect each song element differently	Wheelwright et al. (2008); Chew (1981); Williams et al. (2013)
Sedge warbler	*Acrocephalus schoenobaenus*	Acrocephali-dae	Passeriformes (oscines)	Imitation	Male song	Improvisation and precise imitation of adult tutor songs, open-ended learning, neighboring males share song types; normal song development when raised in acoustic isolation	Nicholson et al. (2007); Leitner et al. 2002
Nightingale	*Luscinia megarhynchos*	Muscicapi-dae	Passeriformes (oscines)	Imitation	Male song	Improvisation and precise imitation of adult tutor songs; open-ended learning; neighboring males share song types but also invent new ones	Todt et al. (1979); Todt and Geberzahn (2003); Hutsch and Todt (1981); Sprau and Mundry (2010)
Brown-headed cowbird	*Molothrus ater*	Icteridae	Passeriformes (oscines)	Imitation	Male song	Male song is copied from adult tutors; improvisation instead of precise imitation in the presence of female conspecifics	King and West (1989)
European starling	*Sturnus vulgaris*	Sturnidae	Passeriformes (oscines)	Mimicry	Songs/calls from sympatric birds, artificial sounds	Mimicry of sympatric bird species and artificial sounds	Hindmarsh 1986)
				Imitation	Male song	Male song is copied from adult tutors; open-ended learning	Eens et al. (1992); Chaiken et al. (1993)

(*continued*)

Table 46.2 (*continued*)
Selected cases of avian VPL

English name	Latin name	Family	Order	VPL type	Vocalization type	Description	References
Common hill mynah	*Gracula religiosa*	Sturnidae	Passeriformes (oscines)	Mimicry	Human speech (English)	Mimicry of human speech, close association with humans	Grosslight et al. (1962); Ginsburg (1963)
				Imitation	Whole call repertoire	Calls are learned by imitating adult neighbors (both sexes); no heterospecific mimicry in the wild	Bertram (1970)
Superb lyrebird	*Menura novaehollandiae*	Menuridae	Passeriformes (oscines)	Mimicry	Songs/calls from sympatric birds, artificial sounds	Mimicry of sympatric bird species and artificial sounds	Zann and Dunstan (2008)
Bare-throated bellbird	*Procnias nudicollis*	Cotingidae	Passeriformes (suboscines)	Mimicry	Calls of chopi blackbird	Mimicry of heterospecific calls, close association with a chopi blackbird (*Gnorimopsar chopi*)	Kroodsma et al. (2013)
Three-wattled bellbird	*Procnias tricarunculata*	Cotingidae	Passeriformes (suboscines)	Imitation	Male song	Regional dialect	Saranathan et al. (2007); Kroodsma et al. (2013)
African gray parrot	*Psittacus erithacus*	Psittacidae	Psittaciformes	Mimicry (referential)	Human speech (English)	Referential use of English words as a result of intensive training	Pepperberg (1981, 1994, 2006, 2010)
				Mimicry	Human speech (German); songs/calls from nine sympatric birds and one bat	Mimicry of human speech, close association with humans; mimicry of sympatric species	Todt (1975); Cruickshank et al. (1993)
Budgerigar	*Melopsittacus undulatus*	Psittacidae	Psittaciformes	Imitation	Contact call	Flock mates imitate each others' calls; males imitate female contact calls after pair bonding	Farabaugh et al. (1994); Hile & Striedter (2000); Hile et al. (2000, 2005)

Yellow-naped amazon	*Amazona auropalliata*	Psittacidae	Psittaciformes	Imitation	Contact call	Regional dialects; imitation of local call types by immigrant birds after dispersal	Wright (1996); Wright et al. (2005, 2008); Salinas-Melgoza and Wright (2012)
Orange-fronted conure/parakeet	*Aratinga canicularis*	Psittacidae	Psittaciformes	Imitation	Contact call	Regional dialect; vocal convergence or divergence prior to flock fusions	Bradbury et al. (2001); Balsby and Scarl (2008); Balsby and Bradbury (2009)
Green-rumped parrotlet	*Forpus passerinus*	Psittacidae	Psittaciformes	Imitation	Contact call	Individual signature character of nestlings' contact calls is learned from both parents	Berg et al. (2012)
Anna's hummingbird	*Calypte anna*	Trochilidae	Apodiformes	Imitation	Male song	Regional dialect; males raised in isolation develop abnormal song	Baptista and Schuchmann (1990)
Sparkling violetear	*Colibri coruscans*	Trochilidae	Apodiformes	Imitation	Male song	Regional dialect	Gaunt et al. (1994)
Green violetear	*Colibri thalassinus*	Trochilidae	Apodiformes	Imitation	Male song	Regional dialect	Gaunt et al. (1994)
Little hermit	*Phaethornis longuemareus*	Trochilidae	Apodiformes	Social modification and imitation	Male song	Regional dialect; open-ended song type learning	Snow (1968); Wiley (1971); Araya-Salas and Wright (2013)

Table 46.3

Occurrence of mammalian VPL

English name	Latin name	Family	Order	VPL type	Vocalization type	Description	References
Common bottlenose dolphin	*Tursiops truncatus*	Delphinidae	Cetacea	Mimicry	Artificial sounds	Referential use of learned acoustic signals (artificial)	Richards et al. (1984); Reiss and McCowan (1993)
				Imitation	Cohesion calls (signature whistles)	Copying of conspecific whistles	Janik (2000); King et al. (2013); King and Janik (2013)
				Social modification	Cohesion calls (signature whistles)	Whistles of young dolphins converge toward whistles of group members	Fripp et al. (2005)
Killer whale	*Orcinus orca*	Delphinidae	Cetacea	Mimicry	Barks of California sea lions	Mimicry of sea lion barks, close association with California sea lions	Foote et al. (2006)
				Mimicry	Human speech (English)	Mimicry of human speech, close association with humans	Abramson et al. (2018)
				Social modification	Different social calls	Regional dialects	Deecke et al. (2000); Weiß et al. (2011); Kremers et al. (2012); Filatova et al. (2012)
Risso's dolphin	*Grampus griseus*	Delphinidae	Cetacea	Mimicry	Whistles	Mimicry of bottlenose dolphin whistles, close association with bottle-nose dolphin	Favaro et al. (2016)
Beluga	*Delphinapterus leucas*	Monodonti-dae	Cetacea	Mimicry	Human speech (English)	Mimicry of human speech, close association with humans	Eaton (1979); Ridgway et al. (2012); Murayama et al. (2014)
Humpback whale	*Megaptera novaeangliae*	Balaenop-teridae	Cetacea	Imitation	Male song	Regional dialects, cultur-ally induced change in song motifs	Noad et al. (2000); Garland et al. (2011)
Harbor seal	*Phoca vitulina*	Phocidae	Carnivora	Mimicry	Human speech (English)	Mimicry of human speech, close association with humans	Ralls et al. (1985)
				Invention	Novel sounds	Novel sound production through reinforcement training	Schustermann (2008)

Northern elephant seal	*Mirounga angustirostris*	Phocidae	Carnivora	Social modification	Male agonistic vocalizations	Regional dialects	Le Boeuf and Peterson (1969); Le Boeuf and Petrinovich (1974)
Southern elephant seal	*Mirounga leonina*	Phocidae	Carnivora	Social modification or imitation	Male agonistic vocalizations	Young adults' vocalizations closely resemble the ones of dominant males present during subadult period of development	Sanvito et al. (2007)
Walrus	*Odobenus rosmarus*	Odobenidae	Carnivora	Invention	Novel sounds	Novel sound production through reinforcement training	Schustermann and Reichmuth (2008)
Egyptian fruit bat	*Rousettus aegyptiacus*	Pteropodidae	Chiroptera	Social modification	Different social calls	Immature vocal repertoire when raised in acoustic isolation; frequency shift to conspecifics' calls	Prat et al. (2015); Prat et al. (2017)
Greater sac-winged bat	*Saccopteryx bilineata*	Emballonuridae	Chiroptera	Imitation	Male territorial songs	Juveniles imitate adult tutor song	Knörnschild et al. (2010)
				Social modification	Juvenile isolation calls	Juvenile isolation calls converge toward fellow juveniles' calls	Knörnschild et al. (2012)
Greater spear-nosed bat	*Phyllostomus hastatus*	Phyllostomidae	Chiroptera	Social modification	Short-distance contact calls (screeches)	Screeches of unrelated group members converge	Boughman (1998)
Pale spear-nosed bat	*Phyllostomus discolor*	Phyllostomidae	Chiroptera	Social modification	Juvenile isolation calls	Isolation calls converge toward maternal directive calls	Esser and Schmidt (1989); Esser (1994)
Greater horseshoe bat	*Rhinolophus ferrumequinum*	Rhinolophidae	Chiroptera	Social modification	Echolocation calls	Resting frequency of juvenile echolocation calls converges toward maternal echolocation calls	Jones and Ransome (1993)
Asian elephant	*Elephas maximus*	Elephantidae	Proboscidea	Mimicry	Human speech (Korean)	Mimicry of human speech, close association with humans	Stoeger et al. (2012)
African elephant	*Loxodonta africana*	Elephantidae	Proboscidea	Mimicry	Artificial sound, chirps of Asian elephants	Mimicry of Asian elephant chirps, close association with Asian elephants	Poole et al. (2005)

(*continued*)

TABLE 46.3 (*continued*)

Occurrence of mammalian VPL

English name	Latin name	Family	Order	VPL type	Vocalization type	Description	References
Common chimpanzee	*Pan troglodytes*	Hominidae	Primates	Social modification (subtle)	Long-distance call (pant hoot)	Adult males modify pant-hoots to be different from their neighbors; novel pant-hoot variant invented by one male spread in its social group	Crockford et al. (2004); Marshall et al. (1999)
				Social modification (subtle)	Food grunt	Call convergence of group members once affiliative social relationships are formed	Watson et al. (2015)
Pygmy marmoset	*Cebuella pygmaea*	Callitrichidae	Primates	Social modification (subtle)	Short-distance contact call (trill)	Convergence in trill structure between adult pair members	Snowdon and Elowson (1999)
Common marmoset	*Callithrix jacchus*	Callitrichidae	Primates	Social modification	Long-distance contact call (phee call)	Contingent vocal responses from parents influence the timing of the transition from infant cries to adult-like phee calls by reinforcing the production of phee calls	Takahashi et al. (2015, 2017)
				Social modification	Different social calls	Immature vocal repertoire when raised in acoustic isolation	Gultekin and Hage (2017)
Domestic goat	*Capra hircus*	Bovidae	Artiodactyla	Social modification (subtle)	Juvenile contact calls	Juvenile contact calls converge toward fellow juveniles' calls	Briefer and McElligott (2012)
Mouse	*Mus musculus (B6 strain)*	Muridae	Rodentia	Social modification (subtle)	Courtship vocalizations	B6 males decrease pitch toward the range of BxD males when housed under competitive social conditions	Arriaga et al. (2012)

of vocalizations. The filter domain of vocal production (also called supralaryngeal/-syringeal articulatory system) is associated with the vocal tract and shapes the dispersion and transition of formants in vocalizations. As vocal tract resonances, formants are concentrations of energy at certain frequencies and thus shape the spectrum of a vocalization. Formants constitute the primary phonetic cue in speech (Lieberman & Blumstein, 1988) and animals can perceive them in con- and heterospecific vocalizations (Fitch, 1997; Fitch & Kelley, 2000; Ohms, Escudero, Lammers, & ten Cate, 2012; Kriengwatana & Beckers, chapter 48 of this volume; Townsend & Manser, 2011). Changing temporal characteristics (duration, interval) is deemed to be easier than changing spectral parameters (pitch, formants), which is why the respiratory domain of VPL is considered to be less cognitively demanding than the phonatory or filter domain of VPL (Boughman & Moss, 2003; Janik & Slater, 1997, 2000). However, these three domains influence one another; rapid amplitude modulations by the respiratory domain, for instance, cause additional harmonics in a vocalization and therefore interact also with the phonatory domain (Fitch, 2006; Fitch, Neubauer, & Herzel, 2002; Wilden, Herzel, Peters, & Tembrock, 1998). The sequences in which the individual vocalization syllables/calls are produced are thought to be controlled by pattern generators in the brain (Mackevicius & Fee, 2017). Complex vocalizations require sophisticated coordination between respiratory, laryngeal/syringeal, and articulatory muscles, and thus involve all domains of production (Beckers, 2011; Fitch, 2000, 2006; Janik & Slater, 2000).

1.4. Diversity in Development of Vocal Learning Strategies The development of extensive VPL has been studied in much greater depth in songbirds and humans than in any other species capable of VPL (Brainard & Doupe, 2002; Catchpole & Slater, 1995; Konishi, 1985; Kroodsma & Miller, 1996). Like human infants, the young of many songbirds listen to conspecific vocalizations and memorize them (sensory or perception phase) before they start imitating them during a babbling phase (sensorimotor or production phase (Brainard & Doupe, 2002; Doupe & Kuhl, 1999; Wilbrecht & Nottebohm, 2003). In songbirds, the initial rehearsal periods (subsong and plastic song) can commence weeks or months after the sensory phase, depending on the species. To fully master adult song, different species need different amounts of time, from a few months to more than a year. In some species, for example, zebra finches, the sensory phase, and the sensory-motor phase overlap completely (Roper & Zann, 2006; Slater, Eales, & Clayton, 1988). Like humans, some avian species are capable of extensive VPL throughout their lives, for example, canaries (Nottebohm, Nottebohm, & Crane, 1986), starlings (Mountjoy & Lemon, 1995), and many parrots (Bradbury & Balsby, 2016; Wright & Dahlin, 2017), whereas VPL only occurs in a sensitive/critical period during development in other species, for example, the white-crowned sparrow (Marler, 1970), zebra finch (Slater et al., 1988), and song sparrow (Beecher, 2017). Diversity also exists in song repertoire size (small vs. large), the accuracy of learning (precise imitation vs. improvisation), the importance of early exposure (development of species-typical song only after early exposure vs. even when raised in isolation), and learning flexibility (imitation within vs. outside species-specific constraints). This diversity of song-learning strategies, both within the same species (e.g., zebra finches: Liu, Gardner, & Nottebohm, 2004) and between different species (Beecher & Brenowitz, 2005; Brenowitz & Beecher, 2005) is an opportunity to understand the evolution and mechanisms of VPL and spoken language. For different aspects of language learning, there is likely to be a suitable songbird species to study.

1.5. Selection for Vocal Learning Avian and mammalian vocalizations are under strong sexual and natural selection, depending on the vocalization types and the sex involved (Catchpole & Slater, 1995; Garland et al., 2011; Gaunt et al., 1994; Knörnschild, 2014; Price, 2015; Sanvito, Galimberti, & Miller, 2007; Sjare, Stirling, & Spencer, 2003). Sexually selected songs and calls facilitate mate attraction and rival deterrence, functioning as advertisement and territorial signals, respectively. VPL can serve to increase song/call complexity, song/call type sharing with neighbors, and local dialects (Janik & Slater, 1997, 2000).

The function of learned vocalizations under natural selection pressures often concerns recognition processes on the individual, group, or population level and facilitates pair/group cohesion, alliance maintenance, cooperation, and mother-offspring reunion (Bradbury & Balsby, 2016; Boughman, 1998; Farabaugh et al., 1994; Janik & Slater, 1997, 2000; King, Sayigh, Wells, Fellner, & Janik, 2013; Nowicki, 1989). Whether VPL in humans is still relevant to sexual and natural selection is a fascinating question from a behavioral ecologist perspective (Lange, Henninghausen, Bril, & Schwab, 2016; Miller, 2013). Behavioral selection for the VPL trait is expected to be associated with molecular selection of genetic changes required for the trait and is the subject of section 2.

2. *Comparative Neurogenetics*

Many misconceptions exist about the relationship among genes, brains, and language. One of those concerns the assumption that because human language is unique, the genes coding for the proteins required to build a "language-ready" brain should also exist solely in humans, and that such brain regions are also unique in humans. The evidence to date has not supported this view. Novel traits can evolve even when the coding sequence of genes remains the same but their regulation changes during the course of evolution (Carroll, 2005; Romero, Ruvinsky, & Gilad, 2012). As a result, the protein may be the same, but where and when it is expressed, and how much there is of it, can vary between species and result in different phenotypes. In addition, many genes that already existed in the primate lineage before humans emerged are extant in humans. Slightly changed versions of genes can lead to altered cellular functions (Castellano et al., 2014). Some of these alterations might also have contributed toward new, human-specific traits. Finally, a small number of genes are truly human unique, for instance as a result of incomplete duplication and fusion with another gene (Dougherty et al., 2017). Together, these changes probably synergized in developing speech and language in the hominin lineage.

Many of the aspects of language mentioned have to come together for it to function. Speech requires the concerted action of respiration, mouth and tongue muscles, and the language regions of the brain, all of which are affected by the activity of genes. However, many of those genes are also involved in processes and behaviors unrelated to language. The same genes that are necessary for development of the brain circuits involved in language perception or production can also contribute to the development of brain regions that process sounds other than language or that control movements of the mouth required for eating. As a result, there are likely many genes required for language, and none thus far have been found to be exclusively associated with it (see section 2.3.2). Of particular interest are genes required for the specialized neural mechanisms that subserve our elaborate form of vocal communication, allowing us to externalize our thoughts through sound. To understand the neurogenetics behind language, we must also understand brain mechanisms of language.

2.1. Brain Pathways for Vocal Learning and Spoken Language Consistent with the widespread auditory comprehension learning, all vertebrate species examined to date have an auditory pathway that reaches the forebrain and is thought to be involved in auditory comprehension learning. This pathway begins at the sensory hair cells inside the ear and connects through the brain stem to the forebrain where it forms a network with the auditory cortex and other connected regions (figure 46.3).

Species with extensive VPL possess a specialized forebrain pathway that controls the learning and production of vocalizations (Jarvis, 2004; Petkov & Jarvis, 2012). This VPL pathway has been studied in songbirds and humans. It consists of an anterior component through the frontal cortex, basal ganglia, and thalamus that is required for vocal imitation, and a posterior motor cortex component that is required to produce learned vocalizations (Jürgens, 2002, 2009; Mackevicius & Fee, 2017; Scharff & Nottebohm, 1991; figure 46.3). The VPL pathway is embedded within a motor pathway that controls nonvocal behaviors, and both VPL and motor pathways receive auditory and other sensory input (Belyk & Brown, 2017; Chakraborty & Jarvis, 2015; Feenders et al., 2008). One specialized feature of the posterior component of the VPL pathway is that cortical neurons in layer 5 and their equivalent neurons in the robust nucleus of the arcopallium (RA) in songbirds project directly onto brain stem vocal motor neurons (figure 46.3). Direct projections from the cortex to brain stem motor neurons correlate with greater fine motor control of the associated muscles that might be necessary for human speech and learned birdsong.

In most species with limited or no VPL, a VPL pathway has not been found, with the possible exception of nonhuman primates, which have been proposed to have a laryngeal motor cortex (LMC) and rudimentary Broca's area (Jürgens, 2009; Simonyan, 2014). Jürgens (2002, 2009) proposed that the main difference between humans and nonhuman primates is the direct human projection from LMC to the vocal motor neurons, as opposed to the indirect connection in nonhuman primates. In contrast, Rilling and colleagues considered the main difference to be the direct projection from the higher auditory cortex (Wernicke's area) to speech-language cortex (Broca's area), which is either indirect or absent in monkeys or weak in great apes (Rilling, 2014; Rilling et al., 2008). Since mice also have a putative LMC region that directly projects, although very sparsely, to brain stem vocal motor neurons, and it receives a direct robust projection from secondary auditory cortex (Arriaga et al., 2012), it seems that this type of connectivity is not a hallmark exclusively associated with extensive VPL. Like in humans and song-learning birds, the putative LMC in mice consists of motor and premotor parts and forms a loop through the basal ganglia and thalamus. However, unlike humans and song-learning birds, in the mouse, the direct projection from

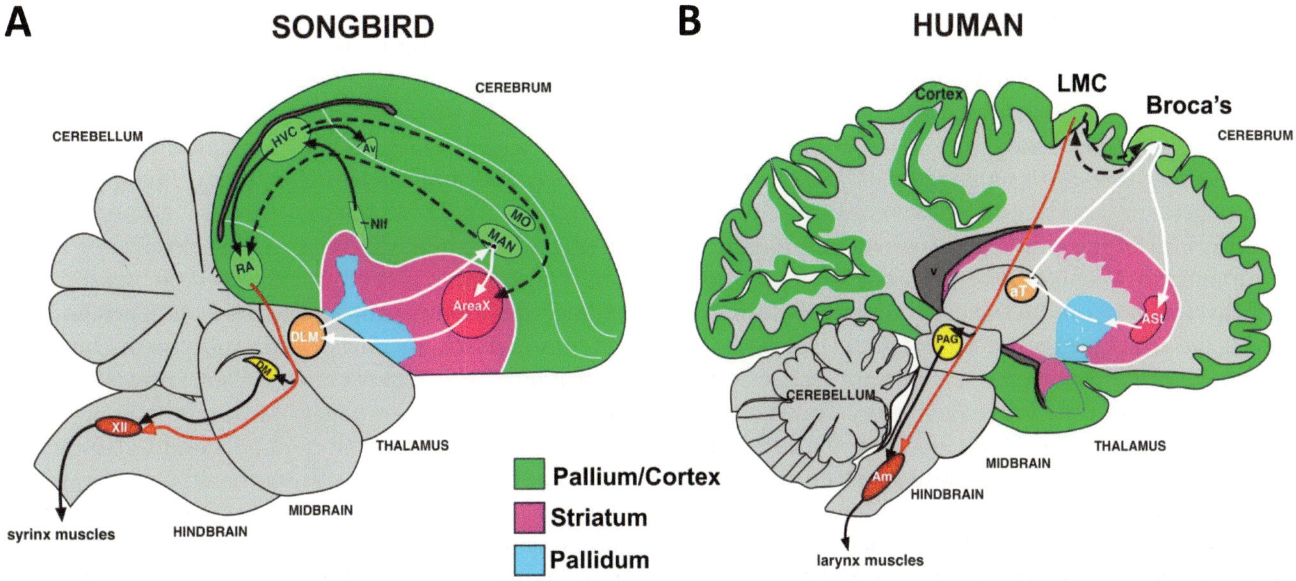

FIGURE 46.3 Brain pathways for vocal learning and spoken language in songbirds and humans. (A) Drawing of a songbird brain section showing connectivity of posterior (a vocal nucleus [HVC], RA, avalanche [Av], interfacial nucleus of the nidopallium [NIf]) and anterior (MAN, Area X, oval nucleus of the anterior mesopallium [MO]) song pathways. (B) Drawing of a human brain section showing proposed vocal pathway connectivity including LMC and part of anterior striatum (ASt) that shows convergence with songbird RA and Area X (Pfenning et al., 2014). Solid black arrows indicate the connections and regions of the posterior vocal motor pathway; white arrows indicate the connections and regions of the anterior vocal pathway. Dashed black arrows indicate the connections between the two pathways. Red arrows indicate the dense direct projection found only in vocal learners, from vocal motor cortex regions to brain stem vocal motor neurons. Am = nucleus ambiguous; aT = anterior thalamus speech area; DLM = dorsolateral nucleus of the thalamus; DM = dorsal medial nucleus of the midbrain; PAG = peri-aqueductal gray; v = ventricle space; XII = 12th vocal motor neurons in birds. Figure from Wang et al. (2015), modified from Arriaga et al. (2012); Petkov and Jarvis (2012); Pfenning et al. (2014).

LMC to vocal motor neurons is very sparse and lacks the specialized gene regulation found in language-relevant human brain regions and their counterparts in song-learning birds (discussed further in section 2.4). Moreover, the mouse LMC modulates vocalizations (Arriaga et al., 2012), but it is not required for their production (Arriaga et al., 2012; Hammerschmidt, Whelan, Eichele, & Fischer, 2015). These findings in primates, mice, and similar findings in a suboscine bird species (Liu, Wada, Jarvis, & Nottebohm, 2013) are consistent with the continuum hypothesis of VPL.

A synthesis of the anatomical and behavioral findings along a continuum of no VPL to extensive VPL is provided by the motor theory of vocal learning origin (Feenders et al., 2008). This theory proposes that in the common ancestor of all vertebrates, a forebrain pathway mediating general motor learning was present but that VPL was absent. The general motor learning pathway was duplicated and the new pathway, still embedded in the evolutionary older one, formed connections with the brain stem vocal motor system, facilitating limited VPL. In extensive VPLers, the new circuit segregated anatomically and functionally from the general motor learning pathway became more specialized and developed robust direct projections controlling the brain stem vocal motor and respiratory systems. In parrots and humans, a second duplication event of the VPL circuit generated two parallel VPL circuits allowing for greater VPL complexity (Chakraborty & Jarvis, 2015). Direct input from the auditory system is proposed to have been already present in the general motor learning pathway before extensive VPL evolved, and the VPL pathway inherited that input, allowing for auditory-vocal motor integration. The trait of VPL and its anatomical underpinnings might be evolutionarily quite plastic, leading to the development and the disappearance of VPL multiple times. For instance, in the majority of songbirds VPL exists in both sexes. However, in songbird species where only males sing, like zebra finches, females apparently lost extensive VPL (Odom, Hall, Riebel, Omland, & Langmore, 2014). In line with this notion, female zebra finches retain a vestigial VPL circuit that can be activated by hormone treatments during development, leading to the development of a VPL pathway and extensive VPL ability (Gurney, 1982). It is within this context of a continuum of neural circuitry and behavior predating speech and the anatomy underlying it, that we discuss how we look for the associated genes.

2.2. Gene Hunting Strategies There are two main ways to identify genes associated with vocal learning and language (see also chapter 41 by Vernes in this volume). First, by searching for the mutation within an individual gene associated with a particular disorder affecting language. Such monogenic disorders are rare. Usually, gene networks, not a single gene, underlie complex traits including language. However, in some cases, a mutation in a single gene can have serious consequences for the production and comprehension of speech, as is the case for the *FOXP2* gene,[1] which we will discuss in section 2.3. In the second approach, one correlates variation in the language faculty with the variation that occurs in genes. Individuals' genomes differ at millions of sites by single nucleotides (SNPs). The totality of those SNP variations is an individual's genetic fingerprint. One can search for associations between specific SNPs and a particular language phenotype, be it an impairment or a talent (see Deriziotis & Fisher, 2017, and chapter 39 by Luciano & Bates in this volume, for more details on methods and findings). This approach can be extended by probing for genes required for specific components of language, such as VPL. The diversity of VPL types among species is a great resource to identify genes involved in different aspects of VPL. Next we discuss some of the most informative discoveries thus far, a single gene approach and multigenic approaches comparing species that exhibit obligate VPL with those that do not.

2.3. The FOXP Gene Family in Humans and Nonhuman Animals

2.3.1. Core phenotype of patients with FOXP2 mutations The first gene associated specifically (but not exclusively) with speech was *FOXP2* (Lai, Fisher, Hurst, Vargha-Khadem, & Monaco, 2001). FOXP2 codes for a "transcription factor," that is, a protein that regulates the transcription of many other genes, called *target genes* (Spiteri et al., 2007). Mutations in FOXP2 cause Developmental Verbal Dyspraxia (also referred to as Childhood Apraxia of Speech), a severe speech disorder (Vargha-Khadem, Gadian, Copp, & Mishkin, 2005). Patients perform poorly in tests of productive aspects of language. Perceptive aspects of language and learning of other motor tasks are less affected (Morgan, Fisher, Scheffer, & Hildebrand, 2017). FOXP2 mutations lead to altered structure and function of corticostriatal and corticocerebellar circuits (Liégeois et al., 2016).

2.3.2. FOXP is evolutionary conserved The association of *FOXP2* with speech led to speculations that it might be "the language gene," unique to humans and the magic bullet to understand language evolution. However, *specifically associated with some aspects of language* does not need to translate to *unique to humans*. In fact, the coding sequence is highly conserved among most vertebrates (Li, Wang, Rossiter, Jones, & Zhang, 2007; Scharff & Haesler, 2005; Zhang, Webb, & Podlaha, 2002). In bats (Li et al., 2007) and teleost fish (Song, Wang, & Tang, 2013), the FoxP2 sequences are also conserved but diverged substantially more during the course of evolution. Importantly, FoxP2 is expressed in many brain regions that are relevant for speech in humans and for VPL in songbirds. Yet, FoxP2 expression is not limited to those neuron populations but it is also expressed in neurons relevant for other behaviors (Geerling et al., 2016; Haesler et al., 2004; Teramitsu, Kudo, London, Geshwind, & White, 2004; Verstegen, Vanderhorst, Gray, Zeidel, & Geerling, 2017). Its expression has been mapped in human embryos (Lai, Gerrelli, Monaco, Fisher, & Copp, 2003; Teramitsu et al., 2004), mice (Ferland, Cherry, Preware, Morrisey, & Walsh, 2003), songbirds (Haesler et al., 2004; Vicario, Mendoza, Abellan, Scharff, & Medina, 2017), fish (Bonkowsky & Chien, 2005), bats (Rodenas-Cuadrado et al., 2018), and even in drosophila and honeybees (DasGupta, Ferreira, & Miesenböck, 2014; Lawton, Wassmer & Deitcher, 2014; Schatton & Scharff, 2017). The fact that FoxP2 is expressed in similar brain circuits in many vertebrates suggests that this transcription factor fulfills important and potentially similar functions across a large variety of animals (Wohlgemuth, Adam, & Scharff, 2014). FoxP2 expression in invertebrates and vertebrates commences during embryogenesis and marks specific neuron populations, consistent with its demonstrated role in neuronal differentiation (Chiu et al., 2014).

2.3.3. FoxP2 manipulations in animal models affect brain development and vocalizations In the developing mouse brain, experimental manipulations of Foxp2 cause abnormal cortical neurogenesis (Tsui, Vessey, Tomita, Kaplan, & Miller, 2013), cerebellar foliation (Groszer et al., 2008), and spinal cord motor neuron delamination (Rousso et al., 2012). FoxP2 continues to function in the adult nervous system, shown by gene function studies of *FoxP2* in songbirds and mice (Adam, Mendoza, Kobalz, Wohlgemuth, & Scharff, 2016; Castellucci, McGinley, & McCormick, 2016; Chabout et al., 2016; Fisher & Scharff, 2009; Gaub, Fisher, & Ehret, 2016; Haesler et al., 2007; Heston & White, 2015; Murugan, Harward, Scharff, & Mooney, 2013; Schreiweis et al., 2014). In zebra finches, during the song-learning phase, both too little and too much FoxP2 in the striatal song nucleus Area X, an anterior forebrain component of the VPL pathway required for song learning (figure 46.3), prevents birds from developing normal song (Haesler et al., 2007; Murugan et al., 2013; Heston & White, 2015). After virus-mediated downregulation in Area X, juveniles developed abnormal song bearing phenotypic parallels to speech features of patients with FOXP2 mutations (Haesler et al., 2007), namely a smaller number of elements, shorter phrases, and inaccurate, variable delivery (Watkins, Dronkers, & Vargha-Khadem, 2002). In mice, a heterozygous Foxp2 missense mutation (Foxp2-R552H $^{+/-}$) that is equivalent to a mutation that causes speech impairments in humans affects their ability to produce complex sequences of sound units when courting a female, but does not affect simple sequences or their acoustic structure (Castellucci et al., 2016; Chabout et al., 2016; Gaub et al., 2016). Furthermore, the layer 5 LMC neurons in these mice are not properly coalesced in the LMC location (Chabout et al., 2016).

Comparing the findings in zebra finches and lab mice with those of human patients are intriguing. In humans carrying a heterozygous FOXP2-R553H point mutation, the speech deficit affects both the acoustic features of individual speech sounds and the sequencing of sound units that make up multisyllable words (Watkins et al., 2002). In zebra finches with reduced levels of FoxP2 in Area X, the acoustic structure of individual song elements was more impaired than their sequential delivery (Haesler et al., 2007). In Foxp2-R552H $^{+/-}$ mice, the acoustic features of the individual sound units are not affected but complex sequencing is. Thus, in species with extensive VPL, humans, and songbirds, it appears that FoxP2 influences circuit control of the acoustic structure of vocalizations, whereas in species with no or limited VPL, it does not play a role in acoustic structure.

2.3.4. FoxP2 affects striatal function in humans and animal models When comparing these results one should bear in mind that in mice and humans, the FoxP2 mutation was in the genome, impacting all cells in the body that express the gene. In contrast, in songbirds, the experimental manipulations were limited to Area X, the brain region important for imitative song learning, thus directly linking the relevance of FoxP for VPL to striatal activity. Follow-up studies showed that the FoxP2 downregulation in Area X alters the spine density of the striatal spiny neurons and glutamatergic transmission, impacting Dopamine 1 Receptor (D1R)–mediated modulation of signal propagation through Area X (Adam et al., 2016; Murugan et al., 2013; Schulz, Haesler, Scharff, & Rochefort, 2010). Concomitantly, D1R and DARPP-32 protein levels in Area X were reduced. In addition, social context-dependent song variability and the associated context-dependent neural variability in the cortical nucleus (lateral magnocellular nucleus of the anterior nidopallium or LMAN) that projects to Area X (figure 46.3) were abolished. Further support for the relevance of FoxP2 in striatal function is provided by in vivo electrophysiological recordings in the striatum of awake-behaving mice that were genetically modified to carry the same mutation occurring in human FOXP2 patients (French et al., 2012). Likewise, the striatum in FOXP2 patients is structurally and functionally affected (Liégeois et al., 2003).

2.3.5. FoxP2 levels can be dynamically regulated by behavior FoxP2 is transiently upregulated in Area X of young zebra finches during the developmental song-learning period and of adult canaries during seasonal song plasticity (Haesler et al., 2004; Teramitsu, Poopatanapong, Torrisi, & White, 2010; Teramitsu & White, 2006). In addition, singing behavior in young and adult zebra finches can acutely downregulate FoxP2 expression in Area X (Miller et al., 2008; Teramitsu et al., 2010; Thompson et al., 2013); intriguingly, in budgerigars (a parrot), FoxP2 is constitutively low in the region equivalent to Area X and does not decrease further with production of learned calls (learned song was not tested) (Hara et al., 2015), possibly related to the life-long vocal plasticity in budgerigars (Farabaugh et al., 1994; Hile, Plummer, & Striedter, 2000). In mice, changes in Foxp2 expression after vocalizing were not addressed, but FoxP2 is upregulated by sound exposure in the auditory thalamic (medial geniculate) nucleus (Horng et al., 2009).

2.3.6. FoxP2 modulates target genes relevant for synaptic function Hundreds of *FoxP2* target genes were discovered in experiments with human fetal brain, human neuron-like cells, and developing mouse brain (Vernes et al., 2007; Vernes et al., 2008; Vernes et al., 2011; Vernes, chapter 41 of this volume). The expression of two of those target genes is positively correlated with FoxP2 expression levels in zebra finch Area X: the very low-density lipoprotein receptor (*VLDLR*), encoding one of the reelin receptors, and Contactin-associated protein-like 2 (*CNTNAP2*), encoding a neurexin, both are important for synaptic function (Rodenas-Cuadrado, Ho, & Vernes, 2014). Zebra finch FoxP2 binds to the promoters of *VLDLR* and *CNTNAP2* (Adam et al., 2016, 2017) and experimental downregulation of FoxP2 in Area X results in reduced expression of *VLDLR* and *CNTNAP2*. Further findings raise the possibility that the regulatory relationship between FoxP2 and VLDLR guides structural plasticity toward the subset of FoxP2-positive medium spiny neurons of Area X in a singing-dependent manner via the reelin pathway (Adam et al., 2016, 2017). This highlights the need to think about the regulation of FoxP2 (Shi et al., 2018) and its targets in an activity-dependent and cell-specific manner associated with moment-to-moment changes in behavior (Adam et al., 2016, 2017; Becker, Devanna, Fisher, & Vernes, 2018; Panaitof, Abrahams, Dong, Geschwind, & White, 2010).

2.3.7. FoxP2 interacts with FoxP1 and FoxP4 FoxP2 and its paralogs, FoxP1 and FoxP4, are co-expressed in Area X, with most medium spiny neurons expressing all three FoxPs, but all other combinations also exist, with different frequencies (Mendoza & Scharff, 2017). In human cell culture experiments, the combinatorial protein-protein interaction among FoxP1, FoxP2, and FoxP4 regulates the transcription of target genes differentially (Sin, Li, & Crawford, 2015). Likewise, in zebra finches, the three FoxPs regulate the CNTNAP2 promoter differentially: FoxP1 activates it, FoxP2 represses it, and FoxP4 neither binds to nor regulates it (Mendoza & Scharff, 2017). Together, these results emphasize the functional importance of the protein-protein interactions among the FoxP family members in regulating their target genes and predict an involvement of FoxP1 and FoxP4 in speech and language. Indeed, FOXP1 mutations also affect speech and language, in addition to a variety of other clinical symptoms (Meerschaut et al., 2017; Siper et al., 2017). A FOXP4 mutation has only been reported in one patient that was developmentally delayed and had larynx problems (Charng et al., 2016).

2.3.8. FoxP1 and FoxP2 expression levels can be sex-specific FOXP2 protein levels in the left-hemispheric Brodmann's area 44 (part of Broca's area) of four-year-old girls (detected by Western blot from postmortem tissue) are higher than those of age-matched boys (Bowers, Perez-Pouchoulen, Edwards, & McCarthy, 2013). This leads to the question of whether the FOXP2 haploinsufficiency (e.g., loss of 50% of functioning protein) in humans with FOXP2 mutations might affect males more than females. However, a phenotypic difference between the sexes in patients with FOXP2 mutations has not been reported so far (Watkins et al., 2002). Sex differences in FoxP1 or FoxP2 expression in Area X were not noted in budgerigars or strawberry finch songbirds (*Amandava amandava*) (Haesler et al. 2004; Hara et al., 2015), whereas the absolute density of FoxP2 expression in Area X neurons was higher in male than in female Black-Capped Chickadees (*Poecile atricapillus*), which is consistent with males being the more vocal sex in this species (Phillmore, MacGillivray, Wilson, & Martin, 2015). Yet, a sex difference in expression levels also exists in multiple brain regions of rat pups, albeit in the opposite direction from the one reported in humans, with higher levels in males than females. Moreover, male pups produce more ultrasound vocalizations when separated from their mother than female pups do, and experimental downregulation of Foxp2 levels in the males leads to more female-like calling (Bowers et al., 2013). Very similar findings were found for Foxpl in mice, linking sexual dimorphic pup separation calls to cortical and subcortical Foxpl expression levels (Fröhlich, Rafiullah, Schmitt, Abele, & Rappold, 2017). This suggests a mechanistic link between the sex specificity of call activity and sexually dimorphic Foxp2 expression levels in rodents (Bowers et al., 2013), with androgens as a mediator (Bowers, Perez-Pouchoulen, Roby, Ryan, & McCarthy, 2014; Hamson, Csupity, Gaspar, & Watson, 2009).

2.3.9. FoxP affects behavioral function in fruit flies During the transition from invertebrates to vertebrates, a single *FoxP* gene gave rise to the four vertebrate FoxP gene paralogs, as a result of two postulated genome duplication events (Santos, Athanasiadis, Leitao, DuPasquier, & Sucena, 2011; Song, Tang, & Wang, 2016). Given the similarity of the DNA-binding domain of the single *FoxP* gene in *Drosophila* and the four vertebrate paralogs prompted studies into the behavioral consequences of FoxP loss of function in fruit flies. *Drosophila* mutants (dFoxPS-SZ-3955) have altered levels of the corresponding messenger RNA compared to wild-type flies and are incapable of learning to fly away from a noxious stimulus if only somatosensory cues are available to them (Mendoza et al., 2014). In a different study, dFoxPS-SZ-3955 mutants were slower to translate decisions based on different odor concentrations into a turning-away movement (DasGupta et al., 2014). A third study using a different mutant found problems with courtship behavior (Lawton et al., 2014). These different behavioral phenotypes in FoxP fly mutants are consistent with the hypothesis that tightly regulated FoxP protein levels in particular neurons are needed to translate stored information (e.g., where the body is in space) into motor programs (Schatton & Scharff 2017). Whether these are superficial similarities to the behavioral effects of FoxP1 and FoxP2 manipulations in vertebrates or whether they are based on true "deep homology" of FoxP and its associated molecular modules requires further scrutiny.

2.3.10. The human version of FOXP2 affects brain development and behavior It seems clear from the preceding summary that FoxP2 was relevant for motor behaviors including vocal communication well before spoken language evolved. But human spoken language only evolved once. Does the fact that the human FOXP2 sequence differs from all other FoxP2 sequences provide a clue to the particular specialization of human speech and language? The human *FOXP2* gene differs from other primates in only two amino acids (Enard et al., 2002) and one of those also differs from all mammals investigated (Zhang et al., 2002). The two human amino acid substitutions (T303N, N325S) were apparently subject to positive selection and became fixed in the human population after the human split from the chimpanzee lineage. This led to the notion that T303N and N325S contributed to the evolution of human language. To address this, the human-specific amino acids were introduced to the mouse Foxp2 gene, Foxp2^{hum} (Enard et al., 2002). Foxp2^{hum} pups produce qualitatively different ultrasonic calls when separated from their mothers. As adults, these mice were less exploratory but better at learning specific labyrinth tasks than their "nonhumanized Foxp2" siblings. In the brain, the Foxp2^{hum} had lower dopamine concentrations and longer dendrites in the cerebral cortex, thalamus, and striatum and increased synaptic plasticity (Long Term Depression; LTD) in the striatal medium spiny neurons (Enard et al., 2009; Reimers-Kipping, Hevers, Pääbo, & Enard, 2011; Schreiweis et al., 2014). These findings underscore the fact that even though the human and the mouse FoxP2 protein differ in only 3 of over 700 amino acids, small changes have the potential to alter neural development and brain function for vocalizations and could have promoted human cognitive development during evolution, for example, enhanced synaptic plasticity and behavioral flexibility.

2.4. Genome- and Transcriptome-wide Changes in VPLers If VPL and its contribution to speech is a multigenic trait, more genes than *FOXP2* should be discoverable. This is supported by recent informative findings of RNA expression of thousands of genes, the transcriptome. Using oligo-microarrays representing the brain transcriptome revealed that the song nucleus RA in songbirds and its analog in parrots and hummingbirds have convergent gene expression specializations with each other and with the human LMC (Pfenning et al., 2014; figure 46.3). These gene expression specializations are not found in birds or nonhuman primates that have little or no VPL. A gene expression specialization is where a gene's RNA or protein product is increased or decreased relative to the surrounding brain region, in this case, the adjacent nonvocal avian motor arcopallium or mammalian primary motor cortex. Of a total of 7,000 genes examined, there are ~55 such specialized genes in RA and the LMC; many of which are related to neural connectivity and neural development. One of the candidate genes is *SLIT1*, a ligand for the ROBO1 receptor, which is important for axon guidance (Pfenning et al., 2014; Wang et al., 2015). Mutations in SLIT1 and ROBO1 loci are associated with language deficits and autism (Wang et al., 2015). Interestingly, *SLIT1* and other genes relevant for neural connectivity were downregulated in avian RA and human LMC. One idea why this might be so is that since SLIT1 is a repulsive molecule, high expression in species without VPL may prevent the direct projection from RA/LMC to brain stem vocal motor neurons typical for VPL from forming (Wang et al, 2015). This can be tested by experimentally downregulating SLIT1 in the motor cortex of limited VPL or upregulating it the cortical RA/LMC of extensive vocal learners. There is precedent for such a possibility. In the study of another repulsive axon guidance ligand, PlexinA1, recapitulating its downregulation in human motor cortex in the mouse brain allowed direct projections to spinal cord motor neurons to remain after infant development (Gu et al., 2017).

Brain regions with convergent specialized transcriptome expression also include songbird Area X and a portion of the anterior human striatum (across the boundary of the caudate-putamen) that is activated during speech production (Pfenning et al., 2014; figure 46.3). The over 70 genes identified with specialized expression are over-represented in neurotransmission and movement disorder functions, consistent with the known function of Area X. These genes include a D2 dopamine receptor that is important for neurotransmission and synaptic plasticity in the striatum and ROBO1. The identified human striatal region overlaps a part of the striatum that is affected in people with a FOXP2 point mutation (Belton, Salmond, Watkins, Vargha-Khadem, & Gadian, 2003). Other avian song learning brain regions showed only weak trends of convergent expression with other human speech regions (e.g., songbird LMAN and human Broca's area), but these also did not have the immediate surrounding brain regions in birds profiled (Pfenning et al., 2014). Preliminary experiments profiling all the surrounding brain regions indicate that most if not all song learning nuclei of song-learning birds have convergent gene expression specializations with human spoken-language brain regions (Gedman, Pfenning, Wirthlin, Audet, & Jarvis, 2017).

One explanation, postulated in the motor theory of vocal learning, for the convergent molecular similarities is that the brain regions surrounding the specialized song-learning and spoken-language areas are homologous and served as a pre-existing substrate out of which similar brain pathways could evolve(Feenders et al., 2008). In this regard, the songbird RA shows an apparent homologous gene expression profile with mammalian motor cortex layer 5 neurons; a vocal nucleus called HVC, which projects to RA, shows molecular similarity to layers 2 and 3 (Pfenning et al., 2014). This finding supports Karten's 1969 hypothesis, originally proposed for the auditory and visual cortices, that different pallial populations of the avian brain are homologous to different cortical layers of the mammalian brain (Jarvis et al., 2005; Wang, Brzozowska-Precht, & Karten, 2010). Interestingly, the auditory regions of both song-learning birds and humans do not show as high a level of gene expression specialization relative to the surrounding sensory cortices, nor do they show convergent expression. Taken together, these findings are consistent with known species differences in the different vocal production learning types. They indicate that the most anatomically and molecularly specialized components of the avian song and human language systems are those involved in song and spoken-language acquisition and production.

In humans, additional molecular changes have occurred that affect the brain. This is the case in the Slit-Robo GTPase 2 gene (*SRGAP2*), which interacts with the speech relevant ROBO1 receptor and modulates its function. In humans, SRGAP2 is partially duplicated, and the duplicated copies act as competitive inhibitors of the parent gene (Charrier et al., 2012; Dennis et al., 2012). Partial inhibition of SRGAP2 and thus of ROBO1 receptor function causes dendrites of cortical neurons in humans to remain in a more immature state throughout life, similar to the finding of Foxp2^{hum} mice. The immature state allows for continued higher levels of learning into adulthood compared to learning in other species without such duplication. Putting studies together, one can imagine how the consequences of specialized expression of SLIT1 and ROBO1, regulated by FoxP2, in some spoken-language VPL brain regions, is in turn enhanced by the SRGAP2 duplication in humans.

The specialized gene regulation in speech brain regions is presumably controlled by changes in regulatory regions of those genes or in the coding sequences of genes that regulate them. Studies have looked for genome-wide sequence changes in VPL (Zhang et al., 2014) as well as in humans with speech disorders (Chen et al., 2017). Such investigations are subject to high rates of false positive sequence changes not associated with the trait, but that are either neutral or associated with another shared trait. To mitigate these false positives, multiple pieces of evidence are necessary. In this regard, noncoding sequences (nucleotide changes) near some of the genes with specialized expression in songbirds evolved faster than in non-VPL species (Zhang et al., 2014). Searching for convergent protein coding sequences among extensive VPL birds, one study identified genes previously found to be associated with speech deficits, including ROBO1, and some overlap with those specialized in speech brain regions (Lei et al., 2017).

In a recent sequencing study of 43 families with individuals that have specific language impairment or dyslexia, rare mutations were identified in dozens of genes (Chen et al., 2017) as well as a common variant (SNP) (Devanna et al., 2018). Of these, 14 genes had previously been associated with language/dyslexia deficits when mutated, including a receptor involved in neurotransmission and plasticity (*GRIN2B*), neurotransmitter release (*ERC1*), and a urokinase receptor ligand (*SPRX2*) involved in promoting synapse formation. Novel candidates included some present in multiple affected families or multiple mutations in the same gene, including genes involved in cellular lipid activity (*STARD9*), potassium channels (*SCN9A*), and histone H3 methyltransferase (*KMT2D*) that modulates gene regulation of the chromatin. Thus far, none of the mutations have been studied functionally in nonhuman animals to address the mechanism that may cause the language deficit. When such studies are done, they will need to consider the type of VPL behavior and whether the brain pathways examined are specialized for song and spoken language.

3. Conclusion

In conclusion, studying nonhuman animals has provided insights into components of human speech and its evolution at the behavioral, neuroanatomical, and genetic levels. As expected, the molecular profiles linked to VPL in nonhuman animals and speech in humans are related to many genes. The challenges for the future will be to determine the remaining molecular players, narrowing down the most important ones, and deciphering their mechanisms of action and their gene networks for spoken language. This will enable scientists to decipher what is ubiquitous among vocally communicating species, what is specialized in different abilities of VPL, and what may be unique to humans. Eventually, this will allow us to disambiguate between the discrete and continuum hypothesis of vocal production learning and thereby spoken language.

NOTE

1. Following standard nomenclature, genes are denoted in italics, proteins in regular font. Uppercase letters denote the human version of the gene (i.e., *FOXP2*), lowercase the mouse version of the gene (i.e., *Foxp2*), for all other species *FoxP2* is used.

REFERENCES

Abramson, J. Z., Hernández-Lloreda, M. V., García L., Colmenares, F., Aboitiz, F., & Call, J. (2018). Imitation of novel conspecific and human speech sounds in the killer whale (*Orcinus orca*). *Proceedings of the Royal Society B: Biological Sciences, 285*(1871), 20172171.

Adam, I., Mendoza, E., Kobalz, U., Wohlgemuth, S., & Scharff, C. (2016). FoxP2 directly regulates the reelin receptor VLDLR developmentally and by singing. *Molecular and Cellular Neuroscience, 74*, 96–105.

Adam, I., Mendoza, E., Kobalz, U., Wohlgemuth, S., & Scharff, C. (2017). CNTNAP2 is a direct FoxP2 target in vitro and in vivo in zebra finches: Complex regulation by age and activity. *Genes, Brain and Behavior, 16*(6), 635–642.

Araya-Salas, M., & Wright, T. (2013). Open-ended song learning in a hummingbird. *Biology letters, 9*, 20130625.

Arbib, M. A., Liebal, K., & Pika, S. (2008). Primate vocalization, gesture, and the evolution of human language. *Current Anthropology, 49*(6), 1053–1076.

Arriaga, G., & Jarvis, E. D. (2013). Mouse vocal communication system: Are ultrasounds learned or innate? *Brain and Language, 124*(1), 96–116.

Arriaga, G., Zhou, E. P., & Jarvis, E. D. (2012). Of mice, birds, and men: The mouse ultrasonic song system has some features similar to humans and song-learning birds. *PLOS ONE, 7*(10), e46610.

Balsby, T. J. S., & Bradbury, J. W. (2009). Vocal matching by orange-fronted conures (*Aratinga canicularis*). *Behavioural processes, 82*(2), 133–139.

Balsby, T. J. & Scarl, J. C. (2008). Sex-specific responses to vocal convergence and divergence of contact calls in orange-fronted conures (*Aratinga canicularis*). *Proceedings of the Royal Society B: Biological Sciences, 275*(1647), 2147–2154.

Baptista, L. F. (1977). Geographic variation in song and dialects of the Puget Sound white-crowned sparrow. *The Condor, 79*, 356–370.

Baptista, L. F., & Schuchmann, K.-L. (1990). Song learning in the Anna hummingbird (*Calypte anna*). *Ethology, 84*(1), 15–26.

Becker, M., Devanna, P., Fisher, S. E., & Vernes, S. C. (2018). Mapping of human FOXP2 enhancers reveals complex regulation. *Frontiers in Molecular Neuroscience, 11*, 47.

Beckers, G. J. (2011). Bird speech perception and vocal production: A comparison with humans. *Human Biology, 83*(2), 191–212.

Beecher, M. D. (2017). Birdsong learning as a social process. *Animal Behaviour, 124*, 233–246.

Beecher, M. D., & Brenowitz, E. A. (2005). Functional aspects of song learning in songbirds. *Trends in Ecology and Evolution, 20*(3), 143–149.

Beecher, M. D., Campbell, S. E. & Stoddard, P. K. (1994). Correlation of song learning and territory establishment strategies in the song sparrow. *Proceedings of the National Academy of Sciences, 91*, 1450–1454.

Beecher, M. D., Stoddard, P. K., Campbell, S. E. & Horning, C. L. (1996). Repertoire matching between neighbouring song sparrows. *Animal Behaviour, 51*, 917–923.

Belton, E., Salmond, C. H., Watkins, K. E., Vargha-Khadem, F., & Gadian, D. G. (2003). Bilateral brain abnormalities associated with dominantly inherited verbal and orofacial dyspraxia. *Human Brain Mapping, 18*(3), 194–200.

Belyk, M., & Brown, S. (2017). The origins of the vocal brain in humans. *Neuroscience and Biobehavioral Reviews, 77*, 177–193.

Berg, K. S., Delgado, S., Cortopassi, K. A., Beissinger, S. R., & Bradbury, J. W. (2011). Vertical transmission of learned signatures in a wild parrot. *Proceedings of the Royal Society B: Biological Sciences, 279*(1728), 585–591.

Bertram, B. (1970). The Vocal Behaviour of the Indian Hill Mynah, Gracula religiosa. *Animal Behaviour Monographs, 3*(2), 79–192, IN1.

Berwick, R. C., Okanoya, K., Beckers, G. J., & Bolhuis, J. J. (2011). Songs to syntax: the linguistics of birdsong. *Trends in Cognitive Sciences, 15*(3), 113–121.

Bettelheim, B. (1959). Feral children and autistic children. *American Journal of Sociology, 64*(5), 455–467.

Blumstein, D. T. (1999). The evolution of functionally referential alarm communication: Multiple adaptations; multiple constraints. *Evolution of Communication, 3*(2), 135–147.

Boeckx, C., & Grohmann, K. K. (2013). *The Cambridge handbook of biolinguistics*. Cambridge: Cambridge University Press.

Böhner, J. (1983). Song learning in the zebra finch (*Taeniopygia guttata*): selectivity in the choice of a tutor and accuracy of song copies. *Animal Behaviour, 31*, 231–237.

Böhner, J. (1990). Early acquisition of song in the zebra finch, *Taeniopygia guttata. Animal Behaviour, 39*, 369–374.

Bolhuis, J. J., Tattersall, I., Chomsky, N., & Berwick, R. C. (2014). How could language have evolved? *PLOS Biology, 12*(8), e1001934.

Bonkowsky, J. L., & Chien, C.-B. (2005). Molecular cloning and developmental expression of foxP2 in zebrafish. *Developmental Dynamics, 234*(3), 740–746.

Boughman, J. W. (1998). Vocal learning by greater spear-nosed bats. *Proceedings of the Royal Society B: Biological Sciences, 265*(1392), 227–233.

Boughman, J. W., & Moss, C. F. (2003). Social sounds: Vocal learning and development of mammal and bird calls. In A. M. Simmons, R. R. Fay, & A. N. Popper (Eds.), *Acoustic communication* (pp. 138–224). New York: Springer.

Bowers, J. M., Perez-Pouchoulen, M., Edwards, N. S., & McCarthy, M. M. (2013). Foxp2 mediates sex differences in ultrasonic vocalization by rat pups and directs order of maternal retrieval. *Journal of Neuroscience, 33*(8), 3276–3283.

Bowers, J. M., Perez-Pouchoulen, M., Roby, C. R., Ryan, T. E., & McCarthy, M. M. (2014). Androgen modulation of Foxp1 and Foxp2 in the developing rat brain: impact on sex specific vocalization. *Endocrinology, 155*(12), 4881–4894.

Bradbury, J. W., & Balsby, T. J. (2016). The functions of vocal learning in parrots. *Behavioral Ecology and Sociobiology, 70*(3), 293–312.

Bradbury, J. W., Cortopassi, K. A. & Clemmons, J. R.(2001). Geographical variation in the contact calls of orange-fronted parakeets. *The Auk, 118*(4), 958–972.

Brainard, M. S., & Doupe, A. J. (2002). What songbirds teach us about learning. *Nature, 417*, 351.

Brenowitz, E. A., & Beecher, M. D. (2005). Song learning in birds: Diversity and plasticity, opportunities and challenges. *Trends in Neurosciences, 28*(3), 127–132.

Briefer, E. F. & McElligott, A. G. (2012). Social effects on vocal ontogeny in an ungulate, the goat, *Capra hircus. Animal Behaviour, 83*(4), 991–1000.

Carroll, S. B. (2005). Evolution at two levels: On genes and form. *PLOS Biology, 3*(7), e245.

Castellano, S., Parra, G., Sanchez-Quinto, F. A., Racimo, F., Kuhlwilm, M., Kircher, M., ... Pääbo, S. (2014). Patterns of coding variation in the complete exomes of three Neandertals. *Proceedings of the National Academy of Sciences, 111*(18), 6666–6671.

Castellucci, G. A., McGinley, M. J., & McCormick, D. A. (2016). Knockout of Foxp2 disrupts vocal development in mice. *Scientific Reports, 6*, 23305.

Catchpole, C., & Slater, P. (1995). *Bird song: Biological themes and variations.* Cambridge: Cambridge University Press.

Chabout, J., Sarkar, A., Patel, S. R., Radden, T., Dunson, D. B., Fisher S. E., & Jarvis, E. D. (2016). A Foxp2 mutation implicated in human speech deficits alters sequencing of ultrasonic vocalizations in adult male mice. *Frontiers in Behavioral Neuroscience, 10*, 197.

Chaiken, M., Boehner, J., & Marler, E. (1993). Song acquisition in European starlings, Sturnus vulgaris: A comparison of the songs of live-tutored, tape-tutored, untutored, and wild-caught males. *Animal Behaviour, 46*, 1079–1090.

Chakraborty, M., & Jarvis, E. D. (2015). Brain evolution by brain pathway duplication. *Philosophical Transactions of the Royal Society B: Biological Sciences, 370*(1684), pii: 20150056.

Chalfin, L., Dayan, M., Levy, D. R., Austad, S. N., Miller, R. A., Iraqi, F. A., ... Kimchi, T. (2014). Mapping ecologically relevant social behaviours by gene knockout in wild mice. *Nature Communications, 5*, 4569.

Charng, W. L., Karaca, E., Coban Akdemir, Z., Gambin, T., Atik, M. M., Gu, S., ... Lupski, J. R. (2016). Exome sequencing in mostly consanguineous Arab families with neurologic disease provides a high potential molecular diagnosis rate. *BMC Medical Genomics, 9*(1), 42.

Charrier, C., Joshi, K., Coutinho-Budd, J., Kim, J. E., Lambert, N., de Marchena, J., ... Polleux, F. (2012). Inhibition of SRGAP2 function by its human-specific paralogs induces neoteny during spine maturation. *Cell, 149*(4), 923–935.

Chen, X. S., Reader, R. H., Hoischen, A., Veltman, J. A., Simpson, N. H., Francks, C., ... Fisher, S. E. (2017). Next-generation DNA sequencing identifies novel gene variants and pathways involved in specific language impairment. *Scientific Reports, 7*, 46105.

Chew, L. (1981). Geographic and individual variation in the morphology and sequential organization of the song of the Savannah sparrow (*Passerculus sandwichensis*). *Canadian Journal of Zoology, 59*, 702–713.

Chiu, Y.-C., Li, M.-Y., Liu, Y.-H., Ding, J.-Y., Yu, J.-Y., & Wang, T.-W. (2014). Foxp2 regulates neuronal differentiation and neuronal subtype specification. *Developmental Neurobiology, 74*(7), 723–738.

Christiansen, M. H., & Kirby, S. (2003). Language evolution: Consensus and controversies. *Trends in Cognitive Sciences, 7*(7), 300–307.

Crockford, C., Herbinger, I., Vigilant, L., & Boesch, C. (2004) Wild chimpanzees produce group-specific calls: a case for vocal learning? *Ethology, 110*, 221–243.

Cruickshank, A. J., Gautier, J. & Chappuis, C. (1993). Vocal mimicry in wild African grey parrot *Psittacus erithacus. Ibis, 135*, 293–299.

DasGupta, S., Ferreira, C. H., & Miesenböck, G. (2014). FoxP influences the speed and accuracy of a perceptual decision in *Drosophila. Science, 344*(6186), 901–904.

Deecke, V. B., Ford, J. K., Spong, P. (2000). Dialect change in resident killer whales: implications for vocal learning and cultural transmission. *Animal Behaviour, 60*, 629–638.

Dennis, M. Y., Nuttle, X., Sudmant, P. H., Antonacci, F., Graves, T. A., Nefedov, M., ... Eichler, E. E. (2012). Evolution of human-specific neural SRGAP2 genes by incomplete segmental duplication. *Cell, 149*(4), 912–922.

Derégnaucourt, S., & Gahr, M. (2013). Horizontal transmission of the father's song in the zebra finch (*Taeniopygia guttata*). *Biology Letters, 9*, 20130247.

Deriziotis, P., & Fisher, S. E. (2017). Speech and language: Translating the genome. *Trends in Genetics, 33*(9), 642–656.

Devanna, P., Chen, X. S., Ho, J., Gajewski, D., Smith, S. D., Gialluisi, A., ... Vernes, S. C. (2018). Next-gen sequencing identifies non-coding variation disrupting miRNA-binding sites in neurological disorders. *Molecular Psychiatry, 23*(5), 1375–1384.

Di Sciullo, A.-M., & Boeckx, C. (2011). *The biolinguistic enterprise: New perspectives on the evolution and nature of the human language faculty.* Oxford: Oxford University Press.

Dougherty, M. L., Nuttle, X., Penn, O., Nelson, B. J., Huddleston, J., Baker, C., ... Eichler, E. E. (2017). The birth of a human-specific neural gene by incomplete duplication and gene fusion. *Genome Biology, 18*(1), 49.

Doupe, A. J., & Kuhl, P. K. (1999). Birdsong and human speech: Common themes and mechanisms. *Annual Review of Neuroscience, 22*(1), 567–631.

Eales, L. A. (1985). Song learning in zebra finches: Some effects of song model availability on what is learnt and when. *Animal Behaviour, 33*(4), 1293–1300.

Eaton, R. L. (1979). A beluga whale imitates human speech. *Carnivore, 2*, 22–23.

Eco, U. (1995). *The search for the perfect language.* Hoboken, NJ: Wiley-Blackwell.

Eens, M., Pinxten, R., & Verheyen, R. F. (1992). Song learning in captive European starlings, *Sturnus vulgaris. Animal Behaviour, 44*(6), 1131–1143.

Enard, W., Gehre, S., Hammerschmidt, K., Hölter, S. M., Blass, T., Somel, M., ... Pääbo, S. (2009). A humanized version of Foxp2 affects cortico-basal ganglia circuits in mice. *Cell, 137*(5), 961–971.

Enard, W., Khaitovich, P., Klose, J., Zöllner, S., Heissig, F., Giavalisco, P., ... Ravid, R. (2002). Intra- and interspecific variation in primate gene expression patterns. *Science, 296*(5566), 340–343.

Esser, K. H. (1994). Audio-vocal learning in a non-human mammal: the lesser spear-nosed bat *Phyllostomus discolor. Neuroreport, 5*(14), 1718–1720.

Esser, K. H., & Schmidt, U. (1989). Mother-Infant Communication in the Lesser Spear-nosed Bat *Phyllostomus discolor* (*Chiroptera, Phyllostomidae*) - Evidence for Acoustic Learning. *Ethology, 82*(2), 156–168.

Farabaugh, S. M., Linzenbold, A., & Dooling, R. J. (1994). Vocal plasticity in budgerigars (*Melopsittacus undulatus*): Evidence for social factors in the learning of contact calls. *Journal of Comparative Psychology, 108*(1), 81–92.

Favaro, L., Neves, S., Furlati, S., Pessani, D., Martin, V., & Janik, V. M. (2016). Evidence suggests vocal production learning in a cross-fostered Risso's dolphin (*Grampus griseus*). *Animal Cognition, 19*(4), 847–853.

Feenders, G., Liedvogel, M., Rivas, M., Zapka, M., Horita, H., Hara, E., ... Jarvis, E. D. (2008). Molecular mapping of movement-associated areas in the avian brain: A motor theory for vocal learning origin. *PLOS ONE, 3*(3), e1768.

Fehér, O., Wang, H., Saar, S., Mitra, P. P., & Tchernichovski, O. (2009). De novo establishment of wild-type song culture in the zebra finch. *Nature, 459*(7246), 564–568.

Ferland, R. J., Cherry, T. J., Preware, P. O., Morrisey, E. E., & Walsh, C. A. (2003). Characterization of Foxp2 and Foxp1 mRNA and protein in the developing and mature brain. *Journal of Comparative Neurology, 460*(2), 266–279.

Ferreira, A. R., Smulders, T. V., Sameshima, K., Mello, C. V., & Jarvis, E. D. (2006). Vocalizations and associated behaviors of the sombre hummingbird (*Aphantochroa cirrhochloris*) and the rufous-breasted hermit (*Glaucis hirsutus*). *Auk, 123*(4), 1129–1148.

Filatova, O. A., Burdin, A. M., & Hoyt, E. (2010). Horizontal transmission of vocal traditions in killer whale (*Orcinus orca*) dialects. *Biology Bulletin, 37*(9), 965–971.

Fischer, J., Wheeler, B. C., & J. P. Higham (2015). Is there any evidence for vocal learning in chimpanzee food calls? *Current Biology, 25*(21), R1028–R1029.

Fisher, S. E., & Scharff, C. (2009). FOXP2 as a molecular window into speech and language. *Trends in Genetics, 25*(4), 166–177.

Fitch, W. T. (1997). Vocal tract length and formant frequency dispersion correlate with body size in rhesus macaques. *Journal of the Acoustical Society of America, 102*(2), 1213–1222.

Fitch, W. T. (2000). The evolution of speech: A comparative review. *Trends in Cognitive Sciences, 4*(7), 258–267.

Fitch, W. T. (2006). Production of vocalizations in mammals. In K. Brown (Ed.), *Encyclopedia of language and linguistics* (2nd ed., pp. 115–121). Oxford: Elsevier.

Fitch, W. T. (2010). *The evolution of language*. Cambridge: Cambridge University Press.

Fitch, W. T., & Kelley, J. P. (2000). Perception of vocal tract resonances by whooping cranes *Grus americana*. *Ethology, 106*(6), 559–574.

Fitch, W. T., Neubauer, J., & Herzel, H (2002). Calls out of chaos: The adaptive significance of nonlinear phenomena in mammalian vocal production. *Animal Behaviour, 63*(3), 407–418.

Foote, A. D., Griffin, R. M., Howitt, D., Larsson, L., Miller, P. J., & Hoelzel, A. R. (2006). Killer whales are capable of vocal learning. *Biology Letters, 2*(4), 509–512.

French, C. A., Jin, X., Campbell, T. G., Gerfen, E., Groszer, M., Fisher, S. E., & Costa, R. M. (2012). An aetiological Foxp2 mutation causes aberrant striatal activity and alters plasticity during skill learning. *Molecular Psychiatry, 17*(11), 1077–1085.

Fripp, D., Own, C., Quintana-Rizzo, E., Shapiro, A., Buckstaff, K., & Jankowski, K. (2005). Bottlenose dolphin (*Tursiops truncatus*) calves appear to model their signature whistles on the signature whistles of community members. *Animal Cognition, 8*, 17–26.

Fröhlich, H., Rafiullah, R., Schmitt, N., Abele, S., & Rappold, G. A. (2017). Foxp1 expression is essential for sex-specific murine neonatal ultrasonic vocalization. *Human Molecular Genetics, 26*(8), 1511–1521.

Fromkin, V., Krashen, S. Curtiss, S., Rigler, D., & Rigler, M. (1974). The development of language in genie: A case of language acquisition beyond the critical period. *Brain and Language, 1*(1), 81–107.

Gahr, M. (2000). Neural song control system of hummingbirds: Comparison to swifts, vocal learning (songbirds) and nonlearning (suboscines) passerines, and vocal learning (budgerigars) and nonlearning (dove, owl, gull, quail, chicken) nonpasserines. *Journal of Comparative Neurology, 426*(2), 182–196.

Garamszegi, L. Z., Eens, M., Pavlova, D. Z., Avilés, J. M., & Møller, A. P. (2007). A comparative study of the function of heterospecific vocal mimicry in European passerines. *Behavioral Ecology, 18*(6), 1001–1009.

Garland, E. C., Goldizen, A. W., Rekdahl, M. L., Constantine, R., Garrigue, C., Hauser, N. D., ... Noad, M. J. (2011). Dynamic horizontal cultural transmission of humpback whale song at the ocean basin scale. *Current Biology, 21*(8), 687–691.

Gaub, S., Fisher, S. E., & Ehret, G. (2016). Ultrasonic vocalizations of adult male Foxp2-mutant mice: Behavioral contexts of arousal and emotion. *Genes, Brain and Behavior, 15*(2), 243–259.

Gaunt, S. L., Baptista, L. F., Sánchez, J. E., & Hernandez, D. (1994). Song learning as evidenced from song sharing in two hummingbird species (*Colibri coruscans* and *C. thalassinus*). *Auk, 111*(1), 87–103.

Gedman, G., Pfenning, A. R., Wirthlin, M., Audet, J.-N., & Jarvis, E. D. (2017). *Further resolution of hypotheses on convergent brain regions for learned song in songbirds and speech in humans*. Washington, DC: Society for Neuroscience.

Geerling, J. C., Kim, M., Mahoney, C. E., Abbott, S. B. G., Agostinelli, L. J., Garfield, A. S., ... Scammell, T. E. (2016). Genetic identity of thermosensory relay neurons in the lateral parabrachial nucleus. *American Journal of Physiology—Regulatory, Integrative and Comparative Physiology, 310*(1), R41.

Ginsburg, N. (1963). Conditioned talking in the Mynah bird. *Journal of comparative and physiological psychology, 56*, 1061–1063.

Grosslight, J. H., Harrison, P. C., & Weiser, C. M. (1962). Reinforcement control of vocal responses in the Mynah bird (*Gracula religiosa*). *The Psychological Record, 12*, 193–201.

Groszer, M., Keays, D. A., Deacon, R. M., De Bono, J. P., Prasad-Mulcare, S., Gaub, S., ... Enard, W. (2008). Impaired synaptic plasticity and motor learning in mice with a point mutation implicated in human speech deficits. *Current Biology, 18*(5), 354–362.

Gu, Z., Kalambogias, J., Yoshioka, S., Han, W., Li, Z., Kawasawa, Y. I., ... Yoshida, Y. (2017). Control of species-dependent cortico-motoneuronal connections underlying manual dexterity. *Science, 357*(6349), 400–404.

Gultekin, Y. B., & Hage, S. R. (2017). Limiting parental feedback disrupts vocal development in marmoset monkeys. *Nature Communications, 8*, 14046.

Gurney, M. E. (1982). Behavioral correlates of sexual differentiation in the zebra finch song system. *Brain Research, 231*(1), 153–172.

Haesler, S., Rochefort, C., Georgi, B., Licznerski, P., Osten, P., & Scharff, C. (2007). Incomplete and inaccurate vocal imitation after knockdown of FoxP2 in songbird basal ganglia nucleus Area X. *PLOS Biology, 5*(12), e321.

Haesler, S., Wada, K., Nshdejan, A., Morrisey, E. E., Lints, T., Jarvis, E. D., & Scharff, C. (2004). FoxP2 expression in avian vocal learners and non-learners. *Journal of Neuroscience, 24*(13), 3164–3175.

Hage, S. R., Gavrilov, N., & Nieder, A. (2013). Cognitive control of distinct vocalizations in rhesus monkeys. *Journal of Cognitive Neuroscience, 25*(10), 1692–1701.

Hage, S. R., & Nieder, A. (2013). Single neurons in monkey prefrontal cortex encode volitional initiation of vocalizations. *Nature Communications, 4*, 2409.

Hammerschmidt, K., Reisinger, E., Westekemper, K., Ehrenreich, L., Strenzke, N., & Fischer, J. (2012). Mice do not require auditory input for the normal development of their ultrasonic vocalizations. *BMC Neuroscience, 13*, 40.

Hammerschmidt, K., Whelan, G., Eichele, G., & Fischer, J. (2015). Mice lacking the cerebral cortex develop normal song: Insights into the foundations of vocal learning. *Scientific Reports, 5*, 8808.

Hamson, D. K., Csupity, A. S., Gaspar, J. M., & Watson, N. V. (2009). Analysis of Foxp2 expression in the cerebellum reveals a possible sex difference. *Neuroreport, 20*(6), 611–616.

Hara, E., Perez, J. M., Whitney, O., Chen, Q., White, S. A., & Wright, T. F. (2015). Neural FoxP2 and FoxP1 expression in the budgerigar, an avian species with adult vocal learning. *Behavioural Brain Research, 283*, 22–29.

Hauser, M. D., Chomsky, N., & Fitch, W. T. (2002). The faculty of language: What is it, who has it, and how did it evolve? *Science, 298*(5598), 1569–1579.

Heston, J. B., & White, S. A. (2015). Behavior-linked FoxP2 regulation enables zebra finch vocal learning. *Journal of Neuroscience, 35*(7), 2885–2894.

Hile, A. G., Burley, N. T., Coopersmith, C. B., Foster, V. S. & Striedter, G. F. (2005), Effects of Male Vocal Learning on Female Behavior in the Budgerigar, *Melopsittacus undulatus. Ethology, 111*: 901–923.

Hile, A. G., Plummer, T. K., & Striedter, G. F. (2000). Male vocal imitation produces call convergence during pair bonding in budgerigars, *Melopsittacus undulatus. Animal Behavior, 59*(6), 1209–1218.

Hile, A. G., & Striedter, G. F. (2000). Call convergence within groups of female budgerigars (*Melopsittacus undulatus*). *Ethology, 106*(12), 1105–1114.

Hindmarsh, A. (1986). The functional significance of vocal mimicry in song. *Behaviour, 99*, 87–100.

Horng, S., Kreiman, G., Ellsworth, C., Page, D., Blank, M., Millen, K., & Sur, M. (2009). Differential gene expression in the developing lateral geniculate nucleus and medial geniculate nucleus reveals novel roles for Zic4 and Foxp2 in visual and auditory pathway development. *Journal of Neuroscience, 29*(43), 13672–13683.

Hultsch, H., & Todt, D. (1981). Repertoire sharing and song-post distance in nightingales (*Luscinia megarhynchos*). *Behavioral Ecology and Sociobiology, 8*, 183–188.

Immelmann, K. (1969). Song development in the zebra finch and other estrildid finches. In R. A. Hinde (Ed.), *Bird Vocalizations* (pp. 61–74). Cambridge: Cambridge University Press.

Janik, V. M. (2000). Whistle matching in wild bottlenose dolphins (*Tursiops truncatus*). *Science, 289*, 1355–1356.

Janik, V. M. (2014). Cetacean vocal learning and communication. *Current Opinion in Neurobiology, 28*, 60–65.

Janik, V. M., & Slater, P. J. B. (1997). Vocal learning in mammals. *Advances in the Study of Behaviour, 26*, 59–100.

Janik, V. M., & Slater, P. J. B. (2000). The different roles of social learning in vocal communication. *Animal Behaviour, 60*(1), 1–11.

Jarvis, E. D. (2004). Learned birdsong and the neurobiology of human language. *Annals of the New York Academy of Sciences, 1016*(1), 749–777.

Jarvis, E. D., Güntürkün, O., Bruce, L., Csillag, A., Karten, H., Kuenzel, W., … Butler, A. B. (2005). Avian brains and a new understanding of vertebrate brain evolution. *Nature Reviews Neuroscience, 6*(2), 151–159.

Jarvis, E. D., Mirarab, S., Aberer, A. J., Li, B., Houde, P., Li, C., … Zhang, G. (2014). Whole-genome analyses resolve early branches in the tree of life of modern birds. *Science, 346*(6215), 1320–1331.

Jones, A. E., ten Cate, C., & Slater, P. J. B. (1996). Early experience and plasticity of song in adult male zebra finches (*Taeniopygia guttata*). *Journal of Comparative Psychology, 110*(4), 354–369.

Jones, G., & Ransome, R. D. (1993). Echolocation calls of bats are influenced by maternal effects and change over a lifetime. *Proceedings of the Royal Society of London. Series B: Biological Sciences, 252*(1334), https://doi.org/10.1098/rspb.1993.0055

Jürgens, U. (2002). Neural pathways underlying vocal control. *Neuroscience and Biobehavioral Reviews, 26*(2), 235–258.

Jürgens, U. (2009). The neural control of vocalization in mammals: A review. *Journal of Voice, 23*(1), 1–10.

Kaminski, J., Call, J., & Fischer, J. (2004). Word learning in a domestic dog: Evidence for fast mapping. *Science, 304*(5677), 1682–1683.

Karten, H. J. (1969). The organization of the avian telencephalon and some speculations on the phylogeny of the amniote telencephalon. *Annals of the New York Academy of Sciences, 167*(1), 164–179.

Kelley, L. A., Coe, R. L., Madden, J. R., & Healy, S. D. (2008). Vocal mimicry in songbirds. *Animal Behaviour, 76*(3), 521–528.

King, A. P., & West, M. J. (1989). Presence of female cowbirds (*Molothrus ater ater*) affects vocal imitation and improvisation in males. *Journal of Comparative Psychology, 103*(1), 39–44.

King, S. L., & Janik, V. M. (2013). Bottlenose dolphins use learned vocal labels to address each other. *Proceedings of the National Academy of Sciences, 110*, 13216–13221

King, S. L., Sayigh, L. S., Wells, R. S., Fellner, W., & Janik, V. M. (2013). Vocal copying of individually distinctive signature whistles in bottlenose dolphins. *Proceedings of the Royal Society of London B: Biological Sciences, 280*(1757), 20130053.

Knörnschild, M. (2014). Vocal production learning in bats. *Current Opinion in Neurobiology, 28*, 80–85.

Knörnschild, M., Nagy, M., Metz, M., Mayer, F., & von Helversen, O. (2010). Complex vocal imitation during ontogeny in a bat. *Biology Letters, 6*(2), 156–159.

Knörnschild, M., Nagy, M., Metz, M., Mayer, F., & von Helversen, O. (2012). Learned vocal group signatures in the polygynous bat *Saccopteryx bilineata. Animal Behaviour, 84*(4), 671–679.

Konishi, M. (1963). The role of auditory feedback in the vocal behavior of the domestic fowl. *Ethology, 20*(3), 349–367.

Konishi, M. (1985). Birdsong: From behavior to neuron. *Annual Review of Neuroscience, 8*(1), 125–170.

Kremers, D., Lemasson, A., Almunia, J., & Wanker, R. (2012). Vocal Sharing and Individual Acoustic Distinctiveness within a Group of Captive Orcas (*Orcinus orca*). *Journal of comparative psychology, 126*, 433–445.

Kroodsma, D. E. (1982). Appendix: A world survey of evidence for vocal learning in birds. In D. E. Kroodsma & E. H. Miller, *Acoustic communication in birds* (pp. 311–337). New York: Academic Press.

Kroodsma, D. E. (1989). Male eastern phoebes (*Sayornis phoebe*; *Tyrannidae, Passeriformes*) fail to imitate songs. *Journal of Comparative Psychology, 103*(3), 227.

Kroodsma, D., Hamilton, D., Sánchez, J. E., Byers, B. E., Fandiño-Mariño, H., Stemple, D. W., ... Powell, G. V. (2013). Behavioral evidence for song learning in the suboscine bellbirds (*Procnias* spp.; Cotingidae). *Wilson Journal of Ornithology, 125*(1), 1–14.

Kroodsma, D. E., Houlihan, P. W., Fallon, P. A., & Wells, J. A. (1997). Song development by grey catbirds. *Animal Behaviour, 54*(2), 457–464.

Kroodsma, D. E., & Konishi, M. (1991). A suboscine bird (eastern phoebe, *Sayornis phoebe*) develops normal song without auditory feedback. *Animal Behaviour, 42*(3), 477–487.

Kroodsma, D. E., & Miller, E. H. (1996). *Ecology and evolution of acoustic communication in birds.* Ithaca, NY: Comstock Publishing Associates.

Lai, C. S. L., Fisher, S. E., Hurst, J. A., Vargha-Khadem, F., & Monaco, A. P. (2001). A forkhead-domain gene is mutated in a severe speech and language disorder. *Nature, 413*, 519.

Lai, C. S. L., Gerrelli, D., Monaco, A. P., Fisher, S. E., & Copp, A. J. (2003). FOXP2 expression during brain development coincides with adult sites of pathology in a severe speech and language disorder. *Brain, 126*(11), 2455–2462.

Lange, B. P. H., Henninghausen, C., Brill, M., & Schwab, F. (2016). Only cheap talk after all? New experimental psychological findings on the role of verbal proficiency in mate choice. *Psychology of Language and Communication, 20*(1), 1–22.

Larson, R. K., Déprez, V., & Yamakido, H. (2010). *The evolution of human language.* Cambridge: Cambridge University Press.

Lawton, K. J., Wassmer, T. L., & Deitcher, D. L. (2014). Conserved role of *Drosophila melanogaster* FoxP in motor coordination and courtship song. *Behavioural Brain Research, 268*, 213–221.

Le Boeuf, B. J., & Peterson, R. S. (1969). Dialects in elephant seals. *Science, 166*, 1654–1656.

Le Boeuf, B. J., & Petrinovich, L. F. (1974). Dialects of northern elephant seals, *Mirounga angustirostris*: Origin and reliability. *Animal Behaviour, 22*(3), 656–663.

Lei, H., Yan, Z., Sun, X., Zhang, Y., Wang, J., Ma, C., ... Sun, Z. (2017). Axon guidance pathways served as common targets for human speech/language evolution and related disorders. *Brain and Language, 174*, 1–8.

Leitner, S., Nicholson, J., Leisler, B., DeVoogd, T. J., & Catchpole, C. K. (2002). Song and the song control pathway in the brain can develop independently of exposure to song in the sedge warbler. *Proceedings of the Royal Society of London B: Biological Sciences, 269*(1509), 2519–2524.

Li, G., Wang, J., Rossiter, S. J., Jones, G., & Zhang, S. (2007). Accelerated FoxP2 evolution in echolocating bats. *PLOS ONE, 2*(9), e900.

Lieberman, P., & Blumstein, S. E. (1988). *Speech physiology, speech perception, and acoustic phonetics.* Cambridge: Cambridge University Press.

Liégeois, F., Baldeweg, T., Connelly, A., Gadian, D. G., Mishkin M., & Vargha-Khadem, F. (2003). Language fMRI abnormalities associated with FOXP2 gene mutation. *Nature Neuroscience, 6*(11), 1230–1237.

Liégeois, F. J., Hildebrand, M. S., Bonthrone, A., Turner, S. J, Scheffer, I. E., Bahlo, M., ... Morgan, A. T. (2016). Early neuroimaging markers of FOXP2 intragenic deletion. *Scientific Reports, 6*, 35192.

Lilly, J. C. (1965). Vocal mimicry in *Tursiops*: Ability to match numbers and durations of human vocal bursts. *Science, 147*(3655), 300–301.

Liu, W.-C., Gardner, T. J., & Nottebohm, F. (2004). Juvenile zebra finches can use multiple strategies to learn the same song. *Proceedings of the National Academy of Sciences, 101*(52), 18177–18182.

Liu, W.-C. & Kroodsma, D. E. (1999). Song development by chipping sparrows and field sparrows. *Animal Behaviour, 57*(6), 1275–1286.

Liu, W.-C. & Kroodsma, D. E. (2006). Song learning by chipping sparrows: When, where, and from whom. *The Condor, 108*(3), 509–517.

Liu, W.-C., & Nottebohm, F. (2007). A learning program that ensures prompt and versatile vocal imitation. *Proceedings of the National Academy of Sciences, 104*(51), 20398–20403.

Liu, W.-C., Wada, K., Jarvis, E. D., & Nottebohm, F. (2013). Rudimentary substrates for vocal learning in a suboscine. *Nature Communications, 4*, 2082.

Mackevicius, E. L., & Fee, M. S. (2017). Building a state space for song learning. *Current Opinion in Neurobiology, 49*, 59–68.

Mahrt, E. J., Perkel, D. J., Tong, L., Rubel, E. W., & Portfors, C. V. (2013). Engineered deafness reveals that mouse courtship vocalizations do not require auditory experience. *Journal of Neuroscience, 33*(13), 5573–5583.

Marler, P. (1970). A comparative approach to vocal learning: Song development in white-crowned sparrows. *Journal of Comparative and Physiological Psychology, 71*(2), 1.

Marler P., & Peters S. (1977). Selective vocal learning in a sparrow. *Science, 198*, 519–521.

Marler P., & Peters S. (1982). Structural changes in song ontogeny in the swamp sparrow *Melospiza georgiana. The Auk, 99*, 446–458.

Marler, P., & Peters, S. (1987). A sensitive period for song acquisition in the song sparrow, *Melospiza melodia*: A case of age-limited learning. *Ethology, 76*, 89–100.

Marler P., & Peters S. (1988). Sensitive periods for song acquisition from tape recordings and live tutors in the swamp sparrow, *Melospiza georgiana. Ethology, 77*, 76–84.

Marler, P., & Tamura, M. (1962). Song dialects in three populations of white-crowned sparrows. *The Condor, 64*, 368–377.

Marshall, A. J., Wrangham, R. W., & Arcadi, A. C. (1999). Does learning affect the structure of vocalizations in chimpanzees? *Animal Behaviour, 58*, 825–830.

Meerschaut, I., Rochefort, D., Revencu, N., Petre, J., Corsello, C., Rouleau, G. A., ... Callewaert, B. (2017). FOXP1-related intellectual disability syndrome: A recognisable entity. *Journal of Medical Genetics, 54*(9), 613–623.

Mendoza, E., Colomb, J., Rybak, J., Pfluger, H. J., Zars, T., Scharff, C., & Brembs, B. (2014). *Drosophila* FoxP mutants are deficient in operant self-learning. *PLOS ONE, 9*(6), e100648.

Mendoza, E., & Scharff, C. (2017). Protein-protein interaction among the FoxP family members and their regulation of two target genes, VLDLR and CNTNAP2 in the zebra finch song system. *Frontiers in Molecular Neuroscience, 10*, 112.

Meredith, R. W., Janecka, J. E., Gatesy, J., Ryder, O. A., Fisher, C. A., Teeling, E. C., ... Murphy, W. J. (2011). Impacts of the cretaceous terrestrial revolution and KPg extinction on mammal diversification. *Science, 334*(6055), 521–524.

47 The Grammatical Abilities of Animals: A Comparative Overview

CAREL TEN CATE AND CHRISTOPHER I. PETKOV

Syntax is a core feature of the human language faculty. It concerns the computational system that structures language. Both the evolution and the development of these computational abilities are debated timely topics. Some researchers defend the position that genetic changes occurring around 100,000 years ago resulted in the sudden emergence of a "universal grammar" ability that guides and constrains language learning (e.g., Berwick & Chomsky, 2015). Others see an evolutionary scenario in which the complex rule learning mechanisms used for language arose gradually from domain general learning and sequence-processing mechanisms, linked to and dependent on cultural transmission (e.g., Tomasello, 2003; Christiansen & Chater, 2015). Also, views differ on whether the processing and acquisition of linguistic structure during infant development results from a single learning mechanism (Aslin & Newport, 2014) or multiple mechanisms (Endress & Bonatti, 2007). Against this background, there is increasing interest in comparative research, which can address what types of sequence learning and processing abilities are present in other species and to what extent they are comparable to those used by humans and relate to language. The comparative approach may provide a window onto what might have been precursors of human language abilities and which of the core elements involved in language are human-specific or of a more general nature.

In this chapter, we provide an overview of this rapidly expanding field of behavioral research. We first consider the structure and level of complexity of natural vocalizations in various groups of animals, couched within theoretical frameworks on the syntactic complexity of language. While this overview provides insight into the "grammar" of animals' vocal output, it does not reveal the full extent of the computational competences that go beyond any given animal's vocal production prowess. Animals may be capable of more complex sequence processing and rule learning but not apply these abilities in their own vocalizations. It could be that these "hidden" competences evolved in contexts other than for vocal communication and may trace an evolutionary trajectory toward human syntactic competences. These competences can be examined in a systematic way with artificial grammar learning (AGL) experiments, which can be designed to emulate different forms of sequencing complexity. The AGL paradigm is used to address whether subjects can detect and learn the rules or patterns underlying structured strings of items and how this compares across species. We overview different types of AGL studies that have been conducted with a range of animal species. We consider what new insights this line of research provides about the grammatical competences of animals or what may be unique about being human.

1. *Structure in Natural Vocalizations*

Speech is the default medium humans use to produce language. Languages are characterized by using a finite variety of speech sounds that can be combined to form a theoretically infinite number of words, which in turn can be used to construct an infinite number of meaningful expressions.

The hierarchical structuring of language (from phonemes to words to sentences, see figure 47.1, *top*) appears to be far removed from the vocalizations present in many mammals. Some primate species do have elaborate vocalizations, like the songs of various gibbon species (e.g., Wanelik, Azis, & Cheyne, 2013), the vocal sequences of the Titi monkey (Robinson, 1979), or the pant-hoots of chimpanzees (e.g., Fedurek, Schel, & Slocombe, 2013), each of which consist of a series of different element types. But at the same time, these vocal strings lack the prominent hierarchical properties of language, such as the nesting of different meaningful phrases in a sentence (Berwick, Beckers, Okanoya, & Bolhuis, 2012; Berwick, Okanoya, Beckers, & Bolhuis, 2011). These vocalizations are used in a particular context, with variations in the phonetic details of calls having the capability to potentially vary call meaning (see chapter 45 by Fischer & Hage in this volume). Occasionally calls, each of which have a meaning on their own, are combined to produce call

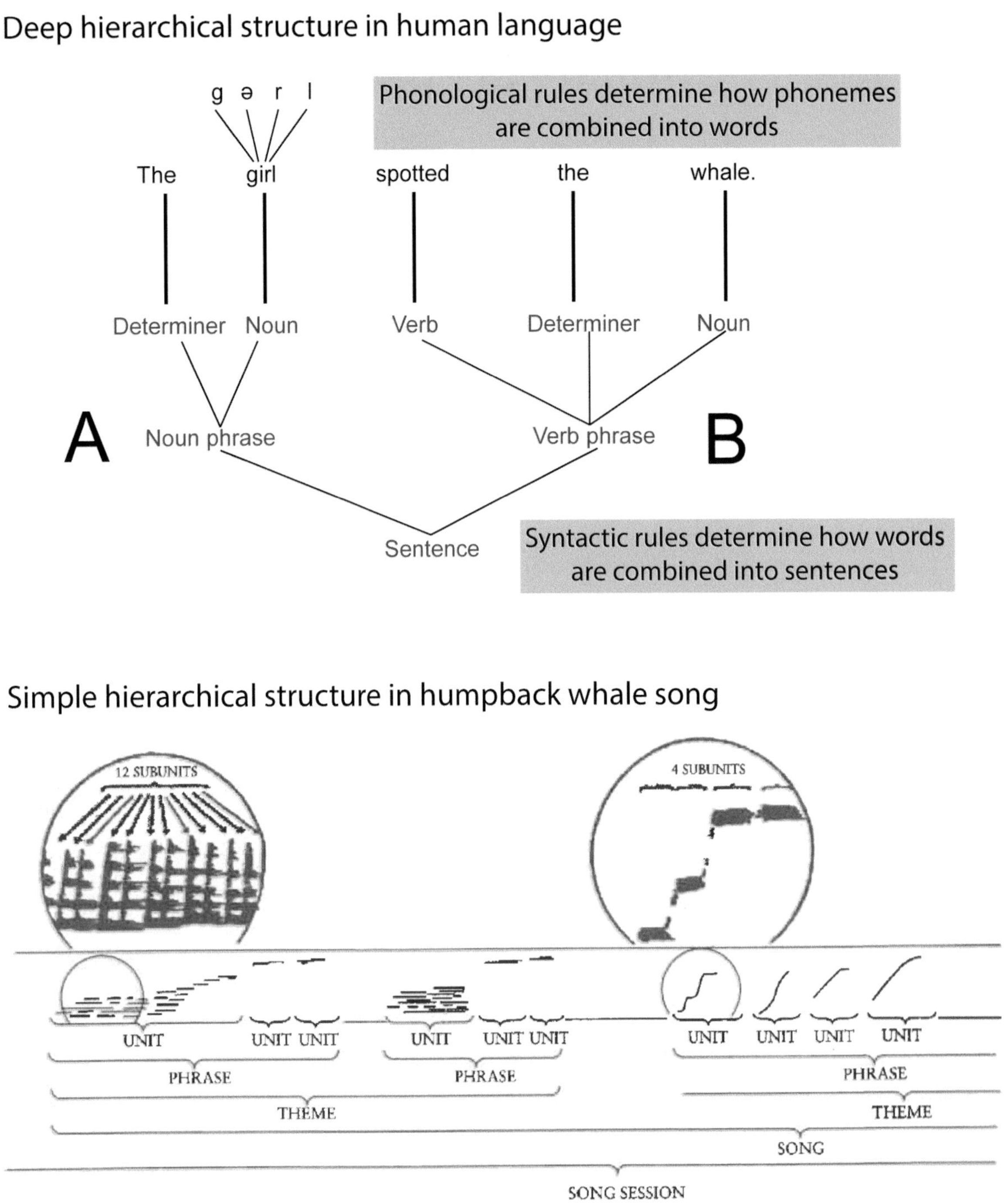

FIGURE 47.1 Hierarchical structure in human language and humpback whale song. (*Top*) Shown is an example sentence, illustrating the deep hierarchical structure of human language. Phonological rules are used to combine phonemes into words and syntactic rules to grammatically combine words in a sentence. Note that at the phrase level there are two phrases (A and B) that define the necessary constituent elements in every sentence: a noun phrase and a verb phrase. (*Bottom*) Illustration of humpback whale song by Payne and McVay (1971) with annotations as shown in Hurford (2012). Note that the humpback whale song has simple but clear hierarchical structure where subunits combine into units, which in turn are combined into phrases. A song typically has no more than two phrases (as shown here) within which the whale elaborates on the numbers of repetitions of units within a phrase.

combinations that result in different behavioral responses and thus presumably have different meanings for the animals. For instance, putty nosed monkeys produce "hack" vocalizations in response to terrestrial predators such as leopards, while "pyows" are produced for avian predators such as eagles (Arnold & Zuberbühler, 2006). Each of these calls gives rise to a different avoidance behavior in the receiver within the vicinity (climb up or down, depending on where the avian or terrestrial predator is likely to be), which might be interpreted as having a different "meaning." When the two calls are produced together, this combination gives rise to a different behavior, making the group move away from the area. This is an example of a simple "combinatorial" or "phonological" syntax (Hurford, 2012), in which sounds are combined in different ways to arrive at a new meaningful utterance. Although these elaborations of primate vocalization usage resemble how humans combine different phonemes to create a novel word, it is not clear whether the order of the combination of monkey calls matters to the animals. In human language, changing the order of a few words in a sentence (such as swapping *girl* with *whale* in the example in figure 47.1) completely changes the meaning of the sentence. Thus human language is referred to as having a "semantically compositional syntax" or "compositional syntax," where the structure of language itself carries meaning (Hurford, 2012). So far, no primate vocalization data has produced clear evidence for the presence of a compositional syntax, although disrupting the order of natural vocalization sequences in playback experiments seems to matter to primates, such as gibbons and Titi monkeys (Mitani & Marler, 1989; Robinson, 1979).

Apart from its semantic and syntactic hierarchical structure, an important difference between primate vocalizations and human language is that learning plays an important role in the development of human language: infants and adults need considerable exposure and practice with any given language in order to be able to produce correct speech sounds and symbols as well as grammatically correct sentences of that language. In contrast, primate vocalizations develop without much prior learning and exposure. Some nonhuman primate vocalizations show acoustic changes during development (marmosets: Takahashi et al., 2015) or with experience (e.g., chimpanzee food grunts: Watson et al., 2015; pant-hoot calls: Marshall, Wrangham, & Arcadi, 1999). However, the overall phonological structure of the calls does not appear to result from vocal production learning, which is defined as the capacity to model one's own vocalization on ones heard from another animal, for example, from a tutor. Thus while some vocal plasticity seems present in nonhuman primates, it is limited, consistent with a continuum of vocal production learning capacity (Petkov & Jarvis, 2012; Arriaga, Zhou, & Jarvis, 2012).

However, there are a number of other mammalian taxa that are capable of vocal production learning: some pinnipeds (seals), cetaceans (dolphins and whales), bats, and elephants. Vocal learning is likely to have evolved independently in each of these groups. In theory, vocal production learning may expand the vocal repertoire of a species and hence also provide more possibilities for combining vocalizations. Of the vocal learning mammal species, some whales produce elaborate vocalizations where long strings of vocalization elements contain a range of distinct elements, including combinations of simple phrases (Hurford, 2012; figure 47.1, *bottom*). One such species is the humpback whale. On a global scale, different populations sing different songs, with one song type dominating in a particular population. At the individual level, a song consists of a long sequence of a variety of sounds that may be repeated for hours (Payne & McVay, 1971; see figure 47.1, *bottom*). A humpback whale song can have a specific nested structure, a simple form of hierarchical organization. The smallest identifiable sound elements (called *subunits*) form fixed "units." Units follow each other in a particular sequence, forming a "phrase" (figure 47.1). Successive renditions of these phrases show slight changes in the structure and repetitions of the underlying units. A succession of particular phrases (which can be repeated a variable number of times, with some slight phonological changes between phrases) form a "theme." Finally, a number of different themes, following each other in a fixed sequence, constitute a song (see Hurford's 2012 analysis based on Payne & McVay, 1971). Therefore, there is certainly some form of simple hierarchical structure in humpback songs, but, as noted by Hurford (2012) and Payne and McVay (1971), the song structure has a high level of predictability, typically consisting of a couple of phrases and can be adequately covered by a state chain diagram (a schematic of the transitions between song components; see Hurford, 2012; figure 47.2).

Mice are a mammalian group that have classically been thought not to learn their vocalizations. Although this view has been challenged (Arriaga et al., 2012), their vocal plasticity seems limited (e.g., Hammerschmidt et al., 2012) comparable to that of most primates. However, with respect to their structure, recent work suggests that the ultrasonic vocalizations of male mice during sexual courting behavior have a number of features indicating a higher structural complexity (Holy & Guo, 2005). Analysis of the ultrasonic mice

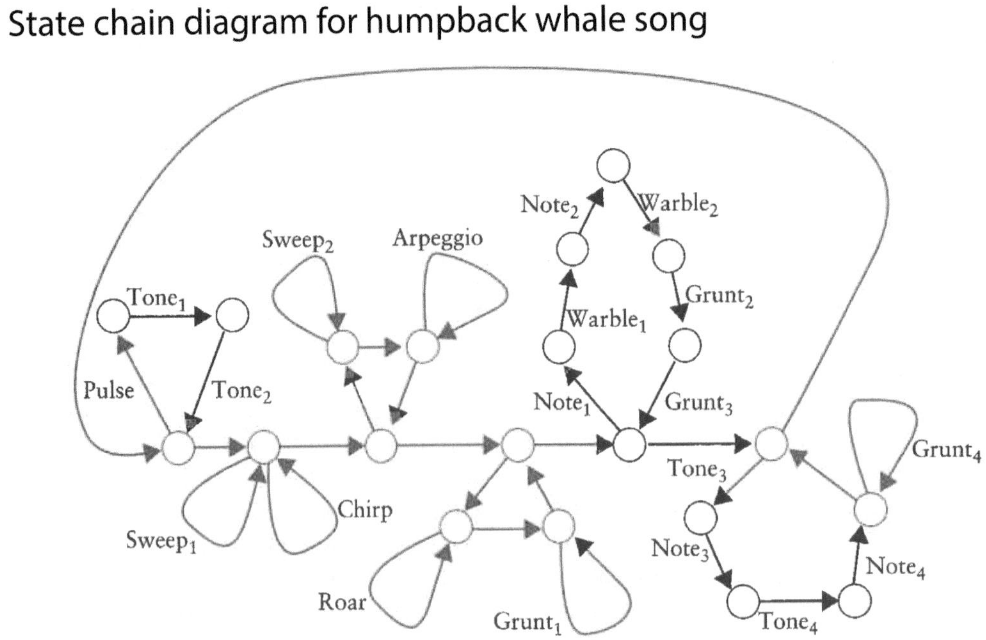

FIGURE 47.2 State chain diagram for humpback whale song. Interpretation, state chain diagram, and annotation of humpback whale song data from Payne and McVay (1971) by James Hurford in his book (for details, see Hurford, 2012). Note that a state chain diagram can well describe the song transitions. Human language, by comparison, cannot be easily described by state chain diagrams.

vocalizations (Chabout, Sarkar, Dunson, & Jarvis, 2015) showed an interesting set of transitions between syllables, which depended on the context that the male animal was in (e.g., with the female, only smelling her). Chabout et al. (2015) identified a common "s" syllable that is often repeated. Also the mouse state transition diagram shows a fair amount of nonlinear transitions where "s" can transition to two or three other syllables with differing probability. It is also interesting to note that female mice respond more favorably to more elaborate male vocalizations, as is seen in many songbirds (Chabout et al., 2015).

Not much is known about the extent to which cetaceans, mice, or other mammal species combine sounds in more complex structures to create vocalizations with a different meaning. As we've described, this is also unknown for primates. Currently the impression, based primarily on a lack of evidence rather than evidence for the ability being absent, is that complex vocalization structures where the ordering relationships between units or phrases carries meaning, as is the case in human language, is not evident in nonhuman mammals, whether the animals are vocal learners or not.

While complex vocalizations and vocal learning are uncommon among mammals, they are more common among birds. Vocal production learning evolved independently in at least three groups of birds: hummingbirds, parrots, and songbirds. This last group consists of some 4,000 of the around 10,000 bird species. Most songbird species produce a variety of different vocalizations, which are often classified as either songs or calls. The term *song* is used for the, often complexly structured, vocalizations made in the context of reproduction—mate attraction and territorial defense. Songs, unlike many calls, require exposure to a model (tutor) to develop properly. However, although this may suggest that songs of a species can develop wide-ranging individual variation in structure, there are often clear constraints on this variation, with each species having its own "structuring rules" and singing styles. Some species have one rather fixed stereotyped song, like the zebra finch (Lachlan, van Heijningen, ter Haar, & ten Cate, 2016), but there are also elaborate singers with a rich repertoire like nightingales and extensive imitators with variable songs like mockingbirds and starlings. These observations raise the question whether the species-specific constraints on song structures are akin to innate species-specific grammar rules and/or whether experience can modify song structuring rules or singing styles. One species where experience can shape the singing style is the nightingale. Nightingales have a song repertoire of up to 200 different songs. Normally, nightingales hardly ever repeat the same song immediately—a singing style known as *immediate variety*. However, young nightingales raised in the laboratory and exposed to "tutor" songs that were always repeated several times before a new one was played showed considerable immediate repetition in their own singing style when they became adults (Hultsch, 1991). So, even though immediate variety is something akin to a universal grammar rule for wild nightingales, experience can affect its development in ways resembling language learning and use during infant development. Experiments on other species have also shown that the structural song characteristics that are universal in the wild can be modified by experience (e.g., Rose et al., 2004). Nevertheless, despite all the variation as well as developmental plasticity there may be within and between songs of a species, there is, as for mammal vocalizations, no clear evidence that the ordering relationships between elements within songs, or songs within a song sequence, carries meaning comparable to the compositional syntax of human language.

The other type of bird vocalizations, calls, are usually produced by both sexes, linked to a specific context (like alarm calls or begging calls). These calls have a simple acoustical structure, usually consisting of a single or just a few elements, and many develop without vocal production learning. While there seems to be no semantical compositionality in songs, two recent studies have provided some evidence for a rudimentary form of semantical compositionality in call sequences of two songbird species. The Japanese tit can use a call combination in which receivers combine the meaning of the separate calls (Suzuki, Wheatcroft, & Griesser, 2016). One call type, consisting of the note sequence *ABC*, on its own elicits scanning behavior by listening birds, while another call type, a *D* note, elicits approach. The combination in the sequence *ABC-D* results in both responses, while the reverse combination does not result in scanning or approach. This finding can be interpreted as sensitivity to the composition of the call combination, a rudimentary form of semantic compositionality. A similar example of sensitivity to call composition has been observed in southern pied babblers (Engesser, Ridley, & Townsend, 2016).

2. *Quantifying the Complexity of Animal Vocalizations*

We've discussed whether variation in the sequential structure of animal vocalizations shows any indication of variation in meaning in a way comparable to the compositional syntax in human language. Another question is how to measure the structural complexity underlying the variation in animal vocalizations and how this complexity relates to linguistic structure.

The relative ease with which bird songs can be recorded and their variability calculated have made them a prime target for formal analyses of their structure, which can be related to the complexity of language using computational theoretic frameworks. The earliest attempts to extract structure from bird songs were based on using transition diagrams, illustrating the transitional probabilities between elements. These analyses showed that the species-specific vocal structures of some species could be described with simple Markov models (reviewed by ten Cate, Lachlan, & Zuidema, 2013). First-order Markov models (or chains) consist of pairwise transitions between elements in a string (bigrams), where one element predicts the next one. Second-order Markov chains are sensitive to triplets in a string (trigrams) and so on. However, when songs become more complex, these models are not sufficient to describe the increase in complexity. For instance, Bengalese finches sing songs with variable song note transitions, where often the same element is present in different parts of the song. Moreover, several elements together in a Bengalese finch song can form chunks (like in the humpback whale's song), and multiple chunks are sung in a variable but somewhat predefined order. Analyses of these songs, and also of the complex vocalizations of other species, led to the conclusion that more powerful algorithms, such as probabilistic or hidden Markov models (Berwick et al., 2011; Hurford, 2012; ten Cate et al., 2013; ten Cate & Okanoya, 2012), state chain processes (Hurford, 2012), or renewal processes (Kershenbaum et al., 2014) provide better descriptions than lower-order Markov models do.

Frameworks that can be useful for understanding the structure of language and the computational power of artificial and biological systems are revised versions of the Formal Language Hierarchy (FLH) (Berwick & Chomsky, 2015; Berwick et al., 2012; Chomsky, 1957; Hurford, 2012; Jäger & Rogers, 2012). Even the most elaborate structure of nonhuman animal songs that we have been considering is currently thought to be well within the realm of regular or finite state grammars in the FLH. Finite state grammars, and regular languages, are at the bottom of the FLH, which has been used as a yardstick to measure the level of generative or syntactic complexity (e.g., Berwick et al., 2011). A formal description of language requires more complex generative capacity that incorporates, for instance, the ability to create deeply nested, hierarchically organized sentences containing multiple levels of structure (figure 47.1, *top*).

To summarize, so far no analysis of vocalizations found naturally in the animal kingdom has shown evidence of a syntactic combinatorial complexity comparable to that of human language. This has been interpreted by some as underlining a fundamental gap in relevant cognitive linguistic abilities between animal vocalizations and language, with humans clearly more advanced than other animals. However, it is also clear that complexity in the forms of call combinations and song structure is increasingly becoming more evident in various animals. Some of this looks, at least on the surface, like rudimentary forms of core human language operations. There is a strong need, though, for more thorough analysis of the depth of nested hierarchical structure in natural vocalizations and songs. Formal analyses also need to be informed by behavioral data on the meaningful units and ordering relationships that matter to the animals—data still lacking in most cases. Thereby, the strong conclusions that have been made regarding where animal computational abilities fall are in our view highly preliminary as they are based on a paucity of data in a handful of species. Additionally, as we consider next, the structure of nonhuman animal vocalizations need not be fully reflective of animals' cognitive abilities in general.

3. Artificial Grammars and Animal Sequence Learning Capabilities

While studying the way that nonhuman animals structure their vocalizations and how this relates to language can be informative, this is but one facet in the language evolution puzzle. Although researchers may differ strongly on their views about how language evolved, a broadly shared view is that the roots of language are likely to be found in computations and combinatorial capacities that did not primarily evolve for vocal communication (e.g., Berwick & Chomsky, 2015; Christiansen & Chater, 2015; Everaert, Huybregts, Chomsky, Berwick, & Bolhuis, 2015; Hauser, Chomsky, & Fitch, 2002; Tomasello, 2003). Therefore, it may well be that many animals are able to detect more complex structural patterns in sound sequences than what they vocally produce (Petkov & Jarvis, 2012; ten Cate, 2017). We next examine the more general computational and rule learning abilities in nonhuman animals and how these compare to those of humans. We ask what types of structured patterns animals can recognize.

The AGL paradigm provides a powerful approach to examine the abilities of animals to process sequential input. This paradigm was initially developed to test the grammar learning abilities of human adults (Reber, 1967). In AGL experiments, arbitrary, meaningless, auditory or visual items are presented to subjects in strings arranged according to particular grammatical rules that define the ordering relationships between particular items. After exposure to "grammatical" sequences, subjects are next tested with new "grammatical" or "ungrammatical" (i.e., "consistent" or "inconsistent") sequences, and their responses to each of these are measured to determine whether the participants are sensitive to sequences consistent with the artificial grammar. The processes involved in AGL need not map one-to-one with natural language learning (Smith et al., 2017). However, AGL experiments have proven to be very successful for analyzing and understanding the rule-based sequence learning abilities of human adults. Moreover, as the strings consist of meaningless items and responses don't need to be verbal but can instead be behavioral or neural, AGL paradigms are suitable for testing young infants and are increasingly being used to address the sequence learning abilities of nonhuman animals (Fitch & Friederici, 2012; Fitch, Friederici, & Hagoort, 2012).

AGL studies of adult humans are often conducted by testing what the participant learned from exposure to a series of sound strings. This is done by asking whether a testing sequence follows or is different from the sequences heard during the prior habituation/exposure period. Animals cannot be instructed as such. In this respect animal studies are similar to studies of human infants. Such studies often use a familiarization or habituation method in which infants are exposed to a string of sounds organized according to one particular algorithm. After hearing this for some time, commonly resulting in a decline in interest to the sounds often measured by looking in the direction of the stimuli, the infants then hear novel test strings organized either according to a pattern like the familiarized strings or strings containing some form of violations of the AG sequencing relationships. A difference in responding to (e.g., looking toward) consistent and violation testing strings indicates that the infants learned something about the sequence structure (e.g., Marcus, Vijayan, Bandi Rao, & Vishton, 1999). The familiarization method can also be used to test nonhuman animal species, as has been done with various monkey species (e.g., tamarins and rhesus macaques) or apes. A second frequently used method for nonhuman animal testing is using operant conditioning in which animals are trained to discriminate sets of strings consistent with or in violation of the AG. Each method has its pros and cons (ten Cate & Okanoya, 2012), but it is beyond the scope of this chapter to go into any considerable detail on each of these methods.

The key question for most AGL experiments concerns whether the ability to learn particular sequencing relationships is specific to humans or more broadly shared. Since animals lack human language, if they can nonetheless process certain relationships that emulate aspects of human language along defined dimensions of complexity, this would support the notion that such processes might have evolved prior to the emergence of language in modern humans. Sequence learning capacities might have been present in human ancestors and been co-opted during language evolution. The next question then becomes the degree of similarity between animal and human AGL abilities. One would expect that humans would outperform other species in various AGL tasks. But what exactly is the level of learning that other species can achieve and how and why do they differ, not only from humans but also from other species or individuals? We consider these questions next, focusing on the behavioral capacities involved rather than the neural mechanisms. First, we overview infant experiments assessing pairwise transitional probabilities and adjacent sequencing dependencies within rule-based patterns. We then consider paradigms with more variable and less predictable adjacent dependencies. This is followed by studies assessing nonadjacent dependency learning, and we finish with studies of learning hierarchically organized structures. Given the already rich and continually growing literature on nonhuman animal AGL, it will not be possible in this chapter to review all available papers. Thus, we focus on a few key observations and studies per section.

4. Adjacent Dependencies—Transitional Probabilities

We'll begin with a roughly historical overview of initial statistical learning experiments in infants that inspired comparative studies in nonhuman animals. In a seminal study, Saffran, Aslin, and Newport (1996) exposed eight-month old infants to a continuous stream of speech syllables, each consisting of a consonant-vowel combination, for example, *bidakupadotigolabidaku* ... , generated with a speech synthesizer. This string contained three three-syllable "words," for example, *bidaku*. The only cues to word boundaries were the transitional probabilities between subsequent syllable pairs, which were high within words (1.0, e.g., *bida*) and low between words (0.33 in all cases, e.g., *kupa*). The infants were first presented with the speech stream for two minutes. Subsequently, in the test phase, they were presented with either the statistical words or "nonwords" consisting of three syllables in a nonfamiliar order. In another experiment, statistical words were contrasted with part-words consisting of syllables that were paired with low probability during familiarization. Infants looked longer to both nonwords and part-words than to the statistically predictable words, hence detected a difference between the experimental conditions. Various studies confirmed and extended these findings, for both human infants and adults, and they also provided evidence that the infants relied on statistical cues rather than just the frequency of co-occurrence of syllables (Aslin, Saffran, & Newport, 1998). Further studies also showed that nonspeech input, such as musical tones (Saffran, Johnson, Aslin, & Newport, 1999), can similarly be segmented based on transitional probabilities, suggesting that the segmentation process may, at least to some extent, be language-independent.

The above-mentioned experiments inspired nonhuman animal studies. Hauser, Newport, and Aslin (2001) exposed cotton-top tamarins to the same strings as those used by the infant study by Saffran et al. (1999). After exposure to the speech stream, the response to the test sequences was measured by examining the orienting behavior toward the loudspeaker from which the sounds were played. The tamarins oriented more to novel than to familiar sequences, indicating that they detected the difference. Toro and Trobalon (2005) showed that rats could also segment a speech stream using the stimuli from Saffran et al. (1996), although the rats' performance was affected by how often particular stimuli occurred. Songbirds have also been examined for their ability to use statistical and prosodic cues to segment a stream of acoustic items. Behavioral experiments on zebra finches suggested that they could recognize triplets from a training string, but only when these were separated by brief pauses (Spierings, de Weger, & ten Cate, 2015). Multielectrode recordings in nuclei in the forebrain (caudal mesopallium, CLM, and caudomedial nidopallium, NCM) of zebra finches exposed to a continuous stream of tones, revealed that the neurons could differentiate "words" consisting of six subsequent tones, from part-words and nonwords (Lu & Vicario, 2014). Overall, while there thus is accumulating evidence that animals are sensitive to adjacent dependencies, it is not clear whether this is based on transitional probabilities or co-occurrence of items.

5. Adjacent Dependencies—Rule Learning and Animal Perceptual Strategies

Historically, another key series of experiments do not depend on transitional probabilities, but rather on whether infants can generalize a rule-based pattern of sequences to novel stimuli following the same or a different pattern. These rule-based learning experiments were inspired by a seminal study by Marcus et al. (1999), in which seven-month-old infants were habituated to *XYX* or *XYY* strings (*X* and *Y* being speech syllables, resulting in strings like *gatiga* or *gatiti*). Afterward, their responses to novel strings showed that they extracted the underlying regularity and generalized it to strings composed of novel syllables. The *X* and *Y* items were all drawn from the same pool and in principle there was no other cue to discriminate the consistent from inconsistent novel strings other than the similarity or dissimilarity in string structure.

The seemingly simple task of discriminating *XYX* and *XXY* structures lends itself very well for comparing how humans and various animal species learn them. Using both habituation and operant conditioning tasks, a number of species have been tested using this paradigm, such as rhesus macaques (Hauser & Glynn, 2009), rats (Murphy, Mondragon, & Murphy, 2008), Bengalese finches (*Lonchura striata domestica*) (Seki, Suzuki, Osawa, & Okanoya, 2013), zebra finches (Chen, van Rossum, & ten Cate, 2015; Spierings & ten Cate, 2016; van Heijningen, Chen, van Laatum, van der Hulst & ten Cate, 2013), and budgerigars (*Melopsittacus undulates*) (Spierings & ten Cate, 2016).

Murphy et al. (2008) trained rats using a discrimination paradigm to distinguish *XYX*, *XXY*, and *YXX* strings where the training stimuli consisted of pure tone sequences, using one tone as an *X* and another one as a *Y* tone. The rats continued to discriminate the patterns when the tones were transposed to different frequencies. Whether this reflects recognition of the abstract pattern is disputed, as it could be more simply interpreted as recognition of the novel tone sequences as transpositions in frequency of familiar ones (Corballis, 2009, but see Mondragon, Murphy, & Murphy, 2009). This issue of whether animals detect an abstract pattern or rely on a physical (acoustical) similarity to recognize and differentiate novel stimuli is a key question in nonhuman animal studies using this type of AGL paradigm (ten Cate, 2014; ten Cate & Okanoya, 2012; van Heijningen et al., 2013).

To overcome the limitations in interpretation of the prior nonhuman animal studies on rule learning, the ability to generalize discrimination to strings consisting of novel sounds was tested in a recent experiment with zebra finches and budgerigars. Both species were first trained to discriminate a set of *XYX* from *XXY* strings, with *X* and *Y* being different zebra finch song elements. Next they were tested with strings consisting of novel arrangements of familiar element types and strings consisting of novel elements that, similar to the study by Marcus et al. (1999), had no acoustical similarity to training sounds. So, these test sounds only shared the string structure (Spierings & ten Cate, 2016). The results were intriguing in that the zebra finches did not differentiate among *XYX* and *XXY* strings consisting of novel items. Nevertheless, strings consisting of novel arrangements of familiar items were differentiated, albeit not by relying on the underlying structure. Instead, the zebra finches seemed to have memorized the positions of the separate acoustical items in all training stimuli and judged new strings by similarity to the training strings with respect to item position and acoustic identity. Earlier experiments also indicated that zebra finches attend to specific parts of training strings and use these to classify novel arrangements of stimuli (Chen et al., 2015; van Heijningen et al., 2013). This strategy was effective for discriminating the training strings, but it fails when encountering novel strings.

In contrast to the zebra finches, budgerigars exposed to the same strings behaved quite differently. They were able to classify strings consisting of fully novel items by their structure. They thus demonstrated a remarkable ability to attend to the abstract structural similarity of training and test strings, similar to the infants in the experiments of Marcus et al. (1999). The cross-species differences in this study and the evidence of more abstract rule learning in budgerigars may stem from the different ways that these birds attend to and use vocal communication signals, possibly relating to the complexity of their vocal communication signal reception and production capabilities. As mentioned earlier, zebra finches have rather stereotyped songs, and recognizing other individuals may depend on whether a particular bird's song has specific song elements in specific positions. Thus, whether the species differences have anything to do with their natural singing styles or other differences in cognitive processing remains to be examined.

Yet, in contrast to what seems to be a common assumption when comparing human and animal computational abilities, humans also do not always use a higher-order regularity as their default strategy. Often they also rely on lower-level or specific local regularities similar to the strategies used by nonhuman animals (e.g., Gerken, 2006). And if humans can apply different strategies depending on the context, should we be surprised if other animals do the same? So while nonhuman animal species may preferentially process salient lower-level regularities as a default, they may in some cases be able to use higher-level ones if needed. Hence the gap between the constraints for humans to use higher-order strategies and for animals to use lower ones may be smaller than it is sometimes claimed to be, requiring more direct comparisons with humans.

6. Adjacent Dependencies with Variable Sequencing Relationships

The original studies by Reber (1967) and Miller (1967) on human AGL used sequences of strings of varying lengths. The strings also contained a variety of transitional probabilities determining which item/element transitioned to which other one or more items (Reber, 1967; also see Miller, 1967). Children also need to be able to manage with unpredictability and variation during language learning (Smith et al., 2017). Using similar Reber-like grammars, infants were exposed to a limited set of items ("words") (e.g., *PEL, VOT, TAM, BIFF, CAV*) arranged in different strings, based on an artificial grammar determining which items could or could not follow each other (e.g., Gómez & Gerken, 1999, 2000; Saffran 2002, Saffran et al, 2008). After exposure to a limited number of strings, infants were able to discriminate novel sequences generated by the grammar from sequences violating that grammar (Gómez & Gerken, 1999; Saffran, 2002). These experiments suggest that infants manage well with variability in transitional probabilities and can extract and apply transitional relationships beyond specific word order.

Some researchers studied the ability of birds to learn artificial grammars of these sorts. Herbranson and Shimp (2003, 2008) trained pigeons in an operant design, in which they were presented with strings of colored letters (*T, X, V, P, S*) on a monitor and were taught to peck one key for one type of sequence and another for the other type. Although it is clear that the pigeons learned some regularities present in the grammar, a detailed analysis of similarities between training and test strings revealed that they responded mainly to specific substrings shared between training and testing stimuli. Therefore, it is not clear what precisely the pigeons learned and whether their behavior would differ from humans.

More recent comparative data with two species of monkeys and adult humans identifies intriguing species differences in behavioral effects and provides insights into the learning strategies and sensitivities of the three primate species. A study with marmosets (New World monkeys) and macaques (Old World monkeys) using a modified version of the Saffran et al. (2008) paradigm, created different types of AG consistent and violation sequences that allowed testing which aspects of the sequencing relationships the animals were able to process (Wilson et al., 2013). The study tested whether the macaques and marmosets could notice a very simple acoustically driven "rule," such as every sequence needs to start with an *A* item (a specific nonsense word). Since the grammar used had multiple levels of complexity, the study also tested whether the animals were sensitive to the other sequencing relationships, such as those defined by adjacent or nonadjacent dependencies between items throughout the sequences. In relation to the start with *A* property, both marmosets and macaques were sensitive to this acoustically driven aspect in the first position of the sequences. However, the macaques showed sensitivity to the local transitions throughout the sequences, which was not the case for the marmosets (Wilson et al., 2013).

In follow-up work, macaque eye-tracking responses were further studied and compared to adult human results, revealing considerable correspondence between the human and macaque sensitivity to adjacent transitional probabilities (Wilson, Smith, & Petkov, 2015). Furthermore, the Saffran et al. (2008) grammar that was used, by design, also contained nonadjacent dependencies. However, in the presence of multiple cues to "grammaticality," the nonadjacent relationships were found not to be salient enough to be perceived by the macaques and many of the humans.

7. Nonadjacent Dependencies

Nonadjacent dependencies are relationships between items in a string that are more distantly separated, arguably an increase in sequencing complexity over strictly local sequencing relationships and a step toward building hierarchical relationships. Newport and Aslin (2004) examined whether adult humans could detect dependencies between syllables when exposed to a continuous stream of syllables as in the studies previously mentioned, but here with fixed nonadjacent dependencies between the first and third syllables of three-syllabic words separated in time by a varying *X* syllable in between (*AXC*). Humans can detect such nonadjacent dependencies (Newport & Aslin, 2004), not just for speech syllables but also for tones (Creel, Newport, & Aslin, 2004). However, these and other experiments (e.g., Gebhart, Newport, & Aslin, 2009; Gómez & Maye, 2005; Lany & Gómez, 2008; Onnis, Monaghan, Richmond, & Chater, 2005) have also shown that the ability to detect nonadjacent dependencies seems constrained to particular contexts. For instance, humans benefit from perceptual cues or acoustical similarities, such as consonants in both *A* and *C* positions over the intervening *X* vowels (Newport & Aslin, 2004).

A first study in nonhuman animals to study nonadjacent dependency learning was by Newport, Hauser, Spaepen, and Aslin (2004), in which tamarin monkeys were exposed to the same strings that Newport and Aslin (2004) used for human adults. Like humans, tamarins could also detect nonadjacent dependencies when the syllables showed predictable relations between vowel segments. However, unlike humans, the tamarins failed to detect nonadjacent dependencies over consonants. A related study in rats failed to find nonadjacent dependency learning (Toro & Trobalon, 2005). However, a later study in which *AAB* or *ABC* patterns were instantiated with nonadjacent regularities over either vowels or consonants showed that rats could detect the nonadjacent dependencies with both vowels and consonants, while humans only detected the dependencies over vowels (de la Mora & Toro, 2013). Another monkey species, the squirrel monkey, could detect nonadjacent dependencies in tone sequences where the first and last tone of a string were low in pitch, separated by one to three tones of a higher frequency (Ravignani, Sonnweber, Stobbe, & Fitch, 2013). Chimpanzees are also able to detect nonadjacent dependencies between different items at the beginning and end (the "edges") of strings both for auditory strings (Endress, Carden, Versace, & Hauser, 2010) and for visual ones (Sonnweber, Ravignani, & Fitch, 2015). However, they seem to do so primarily by identifying a class of items that are at the beginning and another class of items that are at the end of strings rather than by recognizing the linkage between specific items (see also Endress et al., 2010). The only study so far examining the ability of birds to detect nonadjacent dependencies demonstrated that zebra finches could learn edge-based dependencies between two sets of song elements (*A* linked to *B*; *C* linked to *D*) and generalize this over different numbers and types of intervening elements (Chen & ten Cate, 2017).

To summarize, the findings in various animal species indicate that the ability to detect nonadjacent dependencies is present in nonhuman animals. At the same time, detecting such dependencies is clearest when they are the edges and are perceptually salient, as is also the case with humans (e.g., see adult human data in Mueller, Friederici, & Männel, 2012). Thus a key question for future empirical pursuit is whether the ability of nonhuman animals to detect nonadjacent dependencies can go beyond edge-based dependencies, generalize to dependencies in other positions in the string, and whether and how learning nonadjacent dependencies is facilitated by the types of items used.

8. Learning Center-Embedded Hierarchical Structures

The above-mentioned studies dealt with experiments in which the transitions or dependencies can be described by a finite state grammar. However, as we have noted (figure 47.1, *top*), humans are able to form and detect deep, embedded hierarchical structures. Such structures are computationally of higher complexity than the finite state grammars used in the experiments discussed so far. Are animals also able to detect nested hierarchical relationships?

The original experiment testing the ability of nonhuman primates and humans to cope with hierarchical structures was conducted by Fitch and Hauser (2004). In this study, adult humans and tamarins were habituated to speech syllables arranged either in *ABAB* or *ABABAB* ($(AB)^n$) strings or in *AABB* and *AAABBB* strings (A^nB^n). The *A* and *B* indicate sounds belonging to two different sound categories—produced by male (*A*) or female (*B*) voices, respectively. The A^nB^n pattern requires the animal to keep track of the number of *A* items in order to assess whether an equal number of *B* items is present. This is not required for keeping track of the $(AB)^n$ alternation and is suggested to cross the border of what a finite state grammar is expected to deal with (for a discussion, see Fitch et al., 2012). Humans in the study clearly distinguished both string types, irrespective of the habituation string type. However, only those tamarins habituated with the $(AB)^n$ strings detected the difference from the other string type, suggesting that they did not detect the A^nB^n structure.

The tamarin study inspired similar experiments with two songbirds species: starlings (Gentner, Fenn, Margoliash, & Nusbaum, 2006) and zebra finches (van Heijningen, de Visser, Zuidema, & ten Cate, 2009). Both species were able to distinguish $(AB)^n$ from A^nB^n strings. While this might suggest they had learned the rules, van Heijningen et al. (2009) showed that zebra finches discriminate the strings by attending to local regularities of the strings, such as the presence of *AA* bigrams (which are only present in A^nB^n strings), rather than attending to the overall structure. These discrimination strategies differed between individuals (van Heijningen et al., 2009), demonstrating the importance of analyzing the responses to probe strings at the individual level. Thus, zebra finch training with an A^nB^n or $(AB)^n$ grammar seems to result primarily in the animals using reduplicate bigrams to distinguish these patterns, as has also been observed for training with *XYX* and *XXY* patterns (e.g., Chen et al., 2015), rather than the birds extracting and generalizing the key structural regularities underlying the patterns.

Similar experiments on the $(AB)^n$ versus A^nB^n task were conducted with pigeons and keas (*Nestor notabilis*), using strings of visual tiles with either a circle or small squares as different variants of the *A* and *B* classes of stimuli. These species also based their discrimination on the presence or absence of local regularities, rather than on the underlying structure (Ravignani, Westphal-Fitch, Aust, Schlumpp, & Fitch, 2015; Stobbe, Westphal-Fitch, Aust, & Fitch, 2012).

In another study, using a self-training paradigm with touch screens, baboons first learned pairwise associations between several visual symbols, for example, *A1-B1*, *A2-B2* (Rey, Perruchet, & Fagot, 2012). In a later testing phase, the animals were presented with the initial *A* elements of two pairs (e.g., *A1-A2*). They were then allowed to select the following *B* elements. The animals were seen to preferentially pair the *B* partner of the most recent *A* element that was seen (e.g., *A2-B2*), followed by the partner pair of the first element (e.g., *A1-B1*). This resulted in the most often selected pattern, *A1-A2-B2-B1,* which resembles the hierarchical "center-embedded" structure, although a more parsimonious explanation, also noted by Rey et al. (2012) and others (Poletiek, Fitz, & Bocanegra, 2016) is that the baboons relied on an associative memory trace of the element pairs that they were trained to recognize.

We conclude that, while animals may have the ability to detect higher-order sequencing regularities, it is unclear the extent to which any species can recognize hierarchical structures like those that are so prominent in human language. This area of research remains controversial and has not unambiguously been resolved by the prior experiments. Nonetheless, all of the prior studies provide a considerable amount of information to guide future empirical pursuits on this important question. It is also worth noting that without meaning (semantics), even humans can struggle learning hierarchical relationships in AGL tasks that are by their nature designed to address structure devoid of meaning (de Vries, Monaghan, Knecht, & Zwitserlood, 2008; Hochmann, Azadpour, & Mehler, 2008; Uddén, Ingvar, Hagoort, & Petersson, 2012).

9. Overall Conclusions and Empirical Outlook

To conclude, formal analyses of vocalizations in various species point to a range of combinatorial capacities, with a small set of animals to date showing simple hierarchical nesting of vocal units into phrases. Whether this can be considered as resembling, or being at the root of, the deeply nested hierarchical relationships (phonemic, syllabic, syntactic) prominent in language still needs to be explored further.

Our overview of different types of artificial grammars that have been used to examine the rule or pattern learning abilities of both humans and nonhuman animals shows that a number of species appear capable of learning simple adjacent dependencies and the transitional probabilities within sequences of variable length. Once the sequencing relationships become less predictable and more abstract, species differences become more apparent. Some animals are also sensitive to nonadjacent dependencies but benefit, as do human adults and infants, from perceptually salient features shared between the items forming a nonadjacent dependency. Regarding the ability of animals to detect hierarchical sequencing relationships, particularly of the forms that are prominent in language, it is fair to say that the evidence for animals being able to attend to higher-order regularities in strings is equivocal and requires further empirical study (e.g., Beckers, Bolhuis, Okanoya, & Berwick, 2012; Beckers, Berwick, Okanoya & Bolhuis, 2017; Corballis, 2009; Poletiek et al., 2016; van Heijningen et al., 2009). Nevertheless, while many findings can be explained by the use of simpler strategies, this does not imply that animals are unable to use higher-order ones. Nonhuman animals' full potential for grammar or rule learning may not have been revealed by the current experiments (see also ten Cate, 2017). More work using AGL paradigms is thus needed. Furthermore, apart from focusing on a number of model species, the emerging presence of intriguing species differences argues for extending AGL studies to a greater array of more direct cross-species comparisons and relating the findings to the natural behavior and cognitive abilities of these species. If some species do have qualitatively similar rule learning and abstraction abilities to those present in humans, the evidence would lend support to the hypothesis that the uniqueness of language is not due to a single feature of a uniquely human language learning mechanism.

Acknowledgments

We thank Tecumseh Fitch, Ben Wilson, and Erich Jarvis for useful suggestions and discussion on a previous version of the chapter and James (Jim) Hurford for inspiration. This work was supported by NWO-GW (Netherlands Organization for Scientific Research-Social Sciences and Humanities) grant 360.70.452 to C.t.C., Wellcome Trust Investigator Award to C.P. (WT092606AIA), and a Theme Group grant from the Netherlands Institute for Advanced Study in the Humanities and Social Sciences and the Lorentz Institute to C.t.C. and C.P.

REFERENCES

Arnold, K., & Zuberbühler, K. (2006). Language evolution: Semantic combinations in primate calls. *Nature, 441,* 303.

Arriaga, G., Zhou, E. P., & Jarvis, E. D. (2012). Of mice, birds, and men: The mouse ultrasonic song system has some features similar to humans and song-learning birds. *PLOS ONE, 7,* e46610.

Aslin, R. N., & Newport, E. L. (2014). Distributional language learning: Mechanisms and models of category formation. *Language Learning, 64*(s2), 86–105.

Aslin, R. N., Saffran, J. R., & Newport, E. L. (1998). Computation of conditional probability statistics by 8-month-old infants. *Psychological Science, 9,* 321–324.

Beckers, G. J. L., Berwick, R. C., Okanoya, K., & Bolhuis, J. J. (2017). What do animals learn in artificial grammar studies? *Neuroscience and Biobehavioral Reviews, 81,* 238–246.

Beckers, G. J. L., Bolhuis, J. J., Okanoya, K., & Berwick, R. C. (2012). Birdsong neurolinguistics: Songbird context-free grammar claim is premature. *NeuroReport, 23*(3), 139–145.

Berwick, R. C., Beckers, G. J. L., Okanoya, K., & Bolhuis, J. J. (2012). A bird's eye view of human language evolution. *Frontiers in Evolutionary Neuroscience, 4*, 5.

Berwick, R. C., & Chomsky, N. (2015). *Why only us? Language and evolution*. Cambridge, MA: MIT Press.

Berwick, R. C., Okanoya, K., Beckers, G. J. L., & Bolhuis, J. J. (2011). Songs to syntax: The linguistics of birdsong. *Trends in Cognitive Science, 15*, 113–121.

Chabout, J., Sarkar, A., Dunson, D. B., & Jarvis, E. D. (2015). Male mice song syntax depends on social contexts and influences female preferences. *Frontiers in Behavioral Neuroscience, 9*, 76.

Chen, J., & ten Cate, C. (2017). Bridging the gap: Learning of acoustic nonadjacent dependencies by a songbird. *Journal of Experimental Psychology: Animal Learning and Cognition, 43*(3), 295–302.

Chen, J., van Rossum, D., & ten Cate, C. (2015). Artificial grammar learning in zebra finches and human adults: XYX vs XXY. *Animal Cognition*, 18, 151–164.

Chomsky, N. (1957). *Syntactic structures*. The Hague, the Netherlands: Mouton.

Christiansen, M. H., & Chater, N. (2015). The language faculty that wasn't: A usage-based account of natural language recursion. *Frontiers in Psychology*, 6, 1182.

Corballis, M. C. (2009). Do rats learn rules? *Animal Behaviour, 78*, E1–E2.

Creel, S. C., Newport, E. L., & Aslin, R. N. (2004). Distant melodies: Statistical learning of nonadjacent dependencies in tone sequences. *Journal of Experimental Psychology: Learning, Memory, and Cognition, 30*(5), 1119–1130.

de la Mora, D. M., & Toro, J. M. (2013). Rule learning over consonants and vowels in a non-human animal. *Cognition, 126*, 307–312.

de Vries, M. H., Monaghan, P., Knecht, S., & Zwitserlood, P. (2008). Syntactic structure and artificial grammar learning: The learnability of embedded hierarchical structures. *Cognition, 107*, 763–774.

Endress, A. D., & Bonatti, L. L. (2007). Rapid learning of syllable classes from a perceptually continuous speech stream. *Cognition, 105*, 247–299.

Endress, A. D., Carden, S., Versace, E, & Hauser, M. D. (2010). The apes' edge: Positional learning in chimpanzees and humans. *Animal Cognition, 13*(3), 483–495.

Engesser, S., Ridley, A. R., & Townsend, S.W. (2016). Meaningful call combinations and compositional processing in the southern pied babbler. *Proceedings of the National Academy of Sciences, 113*, 5976–5981.

Everaert, M. B. H., Huybregts, M. A. C., Chomsky, N., Berwick, R. C., & Bolhuis, J. J. (2015). Structures, not strings: Linguistics as part of the cognitive sciences. *Trends in Cognitive Sciences, 19*(12), 729–743.

Fedurek, P, Schel, A. M., & Slocombe, K. E. (2013). The acoustic structure of chimpanzee pant-hooting facilitates chorusing. *Behavioral Ecology and Sociobiology, 67*(11), 1781–1789.

Fitch, W. T., & Friederici, A. D. (2012). Artificial grammar learning meets formal language theory: An overview. *Philosophical Transactions of the Royal Society B, 367*(1598), 1933–1955.

Fitch, W. T., Friederici, A. D., & Hagoort, P. (2012). Pattern perception and computational complexity: Introduction to the special issue. *Philosophical Transactions of the Royal Society London B, 367*, 1925–1932.

Fitch, W. T., & Hauser, M. D. (2004). Computational constraints on syntactic processing in a nonhuman primate. *Science, 303*(5656), 377–380.

Gebhart, A. L., Newport, E. L., & Aslin, R. N. (2009). Statistical learning of adjacent and nonadjacent dependencies among nonlinguistic sounds. *Psychonomic Bulletin and Review, 16*(3), 486–490.

Gentner, T. Q., Fenn, K. M., Margoliash, D., & Nusbaum, H. C. (2006). Recursive syntactic pattern learning by songbirds. *Nature, 440*, 1204–1207.

Gerken, L. (2006). Decisions, decisions: Infant language learning when multiple generalizations are possible. *Cognition, 98*(3), B67–B74.

Gómez, R. L, & Gerken, L. (1999). Artificial grammar learning by 1-year-olds leads to specific and abstract knowledge. *Cognition, 70*, 109–135.

Gómez, R. L., & Gerken, L. (2000). Infant artificial language learning and language acquisition. *Trends in Cognitive Sciences, 4*, 178–186.

Gómez, R. L., & Maye, J. (2005). The developmental trajectory of nonadjacent dependency learning. *Infancy*, 7(2), 183–206.

Hammerschmidt, K., Reisinger, E., Westekemper, K. Ehrenreich, L., Strenzke, N., & Fischer, J. (2012). Mice do not require auditory input for the normal development of their ultrasonic vocalizations. *BMC Neuroscience, 13*, 40.

Hauser, M. D., Chomsky, N., & Fitch, W. T. (2002). The faculty of language: What is it, who has it, and how did it evolve? *Science, 298*, 1569–1579.

Hauser, M. D., & Glynn, D. (2009). Can free-ranging rhesus monkeys (*Macaca mulatta*) extract artificially created rules comprised of natural vocalizations? *Journal of Comparative Psychology, 123*, 161–167.

Hauser, M. D., Newport, E. L., & Aslin, R. N. (2001). Segmentation of the speech stream in a non-human primate: Statistical learning in cotton-top tamarins. *Cognition, 78*, B53–B64.

Herbranson, W. T., & Shimp, C.P. (2003). Artificial grammar learning in pigeons: A preliminary analysis. *Learning and Behavior, 31*, 98–106.

Herbranson, W. T., & Shimp, C. P. (2008). Artificial grammar learning in pigeons. *Learning and Behavior*, 36, 116–137.

Hochmann, J.-R., Azadpour, M., & Mehler, J. (2008). Do humans really learn A(n) B(n) artificial grammars from exemplars? *Cognitive Science, 32*, 1021–1036.

Holy, T. E., & Guo, Z. (2005). Ultrasonic songs of male mice. *PLOS Biology, 3*(12), e386.

Hultsch, H. (1991). Early experience can modify singing styles: Evidence from experiments with nightingales, *luscinia-megarhynchos. Animal Behaviour, 42*, 883–889.

Hurford, J. R. (2012). *The origins of grammar: Language in the light of evolution*. Oxford: Oxford University Press.

Jäger, G., & Rogers, J. (2012). Formal language theory: Refining the Chomsky hierarchy. *Philosophical Transactions of the Royal Society of London B: Biological Sciences, 367*(1598), 1956–1970.

Kershenbaum, A., Bowles, A. E., Freeberg, T. M., Jin, D. Z., Lameira, A. R., & Bohn, K. (2014). Animal vocal sequences: Not the Markov chains we thought they were. *Proceedings of the Royal Society B: Biological Sciences, 281*, 20141370.

Lachlan, R. F., van Heijningen, C. A. A., ter Haar, S. M., & ten Cate, C. (2016). Zebra finch song phonology and syntactical structure across populations and continents: A computational comparison. *Frontiers in Psychology, 7,* 980.

Lany, J., & Gómez, R. L. (2008). Twelve-month-old infants benefit from prior experience in statistical learning. *Psychological Science, 19*(12), 1247–1252.

Lu, K., & Vicario, D. S. (2014). Statistical learning of recurring sound patterns encodes auditory objects in songbird forebrain. *Proceedings of the National Academy of Sciences, 111,* 14553–14558.

Marcus, G. F., Vijayan, S., Bandi Rao, S., & Vishton, P. M. (1999). Rule learning by seven-month-old infants. *Science, 283,* 77–80.

Marshall, A. J., Wrangham, R. W., & Arcadi, A. C. (1999). Does learning affect the structure of vocalizations in chimpanzees? *Animal Behaviour, 58,* 825–830.

Miller, G. A. (1967). *Psychology of communication.* New York: Basic Books.

Mitani, J. C., & Marler, P. (1989). A phonological analysis of male gibbon singing behavior. *Behaviour, 109,* 20–45.

Mondragon, E., Murphy, R. A., & Murphy, V. A. (2009). Rats do learn XYX rules. *Animal Behaviour, 78*(4), E3-E4.

Mueller, J. L., Friederici, A. D., & Männel, C. (2012). Auditory perception at the root of language learning. *Proceedings of the National Academy of Sciences, 109,* 15953–15958.

Murphy, R. A., Mondragon, E., & Murphy, V. A. (2008). Rule learning by rats. *Science, 319,* 1849–1851.

Newport, E. L., & Aslin, R. N. (2004). Learning at a distance: I. Statistical learning of non-adjacent dependencies. *Cognitive Psychology, 48*(2), 127–162.

Newport, E. L., Hauser, M. D., Spaepen, G., & Aslin, R. N. (2004). Learning at a distance: II. Statistical learning of non-adjacent dependencies in a non-human primate. *Cognitive Psychology, 49,* 85–117.

Onnis, L., Monaghan, P., Richmond, K., & Chater, N. (2005). Phonology impacts segmentation in online speech processing. *Journal of Memory and Language, 53*(2), 225–237.

Payne, R. S., & McVay, S. (1971). The songs of humpback whales. *Science, 173,* 587–597.

Petkov, C. I., & Jarvis, E. (2012). Birds, primates, and spoken language origins: Behavioral phenotypes and neurobiological substrates. *Frontiers in Evolutionary Neuroscience, 4,* 12.

Poletiek, F. H., Fitz, H., & Bocanegra, B. R. (2016). What baboons can (not) tell us about natural language grammars. *Cognition, 151,* 108–112.

Ravignani, A., Sonnweber, R.-S., Stobbe, N., & Fitch, W. T. (2013). Action at a distance: Dependency sensitivity in a New World primate. *Biology Letters, 9*(6), 20130852.

Ravignani, A., Westphal-Fitch, G., Aust, U., Schlumpp, M. M., & Fitch, W. T. (2015). More than one way to see it: Individual heuristics in avian visual computation. *Cognition, 143,* 13–24.

Reber, A. S. (1967). Implicit learning of artificial grammars. *Journal of Verbal Learning and Verbal Behaviour, 6,* 855–863.

Rey, A., Perruchet, P., & Fagot, J. (2012). Centre-embedded structures are a by-product of associative learning and working memory constraints: Evidence from baboons (*Papio papio*). *Cognition, 123*(1), 180–184.

Robinson, J. G. (1979). An analysis of the organization of vocal communication in the titi monkey *Callicebus moloch. Ethology, 49*(4), 381–405.

Rose, G. J., Goller, F., Gritton, H. J., Plamondon, S. L., Baugh, A. T., & Cooper, B. G. (2004). Species-typical songs in white-crowned sparrows tutored with only phrase pairs. *Nature, 432,* 753–758.

Saffran, J. R. (2002). Constraints on statistical language learning. *Journal of Memory and Language, 47,* 172–196.

Saffran, J. R., Aslin, R. N., & Newport, E. L. (1996). Statistical learning by 8-month-old infants. *Science, 274,* 1926–1928.

Saffran, J. R., Hauser, M., Seibel, R., Kapfhamer, J., Tsao, F., & Cushman, F. (2008). Grammatical pattern learning by human infants and cotton-top tamarin monkeys. *Cognition, 107*(2), 479–500.

Saffran, J. R., Johnson, E. K., Aslin, R. N., & Newport, E. L. (1999). Statistical learning of tone sequences by human infants and adults. *Cognition, 70,* 27–52.

Seki, Y., Suzuki, K., Osawa, A. M., & Okanoya, K. (2013). Songbirds and humans apply different strategies in a sound sequence discrimination task. *Frontiers in Psychology, 4,* 447.

Smith, K., Perfors, A., Fehér, O., Samara, A., Swoboda, K., & Wonnacott, E. (2017). Language learning, language use and the evolution of linguistic variation. *Philosophical Transactions of the Royal Society B, 372,* 20160051.

Spierings, M., de Weger, A., & ten Cate C. (2015). Pauses enhance chunk recognition in song element strings by zebra finches. *Animal Cognition, 18,* 867–874.

Spierings, M., & ten Cate, C. (2016). Budgerigars and zebra finches differ in how they generalize in an artificial grammar learning experiment. *Proceedings of the National Academy of Sciences, 113,* E3977–E3984.

Sonnweber, R., Ravignani, A., & Fitch, W. T. (2015). Non-adjacent visual dependency learning in chimpanzees. *Animal Cognition, 18*(3), 733–745.

Stobbe, N., Westphal-Fitch, G., Aust, U., & Fitch, W. T. (2012). Visual artificial grammar learning: comparative research on humans, kea (*Nestor notabilis*) and pigeons (*Columba livia*). *Philosophical Transactions of the Royal Society London B, 367,* 1995–2006.

Suzuki, T. N., Wheatcroft, D., & Griesser, M. (2016). Experimental evidence for compositional syntax in bird calls. *Nature Communications, 7,* 10986.

Takahashi, D. Y., Fenley, A. R., Teramoto, Y., Narayanan, D. Z. Borjon, J. I., Holmes, P., & Ghazanfar, A. A. (2015). The developmental dynamics of marmoset monkey vocal production. *Science, 349,* 734–738.

ten Cate, C. (2014). On the phonetic and syntactic processing abilities of birds: From song to speech and artificial grammars. *Current Opinion in Neurobiology, 28,* 157–164.

ten Cate, C. (2017). Assessing the uniqueness of language: Animal grammatical abilities take center stage. *Psychonomic Bulletin and Review, 24,* 91–96.

ten Cate, C., Lachlan, R., & Zuidema, W. (2013). Analyzing the structure of bird vocalizations and language: Finding common ground. In J. J. Bolhuis & M. Everaert (Eds.), *Birdsong, speech and language* (pp. 243–260). Cambridge, MA: MIT Press.

ten Cate, C., & Okanoya, K. (2012). Revisiting the syntactic abilities of non-human animals: Natural vocalizations and artificial grammar learning. *Philosophical Transactions of the Royal Society London B, 367,* 1984–1994.

Tomasello, M. (2003). *Constructing a language: A usage based theory of language acquisition.* Cambridge, MA: Harvard University Press.

Toro, J. M., & Trobalon, J. B. (2005). Statistical computations over a speech stream in a rodent. *Perception and Psychophysics, 67,* 867–875.

Uddén, J., Ingvar, M., Hagoort, P., & Petersson, K. M. (2012). Implicit acquisition of grammars with crossed and nested non-adjacent dependencies: Investigating the push-down stack model. *Cognitive Science, 36*(6), 1078–1101.

van Heijningen, C. A., Chen, J., van Laatum, I., van der Hulst, B., & ten Cate, C. (2013). Rule learning by zebra finches in an artificial grammar learning task: Which rule? *Animal Cognition, 16,* 165–175.

van Heijningen C. A., de Visser, J., Zuidema, W., & ten Cate, C. (2009). Simple rules can explain discrimination of putative recursive syntactic structures by a songbird species. *Proceedings of the National Academy of Sciences, 106,* 20538–20543.

Wanelik, K. M., Azis, A., & Cheyne, S. M. (2013). Note-, phrase- and song-specific acoustic variables contributing to the individuality of male duet song in the Bornean southern gibbon (*Hylobates albibarbis*). *Primates, 54*(2), 159–170.

Watson, S. K., Townsend, S. W., Schel, A. M., Wilke, C., Wallace, E. K., Cheng, L., ... Slocombe, K. E. (2015). Vocal learning in the functionally referential food grunts of chimpanzees. *Current Biology, 25*(4), 495–499.

Wilson, B., Slater, H., Kikuchi, Y., Milne, A. E., Marslen-Wilson, W. D., Smith, K., & Petkov, C. I. (2013). Auditory artificial grammar learning in macaque and marmoset monkeys. *Journal of Neuroscience, 33,* 18825–18835.

Wilson, B., Smith, K., & Petkov, C. I. (2015). Mixed-complexity artificial grammar learning in humans and macaque monkeys: Evaluating learning strategies. *European Journal of Neuroscience, 41,* 568–578.

48 Speech Perception: What Do Nonhuman Animals Have to Say?

BUDDHAMAS P. KRIENGWATANA AND GABRIËL J. L. BECKERS

Speech is the most common vehicle through which language is communicated and learned, hence understanding speech perception is of major importance in understanding human behavior and cognition. Although speech perception may seem effortless, it is a deceptively complex process that involves much more than a simple one-to-one mapping of acoustic information with phonemes. This is because the acoustics of speech are highly volatile and never entirely reduplicated. Even a single phoneme repeated by the same speaker may vary considerably in this respect. The difficulty of cracking the speech code is exponentially amplified when one considers that speech signals are susceptible to interference from background noise, distorted by neighboring speech sounds (co-articulation), and modified depending on speaker voice and manner of speaking (speaker variability). Nonetheless, human speech perception is surprisingly robust, as listeners manage to accurately perceive speech even though no consistent acoustic parameters have been identified that invariably characterize different phonemes (lack of invariance problem).

To account for how listeners deal with variability in the speech signal, some researchers have suggested that human brains and auditory systems have evolved to treat speech as different from other acoustic signals. Consequently, comparative studies of speech perception in nonhuman animals (hereafter referred to as *animals*) have been crucial in this "speech is special" debate because they can reveal whether speech perception is possible only through uniquely human mechanisms or may be achieved through mechanisms shared with other animals, in which case they have often been assumed to be domain-general auditory ones. From a broader perspective, this information enriches our knowledge of whether speech perceptual systems evolved especially for language or whether they are largely based on preexisting, and perhaps taxonomically distributed perceptual and cognitive architectures.

This chapter is divided into three sections. First we discuss pioneering work in comparative speech perception that contributed to fundamental changes in theoretical perspectives on the specialization of speech perception mechanisms. Comparative work that relates to more recent developments in theories of human speech perception will also be reviewed. Second, we turn our attention to similarities and differences in vocal production mechanisms in humans and animals, a fundamental topic in speech perception because of potential interdependence of vocal perception and production mechanisms. Third, we synthesize the insights that animal models have provided regarding the extent to which domain-general mechanisms guide speech perception and discuss how close we might be to resolving the debate on the uniqueness of human speech perception.

1. Speech and Vocal Perception

1.1. Theories of Speech Perception

Although the study of speech perception in animals has a long history, it became more intensively and widely studied after the development of the Motor Theory of speech perception (Liberman, Cooper, Shankweiler, & Studdert-Kennedy, 1967; Liberman & Mattingly, 1989). According to the Motor Theory, humans possess specialized mechanisms that perceive speech as a speaker's intended articulatory gestures, which thereby overcomes the problem of speaker variability, co-articulation, and lack of invariance inherent in the acoustic speech signal. One of the most popular pieces of evidence used to support the Motor Theory's claim of a highly specialized and uniquely human speech module was categorical perception: listeners presented speech sounds varying along a continuum did not perceive them as continual, but perceived stimuli as having sharp and distinct boundaries between phonetic categories. Results from comparative studies in birds and mammals, however, posed critical challenges for the idea of a specialized speech processing module. These studies showed that chinchillas can exhibit categorical perception of speech with phoneme boundaries occurring in similar places as humans (Kuhl & Miller, 1975), Japanese quail can form stable phonetic categories without relying on specific invariant acoustic cues and can perceptually compensate for co-articulation (Kluender, Diehl, & Killeen, 1987; Lotto, Kluender, & Holt, 1997), and chinchillas, budgerigars, and zebra finches can maintain discrimination of phonetic categories across different speakers (Burdick & Miller, 1975; Dooling & Brown, 1990; Ohms, Gill, van Heijningen, Beckers, & ten Cate, 2010; Kriengwatana, Escudero, Kerkhoven, & ten Cate, 2015). Studies in humans using nonspeech analogs complement these findings in animals, further weakening the Motor Theory's contention of a human-specific specialized module for speech processing (Lotto & Kluender, 1998; Holt, Lotto, & Kluender, 2000; Pisoni, Carrell, & Gans, 1983; Diehl & Walsh, 1989; Holt, 2005; Sjerps, Mitterer, & McQueen, 2011; Laing, Liu, Lotto, & Holt, 2012; Carbonell & Lotto, 2014).

In light of this evidence against the Motor Theory, two competing hypotheses have been put forward: the Direct Realist theory (Fowler, 1986) and the General Auditory approach (Diehl, Lotto, & Holt, 2004). Both favor the idea that speech perception is not guided by specialized mechanisms specific to speech or humans. However, they differ in what they assume listeners perceive. According to the Direct Realist theory, the objects of speech perception are the speaker's actual articulatory gestures, not the speaker's intended gestures as conceived by Motor Theory (Fowler, 1996). As the acoustic speech signal is physically shaped by articulatory movements and gestures required to produce it, the signal is therefore assumed to contain sufficient information for listeners to be able to recover gestures for speech perception (Best, 1995). However, even though Direct Realist assumes that perceptual systems of animals allow them to perceive human speech through gestures, it has been difficult to convincingly prove that animals can or do perceive articulatory gestures at all. Recent findings in songbirds may provide some of the first but still preliminary support for the idea that gestures may possibly play a role in song perception, as playing back a synthesized version of a bird's own song elicited cortical premotor neuron activity that was precisely timed with the vocal motor gestures (defined as the trajectories of air-sac pressure and syringeal labia tension) required for producing the song (Amador, Perl, Mindlin, & Margoliash, 2013). Still, it is difficult to imagine how songbirds could retrieve articulatory gestures from human speech when the articulatory filters of humans and birds differ so much (e.g., birds have no lips). So, even if songbirds do perceive gestures in their own vocalizations, this may not help them perceive gestures in human speech. It would also imply that perception of own-species' and other species' articulatory gestures are different, even though both speech and song are shaped by vocal tract modulations.

The General Auditory approach advocates that the acoustic signal is the medium for speech perception and that speech is treated like any other auditory signal (Diehl et al., 2004). This approach argues that commonalities in how sounds are processed by the auditory systems of humans and animals account for the sensitivity to acoustic features of speech that humans and animals share (reviewed in Diehl et al., 2004; Holt & Lotto, 2008). For instance, in the experiment that showed that chinchillas and humans exhibit categorical perception of voice onset time in a /da-ta/ continuum, the boundary between /da/ and /ta/ occurred between 30 and 40 ms for both humans and chinchillas (Kuhl & Miller, 1975). Further work showed that chinchillas are most sensitive to changes in voice onset time at around 30–40 ms (Kuhl, 1981), which is likely mediated by the more precise encoding of voice onset times that occur in that time range by the chinchilla auditory nerve compared to shorter or longer voice onset times (Sinex, McDonald, & Mott, 1991). Hence, the chinchilla auditory system could be naturally tuned to preserve temporal information related to voice onset times within the 30–40 ms range. Interestingly, chinchillas produce a bark vocalization that may contain an initial voicing element (Hunyady, 2008), although it was not reported whether the voice onset times of chinchilla barks fall within the time window in which their auditory systems show heightened temporal resolution. If this were the case, it would considerably strengthen the argument that even though the auditory systems of animals did not evolve for speech, they can nonetheless be exploited for speech perception because the demands of their natural communicative environments parallel those of humans (Kriengwatana, Escudero, & ten Cate, 2015; Sinnott, 1989).

1.2. Infant Speech Perception Animal studies have also been important for advancing theories of speech acquisition, especially for understanding whether there are innate and specialized processes that help infants to acquire speech (Weiss & Newport, 2006). Some language acquisition theories posit a combination of human-specific speech processing mechanisms or general biases operating in concert with general auditory mechanisms (reviewed in Best, 1994; Kuhl, 2000; Werker & Curtin, 2005). For example, the Native Language Magnet theory argues that domain-general mechanisms found in humans and animals initially guide infant speech perception, but subsequent phonetic learning is unique to humans (see Kuhl et al.,

2008). Central to the Native Language Magnet theory is the perceptual magnet effect, which describes the perceptual distortion that occurs when tokens around a phonetic prototype (i.e., particularly good exemplar of a phonetic category) are perceived as nearly identical to the prototype, even though the tokens actually vary in their acoustic similarity to the prototype. This indicates that speech categories have an internal and graded structure that human listeners are sensitive to. The perceptual magnet effect is one example used to illustrate phonetic learning, as it was exhibited by human infants and adults, but not rhesus monkeys (Kuhl, 1991).

Although the perceptual magnet effect has been attributed to phonetic learning, a songbird study indicated that it may also be achieved through other general learning processes. Kluender, Lotto, Holt, and Bloedel (1998) trained European starlings to discriminate between multiple tokens of two vowel categories, including those used by Kuhl (1991). When tested on novel vowel tokens, birds exhibited behaviors characteristic of the perceptual magnet effect. Specifically, they responded most strongly to novel tokens surrounding the center of the vowel category, which is where the prototype is thought to be located. This implies that starlings were classifying novel tokens in a way that was not simply based on acoustic distances alone. For tokens that straddled the border between vowel categories, response strength was reduced, indicating that these stimuli were not perceived as clearly belonging to either vowel category. Strikingly, adult human listeners tested on the same stimuli showed almost identical responses. Taken together, this study demonstrates that with sufficient learning experience, animals can create structured internal representations and categorize speech sounds in a fashion similar to humans. The corollary to these findings is that rhesus monkeys in Kuhl's (1991) experiment may not have exhibited the perceptual magnet effect because they lacked the experience required to build structured representations.

Still, it is possible the perceptual magnet effect observed in human infants and European starlings reflect different types of learning. This is because we do not know whether animals would form these structured representations without explicit training, as is the case with human infants (Kuhl, 2000). Information about the existence of auditory prototypes in songbird vocalizations may help to answer this question because songbirds, like humans, must learn their songs, and they perceive their vocalizations as open-ended categories (Sturdy, Phillmore, Price, & Weisman, 1999; Sturdy, Phillmore, & Weisman, 2000; Sturdy, Bloomfield, Farrell, Avey, & Weisman, 2007). In fact, recent work suggests that birdsongs may exhibit internal structures similar to human speech categories, as songbirds will respond more strongly to good exemplars of a song type compared to poor exemplars. From hundreds of recordings of swamp sparrow song syllables, Lachlan, Anderson, Peters, Searcy, and Nowicki (2014) quantified the degree to which each syllable was more or less typical within its syllable category—in other words, how acoustically distant a syllable was to the center of its syllable cluster. When typical songs were presented, male swamp sparrows displayed more aggressive behaviors and female swamp sparrows displayed more copulatory behaviors (Lachlan et al., 2014). Thus, differentiating between good and poor members of an acoustic category is a natural activity for songbirds and is functional because song typicality may be a good indicator of song learning ability, which has been linked to male quality and fitness (Lachlan & Nowicki, 2012; Nowicki, Peters, & Podos, 1998). This highlights exciting opportunities to examine the existence and role of perceptual magnets in animal vocalizations, as well as to investigate the parallels between the development of internal structures in auditory categories in humans and songbirds.

Studies demonstrating perceptual magnet effects have reported finding asymmetries in vowel discrimination, where discrimination of two vowels was enhanced if the second vowel was more peripheral in the acoustic vowel space, with the most peripheral vowels being those such as /i/, /a/, and /u/ (Polka & Bohn, 1996; Polka & Werker, 1994; Kuhl, 1991). The Natural Referent Vowel framework proposed by Polka and Bohn (2011) suggests that this asymmetry exists because peripheral vowels occupy the most extreme articulatory space and have acoustic properties that make them more perceptually salient. Specifically, peripheral vowels are more salient because of the greater convergence of vowel formant frequencies (mainly the first and second formants, but also possibly the second and third formants; Schwartz, Abry, Boe, & Valle, 2005; Polka & Bohn, 2011). Greater convergence of vowel formant frequencies within a vowel (i.e., increased spectral similarity between neighboring formants [F1,F2], [F3,F4], or [F3,F4]; Schwartz et al., 2005) decreases spectral variability and increases acoustic salience, which is why peripheral vowels are thought to act as perceptual anchors that help infants form and organize other vowel categories depending on their language experience. As directional asymmetries in vowel perception may serve a particular role in human language development, Polka and Bohn (2003, 2011) suggested that it should be human-specific. Thus, the Natural Referent

Vowel framework also subscribes to the idea of uniquely human biases that interact with domain-general perceptual systems during phonetic development.

Further studies of native and non-native vowel discrimination by infants and adults, respectively, conformed to patterns of asymmetry as predicted by the Natural Referent Vowel framework (see Polka & Bohn, 2011). Results from animal studies show that the patterns of asymmetries differ from humans, but do not yet conclusively support the Natural Referent Vowel framework because the vowel contrasts tested and experimental procedures used differ, making direct comparisons impossible (see Kriengwatana, Escudero, & ten Cate, 2015). For example, compared to human studies, animal experiments typically involve presentation of few tokens, short interstimulus intervals, and extensive reinforcement training. This approach may reveal an animal's capacity to detect acoustic differences rather than its ability to process phonetic information, which is presumed to occur in human experiments (Polka, Bohn, & Weiss, 2015).

An important issue facing the Natural Referent Vowel framework is that if directional asymmetries in vowel perception are grounded on the concept of focalization (i.e., vowel formant convergence increases saliency and memorability of referent vowels), why should a uniquely human pattern of directional asymmetries in vowel discrimination be expected? That is, what aspect of focalization would render referent vowels available to humans as anchors for discrimination, but not to animals? Sensitivity and ability to perceive formants is not a probable explanation, as most of the birds and nonhuman mammals tested show good vowel discrimination capacities, often comparable to humans (Dooling & Brown, 1990; Hienz, Aleszczyk, & May, 1996; Hienz, Jones, & Weerts, 2004; Ohms, Escudero, Lammers, & ten Cate, 2012; Sinnott, 1989; Sinnott & Kreiter, 1991). Animals also readily attend to formants as they can play important roles in their communication systems (e.g., Fitch & Fritz, 2006; Fitch & Kelley, 2000). Therefore, uncovering explanations for the reasons and functions of perceptual asymmetries in animals would help to develop the Natural Referent Vowel framework into a more comprehensive theory of language acquisition.

1.3. Adapting to Voices In addition to conveying a message, vocal communication signals often carry information about the signaler in the form of voice characteristics. The ability to perceive both types of information is also found in animals, such as discrimination of neighbor versus stranger songs by warblers (Godard, 1991; Godard & Wiley, 1995) and discrimination of highly similar songs of mates and male neighbors by female great tits (Blumenrath, Dabelsteen, & Pedersen, 2007). Although voices provide useful information about the signaler, in human speech they contribute a significant and inescapable source of acoustic variability that can blur the boundaries between phonetic categories, yet human listeners have no problem identifying phonemes produced by different speakers (Potter & Steinberg, 1950; Peterson & Barney, 1952; Hillenbrand, Getty, Clark, & Wheeler, 1995). Signaler characteristics such as body size also cause variation in vocal sounds animals produce (Fitch, 1997, 1999). Thus given the importance of both sources of information, humans and animals have most likely evolved means to partition message and signaler information in vocal communication signals.

Human voices vary in part due to unique individual differences in vocal tract size and shape. For example, voice characteristics are shaped by the length of the vocal cords in the larynx and supralaryngeal vocal tract, which differ between males and females to cause variation in resonance frequencies used for vowel identification. So it is not surprising that traditional speech perception theories see voice information as an interfering property that listeners must remove from the speech signal (i.e., normalizing the signal). The normalization process minimizes acoustic variability of members within a category and maximizes differences between phonetic categories (Magnuson & Nusbaum, 2007), consequently enabling phonetic information to be mapped onto invariant phonetic representations (similar to matching input to entries in a mental dictionary; Johnson, 2005). Normalization based on acoustic parameters correlated with a speaker's vocal tract length or based on differences in the relative patterns of acoustic energy (i.e., formant ratios) are examples of normalization approaches (Johnson, 2005). However, there is evidence that voice information is not wholly discarded and interacts with speech perception (e.g., Newman & Evers, 2007; Nygaard & Pisoni 1998; Pisoni, 1993). More recently, exemplar-based theories of categorization have been applied to speech perception and suggest that listeners store all acoustic information available at the time speech samples are encountered, attaching specific auditory information to specific category labels (e.g., vowel category, speaker identity, and speaker sex). Incoming speech samples are compared to these stored exemplars and classified as a member of a phonetic category based on its resemblance with exemplars (Johnson, 1997; Goldinger, 1996, 1998).

Comparative research in animals can offer meaningful insights into how listeners adjust to different speakers, and perhaps the degree of specialness of normalization and exemplar-based approaches in human speech perception. For instance, do animals create invariant representations of (human) phonemes? Is sensitivity to and use of formant ratios spontaneous in other animals or exhibited only by humans for speech perception? How much experience is required to build a reliable set of exemplars that enable correct classification of speech samples? If animals can perceive human voices as a source of signaler identity, then they may be able to dissociate this information from phonetic information in the speech signal and perceptually adapt their perception of speech to different voices. So far, the methods that animals employ to adapt to voices in speech perception remain essentially unknown.

Nonetheless, experiments in cats, birds, dogs, chinchillas, rats, ferrets, and nonhuman primates indicate that animals can discriminate vowel categories produced by different speakers (Bizley, Walker, King, & Schnupp, 2013; Burdick & Miller, 1975; Dewson, 1964; Dooling & Brown, 1990; Eriksson & Villa, 2006; Griebel & Oller, 2012; Kojima & Kiritani, 1989). In some of these studies, however, it is difficult to know whether animals truly needed to adapt to speaker voices because studies either used vowels that differed maximally in psychoacoustic space or manipulated only one acoustic parameter (e.g., fundamental frequency, which correlates moderately with speaker sex). To the best of our knowledge, only two studies have convincingly shown that animals can adapt to different human voices during vowel discrimination. Ohms, Gill, et al. (2010) demonstrated that zebra finches can differentiate between the psychoacoustically adjacent vowel categories /i/ and /e/ in naturally spoken words (*wit* and *wet*) and generalize this discrimination to voices of multiple novel male and female speakers. Remarkably, zebra finches can generalize discriminations of /i/ and /e/ to novel speakers of the same or different sex even if they have only ever had experience discriminating /i/ and /e/ from one speaker (Kriengwatana, Escudero, Kerkhoven, & ten Cate, 2015). The results from Ohms, Gill, et al. (2010) and Kriengwatana, Escudero, Kerkhoven, and ten Cate (2015) indicate that knowledge of the acoustic correlates of human vocal tract length and shape are not necessary for speaker adaptation and that speaker adaptation in vowels is possible after experience with only a single exemplar.

One explanation for why zebra finches show speaker adaptation of vowels without prior experience is that the auditory-perceptual systems of humans and songbirds share mechanisms or biases that facilitate adaptation to human voices (Kriengwatana, Escudero, Kerkhoven, & ten Cate, 2015; Kriengwatana, Escudero, & ten Cate, 2015). However, the results are more likely attributable to how zebra finches perceive own-species' vocalizations and that this enables them to cope with voices in a human-like way. That is, zebra finch vocalizations exhibit formant-like peaks that vary systematically across vocalization categories and may therefore be important for discrimination of different vocalization types (Elie & Theunissen, 2016). Hence their surprising tendency to focus on vowels rather than voices may have been owed to the importance of formants in their own vocalizations. An equally thorough analysis of quail vocalizations may demystify how Japanese quails solved lack of invariance problem and compensated for co-articulation in Kluender et al. (1987) and Lotto et al. (1997). This latter explanation stresses the need to couple comparative speech perception research with an in-depth knowledge of animal communication systems to gain a broader perspective of why humans and animals might show similar speech perception abilities. In the past, comparative speech perception research has often overlooked the importance of understanding the ecology and communication habits of the animal model used, but we believe that more attention needs to be paid to this aspect in future speech perception research.

We have argued that animals show adaptation to speaker voices, but it is not yet clear that animals can do the reverse and discriminate speakers based on voice information. The ability to identify specific members of their own and other species may serve to help animals know which individuals in their environment represent a threat (or potentially a reliable source of information) and which do not. Unfortunately, there is almost no data on whether animals actually do this, even though there is ample evidence that animals can identify specific members of their own species and respond to these individuals based on the reliability of their signals (Blumstein & Daniel, 2004; Cheney & Seyfarth, 1988). What is currently known is that a range of animals are able to discriminate between familiar and unfamiliar human voices (Kriengwatana, Escudero, & ten Cate, 2015), but these studies do not unambiguously show that animals can recognize individual voices because they do not test the ability to discriminate between two familiar voices or to discriminate novel speech samples from familiar voices. The stimuli used are almost always familiar words or sentences, and animals may be able to recognize patterns in utterances—by using prosodic information for instance—rather than voice characteristics per se. Given that animals seem to be able to show speaker adaptation, demonstrations that animals can recognize individual human voices would suggest that animals, like humans, are flexible in terms of how they perceive and/or represent information available in vocal signals.

fundamental frequency of their song. *Proceedings of the National Academy of Sciences, 103*(14), 5543–5548.

Riede, T., & Zuberbühler, K. (2003). The relationship between acoustic structure and semantic information in Diana monkey alarm vocalization. *Journal of the Acoustical Society of America, 114*(2), 1132–1142.

Schwartz, J.-L., Abry, C., Boe, L.-J., & Valle, N. (2005). Asymmetries in vowel perception, in the context of the Dispersion-Focalisation theory. *Speech Communication, 45*, 425–434.

Sinex, D. G., McDonald, L. P., & Mott, J. B. (1991). Neural correlates of nonmonotonic temporal acuity for voice onset time. *Journal of the Acoustical Society of America, 90*, 2441–2449.

Sinnott, J. M. (1989). Detection and discrimination of synthetic English vowels by Old World monkeys (*Cercopithecus, Macaca*) and humans. *Journal of the Acoustical Society of America, 86*(2), 557–565.

Sinnott, J. M., & Kreiter, N. A. (1991). Differential sensitivity to vowel continua in Old World monkeys and humans. *Journal of Acoustic Society America, 89*(5), 2421–2429.

Sjerps, M. J., Mitterer, H., & McQueen, J. M. (2011). Listening to different speakers: On the time-course of perceptual compensation for vocal-tract characteristics. *Neuropsychologia, 49*(14), 3831–3846.

Stevens, K. N., & House, A. S. (1955). Development of a quantitative description of vowel articulation. *Journal of the Acoustical Society of America, 27*(3), 484–493.

Sturdy, C. B., Bloomfield, L. L., Farrell, T. M., Avey, M. T., & Weisman, R. G. (2007). Auditory category perception as a natural cognitive activity in songbirds. *Comparative Cognition and Behavior Reviews, 2*, 93–110.

Sturdy, C. B., Phillmore, L. S., Price, J. L., & Weisman, R. G. (1999). Song-note discriminations in zebra finches (*Taeniopygia Guttata*): Categories and pseudocategories. *Journal of Comparative Psychology, 113*(2), 204–212.

Sturdy, C. B., Phillmore, L. S., & Weisman, R. G. (2000). Call-note discriminations in black-capped chickadees (*Poecile Atricapillus*). *Journal of Comparative Psychology, 114*(4), 357–364.

Suthers, R. A. (1997). Peripheral control and lateralization of birdsong. *Journal of Neurobiology, 33*(5), 632–652.

Suthers, R. A., Goller, F., & Pytte, C. (1999). The neuromuscular control of birdsong. *Philosophical Transactions of the Royal Society B, 354*, 927–939.

Suthers, R. A., Rothgerber, J. R., & Jensen, K. K. (2016). Lingual articulation in songbirds. *Journal of Experimental Biology, 219*(4), 491–500.

Titze, I. R. (1994). *Principles of voice production.* Upper Saddle River, NJ: Prentice Hall.

Townsend, S. W., Charlton, B. D., & Manser, M. B. (2014). Acoustic cues to identity and predator context in meerkat barks. *Animal Behaviour, 94*, 143–149.

van den Berg, J. (1958). Myoelastic-aerodynamic theory of voice production. *Journal of Speech, Language, and Hearing Research, 1*(3), 227–244.

Warren, D. K., Patterson, D. K., & Pepperberg, I. M. (1996). Mechanisms of American English vowel production in a grey parrot (*Psittacus Erithacus*). *Auk, 113*(1), 41–58.

Weiss, D. J., & Newport, E. L. (2006). Mechanisms underlying language acquisition: Benefits from a comparative approach. *Infancy, 9*(2), 241–257.

Werker, J. F., & Curtin, S. (2005). PRIMIR: A developmental framework of infant speech processing. *Language Learning and Development, 1*(2), 197–234.

49 Key Issues and Future Directions: The Comparative Approach to Language

CAREL TEN CATE AND CONSTANCE SCHARFF

Like humans, many animals use their voice to communicate. Like human speech and language, animals' vocalizations can inform others about identity (what species they are, female or male, which individual). Like humans, animals flirt and fight with their voice, negotiate and manipulate, and can beg or raise alarm. Unlike humans, many animals have a small repertoire of communication sounds, are less flexible in its use, and do not need to imitatively learn their vocalizations. Moreover, there is no evidence that animals can convey complex abstract content the way language can. So, since human language differs in some important ways from most animal vocal communication systems, why then should one be interested in comparisons? What can be learned from the study of other animal species? In this chapter we provide our perspective on some key issues in the comparative approach and identify promising and relevant areas for further study.

1. *Why Comparative Studies?*

As outlined in chapter 44 by Fitch, it is crucial to realize that while human language is unique to our species, it consists of a set of processes that are partly also evident in other animals. Recognizing the multicomponent nature of language invites questions about the nature, development, and evolution of the different constituent components. Comparative studies can help to disentangle the knot of mechanisms underlying human language into components. Widely shared mechanisms indicate that these are unlikely to have evolved specifically for, or in consort with, communication by language. Rather, these mechanisms might have served as evolutionary building blocks for language. Comparative studies may also reveal mechanisms that are derived in humans, that is, for which the human abilities differ from those of other species and the evolutionary origin may not be immediately obvious, for instance, the syntactic structure of language (see ten Cate & Petkov, chapter 47). This invites studies as to what exactly it is that differentiates the human ability from that of other species, for instance, to what extent the learning of syntactical rules is based on general learning mechanisms that may also be present in other species, such as statistical learning (e.g., Santolin & Saffran, 2018). Also, studies on animals provide options for experimental approaches that for ethical or other reasons cannot be done on humans, such as manipulating the expression of *FoxP* genes (see chapter 46 by Scharff, Knörnschild, & Jarvis) or rearing individuals under highly controlled conditions to examine factors affecting vocal learning. Finally, comparative studies can address one of the most challenging questions in science: how human language evolved. Comparing the presence and absence of specific traits in various species, in relation to variation in their ecological and social environment, their cognitive abilities and life history traits may contribute to identifying which selective forces have given rise to separate traits as well as to their linkage. Such studies may provide hypotheses about the stepping stones and processes responsible for our evolution from prelinguistic to linguistic human beings. Altogether, comparative studies can shed light on the question whether the evolutionary emergence of language is the product of one big leap or of an accumulation of smaller steps.

2. *Comparative Studies: How?*

There are two approaches to comparative studies (see Fitch, chapter 44). One is to compare related species, originating from a common ancestor, for example, comparing humans with great apes or other primates. It is based on the principle that evolution of any trait in any species, no matter how complex, is in one way or another based on traits present in an ancestral species. The closer two species are related, the higher the chances that comparing them will make it possible to trace the origin of a trait and to identify in what way the ancestral trait has been modified to gain its current utility. This approach is exemplified in chapter 45 by Fischer and Hage. It focuses on primate vocalizations: their meaning, their link to internal emotional or motivational state, the degree of flexibility in usage and structure, and the plasticity in development. They also address the degree of volitional control that primates may have over their use of vocalizations and how this may be related to neural control mechanisms, such as the monkey homologue of the human Broca's area. However, to get from an ancestral primate vocal production system to the human one also required a series of steps such as the mapping of auditory input to vocal output; coordination of breathing, articulation, and voicing; and the integration between brain areas supporting symbolic and linguistic thought. Fischer and Hage (chapter 45) illustrate how comparative studies help to identify crucial components involved in, and steps toward, language.

The second approach to comparative studies does not focus on closely related species, but on species that share particular traits. Over the course of evolution, analogous traits may have emerged in species not closely related, as a result of adaptations to similar selection pressures, that is, by convergence. A clear example of such a trait, mentioned in several chapters, is the presence of vocal production learning—the ability to learn to produce new vocalizations based on those heard or in responses to social influences by other individuals. This ability is present in several groups of mammals and birds, in which it evolved independently. Comparing species having this ability with those who don't have it, or have it only to some degree, as well as comparing the nature of the learning mechanisms in different vocal learning species can provide a wealth of information on what factors may affect such learning processes, such as predispositions or social interactions between the learner and other individuals. It may also provide animal models to examine neurobiological mechanisms underlying such learning. Additionally, it can offer insights in the social or ecological factors that may have contributed to the evolution of vocal learning.

Both types of comparative studies may thus provide insight into questions about mechanisms (what are components of the language faculty, how do they work) as well as about the evolution of these mechanisms (what was their origin, how did they become the way they are).

3. Key Issues

The various chapters in this part reflect the current topics of comparative studies. They also identify several key issues for further study. We provide a brief discussion of these issues.

3.1. Vocal Production Learning Vocal production learning is a major area of research. This has demonstrated that humans are not the only species that learns most of their vocalizations imitatively from conspecifics. The process has fascinating parallels between such distantly related groups as humans and three orders of birds, which extend to the level of genes involved in the process (see Scharff, Knörnschild, & Jarvis, chapter 46). The considerable research efforts focused on songbirds have made song learning the prime model for the study of vocal learning in nonhuman animals. It has given rise to several prominent theories about the behavioral and neurobiological mechanisms underlying vocal learning (see Scharff, Knörnschild, & Jarvis, chapter 46). However, there are other groups of vocal learners, such as cetaceans (whales and dolphins) and bats. In terms of what we know about the learning mechanisms involved, these groups are much less studied. This is partly a consequence of the logistic constraints: cetaceans cannot readily be studied under controlled laboratory conditions and subjected to rigid experiments. There is a danger that because the outcome of their vocal development is clearly affected by learning, the as yet unknown underlying learning mechanisms and factors affecting the process are too readily assumed to be similar to what is occurring in birds, relying heavily on imitative learning. This may be right, but it may well be that other types of learning are involved. For instance, the example of an elephant producing human words (Stoeger et al., 2012) is striking, but its development is unclear. Is the system involved here the same as the one used for producing conspecific vocalizations? Or might it be the outcome of a separate process, one of skilled operant conditioning in which a trainer has shaped a behavior, in this case one producing a sound, to become similar to a human speech sound? This cautionary note also applies to the cases of "limited vocal production learning" as summarized by Scharff, Knörnschild, and Jarvis (chapter 46). The undisputed observation that there is a graded scale with respect to the occurrence of vocal imitation as outcome of a developmental process need not imply that the underlying process is also one that differs only in degree. It might reflect the presence of different underlying mechanisms. There is still a lot to uncover and the variation in the degree of vocal imitation among both birds and mammals can be used to reveal variation in underlying learning mechanisms. This may provide new hypotheses about human vocal learning, which is not just due to imitation, but also shaped by predispositions and the rewarding responses and guidance by infant caretakers.

3.2. Comprehension Learning, Usage Learning, and Referential Signaling

Linking the hearing of a sound to a particular event or object (comprehension or recognition learning) and learning the appropriate usage of a vocalization (usage learning) are more common than production learning. This raises the question of how they interact and are coupled to production learning: have they become an integral part of the production learning system, or are they to some extent separate? This is another understudied area. For instance, while only juvenile male zebra finches show song production learning, adult males (who sing) and females (who do not sing) can both be readily trained to discriminate different songs (Kriengwatana, Spierings, & ten Cate, 2016). For males this occurs without impacting their own songs. Comprehension learning thus seems at least partly separate from production learning, but it is unclear to what extent.

Usage learning links the production of a vocalization to a specific context. One aspect of this process concerns the degree of voluntary control over the production of such vocalizations. This varies among species. The example of macaques (Fischer & Hage, chapter 45), which after long training became capable of using two different vocalizations depending on context, indicate that for these primates flexibility seems limited. Some bird species show more flexibility. Fork-tailed drongos, for instance, mimic the alarm calls of other species (a case of production learning) and can flexibly use these calls to target particular species (Flower, Gribble, & Ridley, 2014), most likely the result of experience—that is, by usage learning. In this way, the drongos scare other species away from food that they can subsequently eat themselves.

Both comprehension and usage learning are required for referential signaling. Various great apes are able to use gestures or icons to refer to different objects or actions (e.g., Savage-Rumbaugh, McDonald, Sevcik, Hopkins, & Rubert, 1986). They clearly have the cognitive ability for referential communication, but there is no evidence that they link this to using particular vocalizations, let alone vocal production learning. This ability has, however, been demonstrated by some dolphins (Richards, Wolz, & Herman, 1984) and also by grey parrots. In particular one parrot, Alex, was able of vocally expressing labels for many objects and actions (e.g., Pepperberg, 1999). He also mastered more abstract concepts such as *same* and *different* as well as several other cognitively advanced abilities. Here also, the considerable investment in training effort and the logistic constraints on controlled experiments with dolphins, parrots, and corvids are prime factors that are challenging when studying these species. However, the developmental processes underlying referential signaling and their linkage to production learning are topics of great relevance if we want to understand how their nexus as present in language emerged.

3.3. Cultural Evolution

Another aspect of vocal production learning concerns its consequences. In humans, the transfer of language from one generation to the next has given rise to the emergence of the multitude of different languages as a consequence of this process of cultural transmission. This process is not specific to humans. Regional vocal differentiation among members of the same species also occurs in songbirds (e.g., white crowned sparrows; Baker, 1975) and sea mammals (e.g., humpback whales; Garland et al., 2011). The variation among and within species can be studied to assess the impact of social or geographical factors affecting the formation, disappearance, or changes of dialects and language diversity.

3.4. Speech Production and Perception

Over the years, many of the features that were once considered specialized adaptations to the production and perception of speech have been detected in other animal species. The permanently descended larynx in humans, for instance, was thought to be a uniquely human trait linked to the evolution of the ability to produce speech sounds by vocal tract filtering (Lieberman, Klatt, & Wilson, 1969). It certainly is a derived trait when humans are compared with great apes. However, a descended larynx also exists in other species, suggesting that it evolved by selection for producing low-frequency formants, driven by selection to simulate the presence of a larger animal to a receiver (e.g., Charlton & Reby, 2016). Also, many species, including several bird species (see chapter 48 by Kriengwatana & Beckers) are able to produce vocal variation by controlled and flexible filtering of harmonic spectra originating from a vibrating sound source. Comparative studies thus indicate a scenario in which the descended larynx of humans evolved for other reasons and only later became of use for producing speech sounds. Another comparative study (Fitch, Marthur, de Boer, & Ghazanfar, 2016) showed that the vocal tract of macaques is capable of producing a large range of vowel-like sounds, in contrast to what was thought in the past (Lieberman et al., 1969). So, a variety of studies indicate that a source-filter mechanism and the morphology of the sound producing system, and with this the ability to produce a wide range of sounds, are not uniquely human. This shifts the emphasis for studying what is special on human speech production to the way the system is being used, that is, its neural control (Fitch, 2018): what distinguishes the abilities of humans and

Pepperberg, I. M. (1999). *The Alex studies: Cognitive and communicative abilities of grey parrots.* Cambridge, MA: Harvard University Press.

Richards, D. G., Wolz, J. P., & Herman, L. M. (1984). Vocal mimicry of computer-generated sounds and vocal labeling of objects by a bottlenosed dolphin, *tursiops-truncatus. Journal of Comparative Psychology, 98,* 10–28.

Santolin, C., & Saffran, J. R. (2018). Constraints on statistical learning across species. *Trends in Cognitive Sciences, 22,* 52–63.

Savage-Rumbaugh, S., McDonald, K., Sevcik, R. A., Hopkins, W. D., & Rubert, E. (1986). Spontaneous symbol acquisition and communicative use by pygmy chimpanzees (*Pan paniscus*). *Journal of Experimental Psychology: General, 115,* 211–235.

Schlenker, P., Chemla, E., & Zuberbühler, K. (2016). What do monkey calls mean? *Trends in Cognitive Sciences, 20,* 894–904.

Smirnova, A., Zorina, Z., Obozova, T., & Wasserman, E. (2015). Crows spontaneously exhibit analogical reasoning. *Current Biology, 25,* 256–260.

Stoeger, A. S., Mietchen, D., Oh, S., de Silva, S., Herbst, C. T., Kwon, S., & Fitch, W. T. (2012). An Asian elephant imitates human speech. *Current Biology, 22,* 2144–2148.

CONTRIBUTORS

SHANLEY E. M. ALLEN Faculty of Social Sciences, Technische Universität Kaiserslautern, Kaiserslautern, Germany

SALLY ANDREWS School of Psychology, University of Sydney, Sydney, Australia

ANDREAS BARTSCH Oxford Centre for Functional Magnetic Resonance Imaging of the Brain, University of Oxford, Oxford, United Kingdom; Department of Neuroradiology, University of Heidelberg, Heidelberg, Germany; Department of Neuroradiology, University of Würzburg, Würzburg, Germany; Radiologie Bamberg, Bamberg, Germany

TIMOTHY C. BATES Department of Psychology, University of Edinburgh, Edinburgh, United Kingdom

GABRIËL J. L. BECKERS Faculty of Social and Behavioural Sciences, Utrecht University, Utrecht, the Netherlands

CHRISTIAN F. BECKMANN Donders Institute for Brain, Cognition and Behaviour, Radboud University, Nijmegen, the Netherlands; Department of Cognitive Neuroscience, Radboud University Medical Centre, Nijmegen, the Netherlands; Oxford Centre for Functional Magnetic Resonance Imaging of the Brain, University of Oxford, Oxford, United Kingdom

MARINA BEDNY Department of Psychological and Brain Sciences, Johns Hopkins University, Baltimore, Maryland, USA

HEIKE BEHRENS Department of Language and Literature, University of Basel, Basel, Switzerland

SERENA BIANCHI Simons Foundation Autism Research Initiative, New York, New York, USA

EDWARD F. CHANG Department of Neurological Surgery, University of California San Francisco, San Francisco, California, USA; Center for Neural Engineering at University of California Berkeley and San Francisco, Berkeley and San Francisco, California, USA

VASILIKI CHONDROGIANNI School of Philosophy, Psychology and Language Sciences, University of Edinburgh, Edinburgh, United Kingdom

HERBERT H. CLARK Department of Psychology, Stanford University, Stanford, California, USA

DELPHINE DAHAN Department of Psychology, University of Pennsylvania, Philadelphia, Pennsylvania, USA

BART DE BOER Artificial Intelligence Lab, Vrije Universiteit Brussel, Brussels, Belgium

VERA DEMBERG Department of Computer Science, Saarland University, Saarbrücken, Germany

FERNANDA FERREIRA Department of Psychology, University of California Davis, Davis, California, USA

VICTOR S. FERREIRA Department of Psychology, University of California San Diego, La Jolla, California, USA

JULIA FISCHER Cognitive Ethology Laboratory, German Primate Center, Göttingen, Germany; Leibniz Science Campus for Primate Cognition, Göttingen, Germany; Georg-August-University, Göttingen, Germany

SIMON E. FISHER Language and Genetics Department, Max Planck Institute for Psycholinguistics, Nijmegen, the Netherlands; Donders Institute for Brain, Cognition and Behaviour, Radboud University, Nijmegen, the Netherlands

W. TECUMSEH FITCH Department for Cognitive Biology, University of Vienna, Vienna, Austria

HARTMUT FITZ Donders Institute for Brain, Cognition and Behaviour, Radboud University, Nijmegen, the Netherlands; Neurobiology of Language Department, Max Planck Institute for Psycholinguistics, Nijmegen, the Netherlands

ELIA FORMISANO Department of Cognitive Neuroscience, Faculty of Psychology and Neuroscience, Maastricht University, Maastricht, the Netherlands; Maastricht Brain Imaging Center, Maastricht, the Netherlands; Maastricht Center for Systems Biology, Maastricht, the Netherlands

CLYDE FRANCKS Language and Genetics Department, Max Planck Institute for Psycholinguistics, Nijmegen, the Netherlands; Donders Institute for Brain, Cognition and Behaviour, Radboud University, Nijmegen, the Netherlands

STEFAN L. FRANK Centre for Language Studies, Radboud University, Nijmegen, the Netherlands

ALONA FYSHE Department of Computing Science, Department of Psychology, University of Alberta, Edmonton, Canada; Department of Computer Science, University of Victoria, Victoria, Canada

SUSAN GOLDIN-MEADOW Department of Psychology, University of Chicago, Chicago, Illinois, USA

JOACHIM GROSS Institute for Biomagnetism and Biosignalanalysis, University of Münster, Germany; Institute of Neuroscience and Psychology, University of Glasgow, Glasgow, United Kingdom

KOEN V. HAAK Donders Institute for Brain, Cognition and Behaviour, Radboud University, Nijmegen, the Netherlands; Department of Cognitive Neuroscience, Radboud University Medical Centre, Nijmegen, the Netherlands

STEFFEN R. HAGE Neurobiology of Vocal Communication, Werner Reichardt Centre for Integrative Neuroscience, University of Tübingen, Tübingen, Germany

PETER HAGOORT Neurobiology of Language Department, Max Planck Institute for Psycholinguistics, Nijmegen, the Netherlands; Donders Institute for Brain, Cognition and Behaviour, Radboud University, Nijmegen, the Netherlands

GREGORY HICKOK Departments of Cognitive Sciences and Language Science, University of California, Irvine, California, USA

RAY JACKENDOFF Center for Cognitive Studies, Tufts University, Medford, Massachusetts, USA; Department of Brain and Cognitive Sciences, Massachusetts Institute of Technology, Cambridge, Massachusetts, USA

ERICH D. JARVIS Laboratory of Neurogenetics of Language, The Rockefeller University, New York, New York, USA; The Howard Hughes Medical Institute, Chevy Chase, Maryland, USA

ELIZABETH K. JOHNSON Department of Psychology, University of Toronto, Toronto, Canada

FRANK KELLER School of Informatics, University of Edinburgh, Edinburgh, United Kingdom

EVAN KIDD Language Development Department, Max Planck Institute for Psycholinguistics, Nijmegen, the Netherlands; Research School of Psychology, The Australian National University, Canberra, Australia

MIRJAM KNÖRNSCHILD Institute of Biology, Department of Biology, Chemistry, Pharmacy, Freie Universität Berlin, Berlin, Germany; Smithsonian Tropical Research Institute, Panama, Republic of Panama; Museum für Naturkunde, Leibniz Institute for Evolution and Biodiversity Science, Berlin, Germany

EMIEL KRAHMER Tilburg School of Humanities and Digital Sciences, Department of Communication and Cognition, Tilburg University, Tilburg, the Netherlands

BUDDHAMAS P. KRIENGWATANA School of Psychology and Neurosciences, University of St. Andrews, St. Andrews, United Kingdom; College of Medical, Veterinary and Life Sciences, University of Glasgow, Glasgow, United Kingdom

SARAH C. KUCKER Department of Psychology, Oklahoma State University, Stillwater, Oklahoma, USA

PHONG LE Institute for Language, Cognition and Computation, University of Edinburgh, Edinburgh, United Kingdom

STEPHEN C. LEVINSON Language and Cognition Department, Max Planck Institute for Psycholinguistics, Nijmegen, the Netherlands

MICHELLE LUCIANO Department of Psychology, University of Edinburgh, Edinburgh, United Kingdom

MAIRÉAD MACSWEENEY Deafness, Cognition and Language Research Centre, University College London, London, United Kingdom; Institute of Cognitive Neuroscience, University College London, London, United Kingdom

WILLIAM D. MARSLEN-WILSON Center for Speech, Language, and Brain, Department of Psychology, University of Cambridge, Cambridge, United Kingdom; 2MRC Cognition and Brain Sciences Unit, Cambridge, United Kingdom

JAMES M. MCQUEEN Donders Institute for Brain, Cognition and Behaviour, Radboud University, Nijmegen, the Netherlands; Max Planck Institute for Psycholinguistics, Nijmegen, the Netherlands

ANTJE S. MEYER Psychology of Language Department, Max Planck Institute for Psycholinguistics, Nijmegen, the Netherlands; Donders Institute for Brain, Cognition and Behaviour, Radboud University, Nijmegen, the Netherlands

TOM M. MITCHELL School of Computer Science, Machine Learning Department, Carnegie Mellon University, Pittsburgh, Pennsylvania, USA

PADRAIC MONAGHAN Amsterdam Center for Language and Communication, University of Amsterdam, Amsterdam, the Netherlands

ANGELA MORGAN Speech and Language Group, Murdoch Childrens Research Institute, Victoria, Australia; University of Melbourne, Melbourne, Australia

MANTE S. NIEUWLAND Neurobiology of Language Department, Max Planck Institute for Psycholinguistics, Nijmegen, the Netherlands

ASLI ÖZYÜREK Donders Institute for Brain, Cognition and Behaviour, Radboud University, Nijmegen, the Netherlands; Max Planck Institute for Psycholinguistics, Nijmegen, the Netherlands

NICOLA PALOMERO-GALLAGHER Institute of Neuroscience and Medicine, Research Centre Jülich, Jülich, Germany; Department of Psychiatry, Psychotherapy and Psychosomatics, Medical Faculty, Rheinisch-Westfälische Technische Hochschule Aachen, Aachen, Germany

CHRISTOPHER I. PETKOV Institute of Neuroscience and Centre for Behaviour and Evolution, Newcastle University, Newcastle upon Tyne, United Kingdom

DAVID POEPPEL Department of Psychology, New York University, New York, New York, USA; Neuroscience Department, Max Planck Institute for Empirical Aesthetics, Frankfurt am Main, Germany

IZABELA PRZEŹDZIK Donders Institute for Brain, Cognition and Behaviour, Radboud University, Nijmegen, the Netherlands; Department of Cognitive Neuroscience, Radboud University Medical Centre, Nijmegen, the Netherlands

ERIK D. REICHLE Department of Psychology, Macquarie University, Sydney, Australia

ARDI ROELOFS Donders Institute for Brain, Cognition and Behaviour, Radboud University, Nijmegen, the Netherlands

FEDERICO ROSSANO Department of Cognitive Science, University of California San Diego, San Diego, California, USA

CAROLINE F. ROWLAND Language Development Department, Max Planck Institute for Psycholinguistics, Nijmegen, the Netherlands; Department of Psychological Sciences, University of Liverpool, Liverpool, United Kingdom

CONSTANCE SCHARFF Institute of Biology, Department of Biology, Chemistry, Pharmacy, Freie Universität Berlin, Berlin, Germany

KRISTINA SIMONYAN Department of Otolaryngology, Massachusetts Eye and Ear Infirmary, Department of Neurology, Massachusetts General Hospital, Harvard Medical School, Boston, Massachusetts, USA

MATTHIAS J. SJERPS Neurobiology of Language Department, Max Planck Institute for Psycholinguistics, Nijmegen, the Netherlands; Donders Institute for Brain, Cognition and Behaviour, Radboud University, Nijmegen, the Netherlands

MICHAEL A. SKEIDE Department of Neuropsychology, Max Planck Institute for Human Cognitive and Brain Sciences, Leipzig, Germany

ARJEN STOLK Helen Wills Neuroscience Institute, University of California Berkeley, Berkeley, California, USA

CAREL TEN CATE Institute of Biology, Leiden University, Leiden, the Netherlands; Leiden Institute for Brain and Cognition, Leiden University, Leiden, the Netherlands

IVAN TONI Donders Institute for Brain, Cognition and Behaviour, Radboud University, Nijmegen, the Netherlands

CHARA TSOUKALA Centre for Language Studies, Radboud University, Nijmegen, the Netherlands

JOS J. A. VAN BERKUM Utrecht Institute of Linguistics, Department of Languages, Literature and Communication, Utrecht University, Utrecht, the Netherlands

SONJA C. VERNES Neurogenetics of Vocal Communication Group, Max Planck Institute for Psycholinguistics, Nijmegen, the Netherlands; Donders Institute for Brain, Cognition and Behaviour, Radboud University, Nijmegen, the Netherlands

LEILA WEHBE School of Computer Science, Machine Learning department, Carnegie Mellon University, Pittsburgh, Pennsylvania, USA

KATHERINE S. WHITE Department of Psychology, University of Waterloo, Waterloo, Canada

BENCIE WOLL Deafness Cognition and Language Research Centre, University College London, London, United Kingdom

KARL ZILLES Institute of Neuroscience and Medicine, Research Centre Jülich, Jülich, Germany; Department of Psychiatry, Psychotherapy and Psychosomatics, Medical Faculty, Rheinisch-Westfälische Technische Hochschule Aachen, Aachen, Germany; Jülich Aachen Research Alliance–Translational Brain Medicine, Aachen, Germany

WILLEM ZUIDEMA Institute for Logic, Language and Computation, University of Amsterdam, Amsterdam, the Netherlands

INDEX

Note: Figures and tables are indicated by "f" and "t" respectively, following page numbers. Notes are indicated by "n," following page numbers.